FOUNDATIONS OF PARASITOLOGY

"They do certainly give strange and new-fangled names to diseases."

PLATO

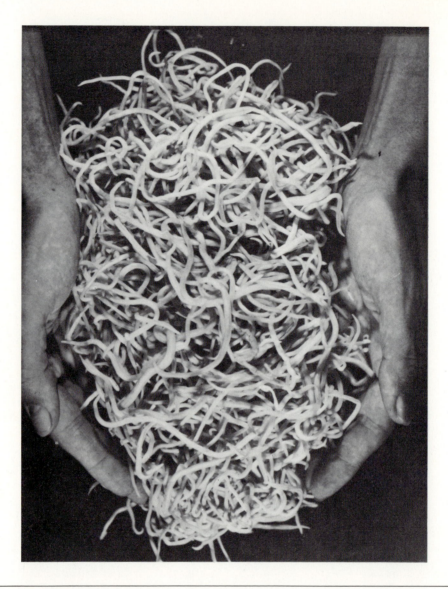

A mass of *Ascaris lumbricoides* from the intestine of a two-year-old African child. This was a fatal infection (see Chapter 27).

Courtesy J.K. Baird, et al., Armed Forces Institute of Pathology, Washington, D.C.

FOUNDATIONS OF PARASITOLOGY

GERALD D. SCHMIDT

Professor of Zoology/Parasitology,
University of Northern Colorado,
Greeley, Colorado

LARRY S. ROBERTS

Professor of Biological Sciences,
Texas Tech University,
Lubbock, Texas

FOURTH EDITION

TIMES MIRROR/MOSBY
COLLEGE PUBLISHING

ST. LOUIS • TORONTO • BOSTON • LOS ALTOS   1989

Editor: David Kendric Brake
Developmental Editor: Mary M. Huggins
Project Manager: Teri Merchant
Production Editor: Deborah Vogel
Book and cover design: Gail Morey Hudson

Cover photos (clockwise from top): *Dermacentor variabilis*
(D.M. Phillips/Taurus Photos); Trematodes on fish gills
(Biological Photo Service); *Prosthogonimus macrorchis*
(Biological Photo Service); *Trypanosoma* (Ed Reschke).

FOURTH EDITION

Schmidt, Gerald D.
    Foundations of parasitology/Gerald D. Schmidt, Larry S. Roberts.—4th ed.
    Includes bibliographies and index.
    ISBN 0-8016-5039-9
    1. Parasitology. I. Roberts, Larry S. II. Title.
    [DNLM: 1. Parasites. 2. Parasitology. QX 4 S351f]
    QL757.S35 1989
    591.5′249—dc19
    DNLM/DLC

C/VH/VH 9 8 7 6 5 4 3 2 1

PREFACE

If the success of the first three editions of *Foundations of Parasitology* is any indication, the science of parasitology is a thriving one. Much of what was new, frontier information only 4 years ago is now standard teaching material. In the meantime still more advances have been made in many exciting and important fields, such as AIDS and malaria research, requiring nearly total revision of the book. This we have done, picking and choosing from the myriad of facts and findings that we have learned from perusing hundreds of journals, reviews, and books. The process was not easy, but we feel that we have succeeded. Much credit, also, should go to the many users of the third edition who have volunteered suggestions or pointed out errors. These people are too numerous to mention; we are grateful to them all.

THE SCOPE OF THIS BOOK

This textbook is designed especially for upper division courses in general parasitology. It emphasizes principles, illustrating them with primary sections on the biology, physiology, morphology, and ecology of the major parasites of humans and domestic animals. We have found that these examples are of most interest to the majority of students. Other parasites are included where they are of unusual biological interest. Through clear phylogenetic development, the student is led from the simplest to the most complex parasites. Different instructors emphasize different materials, so there is more in this book than can be covered adequately in a single semester. But the organization is such that any teacher can easily pick and choose reading assignments for the students. Always, we have strived for readability, the words being enhanced by photographs, drawings, electron micrographs, and tables. The feedback we continue to receive from teachers and students tells us we have hit the mark.

NEW TO THIS EDITION

We have not only read each page carefully for the purpose of eliminating the extraneous and adding new material, but the book is also updated throughout. Material on immunology has been extensively rewritten and updated in response to the rapidly expanding body of knowledge in this area. To cover this trend adequately, the second chapter of the third edition has been divided into two chapters in this new edition, the first explaining definitions and ecology and the second treating immunology and pathology. Other key features of the fourth edition include:

- New treatment of AIDS and its relationship to opportunistic parasites has been added, as well as a new section on immunodiagnosis, including enzyme-linked immunosorbent assays (ELISA).
- The "Form and Function" chapters on protozoa, trematodes, cestodes, nematodes, and arthropods have been significantly rewritten to provide a stronger base of knowledge with which to further investigate each group.
- Chapter 5 on the Kinetoplastida has been revised to include the latest information on molecular biology of trypanosomes and the taxonomy of *Leishmania*.
- Chapter 9 has been revised to include the latest insights in malaria research, including the metabolism of *Plasmodium,* action of drugs (including the Chinese drug *qinghaosu*), and the current status of research on malaria vaccines.

INSTRUCTIVE DESIGN

Students using the fouth edition of *Foundations of Parasitology* are guided to a clear understanding of the topic through the authors' careful use of study aids. Essential terms, which are defined in a complete 500-entry **glossary,** are **boldfaced** in the text to provide emphasis and ease in reviewing. **Numbered references** at the end of each chapter make supportive data and further study easily accessible. Clear **labeling** makes all illustrations approachable and self-explanatory to the student.

The exemplary illustration program that distinguished the third edition has evolved into an even more attractive fourth edition. Primarily we have taken full advantage of modern microscopic and photographic techniques to provide the sharpest and most instructive photographs. Ian Grant of the South Australian Museum and Dr. William Ober have contributed their talents to create new artwork that is both more attractive and more instructive. Through their efforts, twenty-five percent of the artwork has been redrawn. As well, labeling has been added to a large percentage of existing illustrations, on the advice of reviewers.

ACKNOWLEDGMENTS

Many individuals contributed time and effort toward this fourth edition. We especially wish to thank:

Claude G. Alexander, *San Francisco State University*
David T. Clark, *Portland State University*
Peter W. Pappas, *The Ohio State University*
Miles B. Markus
George V. Hillyer
William Coil

Ralph B. Eckerlin
Armand M. Kuris
George L. Stewart, *University of Texas at Arlington*
Leslie Uhazy, *University of Missouri—Columbia*
Janine N. Caira, *University of Connecticut*
J.K. Frenkel, *University of Kansas Medical Center*
Lauritz A. Jensen, *University of Health Sciences—Kansas City*

We thank the staff of The C.V. Mosby Co., especially David K. Brake, Editor; Mary M. Huggins, Developmental Editor; Teri Merchant, Project Manager; and Deborah Vogel, Production Editor.

We hope that teachers, students, and other colleagues will continue to suggest improvements and to ferret out any remaining errors, so that we may continue to improve this book. Our readership is still expanding. We thank you all.

Gerald D. Schmidt
Larry S. Roberts

CONTENTS

CHAPTER 1

INTRODUCTION TO PARASITOLOGY

Big fleas have little fleas
upon their back to bite 'em,
Little fleas have lesser fleas
and so, ad infinitum.

SWIFT

Few persons realize that there are far more kinds of parasitic than nonparasitic organisms in the world. Even if we exclude the viruses and rickettsias, which are all parasitic, and the many kinds of parasitic bacteria and fungi, the parasites are still in the majority. In general the parasitic way of life is highly successful, since it evolved independently in nearly every phylum of animals, from protozoa to arthropods and chordates, as well as in many plant groups. Organisms that are not parasites are usually hosts. Humans, for example, are hosts to more than a hundred kinds of parasites, again not counting viruses, bacteria, and fungi. It is unusual to examine a domestic or wild animal without finding at least one species of parasite on or within it. Even animals reared under strict laboratory conditions are commonly infected with protozoa or other parasites. Often the parasites themselves are the hosts of other parasites. It is no wonder then that the science of parasitology has developed out of efforts to understand parasites and their relationships with their hosts.

RELATIONSHIP OF PARASITOLOGY TO OTHER SCIENCES

Parasitology has passed through a series of stages in its history, each of which, today, is an active discipline in its own right. The first and most obvious stage is the discovery of the parasites themselves. This undoubtedly began in the shadowy eons of prehistory; but the ancient Persians, Egyptians, and Greeks recorded their observations in such a way that later generations could, and often still can, recognize the animals they were writing about. The discovery and naming of unknown parasites, their study, and their arrangement into a classification is an exciting and popular branch of parasitology today.

When people became aware that parasites were troublesome and even serious agents of disease, they began a continuing campaign to heal the infected and eliminate the parasites. The later discovery that other animals could be the **vectors,** or means of dissemination, of parasites opened the door to other approaches for control of parasitic diseases.

Development of better lenses led to basic discoveries in cytology and genetics of parasites that are applicable to all of biology. In the twentieth century refined techniques in physics and chemistry have contributed much to our knowledge of host-parasite relationships. Some of these have added to our understanding of basic biological principles and mechanisms, such as the discovery of cytochrome and the electron transport system by David Keilin in 1925[4] during his investigations of parasitic worms and insects. Today biochemical techniques are widely used in studies of parasite metabolism, immunology, serology, and chemotherapy. The advent of the electron microscope has resulted in many new discoveries at the subcellular level. Molecular biology and recombinant DNA techniques have contributed new diagnostic techniques and new knowledge of relationships between parasites, and they offer much hope in the development of new vaccines. Thus, parasitologists employ the tools and concepts of many scientific disciplines in their research.

Parasitology today usually does not include virology, bacteriology, and mycology because these sciences have developed into disciplines in their own

1

right. Exceptions occur, however, since it is not uncommon for parasitological research to overlap these areas. Medical entomology, too, has branched off as a separate discipline, but it remains a subject of paramount importance to the parasitologist, who must understand the relationships between arthropods and the parasites they harbor and disperse.

PARASITOLOGY AND HUMAN WELFARE

Human welfare has suffered greatly through the centuries because of parasites. Fleas and bacteria conspired to destroy a third of the European population in the seventeenth century, and malaria, schistosomiasis, and African sleeping sickness have sent untold millions to their graves. Even today, after successful campaigns against yellow fever, malaria, and hookworm infections in many parts of the world, parasitic diseases in association with nutritional deficiencies are the primary killers of humans. Recent summaries of the worldwide prevalence of selected parasitic diseases show that there are more than enough existing infections for every living person to have one, were they evenly distributed:[5,6]

Disease category	Number of human infections
All helminths	4.5 billion
Ascaris	1.26 billion
Hookworms	932 million
Trichuris	687 million
Filarial worms	657 million
Schistosomes	271 million
Malaria	300 million

These, of course, are only a few of the many kinds of parasites that infect humans, which points out that parasitic diseases are an important fact of life for many people. The majority of the more serious infections occur in the so-called tropical zones of the earth, so most dwellers within temperate regions are unaware of the magnitude of the problem. For instance, of the approximately 60 million annual worldwide deaths from all causes, 30 million are children under 5 years. Half of these, 15 million, are attributed to the combination of malnutrition and intestinal infection.[3]

However, the notion held by the average person that humans in the United States are free of worms is largely an illusion—an illusion created by the fact that the topic is rarely discussed because of our attitudes that worms are not the sort of thing that refined people talk about, the apparent reluctance of the media to disseminate such information, and the fact that poor people are the ones most seriously affected. Some estimates place the number of children in the United States infected with worms at about 55 million, although this is a gross underestimation if one includes such parasites as pinworms (*Enterobius vermicularis*). Only occasionally is the situation accurately reflected in the popular press: "If I brought in a jar of some child's roundworms, a great many people would be thoroughly nauseated. It is the sort of thing that is left unsaid, undiscussed and unreported throughout the U.S. A good note to close on! Let's not disturb folks. The thought of that jar upsets refined people. Things should be kept in their place, in the . . . well, let's skip it. Sleep well, good people—only a few million kids are affected."[9]

Even though there are many "native-born" parasite infections in the United States, many "tropical" diseases are imported within infected humans coming from endemic areas. After all, one can travel halfway around the world in a day or two. Many thousands of immigrants who are infected with schistosomes, malaria organisms, hookworms, and other parasites—some of which are communicable—currently live in the United States. It is estimated that about 100,000 cases of *Schistosoma mansoni* (Chapter 18) in the continental United States originated in Puerto Rico. Servicemen returning from abroad often bring parasite infections with them. In 1971 the Centers for Disease Control in Atlanta reported 3047 cases of malaria in the United States, about three fourths of which were acquired in Vietnam. There are still viable infections of filariasis and *Strongyloides* in ex-servicemen who contracted the disease in the South Pacific or Asia 45 years ago! A traveler may become infected during a short layover in an airport, and many pathogens find their way into the United States as stowaways on or in imported products. Travel agents and tourist bureaus are reluctant to volunteer information on how to avoid the tropical diseases that a tourist is likely to encounter—they might lose the customer.[2] Small wonder, then, that "exotic" diseases confront the general practitioner with more and more frequency. One family physician claims to have treated virtually every major parasitic disease of humans during the years of his practice in Amherst, Massachusetts. In another example, a survey of intestinal parasites of 776 Southeast Asian immigrants in New Mexico revealed 20 different species of parasites, some of which are not common in the United States.[8]

There are other, much less obvious, ways in which parasites affect all of us, even those in comparatively parasite-free areas. Primary among these is malnutrition, as the result of inefficient use of arable land and of food energy. Only 3.4 billion of the 7.8 billion acres of total potentially arable land in the world are now under cultivation.[3] Much of the remaining 4.4

billion acres cannot be developed because of malaria, trypanosomiasis, schistosomiasis, and onchocerciasis. In Africa alone, an area of land equal to the size of the United States exists in which people cannot live and grow livestock because of trypanosomes. How many starving people could be fed if this land were cultivated? As many as half of the world's population today are undernourished. The population will double again in 35 years. It is impossible to ignore the potentially devastating effects that worldwide famine will have on all humankind.

Even where food is being produced, it is not always used efficiently. Considerable caloric energy is wasted by fevers caused by parasitic infections. Heat production of the human body increases about 7.2% for each degree rise in Fahrenheit. A single, acute day of fever caused by malaria requires approximately 5000 calories, or an energy demand equivalent to 2 days of hard manual labor. To extrapolate, in a population with an average diet of 2200 calories per day, if 33% had malaria, 90% had a worm burden, and 8% had active tuberculosis (conditions that are repeatedly observed), there would be an energy demand equivalent to 7500 tons of rice per month per million people in addition to normal requirements. That is a waste of 25% to 30% of the total energy yield from grain production in many societies.[7]

Another cause of energy loss is malabsorption of digested food. This is a common occurrence in parasitic infections. It is difficult to quantify this loss, but it undoubtedly is highly significant, especially in those who are undernourished to begin with.

People create many of their own disease conditions because of high population density and subsequent environmental pollution. Despite great progress in extending water supplies and sewage disposal programs in developing countries, not more than 10% to 15% of the world population is thus served. Population shifts from rural to urban areas commonly overload the water and sewage capabilities of even major cities. Usually an adequate water supply has first priority, with sewage disposal running a poor second (Fig. 1-1). When one recalls that most parasite infections are caused by ingesting food or water contaminated with human feces, it is easy to understand why 15 million children die of intestinal infections every year.

Parasites are also responsible for staggering financial loss. Malaria, for example, is usually a chronic, debilitating, periodically disabling disease. In situations where it is prevalent the number of hours lost from productive labor multiplied by the number of malaria sufferers yields a figure that can be charged as loss in the manufacture of goods, in the production of crops, or in the earning of a gross national product.

FIG. 1-1.

"Nightsoil" is a logical use of human feces and urine. Here it is applied to a vegetable garden, a technique practiced in much of the world. Although sometimes controlled by government regulations it still serves as a limited means for distribution of eggs of some helminths and certain protozoan cysts.

Photograph by Robert E. Kuntz.

On the basis of estimates this figure is about 2 billion dollars annually. Nations that import goods from countries infected with malaria, schistosomiasis, hookworm, and many other parasitic diseases pay more for these products than they would had the products been produced without the burden of disease. Plant parasites further diminish the productive capacities of all countries.

At first glance it seems incongruous that the nations that suffer the most from disease are also the nations whose populations are undergoing the most rapid growth. The world's population has doubled three times in the past 200 years, and it will double again in the next 35 years, from 4 billion to 8 billion. During this time Latin America will add 400 million, and Asia will double its 1.6 billion to 3.2 billion, which together approximates the total current world population. The efforts at family planning are beginning to be noticed in several countries, especially those in which disease is minimal. But what can we say to a mother who wants to have seven children so that three can survive? A Johns Hopkins University study on family planning motivation confirms the importance of child survival to the sustained practice and acceptance of family planning.[3] The role of the parasitologist then, together with that of other medical disciplines, is to help achieve a lower death rate. However, it is imperative that this be matched with a concurrent lower birth rate. If not, we are faced with the "parasitologist's dilemma," that of sharply increasing a population that cannot be supported by the resources of the country. Dr. George Harrar, president of the Rockefeller Foundation, observed, "It would be a melancholy paradox if all the extraordinary social and technical advances that have been made were to bring us to the point where society's sole preoccupation would of necessity become survival rather than fulfillment." Harrar's paradox is already a fact for half the world. Parasitologists have a unique opportunity to break the deadly cycle by contributing to the global eradication of communicable diseases while making possible more efficient use of the earth's resources.

PARASITES OF DOMESTIC AND WILD ANIMALS

Both domestic and wild animals are subject to a wide variety of parasites that demand the attention of the parasitologist. Although wild animals are usually infected with several species of parasites, they seldom suffer massive deaths, or **epizootics,** because of the normal dispersal and territorialism of most species. However, domesticated animals are usually confined to pastures or pens year after year, often in great numbers, so that the parasite eggs, larvae, and cysts become extremely dense in the soil and the burden of adult parasites within each host becomes devastating.[1] For example, the protozoa known as the coccidia thrive under crowded conditions; they may cause up to 100% mortality in poultry flocks, 28% reduction in wool in sheep, and 15% reduction in weight of lambs.[7] In 1965 the U.S. Department of Agriculture estimated the annual loss in the United States as the result of coccidiosis of poultry alone at about $45 million. Many other examples can be given, some of which are discussed later in this book. Agriculturists, then, are forced to expend much money and energy in combating the phalanx of parasites that attack their animals. Thanks to the continuing efforts of parasitologists around the world, the identifications and life cycles of most parasites of domestic animals are well known. This knowledge, in turn, exposes weaknesses in the biology of these pests and suggests possible methods of control. Similarly, studies of the biochem-

istry of organisms continue to suggest modes of action for chemotherapeutic agents.

Less can be done to control parasites of wild animals. Although it is true that most wild animals tolerate their parasite burdens fairly well, the animals will succumb when crowded and suffering from malnutrition, just as will domestic animals and humans. For example, the range of the big horn sheep in Colorado has been reduced to a few small areas in the high mountains. The sheep are unable to stray from these areas because of human pressure. Consequently, lungworms have so increased in numbers that in some herds no lambs survive the first year of life. These herds seem destined for quick extinction unless a means for control of the parasites can be found in the near future.

A curious and tragic circumstance has resulted in the destruction of large game animals in Africa in recent years. These animals are heavily infected with species of *Trypanosoma,* a flagellate protozoan of the blood. The game animals tolerate infection well but function as **reservoirs** of infection for domestic animals, which quickly succumb to trypanosomiasis. One means of control employed is the complete destruction of the wild animal reservoirs themselves. Hence, the parasites of these animals are the indirect cause of their death. It is hoped that this parasitological quandary will be solved in time to save the magnificent wild animals.

Still another important aspect of animal parasitology is the transmission to humans of parasites normally found in wild and domestic animals. The resultant disease is called a **zoonosis.** Many zoonoses are rare and cause little harm, but some are more common and of prime importance to public health. An example is trichinosis, a serious disease caused by a minute nematode, *Trichinella spiralis* (Chapter 24). This worm exists in a **sylvatic** cycle that involves rodents and carnivores and in an **urban** or **domestic** cycle chiefly among rats and swine. People become infected when they enter either cycle, such as by eating undercooked bear or pork. Another zoonosis is echinococcosis, or hydatid disease, in which humans accidentally become infected with larval tapeworms when they ingest eggs from dog feces (Chapter 22). *Toxoplasma gondii,* which is normally a parasite of felines and rodents, is now known to cause many human birth defects (Chapter 8).

New zoonoses are being recognized from time to time. It is the obligation of the parasitologist to identify, understand, and suggest means of control of such diseases. The first step is always the proper identification and description of existing parasites so that other workers can recognize and refer to them by name in their work. Thousands of species of parasites of wild animals are still unknown and will occupy the energies of taxonomists for many years to come.

CAREERS IN PARASITOLOGY

It can be truly said that there is an area within parasitology to interest every biologist. The field is large and has so many approaches and subdivisions that anyone who is interested in biological research can find a lifetime career in parasitology. It is a satisfying career because one knows that each bit of progress made, however small, contributes to our knowledge of life and to the eventual conquering of disease. As in all scientific endeavor, every major breakthrough depends on many small contributions made, usually independently, by individuals around the world. Previously ignored opportunistic parasites suddenly have become major killers of acquired immune deficiency syndrome (AIDS) patients. Had their identifications and life cycles been better understood, it would have saved much expense and time in recognizing this complex disease.

The training required to prepare a parasitologist is rigorous. Modern researchers in parasitology are well grounded in physics, chemistry, and mathematics, as well as biology from the subcellular through the organismal and populational levels. Certainly they must be firmly grounded in medical entomology, histology, and basic pathology. Depending on their interests, they may require advanced work in physical chemistry, immunology, molecular biology, genetics, and systematics. Most parasitologists hold a Ph.D. or other doctoral degree, but significant contributions have been made by persons with a master's or bachelor's degree. Such intense training is understandable, since parasitologists must be familiar with the principles and practices that apply to over a million species of animals; in addition they need thorough knowledge of their fields of specialty. Once they have received their basic training, parasitologists continue to learn during the rest of their lives. Even after retirement, many remain active in research for the sheer joy of it. Parasitology indeed has something for everyone.

REFERENCES

1. Ershov, V.S. 1956. Parasitology and parasitic diseases of livestock. State Publishing House of Agricultural Literature, Moscow.
2. Grimes, P. 1980. Travelers are warned of increasing danger of malaria. Spread of the disease is blamed on ignorance and the failure of travel agents and tourist bureaus to caution clients. New York Times, May 11, p. 15XX.
3. Howard, L.M. 1971. The relevance of parasitology to the growth of nations. J. Parasitol. 57(sect. 2, part 5):143-147.
4. Keilin, D. 1925. On cytochrome, a respiratory pigment, common to animals, yeast and higher plants. Proc. R. Soc. Lond. (Biol) 98:312-339.
5. Le Riche, W.H. 1967. World incidence and prevalence of the major communicable diseases. In Health of mankind. Little, Brown & Co., Boston, pp. 1-42.
6. Peters, W. 1978. Medical aspects—comments and discussion II. In Taylor, A.E.R., and R. Muller, editors. The relevance of parasitology to human welfare today. Symposia of the British Society for Parasitology, vol. 16. Blackwell Scientific Publications Ltd., Oxford, Eng.
7. Pollack, H. 1968. Disease as a factor in the world food problem. Institute for Defense Analysis, Arlington, Va.
8. Skeels, M.R., et al. 1982. Intestinal parasitosis among Southeast Asian immigrants in New Mexico. Am. J. Public Health 72:57-59.
9. TRB. 1969. Sleep well. The New Republic 160(Mar. 6):6.

SUGGESTED READINGS

Anonymous. 1989. Careers in parasitology, medical zoology, tropical medicine. American Society of Parasitology, Kansas City, Mo.

Baer, J.G. 1971. Parasitology in the world today. J. Parasitol. 57(sect. 2, part 5):136-138.

Cheng, T.C. 1973. The future of parasitology: one person's view. Bios 44:163-171.

Hoekenga, M.T. 1983. The role of pharmaceuticals in the total health care of developing countries. Am. J. Trop. Med. Hyg. 32:437-446.

Mueller, J.F. 1961. From rags to riches, or the perils of a parasitologist. New Physician 272:46-50.

Taylor, A.E.R., and R. Muller, editors. 1978. The relevance of parasitology to human welfare today. Symposia of the British Society for Parasitology, vol. 16. Blackwell Scientific Publications Ltd, Oxford, Eng. (A good discussion of the current problems humans face in medical, veterinary, and agricultural parasitology.)

Tribe, H.T. 1980. Prospects for the biological control of plant parasitic nematodes. Parasitology 81:619-639.

CHAPTER 2

BASIC PRINCIPLES AND CONCEPTS

*The host is an island invaded by strangers with
different needs, different food requirements,
different localities in which to raise their progeny.*

TALIAFERRO

DEFINITIONS
Symbionts

The science of parasitology is largely a study of
symbiosis, especially the form known as parasitism.
Although some authors, especially in Europe, restrict
the term symbiosis to relationships wherein both partners benefit, we prefer to use the term in a wider
sense, as originally proposed by de Bary in 1879. We
consider any two organisms living in close association, commonly one living in or on the body of the
other, as symbiotic, as contrasted with "free living."
With this usage, the effect of one on the other, for example, beneficial or damaging, is not implied. Usually
the **symbionts** are of different species but not necessarily. The study of all aspects of symbiosis is called
symbiology.

We can then subdivide symbiosis into several categories based on the effects of the symbionts on each
other. It should be recognized that not all relationships
fit obviously into one category or another, since they
often overlap each other; furthermore, the exact relationship cannot be determined in some cases. Not all
authors agree on the definitions of these categories,
and some subdivide them further.

■ Phoresis

Phoresis exists when two symbionts are merely
"traveling together." Neither is physiologically dependent on the other. Usually one **phoront** is smaller than
the other and is mechanically carried about by its
larger companion (Fig. 2-1). Examples are bacteria on
the legs of a fly or fungous spores on the feet of a beetle.

■ Mutualism

In this relationship the partners are called **mutuals**
because both members benefit from the association.
Mutualism is usually obligatory, since in most cases
physiological dependence on one another has evolved
to such a degree that one mutual cannot survive without the other. A good example is the termite and its
intestinal protozoan fauna. Termites cannot digest cellulose fibers because they do not secrete the enzyme
cellulase. However, myriad flagellate protozoa, which
dwell within the gut of the termite, synthesize cellulase freely and are able to employ as nutrient the wood
eaten by the termites. The termite is nourished by the
fermentation products excreted by the protozoa. That
the protozoa are necessary to the termite can be shown
by defaunating the insects (killing the protozoa by
subjecting their hosts to elevated temperature or oxygen tension); the termites then die, even with plenty of
choice wood to eat. The protozoa benefit by living in
a stable, secure environment, constantly supplied with
food, and by being provided with a low-oxygen environment, since they are obligate anaerobes. The ter-
mite-flagellate association is the most often cited example of insect-microbe mutualism. A wide variety of
insects have bacteria or yeast-like organisms in their
guts or other organs, and these are physiologically
necessary for the insects in almost all cases studied,
apparently furnishing vitamins or other micronutri-
ents. Some insects even have specialized organs (**mycetomes**) where they "keep" their microbes, and the
symbionts are passed to the progeny transovarially.
What is apparently an evolutionary origin of a case of
mutualism has been discovered in the free-living spe-

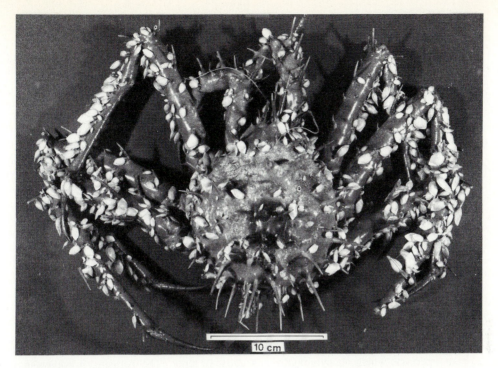

FIG. 2-1

Gooseneck barnacles (*Poecilasma kaempferi*) growing on the legs and carapace of a crab (*Neolithodes grimaldi*). This is an example of phoresis since the two species are merely "traveling together." However, the relationship could grade into commensalism; some advantages probably accrue to the barnacles.

From Williams, R., and J. Moyse. 1988. J. Crust. Biol. 8:177-186.

cies *Amoeba proteus*. A strain of *A. proteus* became infected with a parasitic bacteria, and over a period of several years the amebas became unable to survive without the bacteria.[25] When the bacteria were killed by culturing the amebas at a slightly elevated temperature, the amebas died shortly thereafter, unless reinfected with the bacteria by microinjection.[28] A similar case is known in which a type of the ciliate protozoan *Euplotes* requires the presence of endosymbiotic bacteria to divide, whereas closely related types do not require the bacteria.[19] The nature of the contribution to the ameba or the ciliate by the bacteria is not known.

One form of mutualism that is not obligatory is usually called **cleaning symbiosis.** In this instance certain animals, called cleaners, remove ectoparasites, injured tissues, fungi, and other organisms from a cooperating host. For example, often one or several cleaners establish cleaning stations, and fish to be cleaned visit these locations repeatedly to enjoy the services of the cleaners. The fish may remain immobile at the cleaning station while the cleaners graze its external surface and enter its mouth and branchial cavity with impunity. Evidence exists that such associations may be in fact obligatory; when all cleaners are carefully removed from a particular area of reef, for example, all the other fish leave too. Some terrestrial cleaning associations are known; two examples are the cleaning of a crocodile's mouth by the Egyptian plover and the cleaning of the rhinoceros by tick birds. Some excellent accounts of a wide variety of mutualistic and related associations are found in the texts edited by Henry[20] and Cheng.[6]

■ **Commensalism**

When one symbiont, the **commensal,** benefits from its relationship with the **host,** but the host neither benefits nor is harmed, the condition is known as commensalism. The term means "eating at the same table," and most, but certainly not all, examples of commensalism involve the commensal feeding on unwanted or unusable food captured by the host.

Pilot fish (*Naucrates*) and remoras (Echeneidae) are often cited as examples of commensals. A remora is a slender fish whose dorsal fin is modified into an adhesive organ, with which it attaches to large fish, turtles, and even submarines! It gets free rides this way and perhaps some crumbs left over when its host makes its kill, but in no way does it harm the host or rob it of food. In fact it has now been found that some species of remoras perform important cleaning functions for their host, feeding at least partially on the host's parasitic copepods.[7] Thus remoras, the "classic" commensals, are often mutuals. Depending on the species of echeneid and its degree of specialization, the association may be more or less **facultative,** the remora being able to leave its host at will. Other species are more specialized; for example, *Remoropsis pallida,* at certain stages in its life, lives entirely in the branchial chambers of marlins.

An example of an obligate commensal is *Entamoeba gingivalis,* an ameba that lives in the mouth of humans. Here it feeds on bacteria, food particles, and dead epithelial cells but never harms the tissues of its host. It cannot live anywhere else and in fact has no cyst stage to withstand life outside the buccal cavity; it is transmitted from person to person by direct contact. Humans harbor several species of commensal protozoa; for this reason parasitologists must be able to distinguish between commensal and parasitic species. It is not always easy to determine whether a symbiont is harming its host; adult tapeworms universally are referred to as parasites, yet in some cases they have no known ill effect and might possibly be regarded as commensals.[23]

■ Parasitism

When a symbiont actually harms its host or in some sense lives at the expense of the host, it is then a **parasite.** It may harm its host in any number of ways: by mechanical injury, such as boring a hole into it; by eating or digesting and absorbing its tissues; by poisoning the host with toxic metabolic products; or simply by robbing the host of nutrition. Most parasites inflict a combination of these conditions on their hosts.

If a parasite lives on the surface of its host, it is called an **ectoparasite;** if internal, it is an **endoparasite.** Most parasites are **obligate parasites;** that is, they spend at least a part of their lives as parasites to survive and complete their life cycles. However, many obligate parasites have free-living stages outside any host, including some periods of time in the external environment within a protective egg shell or cyst. **Facultative parasites** are not normally parasitic but can become so, at least for a time, when they are accidentally eaten or enter a wound or other body orifice. Two examples are certain free-living amebas, such as *Naegleria* (p. 110), and free-living nematodes belonging to the genus *Micronema.*[14] Infection of humans with either of these is extremely serious and usually fatal.

When a parasite enters or attaches to the body of a species of host different from its normal one, it is called an **accidental,** or **incidental, parasite.** For instance, it is common for nematodes, normally parasitic in insects, to live for a short time in the intestines of birds or for a rodent flea to bite a dog or human. Accidental parasites usually are unable to stay long on, or live long in, the wrong host. While there, however, they may add a good deal of confusion to the life of a parasite taxonomist! More important, some accidental parasites are extremely pathogenic to their unfamiliar hosts (see *Baylisascaris, Toxocara,* in Chapter 28).

Some parasites live their entire adult lives within or on their hosts and may be called **permanent parasites,** whereas a **temporary,** or **intermittent, parasite,** such as a mosquito or bedbug, only feeds on the host and then leaves. Temporary parasites are often referred to as **micropredators,** in recognition of the fact that they usually "prey" on several different hosts (or the same host at several discrete times). Indeed, predation and parasitism are conceptually similar in that both the parasite and the predator live at the expense of the host or prey. The parasite, however, normally does not kill its host, is small relative to the size of the host, has only one host (or one host at each stage in its life cycle), and is symbiotic. The predator kills its prey, is large relative to the prey, has numerous prey, and is not symbiotic.

Of course, a parasite sometimes kills its host, but it is clearly not a selective advantage, since the life of the parasite also is thereby terminated. To produce few pathological conditions is often regarded as a mark of a well-adapted parasite. Yet, some parasites require pathological changes in host tissue in order to complete their life cycles. The term **parasitoid** is given to a large number of insects whose immature stages feed on their host's body, usually another arthropod, but finally kill the host during or after completion of the development of the parasitoid. They resemble predators in that they kill their hosts (prey) and parasites in that they require only one host (Chapter 40).

Hosts

Hosts also are placed in different categories. The host in which a parasite reaches sexual maturity and reproduction is termed the **definitive host.** If no sexual reproduction occurs in the life of the parasite, such as an ameba or trypanosome, the host believed to be most important is arbitrarily called the definitive host. An **intermediate host** is one in which some development of the parasite occurs but in which it does not reach maturity. Hence, in the case of the malaria organisms, *Plasmodium* spp., the mosquito is the definitive host, and humans or other vertebrates are the intermediate hosts (see Chapter 9).

When a parasite enters the body of a host and does not undergo any development but continues to stay alive and be infective to a definitive host, the host is called a **paratenic,** or **transport, host.** Paratenic hosts are often useful or even necessary for completion of the life cycle of the parasite, since they may bridge an ecological gap between the intermediate and definitive hosts. An owl may be the definitive host of a thorny-headed worm (see Chapter 32), whereas an insect is the intermediate host. The parasite might have little chance of being eaten by the owl while in the insect, but when a shrew eats the insect and the larval worm encysts in the shrew's mesentery, the parasite stands a better chance of a happy life in the alimentary tract of the owl. Furthermore, the shrew might accumulate large numbers of larvae before it becomes an owl's dinner, thereby increasing the chances of both sexes of a dioecious parasite sharing a common host.

Some parasites can live and develop normally in only one or two species of host. These exhibit high **host specificity.** Others have low host specificity or something in between. For example, the pork tapeworm, *Taenia solium,* apparently can mature only in humans, so it has absolute host specificity, whereas the trichina worm, *Trichinella spiralis,* seems to be able to mature in almost any warm-blooded vertebrate. Any animal, including humans, that harbors an infection that can be transmitted to people is called a **reservoir host,** even if the animal is a normal host of the parasite; it is a reservoir for a zoonotic infection of people. Examples are the rat with the trichina and dogs, cats, and armadillos with the agent of Chagas' disease, *Trypanosoma cruzi* (see Chapter 5).

Finally, many parasites host other parasites, a condition known as **hyperparasitism.** Examples are *Plasmodium* spp. in a mosquito, a tapeworm juvenile in a flea, an ameba within an opalinid protozoan, a monogenetic trematode *(Udonella)* on a copepod parasite of fish, and the many insect hyperparasites of primary parasites of other insects.

In nearly all cases of parasitism the host is a different species from the parasite. Exceptions occur, however, as in the case of the nematode parasite of rats, *Trichosomoides,* where the male lives its mature life within the uterus of the female worm, obtaining its nourishment from her tissues. This is also the case with *Gyrinicola japonicus,* a nematode parasite of frogs, and with the echiurid worm, *Bonellia viridis,* the female of which is free living. An even stranger relationship has evolved in some species of anglerfish in which the male bites the skin of the female and

sucks her blood and tissue fluids for nourishment. Eventually they grow together, and he shares her bloodstream! *Syngamus trachea,* a common hookworm of birds, is known to pierce the body wall and feed on the fluids of other worms of the same species. This might better be called predation, but specimens are found with healed wounds and so were not killed by the attack.

The ultimate in intraspecific parasitism would be that of a tumor. Here self (tumor) parasitizes self (host), often very much to the detriment of the host, such as in a malignancy, or with only minor inconvenience to the host, such as with a wart. The tumor may be stimulated by a virus, but the tumor itself harms the host.

Modern authors increasingly have realized the utility of de Bary's distinction, but for many people symbiosis continues to mean mutualism. Nevertheless, symbiosis has evolutionary, morphological, ecological, and physiological implications, whether mutualistic, commensalistic, or parasitic, that a free-living habitus does not.

We must stress that definitions are always arbitrary, and when we construct pigeonholes to receive descriptions of situations in the real world, we must not be dismayed to find one or another situation that does not quite fit our assignment. Thus we may cite symbiotic associations with greater or lesser dependence, of greater or lesser duration, or with some grading over into free living. Certainly many symbiotic associations cannot be specified with certainty as to the effects on the hosts; an apparent case of commensalism may have damaging effects on the host that are subtle, have not been observed, or are present only in certain circumstances. Conversely, a case of assumed parasitism may, on closer inspection, turn out to be commensalism. Definitions are necessary to communication; therefore we must strive for definitions that are concise, meaningful, and, above all, useful.

EVOLUTION OF PARASITISM

The question of how parasitism originates or has originated is impossible to answer definitely. However, indirect indications that are impossible to ignore suggest several factors involved in adaptations to a parasitic mode of life. Particularly important among these is the concept of **preadaptation.** A preadapted organism has a set of characteristics or adaptations that have evolved as the result of the selective pressures present in its environment and that coincidentally give it the *potential* of surviving in a different envionment. If the organism finds itself in an environment for which it is preadapted, it may successfully survive and reproduce there and give rise to generations of progeny that become progressively more adapted to the environment. Preadaptations may be morphological or physiological and may also allow such organisms to pass from a generalized type of environment to a more specialized one. For example, arthropods of several types may have fed on carrion that was also found and eaten by primitive humans. Any arthropod eaten along with the carrion may well have been digested, but if the pinworm nematodes in its hindgut were preadapted to survive at a higher temperature, they might have then colonized the hindgut or cecum of the human. This isolated population could no longer share the gene pool with its arthropod-dwelling relatives and would thus be free to undergo genetic drift and certainly would experience different selective pressures. Although the likelihood of such an event happening seems chancy, at best, it has been such a successful process that nearly every species of living organism has several kinds of parasites peculiar to it. The result is more species of parasites than of nonparasites.

Of course, the immense number of parasites today is the result of other processes as well. Once a parasite is in the relatively stable environment of a host, selective pressure by the external environment is decreased. Genetic drift is favored in such circumstances. Both beneficial and nonharmful mutations could accumulate in the gene pool, preadapting the species to invade a new host, organ, or tissue successfully should the opportunity arise. It is also conceivable that a parasite could become nonparasitic, but this probably never occurs, because the organism will have become too specialized, and evolution generally does not progress from the more specialized to the more generalized. The parasite could not compete with well-adjusted organisms already occupying external niches.

The probability that two organisms will establish a host-parasite relationship depends largely on ecological and behavioral factors. For a prospective parasite to succeed, it must come into frequent contact with a potential host. The behavior of both must favor such a contact. Ultimately the distribution of closely related parasites reflects the opportunities available to them when they became parasitic. This can lead to similar species parasitizing phylogenetically unrelated hosts. For instance, the nematode genus *Molineus* has species in primates, insectivores, rodents, and carnivores, the ancestors of which were available to the ancestors of the worms at the time of their radiation.

Undoubtedly parasitism arose in many different ways at many different times. Ectoparasites may well have evolved from nonparasitic omnivores or predators or from ancestors that sought body secretions from nearby animals. A hypothetical example of the latter is seen in certain moths of the family Noctuidae in Southeast Asia. Some species feed on plant secretions, whereas others, such as *Calpe thalicteri*, can pierce ripe fruit to suck the sweet juices inside.[4] One species, *Lobocraspis griseifusa*, gathers around the eyes of a large mammal, such as a cow, and sucks tears from its orbits (Fig. 2-2). It seems a natural step from such lachryphagous species to the skin-piercing bloodsucking noctuid *Calpe eustrigata* (Fig. 2-3). Other ectoparasitic arthropods may have followed a similar road to parasitism, or they may have accomplished the transition in a single step.

Endoparasites probably found entrance into a host easiest through the alimentary tract. Cyst-forming protozoa and nematodes protected by a tough cuticle most readily survived maceration and gastric juices. Species adapted to endocommensalism might have found it possible to go a step further and become parasitic, but once again, it was probably a one-step process in many cases. We will not discuss individual groups of parasites; the interested reader is referred to the useful work edited by Taylor.[39] Parasites in the fossil record are the subject of an interesting review by Morris.[33]

Whatever the mode of attack by incipient parasites, they are confronted by numerous barriers that protect the host from such incursions. The temperature of the host places severe restrictions on potential parasites of birds and mammals, since only those preadapted to survive the higher body temperatures can become established. The size of the parasite can prevent its effective entrance into a smaller host or one that is a filter feeder. A mechanism that enables the invader to maintain its position in the gut of the host would seem to be a necessity in most cases. This is provided by suckers or hooks in some species or by a strong swimming action, as with many nematodes.

When free-living organisms enter the gut or tissues of a prospective host, they move into a medium of higher osmotic pressure. Only those which can survive this challenge are successful. The high hydrogen ion concentration of the stomach is a formidable barrier to invasion. However, not only do hordes of parasites survive this pitfall but also some live, mature, and reproduce bathed in the hydrochloric acid medium of the vertebrate stomach. The digestive enzymes of the host are sufficient to destroy most would-be colonists, but successful endoparasites secrete a tegument as fast as it is digested or have some other specialization that protects them.

Oxygen is in short supply in the digestive tract, but immediately adjacent to the gut mucosa oxygen concentration may approach that of the blood. Oxygen is

FIG. 2-2

Eight *Lobocraspis griseifusa* (Lepidoptera, Noctuidae) suck tears from the eye of a banteng, *Bos banteng*, in northern Thailand. Note the proboscis of each moth extended to feed at the eye perimeter.

Photograph by Hans Bänziger.

FIG. 2-3

A noctuid moth, *Calpe eustrigata*, piercing the skin and sucking the blood of a Malayan tapir. This is the only known bloodsucking moth.

Photograph by Hans Bänziger.

much lower in the gut lumen, and in deep tissues of large parasites or in areas of heavy bacterial growth, such as the colon, oxygen is consumed faster than it can be diffused, and its concentration approaches zero. A facultative anaerobe, such as a saprophytic protozoan or a nematode, would not find such a low oxygen concentration to be a major barrier; but organisms that rely on the Krebs cycle and classical oxidative phosphorylation could not survive. Hence endoparasites have either been derived from facultatively anaerobic ancestors or have evolved metabolic adaptations to tolerate low oxygen tension. Interestingly, to a considerable degree this generalization applies also to parasites in the blood and other tissues. Whatever their origin, metabolic adaptations to low oxygen have involved a shortening or modification of oxidative metabolism to the point where glycolysis assumes major importance in energy derivation. End products of the parasite's energy metabolism, such as lactic, pyruvic, and succinic acids, excreted by many parasites, can then be catabolized by the host in its own energy metabolism.

A major barrier to incursion by parasites is the immune response of the host. This will be discussed in more detail in Chapter 3.

After its establishment as a parasite, the organism is subjected to the strong selective pressures composed of the conditions in the host's body. The host continues to evolve, being selected for by the pressures in its environment. Thus each member—the host and the parasite—evolves independently, but also the partners evolve together as a unit. In time a given parasite may be so adapted to its host species that it cannot mature and survive in any other. Such host specificity may be nearly absolute, or it may be slight, depending on the degree of adaptation by the symbionts. As each host moves through time and accumulates its infinite series of minute genetic changes, its parasites move with it, adapting to their host's mutations or else perishing.

When considering the morphological changes associated with parasitism, we are confronted with two related but opposite phenomena, those of progressive specialization and regressive elimination of specializations. In the former we find structures such as the hooks of tapeworms and the Acanthocephala, which cannot be other than adaptations to parasitism. Other examples of favorable morphological adaptations are acetabula, tribocytic organs, and tegumentary spines in digenetic trematodes; posterior clamps and suckers in monogeneans; and teeth and cutting plates in nematodes.

In contrast, structures may be lost or diminished in parasitic animals because they have no selective value. The digestive system is completely absent in tapeworms and thorny-headed worms, although they may have evolved from fully equipped ancestors. Fleas and lice have lost their wings, and in some cases their eyes, and in parasitic nematodes the sensory structures called amphids are much reduced in size. In fact the loss of sense organs is a common feature of parasitism, although specialized receptors are still found on all metazoan species.

As parasite and host coevolve, it appears that a decrease in pathogenesis would be a selective advantage for the parasite. Common sense tells us that a longer life for the host maximizes reproductive potential for the parasite. Thus it has become dogma that the least pathogenic parasite is the most successful. The dogma has been questioned by Anderson and May,[3] who argue that lack of virulence does not necessarily coincide with maximal production of progeny. They concluded that "the co-evolutionary trajectory followed by any particular host-parasite association will ultimately depend on the way the virulence and the production of transmission stages of the parasite are linked together: depending on the specifics of this linkage, the co-evolutionary course can be toward essentially zero virulence, or to very high virulence, or to some intermediate grade." There is no doubt that an important adaptation for parasitism is an increase in reproductive potential, discussed in more detail in the next section. This adaptation compensates for the massive mortality of offspring while enacting their complex life cycles.

REPRODUCTIVE POTENTIAL OF PARASITES

All organisms must reproduce successfully or else they will join the legions of the extinct. Most parasites, especially endoparasites, are faced with special problems, since they must survive not only the concerted defense efforts of the host but also the danger-fraught interludes between hosts. All parasites must produce offspring that can infect the next host, and many species have complex life cycles that involve a series of intermediate hosts, as well as a definitive host. When one considers that chance governs the successful completion of much of the life cycle of any given parasite, it becomes apparent that the odds against success are nearly overwhelming. Parasites, more than most other groups of organisms, beat the odds by producing enormous numbers of offspring. Most of these perish, but enough have survived to maintain extant species.

Different groups have evolved various methods to solve the problem of mass reproduction. For example, **multiple fission,** or **schizogony,** is found in many parasitic protozoa. In this case, instead of a single mitotic division that produces two daughter cells, the nucleus redivides several times before the cytoplasm is divided among many daughter cells. These daughter cells are produced simultaneously, flooding the microenvironment with so many individuals that one or more is likely to infect the next host and initiate the next stage of development. A more detailed discussion of schizogony and similar processes and illustrations of their functions are given in the chapters on protozoa.

When an individual possesses both male and female reproductive systems (**hermaphroditism),** it has solved the problem of finding a mate. Many tapeworms and flukes fertilize their own eggs (selfing); this method, although not likely to produce anything original, certainly is efficient in guaranteeing offspring. However, some hermaphrodites must rely on cross-fertilization, whereas others can swing either way.

Most tapeworms undergo a type of continuous replication of reproductive organs, called **strobilation,** in which segments differentiate at a zone near the holdfast organ, or **scolex.** The resulting chain of segments, or **proglottids,** consists of a linear series of units, in most cases each with functioning sets of male and female reproductive systems. Thus, instead of an animal with a single set of reproductive systems, a veritable factory of reproduction exists. The ultimate must be *Hexagonoporus,* a tapeworm of whales, which consists of about 45,000 proglottids, each with 5 to 14 sets of male and female reproductive systems. The reproductive potential of this 100-foot monster is staggering. However, there are few whales, and the ocean is large; thus the chances of survival of even this species remain slim.

One of the most common means of increasing the biotic potential of species is the production of numerous eggs. Thus the common rat tapeworm, *Hymenolepis diminuta,* produces up to 250,000 eggs per day or 100×10^6 during the life of its host. If all of these could reach maturity in new hosts, they would represent more than 20 tons of tapeworm tissue. A female *Ascaris* will produce more than 200,000 eggs per day for several months, and the filarial nematode *Wuchereria* produces several million offspring during her lifetime.

Asexual reproduction is common in several groups of parasites. Simple mitotic fission is especially important among the protozoa and, indeed, is the only form of reproduction in the Sarcomastigophora, except in foraminiferans. Rapid fission often results in millions of offspring in a matter of only a few days.

More complex forms of asexual reproduction are found in the flatworms. The juveniles (**metacestodes**) of several tapeworms are capable of external or internal budding of more metacestodes. The cysticercus juvenile of *Taenia crassiceps,* for instance, can bud off as many as a hundred small bladderworms while in the abdominal cavity of the mouse intermediate host. Each new metacestode develops a scolex and neck, and when the mouse is eaten by a carnivore, each develops into an adult cestode. The hydatid metacestode of the tapeworm *Echinococcus granulosus* is capable of budding off hundreds of thousands of new scolices within itself. When such a structure is eaten by a dog, vast numbers of adult tapeworms are produced.

Perhaps the most bizarre and astonishing asexual reproduction in all zoology is found among the digenetic trematodes, a large, successful group of parasites commonly called the "flukes." These animals undergo a series of reproductive stages, each of which produces the next generation of larvae, rapidly building up a huge population (see Chapter 17). Although there are many variations, the basic scheme is egg (miracidium) → sporocyst → redia → cercaria → adult. The microscopic miracidium that hatches from an egg enters the first intermediate host, usually a mollusc, and becomes a saclike sporocyst. By a form of internal budding[41] (which still is not well understood), the next larval stage, the redia, develops within it. Several rediae are usually produced by a single sporocyst. The rediae break out of the sporocyst and grow, feeding actively on host tissues. Several cercariae form within each redia by a process apparently similar to the formation of the redia. The cercariae pass out of the redia and the intermediate host to seek their fortunes elsewhere. There may be more than one generation of

sporocysts or rediae, or neither of these may be present. The point is that each egg has the potential of producing hundreds of offspring by these successive asexual stages. When one observes that most flukes give birth to thousands of eggs each day, the biotic potential of these worms is staggering. Parasites are, indeed, masters of reproduction.

PARASITE DISTRIBUTION WITHIN THE HOST

Species of parasites have adapted to virtually every tissue, organ, and space in the body. Parasites that live within tissues are called **histozoic,** whereas those inhabiting the lumen of the intestine or other hollow organs are said to be **coelozoic.** Most endoparasites of vertebrates live in the digestive system. This cannot be considered a single niche since many different environments exist between the mouth and the anus. Furthermore, digestive systems vary greatly between species (compare the stomachs of human and ox) and even between different stages in the life of the host, such as a tadpole and a frog. (For a review of variations between vertebrate intestinal tracts see Crompton.[9]) Even a given level along the length of the alimentary tract cannot be considered a single niche because subtle differences occur in oxygen and carbon dioxide tension, pH, and other chemical and physical factors between the mucosa and the center of the lumen. Such differences occur even between the tip of a villus and its base, making at least two different niches available for colonization by parasites of suitable sizes. Obviously, then, when two species of parasites are found in the same region of the intestine, one cannot state that they are occupying the same niche, since although in proximity to one another, they may be in entirely different microenvironments. Thus Schad[35] found eight species of the nematode *Tachygonetria* in the large intestine of the turtle *Testudo graeca,* all apparently occupying different habitats. Obviously, though, one large parasite may occupy several microenvironments simultaneously.

If two species cannot occupy the same niche simultaneously, there must be *competition* between them when they enter a host. As is usually the case in nature, the species that is first established normally will not be displaced by the intruder. How it manages to prevent establishment by the interloper is not well understood.

The principles outlined previously apply to all areas of the host's body, not just the alimentary tract. Adult parasites are found in blood, skin, nervous tissues, and virtually all other parts of the body; and juvenile stages often undergo elaborate migrations through distant regions of the body before arriving at their definitive sites. Parasites that enter a host in which they are incapable of maturing often wander about until they die, or become dormant for long periods of time within the accidental host's tissues. It has been said that if a host were infected with all the parasites capable of infecting it and the host tissues were then removed, leaving only the parasites, the host could still be recognized! Although this may be an exaggeration, it is only slightly so. Consider the human eye, for example, an organ not particularly suited to infection by parasites. However, the retina may be infected by the protozoan *Toxoplasma gondii* and juveniles of the nematode *Onchocerca volvulus;* the chamber may harbor the bladderworm metacestodes of the tapeworms *Taenia solium, T. crassiceps, T. multiceps,* or *Echinococcus granulosus;* the conjunctiva may host a wandering nematode, *Loa loa;* and the orbit may be the home of nematodes of the genus *Thelazia.* Parasitologically the vertebrate body can be considered as a great mass of ecological niches, which have been colonized by a great variety of parasite species.

ECOLOGY OF PARASITIC INFECTIONS

The general principles of ecology apply equally well to parasitic organisms as to others, but we will mention a few ecological topics of particular interest to parasitologists. Two different although related approaches to the ecology of parasites are epidemiology and population ecology.

Epidemiology

Epidemiology is the study of the factors affecting the transmission and distribution of any disease entity, one of the most important of which is the presence and characteristics of the **vector.** A vector is the means of transmission of a disease organism from one host to another. Thus water may be the vector for *Entamoeba* species, aerosol sputum droplets for tuberculosis, and wind for the fungal disease coccidioidomycosis. Mosquitoes, flies, ticks, and other arthropods are vectors for many parasitic diseases; their biology and habits must be well understood before prevention and control of the diseases are possible. Intermediate hosts of parasites may be their vectors. Another important consideration in the epidemiology of an infection is the presence of reservoir hosts. The concept of the reservoir commonly, although not always, is used with reference to parasitic infections of humans. The reservoir is the host that maintains the disease in nature and provides a source for infection of humans. For example, rodents are reservoirs of oriental sore and kala-azar, and the house mouse is a reservoir of

the dwarf tapeworm, *Vampirolepis nana.* Other extremely important epidemiological factors in parasitic infections include the habits and sanitary levels of the hosts and the density of their population. A renewed attempt at developing safe water supplies on a global scale is expected to yield a rich harvest in human health.[5]

The behavior of the host should not be underestimated in the epidemiology of parasitic diseases. For instance, much of human suffering from parasites is brought on by patterns, and sometimes quirks, of human behavior.[16]

Between World Wars I and II it was noted by the Russian school of Pavlovsky that certain parasitic diseases occur in certain ecosystems and not in others and that factors making up these ecosystems can be categorized such that they can be recognized wherever they are encountered. Thus, each disease has a natural focus, or **nidus.** Discovery of this *natural nidality of infection* was a landmark in the history of parasitology because it enabled the epidemiologist to recognize "landscapes" where certain diseases could be expected to exist or, equally, where the possibility of their presence could be eliminated. Such **landscape epidemiology** requires thorough knowledge of all parameters that influence the infection, such as climate, plant and animal populations and their densities, geology, and human activities within the nidus. This holistic approach is best applied to parasitic diseases of wild and domestic animals and to the **zoonoses,** diseases of animals transmissable to humans. However, the principles of landscape epidemiology can be applied equally to whipworm infections in mental institutions or *Giardia lamblia* outbreaks in swank vacation communities.

Recent developments in biotechnology, such as the use of monoclonal antibodies and recombinant DNA, have resulted in new methods of detection of certain parasitic diseases. These new methods are becoming important in epidemiological studies on prevalence and transmission of parasites.[37,40]

Population ecology

Application of the methods of modern **population ecology** helps us come to grips with the epidemiology of infections on more precise terms. In population ecology we seek to explain the numbers of organisms that occur in natural populations. Thus we need to understand the nature of the changes in numbers and relate these to biotic and abiotic factors in the environment,[12] that is, how the numbers in a population are regulated. Although many of the problems in understanding regulation of free-living populations also are pertinent to populations of parasites, the nature of parasitism presents additional difficulties in understanding regulatory phenomena. Among these are the fact that the environment (host) of the parasite is alive and can respond with defense reactions and that in some circumstances a parasite can eliminate (kill) its own environment, preventing further recruitment. Even the definition of "population" is not so straightforward in parasites as in free-living organisms. Whether we consider all parasites in a single host as the population or all members of the species in all hosts in the ecosystem, the population will certainly affect our analysis. Esch et al.[11] designated the former as the parasite **infrapopulation** and the latter as the **suprapopulation.** A variety of factors may influence the regulation of parasite infrapopulations: host diet, age, sex and sexual maturity, behavior, environmental temperature, and dispersion of the parasites among available hosts. Parasite infrapopulations are characteristically not dispersed randomly among hosts, but rather a minority of hosts will harbor a majority of parasites. This phenomenon is described as **overdispersion,** and complete overdispersion would occur if all of the parasites were habored by a single host individual. Complete **underdispersion** would exist if all hosts harbored an equal number of parasites, and the midpoint of the dispersion spectrum would be a completely random distribution of parasites in the host population.[2] Overdispersion has been shown in numerous studies of parasite prevalence. For example, a study of several nematode parasites in villages in Iran showed that 1% to 3% of the people carried 11%, 16%, 30%, 38%, or 84% of all worms collected.[8] Kennedy[26] studied the population dynamics of *Tylodelphys podicipina,* an eye fluke of fish, during the years after the fluke was introduced into a lake. He found that initially, when infection levels were low, the dispersion was random, but overdispersion became marked when the parasite population in the lake increased.

A concept useful in understanding the regulation of suprapopulations is that of **r** and **K strategies.** Although sometimes controversial, this concept has been applied to parasites by several authors in recent years.[12,13,24] "Strategy" indicates a genetically fixed adaptation or characteristic, an outcome of selective pressures in the environment. Therefore, r-strategists and K-strategists are said to be r-selected and K-selected, respectively. Factors selecting for r-strategists are unstable, variable environmental conditions, whereas those for K-strategists are environmental conditions that are relatively stable over a period of time. Species of r-strategists are characterized by high fecundity, high (often catastrophic) mortality, short life span, and population sizes variable in time, usually well below the carrying capacity of the

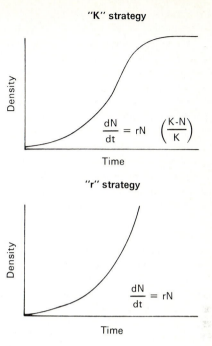

"K" strategy

$$\frac{dN}{dt} = rN \left(\frac{K-N}{K} \right)$$

"r" strategy

$$\frac{dN}{dt} = rN$$

FIG. 2-4

Population growth curves for a K-strategist and an r-strategist. In essence the equation says that the rate of population increase $\left(\frac{dN}{dt} \right)$ is equal to the maximal growth rate times the number in the population (rN) times the degree of realization of the maximal rate $\left(\frac{K-N}{K} \right)$. *K* represents the upper asymptote, or the carrying capacity of the environment. The population growth curve of a K-strategist is S shaped and is described by a logistic equation. The growth curve of the r-strategist is the exponential phase of the logistic equation, and *r* represents the intrinsic rate of natural increase. The population does not reach the carrying capacity of the environment and must be controlled by a high, often calamitous, mortality rate at some point in the population growth.

From Esch, G.W., T.C. Hazen, and J.M. Aho. 1977. In Esch, G.W., editor. Regulation of parasite populations. Academic Press, Inc., New York.

environment.[34] K-strategists are characterized by the opposite characteristics. Populations of r-strategists are controlled by density-independent factors, and populations of K-strategists are controlled by density-dependent factors and tend to be resource limited. Idealized population growth curves are shown in Fig. 2-4. An r-K continuum exists, and particular species are often not at the extremes of the continuum but are somewhere between.[34] Furthermore, the concept is a relative one; that is, species A may be an r-strategist compared with species B but a K-strategist with respect to species C.[12] A good example would be the digenetic trematode as an r-strategist compared with a free-living flatworm. The typical digenean has a high fecundity, including reproduction in an intermediate host, but a high mortality rate and a highly variable environment (as many as three or more different hosts, plus two free-swimming sojourns in water to complete its life cycle). Esch et al.[12] concluded that, on balance, most parasites are r-strategists. On the other hand, Keymer[27] discussed possible density-dependent factors in control of helminth parasites, such as parasite mortality and fecundity, overdispersion, parasite-induced host mortality, and transmission by predator-prey links.

Dobson[10] described two kinds of competition among parasites that affect their population dynamics: (1) in **exploitation competition** there is a joint exploitation of a limiting resource within a host species by two or more parasite species; and (2) in **interference competition** antagonistic mechanisms are used by one species to reduce survival of another or to displace it from a preferred site of attachment.

Parasite-induced behavioral changes in a host population may have a direct impact on parasite populations, either to the benefit or detriment of the parasite.[30,31] For example, the acanthocephalan *Polymorphis marilis* often castrates its crustacean intermediate host, *Gammarus lacustris,* but only in the spring, because the parasite's development goes into diapause in cold winter water. In response, the crustacean has altered its reproduction time to late summer and fall.

On the other hand, many examples are known where changes in intermediate host behavior increase parasite success rate. Moore[32] has shown that when isopod crustaceans are infected with the acanthocephalan *Plagiorhynchus cylindraceus* they lose their light-avoiding behavior and become more available to the predatory bird that is the parasite's definitive host. Many other examples are described in the following chapters.

Construction of **mathematical models** to describe population dynamics can be valuable, both to predict population behavior over periods of time and the outcome of artificial manipulations of the populations and to discover heretofore unknown biological principles through extrapolations or combinations of models.[21] An interesting example of the latter is that of Anderson,[1] whose model suggests that parasitic species that cause little harm or stress to the host are the most effective regulators of host population growth. If a parasite is too harmful, it will fail to regulate the host population when it causes the death of too many subpopulations of parasites within individual hosts, which, incidentally, supports the dogma mentioned

earlier that the most successful parasite is one that causes little harm to its host. A number of examples now show the predictive value of mathematical models. In one of the earliest uses of models in parasitology, Hairston[17,18] examined the population dynamics of *Schistosoma japonicum* in the Philippines. *S. japonicum* is an important parasite of humans; but it also uses dogs, pigs, and field rats as definitive hosts. Hairston's model suggests that rats alone could support the suprapopulation of *S. japonicum* because of high rat population and high rates of rat-snail contact. Thus, if all infections of humans were removed (cured) at once, only a few years would be required for the infrapopulations in humans to reach the previous level. Gettinby's model[15] predicts that control of *Fasciola hepatica,* an economically important trematode parasite of sheep and cattle, would be more effective by killing the intermediate hosts with a molluscicide than by treating the definitive hosts with an anthelmintic. Smith[38] provided a model explaining density-dependent mechanisms in the regulation of *Fasciola hepatica* populations in sheep.

Holmes et al.[22] modeled the population dynamics of an acanthocephalan, *Metechinorhynchus salmonis,* parasitizing a number of fish species in an Alberta lake. They concluded that the parasite population could be regulated through a mechanism operating on infrapopulations in only one of several species of hosts. The host species would neither have to be the one in which the parasite was most abundant, nor even the one through which there was the greatest flow of parasites. The only requirement was that the combined flow through the other host species not be adequate to maintain the suprapopulation.

Ecological terms in parasitology

Various authors have been inconsistent in their usage of ecological terms, leading to a certain amount of confusion. In an effort to standardize these usages the American Society of Parasitologists has adopted the following terms and definitions[29]:

1. Term: **Prevalence** (usually expressed as a percentage).

   Concept or definition: Number of individuals of a host species infected with a particular parasite species ÷ Number of hosts examined.

   Remarks: **Incidence** is commonly misused for this concept. In Russian and other eastern European literature the term **extensity** is used for this concept.

2. Term: **Incidence** (usually expressed as a percentage).

   Concept or definition: Number of new cases of a disease or infection appearing in a population within a given period of time ÷ Number of uninfected individuals in the population at the beginning of the time period.

Remarks: **Incidence** is not likely to be applicable when studying populations of feral animals because the number of uninfected individuals at the beginning of the time period is rarely known.

3. Term: **Intensity** (commonly expressed as a numerical range).

Concept or definition: Number of individuals (determined directly or indirectly) of a particular parasite species in each infected host (i.e., in an **infrapopulation**—see Term 7) in a sample.

4. Term: **Mean intensity.**

Concept or definition: Total number of individuals of a particular parasite species in a sample of a host species ÷ Number of infected individuals of the host species in the sample (= Mean number of individuals of a particular parasite species per infected host in a sample).

5. Term: **Density.**

Concept or definition: Number of individuals of a particular parasite species per unit area, volume, or weight of infected host tissue or organ. (The units should always be clearly specified.)

6. Term: **Relative density** or **Abundance.**

Concept or definition: Total number of individuals of a particular parasite species in a sample of hosts ÷ Total number of individuals of the host species (infected + uninfected) in the sample (= Mean number of individuals of a particular parasite species per host examined). It equals **mean intensity** (Concept 4) × **prevalence** (Concept 1).

Remarks: This concept seems to cause the greatest problem with respect to its terminology. **Mean intensity,** which should be restricted in its use to Concept 4 above, has been misused for this concept. The term **density** has also been used by some authors for this concept. However, **density** should be reserved for Concept 5 above, which is in keeping with the use of this term in general ecology. **Abundance** has been recommended for this concept, but it has the disadvantage of having been widely used as a more general term without a specific quantitative meaning. As an alternative the committee considered **relative density,** which is used in ecology for a similar concept, but it has the disadvantage of possibly being confused with **density** (Concept 5) or being considered to have been derived from **density.**

It is appropriate to comment here on the term **density-dependent.** In general ecology this term is used to describe factors or mechanisms whose effects on, or regulation of, populations are a function of the **density** (in the sense of Concept 5) of the population. In parasitology it has been used with reference to factors affecting **infrapopulations** (see Term 7). However, since **intensity,** rather than **density,** is the term applied to the number of individuals in a parasite **infrapopulation, intensity-dependent** would be the appropriate term when parasite **infrapopulations** are concerned. It is also improper to use **density-dependent** in relation to the synonymous terms **relative density** or **abundance** (**density** of some authors) because their calculation uses **prevalence** and **intensity,** parasite population parameters that are controlled or regulated by different mechanisms.

7. Term: **Infrapopulation.**

Concept or definition: All individuals of a species of parasite occurring in an individual host.

8. Term: **Suprapopulation.**

Concept or definition: All individuals of a species of parasite in all stages of development within all hosts in an ecosystem.

Remarks: This term has at times been misused in two ways: (1) to denote all of the individuals of a species of parasite occurring in only one of its host species in an ecosystem, and (2) to indicate all of the individuals of a particular stage in an ecosystem (e.g., all adults of a cestode species in all definitive hosts in the system).

9. Term: **Site or location.**

Concept or definition: The tissue, organ, or part of the host in which a parasite was found.

Remarks: These two terms are synonymous and essentially interchangeable. **Site** is preferable because of the possibility of confounding **location** and **locality,** the latter denoting the geographic place of capture or collection of the host. **Site** has a further advantage in that the compound words derived from it, such as **site-selection** and **site-specific,** are more euphonius than the equivalents formed from **location.** The use of **niche** should generally be avoided. **Habitat** refers not only to the spatial location of the parasites but to their physical and chemical environment as well, and, therefore, should not be used synonymously with **site** and **location.**

17

THE SPECIES PROBLEM
AND CLASSIFICATION

The delineation of any species of organism is a difficult task. This applies also to parasitic forms, which in many cases are even more difficult to recognize than free-living species. Many factors contribute to the confusion. Probably the single most important factor is the paucity of known specimens. Many species of parasites, even some that infect humans, are known from only a few individuals. Obviously nothing can be known about infraspecific variation of morphological and physiological characteristics of such animals.

However, even when parasites are common and easily obtained, their uniqueness is often in question. For example, in well-known forms such as *Trypanosoma brucei, Trichinella spiralis,* and *Entamoeba histolytica,* the more we learn about each species, the more we become aware that it is not simply a kind of animal different and reproductively isolated from all other types, but rather it is a complex of strains, or races, with slightly different characteristics from each other. Even more sophisticated techniques are then required to separate members of **species flocks**—true species that are morphologically indistinguishable.

Most definitions of species include some consideration of reproductive isolation. This is a valid concept, but there are so many exceptions that it is useful only as a general principle. For example, sexuality and biparental reproduction are completely unknown for many species. How can the concept of genetic incompatibility apply to organisms that reproduce only by mitotic fission? Where would parthenogenesis fit into such a scheme? The series of asexual reproduction exhibited by digenetic trematodes has nothing in common with the usual concept of gamete exchange with resultant diploid recombinations.

Another pitfall in our attempt to recognize species is the tendency among many parasites to alter their form according to age, host, or nutrition. Juvenile parasites often bear no superficial resemblance to their parents. Who would guess that a cysticercus is a tapeworm, that a miracidium is a fluke, or, for that matter, that what hatches from a flea's egg is a flea? Some parasites have altered forms, according to the host in which they find themselves. Digenetic trematodes, for example, are notorious for assuming different sizes and shapes when in different host species. This has accounted for many redundant species names, some of which may never be reconciled. The nutrition available to the parasite, the size and age of the host, and the effectiveness of the parasite's and host's defense mechanisms also tend to alter the morphology of the specimen in question. Even as adults some species have alternating parasitic and free-living phases that are quite unlike one another. The parasite taxonomist must be able to recognize all these variations.

Surprisingly the taxonomist usually does recognize the variations. More than 200 years of taxonomy, based mainly on adult morphology, have established the present concept of species delineation. As new information on life cycles, genetics, and other cryptic aspects of the species has become available, it has, in the majority of instances, borne out the conclusions previously reached by an experienced taxonomist, sometimes hundreds of years previously. Recent advances in numerical taxonomy, chemotaxonomy, and serotaxonomy provide useful tools for taxonomic research in some groups. The taxonomist welcomes these additional tools to apply to the perplexing problem of defining a species population.[36]

The names given to parasites reflect the vagueness, eruditeness, and imagination of their discoverers. To the beginning student of parasitology such tongue twisters as *Macracanthorhynchus hirudinaceus* and *Leucochloridium macrostomum* may seem to be insurmountable barriers to true knowledge. However, when these names are understood as being symbolic of discrete populations of animals and as being used by scientists in every nation in the world regardless of their native tongue, their usefulness becomes apparent. Every described species must have a name so that we can refer to it and retrieve published knowledge about it in an efficient way. True, the names could just as easily be numerals or other symbols, but the Latin alphabet has been accepted for hundreds of years and has the advantage of possessing a wry charm. Consequently, it is much easier to remember names like *Trichinella spiralis* and *Entamoeba histolytica* than coded groups of numbers. Our system of classification has enabled us to catalog 1.5 million species of organisms in such a way that we can retrieve all published information about any one of them in a very short time, regardless of the language we speak. Scientific names, then, should not be dreaded or avoided but should be welcomed as a simple scheme that avoids unimaginable chaos.

In this book we have included brief summaries of the classification of each major group. These can be referred to whenever the reader wishes to know where a given taxon is placed in the taxonomic hierarchy.

REFERENCES

1. Anderson, R.M. 1978. The regulation of host population growth by parasitic species. Parasitology 76:119-157.
2. Anderson, R.M., and D.M. Gordon. 1982. Processes influencing the distribution of parasite numbers within host populations with special emphasis on parasite-induced host mortalities. Parasitology 85:373-398.

3. Anderson, R.M., and R.M. May. 1982. Coevolution of hosts and parasites, Parasitology. 85:411-426.

4. Bänziger, H. 1970. The piercing mechanism of the fruit-piercing moth *Calpe (Calyptra) thalicteri* Bkh. (Noctuidae) with reference to the skin-piercing blood sucking moth *C. eustrigata* Hmps. Acta Trop. 27:54-88.

5. Birley, M.H. 1985. Forecasting the vector-borne disease implications of water development. Parasitol. Today 1:34-36.

6. Cheng, T.C. 1971. Aspects of the biology of symbiosis. University Park Press, Baltimore.

7. Cressey, R.F., and E.A. Lachner. 1970. The parasitic copepod diet and life history of diskfishes (Echeneidae). Copeia. No. 2:310-318.

8. Croll, N.A., and E. Ghadirian. 1981. Wormy persons: contributions to the nature and patterns of overdispersion with *Ascaris lumbricoides, Ancylostoma duodenale, Necator americanus,* and *Trichuris trichiura.* Trop. Geogr. Med. 33:241-248.

9. Crompton, D.W.T. 1973. The sites occupied by some parasitic helminths in the alimentary tract of vertebrates. Biol. Rev. 48:27-83.

10. Dobson, A.P. 1985. The population dynamics of competition between parasites. Parasitology 91:317-347.

11. Esch, G.W., J.W. Gibbons, and J.E. Bourque. 1975. An analysis of the relationship between stress and parasitism. Am. Midl. Nat. 93:339-353.

12. Esch, G.W., T.C. Hazen, and J.M. Aho. 1977. Parasitism and r- and K-selection. In Esch, G.W., editor. Regulation of parasite populations. Academic Press, Inc. New York, pp. 9-62.

13. Force, D.C. 1975. Succession of r and K strategists in parasitoids. In Price, P.W., editor. Evolutionary strategies of parasitic insects and mites. Plenum Publishing Corp., New York, pp. 112-129.

14. Gardiner, C.H., D.S. Koh, and T.A. Cardella. 1981. *Micronema* in man: third fatal infection. Am. J. Trop. Med. Hyg. 30:586-589.

15. Gettinby, G. 1974. Assessment of the effectiveness of control techniques for liver fluke infection. In Usher, M.B., and M.H. Williamson, editors. Ecological stability, Chapman & Hall Ltd., London.

16. Gillett, J.D. 1985. The behaviour of *Homo sapiens,* the forgotten factor in the transmission of tropical disease. Trans. R. Soc. Trop. Med. Hyg. 79:12-20.

17. Hairston, N.G. 1962. Population ecology and epidemiological problems. In Wolstenholme, G.E.W., and M. O'Connor, editors. Bilharziasis. Ciba Foundation Symposium. J. & A. Churchill Ltd., London.

18. Hairston, N.G. 1965. On the mathematical analysis of schistosome populations. Bull. WHO 33:45-62.

19. Heckmann, K., R.T. Hagen, and H.-D. Görtz. 1983. Freshwater *Euplotes* species with a 9 type 1 cirrus pattern depend upon endosymbionts. J. Protozool. 30:284-289.

20. Henry, S.M., editor. 1966, 1967. Symbiosis, vols. 1 and 2. Academic Press, Inc., New York.

21. Hirsch, R.P. 1977. Use of mathematical models in parasitology. In Esch, G.W., editor. Regulation of parasite populations. Academic Press, Inc., New York, pp. 169-207.

22. Holmes, J.D., R.P. Hobbs, and T.S. Leong. 1977. Populations in perspective: community organization and regulation of parasite populations. In Esch, G.W., editor. Regulation of parasite populations. Academic Press, Inc., New York, pp. 209-245.

23. Insler, G.D., and L.S. Roberts. 1976. *Hymenolepis diminuta:* lack of pathogenicity in the healthy rat host. Exp. Parasitol. 39:351-357.

24. Jennings, J.B., and P. Calow. 1975. The relationship between high fecundity and the evolution of entoparasitism. Oecologia 21:109-115.

25. Jeon, K.W. 1980. Symbiosis of bacteria with Amoeba. In Cook, C.B., P. Pappas, and E. Rudolph, editors. Cellular interactions in symbiotic and parasitic relationships. Ohio State University Press, Columbus, pp. 245-262.

26. Kennedy, C.R. 1981. The establishment and population biology of the eye-fluke *Tylodelphys podicipina* (Digenea: Diplostomatidae) in perch. Parasitology 82:245-255.

27. Keymer, A. 1982. Density-dependent mechanisms in the regulation of intestinal helminth populations. Parasitology 84:573-587.

28. Lorch, I.J., and K.W. Jeon. 1980. Resuscitation of amebae deprived of essential symbiotes: micrurgical studies. J. Protozool. 27:423-426.

29. Margolis, L., G.W. Esch, J.C. Holmes, A.M. Kuris, and G.A. Schad.1982. The use of ecological terms in parasitology (report of an ad hoc committee of The American Society of Parasitologists). J. Parasitol. 68:131-133.

30. Minchella, D.J. 1985. Host life-history variation in response to parasitism. Parasitology 90:205-216.

31. Molyneaux, D.H., and D. Jefferies. 1986. Feeding behaviour of pathogen-infected vectors. Parasitology 92:721-736.

32. Moore, J. 1983. Responses of an avian predator and its isopod prey to an acanthocephalan parasite. Ecology 64:1000-1015.

33. Morris, S.C. 1981. Parasites and the fossil record. Parasitology 82:489-509.

34. Pianka, E.R. 1970. On r- and K-selection. Am. Naturalist 104:592-597.

35. Schad, G.A. 1963. Niche diversification in a parasitic species flock. Nature 198:404-406.

36. Schmidt, G.D., editor. 1969. Problems in systematics of parasites. University Park Press, Baltimore.

37. Simpson, A.J.G., editor. 1986. Parasites and molecular biology: applications of new techniques. Parasitology 92 (suppl. Symposia of the British Society for Parasitology 23: 1-174.)

38. Smith, G. 1984. Density-dependent mechanisms in the regulation of *Fasciola hepatica* populations in sheep. Parasitology 88:449-461.

39. Taylor, A.E.R., editor. 1965. Evolution of parasites. Blackwell Scientific Publications Ltd., Oxford, Eng.

40. Wirth, D.F., W.O. Rogers, R. Barker, Jr., H. Dourado, L. Suesebang, and B. Albuquerque. 1986. Leishmaniasis and malaria: new tools for epidemiologic analysis. Science 234:975-979.

41. Whitfield, P.J., and N.A. Evans. 1983. Parthenogenesis and asexual multiplication among parasitic platyhelminths. Symposia of the British Society for Parasitology, vol. 20. Parasitology 86:121-160.

SUGGESTED READINGS

Admadjian, V., and S. Paracer. 1986. Symbiosis. An introduction to biological associations, University Press of New England, Hanover, New Hampshire.

Campbell, W.C., and R.S. Rew, editors. 1986. Chemotherapy of parasitic diseases. Plenum, New York.

Futuyma, D.J., and M. Slatkin, editors. 1983. Coevolution. Sinauer Associates, Inc., Sunderland, Mass.

Keymer, A.E., and A.F.G. Slater. 1987. Helminth fecundity: density dependence or statistical illusion? Parasitol. Today 3:56-58.

MacInnis, A.J. 1976. How parasites find hosts: some thoughts on the inception of host-parasite integration. In Kennedy, C.R., editor. Ecological aspects of parasitology. North-Holland Publishing Co., Amsterdam.

Pavlovsky, E.N. 1966. Natural nidality of transmissible diseases, with special reference to the landscape epidemiology of zooanthroponoses (English translation). University of Illinois Press, Urbana.

Price, P.W. 1980. Evolutionary biology of parasites. Monographs in population biology 15. Princeton University Press, Princeton, New Jersey.

Smith, T. 1963. Parasitism and disease. Hafner Publishing Co., New York. (Another classic, originally published in 1934.)

Soulsby, E.J.L., editor. 1976. Pathophysiology of parasitic infection. Academic Press, Inc., New York.

Trager, W. 1986. Living together. The biology of animal parasitism. Plenum, New York.

CHAPTER 3

BASIC PRINCIPLES AND CONCEPTS II: IMMUNOLOGY AND PATHOLOGY

As a symbiont becomes progressively more specialized, it increasingly limits its potential host species; that is, it increases its host specificity. A vital component in the process is the habitat (host), which is a dynamic, living, and evolving partner in the relationship. It reacts to the presence of the symbiont in defense against a foreign invader, and the successful symbiont must evolve strategies to evade the host defenses. Parasitologists have come to recognize that not only is host specificity determined in great degree by which host individuals can mount an effective defense and which parasites can evade that defense, but also that much of the disease caused by the parasite is directly related to the host's defense mechanisms. Therefore this chapter will introduce concepts related to host defenses, to the evasion of host defenses by parasites, and to how parasites cause disease in the host.

SUSCEPTIBILITY AND RESISTANCE

Before discussing host defense mechanisms, we will distinguish certain terms that often are used with other than their strict biological meanings. A host is **susceptible** to a parasite if the host is in a physiologic state such that it will not eliminate the parasite before the parasite can become established in the host. The host is **resistant** if the host's physiological status prevents the establishment and survival of the parasite. A corresponding term from the view point of the parasite would be **infectivity.**

These terms deal only with the ease or difficulty of infection, not with the mechanisms producing the result. The mechanisms that increase resistance (and correspondingly reduce the susceptibility and infectivity) may involve either attributes of the host not related to active defense mechanisms or specific defense mechanisms mounted by the host in response to a foreign invader. Furthermore, the terms are relative, not absolute; for example, one individual organism may be more or less resistant than another.

The term **immunity** has often been used on the one hand as synonymous with resistance and on the other hand has been associated with the sensitive and specific immune response exhibited by vertebrates. However, since many invertebrates can be immune to infection with various agents, a more general yet concise statement would be that an animal demonstrates immunity if it possesses tissues capable of recognizing and protecting the animal against nonself.[14] The type of immunity shown by most invertebrates may be described as **innate.** In addition to innate immunity, vertebrates develop **acquired immunity,** which is specific to the particular nonself material, requires time for its development, and occurs more quickly and vigorously on secondary response.

DEFENSE MECHANISMS
Phagocytosis

Most animals have one or more innate mechanisms to protect themselves against invasion of a foregin body or infectious agent. These may be coincidental attributes of certain structures (e.g., a tough skin or high stomach acidity), or they may be characteristics evolved as adaptations for defense. For defense against an invader, the cells in an animal must "know" when a substance does not belong in that animal; they must recognize "nonself." **Phagocytosis** illustrates nonself recognition, and it is found in almost all metazoa and is a feeding mechanism in protozoa. A cell that has this ability is a **phagocyte.** Phagocytosis is a process of engulfment of the invading particle within an invagination of the phagocyte's cell membrane. The invagination becomes pinched off, and the particle is thereby enclosed in an intracellular vacuole. **Lysosomes** empty digestive enzymes (**lysozymes**) into the vacuole to destroy the particle.

In vertebrates there are **fixed** and **mobile** phagocytes. The fixed phagocytes taken together form the **reticuloendothelial (RE) system** and are spread through a variety of tissues. The RE system filters out and destroys particles and spent red blood cells from the blood that passes throughout the organs where the cells are located. Cells of the RE system of humans include the **Kupffer cells** in sinusoids of the liver; the phagocytic reticular cells of lymphatic tissue, myeloid tissue, and spleen; and the "dust cells" of the lungs. **Macrophages** in the lymph nodes help remove foreign particles from the lymph.

The most numerous of the mobile or circulating phagocytes are the **polymorphonuclear leukocytes,** or **granulocytes.** The first name refers to the fact that the nucleus is highly variable in shape; the second name refers to the many small granules that can be seen in their cytoplasm, especially when stained with a Romanovsky type of stain. According to their staining properties, granulocytes are further subdivided into **neutrophils, eosinophils,** and **basophils.** Neutrophils are the most abundant of the white cells in the blood, and they provide the first line of phagocytic defense in an infection. Eosinophils in normal blood account for about 2% to 5% of the total leukocytes, and basophils are the least numerous at about 0.5%. A high **eosinophilia** (eosinophil count in the blood) is often associated with allergic diseases and parasitic infections. The function of eosinophils has long been obscure. Some recent discoveries on their role in parasitic infections will be described later.

Another circulating white cell important in phagocytosis is the **monocyte.** The cell is not considered a granulocyte, though its cytoplasm may contain fine granules. When monocytes move into the tissue, they differentiate into macrophages, which are active phagocytes. Macrophages also have important roles in the specific immune response of vertebrates.

Immunity in invertebrates

Our presentation will be brief; for further information the reader should consult the book edited by Maramorosch and Shope[19] and especially the review by Lackie.[14]

Many invertebrates have specialized cells that function as itinerant troubleshooters within the body, acting to engulf or wall off foreign material (Table 3-1) and to repair wounds. The cells are variously known as amebocytes, hemocytes, coelomocytes, and so on, depending on the animals in which they are found. If the foreign particle is small, it is engulfed by phagocytosis; but if it is larger than about 10 μm, it is usually encapsulated. Arthropods can wall off the foreign object also by deposition of melanin around it, either from the cells of the capsule or by precipitation from the **hemolymph** (blood).

One of the principal tests of the ability of invertebrate tissues to recognize nonself is by grafting a piece

TABLE 3-1

Some invertebrate leukocytes and their functions

Group	Cell types and functions	Phagocytosis	Encapsulation	Transplantation reaction
Sponges	Archaeocytes (wandering cells that differentiate into other cell types and can act as phagocytes)	+	+	?*
Cnidarians	Amebocytes; "lymphocytes"	+		?
Nemertines	Agranular leukocytes; granular macrophage-like cells	+		+
Annelids	Basophilic amebocytes (accumulate as "brown bodies"); acidophilic granulocytes	+	+	+
Sipunculids	Several types	+	+	+
Insects	Several types, depending on family; e.g., plasmatocytes, granulocytes, spherule cells, coagulocytes—blood clotting	+	+	+
Crustaceans	Granular phagocytes; refractile cells that lyse and release contents	+	+	+
Molluscs	Amebocytes	+	+	+
Echinoderms	Amebocytes, spherule cells, pigment cells, vibratile cells—blood clotting	+	+	+
Tunicates	Many types, including phagocytes; "lymphocytes"	+	+	+

Modified from Lackie, A.M. 1980. Parasitology 80:393-412; published by Cambridge University Press, Cambridge, Eng. (See Lackie's article for references.)

*Transplantation reactions occur, but the extent to which the leukocytes are involved is unknown.

of tissue from another individual of the same species (**allograft**) or a different species (**xenograft**) onto the host. If the graft grows in place with no host response, the host tissue is treating it as self, but if cell response and rejection of the graft occur, the host exhibits immune recognition. Most invertebrates tested reject xenografts; and sponges, cnidarians, annelids, sipunculids, echinoderms, and tunicates can reject allografts.[14] Interestingly, nemertines, arthropods, and molluscs apparently do not reject allografts. It is curious that such primitive phyla as Porifera and Cnidaria show the mechanism of allograft rejection; Lackie[14] suggested that this may be an adaptation to avoid loss of integrity of the individual sponge or colony under conditions of crowding, with its danger of overgrowth or fusion with other individuals.

Lysozyme is released from the hemocytes of molluscs during phagocytosis and encapsulation,[4,10] and bactericidal substances have been found in the body fluids of a variety of invertebrates,[14] evidence of humoral components in immunity. Substances that can agglutinate vertebrate erythrocytes in vitro have been found in many invertebrates, although the function of these lectin-like molecules in vivo is unclear. A number of reports suggest that these molecules may act as opsonins, enhancing or facilitating phagocytosis, in molluscs and arthropods.

Generally, invertebrates do not seem able to acquire a specific immune response. They will either respond at first exposure to a foreign material or not at all; repeated exposure does not elicit immunity. Numerous reports of memory in the immune response of invertebrates, that is, enhanced response on second and subsequent exposures, have not been substantiated by other investigators. Some of the problems in duplicating results may derive from reactions of hosts to differing species of parasites or various reactions to the same parasite by different strains of hosts. These interactions have been studied by Lie[18] in the snail *Biomphalaria glabrata* and the trematodes *Echinostoma* spp. and *Schistosoma mansoni*. He found varying degrees of resistance among different strains of *B. glabrata*, accelerated response on subsequent exposure, and species specificity of the immune response.

Acquired immune response in vertebrates

Vertebrates have a specialized system of nonself recognition that results in increased resistance to a *specific* foreign substance or invader on repeated exposures. Investigations on the mechanisms involved currently are intense, and our knowledge of them is increasing rapidly. For an excellent account of the current status, the reader should consult Tonegawa.[26]

The immune response is stimulated by the specific foreign substance called an *antigen,* and, circularly, an antigen is any substance that will stimulate an immune response. Antigens may be any of a variety of substances with a molecular weight of over 3,000, most commonly proteins, and are usually (but not always) foreign to the host. There are two arms of the immune response, known as **humoral** and **cellular.** Humoral immunity is based on **antibodies,** which are dissolved in and circulate in the blood, whereas cellular immunity is associated with cell surfaces. Although the two types interact, humoral immunity seems to be more important in a variety of bacterial infections, whereas the cellular response is of particular importance in tissue rejection reactions and a variety of viral, fungal, and parasitic infections.

■ Basis of self and nonself recognition

It has been known for many years that nonself recognition was very specific. If tissue from one individual was transplanted into another individual in the same species, the graft would grow for a time and then die as immunity against it arose. Tissue grafts would only grow successfully if they were between identical twins or between individuals of highly inbred strains of animals. It has been found in recent years that the molecular basis for this nonself recognition depends on certain proteins imbedded in the cell surface. These proteins are coded by certain genes, now known as the **major histocompatibility complex (MHC),** because they were discovered in tissue graft experiments. The MHC proteins are among the most variable known, and unrelated individuals almost always have different genes. There are two types of MHC proteins: class I and class II. Class I proteins are found on the surface of virtually all cells, whereas class II MHC proteins are found only on certain cells participating in the immune responses, such as certain lymphocytes and macrophages.

■ Antibodies

Antibodies are proteins called **immunoglobulins.** The basic antibody molecule consists of four polypeptide strands: two identical light chains and two identical heavy chains, held together in a Y-shape by disulfide bonds and hydrogen bonds (Fig. 3-1). The amino acid sequence toward the ends of the Y varies in both the heavy and light chains, according to the specific antibody molecule (the **variable region**), and this determines with which antigen the antibody can bind. Each of the ends of the Y forms a cleft that acts as the antigen-binding site (see Fig. 3-1), and the specificity of the molecule depends on the shape of the cleft and the properties of the chemical groups that line its walls. The remainder of the antibody is known as the

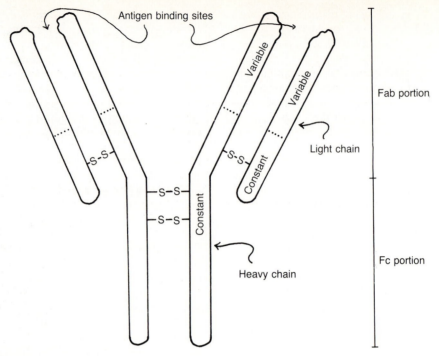

FIG. 3-1

Diagram of an immunoglobulin molecule. The bars represent the polypeptide chains bound together by disulfide (—S—S—) and hydrogen bonds. The light chains may be either of two types, kappa or lambda. The class of antibody is determined by the type of heavy chain: mu (IgM), gamma (IgG), alpha (IgA), delta (IgD), or epsilon (IgE). The constant portion of each chain does not vary for a given type or class, and the variable portion varies with the specificity of the antibody. Antigen-binding sites are in clefts formed in the variable portions of the heavy and light chains. IgM normally occurs as a pentamer, five of the structures illustrated being bound together by another chain. IgA may occur as a monomer, dimer, or trimer.

constant region, but the "constant" region also varies to some extent. The variable end of the antibody molecule is referred to as **Fab** (for antigen-binding fragment), and the constant end is the **Fc** (for crystallizable fragment) (see Fig. 3-1). The constant region of the light chains can be either of two types: kappa or lambda; the heavy chains may be any of five types: mu, gamma, alpha, delta, or epsilon. The latter determines the **class** of the antibodies, referred to as IgM, IgG (now familiar to many people as "gamma globulin"), IgA, IgD, and IgE, respectively. The class of the antibody determines the role of the antibody in the immune response but not the antigen it recognizes, for example, whether the antibody is secreted or held on a cell surface.

■ **Generation of a humoral response**

The immune response is caused primarily by a type of leukocyte called **lymphocytes,** of which there are two broad categories: T **lymphocytes (T cells)** and **B lymphocytes (B cells).** B cells have antibody molecules in their surface and give rise to cells that actively secrete antibodies into the blood. T cells have surface receptors that bind antigens, but the receptors are somewhat different in structure from antibodies. There are a vast number of different kinds of B cells, each bearing on its surface molecules of antibody that will bind with one particular antigen, even though that antigen has never been present in the body previously. There are probably an equally great number of different T cells with receptors for specific antigens.

When an antigen is introduced into the body, it binds to specific antibody on the surface of the appropriate B cell, but this is usually insufficient to activate the B cell to multiply. Some of the antigen is taken up by **antigen-presenting cells (APCs),** such as macrophages,[27] that partially digest the antigen. The APCs then incorporate portions of the antigen into

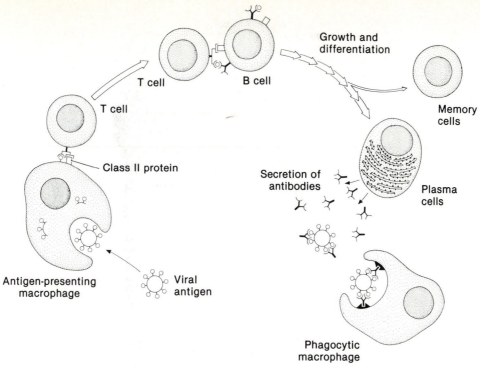

FIG. 3-2

Stimulation of a humoral immune response by an antigen. *1,* Macrophage consumes antigen, partially digests it, and displays epitope on its surface, along with class II MHC protein. *2,* T helper cell recognizes epitope and class II protein on macrophage and is activated. *3,* T helper then activates B cell, which carries same antigen and class II protein on its surface. *4,* Activated B cell multiples, finally producing many plasma cells which secrete antibody. *5,* Some of B cell progeny become memory cells. *6,* Antibody produced by plasma cells binds to antigen and stimulates macrophages to consume antigen (opsonization).

From Hickman, C.P., Jr., L.S. Roberts, and F.M. Hickman, 1988. Integrated principles of zoology, ed. 8. The C.V. Mosby Co., St. Louis.

their own cell surface (Fig. 3-2). That portion of the antigen presented on the surface of the macrophage or other APC is called the **epitope** (or **determinant**). The macrophages also secrete a substance known as **inter-leukin-1.** The epitopes on the surface of the macrophages are recognized by a subset of T cells called **T helper cells (T_H),** in conjunction with the class II MHC protein on the macrophage surface. (Both the class II protein and the epitope must be present; neither is effective alone.) The T_H cells then activate the B cell that has the same epitope and a class II MHC protein on its surface. Interleukin-1 is also necessary for this activation. The B cell multiplies rapidly and produces many **plasma cells,** which secrete large quantities of antibody for a period of time, then die. Thus if the amount of the antibody **(titer)** is measured soon after the antigen is injected, little or none can be detected. The titer rises rapidly as the plasma cells secrete antibody, then it may decrease somewhat as they

die and the antibody is degraded (Fig. 3-3). However, if another dose of antigen (the **challenge**) is given, there is no lag, and the antibody titer rises quickly to a higher level than after the first dose. This is the **secondary** or **anamnestic response,** and it occurs because some of the activated B cells gave rise to long-lived **memory cells.** There are many more memory cells present in the body than the original B lymphocyte with the appropriate antibody on its surface, and they rapidly multiply to produce additional plasma cells.

Antibodies can mediate destruction of an invader (antigen) in a number of ways. A foreign particle, for example, becomes coated with antibody molecules as their Fab regions become bound to it. Phagocytic macrophages recognize the projecting Fc regions and are stimulated to engulf the particle. This is the process of **opsonization.**

Another important process, particularly in the destruction of bacterial cells, is the interaction with **com-**

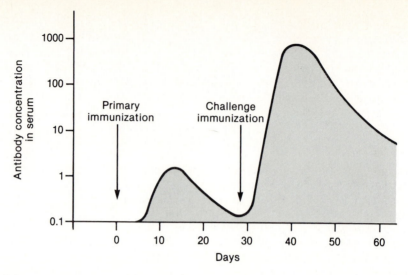

FIG. 3-3

Typical immunoglobulin response after primary and challenge immunizations. Secondary response is result of large numbers of memory cells produced after primary B-cell activation.

From Hickman, C.P., Jr., L.S. Roberts, and F.M. Hickman, 1988. Integrated principles of zoology, ed. 8. The C.V. Mosby Co., St. Louis.

plement. Complement is a series of 12 enzymes that are activated by bound antibody, and they actually punch holes in the bacterial cell surface. Complement may also play a role in opsonization.

Antibody bound to the surface of an invader may trigger contact killing of the invader by host cells (**antibody-dependent, cell-mediated cytotoxicity [ADCC]**). Eosinophils can be effector cells in ADCC. They, as well as lymphoid cells and neutrophils, can destroy bloodstream forms of *Trypanosoma cruzi* in the presence of antibody against this organism.[11,12] The parasites are phagocytosed, and the granules in the eosinophils and neutrophils fuse with the phagosome and kill the *T. cruzi*.[22] Interestingly, *T. cruzi* phagocytosed by nonactivated macrophages are not killed but multiply within the host cell. Purified protein from the granules of eosinophils (major basic protein) can damage newborn *Trichinella spiralis* and *Schistosoma mansoni* schistosomula (juveniles)[3,29] When eosinophils are incubated with schistosomula in vitro in the presence of complement and antibody to the worms, they adhere to the worm's surface, degranulate (expel the contents of their granules), and damage the worm's tegument.[20] Neutrophils are also active in this process.[9]

■ **Types of T cells**

In addition to the T helper cells, three other T cell types are known. **T inducer cells** trigger the maturation of T lymphocytes from precursors into function-

ally distinct cells. **Cytotoxic T cells** are important in the defense against viruses. They recognize class I MHC protein on the cell surface of infected cells in conjunction with viral protein that was left on the surface when the virus invaded the cell, and they kill the infected cell. The action of cytotoxic T cells is mediated by T_H cells. **T suppressor cells,** the fourth type, interact with T_H cells in complex ways but in general have effects on cytotoxic cells opposite those of T_H cells. They turn off the T_H cells and dampen down the immune response. T suppressor cells are very important in the regulation of the immune response.

Certain biochemical markers on the surfaces of the T cells allow classification of the four types into two: T inducer and T helper cells are known as $CD4^+$ (or T4) because a receptor protein designated CD4 is on their surfaces, and T suppressor and cytotoxic T cells are known as $CD8^+$ (or T8) because they bear CD8 receptors. $CD4^+$ cells recognize class II MHC proteins in association with antigenic proteins in other cell surfaces, and $CD8^+$ cells recognize class I MHC proteins along with foreign proteins.

■ **Acquired immune deficiency syndrome (AIDS)**

AIDS is an extremely serious disease in which the ability to mount an immune response is crippled. The first case was recognized in 1981, and by the end of 1987, over 45,000 individuals had contracted AIDS in the United States alone, of which almost 25,000 had died.[25] AIDS patients are continuously plagued by infections with agents (often parasites) that cause insignificant problems in persons with normal immune responses. The disease is caused by a virus that, although a variety of cell types are infected, preferentially invades and destroys CD4$^+$ lymphocytes. Normally, CD4$^+$ cells make up 60% to 80% of the T-cell population; in AIDS they can become too rare to be detected.[16]

■ The cell-mediated response

Some immune responses involve little, if any, antibody and depend on the action of cells only. In cell-mediated immunity (CMI) the epitope of the antigen is also presented by macrophages, but only T cells respond. The T cells with the specific receptors for the particular antigen are activated and proliferate. They produce soluble effector molecules called **lymphokines.** Examples of lymphokines are the **interferons, interleukin-2,** and **macrophage migration inhibitory factor (MMI).** Interferons are a family of 20 to 25 proteins that act on other cells to increase their resistance to viral infection. They affect infectious agents and tumors indirectly by activation of neutrophils and natural killer cells. MMI inhibits the migration of macrophages from the vicinity of the local reaction; that is, any that arrive there are inhibited from migrating, and they are therefore present to perform their phagocytic activity. Interleukin-2 causes proliferation of enlarged populations of mature cytotoxic, suppressor, and helper T cells. It also bolsters the **natural killer cells,** which are lymphocyte-like cells that can kill virus-infected and tumor cells in the absence of antibody. They do not directly interact with lymphocytes or recognize antibody. Whether natural killer cells have any important role in protozoan or helminth infections is at present unknown. Other lymphokines stimulate phagocytic activity in macrophages, and still others have a chemotactic effect, attracting macrophages and other white cells to the vicinity. The specific interaction of lymphocyte and antigen in the CMI greatly influences subsequent events in a nonspecific response, inflammation. The immune response elicited by at least several parasitic worms is not directly responsible for their expulsion from the host. Rather, the expulsion is caused by the inhospitable chemical conditions in the ensuing inflammation that has been enhanced by the specific CMI.[15]

Like humoral immunity, CMI shows a secondary response due to large numbers of memory T cells that were produced from the original activation. For example, a second tissue graft (challenge) between the same donor and host will be rejected much more quickly than the first.

A manifestation of CMI, the **delayed type hyper-** sensitivity (DTH) reaction, once thought to produce only tissue damage is now understood to have a basically protective role. The term **delayed type hypersensitivity** is derived from the fact that a period of 24 hours or more elapses between the time of antigen injection and the response to it in an immunized subject. This is in contrast to **immediate hypersensitivity** reactions, mediated by antibodies, in which maximal response is reached within a few minutes or hours. An example of DTH that is familiar to most people is the tuberculin diagnostic test. When the nonimmunogenic tuberculin antigen is injected into a person who has had no contact with tuberculosis, there is no reaction. However, if the person has had even a slight tuberculosis infection in the past, after about 4 hours a small swelling appears at the site of injection, increasing to a maximal size and redness after 24 to 48 hours and then gradually subsiding. The mechanism of this reaction, as in other CMI reactions, seems to involve comparatively small numbers of specifically sensitized lymphocytes that are circulating in the blood and arrive randomly at the injection site. This is a *specific* interaction of antigen with lymphocyte, the presence of the sensitized lymphocyte being the result of a prior immunizing dose of antigen, just as in the humoral antibody response.

Inflammation

Inflammation is a vital process in the mobilization of the body defenses against an invading organism or other tissue damage and in the repair of damage thereafter. Three categories of inflammation can be recognized: (1) simple acute, (2) chronic, and (3) immunologically mediated.[8] Acute inflammation normally follows a time course of 8 to 10 days between initiation and final healing of the inflamed site, whereas inflammation that continues over a period of weeks or even years is considered chronic. In addition, chronic inflammation may be immunologically mediated, and such is usually the case in parasitic infections. Systemic effects of inflammation often include fever, general malaise, and swelling and soreness of the lymph nodes. Lymph nodes have important roles in the body's defense against disease. They contain large numbers of macrophages and both T and B lymphocytes. The lymph nodes and the spleen are major locations in which antigens are taken up by macrophages and lymphocytes are activated and proliferate to generate the immune response.

Tissue damage initiates degranulation of **mast cells** in the area. Mast cells are basophil-like cells found in the dermis and other tissues. Their surfaces bear receptors for the Fc portions of IgE and IgG. Occupation of these sites by antigen-specific antibodies enhances degranulation of the mast cells when the Fab portions bind the particular antigen. For a review of the role of mast cells in helminth parasite infections, see Lee et al.[17]

Mast cells release a wide variety of substances, including histamine; serotonin; enzymes; neutrophil and eosinophil chemotactic factors, prostaglandins and leukotrienes.[17] Some of these substances increase the diameter and permeability of nearby small blood vessels. Thus, redness and warmth result from the increased amount of blood in the area (hence "inflammation"), and more proteins and fluid escape into the tissue, causing swelling. The first phagocytic line of defense is the neutrophils, which may last only a few days, then macrophages (either fixed or differentiated from monocytes) become predominant. "Pus" of an infection is formed principally from exuded tissue fluid and spent phagocytes. Activation of complement by fixed antibody stimulates opsonization by macrophages and increased release of substances from mast cells. As appropriate T cells arrive at the site, binding of antigen to their surface stimulates them to release lymphokines, such as MMI. Some degree of cell death (**necrosis**) always occurs, but necrosis may not be prominent in minor inflammation. If the necrotic debris is confined within a localized area, the pus may increase in hydrostatic pressure, forming an **abscess.** An area of inflammation that opens out to a skin or mucous surface is an **ulcer.**

Healing of the inflamed area involves regeneration or replacement, depending on the extent of the lesion and its location. Some cells, such as epithelial cells of the skin or cells of the lining of the digestive tract, are capable of division and may regenerate the damaged area without scar formation. Healing of other lesions involves migration by fibroblasts, collagen deposition, and organization of fibrous connective tissue (scar).

When the source of irritation or damage is not resolved over a period of weeks, the inflammation becomes chronic. Chronic inflammation can be damaging to the long-term well being of the host. It is characterized by continued residence of macrophages and other white cells, as well as infiltration with fibroblasts as the system attempts to wall off or heal the lesion (**granuloma** formation). When chronic inflammation is immune mediated, as it usually is in parasitic infections, the reactions are more severe.

Innate immunity in vertebrates

Under this heading one may place the "accidents" of a host's structural and physiological characteristics that reduce its susceptibility to certain parasites. This includes the physiological adaptations possessed by a wide range of hosts, evolved as adaptations for defense and effected on first encounter with a wide range

of invading organisms. Examples of the first category include physical barriers, such as a thick, cornified epidermis or other protective external covering; the ability to repair damaged tissues rapidly; and high acidity in the stomach. Such mechanisms are not always easy to distinguish from some "natural" defense adaptations, and the distinction may not be useful. In any case a variety of parasiticidal substances is known to be present in such animal body secretions as tears, mucus, saliva, and urine. In fact at least one of what were previously thought to be nonspecific parasiticidal substances is now known to be a class of antibody, IgA. IgA can cross cellular barriers easily. It seems to be an important protective agent in the mucus of the intestinal epithelium, and it is present in mucus in the respiratory tract, in tears, in saliva, and in sweat. Substances have been found in normal human milk that can kill intestinal protozoa such as *Giardia lamblia* and *Entamoeba histolytica,* and these substances may be important in protection of infants against such infections.[7] The nature of the substances is not yet known, but they seem not to be IgA or other antibody.

As indicated previously, the specific immune responses of vertebrates can greatly enhance or modify the nonspecific processes, phagocytosis, inflammation, and the action of complement. Furthermore, the immune response elicited by one infective agent may, by coincidence, provide nonspecific resistance to another. An example is the **allergic klendusity** (disease-escaping ability) conferred on rabbits against the tularemia organism *Francisella tularensis* by a hypersensitivity to the bites of the tick vector *Dermacentor andersoni.*[2] The immunity is stimulated by previous bites of uninfected ticks.

IMMUNODIAGNOSIS

Because the body produces specific antibodies in response to the stimulus of any particular antigen, many methods have been developed to detect the presence of the antibodies. If specific antibodies are present, the host either is infected with the agent that stimulated them or has had an immunizing experience in the past with that agent. The tuberculin test, previously described, is an example of an immunological skin test. In this section we will explain two additional diagnostic tests. Space does not permit a more thorough treatment, but the parasitology student should be aware of these extremely valuable diagnostic tools. You should also be aware of some difficulties. For example, false positives will arise when two related agents have antigens in common or similar enough to cross-react with antibodies raised against the other.

Complement, the series of enzymes activated by bound antibody, has already been described. Comple-

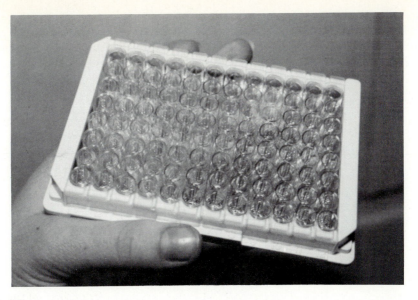

FIG. 3-4

A microplate for an ELISA test. Wells are available on the plate for testing several individuals in addition to positive and negative controls. Positive controls are wells in which antibody is known to be present, and negative controls omit the enzyme-linked anti-Ig.

Photograph by L.S. Roberts.

ment participates in a variety of antigen-antibody reactions, not all of which produce a visible result, such as lysis of cells. This has led to the development of a means of diagnosis called the **complement fixation test.** This test generally can be adapted to diagnose any infection that stimulates antibodies that bind to complement, including many parasitic infections. The most widely known complement fixation test is the Wassermann test for syphilis. In preparation of reagents for such a test, *rabbits* are immunized against *sheep* erythrocytes. Antigen is derived from the parasite (or a related species) to be diagnosed, and standard complement is prepared, usually from guinea pig blood. To perform the test, a small amount of serum from the patient is mixed with the known antigen and a quantity of complement. The rabbit antibody (against sheep erythrocytes) and the sheep red blood cells are then added to the mixture. If antibodies to the specific antigen were present in the patient's serum, complement will have been fixed before the rabbit antibody and sheep cells are added, and *no lysis* results; therefore the test is *positive.* In contrast, if antibodies to the infection were not present in the patient's serum, complement will still be available, and the sheep cells will be disrupted; therefore *lysis* of the sheep cells indicates a *negative* result.

The **enzyme-linked immunosorbent assay (ELISA)** is a type of immunodiagnostic test that has become popular in recent years for a number of parasite infections. It is simple to perform and does not require sophisticated equipment. A small quantity of antigen is adsorbed to the bottom of a small cup in a plastic microplate (Fig. 3-4). Next a portion of the serum to be tested is added to the cup (Fig. 3-5). The serum is removed, and the cup is rinsed several times. If the serum contained antibodies to the antigen, they will have bound to the antigen and would not have been removed by the rinses. Then a solution containing antibodies to human Ig (anti-Ig) is added. The anti-Ig must be prepared beforehand and linked covalently to an enzyme. The enzyme can be any one of several whose reaction product is colored. This solution is then removed from the cup, the cup is rinsed again, and the substrate for the enzymatic reaction is added. If the tested serum contained antibodies against the antigen, the anti-Ig will be bound to them, and the enzymatic reaction will occur, producing a color.

PATHOGENESIS OF PARASITIC INFECTIONS

The pathogenic effects of a parasitic infection may be so subtle as to be unrecognizable, or they may be strikingly obvious. An apparently healthy animal may be host to hundreds of parasitic worms and yet show no signs of distress, at least none that is detectable. On the other hand, another host may be so anemic,

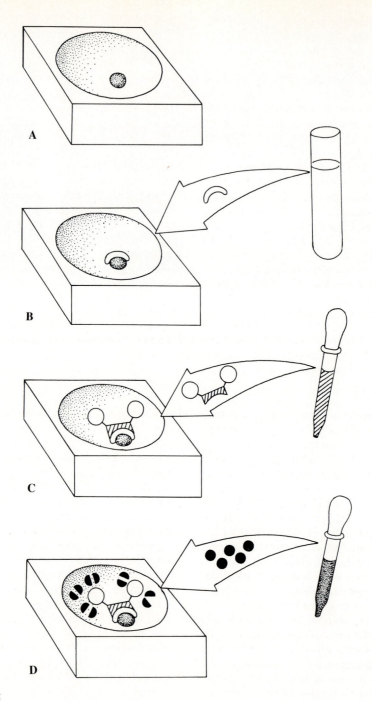

FIG. 3-5

Sequence of steps in performance of an ELISA test. **A,** Known antigen is adsorbed to bottom of microplate well. **B,** Serum from patient is added, then well is rinsed. **C,** Enzyme-linked antibody against human immunoglobulin is added, then well is rinsed again. **D,** Enzyme substrate is added. If colored products of enzyme reaction are observed, this indicates presence of bound anti-Ig, which in turn indicates presence of antibody against antigen. Thus test is positive.

unthrifty, and stunted that parasites are undoubtedly the reason for its sad state. The pathogenic effects of parasites are many and varied but for the sake of convenience can be discussed under the headings of trauma, nutrition robbing, and interactions of the host immune responses.

Physical trauma, or destruction of cells, tissues, or organs by mechanical or chemical means, is common in parasite infections. When an *Ascaris* or hookworm larva penetrates a lung capillary to enter an air space, it damages the blood vessel and causes hemorrhage and possible infection by bacteria that may have been inhaled. The hookworm, after completing its migration to the small intestine, feeds by biting deeply into the mucosa and sucking blood and tissue fluids. The dysentery ameba *Entamoeba histolytica* digests away the mucosa of the large intestine, forming ulcers and abscessed pockets that can cause severe disease. These are but a few examples of known physical trauma caused by parasites. Many are discussed in later chapters in conjunction with the particular parasite involved.

A less obvious but often pernicious pathogenic situation is diversion of the host's nutritive substances. Tapeworms and acanthocephalans, for instance, lack digestive systems and rely on the host's daily intake for their own food. Although most tapeworms absorb so little food in proportion to the amount eaten by the host that the host still manages very well, when the level of subsistence of the host is low, one or two large tapeworms may absorb enough to cause serious deficiencies in the host's diet. The broad fish tapeworm *Diphyllobothrium latum* has such strong affinity for vitamin B_{12} that it absorbs large amounts from the intestinal wall and contents of its host. Since B_{12} is necessary for erythrocyte production, a severe anemia may result. The giant nematode *Ascaris lumbricoides* inhabits the small intestine—often in large numbers—and consumes a good deal of food the host intends for itself. One study showed 21% greater weight gain in children treated for *Ascaris* infection compared with those whose infections went untreated.[30] The tiny protozoan *Giardia* robs its host in a different way. It is concave on its ventral surface and applies this suction cup to the surface of an intestinal epithelial cell. When many of these parasites are present, they cover so much intestinal absorptive surface that they interfere with the host's absorption of nutrients. The unused nutrients then pass uselessly through the intestine and are wasted.

In recent years we have come to realize that a great deal, perhaps the most serious and most pervasive, of pathogeneses are actually caused by the host's own defense system: the immune response and inflammation.[28] A number of cases previously thought due to toxins released by the parasite are now understood as caused by the host's reaction to parasite products. For example, the protozoan *Trypanosoma cruzi* develops clusters of cells in the smooth and cardiac muscle cells of its host, and when the parasites degenerate—sometimes years later—the inflammatory response damages the supporting cells of the nerve ganglia that control peristalsis and heart contraction. Parasite antigens on the host's own cells, particularly in the endocardium, cause autoimmune reactions, and the host's cells are attacked as foreign by the immune system. Some of the large amount of antigen-antibody complex formed in infections with the African trypanosomes (*T. brucei rhodesiense* and *T. b. gambiense*) adsorbs to the host's red blood cells, activating complement and causing lysis with resulting anemia. Many of the eggs laid by schistosomes are carried in the blood to the liver where they lodge, leaking antigen and initiating a cell-mediated, delayed-type hypersensitivity reaction. The formation of granulomas around the eggs eventually impedes blood flow through the liver, causing cirrhosis and portal hypertension. A similar reaction is caused by the filarial nematode *Onchocerca volvulus*. When the females located in the skin of the head or neck release juveniles, many are likely to wander into the retinas of the eyes, eliciting a powerful immune reaction. The invasion of host defense cells and granuloma formation often destroys the retina, causing permanent blindness. Today there are villages in Africa and Central America where the majority of adults are blind because of this parasite. Long-time infections with another filarial nematode, *Wuchereria bancrofti,* can cause swelling and thickening of the lower legs, scrotum, vulva, and breasts, sometimes achieving the horrible dimensions of elephantiasis. This and many other diseases to be discussed in context are examples of the immune response gone wrong. We could scarcely do without the defenses of our immune system, but some manifestations of the immune response are responsible for much of the pathogenesis we observe.

ACCOMMODATION AND TOLERANCE IN THE HOST-PARASITE RELATIONSHIP

Successful parasites of vertebrates have had to evolve one or more tactics to avoid a protective immunity in a given host. Otherwise the host is simply not susceptible. In recent years an astonishing array of such tactics has been found (Table 3-2), and most parasites appear to have more than one mechanism, sometimes many, for evading the host's immune response. We will describe only a few exam-

TABLE 3-2

Mechanisms favoring immune evasion in human parasitic infections

Disease	Parasite antigens						Modification of host immune responsiveness				
	Anatomical seclusion	Stage specificity	Antigenic variation	Shedding and renewal	Antigenic disguise	Antibody cleavage	Complement consumption	Modified leukocyte function	Immuno- suppression	Polyclonal lymphocyte activation	Circulating immune complexes
Amebiasis				+							
Giardiasis	+										
African trypanosomiasis	+		+						+	+	+
South American trypanosomiasis	+			+		+			+		
Leishmaniasis	+								+	+	+
Toxoplasmosis	+		+	+					+		
Malaria	+	+	+	+				+	+	+	+
Babesiosis			+		+			+	+	+	
Schistosomiasis	+			+	+	+	+	+	+	+	+
Fascioliasis	+			+	+			+			
Filariasis									+		
Onchocerciasis				+			+		+		
Trichinosis	+	+							+		
Cestode infections									+		
Nematode infections		+							+		

From Cohen, S. 1982. Survival of parasites in the immunocompetent host. In Cohen, S., and K.S. Warren, editors. Immunology of parasitic infections, ed. 2. Blackwell Scientific Publications Ltd. Oxford, Eng., pp. 138-161. (See Cohen's article for references.)

ples; for further information the reader should consult Parkhouse.[23]

The location of the parasite may provide some protection against host defenses. The lumen of the intestine is one such site. Although IgA is secreted into the intestine, IgA is not a very potent effector molecule against worms, and complement and phagocytic cells are normally not found in the intestine. However, the rat nematode *Nippostrongylus brasiliensis* can be expelled because inflammation and an immediate hypersensitivity reaction change the permeability of the mucosa and allow IgG to leak into the lumen. Many other intestinal parasites, not provoking such inflammation, are relatively long-lived. Numerous parasites achieve protection from the host response by the development of a cyst wall, such as juvenile tapeworms (cestodes) in various tissues and a nematode, *Trichinella spiralis,* in muscle. Others are shielded by their location within a host cell. Recognition of the infected cell by the host's cell-mediated effector systems is precluded if no parasite antigens are present in the outer membrane of the infected cell, as seems to be the case in liver cells infected with malaria parasites.

Parasites that are constantly or frequently bathed in blood would seem particularly vulnerable to the range of host defenses, but they have evolved fascinating mechanisms for evasion. The protozoa causing African trypanosomiasis display a "moving target," that is, a continuing succession of variant antigenic types, so that just as the host mounts an antibody response to one, another type proliferates. This phenomenon is described further in Chapter 5. Other important mechanisms of evasion are present in these infections as well. Antibody and cell-mediated responses are suppressed, apparently by some substance secreted by the trypanosomes. Suppression may be achieved by polyclonal B cell activation early in the infection; many subtypes of B cells are stimulated to divide, leading to the production of nonspecific IgG and autoantibodies.[13] Eventually the immune system is exhausted, and there is loss of B and T cell memory.[1]

In addition to immunosuppression, polyclonal lymphocyte activation, and other mechanisms, the blood fluke *Schistosoma* actually adsorbs host antigen so that the host immune system "sees" only self, not recognizing the parasite as foreign.[21,24] For example, if adult worms are removed from mice and transferred surgically to monkeys, the worms stop producing eggs for a time but then recover and resume normal egg production. However, if the worms from mice are transferred to a monkey that has been previously immunized against mouse red blood cells, the worms are promptly destroyed. Early in the infection, however, just after the cercariae (juvenile flukes) have penetrated the host skin, adsorption of host antibody does not seem to be the explanation of the protection. Although the cercariae themselves are killed by host immune effectors, they begin to acquire protection during the transit through the skin, and in time progressively fewer of the young worms are damaged. Current opinion is that unknown substances in the host skin and blood stimulate protective structural changes in the worm's surface tegument.[20,21]

Damian[6] suggested that in some instances, parasites may not only evade the host immune response but actually exploit the immune response to their own advantage. For example, because schistosomes live in the blood vessels adjacent to the host intestine or urinary bladder, how their eggs manage to traverse the intestinal wall to the lumen (and thereby gain access to the external environment via the feces or urine) has long been a mystery. There is evidence that the granuloma that forms around each egg expedites this transit. Immunosuppression of the host considerably reduces the number of eggs that reach the lumen. Other possible examples of exploitation were cited by Damian.[6]

REFERENCES

1. Askonas, B.A., A.C. Corsini, C.E. Clayton, and B.M. Ogilvie. 1979. Functional depletion of T- and B-memory cells and other lymphoid cell subpopulations during trypanosomiasis. Immunology 36:313-321.
2. Bell, J.R., S.J. Stewart, and S.K. Wikel. 1979. Resistance to tick-borne *Francisella tularensis* by tick-sensitized rabbits: allergic klendusity. Am. J. Trop. Med. Hyg. 28:876-880.
3. Butterworth, A.E., D.L. Wassom, G.J. Gleich, D.A. Loegering, and J.R. David. 1979. Damage to schistosomula of *Schistosoma mansoni* induced directly by eosinophil major basic protein. J. Immunol. 122:221-229.
4. Cheng, T.C., M.J. Chorney, and T.P. Yoshino. 1977. Lysosome-like activity in the hemolymph of *Biomphalaria glabrata* challenged with bacteria. J. Invertebr. Pathol. 29:170-174.
5. Cohen, S. 1982. Survival of parasites in the immunocompetent host. In Cohen, S., and K.S. Warren, editors. Immunology of parasitic infections, ed. 2. Blackwell Scientific Publications Ltd., Oxford, Eng., pp. 138-161.
6. Damian, R.T. 1987. The exploitation of host immune responses by parasites. J. Parasitol. 73:1-13.
7. Gillen, F.D., D.S. Reiner, and Ch.-S. Wang. 1983. Human milk kills parasitic intestinal protozoa. Science 221:1290-1291.
8. Groër, M.W., and M.E. Shekleton. 1983. Basic pathophysiology. A conceptual approach. The C.V. Mosby Co., St. Louis.
9. Incani, R.N., and D.J. McLaren. 1983. Ultrastructural observations on the in vitro interaction of rat neutrophils with schistosomula of *Schistosoma mansoni* in the presence of antibody and/or complement. Parasitology 86:345-357.
10. Kassim, O.O., and C.S. Richards. 1978. *Biomphalaria glabrata:* lysozyme activities in the hemolymph, digestive gland and headfoot of the intermediate host of *Schistosoma mansoni.* Exp. Parasitol. 46:218-224.

11. Kierszenbaum, F. 1979. Antibody-dependent killing of blood-stream forms of *Trypansoma cruzi* by human peripheral blood leukocytes. Am. J. Trop. Med. Hyg. 28:965-968.

12. Kierszenbaum, F., S.J. Ackerman, and G.J. Gleich. 1981. Destruction of bloodstream forms of *Trypansoma cruzi* by eosinophil granule major basic protein. Am. J. Trop. Med. Hyg. 30:775-779.

13. Kobayakawa, T., J. Louis, S. Isui, and P.H. Lambert. 1979. Autoimmune response to DNA, red blood cells and thymocyte antigens in association with polyclonal antibody synthesis during experimental African trypanosomiasis. J. Immunol. 122:296-301.

14. Lackie, A.M. 1980. Invertebrate immunity. Parasitology 80:393-412.

15. Larsh, J.E., Jr., and N.F. Weatherly. 1975. Cell-mediated immunity against certain parasitic worms. In Dawes, B., editor. Advances in parasitology, vol. 13. Academic Press, Inc., New York, pp. 183-222.

16. Laurence, J. 1985. The immune system in AIDS. Sci. Am. 253:84-93 (Dec.).

17. Lee, T.D.G., M. Swieter, and A.D. Befus. 1986. Mast cell responses to helminth infection. Parasitol. Today 2:186-191.

18. Lie, K.J. 1982. Swellengrebel lecture. Survival of *Schistosoma mansoni* and other trematode larvae in the snail *Biomphalaria glabrata*. A discussion of the interference theory. Trop. Geogr. Med. 34:111-122.

19. Maramorosch, K., and R.E. Shope, editors. 1975. Invertebrate immunity. Mechanisms of invertebrate vector-parasite relations. Academic Press, Inc., New York.

20. McLaren, D.J., F.J. Ramalho-Pinto, and S.R. Smithers. 1978. Ultrastructural evidence of complement and antibody-dependent damage to schistosomula of *Schistosoma mansoni* by rat eosinophils *in vitro*. Parasitology 77:313-324.

21. McLaren, D.J., and R.J. Terry. 1981. The protective role of acquired host antigens during schistosome maturation, Parasite Immunol. 4:129-148.

22. Okabe, K., T.L., Kipnis, V.L.G. Calish, and W.D. daSilva. 1980. Cell-mediated cytotoxicity to *Trypansoma cruzi*. I. Antibody-dependent cell mediated cytotoxicity to trypomastigote bloodstream forms. Clin. Immunol. Immunopathol. 16:344-353.

23. Parkhouse, R.M.E., editor. 1984.: Parasite evasion of the immune response, Symposium of the British Society Parasitology, vol 21, Parasitology 88:571-682.

24. Rasmussen, K.R., and W.M. Kemp. 1987. *Schistosoma mansoni:* demonstration of homospecific antibody adsorbed to the tegumental surfaces of adult male parasites from mice. J. Parasitol. 73:448-451.

25. Thompson, L. 1988. AIDS diary. Discover 9:36-38.

26. Tonegawa, S. 1985. The molecules of the immune system. Sci. Am. 253:122-131 (Oct.).

27. Unanue, E.R., and P.M. Allen. 1987. The basis of the immunoregulatory role of macrophages and other accessory cells. Science 236:551-557.

28. Warren, K.S. 1982. Mechanisms of immunopathology in parasitic infections. In Cohen, S., and K.S. Warren, editors. Immunology of parasitic infections, ed. 2. Blackwell Scientific Publications Ltd., Oxford, Eng. pp. 116-137.

29. Wassom, D.L., and G.J. Gleich. 1979. Damage to *Trichinella spiralis* newborn larvae by eosinophil major basic protein. Am. J. Trop. Med. Hyg. 28:860-863.

30. Willett, W.C., W.L. Kilama, and C.M. Kihamia. 1979. *Ascaris* and growth rates: a randomized trial of treatment. Am. J. Public Health. 69:987-991.

SUGGESTED READINGS

Beutler, B., and A. Cerami. 1987. Cachectin—tumour necrosis factor: a cytokine that mediates injury initiated by invasive parasites. Parasitol. Today 3:345-346.

Capron, A., J.-P. Dessaint, M. Capron, M. Joseph, and G. Torpier. 1982. Effector mechanisms of immunity to schistosomes and their regulation. Immunol. Rev. 61:41-66.

Clark, I.A. 1987. Cell-mediated immunity in protection and pathology of malaria. Parasitol. Today 3:300-305.

Cohen, S., and K.S. Warren, editors. 1982. Immunology of parasitic infections, ed. 2. Blackwell Scientific Publications Ltd., Oxford Eng. (Excellent treatment of parasite immunology. Has general chapters and chapters on each group of medical importance.)

Kennedy, R.C., J.L. Melnick, and G.R. Dreesman. 1986. Anti-idiotypes and immunity. Sci. Am. 255:48-56 (July). (Certain portions of antibodies [idiotypes] stimulated by an antigen themselves stimulate production of antibodies against the idiotypes [anti-idiotypes]. These ramify into an intricate network of antibodies against antibodies that help regulate the immune system.)

Mitchell, G.F. 1987. Injection versus infection: the cellular immunology of parasitism. Parasitol. Today 3:106-111.

Playfair, J.H.L. 1982. Immunology at a glance. Oxford, England, Blackwell Scientific Publications. (Good diagrams with an excellent summary of modern immunology, but it takes more than a "glance.")

Rose, M.E., and D.J. McLaren, editors. 1986. Pathophysiological responses to parasites, Symposium of the British Society of Parasitology, vol. 24. Parasitology 94:S1-S182.

Soulsby, E.J.L., editor. 1976. Pathophysiology of parasitic infection. Academic Press, Inc., New York.

Wakelin, D. 1985. Genetic control of immunity to helminth infections. Parasitol. Today 1:17-23.

Woo, P.T.K. 1986. Immune response of fish to parasitic protozoa. Parasitol. Today 3:186-188.

Young, J.D., and Z.A. Cohn. 1988. How killer cells kill. Sci. Am. 258:38-44 (Jan.). (The mechanism by which cytotoxic T and natural killer cells kill their target cells is being studied at the molecular level. They punch holes in the target cells in a manner similar to the action of complement.)

CHAPTER 4

PARASITIC PROTOZOA: FORM, FUNCTION, AND CLASSIFICATION

Because of the small size of most of them, protozoa were not detected until Leeuwenhoek invented his magnifying lenses in the seventeenth century. He recounted his discoveries to the Royal Society of London in a series of letters covering a period between 1674 and 1716. Among his observations were oocysts of a parasite of rabbit livers, the species known today as *Eimeria stiedai*. Another 154 years passed before the second sporozoan was found, when in 1828 Delfour described gregarines from the intestines of beetles. Leeuwenhoek also found *Giardia intestinalis* in his own diarrheic stools, and he found *Opalina* and *Nyctotherus* in the intestines of frogs. By the middle of the eighteenth century other parasitic protozoa were being reported at a rapid rate, and such discoveries have continued unabated to the present. At least 45,000 species of protozoans have been described to date, many of which are parasitic. Parasitic protozoans still kill, mutilate, and debilitate more people in the world than any other group of disease organisms. Because of this, studies on protozoa occupy a prominent place in parasitology and are covered in some detail in this book. We will begin their study with a review of protozoan form and function.

FORM AND FUNCTION

Every protozoan consists of a single cell, although many species contain more than one nucleus during all or portions of their life cycles. Phenomenal adaptations to wide varieties of ecological niches have evolved, a great many of which resulted in parasitic or other symbiotic associations. The success of protozoa is, to a large extent, the result of their remarkable development of organelles, which perform the same functions as do organs in higher life forms.

Protozoa have traditionally been considered a single phylum of animals, although it was recognized that the group was a large, heterogenous assemblage that was almost certainly not monophyletic. A revision of

classification published in 1980 by a committee of the Society of Protozoologists recognizes seven phyla (p. 49).[13] We will follow the Society of Protozoologists' classification in this book. In spite of their diversity, studies of the fine structure of protozoa have shown that most of their organelles do not differ in any basic way in any of the phyla or from those in metazoan cells. Indeed, Pitelka[17] concluded "that the fine structure of protozoa is directly and inescapably comparable with that of cells of multicellular organisms," and the "morphologist has to start out by admitting that protozoa are, at the least, cells." In the following discussion we shall attempt to emphasize the basic similarities of organelles in protozoa and those in other kinds of cells and to use terminology that is consistent with modern knowledge of ultrastructure.

Nucleus and cytoplasm

Together with metazoa, fungi, and plants, protozoa are described as **eukaryotes;** that is, their genetic material (**deoxyribonucleic acid,** or **DNA**) is carried on well-defined **chromosomes** combined with basic proteins called **histones,** and the chromosomes are contained within a membrane-bound **nucleus.** In contrast, the DNA of bacteria and cyanobacteria (blue-green algae)—**prokaryotes**—is a long, coiled, single molecule, lying free in the cytoplasm. In addition to their simple chromosome, prokaryotes do not have the elaborate differentiation of membranous organelles characteristic of eukaryotes.

Like all cells, the bodies of protozoa are covered by a **plasma membrane.** The membrane appears three layered in electron micrographs because the central lipid portion looks light or clear (**electron lucent**) and is enclosed by the darker (**electron dense**) protein layers. As in other eukaryotes, the nuclei of protozoa are bound by a double membrane with pores. The several other membranous organelles characteristic of eukaryotes, such as endoplasmic reticulum, mitochondria, various membrane-bound vesicles, and Golgi bodies,

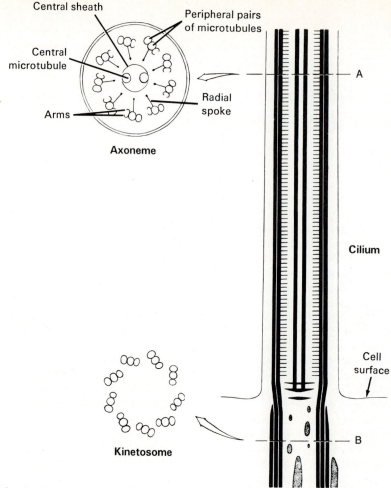

FIG. 4-1

Structure of a cilium showing a section through the axoneme within the cell membrane (**A**) and a section through the kinetosome (**B**). The nine pairs of filaments plus the central pair make up the axoneme. The central pair ends at about the level of the cell surface in a basal plate (axosome). The peripheral filaments continue inward for a short distance to comprise two of each of the triplets in the kinetosome.

From Hickman, C.P., Jr., L.S. Roberts, and F.M. Hickman. 1988. Integrated principles of zoology, ed. 8. The C.V. Mosby Co., St. Louis.

are usually found in protozoa. **Mitochondria,** the organelles that bear the enzymes of oxidative phosphorylation and the tricarboxylic acid cycle, often have tubular rather than lamellar cristae in protozoa, although they may be absent altogether. As in many other eukaryotic cells, protozoa possess membrane-bound bodies called **microbodies.**[7] These are usually spherical structures with a dense, granular matrix. In most animal and many plant cells the microbodies contain oxidases and catalase. The oxidases reduce oxygen to hydrogen peroxide, and the catalase decomposes the hydrogen peroxide to water and oxygen. Thus the microbodies in these cells are called **peroxisomes** because of their biochemical activity. Peroxisomes are found in many aerobic protozoa,[14] that is, protozoa in which oxygen is a terminal electron acceptor in their metabolism (p. 46). In at least some anaerobic protozoa, the microbodies produce molecular hydrogen and are called **hydrogenosomes** (p. 88, Figs. 4-2 and 4-3). The microbodies in at least some Kinetoplastida have been characterized as **glycosomes**[16] (p. 60).

The cytoplasmic matrix consists of very small gran-

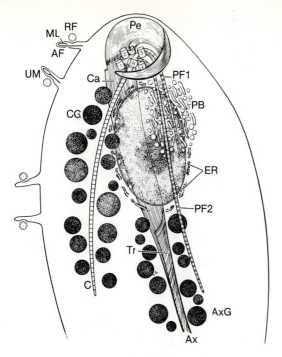

FIG. 4-2

A composite schematic diagram of a trichomonad flagellate seen from a dorsal and slightly right view. *AF*, Accessory filament; *AxG*, paraxostylar granules (hydrogenosomes): *C*, costa; *Ca*, capitulum of the axostyle; *CG*, paracostal granules (hydrogenosomes); *ER*, endoplasmic reticulum; *ML*, marginal lamella; *Pe*, pelta; *PB*, parabasal body; *PF1* and *PF2*, parabasal filaments; *R*, kinetosome of recurrent flagellum; *RF*, recurrent flagellum; *Tr*, trunk of the axostyle; *UM*, undulating membrane; *1* to *4*, kinetosomes of the anterior flagella.

From Mattern, C.F.T., B.M. Honigberg, and W.A. Daniel. 1967. J. Proto-zool. 14:321. Reprinted with permission of the Society of Protozoologists.

ules and filaments suspended in a low-density medium with the physical properties of a colloid. Central and peripheral zones of cytoplasm can often be distinguished as the **endoplasm** and the **ectoplasm.** The endoplasm is in the **sol state** of the colloid, and it bears the nucleus, mitochondria, Golgi bodies, and so on. The ectoplasm is often in the **gel state;** it appears more transparent under the light microscope, and it helps give structural rigidity to the protozoan's body. The bases of the flagella or cilia and their associated fibrillar structures, which may be very complex, are embedded in the ectoplasm. The outer membrane and structures immediately beneath it often are referred to as the **pellicle.** Pellicular microtubules or fibrils may course just beneath the unit membrane, presumably to contribute structural integrity.

Nuclei of protozoa exhibit a wide variety of appearances, particularly under the light microscope. The most common type of nucleus in protozoa other than ciliates is described as **vesicular.** These nuclei are characterized by such an irregular distribution of chromatin material that "clear" areas are apparent in the nuclear sap. Condensations of chromatin within the nucleus may be peripheral or internal. One or more nucleoli may be present. **Endosomes,** conspicuous internal bodies, are thought to be analogous to nucleoli, although they do not disappear during mitosis. Parasitic amebas, trypanosomes, and phytoflagellates have endosomes. **Compact** or **condensed nuclei** are exemplified in the Ciliophora. Ciliates have two types of nuclei, **micronuclei** and **macronuclei.** Micronuclei, as the name implies, are much smaller than macronuclei. Their major function seems to be the sequestration of genetic material for exchange during **conjugation,** the process of sexual reproduction in ciliates. During conjugation the micronuclei undergo meiosis, and haploid micronuclei are exchanged between two fused individuals. During conjugation the macronuclei have been resorbed and are subsequently reformed by division from the micronuclei. Macronuclei take a variety of forms, according to species, but their common function is the genetic direction of the phenotypic expression of the organism (feeding, digestion, locomotion, excretion, and so on). Macronuclei divide amitotically and are hyperpolyploid. They appear "compact" by light microscopy because clear areas of nucleoplasm are not observable, although present. On the electron microscope level one can distinguish a large number of granules, apparently chromatin, randomly scattered throughout a fine fibrogranular reticulum. Nucleoli are large and sometimes numerous in ciliates.

Locomotory organelles

Protozoa move by four basic types of organelles: flagella, cilia, pseudopodia, and undulating ridges. **Flagella** are slender, whiplike structures composed of a central **axoneme** and an outer sheath that is a continuation of the cell membrane. The axoneme consists of nine peripheral and one central pair of microtubules that are enclosed in an inner sheath. The central two microtubules are bilateral, and the plane of the flagellar beat is associated with their orientation. The entire unit—the shaft and its basal fibrils and organelles—is called the **kinetid** or **mastigont.** The flagellum may be buried in the cell membrane along much of its length, forming a finlike **undulating membrane.** A flagellum is capable of a variety of movements, which may be fast or slow, forward, backward, lateral, or spiral. The stroke may originate at the base, thereby propelling the rest of the cell ahead of it, or it may begin at its tip, effecting a force that pulls the cell behind it.

The base of the axoneme terminates in a complicated **root system** that varies greatly in complexity in different flagellates. The entire flagellar base is sunken into an elongated, blind pouch, the **flagellar reservoir.** The axoneme arises from a small centriole, called a **basal body, kinetosome** or **blepharoplast,** in the cytoplasm. (A second, nonfunctional basal body may be found nearby.) In the order Kinetoplastida, including the trypanosomes (Chapter 5), a large, dark-staining body called a **kinetoplast** is found near the kinetosome. It varies in structure in different species but consists of a double membrane enclosing DNA that has different genetic properties from the nucleus. The kinetoplast is part of a mitochondrion that may run most of the length of the animal's body. It divides by binary fission at mitosis. Under light microscopy the kinetosome and kinetoplast are often too close together to be differentiated.

The structure of the kinetosome varies in detail over the wide range of flagellated or ciliated cells, but the basic structure is always similar. Fig. 4-1 illustrates the structure of a typical kinetosome.[10] A short cylinder, with a constant diameter throughout its length, is formed by nine groups of three microtubules and the microtubules in each triplet are so close together that they share a common wall where they touch. If one views a kinetosome from the base distally (Fig. 4-1), *B*) it is clear that the triplets are skewed inward in a clockwise direction, with fine filaments projecting in a cartwheel-like fashion toward a central hub from each triplet. The "lumen" of the kinetosome is open to and appears continuous with the rest of the cytoplasm in the cell. More distally in the kinetosome, the skewing becomes less, and the cartwheel disappears. Finally, at the distal end *one* microtubule in each triplet tapers to an end, and the kinetosome is closed by a somewhat more electron-dense, discoid structure called the **terminal plate.** This plate is often positioned very close to the cell surface where the shaft of the flagellum begins its protrusion. Distal to the terminal plate is a disc-shaped **axosome** from which two central microtubules originate and continue throughout the flagellar shaft. Thus, the flagella of protozoa, like those of other eukaryotes, have the familiar 9 + 2 structure: a circle of nine pairs of microtubules and two central microtubules. Exceptions to the 9 + 2 pattern in motile flagella of sperm in a number of groups are known.[20] Examples are the 9 + 1 flagella of sperm tails of flatworms (Chapter 15) and the 6 + 0 and 3 + 0 patterns of some gregarines.[18] The central and peripheral microtubules, plus the cytoplasm within the cylinder, constitute the **axoneme** of the flagellum. Regularly spaced, short arms extend from one of the fibrils in each pair in the direction of the next pair (Fig. 4-1, *A*); radial spokes extend from each peripheral pair of microtubules toward the central two.

The kinetosome is of critical importance in the formation and operation of the individual flagellum. Not only are the nine peripheral pairs of microtubules in the axoneme direct extensions of two microtubules in each triplet in the kinetosome, but the kinetosome is responsible for actual formation of the axoneme. Furthermore, when the cells divide, the kinetosomes replicate themselves (or serve as organizing centers for such replication) before forming the new flagella. Thus, the kinetosome closely resembles a structure found in many other animal cells, the **centriole,** which is so important in organizing fibrillar structures (for example, the spindle) during cell division. A more complex flagellar root system occurs in a number of species with multiple flagella (Figs. 4-2 and 4-3). Some species have a prominent, striated rod, the **costa,** which courses from one of the kinetosomes along the margin of the organism just beneath the recurrent flagellum and undulating membrane. A tube-like **axostyle,** formed by a sheet of microtubules, may run from the area of the kinetosomes to the posterior end, where it may protrude. A Golgi body may be present, and if a periodic fibril, the **parabasal filament,** runs from the Golgi body to contact a kinetosome, the Golgi body is referred to as a **parabasal body.** The function of the parabasal body is probably similar to that of the Golgi body in other cells. A fibril running from a kinetosome to a point near the surface of the nuclear membrane is called a **rhizoplast,** and the entire complex of organelles of the mastigont and nucleus may be referred to as the **karyomastigont.**

Some symbionts, such as the flagellates found in cockroaches and ruminants, have dense coats of flagella covering their entire body. The root systems are correspondingly complex but will not be discussed here.

One feature that differentiates flagellates with numerous flagella from the ciliates is that division of the body at fission occurs between the rows of flagella **(symmetrogenic fission),** but in ciliates the body divides across the rows of cilia **(homothetogenic).**

Cilia are structurally similar to flagella, with a kinetosome and an axoneme composed of two central and nine peripheral microtubules. Cilia differ from flagella only in the pattern of their beat; cilia propel fluid parallel to the surface of the cell, whereas flagella propel fluid parallel to the long axis of the flagellum. Because cilia are usually numerous on ciliate protozoa, their root systems are complex. A fiber, the kinetodesma (plural: kinetodesmata), arises from each kinetosome and joins a similar fiber from the adjoin-

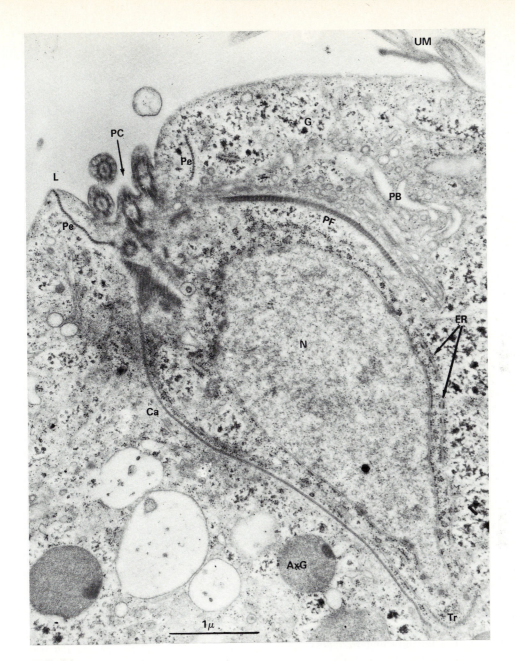

FIG. 4-3

Section through the anterior portion of *Trichomonas gallinae* seen in dorsal and slightly left view. The capitulum *(Ca)* of the axostyle and a few paraxostylar granules (hydrogenosomes) *(AxG)* are located ventral to the nucleus *(N);* the parabasal body *(PB)* and its accompanying filament *(PF)* are dorsal and to the right of the nucleus. The pelta *(Pe)* extends to the extreme anterior end of the organism, terminating at the cell membrane in the area of the periflagellar lip *(L)*. The proximal segments of the flagella are seen within the periflagellar canal *(PC)*. The undulating membrane *(UM)* with its recurrent flagellum is located dorsally. Near the lower right corner note the proximal part of the axostylar trunk *(Tr)*. Two additional noteworthy features of this micrograph are the absence of mitochondria and the presence of large numbers of dense granules presumed to be glycogen *(G)*. *ER,* Endoplasmic reticulum. (×32,600.)

From Mattern, C.F.T., B.M. Honigberg, and W.A. Daniel. 1967. J. Protozool. 14:322. Reprinted with permission of The Society of Protozoologists.

ing cilium in the same row. The resulting compound fiber of kinetodesmata is called a **kinetodesmose.** The row of kinetosomes and their kinetodesmose is a **kinety,** and all the kineties and associated fibrils constitute the **infraciliature.** The mechanism of coordination appears to be by waves of depolarization of the cell membrane, similar to the phenomenon in a nerve impulse.

The mechanism by which flagella and cilia move requires ATP and involves the interaction of the arms of each microtubule pair (Fig. 4-1) with the neighboring pair of microtubules. This causes one member of a pair to slide lengthwise relative to the other microtubule in the pair ("sliding microtubule model"). For a more complete explanation, refer to Satir.[20]

Several varieties of ciliary specialization have evolved within the ciliate protozoa. One specialization involves the reduction of somatic ciliature across parts of the body. In many cases cilia fuse to form **ciliary organelles** (Fig. 4-4). Fused somatic cilia form tuftlike brushes of cilia, called **cirri,** in some species. These function like tiny legs, enabling their owner to walk about. Most ciliary specializations occur in the oral region. If a longitudinal row of cilia fuses along its base, it forms an **undulating membrane** (which must not be confused with the undulating membrane of the flagellates). This functions in moving food particles into the oral groove. Short, transverse rows of cilia, fused at their bases to form a triangular flap, are called **membranelles** (Fig. 4-5). These also are organelles that serve to move food particles toward the cytostome. In the heterotrichs and other ciliates of the class Polymenophorea, there is a prominent **adoral zone of membranelles (AZM)** (Fig. 4-4). A group of kinetosomes forming a tuft of ciliary organelles in the aboral region of peritrich ciliates is called the **scopula.** It is involved in stalk formation.

Pseudopodia are temporary organelles found in the Sarcodina (and other organisms) that cause the organism to move and aid it in capturing food. They do not occur in all sarcodines; some amebas flow along with no definite body extensions (**limax forms,** named after the slug *Limax*). Four types of pseudopodia are found among the rest of the Sarcodina (Fig. 4-6). Most of the amebas have **lobopodia,** which are finger-shaped, round-tipped pseudopodia that usually contain both ectoplasm and endoplasm. All parasitic and commensal amebas of humans have this kind of pseudopodium. **Filopodia** are slender, sharp-pointed organelles, composed only of ectoplasm. They are not branched like **rhizopodia,** which branch extensively and fuse together to form netlike meshes. **Axopodia** are like filopodia but contain a slender axial filament composed of microtubules that extend into the interior of the cell.

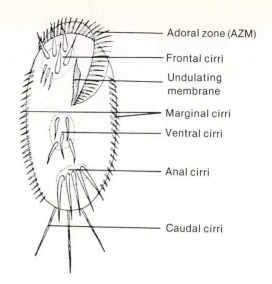

FIG. 4-4

Stylonychia, a ciliate protozoan, showing ciliary organelles.
From Kudo, R. 1966. Protozoology, ed. 5. Charles C Thomas, Publisher, Springfield, Ill.

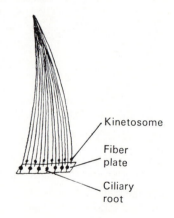

FIG. 4-5

A ciliary membranelle.
From Kudo, R. 1966. Protozoology, ed. 5. Charles C. Thomas, Publisher, Springfield, Ill.

Movement by means of pseudopodia is a complex form of protoplasmic streaming. Current evidence is that the mechanism, like most other forms of cell movement (except in cilia and flagella), involves the interaction of microfilaments of actin with myosin. Nachmias[15] gives a good review of cell movement by this mechanism.

In most sporozoan parasites locomotion is accomplished by **undulating ridges.** The merozoites, ookinetes, and sporozoites appear to glide through fluids with no subcellular motion whatever, but electron microscope studies reveal tiny undulatory waves that form in the cell membrane and pass posteriad. This ef-

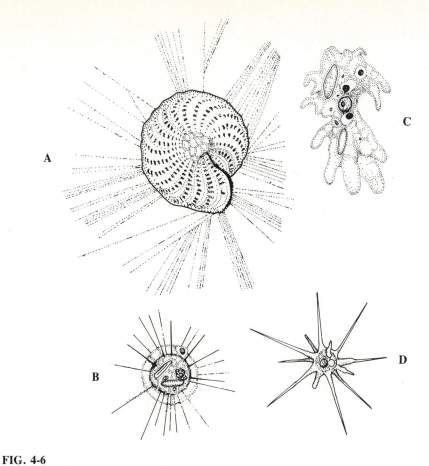

FIG. 4-6

Types of pseudopodia. **A,** Rhizopodia. **B,** Axopodia. **C,** Lobopodia. **D,** Filopodia.

Redrawn from Kudo, R. 1966. Protozoology, ed. 5. Charles C. Thomas, Publisher, Springfield, Ill.

fectively propels the cell forward, albeit at a slow rate. Subpellicular microtubules may aid this action in some species. The mechanism is unknown, but it probably also involves actin and myosin.

Nutrition

Several types of nutrition are found in protozoa. In **holophytic nutrition** (also known as **photoautotrophic nutrition**) carbohydrates are synthesized by chloroplasts, the organelles of the "typical" plant. Zooxanthellae (order Dinoflagellida) are very important mutuals living in the cells of reef-forming corals and other invertebrates (including some other protozoa). They contribute a substantial amount of nutrient to their hosts.

Holozoic nutrition is typical of many parasitic protozoa, which feed by ingesting entire organisms or particles thereof. Their mouth openings may be temporary, as in amebas, or permanent **cytostomes,** as in most ciliates. Particulate food passes into a food vac-

uole, which is a digestive organelle that forms around any food thus ingested. Indigestible material is voided either through a temporary opening or through a permanent **cytopyge,** which is found in many ciliates. **Pinocytosis** is an important activity in many protozoa, as is **phagocytosis.** Both pinocytosis and phagocytosis are examples of **endocytosis,** differing only in that pinocytosis deals with droplets of fluid, whereas phagocytosis is the process of internalizing particulate matter. Both may be important in nutrition of protozoa, although the extent of each is not always clear.[1,9] A submicroscopic micropore is present in *Eimeria* and *Plasmodium* and, in certain stages, is involved in taking in nutrients (Fig. 4-7).

In **saprozoic nutrition,** nutrients are assimilated either by diffusion through the cell membrane or by pinocytosis.

Permeation, the passing of molecules directly through the outer cell membrane, may be one of three different types. **Diffusion** is possible when the cell

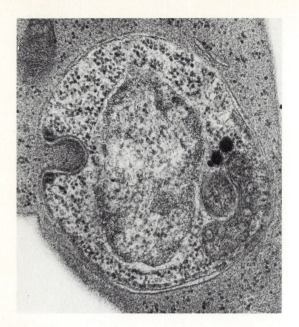

FIG. 4-7

A uninucleate trophozoite of *Plasmodium cathemerium* ingesting host cell cytoplasm through a cytostome (micropore). (×52,000.)

From Aikawa, M.P.K. Hepler, C.G. Huff, and H. Sprinz. 1966. J. Cell Biol. 28:362.

membrane is permeable to a particular molecule and when the concentration of that molecule is lower inside the cell than outside. Few molecules satisfy these requirements; the diffusion component of nutrition for most protozoa must be negligible, although it may be more important for some parasites. A carrier molecule that has binding sites for the nutrient may be present in the cell membrane. The membrane itself might not be permeable to the carrier, which picks up the nutrient molecule at the outer membrane surface and then releases it into the cytoplasm. This mechanism is known as **facilitated diffusion.** However, like free diffusion, facilitated diffusion cannot operate against a concentration gradient. The accumulation of molecules against a concentration gradient requires the expenditure of energy and is called **active transport.** Active transport appears to operate through a carrier in the cell membrane, as in facilitated diffusion, but the action of the carrier must be coupled with an energy-yielding metabolic reaction. Obviously the ultimate value of the molecule or the energy derived from it must be greater than the energy expended in acquiring it. Some important food molecules such as glucose are brought into the cell by active transport.

The nutrition of intracellular parasitic protozoa is so intimately bound to the metabolic activities of the host cell that it almost appears as if the parasitized cell willingly contributes to the welfare of its guest. In some, entry into the host cell is by phagocytosis of the parasite. An example is *Leishmania donovani,* which is eaten by reticuloendothelial cells. The host cell forms a membrane-bound vacuole around the parasite, but instead of killing the parasite with digestive enzymes, as might be expected, the host cell provides it with nutrients. The host is controlling the flow of materials to the parasite that will kill it.

A different mode of entry into a host cell is employed by members of the phylum Microspora (Chapter 10). The cyst stage of these parasites contains a coiled, hollow filament that apparently is under great pressure. When eaten by the host, which is usually an arthropod, the tubule is forcibly extruded from the cyst and penetrates an adjoining host cell. The organism within the spore (**sporoplasm**) crawls through the tube and enters its host. In this case the membrane of the parasite is in direct contact with the cytoplasm of the host, with no vacuole being formed around it. The protozoan is then free to assimilate the nutrients it needs.

Active invasion of host cells by motile infective stages of protozoa is another means by which these tiny animals become intracellular parasites. Several of the Coccidia, such as *Toxoplasma, Eimeria,* and *Cryptosporidium* (Fig. 8-22), have been found to penetrate host cells actively, not by injection as in the Microspora but by a boring action, probably aided by digestive secretions.[21] Once within the cell, the parasite is surrounded by layers of host endoplasmic reticulum, forming a **parasitophorous vacuole** (Fig. 4-8). The host cell then proceeds to provide nourishment for the parasite, as in the case of *Leishmania.*

Whether an intracellular parasite is bound by one membrane or by two, it obtains its nourishment mainly by endocytosis. When a parasitophorous vacuole surrounds the parasite, the host appears to extrude material into it, and the protozoan then takes it up by phagocytosis.[9] How the parasite thus manipulates its host to provide room and board is unknown.

Excretion and osmoregulation

Most protozoa appear to be **ammonotelic;** that is, they excrete most of their nitrogen as ammonia, most of which readily diffuses directly through the cell membrane into the surrounding medium. Other, sometimes unidentified waste products are also produced, at least by intracellular parasites. These substances are secreted and accumulated within the host cell and, on the death of the infected cell, have toxic effects on the host. Carbon dioxide, lactate, pyruvate, and short-

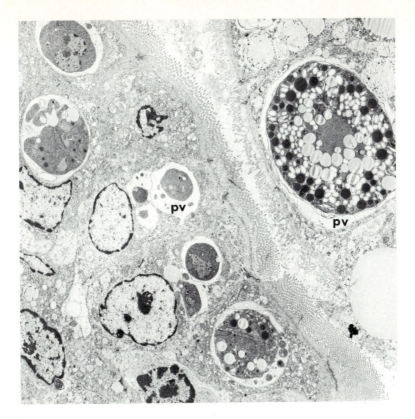

FIG. 4-8

Electron micrograph of several sexual stages of *Eimeria magna* in epithelial cells of the lower two thirds of the small intestine of a domestic rabbit. Each parasite is surrounded by a parasitophorous vacuole *(pv),* two of which are labeled, 5½ days after experimental inoculation of rabbit with 200,000 oocysts. (×4100.)

Courtesy Clarence Speer.

chain fatty acids are also common waste products.

Contractile vacuoles are probably more involved with osmoregulation than with excretion per se. Because free-living, freshwater protozoa are hypertonic to their environment, they imbibe water continuously by osmosis. The action of contractile vacuoles effectively pumps out the water. Marine species and most parasites do not form these vacuoles, probably because they are more isotonic to their environment. However, *Balantidium* (Chapter 11) contains contractile vacuoles.

Reproduction

Reproduction in protozoa may be either asexual or sexual, although many species alternate types in their life cycles. Most often, **asexual reproduction** is by **binary fission,** in which the individual divides into two. The plane of fission is random in Sarcodina, longitudinal in flagellates (between kineties, symmetrogenic),

and transverse in ciliates (across kineties, homothetogenic). The sequence of division is kinetosome(s), kinetoplast (if present), nucleus, and cytokinesis. With the exception of the macronucleus of ciliates, nuclear division during asexual reproduction of protozoa is by mitosis. However, patterns of mitosis are much more diverse among the protozoa than among the metazoa. Details of the patterns of diversity are beyond the scope of this book but include the fact that the nuclear membrane often retains its identity through mitosis, that spindle fibers may form within the nuclear membrane, that centrioles may not be present, or that the chromosomes may not go through a well-defined cycle of condensation and decondensation. Nevertheless, the essential features of mitosis—replication of the chromosomes and regular distribution of the daughter chromosomes to the daughter nuclei—are always present.

Multiple fission, or **merogony, schizogony,** occurs

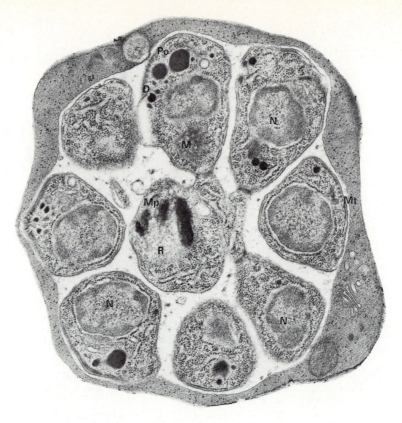

FIG. 4-9

A late stage in the development of *Plasmodium cathemerium* within the host erythrocyte. The segmentation has been almost completed, and paired organelles *(Po)*, dense bodies *(d)*, nucleus *(n)*, mitochondrion *(M)*, pellicular complex with microtubules *(Mt)*, and ribosomes are observed in the new merozoites. A residual body *(R)* surrounded by a rim of cytoplasm of the mother schizont contains a cluster of malarial pigment *(Mp)* granules. (×30,000.)

From Aikawa, M. 1966. Am. J. Trop. Med. Hyg. 15:467.

in the Sarcodina and Sporozoea. In this type of division the nucleus and other essential organelles divide repeatedly before cytokinesis; thus a large number of daughter cells are produced almost simultaneously. During schizogony the cell is called a **schizont, meront,** or **segmenter.** The daughter nuclei in the schizont arrange themselves peripherally, and the membranes of the daughter cells form beneath the cell surface of the mother cell, bulging outward (Fig. 4-9). The daughter cells are **merozoites,** and they finally break away from a small residual mass of protoplasm remaining from the mother cell to initiate another phase of merogony or begin gametogony. Schizogony to produce merozoites may be referred to as **merogony.** Another type of multiple fission often recognized is **sporogony,** which is multiple fission after the union of gametes (see sexual reproduction).

Several forms of **budding** can be distinguished. **Plasmotomy,** sometimes regarded as budding, is a phenomenon in which a multinucleate individual divides into two or more smaller, but still multinucleate, daughter cells. Plasmotomy itself is not accompanied by mitosis. **External budding** is found among some complex ciliates, such as the Suctoria. Here nuclear division is followed by unequal cytokinesis, resulting in a smaller daughter cell, which then grows to its adult size.

Internal budding, or **endopolyogeny,** differs from schizogony only in the location of the formation of daughter cells. In this process the daughter cells begin forming within their cell membranes, distributed throughout the cytoplasm of the mother cell rather than at the periphery. The process occurs in some stages of the schizonts of the Eimeriina. **Endodyogeny**

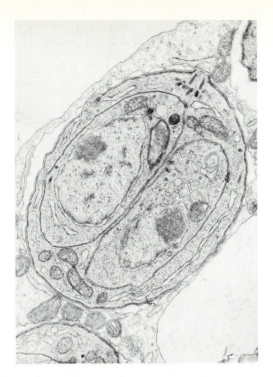

FIG. 4-10

Toxoplasma gondii exhibiting two daughter cells in a mother cell, formed by endodyogeny.

From Vivier, E., and A. Petitprez. Reproduced from *The Journal of Cell Biology,* 1968, vol. 43, p. 337, by copyright permission of The Rockefeller University Press.

is endopolyogeny in which only two daughter cells are formed (Fig. 4-10).

Sexual reproduction is of two basic types in parasitic protozoa. Sexual reproduction involves reductional division in meiosis, causing a change from diploidy to haploidy, with a subsequent union of two cells to restore diploidy by amphimixis. The cells that join to restore diploidy are the **gametes,** and the process of producing the gametes is **gametogony.** Cells responsible for the production of gametes are **gamonts.**

Reproduction may be **amphimictic,** involving the union of gametes from two parents, or **automictic,** in which one parent gives rise to both gametes. Uniting gametes may be entire cells or only nuclei. When they are whole cells, the union is called **syngamy.** When only nuclei unite, the process is termed **conjugation.** Conjugation is found only among the ciliates, whereas syngamy occurs in all other groups in which sexual reproduction is found. Meiosis is known in both types of sexual reproduction.

In the majority of the protozoa, including all of the Sporozoea, meiosis occurs in the first division of the zygote (**zygotic meiosis**),[8] and all other stages are haploid. **Intermediary meiosis,** which occurs only in the Foraminiferida among the protozoa but which is widespread in plants, exhibits a regular alternation of haploid and diploid generations.

In syngamy the gametes may be outwardly similar (**isogametes**) or dissimilar (**anisogametes**). Isogamy is most common in the more primitive groups and, therefore, is considered more primitive than anisogamy. Although isogametes look similar, they will fuse only with isogametes of another "mating type," thus avoiding inbreeding. Differences between these "sexes" in anisogametes vary from slight size differences to marked dimorphism. The larger, more quiescent of the two is the **macrogamete,** which corresponds to the ovum of metazoa. The smaller, more active gamete is the **microgamete,** which corresponds to the spermatozoon, although it is debatable whether "male" and "female" sexes can be distinguished in protozoa or whether such a distinction is even useful. Fusion of the microgamete and macrogamete produces the **zygote,** which is often a resting stage that overwinters or forms spores that enable survival between hosts.

Conjugation occurs only in ciliates and varies somewhat in details among species. Two individuals ready for conjugation unite, fusing their pellicles at the point of contact. The macronucleus in each disintegrates, and the micronuclei undergo divisions involving meiosis. A pronucleus from each passes into the other conjugant, there to fuse with a remaining pronucleus and restore diploidy. The cells separate, and subsequent nuclear divisions produce one or more macronuclei. The resultant gene recombination lends renewed vigor to the exconjugants, which then actively reproduce by fission. Variations of conjugation are **cytogamy,** in which two individuals fuse but do not exchange pronuclei, with two pronuclei in each cell rejoining to restore diploidy, and **autogamy,** in which haploid pronuclei from the same cell fuse but there is no cytoplasmic fusion with another individual.

Encystment

Many protozoa can secrete a resistant covering and go into a resting stage called as **cyst.** Cyst formation is particularly common among free-living protozoa found in temporary bodies of water that are subject to drying or other harsh conditions and among parasitic forms that must survive transferral to new hosts.[22] In addition to protection against unfavorable conditions, cysts may serve as sites for reorganization and nuclear division, followed by multiplication after excystation. In a few forms, such as *Ichthyophthirius,* a ciliate parasite of fish, the cyst falls from the host to the substrate and sticks there until excystation occurs

(Chapter 10). Cellulose has been found in the cyst walls of some amebas, and others contain chitin.[2]

The conditions favoring encystment are not fully understood, but they are thought in most cases to involve some adverse change in the environment, such as food deficiency, desiccation, increased tonicity, decreased oxygen concentration, or pH or temperature change. In parasitic species the normal feeding form (**trophozoite,** also sometimes referred to as the **vegative stage**) often cannot infect a new host or is too fragile to survive the transfer. Human amebiasis, caused by *Entamoeba histolytica,* is spread by persons who often have no clinical symptoms but who pass cysts in their feces (Chapter 9). Therefore understanding the elusive factors that induce cyst formation within the host is important.

During encystment the cyst wall is secreted, and some food reserves, such as starch or glycogen, are stored. Projecting portions of locomotor organelles are partially or wholly resorbed, and certain other structures, such as contractile vacuoles, may be dedifferentiated. During the process or following soon therafter, one or more nuclear divisions give the cyst more nuclei than the trophozoite. In the flagellates and amebas, cytokinesis occurs in a characteristic division pattern after excystation. In Sporozoea the cystic form is the **oocyst,** which is formed after gamete union and in which multiple fission (**sporogony**) occurs with cytokinesis to produce **sporozoites.** In the Eimeriina the oocyst containing the sporozoites serves as the resistant stage for transferral to a new host, whereas in the Haemosporina (containing the causative agent of malaria, *Plasmodium* spp.) the oocyst merely serves as a developmental capsule for the sporozoites within the insect host.

In species in which the cyst is a resistant stage, a return of favorable conditions stimulates excystation. In parasitic forms some degree of specificity in the requisite stimuli provides that excystation will not take place except in the presence of conditions found in the host gut. Mechanisms for excystation may include absorption of water with consequent swelling of the cyst, secretion of lytic enzymes by the protozoan, and action of host digestive enzymes on the cyst wall. Excystation must include reactivation of enzyme pathways that were "turned off" during the resting stage, internal reorganization, and redifferentiation of cytoplasmic and motor organelles.

Metabolism

Because the phyla in the chapters to follow are so diverse and because metabolic studies have been concentrated on so few parasitic species, few generalizations are possible or practical. Therefore, the following considerations will be rather simplified and limited, and we will comment appropriately on specific groups in subsequent chapters.

The main energy in protozoa, as in other cells, is in the form of **high-energy phosphate bonds,** primarily in **adenosine triphosphate (ATP).** Energy is released in the step-by-step, enzymatic oxidation of food molecules; part of this energy is conserved by coupling these oxidations with the phosphorylation of **adenosine diphosphate (ADP)** to ATP. Subsequent hydrolysis of the high-energy phosphate bond yields energy to drive other endergonic reactions in the cell. Conceptually, oxidation of the main energy-source molecule, **glucose,** can be divided into three phases: **glycolysis,** the **Krebs** or **tricarboxylic acid cycle,** and **electron transport.** Glycolysis, sometimes called the Embden-Meyerhof pathway, is the degradation of the six-carbon compound, glucose, to the three-carbon pyruvate (one molecule glucose → two molecules pyruvate) (Fig. 4-11). The pyruvate then may be reduced to lactate, or it may enter a mitochondrion and be routed into the tricarboxylic acid cycle. To enter the tricarboxylic acid cycle, pyruvate is decarboxylated to a two-carbon group and joined to the acyl carrier, **coenzyme A (CoA),** a low-molecular weight compound containing the vitamin pantothenic acid. The resulting compound, **acetyl-CoA,** is condensed with oxaloacetate to form citrate. Through the series of reactions constituting the tricarboxylic acid cycle, the citrate is decarboxylated twice, finally producing another molecule of oxaloacetate to condense with another molecule of acetyl-CoA (Fig. 4-12). Since one molecule of glucose produces two of pyruvate, the tricarboxylic acid cycle goes through two cycles per molecule of glucose.

In overview the reactions of glycolysis and the tricarboxylic acid cycle may be considered the oxidation of the carbons in glucose to carbon dioxide. Of course, in any given oxidation reaction, one compound is always **oxidized (electron donor),** and another is always **reduced (electron acceptor).** The tendency of a compound to give up or gain electrons determines whether it is an oxidizing or reducing agent and also the final distribution of electrons at equilibrium. In the oxidation-reduction reactions of glycolysis and then in the tricarboxylic acid cycle, the principal electron acceptor is **nicotinamide-adenine dinucleotide (NAD).** The reduced NAD is reoxidized by transferring its electrons to acceptors in the **electron transport chain,** which is a series of carriers that are alternately reduced and oxidized as they accept electrons and then donate them to the next compound in the chain. Several components in the chain are heme-containing proteins called **cytochromes;** hence the

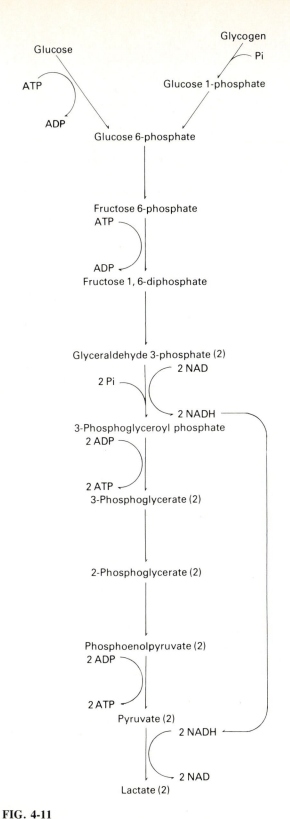

FIG. 4-11

Glycolytic pathway.

chain is sometimes referred to as the **cytochrome system.** The last electron acceptor to be reduced is molecular oxygen, producing water.

As noted, part of the energy released in the oxidation of glucose is conserved by the synthesis of high-energy phosphate bonds in ATP. In glycolysis four molecules of ATP are produced per molecule of glucose, but two ATPs are used in phosphorylation "priming" reactions; hence there is a net gain of two ATPs. Another direct, **substrate level** phosphorylation reaction takes place in the tricarboxylic acid cycle, but a much larger proportion of the energy released by the cycle results from the passage of the reducing power it forms along the electron transport sequence by **oxidative phosphorylation.** Specifically, four ATPs are produced by substrate level (two in glycolysis and two in the Krebs cycle) and 32 by oxidative phosphorylation. Glycolysis can proceed in both the presence or absence of molecular oxygen, that is, either **aerobically** or **anaerobically.** The tricarboxylic acid cycle and electron transport are essentially aerobic processes, oxygen being the necessary, ultimate acceptor of all the electrons initially accepted by NAD. When glycolysis is proceeding under aerobic conditions, the two reduced NAD produced in that path may enter the mitochondria and the electron transport system, but under anaerobic conditions the reduced NAD must be reoxidized in another manner. In vertebrate muscle and some other tissues, the NAD is reoxidized by reduction of the pyruvate from glycolysis to lactate.

Most of our knowledge of the preceding classical system of energy metabolism has been derived from a few species of mammals and microorganisms. Much of it has evolved from discovery to dogma in the space of a few years. It is now becoming clear, however, that parasites, both protozoan and metazoan, are unexpectedly variable in their energy metabolism, particularly in the portions after glycolysis. Some important biological factors to consider are that many parasites must survive in locations in which the oxygen supply is quite limited and that, even in many cases in which oxygen is not limited, neither is glucose; therefore there is no advantage in completely oxidizing glucose. If glucose is in plentiful supply, the organism can live on little more than the energy derived from glycolysis, simply by consuming more glucose, and the partially oxidized products can be excreted as waste. The complete Krebs cycle and cytochrome system then become so much excess metabolic machinery, at least in terms of energy production. However, the problem of reoxidation of the net accumulation of reduced NAD remains, since even without the need for the energy obtainable in subsequent electron transfer, the oxidized compounds must be available for

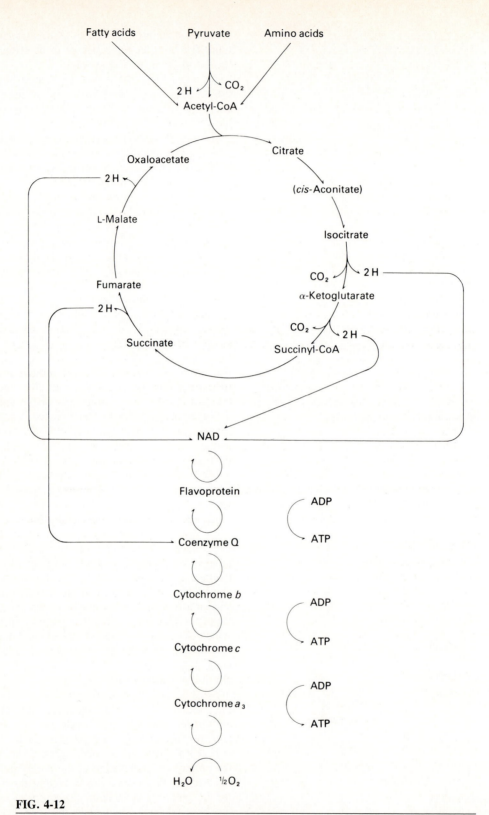

FIG. 4-12

Tricarboxylic acid cycle and classical electron transport system.

continuous functioning of glycolysis. In some parasites the electrons are transferred to the pyruvate, and lactate is excreted, as in the classical model; however, many organisms excrete such compounds as succinate, acetate, and short-chain fatty acids as end products of glycolysis. Some metabolic solutions to the problem of NAD oxidation will be mentioned in subsequent chapters.

Endosymbionts

Just as many protozoa live symbiotically in the bodies of larger animals, many examples are known of organisms living within the bodies of protozoa. Zooxanthellae were mentioned earlier. Many biologists now believe that chloroplasts and mitochondria, and perhaps flagella, arose from prokaryotes that came to live inside other cells. Corliss[5] proposed that all such structures or organisms be referred to as **xenosomes,** which is any body or constituent organelle that contains DNA and is bounded by at least one membrane, living within a cell and capable of reproducing itself. The term implies that "the symbiont once functioned as a free-living organism outside its present residence." In addition to the previous examples, zoochlorellae (green algal cells in protozoa and some multicellular animals), a variety of prokaryotes in protozoa, many intracellular protozoan parasites of multicellular animals, many hyperparasites of parasitic protozoa, and even nuclei of eukaryotic cells would be included. Nonetheless, endosymbionts living within protozoa are numerous and have been discussed by several authors.[3,11,12,19] Their contribution to and interaction with the metabolism of their hosts undoubtedly varies, but in many cases it is critical.

CLASSIFICATION OF THE PROTOZOAN PHYLA

All of the following groups (plus some others) may be properly encompassed in the kingdom Protista, of which they comprise the subkingdom Protozoa. The classification is based on that of Levine et al.[13] Although this is the most authoritative classification currently available, some authorities feel strongly that it does not yet fully reflect true phylogenetic relationships.[4,6] Most of the terminology used has been covered in this chapter; some terms will be defined in the chapters to follow.

PHYLUM SARCOMASTIGOPHORA

Single type of nucleus, except in some Foraminiferida; sexuality, when present, essentially syngamy; flagella, pseudopodia, or both types of locomotor organelles.

Subphylum Mastigophora
One or more flagella typically present in trophozoites; asexual reproduction basically by symmetrogenic binary fission; sexual reproduction known in some groups.

Class Phytomastigophorea
Typically with chloroplasts; if chloroplasts lacking, relationship to pigmented forms clearly evident; mostly free living.

Order Dinoflagellida
Two flagella, typically one transverse and one trailing; body usually grooved transversely and longitudinally, forming a girdle and sulcus, each containing a flagellum; chromatophores usually yellow or dark brown, occasionally green or blue-green; nucleus unique among eukaryotes in having chromosomes that consist primarily only of nonprotein complexed DNA; mitosis intranuclear; flagellates, coccoid unicells, colonies, and simple filaments; sexual reproduction present; few parasites of invertebrates; one or more species (*Zooxanthella microadriatica)* very important mutuals in tissues of various marine invertebrates, especially cnidarians.[6]

Other Orders
Cryptomonadida, Euglenida, Chrysomanadida, Heterochlorida, Chloromonadida, Prymnesiida, Volvocida, Prasinomonadida, Silicoflagellida

Class Zoomastigophorea
Chloroplasts absent; one to many flagella; ameboid forms, with or without flagella in some groups; sexuality known in few groups; polyphyletic.

Order Kinetoplastida
One or two flagella arising from depression; flagella typically with paraxial rod in addition to axoneme; single mitochondrion (nonfunctional in some forms) extending length of body as single tube, hoop, or network of branching tubes, usually containing conspicuous, DNA-containing kinetoplast located near flagellar kinetosomes; Golgi apparatus typically in region of flagellar depression, not connected to kinetosomes and flagella; majority parasitic, some free-living.

Suborder Bodonina
Typically two unequal flagella, one directed anteriorly, the other posteriorly; no undulating membrane; kinetoplastic DNA in several discrete bodies in some, dispersed throughout mitochondrion in some; free-living and parasitic. *Bodo, Cryptobia, Rhynchomonas.*

Suborder Trypanosomatina
Single flagellum either free or attached to body by undulating membrane; kinetoplast relatively small and compact; parasitic. *Blastocrithidia, Leishmania, Trypanosoma.*

Order Proteromonadida
One or two pairs of unequal flagella without paraxial rods; single mitochondrion, distant from kinetosomes, curling around nucleus, not extending length of body, without kinetoplast; Golgi apparatus encircling band-shaped rhizoplast passing from kinetosomes near surface of nucleus to mitochondrion; cysts present; parasitic. *Karotomorpha, Proteromonas.*

Order Retortamonadida
Two to four flagella, one turned posteriorly and associated with ventrally located cytostomal area bordered by fibril; mitochondria and Golgi apparatus absent; intranuclear division spindle; cysts present; parasitic. *Chilomastix, Retortamonas.*

Order Diplomonadida
One or two karyomastigonts; individual mastigonts with one to four flagella, typically one of them recurrent and associated with cytostome or with organelles forming cell axis; mitochondria and Golgi apparatus absent; intranuclear division spindle; cysts present; free living or parasitic.

Suborder Entermonadina

Single karyomastigont containing one to four flagella; one recurrent flagellum in genera with more than single flagellum; frequent transitory forms with two karyomastigonts; parasitic. *Enteromonas, Trimitus.*

Suborder Diplomonadina

Two karyomastigonts; body with twofold rotational symmetry; each mastigont with four flagella, one recurrent; with variety of microtubular bands; free living or parasitic. *Giardia, Hexamita.*

Order Oxymonadida

One of more karyomastigonts, each containing four flagella typically arranged in two pairs in motile stages; one to many axostyles per organism; mitochondria and Golgi apparatus absent; division spindle intranuclear; cysts in some; sexuality in some; parasitic. *Monocercomonoides, Oxymonas.*

Order Trichomonadida

Typically at least some kinetosomes associated with rootlet filaments characteristic of trichomonads, parabasal body present; mitochondria absent; division spindle extranuclear; karyomastigonts with four to six flagella, but only one flagellum in one genus and no flagella in another; pelta and noncontractile axostyle in each mastigont, except for one genus; hydrogenosomes present; true cysts rare; parasitic. *Dientamoeba, Histomonas, Monocercomonas, Trichomonas.*

Order Hypermastigida

Mastigont system with numerous flagella and multiple parabasal bodies, resembling in arrangement kinetosomes of trichomonads and associated with rootlet filaments characteristic of these flagellates, present in many genera; flagella-bearing kinetosomes distributed in complete or partial circle, in plate or plates, or in longitudinal or spiral rows meeting in a centralized structure; one nucleus per cell; mitochondria absent; division spindle extranuclear; cysts in some; sexuality in some; parasitic.

Suborder Lophomonadina

Extranuclear organelles arranged in one system; typically all old structures resorbed in division and new organelles formed in daughter cells. *Lophomonas, Microjoenia.*

Suborder Trichonymphina

Two or occasionally four mastigont systems; typically equal separation of mastigont systems in division, with total or partial retention of old structures when new systems are formed. *Barbulanympha, Spirotrichonympha, Trichonympha.*

Other Order

Choanoflagellida

Subphylum Opalinata

Numerous cilia in oblique rows over entire body; cytostome absent; binary fission generally symmetrogenic; known life cycles involve syngamy with anisogamous flagellated gametes; parasitic.

Class Opalinatea

With characters of the subphylum.

Order Opalinida

With characters of the class. *Opalina.*

Subphylum Sarcodina

Pseudopodia, or locomotive protoplasmic flow without discrete pseudopodia; flagella, when present, usually restricted to developmental or other temporary stages; body naked or with external or internal test or skeleton; asexual reproduction by fission; sexuality, if present, associated with flagellate or, more rarely, ameboid gametes; most free-living.

SUPERCLASS RHIZOPODA

Locomotion by lobopodia, filopodia, or reticulopodia or by protoplasmic flow without production of discrete pseudopodia.

Class Lobosea

Pseudopodia lobose or more or less filiform but produced from broader hyaline lobe; usually uninucleate; no sporangia or similar fruiting bodies.

Subclass Gymnamoebia
Without test.

Order Amoebida

Typically uninucleate; mitochondria typically present; no flagellate stage.

Suborder Tubulina
Body branched or unbranched cylinder. *Entamoeba.*

Suborder Acanthopodina
More or less finely tipped, sometimes filiform, often branched hyaline subpseudopodia, produced from a broad hyaline lobe.

Other Suborders
Thecina, Flabellina, Conopodina

Order Schizopyrenida

Body with shape of monopodial cylinder, usually moving with more or less eruptive, hyaline, hemispherical bulges; typically uninucleate; temporary flagellate stages in most species. *Naegleria.*

Other Order
Pelobiontida
Other Subclass
Testacealobosia

Class Plasmodiophorea

Obligate intracellular parasites with minute plasmodia; zoospores produced in sporangia and bearing anterior pair of unequal flagella; sexuality reported in some species.

Order Plasmodiophorida

With characters of the class. *Plasmodiophora.*

Other Classes

Acarpomyxea, Acrasea, Eumycetozoea, Filosea, Granuloreticulosea, Xenophyophorea

OTHER SUPERCLASS

Actinopoda. **Classes.** Acantharea, Polycystinea, Phaeodarea, Heliozoea

PHYLUM LABYRINTHOMORPHA

Trophic stage as ectoplasmic network with spindle-shaped or spherical, nonameboid cells; in some genera ameboid cells move within network by gliding; with sagenogenetosome (unique cell-surface organelle, associated with ectoplasmic network); saprozoic and parasitic on algae, mostly marine and estuarine.

Class Labyrinthulea

With characters of the phylum.

Order Labyrinthulida
With characters of the class. *Labyrinthula*.

PHYLUM APICOMPLEXA

Apical complex generally consisting of polar ring, micronemes, rhoptries, subpellicular tubules, and conoid present at some stage; micropore(s) usually present; cilia and flagella absent except for flagellated microgametes in some groups; sexuality by syngamy; all parasitic.

Class Perkinsea

Conoid forming incomplete cone; "zoospores" (sporozoites?) flagellated, with anterior vacuole; no sexual reproduction; homoxenous.

Order Perkinsida
With characters of the class. *Perkinsus*.

Class Sporozoea

Conoid, if present, forming complete cone; reproduction generally both sexual and asexual; oocysts generally containing infective sporozoites resulting from sporogony; locomotion of mature organisms by body flexion, gliding, or undulation of longitudinal ridges; flagella present only in microgametes of some groups; pseudopods ordinarily absent, if present used for feeding, not locomotion.

Subclass Gregarinia
Mature gamonts large, extracellular; mucron or epimerite in mature organism; mucron formed from conoid; generally syzygy of gamonts; gametes usually isogamous or nearly so; zygotes forming oocysts within gametocytes; in digestive tract or body cavity of invertebrates; generally homoxenous.

Order Archigregarinida
Apparently primitive life cycle, usually with merogony, gametogony, sporogony; in annelids, sipunculids, hemichordates, or ascidians. *Exoschizon, Selenidioides*.

Order Eugregarinida
Merogony absent; gametogony and sporogony present; typically parasites of arthropods and annelids.

Suborder Blastogregarinina
Gametogony by gamonts while still attached to intestine; no syzygy; gametocysts absent; gamont of single compartment with mucron, without definite protomerite and deutomerite; in polychaete annelids. *Siedleckia*.

Suborder Aseptatina
Gametocysts present; gamont of single compartment, without definite protomerite and deutomerite but with mucron in some species; syzygy present. *Lecudina, Monocystis, Selenidium*.

Suborder Septatina
Gametocysts present; gamont divided into protomerite and deutomerite by septum; with epimerite; in alimentary canal of invertebrates, especially arthropods. *Actinocephalus, Cephaloidophora, Gregarina*.

Order Neogregarinida
Merogony, presumably acquired secondarily; in Malpighian tubules, intestine, hemocoel, or fat tissues of insects. *Caulleryella, Gigaductus*.

Subclass Coccidia
Gamonts ordinarily present; mature gamonts small, typically intracellular, without mucron or epimerite; syzygy generally absent; life cycle characteristically consisting of merogony, gametogony, and sporogony; most species in vertebrates.

Order Agamococcidiida
Merogony and gametogony absent. *Rhytidocystis*.

Order Protococcidiida
Merogony absent; in invertebrates. *Eleutheroschizon, Grellia*.

Order Eucoccidiida
Merogony present; in vertebrates and/or invertebrates.

Suborder Adeleina
Macrogamete and microgamont usually associated in syzygy during development; microgamont producing one to four microgametes; sporozoites enclosed in envelope. *Adelea, Haemogregarina, Klossiella*.

Suborder Eimeriina
Macrogamete and microgamont developing independently; no syzygy; microgamont typically producing many microgametes; zygote not motile; sporozoites typically enclosed in sporocyst within oocyst; homoxenous or heteroxenous. *Aggregata, Eimeria, Isospora, Sarcocystis, Toxoplasma*.

Suborder Haemosporina
Macrogamete and microgamont developing independently; no syzygy; conoid ordinarily absent; microgamont producing eight flagellated microgametes; zygote motile (ookinete); sporozoites naked, with three-membraned wall; heteroxenous, with merogony in vertebrates and sporogony in invertebrates; transmitted by blood-sucking insects. *Haemoproteus, Leucocytozoon, Plasmodium*.

Subclass Piroplasmia
Piriform, round, rod-shaped, or ameboid; conoid absent; no oocysts, spores and pseudocysts; flagella absent; usually without subpellicular microtubules, with polar ring and rhoptries; asexual and probably sexual reproduction; parasitic in erythrocytes and sometimes also in other circulating and fixed cells; heteroxenous, with merogony in vertebrates and sporogony in invertebrates; sporozoites with single-membraned wall; known vectors are ticks. *Babesia, Theileria*.

PHYLUM MICROSPORA

Unicellular spores, each with imperforate wall, containing one uninucleate or dinucleate sporoplasm and simple or complex extrusion apparatus always with polar tube and polar cap; without mitochondria; intracellular parasites in nearly all major animal groups.

Class Rudimicrosporea

Spore with simple extrusion apparatus; polaroplast and posterior vacuole absent; sporulation sequence with dimorphism, occurring either in parasitophorous vacuole or in thick-walled cyst; hyperparasites of gregarines in annelids.

Order Metchnikovellida
With characters of the class. *Amphiacantha, Metchnikovella*.

Class Microsporea

Spore with complex extrusion apparatus of Golgi origin, often including polaroplast and posterior vacuole; spore wall with three layers; sporocyst present or absent; often dimorphic in sporulation sequence.

Order Minisporida

Tendency toward minimal development of accessory spore organelles and toward maximal development of sporocysts; spore without well-developed polaroplast; relatively short polar filament; some genera dimorphic in sporulation sequence. *Burkea, Chytridiopsis, Hessea.*

Order Microsporida

Tendency toward maximal development and specialization of accessory spore organelles, with accompanying reduction of sporocysts; sporocysts inside host cell present or absent; merogony present; often dimorphic in sporulation sequence.

Suborder Pansporoblastina

Sporulation sequence occurring within more or less persistent intracellular (in host cell) sporocyst (pansporoblastic membrane); often dimorphic, with another sporulation sequence not involving such membrane. *Amblyospora, Pleistophora, Thelohania.*

Suborder Apansporoblastina

Pansporoblastic membrane absent or vestigial, never persisting as sporophorous vesicle. *Encephalitozoon, Glugea, Nosema.*

PHYLUM ASCETOSPORA

Spore multicellular (or unicellular?); with one or more sporoplasms; without polar capsule or polar filaments; all parasitic in invertebrates.

Class Stellatosporea

Haplosporosomes present; spore with one or more sporoplasms.

Order Occlusosporida

Spore with more than one sporoplasm; spore wall entire. *Marteilia.*

Order Balanosporida

Spore with one sporoplasm; spore wall with orifice. *Haplosporidium, Urosporidium.*

Class Paramyxea

Spore bicellular, consisting of parietal cell and one sporoplasm; spore without orifice.

Order Paramyxida

With characters of the class. *Paramyxa.*

PHYLUM MYXOZOA

Spores of multicellular origin, with one or more polar capsules (enclosing polar filaments) and sporoplasms; with one, two, or three (rarely more) valves; all parasitic.

Class Myxosporea

Spore with one or two sporoplasms and one to six (typically two) polar capsules; spore membrane generally with two, occasionally up to six, valves; trophozoite stage well developed, main site of proliferation; coelozoic or histozoic in cold-blooded vertebrates.

Order Bivalvulida

Spore wall with two valves.

Suborder Bipolarina

Spores with polar capsules at opposite ends of spore or with widely divergent polar capsules located in sutural plane or sutural zone. *Myxidium, Sphaeromyxa.*

Suborder Eurysporina

Spores with two to four polar capsules at one pole in plane perpendicular to sutural plane. *Ceratomyxa, Sphaerospora.*

Suborder Platysporina

Spores with two polar capsules at one pole in sutural plane; spores bilaterally symmetrical, unless with single polar capsule. *Henneguya, Myxobolus, Thelohanellus.*

Order Multivalvulida

Spore wall with three or more valves. *Hexacapsula, Kudoa.*

Class Actinosporea

Spores with three polar capsules; membrane with three valves; several to many sporoplasms; trophozoite stage reduced, proliferation mainly during sporogenesis; in invertebrates, especially annelids.

Order Actinomyxida

With characters of the class. *Triactinomyxon.*

PHYLUM CILIOPHORA

Simple cilia or compound ciliary organelles typical in at least one stage of life cycle; with subpellicular infraciliature present even when cilia absent; two types of nuclei, with rare exception; binary fission transverse, basically homothetogenic, but budding and multiple fission also occur; sexuality involving conjugation, autogamy, and cytogamy; contractile vacuole typically present; most species free-living, but many commensal, some parasitic. (Numerous taxa with commensal and some with parasitic species will not be characterized here because they are not covered in the text, and the morphological terminology is beyond the scope of this book.)

Class Kinetofragminophorea

Oral infraciliature only slightly distinct from somatic infraciliature and differentiated from anterior parts, or other segments, of all or some somatic kineties; cytostome often apical (or subapical or midventral) on surface of body or at bottom of atrium or vestibulum; compound ciliature typically absent.

Subclass Vestibulifera

Apical or near apical vestibulum commonly present, equipped with cilia derived from anterior parts of somatic kineties and leading to cytostome; free-living or parasitic, especially in digestive tract of vertebrates and invertebrates.

Order Trichostomatida

No reorganization of somatic kineties at level of vestibulum other than more packed alignment of kinetosomes or addition of supernumerary segments of kineties; many species endocommensals in vertebrate hosts.

Suborder Trichostomatina

Somatic ciliature not reduced. *Balantidium, Isotricha, Sonderia.*

Suborder Blepharocorythina

Somatic ciliature markedly reduced; in herbivorous mammals, especially equids. *Blepharocorys, Ochoterenaia.*

Order Entodiniomorphida

Somatic ciliature in form of unique ciliary tufts or bands, otherwise body naked; pellicle generally firm, sometimes drawn out into processes; skeletal plates in many species; commensals in mammalian herbivores, including anthropoid apes. *Entodinium, Ophryoscolex.*

Other Order

Colpodida

Other Subclasses

Gymnostomatia, Hypostomatia, Suctoria

Class Oligohymenophorea

Oral apparatus, at least partially in buccal cavity, generally well defined, although absent in one group; oral ciliature, clearly distinct from somatic ciliature, consisting of paroral membrane on right side and small number of compound organelles on left side; cytostome usually ventral and/or near anterior end, present at bottom of buccal or infundibular cavity.

Subclass Hymenostomatia

Body ciliation often uniform and heavy; buccal cavity, when present, ventral; kinetodesmata regularly present, usually conspicuous; sessile forms and stalk, colony and cyst formation relatively rare; freshwater forms predominant.

Order Hymenostomatida

Buccal cavity well defined; oral area on ventral surface, usually in anterior half of body.

Suborder Ophryoglenina

Large, primarily freshwater, histophagous forms; life cycle with cyst stage; several species causing white spot disease in marine and freshwater fishes. *Ichthyophthirius, Ophryoglena.*

Other Suborders
Tetrahymenina, Peniculina

Other Orders

Scuticociliatida, Astomatida

Subclass Peritrichia

Oral ciliary field prominent, covering apical end of body and dipping into infundibulum; paroral membrane and adoral membranelles present; somatic ciliature reduced to temporary posterior circlet of locomotor cilia; many stalked and sedentary, others mobile, all with aboral scopula; conjugation total, involving fusion of microconjugants and macroconjugants.

Order Peritrichida

With characters of the subclass.

Suborder Mobilina

Mobile forms, usually conical or cylindrical (or discoidal and orally-aborally flattened), with permanently ciliated trochal band (ciliary girdle); complex thigmotactic apparatus at aboral end, often with highly distinctive denticulate ring; all ectoparasite or endoparasites of freshwater or marine vertebrates and invertebrates. *Trichodina, Urceolaria.*

Other Suborder
Sessilina

Class Polymenophorea

Dominated by well-developed, conspicuous adoral zone of numerous buccal or peristomial membranelles (AZM), often extending out onto body surface; somatic ciliature complete or reduced or appearing as cirri; cytostome at bottom of buccal cavity or infundibulum; somatic infraciliature rarely including kinetodesmata; cytoproct often absent; cysts common in some groups.

Subclass Spirotrichia

With characters of the class.

Order Heterotrichida

Generally large to very large forms, often highly contractile, sometimes pigmented; body dominated by AZM but often bearing heavy ciliation; macronucleus oval or, often, beaded; parasitic and free-living species.

Suborder Clevelandellina

Somatic ciliature well developed, sometimes separated into distinct areas by well-defined suture lines; several specialized unique fibers associated with kinetosomes; macronuclear karyophore (region of cytoplasm apparently supporting nucleus) and/or conspicuous dorsoanterior sucker characteristic of many species; endoparasitic in digestive tract of insects and other arthropods or lower vertebrates, occasionally in oligochaetes or molluscs. *Clevelandella, Nyctotherus.*

Other Suborders

Heterotrichina, Armophorina, Coliphorina, Plagiotomina, Licnophorina

Other Orders

Odontostomatida, Oligotrichida, Hypotrichida

REFERENCES

1. Aikawa, M. 1971. Parasitological review: *Plasmodium:* the fine structure of malarial parasites. Exp. Parasitol. 30:284-320.
2. Arroyo-Begovich, A., A. Cárabez-Trejo, and J. Ruíz-Herrera. 1980. Identification of the structural component in the cyst wall of *Entamoeba invadens.* J. Parasitol. 66:735-741.
3. Cavalier-Smith, T., and J.J. Lee. 1985. Protozoa as hosts for endosymbioses and the conversion of symbionts into organelles, J. Protozool. 32:376-379.
4. Corliss, J.O. 1981. What are the taxonomic and evolutionary relationships of the protozoa to the Protista? Biosystems 14:445-459.
5. Corliss, J.O. 1985. Concept, definition, prevalence, and host-interactions of xenosomes (cytoplasmic and nuclear endosymbionts). J. Protozool. 32:373-376.
6. Dawes, C.J. 1981. Marine botany. John Wiley & Sons, Inc., New York.
7. de Duve, C. 1983. Microbodies in the living cell. Sci. Am. 248:74-84 (May).
8. Grell, K.G. 1973. Protozoology. Springer-Verlag New York, Inc., New York.
9. Hammond, D.M., E. Scholtyseck, and B. Chobotar. 1967. Fine structure associated with nutrition of the intracellular parasite *Eimeria auburnensis.* J. Protozool. 14:678-683.
10. Hickman, C.P., Jr., L.S. Roberts, and F.M. Hickman. 1988. Integrated principles of zoology, ed. 8. The C.V. Mosby Co., St. Louis.
11. Lee, J.J., M.J. Lee, and D.S. Weis. 1985. Possible adaptive value of endosymbionts to their protozoan hosts. J. Protozool. 32:380-382.
12. Lee, J.J., A.T. Soldo, W. Reisser, M.J. Lee, K.W. Jeon, and H.-D. Görtz. 1985. The extent of algal and bacterial endosymbioses in protozoa. J. Protozool. 32:391-403.
13. Levine, N.D., et al. 1980. A newly revised classification of the protozoa. J. Protozool. 27:37-58.
14. Müller, M. 1975. Biochemistry of protozoan microbodies: peroxisomes, α-glycerophosphate oxidase bodies, hydrogenosomes. Ann. Rev. Microbiol. 29:467-483.
15. Nachmias, V.T. 1984. Microfilaments. Carolina Biology Reader, No. 130. Carolina Biol. Supply. Co., Burlington, North Carolina.
16. Opperdoes, F.R., and P. Borst. 1977. Localization of nine glycolytic enzymes in a microbody-like organelle in *Trypanosoma brucei:* The glycosome. FEBS Letters 80:360-364.

17. Pitelka, D.R. 1963. Electron-microscopic structure of Protozoa. Pergamon Press, Inc., Elmsford, N.Y.

18. Prensier, G., E. Vivier, S. Goldstein, and J. Schrével. 1980. Motile flagellum with a "3 + 0" ultrastructure. Science 207:1493-1494.

19. Reisser, W., R. Meier, H.-D. Görtz, and K.W. Jeon. 1985. Establishment, maintenance, and integration mechanisms of endosymbionts in protozoa. J. Protozool. 32:383-390.

20. Satir, P. 1983. Cilia and related organelles. Carolina Biology Reader No. 123. Carolina Biological Supply Co., Burlington, NC.

21. Sheffield, H.J., and M.L. Melton. 1968. The fine structure and reproduction of *Toxoplasma gondii*. J. Parasitol. 54:209-226.

22. van Wagtendonk, W.J. 1955. Encystment and excystment of Protozoa. In Hutner, S.H., and A. Lwoff, editors. Biochemistry and physiology of Protozoa, vol. 2. Academic Press, Inc., New York.

SUGGESTED READINGS

Farmer, J.N. 1980. The Protozoa: introduction to protozoology. The C.V. Mosby Co., St. Louis.

Hyman, L.H. 1940. The invertebrates, vol. 1. Protozoa through Ctenophora. McGraw-Hill Book Co., New York. (An excellent reference to general aspects of the Protozoa.)

Jahn, T.L., E.C. Bovee, and F.F. Jahn. 1979. How to know the Protozoa, ed. 2. William C. Brown Co., Publishers, Dubuque, Iowa (Identification keys to the common Protozoa.)

Kreier, J.P., and J.R. Baker. 1987. Parasitic protozoa. Allen and Unwin, Boston.

Lee, J.J., S.H. Hutner, and E.C. Bovee (editors). 1986. An illustrated guide to the protozoa. Society of Protozoologists, Lawrence, Kansas.

Levine, N.D. 1973. Protozoan parasites of domestic animals and man, ed. 2. Burgess Publishing Co., Minneapolis. (A very useful reference to the parasites indicated by the title.)

Margulis, L. 1981. Symbiosis in cell evolution. W.H. Freeman and Co., Publishers, San Francisco. (Presents the case for the symbiotic origin of the eukaryotes.)

Scholtyseck, E. 1979. Fine structure of parasitic protozoa. Springer-Verlag, Berlin. (Atlas of electron micrographs accompanied by labeled diagrams. Heavy on Apicomplexa.)

Trager, W. 1986. Living together. The biology of animal parasitism. Plenum Press, New York.

Whittaker, R.H. 1977. Broad classification: the kingdoms and the protozoans. In Kreier, J.P., editor. Parasitic protozoa, vol. 2. Academic Press, Inc., New York.

CHAPTER 5

ORDER KINETOPLASTIDA: TRYPANOSOMES AND THEIR KIN

FAMILY TRYPANOSOMATIDAE

Most members of the family Trypanosomatidae are **heteroxenous:** during one stage of their lives they live in the blood and/or fixed tissues of all classes of vertebrates, and during other stages they live in the intestines of bloodsucking invertebrates. Thus they usually are called hemoflagellates. All species are either elongate with a single flagellum or rounded with a very short, nonprotruding flagellum. All forms have a single nucleus. Sexual phenomena have not been observed directly in these organisms, but there is indirect evidence of sexuality.[66]

The flagellum arises from the kinetosome and, when well developed, propels the organism (Fig. 5-1). The flagellum may also attach the organism to an insect host's gut wall or salivary gland epithelium.[74] Closely associated with and usually posterior to the kinetosome is a structure unique to the order, the **kinetoplast.** The sausage- or disc-shaped kinetoplast contains the mitochondrial DNA, and the single mitochondrion arises from it. Electron micrographs reveal the DNA fibers running in an anterior-posterior direction within the kinetoplast. The kinetoplast DNA (K-DNA) is organized into a network of linked circles,[5] quite unlike the organization of DNA in a chromosome. There are up to 20,000 tiny circles (minicircles) and 20 to 50 larger circles (maxicircles) in the kinetoplast network. Evidence indicates that the maxicircles are the equivalent of mitochondrial DNA of other species and code for similar information, but the function of the K-DNA in the minicircles remains obscure. One suggestion is that the minicircles are elements that regulate the replication rate of the mitochondrion during the cell cycle.[5]

The kinetoplast and the kinetosome are always closely associated, sometimes so close that they appear as a single body under the light microscope. Most electron micrographs show no physical connection between the kinetosome and kinetoplast.

The family originally parasitized the digestive tract

of insects and, possibly, annelids. Many species are still only parasitic within a single arthropod host **(monoxenous).**[79] Some species pass through different morphological stages, depending on the phase of their life cycle and the host they are parasitizing. In the past these stages were named after the genera they most resembled, such as *Crithidium* or leptomonad (for *Leptomonas*), but currently a nomenclature referring to the flagellum prevails.

Before proceeding to the general characteristics of these stages, we will describe the morphology of one of them, the **trypomastigote** stage, more fully. In this form the kinetoplast and kinetosome are near the posterior end of the body, and the flagellum runs along the surface, usually continuing as a free whip anterior to the body. Running along the axoneme within the flagellar membrane is a **paraxial rod** (Fig. 5-2). The paraxial rod has a lattice-like structure, and short projections connect it to the axonemal microtubules of the flagellum.[12] The flagellar membrane is closely apposed to the body surface, and when the flagellum beats, this area of the pellicle is pulled up into a fold; the fold and the flagellum constitute the **undulating membrane.** A second, "barren" kinetosome, without a flagellum, is usually found near the flagellar kinetosome. In the typical bloodstream form of trypomastigote, a simple mitochondrion with or without tubular cristae runs anteriorly from the kinetoplast. The mitochondrion is much larger and more complex, with lamellar cristae, in the insect stage of the organism. At the base of the flagellum and surrounding the kinetosome is a **flagellar pocket** or reservoir. A system of **pellicular microtubules** spirals around the body just beneath the cell membrane (Fig. 5-2). These give supportive resistance to the deformation of the body caused by the beating flagellum.[78] Rough endoplasmic reticulum is well developed, and a Golgi body is found between the nucleus and kinetosome.

The trypomastigote is the definitive stage of the ge-

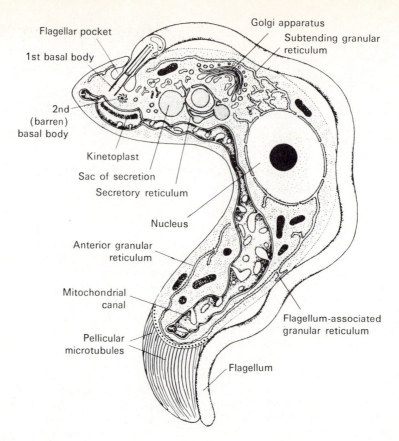

FIG. 5-1

Diagram to show principal structures revealed by the electron microscope in the bloodstream try-pomastigote form of the salivarian trypanosome, *Trypanosoma congolense*. It is shown cut in sagittal sections, except for most of the shaft of the flagellum and the anterior extremity of the body.

From Vickerman, K. 1969. J. Protozool. 16:54-69.

nus *Trypanosoma*. The other forms differ in body shape, position of the kinetosome and kinetoplast, or development of the flagellum (Fig. 5-3). A spheroid **amastigote** occurs in the life cycle of some species and is definitive in the genus *Leishmania*. The flagellum is very short, projecting only slightly beyond the flagellar pocket.

In the **promastigote** stage the elongate body has the flagellum extending forward as a functional organelle. The kinetosome and kinetoplast are located in front of the nucleus, near the anterior end of the body. The promastigote form is found in the life cycles of several species while they are in their insect hosts. It is the mature form in the genus *Leptomonas*. If the flagellum emerges through a wide, collar-like process, the type is termed a **choanomastigote**, which is found in some species of *Crithidia* (parasitic in insects).

The **epimastigote** form is encountered in some life cycles. Here the kinetoplast and kinetosome are still located between the nucleus and the anterior end, but a short undulating membrane lies along the proximal part of the flagellum. The genera *Crithidia* and *Blastocrithidia*, both parasites of insects, exhibit this form during their life spans. Finally, the **opisthomastigote** form is found in *Herpetomonas*, a widespread group of insect parasites. The kinetosome and kinetoplast are located between the nucleus and posterior end, but there is no undulating membrane. The flagellum pierces a long reservoir that passes through the entire length of the body and opens at the anterior end.

Genus *Trypanosoma*

All trypanosomes (except *T. equiperdum*) are **hexteroxenous** or at least are transmitted by an animal vector. Various species pass through amastigote, promastigote, epimastigote, and/or trypomastigote stages,

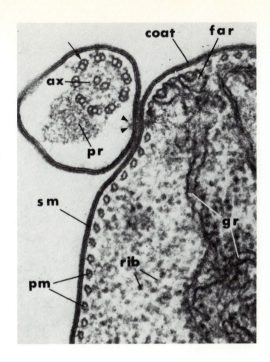

FIG. 5-2

Trypanosoma congolense. Transverse section of shaft of flagellum and adjacent pellicle in region of attachment. Both flagellum and body surface have a limiting unit membrane *(sm)* covered by a thick coating *(coat)* of dense material. The axoneme *(ax)* of the flagellum shows the partition *(arrow)* dividing one of the tubules of each doublet; alongside the axoneme lies the paraxial rod *(pr)*. Pellicular microtubules *(pm)* underlie the surface membrane of the body, and a diverticulum *(far)* of the granular reticulum *(gr)* is always found embracing three or four of these microtubules close to the flagellum. Note the fibrous condensations *(arrowheads)* on either side of the opposed surface membranes, apparently "riveting" the flagellum to the body. A row of these "rivets" replaces a microtubule along the line of adherence. *rib,* Ribosomes. (×66,000.)

From Vickerman, K. 1969. J. Protozool. 16:54-69.

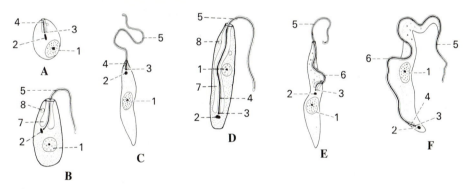

FIG. 5-3

Genera of Trypanosomatidae. **A,** *Leishmania* (amastigote form); **B,** *Crithidia* (choanomastigote); **C,** *Leptomonas* (promastigote); **D,** *Herpetomonas* (opisthomastigote); **E,** *Blastocrithidia* (epimastigote); **F,** *Trypanosoma* (trypomastigote); *1,* nucleus; *2,* kinetoplast; *3,* kinetosome; *4,* and *5,* axoneme and flagellum; *6,* undulating membrane; *7,* flagellar pocket; *8,* contractile vacuole.

From Olsen, O.W. 1974. Animal parasites, their biology and life cycles. University Park Press, Baltimore.

with the other forms developing in the invertebrate hosts. Much research has been conducted on this genus because of its extreme importance to the health of humans and domestic animals. Reviews are available dealing with various aspects of the group: host susceptibility,[51] physiology and morphology,[73,74] chemotherapy,[82] taxonomy,[28,54] immunology,[8,44,64,76,77] evolution,[20,30] and vector relationships.[47] Wallace[79] outlined the trypanosomatids of arachnids and insects.

It is possible that some species now known only from invertebrates may also be parasites of vertebrates.[40]

Members of the genus *Trypanosoma* are parasites of all classes of vertebrates. Most live in the blood and tissue fluids, but some important ones, such as *T. cruzi,* occupy intracellular habitats as well. Although other means of transmission exist, the majority are transmitted by blood-feeding invertebrates. Most trypanosomes parasitize animals of no particular impor-

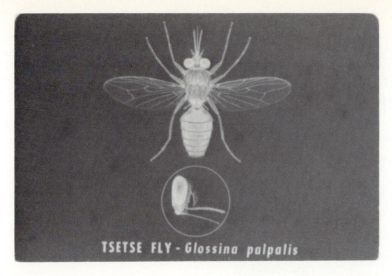

FIG. 5-4

Glossina palpalis, a tsetse fly. Species in this genus are vectors of some species of *Trypanosoma*. *Glossina* spp. are moderate in size, ranging from 7.5 to 14 mm in length.

tance to humans, but a few species are responsible for misery and privation of enormous proportions. In Africa alone 4.5 million square miles, an area larger than the United States, are incapable of supporting agriculture—not because the land is poor, since it is much like the grasslands of the American West, but because domestic livestock are killed by trypanosomes that are nonfatal parasites of native grazing animals. Thus semiarid lands that otherwise could support agronomy are denied to millions of persons who most need the protein affordable by the rich soil. More directly affected by trypanosomes are millions of people in South America who have never known a day of good health because of trypanosome infections.

Trypanosomes are divided into two broad groups, or "sections," based on characteristics of their development in their invertebrate hosts. If a species develops in the front portions of the digestive tract, it is said to undergo **anterior station** development and is relegated to the section Salivaria, which contains several subgenera. When a species develops in the hindgut of its invertebrate host, it is said to undergo **posterior station** development and is placed in the section Stercoraria. Other developmental and morphological criteria separate the two sections and further aid in placement in the proper section of species that do not require development in an intermediate host *(T. equiperdum, T. equinum)*.[28] Classification of the various species into subgenera is based on their physiology, morphology, and biology.

■ **Section Salivaria**

Trypanosoma (Trypanozoon) brucei **subspecies:** *T. b. brucei, T. b. gambiense* and *T. b. rhodesiense.* The three subspecies of *Trypanosoma brucei* are morphologically indistinguishable and traditionally have been considered separate species. They vary in infectivity for different species of hosts and produce somewhat different pathological syndromes. It seems clear that the ancestral form in the complex is represented by *T. b. brucei,* and the other two should be considered subspecies of *T. brucei.*[54] These typanosomes are widely distributed in tropical Africa between latitude 15 degrees N and 25 degrees S, corresponding in distribution with their vectors, tsetse flies (*Glossina* spp.) (Fig. 5-4). An interesting history of the classification was presented by Gibson.[19]

Trypanosoma brucei brucei is fundamentally a parasite of the bloodstream of native antelopes and other African ruminants, causing a disease called **nagana.** Unfortunately, the parasite also infects introduced livestock, including sheep, goats, oxen, horses, camels, pigs, dogs, donkeys, and mules. It is pathogenic to these animals, as well as to several native species. Humans, however, are not susceptible.

Trypanosoma brucei gambiense and *T. b. rhodesiense* are the etiological agents of African sleeping sickness. There are physiological differences between them, and they differ in pathogenesis, growth rate, and biology. As early as 1917 a German researcher named Taute showed that the nagana trypanosome would not infect humans. He repeatedly inoculated

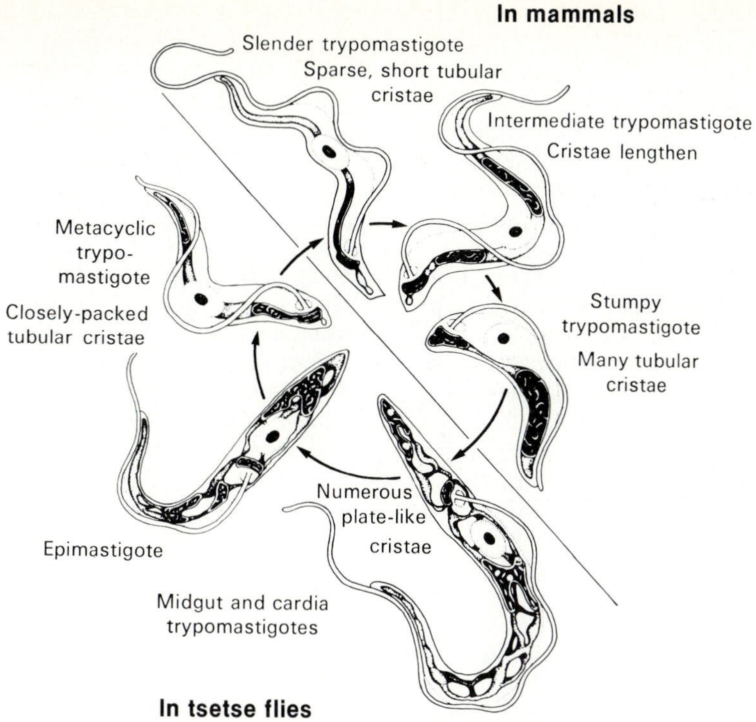

FIG. 5-5

Diagram to show changes in form and structure of the mitochondrion of *Trypanosoma brucei* throughout its life cycle. The slender bloodstream form lacks a functional Krebs cycle and cytochrome chain. Stumpy forms have a partially functional Krebs cycle but still lack cytochromes. The glycerophosphate oxidase system functions in terminal respiration of bloodstream forms. The fly gut forms have a fully functional mitochondrion with active Krebs cycle and cytochrome chain. Cytochrome oxidase may be associated with the distinctive plate-like cristae of these forms. Reversion to tubular cristae in the salivary gland stages may therefore indicate loss of this electron transfer system.

From Vickerman, K. 1971. In Fallis, A.M., editor. Ecology and physiology of parasites. University of Toronto Press, Toronto.

himself and native "volunteers" with nagana-ridden blood; none of them acquired sleeping sickness. His work was largely discounted by British experts for reasons that can only be guessed.

Trypanosoma brucei gambiense causes a chronic form of the disease and is found in west central and central Africa, whereas *T. b. rhodesiense* occurs in central and east central Africa and causes a more acute type of infection. Native game animals are thought to serve as reservoirs for Rhodesian trypanosomiasis but not for Gambian trypanosomiasis.

MORPHOLOGY AND LIFE HISTORY. In recent years correlated findings from the fields of cytology, biochemistry, and immunology have combined to make the life history of these trypanosomes one of the most fascinating stories of development in parasitology. We will trace the sequence of events in the life history itself and then briefly consider the physiological and ultrastructural changes associated with these events.

Trypanosoma brucei in natural infections tends to be **pleomorphic** (polymorphic) in its vertebrate host, ranging from long, slender trypomastigotes with a long free flagellum through intermediate forms to short, stumpy individuals with no free flagellum (Fig. 5-5). The small kinetoplast is usually very near the posterior end, and the undulating membrane is conspicuous.

The insect vectors of *T. b. brucei* and *T. b. rhodesiense* are *Glossina morsitans*, *G. pallidipes*, and *G. swynnertoni*, whereas those of *T. b. gambiense* are *G.*

palpalis and *G. tachinoides*. At least 90% of the flies are refractive to infection. When eaten by a susceptible fly, along with a blood meal, *T. brucei* locates in the posterior section of the midgut of the insect, where it multiplies in the trypomastigote form for about 10 days. At the end of this time the slender individuals produced migrate forward into the foregut, where they are found on the twelfth to twentieth days. They then migrate farther forward into the esophagus, pharynx, and hypopharynx and enter the salivary glands. Once in the salivary glands they transform into the epimastigote form and attach to host cells or lie free in the lumen. After several asexual generations they transform into the **metacyclic trypomastigote** form, which is small, stumpy, and lacks a free flagellum. The metacyclics are the only stage in the vector that is infective to the vertebrate host. When feeding, the tsetse fly may inoculate a host with up to several thousand protozoa with a single bite. The entire cycle within the fly can be completed in 15 to 35 days.

Once within a vertebrate, the trypanosomes multiply as trypomastigotes in the blood and lymph. Amastigote forms have been reported from the liver and spleen of experimentally infected mice and from the myocardium of monkeys.[53,83] In the chronic form of the disease many of the trypanosomes invade the central nervous system, multiply, and enter the intercellular spaces within the brain.

Biochemical, ultrastructural, and immunological studies have added greatly to our understanding of trypanosomes.[6,73] Of considerable value is the fact that essentially pure preparations of certain morphological stages can be obtained. When *T. brucei* is passed by syringe from one vertebrate host to another, the strain tends to become monomorphic after a period of time, consisting only of slender trypomastigotes that are no longer infective to tsetse flies, nor can they be cultivated in vitro. Their morphology and metabolism correspond to the slender trypomastigote in natural infections. In contrast, when *T. brucei* is placed in certain in vitro culture systems, its morphology and metabolism revert to that found in the fly midgut, with the kinetoplast further from the posterior end and close to the nucleus. It has been found that the monomorphic, syringe-passed strain depends entirely on glycolysis for its energy production, degrading glucose only as far as pyruvate and having no tricarboxylic acid cycle or oxidative phosphorylation by way of the classical cytochrome system. The reduced NAD produced in glycolysis is reoxidized by a nonphosphorylating glycerophosphate oxidase system, which, although it requires oxygen, is not sensitive to cyanide. This respiratory system is inhibited by **suramin,** an antitrypanosomal drug, and it is apparently localized in membrane-bound microbodies called α-glycero-

phosphate oxidase bodies, or **glycosomes.**[26,49] The long, slender trypomastigote is very active, and it consumes substantial quantities of both glucose and oxygen in its inefficient energy production. However, since the blood and lymph have a plentiful supply of both, no selective value is attached to the conservation of either. The situation is quite different when the trypanosome finds itself in a blood clot in its vector's midgut; interestingly enough, in this case the form completely degrades glucose via glycolysis, the tricarboxylic acid cycle, and the cyanide-sensitive cytochrome system. The oxygen and glucose consumption of the midgut (or culture) form is only one tenth that of the bloodstream form. Culture forms have two oxidase systems with a high affinity for oxygen, one containing cytochrome aa_3, which accounts for more than 50% of the respiration, and the other apparently containing cytochrome o, accounting for 25% to 35% of oxygen consumption.[26] The glycerophosphate oxidase system is also present in the culture forms, but its activity now is sensitive to mitochondrial inhibitors.[49]

Ultrastructural observations on the mitochondria in the respective forms correlate beautifully with the biochemical findings. The long, slender trypomastigote has a single, simple mitochondrion extending anteriorly from its kinetoplast, and the cristae are few, short, and tubular. The midgut stage has an elaborate mitochondrion extending both posteriorly and anteriorly from the kinetoplast, and the cristae are numerous and platelike. The curious movement of the kinetoplast away from the posterior end in the midgut trypomastigote and anterior to the nucleus in the epimastigote can now be understood as reflecting the elaboration of the posterior section of mitochondrion; it "pushes" the kinetosome forward. Furthermore, current evidence indicates that the short, stumpy form is the only one infective to the tsetse fly and that the intermediate form is transitional from the long, slender noninfective form (Fig. 5-5). Correspondingly, electron microscopy has shown that this transition is marked by increasing elaboration of the mitochondrion; synthesis of mitochondrial enzymes has been shown by cytochemical means. Similarly the metacyclic form, as though "gearing down" for infection of the vertebrate, has a mitochondrion much like the bloodstream form.

IMMUNOLOGY. An important remaining question is what triggers the bloodstream trypomastigotes to become short, stumpy forms infective to flies? A possibility is suggested by some interesting immunological observations. The clinical course characteristic of the infection varies according to the host infected, but in certain hosts (guinea pig, dog, and rabbit) repeated remissions alternate with very high levels of parasitemia. That is, periods with few trypanosomes (and disease symptoms) evident are followed by a large increase in parasite population. This cycle tends to repeat itself until the host dies.[61] The mechanism of the phenomenon lies in the successive dominance of each of a series of variant antigenic types (VATs) over time. The remissions seem to result from the generation of protective antibodies by the host that destroy the homologous trypanosomes. The parasites have evolved an amazing subterfuge to escape obliteration by the host's defenses: each time the host's antibodies are almost successful in eliminating the infection, the trypanosomes elude destruction by rapid multiplication of protozoa of a different antigen type! Furthermore, the only apparent limit to the number of VATs that can develop in a clone strain of trypanosomes during an untreated infection is the host's life span; the maximal number shown so far for a trypanosome stock is 101.[6]

The means by which the parasites achieve this succession of antigenic types is a subject of intensive investigation and is a fascinating story of gene expression.[24,25,34,75] A much better understanding has been achieved by the methods of modern molecular biology.[6,56] The antigen recognized by the host's immune system is a variant-specific surface glycoprotein (VSG) released through the flagellar reservoir of the trypanosome and completely covering the organism as a surface coat (Fig. 5-2). Each *T. brucei* individual possesses a large number of basic copy (BC) genes coding for VSGs, possibly up to 2000; but when an organism is switching from one VSG to another,[15] only one VSG gene is expressed at any time. The others are transcriptionally silent. There are apparently two mechanisms of rearrangement that result in expression of a gene.[87] In one of these the BC gene is duplicated and transferred to another position near the end (telomere) of a chromosome. This becomes the expression-linked copy (ELC). It is located downstream from a promoter on the chromosome and is thus transcribed. The other mechanism involves VSG genes that do not have to be duplicated for transcription (nonduplication associated [NDA] genes).[39] What causes the NDA genes to be turned on is not understood, but they are also telomeric in position. Furthermore, some evidence suggests that there may be two or more ELC expression sites.[50] Expression of the genes occurs in an imprecisely predictable order; that is, expression of a given VSG more commonly occurs after the expression of another, particular VSG, but not invariably. Thus the VSG genes in a population of trypanosomes are heterogeneous at any time in a chronic infection, but there is a single VAT that is predominant in the blood, and against which the host mounts its antibody defense. Other dominant VATs can be found in such places as the brain and liver.[65]

Adding to the complexity of the system, trypanosomes can lose VSG genes and add new genes to their repertoire by a mechanism of segmental gene conversion (nonreciprocal crossing over).[57,58]

When the trypanosome is ingested by *Glossina,* it loses its VSG surface coat. This implies that there is yet another mechanism for activating and deactivating VSG genes. Expression resumes when the trypanosomes reach the metacyclic state and are then able to infect the mammalian host. A much smaller number of VATs characterize the metacyclic trypanosomes; only 8 VATs comprised 60% to 80% of the population in one *T. b. rhodesiense* clone.[70] Although the first VSG gene expressed after infection of the mammalian host is one of the metacyclic VATs, within a few days the VSG found on the trypanosome that had been ingested by the fly will be expressed. This reexpression of the ingested VAT is referred to as **anamnestic** expression. While the promastigotes are without a surface coat in the tsetse fly (a period of several weeks), they are vulnerable to antibodies in subsequent blood meals of the vector. This points to a possible means of protection in cattle ranches if stock have a level of immunity, since the flies have little else to eat.[4]

Although the VAT story is best known for *T. brucei,* antigen switching is also found in at least some other trypanosomes, such as *T. vivax.*[3,17]

PATHOGENESIS. In their vertebrate hosts these trypanosomes live in the blood, lymph nodes and spleen, and cerebrospinal fluid. They do not invade or live within cells but inhabit connective tissue spaces within various organs and the reticular tissue spaces of the spleen and lymph nodes. They are particularly abundant in the lymph vessels and intercellular spaces in the brain.

The clinical course in *T. b. brucei* infections depends on the susceptibility of the host species. Horses, mules, donkeys, some ruminants, and dogs suffer acutely, and they usually survive only 15 days to 4 months. Symptoms include anemia, edema, watery eyes and nose, and fever. Within a few days the animals become emaciated, uncoordinated, and paralyzed and die shortly afterward. Blindness as the result of the infection is common in dogs. Cattle are somewhat more refractory to the disease, often surviving for several months after onset of symptoms. Swine usually recover from the infection.

In human infections with *T. b. rhodesiense* and *T. b. gambiense* a small sore often develops at the site of inoculation of metacyclic trypansomes. This disappears after 1 to 2 weeks, while the protozoa gain entrance to the blood and lymph channels. Reproducing rapidly, they produce a parasitemia and invade nearly all organs of the body. *Trypansoma b. rhodesiense* rarely invades the nervous system as does *T. b. gam-*

biense but usually causes a more rapid course toward death. The lymph nodes become swollen and congested, especially in the neck, groin, and legs. Swollen nodes at the base of the skull were recognized by slave traders as signs of certain death, and slaves who developed them were routinely thrown overboard by slavers bound for the Caribbean markets. Today such swollen lymph nodes are called **Winterbottom's sign,** named after the British officer who first described the symptom. The symptoms of illness usually are more marked in nonnative than in native peoples. Intermittent periods of fever accompany the early stages of the disease, and the number of trypanosomes in the circulating blood greatly increases at these times. As previously noted, the successive parasite populations represent different antigenic types. With fever there is an increase in swelling of lymph nodes, generalized pain, headache, weakness, and cramps. Infection of *T. b. rhodesiense* causes rapid weight loss and heart involvement. Death may occur within a few months after infection, but *T. b. rhodesiense* causes no somnambulism or other protracted nervous disorders found with *T. b. gambiense* because the host usually dies before these can develop.

When the trypanosomes of *T. b. gambiense* invade the central nervous system, they initiate the chronic, sleeping-sickness stage of infection. Increasing apathy, a disinclination to work, and mental dullness accompany disturbances of coordination. Tremor of the tongue, hands, and trunk is common, and paralysis or convulsions usually follow. Sleepiness increases, with the patient falling asleep even while eating or standing. Finally, coma and death ensue. Actually, death may result from any one of a number of causes, including malnutrition, pneumonia, heart failure, other parasitic infections, or a severe fall.

The mechanism of pathogenesis is unclear. In the acute infection of small mammals, in which death occurs rapidly at a time of a high level of parasitemia, mortality probably is a result of overall disruption of normal physiological processes.[42] In the case of mammals, including humans, present evidence suggests that pathogenesis may be caused in part by the antigenic "performance" of the trypanosomes and the host's immune reactions. Because of the repeated changes in the surface antigens of the parasites and the fact that these surface antigens are being released into the blood almost constantly (exoantigens), the immune system of the host is greatly stimulated, and huge amounts of immunoglobulins are produced. It has been shown that some of the trypanosome antigens can adsorb to the surface of some host cells, and binding of the specific antibody to the adsorbed antigen, in conjunction with complement, leads to lysis of the *host's* cells. Lysis of red blood cells by this mech-

anism may account for the anemia of trypanosomiasis. Action of immunoconglutinin against the antigen-antibody-complement complex on the red cells also may be an important factor.[59] Lysis of mast cells releases pharmacologically active kinins, with resultant reactions in susceptible tissues.

It has been reported that the free tyrosine levels in the brain of voles *(Microtus)* infected with *T. brucei* were only about 50% of the level in controls, possibly caused by nutritional requirements of the trypanosomes. Resulting interference with protein and neurotransmitter synthesis in the brain may help account for neurological symptoms.[52]

DIAGNOSIS AND TREATMENT. Demonstration of the parasite in the blood, bone marrow, or cerebrospinal fluid establishes diagnosis. In native populations in which early infections are usually symptomless, a serological test is available.[1]

Arsenical drugs historically have been used in the treatment of African trypanosomiases, but these drugs have severe drawbacks.[21] They cause eye damage and are best administered intravenously; furthermore, trypanosomes rapidly become tolerant to them. Other drugs (suramin, pentamidine, and Berenil) have been developed in recent years and have proved to be satisfactory in most early cases. Prognosis, however, is poor if the nervous system has become involved. A combination of Berenil and certain nitroimidazoles has been shown to be effective experimentally in cases with nervous system involvement.[34]

More recently difluoromethylornithine (DFMO) has been found to be very efficacious in the treatment of African trypanosomiasis, especially in brain infections, and is the drug of choice at this writing.[63]

Nutrition of the vertebrate host can also affect the course of the disease. Adequate dietary lipid has been shown to limit the infectivity of *T. brucei* in rats and probably also protect people against African sleeping sickness.[55]

EPIDEMIOLOGY AND CONTROL. Tsetse flies occupy 4.5 million square miles of Africa, making much of that area impractical for human habitation. Trypanosomes of the brucei group do not occur throughout the entire range of tsetse flies, and not all species of *Glossina* are vectors for them. Therefore transmission varies locally, depending on coincidence of the trypanosome and the proper fly species. Furthermore, there is an inheritance of susceptibility to trypanosomes in tsetse flies.[43]

Glossina vectors of *T. b. brucei* and *T. b. rhodesiense* occur in open country, pupating in dry, friable earth. Vectors of *T. b. gambiense* are riverine flies, breeding in shady, moist areas along rivers.

Control of trypanosomiasis brucei is conducted along several lines, most of which involve the vectors.

Tsetse flies are larviparous, and they deposit their young on the soil under brush. Because of this, and because the adults rest in bushes at certain heights above the ground and no higher, brush removal and trimming are very successful means of control. When wide belts of land are thus cleared, the flies seldom cross them and can be more easily controlled. However, this method is expensive and must be followed up every year to remove new growth.

Elimination of the wild game reservoirs has been proposed and practiced in some regions, stimulating an outcry among conservation-minded people all over the world.

Programs have been established in which people simply sit and catch flies that try to bite them. Because the flies feed only during the day, some farmers graze their livestock at night, moving them into enclosures during the day and protecting them from flies with switches.

The most satisfactory means of control is by spraying insecticides by aircraft. DDT and benzene hexachloride are inexpensive and highly effective for this purpose. *Glossina pallidipes* was eradicated from Zululand in this manner at a cost of about 40¢ per acre. The possibilities of harmful side effects of DDT must be carefully weighed against the benefits gained by its use.

So far, 80 years of tsetse eradication have had little impact on tsetse distribution. Development and use of trypanotolerant cattle stocks may offer a practical alternative to control of the vectors or trypanosomes.[51] Breeds of cattle that have survived in trypanosome-infested regions of Africa show great promise as trypanotolerant animals. Three generations of cattle breeding using trypanotolerant bulls on a trypanosensitive breed will produce an 87.5% trypanotolerant genotype.[14]

The political situation in Africa, often with decreased cooperation between adjacent nations and tribes, may contribute to new epidemics. At the same time increased mobility of the human population contributes to wider and faster spread of the disease.

Trypanosoma (Nannomonas) congolense **and** *Trypanosoma (Duttonella) vivax.* Nagana also is caused by *Trypanosoma congolense,* which is similar to *T. brucei* but lacks a free flagellum.[72] It occurs in South Africa, where it is the most common trypanosome of large mammals. The life cycle, pathogenesis, and treatment are as for *T. brucei.* The vascular damage reported in chronic *T. congolense* infections may be a result of the propensity of the trypanosomes to attach to the walls of small blood vessels by their anterior ends.[2]

Trypanosoma vivax is also found in the tsetse fly belt of Africa and has spread to the western hemisphere and Mauritius. Very similar to *T. brucei,* it causes a like disease in the same hosts. In the New World, transmission is mechanical and involves tabanid flies. Pathogenesis and control are as for *T. brucei.* Also, the changes in the mitochondrion through the life cycle in *T. vivax* and *T. congolense* are similar to those in *T. brucei,* described previously. However, these two species appear to retain some mitochondrial function in the bloodstream form. A good review of *T. vivax* was given by Gardiner and Wilson.[18]

Trypanosoma (Trypanozoon) evansi **and** *Trypanosoma (Trypanozoon) equinum.* *Trypanosoma evansi* causes a widespread disease of camels, horses, elephants, deer, and many other mammals. The disease goes by many different names in different languages and countries but is most often called **surra.** *T. evansi* probably was originally a parasite of camels.[27] Today it is distributed throughout the northern half of Africa, Asia Minor, southern Russia, India, southwestern Asia, Indonesia, Philippines, and Central and South America. The Spaniards introduced the disease to the western hemisphere by way of infected horses in the sixteenth century.

This trypanosome is morphologically indistinguishable from *T. brucei.* Typically it is 15 to 34 μm long. Most are slender in shape, but stumpy forms occasionally appear. However, the biology of *T. evansi* is quite different from that of *T. brucei.* The life cycle does not involve *Glossina* spp. or development within an arthropod vector. In most areas contaminated mouthparts of horseflies (*Tabanus* spp.) mechanically transmit the disease, but *Stomoxys, Lyparosia,* and *Haemotopota* spp. can also transmit it. In South America vampire bats are common vectors of the disease, known there as **murrina.**[29]

The disease is most severe in horses, elephants, and dogs, with nearly 100% fatalities in untreated cases. It is less pathogenic to cattle and buffalo, which may be asymptomatic for months. In camels surra is serious, but it tends to remain chronic. Pathogenesis, symptoms, and treatment are the same as for *T. brucei.*

Trypanosoma evansi probably originated from *T. brucei,* when camels were brought into the tsetse fly belt. Subsequently, the organism apparently lost its requirement for development within an insect.

Trypanosoma (Trypanozoon) equinum occurs in South America, where it causes a disease in horses similar to surra. The condition is known as **mal de caderas.** *Trypanosoma equinum* is similar to *T. evansi* except that it appears to lack a kinetoplast. Actually a vestigial kinetoplast can be seen in electron micrographs, but it does not function in activation of the mitochondrion. The condition is known as **dyskinetoplasty.** *Trypanosoma brucei* and *T. evansi* can be rendered dyskinetoplastic with certain drugs, and the character is inherited as a mutation. Such organisms can survive as blood-stream parasites but no longer can infect flies. *Trypanosoma equinum* most likely evolved from *T. evansi*. It also is transmitted mechanically by tabanid flies. Pathogenesis, symptoms, and treatment are as for *T. evansi*.

Trypanosoma (Trypanozoon) equiperdum. Another trypanosome, *T. equiperdum,* also morphologically indistinguishable from *T. brucei,* causes a venereal disease called **dourine** in horses and donkeys. The organisms are transmitted during coitus, and no arthropod vector is known. No doubt this trypanosome also originated from *T. brucei*.

The disease is found in Africa, Asia, southern and eastern Europe, Russia, and Mexico. It was once common in western Europe and North America but has been eradicated from these areas.

Dourine exhibits three stages. In the first the genitalia become edematous, with a discharge from the urethra and vagina. Areas of the penis or vulva may become depigmented. In the second stage a prominent rash appears on the sides of the body, remaining for 3 or 4 days. The third stage produces paralysis, first of the neck and nostrils and then the hind body; the paralysis finally becomes general. Dourine is usually fatal unless treated.

Diagnosis depends on finding trypanosomes in the blood, genital secretions, or fluids from the large urticarious patches of the skin during the second stage. A complement fixation test is very reliable and was used by U.S. Department of Agriculture (USDA) personnel to ferret out infective horses during their successful campaign to eradicate the disease in the United States. All horses now entering the United States must be tested for dourine before being admitted.

■ Section Stercoraria

Trypansoma (Schizotrypanum) cruzi. *Trypanosoma cruzi* carries the unusual distinction of having been discovered and studied several years before it was found to cause a disease. In 1910 a 40-year-old Brazilian, Carlos Chagas, dissected a number of cone-nosed bugs (Hemiptera, family Reduviidae, subfamily Triatominae) and found their hindguts swarming with trypanosomes of the epimastigote type. The biology and habits of triatomines are discussed in Chapter 37.

Chagas sent a number of the bugs to the Oswaldo Cruz Institute, where they were allowed to feed on marmosets and guinea pigs. Trypanosomes appeared in the blood of the animals within a month. Chagas thought the parasites went through a type of schizogony in the lungs, so he named them *Schizotrypanum*

cruzi. The name *Schizotrypanum* still is employed by some workers, although most prefer to use it as a subgenus of *Trypanosoma*. By 1916 Chagas demonstrated that an acute, febrile disease, common in children throughout the range of cone-nosed bugs, was always accompanied by the trypanosome. Unfortunately, he thought that goiter and cretinism also were caused by this parasite. When that was disproved, suspicion was cast on the rest of his work. Also, Chagas maintained to near the end of his life that transmission of the disease, which now bears his name, was through the bite of the insect. It was not until the early 1930s that Chagas' disease was proved to be transmitted by way of the feces of the cone-nosed bug.

Trypanosoma cruzi is distributed throughout most of South and Central America, where it infects 12 to 19 million persons. Another 35 million are exposed to infection.[85] In some surveys in Brazil it has been reported that 30% of all adults die of *T. cruzi* infection. Many kinds of wild and domestic mammals serve as reservoirs. Animals that live in proximity to humans, such as dogs, cats, opossums, armadillos, and wood rats, are particularly important in the epidemiology of Chagas' disease.[64]

In the United States *T. cruzi* has been found in Maryland, Georgia, Florida, Texas, Arizona, New Mexico, California, Alabama, and Louisiana. Fourteen species of infected mammals have been found in the United States.[35] The first indigenous infection in a human in the United States was reported in 1955.[84] Since then several cases have been reported in Arizona, mainly in Indian reservations. Several North American strains have been isolated. They are morphologically indistinguishable from any other *T. cruzi,* but they seem to be much less pathogenic. A recent survey has shown 0.8% positive in immunological tests of a random sample of 500 persons in the lower Rio Grande Valley of Texas.[8] It is possible that this disease in humans is more widespread in the United States than is now known.

MORPHOLOGY. The trypomastigote form is found in the circulating blood. It is slender, 16 to 20 μm long, and its posterior end is pointed. The free flagellum is moderately long, and the undulating membrane is narrow, with only two or three undulations at a time along its length. The kinetoplast is subterminal and is the largest of any trypanosome; it sometimes causes the body to bulge around it. The protozoan commonly dies in a question mark shape, the appearance it retains in stained smears (Fig. 5-6.)

Amastigotes develop in muscles and other tissues. They are spheroid, 1.5 to 4.0 μm wide, and occur in clusters composed of many organisms. Intermediate forms are easily found in smears of infected tissues.

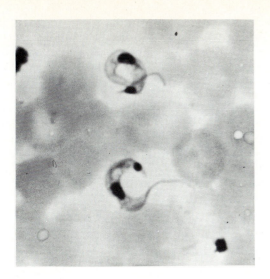

FIG. 5-6

Trypanosoma cruzi: trypomastigote form in a blood film.

Courtesy Ann Arbor Biological Center.

BIOLOGY. (Fig. 5-7). When reduviid bugs feed, they often defecate on the skin of their host. Their feces may contain metacylic trypanosomes, which gain entry into the body of the vertebrate host through the bite, through scratched skin, or, most often, through mucous membranes that are rubbed with fingers contaminated with the insect's feces. Also reservoir mammals can become infected by eating infected insects.[86] Although trypomastigotes are abundant in the blood in early infections, they do not reproduce until they have entered a cell and have transformed into amastigotes. The cells most frequently invaded are reticuloendothelial cells of the spleen, liver, and lymphatics and cells in cardiac, smooth, and skeletal muscles. The nervous system, skin, gonads, intestinal mucosa, bone marrow, and placenta also are infected in some cases (Fig. 5-8). There is some evidence that the trypanosomes can actively penetrate host cells, but they may also enter through phagocytosis by the host macrophages.

The undulating membrane and flagellum disappear soon after the parasite enters a host cell. Repeated bi-

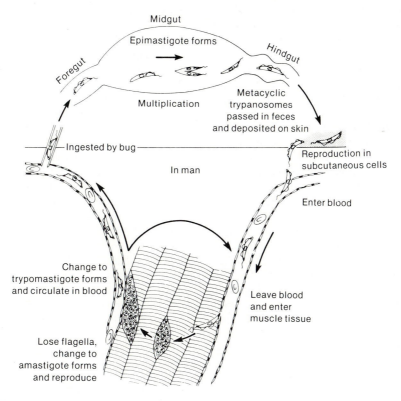

Midgut

Epimastigote forms

Foregut

Hindgut

Multiplication

Metacyclic trypanosomes passed in feces and deposited on skin

Ingested by bug

Reproduction in subcutaneous cells

In man

Enter blood

Change to trypomastigote forms and circulate in blood

Leave blood and enter muscle tissue

Lose flagella, change to amastigote forms and reproduce

FIG. 5-7

Life cycle of *Trypanosoma cruzi.*

From Adam, K.M.G., J. Paul, and V. Zaman. 1971. Medical and veterinary protozoology. An illustrated guide. Churchill Livingstone, Edinburgh.

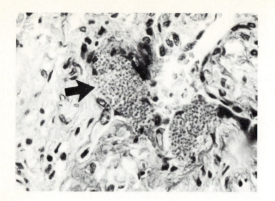

FIG. 5-8

Clusters of *Trypanosoma cruzi* amastigotes *(arrow)* in placenta.

AFIP photograph neg. no. 63-5589.

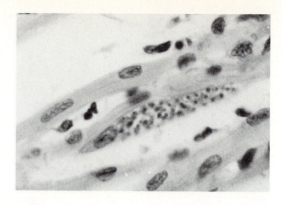

FIG. 5-9

Trypanosoma cruzi pseudocyst in cardiac muscle. (×780.)

Photograph by S.S. Desser.

nary fission produces so many amastigotes that the host cell soon is killed and lyses. When released, the protozoa attack other cells. Cystlike pockets of parasites, called **pseudocysts,** form in muscle cells (Fig. 5-9). Intermediate forms (promastigotes and epimastigotes) can be seen in the interstitial spaces. Some of these complete the metamorphosis into trypomastigotes and find their way into the blood. *Trypanosoma cruzi* is a "partial aerobic fermenter;" some of the glucose carbon it consumes is degraded completely to carbon dioxide, but a substantial portion is excreted as relatively reduced products such as succinate and acetate.[9] The oxygen consumption of blood and intracellular forms is the same as that of culture forms, and bloodstream forms apparently have a Krebs cycle and classical cytochrome system.[13,23] Regulation of this "partial aerobic fermentation" may take place via the differing regulatory properties of two forms of "malic" enzyme (malic dehydrogenase [decarboxylating], p. 338).[9]

Trypomastigotes that are ingested by triatomine bugs pass through to the posterior portion of the insect's midgut, where they become short epimastigotes. These multiply by longitudinal fission to become long, slender epimastigotes. Short metacyclic trypomastigotes appear in the insect's rectum 8 to 10 days after infection. These are passed with the feces and can infect a mammal if rubbed into a mucous membrane or wound in the skin. It has been observed that the first-generation amastigotes in the insect's stomach group together to form aggregated masses.[7] These fuse and may represent a primitive form of sexual reproduction, although Tibayrenc et al. present evidence that disputes this interpretation.[68]

PATHOGENESIS. Entrance of the metacyclic trypanosomes into cells in the subcutaneous tissue produces an acute local inflammatory reaction. Within 1 to 2 weeks after infection, they spread to the regional lymph nodes and begin to multiply in the cells that phagocytose them. The intracellular amastigote undergoes repeated divisions to form large numbers of parasites, producing the so-called pseudocyst. After a few days some of the organisms retransform into trypomastigotes and burst out of the pseudocyst, destroying the cell that contains them. A generalized parasitemia occurs then, and almost every type of tissue in the body can be invaded, although the parasites show a particular preference for muscle and nerve cells (Fig. 5-10). Reversion to amastigote, pseudocyst formation, retransformation to trypomastigote, and pseudocyst rupture are repeated in the newly invaded cells; then the process begins again. Rupture of the pseudocyst is accompanied by an acute, local inflammatory response, with degeneration and necrosis (cell or tissue death) of nerve cells in the vicinity, especially ganglion cells. This is the most important pathological change in Chagas' disease, and it appears to be the indirect result of parasitism of supporting cells, such as glial cells and macrophages, rather than invasion of neurons themselves.[67]

Chagas' disease manifests *acute* and *chronic* phases. The acute phase is initiated by inoculation into the wound of the trypanosomes from the bug's feces. The local inflammation produces a small red nodule, known as a **chagoma,** with swelling of the regional lymph nodes. In about 50% of the cases the trypanosomes enter through the conjunctiva of the eye, causing edema of the eyelid and conjunctiva and swelling

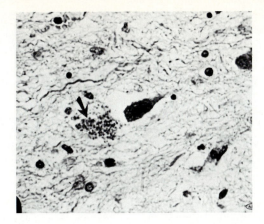

FIG. 5-10

Pseudocyst of *Trypanosoma cruzi* in brain tissue.

AFIP photograph neg. no 67- 5313.

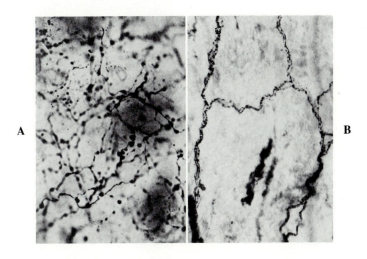

FIG. 5-11

Diaphanised tricuspid valves with zinc-osmium impregnation of nerve fibers (dark lines) **A,** Normal heart; **B,** Chagas' cardiopathy with marked reduction of nerve fibers.

From Hutt, M.S.R., F. Köberle, and K. Salfelder, 1973. In Spencer, H., editor. Tropical pathology. Springer- Verlag, Berlin.

of the preauricular lymph node. This symptom is known as **Romaña's sign.** As the acute phase progresses, pseudocysts may be found in almost any organ of the body, although the intensity of attack varies from one patient to another. The heart muscle usually is invaded, with up to 80% of the cardiac ganglion cells being lost. Symptoms of the acute phase include anemia, loss of strength, nervous disorders, chills, muscle and bone pain, and varying degrees of heart failure. Death may ensue 3 to 4 weeks after infection. The acute stage is most common and severe among children less than 5 years old.

The chronic stage is most often seen in adults. Its spectrum of symptoms is primarily the result of central and peripheral nervous dysfunction, which may last for many years. Some patients may be virtually asymptomatic and then suddenly succumb to heart failure. Chagas' disease accounts for about 70% of cardiac deaths in young adults in endemic areas. Part of the inefficiency in heart function is caused by loss of muscle tone resulting from the destroyed nerve ganglia. (Fig. 5-11). The heart itself becomes greatly enlarged and flabby.

In some regions of South America it is common for

the autonomic ganglia of the esophagus or colon to be destroyed. This ruins the tonus of the muscularis, resulting in deranged peristalsis and gradual flabbiness of the organ, which may become huge in diameter and unable to pass materials within it. This advanced condition is called **megaesophagus** or **megacolon,** depending on the organ involved (Fig. 5-12). Advanced megaesophagus may be fatal when the patient can no longer swallow. It has been experimentally demonstrated that testis tubules and epididymis atrophy in chronic cases.[16]

EPIDEMIOLOGY. The principal vectors of *T. cruzi* in Brazil are *Panstrongylus megistus, Triatoma sordida,* and *T. brasiliensis;* in Uruguay, Chile, and Argentina, *Triatoma infestans* is the primary culprit. *Rhodnius prolixus* is the main vector in northern South America and *Triatoma dimidiata* in Central America, whereas species of the *Triatoma protracta* group serve as vectors in Mexico. Several other species of triatomines have been found naturally infected throughout this range. Natural infections in *Triatoma sanguisuga* have been found in the United States. The insects can become infected as nymphs or adults. Triatomines can infect themselves when they feed on each other, presumably by sucking the contents of the intestine. Ticks, sheep keds, and bedbugs have been experimentally infected, but no evidence that they serve as natural vectors has been found. Natural mammalian reservoirs of infection have been mentioned, but domestic dogs and cats probably are the most important to human health.

Transmission from human to human during coitus or through breast milk may be possible, although this has yet to be documented. It has been shown conclusively that *T. cruzi* can and does cross the placental barrier from mother to fetus (Fig. 5-8). Newborn infants with advanced cases of Chagas' disease, including megaesophagus, have been described in Chile. In some villages in Mexico triatomines are thought to be aphrodisiac; therefore, they are eaten, and the trypanosomes gain access through the oral mucosa.[62]

Finally, the hazard of transmission by blood transfusion from donors with cryptic infection should not be underestimated. The frequency of this mode of transmission now ranks second only to natural vector transmission in endemic areas.[60]

The age of the victim is important in the epidemiology of Chagas' disease. Most new infections are in children from a few weeks to 2 years of age. The acute phase is most often fatal in this age group.

Because the bugs hide by day, primitive or poor-quality housing favors their presence. Thatched roofs, cracked walls, and trash-filled rooms are ideal for the

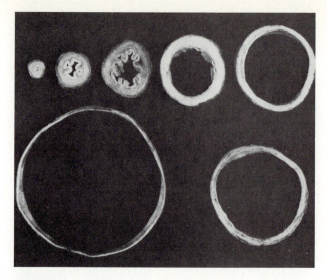

FIG. 5-12

Different stages of chagasic esophagopathy beginning with a normal organ, passing through hypertrophy and dilatation to the final megaesophagus.

From Hutt, M.S.R., F. Köberle, and K. Salfelder. 1973. In Spencer. H., editor. Tropical pathology. Springer-Verlag. Berlin.

breeding and survival of the insects. Misery compounds itself.

DIAGNOSIS AND TREATMENT. Diagnosis usually is by demonstration of trypanosomes in blood, cerebrospinal fluid, fixed tissues, or lymph. Trypomastigotes are most abundant in peripheral blood during periods of fever; they may be difficult to find at other times or in cases of chronic infection. In these cases blood can be inoculated into guinea pigs, mice, or other suitable hosts, and the animals in turn can be examined by heart smear or spleen impression. Another method that is widely employed is **xenodiagnosis.** Laboratory-reared triatomines are allowed to feed on the patient; then after a suitable period of time (10 to 30 days) they are examined for intestinal flagellates. This technique can detect cases in which trypanosomes in the blood are too few to be found by ordinary examination of blood films.

Complement fixation or other immunodiagnostic tests are extremely effective in demonstrating chronic cases, although they may give false positive reactions if the patient is infected with *Leishmania* or another species of trypanosome. In experiments with infected opossums, *Didelphis marsupialis,* an indirect fluorescent antibody test (IFAT) was the most sensitive test for *T. cruzi,* followed by xenodiagnosis.[33]

There is some suggestion that autoimmunity exists in some cases, but the topic is still being debated.[7a,32,36]

Unlike the other trypanosomes of humans, *T. cruzi* does not respond well to chemotherapy. The best drugs kill only the extracellular protozoa, but the intracellular forms defy the best efforts at eradication. This seems to be because the reproductive stages, inside living host cells, are shielded from the drugs. The lives and strength of millions of Latin American people depend on the discovery of a drug or vaccine that is effective against *T. cruzi.* One hope is the drug ketoconazole, which completely cured 78.5% of otherwise fatally infected mice.[45]

Trypanosoma (Herpetosoma) rangeli. Trypanosoma rangeli first was found, as was *T. cruzi,* in a triatomine bug in South America. *Rhodnius prolixus* is the most common vector, but *Triatoma dimidiata* and other species will also serve. Development is in the hindgut, and the epimastigote stages that result are from 32 to more than 100 μm long. The kinetoplast is minute, and the species can thereby be differentiated from *T. cruzi,* with which it often coexists.

Trypanosoma rangeli is common in dogs, cats, and humans in Venezuela, Guatemala, Chile, El Salvador, and Colombia. It has been found in monkeys, anteaters, opossums, and humans in Colombia and Panama. Trypomastigotes, 26 to 36 μm long, are larger than those of *T. cruzi.* The undulating membrane is large and has many curves. The nucleus is preequatorial, and the kinetoplast is subterminal.

The method of transmission is unclear. Although development is by posterior station, transmissions both by fecal contamination and by feeding inoculation have been reported.[69] *Trypanosoma rangeli* multiplies by binary fission in the mammalian host's blood. No intracellular stage is known, and the organism is apparently not pathogenic in humans.

However, infections with *T. rangeli* or mixed infections with *T. rangeli* and *T. cruzi* are potential problems for diagnosis.[22] Conventional immunofluorescence and ELISA assays, reinforced by immunoprecipitation and Western blot analysis diagnoses either or both infections.

Trypanosoma (Herpetosoma) lewisi. (Fig. 5-13). *Trypanosoma lewisi* is a cosmopolitan parasite of *Rattus* spp. Other rodents apparently are not susceptible, not even mice. The vector is the northern rat flea, *Nosopsyllus fasciatus,* in which the parasite develops inside cells of the posterior midgut. Metacyclic trypomastigotes appear in large numbers in the rectum of the insect, infecting rats that have eaten a flea or its feces. The parasite seems to be nonpathogenic in most

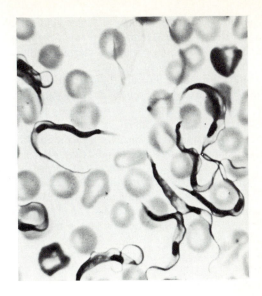

FIG. 5-13

Trypanosoma lewisi trypomastigotes in the blood of a rat.
Courtesy Turtox/Cambosco.

cases, perhaps even promoting the growth of its rat host.[41] However, infection may contribute to abortion and arthritis.

Much research has been conducted on this species because of the ease of maintaining it in the laboratory rat. One fascinating subject of this research is the "ablastin" phenomenon.[73] **Ablastin** is an antibody that arises during the course of an infection. After a rat is infected by the metacyclic trypomastigotes the parasite begins reproducing in the epimastigote form in the visceral blood capillaries. After about 5 days the trypanosome appears in the peripheral blood as a rather "fat" form, and shortly thereafter a crisis occurs in which most of the trypanosomes are killed by a trypanocidal antibody. A small population of slender trypomastigotes remains, which are infective for the flea but do not reproduce further while in the rat. After a few weeks the host produces another trypanocidal antibody, which clears the remaining trypanosomes, and the infection is cured. The slender trypomastigotes are sometimes known as "adults," and it has been shown that their reproduction is inhibited by the ablastin. Ablastin is a globulin with many characteristics of a typical antibody, but its action inhibits reproduction. Nucleic acid and protein synthesis by the trypanosome are inhibited, as is uptake of nucleic acid precursors. This seems to be an interesting adaptive accommodation of parasite and host: the parasite eludes the first onslaught of its host's immune defenses, but the host "allows" the parasite to remain available for the intermediate host for a longer period, meanwhile holding its reproduction in check. A somewhat less effective ablastin is produced by mice, when infected by the mouse trypanosome *T. musculi.* Interestingly, the two ablastins are cross-reactive: that is, anti-*lewisi* ablastin from a rat inhibits *T. musculi* when injected into a mouse, and anti-*musculi* ablastin inhibits *T. lewisi;* neither trypanosome is infective for the alternate host.

Trypanosoma (Megatrypanum) theileri. Trypanosoma theileri is a cosmopolitan parasite of cattle. The vectors are horseflies of the genera *Tabanus* and *Haematopota.* The trypanosome reproduces in the fly gut as an epimastigote.

The size of *T. theileri* varies with the strain—from 12 to 46 μm, 60 to 70 μm, and even 120 μm in length. The posterior end is pointed, and the kinetoplast is considerably anterior to it. Both trypomastigote and epimastigote forms can be found in the blood. Reproduction in the vertebrate host is in the epimastigote form and apparently occurs extracellularly in the lymphatics.

Trypanosoma theileri is usually nonpathogenic, but under conditions of stress it may become quite virulent. When cattle are stressed by immunization against another disease, undergo physical trauma, or become pregnant, the parasite may cause serious disease.

This parasite is rarely found in routine blood films. Detection usually depends on in vitro cultivation from blood samples. In fact, during tissue culture of bovine blood or cells, *T. theileri* is the most commonly found contaminant. Strong evidence points to transplacental transmission. In the United States a similar trypanosome is also common in deer and elk.

• • •

Other species of *Trypanosoma* are common in other classes of vertebrates, for example, *T. percae* in perch, *T. granulosum* in eels, *T. rotatorium* in frogs, *T. avium* in birds, and incompletely known species in turtles and crocodiles. Trypanosomes are commonly found in a variety of marine fishes.

Genus *Leishmania*

Like the trypanosomes, the leishmanias are heteroxenous. Part of their life cycle is spent in the gut of a fly, where they assume the form of a promastigote; the remainder of their life cycle is completed in vertebrate tissues, where only the amastigote form is found. Traditionally the amastigote is also known as a **Leishman-Donovan (L-D) body.** Species in humans are widely distributed (Fig. 5-14).

The vertebrate hosts of *Leishmania* spp. are primarily mammals. Although nearly a dozen species have been reported from lizards, they now are considered to be species of *Trypanosoma.*[10] The mammals most commonly infected are humans, dogs, and several species of rodents, in which they cause a complex of diseases called *Leishmaniasis.* Several of these are infective to humans. The intermediate hosts and vectors of leishmaniasis are **sandflies,** (Fig. 5-15) small bloodsucking insects in the family Psychodidae, subfamily Phlebotominae (see Chapter 39). There are over 600 species of sandflies divided into five genera: *Phlebotomus* and *Sergentomyia* in the Old World, and *Lutzomyia, Brumptomyia,* and *Warileya* in the New World. When they suck the blood of an infected animal, they ingest amastigote forms. These pass to the midgut, or hindgut, where they transform into promastigotes and multiply by binary fission. The parasites attach to the walls of the fly's gut and replicate. By the fourth or fifth day after feeding, promastigotes move forward to the esophagus and pharynx. When leishmanias begin to clog up the esophagus, the feeding sandfly pumps its esophageal contents in and out to clear the obstruction, thereby inoculating promastigotes into the skin of a luckless victim. Transmission also can occur when infected sandflies are crushed into the skin or mucous membrane.

All amastigotes in vertebrate tissues look similar (Fig. 5-16). They are spheroid to ovoid, usually 2.5 to 5 μm wide, although smaller ones are known. This makes them among the smallest nucleated cells known. In stained preparations only the nucleus and a very large kinetoplast can be seen, and the cytoplasm appears vacuolated. Exceptionally, a short axoneme is visible within the cytoplasm under the light microscope.

Although all *Leishmania* spp. exhibit similar morphology, they differ clinically, biologically, and serologically.[81] Even so, these characteristics often overlap, so distinction between species are not clear-cut. Leishmaniases that normally are visceral may become dermal; dermal forms can become mucocutaneous; and an immunodiagnostic test derived from the antigens of one species may give positive reactions in the presence of other species of *Leishmania,* or even *Trypanosoma.* These populations are closely related and are in the process of speciation as the result of recent geographical isolation. It is likely that the transport of slaves to the western world from Africa through the Middle East and Asia spread *Leishmania* into previously uncontaminated areas, where they are now rapidly speciating.

The result of the difficulty in species definition within the genus is that several schemes of classification have been proposed, nearly all of them having

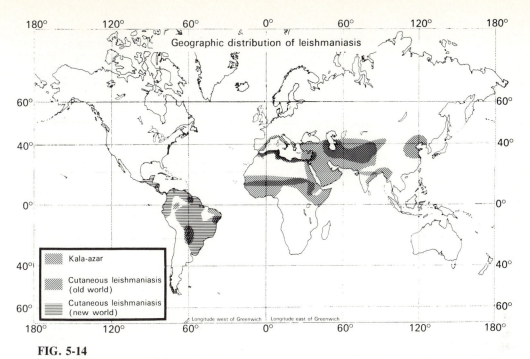

FIG. 5-14

Geographical distribution of leishmaniasis.

AFIP photograph neg. no. 68-1805- 2.

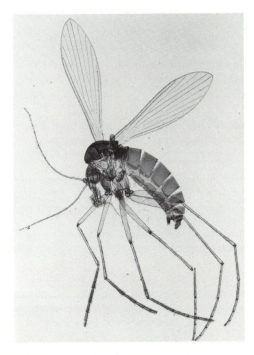

FIG. 5-15

The sandfly *Phlebotomus*, a vector of *Leishmania* spp. Sandflies are about 3 mm long.

Photograph by Jay Georgi.

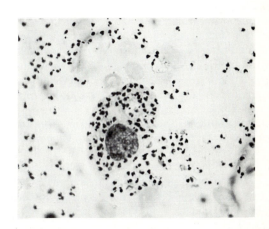

FIG. 5-16

Spleen smear showing numerous intracellular and extracellular amastigotes of *Leishmania donovani*.

AFIP photograph neg. no. 55-17580.

some acceptance. The published literature is confusing, since some researchers refer to several species and others consider the same organisms as a single, widespread species with slightly different clinical manifestations but similar or identical immunological properties. Although this is rather a nuisance to the student, it is a good demonstration of evolution. The more similarities two or more populations exhibit, the more recently they have diverged from one another.

Species of flagellate that develop in the sandfly's midgut before moving anteriad are placed in **section Suprapylaria.** Those that develop in the hindgut first are placed in **section Peripylaria.**[37] Species and subspecies of *Leishmania* currently recognized are listed in Table 5-1. These taxa are in general agreement with species separation on the basis of biochemical criteria.[10] Of those species complexes we will consider the five most important to human welfare: *L. tropica, L. major, L. donovani, L. mexicana,* and *L. braziliensis.*

Leishmania tropica and L. major. These two species have similar life cycles and clinical symptoms. However, they are found in different localities, have different reservoir and intermediate hosts, and the lesions they cause are somewhat different. Further, they can be differentiated biochemically.

These parasites produce a cutaneous ulcer variously known as **oriental sore, cutaneous leishmaniasis, Jericho boil, Aleppo boil,** or **Delhi boil.** They are found in west-central Africa, the Middle East, and Asia Minor into India.

MORPHOLOGY AND LIFE CYCLE. The appearance of their amastigotes is similar to that of the other leishmanias of humans (Fig. 5-16). Sandflies of the genus *Phlebotomus* are the intermediate hosts and vectors. When the fly takes a blood meal containing amastigotes, the parasites multiply in the midgut, move to the pharynx, and are inoculated into the next mammalian victim. There they multiply in the reticuloendothelial system and lymphoid cells of the skin. Few amastigotes are found except in the immediate vicinity of the site of infection, so the sandflies must feed there to become infected.

PATHOGENESIS. The incubation period lasts from a few days to several months. The first symptom of infection is a small, red papule at the site of the bite. This may disappear in a few weeks, but usually it develops a thin crust that hides a spreading ulcer underneath. Two or more ulcers may coalesce to form a large sore (Fig. 5-17). In uncomplicated cases the ulcer will heal within 2 months to a year, leaving a depressed, unpigmented scar. It is common, however, for secondary infection to occur, including, for exam-

TABLE 5-1

Species and subspecies of *Leishmania*

Parasite	Locality
Section SUPRAPYLARIA	
L. tropica	Urban areas of Middle East and India
L. major	Africa, Middle East, Soviet Asia
L. donovani donovani	China, India, Bangladesh
L. donovani infantum	North Central Asia, North West China, Middle East, Southern Europe, North West Africa
L. donovani chagasi	South and Central America
L. donovani archibaldi	Sudan, Ethiopia
L. mexicana mexicana	Belize, Guatemala, Mexico, South Central United States
L. mexicana aristedesi	Panama
L. mexicana garnhami	Venezuela
L. mexicana pifanoi	Venezuela
L. mexicana venezuelensis	Venezuela
L. mexicana amazonensis	Amazon Basin, Brazil
L. mexicana ssp.	Trinidad, Brazil
L. enriettii	Brazil
L. hertigi hertigi	Panama, Costa Rica
L. hertigi deanei	Brazil
L. aethiopica	Ethiopia, Kenya
L. gerbilli	China, Mongolia
Section PERIPYLARIA	
L. braziliensis braziliensis	Brazil
L. braziliensis guyanensis	French Guiana, Guyana, Suriname
L. braziliensis panamensis	Panama, Costa Rica
L. braziliensis peruviana	Western Andes
L. braziliensis ssp.	Belize, Panama, Costa Rica

ple, yaws (disfiguring disease caused by a spirochete) and myiasis (infection with fly maggots, see Chapter 39).

Leishmania tropica is found in more densely populated areas. Its lesion is dry, persists for months before ulcerating, and has numerous amastigotes within it. By contrast, *L. major* is found in sparsely inhabited regions. Its papule ulcerates quickly, is of short duration, and contains few amastigotes.

Most species and subspecies of *Leishmania* also can produce cutaneous lesions. There is an astonishing va-

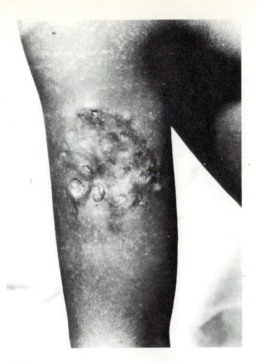

FIG. 5-17

Oriental sore: a complicated case with several lesions.
AFIP photograph neg. no. A-43418-1.

riety of forms of such lesions, from tiny, pin–prick-like sores to massive, diffused ulcers. Some even have been misdiagnosed as leprosy or tuberculosis. Diagnosis, then, becomes difficult at times, especially when two species occur in the same locality.

DIAGNOSIS. Diagnosis of infection is greatly facilitated by finding amastogotes. Scrapings from the side or edge of the ulcer, smeared on a slide and stained with Wright's or Giemsa's stain, will show the parasites in endothelial cells and monocytes, even though they cannot be found in the circulating blood. Cultures should be made in case amastigotes go undetected.

TREATMENT AND CONTROL. Lesions caused by *L. tropica* and *L. major* usually heal spontaneously in a few months, so treatment may not be required. If the site of infection is cosmetically important treatment might be desirable. Systemic therapy with pentavalent antimonials is the treatment of choice. In earlier years trivalent antimonials were the only drugs available but they were so toxic as to be downright dangerous. Newer drugs are much less so, but toxic effects still are common. Two preparations are available: Pentostam and Glucantime; only Pentostam is available in the United States, through the Centers for Disease Control

parasite drug service. More recently a different drug, ketoconazole has shown much promise in cutaneous lesions.[71]

Protective immunity following medical treatment seems to be absolute, and immunity as a result of the natural course of the disease is 97% to 98% effective. Recognizing this, some native peoples deliberately inoculate their children on a part of their body normally hidden by clothes, which prevents their later developing a disfiguring scar on an exposed part of the body. In Lebanon and Russia attempts at mass vaccination with promastigotes show promising results. Control, however, ultimately depends on eliminating the sandflies and reducing the population of rodent reservoirs.

Leishmania donovani. In 1900 Sir William Leishman discovered *L. donovani* in spleen smears of a soldier who died of a fever at Dum-Dum, India. The disease was known locally as **Dum-Dum fever** or **kala-azar.** Leishman published his observations in 1903, the year that Charles Donovan found the same parasite in a spleen biopsy. The scientific name is in honor of these men, as is the common name of the amastigote forms, Leishman-Donovan (L-D) bodies. The Indian Kala-azar Commission (1931 to 1934) demonstrated the transmission of *L. donovani* by *Phlebotomus* spp.

Leishmania donovani has a wide distribution, and several varieties are distinguished on epidemiological and clinical grounds (Table 5-1).

MORPHOLOGY. *Leishmania donovani* ssp. cannot be differentiated from other species of *Leishmania* on the basis of morphology. Its rounded or ovoid body measures 2 to 3 μm, with a large nucleus and kinetoplast. It lives within cells of the reticuloendothelial system of the viscera, including spleen, liver, mesenteric lymph nodes, intestine, and bone marrow. Amastigotes have been found in nearly every tissue and fluid of the body.

LIFE CYCLE. The life cycle parallels that of *L. tropica* except that *L. donovani* is primarily a visceral infection. When a sandfly of the genus *Phlebotomus* ingests amastigote forms along with its blood meal, the parasites lodge in the midgut and begin to multiply. They transform into slender promastigotes and quickly block the gut of the insect. Soon they can be seen in the esophagus, pharynx, and buccal cavity, where they are injected into a new host with the fly's bite. Not all strains of *L. donovani* are adapted to all species and strains of *Phlebotomus*. Once in the mammalian host, the parasite is immediately engulfed by a macrophage, in which it divides rapidly by binary fission, killing the host cell. Escaping the dead macrophage, the protozoa are engulfed by other macrophages, which they also kill, and by this means eventually severely damage the RE system, one of the host's primary defense mechanisms against disease. Interestingly, amastigotes engulfed by neutrophils are killed.[11] Although less effective as a phagocyte, eosinophils also kill the *L. donovani* they consume.

PATHOGENESIS. The incubation period in humans may be as short as 10 days or as long as a year, but usually is 2 to 4 months. The disease usually begins slowly with low-grade fever and malaise and is followed by progressive wasting and anemia, protrusion of the abdomen from enlarged liver and spleen (Fig. 5-18), and finally by death (in untreated cases) in 2 to 3 years. In some cases the symptoms may be more acute in onset, with chills, fever up to 104° F, and vomiting; death may occur within 6 to 12 months. Accompanying symptoms are edema, especially of the face, bleeding of the mucous membranes, breathing difficulty, and diarrhea. The immediate cause of death often is the invasion of secondary pathogens that the body is unable to combat. A certain proportion of cases, especially in India, recover spontaneously, and post–kala-azar dermal leishmanoid (see below) also is more common in India.

Visceral leishmaniasis may be viewed essentially as a disease of the reticuloendothelial system. The phagocytic cells, which are so important in defending the host against invasion, are themselves the habitat of the parasites. Blood-forming organs, such as the spleen and bone marrow, undergo compensatory production of macrophages and other phagocytes (hyperplasia) to the detriment of red cell production. Thus the spleen and the liver become greatly enlarged (hepatosplenomegaly) (Fig. 5-19), while the patient becomes severely anemic and emaciated.

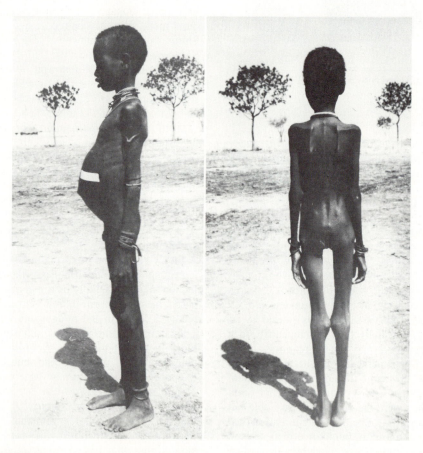

FIG. 5-18

Advanced kala-azar. Boy, about 6 years old, from Sudan, showing extreme hepatosplenomegaly and emaciation typical of advanced kala-azar.

From Hoogstraal, H., and D. Heyneman. 1969. Am. J. Trop. Med. Hyg. 18:1091- 1210.

A condition known as **post–kala-azar dermal leishmanoid** develops in some cases.[48] It is rare in the Mediterranean and Latin American areas but develops in 5% to 10% of cases in India. The condition usually becomes apparent about 1 to 2 years after inadequate treatment for kala- azar. It is marked by reddish, depigmented nodules in the skin (Fig. 5-20)

EPIDEMIOLOGY. Transmisson of visceral leishmaniasis is related to the activities of humans and the biology of sandflies. *Phlebotomus* spp. exist mainly at altitudes under 2000 feet, most commonly in flat plains areas. Even in desert areas such as in the Sudan, the flies rest in and are protected by cracks in the parched earth and under rocks. In such conditions the flies are active only during certain hours of the day. For humans to become infected, they must be in sandfly areas at these times.

Age of the victim is a factor in the course of the disease, and fatal outcome is most frequent in infants and small children. Males are more often infected than are females, most likely as the result of more exposure to sandflies. Poor nutrition, concomitant infection with other pathogens, and other stress factors predispose the patient to lethal consequences.

A wide variety of animals can be infected experimentally, although dogs are the main important reservoir in most areas. Canine infection is less common in India, where it is believed that a fly-to-human relationship is maintained.

DIAGNOSIS. As in *L. tropica,* diagnosis depends on finding L-D bodies in tissues or secretions. Spleen punctures, blood or nasal smears, bone marrow, and other tissues should be examined for the characteristic parasites, and cultures from these and other organs should be attempted. Immunodiagnostic tests are sensitive but cannot differentiate between species of *Leishmania* or between current and cured cases. The tests most frequently used are the enzyme-linked immunosorbent assay (ELISA) and the indirect fluorescent antibody test (IFA). Until recently the main limitation of these tests was a cross-reacting positive response to *Trypanosoma cruzi.* This problem has been resolved by improved techniques of antigen purification, special antigen polymerization, and antigen se-

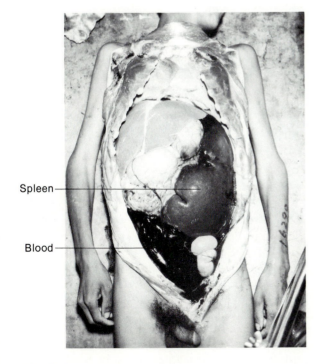

FIG. 5-19

A patient with kala-azar who died of hemorrhage after a spleen biopsy. Note the greatly enlarged spleen. (The dark matter in the lower abdominal cavity is blood.)

AFIP neg. no. A-45364.

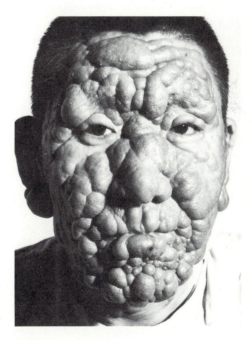

FIG. 5-20

Post kala-azar dermal leishmanoid. This patient responded very well to treatment, regaining a nearly normal appearance.

Photograph by Robert E. Kuntz.

75

lection by use of monoclonal antibodies.[44] Other diseases that might have symptoms similar to kala-azar are typhoid and paratyphoid fevers, malaria, syphilis, tuberculosis, dysentery, and relapsing fevers. Each must be eliminated in the diagnosis of kala-azar.

TREATMENT AND CONTROL. Treatment consists of injections of various antimony compounds, as previously described for *L. tropica*, and good nursing care. An adequate diet with high protein and vitamins is essential to combat nutritional deficiency. These antimonial drugs are toxic and must be used with great care. Furthermore, relapses and post–kala-azar dermal leishmanoid may follow insufficient treatment.

Control of sandflies and reservoirs is urgently required in endemic areas.

Leishmania braziliensis. Leishmania braziliensis produces a disease in humans variously known as **espundia, uta,** or **mucocutaneous leishmaniasis.** It is found throughout the vast area between central Mexico and northern Argentina, although its range does not extend into the high mountains, except for the South slope of the Andes. Clinically similar cases have been reported in northwest Africa, due to *L. donovani* ssp. The clinical manifestations of the disease vary along its range, which has led to confusion regarding the identity of the organisms responsible. Several species names have been proposed for different clinical and serological types, (see Table 5-1). Once again, it appears that the parasite is rapidly evolving into groups that are adapting to local populations of humans and flies. Morphologically, *L. braziliensis* cannot be differentiated from *L. tropica, L. mexicana,* or *L. donovani.* An interesting historical account of this disease, with evidence of its pre-Columbian existence in South America is given by Hoeppli.[31]

LIFE CYCLE AND PATHOGENESIS. The life cycle and methods of reproduction are identical to those of *L. donovani* and *L. tropica* except that the promastigotes reproduce in the hindgut of the sandfly, with several species of *Lutzomyia* serving as vectors. Inoculation of promastigotes by the bite of a sandfly causes a small, red papule on the skin. This becomes an itchy, ulcerated vesicle in 1 to 4 weeks and is similar at this stage to oriental sore. This primary lesion heals within 6 to 15 months. The parasite never causes a visceral disease but often develops a secondary lesion on some region of the body.

In Venezuela and Paraguay the lesions more often appear as flat, ulcerated plaques that remain open and oozing. The disease is called **pian bois** in that area. Sloths and anteaters are the primary reservoirs of pian bois in northern Brazil.[38]

In the more southerly range of *L. braziliensis* the parasites have a tendency to metastasize, or spread directly from the primary lesion to mucocutaneous zones. The secondary lesion may appear before the primary has healed, or it may be many years (up to 30) before secondary symptoms appear.[80]

The secondary lesion often involves the nasal system and buccal mucosa, causing degeneration of the cartilaginous and soft tissues (Fig. 5-21). Necrosis and secondary bacterial infection are common. **Espundia** and **uta** are the names applied to these conditions. The ulceration may involve the lips, palate, and pharynx, leading to great deformity. Invasion of the infection into the larynx and trachea destroys the voice. Rarely the genitalia may become infected. The condition may last for many years, and death may result from secondary infection or respiratory complications. A similar condition is known to occur in Ethiopia.

DIAGNOSIS AND TREATMENT. Diagnosis is established by finding L-D bodies in affected tissues. Espundia-like conditions are also caused by tuberculosis, leprosy, syphilis, and various fungal and viral dis-

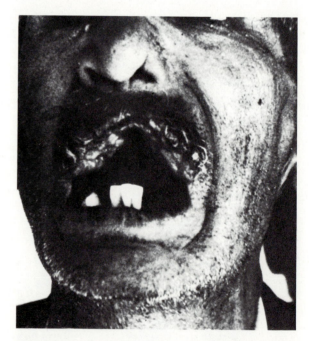

FIG. 5-21

Espundia of 2 years' development after 24 years' delay in onset. The upper lip, gum, and palate are destroyed.

From Walton, B.C., L.C. Chinel, and O. Eguiay E. 1973. Am. J. Trop. Med. Hyg. 22:696-698.

eases, and these must be differentiated in diagnosis. Skin tests are available for diagnosis of occult infections. Culturing the parasite in vitro is also a valuable technique when L-D bodies cannot be demonstrated in routine microscope preparation.

Treatment is similar to that for kala-azar and tropical sore: antimonial compounds applied on the lesions or injected intravenously—or intramuscularly. Secondary bacterial infections should be treated with antibiotics. Mucocutaneous lesions are particularly refractory to treatment and require extensive chemotherapy. Relapse is common, but once cured, a person usually has lifelong immunity. However, if the infection is not cured but merely becomes occult, there may be a relapse with onset of espundia many years later. Because this is primarily a sylvatic disease, there is little opportunity for its control.

Leishmania mexicana. This parasite is found in northern Central America, Mexico, Texas, and possibly the Dominican Republic and Trinidad. Primarily a cutaneous form, it infects several thousand persons a year, especially agricultural or forest laborers. The disease is on the increase due to increased clearing of forests and concomitant expansion of farm land. Three clinical manifestations are found: cutaneous, nasopharyngeal mucosal, and visceral, although some records probably are due to *L. braziliensis.* Traditionally the cutaneous form of disease has been called "chiclero ulcer" because it is so common in "chicleros," forest-dwelling people who glean a living by harvesting the gum of chicle trees. In Belize, an English-speaking country, it is called "bay sore."

LIFE CYCLE AND PATHOGENESIS. Like the other species of *Leishmania,* sandflies are the vectors of *L. mexicana.* Several species of *Lutzomyia* are involved. The disease is a zoonosis and the main reservoirs are rodents. The most important reservoirs are those that live at, or travel to ground level. Obviously, arboreal reservoirs are less efficient sources of infection to humans. No domestic reservoir is known for chiclero ulcer.

Cutaneous leishmaniasis due to the *L. mexicana* complex usually heals spontaneously in a few months except when the lesions are in the ear. Ear cartilage is poorly vascularized so immune responses are weak. Chronic lesions are known with a duration of up to 40 years. Considerable mutilation may result. Mucocutaneous and visceral manifestations are rare.

At least eight cases of autochthonous infections of *L. mexicana* in Texas are known in humans, with another on the ear of a cat. Visceral cases in dogs in Oklahoma suggest that canine leishmaniasis may have become endemic in the United States.

DIAGNOSIS AND TREATMENT. The diagnosis and treatment of *L. mexicana* is the same as for *L. tropica.*

Genus *Leptomonas*

Leptomonas organisms are parasitic in invertebrates and are of no medical importance. *Leptomonas* is variously a promastigote and an intracellular amastigote throughout its monoxenous life cycle. Species are found in molluscs, nematodes, insects, and other protozoa.

Genus *Herpetomonas*

Members of the genus *Herpetomonas* also are characteristically monoxenous in insects. They pass through amastigote, promastigote, opisthomastigote, and possibly epimastigote stages in their life cycles. In the opisthomastigote the flagellum arises from a reservoir that runs the entire length of the body.

Genus *Crithidia*

Crithidia spp. are tiny (4 to 10 μm) choanoflagellates of insects. They are often clustered together against the intestine of their host. They can assume the amastigote form and are monoxenous.

Genus *Blastocrithidia*

Blastocrithidia organisms are monoxenous insect parasites, usually found as epimastigotes and amastigotes in the intestines of their hosts. Species are common in water striders (family Gerridae).

Genus *Phytomonas*

Phytomonas is a parasite of milkweeds and related plants. It passes through promastigote and amastigote phases in the intestines of certain beetles and appears as promastigotes in the sap (latex) of its plant hosts.

REFERENCES

1. Bailey, N.M. 1967. Recent development in the screening of populations for human trypanosomiasis and their possible application in other immunizing diseases. East Afr. Med. J. 44:475-481.
2. Banks, K.L. 1978. Binding of *Trypanosoma congolense* to the walls of small blood vessels. J. Protozool. 25:241-245.
3. Barry, J.D. 1986a. Antigenic variation during *Trypanosoma vivax* infections of different host species. Parasitology 92:51-65.
4. Barry, J.D. 1986b. Surface antigens of African trypanosomes in the tsetse fly. Parasitol. Today 2:143-145.
5. Battaglia, P.A., M. del Bue, M. Ottaviano, and M. Ponzi. 1983. A puzzle genome: kinetoplast DNA. In Guardiola, J., L. Luzzatto, and W. Trager, editors: Molecular biology of parasites. Raven Press, New York, pp. 107-124.

6. Borst, P., and G.A.M. Cross. 1982. Molecular basis for trypanosome antigenic variation. Cell 29:291-303.

7. Brener, Z. 1972. A new aspect of *Trypanosoma cruzi* life cycle in the intermediate host. J. Protozool. 19:23-27.

7a. Brener, Z. 1980. Immunity to *Trypanosoma cruzi*. In Lumsden, W.H.R., R. Muller, and J. R. Baker, editors. Advances in parasitology, vol. 18. Academic Press, Inc., New York, pp. 247-292.

8. Burkholder, J.E., T.C. Allison, and V.P. Kelly. 1980. *Trypanosoma cruzi* (Chagas) (Protozoa: Kinetoplastida) in invertebrate, reservoir, and human hosts of the lower Rio Grande Valley of Texas. J. Parasitol. 66:305-311.

9. Cannata, J.J.B., A.C.C. Frasch, M.A. Cataldi de Flombaum, E.L. Segura, and J.J. Cazzulo. 1979. Two forms of "malic" enzyme with different regulatory properties in *Trypanosoma cruzi*. Biochem. J. 184:409-419.

10. Chance, M.C. 1985. The biochemical and immunological taxonomy of *Leishmania*. In Chang, K.P., and R.S. Bray, editors. Human parasitic diseases, vol. 1. Leishmaniasis. Elsevier, New York, pp. 93-110.

11. Chang, K.-P. 1981. Leishmanicidal mechanisms of human polymorphonuclear phagocytes. Am. J. Trop. Med. Hyg. 30:322-333.

12. De Souza, W., and T. Souto-Padrón. 1980. The paraxial structure of the flagellum of Trypanosomatidae. J. Parasitol. 66:229-235.

13. Docampo, R., F.S. Cruz, W. Leon, and G.A. Schmunis. 1979. Acetate oxidation by bloodstream forms of *Trypanosoma cruzi*. J. Protozool. 26:301-303.

14. Dolan, R.B. 1987. Genetics and trypanotolerance. Parasitol. Today 3:137-143.

15. Esser, K.M., and M.J. Schoenbechler. 1985. Expression of two variant surface glycoproteins on individual African trypanosomes during antigen switching. Science 229:190-193.

16. Ferreira, A.L., and M.A. Rossi. 1973. Pathology of the testis and epididymis in the late phase of experimental Chagas' disease. Am. J. Trop. Med. 22:699-704.

17. Gardiner, P.R., T.W. Pearson, M.W. Clarke, and L.M. Mutharia. 1987. Identification and isolation of a variant surface glycoprotein from *Trypanosoma vivax*. Science 235:774-777.

18. Gardiner, P.R., and A. J. Wilson. 1987. *Trypanosoma (Duttonella) vivax*. Parasitol. Today 3:49-52.

19. Gibson, W.C. 1986. Will the real *Trypanosoma b. gambiense* please stand up. Parasitol. Today 2:255-257.

20. Gibson, W.C., T.F. de C. Marshall, and D.G. Godfrey. 1980. Numerical analysis of enzyme polymorphism: a new approach to the epidemiology and taxonomy of trypanosomes of the subgenus *Trypanozoon*. In Lumsden, W.H.R., R. Muller, and J.R. Baker, editors. Advances in parasitology, vol. 18. Academic Press, Inc., New York, pp. 175-246.

21. Goodwin, L.G. 1964. The chemotherapy of trypanosomiasis. In Hutner, S.M., editor. The biochemistry and physiology of protozoa, vol. 3. Academic Press, Inc., New York, pp. 495-524.

22. Guhl, F., L. Hudson, C.J. Marinkelle, C.A. Jaramillo, and D. Bridge. 1987. Clinical *Trypanosoma rangeli* infection as a complication of Chagas' disease. Parasitology 94:475-484.

23. Gutteridge, W.E., B. Cover, and M. Gaborak. 1978. Isolation of blood and intracellular forms of *Trypanosoma cruzi* from rats and other rodents and preliminary studies of their metabolism. Parasitology 76:159-176.

24. Hajduk, S.L., C.R. Cameron, J.D. Barry, and K. Vickerman. 1981. Antigenic variation in cyclically transmitted *Trypanosoma brucei*. Variable antigen type composition of metacyclic trypanosome populations from the salivary glands of *Glossina morsitans*. Parasitology 83:595-607.

25. Hajduk, S.L., and K. Vickerman. 1981. Antigenic variation in cyclically transmitted *Trypanosoma brucei*. Variable antigen type composition of the first parasitaemia in mice bitten by trypanosome-infected *Glossina morsitans*. Parasitology 83:609-621.

26. Hill, G.C. 1976. Characterization of the electron transport systems present during the life cycle of African trypanosomes. In Van den Bossche, H., editor. Biochemistry of parasites and host-parasite relationships. North-Holland Publishing Co., Amsterdam, pp. 31-50.

27. Hoare, C.A. 1956. Morphological and taxonomic studies on the mammalian trypanosomes. VIII. Revision of *Trypanosoma evansi*. Parasitology 46:130-172.

28. Hoare, C.A. 1964. Morphological and taxonomic studies on mammalian trypanosomes. X. Revision of the systematics. J. Protozool. 11:200-207.

29. Hoare, C.A. 1965. Vampire bats as vectors and hosts of equine and bovine typanosomes. Acta Trop. 22:204-216.

30. Hoare, C.A. 1967. Evolutionary trends in mammalian trypanosomes. In Dawes, B., editor. Advances in parasitology, vol. 5 Academic Press, Inc., New York, pp. 47-91.

31. Hoeppli, R. 1969. Parasitic diseases in Africa and the Western Hemisphere. Early documentation and transmission by the slave trade. Verlag für Recht and Gesellschaft AG, Basel.

32. Hudson, L. 1985. Autoimmune phenomena in chronic chagasic cardiopathy. Parasitol. Today 1:6-7.

33. Jansen, A.M., P.L. Moriearty, B.G. Castro, and M.P. Deane. 1985. *Trypanosoma cruzi* in the opossum *Didelphis marsupialis:* an indirect fluorescent antibody test for the diagnosis and follow-up of natural and experimental infections. Trans. R. Soc. Trop. Med. Hyg. 79:474-477.

34. Jennings, F.W., G.M. Urquhart, P.K. Murray, and B.M. Miller, 1980. "Bernil" and nitroimidazole combinations in the treatment of *Trypanosoma brucei* infection with central nervous system involvement. Int. J. Parasitol. 10:27-32.

35. Kagan, I., L. Norman, and D.S. Allain. 1966. Studies on *Trypanosoma cruzi* isolated in the United States: a review. Rev. Biol. Trop. 14:55-73.

36. Kierzenbaum, F. 1985. Is there autoimmunity in Chagas' disease? Parasitol. Today 1:4-6.

37. Lainson, R. 1982. In Symposium of the Zoological Society of London 50:137-179.

38. Lainson, R., J.J. Shaw., and M. Póvoa. 1981. The importance of edentates (sloths and anteaters) as primary reservoirs of *Leishmania braziliensis guyanensis,* causative agent of "pianbois" in north Brazil. Trans. R. Soc. Trop. Med. Hyg. 75:611-612.

39. Laurent, M., E. Pays, K. Delinte, E. Magnus, N. Van Meirvenne, and M. Steinert. 1984. Evolution of a trypanosome surface antigen gene repertoire linked to non-duplicative gene activation. Nature 308:370-373.

40. Levine, N.D. 1973. Protozoan parasites of domestic animals and of man, ed. 3. Burgess Publishing Co., Minneapolis.

41. Lincicome, D.R., R.N. Rossan, and W.C. Jones. 1963. Growth of rats infected with *Trypanosoma lewisi*. Exp. Parasitol. 14:54-65.

42. Lumsden, W.H.R. 1971. Pathobiology of trypanosomiasis. In Gaafar, S.M., editor. Pathology of parasitic diseases. Purdue Research Foundation, Lafayette, Ind., pp. 1-14.

43. Maudlin, I. 1985. Inheritance of susceptibility to trypanosomes in tsetse flies. Parasitol. Today 1:59-60.

44. Mauel, J., and R. Behin. 1982. Leishmaniasis: immunity, immunopathology and immunodiagnosis. In Cohen, S., and K.S. Warren editors. Immunology of parasitic infections. Blackwell Scientific Publications Ltd., Oxford, Eng., pp. 299-355.

45. McCabe, R.E., J.S. Remington, and F.G. Araujo. 1987. Ketoconazole promotes parasitological cure of mice infected with *Trypanosoma cruzi*. Trans. R. Soc. Trop. Med. Hyg. 81:613-615.

46. Miller, E.N., and M.J. Turner. 1981. Analysis of antigenic types appearing in first relapse populations of clones of *Trypanosoma brucei*. Parasitology 82:63-80.

47. Molyneux, D.H. 1977. Vector relationships in the Trypanosomatidae. In Dawes, B., editor. Advances in parasitology, vol. 15. Academic Press, Inc., New York. pp. 1-82.

48. Morgan, F.M., R.H. Watten, and R.E. Kuntz. 1962. Post—kala-azar dermal leishmaniasis. A case report from Taiwan (Formosa). J. Formosa Med. Assoc. 61:282-291.

49. Müller, M. 1975. Biochemistry of protozoan microbodies: peroxisomes, α-glycerophosphate oxidase bodies, hydrogenosomes. Ann. Rev. Microbiol. 29:467-483.

50. Murphy, W.J., S.T. Brentano, A.C. Rice-Ficht, D.M. Dorfman, and J.E. Donelson. 1984. DNA rearrangements of the variable surface antigen genes of the trypanosomes. J. Protozool. 31:65-73.

51. Murray, M., W.I. Morrison, and D.D. Whitelaw. 1982. Host susceptibility to African trypanosomiasis: trypanotolerance. In Baker, J.R., and R. Muller, editors. Advances in parasitology, vol. 21. Academic Press, Inc. New York. pp. 1-68.

52. Newport, G.R., and C.R. Page, III. 1977. Free amino acids in brain, liver, and skeletal muscle tissue of voles infected with *Trypanosoma brucei gambiense*. J. Parasitol. 63:1060-1065.

53. Noble, E.R. 1955. The morphology and life cycles of trypanosomes. Quart. Rev. Biol. 30:1-28.

54. Ormerod, W.E. 1967. Taxonomy of the sleeping sickness trypanosomes. J. Parasitol. 53:824-830.

55. Ormerod, W.E. 1985. How do lipids affect African trypanosomes? Parasitol. Today 1:86-87.

56. Pays, E., M. Lheureux, and M. Steinert. 1982. Structure and expressions of a *Trypanosoma brucei gambiense* variant specific antigen gene. Nucleic Acids Res. 10:3149-3163.

57. Pays, E., S. Van Assel, M. Laurent, M. Darville, T. Vervoort, N. Van Meirvenne, and M. Steinert. 1983a. Gene conversion as a mechanism for antigenic variation in trypanosomes. Cell 34:371-381.

58. Pays, E., M.-F. Delauw, S. Van Assel, M. Laurent, T. Vervoort, N. Van Meirvenne, and M. Steinert. 1983b. Modifications of a *Trypanosoma b. brucei* antigen gene repertoire by different DNA recombinational mechanisms. Cell 35:721-731.

59. Rickman, W.J., and H.W. Cox. 1983. Trypanosome antigen-antibody complexes and immunoconglutinin interactions in African trypanosomiasis. Int. J. Parasitol. 13:389-392.

60. Rohwedder, R. 1965. Chagas' infection in blood donors and the possibilities of its transmission by means of transfusion. Bull. Chil. Parasitol. 24:88-93.

61. Ross, R., and D. Thompson. 1910. A case of sleeping sickness studied by precise enumerative methods: regular periodical increase of the parasites disclosed. Proc. R. Soc. Lond. (Biol.) 82:411-415.

62. Salazar-Schettino, P.M. 1983. Customs which predispose to Chagas' disease and cyticercosis in Mexico. Am. J. Trop. Med. Hyg. 32:1179-1180.

63. Schechter, P.J., and A. Sjoerdsma. 1986. Difluoromethylornithine is the treatment of African trypanosomiasis. Parasitol. Today 2:223-224.

64. Scott, M.T., and D. Snary. 1982. American trypanosomiasis (Chagas' disease). In Cohen, S., and K.S. Warren, editors. Immunology of parasitic infections. Blackwell Scientific Publications Ltd., Oxford, Eng., pp. 261-298

65. Seed, J.R., R. Edwards, and J. Sechelski. 1984. The ecology of antigenic variation. J. Protozool. 31:48-53.

66. Tait, A. 1983. Sexual processes in the Kinetoplastida. Symposia of the British Society for Parasitology, vol. 20. Parasitology 86(4):29- 57.

67. Tanowitz, H.B., C. Brosnan, D. Guastamacchio, G. Baron, C. Raventos-Suarez, M. Bornstein, and M. Wittner. 1982. Infection of organotypic cultures of spinal cord and dorsal root ganglia with *Trypanosoma cruzi*. Am. J. Trop. Med. Hyg. 31:1090-1097.

68. Tibayrenc, M., L. Echalar, J.P. Dujardin, O. Poch, and P. Desjeux. 1984. The microdistribution of isoenzymic strains of *Trypanosoma cruzi* in southern Bolivia; new isoenzyme profiles and further arguments against Mendelian sexuality. Trans. R. Soc. Trop. Med. Hyg. 78:519-525.

69. Tobie, E.J. 1965. Biological factors influencing transmission of *Trypanosoma rangeli* by *Rhodnius prolixus*. J. Parasitol. 51:837-841.

70. Turner, C.M.R., J.D. Barry, and K. Vickerman. 1986. Independent expression of the metacyclic and bloodstream variable antigen repertoires of *Trypanosoma brucei rhodesiense*. Parasitology 92:67-73.

71. Viallet, J., J.D. MacLean, and H. Robson. 1986. Response to ketoconazole in two cases of longstanding cutaneous leishmaniasis. Am. J. Trop. med. Hyg. 35:491-495.

72. Vickerman, K. 1969. The fine structure of *Trypanosoma congolense* in its bloodstream phase. J. Protozool. 16:54-69.

73. Vickerman, K. 1971. Morphological and physiological considerations of extracellular blood protozoa. In Fallis, A.M., editor. Ecology and physiology of parasites. University of Toronto Press, Toronto, pp. 58-91.

74. Vickerman, K. 1972. The host-parasite interface of parasitic Protozoa. Some problems posed by ultrastructural studies. In Taylor, A.E.R., and R. Muller, editors. Functional aspects of parasite surfaces. Blackwell Scientific Publications Ltd., Oxford, England, pp. 71- 91.

75. Vickerman, K. 1974. Antigenic variation in African trypanosomes. In Parasites in the immunized host: mechanisms of survival (Ciba Foundation Symposium 25, new series). Elsevier, Amsterdam, pp. 53-80.

76. Vickerman, K. 1978. Antigenic variation in trypanosomes. Nature 273:613-617.

77. Vickerman, K., and J.D. Barry. 1982. African trypanosomiasis. In Cohen, S., and K.S Warren, editors. Immunology of parasitic infections. Blackwell Scientific Publications Ltd. Oxford, Eng., pp. 204-260.

78. Vickerman, K., and F.E.G. Cox. 1967. The Protozoa. Houghton Mifflin Co., Boston.

79. Wallace, F.G. 1966. The trypanosomatid parasites of insects and arachnids.Exp. Parasitol. 18:124-193.

80. Walton, B.C., L.V. Chinel, and O. Eguia y E. 1973. Onset of espundia after many years of occult infection with *Leishmania braziliensis*. Am. J. Trop. Med. Hyg. 22:696-698.

81. Williams, P., and M. de Vasconcellos Coelho. 1978. Taxonomy and transmission of Leishmania. In Lumsden, W.H.R., R. Muller and J.R. Baker, editors. Advances in parasitology, vol. 16, Academic Press, Inc., New York, pp. 1-42.

82. Williamson, J. 1962. Chemotherapy and chemoprophylaxis in African trypanosomiasis. Exp. Parasitol. 12:274-367.

83. Woo, P.T.K., and M.A. Soltys. 1970. Animals as reservoir hosts of human trypanosomes. J. Wildl. Dis. 6:313-322.

84. Woody, N.C., and H.B. Woody. 1955. American trypanosomiasis (Chagas' disease). First indigenous case in the United States. J.A.M.A. 159:476-477.

85. World Health Organization. 1960. Chagas' disease. Report of a study group. Technical Report Series no. 202, Geneva.

86. Yaeger, R.G. 1971. Transmission of *Trypanosoma cruzi* infection to opossums via the oral route. J. Parasitol. 57:1375-1376.

87. Young, J.R., J.S. Shah, G. Matthyssens, and R.O. Williams. 1983. Relationship between multiple copies of a *T. brucei* variable surface glycoprotein gene whose expression is not controlled by duplication. Cell 32:1149-1159.

SUGGESTED READINGS

Adler, S. 1964. Leishmania. In Dawes, B., editor. Advances in parasitology, vol. 2. Academic Press, Inc., New York, pp. 1-34. (An advanced treatise on the subject. Recommended reading for all who are interested in the Trypanosomatidae.)

Barker, D.C. 1987. DNA diagnosis of human leishmaniasis. Parasitol. Today 3:177-184.

Blackwell, J., et al. 1986. Molecular biology of *Leishmania*. Parasitol. Today 2:45-53.

Chang, K.-P., and R.S. Bray, editors. 1985. Human parasitic diseases. vol 1. Leishmaniasis. Elsevier Publications, Amsterdam.

Evans, D.A. 1985. *Leishmania* reference strains. Parasitol. Today 1:172-173.

Foster, W.D. 1965. A history of parasitology. E. & S. Livingstone, Edinburgh. (Chapter 10, "The Trypanosomes," is a very interesting account of the history of knowledge about this group.)

Grimaldi, G., Jr., J.R. David, and D. McMahon-Pratt. 1987. Identification and distribution of New World *Leishmania* species characterized by serodeme analysis using monoclonal antibodies. Am. J. Trop. Med. Hyg. 36:270-287.

Hoogstraal, H., and D. Heyneman. 1969. Leishmaniasis in the Sudan Republic. 30. Final epidemiological report. Am. Trop. Med. Hyg. 18:1089-1210. (An extensive account of the aspects of leishmaniasis by two men who have an unashamed love for humanity. It should be required reading for all students of parasitology, and it will stand by itself as an example of what scientific writing should be.)

Jordan, A.M. 1985. Tsetse eradication plans for southern Africa. Parasitol. Today 1:121-123.

Marsden, P.D. 1985. Clinical presentations of *Leishmania braziliensis braziliensis*. Parasitol. Today 1:129-133. (An outstanding review of the subject with excellent illustrations).

Marsden, P.D. 1986. Mucosal leishmaniasis ("espundia" Escomel 1911). Trans. R. Soc. Trop. Med. Hyg. 80:859-876.

McCabe, R.E., J.S. Remington, and F.G. Araujo. 1984. Mechanisms of invasion and replication of the intracellular stage in *Trypanosoma cruzi*. Infect. Immun. 46:372-376.

Nantulya, V.M. 1986. Immunological approaches to the control of animal trypanosomiasis. Parasitol. Today 2:168-175.

Steinart, M., and E. Pays. 1986. Selective expression of surface antigen genes in African trypanosomes. Parasitol. Today 2:15-19.

Vickerman, K. 1985. Leishmaniasis—the first centenary. Parasitol. Today 1:149, 172.

Walters, L.L., G.B. Modi, R.B. Tesh, and T. Burrage. 1987. Host-parasite relationship of *Leishmania mexicana mexicana* and *Lutzomyia abonnenci* (Diptera: Psychodidae). Am. J. Trop. Med. Hyg. 36:294-314

CHAPTER 6

OTHER FLAGELLATE PROTOZOA

Several groups of flagellate protozoa have members that are parasitic in or on invertebrate and vertebrate animals. Certain dinoflagellates, for example, parasitize copepods, diatoms, and pelagic invertebrates. Although such forms are interesting, limited space allows consideration of only a few examples. Representative species are drawn from four orders.

ORDER RETORTAMONADIDA
Family Retortamonadidae

Two species in the family Retortamonadidae are commonly found in humans. Although they are apparently harmless commensals, they are worthy of note because they can easily be mistaken for highly pathogenic species.

■ *Chilomastix mesnili* (Fig. 6-1)

Chilomastix mesnili infects about 3.5% of the population of the United States and 6% of the world population.[1] It lives in the cecum and colon of humans, chimpanzees, orangutans, monkeys, and pigs. Other species are known in other mammals, birds, reptiles, amphibians, fish, leeches, and insects.

The living trophozoite is pyriform, with the posterior end drawn out into a blunt point, and is 6 to 24 μm by 3 to 10 μm. A longitudinal **spiral groove** occurs in the surface of the middle of the body, but this is usually visible only on living specimens. A sunken **cytostomal groove** is prominent near the anterior end. Along each side of the cytostome runs a cytoplasmic **cytostomal fibril,** presumably strengthening the lips of the cytostome. The cytostome leads into the cytopharynx, where endocytosis takes place. Four flagella, one longer than the others, emerge from kinetosomes on the anterior end of the body, and the kinetosomes are interconnected by microfibrillar material.[4] One of the flagella is very short and delicate, curving back into the cytostome, where it undulates. The large nucleus is near the front end.

A cyst stage occurs, especially in formed stools (Fig. 6-2). A typical cyst is thick walled, 6.5 to 10 μm long, and pear or lemon shaped. It has a single nucleus and retains all the cytoplasmic organelles, including cytostomal fibrils, kinetosomes, and axonemes.

Transmission is by ingestion of cysts, since trophozoites cannot survive stomach acid. Fecal contamination of drinking water is the most important means of transmission.

Chilomastix mesnili usually is considered to be non-pathogenic, although Mueller[29] suggested that it might cause a watery stool in some instances.

■ *Retortamonas intestinalis* (Fig. 6-3)

Retortamonas intestinalis is a tiny protozoan that is basically similar to *C. mesnili,* but the trophozoite is only 4 μm to 9 μm long. Furthermore, it has only two flagella, one of which extends anteriad, whereas the other emerges from the cytostomal groove and trails posteriad. The living trophozoite usually extends into a blunt point at its posterior end, but it bends to round up in fixed specimens. The ovoid to pear-shaped cysts contain a single nucleus.

Like *C. mesnili,* this species is probably a harmless commensal. It lives in the cecum and large intestine of monkeys and chimpanzees, as well as humans, and apparently is not a common symbiont anywhere in the world.

ORDER DIPLOMONADIDA
Family Hexamitidae

Members of the Hexamitidae are easily recognized because they have two equal nuclei lying side by side. There are several species in five genera, most of which are parasitic in vertebrates or invertebrates. One species is a parasite of humans. It will serve to illustrate the genus *Giardia,* whereas *Hexamita meleagridis* is an example of a related species in domestic animals.

■ *Giardia intestinalis*

Giardia intestinalis was first discovered in 1681 by Leeuwenhoek, who found it in his own stools. The taxonomy of the species was confused in the nine-

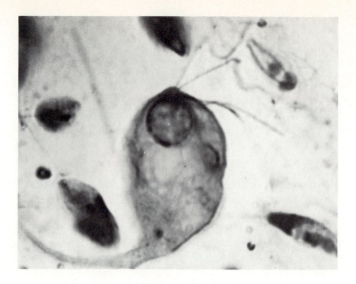

FIG. 6-1

Trophozoite of *Chilomastix caulleryi*, which is similar morphologically to *C. mesnili*. Note the four flagella and the cytostomal fibrils. It is 6 to 24 μm long.

Photograph by L.S. Roberts

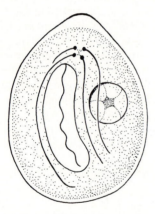

FIG 6-2

Cyst of *Chilomastix mesnili* from a human stool, showing the characteristic lemon or pear shape. Also visible are the large, irregular karyosome and the cytostomal fibrils.

Drawing by William C. Ober.

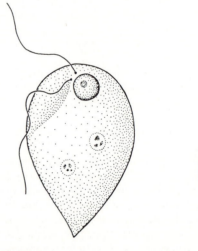

FIG 6-3

Retortamonas intestinalis trophozoite and cyst.

Drawing by William C. Ober.

teenth century, and even today it is often called *Giardia lamblia*. The species is cosmopolitan in distribution but occurs most commonly in warm climates, and children are especially susceptible. *G. intestinalis* is the most common flagellate of the human digestive tract.

Morphology. The trophozoite (Figs. 6-4 and 6-5), is 12 to 15 μm long, rounded at the anterior end, and

pointed at its posterior end. The organism is dorsoventrally flattened and is convex on the dorsal surface. The flattened ventral surface bears a concave, bilobed **adhesive disc** (Figs. 6-6 and 6-7). The so-called adhesive disc actually is a rigid structure, reinforced by microtubules and fibrous ribbons, surrounded by a flexible, apparently contractile, striated rim of cytoplasm. Application of this flexible rim to the host's in-

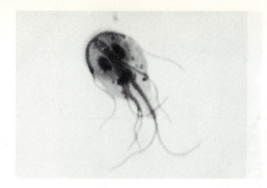

FIG 6-4

Giardia intestinalis trophozoite in a human stool. It is 12 to 15 μm long.

Photograph by Sherwin Desser.

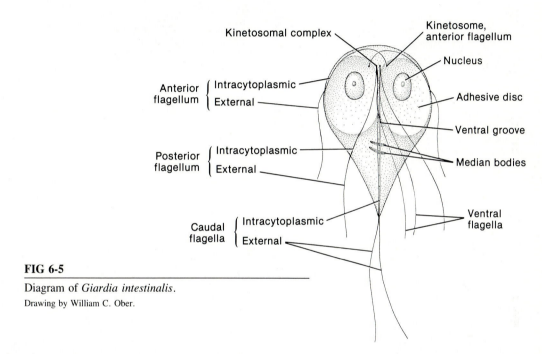

FIG 6-5

Diagram of *Giardia intestinalis*.

Drawing by William C. Ober.

testinal cell, working in conjunction with the **ventral flagella,** found in the **ventral groove,** is responsible for the organism's remarkable ability to adhere to the host cell (Fig. 6-8). The pair of ventral flagella, as well as three more pairs of flagella—arises from kinetosomes located between the anterior portions of the two nuclei (Fig. 6-5). The axonemes of all flagella course through the cytoplasm for some distance before emerging from the cell body; those of the anterior flagella actually cross and emerge laterally from the adhesive disc area on the side opposite from their respective kinetosomes. A pair of large, curved, transverse, dark-staining **median bodies** lies behind the adhesive disc. These bodies are unique to the genus *Giardia.* Various authors have regarded them as parabasal bodies, kinetoplasts, or chromatoid bodies, but ultrastructural studies have shown they they are none of these.[8,13] Their function is obscure, although it has

been suggested that they may help support the posterior end of the organism, or they may be involved in its energy metabolism. There is no true axostyle; the structure so described by previous authors is formed by the intracytoplasmic axonemes of the ventral flagella and associated groups of microtubules. Interestingly, no mitochondria, smooth endoplasmic reticulum, Golgi bodies, or lysosomes have been found.[13]

The overall effect of the two nuclei behind the lobes of the adhesive disc and the median bodies is that of a wry little face that seems to be peering back at the observer.

Life cycle. *Giardia intestinalis* lives in the duodenum, jejunum, and upper ileum of humans, with the adhesive disc fitting over the surface of an epithelial cell. In severe infections the free surface of nearly every cell is covered by a parasite. The protozoa can swim rapidly using their flagella.

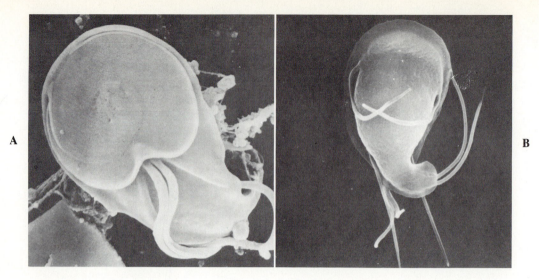

FIG 6-6

Scanning electron micrograph of *Giardia*. **A,** The ventral view shows the flat adhesive disc and the relationship of the ventral and posterior flagella and ventral groove, but the caudal flagella curve around to the other side in this photograph. **B,** The dorsal view shows these flagella, as well as the anterior flagella. The organism is 12 to 15 μm long.

Photographs by Dennis Feely.

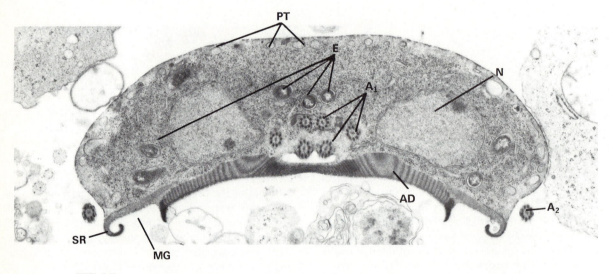

FIG 6-7

Transmission electron micrograph of a transverse section of a *Giardia muris* trophozoite found in the small bowel of an infected mouse. The marginal groove is the space between the striated rim of cytoplasm and the lateral ridge of the adhesive disc. The beginning of the ventral groove can be seen dorsal to the central area of the adhesive disc. This specimen bears endosymbionts, which are apparently bacteria. *PT,* Peripheral tubules; *E,* endosymbionts; *N,* nucleus; *A₁,* axonemes of posterior, ventral, and caudal flagella; *A₂,* axoneme of anterior flagellum; *AD,* adhesive disc; *MG,* marginal groove; *SR,* striated rim of cytoplasm. (×15,350.)

Reprinted from J. Infect. Dis. 140:222-228 by Nemanic, P.C., R.L. Owen, D.P. Stevens, and J.C. Mueller by permission of The University of Chicago Press. © 1979 by The University of Chicago.

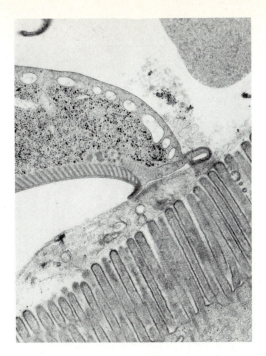

FIG 6-8

Periphery of *Giardia muris* in contact with the mucous stream covering the microvilli of a duodenal epithelial cell. It appears that the peripheral flange of striated cytoplasm is the grasping organelle of the ventral surface. (×33,000.)

From Friend, D.S. Reproduced from The Journal of Cell Biology, 1966, vol. 29, pp. 317-332, by copyright permission of The Rockefeller University Press.

Trophozoites divide by binary fission. First the nuclei divide, then the locomotor apparatus and the sucking disc, and finally the cytoplasm. Enormous numbers can build up rapidly in this way. It has been calculated that a single diarrheic stool can contain 14 billion parasites, whereas a stool in a moderate infection may contain 300 million cysts.[7]

In the small intestine and in watery stools only the trophic stage can be found. However, as the feces enter the colon and begin to dehydrate, the parasites become encysted. First the flagella shorten and no longer project. The cytoplasm condenses and secretes a thick, hyaline cyst wall. The ovoid cysts (Fig. 6-9) are 8 to 12 μm by 7 to 10 μm in size. Newly formed cysts have two nuclei, but older ones have four. Soon the sucking disc and the locomotor apparatus are doubled, and the Siamese-twinned flagellates are ready to emerge. When swallowed by the host, they pass safely through the stomach and excyst in the duodenum, immediately completing the division of the cytoplasm. The flagella grow out, and the parasites are once again at home.

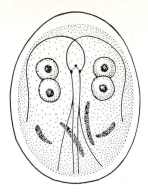

FIG. 6-9

Cyst of *Giardia intestinalis* in the human stool. It is 8 to 12 μm long. The karyosomes of all four cyst nuclei, as well as several intracytoplasmic axonemes, are visible.

Drawing by William C. Ober.

Metabolism. *G. intestinalis* is an aerotolerant anaerobe.[21,37] As mentioned earlier, these protozoa have no mitochondria. The tricarboxylic acid cycle and cytochrome system are absent, but the organisms avidly consume oxygen when it is present. Glucose is apparently the primary substrate for respiration, and they store glycogen. The principal end products are ethanol, acetate, and CO_2, both aerobically and anaerobically. In the absence of oxygen, more reducing equivalents are transferred to acetaldehyde to produce ethanol. When oxygen is present, they produce more acetate and less ethanol. All their energy is produced by substrate-level phosphorylation via a flavin, iron-sulfur, protein-mediated fermentative pathway.[21] This pathway is blocked by the flavoantagonists quinacrine and chloroquine.[37]

Pathogenesis. Many cases of infection show no evidence of disease. Apparently some persons are more sensitive to the presence of *G. intestinalis* than are others, and considerable evidence suggests that some protective immunity can be acquired. In other cases there is a marked increase of mucus production, diarrhea, dehydration, intestinal pain, flatulence, and weight loss. The stool is fatty but never contains blood. The protozoan does not lyse host cells but appears to feed on mucous secretions. A dense coating of flagellates on the intestinal epithelium interferes with the absorption of fats and other nutrients, which probably triggers the onset of disease. The gallbladder may become infected, which can cause jaundice and colic. The disease is not fatal but can be intensely discomforting.

Epidemiology. Giardiasis is highly contagious. If one member of a family catches it, others will usually become infected. Transmission depends on the swallowing of mature cysts; prevention, therefore, depends on a high level of sanitation.

A summary of surveys of 134,966 persons throughout the world showed that the prevalence of the infection ranged from 2.4% to 67.5%.[1] In 1984, 26,560 cases of giardiasis were reported in the United States.[28]

Outbreaks continue to flare up in the United States, often without regard for the affluence of the people involved. For instance, an epidemic occurred in Aspen, Colorado, during the 1965-1966 ski season, with at least 11% of 1094 skiers infected.[26] Of the permanent population, 5% remained infected after the epidemic.[14]

Faunal surveys in watersheds that were known sources of infections to people have shown that several animals, including beavers, dogs, cats,[19] and sheep, serve as reservoirs of infection. Beavers, in particular, are epidemiologically significant in human giardiasis. When one has hiked for miles in the wild on a hot day, it can be very tempting to fill a canteen and drink from a crystal-clear beaver pond. Many cases have been acquired in just that way, including that of a son of one of the authors. In 1980 numerous cases of giardiasis were diagnosed in the resort village of Estes Park, Colorado. Surprisingly, all were in one half of the town, with the other half remaining parasite free. Each half is served with water from a different river. Both rivers have beavers in abundance, but the municipal water filtration system had broken down for one source but not the other. In nearby Rocky Mountain National Park, Monzingo and Hibler found a prevalence of 20% to 60% in beavers from one valley.[25] Muskrats were also infected.

It now appears that many nominal "species" described from mammals may be identical to *G. intestinalis*. In at least some cases, *Giardia* from beavers can be established in gerbils,[36] but *G. muris* from microtine rodents may be distinct. *G. agilis* from amphibians may be another species.

Diagnosis and treatment. Recognition of trophozoites or cysts in stained fecal smears is adequate for diagnosis. However, an otherwise benign infection with the flagellates may coexist with a peptic ulcer, enteritis, tumor, or strongyloidiasis, any of which could actually be causing the symptoms. Treatment with quinacrine or metronidazole (Flagyl) usually effects complete cure within a few days; both are recommended drugs.[27] All members of a family should be treated simultaneously to avoid reinfection of the others. In a small percentage of cases cysts are not passed or are passed sporadically. Duodenal aspiration may be necessary for diagnosis by demonstrating trophozoites.

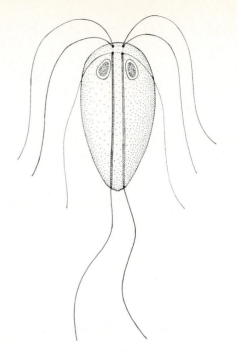

FIG. 6-10

Diagram of a trophozoite of *Hexamita meleagridis*. It is 6 to 12 μm long.

Drawing by William C. Ober.

■ *Hexamita meleagridis*

Hexamita meleagridis is a parasite of the small intestine of young galliform birds, including turkey, quail, pheasant, partridge, and peafowl. It is known from the United States, Great Britain, and South America, although it probably is common elsewhere. In the United States, at least, it causes millions of dollars in loss to the turkey industry every year.

Morphologically, *Hexamita* is quite similar to *Giardia*, being elongate, with two nuclei and four pairs of flagella (Fig. 6-10). However, it is smaller, has no sucking disc, possesses karyosomes two thirds the size of the nuclei, and has no median bodies. As in *Giardia* spp., the kinetosomes are grouped anterior to and between the nuclei, but three pairs of axonemes emerge anteriorly, whereas one pair courses intracytoplasmically. The intracytoplasmic axonemes run posteriorly along granular lines and emerge to become the posterior flagella.

The life cycle is essentially the same as for *Giardia* spp., except that birds are the normal hosts rather than mammals.

Like *G. intestinalis* infection, hexamitosis is mainly

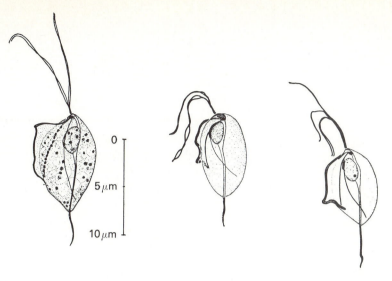

0

5 μm

10 μm

FIG. 6-11

Typical trophozoites of *Trichomonas tenax*.

From Honigberg, B.M., and J.J. Lee. 1959. Am. J. Hyg. 69:183.

a disease of young animals. Symptomless adults are reservoirs of infection.

Mortality in a flock may range from 7% to 80% in very young birds. Survivors are somewhat immune but commonly are stunted in size. They become a ready source of infection for new broods. No completely satisfactory treatment is available, but prevention in domestic flocks is possible by proper management and sanitation. Separation of chicks from adult birds is mandatory.

ORDER TRICHOMONADIDA
Family Trichomonadidae

The many members of this family are rather similar in structure. They are easily recognized because they have an anterior tuft of flagella, a stout median rod (the **axostyle**), and an **undulating membrane** along the recurrent flagellum. They are found in intestinal or reproductive tracts of vertebrates and invertebrates, with one group occurring exclusively in the gut of termites. Unlike other protozoa covered in this chapter, most members of this order do not form cysts. Three species are common in humans, and one is of extreme importance in domestic ruminants. These will serve to illustrate the order.

The three trichomonads of humans, *Trichomonas tenax*, *T. vaginalis*, and *Pentatrichomonas hominis*, are similar enough morphologically to have been con-

sidered conspecific by many taxonomists. More recently there has been a wide recognition of the differences between *P. hominis* and the other two. As currently defined, the genus *Trichomonas* contains only three species, *T. tenax*, *T. vaginalis*, and a species found in birds, *T. gallinae*, which is more like *T. tenax* than is *T. vaginalis*.[15]

■ *Trichomonas tenax* (Fig. 6-11)

Trichomonas tenax was first discovered by O.F. Müller in 1773, when he examined an aqueous culture of tartar from teeth. *T. tenax* is now known to have worldwide distribution.

Morphology (Fig. 6-11). Like all species of *Trichomonas*, *T. tenax* has only a trophic stage. It is an oblong cell 5 to 16 μm long by 2 to 15 μm wide, with size varying according to strain. There are four anterior, free flagella, with a fifth flagellum curving back along the margin of an undulating membrane and ending posterior to the middle of the body[24] (see Fig. 4-3). The recurrent flagellum is not enclosed by the undulating membrane but is closely associated with it in a shallow groove. A densely staining lamellar structure (**accessory filament**) courses within the undulating membrane along its length. A **costa** arises in the kinetosome complex and runs superficially beneath and generally parallel to the serpentine path of the undulating membrane. The costa distinguishes the Trichomonadidae from other families in its order. It is a rod-

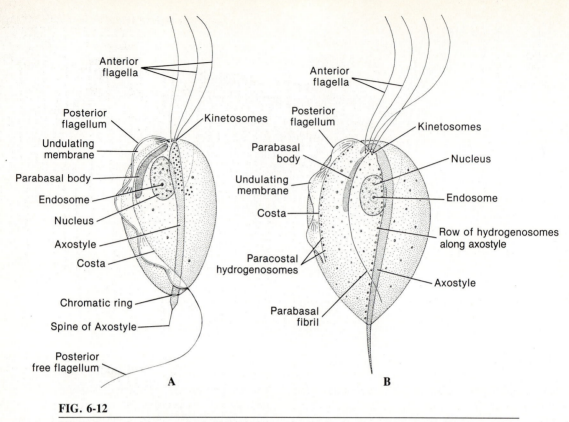

FIG. 6-12

Morphology of trichomonads **A,** *Tritrichomonas foetus.* **B,** *Trichomonas vaginalis.* The hydrogenosomes are not always in a definite row.

Modified from Wenrich, D.H., and M.A. Emmerson, 1933. J. Morphol. 55:195.

Drawing by William C. Ober.

like structure with complex cross-striations, which probably serve as a strong, flexible support in the region of the undulating membrane.

A parabasal body (Golgi body) lies near the nucleus with the parabasal filament running from the kinetosome complex, through or very near the parabasal body, and ending in the posterior portion of the body. A small, "minor" parabasal filament, which is inconspicuous in light microscope preparations, has been shown in other trichomonads, and it probably is present in *T. tenax* as well. The tubelike axostyle extends from the area of the kinetosomes posteriorly to protrude from the end of the body (covered by cell membrane). The axostylar tube is formed by a sheet of microtubules, and its anterior, middle, and posterior parts are known as **capitulum, trunk,** and **caudal tip,** respectively. Toward the capitulum, the tubular trunk opens out to curve around the nucleus, and the

microtubules of the capitulum slightly overlap the curving, collar-like **pelta.** The pelta also comprises a sheet of microtubules and appears to function in supporting the "periflagellar canal," a shallow depression in the anterior end from which all the flagella emerge (see Fig. 4-2). A cytostome is not present. *Trichomonas tenax* has concentrations of microbodies traditionally called **paracostal granules** along its costa, and other species of *Trichomonas* have **paraxostylar granules** along the axostyle. These bodies are now referred to as **hydrogenosomes** on the basis of their biochemical characteristics (see Fig. 4-2). The metabolic functions of hydrogenosomes have been investigated in *T. vaginalis* and *Tritrichomonas foetus,* but most characteristics are probably shared by all the trichomonads. Metabolism will be described below.

Biology. *Trichomonas tenax* can live only in the mouth and, apparently, cannot survive passage through the digestive tract. Transmission, then, is direct, usually by kissing or common use of eating or drinking utensils. Trophozoites divide by binary fission. They are harmless commensals, feeding on microorganisms and cellular debris. They are most abundant between the teeth and gums and in pus pockets, tooth cavities, and crypts of the tonsils, but they also have been found in the lungs and trachea. *T. tenax* is resistant to changes in temperature and will live for several hours in drinking water. Thus the "communal dipper" may be a route of infection in some situations.

Although good oral hygiene is said to decrease or eliminate the infection, in one survey 15.7% of patients in a clinical practice in New York were positive, and none had oral hygiene rated as poor.[3]

■ *Trichomonas vaginalis* (Figs. 6-12, 6-13)

This species was first found by Donné in 1836 in purulent vaginal secretions and in secretions from the male urogenital tract. In 1837 he named it *Trichomonas vaginalis,* thereby creating the genus. It is a cosmopolitan species, found in the reproductive tracts of both men and women the world over. Donné thought the organism was covered with hairs, which is what prompted the generic name (Greek *thrix,* hair).

Morphology. *Trichomonas vaginalis* is very similar to *T. tenax* but differs in the following ways: it is somewhat larger, 7 to 32 μm long by 5 to 12 μm wide; its undulating membrane is relatively shorter; and there are more granules along the axostyle and costa. In living and appropriately fixed and stained specimens, the constancy in presence and arrangement of the hydrogenosomes is the best criterion for distinguishing *T. vaginalis* from other *Trichomonas* spp.[16] *Trichomonas vaginalis* frequently produces pseudopodia.

Biology. *Trichomonas vaginalis* lives in the vagina and urethra of women and in the prostate, seminal vesicles, and urethra of men. It is transmitted primarily by sexual intercourse,[18] although it has been found in newborn infants. Its presence occasionally in very young children, including virginal females, suggests that the infection can be contracted from soiled washcloths, towels, and clothing. Viable cultures of the organism have been obtained from damp cloth as long as 24 hours after inoculation. The acidity of the normal vagina (pH 4 to 4.5) ordinarily discourages infection, but once established, the organism itself causes a shift toward alkalinity (pH 5 to 6), which further encourages its growth.

Metabolism. Like *Giardia,* trichomonads are aerotolerant anaerobes, degrading carbohydrates incompletely to short-chain organic acids (principally acetate

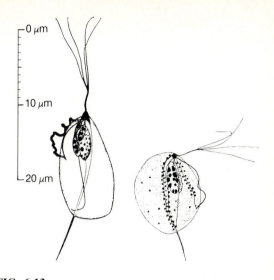

FIG. 6-13

Typical trophozoites of *Trichomonas vaginalis.*

From Honigberg, B.M., and V.M. King. 1964. J. Parasitol. 50:345-364.

and lactate) and carbon dioxide, regardless of whether oxygen is present.[22] Unlike *Giardia,* however, trichomonads produce molecular hydrogen in the absence of oxygen. These reactions take place in the hydrogenosomes, hence the name of the organelle. Hydrogenosomes are analogous to mitochondria (which are absent in trichomonads) in other eukaryotes; but their distinctness is shown by their morphology, the absence of DNA, and the absence of cardiolipin, which is present in the membranes of mitochondria.[31,35] Hydrogenosomes are surrounded by two, closely apposed 6 nm membranes.[2] Similar organelles have now been reported in certain rumen ciliates.[30]

Pyruvate is produced in the cytoplasm by glycolysis (see Fig. 4-11). Part of the pyruvate is reduced to lactate by lactic dehydrogenase and excreted. Part of the pyruvate enters the hydrogenosomes where it is oxidatively decarboxylated, and the electrons are accepted by **ferredoxin** (Fig. 6-14).[23] Under anaerobic conditions, the electrons are then transferred to protons by a hydrogenase to form molecular hydrogen. When oxygen is present, it apparently accepts the electrons and, along with H^+, forms water. The oxidation of pyruvate to acetate is coupled to substrate level generation of ATP (Fig. 6-14); therefore, the hydrogenosomes participate in energy production in the cell. The drug metronidazole is reduced by ferredoxin to form toxic products, thus explaining the effectiveness of this drug in chemotherapy for trichomoniasis.

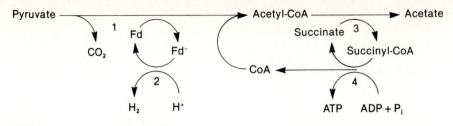

FIG. 6-14

Pathway of pyruvate degradation in *Tritrichomonas foetus* hydrogenosomes under anaerobic conditions. Step 1 is catalyzed by pyruvate: ferredoxin oxidoreductase; step 2 by hydrogenase; step 3, acetate:succinate-CoA transferase; and step 4, succinate thiokinase. *Fd,* ferredoxin.

Modified from Marczak, R., T.E. Gorrell, and M. Müller. 1983. J. Biol. Chem. 258:12427-12433.

Pathogenesis. Most strains are of such low pathogenicity that the infected person is virtually asymptomatic. However, other strains cause an intense inflammation, with itching and a copious white discharge (**leukorrhea**) that is swarming with trichomonads. They feed on bacteria, leukocytes, and cell exudates and are themselves ingested by monocytes. Like all mastigophorans, *T. vaginalis* divides by longitudinal fission, and like most trichomonads, it does not form cysts.

A few days after infection there is a degeneration of the vaginal epithelium followed by leukocytic infiltration. The vaginal secretions become abundant and white or greenish, and the tissues become intensely inflamed. An acute infection will usually become chronic, with a lessening of symptoms, but will occasionally flare up again. It should be noted, however, that leukorrhea is not symptomatic of trichomoniasis; indeed, at least half of patients even with severe leukorrhea are negative for *T. vaginalis.*[12] In men the infection is usually asymptomatic, although there may be an irritating urethritis or prostatitis.

Diagnosis depends on recognizing the trichomonad in a secretion or from an in vitro culture made from a vaginal irrigation. Cultivation is recommended to detect low numbers of organisms.[32] Oral drugs, such as metronidazole, usually cure infection in about 5 days. Some apparently recalcitrant cases may be caused by reinfection by the sexual partner. Suppositories and douches are useful in promoting an acid pH of the vagina. Sexual partners should be treated simultaneously to avoid reinfection.

■ *Pentatrichomonas hominis* (Fig. 6-15)

The third trichomonad of humans is a harmless commensal of the intestinal tract. It was first found by Davaine, who named it *Cercomonas hominis* in 1860. Traditionally it has been called *Trichomonas hominis,* but since most specimens actually bear five anterior flagella, the organism has been assigned to the genus *Pentatrichomonas.* Next to *Giardia intestinalis* and *Chilomastix mesnili,* this is the most common intestinal flagellate of humans. It is also known in other primates and in various domestic animals. The prevalence among 13,517 persons examined in the United States was 0.6%.[1]

Morphology. This species is superficially similar to *T. tenax* and *T. vaginalis* but differs in several respects. Its size is 8 μm to 20 μm by 3 μm to 14 μm. Five anterior flagella are present in most specimens, although individuals with fewer flagella are sometimes found. The arrangement is referred to as "four-plus-one," since the fifth flagellum originates and beats independently of the others.[15] A recurrent (sixth) flagellum is aligned alongside the undulating membrane, as in *T. tenax* and *T. vaginalis,* but in contrast to these two species, the recurrent flagellum in *P. hominis* continues as a long, free flagellum past the posterior end of the body. Axostyle, pelta, parabasal body, "major" and "minor" parabasal filaments, costa, and paracostal hydrogenosomes are present. Paraxostylar hydrogenosomes are absent.

Biology. *Pentatrichomonas hominis* lives in the large intestine and cecum, where it divides by binary fission, often building up incredible numbers. It feeds on bacteria and debris, probably taking them in with active pseudopodia. The organism often is present in routine examinations of diarrheic stools, but no indication exists that it contributes to this or other disease conditions. In formed stools the flagellates are rounded and dormant but not encysted. They are difficult to identify at this stage because they do not move,

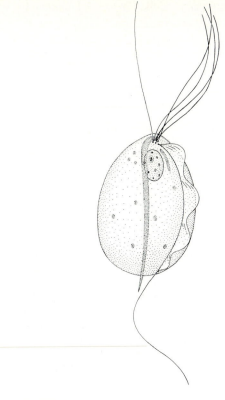

FIG. 6-15

Pentatrichomonas hominis trophozoite. It ranges from 8 to 20 μm long.

Drawing by William C. Ober.

and the structures normally characteristic for the species cannot be distinguished.

The organism apparently can survive acidic conditions of the stomach, and transmission occurs by contamination. Filth flies can serve as mechanical vectors. Higher prevalence is correlated with unsanitary conditions.

Diagnosis depends on identification of the animal in fecal preparations, and prevention depends on personal and community sanitation. The organism cannot establish in the mouth or urogenital tract.

■ *Tritrichomonas foetus* (Fig. 6-12)

Tritrichomonas foetus is responsible for a serious genital infection in cattle, zebu, and possibly other large mammals. Probably the third leading cause of abortion in cattle (after brucellosis and leptospirosis), *T. foetus* is especially common in Europe and the United States. The USDA estimated losses as the re-

sult of *T. foetus* in the United States between 1951 and 1960 at $8.04 million.

Morphology. The cell is spindle to pear shaped, 10 to 25 μm long by 3 to 15 μm wide. There are three anterior flagella, and the fourth, the recurrent flagellum, extends free from the posterior end of the body about the length of the anterior flagella. The mastigont system is generally similar in organization to the trichomonads described previously, but it is even more complex and will not be detailed here.[17] The costa is prominent and, although similar in position and function to those of the other trichomonads, differs in ultrastructural detail, resembling a parabasal filament in this respect. The structure of the undulating membrane is curious, consisting of two parts. The proximal part is a foldlike differentiation of the dorsal body surface, and the distal part, which contains the axoneme of the recurrent flagellum, courses along the rim of the proximal part with no obvious physical connection to it. The thick axostyle protrudes from the posterior end of the body. Numerous paraxostylar hydrogenosomes are present in the posterior part of the organism, just anterior to the point of the axostyle, and these are apparent in the light microscope preparations as the "chromatic ring."

Biology. These trichomonads live in the preputial cavity of the bull, although the testes, epididymis, and seminal vesicles also may be infected. In the cow the flagellates first infect the vagina, causing a vaginitis, and then move into the uterus. After establishing in the uterus, they may disappear from the vagina or remain there as a low-grade infection. Bovine genital trichomoniasis is a venereal disease transmitted by coitus, although transmission by artificial insemination is possible. Trichomonads multiply by longitudinal fission and form no cyst.

Pathogenesis. The most characteristic sign of bovine trichomoniasis is early abortion, which usually happens 1 to 16 weeks after insemination. Because of the small size of the fetus, the owner may not notice that the cow has aborted and therefore may believe she did not conceive. The cow may recover spontaneously if all fetal membranes are passed after abortion; however, if they remain, she usually develops chronic endometritis, which may cause permanent sterility. Normal gestation and delivery occasionally occur with an infected animal. Cows that recover from trichomoniasis are usually immune to further infection.

Pathogenesis is not observable in bulls, but an infected bull is worthless as a breeding animal; unless treated, it usually remains infected permanently. Treatment is expensive, difficult, and not always effective. Because of the immense prices paid for top-quality bulls, the loss of a single animal may bankrupt the breeder.

Epidemiology. It has not been determined definitely whether *T. foetus* can infect animals other than cattle and closely related species. Experimental infections have been established in rabbits, guinea pigs, hamsters, dogs, goats, sheep, and pigs. Trichomonads similar to *T. foetus* have been found as natural infection in pigs and horses. Whether these can be transmitted to cattle by contamination is not known, but experiments demonstrate that it is likely.

Trichomonads can survive freezing in semen ampules, although some media are more detrimental than are others. This precludes use, by artificial methods, of semen from infected bulls.

Diagnosis, treatment, and control. Direct identification of protozoa from smears or culture remains the only sure means of diagnosis, although a mucus agglutination test is available. In light infections a direct smear of mucus or exudate is sufficient. Smears can be obtained from amniotic or allantoic fluid, vaginal or uterine exudates, placenta, fetal tissues or fluids, or preputial washings from bulls. Flagellates fluctuate in numbers in bulls; in cows they are most numerous in the vagina 2 or 3 weeks after infection.

No satisfactory treatment is known for cows, but the infection is usually self-limiting in them, with subsequent, partial immunity. Bulls can be treated if the condition has not spread to the inner genital tubes and testes. Treatment is usually attempted only on exceptionally valuable animals, since it is a tedious, expensive task. Preputial infection is treated by massaging antitrichomonal salves or ointments into the penis, after it has been let down by nerve block or by injection of a tranquilizer into the penis retractor muscles. Repeated treatment is usually necessary. Systemic drugs show promise of becoming the standard method of treatment.

Control of bovine genital trichomoniasis depends on proper herd management. Cows that have been infected should be bred only by artificial insemination to avoid infecting new bulls. Bulls should be examined before purchase, with a wary eye for infection in the resident herd. Unless they are extremely valuable, infected bulls should be killed. Like any venereal disease, trichomoniasis can be controlled and eventually eliminated with proper treatment and reporting, but the disease is likely to remain a problem for some time.

Family Monocercomonadidae

The Monocercomonadidae show affinities with the Sarcodina, since pseudopodia are well developed, an undulating membrane is absent, and flagella tend to be reduced. Most species are parasites of insects, but three genera infect domestic animals. One of these is economically important and has evolved a unique mode of transmission: in the egg of a nematode.

■ *Histomonas meleagridis* (Fig. 6-16)

Histomonas meleagridis, a cosmopolitan parasite of gallinaceous fowl, including chickens, turkeys, peafowl, and pheasant, causes a severe disease known variously as blackhead, infectious enterohepatitis, and histomoniasis. The disease is more virulent in some species of host than others: chickens show disease less often than do turkeys, for example. The USDA estimated that the economic loss in the United States as the result of histomoniasis in chickens and turkeys during 1951 to 1960 amounted to $9.3 million.

The taxonomic history of *H. meleagridis* has been very confused because of its polymorphism in different situations. At various times it has been confused with amebas, coccidia, fungi, and *Trichomonas* spp. Even the disease that it causes has been attributed to different organisms, from amebas to viruses. Today much is known about the organism, and its biology and pathogenesis are less mysterious.

Morphology. *Histomonas meleagridis* is pleomorphic; its stages change size and shape in response to environmental factors. There is no cyst in the life cycle, only various trophic stages. When they are found in the lumen of the cecum (which is rare) or in culture, the stages are ameboid, 5 to 30 μm in diameter, and almost always with only one flagellum. However, there are usually four kinetosomes, the basic number for trichomonads, although this condition has been attributed to duplication of the kinetic apparatus in preparation for mitosis.[33] More likely they are vestiges of an ancestral type with four flagella. The nucleus is vesicular and often has a distinct endosome. One can usually discern a clear ectoplasm and a granular endoplasm. Food vacuoles may contain host blood cells, bacteria, or starch granules. Electron microscope studies have revealed a pelta, a V-shaped parabasal body, a parabasal filament, and a structure resembling an axostyle (Fig. 6-17). These cannot be seen with light microscopy, but their presence supports placement of *Histomonas* spp. in the order Trichomonadida. No mitochondria have been observed. The forms within the tissues have no flagella, although kinetosomes are present near the nucleus.

Biology and epidemiology. Like other flagellates, *H. meleagridis* divides by binary fission. No cysts or sexual stages are found in the life cycle.

Trophozoites are fragile and cannot long survive in

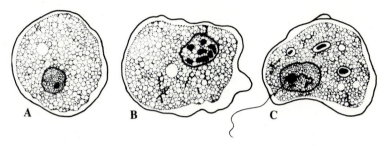

FIG. 6-16

Examples of *Histomonas meleagridis*. **A,** Tissue type of *H. meleagridis* in fresh preparation from liver lesion; viewed with phase contrast. **B,** *H. meleagridis* in transitional stage in lumen of the cecum. Pseudopodia have been formed, and the distribution of chromatin suggests that binary fission is approaching. However, the flagellum has not yet appeared. **C,** An organism in same cecal preparation as **B,** but this one completely adapted as a lumen dweller.

From Lund, E.E. 1969. In Brandly, C.A., and C.E. Cornelius, editors. Advances in veterinary science and comparative medicine. Academic Press, Inc., New York.

the external environment or the host's stomach acids. Certain factors can, and sometimes do, conspire to allow infection by trophozoites. If trophozoites are eaten with certain foods that raise the stomach pH, they may survive to initiate a new infection. This can be the means of an epizootic in a dense flock of birds.

The most important, and by far the most interesting, mode of transmission is within the egg of the cecal nematode, *Heterakis gallinarum*. Since the protozoan undergoes development and multiplication in the nematode, the worm can be considered a true intermediate host.[20] After being ingested by the worm, the flagellates enter the nematode's intestinal cells, multiply, and then break out into the pseudocoel and invade the germinative area of the nematode's ovary. There they feed and multiply extracellularly and move down the ovary with the developing oogonia, then penetrate the oocytes (Fig. 6-18). Feeding and multiplication continue in the oocytes and newly formed eggs. Passing out of the mother worm and out of the bird with its feces, the protozoan divides rapidly, invading the tissues of the juvenile nematode, especially those of the digestive and reproductive systems. Interestingly, *H. meleagridis* also parasitizes the reproductive system of the male nematodes.[20] Presumably it could be transmitted to the female during copulation, thus con-

stituting a venereal infection of nematodes!

Infected eggs can survive for at least 2 years in the soil. If the worm eggs are eaten by an appropriate bird, they hatch in the intestine, and the juvenile *Heterakis* passes down into the cecum, where *Histomonas* is free to leave its temporary host to begin residence in a more permanent one.

Earthworms are important paratenic hosts of both the *Heterakis* and its contained *Histomonas*. When eaten by an earthworm, the nematode eggs will hatch, releasing second-stage juveniles that become dormant in the earthworm's tissues. When the earthworm is eaten by a gallinaceous fowl, the *Heterakis* juveniles are released, and the bird becomes infected by two kinds of parasites at once. Earthworms can serve to maintain the parasites in the soil for long periods of time. Chickens are the most important reservoirs of infection because they are less often affected by *Histomonas* than are turkeys. Because *Heterakis* eggs and infected earthworms can survive for such long periods in the soil, it is almost impossible to raise uninfected turkeys in the same yards in which chickens have lived.

Pathogenesis. Turkeys are most susceptible between the ages of 3 and 12 weeks, although they can become infected as adults. In very young poults losses

FIG. 6-17

Histomonas meleagridis. **A,** Composite, schematic diagram of the mastigont system and nucleus as seen from a dorsal and somewhat right view. **B,** Composite diagram of an organism, with the mastigont system seen in the same view as in **A.** The flagellum arises from the kinetosomal complex just anterior to the V-shaped parabasal body. The cytoplasm appears highly vacuolated and contains ingested bacteria and rice starch. *Ax,* axostyle; *Ca,* capitulum; *F,* flagellum; *K,* kinetosomal complex; *N,* nucleus; *Pe,* pelta; *PB,* parabasal body; *PF,* parabasal fibril; *Tr,* trunk of axostyle. (×4270.)

From Honigberg, B.M., and C.J. Bennett. 1971. J. Protozool. 18:688.

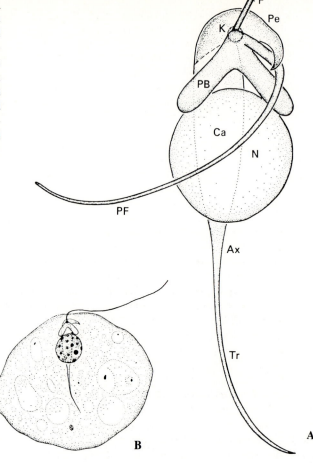

FIG. 6-18

Electron micrograph of section through the growth zone of the ovary of *Heterakis gallinarum* to show *Histomonas meleagridis* in the process of entering an oocyte *(arrow).* (×13,800.)

From Lee, D.L. 1971. In Fallis, A.M. editor. Ecology and physiology of parasites. University of Toronto Press, Toronto.

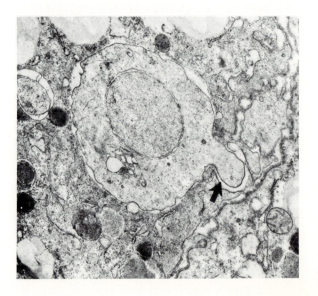

may approach 100% of the flock. Chickens are less prone to the disease, but outbreaks among young birds have been reported. Quails and partridges show varying degrees of susceptibility.

The principal lesions of histomoniasis are found in the cecum and liver. At first, pinpoint ulcers are formed in the cecum. These may enlarge until nearly the entire mucosa is involved. The ceca often become filled with cheesy, foul-smelling plugs that adhere to the cecal walls. Complete perforation of the cecum, with peritonitis and adhesions, can occur. The ceca are usually enlarged and inflamed. Liver lesions are rounded, with whitish or greenish areas of necrosis. Their size varies, and they penetrate deep into the parenchyma.

Infected birds show signs of droopiness, ruffled feathers, and hanging wings and tail. The skin of the head turns black in some cases, giving the disease the name **blackhead**. Other diseases can cause this symptom, however. Yellowish diarrhea usually occurs.

It has been shown that *H. meleagridis* by itself is incapable of causing blackhead but does so only in the presence of intestinal bacteria of several species, particularly *Escherichia coli* and *Clostridium perfringens*. Birds that survive are immune for life. A related histomonad, *Parahistomonas wenrichi*, also is transmitted by *Heterakis* but is not pathogenic.

Diagnosis, treatment, and control. Cecal and liver lesions are diagnostic. Scrapings of these organs will reveal histomonads, thereby distinguishing the disease from coccidiosis.

Several types of drugs are used in prevention and treatment, including nitrofurans, nitroimidazoles and phenylarsonic acid derivatives. These successfully inhibit, suppress, or cure the disease, but some have undesirable side effects, such as delaying sexual maturity of the bird. Treatment of birds with nematocides, such as mebendazole, cambendazole, and levamisole, to eliminate *Heterakis* is effective in preventing future outbreaks, since *H. meleagridis* cannot survive in the soil by itself.

Control depends on effective management techniques, such as rearing young birds on hardware cloth above the ground, keeping young birds on dry ground, and controlling *Heterakis*. Pasture rotation of *Heterakis*-free flocks is also successful.

■ *Dientamoeba fragilis*

Dientamoeba fragilis has traditionally been considered a member of the ameba family Endamoebidae, but it has long been recognized as being unlike other members of this family. For example, a large proportion of individuals have two nuclei, the nuclear structure is rather unlike other Endamoebidae, an extranuclear spindle is present during division, and cysts are

not formed. The last is a characteristic shared with a more typical member of the family, *Entamoeba gingivalis*. More than 35 years ago Dobell believed that *D. fragilis* was closely related to the ameboflagellate *Histomonas*.[9] On the basis of ultrastructural and immunological evidence, Honigberg placed the genus *Dientamoeba* in a subfamily of the Monocercomonadidae in the flagellate order Trichomonadida.[6] This seems to reflect the phylogenetic relationship of the organism rather than the fact that it moves by pseudopodia instead of flagella, and we will use it in this text. *Dientamoeba fragilis*, infecting about 4% of the human population, is the only species known in the genus.

Morphology. Only trophozoites are known in this species; cysts are not formed. The trophozoites (Fig. 6-19) are very delicate and disintegrate rapidly in feces or water. They are 6 to 12 μm in diameter, and the ectoplasm is somewhat differentiated from the endoplasm. A single, broad pseudopodium usually is present. The food vacuoles contain bacteria, yeasts, starch granules, and cellular debris. About 60% of the amebas contain two nuclei, which are connected to each other by a filament, observable by light microscopy; the rest have only one nucleus. By electron microscopy one can discern that the filament connecting the nuclei is a division spindle composed of microtubules; the binucleate individuals are, in reality, in an arrested telophase. The endosome is eccentric, sometimes fragmented or peripheral in the nucleus, and concentrations of chromatin are usually apparent. A filament and Golgi apparatus are present, which are reminiscent of the parabasal fibers and parabasal bodies found in *Histomonas* and trichomonads. There are no kinetosomes or centrioles.

Biology. *Dientamoeba fragilis* lives in the large intestine, especially in the cecal area. It feeds mainly on debris and traditionally has been considered a harmless commensal. However, a study of 43,029 people in Ontario showed a high percentage of intestinal problems in those infected with *D. fragilis*.[38] Symptoms included diarrhea, abdominal pain, anal pruritus, abnormal stools, and other indications of abdominal distress. It seems probable that *D. fragilis* is responsible for many such cases of unknown etiology, especially in small children.

The mode of transmission is unknown, since the parasite does not form cysts, and it cannot survive the upper digestive tract. It is possible that the organism survives transmission in the eggs of a parasitic nematode, as does its relative, *Histomonas meleagridis*. Small, ameboid organisms resembling *D. fragilis* have been found in the eggs of the common human pinworm, *Enterobius vermicularis*, and there is overwhelming epidemiological evidence that the nematode may be the vector of the protozoan.[5,38]

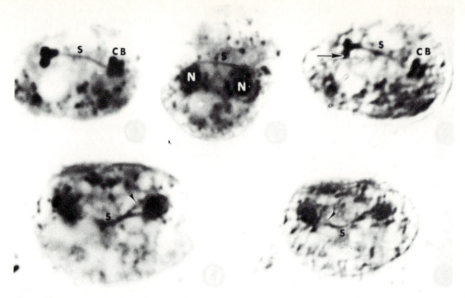

FIG. 6-19

Dientamoeba fragilis: photomicrographs of binucleate organisms. Four chromatin bodies *(CB)* can be resolved within the telophase nucleus of the organism, shown in the first and third figures. The extranuclear spindle *(S)* extends between the nuclei *(N)* in all figures. Note the branching of the spindle *(arrowheads)* near the nucleus in the fourth and fifth figures. (Bouin's fixative. First, second, and fourth figures: bright field [×4950]; third and fifth figures: Nomarski differential interference [×3650].)

From Camp, R.R., C.F.T. Mattern, and B.M. Honigberg. 1974. J. Protozool. 21:69-82.

Subphylum Opalinata

ORDER OPALINIDA
Family Opalinidae

There are about 150 species of opalinids, most of which live in the intestines of amphibians. They are of no economic or medical importance but are of zoological interest because of their peculiar morphology and the fact that their reproductive cycles apparently are controlled by host hormones.[11] Also, study of opalinids has contributed evidence to support the theory of continental drift, and further investigation is likely to provide much better understanding of amphibian zoogeography and evolution.[10] Finally, they are commonly encountered in routine dissections of frogs in teaching laboratories.

Numerous oblique rows of cilia occur over the entire body surface of opalinids, giving them a strong resemblance to ciliates (Fig. 6-20), and they traditionally have been classified with the Ciliophora. However, opalinids have several important differences from ciliates and are now placed as a separate subphylum in the phylum Sarcomastigophora. For example, opalinids possess two to many nuclei of similar structure and reproduce sexually by anisogamous syngamy. Asexually they undergo binary fission between kineties.

Adult opalinids reproduce asexually by binary fission in the rectum of frogs and toads during the summer, fall, and winter. In the spring, which is their host's breeding season, they accelerate divisions and produce small, precystic forms, which then form cysts and pass out with the feces of the host. When the cysts are eaten by tadpoles, male and female gametes excyst and fuse to form the zygote, which resumes asexual reproduction. The exact chemical identity of the compound(s) that stimulates encystment is not known, but present evidence indicates that it is one or more breakdown products of steroid hormones excreted in the frog's urine. This is an interesting example of a physiological adaptation to ensure the production of infective stages at the time and place of new host availability. The effectiveness of the adaptation is attested to by the prevalence of opalinids in frogs and toads.

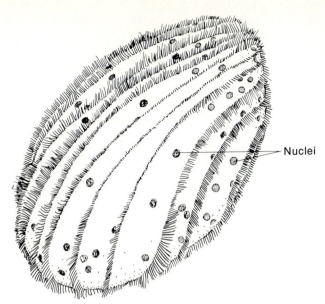

Nuclei

FIG. 6-20

Opalina sp. from the rectum of a frog. Note the numerous nuclei.

Drawing by Ian Grant.

A curious symbiosis is found in *Zelleriella opisthocarya,* a parasite of toads, and *Entamoeba* sp., in which more than 200 cysts of the ameba were found in one opalinid.[34]

REFERENCES

1. Belding, D.L. 1965. Textbook of clinical parasitology, ed. 3 Appleton-Century-Crofts, New York.
2. Benchimol, M., and W. De Souza. 1983. Fine structure and cytochemistry of the hydrogenosome of *Tritrichomonas foetus.* J. Protozool. 30:422-425.
3. Brooks, B., and F.L. Schuster. 1984. Oral protozoa: survey, isolation, and ultrastructure of *Trichomonas tenax* from clinical practice. Trans. Am. Micr. Soc. 103:376-382.
4. Brugerolle, G. 1973. Etude ultrastructural du trophozoite et du kyste chez le genre *Chilomastix* Alexeieff, 1910 (Zoomastigophorea, Retortamonadida Grassé). J. Protozool. 20:574-585.
5. Burrows, R.B., and M.A. Swerdlow. 1956. *Enterobius vermicularis* as a probable vector of *Dientamoeba fragilis.* Am. J. Trop. Med. Hyg. 5:258-265.
6. Camp, R.R., C.F.T. Mattern, and B.M. Honigberg. 1974. Study of *Dientamoeba fragilis* Jepps and Dobell. I. Electronmicroscopic observations of the binucleate stages. II. Taxonomic position and revision of the genus. J. Protozool. 21:69-82.
7. Chandler, A.C., and C.P. Read. 1961. Introduction to parasitology, ed. 10, John Wiley & Sons, Inc., New York.
8. Cheissin, E.M. 1964. Ultrastructure of *Lamblia duodenalis.* I. Body surface, sucking disc, and median bodies. J. Protozool. 11:91-98.
9. Dobell, C. 1940. Researches on the intestinal protozoa of monkeys and man. X. The life history of *Dientamoeba fragilis*—observations, experiments and speculations. Parasitology 32:417-459.
10. Earl, P.R. 1979. Notes on the taxonomy of the opalinids (Protozoa), including remarks on continental drift. Trans. Am. Microsc. Soc. 98:549-557.
11. El Mofty, M.M., and I.A. Sadek, 1973. The mechanism of action of adrenaline in the induction of sexual reproduction (encystation) in *Opalina sudafricana* parasitic in *Bufo regularis.* Int. J. Parasitol. 3:425-431.
12. Fouts, A.C., and S.J. Kraus. 1980. *Trichomonas vaginalis:* reevaluation of its clinical presentation and laboratory diagnosis. J. Infect. Dis. 141:137-143.
13. Friend, D.S. 1966. The fine structure of *Giardia muris.* J. Cell Biol. 29:317-332.
14. Gleason, N.N., M.S. Horwitz, L.H. Newton, and G.T. Moore. 1970. A stool survey for enteric organisms in Aspen, Colorado. Am. J. Trop. Med. Hyg. 19:480-484.
15. Honigberg, B.M. 1963. Evolutionary and systematic relationships in the flagellate Order Trichomonadida Kirby. J. Protozool. 10:20-63.
16. Honigberg, B.M., and V.M. King. 1964. Structure of *Trichomonas vaginalis* Donné. J. Parasitol. 50:345-364.
17. Honigberg, B.M., C.F.T. Mattern, and W.A. Daniel. 1971. Fine structure of the mastigont system in *Tritrichomonas foetus* (Riedmüller). J. Protozool. 18:183-198.
18. Jírovic, O. 1965. Neuere Forschungen über *Trichomonas vaginalis* und vaginale Trichomonosis. Angew. Parasitol. 6:202-210.
19. Kirkpatrick, C.E., and J.P. Farrell. 1984. Feline giardiasis: observations on natural and induced infections. Am. J. Vet. Res. 45:2182-2188.
20. Lee, D.L. 1971. Helminths as vectors of micro-organisms. In Fallis, A.M., editor. Ecology and physiology of parasites. University of Toronto Press, Toronto, pp. 104-122.
21. Lindmark, D.G. 1980. Energy metabolism of the anaerobic protozoon *Giardia lamblia.* Mol. Biochem. Parasitol. 1:1-12.
22. Mack, S.R., and M. Müller. 1980. End products of carbohydrate metabolism in *Trichomonas vaginalis.* Comp. Biochem. Physiol. 67B:213-216.
23. Marczak, R., T.E. Gorrell, and M. Müller. 1983. Hydrogenosomal ferredoxin of the anaerobic protozoon, *Tritrichomonas foetus.* J. Biol. Chem. 258:12427-12433.

24. Mattern, C.F.T., B.M. Honigberg, and W.A. Daniel. 1967. The mastigont system of *Trichomonas gallinae* (Rivolta) as revealed by electron microscopy. J. Protozool. 14:320-339.

25. Monzingo, D.L., Jr., and C.P. Hibler. 1987. Prevalence of *Giardia* sp. in a beaver colony and the resulting environmental contamination. J. Wildl. Dis. 23:576-585.

26. Moore, G.T., W.M. Cross, D. McGuire, et al. 1970. Epidemic giardiasis at a ski resort. N. Engl. J. Med. 281:402-407.

27. Morbidity and Mortality Weekly Report. 1985. 1985 STD treatment guidelines. 34(4S):75S-108S.

28. Morbidity and Mortality Weekly Report. 1986. Annual Summary 1984. 33(54):1-135.

29. Mueller, J.F. 1959. Is *Chilomastix* a pathogen? J. Parasitol. 45:170.

30. Müller, M. 1985. Search for cell organelles in protozoa. J. Protozool. 32:559-563.

31. Paltauf, F., and J.G. Meingassner. 1982. The absence of cardiolipin in hydrogenosomes of *Trichomonas vaginalis* and *Tritrichomonas foetus*. J. Parasitol. 68:949-950.

32. Peterson, K.M., and J.F. Alderete. 1984. Selective acquisition of plasma proteins by *Trichomonas vaginalis* and human lipoproteins as a growth requirement for this species. Mol. Biochem. Parasitol. 12:37-48.

33. Schuster, F.L. 1968. Ultrastructure of *Histomonas meleagridis* (Smith) Tyzzer, a parasitic amebo-flagellate. J. Parasitol. 54:725-737.

34. Stabler, R.M., and T. Chen. 1936. Observations on an *Endamoeba* parasitizing opalinid ciliates. Biol. Bull. 70:56-71.

35. Turner, G., and M. Müller. 1983. Failure to detect extranuclear DNA in *Trichomonas vaginalis* and *Tritrichomonas foetus*. J. Parasitol. 69:234-236.

36. Wallis, P.M., J.M. Buchanan-Mappin, G.M. Fauber, and M. Belosevic. 1984. Reservoirs of *Giardia* spp. in southwestern Alberta. J. Wildl. Dis. 20:279-283.

37. Weinbach, E.C., C.E. Claggett, D.B. Keister, L.S. Diamond, and H. Kon. 1980. Respiratory metabolism of *Giardia lamblia*. J. Parasitol. 66:347-350.

38. Yang, J., and T. Scholten. 1977. *Dientamoeba fragilis:* a review with notes on its epidemiology, pathogenicity, mode of transmission, and diagnosis. Am. J. Trop. Med. Hyg. 26:16-22.

SUGGESTED READINGS

Honigberg, B.M. 1978a. Trichomonads of importance in human medicine. In Kreier, J.P., editor. Parasitic protozoa, vol. 3. Academic Press, Inc., New York.

Honigberg, B.M. 1978b. Trichomonads of veterinary importance. In Kreier, J.P., editor. Parasitic protozoa, vol. 3. Academic Press, Inc., New York.

Kulda, J., and E. Nohynkova. 1978. Flagellates of the human intestine and intestines of other species. In Kreier, J.P., editor. Parasitic protozoa, vol. 3. Academic Press, Inc., New York.

McDougald, I.R., and W.M. Reid 1978. *Histomonas meleagridis* and its relatives. In Kreier, J.P., editor. Parasitic protozoa, vol. 3. Academic Press, Inc., New York.

Meyer, E.A., and S. Radulescu. 1979. *Giardia* and giardiasis. In Lumsden, W.H.R., editor. Advances in parasitology, vol. 17. Academic Press, Inc., New York.

Wessenberg, H. 1978. Opalinata. In Kreier, J.P., editor. Parasitic protozoa, vol. 3. Academic Press, Inc., New York.

CHAPTER 7

SUBPHYLUM SARCODINA: AMEBAS

Students of biology are introduced to amebas (subclass Gymnamoebia) early in their careers. Most are left with the impression that amebas are harmless, microscopic creatures that spend their lives aimlessly wandering about in mud, water, and soil, occasionally catching a luckless ciliate for food and unemotionally reproducing by binary fission. Actually this is a pretty fair account of most amebas. However, a few species are parasites of other organisms, and one or two are responsible for much misery and death of humans. Still others are commensals. These must be recognized, however, to differentiate them from the pathogenic species.

The subphylum Sarcodina evolved from the Mastigophora, but as traditionally conceived, the group is polyphyletic. One line passes from the flagellate *Tetramitus,* which includes flagellate and ameboid stages (the flagellate stage has a permanent cytostome), through *Naegleria* to *Vahlkampfia.* The life cycle of *Naegleria* also includes flagellate and ameboid stages (Chapter 4), but no permanent cytostome is found in them. *Vahlkampfia* has no flagellate stage, but its ameboid stage is like that of *Naegleria.* Another line passes from the ameboid-flagellate *Histomonas* to the related *Dientamoeba,* which itself is still considered an ameba by many parasitologists. No doubt other lines of evolution have evolved in the subphylum.

Of the many families of amebas, only the Endamoebidae has species of great medical or economic importance. Two other families, Schizopyrinidae and Hartmannelidae, have species that can become facultatively parasitic in humans.

FAMILY ENDAMOEBIDAE

Species in the Endamoebidae are parasites or commensals of the digestive systems of arthropods and vertebrates. The genera and species are differentiated on the basis of nuclear structure. Three genera contain known parasites or commensals of humans and domestic animals: *Entamoeba, Endolimax,* and *Iodamoeba.*

Genus *Entamoeba*

Species of *Entamoeba* possess a nucleus that is vesicular and that has a small endosome at or near the center. Chromatin granules are arranged around the periphery of the nucleus and, in some species, also around the endosome. The cytoplasm contains a variety of food vacuoles, often containing particles of food being digested, usually bacteria or starch grains.[12] On the ultrastructural level both lysosomes and some endoplasmic reticulum are seen, and ribosomes are abundant. Golgi bodies and mitochondria apparently are absent. Curious, small **helical bodies** can be seen widely distributed in the cytoplasm of some trophozoites. These bodies are 0.3 to 1 μm in length and are ribonucleoprotein, perhaps some form of "packaged" messenger RNA. These become aggregated in a crystalline array in some species[16] and then are visible by light microscopy as **chromatoidal bars** (Figs. 7-1 and 7-2). These bodies stain darkly with basic dyes and have been known by parasitologists for many years, although their RNA character has only recently been discovered. The chromatoidal bars may be blunt rods or splinter shaped, according to species, and in some species they are noticeable only in young cysts. As the cyst ages, the bars apparently are disassembled and disappear.

Species of *Entamoeba* are found in both vertebrate and invertebrate hosts. Four species are common in humans *(E. histolytica, E. hartmanni, E. coli,* and *E. gingivalis)* and will be considered here in some detail. *Entamoeba polecki* is mentioned in passing.

■ *Entamoeba histolytica* (Fig. 7-3)

Dysentery, both bacterial and amebic, has long been known as a handmaiden of war, often inflicting more casualties than bullets and bombs. Accounts of epidemics of dysentery accompany nearly every thorough account of war, from antiquity to the prison camp horrors of World War II and Vietnam. Captain James Cook's first voyage met with amebic disaster in Batavia, Java, and modern tourists, too, often find themselves similarly afflicted on visiting foreign ports.

FIG. 7-1

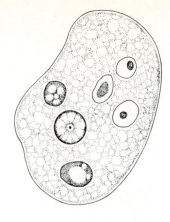

Entamoeba histolytica trophozoite and cyst.
Drawing by Jeanne Robertson.

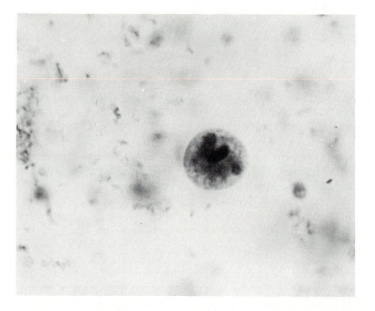

FIG. 7-2

Young cyst of *Entamoeba histolytica,* containing two nuclei and a prominent chromatoidal bar. Such a cyst usually is 10 to 20 μm wide.

This organism is the third most common cause of parasitic death in the world. Close to 500 million people are infected at any one time, with up to 100,000 deaths per year. These numbers may increase as urban migration and deteriorating economies of some developing countries result in unhygienic conditions. In addition, high rates of infection exist in certain high-risk groups, such as homosexuals, where infections have reached epidemic levels.

The history of acquired knowledge of the parasite *E. histolytica* is rampant with confusion and false conclusions. An interesting account has been prepared by Foster.[6]

The ameba was first discovered by a clinical assistant, D.F. Lösch, in St. Petersburg (now Leningrad), Russia, in 1873. The patient, a young peasant with bloody dysentery, was passing large numbers of amebas in his stools. Many of these, Lösch observed, contained erythrocytes in their food vacuoles. He successfully infected a dog by injecting amebas from his patient into the dog's rectum. On dissection Lösch found the dog's colonic mucosa riddled with ulcers that contained amebas. His human patient soon died, and at autopsy Lösch found identical ulcers in the intestinal mucosa. Despite these clear-cut observations, Lösch concluded that the ulcers were caused by some other

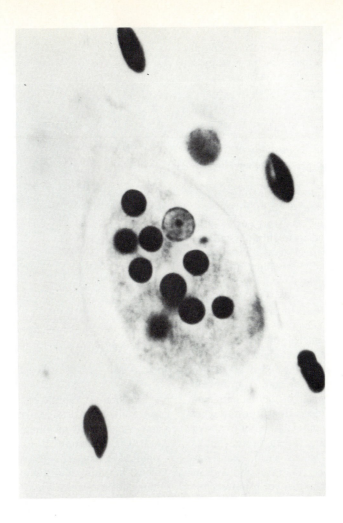

FIG. 7-3

Trophozoite of *Entamoeba histolytica* with several erythrocytes in food vacuoles.

From Kenney, M., and L.K. Eveland. 1981. Bull. N.Y. Acad. Med. 57:234-239.

agent and that the amebas merely interfered with their healing. It was nearly 40 years before it was generally accepted that an intestinal ameba can cause disease.

A major part of the problem was the then unrecognized fact that several species of amebas are found in the human intestine. Once this was established and nonpathogenic species were delineated, only one species complex remained that appeared to cause disease, and only occasionally at that. In 1903 Schaudinn named this group *Entamoeba histolytica*,[20] although the epithet "coli" was already applied to it by Lösch (as *Amoeba coli*). Schaudinn applied the latter name to a nonpathogenic species that he named *Entamoeba coli*.

Through the years it became obvious that *E. histolytica* occurs in two sizes. The smaller-sized amebas

have trophozoites 12 to 15 μm in diameter and cysts 5 to 9 μm wide. This form is encountered in about a third of those who harbor amebas and is not associated with disease. The larger form has trophozoites 20 to 30 μm in diameter and cysts 10 to 20 μm wide. The larger form may actually consist of two races, one sometimes pathogenic and the other always a commensal.

The small, nonpathogenic type is considered here as a separate species called *E. hartmanni*. Its life cycle, general morphology, and overall appearance, with the exception of size, are identical to those of *E. histolytica*. The task of proper identification is placed on the diagnostician, whose diagnosis may save the life of the patient or add the burden of unnecessary medication.

A third species, *E. moshkovskii,* is identical in morphology to *E. histolytica,* but it is not a symbiont. It dwells in sewage and is often mistaken for a parasite of humans. Indeed, it may be a strain that recently derived from one of the symbionts of humans.

There are several pathogenic strains (zymodemes) of *E. histolytica.*

Morphology and life cycle. Several successive stages occur in the life cycle of *E. histolytica:* the **trophozoite, precyst, cyst, metacyst,** and **metacystic trophozoite.**

Although the diameter of most trophozoites (Figs. 7-3 and 7-1) falls into the range of 20 to 30 μm, occasional specimens are as small as 10 μm or as large as 60 μm. In the intestine and in freshly passed, unformed stools, the parasites actively crawl about, their short, blunt pseudopodia rapidly extending and withdrawing. They also have filopodia, which are usually not discernible by light microscopy.[13] The clear ectoplasm is rather thin but is clearly differentiated from the granular endoplasm. The nucleus is difficult to discern in living specimens, but nuclear morphology may be distinguished after fixing and staining with iron-hematoxylin. The nucleus is spherical and is about one sixth to one fifth the diameter of the cell. A prominent endosome is located in the center of the nucleus, and delicate, achromatic fibrils radiate from it to the inner surface of the nuclear membrane. Chromatin is absent from a wide area surrounding the endosome but is concentrated in granules or plaques on the inner surface of the nuclear membrane. This gives the appearance of a dark circle with a bull's-eye in the center. The nuclear membrane itself is quite thin.

Food vacuoles are common in the cytoplasm of active trophozoites and may contain host erythrocytes in samples from diarrheic stools (Fig. 7-3). Granules typical for all amebas are numerous in the endoplasm. Chromatoidal bars are not found in this stage.

In a normal, asymptomatic infection, the amebas are carried out in formed stools. As the fecal matter passes posteriad and becomes dehydrated, the ameba is stimulated to encyst. Cysts are neither found in the stools of patients with dysentery nor formed by the amebas when they have invaded the tissues of the host. Trophozoites passed in stools are unable to encyst. At the onset of encystment, the trophozoite disgorges any undigested food it may contain and condenses into a sphere, called the **precyst.** A precyst is so rich in glycogen that a large glycogen vacuole may occupy most of the cytoplasm in the young cyst. The chromatoidal bars that form typically are rounded at the ends. The bars may be short and thick, thin and curved, spherical, or very irregular in shape, but they do not have the splinter-like appearance found in *E. coli.*

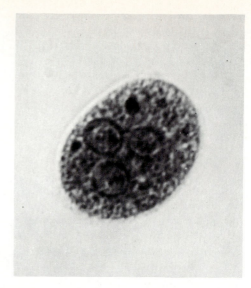

FIG. 7-4

Three of the four nuclei are in focus in the metacyst of *Entamoeba histolytica,* and two small chromatoidal bodies can be seen.

The precyst rapidly secretes a thin, tough hyaline **cyst wall** around itself to form a **cyst.** The cyst may be somewhat ovoid or elongate, but it usually is spheroid. It is commonly 10 to 20 μm wide but may be as small as 5 μm. The young cyst has only a single nucleus, but this rapidly divides twice to form two- and four-nucleus stages (Fig. 7-4). As the nuclear division proceeds and the cyst matures, the glycogen vacuole and chromatoidal bodies disappear. In semiformed stools one can find precysts and cysts with one to four nuclei, but quadrinucleate cysts (**metacysts**) are most common in formed stools (Figs. 7-1 and 7-4). This stage can survive outside the host and can infect a new one. After excysting in the small intestine, both the cytoplasm and nuclei divide to form eight small amebulae, or **metacystic trophozoites.** These are basically similar to mature trophozoites except in size.

Biology. Trophozoites may live and multiply indefinitely within the crypts of the mucosa of the large intestine, apparently feeding on starches and mucous secretions and interacting metabolically with enteric bacteria. However, such trophozoites commonly initiate tissue invasion when they hydrolyse mucosal cells and absorb the predigested product. At this stage they no longer require the presence of bacteria to meet their nutritional requirements. It has been shown that under optimal conditions of pH and ionic concentrations, nonvirulent *E. histolytica* becomes virulent when it contains a certain ratio of starch to cholesterol.[21] It has been suggested that the tiny cytoplasmic extensions from the surface (seen in some electron micrographs) could be "triggers" for "surface-active lysosomes" and function in cytolysis of host cells, although there is no evidence for this at present.[12] Lushbaugh and Pittman[13] showed that these "triggers" were actually filopodia, and they speculated that any or all of the following could be functions of the filopodia: (1) endocytosis or pinocytosis, (2) exocytosis, (3) attachment to the substrate, (4) penetration of tissue, (5) release of cytotoxic substances, or (6) contact cytolysis of host cells. Both pathogenic and nonpathogenic strains can possess proteolytic enzymes that presumably would make tissue invasion possible. Virulence of a particular strain can be attenuated by in vitro cultivation and sometimes restored by passage through certain experimental hosts. Some evidence exists that viral infection of the amebas may affect virulence.[14] The complex of factors involved in the environmental conditions in the host are even more difficult to untangle because the conditions mutually interact. The oxidation-reduction potential and the pH of the gut contents influence invasiveness, but these conditions are determined largely by the bacterial flora, which is in turn influenced by the host's diet and perhaps even its overall nutritional state. It is believed that one reason newcomers to areas of endemicity suffer more than the local population may be the differences in their bacterial flora.

Invasive organisms erode ulcers into the intestinal wall, eventually reaching the submucosa and underlying blood vessels. From there, they may travel with the blood to other sites in the body, such as the liver, lungs, or skin. Although these endogenous forms are active, healthy amebas that multiply rapidly, they are on a dead-end course. They cannot leave the host and infect others and so must perish with their luckless benefactor.

Mature cysts in the large intestine, on the other hand, leave the host in great numbers. The host that produces such cysts is usually asymptomatic or only mildly afflicted. Cysts of *E. histolytica* can remain viable and infective in a moist, cool environment for at least 12 days, and in water they can live up to 30 days. They are rapidly killed by putrefaction, desiccation, and temperatures below −5° C and above 40° C. They can withstand passage through the intestines of flies and cockroaches. The cysts are resistant to levels of chlorine normally used for water purification.

When swallowed, the cyst passes through the stomach unharmed and shows no activity while in an acidic environment. When it reaches the alkaline medium of the small intestine, the metacyst begins to move within the cyst wall, which rapidly weakens and tears. The quadrinucleate ameba emerges and divides into amebulae that are swept downward into the cecum.

This is the first opportunity of the organism to colonize, and its success depends on one or more metacystic trophozoites making contact with the mucosa. Obviously chances for establishment are improved when large numbers of cysts are swallowed.

Pathogenesis. To quote Elsdon-Dew, "Were one tenth, nay, one hundredth, of the alleged carriers of this parasite to suffer even in minor degree, then the ameba would rank as the major scourge of mankind."[5] Obviously not every infected person shows symptoms of disease.

Entamoeba histolytica is almost unique among the amebas of humans in its ability to hydrolyse host tissues. Once in contact with the mucosa, the amebas secrete proteolytic enzymes, which enable them to penetrate the epithelium and begin moving deeper. The **intestinal lesion** (Fig. 7-5) usually develops initially in the cecum, appendix, or upper colon and then spreads the length of the colon. The number of parasites builds up in the ulcer, increasing the speed of mucosal destruction. The muscularis mucosae is somewhat of a barrier to further progress, and pockets of amebas form, communicating with the lumen of the intestine through a slender, ductlike ulcer. The lesion may stop at the basement membrane or at the muscularis mucosae and then begin eroding laterally, causing broad, shallow areas of necrosis. The tissues may heal nearly as fast as they are destroyed, or the entire mucosa may become pocked. These early lesions usually are not complicated by bacterial invasion, and there is little cellular response by the host. In older lesions the amebas, assisted by bacteria, may break through the muscularis mucosae, infiltrate the submucosa, and even penetrate the muscle layers and serosa. This enables trophozoites to be carried by blood and lymph to ectopic sites throughout the body where secondary lesions then form. A high percentage of deaths result from perforated colons with concomitant peritonitis. Surgical repair of perforation is difficult because a heavily ulcerated colon becomes very delicate.

Sometimes a granulomatous mass, called an **ameboma,** forms in the wall of the intestine and may obstruct the bowel. It is the result of cellular responses to a chronic ulcer and often still contains active trophozoites. The condition is rare except in Central and South America.

Secondary lesions have been found in nearly every organ of the body, but the liver is most commonly affected (about 5% of all cases). Regardless of the secondary site, the initial infection was an intestinal abscess, even though it may have gone undetected. **Hepatic amebiasis** results when trophozoites enter the mesenteric venules and travel to the liver through the hepatoportal system. They digest their way through the portal capillaries and enter the sinusoids, where

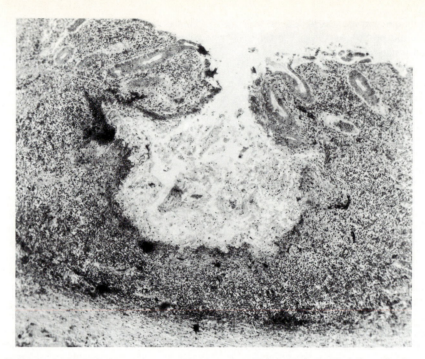

FIG. 7-5

Typical flask-shaped amebic ulcer of the colon. Extensive tissue destruction has resulted from invasion by *Entamoeba histolytica*.

AFIP neg. no. N- 44718.

they begin to form abscesses. The lesions thus produced may remain pinpoint size, or they may continue to grow, sometimes reaching the size of a grapefruit. The center of the abscess is filled with necrotic fluid, a median zone consists of liver stroma, and the outer zone consists of liver tissue being attacked by amebas, although it is bacteriologically sterile. The abscess may rupture, pouring debris and organisms into the body cavity, where they attack other organs.

Pulmonary amebiasis is the next most common secondary lesion. It usually develops by metastasis from a hepatic lesion but may originate independently. Most cases originate when a liver abscess ruptures through the diaphragm. Other ectopic sites occasionally encountered are the brain, skin and penis (possibly acquired venereally). Rare ectopic sites are kidneys, adrenals, spleen, male and female genitalia, pericardium, and others. As a rule, all ectopic abscesses are bacteriologically sterile.

Symptoms. Symptoms of infection vary greatly between cases. The strain of *E. histolytica* present, the host's natural or acquired resistance to that strain, and the host's physical and emotional condition when challenged all affect the course of the disease in any individual. When conditions are appropriate, a highly pathogenic strain can cause a sudden onset of severe disease. This usually is the case with waterborne epidemics. More commonly the disease develops slowly, with intermittent diarrhea, cramps, vomiting, and general malaise. Infection in the cecal area may mimic the symptoms of appendicitis. Some patients tolerate intestinal amebiasis for years with no sign of colitis (but are passing cysts) and then suddenly succumb to an ectopic lesion. Depending on the number and distribution of intestinal lesions, the patient might develop pain in the entire abdomen, fulminating diarrhea, dehydration, and loss of blood. Amebic diarrhea is marked by bouts of abdominal discomfort with four to six loose stools per day but little fever.

Acute amebic dysentery is a less common condition, but the sufferer from this affliction can best be described as miserable. The onset may be sudden after an incubation period of 8 to 10 days or after a long period as an asymptomatic cyst passer. In acute onset there may be headache, fever, severe abdominal cramps, and sometimes prolonged, ineffective straining at stool. An average of 15 to 20 stools are passed per day, consisting of liquid feces flecked with bloody mucus. Death may occur from peritonitis, resulting from gut perforation, or from cardiac failure and exhaustion. Bacterial involvement may lead to extensive scarring of the intestinal wall, with subsequent loss of peristalsis. Symptoms arising from ectopic lesions are typical for any lesion of the affected organ.

Epidemiology. *Entamoeba histolytica* is found throughout the world. Approximately 500 million persons are infected, of whom about 100 million suffer acute or chronic effects of the disease. Although clinical amebiasis is most prevalent in tropical and subtropical areas, the parasite is well established from Alaska to the southern tip of Argentina. The prevalence of infection varies widely, depending on local conditions, from less than 1% in Canada and Alaska to 5% in the contiguous United States to 40% in many tropical areas. The prevalence in the United States may be much higher among particular groups, such as persons in mental hospitals or orphanages. Age influences the prevalence of infection: Children under 5 years old have a *lower* rate than other age groups. In the United States the greatest prevalence occurs in the age group 26 to 30. The higher prevalence in the tropics results from lower standards of sanitation and the greater longevity of cysts in a favorable environment. The onset of the disease in persons who travel from temperate regions to endemic tropical areas may partly be the result of lessened resistance from the stress of travel and unaccustomed heat, in addition to the change in bacterial flora in the gut, as mentioned previously. All races are equally susceptible.

In 1977-1978 amebiasis was recognized as a sexually transmitted disease of increasing prevalence in New York City and a major health problem, particularly among homosexual men. In a study of 126 homosexual volunteers who participated in a gay men's health project, 39.7% were infected with *E. histolytica* and 18.3% with *Giardia intestinalis*, both fecal-borne organisms.[9] The authors believed that if multiple stools were examined, the figure of 39.7% could have been increased at least to 50%. Clearly the primary mode of infection in these cases was by oral to anal contact, and certainly the situation is not restricted to New York City. Thus a "new" health problem was discovered that probably has been fairly common for thousands of years.

The manner of disposal of human wastes in a given area is the most important factor in the epidemiology of this organism. Transmission depends heavily on contaminated food and water. Filth flies, particularly *Musca domestica,* and cockroaches also are important mechanical vectors of cysts. Their sticky, bristly appendages can easily carry cysts from a fresh stool to the dinner table, and the habit of the housefly to vomit and defecate while it feeds has been shown to be an important means of transmission. Polluted water sup-

plies, such as wells, ditches, and springs, are common sources of infection. Instances of careless plumbing have been known, in which sanitary drains were connected to freshwater pipes with resultant epidemics. Carriers (cyst passers) handling food can infect the rest of their family group or hundreds of people if the carrier works in a restaurant. The use of human feces as fertilizer in Asia, Europe, and South America contributes heavily to transmission.

Although humans are the most important reservoir of this disease, dogs, pigs, and monkeys are also implicated.

A bizarre event occurred in Colorado in 1980, when an epidemic of amebiasis was caused by colonic irrigation with a contaminated enema machine in a chiropractic clinic. Ten patients had to have a colectomy; seven of them died.[22]

Diagnosis and treatment. Demonstration of trophozoites or cysts is necessary for the accurate diagnosis of *E. histolytica*. However, a large proportion of patients with extraintestinal amebiasis have no concurrent intestinal infection; therefore diagnosis in such cases must be primarily by clinical and immunological means. X-ray examination and other means of scanning the liver may be useful in diagnosing abscesses. Examination of stool samples is the most effective means of diagnosis of gut infection. A direct smear examined either as a wet mount or fixed and stained will usually detect heavy infections. Even so, repeated examinations may be necessary; one of us found abundant trophozoites in the stool of a hospital patient after negative findings on 3 previous days. Lighter infections of cyst passers may be detected with concentration techniques, such as zinc sulfate flotation.

Immunological diagnosis is promising but has yet to be perfected. Fluorescent antibody technique has some value in diagnosis but cannot differentiate *E. histolytica* from *E. hartmanni*. Serological procedures that have been adopted widely are the hemagglutination test and agar gel diffusion. Many other diseases can easily be confused with amebiasis; on the hospital chart of the aforementioned patient, a dozen possible explanations for his persistent diarrhea, *other than amebiasis*, had been listed. Hence demonstration of the organism is nearly mandatory in diagnosis.

Several drugs have a high level of efficacy against colonic amebiasis. Most fall into the categories of arsanilic acid derivatives, iodochlorhydroxyquinolines, and other synthetic and natural chemicals. Antibiotics, particularly tetracycline, are useful as bactericidal adjuvants. These drugs are not as effective in ectopic infections, for which chloroquine phosphate and niridazole show promise of efficacy. Metronidazole (a 5-nitroimidazole derivative) has become the preferred

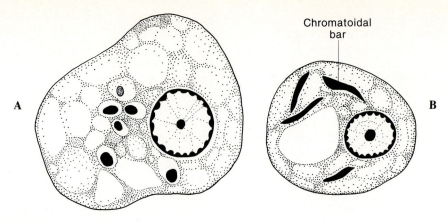

FIG. 7-6

Entamoeba polecki. **A,** Trophozoite. **B,** Cyst. Note the angular, tapering chromatoidal bars and single nucleus in the cyst.

Drawing by Ian Grant.

drug in treatment of amebiasis. It is low in toxicity and is effective against both extraintestinal and colonic infections, as well as cysts. However, metronidazole has been reported as being mutagenic in bacteria and carcinogenic in mice at doses not much higher than those given for the treatment of amebiasis. Furthermore, patients must be warned that the drug cannot be taken with alcohol because of its Antabuse-like effect. Finally, its efficacy may not be as high as originally reported. Tetracycline in combination with diiodohydroxyquin results in a high rate of cures. Two new 5-nitroimidazole derivatives, ornidazole and tinidazole, can cure amebic liver abscess with a single dose.[10]

Metabolism. The metabolism of *E. histolytica* has received some attention, but a detailed picture has yet to emerge.[1,2,15] It had been believed that the organism was an obligate anaerobe because it requires a low oxidation-reduction potential for optimal growth. It has neither a cytochrome system nor mitochondria. Band and Cirrito[1] concluded that the organism is an anaerobe with limited ability to detoxify products of oxygen reduction. During electron transport to oxygen as a final electron acceptor and in various hydroxylation and oxygenation reactions in many cells, toxic partial oxidation products of oxygen are formed such as O_2^- and hydrogen peroxide. In aerobic cells these are removed by superoxide dismutase and catalase or peroxidase. *Entamoeba histolytica* has a superoxide dismutase but not catalase or peroxidase.[23] Its ability to detoxify O_2^- but not hydrogen peroxide is consistent with its life in aerobic host tissue but not with the use of oxygen as a terminal electron acceptor.[1] In vivo the hy-

drogen peroxide may be detoxified by the host for the ameba, but in vitro the availability of sulfhydryl groups (as in cysteine) in the medium is necessary for optimal growth of the organisms in the presence of 5% oxygen. Growth is inhibited in the presence of 10% oxygen. Like many other gut-dwelling organisms, *E. histolytica* requires carbon dioxide.

Anaerobically the organism ferments glucose to ethanol and acetate in a ratio of 3:1 and evolves carbon dioxide and molecular hydrogen. Aerobically the ratio of ethanol and acetate is reversed, and hydrogen is not produced. It is reasonable to suppose that the ethanol is produced by a decarboxylation and then a reduction of the pyruvate from glycolysis, reoxidizing the NAD, as is the case in many bacteria and yeasts.

■ *Entamoeba polecki*

Entamoeba polecki (Fig. 7-6) is usually a parasite of pigs and monkeys, although on rare occasions it occurs in humans. It is generally nonpathogenic in humans, but symptomatic cases may be difficult to treat.[19] It can be distinguished from *E. histolytica* by several morphological criteria, including the fact that cysts of *E. polecki* have just one nucleus, with only about 1% of cysts ever reaching the binucleate stage. Uninucleate cysts of *E. histolytica* are infrequent.

■ *Entamoeba coli*

Entamoeba coli often coexists with *E. histolytica* and, in the living trophozoite stage, is difficult to differentiate from it. Unlike *E. histolytica,* however, *E. coli* is a commensal that never lyses its host's tissues. It feeds on bacteria, other protozoa, yeasts, and, occa-

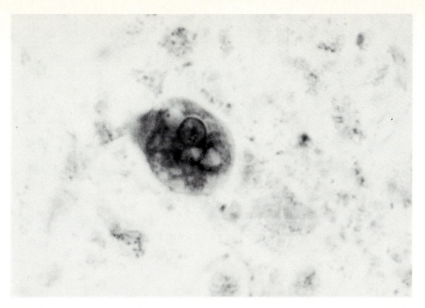

FIG. 7-7

Trophozoite of *Entamoeba coli,* a commensal in the human digestive tract. Note the characteristic, eccentrically located endosome. The size is usually 20 to 30 μm.

Photograph by Sherwin Desser.

sionally, blood cells that may be casually available to it. The diagnostician must identify this species correctly; if it is incorrectly diagnosed as *E. histolytica,* the patient may be submitted to unnecessary drug therapy.

Entamoeba coli is more common than *E. histolytica,* partly because of its superior ability to survive in putrefaction.

Morphology. The trophozite of *E. coli* (Fig. 7-7) is 15 to 50 μm (usually 20 to 30 μm) in diameter and is superficially identical to that of *E. histolytica.* However, their nuclei differ. The endosome of *E. coli* is usually eccentrically placed, whereas that of *E. histolytica* is central. Also, the chromatin lining the nuclear membrane is ordinarily coarser, with larger granules, than that of *E. histolytica.* The food vacuoles of *E. coli* are more likely to contain bacteria and other intestinal symbionts than are those of *E. histolytica,* although both may ingest available blood cells.

Encystment follows the same pattern as for *E. histolytica.* A precyst is formed, which rapidly secretes the cyst wall. The young cyst usually has a dense mass of chromatoidal bars that are splinter shaped, rather than blunt as in *E. histolytica.* As the cyst matures, the nucleus divides repeatedly to form eight nuclei (Fig. 7-8). Rarely, as many as 16 nuclei may be produced. The cysts vary in diameter from 10 to 33 μm.

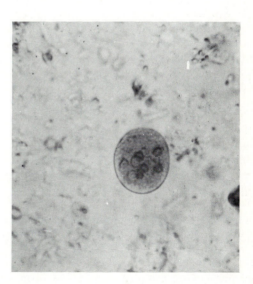

FIG. 7-8

Metacyst of *Entamoeba coli,* showing eight nuclei. The size is 10 to 33 μm.

Photograph by David Oetinger.

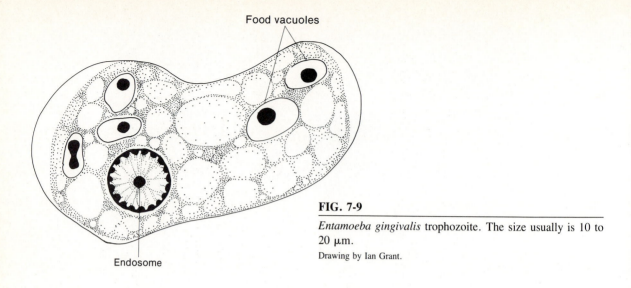

Food vacuoles

Endosome

FIG. 7-9

Entamoeba gingivalis trophozoite. The size usually is 10 to 20 μm.

Drawing by Ian Grant.

Biology. Infection and migration to the large intestine are identical to that of *E. histolytica*. The octanucleate metacyst produces 8 to 16 metacystic trophozoites, which first colonize the cecum and then the general colon. Infection is by contamination, in some areas of the world reaching nearly 100%. Obviously, this is a reflection of the level of the sanitation and water treatment. Because *E. coli* is a commensal, no treatment is required. However, infection with this protozoan indicates that opportunities exist for ingestion of *E. histolytica*.

■ *Entamoeba gingivalis*

Entamoeba gingivalis was the first ameba of humans to be described. It is present in all populations, dwelling only in the mouth. Like *E. coli*, it is a commensal and is of interest to parasitologists as another example of niche location and speciation.

Morphology. Only the trophozoite has been found, and encystment probably does not occur. The trophozoite (Fig. 7-9) is 10 to 20 μm (exceptionally 5 to 35 μm) in diameter and is quite transparent in life. It moves rather quickly, by means of numerous blunt pseudopodia. The spheroid nucleus is 2 to 4 μm in diameter and has a small, nearly central endosome. As in all members of this genus, the chromatin is concentrated on the inner surface of the nuclear membrane. Food vacuoles are numerous and contain cellular debris, bacteria, and, occasionally blood cells.

Biology. *Entamoeba gingivalis* lives on the surface of the teeth and gums, in the gingival pockets near the base of the teeth, and sometimes in the crypts of the tonsils. The organisms often are abundant in cases of gum or tonsil disease, but no evidence shows that they cause these conditions. More likely, the protozoa multiply rapidly with the increased abundance of food. They even seem to fare well on dentures, if the devices are not kept clean. The commensal also infects other primates, dogs, and cats.

Because no cyst is formed, transmission must be direct from one person to another, by kissing, by droplet spray, or by sharing eating utensils. Up to 95% of persons with unhygienic mouths may be infected, and up to 50% of persons with healthy mouths may harbor this ameba.[7]

Genus *Endolimax*

Members of the genus *Endolimax* live in both vertebrates and invertebrates. These amebas are small, with a vesicular nucleus. The endosome is comparatively large and irregular and is attached to the nuclear membrane by achromatic threads. Encystment occurs in the life cycle.

■ *Endolimax nana*

Endolimax nana lives in the large intestine of humans, mainly at the level of the cecum, and feeds on bacteria. Like *E. coli*, it is a commensal.

Morphology. The trophozoite of this tiny ameba (Fig. 7-10) measures 6 to 15 μm in diameter, but it is usually less than 10 μm. The ectoplasm is a thin layer surrounding the granular endoplasm. The pseudopodia

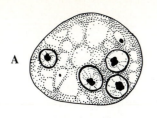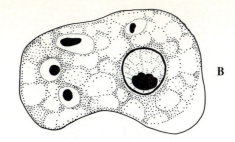

FIG. 7-10

Endolimax nana. **A,** Cyst. **B,** Trophozoite. Note the large karyosome and absence of chromatin granules on the nuclear membrane.

Drawings by Ian Grant.

are short and blunt, and the ameba moves very slowly, characteristics from which its name is derived: "dwarf internal slug." The nucleus is small and contains a large centrally or eccentrically located endosome. The marginal chromatin is a thin layer. Large glycogen vacuoles are often present, and food vacuoles contain bacteria, plant cells, and debris.

Encystment follows the same pattern as in *E. coli* and *E. histolytica.* The precyst secretes a cyst wall, and the young cyst thus formed includes glycogen granules and, occasionally, small curved chromatoidal bars. The mature cyst (Fig. 7-10) is 5 to 14 μm in diameter and contains four nuclei.

Biology. As with other cyst-forming amebas that infect humans, the mature cyst must be swallowed. The metacyst excysts in the small intestine, and colonization begins in the upper large intestine. Incidence of infection parallels that of *E. coli* and reflects the degree of sanitation practiced within a community. The cyst is more susceptible to putrefaction and desiccation than is that of *E. coli.* Although the protozoan is not a pathogen, its presence indicates that opportunities exist for infection by disease-causing organisms.

Genus *Iodamoeba*
■ *Iodamoeba buetschlii*

The genus *Iodamoeba* has only one species, and it infects humans, other primates, and pigs. Its distribution is worldwide. *Iodamoeba buetschlii* is the most common ameba of swine, which probably is its original host. The prevalence of *I. buetschlii* in humans is 4% to 8%, considerably lower than that of *E. coli* or *E. nana.*

Morphology. The trophozoite (Fig. 7-11) is usually 9 to 14 μm long but may range from 4 to 20 μm. It moves slowly by means of short, blunt pseudopodia. The ectoplasm is not clearly demarcated from the

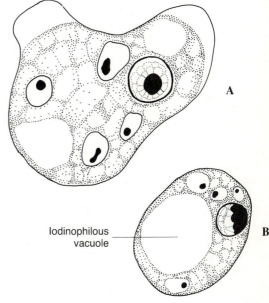

Iodinophilous vacuole

FIG. 7-11

Iodamoeba buetschlii. **A,** Trophozoite. **B,** Cysts. Note the persistance of glycogen mass in cyst, large eccentric karyosome.

Drawings by Ian Grant.

granular endoplasm. The nucleus is relatively large and vesicular, containing a large endosome that is surrounded by lightly staining granules about midway between it and the nuclear membrane. Achromatic strands extend between the endosome and the nuclear membrane, which has no peripheral granules. Food vacuoles usually contain bacteria and yeasts.

The precyst is usually oblong and contains no undigested food. It secretes the cyst wall that also is usu-

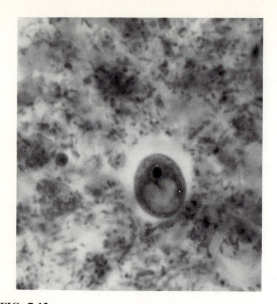

FIG. 7-12

Iodamoeba buetschlii cyst in human feces. Note the large iodinophilous vacuole. The size is 6 to 15 μm.

Photograph by James Jensen.

ally oblong, measuring 6 to 15 μm long. The mature cyst (Figs. 7-11 and 7-12) nearly always has only one nucleus. A large conspicuous glycogen vacuole stains deeply with iodine, hence the generic name.

Biology. *Iodamoeba buetschlii* lives in the large intestine, mainly in the cecal areas, where it feeds on intestinal flora. Infection spreads by contamination, since mature cysts must be swallowed to induce infection. It is possible that humans become infected through pig feces, as well as human feces. A few reports of *I. buetschlii* causing ectopic abscesses such as those of *E. histolytica* probably were actually misidentifications of *Naegleria fowleri.*

FAMILY SCHIZOPYRENIDAE

Schizopyrenidae are aerobic inhabitants of soil and water and mainly are bacteriophagous. They show affinities with the Mastigophora, since they possess a flagellated stage and an ameboid form. Binary fission seems to take place only in the ameboid form; thus these are diphasic amebas, with the ameboid stage predominating over the flagellated stage. Although the several genera and species in this family live in stagnant water, soil, sewage, and the like, a few are able to become facultative parasites in vertebrates. There has been confusion in the taxonomy of the genera *Naegleria, Hartmannella,* and *Acanthamoeba,* with

reports of all three as facultative parasites of humans. We take the view that *Naegleria* is placed in this family, whereas the other two genera belong in Hartmannellidae (see p. 112).

■ *Naegleria fowleri* (Fig. 7-13)

This species is also known in some of the literature as *N. aerobia.* It is the major cause of a disease called **primary amebic meningoencephalitis (PAM).** Other known species, *N. gruberi, N. lovaniensis,* and *N. australiensis,* appear to be harmless.

The flagellated stage of *N. fowleri* bears two long flagella at one end, is rather elongate, and does not form pseudopodia; the ameboid stage usually has a single blunt pseudopodium. Pointed tips of the pseudopodia are visible with the scanning electron microscope (Fig. 7-14). Transformation of the ameboid form to the flagellated form is quite rapid; once the flagella develop, the organism can swim rapidly, like a mastigophoran. The nucleus is vesicular and has a large endosome and peripheral granules. Dark polar masses are formed at mitosis, and Feulgen-negative **interzonal bodies** are present during late stages of nuclear division. A contractile vacuole is conspicuous in free-living forms. Food vacuoles contain bacteria in free-living stages but are filled with host cell debris in parasitic forms. Sucker-like structures called **amebastomes** are present, at least in culture forms; the amebastomes function in phagocytosis (Fig. 7-14).[8] The cyst has a single nucleus.

Primary amebic meningoencephalitis (PAM). This is an acute, fulminant, rapidly fatal illness usually affecting children and young adults who have been exposed to water harboring free-living *N. fowleri.* Most cases have been contracted in lakes or swimming pools. Probably the flagellated trophozoites are forced deep into the nasal passages when the victim dives into the water. One well-documented case involved ritual washing before prayers by a Nigerian Muslim farmer.[11] The washing occurred five times a day and included sniffing water up the nose. After entrance to the nasal passages, the amebas migrate along the olfactory nerves, through the cribriform plate, and into the cranium. Death from brain destruction is rapid, and few cures have been reported. *Naegleria fowleri* has been isolated and cultured from many fatal cases. These amebas kill a variety of laboratory animals when injected intranasally, intravenously, or intracerebrally.[3] They do not form cysts in the host. They have even been isolated from bottled mineral water in Mexico.[18]

As of 1962, only 92 cases of PAM had been reported. Since then, over 100 cases have been recorded in widely separated parts of the world, for example,

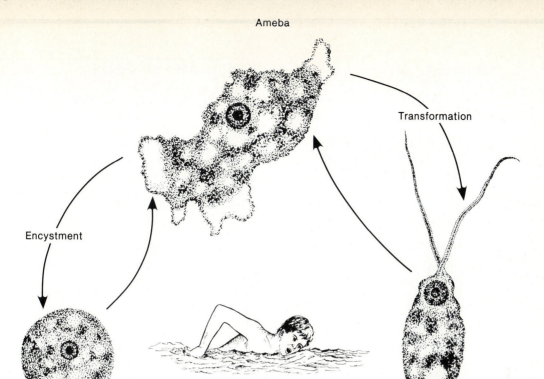

Ameba

Transformation

Encystment

[Human infection]

Cyst

Flagellate

FIG. 7-13

Life cycle of *Naegleria fowleri*.

From John, T.D. 1985. Amebas. In McGraw-Hill yearbook of science and technology 1986. McGraw-Hill Book Co., New York.

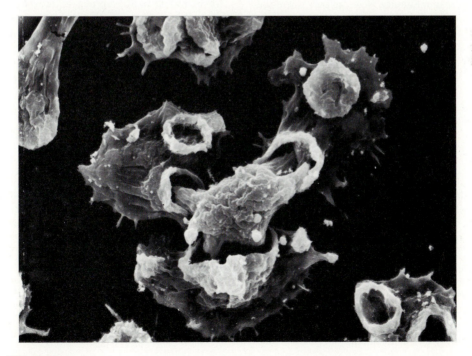

FIG. 7-14

Three *Naegleria fowleri* from axenic culture, attacking and beginning to devour or engulf a fourth, presumably dead, ameba with their amebastomes. (×2160.)

From John, D.T., T.B. Cole, Jr., and F.M. Marciano-Cabral. 1984. Appl. Environ. Microbiol. 47:12-14.

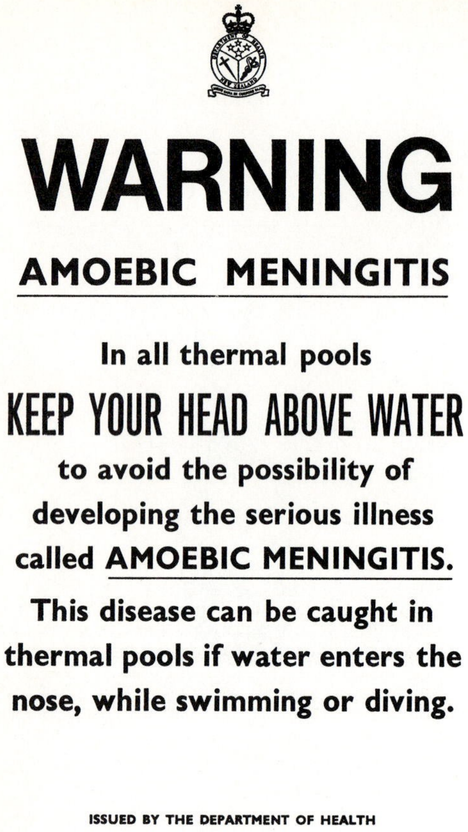

WARNING

AMOEBIC MENINGITIS

In all thermal pools

KEEP YOUR HEAD ABOVE WATER

to avoid the possibility of developing the serious illness called AMOEBIC MENINGITIS.

This disease can be caught in thermal pools if water enters the nose, while swimming or diving.

ISSUED BY THE DEPARTMENT OF HEALTH

FIG. 7-15

Warning encountered in Rotorua, New Zealand, where several cases of primary amebic meningoencephalitis have been contracted in hot pools.

Courtesy New Zealand Department of Health.

from the United States, Czechoslovakia, Mexico, Africa, New Zealand, and Australia. Undoubtedly, many cases remain undiagnosed. *Naegleria* amebas proliferate rapidly as water temperature rises, so thermal pools that are contaminated by rainwater run-off are particularly at risk (Fig. 7-15). Although these amebas are ubiquitous, the risk of acquiring this infection fortunately is small.

Treatment. Unfortunately most cases of PAM are diagnosed at autopsy. The disease is so rare and its course of brain destruction so rapid that only seldom has it been diagnosed in time for treatment to be attempted. Amphotericin B kills *Naegleria* in vitro and has been used successfully in at least two human cases. Recently *N. fowleri* was shown to be sensitive to qinghaosu in vitro. The lack of toxicity of this drug makes it potentially useful for therapy of PAM.[4]

FAMILY HARTMANNELLIDAE

A single genus, *Acanthamoeba*, is a facultative parasite of humans in much the same manner as *Naegleria*. The species *A. culbertsoni*, *A. polyphaga*, *A. hatchetti*, *A. castellanii* and *A. rhysodes* have been identified in human tissues. Some of these have been reported as species of *Hartmannella* but that genus is not pathogenic.[24] Biology of the free-living forms is similar to that of *Naegleria* except that flagella are not known to be produced and *Acanthamoeba* cannot tolerate water as hot as can *Naegleria*. *Acanthamoeba* spp. usually cause chronic infection of the skin or central nervous system in immunocompromised persons, although immunocompetent victims may suffer corneal ulcers and keratitis.

Live trophozoites of *Acanthamoeba* are easily differentiated from *Naegleria*, since they have small spiky acanthopodia and move very slowly. By contrast, *Naegleria* has a single, blunt lobopodium and moves rapidly (more than two body lengths per minute).

Twenty four cases of *Acanthamoeba* keratitis were reported to Centers for Disease Control in a 9-month period during 1985-1986.[17] In two patients the infected eye was enucleated; 12 patients underwent corneal transplantation. Twenty-two of the patients were initially diagnosed as having corneal herpes simplex virus infections. Most wore contact lenses and were found to use home-made saline washes containing live *Acanthamoeba*. Other recorded cases usually involve some trauma to the cornea before exposure to parasites. When one notes that *Acanthamoeba* is the most common ameba in fresh water and soil, it is surprising that more infections do not occur.

Treatment is difficult, although some cases have been treated successfully with ketoconazole, miconazole and propamidine isethionate.[17]

REFERENCES

1. Band, R.N., and H. Cirrito. 1979. Growth response of axenic *Entameoba histolytica* to hydrogen, carbon dioxide, and oxygen. J. Protozool. 26:282-286.
2. Bryant, C. 1970. Electron transport in parasitic helminths and protozoa. In Dawes, B., editor. Advances in parasitology, vol. 8. Academic Press, Inc., New York, pp. 139-172.
3. Chang, S.H. 1974. Etiological, pathological, epidemiological, and diagnostical consideration of primary amoebic meningoencephalitis. CRC Crit. Rev. Microbiol. 3:135-159.
4. Cooke, D.W., G.J. Lalliger, and D.T. Durack. 1987. *In vitro* sensitivity of *Naegleria fowleri* to qinghaosu and dihydroqinghaosu. J. Parasitol. 73:411-413.
5. Elsdon-Dew, R. 1964. Amoebiasis. Exp. Parasitol. 15:87-96.
6. Foster, W.D. 1965. A history of parasitology. E. & S. Livingstone, Edinburgh.
7. Jaskoski, B.J. 1963. Incidence of oral Protozoa. Trans. Am. Microsc. Soc. 82:418-420.

8. John, D.T., T.B. Cole, Jr., & R.A. Bruner. 1985. Amebastomes of *Naegleria fowleri.* J. Protozool. 32:12-19.

9. Kean, B.H., D.C. William, and S.K. Luminais. 1979. Epidemic of amebiasis and giardiasis in a biased population. Br. J. Vener. Dis. 55:375-378.

10. Lasserre, R., et al. 1983. Single-day drug treatment of amebic liver abcess. Am. J. Trop. Med. Hyg. 32:723-726.

11. Lawande, R.V., J.T. Macfarlane, W.R.C. Weir, and C. Awunor-Renner. 1980, A case of primary amebic meningoencephalitis in a Nigerian farmer. Am. J. Trop. Med. Hyg. 29:21-25.

12. Ludvik, J., and A.C. Shipstone. 1970. The ultrastructure of *Entamoeba histolytica.* Bull. WHO 43:301-308.

13. Lushbaugh, W.B., and F.E. Pittman. 1979. Microscopic observations on the filopodia of *Entamoeba histolytica.* J. Protozool. 26:186-195.

14. Mattern, C.F.T., D.B. Keister, and L.S. Diamond. 1979. Experimental amebiasis. IV. Amebal viruses and virulence of *Entamoeba histolytica.* Am. J. Trop. Med. Hyg. 28:653-657.

15. Montalvo, R.E., R.E. Reeves, and L.G. Warren. 1971. Aerobic and anaerobic metabolism in *Entamoeba histolytica.* Exp. Parasitol. 30:249-256.

16. Morgan, R.S., and B.G. Uzman, 1966. Nature of the packing of ribosomes within chromatoid bodies. Science 152:214-216.

17. Newton, C. et al. 1986. *Acanthamoeba* keratitis associated with contact lenses—United States. Morbid. Mortal. Weekly Rep. 35:405-408.

18. Rivera, F., et al. 1981. Bottled mineral waters polluted by protozoa in Mexico. J. Protozool. 28:54-56.

19. Salaki, J.S., J.L. Shirey, and G.T. Strickland. 1979. Successful treatment of *Entamoeba polecki* infection. Am. J. Trop. Med. Hyg. 28:190-193.

20. Schaudinn, F. 1903. Untersuchungen über die Fortpflanzung einiger Rhizopoden. Arb. Kaiserl. Gesundh.-Amte 19:547-576.

21. Sharma, R. 1959. Effect of cholesterol on the growth and virulence of *Entamoeba histolytica.* Trans. R. Soc. Trop. Med. Hyg. 53:278-281.

22. Simmons, R., et al. 1981. Amebiasis associated with colonic irrigation—Colorado. Morbid. Mortal. Weekly Rep. 30:101-102.

23. Sykes, D.E., and R.N. Band. 1977. Superoxide dismutase activity of *Acanthamoeba* and two anaerobic *Entamoeba* species. J. Cell Biol. 75:86a.

24. Warhurst, D.C. 1985. Pathogenic free-living amoebae. Parasitol. Today 1:24-28.

SUGGESTED READINGS

Albach, R.A., and T. Booden. 1978. Amoebae. In Kreier, J.P. editor. Parasitic protozoa, vol. II. Academic Press, Inc., New York.

Band, R.N., et al. 1983. Symposium—the biology of small amoebae. J. Protozool. 30:192-214.

Chang, S.L. 1971. Small, free-living amebas: cultivation, quantitation, identification, classification, pathogenesis, and resistance. In Cheng, T.C., editor. Current topics in comparative pathobiology, vol. 1. Academic Press, Inc., New York, pp. 202-254. (A review of the facultatively parasitic amebas.)

Connor, D.H., R.C. Neafie, and W.M. Meyers. 1976. Amebiasis. In Binford, C.H., and D.H. Connor, editors. Pathology of tropical and extraordinary diseases. Armed Forces Institute of Pathology, Washington, D.C.

Culbertson, C.G. 1976. Amebic meningoencephalitides. In Binford, C.H., and D.H. Connor, editors. Pathology of tropical and extraordinary diseases. Armed Forces Institute of Pathology, Washington, D.C.

Cursons, R.T.M., and T.J. Brown. 1976. Identification and classification of the aetiological agents of primary amoebic meningoencephalitis. N.Z. J. Mar. Freshwater Res. 10:245-262.

Griffin, J. 1978. The pathogenic free living amoebae. In Kreier, J.P., editor. Parasitic protozoa, vol. 2. Academic Press, Inc., New York.

Hoare, C.A. 1958. The enigma of host-parasite relations in amebiasis. Rice Inst. Pamphlet 45:23-35. (Very interesting reading.)

Lösch, F.A. 1875. Massive development of amebas in the large intestine. (Translated by Kean, B.H., and K.E. Mott, 1975.) Am. J. Trop. Med. Hyg. 24:383-392.

Martínez-Paloma, A., editor. 1986. Human parasitic diseases, vol. 2. Elsevier Science Publishers, Amsterdam.

Martinez-Paloma, A. 1987. The pathogenesis of amoebiasis. Parasitol. Today 3:111-118.

McLaughlin, J., and S. Aley. 1985. The biochemistry and functional morphology of the *Entamoeba.* J. Protozool. 32:221-240.

Neal, R.A. 1983. Experimental amoebiasis and the development of anti-amoebic compounds. Parasitology 86:175-191.

Singh, B.N. 1975. Pathogenic and non-pathogenic amoebae. John Wiley & Sons, Inc., New York.

Ungar, B.L.P., R.H. Yolken, and T.C. Quinn. 1985. Use of a monoclonal antibody in an enzyme immunoassay for the detection of *Entamoeba histolytica* in fecal specimens. Am. J. Trop. Med. Hyg. 34:465-472.

CHAPTER 8

PHYLUM APICOMPLEXA: GREGARINES, COCCIDIA, AND RELATED ORGANISMS

Protozoologists had become increasingly dissatisfied with the grouping of parasitic protozoa traditionally assigned to the class or subphylum Sporozoa. Levine[24] proposed a new subphylum, now considered a phylum, the Apicomplexa, to comprise organisms that possess a certain combination of structures, the **apical complex,** distinguishable with the electron microscope.[24] The structures that make up the apical complex will be described further, but they typically include a polar ring, micronemes, rhoptries, subpellicular tubules, micropore(s) (cytostome), and a conoid. Members of the Apicomplexa have a single type of nucleus and no cilia or flagella, except for the flagellated microgametes in some groups. The phylum contains two classes, Perkinsea and Sporozoea. Perkinsea consists of two species *Perkinsus,* parasites of oysters and abalones, and will not be considered further here. Sporozoea is divided into three subclasses: Gregarinia, Coccidia, and Piroplasmia.

Class Sporozoea

Members of the class Sporozoea are, without exception, parasites. All classes of vertebrates and most invertebrates harbor Sporozoea. Several species are of exceptional importance to humans, since they kill and weaken most of the domestic animal species and thereby cause millions of dollars of loss to agriculture every year. Furthermore, Sporozoea are responsible for malaria, the most important infectious disease of humans in the world today.

Most Sporozoea produce a resistant spore, or **oocyst** (containing sporozoites), which survives the elements between hosts (Fig. 8-1). In some the spore wall has been eliminated, and the development of the sporozoites is completed within an invertebrate vector.

Locomotor organelles are not as obvious as they are in the other phyla of protozoa. Pseudopodia are found only in some tiny, intracellular forms; flagella occur only on gametes of a few species, and a very few have

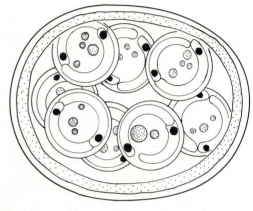

FIG. 8-1

Oocyst of *Adelina* sp., containing numerous sporocysts, each with two sporozoites.

Drawing by Ian Grant.

cilia-like appendages. Various species have sucker-like depressions, knobs, hooks, myonemes, and/or internal fibrils that aid in limited locomotion. The myonemes and fibrils form tiny waves of contraction across the body surfaces; these can propel the animal slowly through a liquid medium.

Both asexual and sexual reproduction are known in many Sporozoea. Asexual reproduction is either by binary or multiple fission or by endopolyogeny. Sexual reproduction is by isogamous or anisogamous fusion; in many cases this stage marks the onset of spore formation.

SUBCLASS GREGARINIA

Members of the large group of Sporozoea known as Gregarinia parasitize only invertebrates. Because gregarines are widespread, common, and may be large in

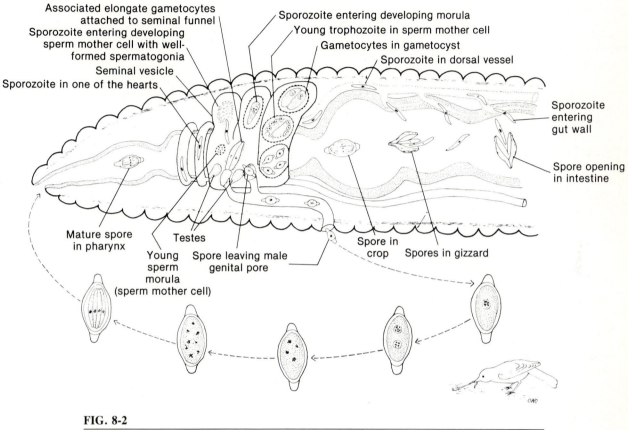

FIG. 8-2

Endogenous stages of *Monocystis lumbrici,* an aseptatine gregarine of earthworms.

From Olsen, O.W. 1974. Animal parasites. Their life cycles and ecology. University Park Press, Baltimore, Md.

size, they are often used as models in elementary and advanced zoology laboratories. Accordingly, we will discuss some examples to outline the characteristics of the group.

Multiple fission, or **schizogony** (or **merogony**) occurs in a few families of gregarines (orders Archigregarinida and Neogregarinida). However, these are not commonly encountered and so will not be discussed. Most gregarines (order Eugregarinida) have no schizogony and produce spores in their life cycles. They range in size from only a few micrometers in diameter to at least 10 mm long. Some are so large that nineteeth century zoologists placed them among the worms!

In **acephaline** gregarines the body consists of a sin-

gle unit that may have an anterior anchoring device, the **mucron.** In the **cephaline** species the body is divided by a septum into an anterior **protomerite** and a posterior **deutomerite** that contains the nucleus. Sometimes the protomerite bears an anterior anchoring device, the **epimerite.** Mucrons and epimerites are considered modified conoids (p. 117). In both the cephalines and acephalines the host becomes infected by swallowing spores. Most species parasitize the body cavity, intestine, or reproductive system of their hosts.

Order Eugregarinida
■ Suborder Aseptatina

Monocystis lumbrici (Fig. 8-2). *Monocystis lumbrici* lives in the seminal vesicle of *Lumbricus terres-*

115

tris and related earthworms, and it is easily demonstrated in the laboratory.

The worm becomes infected when it ingests a spore containing several **sporozoites.** These hatch in the gizzard, and the released sporozoites penetrate the intestinal wall, enter the dorsal vessel, and move forward to the hearts. They then leave the circulatory system and penetrate the seminal vesicles, where they enter into the sperm-forming cells (blastophores) in the vesicle wall. After a short period of growth, during which the parasites destroy the developing spermatocytes, the sporozoites enter the lumen of the vesicle where they become mature trophozoites, or **sporadins (gamonts),** measuring about 200 μm long by 65 μm wide. They attach to cells in the region of the sperm tunnel and undergo a form of union called **syzygy,** in which two or more sporadins connect with one another. The anterior organism is called the **primite** and the posterior the **satellite.** The differences between the two are unclear, but they exhibit different staining reactions. After syzygy the gamonts flatten against each other and surround themselves with a common cyst envelope, forming the **gametocyst.** Although "united" in the same cyst, the gamonts are still morphologically distinct, and each undergoes numerous nuclear divisions. The many small nuclei move to the periphery of the cytoplasm and, taking a small portion of the cytoplasm with them, bud off to become gametes. Some of the cytoplasm of each gamont remains and fuses to become the **residual body.** The gametes from each gamont are morphologically distinguishable and are thus anisogametes. Many species of gregarines have isogamy, however.[15] The fusion of a pair of gametes to form a zygote is followed by the secretion of the **spore** or **oocyst membrane** around that zygote and three cell divisions (sporogony) to form eight **sporozoites.** Thus each gametocyst now contains many oocysts, and the new host may become infected by eating a gametocyst or, if that body ruptures, an oocyst. As in other sporozoeans, meiosis is zygotic: only the zygote is diploid, and reductional division in sporogony returns the sporozoites to the haploid condition. The gametocyst or oocyst passes from the host through the sperm duct to be ingested by another worm.

■ Suborder Septatina

Gregarina polymorpha. Gregarina polymorpha is a common parasite of the mealworm *Tenebrio molitor* that usually infects colonies of beetles maintained in the laboratory. The sporadins are cylindrical, up to 350 μm long by 100 μm wide, with a small, globular epimerite that is inserted into a host cell. The entire life cycle takes place within the midgut of the mealworm larva or adult. The sporadins undergo syzygy,

having detached from the host intestinal epithelium, leaving the epimerite behind. The protomerite is dome shaped, and the deutomerite is cylindrical, is rounded posteriorly, and has a nucleus with an endosome. A gametocyst is formed that in turn produces gametes and zygotes. The cysts pass out with the feces of the host, to be eaten by the next mealworm.

When the mealworms are well fed and healthy, infection with *G. polymorpha* causes no obvious ill effects. However, when the beetles are stressed by malnutrition, the large gregarines take their toll: the pupae are smaller than normal, and fewer of them survive metamorphosis.

SUBCLASS COCCIDIA

Unlike members of the subclass Gregarinia, those of the subclass Coccidia are small, with a prevalent intracellular development and no epimerite or mucron. Some species are monoxenous, wheras others require two hosts to survive. Coccidia live in the digestive tract epithelium, liver, kidneys, blood cells, and other tissues of vertebrates and invertebrates.

The typical coccidian life cycle (see Fig. 8-6) has three major phases: merogony, gametogony, and sporogony. The infective stage is a rod- or banana-shaped sporozoite that enters a host cell and begins to develop. The organism becomes an ameboid trophozoite that multiplies by merogony to form more rod- or banana-shaped **merozoites,** which then escape from the host cell. These enter other cells to initiate further merogony or transform into a gamont (**gametogony**). Gamonts produce "male" **microgametocytes** or "female" **macrogametocytes.** Most species are thus anisogamous, and the macrogametocyte develops directly into a comparatively large, rounded macrogamete, which is an ovoid body filled with globules of a refractile material and has a central nucleus. The microgametocyte undergoes multiple fission to form tiny, biflagellated microgametes. Fertilization produces a zygote. Multiple fission of the zygote (sporogony) produces the sporozoite-filled oocyst. In monoxenous life cycles all stages occur in a single host, although the oocyst matures in the oxygen-rich, lower-temperature environment outside a host. The sporozoites are then released when the sporulated oocyst is eaten by another host. In heteroxenous life cycles, merogony and a part of gametogony occur in a vertebrate host, whereas sporogony occurs in an invertebrate, and the sporozoites are transmitted by the bite of the invertebrate. In still other heteroxenous life cycles, sporozoites are infective to a vertebrate intermediate host, which in turn produces zoites that are infective to the definitive host when it practices carnivorism.

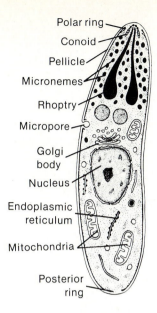

FIG. 8-3

An apicomplexan sporozoite or merozoite, illustrating the apical complex.

From Hickman, C.P., Jr., L.S. Roberts, and F.M. Hickman, 1988. Integrated principles of zoology, ed. 8. The C.V. Mosby Co., St. Louis.

The ultrastructure of the sporozoites and merozoites in the subclass is typical of the Apicomplexa.[24] These banana-shaped organisms (Fig. 8-3) are somewhat more attenuated at the anterior, apical complex end. A constant feature of the apical complex is one or two **polar rings,** electron-dense structures immediately beneath the cell membrane, which encircle the anterior tip. A **conoid** is found in members of the suborder Eimeriina. This structure is a truncated cone of spirally arranged fibrillar structures just within the polar rings. **Subpellicular microtubules** radiate from the polar rings and run posteriorly, parallel to the axis of the body. These organelles probably serve as structural elements and may be involved with locomotive function. Two to several elongate, electron-dense bodies, the **rhoptries,** extend to the cell membrane within the polar rings (and conoid, if present). Smaller, more convoluted elongate bodies, the **micronemes,** also extend posteriorly from the apical complex. The ducts of the micronemes apparently run anteriorly into the rhoptries or join a common duct system with the rhoptries to lead to the cell surface at the apex. The contents of the rhoptries and micronemes seem similar in electron micrographs, and it is thought that this material is secreted during entry into the host cell and is of aid in that process. Host recognition and invasion by various apicomplexan stages are reviewed by Sinden.[44] Along the side of the organism are one or more **micropores,** which function in ingestion of food material during the intracellular life of the parasite. The edges of the micropore are marked by two concentric, electron-dense rings, located immediately beneath the cell membrane. As host cytoplasm or other food matter within the parasitophorous vacuole is pulled through the rings, the parasite's cell membrane invaginates accordingly and finally pinches off to form a membrane-bound food vacuole. With the exception of the micropore, the structures described previously dedifferentiate and disappear after the sporozoite or merozoite penetrates the host cell to become a trophozoite. Members of the phyla Microspora and Myxozoa, which formerly were included in the subphylum Sporozoa, lack these structures at all stages of their life cycles.

Order Eucoccida
■ Suborder Adeleina

In the Adeleina the macrogametocyte and microgametocyte are associated in syzygy during development; therefore this is considered the most primitive suborder in the Eucoccida. The microgametocyte produces only one to four microgametes, the sporozoite is surrounded by a membrane, and endodyogeny is absent. The life cycles are either monoxenous or heteroxenous.

Family Haemogregarinidae

HAEMOGREGARINA STEPANOWI (Fig. 8-4). *Haemogregarina stepanowi* is a parasite of a European turtle, *Emys orbicularis,* and a leech, *Placobdella catenigra.* Similar species are common in turtles and frogs in the United States, and the following description of *H. stepanowi* essentially can be applied to them. Sporogony occurs in the leech; schizogony occurs in the turtle.

Biology. Trophozoites live in the circulating erythrocytes of the turtle. They become U shaped as they grow, and the unequal arms of the U finally fuse together to form an ovoid body. This becomes a macroschizont, and the erythrocyte that bears it lodges in the bone marrow. Meronts of this generation produce 13 to 24 large merozoites that enter other erythrocytes and become microschizonts. The microschizonts produce only six smaller merozoites. When these enter erythrocytes, they become gametocytes, thus ending the merogonous cycle. The gametocytes are elongate. The macrogamont has a small nucleus, and the microgamont has a large nucleus and dark-staining transverse bands at the anterior end. No further development takes place in the turtle.

When a leech ingests infected erythrocytes, the gametocytes are released. A macrogametocyte fuses in

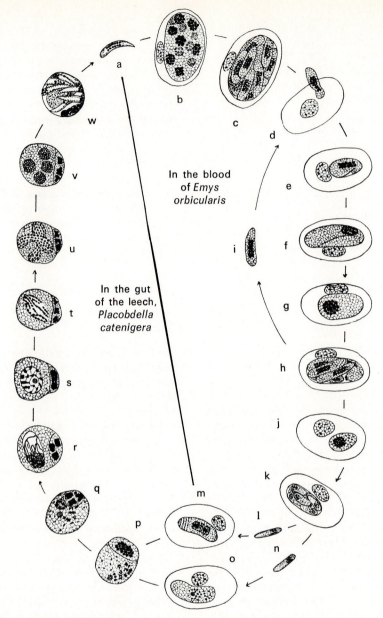

In the blood
of *Emys
orbicularis*

In the gut
of the leech,
*Placobdella
catenigera*

FIG. 8-4

Life cycle of *Haemogregarina stepanowi*. *a.* Sporozoite; *b* to *i,* schizogony; *j* and *k,* gametocyte formation; *l* and *m,* microgametocytes; *n* and *o,* macrogametocytes; *p* and *q,* association of gametocytes; *r,* fertilization; *s* to *w,* division of the zygote nucleus to form eight sporozoites.

From Kudo, R.R. Protozoology, 5th ed., 1966. Courtesy Charles C Thomas, Publisher, Springfield, Ill.

syzygy with a microgametocyte, and the pair becomes surrounded by a thin oocyst membrane. The nucleus of the microgametocyte divides, and the cell forms four microgametes, one of which fertilizes the macrogamete. Eight sporozoites develop from the zygote and, when mature, break out of the thin-walled oocyst into the intestinal lumen. The sporozoites then enter the circulatory system and migrate to the salivary glands, where they are injected into a new turtle host when the leech feeds.

Haemogregarines of terrestrial reptiles apparently are transmitted by mites. The means of transmission of haemogregarines of fish is unknown, but leeches probably are the vectors.

■ Suborder Eimeriina

In this suborder the macrogamete and microgamete develop independently without syzygy. The microgametocyte produces many microgametes, and the sporozoites are enclosed in a sporocyst. This is a very large group with many families and thousands of species. Very few species are found in humans, but many parasitize domestic animals, making this suborder of Eucoccida one of the most important group of parasites in agriculture. The history and taxonomy of the Eimeriina have been reviewed by Levine.[25]

Family Eimeriidae

This family contains numerous genera, of which several are of medical and veterinary importance. The taxonomy of the organisms we will consider in this family has undergone a great flux in recent years with the accumulation of much new knowledge about them, particularly about their life cycles. Levine[27] characterized the family as follows. The organisms develop in the host cell proper, and the gamonts and meronts (schizonts) do not have an attachment organelle. The oocysts have zero, one, two, four, or more sporocysts, each with one or more sporozoites. Merogony and gametogony occur within a host; sporogony typically, although not necessarily, occurs outside. Microgametes have two or three flagella, and particular species may be monoxenous or heteroxenous.

A typical oocyst (*Eimeria* sp.) is diagrammed in Fig. 8-5. The oocyst wall is of two layers, and with the electron microscope a membrane surrounding the outer wall may be distinguished.[42] In many species there is a tiny opening at one end of the oocyst, the **micropyle,** and this may be covered by the **micropylar cap.** A refractile **polar granule** may lie somewhere within the oocyst. The oocyst wall (and probably the sporocyst wall, too) is of a resistant material that helps the organism to survive harsh conditions in the external environment. Wilson and Fairbairn showed that the composition of the wall was of a chitin-like substance, but not chitin, since it does not contain N-acetyl-glucosamine.[52] The chitin-like substance is probably found in the inner layer of the wall because that layer is resistant to sodium hypochlorite.

Most species form sporocysts, which contain the sporozoites, within the oocyst. During sporogony to form the sporozoites, the cytoplasmic material not incorporated into the sporozoites forms the **oocyst residuum.** In like manner, some material may be left over within the sporocysts to become the **sporocyst residuum.** However, it appears that the sporocyst residuum is more than a depository for waste. It contains a large amount of lipid that seems to be an important source of energy for the sporozoites during their sojourn outside a host.[52] The sporocyst wall consists of a thin

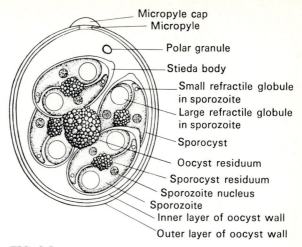

Micropyle cap
Micropyle
Polar granule
Stieda body
Small refractile globule in sporozoite
Large refractile globule in sporozoite
Sporocyst
Oocyst residuum
Sporocyst residuum
Sporozoite nucleus
Sporozoite
Inner layer of oocyst wall
Outer layer of oocyst wall

FIG. 8-5

Structure of sporulated *Eimeria* oocyst.

From Levine, N.D. 1961. Protozoan parasites of domestic animals and of man, ed. 2. Burgess Publishing Co., Minneapolis.

outer granular layer surrounded by two membranes and a thick, fibrous inner layer. At one end of the sporocyst, a small gap in the inner layer is plugged with a homogeneous **Stieda** body. In some species additional plug material underlies the Stieda body and is designated the **substiedal body.** When the sporocysts reach the intestine of a new host or are treated in vitro with trypsin and bile salt, the Stieda body is digested, the substiedal body pops out, and the sporozoites wriggle through the small opening thus created.[42] In addition to the apical complex, nucleus, and so on, the sporozoites themselves may contain one or more prominent **refractile bodies** of unknown function.

The size and shape of the oocyst and its contents, the presence or absence of several of the aforementioned structures, and the texture of the outer wall are all useful taxonomic characters. Identification of a coccidian usually can be accompished by examining its oocyst, which is remarkably constant in its characters in a given species.

Host specificity is more rigid in the genus *Eimeria* than in most other invasive organisms. Not only are *Eimeria* spp. often restricted to a certain host species, but a given species of *Eimeria* may be limited to certain organ systems, narrow zones in that system, specific kinds of cells in that zone, and even specific locations within the cells.[38] One species may be found only at the tips of the intestinal villi, another in the crypts at the bases of the villi, and a third in the interior of the villi, all in the same host. Some species develop below the nucleus of the host cell, others above it, and a few within it. Most coccidia inhabit the di-

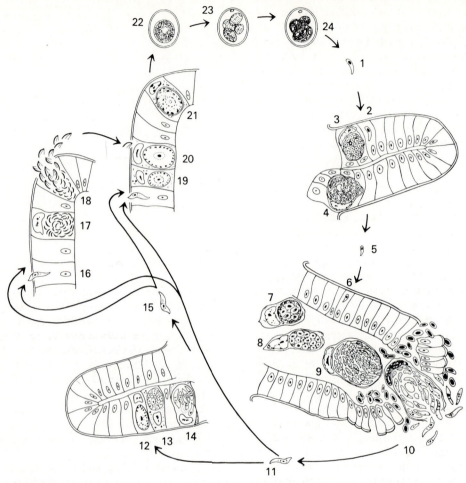

FIG. 8-6

Life cycle of the chicken coccidian *Eimeria tenella*. A sporozoite *(1)* enters an intestinal epithelial cell *(2)*, rounds up, grows, and becomes a first-generation schizont *(3)*. This produces a large number of first-generation merozoites *(4)*, which break out of the host cell *(5)*, enter new intestinal epithelial cells *(6)*, round up, grow, and become second-generation schizonts *(7 and 8)*. These produce a large number of second-generation merozoites *(9 and 10)*, which break out of the host cell *(11)*. Some enter new host intestinal epithelial cells and round up to become third-generation schizonts *(12 and 13)*, which produce third-generation merozoites *(14)*. The third-generation merozoites *(15)* and the great majority of second-generation merozoites *(11)* enter new host intestinal epithelial cells. Some become microgametocytes *(16 and 17)*, which produce a large number of microgametes *(18)*. Others turn into macrogametes *(19 and 20)*. The macrogametes are fertilized by the microgametes and become zygotes *(21)*, which lay down a heavy wall around themselves and turn into young oocysts. These break out of the host cell and pass out in the feces *(22)*. The oocysts then sporulate. The sporont throws off a polar body and forms four sporoblasts *(23)*, each of which forms a sporocyst containing two sporozoites *(24)*. When the sporulated oocyst *(24)* is ingested by a chicken, the sporozites are released *(1)*.

From Levine, N.D. 1961. Protozoan parasites of domestic animals and of man, ed. 2. Burgess Publishing Co., Minneapolis.

gestive tract, but a few are found in other organs, such as the liver and kidneys.

The number of species of coccidia is staggering. Levine and Ivens[32] recognized 204 species of *Eimeria* in rodents, but they estimated that there must be at least 2700 species of *Eimeria* in rodents alone. It has been calculated that *Eimeria* has been described from only 1.2% of the chordate species and 5.7% of the mammals in the world.[23] If all chordates were examined, Levine estimated that 34,000 species of *Eimeria* would be found—3500 of them in mammals—and if all animals were examined, a total of 45,000 species would be found of this single genus.[22] That is nearly equivalent to the number of all living and fossil protozoa described so far.

EIMERIA TENELLA (Fig. 8-6). *Eimeria tenella* lives in the epithelium of the intestinal ceca of chickens, where it produces considerable destruction of tissues, causing a high mortality rate in young birds. This and related species are of such consequence that all commercial feeds for young chickens now contain anticoccidial agents.

Biology. A chicken becomes infected when it swallows food or water that is contaminated with sporulated oocysts. The oocyst of this species is ovoid, smooth, and 14 to 31 μm long by 9 to 25 μm wide. When the micropyle ruptures in the bird's gizzard, the activated sporozoites escape the sporocyst in the small intestine. Once in a cecum the sporozoites first enter the cells of the surface epithelium and pass through the basement membrane into the lamina propria. There they are engulfed by macrophages that carry them to the glands of Lieberkühn. They then escape the macrophages and enter into a glandular epithelial cell of the crypt, where they locate between the nucleus and the basement membrane.

Within the epithelial cell the sporozoite becomes a trophozoite, feeding on the host cell and enlarging to become a meront. During merogony the meront separates into about 900 first-generation merozoites, each about 2 to 4 μm long. They break out into the lumen of the cecum about 2½ to 3 days after infection, destroying the host cell. Each surviving first-generation merozoite enters another cecal epithelial cell to initiate the second endogenous generation. The merozoite develops into a meront that lives between the nucleus and the free border of the host cell. A great many merozoites will form meronts in the lamina propria under the basement membrane.

About 200 to 350 second-generation merozoites, each about 16 μm long, are then formed by merogony. These rupture the host cell and enter the lumen of the cecum about 5 days after infection. Some of these merozoites enter new cells to initiate a third gen-

eration of merogony below the nucleus, producing 4 to 30 third-generation merozoites, each about 7 μm long. Many merozoites are engulfed and digested by macrophages during these cycles of merogony.

Some of the second-generation merozoites enter new epithelial cells in the cecum to begin gametogony. Most develop into macrogametocytes. Both male and female gamonts lie between the host cell nucleus and the basement membrane. The microgametocyte buds to form many slender, biflagellated microgametes that leave the host cell and enter cells containing macrogametes; there fertilization takes place.

The macrogamete has many granules of two types. Immediately after fertilization these granules pass peripherally toward the surface of the zygote, flatten out, and coalesce to form first the outer, then the inner layer of the oocyst wall. This coalescence takes place within the cell membrane of the zygote, and the membrane becomes the covering of the outer wall. The oocyst then is released from the host cell and moves with the cecal contents into the large intestine to be passed out of the body with the feces. Oocysts appear in the feces less than 6 days (138 hours) after infection. Oocysts are passed for several days because not all second-generation merozoites reenter host cells at the same time; furthermore, oocysts often will remain in the lumen of the cecum for some time before moving to the large intestine.

The freshly passed oocyst contains a single cell, the **sporont.** Sporogony (often called sporulation), or development of the sporont into sporocysts and sporozoites, is exogenous. The sporont is diploid, and the first division is reductional, a polar body being expelled. The haploid number of chromosomes is two. The sporont divides into four **sporoblasts,** each of which forms a sporocyst containing two sporozoites. Sporulation takes 2 days at summertime temperatures, whereupon the oocysts are infective.

Although the organisms can survive anaerobic conditions, as might be found in freshly passed feces, the metabolism of sporulation is an aerobic process and will not proceed in the absence of oxygen.[52] Development also is strongly and reversibly inhibited by cyanide, indicating that the cytochrome system is probably very important in the energetics of sporulation. Oxygen consumption is high at first but falls steadily as sporulation is completed. The organisms have large amounts of glycogen, which is rapidly consumed, and measurements of the respiratory quotient indicate that they depend primarily on carbohydrate oxidation for energy during sporoblast formation, then change over to lipid for energy as sporulation is completed. Thus the biochemistry suggests an interesting developmental control in metabolism: first a rapid burst of energy

FIG. 8-7

Cecum of chicken, opened to show patches of hemorrhage caused by *Eimeria tenella*.

Photograph by James Jensen.

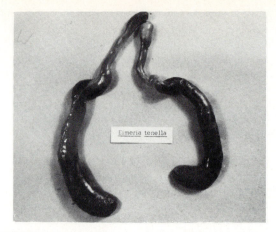

FIG. 8-8

Ceca of chicken infected by *Eimeria tenella*. Note distention caused by clotted blood and debris and dark color from hemorrhage.

Photograph by James Jensen.

fuels sporulation, and then a shift to a low level of maintenance metabolism conserves resources until a new host is reached.

Eimeria infections are self-limiting; that is, asexual reproduction does not continue indefinitely. If the chicken survives through oocyst release, it recovers. It may become reinfected, but a primary infection usually imparts some degree of protective immunity to the host.

The number of oocysts produced in any infection can be astounding. Theoretically one oocyst of *E. tenella,* containing eight sporozoites, can produce 2.52 million second-generation merozoites, most of which will become macrogametes and thereby oocysts. The actual numbers of oocysts produced are far fewer than their theoretical potential, however. Many merozoites and sporozoites are discharged with the feces before they can penetrate host cells, and many are destroyed by host defenses. A complete replacement of cecal epithelium normally occurs about every 2 days, so any merozoite or sporozoite that invades a cell which is about to be sloughed is out of luck. Young chickens are more susceptible to infection and discharge more oocysts than do older birds.

Pathogenesis. Cecal coccidiosis is a serious disease that causes a bloody diarrhea, sloughing of patches of epithelium, and, commonly, death of the host. When merozoites emerge, especially from the lamina propria, host tissues are disrupted. The large schizonts, especially when packed close together, disrupt the delicate capillaries that service the epithelium, further altering normal physiology of the tissues and also causing hemorrhage (Fig. 8-7). A hard core of the clotted blood and cell debris often plugs up the cecum, causing necrosis of that organ (Fig. 8-8). Birds that are not killed outright by the infection become unthrifty, listless, and susceptible to predation and other diseases.

The USDA estimated that loss to poultry farmers in the United States alone in 1986 was $80 million counting the extra cost of medicated feeds and added labor. Annual broiler production in the USA is about 4.2 trillion birds.[34]

Many useful drugs are available as prophylaxes against coccidiosis. However, once infection is established, there is no effective chemotherapy. Therefore a coccidiostat must be administered continuously in food or water to prevent an outbreak of disease. These compounds affect the schizont primarily, so the host can still build up an immunity in response to invading sporozoites.

OTHER *EIMERIA* SPECIES. The number of species of *Eimeria* is so large that it is impossible to discuss more than a representative species, as we have done with *E. tenella*. Levine[26] has summarized the species that parasitize domestic animals, some of the most common of which are *E. auburnensis* and *E. bovis* in cattle, *E. ovina* in sheep, *E. debliecki* and *E. porci* in pigs, *E. stiedai* in rabbits, *E. necatrix* and *E. acervulina* in chickens, *E. meleagridis* in turkeys, and *E. anatis* in ducks. All have life cycles similar to that of *E. tenella* but differing in details.

***ISOSPORA* GROUP.** Confusion about this group of coccidia has been fostered for many years by the discription of stages in intermediate hosts and in definitive hosts as different genera through ignorance of their life cycles. Some clarification of their relationships has begun to emerge only within the last 20 years.

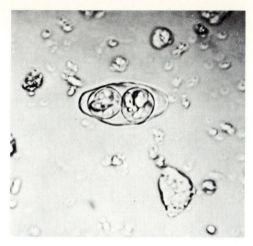

FIG. 8-9

Isospora belli oocyst. It averages 35 by 9 μm.

From Beck, J.W., and J.E. Davies. 1981. Medical parasitology, ed. 3. The C.V. Mosby Co., St. Louis.

The oocyst of *Isospora* contains two sporocysts, each with four sporozoites. Oocysts of the genera *Toxoplasma*, *Sarcocystis*, *Levineia*, *Besnoitia*, *Frenkelia*, and *Arthrocystis* have similarly constructed oocysts, but these parasites are heteroxenous, with vertebrate intermediate hosts. For this reason they are placed in the family Sarcocystidae, discussed later. For a discussion on the taxonomy of these genera see Frenkel et al.[13]

Isospora contains far fewer species than *Eimeria*, most of them in birds, but it includes a human parasite, *I. belli* (Fig. 8-9). Infection with *I. belli* is rather uncommon, and most cases have been reported from the tropics. It can cause severe disease with fever, malaise, persistent diarrhea, and even death, especially in AIDS patients.[8,10] The species previously known as *Hammondia heydorni* is now known to be synonymous with *I. bahiensis*.

Atoxoplasmatidae

Species of *Atoxoplasma*, the only genus in this family, are parasites of birds. They produce *Isospora*-like oocysts and have long been confused with that genus. For a review see Levine.[28] They have a monoxenous life cycle with merogony in blood and intestinal cells and gametogony in intestinal cells of the same host. Sporogony is outside the host; infection is by swallowing the oocyst.

Family Sarcocystidae

Members of this family are very closely related to Eimeriidae, differing principally in having heteroxenous life cycles. Asexual development occurs in ver-

tebrate intermediate hosts, whereas other vertebrates, mainly carnivorous mammals and birds, are definitive hosts. The oocyst contains two sporocysts, each with four sporozoites. There are about 122 species of *Sarcocystis*, 9 of *Toxoplasma*, 7 of *Besnoitia* and 1 of *Arthrocystis*.[2,30] Their classification has been changing rapidly as new information becomes available. One genus, *Toxoplasma*, is known to be of extreme importance to human and animal health. Others are of less known importance and will be mentioned briefly.

The **basic life cycle** found throughout this family is as follows: oocysts from a definitive host sporulate and are swallowed by an intermediate host. Sporozoites released from the cysts infect various tissues and rapidly undergo endodyogeny to form merozoites, also known as **tachyzoites** or **endozoites.** These are infective to other tissues in the body, such as muscles, fibroblasts, liver and nerves. Asexual reproduction in these tissues is much slower, developing large, cyst-like accumulations of merozoites that are now called **bradyzoites.** The cyst itself is called a **zoitocyst,** or simply a **tissue cyst.** A definitive host is infected when it eats meat containing bradyzoites or tachyzoites (rarely), or in some cases when it swallows a sporulated oocyst. Tachyzoites and bradyzoites are known to have antigenic differences.[35]

TOXOPLASMA GONDII (Fig. 8-10). Like *Trypanosoma cruzi*, *Toxoplasma gondii* was discovered before it was known to cause disease in humans. It was first discovered in 1908 in a desert rodent, the gondi, in a colony maintained in the Pasteur Institute in Tunis. Since then, the parasite has been found in almost every country of the world in many species of carnivores, insectivores, rodents, pigs, herbivores, primates, and other mammals, as well as in birds. We now realize that it is cosmopolitan in the human population and can cause disease. The importance of the organism as a human pathogen has stimulated a huge amount of research in recent years. A 1963 bibliography on the subject contained 3706 references, and Jacobs[18] reported more than 2000 references for the years 1967 through 1972. Between 1968 and 1975 an additional 12,500 references were added. Thus the onetime obscure protozoan parasite of an obscure African rodent has become one of the most exciting subjects in parasitology.

Biology. *Toxoplasma* is an intracellular parasite of many kinds of tissues, including muscle and intestinal epithelium. In heavy acute infections the organism can be found free in the blood and peritoneal exudate. It may inhabit the nucleus of the host cell but usually lives in the cytoplasm. The life cycle includes intestinal-epithelial (**enteroepithelial**) and **extraintestinal** stages in domestic cats and other felines, but extrain-

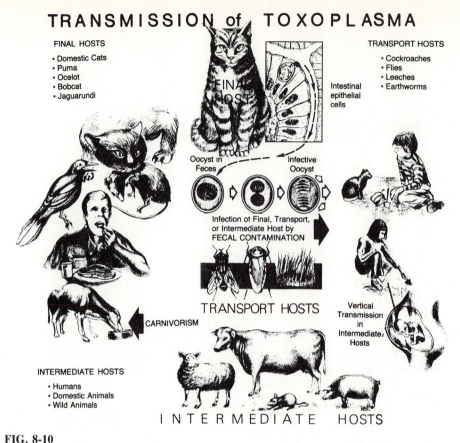

FIG. 8-10

Transmission of *Toxoplasma gondii*.

Courtesy Ronald Fayer, 1976. Natl. Wool Grower 66:22; drawing by R.B. Ewing.

testinal stages only in other hosts. Sexual reproduction of *Toxoplasma* occurs while in the cat, and only asexual reproduction is known while in other hosts.

Extraintestinal stages begin when a cat or other host ingests bradyzoites. Ingested tachyzoites or sporocysts also sometimes are infective. Intrauterine infection is possible (see discussion of pathogenesis). The oocyst (Fig. 8-11) is 10 to 13 μm by 9 to 11 μm and is basically similar in appearance to those of isosporan species. There is no oocyst residuum or polar granule, and the sporocysts have a sporocyst residuum but no Stieda body. The sporozoites escape from the sporocysts and the oocyst in the small intestine. In cats some of the sporozoites enter epithelial cells and remain to initiate the enteroepithelial cycle, whereas others penetrate through the mucosa to begin development in the lamina propria, mesenteric lymph nodes and other distant organs, and white blood cells. In hosts other than cats there is no enteroepithelial development; the sporozoite enters a host cell and begins multiplying by endodyogeny. These rapidly dividing

cells in acute infections are called tachyzoites (Fig. 8-12). Eight to 32 tachyzoites accumulate within the host cell's parasitophorous vacuole before the cell disintegrates, releasing the parasites to infect new cells. These accumulations of tachyzoites in a cell are called **groups.** Tachyzoites apparently are less resistant to stomach secretions; therefore they are less important sources of infection than are other stages.

As infection becomes chronic, the zoites that affect brain, heart, and skeletal muscles multiply much more slowly than in the acute phase. They are now called bradyzoites, and they accumulate in large numbers within a host cell. They become surrounded by a tough wall and are called cysts or zoitocysts (Fig. 8-13). Previous authors have referred to the cysts, as well as the groups of tachyzoites, as "pseudocysts" on the basis that the cyst wall is of host origin; however, since the origin of the cyst wall has not been firmly established, "tissue cyst" is preferred. Cysts may persist for months or even years after infection, particularly in nervous tissue. Cyst formation coincides with

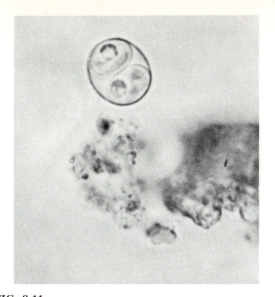

FIG. 8-11

Oocyst of *Toxoplasma gondii* from cat feces. It is 10 to 13 by 9 to 11 μm.

Photograph by Harley Sheffield.

FIG. 8-12

Tachyzoites of *Toxoplasma gondii*. They are about 7 by 2 μm.

the time of development of immunity to new infection, which is usually permanent. If immunity wanes, released bradyzoites can boost the immunity to its prior level. This protection against superinfection by the presence of the infectious agent in the body is called **premunition.** Immunity to *Toxoplasma* is of both the antibody and cell-mediated types; the latter is more important. The tough, thin cyst wall, except when the cyst breaks down, effectively separates the parasite from the host, and an inflammatory reaction is not elicited. The cyst wall and its bradyzoites develop intracellularly, but they may eventually become extracellular because of distention and rupture of the host cell. Bradyzoites are resistant to digestion by pepsin and trypsin, and when eaten, they can infect a new host.

Enteroepithelial stages are initiated when a cat ingests zoitocysts containing bradyzoites, oocysts containing sporozoites, or, occasionally, tachyzoites. Another possible means of epithelial infection is by migration of extraintestinal zoites into the intestinal lining within the cat. Once inside an epithelial cell of the small intestine or colon, the parasite becomes a trophozoite that grows and prepares for merogony. At least five different strains have been studied well enough to allow characterization of the enteroepithelial stages.[11] These strains differ in duration of stages, number of merozoites produced, shape, and other de-

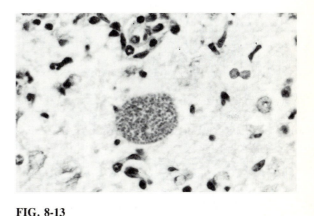

FIG. 8-13

Zoitocyst of *Toxoplasma gondii* in the brain of a mouse.

Photograph by Sherwin Desser.

tails. Basically, from 2 to 40 merozoites are produced by merogony, endopolyogeny, or endodyogeny, and these initiate subsequent asexual stages. The number of merogonous cycles is variable, but gametocytes are produced within 3 to 15 days after cyst-induced infection. Gametocytes develop throughout the small intestine but are more common in the ileum. From 2% to 4% of the gametocytes are male, and each produces about 12 microgametes. Oocysts appear in the cat's

125

feces from 3 to 5 days after infection by cysts, with peak production occurring between days 5 and 8. Oocysts require oxygen for sporulation and sporulate in 1 to 5 days. The extraintestinal development can proceed simultaneously with the enteroepithelial in the cat. Ingested bradyzoites penetrate the intestinal wall and multiply as tachyzoites in the lamina propria. They may disseminate widely in the extraintestinal tissues of the cat within a few hours of infection.[8]

One final note of interest is that in the Pasteur Institute in Tunis in 1908, when gondis were brought in from the field and died, the source of their infection was never established. However, it is known that a cat was roaming the laboratory.[18]

Pathogenesis. In view of the fact that antibody to *Toxoplasma* is widely prevalent in humans throughout the world yet clinical toxoplasmosis is less common, it is clear that most infections are asymptomatic or mild. Several factors influence this phenomenon: the virulence of the strain of *Toxoplasma,* the susceptibility of the individual host and of the host species, the age of the host, and the degree of acquired immunity of the host. Pigs are more susceptible than cattle; white mice are more susceptible than white rats; chickens are more susceptible than most carnivores. The reasons for natural resistance or susceptibility to infection are not known.

Occasionally, circumstances conspire to make a mild case important, as when Martina Navratilova lost the U.S. Open tennis championship and $500,000 in 1982, when she had toxoplasmosis.

About 13% of the world population is infected. In 1976 the global prevalence of toxoplasmosis was estimated to be over 500 million.[17] In countries like France, where raw meat is popular, the infection rate may be high; in Paris, for example, it may be as high as 50%. In the United States, it is estimated that about 3500 infants are born each year with severe infections, costing $300 million annually in care and treatment.[41].

Tachyzoites proliferate in many tissues and tend to kill host cells at a faster rate than does the normal turnover of such cells. Enteroepithelial cells, on the other hand, normally live only a few days, especially at the tips of the villi. Therefore, the extraintestinal stages, particularly in sites such as the retina or brain, tend to cause more serious lesions than do those in the intestinal epithelium.

Since there seems to be an age resistance, infections of adults or weaned juveniles are asymptomatic, although exceptions occur. Asymptomatic infections can suddenly become fulminating if immunosuppressive drugs such as corticosteroids are employed for other conditions. Symptomatic infections can be classified as acute, subacute, and chronic.

In most **acute infections** the intestine is the first site of infection. Cats infected by oocysts usually show little disease beyond loss of individual epithelial cells, and these are rapidly replaced. Actually, cysts probably are of little importance in infecting cats. In massive infections, however, intestinal lesions can kill kittens in 2 to 3 weeks. The first extraintestinal sites to be infected in both cats and other hosts, including humans, are the mesenteric lymph nodes and the parenchyma of the liver. These, too, have rapid regeneration of cells and perform an effective preliminary screening of the parasites. The most common symptom of acute toxoplasmosis is painful, swollen lymph glands in the cervical, supraclavicular, and inguinal regions. This symptom may be associated with fever, headache, muscle pain, anemia, and sometimes lung complications. This syndrome can be mistaken easily for the flu. Acute infection can, although rarely does, cause death. If immunity develops slowly, the condition can be prolonged and is then called subacute.

In **subacute infections** pathogenic conditions are extended. Tachyzoites continue to destroy cells, causing extensive lesions in the lung, liver, heart, brain, and eyes. Damage may be more extensive in the central nervous system than in unrelated organs because of lower immunocompetence in these tissues.

Chronic infection results when immunity builds up sufficiently to depress tachyzoite proliferation. This coincides with the formation of cysts. These cysts can remain intact for years and produce no obvious clinical effect. Occasionally a cyst wall will break down, releasing bradyzoites; most of these are killed by host reactions, although some may form new cysts. Death of the bradyzoites elicits an intense hypersensitive inflammatory reaction, the area of which, in the brain, is gradually replaced by nodules of glial cells. If many such nodules are formed, the host may develop symptoms of chronic encephalitis, with spastic paralysis in some cases. Chronic active or relapsing infections of retinal cells by tachyzoites causes blind spots and extensive infection of the central, macular area, which may lead to blindness. Cysts and cyst rupture in the retina can also lead to blindness. Other kinds of extensive pathological conditions can occur in chronic toxoplasmosis, for example, myocarditis, with permanent heart damage and with pneumonia.

In the immunocompetent person *T. gondii* ordinarily is kept at bay by cell-mediated immunity. When an infected person becomes immunosuppressed, the organism will disseminate rapidly, which may lead to ocular toxoplasmosis and to fatal CNS disorders such as encephalitis. Any long-term steroid therapy, such as is given to some cancer patients, can result in disseminated toxoplasmosis. Presently, *T. gondii* is a serious opportunistic infection in AIDS. Death usually results from cyst rupture with continued multiplication of tachyzoites.

The most tragic form of this disease is **congenital toxoplasmosis.** If a mother contracts acute toxoplasmosis at the time of her child's conception or during pregnancy, the organisms will often infect her developing fetus. Fortunately most neonatal infections are asymptomatic, but a significant number cause death or disability to newborns. It is generally assumed that *Toxoplasma* crosses the placental barrier from the mother's blood; however, because the uterus itself is commonly heavily infected, *at least in mice,* direct transmission cannot be ruled out.

The transmission rate to the fetus from a maternal infection is about 45%. Of those infected, about 60% will be subclinical, 9% may die, and 30% may suffer severe damage such as hydrocephalus, intracerebral calcification, retinochoroiditis, and mental retardation. However, even subclinical cases may develop into ocular toxoplasmosis later in life.

Stillbirths and spontaneous abortions may result from fetal infection with *Toxoplasma* in humans and other animals. Sheep seem to be particularly susceptible, and *Toxoplasma*-caused abortions in this host often reach epidemic proportions. Congenital toxoplasmosis is said to account for half of all ovine abortions in England and New Zealand.[4]

In a study of more than 25,000 pregnant women in France, no case of congenital toxoplasmosis was found whenever maternal infection occurred before pregnancy.[5] However, of 118 cases of maternal infection near the time of or during pregnancy, there were 9 abortions or neonatal deaths without confirmation by examination of the fetus, 39 cases of acute congenital toxoplasmosis with 2 deaths, and 28 cases of subclinical infection. The remainder of fetuses were free of infection. Maternal infection in the first 3 months of pregnancy results in more extensive pathogenesis, but transmission to the fetus is more frequent if the maternal infection occurs in the third trimester.

In cases of twins one may have severe symptoms, and the other no overt evidence of infection. In children who survive infection there is often congenital damage to the brain, manifested as mental retardation and retinochoroiditis. Thus toxoplasmosis is a major cause of human birth defects, probably causing more congenital abnormalities in the United States than rubella, herpes, and syphilis combined.

Epidemiology. In the United States the prevalence of chronic, asymptomatic toxoplasmosis is age related, increasing ½% to 1% per year of age in the United States.[21] Although clinical toxoplasmosis usually affects only scattered individuals, small epidemics occur from time to time.[36] For example, several medical students were infected simultaneously by wolfing down undercooked hamburgers between classes.[19] In 1969, at a university in São Paulo, Brazil, 110 persons were diagnosed with acute toxoplasmosis in a 3-month period. Most admitted to eating undercooked meat. Therefore raw meat seems to be an important source of infection. One large sheep or pig might conceivably be the source of an epidemic at any time. One looks at the fad of backyard cooking and Americans' fondness for rare beef and wonders how many cases of toxoplasmosis are thus acquired every day.

Although beef is certainly a potential source of infection, pork and lamb are much more likely to be contaminated. Freezing at $-14°$ C for even a few hours apparently will kill most cysts. To avoid a multitude of parasites, persons who insist on eating undercooked meat would do well to see that it has been hard frozen.

Feral and domestic cats will continue to be a source of infection of humans. Stray cats lead to problems of several kinds and are reservoirs of several diseases; efforts should be made to keep their numbers down. A more difficult problem to resolve is the household pet, the tabby that spends most of its time in a close, symbiotic relationship with its owners. Any cat, no matter how well fed and protected, may be passing oocysts of *Toxoplasma,* although for only a few days after infection. The possibilities are particularly alarming if someone in the house becomes pregnant. Certainly, a woman who knows she is pregnant should never empty the litterbox or clean up after the cat's occasional indiscretion. (Emptying the box every 2 days should help, but since cysts require 1 to 3 days to sporulate, it is better to have the husband do the job.) Having a cat tested for antibodies is impractical, for their presence does not correlate with shedding of oocysts. Also, because children's sandboxes become a haven for neighborhood cats, they should have tightly fitting covers. This also will protect children from larva migrans from hookworms and ascaridoid juveniles. Any soil reservoir of oocysts is a most important source of infection of humans.

Filth flies and cockroaches are capable of carrying *Toxoplasma* oocysts from cat feces to the dinner table.[50] Earthworms may serve to move oocysts from where cats have buried them to the surface of the ground.

Toxoplasma tachyzoites have been isolated in humans from nasal, vaginal, and eye secretions; milk; saliva; urine; seminal fluid; and feces. The role of any of these in spreading infection is unknown, but it seems reasonable that any or all may be involved. Whole blood or leukocyte transfusions and organ transplants are also potential sources of infection, made more important because the recipient may be immunodeficient because of disease or treatment.

FIG. 8-14

Cross section of zoitocyst of *Sarcocystis tenella* in muscle of experimentally infected sheep. (×6600.)

From Dubey, J.P., C.A. Speer, G. Callis, and J.A. Blixt. 1982. Can. J. Zool. 60:2464- 2477.

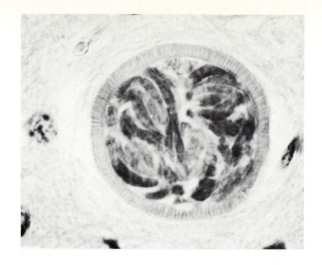

FIG. 8-15

Transmission electron micrograph of *S. tenella* sarcocyst with fully formed wall composed of cytophaneres *(Cw);* note indistinct granular septum *(Se)*, bradyzoites *(Bz)*, metrocytes *(Mc)*, amylopectin *(Am)*, and lipid bodies *(Lb)*. (× 6600.)

From Dubey, J.P., C.A. Speer, G. Callis, and J.A. Blixt, 1982. Can. J. Zool. 60:2464- 2477.

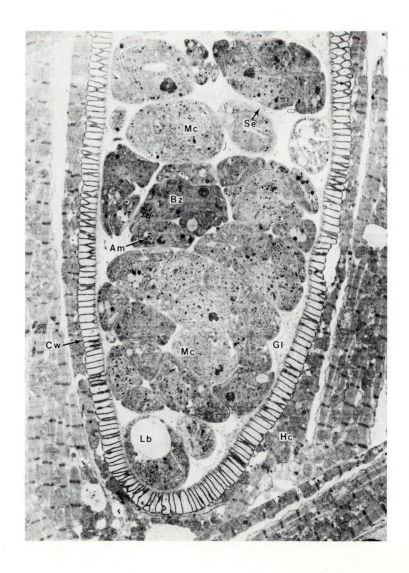

Diagnosis and treatment. Specific diagnosis in humans is based on one or more laboratory tests. Demonstration of the organism at necropsy or biopsy is definitive. Intraperitoneal inoculation of a biopsy of lymph node, liver, or spleen into mice is useful and accurate. Demonstration of specific antibody, using an enzyme-linked, immunosorbent assay (ELISA), is easy and accurate.

Pyrimethamine and sulfonamides given together are widely used drugs against *Toxoplasma*. They act synergistically by blocking the pathway involving *p*-aminobenzoic acid and the folic-folinic acid cycle, respectively. Possible side effects of this treatment are thrombocytopenia and/or leukopenia, but these can be avoided by administration of folinic acid and yeast to the patient. Vertebrates can employ presynthesized folinic acid, whereas *Toxoplasma* cannot.

SARCOCYSTIS SPECIES. *Sarcocystis* spp. have been known from their zoitocysts (Figs. 8-14 and 8-15) in muscle of reptiles, birds, and mammals since the late nineteeth century; but the life cycle remained obscure until 1972 when it was discovered that the bradyzoites would lead to the development of coccidian gametes in cell culture and to oocysts after being fed to cats.[9,43] Since then, it has been found that some species of what was called *Isospora* were in fact stages of *Sarcocystis* in their definitive hosts (for example, *S. bigemina* and *S. hominis*),[27] and what had been considered single species of *Sarcocystis* from particular hosts comprised several species in each (Table 8-1).

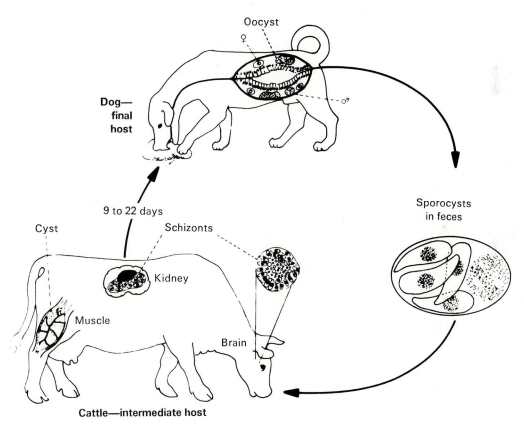

FIG. 8-16

Life cycle of *Sarcocystis cruzi* of cattle with the dog exemplifying the definitive host. Dogs, wolves, coyotes, raccoons, and foxes shed sporulated oocysts or sporocysts in their feces after eating infected bovine musculature. Cattle become infected by ingesting sporocysts from the feces of carnivores. Generalized infection occurs in bovine tissues, and schizonts are formed in many tissues, especially in the kidneys and brain. After schizogonic cycles, cysts are formed in the musculature in 2 months. Current evidence indicates that canines become infected by ingesting only mature cysts. Sporocysts are noninfectious to definitive hosts.

From Dubey, J.P. 1976. J. Am. Vet. Med. Assoc. 169:1061-1078.

TABLE 8-1

Features of *Sarcocystis* spp. of ox, sheep, pig, and horse

Old names → Current names →	Ox *S. fusiformis*			Sheep *S. tenella*		Pig *S. miescheriana*			Horse *S. bertrami*	
	S. cruzi	*S. hirsuta*	*S. hominis*	*S. ovicanis*	*S. tenella*	*S. miescheriana*	*S. porcifelis*	*S. suihominis*	*S. bertrami*	*S. fayeri*
Cyst wall	Thin (0.5 μm)	Thick (6.0 μm) striated	Thick (5.9 μm) striated	Thick, radially striated	Thin	Not known	Not known	Not known	Thin	Thin
Pathogenicity	Pathogenic	Nonpathogenic or slightly pathogenic	Nonpathogenic or slightly pathogenic	Pathogenic	Nonpathogenic	Not known	Pathogenic	Not known	Not known	Not known
Definitive hosts	Dog Coyote Wolf Fox Raccoon	Cat	Man Rhesus monkey Baboon	Dog	Cat	Dog	Cat	Man	Dog	Dog
Prepatent period (days)	9-10	7-9	9-10	8-9	11-14	9-10	5-10	10-17	8	12-15
Sporocysts (μm)*	16 × 11	12 × 8	15 × 9	15 × 10	12 × 8	13 × 10	13 × 8	13 × 9	15 × 10	12 × 8

Species and intermediate hosts

From Dubey, J.P. 1977. In Kreier, J.P., editor. Parasitic protozoa, vol. 3. Academic Press, Inc., New York.
*Mean dimensions (Dubey, J.P. 1976, J. Am. Vet. Med. Assoc. 169:1061-1078).

For example, oocysts of the three following species cannot be distinguished morphologically: *Sarcocystis cruzi* (syn. *S. bovicanis*), *S. tenella* (syn. *S. ovicanis*), and *S. meischeriana* (syn. *S. suicanis*). For a review of the taxonomy of the genus see Levine.[30]

Sarcocystis spp. are obligately heteroxenous, with a herbivorous intermediate host and a carnivorous definitive host (Fig. 8-16). Humans are normally definitive hosts for some species *(S. hominis, S. suihominis),* but zoitocysts of several unidentified species occasionally are found in human muscle[3] (Fig. 8-17). Intermediate hosts of various species include reptiles, birds, small rodents, and hoofed animals.

When sporozoites are released from sporocysts consumed by the intermediate host, they penetrate the intestinal epithelium, are distributed through the body, and invade the endothelial cells of blood vessels in many tissues (Fig. 8-18). There they undergo merogony, and additional merogonous generations may ensue. Zoitocysts (tissue cysts) are then formed in skeletal and cardiac muscle and occasionally the brain. The cysts are also known as **sarcocysts** or **Miescher's tubules,** and some species are large enough to see with the unaided eye. They usually have internal septa and compartments. They are elongate, cylindroid, or spindle shaped, but they may be irregularly shaped. They lie within a muscle fiber, in the same plane as the muscle bundle. The overall size varies, reaching 1 cm in diameter in some cases, but they usually are 1 to 2 mm in diameter and 1 cm or less long. The structure of the cyst wall varies among "species" and in different stages of development of the parasite. In some cases the outer wall is smooth; in others it has an outer layer of fibers, the **cytophaneres,** which radiate out into the muscle (Fig. 8-15). The origin of the cyst wall is controversial: Some authors conclude that it is of host origin; others maintain that it is of parasite origin. It may well be derived from both sources. Two distinct regions can be distinguished in the cyst. The peripheal region is occupied by globular **metrocytes.** After several divisions the metrocytes give rise to the more elongated bradyzoites. The bradyzoites resemble typical coccidian merozoites except that they have a larger number of micronemes; the metrocytes are also structurally similar but lack rhoptries and micronemes. Only bradyzoites are infective to definitive hosts.

When the zoitocyst is consumed by the definitive host, its wall is digested away, and the bradyzoites penetrate the lamina propria of the small intestine. There they undergo gamogony without an intervening merogonic generation. The male gametes penetrate the female gametes, and the oocyst sporulates in the lamina propria. The oocyst wall is thin and is usually broken during passage through the intestine; thus sporo-

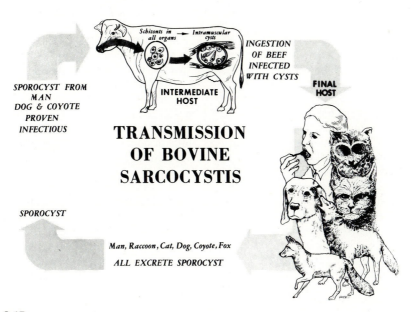

FIG. 8-17

Transmission of bovine *Sarcocystis* spp., illustrating how humans can become infected.

Courtesy Ronald Fayer; drawing by R.B. Ewing.

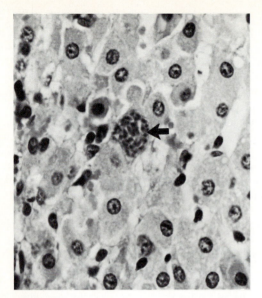

FIG. 8-18

Sarcocystis cruzi meront *(arrow)* in adrenal gland of experimentally infected calf.

From Fayer, R., and A.J. Johnson. 1973. J. Parasitol. 59:1135-1137.

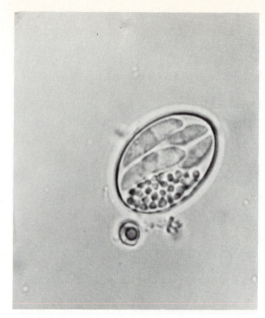

FIG. 8-19

Sporocyst of *Sarcocystis cruzi* from dog. It is 16 by 11 μm.

Photograph by Ronald Fayer.

cysts rather than oocysts are normally passed in the feces (Fig. 8-19). Sporocysts can infect the intermediate hosts but not the definitive hosts.

It is estimated that more than 50% of adult swine, cattle, and sheep are infected with *Sarcocystis* spp.[8] Some of these parasites are nonpathogenic (Table 8-1), but some may cause serious symptoms, which may include loss of appetite, fever, lameness, anemia, weight loss, and abortion in pregnant animals. Heavily infected animals may die. Flies may act as transport hosts.[37] In vitro cultivation may lead the way to immunization and more accurate diagnosis.[45]

FRENKELIA SPECIES. These organisms are poorly known but appear similar structurally and biologically to *Sarcocystis* spp. Very small cysts containing metrocytes and bradyzoites occur in the brains of voles and muskrats. The definitive host of *F. buteonis* is the hawk *Buteo buteo*.

BESNOITIA SPECIES. This is another small group of poorly known species. They are apparently related to *Toxoplasma gondii* but differ in several respects. At least some species are facultatively heteroxenous, and they produce characteristic cysts with a very thick wall, mostly in the connective tissue of their intermediate hosts. The cyst wall contains several flattened giant host cell nuclei. The natural intermediate hosts of *Besnoitia besnoiti* are reported to be cattle and wild

hoofed animals in Africa, the Mediterranean countries, China, and the U.S.S.R. The definitive hosts possibly are cats. The organisms can cause considerable economic losses in cattle, and death may occur in severe infections.

LEVINEIA SPECIES. This genus contains several forms from cats and dogs, which had been considered species of *Isospora* until recently. Based on the discovery that they are facultatively heteroxenous and on the structure of the cysts, Dubey[8] erected the new genus *Levineia* to receive them. Their cysts in the intermediate hosts contain but a single zoite. They show little or no pathogenicity in the definitive hosts and are apparently not pathogenic in the intermediate hosts.[8,33]

The definitive hosts of *L. felis* are cats, whereas rats, mice, hamsters, dogs, and chickens may serve as intermediate hosts. Both cats and the intermediate hosts can become infected by ingestion of oocysts (Fig. 8-20). In cats three generations of meronts and then gametocytes are produced in the epithelium of the small intestine. The prepatent period is 7 to 8 days after oocyst ingestion and 5 to 8 days after ingestion of an infected mouse.[12] The oocysts are passed unsporulated, and the sporozoites develop within 12 hours under optimal conditions. After mice ingest oocysts, cystlike stages are produced in the tissues,

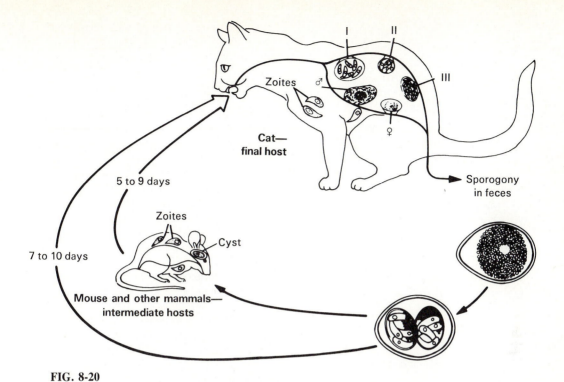

FIG. 8-20

Life cycle of *Levineia felis*. Cats are the definitive hosts, and nonfeline hosts may act as intermediate or transport hosts. Unsporulated oocysts are shed in feline feces. *Levineia felis* can sporulate within 12 hours under optimal conditions. Three generations of schizonts *(I to III)* develop in the epithelium of the small intestine before the formation of gametes and oocysts. The life cycle of *L. rivolta* is essentially similar to that of *L. felis*. In nature *L. felis* and *L. rivolta* are probably transmitted to cats mainly through oocysts.

From Dubey, J.P. 1976. J. Am. Vet. Med. Assoc. 169:1061-1078.

chiefly in the mesenteric lymph nodes. These have a thick wall and contain a single zoite. Extraintestinal stages have been found in cats, although they are usually difficult to demonstrate.

Levineia rivolta also occurs in cats, and *L. canis* (Fig. 8-21) and *L. ohioensis* parasitize dogs. *L. canis* occurs in about 11% of dogs in North America.

Family Cryptosporidiidae. This family contains the single genus *Cryptosporidium,* parasites of the brush borders of epithelia of many tissues of mammals, birds, reptiles and fishes. About 19 species names have been proposed but Levine[29] considers only *C. muris* in mammals, *C. meleagris* in birds, *C. crotali* in reptiles, and *C. nasorum* in fishes to be valid. Lack of host specificity is one of the major characteristics that sets *Cryptosporidium* apart from the rest of the coccidia. For an excellent review of this organism see Tzipori.[46] Until recently cryptosporidiosis was considered to be an infection in animals other than humans. Since 1907, when *Cryptosporidium* was

first described by Tyzzer, until 1975, 15 reports describing *Cryptosporidium* infection in eight species of animals were published. Since 1975, over 100 reports on cryptosporiodiosis have appeared in the literature, most having been published since 1980. Earlier reports have been on enteritis of guinea pigs,[49] turkeys,[16] chickens,[7] calves,[40,48] and lambs,[47] among others. Now it is known to be an opportunistic parasite of humans, both those who are immunodeficient and those who are immunocompetent. As of 1982, the Centers for Disease Control reported 21 cases of cryptosporidiosis in patients with the recently recognized acquired immunodeficiency syndrome (AIDS), and it is an important contributory factor in the deaths of some infected patients who have AIDS.[51]

These coccidians are very small (2 to 6 μm), living in the brush border or just under the free-surface membrane of the host gastrointestinal or respiratory epithelial cell. Oocysts are seen only in the feces, either after treatment with Giemsa's stain or by sugar

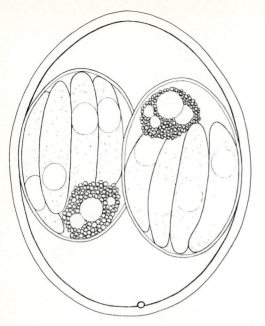

FIG. 8-21

Levineia canis, sporulated oocyst. It measures 35 to 42 by 27 to 33 μm.

From Levine, N.D., and V. Ivens. 1965. J. Parasitol. 51:859-864.

flotation of feces viewed with phase-contrast microscopy. The spherical oocysts are 4 to 5 μm wide, highly refractile, and contain one to eight prominent granules, usually in a small cluster near the margin of the cell. Sporocysts are absent. Each oocyst contains four slender, fusiform sporozoites.

Biology. When swallowed, the sporozoites excyst in the intestine and invade either epithelial cells of the respiratory system or the intestine (from the ileum to the colon) of the next host (Fig. 8-22). The meronts are about 7 μm wide and produce eight banana-shaped merozoites and a small residuum. Microgamonts produce 16 rod-shaped, nonflagellated microgametes that are 1.5 to 2 μm long. Oocysts are passed as early as 5 days after infection. Improved methods for purification of oocysts and sporozoites have been described.[1,20]

Epidemiology. Infection is by fecal-oral contamination. A number of animals can serve as reservoirs of infection. Current et al. experimentally infected kittens, puppies, and goats with oocysts from an immunodeficient person.[6] They also infected calves and mice using oocysts from infected calves and humans. Finally, they diagnosed 12 infected immunocompetent persons who worked closely with calves that were infected with *Cryptosporidium*. Thus cryptosporidiosis

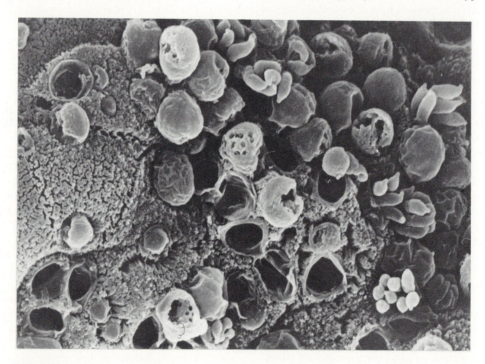

FIG. 8-22

Oocysts of *Cryptosporidium* in various stages of development in intestinal epithelium. The slender, elongated bodies are emerging sporozoites.

SEM by S. Tzipori.

is a zoonosis and in fact may be a fairly common cause of short-term diarrhea in the population at large.

Pathogenesis and treatment. In patients with AIDS, the parasites cause profuse, watery diarrhea lasting for several months. Bowel-movement frequency ranges from 6 to 25 per day, and the maximal stool volume ranges from 1 to 17 liters per day. Despite intensive trials in humans and animals, no effective drug treatment for cryptosporidiosis has been found. Some efficacy has been demonstrated by the use of spiromycin.[51] Fortunately the infection is much less severe in immunocompetent patients, with no symptoms in some and with a self-limiting diarrhea and abdominal cramps lasting from 1 to 10 days in others. It is a common cause of diarrhea (scours) in domestic animals such as calves, lambs, and others and may become a major problem in the cattle industry.

PNEUMOCYSTIS CARINII. The taxonomic position of this organism is undetermined. It is considered to be a fungus by some workers and a protozoan by others. It has ultrastructural properties similar to some fungi and stains with certain fungal stains, such as methenamine silver. On the other hand it has other structural characteristics resembling *Toxoplasma* and *Plasmodium*[53] and is sensitive to antiprotozoal agents such as pentamidine isethionate, pyrimethamine, sulfadiazine, and trimethoprimsulfamethoxazole.[14] Its membrane properties, including formation of small pseudopodia, resemble the amebas. We place *Pneumocystis* in this chapter only for the sake of convenience.

Pneumocystis carinii is an obligate, extracellular parasite that causes interstitial plasma cell pneumonia, especially in the immunosuppressed host, either human or rodent. It occurs commonly in humans in all age groups and is particularly important in the elderly and in infants and children with primary immune deficiency disorders; it is also critical in patients receiving cytotoxic or immunosuppressive drugs for lymphoreticular cancers, organ transplantation, and a variety of other disorders. Patients with AIDS are particularly susceptible, and pneumocystis pneumonia is a major cause of death in that population. In properly prepared laboratory rats the organism appears to respond identically as in humans, and thus a laboratory model for experimentation is available. In nature the organism is widespread in mammals. Many human infections may be acquired from pets.

Morphology and biology. *Pneumocystis carinii* assumes three morphological forms while in the lung: the **tropic stage,** the **precystic stage,** and the **cystic stage.** The trophic stage is pleomorphic, 1 to 5 μm wide, and has small filopodia. The precystic stage (Fig. 8-23) is oval, with few filopodia, has a clump of mitochondria in its center, and has a nucleus that is

seldom seen. The cyst stage (Fig. 8-24) is spherical, has a thick cell membrane, and has one to four nuclei. There are also some cysts that are crescent shaped (Fig. 8-25). Cysts contain several uninuclear, pear-shaped, intracystic bodies about 1 to 2 μm long, which sometimes are called sporozoites. The bodies often are in groups of eight and form a rosette.

All three stages live in the interstitial tissues of the lungs and are not normally found in the alveoli. Their mode of reproduction is unknown, although there is evidence of both sexual and asexual reproduction.[39] Occasionally they do break into air spaces, for they are abundant in pulmonary exudate of an infected person. Human-to-human transmission probably is by aerosol droplets and direct contact. Congenital infection is possible, as it has been found in stillborn infants, newborn germ-free rats, and 3-day-old children.[39]

Pathogenesis. In the infected lung the alveolar septa become thickened and infiltrated with plasma

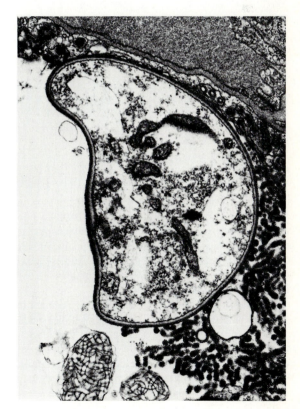

FIG. 8-23

Transmission electron micrograph of a precyst *Pneumocystis carinii.* Note tightly packed outer layer of cell membrane.

From Yoneda, K., P.D. Walzer, C.S. Richey, and M.G. Birk. 1982. Exp. Parasitol. 53:68-76.

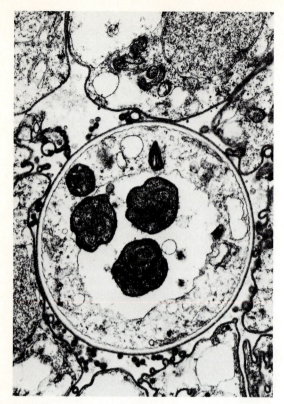

FIG. 8-24

Transmission electron micrograph of a *Pneumocystis carinii* cyst. Note the intracystic bodies.

From Yoneda, K., P.D. Walzer, C.S. Richey, and M.G. Birk. 1982. Exp. Parasitol. 53:68-76.

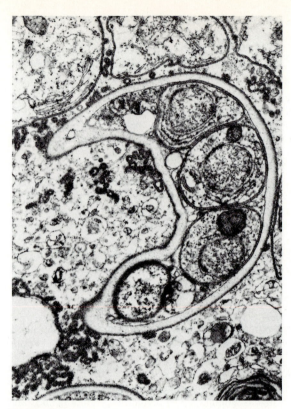

FIG. 8-25

Transmission electron micrograph of a crescent-shaped cyst of *Pneumocystis carinii*. Note intracystic bodies.

From Yoneda, K., P.D. Walzer, C.S. Richey, and M.G. Birk. 1982. Exp. Parasitol. 53:68-76.

cells. The alveolar epithelium becomes partly desquamated, and the alveoli fill with foamy exudate containing parasites. The disease, **interstitial plasma cell pneumonitis,** has a rapid onset associated with fever, cough, rapid breathing, and cyanosis. Death is caused by asphyxia. The mortality rate with this disease is virtually 100% in untreated patients.

Diagnosis. Only by demonstrating the organism by special staining can a positive diagnosis be made. Lung biopsy or bronchial lavage or brushing will yield infected material. Toluidine blue or methenamine silver stains are reported to be reliable, and the Gram-Weigert stain is accurate in demonstrating the cysts.

Treatment. Even with treatment the mortality rate is high in patients with this disease because the immune system is deficient to begin with. The treatment of choice is the combination of trimethoprim-sulfamethoxazole (Bactrim). Equally effective but highly toxic is pentamidine isethionate.

REFERENCES

1. Arrowood, M.J., and C.R. Sterling. 1987. Isolation of *Cryptosporidium* oocysts and sporozoites using discontinuous sucrose and isopycnic percoll gradients. J. Parasitol. 73:314-319.
2. Baker, J.R. 1987. The *Toxoplasma* tangle. Parasitol. Today 3:103-105.
3. Beaver, P.C., R.K. Gadgil, and P. Morera, 1979. *Sarcocystis* in man: a review and report of five cases. Am. J. Trop. Med. Hyg. 28:819-844.
4. Beverly, J.K.A., W.A. Watson, and J.B. Spence. 1971. The pathology of the foetus in ovine abortion due to toxoplasmosis. Vet. Rec. 88:174-178.
5. Couvreur, J. 1971. Prospective study in pregnant women with a special reference to the outcome of the foetus. In Hentsch, D., editor. Toxoplasmosis. Hans Huber Medical Publisher, Bern, pp. 119-135.
6. Current, W.L., N.C. Reese, J.V. Ernst, W.S. Bailey, M.B. Heyman, and W.M. Weinstein. 1983. Human cryptosporidiosis in immunocompetent and immunodeficient persons. Studies of an outbreak and experimental transmission. N. Engl. J. Med. 308:1252-1257.

7. Dhillon, A.S., H.L. Thacker, V. Dietzel, and R.W. Winterfield. 1981. Respiratory cryptosporidiosis in broiler chickens. Avian Dis. 25:747-751.

8. Dubey, J.P. 1977. *Toxoplasma, Hammondia, Besnoitia, Sarcocystis,* and other tissue cyst-forming coccidia of man and animals. In Kreier, J.P., editor. Parasitic protozoa, vol. III. Academic Press, Inc., New York.

9. Fayer, R. 1972. Gametogony of *Sarcocystis* sp. in cell culture. Science 175:65-67.

10. Forthal, D.N., and S.S. Guest. 1984. *Isospora belli* enteritis in three homosexual men. Am. J. Trop. Med. Hyg. 33:1060-1064.

11. Frenkel, J.K. 1973. Toxoplasmosis: parasite life cycles, pathology, and immunology. In Hammond, D.M., and P.L. Long, editors. The Coccidia. *Eimeria, Isospora, Toxoplasma,* and related genera. University Park Press, Baltimore, pp. 343-410.

12. Frenkel, J.K., and J.P. Dubey. 1972. Rodents as transport hosts for cat coccidia, *Isopora felis* and *I. rivolta.* J. Infect. Dis. 125:69-72.

13. Frenkel, J.K., H. Mehlorn, and A.O. Heydorn. 1987. Beyond the oocyst: over the molehills and mountains of coccidialand. Parasitol. Today 3:250-251.

14. Gajdusek, D.C. 1957. *Pneumocystis carinii*—etiologic agent of interstitial plasma cell pneumonia of premature and young infants. Pediatrics 19:543-545.

15. Grell, K.G. 1973. Protozoology. Springer-Verlag, Heidelberg.

16. Hoerr, F.J., F.M. Ranck, and T.F. Hastings. 1978. Respiratory cryptosporidiosis in turkeys. J. Am. Vet. Med. Assoc. 173:1591-1593.

17. Hughes, H.P.A. 1985. Toxoplasmosis—a neglected disease. Parasitol. Today 1:41- 44.

18. Jacobs, L. 1973. New knowledge of *Toxoplasma* and toxoplasmosis. In Dawes, B., editor. Advances in parasitology, vol. 11. Academic Press, Inc., New York, pp. 631-669.

19. Kean, B.H., A.C. Kimball, and W.N. Christenson, 1969. An epidemic of acute toxoplasmosis. J.A.M.A. 208:1002-1004.

20. Kilani, R.T., and L. Sekla. 1987. Purification of *Cryptosporidium* oocysts and sporozoites by cesium chloride and percoll reagents. Am. J. Trop. Med. Hyg. 36:505-508.

21. Krick, J.A., and J.S. Remington. 1978. Toxoplasmosis in the adult. N. Engl. J. Med. 298:550-553.

22. Levine, N.D. 1962. Protozoology today. J. Protozool. 9:1-6.

23. Levine, N.D. 1963. Coccidiosis. Ann. Rev. Microbiol. 17:179-198.

24. Levine, N.D. 1970. Taxonomy of the Sporozoa. J. Parasitol. 56(sect. II, part 1):208-209.

25. Levine, N.D. 1973a. Introduction, history and taxonomy. In Hammond, D.M., and Long, P.L., editors. The Coccidia. *Eimeria, Isospora, Toxoplasma,* and related genera. University Park Press, Baltimore, pp. 1-22.

26. Levine, N.D. 1973b. Protozoan parasites of domestic animals and of man, ed. 2. Burgess Publishing Co., Minneapolis.

27. Levine, N.D. 1977. Nomenclature of *Sarcocystis* in the ox and sheep and of fecal coccidia of the dog and cat. J. Parasitol. 63:36-51.

28. Levine, N.D. 1982. The genus *Atoxoplasma* (Protozoa, Apicomplexa). J. Parasitol. 68:719-723.

29. Levine, N.D. 1984. Taxonomy and review of the coccidian genus *Cryptosporidium* (Protozoa, Apicomplexa). J. Protozool. 31:94-98.

30. Levine, N.D. 1986. The taxonomy of *Sarcocystis* (Protozoa, Apicomplexa) species. J. Parasitol. 72:372-382.

31. Levine, N.D. 1987. Whatever became of *Isopora bigemina?* Parasitol. Today 3:101-103.

32. Levine, N.D., and V. Ivens. 1965. The coccidian parasites (Protozoa, Sporozoa) of rodents. Ill. Biol. Monogr. 33. University of Illinois Press, Urbana.

33. Long, P.L. 1973. Pathology and pathogenicity of coccidial infections. In Hammond, D.L., and P.L. Long, editors. The Coccidia. *Eimeria, Isospora, Toxoplasma,* and related genera. University Park Press, Baltimore.

34. Long, P.L., and T.K. Jeffers. 1986. Control of chicken coccidiosis. Parasitol. Today 2:236-240.

35. Lunde, M.N., and L. Jacobs. 1983. Antigenic differences between endozoites and cystozoites of *Toxoplasma gondii.* J. Parasitol. 69:806-808.

36. Maddison, S.E., et al. 1979. Lymphocyte proliferative responsiveness in 31 patients after an outbreak of toxoplasmosis. Am. J. Trop. Med. Hyg. 28:955-961.

37. Markus, M.B. 1980. Flies as natural transport hosts of *Sarcocystis* and other coccidia. J. Parasitol. 66:361-362.

38. Marquardt, W.C. 1973. Host and site specificity in the Coccidia. In Hammond, D.M., and P.L. Long, editors. The Coccidia. *Eimeria, Isospora, Toxoplasma* and related genera. University Park Press, Baltimore. pp. 23-43.

39. Matsumoto, Y., and Y. Yoshida. 1986. Advances in *Pneumocystis* biology. Parasitol. Today 2:137-142.

40. Polenz, J., H.W. Moon, N.F. Cheville, and W.J. Bemrick. 1978. Cryptosporidiosis as a probable factor in neonatal diarrhea of calves. J. Am. Vet. Med. Assoc. 172:452-457.

41. Remington, J.S., and G. Desmonts. 1976. In Remington, J.S., and J.O. Klein, editors. Infectious diseases of the fetus and newborn infant. W.B. Saunders, Philadelphia. pp. 191-332.

42. Roberts, W.L., C.A. Speer, and D.M. Hammond. 1970. Electron and light microscope studies of the oocyst walls, sporocysts, and encysting sporozoites of *Eimeria callospermophili* and *E. larimerensis.* J. Parasitol. 56:918-926.

43. Rommel, M., A.O. Heydorn, and F. Gruber. 1972. Beiträge zum Lebenzyklus der Sarkosporidien. I. Die Sporozyste von *S. tenella* in der Fäzes der Katze. Berl. Münch. Tierärztl. Wochenschr. 85:101-105.

44. Sinden, R.E. 1985. A cell biologist's view of host cell recognition and invasion by malarial parasites. Trans. R. Soc. Trop. Med. Hyg. 79:598-605.

45. Speer, C.A., and D.E. Burgess. 1987. *In vitro* cultivation of *Sarcocystis* merozoites. Parasitol. Today 3:2-3.

46. Tzipori, S. 1985. *Cryptosporidium:* notes on epidemiology and pathogenesis. Parasitol. Today 1:159-165.

47. Tzipori, S., K.W. Angus, E.W. Gray, I. Campbell, and F. Allen. 1981. Diarrhea in lambs experimentally infected with *Cryptosporidium* isolated from calves. Am. J. Vet. Res. 42:1400-1404.

48. Tzipori, S., I. Campbell, D. Sherwood, and D.R. Snodgrass. 1980. An outbreak of calf diarrhea attributed to cryptosporidial infection. Vet. Rec. 107:579-580.

49. Vetterling, J.M., H.R. Jervis, T.G. Merrill, and H. Sprinz. 1971. *Cryptosporidium wrairi* sp.n. From the guinea pig *Cavia porcellus,* with an emendation of the genus. J. Protozool. 18:243-247.

50. Wallace, G.D. 1971. Experimental transmission of *Toxoplasma gondii* by filth flies. Am. J. Trop. Med. Hyg. 20:411-413.

51. Whiteside, M.E., J.S. Barkin, R.G. May, S.D. Weiss, M.A. Fischl, and C.L. MacLeod. 1984. Enteric coccidiosis among patients with the acquired immunodeficiency syndrome. Am. J. Trop. Med. Hyg. 33:1065-1072.

52. Wilson, P.A.G., and D. Fairbairn. 1961. Biochemistry of sporulation in oocysts of *Eimeria acervulina*. J. Protozool. 8:410-416.

53. Yoneda, K., P.D. Walzer, C.S. Richey, and M.G. Birk. 1982. *Pneumocystis carinii:* freeze-fracture study of stages of the organism. Exp. Parasitol. 53:68-76.

SUGGESTED READINGS

Brandberg, L.L., S.B. Goldberg, and W.C. Breidenbach. 1970. Human coccidiosis—a possible cause of malabsorption. The life cycle in small-bowel mucosal biopsies as a diagnostic feature. N. Engl. J. Med. 283:1306-1313. (A study of six cases and a review of previous reports. Endogenous stages are illustrated for the first time.)

Desmonts, G., and J. Couveur. 1974. Congenital toxoplasmosis. N. Engl. J. Med. 290:1110-1116. (A study of 378 pregnancies.)

Feldman, H.A. 1974. Congenital toxoplasmosis, at long last . . . N. Engl. J. Med. 290:1138-1140. (A short summary of the discovery of congenital toxoplasmosis.)

Long, P.L., editor. 1982. The biology of the Coccidia. University Park Press, Baltimore, Md.

Ma, P. 1987. Protozoa and acquired immune deficiency syndrome (AIDS). ATCC Quart. Newsletter 7:1-2, 7.

Markus, M.B. 1978. *Sarcocystis* and sarcocystosis in domestic animals and man. Adv. Vet. Sci. Comp. Med. 22:159-193.

Ryley, J.F. 1980. Recent developments in coccidian biology: where do we go from here? Parasitology 80:189-209.

CHAPTER 9

PHYLUM APICOMPLEXA: MALARIA AND PIROPLASMS

SUBORDER HAEMOSPORINA

This suborder contains the family Plasmodiidae, including the genera *Plasmodium, Haemoproteus,* and *Leucocytozoon,* which are the malaria and malaria-like organisms.[43] When in host cells, *Plasmodium* and *Haemoproteus* usually produce a pigment called **hemozoin** from host hemoglobin, distinguishing them from the closely related *Leucocytozoon.* Studies with the electron microscope have revealed the basic similarity of these parasites to the coccidia, except that they lack a conoid. Syzygy is absent, and the macrogametocyte and microgametocyte develop independently. The microgametocyte produces about eight flagellated gametes. The zygote is motile and is called an **ookinete;** the sporozoites are not enclosed within sporocysts. They are heteroxenous, with merozoites produced in the vertebrate host and sporozoites developing in the invertebrate host. It is possible that these parasites evolved from the coccidia of vertebrates rather than invertebrates, with mites or other bloodsuckers initiating the cycle in arthropods.

Although most species of Haemosporina are parasites of wild animals and appear to cause little harm in most cases, a few cause diseases that are among the worst scourges of humankind. Indeed, malaria has played an important part in the rise and fall of nations and has killed untold millions the world over. John F. Kennedy said in 1962:

For centuries, malaria has outranked warfare as a source of human suffering. Over the past generation it has killed millions of human beings and sapped the strength of hundreds of millions more. It continues to be a heavy drag on man's efforts to advance his agriculture and industry.[35]

Despite the combined efforts of 90 countries to eradicate malaria, it remains the most important disease in the world today in terms of lives lost and economic burden. Progress has been made, however. Between 1948 and 1965 the number of cases was cut from a worldwide total of 350 million to fewer than 100 million. In some countries, such as the United States, eradication of endemic malaria is complete.

At this writing approximately 1472 million persons live in malarious areas of the world. This unprotected population lives in countries without the administrative, financial, and human resources necessary for control.

Genus *Plasmodium*
■ History

History is, after all, a review of past experiences which influence present events.

ELVIO H. SADUN

Because malaria is still the most important disease of humankind, we think it is of value to relate the history of its conquest in considerable detail.

Malaria has been known since antiquity, with recognizable descriptions of the disease recorded in various Egyptian papyri. The Ebers papyrus (1550 BC) mentions fevers, splenomegaly, and the use of oil of the Balamites tree as a mosquito repellent. Hieroglyphs on the walls of the ancient Temple of Denderah in Egypt describe an intermittent fever following the flooding of the Nile.[25] Hippocrates studied medicine in Egypt and clearly described quotidian, tertian, and quartan fevers with splenomegaly. He believed that bile was the cause of the fevers. Greek states built beautiful cities in the lowlands only to see them devastated by the disease, and wealthy Greeks and Romans traditionally summered in the highlands to escape the heat, mosquitoes, and mysterious fevers. Herodotus (c. 500-424 BC) states that Egyptian fishermen slept with their nets arranged around their beds so that mosquitoes could not reach them. Homer also noted that malaria is most prevalent in the later summer: in the *Iliad* (XXII, 31) we read ". . . like that star which comes on in the autumn . . ., the star they give the name of Orion's dog which is brightest among the stars, and yet is wrought as a sign of evil and brings on the great fever for unfortunate mortals." Medieval England saw crusaders falter and fail as they encountered malaria. As had happened before and has happened since, malaria killed more warriors than did warfare. When Europeans imported slaves and returned their colonial armies to their continent, they brought malaria with them, increasing the concentration of the disease with devastating results.

Throughout history a connection between swamps and fevers has been recognized. It was commonly concluded that the disease was contracted by breathing "bad air," or "malaria." This belief flourished until near the end of the nineteenth century. Another name for the disease, "paludism" (marsh disease), is still in common use in the world.

There has been much speculation as to whether malaria existed in the western hemisphere before the Spanish conquest. It seems inconceivable that the great Olmec and Mayan civilizations could have developed in regions that are now highly malarious. The Spanish Conquistadores made no mention of fevers during the early years of the conquest, and in fact they holidayed in Guayaquil and the coastal area near Veracruz, regions that soon after became very unhealthy because of malaria. Even Balboa, while traversing the Isthmus of Panama, did not mention any encounters with malaria. It therefore seems likely that malaria was introduced into the New World by the Spaniards and their African slaves. However, nagging evidence that Africans reached South America during pre-Columbian times suggests that, while improbable, it is not impossible that malaria existed in localized areas of the continent before the Spanish conquest,[29] could have been brought from Oceania or from Asia by way of the Bering Strait, or could have been introduced by the Vikings.

No progress was made in the etiology of malaria until 1847, when Meckel observed black pigment granules in the blood and spleen of a patient who died of the disease. He even stated that the granules lay within protoplasmic masses. Was he the first to actually see the parasite? In 1879 Afanasiev suggested that the granules caused the disease.

During the next 30 years physicians and scientists of high stature searched diligently for the cause of the disease and its means of transmission to people. It remained for two obscure army medical officers, working in their spare time, under primitive and difficult circumstances, to make these cardinal discoveries.

Most research was directed toward finding an infective organism in water or in the air. Many false hopes were generated when a previously unknown ameba or fungus was discovered, and, when *Bacillus malariae* was declared to be the causative organism by Edwin

Klebs (German) and equally prestigious Corroado Tomasi-Crudelli (Italian), few doubted the truth of their momentous discovery.

Meanwhile, in North Africa, far from academic circles, a young French Army physician named Louis Alphonse Laveran decided that the mysterious pigment in his malarious patients would be a good starting point for further research. He observed the pigment not only free in the plasma but also within leukocytes, and he saw clear bodies within erythrocytes. As the hyaline bodies of irregular shape grew, he saw the erythrocytes grow pale and pigment form within them. He little doubted the parasitic nature of the organisms he saw. Then, on November 6, 1880, he witnessed one of the most dramatic events in protozoology: the formation of male gametes by the process of exflagellation. He quickly wrote of his discovery, reporting on November 23, 1880, to the Academy of Medicine of Paris, where much skepticism was offered his report. Most scientists were loath to abandon the Klebs/Tomasi-Crudelli bacillus in favor of a protozoan that an army physician claimed to have discovered in Algeria. His "organisms" were assumed to be degenerating blood cells. In addition to the prestige of Klebs and Tomasi-Crudelli, other factors influenced this attitude.[76] Opportunities to study the malarial fevers in the academic medical centers of Europe were limited, and there were real limitations and difficulties in interpreting microscopic observations. Technical progress in microscopy was rapid in the decade of 1880-1890, however, and Ettore Marchiafava (a favorite student of Tomasi-Crudelli) and Angelo Celli (his longtime collaborator), who originally favored the bacillus hypothesis, became convinced that Laveran was correct. More strong support came in 1885, when Camillo Golgi differentiated between species of *Plasmodium* and demonstrated the synchronism of the parasite in relation to paroxysm.

Laveran had accurately described the male and female gametes, the trophozoite, and the schizont while working with a poor, low-power microscope and with unstained preparations. By 1890 several scientists in different parts of the world verified his findings. In 1891 Romanovsky, in Russia, developed a new method of staining blood smears based on methylene blue and eosin. Modifications of his stain remain in wide use.

The mode of transmission of malaria was, however, still unknown. Although ideas were rampant ("bad night air" was still a popular candidate), few were as well thought out as that of Patrick Manson, who favored the possibility of transmission by mosquitoes. True, he was conditioned by the proof of mosquitoes as vectors of filariasis, which gave him some insight.

Surgeon-Major Ronald Ross was 38 years old when he met Manson for the first time, while on leave from the Indian Medical Service. Finding in Ross a man who was interested in malaria and who could test his ideas for him, Manson lost no time in convincing Ross that malaria was caused by a protozoan parasite. For the next several years, in India, Ross worked during every spare minute, searching for the mosquito stages of malaria that he was certain existed. Dissecting mosquitoes at random and also after allowing them to feed on malarious patients, he found many parasites, but none of them proved to be what he searched for. During this time he had a steady correspondence with Manson, who encouraged him and brought his discourses to the learned societies of England. Ross left a wonderful record of his moods of excitement, frustration, disappointment, and triumph; in addition, he was a sensitive poet. His journals also contain long quotations from Manson's letters written to him at that time.

Ross' first significant observation was that exflagellation normally occurs in the stomach of a mosquito, rather than in the blood as was currently thought. At this time he was posted to Bangalore to help fight a cholera epidemic, the first in a series of frustrating interruptions by superiors who had no concept of the importance of the work Ross was doing in his spare time. Returning from Bangalore, he continued the search for further development of the parasite within the mosquito. Failing this, he concluded that he had been working with the wrong kinds of mosquitoes (*Culex* and *Stegomyia*). He tried other kinds and was led astray time after time by gregarines and other mosquito parasites, each of which had to be eliminated as possible malaria organisms by laborious experimentation. After 2 years of work, which his superior officers ignored as harmless lunacy, he seemed to have reached an impasse. He was eligible for retirement soon and was determined to try "one more desperate effort to solve the Great Problem." He toiled far into the nights, dissecting mosquitoes, in a hot little office. He could not use the overhead fan lest it blow his mosquitoes away, and swarms of gnats and mosquitoes avenged themselves "for the death of their friends." At last, late in the night of August 16, 1897, he dissected some "dappled-winged" mosquitoes (*Anopheles* spp.) that had fed on a malaria patient, and he found some pigmented, spherical bodies in the walls of the insects' stomachs. The next day he dissected his last remaining specimen and found the spheroid cells had grown. They were most certainly the malaria parasites! That night he penned in a notebook:

This day designing God
Hath put into my hand
A wondrous thing. And God
Be praised. At this command
I have found they secret deeds
Oh million-murdering Death

I know that this little thing
A myriad men will save—
Oh death where is they sting?
Thy victory oh Grave?

He reported his discovery to Manson and immediately set about breeding the correct kind of mosquito in preparation for the first step of transmitting the disease from the insect to humans. He was immediately posted to Bombay, where he could do no further research on human malaria but found similar organisms (*Plasmodium relictum*) in birds. He repeated his feeding experiments with mosquitoes and found similar parasites, when they fed on infected birds. He also found that the spheroid bodies ruptured, releasing thousands of tiny bodies that dispersed throughout the insect's body, including into the salivary glands. Through Manson he reported to the world how malaria is transmitted by mosquitoes. It remained only for a single experiment to prove the transmission to humans. Ross never did it. The authorities were so impressed they ordered Ross to work out the biology of kala-azar in another part of India. This seems to have broken his spirit, for he never really tried again to finish the study of malaria. The concentration had made him ill, his eyes were bothering him, and his microscope had rusted tight from his sweat. Anyway, he was a physician, not a zoologist, and was only interested in learning how to prevent the disease, not in the finer points of the parasite's biology. This he considered to be done, and he retired from the Army. He was awarded the Nobel Prize in Medicine in 1902 and was knighted in 1911. He died in 1932 after a distinguished postarmy career in education and research.

Unfortunately, the history of malariology is tarnished by strife and bitterness. Several persons who were working on the life cycle of the parasite claimed credit for the discovery that pointed to the means of control for malaria. Italian, German, and American scientists all made important contributions to the solution of the problem of malaria transmission. Several of these, including Ross, spent a good portion of their lives quibbling about priorities in the discoveries. Manson-Bahr[49] and Harrison[27] give fascinating accounts of the personalities of the men who conquered the life cycle of malaria. Credit for completing this life cycle should go to Amigo Bignami and Giovanni

Grassi, who experimentally transmitted the malaria parasite from mosquito to human in 1898. Although the life cycle of malaria was thought to be known after Ross' work, it remained to be found that stages occur in the liver.

In the early twentieth century the cycle was thought to progress from the blood to the mosquito back to the blood. This concept gained support from the published work of Fritz Schaudin, who claimed to have seen sporozoites penetrating red blood cells and transforming into trophozoites.

Schaudin's work remained unchallenged until World War I, when a fact began to emerge that could not be explained by the direct cycle between mosquito and blood. Quinine is a well-known antimalarial drug. Its effect is only on the erythrocytic forms. However, it was found that soldiers treated with the drug were cured, that is, there were no blood parasites, but when the treatment stopped and the patients moved to a nonmalarious area, parasites would return to the blood at certain time intervals.

In 1917 Julius von Wagner-Jauregg discovered that the high fevers of malaria could be used to treat neurosyphilis. From this work two additional facts emerged. When a patient was infected by injection with parasitized blood, the incubation period could be shortened or lengthened by changing the number of parasites injected. However, the bite of 1 or 200 infected mosquitoes did not alter the incubation period.

In 1938 James and Tate discovered the exoerythrocytic stages of *P. gallinaceum*. After this discovery, large-scale work began in order to find the exoerythrocytic stages of human malaria parasites. Finally in 1948 Shortt and Garnham demonstrated the exoerythrocytic stages of *P. cynomolgi* in monkeys and *P. vivax* in humans.

These historical notes cannot be concluded without mention of a man who managed to apply these early discoveries for the immense benefit of his country and humankind: William C. Gorgas. Gorgas was the medical officer placed in charge of the Sanitation Department of the Canal Zone, when the United States undertook to build the Panama Canal; were it not for his mosquito control measures, malaria and yellow fever would have defeated American attempts to build the Canal, just as they had the French. In July 1906 the malaria rate in the Canal Zone was 1263 hospital admissions per 1000 population![68] Gorgas' work reduced the rate to 76 hospital admissions per 1000 in 1913, saving his country $80 million and the lives of 71,000 fellow humans. Gorgas became a hero in his lifetime: the President made him Surgeon General, Congress promoted him, Oxford University made him an honorary Doctor of Science, and the King of En-

gland made him a knight. Sir William Osler stated, "There is nothing to match the work of Gorgas in the history of human achievement." It is a sad commentary on our cultural memory that the name of Gorgas is now known by so few, whereas we find it easy to remember the names of generals and tyrants who caused great bloodshed.

For students interested in more details about humanity's fight against malaria, the book by Harrison[27] is likely to become a classic and is highly recommended.

■ Life cycle and general morphology (Fig. 9-1)

Following is a general account of the development and structure of malaria parasites, without reference to particular species. Specific morphological details for each species will be found in the pages to follow. *Plasmodium* spp. require two types of hosts: an invertebrate (mosquito) and a vertebrate (reptile, bird, or mammal). Technically the invertebrate can be considered the definitive host because sexual reproduction occurs there. Asexual reproduction takes place in the tissues of a vertebrate, which thus can be called the intermediate host. However, it has been pointed out that the gametocytes actually form in the blood of the vertebrate, and fertilization occurs while still in this medium in the stomach of the mosquito. By this reasoning the vertebrate is the definitive host.[13] We should also observe that *Plasmodium* spp. were probably derived from an ancestral coccidian whose asexual and sexual reproduction took place in the same (presumably vertebrate) host.

Vertebrate phases. When an infected mosquito takes blood from a vertebrate, she injects saliva containing tiny, elongate sporozoites into the bloodstream. The sporozoite basically is similar in morphology to that of *Eimeria* and other coccidia. It is about 10 to 15 μm long by 1 μm in diameter and has a pellicle composed of a thin outer membrane, a doubled inner membrane, and a layer of subpellicular microtubules. There are three polar rings. The rhoptries are long, extending to the midportion of the organism, and much of the rest of the anterior cytoplasm is taken up by the micronemes. An apparently nonfunctional cytostome is present, and there is a mitochondrion in the posterior end of the sporozoite.[2]

After being injected into the bloodstream, the sporozoites quickly disappear (within an hour) from the circulating blood. Their immediate fate was a great mystery until the mid-1940s, when it was shown that within 1 or 2 days they enter the parenchyma of the liver or other internal organ, depending on the species of *Plasmodium*. Where they are the first 24 hours still is unknown. Entry into the liver initiates a series

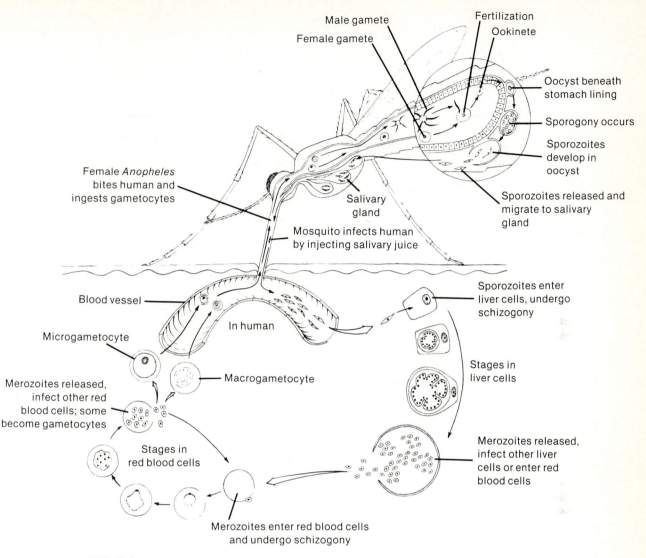

FIG. 9-1

Life cycle of *Plasmodium vivax*.

From Hickman, C.P., Jr., L.S. Roberts, and F.M. Hickman. 1988. Integrated principles of zoology, ed. 8. The C.V. Mosby Co., St. Louis.

of asexual reproductions known as the **preerythrocytic cycle** or **primary exoerythrocytic schizogony,** often abbreviated as the **PE** or **EE** stage. Once within a hepatic cell, the parasite metamorphoses into a feeding trophozoite. The organelles of the apical complex disappear, and the trophozoite feeds on the cytoplasm of the host cell by way of the cytostome and, in the species in mammals, by pinocytosis.

After about a week, depending on the species, the trophozoite is mature and begins schizogony. Numer-

ous daughter nuclei are first formed, transforming the parasite into a schizont (Fig. 9-2), also known as a **cryptozoite.** During the nuclear divisions the nuclear membranes persist, and the microtubular spindle fibers are formed within the nucleus. The mitochondrion becomes larger during the growth of the trophozoite, forms buds, and then breaks up into many mitochondria. Elements of the apical complex form subjacent to the outer membrane, and schizogony proceeds as previously described. The merozoites thus formed af-

143

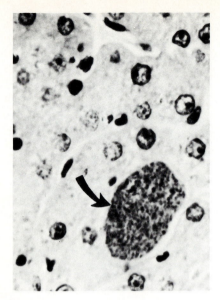

FIG. 9-2

Preerythrocytic schizont of *Plasmodium (arrow)* in liver tissue.

Photograph by Peter Diffley.

ter cytokinesis are referred to in the EE stage as **metacryptozoites.** The merozoites are much shorter than sporozoites—2.5 μm long by 1.5 μm in diameter— and have small, teardrop-shaped rhoptries and small, oval micronemes.

What happens next is a subject of lively debate. For many years it was believed that the merozoites entered new hepatocytes to form new schizonts and then merozoites, at least in the species of *Plasmodium* that are capable of causing a relapse.[73] However, as early as 1913 it was postulated that some sporozoites become dormant for an indefinite time after entering the body.[6] Such dormant cells, now called **hypnozoites,** apparently have been demonstrated.[40,41] They are discussed under relapse in malaria (p. 152).

Eventually, merozoites leave liver cells to penetrate erythrocytes in the blood, initiating the **erythrocytic cycle.** Some of the merozoites may be phagocytized by Kupffer cells in the liver, which may be an important host defense mechanism.[80] On entry into an erythrocyte, the merozoite again transforms into a trophozoite. The host cytoplasm ingested by the trophozoite forms a large food vacuole, giving the young *Plasmodium* the appearance of a ring of cytoplasm with the nucleus conspicuously displayed at one edge (Plate, 1, *1* and *2*). The distinctiveness of the "signet-ring stage" is accentuated by the Romanovsky stains:

the parasite cytoplasm is blue, and the nucleus is red. As the trophozoite grows (Plate 1, *3* to *15*), its food vacuoles become less noticeable by light microscopy, but pigment granules of hemozoin in the vacuoles may become apparent. Hemozoin is the end product of the parasite's digestion of the host's hemoglobin but is not a partially degraded form of hemoglobin. It contains insoluble dimers and monomers of hematin, some ferriprotoporphyrin coupled to a plasmodial protein, and insoluble methemoglobin.[71]

The parasite rapidly develops into a schizont (Plate 1, *16* to *20*). The stage in the erythrocytic schizogony at which the cytoplasm is coalescing around the individual nuclei, before cytokinesis, is called the **segmenter.** When development of the merozoites is completed, the host cell ruptures, releasing parasite metabolic wastes and residual body, including hemozoin. The metabolic wastes thus released are one factor responsible for the characteristic symptoms of malaria, although hemozoin itself is nontoxic. A great many of the merozoites are ingested and destroyed by reticuloendothelial cells and leukocytes, but, even so, the number of parasitized host cells may become astronomical because erythrocytic schizogony takes only from 1 to 4 days, depending on the species.

After an indeterminate number of asexual generations, some merozoites enter erythrocytes and become **macrogamonts (macrogametocytes)** and **microgamonts (microgametocytes)** (Plate 1, *21* to *24*). The size and shape of these cells are characteristic for each species; they also contain hemozoin. Unless they are ingested by a mosquito, gametocytes soon die and are phagocytized by the reticuloendothelial system.

Invertebrate stages. When erythrocytes containing gametocytes are imbibed by an unsuitable mosquito, they are digested along with the blood. However, if a susceptible mosquito is the diner, the gametocytes develop into gametes. Although this development would take place only in a female mosquito in nature, since only females feed on vertebrate blood, males of appropriate species can support development after exprimental infection with the parasite in the laboratory. Suitable hosts for the *Plasmodium* spp. of humans are a wide variety of *Anopheles* spp. (see Fig. 39-10). After release from its enclosing erythrocyte, maturation of the macrogametocyte to the macrogamete involves little obvious change other than a shift of the nucleus toward the periphery. In contrast, the microgametocyte displays a rather astonishing transformation, **exflagellation.** As the microgametocyte becomes extracellular, within 10 to 12 minutes its nucleus divides repeatedly to form six to eight daughter nuclei, each of which is associated with the elements of a developing axoneme. The doubled outer membrane of

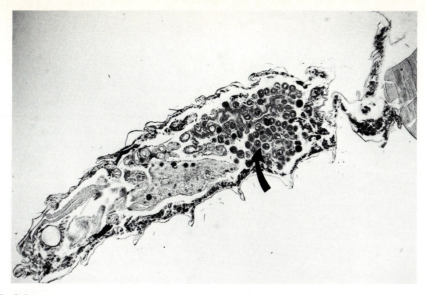

FIG. 9-3

Longitudinal section of a mosquito, with numerous oocysts of *Plasmodium* sp. in the hemocoel *(arrow).*

Photograph by Warren Buss.

the microgametocyte becomes interrupted; the flagellar buds with their associated nuclei move peripherally between the interruptions and then continue outward covered by the outer membrane of the gametocyte. These break free and are the microgametes. The stimulus for exflagellation is an increase in pH caused by escape of dissolved carbon dioxide from the blood.[59] The life span of the microgametes is short, since they contain little more than the nuclear chromatin and the flagellum covered by a membrane. The microgamete swims about until it finds a macrogamete, which it penetrates and fertilizes. The resultant diploid zygote quickly elongates to become a motile **ookinete.** The ookinete is reminiscent of a sporozoite and merozoite in morphology. It is 10 to 12 μm in length and has polar rings and subpellicular microtubules but no rhoptries or micronemes.

The ookinete penetrates the peritrophic membrane in the mosquito's gut, migrates to the hemocoel side of the gut, and begins its transformation into an oocyst. The oocyst (Fig. 9-3) is covered by an electron-dense capsule and soon extends out into the insect's hemocoel. The initial division of its nucleus is reductional; meiosis takes place immediately after zygote formation as in other Sporozoea.[74,75] The oocyst reorganizes internally into a number of haploid nucleated masses called **sporoblasts,** and the cytoplasm contains many ribosomes, endoplasmic reticulum, mi-

tochondria, and other inclusions. The sporoblasts in turn divide repeatedly to form thousands of sporozoites (Fig. 9-4). These break out of the oocyst into the hemocoel and migrate throughout the mosquito's body. On contacting the salivary gland, sporozoites enter its channels and can be injected into a new host at the next feeding.

Sporozoite development takes from 10 days to 2 weeks, depending on the species of *Plasmodium* and the temperature. Once infected, a mosquito remains infective for life, capable of transmitting malaria to every susceptible vertebrate it bites.

Plasmodium sometimes is transmitted by means other than the bite of a mosquito. The blood cycle may be initiated by blood transfusion, by malaria therapy of certain paralytic diseases, by syringe-passed infection among drug addicts, or, rarely, by congenital infection.

■ **Classification of *Plasmodium***

The genus *Plasmodium* can conveniently be divided into nine subgenera, of which three occur in mammals, four in birds, and two in lizards. These are listed and characterized by Garnham,[18] who discusses 127 species of *Plasmodium*. A few of these are of doubtful status, but several can be separated into well-defined subspecies.

Species of *Plasmodium* are not difficult to distin-

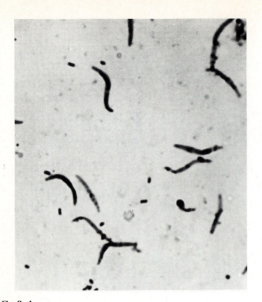

FIG. 9-4

Plasmodium sporozoites.

Photograph by Peter Diffley.

guish after training. Most are parasites of birds; others occur in such animals as rodents, primates, and reptiles. Some species are very useful in laboratory studies of immunity, physiology, and so forth, such as the rodent parasite *Plasmodium berghei* and the chicken parasite *P. gallinaceum*. Still other species, normally parasitic in nonhuman primates, occasionally infect humans as zoonoses or can be acquired by humans when infected experimentally. Such are *P. schwetzi* of chimpanzees and gorillas, *P. eylesi* of Malayan gibbons, *P. cynomolgi, P. knowlesi* and *P. inui* of oriental monkeys, *P. simium* and *P. brasilianum* of New World monkeys, and *P. shortii* of Indian and Ceylonese monkeys. The importance of these species to human medicine is for the most part unassessed; surely they are potential disease agents, at least to the individual who may be exposed under unusual circumstances. In fact some of the foregoing (*P. simium* and *P. brasilianum*) may be conspecific with some species usually considered parasites of humans.[42] Humans are normal hosts for four species of *Plasmodium,* and these will be treated in more detail.

■ *Plasmodium* species parasitic in humans

Plasmodium (Plasmodium) vivax (Plates 1 and 2). *Plasmodium vivax* is the cause of **benign tertian malaria,** also known as **vivax malaria** or **tertian ague.** When early Italian investigators noted the actively motile trophozoites of the organism within host cor-

puscles, they nicknamed it "vivace," foreshadowing the Latin name "vivax," which later was accepted as its epithet. The designation "tertian" is based on the fact that fever paroxysms typically recur every 48 hours, and the name is derived from the ancient Roman custom of calling the day of an event the first day, 48 hours later hence being the third. The species flourishes best in temperate zones, rarely as far north as Manchuria, Siberia, Norway, and Sweden and as far south as Argentina and South Africa. Because malaria eradication campaigns have been so successful in many of the temperate areas of the world, however, the disease has practically disappeared from them. Most vivax malaria today is found in Asia; about 40% of malaria among United States military personnel in Vietnam resulted from *P. vivax*.[9] It is common in North Africa but drops off in tropical Africa to very low levels, partly because of a natural resistance of black persons to infection with this species (see Duffy blood groups, below). About 43% of malaria in the world is caused by *P. vivax*.

Sporozoites that are 10 to 14 μm long invade cells of the liver parenchyma within 1 or 2 days after injection with the mosquito's saliva. By the seventh day the exoerythrocytic schizont is an oval body about 40 μm long, has blue-staining cytoplasm, a few large vacuoles, and lightly staining nuclei. On maturity, the vacuoles disappear, and about 10,000 merozoites are produced. The fate of these merozoites is a subject of debate. Certainly, many of them are killed outright by host defenses. Others invade erythrocytes to initiate the erythrocytic stages of development. Still others may possibly reinfect hepatic cells (see relapse of malaria, p. 152). Relapses up to 8 years after initial infection are characteristic of vivax malaria.

The patient is in normal health during the intervening periods of latency. The relapses are believed to result from genetic differences in the original sporozoites; that is, some give rise to tissue schizonts that take much longer to mature.[12] However, occurrence of relapses may also be related to the immune state of the host (see discussion of immunity later in the chapter).

Plasmodium vivax merozoites invade only young erythrocytes, the reticulocytes, and apparently are unable to penetrate mature red cells. It now is known that merozoites can only penetrate erythrocytes with mediated receptor sites, such sites being genetically determined.[26] Known as the **Duffy blood groups,** there are two codominant alleles, Fy^a and Fy^b, recognized by their different antigens. A third allele, Fy, has no corresponding antigen. The Fy/Fy genotype is common in African and in American black people (40% or more) and rare in white people (about 0.1%). It has been shown that Fy^a and Fy^b are receptors for

P. vivax and *P. knowlesi*[56]; hence *Fy/Fy* is refractory to infection. This explains the natural resistance of black people to vivax malaria. The Duffy negative genotype may represent the original, rather than the mutant, condition in tropical Africa.[51]

Soon after invasion of the erythrocyte and formation of the ring stage, the cytoplasm becomes actively ameboid, throwing out pseudopodia in all directions and fully justifying the name "vivax." Infection of the same erythrocyte with more than one trophozoite may occur but not commonly. As the trophozoite grows, the red cell enlarges, loses its pink color, and develops a peculiar stippling known as **Schüffner's dots** (Plate 1, *5*). These dots are visible by light microscopy after Romanovsky staining. By electron microscopy they can be seen as small surface invaginations (**caveolae**), surrounded by small vesicles.[72] Ring stages occupy approximately one third to one half of the erythrocyte, and the trophozoite occupies about two thirds of the red cell after 24 hours. The vacuole disappears, the organism becomes more sluggish, and hemozoin granules accumulate as the trophozoite grows. By 36 to 42 hours after infection, nuclear division begins and is repeated four times, yielding 16 nuclei in the mature schizont. Fewer nuclei may be produced, especially in older infections or those interfered with by host immunity or chemotherapy. Once schizogony has begun, the pigment granules accumulate in two or three masses in the parasite, ultimately to be left in the residual body and engulfed by the host's reticuloendothelial system. The rounded merozoites, about 1.5 µm in diameter, immediately attack new erythrocytes. Erythrocytic schizogony takes somewhat less than 48 hours, although early in the disease there are usually two populations, each maturing on alternate days, resulting in a daily, or **quotidian,** periodicity (refer to discussion of pathology later in the chapter).

Some merozoites develop into gametocytes rather than into schizonts. The factors determining the fate of a given merozoite are not known, but since the gametocytes have been found as early as the first day of parasitemia in rare instances, it may be possible for exoerythrocytic merozoites to produce gametocytes. The stained macrogametocyte has bright blue, rounded cytoplasm and a nucleus that is compact and dark staining by Romanovsky methods. Dark-brown hemozoin granules are abundant throughout the cytoplasm. The mature macrogametocyte fills most of the enlarged erythrocyte and measures about 10 µm wide. The rounded microgametocyte is more gray than blue with Romanovsky stains. Its nucleus is more diffuse and is much larger, sometimes half the diameter of the entire parasite, and the pigment granules are coarser and more unevenly distributed than in the macrogametocyte. The mature microgametocyte is smaller than the macrogametocyte and usually does not fill the erythrocyte.

Gametocytes take 4 days to mature, twice the length of time for schizonts. Macrogametocytes often outnumber microgametocytes by two to one. A single host cell may contain both a gametocyte and a schizont.

Formation of zygote, ookinete, and oocyst are as described previously. The oocyst may reach a size of 50 µm and produce up to 10,000 sporozoites. The mature oocyst ruptures after 9 days at 25° C. If the ambient temperature is too high or too low, the oocyst blackens with pigment and degenerates, a phenomenon noted by Ross. Too many developing oocysts kill the mosquito before the sporozoites are developed.

Plasmodium (Laverania) falciparum (Plates 3 and 4). Malaria known as **malignant tertian, subtertian, or estivoautumnal (E-A)** is caused by *P. falciparum,* the most virulent of *Plasmodium* spp. in humans. It was nearly cosmopolitan at one time, with a concentration in the tropics and subtropics. It still extends into the temperate zone in some areas, although it has been eradicated in the United States, the Balkans, and around the Mediterranean. Nevertheless, falciparum malaria reigns supreme as the greatest killer of humanity in the tropical zones of the world today, accounting for about 50% of all malaria cases.

Among the many cases studied by Laveran, persons suffering from "malignant tertian malaria" interested him the most. He had long noticed a distinct darkening of the gray matter of the brain and abundant pigment in other tissues of his deceased patients. When, in 1880, he saw crescent-shaped bodies in the blood and watched them exflagellate, he knew he had found living parasites. The confusion that surrounded the correct name for this species was great until 1954, when the International Commission of Zoological Nomenclature validated the epithet *falciparum*.

Malignant tertian malaria is usually blamed for the decline of the ancient Greek civilization, the halting of Alexander the Great's progress to the East, and the destruction of some of the Crusades. In more modern times, the Macedonian campaign of World War I was destroyed by falciparum malaria, and the disease caused more deaths than did battles in some theaters of World War II.

As in other species, the exoerythrocytic schizont of *P. falciparum* grows in liver cells. It is more irregularly shaped than that of *P. vivax,* with projections extending in all directions by the fifth day. The schizont ruptures in about 5½ days, releasing about 30,000 merozoites. There seems to be no second exoerythrocytic cycle, and true relapses do not occur. However, recrudescences of the disease may follow remissions up to a year, occasionally 2 or 3 years, after initial infection, apparently because of small populations of the parasites remaining in the red cells.

Merozoites can invade erythrocytes of any age, including reticulocytes; therefore falciparum malaria is characterized by much higher levels of parasitemia than are the other types. Soon after invasion of the erythrocyte, the trophozoite produces a protein that is inserted into the plasma membrane of the erythrocyte and causes the appearance of "knobs" on the surface of the host cell.[31] The knobs form focal junctions with endothelial cell membranes or with knobs of other erythrocytes, resulting in sequestration of infected erythrocytes along the vascular endothelium of deep tissues,[1,3] including the brain, spleen, and bone marrow. Gametocyte-infected erythrocytes have no knobs and do not stick to the endothelium.[72] Hence one usually observes only ring stages and/or gametocytes in blood smears from patients with falciparum malaria. If schizogony is well synchronized, parasites may be practically absent from peripheral blood toward the end of the 48-hour cycle.

The ring-stage trophozoite is the smallest of any *Plasmodium* spp. of humans: about 1.2 μm. The following are observed more frequently in *P. falciparum* than in other species: (1) **accolé,** or **appliqué,** forms (Plate 3, *2*); (2) ring stages with two chromatin dots (nuclei) (Plate 3, *3* and *4*); and (3) infection of a single erythrocyte with more than one parasite (Plate 3, *2, 8, 10,* and *11*). "Appliqué," or "accolé," forms are ring stages that lie very close to the surface of the red cell, appearing to be "applied" to the cell. The frequency of multiple infections in the same cell has led some parasitologists to believe that the ring stages divide and that the binucleate rings are division stages. As it grows, the protozoan extends wispy pseudopodia, but it is never as active as *P. vivax.* The infected erythrocyte develops irregular blotches known as **Maurer's clefts** (Plate 3, *9*). These are much larger than the fine Schüffner's dots found in *P. vivax* infections. They are caused by extension of the parasitophorous vacuole within the host cytoplasm.[38]

The mature schizont is less symmetrical than are those of the other species infecting humans. It develops 8 to 32 merozoites, with 16 being the usual number. In contrast to the normal situation, schizonts may be fairly common in peripheral blood in some geographical areas. This may reflect strain differences. The erythrocytic cycle takes 48 hours, but the periodicity is not as marked as in *P. vivax,* and it may vary considerably with the strain of parasite. Extremely high levels of parasitemia may occur, with more than 65% of the erythrocytes containing parasites; a density of 25% is usually fatal. Two or three parasites per milliliter of blood may be sufficient to cause disease symptoms.

In *P. vivax* the gametocytes may appear in the peripheral blood almost at the same time as the trophozoites, but in *P. falciparum* the sexual stages require nearly 10 days to develop and then appear in large numbers. They develop in the blood spaces of the spleen and bone marrow, first assuming bizarre, irregular shapes, then becoming round, and finally changing into the crescent shape so distinctive of the species (Plate 3, *26* and *27*). The mature microgametocyte is 9 to 11 μm long, has blunt ends, and has a diffuse nucleus extending over half the length of the organism. Hemozoin granules cluster in the nuclear zone. The cell stains light blue to pinkish with Romanovsky stains. The macrogametocyte is more slender and has slightly pointed ends. It is 12 to 14 μm long, stains a darker blue, and has a more compact nucleus. Pigment granules also cluster around the nucleus of the female cell. This differs from *P. vivax,* in which pigment is diffuse throughout the cytoplasm.

Plasmodium (Plasmodium) malariae (Plates 5 and 6). **Quartan malaria,** with paroxysms every 72 hours, is caused by *P. malariae.* It was recognized by the early Greeks because the timing of the fevers differed from that of the tertian malaria parasites. Although Laveran saw and even illustrated the characteristic schizonts of this parasite, he refused to believe it was different from *P. falciparum.* In 1885 Golgi differentiated the tertian and quartan fevers and gave an accurate description of what is now known as *P. malariae.*

Plasmodium malariae is a cosmopolitan parasite but does not have a continuous distribution anywhere. It is common in many regions of tropical Africa, Burma, parts of India, Sri Lanka, Malaya, Java, New Guinea, and Europe. It is also distributed in the New World, including Guadeloupe, Guyana, Brazil, Panama, and at one time the United States. The peculiar distribution of this parasite has never been satisfactorily explained. Two likely, but opposite, possibilities are that either it was recently a parasite of simian primates, and with the decline of simian populations it too is in the decline, or that it was originally a parasite of ancient human populations and is declining with the improvement and migration of peoples. It may be the only species of human malaria organism that also regularly lives in wild animals. Chimpanzees are infected at about the same rate as humans but are unimportant as reservoirs, since they do not live side by side with people. Some workers believe that *P. brasilianum* is really *P. malariae* in New World monkeys.[42] This species accounts for about 7% of malaria cases in the world.

Exoerythrocytic schizogony is completed in 13 to 16 days.

Erythrocytic forms build up slowly in the blood; the characteristic symptoms of the disease may appear before it is possible to find the parasites in blood smears. The ring forms are less ameboid than those of *P. vivax,* and the cytoplasm is somewhat thicker. Rings often retain their shape for as long as 48 hours, finally transforming into an elongate "band form," which begins to collect pigment along one edge (Plate 3, *6* and *10*). The nucleus divides into 6 to 12 merozoites at 72 hours. The segmenter is strikingly symmetrical and is called a "rosette" or "daisy-head" (Plate 5, *20*). Parasitemia levels are characteristically low, with one parasite per 20,000 red cells, representing a high figure for this species. This low density is accounted for by the fact that the merozoites apparently can invade only aging erythrocytes, which are soon to be removed from circulation by the normal process of blood destruction.

Gametocytes probably develop in the internal organs, since immature forms are rare in peripheral blood. They are slow to develop in sporozoite-induced infections. The microgametocyte fills the entire host cell. It has a nucleus that occupies at least half the volume of the parasite. The remaining cytoplasm stains a grayish green color, mainly because of the diffuse hemozoin granules within it. Macrogametocytes are nearly impossible to identify, appearing identical to large, uninucleate asexual stages. Recrudescences of quartan malaria can occur up to 53 years after initial infection.[19] Because it can live in the blood so long, it is the most important cause of transfusion malaria.

Plasmodium (Plasmodium) ovale (Plate 7)**.** This species causes **ovale,** or **mild tertian, malaria** and is the rarest of the four malaria parasites of humans. It is confined mainly to the tropics, although it has been reported from Europe and the United States. Although common on the west coast of Africa, which may be its original home, the species is scarce in central Africa and present but not abundant in eastern Africa. It is known also in India, the Philippine Islands, New Guinea, and Vietnam. *Plasmodium ovale* is difficult to diagnose because of its similarity to *P. vivax*.

The youngest ring stages have a large, round nucleus and a rather small vacuole that disappears early. The mature schizont is oval or spheroid and is about half the size of the host cell. Eight merozoites are usually formed, with a range of 4 to 16. Schüffner's dots appear early in the infected blood cells. They are very numerous and larger than those in *P. vivax* infections and stain a brighter red color. As in *P. vivax,* the Schüffner's dots are due to caveolae.

Gametocytes of *P. ovale* take longer to appear in the blood than do those of other species. They are numerous enough 3 weeks after infection to infect mosquitoes regularly. Gametocytes of both sexes are about 9 μm in diameter and contain pigment granules in concentric rings or irregular nodes. The macrogametocyte has purplish cytoplasm and a small nucleus, usually on one side. The stained microgametocyte has bluer cytoplasm and a large nucleus nearly half the size of the parasite.

■ Malaria: the disease

Certain disease aspects of *Plasmodium* spp. have been mentioned in the preceding pages; following is a brief consideration of the subject, particularly in relation to pathogenesis and public health. We urge the reader to consult other references for more complete treatment.[8,47,78,85]

Diagnosis. Diagnosis depends to some extent on the clinical manifestations of the disease, but most important is demonstration of the parasites in stained smears of peripheral blood. Technical details can be found in many texts and laboratory manuals of medical parasitology. Characteristic morphology of the respective species has been noted, but the most useful criteria for differential diagnosis are summarized in Table 9-1.

A DNA hybridization technique for diagnosis of *P. falciparum* has been described that is rapid and simple.[64] The technique is now undergoing field trials[30] and should be of great value in screening blood for blood banks and epidemiological surveys, as well as blood of immigrants and returning tourists. A radioimmunoassay using a monoclonal antibody has also been described.[4]

Pathology. The major clinical manifestations of malaria may be attributed to two general factors: (1) the host inflammatory response, which produces the characteristic chills and fever, as well as other related phenomena; and (2) anemia, arising from the enormous destruction of red blood cells. Severity of the disease is correlated with the species producing it: falciparum malaria is most serious and vivax and ovale the least dangerous.

The main causes of the anemia are destruction of both parasitized and nonparasitized erythrocytes, inability of the body to recycle the iron bound in the insoluble hemozoin, and an inadequate erythropoietic response of the bone marrow. Why such large num-

TABLE 9-1.

Criteria for differential diagnosis of *Plasmodium* spp. in humans

P. vivax	*P. falciparum*	*P. ovale*	*P. malariae*
Trophozoites ameboid	Larger trophozoites and	Trophozoites not ameboid	Trophozoites often band
Segmenters form about 16	schizonts not usually in	Segmenters usually form	form
merozoites	peripheral blood	eight merozoites	Segmenters usually form
Host cell enlarged,	Ring stages small, often	Host cells somewhat	eight merozoites
decolorized, frequently	with two chromatin dots	enlarged, sometimes	Host cells not enlarged,
with Schüffner's dots	Appliqué forms frequent	decolorized with oval	decolorized, or stippled
Parasites relatively large	Multiple infections	distortion, Schüffner's	Hemozoin granules large,
	frequent	dots heavy	abundant
	Gametocytes crescent		
	shaped		

Modified from Russell, P.F., et al. 1946. Practical malariology. W.B. Saunders Co., Philadelphia.

bers of nonparasitized red cells are destroyed is still not understood, but some evidence has indicated autoimmune hemolysis. Other reports have suggested increased phagocytosis of erythrocytes by the reticuloendothelial system.[82] The defective bone marrow response may be due in part to limitation in iron supply and in falciparum malaria it may be due to blockage of the capillaries by parasitized erythrocytes. Destruction of erythrocytes leads to an increase in blood bilirubin, a breakdown product of hemoglobin. When excretion cannot keep up with formation of bilirubin, jaundice yellows the skin. The hemozoin is taken up by circulating leukocytes and is deposited in the reticuloendothelial system. In severe cases the viscera, especially the liver, spleen, and brain, become blackish or slaty as the result of pigment deposition (Fig. 9-5, Plate 8).

Fever is a common, nonspecific reaction of the body to infection, functioning at least in part to increase the rate of metabolic reactions important in host defenses. Fever in malaria is correlated with the maturation of a generation of merozoites and the rupture of the red blood cells that contain them. It is widely believed that fever is stimulated by the excretory products of the parasites, released when the erythrocytes lyse, but the exact nature of such substances is not known. There is evidence of production of cytotoxic factors by the parasites: oxidative phosphorylation and respiration are inhibited in mitochondria from infected animals, and damage to liver cells can be observed on the ultrastructural level.[47]

A few days before the first paroxysm, the patient may feel malaise, muscle pain, headache, loss of appetite, and slight fever; or the first paroxysm may occur abruptly, without any prior symptoms. A typical attack of benign tertian or quartan malaria begins with

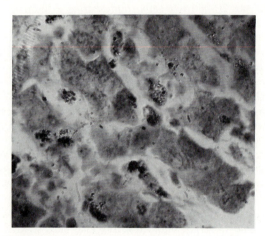

FIG. 9- 5

Section of liver tissue with numerous deposits of malarial pigment.
Photograph by L.S. Roberts.

a feeling of intense cold as the hypothalamus, the body's thermostat, is activated, and the temperature then rises rapidly to 104° to 106°F. The teeth chatter, and the bed may rattle from the victim's shivering. Nausea and vomiting are usual. The hot stage begins after ½ to 1 hour, with intense headache and feeling of intense heat. Often a mild delirium stage lasts for several hours. As copious perspiration signals the end of the hot stage, the temperature drops back to normal within 2 to 3 hours, and the entire paroxysm is over within 8 to 12 hours. The person may sleep for a while after an episode and feel fairly well until the next paroxysm. The foregoing time periods for the stages are usually somewhat shorter in quartan malaria, and the paroxysms recur every 72 hours. In vivax malaria the periodicity is often quotidian early in the infection, since two populations of merozoites usually mature on alternate days. "Double" and "triple" quartan infections also are known. Only after one or more groups drop out does the fever become tertian or quartan, and the patient experiences the classical good and bad days.

Because the synchrony in falciparum malaria is much less marked, the onset is often more gradual, and the hot stage is extended. The fever episodes may be continuous or fluctuating, but the patient does not feel well between paroxysms, as in vivax and quartan malaria. In cases in which some synchrony develops each episode lasts 20 to 36 hours, rather than 8 to 12, and is accompanied by much nausea, vomiting, and delirium. Concurrent infections with *P. vivax* and *P. falciparum* are not uncommon.

Falciparum malaria is always serious, and sometimes severe complications are produced. The most common of these is **cerebral malaria,** which may account for 10% of falciparum malaria admitted to the hospital and 80% of such deaths.[85] Cerebral malaria may be gradual in onset, but it is commonly sudden; a progressive headache may be followed by a coma, an uncontrollable rise in temperature to above 108° F, and psychotic symptoms or convulsions, especially in children. Death may ensue within a matter of hours. Initial stages of cerebral malaria are easily mistaken for a variety of other conditions, including acute alcoholism, usually with disastrous consequences. Another grave and usually fatal complication of severe falciparum malaria is **pulmonary edema,** which in some cases may be a result of overadministration of intravenous fluids. Difficulty in breathing increases and death may ensue in a few hours.[85] A combination of other severe manifestations leads to a condition known as **algid malaria.** This is associated with a bacterial infection of the blood **(septicemia)** with toxemia and massive gastrointestinal hemorrhage.[85] There is a circulatory collapse with markedly low blood pressure. The skin is cold and clammy; peripheral veins are constricted.

The direct cause of these severe complications has traditionally been cited as a "plugging" of the capillaries in the affected organs by clots (Fig. 9-6). Some evidence has suggested that the conditions are caused by a manifestation of the inflammatory response: an increase in vascular permeability, with accompanying water and protein lost from the blood to the tissues, leading to circulatory stasis and hypoxia. However, recent investigations have led to the conclusion that symptoms of cerebral malaria are not due to edema but rather that the dysfunction is a consequence of

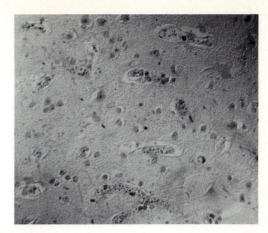

FIG. 9-6

Section of cerebral tissue, demonstrating capillaries plugged with erythrocytes infected by *Plasmodium falciparum.* Infected red cells are marked by pigment; the parasites themselves are transparent.

Photograph by L.S. Roberts.

stagnant hypoxia caused by adherence of the parasitized erythrocytes to the endothelium of cerebral venules and capillaries.[82,83] Edema seen at autopsy is probably a condition that developed at the death of the patient. The symptoms of algid malaria appear to result from circulatory stasis in the gastrointestinal tract due to the same mechanism.

Blackwater fever is another grave condition associated with falciparum malaria, but its clinical picture is distinct from the foregoing. It is an acute, massive lysis of erythrocytes, marked by high levels of free hemoglobin and its breakdown products in the blood and urine and by renal insufficiency. Because of the presence of hemoglobin and its products in the urine, the fluid is quite dark, hence the name of the condition. A prostrating fever, jaundice, and persistent vomiting occur. Renal failure is usually the immediate cause of death. Damage to the kidney is now thought to result from renal anoxia, reducing efficiency of the glomerular filtration and tubular resorption. The massive hemolysis is not directly attributable to the parasites; the organisms frequently cannot be demonstrated. However, the condition is almost always associated with areas of *P. falciparum* hyperendemicity, found in persons with prior falciparum malaria, and very frequently with irregular or inadequate treatment for the infection. Inadequate suppressive or therapeutic doses of quinine most often have been implicated, but many cases have been reported following treatment with quinacrine and pamaquine and can occur in persons who have not been treated at all. It is now believed that blackwater fever is an autoimmune phenomenon and is triggered by some stimulus that results in release of large amounts of antibodies, which act as hemolysins, into the circulation. Mortality is 20% to 50%. The incidence of blackwater fever has declined in recent years, perhaps due to the use of drugs other than quinine for prophylaxis.

Hypoglycemia (reduced concentration of blood glucose) is a common symptom in falciparum malaria. It is usually found in women with uncomplicated or severe malaria who are pregnant or have recently delivered, as well as other cases of severe falciparum malaria.[85] Coma produced by hypoglycemia has commonly been misdiagnosed as cerebral malaria. This condition is usually associated with quinine treatment. For reasons still unclear, the pancreatic islet cells are stimulated to increase insulin secretion, thus lowering blood glucose.[82]

Immunity. Despite the fact that much of the disease results from the inflammatory and immune responses of the host, host defenses are vital in limiting the infection. One vivax segmenter producing 24 merozoites every 48 hours would give rise to 4.59 billion parasites within 14 days, and the host would soon be destroyed if the organisms continued reproducing unchecked.[42] The development of some protective immunity is evident in malaria, and we will consider only briefly some of the practical effects. Relapses and recrudescences may be associated with lowered antibody titers or increased ability of the parasite to deal with the antibody, but they may depend on genetic differences in sporozoite populations. Symptoms in a relapse are usually less severe than those in the primary attack, but the level of parasitemia is higher. After the primary attack and between relapses, the patient may have a **tolerance** to the effects of the organisms and in fact may have as high a circulating parasitemia level as during the primary attack, although remaining asymptomatic. Such tolerant carriers are very important in the epidemiology of the disease. The protective immunity is primarily a premunition, that is, resistance to superinfection. It is effective only as long as a small, residual population of parasites is present; if the person is completely cured, susceptibility returns. Thus, in highly endemic areas, infants are protected by maternal antibodies, and young children are at greatest risk after weaning. The immunity of children who survive a first attack will be continuously stimulated by the bites of infected mosquitoes as long as the children live in the malarious area. Nonimmune adults are highly susceptible. Immunity is species specific and to some degree strain specific so that a person may risk a new

infection by migrating from one malarious area to another. Falciparum malaria is unmitigated in its severity to a person who is immune to vivax malaria.

Black persons are much less susceptible to vivax malaria than are whites, and falciparum malaria in blacks is somewhat less severe. The genetic basis for this phenomenon is explained by the inheritance of Duffy blood groups (p. 146). Other factors that can contribute to genetic resistance are certain heritable anemias: sickle cell, favism, and thalassemia. Although these conditions are of negative selective value in themselves, they have been selected for in certain populations because they confer resistance to falciparum malaria. The most well known of these is **sickle cell anemia.** In persons homozygous for this trait a glutamic acid residue in the amino acid sequence of hemoglobin is replaced by a valine, interfering with the conformation of the hemoglobin and oxygen-carrying capacity of the erythrocytes. Persons with sickle cell anemia usually die before the age of 30. In heterozygotes some of the hemoglobin is normal, and these persons can live relatively normal lives, but the presence of the abnormal hemoglobin inhibits growth and development of *P. falciparum* in their erythrocytes. The selective pressure of malaria in Africa has led to maintenance of this otherwise undesirable gene in the population. This legacy has unfortunate consequences when the people are no longer threatened by malaria, as in the United States, where 1 in 10 Americans of African ancestry is heterozygous for the sickle cell gene, and 1 in 400 is homozygous.

Relapse in malarial infections. Since the advent of an antimalarial drug (quinine) in the sixteenth century, it has been noted that some persons, who have been treated and seemingly recovered, relapse back into the disease weeks, months, or even years after the apparent cure.[24] An interesting history of the phenomenon was given by Coatney.[12] Malarial relapse has engendered much speculation and research for many years. The discovery of preerythrocytic schizogony in the liver by Shortt and Garnham in 1948 seemed to have solved the mystery. It appeared most reasonable to assume that preerythrocytic merozoites simply reinfected other hepatocytes, with subsequent reinvasion of red blood cells. This would explain why relapse occurred after erythrocytic forms were eliminated by erythrocytic schizontocides, such as quinine and chloroquine.

However, not all species of *Plasmodium* cause relapse. Among the parasites of primates, only *P. vivax* and *P. ovale* of humans and *P. cynomolgi, P. fieldi,* and *P. simiovale* of simians cause true relapse. If

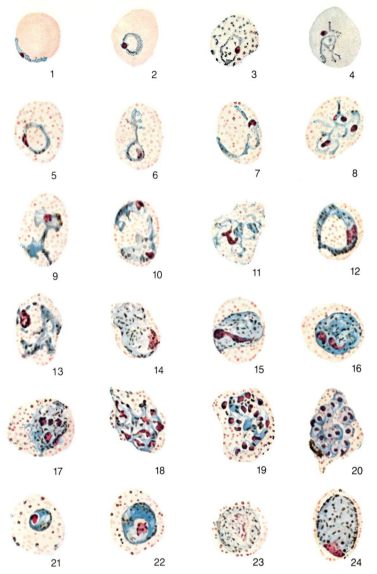

PLATE 1

Plasmodium vivax. **1,** Normal-sized red cell with marginal ring-form trophozoite. **2,** Young signet-ring form trophozoite in macrocyte. **3,** Slightly older ring-form trophozoite in red cell showing basophilic stippling. **4,** Polychromatophilic red cell containing young tertian parasite with pseudopodia. **5,** Ring-form trophozoite showing pigment in cytoplasm, in enlarged cell containing Schüffner's stippling *(dots).* (Schüffner's stippling does not appear in all cells containing growing and older forms of *P. vivax,* as would be indicated by these pictures, but it can be found with any stage from fairly young ring form onward.) **6** and **7,** Very tenuous medium trophozoite forms. **8,** Three ameboid trophozoites with fused cytoplasm. **9** and **11** to **13,** Older ameboid trophozoites in process of development. **10,** Two ameboid trophozoites in one cell. **14,** Mature trophozoite. **15,** Mature trophozoite with chromatin apparently in process of division. **16** to **19,** Schizonts showing progressive steps in division (presegmenting schizonts). **20,** Mature schizont. **21** and **22,** Developing gametocytes. **23,** Mature microgametocyte. **24,** Mature macrogametocyte.

From Wilcox, A. 1960. Manual for the microscopical diagnosis of malaria in man. Department of Health, Education, and Welfare, Public Health Service, U.S. Government Printing Office, Washington, D.C.

PLATE 2

Plasmodium vivax in thick smear. **1,** Ameboid trophozoites, **2,** Schizont—two divisions of chromatin. **3,** Mature schizont. **4,** Microgametocyte. **5,** Blood platelets. **6,** Nucleus of neutrophil. **7,** Eosinophil. **8,** Blood platelet associated with cellular remains of young erythrocytes.

From Wilcox, A. 1960. Manual for the microscopical diagnosis of malaria in man. U.S. Government Printing Office, Washington, D.C.

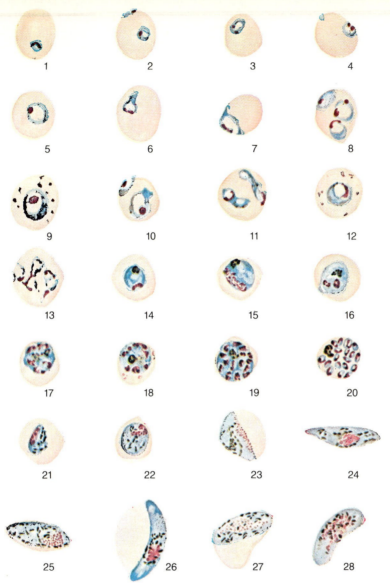

PLATE 3

Plasmodium falciparum. **1,** Very young ring-form trophozoite. **2,** Double infection of single cell with young trophozoites, one a marginal form, the other a signet-ring form. **3** and **4,** Young trophozoites showing double chromatin dots. **5** to **7,** Developing trophozoite forms. **8,** Three medium trophozoites in one cell. **9,** Trophozoite showing pigment, in cell containing Maurer's dots. **10** and **11,** Two trophozoites in each of two cells, showing variations of forms that parasites may assume. **12,** Almost mature trophozoite showing haze of pigment throughout cytoplasm. Maurer's dots in cell. **13,** Estivoautumnal slender forms. **14,** Mature trophozoite showing clumped pigment. **15,** Parasite in process of initial chromatin division. **16** to **19,** Various phases of development of schizont (presegmenting schizonts). **20,** Mature schizont. **21** to **24,** Successive forms in development of gametocyte—usually not found in peripheral circulation. **25,** Immature macrogametocyte. **26,** Mature macrogametocyte. **27,** Immature microgametocyte. **28,** Mature microgametocyte.

From Wilcox, A. 1960. Manual for the microscopical diagnosis of malaria in man. U.S. Government Printing Office, Washington, D.C.

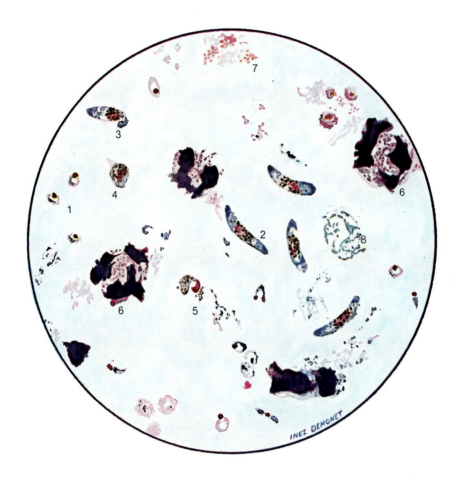

PLATE 4

Plasmodium falciparum in thick film. **1,** Small trophozoites. **2,** Gametocytes—normal. **3,** Slightly distorted gametocyte. **4,** "Rounded-up" gametocyte. **5,** Disintegrated gametocyte. **6,** Nucleus of leukocyte. **7,** Blood platelets. **8,** Cellular remains of young erythrocyte.

From Wilcox, A. 1960. Manual for the microscopical diagnosis of malaria in man. U.S. Government Printing Office, Washington, D.C.

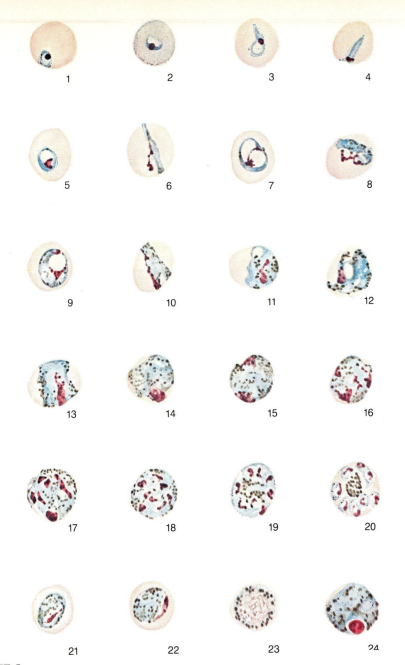

PLATE 5

Plasmodium malariae. **1,** Young ring-form trophozoite of quartan malaria. **2 to 4,** Young trophozoite forms of parasite showing gradual increase of chromatin and cytoplasm. **5,** Developing ring-form trophozoite showing pigment granule. **6,** Early band-form trophozoite—elongate chromatin, some pigment apparent. **7 to 12,** Some forms that developing trophozoite of quartan may take. **13** and **14,** Mature trophozoites—one a band form. **15 to 19,** Phases in development of schizont (pre-segmenting schizonts). **20,** Mature schizont. **21,** Immature microgametocyte. **22,** Immature macrogametocyte. **23,** Mature microgametocyte. **24,** Mature macrogametocyte.

From Wilcox, A. 1960. Manual for the microscopical diagnosis of malaria in man. U.S. Government Printing Office, Washington, D.C.

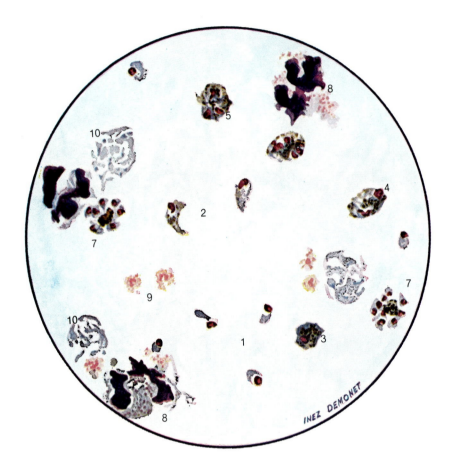

PLATE 6

Plasmodium malariae in thick smear. **1,** Small trophozoites. **2,** Growing trophozoites. **3,** Mature trophozoites. **4** to **6,** Schizonts (presegmenting) with varying numbers of divisions of chromatin. **7,** Mature schizonts. **8,** Nucleus of leukocyte. **9,** Blood platelets. **10,** Cellular remains of young erythrocytes.

From Wilcox, A. 1960. Manual for the microscopical diagnosis of malaria in man. U.S. Government Printing Office, Washington, D.C.

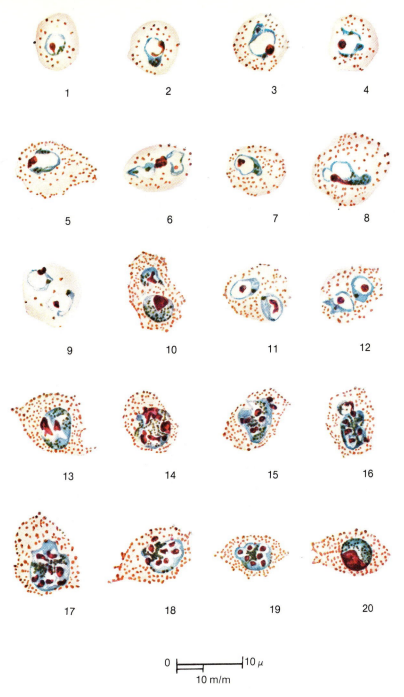

0 ⊢——————⊣ 10 μ
10 m/m

PLATE 7

Plasmodium ovale. **1,** Young ring-shaped trophozoite. **2** to **5,** Older ring-shaped trophozoites. **6** to **8,** Older ameboid trophozoites. **9, 11,** and **12,** Doubly infected cells, trophozoites. **10,** Doubly infected cell, young gametocytes. **13,** First stage of the schizont. **14** to **19,** Schizonts, progressive stages. **20,** Mature gametocyte.

From Wilcox, A. 1960. Manual for the microscopic diagnosis of malaria in man, ed. 2. National Institutes of Health Bulletin No. 180 (Revised). U.S. Government Printing Office, Washington, D.C.

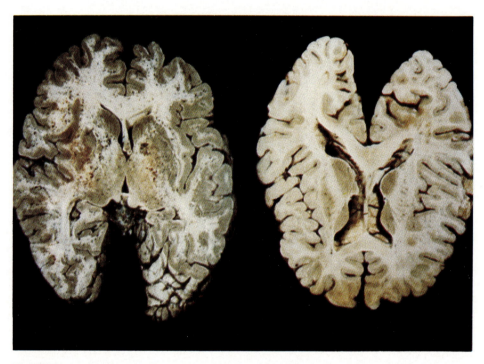

PLATE 8

Cut section of brain from cerebral malaria victim *(left)* compared with normal brain *(right)*. The cortex shows slate-gray color from hemozoin, and tiny hemorrhages (petechiae) around blood vessels in white matter can be seen. The degree to which there is swelling from fluid in the nervous tissue (edema) can be observed by comparing to ventricles and sulci of the normal brain.

Courtesy Toro González, G., G. Román-Campos, and L. Navarro de Roman, 1983. Neurologia tropical: aspectos neuropatológicos de la medicina tropical. Editorial Printer Columbiana Ltda., Columbia.

preerythrocytic merozoites reinvaded hepatocytes, then relapse should occur in all species.

Two populations of exoerythrocytic forms have now been shown.[40,41] One develops rapidly into schizonts, as previously described, but the other remains dormant as **hypnozoites** ("sleeping animalcules").[50] These have been demonstrated for *P. vivax, P. ovale,* and *P. cynomolgi,* but they have not been found in any species that does not cause relapse. How long the hypnozoite can remain capable of initiating schizogony and what triggers it to do so are unknown. Primaquine has been shown to be an effective hypnozoiticide.

It was long thought that *P. malariae,* a dangerous species in humans, also exhibited relapse, but it has been shown that this species can remain in the blood for years, possibly for the lifetime of the host, without showing signs of disease and then suddenly can initiate a clinical condition. This is more correctly known as a recrudescence, since preerythrocytic stages are not involved. The danger of transmission of this parasite in blood transfusion is evident. Treatment of this species with primaquine is unnecessary.

Epidemiology, control, and treatment. In light of the prevalence and seriousness of the disease, epidemiology and control are extremely important, and thorough consideration is far beyond the scope of this book. Some aspects of these subjects have been touched on in the preceding pages, and the following will give the reader additional insight into the problem involved (see also Chapter 39; Spencer and Strickland[78]; Bruce-Chwatt[8]).

In addition to natural or biological transmission, discussed below, malaria can be transmitted from human to human. Accidental transmission can occur by blood transfusion and by the sharing of needles by drug addicts. Although rare, infection of the newborn from an infected mother also occurs.[8] Neurosyphilis was formerly treated by deliberate infection with malaria. (A great deal of knowledge about malaria was gained during these treatments, but we still do not understand why infection with malaria alleviated the symptoms of the terrible disease of neurosyphilis.[10])

A variety of interrelated factors contributes to the level of natural transmission of the disease in a given area (Fig. 9-7). Following (modified from Young[87]) are the most important:

1. Reservoir—the prevalence of the infection in humans, and in some cases other primates, with high enough levels of parasitemia to infect mosquitoes; this would include persons with symptomatic disease and tolerant individuals
2. Vector—suitability of the local anophelines as hosts; their breeding, flight, and resting behavior; feeding preferences; and abundance

3. New hosts—availability of nonimmune hosts
4. Local climatic conditions
5. Local geographical and hydrographical conditions and human activities that determine availability of mosquito breeding areas

One must thoroughly study and understand all these factors before undertaking a malaria control program with any hope of success.

Of the approximately 390 species of *Anopheles,* some are more suitable hosts for *Plasmodium* than are others. Of those which are good hosts, some prefer animal blood other than human; therefore transmission may be influenced by the proximity with which humans live to other animals. The preferred breeding and resting places are very important. Some species breed only in fresh water, others in brackish; some like standing water around human habitations, such as puddles, or trash that collects water, such as bottles and broken coconut shells. Water, vegetation, and amount of shade are important, as are whether the species enters dwellings and rests there after feeding and whether the species flies some distance from breeding areas. As shown in the table given by Young,[87] *Anopheles* spp. exhibit an astonishing variety of such preferences; two specific examples can be cited for illustration. *Anopheles darlingi* is the most dangerous vector in South America, extending from Venezuela to southern Brazil, breeding in shady, fresh water among debris and vegetation. It invades houses and prefers human blood. *Anopheles bellator* is an important vector in cocoa-growing areas of Trinidad and coastal states of southern Brazil, breeding in partial shade in the "vases" of epiphytic bromeliads (plants that grow attached to trees and collect water in the center of their leaf rosettes). It prefers humans but enters dwellings only occasionally and returns to the forest. The importance of thorough investigation of such factors is demonstrated by cases in which swamps have been flooded with seawater to destroy the breeding habitat of the species, only to create extensive breeding areas for a brackish water species that turned out to be just as effective a vector.

Sometimes deliberate government policy exacerbates transmission of malaria. During the 1970s and 1980s, Brazil has been attempting to open up the Amazon regions to farming. This caused a great influx of nonimmune people from the cities into the newly cleared land, and malaria incidence jumped from 6/1000 per year in 1971 to 30/1000 per year in 1986.[14] As the soils of the former rain forest are poorly suited for agriculture, return of the failed farmers (often infected with malaria) to their old homes was inevitable. This has created many new foci of transmission outside of the Amazon region.[14]

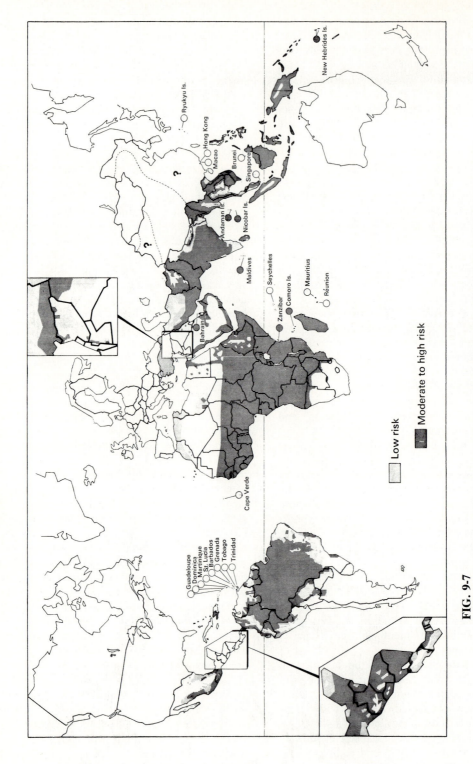

FIG. 9-7

Areas of risk for malaria transmission, December 1977.

WHO Weekly Epidemiological Record No. 22, 1979.

Valuable actions in mosquito control include destruction of breeding places when possible or practical, introduction of mosquito predators such as the mosquito-eating fish *Gambusia affinis,* and judicious use of insecticides. The efficacy and economy of DDT have been a boon to such efforts in underdeveloped countries. Although we now seem to be aware of the supposed environmental dangers of DDT, we consider these dangers preferable and minor compared with the miseries of malaria. Unfortunately, reports of DDT-resistant strains of *Anopheles* are increasing, and this phase of the battle will become more difficult in coming years. For exterminating susceptible *Anopheles* spp. that enter dwellings and rest there after feeding, spraying the insides of houses with residual insecticides can be effective and cheap, without incurring any environmental penalty. Unfortunately, some *Anopheles* rest in houses only briefly before or after feeding, and sufficient quantities of DDT are becoming difficult to obtain on the world market.[14]

Appropriate drug treatment of persons with the disease, as well as prophylactic drug treatment of newcomers to malarious areas, is an integral part of malaria control. Centuries ago the Chinese used extracts of certain plants, such as *chang shan* and *shun qi* (the roots and leaves of *Dichroa febrifuga,* family Saxifragaceae) and *qing hao* (the annual *Artemisia annua,* family Compositae) (Fig. 9-8), that actually had antimalarial properties.[36,67] In the meantime, Europeans were medically powerless and depended on absurd and superstitious remedies until **quinine** was discovered in the sixteenth century. Extracts of bark from Peruvian trees had been used with varying success to treat malaria, but alkaloids from the bark of the Peruvian tree *Cinchona ledgeriana* proved to be dependable and effective. The most widely used of these alkaloids has been quinine. The alkaloid of *D. febrifuga,* febrifugine, is now considered too toxic for human use, but the terpene from *A. annua,* called *qinghaosu,* has been recently "rediscovered" and promises to be a valuable drug.[36]

Only two synthetic antimalarials were discovered before World War II. Japanese capture of cinchona plantations early in the war created a severe quinine shortage in the United States, stimulating a burst of investigation that produced a number of important drugs. The most important of these was **chloroquine.** Subsequently a number of valuable drugs have been developed, including **primaquine, mefloquine, pyrimethamine, proguanil,** sulfonamides such as **sulfadoxine,** and antibiotics such as **tetracycline.** Only primaquine is effective against all stages of all species; the others vary in efficacy according to stages and species, with the erythrocytic stages being most suscepti-

FIG. 9-8

Artemisia annua, the source of the antimalarial drug qinghaosu, being grown in the herb garden of the College of Traditional Medicine, Guangzhou, China.

Photograph by L.S. Roberts.

ble. The drugs of choice are chloroquine and primaquine for *P. vivax* and *P. ovale* malarias and chloroquine alone for *P. malariae* infections. Chloroquine is still recommended for strains of *P. falciparum* sensitive to that drug.[20]

Resistance of *P. falciparum* to chloroquine has now spread through Asia, Africa, and South America,[63] and resistance to other drugs is often present. A combination of sulfadoxine and pyrimethamine (Fansidar) has been in use for chloroquine-resistant falciparum malaria, but Fansidar-resistant *P. falciparum* is now present in a number of areas. For multidrug resistant *P. falciparum,* mefloquine is still effective (although there are reports of mefloquine resistance), or quinine and tetracycline can be given over a period of 7 days.[52,84] Resistance to qinghaosu has not been reported in the field, but resistant strains have been produced in the laboratory.[32] Current recommendations for travelers in malarious regions are to take prophylactic doses of chloroquine regularly and to take a therapeutic dose of Fansidar only if a febrile illness develops. Prophylactic doses of Fansidar are not recommended because these drugs occasionally (1 in 5000 to 8000) cause a severe cutaneous reaction.[54,57,58]

Because of the ominous and dangerous multidrug

resistance in various strains of *P. falciparum,* it is clear that the search for satisfactory malaria treatments must continue; perhaps the answer lies in the development of vaccines. This area is the subject of intensive investigation at present, and much progress is evident.[21,55,60] The current thrust has been made possible by two major advances: development of methods whereby *P. falciparum* could be cultured in vitro,[81] thus making a large supply of organisms available; and the development of recombinant DNA technology. However, difficulties have been numerous. Different stages of the parasite have different antigens on their surface; therefore, antibodies against merozoites will not affect sporozoites or gametocytes. Also, as we already noted, the immunity produced in natural infections is only partial. Sporozoites continually slough off their outer coat and restore it with newly synthesized protein, and they have multiple epitopes.[21] Thus, they evade the host immune response by producing new binding sites and producing "decoys" in the form of sloughed molecules.

Most effort for a vaccine has centered on the sporozoites, since killing sporozoites would prevent initial infection. The gene for a surface protein in sporozoites (circumsporozoite protein, or CS protein) of *P. falciparum* has been cloned in bacteria.[60] The protein produced has been used to immunize volunteers with mixed results.[5,22] Protection from infection with *P. berghei* in mice was greater using radiation-attenuated sporozoites rather than a CS protein antigen.[15] It was found that the CS vaccine immunization was predominantly antibody mediated, while that with the attenuated sporozoites was cell-mediated. These results suggest that strategies designed to induce cell-mediated immunity will be required for efficacious sporozoite vaccine.[11,15]

In the 1950s it was widely thought that, because technical knowledge was adequate, a modicum of effort and money could achieve the eradication of malaria from large areas of the globe: its scourge would be only history. Such views were naively optimistic. Not only did we not anticipate insecticide-resistant *Anopheles* spp., drug-resistant *Plasmodium* and animal reservoirs but also insufficient account was taken of the enormous logistical problems of control in wilderness areas and of dealing with primitive peoples; neither were the disruptive effects of wars and political upheavals on control programs considered. Malaria will be with us for a long time, probably as long as there are people.

■ Metabolism of *Plasmodium* species

Selected features of *Plasmodium* metabolism follow, and we direct the reader to reviews by Fletcher and Maegraith[16] and Sherman[71] for more information.

Energy metabolism. The presence and importance of glycolysis in the degradation of glucose by *Plasmodium* spp. are well established, although subsequent steps are unclear. This is complicated by the fact that malaria species from birds have recognizable mitochondria, whereas unequivocal mitochondria have been demonstrated in very few species from mammals.[79] The bird plasmodia apparently have a functional tricarboxylic acid cycle, but the existence of the complete cycle in the erythrocytic stages of the mammalian parasites is doubtful. Membranous structures in some of the mammalian species may represent mitochondria because of certain mitochondrial enzymes demonstrated in them cytochemically (NADH- and NADPH-dehydrogenases and cytochrome oxidase). Interestingly, the sporogonic stages of these organisms in the mosquito possess prominent, cristate mitochondria, reflecting perhaps a developmental change in metabolic pattern analogous to that observed in trypanosomes. Treatment of the host with qinghaosu leads to swelling of the mitochondria of *P. inui* (a mammalian species with prominent mitochondria) within 2.5 hours.[34] Host mitochondria are unaffected. Similar reactions have been observed after primaquine treatment, leading to the suggestion that these drugs act via inhibition of mitochondrial metabolic reactions.

The erythrocytic forms of *Plasmodium* appear to be facultative anaerobes, consuming oxygen when it is available. Infected red cells take up considerably more oxygen than do uninfected ones when incubated with various substrates. It has been suggested that *Plasmodium* uses oxygen for biosynthetic purposes, especially synthesis of nucleic acids. Also, a branched electron transport system has been proposed, analogous to that suggested for some helminths (p. 337), but a classical cytochrome system has not been demonstrated. Although the bird plasmodia have cristate mitochondria, they nevertheless depend heavily on glycolysis for energy. They convert four to six molecules of glucose to lactate for every one they oxidize completely. A limiting factor may be the parasite's inability to synthesize coenzyme A, which it must obtain from its host; this cofactor is necessary to introduce the two-carbon fragment into the tricarboxylic acid cycle (see Fig. 4-13). Supplies of CoA in the mammalian erythrocyte may be even more limited and may impose restrictions on any CoA-dependent reaction.

The end products of glucose metabolism of the mammalian plasmodia are lactate and some volatile compounds, especially acetate and formate. The bird malaria parasites oxidize glucose more completely, producing some carbon dioxide and organic acids.

Both bird and mammal plasmodia "fix" carbon dioxide into phosphoenolpyruvate, as do numerous other parasites (see Fig. 21-33). In plasmodia the carbon dioxide-fixation reaction can be catalyzed by either phosphoenolpyruvate carboxykinase or phosphoenolpyruvate carboxylase. Chloroquine and quinine inhibit both enzymes, possibly accounting for the antimalarial activity of these drugs. The significance of the carbon dioxide fixation is not clearly understood; it may be to reoxidize NADH produced in glycolysis, or its reactions may function to maintain levels of intermediates for use in other cycles.

The **pentose phosphate pathway** is an important and interesting metabolic pathway in *Plasmodium*. This path has several known functions in various systems, and its importance to plasmodia is probably twofold; to furnish pentoses from hexoses for use in synthesis of nucleic acids (however, *Plasmodium* apparently lacks a full complement of enzymes for nucleic acid synthesis, which will be discussed further) and to provide reducing power in the form of NADPH. The first steps in the path are the dehydrogenation and then hydrolysis of glucose 6-phosphate to 6-phosphogluconate by the enzymes glucose 6-phosphate dehydrogenase (G6PDH) and lactonase, and the next reactions are oxidation, decarboxylation, and isomerization of the 6-phosphogluconate to D-ribose-5-phosphate (a pentose) by an isomerase and 6-phosphogluconate dehydrogenase (6PGDH). Current evidence indicates that the plasmodia are entirely dependent on G6PDH and possibly 6PGDH and the entire pathway from the host cell.[16,71] This dependency becomes even more interesting when it is observed that persons with a genetic deficiency in erythrocytic G6PDH, or **favism,** are more resistant to malaria. Favism is a sex-linked trait in which ingestion of various substances such as aspirin, the antimalarial drug primaquine, sulfonamides, or the broad bean *Vicia favia* brings on a hemolytic crisis in the female homozygote or male hemizygote. The gene is relatively frequent in blacks and some Mediterranean white people.[44] Over 5% of Southeast Asian refugees entering the United States have had a G6PDH deficiency.[70] Since the trait is expressed as a mosaic, even heterozygotes have some red cells deficient in the enzyme. Therefore all conditions—heterozygous, homozygous, and hemizygous—are protected to some extent against *P. falciparum*.[46] However, presence of the deficiency should be determined before treatment with primaquine to avoid a hemolytic crisis.[70]

Digestive metabolism. That the parasites digest host hemoglobin, leaving the iron-containing residue (hemozoin), deserves further comment. The plasmodia depend heavily on this protein source; the trophozoites substantially reduce the hemoglobin content of the erythrocyte. The parasites ingest a portion of host cytosol via the cytostome, and the vesicle thus formed migrates to and joins the central food vacuole, where the hemoglobin is rapidly degraded.[86] Chloroquine is a dibasic amine (a weak base) and increases the pH in the food vacuole to prevent the digestion of hemoglobin. Krogstad and Schlesinger[39] have suggested that the increase in pH is due both to the weak base properties of chloroquine and to some other, nonweak base effects. Chloroquine is a very safe drug because it has no nonweak base effects on mammalian cells, but the basis of chloroquine resistance in *P. falciparum* is due to interference with the nonweak base mechanism. The explanation for the nonweak base effects is unknown. Mefloquine also affects the food vacuoles,[33] and it is believed that quinine acts by a similar mechanism.[45]

Resistance to *P. falciparum* by persons homozygous and heterozygous for sickle cell hemoglobin (HbS) may involve several mechanisms, partly involving feeding and digestion by the protozoa. The parasite develops normally in cells with HbS until those cells are sequestered in the tissues.[17] Kept in this low oxygen environment for several hours, the cells have more of a tendency to sickle than cells that pass through at a normal rate. When sickling occurs, HbS forms filamentous aggregates. The filamentous aggregates actually pierce the *Plasmodium,* apparently releasing digestive enzymes that lyse both parasite and host cell. Furthermore, K^+ leaks out of the sickled cell, depriving the parasite of this ion. Sickled cells also may block capillaries, further decreasing local oxygen concentration. Other workers have shown that sickling denatures hemoglobin and releases ferriprotoporhyrin IX (FP, hemin), which has a membrane toxicity.[61] They suggested that the FP lyses the parasites.

Synthetic metabolism. As a specialized parasite, *Plasmodium* appears to depend on its host cell for a variety of molecules other than the strictly nutritional ones. Specific requirements for maintenance of the parasites free of host cells are pyruvate, malate, NAD, ATP, CoA, and folinic acid. The inability of the organisms to synthesize CoA has been mentioned. They are unable to synthesize the purine ring de novo, thus requiring an exogenous source of purines for DNA and RNA synthesis. The purine source seems to be hypoxanthine "salvaged" from the normal purine catabolism of the host cell.[48]

Several aspects of synthetic metabolism in *Plasmodium* have offered opportunities for attack with antimalarial drugs. Although plasmodia have cytoplasmic ribosomes of the eukaryotic type, several antibiotics

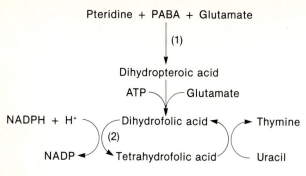

FIG. 9-9

Metabolism of folate in *Plasmodium*. (1) Site of action of PABA analogs, such as sulfadoxine, which inhibit the synthesis of dihydropteroic acid from PABA and pteridine. (2) Site of action of pyrimethamine, which inhibits synthesis of tetrahydrofolic acid from dihydrofolic acid, which prevents the synthesis of thymine required for DNA synthesis.

Redrawn from Looker, D.L., J.J. Marr, and R.L. Stotish. 1986. Modes of action of antiprotozoal agents. In Campbell, W.C., and R.S. Rew, editors. Chemotherapy of parasitic diseases. Plenum Press, New York.

that specifically inhibit prokaryotic (and mitochondrial) protein synthesis; for example, tetracycline and tetracycline derivatives, have a considerable antimalarial potency. It has been shown that tetracycline inhibits protein synthesis in *P. falciparum,* as well as growth in vitro.[7] Antibiotics have only recently been used extensively in malaria therapy because they are effective less rapidly than conventional antimalarials and because of apprehensions relative to development of resistant bacteria.

Tetrahydrofolate is a cofactor that is very important in the transfer of one-carbon groups in various biosynthetic pathways in both prokaryotes and eukaryotes. Mammals require a precursor form, **folic acid,** as a vitamin, and dietary deficiency in this vitamin inhibits growth and produces various forms of anemia, particularly because of impaired synthesis of purines and the pyrimidine thymine. In contrast, *Plasmodium* (in common with bacteria) synthesizes tetrahydrofolate from simpler precursors, including *p*-aminobenzoic acid, glutamic acid, and a pteridine (Fig. 9-9); the organisms are apparently unable to assimilate folic acid. Analogs of *p*-aminobenzoic acid such as **sulfones** and **sulfonamides** block its incorporation, and some of these (for example, sulfadoxine and dapsone) are effective antimalarials. In both the mammalian pathway and the plasmodial-bacterial pathway an intermediate product is dihydrofolate, which must be reduced to tetrahydrofolate by the enzyme **dihydrofolate reductase.** Also, this enzyme is necessary for tetrahydrofo-

late regeneration from dihydrofolate, which is produced in a vital reaction for which tetrahydrofolate is a cofactor: thymidylic acid synthesis. Thus the enzyme is vital to both parasite and host, but fortunately the dihydrofolate reductases from the two sources vary in several respects. These differences include pH optima, molecular weight, and, most important, affinity for certain inhibitors.[45] A concentration of the antimetabolites pyrimethamine and trimethoprim greater than 1000 times is required to produce 50% inhibition of the mammalian enzyme as contrasted with the plasmodial one. Therefore these drugs have been used as potent antimalarials.

Genus *Haemoproteus*

Protozoa belonging to the genus *Haemoproteus* are primarily parasites of birds and reptiles and have their sexual phases in insects other than mosquitoes. Exoerythrocytic schizogony occurs in endothelial cells; the merozoites produced enter erthrocytes to become pigmented gametocytes in the circulating blood (Fig. 9-10).

Haemoproteus columbae is a cosmopolitan parasite of pigeons. The definitive hosts and vectors of this parasite are several species of ectoparasitic flies in the family Hippoboscidae (Chapter 39). Sporozoites are injected with their bite. Exoerythrocytic schizogony is completed in about 25 days in the capillary endothelium of the lungs, with thousands of merozoites produced from each schizont. Merozoites presumably can develop directly from a schizont, or the schizont can break into numerous multinucleate "cytomeres." In this case the host endothelial cell breaks down, releasing the cytomeres, which usually lodge in the capillary lumen, where they grow, become branched, and rupture, producing many thousands of merozoites. A few of these may attack other endothelial cells, but most enter erythrocytes and develop into gametocytes. At first they resemble ring stages of *Plasmodium,* but they grow into mature microgametocytes or macrogametocytes in 5 or 6 days. Multiple infections of young forms in a single red blood cell are common, but one rarely finds more than one mature parasite per cell.

The mature macrogametocyte is 14 μm long and grows in a curve around the nucleus. Its granular cytoplasm stains a deep blue color and contains about 14 small, dark-brown pigment granules. The nucleus is small. The microgametocyte is 13 μm long, is less curved, has lighter-colored cytoplasm, and has six to eight pigment granules. The nucleus is diffuse.

Exflagellation occurs in the stomach of the fly, producing four to eight microgametes. The ookinete is like that of *Plasmodium* except there is a mass of pig-

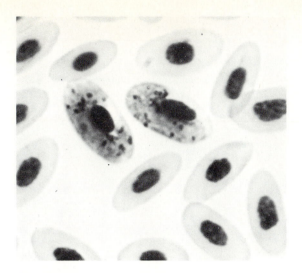

FIG. 9-10

Haemoproteus gametocytes in blood of a mourning dove. They are about 14 μm long.

Photograph by Sherwin Desser.

ment at its posterior end. It penetrates the intestinal epithelium and encysts between the muscle layers. The oocyst grows to maturity by 9 days, measuring 40 μm in diameter. Myriad sporozoites are released when the oocyst ruptures. Many of the sporozoites reach the salivary glands by the following day. Flies remain infected throughout the winter and can transmit infection to young squabs the following spring.

The pathogenesis in pigeons is slight, and infected birds usually show no signs of disease. Exceptionally, birds appear restless and lose their appetite. The air spaces of the lungs may become congested, and some anemia may result from loss of functioning erythrocytes. The spleen and liver may be enlarged and dark with pigment.

More than 80 species of *Haemoproteus* have been named from birds, mainly Columbiformes. The actual number may be much less than that, since life cycles of most of them are unknown.

Of the related genera, *Hepatocystis* spp. parasitize African and oriental monkeys, lemurs, bats, squirrels, and chevrotains; *Nycteria* and *Polychromophilus* are in bats; *Simondia* occurs in turtles; *Haemocystidium* lives in lizards; and *Parahaemoproteus* is common in a wide variety of birds.

Genus *Leucocytozoon* (Fig. 9-11)

Species of *Leucocytozoon* are parasites of birds. Schizogony is in fixed tissues, gametogony is in both leukocytes and immature erythrocytes of the verte-

brate, and sporogony occurs in insects other than mosquitoes. Pigment is absent from all phases of the life cycles. A related genus, *Akiba,* with only one species, occurs in chickens. *Leucocytozoon* has about 60 species in various birds. These are the most important blood protozoa of birds, since they are pathogenic in both domestic and wild hosts.

Leucocytozoon simondi is a circumboreal parasite of ducks, geese, and swans. The definitive hosts and vectors are black flies, family Simuliidae (see Fig. 39-1, *B*). The sporozoite, which is about 9 μm long, is injected into the avian host when the black fly feeds. After the sporozoites enter hepatocytes they develop into small schizonts, 11 to 18 μm in diameter, which produce merozoites in 4 to 6 days. Merozoites that enter red blood cells become round gametocytes (Fig. 9-12). If, however, the merozoite is ingested by a macrophage in the brain, heart, liver, kidney, lymphoid tissues, or other organ, it develops into a huge megaloschizont 100 to 200 μm in diameter. The large form is more abundant than the small hepatic schizont.

The megaloschizont divides internally into primary cytomeres, which in turn multiply in the same manner. Successive cytomeres become smaller and finally multiply by schizogony into merozoites. Up to a million merozoites may be released from a single megaloschizont.

Merozoites penetrate leukocytes or developing erythrocytes to become elongate gametocytes (Fig. 9-12). Gametocytes of both sexes are 12 to 14 μm in diameter in fixed smears and may reach 22 μm in living cells. The macrogametocyte has a discrete, red-staining nucleus. The male cell is pale staining and has a diffuse nucleus that takes up most of the space within the cell. The diffuse nucleus of the macrogametocyte has large numbers of ribosomes.[37] As the gametocytes mature, they cause their host cells to become elongate and spindle shaped.

Exflagellation produces eight microgametes and begins only 3 minutes after the organism is eaten by the fly. A typical ookinete entering an intestinal cell becomes a mature oocyst within 5 days. Only 20 to 30 sporozoites form and slowly leave the oocyst. Rather than entering the salivary glands of the vector, they enter the proboscis directly and are transmitted by contamination or are washed in by saliva.

Leucocytozoon simondi is highly pathogenic for ducks and geese, especially young birds. The death rate in ducklings may reach 85%; older ducks are more resistant, and the disease runs a slower course in them, but they still may succumb. Anemia is a prominent symptom of leukocytozoonosis, as are elevated numbers of leukocytes. The liver enlarges and be-

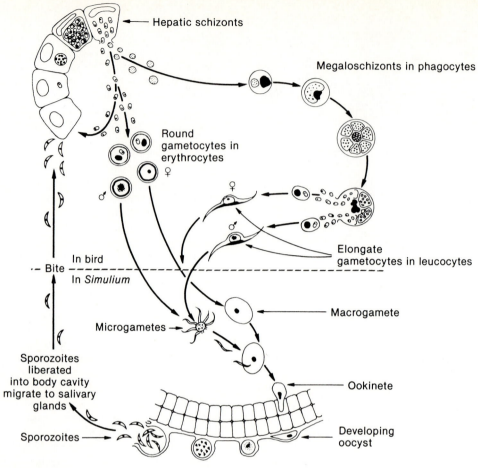

FIG. 9-11

Life cycle of *Leucocytozoon simondi*.

From Adam, K.M.G., J. Paul, and V. Zaman. 1971. Medical and veterinary protozoology. An illustrated guide. Churchill Livingston, Edinburgh.

comes necrotic, and the spleen may increase to as much as twenty times the normal size. *Leucocytozoon simondi* probably kills the host by destroying vital tissues, such as brain and heart. An outstanding feature of an outbreak of leukocytozoonosis is the suddenness of its onset. A flock of ducklings may appear normal in the morning, become ill in the afternoon, and be dead by the next morning. Birds that survive are prone to relapses but, as the result of premunition, are generally immune to reinfection.

Another species of importance is *L. smithi,* which can devastate domestic and wild turkey flocks. Its life cycle is similar to that of *L. simondi.*

SUBCLASS PIROPLASMEA

Members of the subclass Piroplasmea are small parasites of ticks and mammals. They do not produce spores, flagella, cilia, or true pseudopodia; their locomotion, when necessary, is accomplished by body flexion or gliding. No stages produce intracellular pigment. Asexual reproduction is in the erythrocytes or other blood cells of mammals by binary fission or schizogony. Sexual reproduction apparently occurs, at least in some species.[53] The components of the apical complex are reduced but warrant placement in the phylum Apicomplexa.

The single order Piroplasmida contains the two

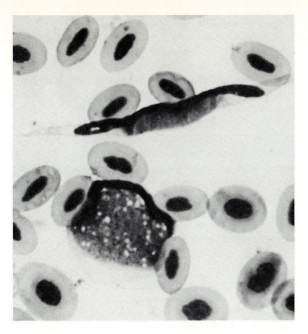

FIG. 9-12

Avian blood cells infected with elongate and round gametocytes of *Leucocytozoon simondi*. The elongate form is up to 22 μm long.

Photograph by Sherwin Desser.

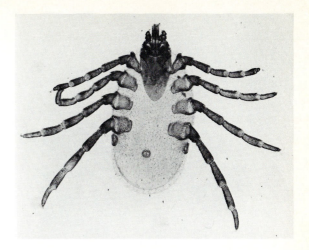

FIG. 9-13

Boophilus annulatus, the vector of *Babesia bigemina.*

Photograph by Jay Georgi.

families Babesiidae and Theileriidae, both of which are of considerable veterinary importance.

Family Babesiidae

Babesiids are usually described from their stages in the red blood cells of vertebrates. They are pyriform, round, or oval parasites of erythrocytes, lymphocytes, histiocytes, erythroblasts, or other blood cells of mammals and of various tissues of ticks. The apical complex is reduced to a polar ring, rhoptries, micronemes, and subpellicular microtubles. A cytostome is present in at least some species. Schizogony occurs in ticks. By far the most important species in America is *Babesia bigemina,* the causative agent of **babesiosis,** or **Texas redwater fever,** in cattle.

■ *Babesia bigemina*

By 1890 the entire southeastern United States was plagued by a disease of cattle, variously called Texas cattle fever, red-water fever, or hemoglobinuria. Infected cattle usually had bloody urine resulting from massive destruction of erythrocytes, and they often died within a week after symptoms first appeared. The death rate was much lower in cattle that had been reared in an enzootic area than in northern animals that were brought south. Also, it was noticed that, when southern herds were driven or shipped north and penned with northern animals, the latter rapidly succumbed to the disease. The cause of red-water fever and its mode of dissemination were a mystery when Theobald Smith and Frank Kilbourne began their investigations of this disease in the early 1880s. In a series of intelligent, painstaking experiments, they showed that the tick *Boophilus annulatus* (Fig. 9-13) was the vector and alternate host of a tiny protozoan parasite that inhabited the red blood cells of cattle and killed these relatively immense animals.[77] This not only pointed the way to an effective means of control but was also the first demonstration that a protozoan parasite could develop in and be transmitted by an arthropod. This book is replete with other examples of this phenomenon, as we have already seen.

Babesia bigemina infects a wide variety of ruminants, such as deer, water buffalo, and zebu, in addition to cattle. When in the erythrocyte of the vertebrate host, the parasite is pear shaped, round, or, occasionally, irregularly shaped and is 4 μm long by 1.5 μm wide. The organisms usually are seen in pairs within the erythrocyte (hence the name "bigemina," the twins) and are often united at the pointed tips (Fig. 9-14). At the light microscope level, they appear to be undergoing binary fission, but the electron microscope has revealed that the process is a kind of binary

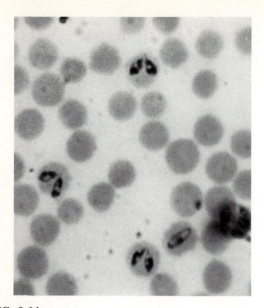

FIG. 9-14

Babesia bigemina trophozoites in the erythrocytes of a cow.
Photograph by Warren Buss.

schizogony, a budding analogous to that occurring in the Haemosporina, with redifferentiation of the apical complex and merozoite formation.[2]

Biology. The infective stage of *Babesia* sp. in the tick is called a **vermicle.** It is about 2 μm long and is pyriform, spherical, or ovoid. After completing development, the vermicles reach the salivary glands of the tick and are injected with its bite. There is no exoerythrocytic schizogony in the vertebrate. The parasites immediately enter erythrocytes, where they become trophozoites, undergo binary schizogony, and ultimately kill their host cell. The merozoites attack other red blood cells, building up an immense population in a short time. Trophozoites lie in parasitophorous vacuoles that appear as clear areas in the host cell by light microscopy. This asexual cycle continues indefinitely or until the host succumbs. Erythrocytic phases are reduced or apparently absent in resistant hosts.

Babesia bigemina is transmitted by ticks of the genus *Boophilus,* and the distribution of babesiosis is limited by the distribution of the tick. *Boophilus annulatus* is the vector in the Americas. It is a one-host tick, feeding, maturing, and mating on a single host. After engorging and mating, the female tick drops to the ground, lays her eggs, and dies. The larval, six-legged ticks that hatch from the eggs climb onto vegetation and attach to animals that brush by the plants.

One would think that a one-host tick would be a poor vector, since if they do not feed on successive hosts, how can they transmit pathogens from one animal to another? This question was answered when it was discovered that the protozoan infects the developing eggs in the ovary of the tick, a phenomenon called **transovarian transmission.**

When ingested by a feeding tick, the parasites are freed from their dead host cells by digestion. They become polymorphic at this stage, assuming a variety of shapes. It is possible, but not certain, that fertilization of one type of cell by another occurs at this time. At any rate, after 24 hours, the parasites are cigar-shaped bodies 8 to 10 μm long. They penetrate into the tick's intestinal epithelium and transform into spheroid bodies up to 16 μm in diameter within 2 days. At this time the nucleus undergoes schizogony, and the resultant vermicles, or merozoites, which are 9 to 13 μm long, migrate into the hemocoel of their host. There they attack the cells of the Malpighian tubules, where they once again undergo multiple fission. The merozoites of this generation enter the ovaries and the eggs they contain. At first they are scattered randomly throughout the egg, but after the egg is laid and begins developing, the parasites migrate to the intestinal epithelium of the larva, where schizogony takes place. Merozoites enter the hemocoels of the embryonic ticks, migrate to the salivary glands, and again undergo multiple fission in host cells. This generation of vermicles consists of enormous numbers of minute, pyriform bodies about 2 to 3 μm long by 1 to 2 μm wide. They enter the channels of the salivary glands and are injected into the vertebrate host by the feeding tick.[65]

Although this is the life cycle as it occurs in a one-host tick, two- and three-host ticks serve as hosts and vectors of *B. bigemina* in other parts of the world. In these cases transovarian transmission is not required and may not occur. All instars of such ticks can transmit the disease.

Pathology. *Babesia bigemina* is unusual in that the disease it causes is more severe in adult cattle than in calves. Calves less than a year old are seldom seriously affected, but the mortality rate in acute cases in untreated adult cattle is as high as 50% to 90%. The incubation period is 8 to 15 days, but an acutely ill animal may die only 4 to 8 days after infection. The first symptom is a sudden rise in temperature to 106° to 108° F; this may persist for a week or more. Infected animals rapidly become dull and listless and lose their appetite. Up to 75% of the erythrocytes may be destroyed in fatal cases, but even in milder infections so many erythrocytes are destroyed that a severe anemia results. Mechanisms for clearance of hemoglobin and its breakdown products are overloaded, producing jaundice, and much excess hemoglobin is excreted by the kidneys, giving the urine the red color mentioned earlier. Chronically infected animals remain thin, weak, and out of condition for several weeks before recovering. Damage to internal organs is described by Levine.[42]

Cattle that recover are usually immune for life with a sterile immunity or, more commonly, premunition. There are strain differences in the degree of immunity obtained; furthermore, little cross-reaction occurs between *B. bigemina* and other species of *Babesia*.

For unknown reasons drugs that are effective against trypanosomes are also effective against *Babesia* spp. A number of chemotherapeutic agents are available, some allowing recovery but leaving latent infection, others effecting a complete cure. It should be remembered that elimination of all parasites also eliminates premunition.

Infection can be prevented by tick control, the means by which red-water fever was eliminated from the United States. Regular dipping of cattle in a tickicide effectively eliminates the tick vector, especially if it is a one-host species. Another method that has been used is artificial premunizing of young animals with a mild strain of *Babesia* before shipping them to enzootic areas.

■ Other species of Babesiidae

Cattle seem particularly suitable as hosts to piroplasms. Other species of *Babesia* in cattle are *B. bovis,* in Europe, Russia, and Africa; *B. berbera,* in Russia, north Africa, and the Middle East; *B. divergens,* in western and central Europe; *B. argentina,* in South America, Central America, and Australia; and *B. major,* in north Africa, Europe, and Russia. Several other species are known from deer, sheep, goats, dogs, cats, and other mammals, as well as birds. Their biology, pathogenesis, and control are generally the same as for *B. bigemina*. In humans 20 to 25 cases of babesiosis have been reported: two were caused by *B. divergens,* a cattle parasite, and the others were caused by species normally parasitic in rodents. In several of the cases, three of which were fatal, the patients had been splenectomized some time before infection, and it was believed that the disabling of the immune system by splenectomy rendered the humans susceptible. However, human infection in a nonsplenectomized patient was reported from Nantucket Island off the coast of Massachusetts in 1969, another in 1973, and five more in 1975.[28] Since then, additional cases have been reported from Nantucket Island, Martha's Vineyard, Shelter Island near Long Island, New York, and eastern Long Island itself.[23,62,69] All have

been caused by *B. microti,* a parasite of meadow voles and other rodents but which can also infect pets. The vector is *Ixodes dammini,* whose adults feed on deer. Deer are refractory to infection with *B. microti,* and the infection is transmitted among rodents and to humans by nymphs of *I. dammini* and among rodents by *I. muris,* which does not feed on humans.[66] It is unclear why this supposedly rare infection has now become almost common with such a restricted distribution. However, *I. dammini* may have been recently introduced on Nantucket Island and only became abundant enough to pose a threat to humans in the last 10 to 15 years.[66]

Family Theileriidae

Like the Babesiidae, the members of this family lack a conoid. The rhoptries, micronemes, subpellicular tubules, and polar ring are well demonstrated in the tick stages. The Theileriidae parasitize blood cells of mammals, and the vectors are hard ticks of the family Ixodidae. Gamogony occurs in the gut of the nymphal tick, resulting in the formation of **kinetes,** which are very similar to the ookinetes of Haemosporina.[53] The kinetes grow in the gut cells of the tick for a time, then leave and penetrate the cells of the salivary glands, where sporogony takes place. Several members of this family infect cattle, sheep, and goats, causing a disease called **theileriosis,** which results in heavy losses in Africa, Asia, and southern Europe.

■ *Theileria parva*

Theileria parva causes a disease called **East Coast fever** in cattle, zebu, and Cape buffalo. It has been one of the most important diseases of cattle in southern, eastern, and central Africa, although it has been eliminated from most of southern Africa. The forms within erythrocytes have blue cytoplasm and a red nucleus in one end, after Romanovsky staining. At least 80% of them are rod shaped, about 1.5 to 2 μm by 0.5 to 1 μm in size. Oval and ring- or comma-shaped forms are also found.

Biology. East Coast fever, like red-water fever, is a disease of ticks and cattle, flourishing in both. The principal vector is the brown cattle tick *Rhipicephalus appendiculatus,* a three-host species. Other ticks, including one- and two-host ticks, can also serve as hosts for this parasite.

When the tick feeds, it injects all piroplasms present in its salivary glands into the next host; there they enter lymphocytes within lymphoid tissue, grow, and undergo schizogony. Schizonts, called **Koch's blue bodies,** can be seen in circulating lymphocytes within 3 days after infection. Two types of schizonts are recognized. The first generation in lymph cells comprises **macroschizonts** and produces about 90 macromerozoites, each 2 to 2.5 μm in diameter. Some of these enter other lymph cells, especially in fixed tissues, and initiate further generations of macroschizonts. Others enter lymphocytes and become **microschizonts,** producing 80 to 90 micromerozoites, each 0.7 to 1 μm wide. If microschizonts rupture while in lymphoid tissues, the micromerozoites enter new lymph cells, maintaining the lymphatic infection. However, if they rupture in the circulating blood, the micromerozoites enter erythrocytes to become the "piroplasms," typical of the disease. Apparently the parasites do not multiply in erythrocytes.

Ticks of all instars can aquire infection when they feed on blood containing piroplasms. However, because three-host ticks drop off the host to ecdyse immediately after feeding, only the nymph and adult are infective to cattle. Transovarian transmission does not occur, as it does in *Babesia* spp.

Ingested erythrocytes are digested, releasing the piroplasms that undergo gamogony and kinete formation, as described before. Sexual reproduction has been described in *T. parva, T. annulata,* and *T. ovis.*

Pathogenesis. As in babesiosis, calves are more resistant to *T. parva* than are adult cattle. Nevertheless, *T. parva* is highly pathogenic: strains with low pathogenicity kill around 23% of infected cattle, whereas highly pathogenic strains kill 90% to 100%. Symptoms such as high fever first appear 8 to 15 days after infection. Other signs are nasal discharge, runny eyes, swollen lymph nodes, weakness, emaciation, and diarrhea. Hematuria and anemia are unusual, although blood is often present in feces.

Animals that recover from theileriosis are immune from further infection, without premunition. Diagnosis depends on finding the parasites in blood or lymph smears. No drug is known to be effective once symptoms appear; however, some of the tetracyclines prevent clinical disease if given during the incubation period. Control depends on tick control and quarantine rules.

Other species of *Theileria* are *T. annulata, T. mutans, T. hirei, T. ovis,* and *T. camelensis,* all parasites of ruminants. Other genera in the family are *Haematoxenus* in cattle and zebu and *Cytauxzoon* in antelope, both in Africa.

REFERENCES

1. Aikawa, M., J.R. Rabbege, I. Udeinya, and L.H. Miller. 1983. Electron microscopy of knobs in *Plasmodium falciparum*-infected erythrocytes. J. Parasitol. 69:434-437.
2. Aikawa, M., and C.R. Sterling. 1974. Intracellular parasitic Protozoa. Academic Press, Inc., New York.
3. Aikawa, M., I.J. Udeinya, J. Rabbege, M. Dayan, J.H. Leech, R.J. Howard, and L.H. Miller. 1985. Structural alteration of the membrane of erythrocytes infected with *Plasmodium falciparum.* J. Protozool. 32:424-429.
4. Avidor, B., J. Golenser, C.H.J. Schutte, G.A. Cox, M. Isaacson, and D. Sulitzeanu. 1987. A radioimmunoassay for the diagnosis of malaria. Am. J. Trop. Med. Hyg. 37:225-229.
5. Ballou, W.R., J.A. Sherwood, F.A. Neva, D.M. Gordon, R.A. Wirtz, G.F. Wasserman, C.L. Diggs, S.L. Hoffman, M.R. Hollingdale, W.T. Hockmeyer, I. Schneider, J.F. Young, P. Reeve, and J.D. Chulay. 1987. Safety and efficacy of a recombinant DNA *Plasmodium falciparum* sporozoite vaccine. The Lancet 6 June 1987:1277-1281.
6. Bignami, A. 1913. Concerning the pathogenesis of relapses in malarial fevers. South. Med. J. 6:79-88.
7. Blum, J.J., A. Yayon, S. Friedman, and H. Ginsberg. 1984. Effects of mitochondrial protein synthesis inhibitors on the incorporation of isoleucine into *Plasmodium falciparum* in vitro. J. Protozool. 31:475-479.
8. Bruce-Chwatt, L.J. 1980. Essential malariology. William Heinemann Medical Books Ltd., London.
9. Canfield, C.J. 1972. Malaria in U.S. military personnel 1965-1971. In Sadun, E.H., editor. Basic research in malaria. Special issue, Proc. Helm. Soc. Wash. 39:15-18.
10. Chernin, E. 1984. The malariatherapy of neurosyphilis. J. Parasitol. 70:611-617.
11. Clark, I.A. 1987. Cell-mediated immunity in protection and pathology of malaria. Parasitol. Today 3:300-305.
12. Coatney, G.R. 1976. Relapse in malaria—an enigma. J. Parasitol. 62:3-9.
13. Corradetti, A. 1950. Ospite definitive e ospite intermedio di parassiti della malaria. Riv. Parasitol. 11:89.
14. Cruz Marques, A. 1987. Human migration and the spread of malaria in Brazil. Parasitol. Today 3:166-170.
15. Egan, J.E., J.L. Weber, W.R. Ballou, M.R. Hollingdale, W.R. Majarian, D.M. Gordon, W.L. Maloy, S.L. Hoffman, R.A. Wirtz, I. Schneider, G.R. Woollett, J.F. Young, and W.T. Hockmeyer. 1987. Efficacy of murine malaria sporozoite vaccines: implications for human vaccine development. Science 236:453-456.
16. Fletcher, A., and B. Maegraith. 1972. The metabolism of the malaria parasite and its host. In Dawes, B., editor. Advances in parasitology, vol. 10. Academic Press, Inc., New York, pp. 31-48.
17. Friedman, M.J., and W. Trager. 1981. The biochemistry of resistance to malaria. Sci. Am. 244:154-164 (Mar.).
18. Garnham, P.C.C. 1966. Malaria parasites and other *Haemosporidia.* Blackwell Scientific Publications Ltd., Oxford, Eng.
19. Garnham, P.C.C. 1977. The continuing mystery of relapses in malaria. Protozool. Abstr. 1:1-12.
20. Geary, T.G., and J.B. Jensen. 1986. Protozoan infections of man: malaria. In Campbell, W.C., and R.S. Rew, editors. Chemotherapy of parasitic diseases. Plenum Press, New York.
21. Godson, G.N. 1985. Molecular approaches to malaria vaccines. Sci. Am. 252:52-59 (May).
22. Greenwood, B.M. 1987. Asymptomatic malaria infections—do they matter? Parasitol. Today 3:206-214.
23. Grunwaldt, E. 1977. Babesiosis on Shelter Island, N.Y. State J. Med. 77:1320-1321.
24. Guazzi, M., and S. Grazi. 1963. Consideratione sa un caso di malaria quartana recidivante dopo se anni di latenza. Riv. Malar. 42:55-59.

25. Halawani, A., and A.A. Shawarby. 1957. Malaria in Egypt. J. Egypt. Med. Assoc. 40:753-792.

26. Hall, A.P., and C.J. Canfield. 1972. Resistant falciparum malaria in Vietnam: its rarity in Negro soldiers. In Sadun, E.H., editor. Basic research in malaria. Special issue, Proc. Helm. Soc. Wash. 39:66-70.

27. Harrison, G. 1978. Mosquitoes, malaria and man: a history of the hostilities since 1880. E.P. Dutton & Co., Inc., New York.

28. Healy, G.R., A. Spielman, and N. Gleason. 1976. Human babesiosis: reservoir of infection on Nantucket Island. Science 192:479-480.

29. Hoeppli, R. 1969. Parasitic diseases in Africa and the western hemisphere. Early documentation and transmission by the slave trade. Verlag für Recht und Gesellschaft AG, Basel.

30. Holmberg, F.C., Shenton, L. Franzen, K. Janneh, R.W. Snow, U. Pettersson, H. Wigzell, and B.M. Greenwood. 1987. Use of a DNA hybridization assay for the detection of Plasmodium falciparum in field trials. Am. J. Trop. Med. Hyg. 37:230-234.

31. Igarashi, I., M.M. Oo, H. Stanley, R. Reese, and M. Aikawa. 1987. Knob antigen deposition in cerebral malaria. Am. J. Trop. Med. Hyg. 37:511-515.

32. Inselberg, J. 1985. Induction and isolation of artemisinine-resistant mutants of Plasmodium falciparum. Am. J. Trop. Med. Hyg. 34:417-418.

33. Jacobs, G.H., M. Aikawa, W.K. Milhous, and J.R. Rabbege. 1987. An ultrastructural study of the effects of mefloquine on malaria parasites. Am. J. Trop. Med. Hyg. 36:9-14.

34. Jiang, J.-B., G. Jacobs, D.-S. Liang, and M. Aikawa. 1985. Qing-haosu-induced changes in the morphology of Plasmodium inui. Am. J. Trop. Med. Hyg. 34:424-428.

35. Kennedy, J.F. 1962. Message of first day of issue of U.S. malaria eradication stamp.

36. Klayman, D.L. 1985. Qinghaosu (artemisinin): an antimalarial drug from China. Science 228:1049-1055.

37. Kocan, A.A., and K.M. Kocan. 1978. The fine structure of elongate gametocytes of Leucocytozoon ziemanni (Laveran). J. Parasitol. 64:1057-1059.

38. Kreier, J.P., and J.R. Baker. 1987. Parasitic protozoa. Allen and Unwin, Boston.

39. Krogstad, D.J., and P.H. Schlesinger. 1987. The basis of antimalarial action: non-weak base effects of chloroquine on acid vesicle pH. Am. J. Trop. Med. Hyg. 36:213-220.

40. Krotoski, W.A., W.E. Collins, R.S. Bray, P.C.C. Garnham, F.B. Cogswell, R.W. Gwadz, R. Killick-Kendrick, R. Wolf, R. Sinden, L.C. Koontz, and P.S. Stanfill. 1982. Demonstration of hypnozoites in sporozoite-transmitted Plasmodium vivax infection. Am. J. Trop. Med. Hyg. 31:1291-1293.

41. Krotoski, W.A., P.C.C. Garnham, R.S. Bray, D.M. Krotoski, R. Killick-Kendrick, C.C. Draper, G.A.T. Targett, and M.W. Guy. 1982. Observations on early and late post-sporozoite tissue stages in primate malaria. I. Discovery of a new latent form of Plasmodium cynomolgi (the hypnozoite), and failure to detect hepatic forms within the first 24 hours after infection. Am. J. Trop. Med. Hyg. 31:24-35.

42. Levine, N.D. 1973. Protozoan parasites of domestic animals and of man, ed. 2. Burgess Publishing Co., Minneapolis.

43. Levine, N.D. 1985. Phylum II. Apicomplexa Levine, 1970. In Lee, J.J., S.H. Hutner, and E.C. Bovee, editors. An illustrated guide to the protozoa. Society of Protozoologists, Lawrence, Kansas.

44. Levitan, M., and A. Montagu, 1971. Textbook of human genetics. Oxford University Press, New York.

45. Looker, D.L., J.J. Marr, and R.L. Stotish. 1986. Modes of action of antiprotozoal agents. In Campbell, W.C., and R.S. Rew, editors. Chemotherapy of parasitic diseases. Plenum Press, New York.

46. Luzzatto, L., E.A. Usanga, and S. Reddy. 1969. Glucose-6-phosphate dehydrogenase deficient red cells: resistance to infection by malarial parasites. Science 164:839-842.

47. Maegraith, B.G., and A. Fletcher. 1972. The pathogenesis of mammalian malaria. In Dawes, B., editor. Advances in parasitology, vol. 10. Academic Press, Inc., New York, pp. 49-75.

48. Manandhar, M.S.P., and K. van Dyke. 1975. Detailed purine salvage metabolism in and outside the free malarial parasite. Exp. Parasitol. 37:138-146.

49. Manson-Bahr, P. 1963. The story of malaria: the drama and the actors. Int. Rev. Trop. Med. 2:329-390.

50. Markus, M.B. 1976. Possible support for the sporozoite hypothesis of relapse and latency in malaria. Trans. R. Soc. Trop. Med. Hyg. 70:535.

51. Mathews, H.M., and J.C. Armstrong. 1981. Duffy blood types and vivax malaria in Ethiopia. Am. J. Trop. Med. Hyg. 30:299-303.

52. Meek, S.R., E.B. Doberstyn, B.A. Gaüzere, C. Thanapanich, E. Nordlander, and S. Phuphaisan. 1986. Treatment of falciparum malaria with quinine and tetracycline or combined mefloquine/sulfadoxine/pyrimethamine on the Thai-Kampuchean border. Am. J. Trop. Med. Hyg. 35:246-250.

53. Mehlhorn, H., E. Schein, and M. Warnecke. 1979. Electron-microscopic studies on Theileria ovis Rodhain, 1916: Development of kinetes in the gut of the vector tick, Rhipicephalus evertsi evertsi Neumann, 1897, and their transformation within the cells of the salivary glands. J. Protozool. 26:377-385.

54. Miller, K.D., H.O. Lobel, R.F. Satriale, J.N. Kuritsky, R. Stern, and C.C. Campbell. 1986. Severe cutaneous reactions among American travelers using pyrimethamine-sulfadoxine (Fansidar) for malaria prophylaxis. Am. J. Trop. Med. Hyg. 35:451-458.

55. Miller, L.H., R.J. Howard, R. Carter, M.F. Good, V. Nussenzweig, and R.S. Nussenzweig. 1986. Research toward malaria vaccines. Science 234:1349-1356.

56. Miller, L.H., S.J. Mason, D.F. Clyde, and M.H. McGinnis. 1976. The resistance factor to Plasmodium vivax in blacks. The Duffy bloodgroup genotype FyFy. N. Engl. J. Med. 295:302-304.

57. Morbidity and Mortality Weekly Report. 1986a. Outbreak of malaria imported from Kenya. M.M.W.R. 35:567-568, 573.

58. Morbidity and Mortality Weekly Report. 1986b. Need for malaria prophylaxis by travelers to areas with chloroquine-resistant Plasmodium falciparum. M.M.W.R. 35:21-22, 28.

59. Nijhout, M.M., and R. Carter, 1978. Gamete development in malaria parasites: bicarbonate-dependent stimulation by pH in vitro. Parasitology 76:39-53.

60. Nussenzweig, V., and R.S. Nussenzweig. 1986. Development of a sporozoite malaria vaccine. Am. J. Trop. Med. Hyg. 35:678-688.

61. Orjih, A.U., R. Chevli, and C.D. Fitch. 1985. Toxic heme in sickle cells: an explanation for death of malaria parasites. Am. J. Trop. Med. Hyg. 34:223-227.

62. Parry, M.F., M. Fox, S.A. Burka, and W.J. Richar. 1977. *Babesia microti* infection in man. J.A.M.A. 238:1282-1283.

63. Payne, D. 1987. Spread of chloroquine resistance in *Plasmodium falciparum*. Parasitol. Today 3:241-246.

64. Pollack, Y., S. Metzger, R. Shemer, D. Landau, D.T. Spira, and J. Golenser. 1985. Detection of *Plasmodium falciparum* in blood using DNA hybridization. Am. J. Trop. Med. Hyg. 34:663-667.

65. Rick, R.F. 1964. The life cycle of *Babesia bigemina* (Smith and Kilbourne, 1893) in the tick vector *Boophilus microplus* (Canastrini). Aust. J. Agr. Res. 15:802–821.

66. Ruebush, T.K., II, D.D. Juranek, A. Spielman, J. Piesman, and G.R. Healy. 1981. Epidemiology of human babesiosis on Nantucket Island. Am. J. Trop. Med. Hyg. 30:937-941.

67. Russell, P.F. 1955. Man's mastery of malaria. Oxford University Press, London.

68. Russell, P.F., L.S. West, and R.D. Manwell. 1946. Practical malariology. W.B. Saunders Co., Philadelphia.

69. Scharfman, W.B., and E.G. Taft. 1977. Nantucket fever. An additional case of babesiosis. J.A.M.A. 238:1281-1282.

70. Schwartz, I.K., W. Chin, J. Newman, and J.M. Roberts. 1984. Glucose-6-phosphate dehydrogenase deficiency in Southeast Asian refugees entering the United States. Am. J. Trop. Med. Hyg. 33:182-184.

71. Sherman, I.W. 1979. Biochemistry of *Plasmodium* (malarial parasites). Microbiol. Rev. 43:453-495.

72. Sherman, I.W. 1985. Membrane structure and function of malaria parasites and the infected erythrocyte. Parasitology 91:609-645.

73. Shortt, H.C., and P.C.C. Garnham. 1948. Demonstration of a persisting exoerythrocytic cycle in *Plasmodium cynomolgi* and its bearing on the production of relapses. Br. Med. J. 1:1225-1232.

74. Sinden, R.E., and R.H. Hartley. 1985. Identification of the meiotic division of malarial parasites. J. Protozool. 32:742-744.

75. Sinden, R.E., R.H. Hartley, and L. Winger. 1985. The development of *Plasmodium* ookinetes *in vitro:* an ultrastructural study including a description of meiotic division. Parasitology 91:227-244.

76. Smith, D.C., and L.B. Sanford. 1985. Laveran's germ: the reception and use of a medical discovery. Am. J. Trop. Med. Hyg. 34:2-20.

77. Smith, T., and F.L. Kilbourne, 1893. Investigations into the nature, causation, and prevention of Texas or southern cattle fever. U.S. Dept. Agr. Bur. Anim. Indust. Bull. 1.

78. Spencer, H.C., and G.T. Strickland. 1984. Malaria. In Strickland, G.T., editor. Hunter's tropical medicine, ed. 6. W.B. Saunders Co., Philadelphia.

79. Sterling, C.R., M. Aikawa, and R.S. Nussenzweig. 1972. Morphological divergence in a mammalian malarial parasite: the fine structure of *Plasmodium brasilianum*. In Sadun, E.H., editor. Basic research in malaria. Special issue, Proc. Helm. Soc. Wash. 39:109-128.

80. Terzakis, J.A., J.P. Vanderberg, D. Foley, and S. Shustak. 1979. Exoerythrocytic merozoites of *Plasmodium berghei* in rat hepatic Kupffer cells. J. Protozool. 26:385-389.

81. Trager, W., and J.B. Jensen. 1976. Human malaria parasites in continuous culture. Science 193:673-675.

82. Warrell, D.A. 1987. Pathophysiology of severe falciparum malaria in man. Symposia of the British Society of Parasitology, vol. 24. Parasitology 94:S53-S76.

83. Warrell, D.A., S. Looareesuwan, R.E. Phillips, N.J. White, M.J. Warrell, H.M. Chapel, S. Areekul, and S. Tharavanij. 1986. Function of the blood-cerebrospinal fluid barrier in human cerebral malaria: rejection of the permeability hypothesis. Am. J. Trop. Med. Hyg. 35:882-889.

84. White, N.J. 1988. The treatment of falciparum malaria. Parasitol. Today 4:10-14.

85. World Health Organization Malaria Action Programme. 1986. Severe and complicated malaria. Trans. R. Soc. Trop. Med. Hyg. 80(suppl.):1-50.

86. Yayon, A.R., Timberg, S. Friedman, and H. Ginsberg. 1984. Effects of chloroquine on the feeding mechanism of the intraerythrocytic human malarial parasite *Plasmodium falciparum*. J. Protozool. 31:367-372.

87. Young, M.D. 1976. Malaria. In Hunter, G.W., III, J.C. Swartzwelder, and D.F. Clyde. Tropical medicine, ed. 5. W.B. Saunders Co., Philadelphia.

SUGGESTED READINGS

Brown, A.W.A., J. Haworth, and A.R. Zahar. 1976. Malaria eradication and control from a global standpoint. J. Med. Entomol. 13:1-25.

Bruce-Chwatt, L.J. 1980. Essential malariology. William Heinemann Medical Books, London.

Bruce-Chwatt, L., and J. de Zulueta. 1980. The rise and fall of malaria in Europe. Oxford University Press, New York.

Faust, E.C. 1951. The history of malaria in the United States. American Scientist 39:121-129.

Harrison, G. 1978. Mosquitoes, malaria and man: a history of the hostilities since 1880. E.P. Dutton & Co., Inc., New York.

Joyner, L.P., and J. Donnelly. 1979. The epidemiology of babesial infections. In Lumsden, W.H.R., R. Muller, and J.R. Baker, editors. Advances in parasitology, vol. 17. Academic Press, inc., New York, pp. 115-140.

Kreier, J.P., editor. 1980. Malaria, vols. 1, 2, and 3. Academic Press, Inc., New York.

Mattingly, P.F. 1976. Evolution of the malarias: the problems of origins. Parassitologia 18:1-8.

Sinden, R.E. 1983. Sexual development of malarial parasites. In Baker, J.R., and R. Muller, Advances in parasitology, vol. 22. editors. Academic Press, London.

Walliker, D. 1983. The genetic basis of diversity in malaria parasite. In Baker, J.R., and R. Muller, editors, Advances in parasitology, vol. 22. Academic Press, London.

CHAPTER 10

PHYLA MYXOZOA AND MICROSPORA: PROTOZOA WITH POLAR FILAMENTS

Members of these two phyla formerly were placed in a class (Cnidosporidea) of the Sporozoa because they form spores. However, it is now recognized that they are quite different from the apicomplexans and, indeed, bear little if any relationship to each other. In fact some workers do not consider the Myxozoa to be protozoa because they are composed of more than one cell. In both groups the polar filaments are tubelike and held coiled within the spores. When eaten, the polar filaments are expelled. In the Myxozoa the filaments are contained within **polar capsules** and apparently serve an anchoring function after expulsion. In the Microspora the polar filament pierces the intestinal epithelium of the host, and the ameba-like **sporoplasm** passes through the filament into the host cell. The Myxozoa attack lower vertebrates, and a few are known from invertebrates.[16] Microsporans are mostly parasites of invertebrates, but some are found in lower vertebrates and rarely humans. So far as is known, most life cycles of both groups are direct, but some microsporans require a second host. They probably are a major natural control of some insect populations.

PHYLUM MYXOZOA

In the phylum Myxozoa the spores are of multicellular origin and are surrounded by two, three, or rarely more valves of various shapes (Fig. 10-1). They have one or more polar capsules and are mostly parasites of fish. A few are reported from amphibians and reptiles, but none is known from birds or mammals. From one to four polar capsules can be found at one end of the spore, except in the suborder Bipolarina, where one capsule is located at each end of the spore. Next to the polar capsules is an ameboid sporoplasm that is infective to the host. Some species have large vacuoles in the sporoplasm that stain readily with iodine and are therefore called **iodinophilous vacuoles.** The valves join at a **sutural plane** that is either twisted or straight. The valves may bear various markings and often are extended as pointed processes at the "posterior" end. More than 1200 species in 46 genera are described in this phylum. Most are host and tissue specific.

Family Myxosomatidae

Of the many families of myxozoans, few are more striking in appearance and importance than are the Myxosomatidae. Fish parasites, they have two or four polar capsules in the spore stage, and their sporoplasm lacks iodinophilous vacuoles.[5] One species is of circumboreal importance to salmonid fish, including trout.

■ *Myxobolus cerebralis*

Myxobolus cerebralis causes **whirling disease** in salmonids, so called because fish with the disease swim in circles when disturbed or feeding. The parasite appears to have been endemic in the brown trout, *Salmo trutta,* from central Europe to southeast Asia, and it causes no disease symptoms in that host. The disease was first noticed in 1900 after the introduction of the rainbow trout, *Salmo gairdneri,* to Europe. Since then it has spread to other localities in Europe, including Sweden and Scotland; to the United States; to South Africa; and to New Zealand.[4] Whirling disease results in a high mortality rate in very young fish and causes corresponding economic loss, especially in hatchery-reared brook and rainbow trout. If a fish survives, damage to the cranium and vertebrae can cause crippling and malformation.

Morphology. The mature spore of *M. cerebralis* (Fig. 10-1, *D*) is broadly oval, with thick sutural ridges on the edges of the valves. It measures 7.4 to 9.7 μm long by 7 to 10 μm wide. The entire spore is covered with a mucoid-like envelope. Two polar capsules are found at the anterior end, each with a filament twisted into five or six coils. During development each polar capsule lies within a polar cell that also contains a nucleus, and the nuclei of the two valvogenic cells may be seen lying adjacent to the inner surface of each valve. The sporoplasm contains two

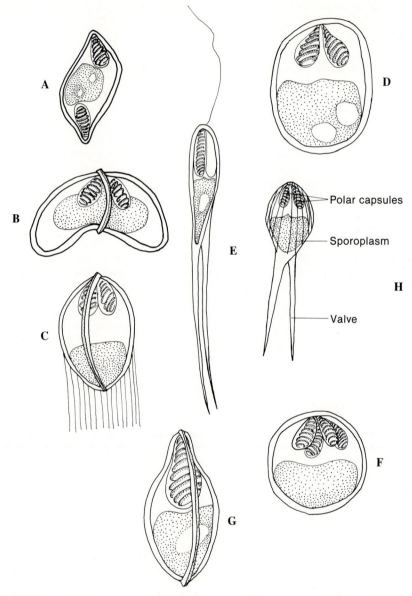

FIG. 10-1

Spores of representative genera of myxosporans. **A,** *Myxidium,* with a polar capsule at each end. **B,** *Ceratomyxa.* **C,** *Sphaerospora.* **D,** *Myxobolus.* **E,** *Henneguya.* **F,** *Chloromyxum,* with four polar capsules. **G,** *Thelohanellus,* with a single polar capsule. **H,** *Myxobilatus.*

Drawings by Ian Grant.

nuclei (presumably haploid), numerous ribosomes, mitochondria, and other typical organelles.[13]

Biology. The life history of *M. cerebralis* has not been fully described, and its details have been quite elusive. Spores from the bottom mud of ponds that were left aging for at least 4 months were reported to be infective when eaten by trout fry.[22] Other attempts to duplicate the experiment have been unsuccessful. Subsequently it has been reported that the spores must first be eaten by tubificid oligochaete annelids, which in turn are eaten by fish.[15] More recently, Wolf and Markiw[26] reported results of their experiments indicating that tubificids are obligatory alternate hosts in the life cycle of *Myxobolus cerebralis*. When the spore is ingested by *Tubifex tubifex* its sporoplasm develop into a *Triactinomyxon,* a myxozoan parasite of annelids, that belongs to a different class, Actinosporea. *Triactinomyxon* has its own complex life cycle in the worm, involving sexual reproduction and the production of spores with three valves (contrasted with two in *Myxobolus*). According to the experiments of Wolf and Markiw, these spores are released from the tubificid and are then infective to trout. If true, this life cycle would be unparalled in having two hosts, each with sexual and asexual stages of development and with spores of different construction. Another difficulty with this theory is that there are many more known species of myxosporeans than actinosporeans. One would expect the known numbers to be more equal. So far, the findings of Wolf and Markiw have not been duplicated.

At any rate, on the basis of knowledge of related species, we would presume that the valves open once they gain entrance to the fish host, and the polar capsules shoot out their filaments in a manner analogous to eversion of the finger of a glove. The polar filament creates a wound in the intestinal epithelium through which the sporoplasm can enter. Either before infection or soon thereafter, the two nuclei of the sporoplasm fuse; it is believed that the nuclei are haploid, and the fusion is autogamous. The sporoplasm makes its way to its preferred site, the cartilage of the head and spine. There it begins to grow, the nuclei dividing repeatedly. There is no corresponding cytokinesis, and by 4 months the multinucleate trophozoite will have reached a diameter of 1 mm (some species can reach a size of several millimeters). It apparently feeds by digesting the surrounding cartilage, thus creating the cavity within which it lies. During the course of the nuclear divisions, two types of nuclei can be distinguished, **generative** and **somatic** (Fig. 10-2). As development proceeds, a certain amount of cytoplasm becomes segregated around each generative nucleus to form a separate cell within the trophozoite. These cells will produce the spores; hence they are called **sporoblasts.** Because in most species each will give rise to more than one spore, they are called **pansporoblasts.** Each pansporoblast in *M. cerebralis* will produce two spores. The generative nucleus for each spore will divide four times, one of the daughter nuclei of each division remaining generative, the other becoming somatic. The first somatic daughter nucleus will form the outer envelope of the spore; the second will divide again to give rise to the valvogenic cells, and the third nucleus will divide to produce the nuclei of the polar cells. Thus the spore of the Myxozoa is of multicellular origin. The fourth division of the generative nucleus produces the two nuclei of the sporoplasm, and this (or one of the preceding divisions) is reductional so that the nuclei of the sporoplasm are haploid.

The cavities within the cartilage become packed with trophozoites and spores by 8 months after infection. Spores may live in the fish for 3 or more years. How they escape into the water is speculative, but it seems reasonable to assume that, when the host is devoured by a larger fish or other piscivorous predator, such as a kingfisher or heron, the spores are released by digestion of their former home. The crippling effect of the parasite can make the host especially vulnerable to predation. The feces of birds that have been fed fish infected with *M. cerebralis* can carry the disease organism and subsequently infect fish held in previously uncontaminated water.[22]

Pathogenesis. The main pathogenic effects of this disease can be attributed to damage to the cartilage in the axial skeleton of young fish, consequent interference with function of adjacent neural structures, and subsequent granuloma formation in healing of the lesions. Invasion of the cartilaginous capsule of the auditory-equilibrium organ behind the eye interferes with coordinated swimming; thus, when the fish is disturbed or tries to feed, it begins to whirl frantically, as if chasing its tail. It may become so exhausted by this futile activity that it sinks to the bottom and lies on its side until it regains strength. Predation most likely occurs at this stage.[7] Often the cartilage of the spine is invaded, especially posterior to the twenty-sixth vertebra. Function of the sympathetic nerves controlling the melanocytes is interfered with, and the posterior part of the fish becomes very dark, producing the "black tail." If the fish survives, granulomatous tissue infiltration of the skeleton may produce permanent deformities: misshapen head, permanently open or twisted lower jaw, or severe spinal curvature (scoliosis) (Fig. 10-3).

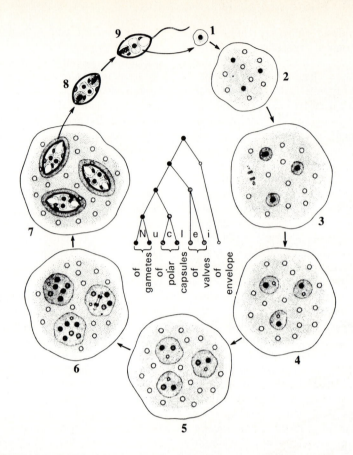

FIG. 10-2

Diagram of the development of a myxosporidian. **1,** Uninucleate amebula. **2,** Multinucleate plasmodium: differentiation into generative *(dark)* and somatic *(light)* nuclei. **3,** Segregation of the sporoblasts. **4** to **6,** Different stages of nuclear multiplication in the sporoblasts, corresponding to the divisional sequence shown in the middle of the diagram (envelope nucleus, *white:* valve nuclei, *cross-hatched;* nuclei of the polar capsule, *dotted;* gamete nuclei [germ line!], *black*). **7,** Plasmodium with spores. **8,** Single spore with binucleate amebula. **9,** Single spore with uninucleate amebula and a polar capsule with discharged polar filament.

From Grell, K.G. 1973. Protozoology. Springer-Verlag, New York.

Epidemiology and prevention. In ponds in which infected fish are held it seems clear that spores can accumulate, whether by release from dead and decomposing fish, passage through predators, or some kind of escape from the tissue of infected living fish. Severity of an outbreak depends on the degree of contamination of a pond, and light infections cause little or no overt disease. Spores are resistant to drying and freezing, surviving for a long period of time, up to 18 days at $-20°$ C.[8]

No effective treatment for infected fish is known, and such fish should be destroyed by burial or incineration. Great care should be exercised to avoid transferring spores to uncontaminated hatcheries or streams, either by live fish that might be carriers or by feeding possibly contaminated food materials, for example, tubificids, to hatchery fish. Earthen and concrete ponds in which infected fish have been held can be disinfected by draining and treating with calcium cyanamide or quicklime. Application of ultraviolet irradiation can disinfect contaminated water.[6] Clearly, we need detailed data on myxozoan life cycles before we can plan effective measures for control.

Extracellular and intracellular development of Myxozoa. In the case of *Myxobolus cerebralis,* described earlier, development of the parasite is intercellular, occurring in the matrix between host cells. This pattern is found in many species. In many others de-

FIG. 10-3

Axial skeleton deformities in living rainbow trout that have recovered from whirling disease *(Myxosoma cerebralis)*. **A,** Note bulging eyes, shortened operculum, and both dorso-ventral and lateral curvature of the spinal column (lordosis and scoliosis). **B,** Note gaping, underslung jaw, and grotesque cranial granuloma.

Photographs by L.S. Roberts.

velopment is intracellular, with the host cell becoming greatly enlarged, or hypertrophic. The hypertrophic host cell, together with its developing parasites, forms a nodule, sometimes several millimeters wide, called a **xenoparasitic complex** or **xenoma.** Less commonly hypertrophy of a host cell is induced by contact with the parasite without actual infection of the cell. Often the xenomas die and are resorbed by the host.[12]

Extrasporogonic phases of the life cycle. Several species of Myxozoa, including the commercially important *Sphaerospora renicola* in carp, have an asexually proliferating phase in the blood of the host. This stage only increases the number of parasites; it does not develop directly into spores. A second extrasporogonic phase invades the swim bladder of carp fry, causing swimbladder inflammation that results in high mortality or growth retardation. Some small plasmodia (ameboid forms) reach the renal tubules where they either produce spores (seasonally) or are destroyed by host reactions.[12]

Some other species, belonging to different genera and families, also commonly occur in fish, amphibians, and reptiles. For general reviews and keys see Hoffman,[5] Hoffman et al.,[7,9] Sprague,[21] and Lom.[12]

PHYLUM MICROSPORA

The phylum Microspora includes fewer than 1000 described species of intracellular parasites of invertebrates and lower (rarely higher) vertebrates.[20] They have been found in protozoa, platyhelminths, nematodes, bryozoa, rotifers, annelids, all classes of arthropods, fish, amphibians, reptiles, birds and some mammals. A few species are known to infect humans. Numerous species are pathogenic, and several are of economic importance. The spores are unicellular and have a single sporoplasm; the spore walls are complete, without suture lines, pores, or other openings. They have a simple or complex extrusion apparatus with polar tube and polar cap. The extrusion apparatus of the class Microsporea is described below; the class Rudimicrosporea has a simple extrusion apparatus without a polaroplast and posterior vacuole. The rudimicrosporeans are all hyperparasites of gregarines (p. 114), which in turn are mostly parasites of marine annelids.

The Microspora formerly included the class Haplosporea. The haplosporideans and some newly described and lesser known forms have been placed in the phylum Ascetospora.[11] The spores are often, or always, multicellular and complex, with one or more sporoplasms but without polar capsules or filaments. They are parasites of a variety of invertebrates; *Haplosporidium* spp. (formerly *Minchinia* spp.) are pathogens in the economically important oyster *Crassostrea virginica.*

For techniques of study see Canning and Lom.[2]

Class Microsporea

The spore is the most conspicuous and morphologically distinctive stage in the life cycle of microsporeans. Spores are ovoid, spheroid, or cylindroid. The spore wall is trilaminar, consisting of an outer, dense exospore; an electron-lucent middle layer (endospore); and a thin membrane surrounding the cytoplasmic contents. Some species have two to five layers of exospore. The wall is dense and refractile; its resistant properties contribute greatly to the survival of the spore. Spores are usually about 3 to 6 μm in length, and little structure can be discerned under the light microscope other than an apparent vacuole at one or both ends. The smallest known spore is *Encephalitozoon* from mammals (2.5 × 1.5 μm), while the largest is *Mrazekia piscicola* from cod (20 × 6 μm). It has long

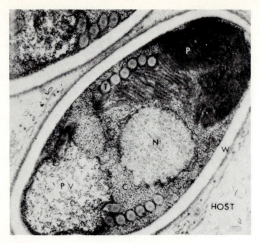

FIG. 10-4

Nosema lophii spore displaying polaroplast *(P)*, nucleus *(N)*, ribosome-rich cytoplasm *(C)*, polar tube *(T)*, posterior vacuole *(PV)*, and wall *(W)*.

From Weidner, E. 1970. Z. Parasitenkd. 40:230-234.

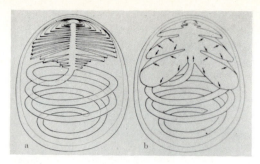

FIG. 10-5

Nosema lophii: diagrammatic interpretation of polaroplast membranes before *(a)* and during *(b)* collapse of polar sac and polar aperture before extrusion.

From Weidner, E. 1970. Z. Parasitenkd. 40:230-234.

been known that microsporeans possess a **polar tube,** or **filament,** since the structure can be stimulated to extrude artificially.

Possession of this structure has been the basis for uniting the microsporans with the myxozoans in the Cnidospora. However, use of the electron microscope has shown that the polar filaments of the two groups are basically dissimilar. There is no polar capsule in the microsporans, and neither is the polar filament formed by a separate capsulogenic cell. At the ultrastructural level one can see a small **polar cap** or **sac** covering the attached end of the filament and just overlying the **polaroplast** (the apparent anterior vacuole) (Fig. 10-4). The ameboid **sporoplasm** surrounds the extrusion apparatus, with its nucleus and most of its cytoplasm lying within the coils of the filament. A **posterior vacuole** may be found at the end opposite the polaroplast. The cytoplasm of the sporoplasm has many free ribosomes and some endoplasmic reticulum but no mitochondria, peroxisomes, or typical Golgi membranes. The membrane and matrix of the polar cap are continuous, with a highly pleated membrane comprising the polaroplast. This in turn is continuous with the anchoring disc or base of the polar filament.[25] When polar filament extrusion is stimulated in the host, a permeability change in the polar cap apparently allows water to enter the spore, and the filament is expelled explosively, simultaneously turning "inside out." The stacked membrane in the polaroplast is unfolded as the filament apparently is dis-

charged and contributes to the expelled filament so that it is much longer than when it is coiled within the spore (Fig. 10-5). The force with which the filament is extruded causes it to penetrate any cell in its path, and the sporoplasm flows through the tubular filament, thereby gaining access to its host cell. The end of the filament within the host cell expands to enclose the sporoplasm and becomes the parasite's new outer membrane.

The nuclei of the intracellular trophozoite may divide repeatedly, and the organism may become a large, multinucleate plasmodium. Finally cytokinesis takes place, and the process may then be repeated. In some species the nuclei may be associated in pairs **(diplokarya),** but such association apparently is not involved with sexual reproduction. The multiple fission of the trophozoites (or meronts) is usually regarded as schizogony, but the process may not be strictly analogous to the schizogony found in the Apicomplexa.

Sporogenesis occurs when the nuclear divisions of the monokaryotic or dikaryotic trophozoites give rise to nuclei destined to become spore nuclei. In a number of genera, *not* including *Nosema,* the nuclear division preceding sporogony is meiotic (reductional), giving rise to haploid spores.[3,14] In these genera the spores are not directly infective to new hosts, leading to the suggstion that there is an alternate (intermediate?) host in which restoration of the diploid condition occurs. For example, see Canning and Hollister[1] for *Amblyospora* in copepods and mosquitoes. Sexual reproduction seems to be restricted to plasmogamy, not karyogamy.

During sporogony the organism becomes a multinucleate, sporogonial plasmodium. This can occur either by internal segregation of cytoplasm around the nuclei to become sporont-determinate areas or by the formation of an envelope at the sporont surface and subsequent separation from developing sporoblasts, leaving a vacuolar space.[17] The spores then differentiate and mature within the pansporoblast. A mass of tubules forms in each sporoblast, which becomes the polar tube and polaroplast.[10] Mitochondria are not present at any stage. A xenoma of considerable size is developed in some species.[2]

■ Family Nosematidae

The genera of Nosematidae are separated on the basis of the number of spores produced by each sporoblast mother cell during the life cycle (from 1 to 16 or more).

Nosema apis. *Nosema apis* is a common parasite of honeybees in many parts of the world, causing much loss annually to beekeepers. It infects the epithelial cells in the midgut of the insect. Infected bees lose strength, become listless, and die. Although the ovaries of the queen are not directly infected, they degenerate when her intestinal epithelium is damaged, an example of parasitic castration. The disease is variously known as nosema disease, spring dwindling, bee dysentery, bee sickness, and May sickness.

The spore of *N. apis* is oval, measuring 4 to 6 μm long by 2 to 4 μm wide. The extended filament is 250 to 400 μm long. Infected bees defecate spores that are infective to other bees when the spores are eaten. Swallowed spores enter the midgut and lodge on the peritrophic membrane. Extruded filaments pierce the peritrophic membrane and intestinal epithelium, and the sporoplasm enters the epithelial cell. The entire process is accomplished within 30 minutes. Sporogony takes place in the second multiple fission generation, and the spores rupture the host cell to be passed with the feces. The entire life history in the bee is completed in 4 to 7 days. Destruction of the intestinal epithelium may kill the host.

OTHER *NOSEMA* SPECIES. Besides *N. apis,* a few other of the many species in this genus are of known direct importance to humans, although many additional ones may be important biological controls of insect populations. *N. bombycis* is a parasite of silk moth larvae, flourishing in the unnaturally crowded conditions of silkworm culture. The parasite affects nearly all tissues of the insect's body, including the intestinal epithelium. Parasitized larvae show brown or black spots on their bodies, giving them a peppered appearance. There is a high rate of mortality. Pasteur, in 1870, devoted considerable effort to understanding and controlling this disease and is credited with saving the silk industry in the French colonies. This also was one of the first "germs" proved to cause disease. The life cycle of *N. bombycis* is basically similar to that of *N. apis* and can be completed in 4 days.

Ameson michaelis is a parasite of the economically important blue crab, *Callinectes sapidus,* which it may kill 15 to 29 days after infection.[24] *Ameson michaelis* undergoes schizogony, probably in the hemocytes, and then undergoes sporogony in the striated muscle, which is extensively damaged.

Species of *Glugea, Pleistophora, Nosema,* and other genera parasitize fish, including several economically important groups, and serious epizootics have been reported.

Encephalitozoon cuniculi is found in laboratory mice and rabbits, and it is also known in monkeys, dogs, rats, birds, guinea pigs, and other mammals, including humans, usually in the brain.[18] It may be transmitted in body exudates or transplacentally. Although damage is usually minimal, the infection can be fatal. It is the most extensively studied of all Microsporidia. At various early times it was thought to be the cause of rabies and polio.

At least two infections in immunocompetent humans are known.[1] High levels of anti-*E. cuniculi* antibodies are common in immunodeficient patients but low in uncompromised people, suggesting that it is yet another opportunistic parasite.

Other microsporidian species have been isolated from AIDS patients and others who were unable to rally their lymphocytic defenses. *Enterocytozoon bieneusi, Microsporidium ceylonensis, M. africanum,* and *Nosema connori* have all been reported in humans. There are no effective drugs for the treatment of microsporidiosis.

REFERENCES

1. Canning, E.U., and W.S. Hollister. 1987. Microsporidia of mammals—widespread pathogens or opportunistic curiosities? Parasitol. Today 3:267-273.
2. Canning, E.U., and J. Lom. 1986. The microsporidia of vertebrates. Academic Press, London.
3. Hazard, E.I., T.G. Andreadis, D.J. Joslyn, and E.A. Ellis. 1979. Meiosis and its implications in the life cycles of *Amblyospora* and *Parathelohania* (Microspora). J. Parasitol. 65:117-122.
4. Hewitt, G.C., and R.W. Little. 1972. Whirling disease in New Zealand trout caused by *Myxosoma cerebralis* (Hofer, 1903) (Protozoa; Myxosporida). N. Z. J. Mar. Freshwater Res. 6:1-10.
5. Hoffman, G.L. 1967. Parasites of North American freshwater fishes. University of California Press, Berkeley.
6. Hoffman, G.L. 1975. Whirling disease *(Myxosoma cerebralis):* control with ultraviolet irradiation and effect on fish. J. Wildlife Dis. 11:505-507.
7. Hoffman, G.L., C.E. Dunbar, and A. Bradford. 1969. Whirling disease of trouts caused by *Myxosoma cerebralis* in the

United States. U.S. Department of Interior, Fish and Wildlife Service, Special Scientific Report, Fisheries No. 427 (1962 report issued with addendum, 1969).

8. Hoffman, G.L., and R.E. Putz. 1969. Host susceptibility and the effect of aging, freezing, heat, and chemicals on spores of *Myxosoma cerebralis*. Progressive Fish-Culturist 31:35-37.

9. Hoffman, G.L., R.E. Putz, and C.E. Dunbar. 1965. Studies on *Myxosoma cartilaginis* n. sp. (Protozoa: Myxosporidea) of centrarchid fish and a synopsis of the *Myxosoma* of North American freshwater fishes. J. Protozool. 12:319-332.

10. Krinsky, W.L., and S.F. Hayes, 1978. Fine structure of the sporogonic stages of *Nosema parkeri*. J. Protozool. 25:177-186.

11. Levine, N.D., et al. 1980. A newly revised classification of the Protozoa. J. Protozool. 27:37-58.

12. Lom, J. 1987. Myxosporea: a new look at long-known parasites of fish. Parasitol. Today 3:327-332.

13. Lom, J., and P. dePuytorac. 1965. Studies on the myxosporidean ultrastructure and polar capsule development. Protistologica 1:53-65.

14. Loubès, C. 1979. Recherches sur la méiose chez les microsporidies: conséquences sur les cycles biologiques. J. Protozool. 26:200-208.

15. Markiw, M.E., and K. Wolf. 1983. *Myxosoma cerebralis* (Myxozoa: Myxosporea) etiologic agent of salmonid whirling disease requires tubificid worm (Annelida: Oligochaeta) in its life cycle. J. Protozool. 30:561-564.

16. Overstreet, R.M. 1976. *Fabespora vermicola* sp. n., the first myxosporidan from a platyhelminth. J. Parasitol. 62:680-684.

17. Overstreet, R.M., and E. Weidner. 1974. Differentiation of microsporidian spore-tails in *Inodosporus spraguei* gen. et sp. n. Z. Parasitenkd. 44:169-186.

18. Shadduck, J.A., W.T. Watson, S.P. Pakes, and A. Cali. 1979. Animal infectivity of *Encephalitozoon cuniculi*. J. Parasitol. 65:123-129.

19. Sprague, V. 1982. Ascetospora. In Parker, S.P., editor. Synopsis and classification of living organisms, vol. 1. McGraw-Hill, Inc., New York, pp. 599-601.

20. Sprague, V. 1982. Microspora. In Parker, S.P., editor. Synopsis and classification of living organisms, vol. 1. McGraw-Hill, Inc., New York, pp. 589-594.

21. Sprague, V. 1982. Myxozoa. In Parker, S.P., editor. Synopsis and classification of living organisms, vol. 1. McGraw-Hill, Inc., New York, pp. 595-597.

22. Taylor, R.L., and M. Lott. 1978. Transmission of salmonid whirling disease by birds fed trout infected with *Myxosoma cerebralis*. J. Protozool. 25:105-106.

23. Uspenskaya, A.V. 1984. The cytology of Myxosporidia. Izd. Nauka, Moscow.

24. Weidner, E. 1970. Ultrastructural study of microsporidian development. 1. *Nosema* sp. Sprague, 1965, in *Callinectes sapidus* Rathbun. Z. Zellforsch. 105:33-54.

25. Weidner, E. 1972. Ultrastructural study of microsporidian invasion into cells. Z. Parasitenkd. 40:227-242.

26. Wolf, K., and M.E. Markiw. 1984. Biology contravenes taxonomy in the Myxozoa: new discoveries show alternation of invertebrate and vertebrate hosts. Science 225:1449-1452.

SUGGESTED READING

Hamilton, A.J., and E.U. Canning. 1985. Transmission of *Myxosoma cerebralis* in trout. Proceedings EAFP, Montpellier, p. 21.

CHAPTER 11

PHYLUM CILIOPHORA: CILIATED PROTOZOA

The possession of simple cilia or compound ciliary organelles in at least one stage of their life cycle is the most conspicuous feature of the Ciliophora. A compound subpellicular infraciliature is universally present, even when cilia are absent. Most species have one or more macronuclei and micronuclei, and fission is homothetogenic. Some species exhibit sexual reproduction involving conjugation, autogamy, and cytogamy. Although each cilium has a kinetosome, centrioles functioning as such are absent. Most ciliates are free living, but many are commensals of vertebrates and invertebrates, and a few are parasitic. The following examples will illustrate the phylum.

CLASS KINETOFRAGMINOPHOREA
Subclass Vestibulifera

The class is named for structures typically found in the oral region of its members: "kinetofragments." These are patches or short files of basically somatic kinetids, only some of which may bear cilia. As now conceived, the class includes some very primitive ciliates, as well as some very specialized ones. The subclass Vestibulifera is composed of those that have a **vestibulum** at the apical or near-apical end of the body. The vestibulum is a depression or invaginated area that leads directly to the cytostome; it is lined with cilia predominantly somatic in nature and origin. In the order Trichostomatida the somatic ciliature is typically distributed uniformly over the body, and there is little reorganization of the somatic kineties at the level of the vestibulum. The curiously appearing entodiniomorphids (order Entodiniomorphida) have unique tufts of cilia on an otherwise naked body and a generally firm pellicle. They are commensals in mammalian herbivores. The remaining order (Colpodida) has a highly reorganized vestibular ciliature. In our consideration of this ciliate class we will discuss in detail only the trichostomatid species *Balantidium coli*, which is an important parasite of humans.

■ Order Trichostomatida

Family Balantidiidae. The family Balantidiidae has the single genus *Balantidium,* species of which are found in the intestines of crustaceans, insects, fish, amphibians, and mammals. The vestibulum leading into the cytostome is at the anterior end, and a cytopyge is present at the posterior tip.

BALANTIDIUM COLI. Balantidium coli is the largest protozoan parasite of humans. It is most common in tropical zones but is present throughout the temperate climes as well. The epidemiology and effects on the host are similar to those of *Entamoeba histolytica.* The organism appears to be basically a parasite of pigs, with strains adapted to various other hosts.

Morphology. Trophozoites (Fig. 11-1) of *B. coli* are oblong, spheroid, or more slender, 30 to 150 μm long by 25 to 120 μm wide. Encysted stages (Fig. 11-2), which are most commonly found in stools, are spheroid or ovoid, measuring 40 to 60 μm in diameter. The macronucleus is a large, sausage-shaped structure. The single micronucleus is much smaller and often hidden from view by the macronucleus. There are two contractile vacuoles, one near the middle of the body and the other near the posterior end. The cytostome is at the anterior end. Food vacuoles contain erythrocytes, cell fragments, starch granules, and fecal and other debris. Living trophozoites and cysts are yellowish or greenish.

Biology. The ciliate *B. coli* lives in the cecum and colon of humans, pigs, guinea pigs, rats, and many other mammals. It is not readily transmissible from one species of host to another, since it seems to require a period of time to adjust to the symbiotic flora of a new host. However, when adapted to a host species, the protozoan flourishes and can become a serious pathogen, particularly in humans. In animals other than primates the organism is unable to initiate a lesion by itself, but it can become a secondary invader if the mucosa is breached by other means.

The trophozoite multiplies by transverse fission. Conjugation has been observed in culture but may oc-

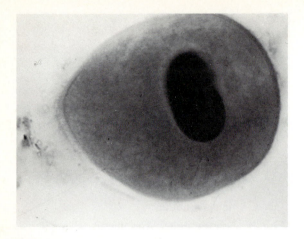

FIG. 11-1

Trophozoite of *Balantidium coli*. Trophozoites range from 30 to 150 μm long by 25 to 120 μm wide.

Photograph by James Jensen.

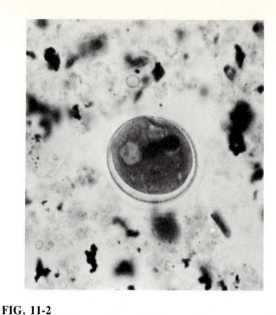

FIG. 11-2

Encysted form of *Balantidium coli*. Cysts are 40 to 60 μm in diameter.

cur only rarely, if at all, in nature. Encystment is instigated by dehydration of feces as they pass posteriad in the rectum. These protozoa can encyst after being passed in stools—an important factor in the epidemiology of the disease. Infection occurs when the cyst is ingested, usually in contaminated food or water. Unencysted trophozoites may live up to 10 days and may possibly be infective if eaten, although this is unlikely under normal circumstances. Since *B. coli* is destroyed by a pH lower than 5, infection is most likely to occur in malnourished persons with low stomach acidity.

Pathogenesis. Under ordinary conditions the trophozoite feeds much like a *Paramecium,* ingesting particles with the vestibulum and cytostome. However, sometimes it appears that the organisms can produce proteolytic enzymes that digest away the intestinal epithelium of the host. Production of hyaluronidase has been detected, and this enzyme could help enlarge the ulcer. The ulcer usually is flask shaped, like an amebic ulcer, with a narrow neck leading into an undermining saclike cavity in the submucosa. The colonic ulceration produces lymphocytic infiltration with few polymorphonuclear leukocytes, and hemorrhage and secondary bacterial invasion may follow. Fulminating cases may produce necrosis and sloughing of the overlying mucosa and occasionally perforation of the large intestine or appendix, as in amebic dysentery. Death often follows at this stage. Secondary foci, such as the liver or lung, may become infected.[1] Urogenital organs are sometimes attacked after contamination, and

vaginal, uterine, and bladder infections have been discovered.

Epidemiology. Balantidiasis in humans is most common in the Philippines but can be found almost anywhere in the world, especially among those who are in close contact with swine. Generally the disease is considered rare and is found in less than 1% of the human population. Higher infection rates have been reported among institutionalized persons. However, in pigs the infection rate may be 20% to 100%. Primates other than humans sometimes are infected and may represent a reservoir of infection to humans, although the reverse is probably more likely. The cysts can remain alive for weeks in pig feces, if the feces do not dry out. The pig is probably the usual source of infection for humans, but the relationship is not clear. The protozoa in swine are essentially nonpathogenic and are considered by some a separate species, *B. suis.* There may be differing strains of *B. coli* that vary in their adaptability to humans.

Treatment and control. Several drugs are used to combat infections of *B. coli,* including carbarsone, diiodohydroxyquin, and tetracycline. The infection often disappears spontaneously in healthy persons, or it can become symptomless, making the person a carrier. Prevention and control duplicate those of *Entamoeba histolytica,* except that particular care should be taken by those who work with pigs.

Other species of *Balantidium* are *B. praenuleatum,* common in the intestines of American and oriental cockroaches; *B. duodeni* in frogs; and *B. caviae* in guinea pigs.

CLASS OLIGOHYMENOPHOREA

Members of this class have a "buccal cavity" bearing a well-defined oral ciliary apparatus. However, the oral ciliary apparatus may be rather inconspicuous. It is composed of only three or four specialized membranelles.

Subclass Hymenostomatia, Order Hymenostomatida

The body ciliature in members of this subclass is often uniform and heavy, and conspicuous kinetodesmata are regularly present. In contrast to the following subclass, sessile and stalked forms are rare. The Hymenostomatida have a well-defined buccal cavity on their ventral surface. Most species are small, but *Ichthyophthirius multifiliis* is a very large one.

■ Family Ophryoglenidae

The family Ophryoglenidae contains one genus of parasites, most of which are unimportant to humans. One species, however, is a common pest in freshwater aquaria and in fish farming, causing much economic loss.

Ichthyophthirius multifiliis (Fig. 11-3). *Ichthyophthirius multifiliis* causes a common disease in aquarium and wild freshwater fish, known as **ick** to many fish culturists. It attacks the epidermis, cornea, and gill filaments.

MORPHOLOGY. Adult trophozoites are as large in diameter as 1 mm. The macronucleus is a large, horseshoe-shaped body that encircles the tiny micronucleus. Each of several contractile vacuoles has its own micropore in the pellicle. A permanent cytopyge is located at the posterior end of the animal.

BIOLOGY. Mature trophozoites form pustules in the skin of their fish hosts (Fig. 11-4). They are set free and swim feebly about when the pustules rupture, finally settling on the bottom of their environment or on vegetation. Within an hour the ciliate secretes a thick, gelatinous cyst about itself and begins a series of transverse fissions, producing up to 1000 infective cells. The daughter trophozoites, or **tomites,** also termed **theronts** or **swarmers,** represent the infective stage and can survive about 96 hours without a host. The tomite is about 40×15 μm. Its narrowed anterior end carries a characteristic long filament that emerges from a conical depression in the pellicle.[2] Apparently, the parasite burrows into the fish's skin

with its pointed end and filament. There it becomes a trophozoite within 3 days, ingesting debris of host cells and forming a pustule that reaches over 1 mm in diameter.

PATHOGENESIS. Grayish pustules form wherever the parasites colonize in the skin. Epidermal cells combat the irritation by producing much mucus, but many die and are sloughed. When many parasites attack the gill filaments, they so interfere with gas exchange that the fish may die. Catfish that recover show protective immunity for up to 8 months, which shows promise for the development of vaccines.[4]

Aquarium fish can be treated successfully with very dilute concentrations of formaldehyde, malachite green, or methylene blue.

Subclass Peritrichia, Order Peritrichida

Although the oral ciliary field is prominent in members of this subclass, the somatic ciliature is much reduced. There is a temporary posterior circlet of locomotor cilia, and many are stalked and sessile. All possess an aboral **scopula,** a structure at the aboral pole composed of a field of kinetosomes with immobile cilia. It functions as a holdfast or may be involved in the formation of the stalk.

■ Family Trichodinidae

Species in the family Trichodinidae lack stalks and are mobile. The oral-aboral axis is shortened, with a prominent basal disc usually at the aboral pole. A protoplasmic fringe, or velum, lies on the margin of the basal disc, and a circle of strong cilia lies underneath. A second circle of cilia, above the disc, cannot always be found. The biology of the group is poorly known. The family contains seven genera, with *Trichodina* being a typical example.[3]

Trichodina species. Members of this genus parasitize a wide variety of aquatic invertebrates, fish, and amphibians. The basal disc contains a corona of hard, pointed "teeth" that aid the parasite in attaching to its host (Fig. 11-5). The number, arrangement, and shapes of these teeth are useful taxonomic characters. The buccal ciliary spiral makes more than one, but fewer than two, complete turns. Species of *Trichodina* may cause some damage to the gills of fish, but most produce little pathogenic effect and are of interest only as beautiful examples of highly evolved protozoa with incredibly specialized organelles. Typical examples are *T. californica* on the gills of salmon, *T. pediculus* on *Hydra,* and *T. urinicola* in the urinary bladder of amphibians.

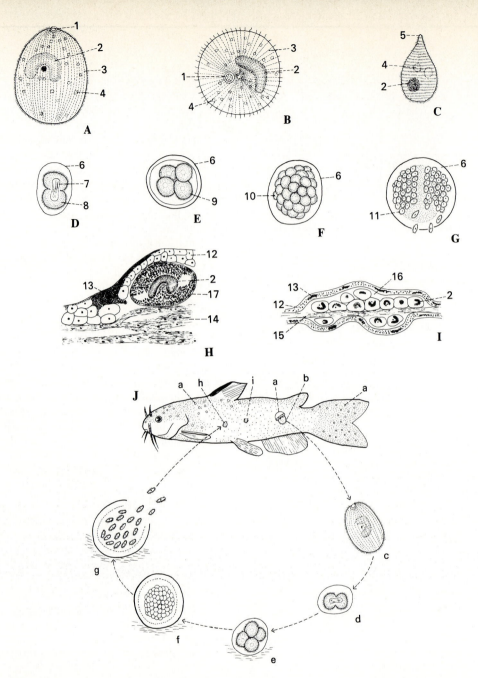

FIG. 11-3

Life cycle of *Ichthyophthirius multifiliis*. **A,** Fully developed trophozoite from pustule. **B,** Anterior end of fully developed trophozoite. **C,** Tomite from cyst. **D** and **E,** First and second divisions of encysted trophozoite. **F,** Later stage of cystic multiplication. **G,** Cyst filled with tomites, some of which are escaping into water. **H,** Section of skin of fish, showing full-grown trophozoite embedded in it. **I,** Section of tail of carp, showing ciliates developing in pustule. **J,** Infected bullhead (*Ameiurus melas*). *1,* Cytostome; *2,* macronucleus with nearby micronucleus; *3,* longitudinal rows of cilia; *4,* contractile vacuoles; *5,* boring or penetrating apparatus; *6,* cyst; *7,* dividing of macronucleus; *8,* two daughter cells formed by first division; *9,* four daughter cells formed by second division in cyst; *10,* numerous daughter cells; *11,* tomites; *12,* epidermis of fish skin; *13,* pigment cell in epidermis; *14,* dermis; *15,* cartilaginous skeleton of tail of carp; *16,* pustule containing trophozoites; *17,* trophozoite under skin; *a,* pustules; *b,* trophozoite escaping from pustule into water; *c,* trophozoite free in water; *d,* encysted trophozoite on bottom of pond in first division, showing two daughter cells; *e,* cyst in second division with four daughter cells; *f,* cyst with many daughter cells; *g,* ruptured cyst liberating tomites; *h,* tomite attached to skin; *i,* tomite partially embedded in skin.

From Olsen, O.W. 1974. Animal parasites, their life cycles and ecology. © 1974 University Park Press, Baltimore.

FIG. 11-4

Sunfish infected with *Ichthyophthirius multifiliis*. Note the light-colored pustules in the skin.

From Hoffman, G. 1977. In Kreier, J., editor. Protozoa of medical and veterinary interest. Academic Press, Inc., New York.

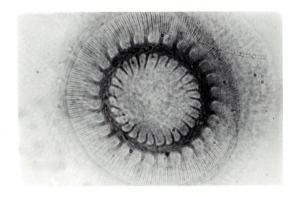

FIG. 11-5

Trichodina sp. from the gill of a fish. *Trichodina* are 35 to 60 μm in diameter, with a height of 25 to 55 μm.

Photograph by Warren Buss.

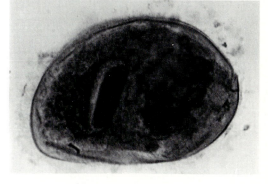

FIG 11-6

Nyctotherus cordiformis trophozoite from the colon of a frog. These protozoa range from 60 to 200 μm in length.

Photograph by Warren Buss.

CLASS POLYHYMENOPHOREA

The Polyhymenophorea have a well-developed, conspicuous system of membranelles in and around their buccal cavity (**adoral zone of membranelles [AZM]**) (Fig. 4-4). The body ciliature may be reduced, or the cilia may be joined into compound organelles called **cirri** (Fig. 4-4).

Order Heterotrichida
■ Family Plagiotomidae

Somatic ciliature is sparse in most species in this order but, when present, is usually uniform. Buccal ciliature is conspicuous, with the AZM typically composed of one to many membranelles or undulating membranes that wind clockwise to the cytosome. Most species are quite large.

179

The Plagiotomidae are robust parasites of the intestine of vertebrates and invertebrates. The entire body has tiny cilia arranged in longitudinal rows. A single undulating membrane extends from the anterior end to deep within the cytopharynx.

The most common genus is *Nyctotherus,* which is easily obtained for laboratory use. These ciliates (Fig. 11-6) are ovoid to kidney shaped, with the cytostome on one side. The anterior half contains a massive macronucleus, with a small micronucleus nearby. The genus has numerous species, some of which are useful in routine laboratory exercises. Common species are *N. ovalis* in cockroaches and *N. cordiformis* in the colon of frogs and toads.

REFERENCES

1. Dorfman, S., O. Rangel, and L.G. Bravo. 1984. Balantidiasis: report of a fatal case with appendicular and pulmonary involvement. Trans. Soc. Trop. Med. Hyg. 78:833-834.

2. McCartny, J.B., et al. 1985. Scanning electron microscopic studies of the life cycle of *Ichthyophthirius multifiliis.* J. Parasitol. 71:218-226.

3. Van As, J.G., and L. Basson. 1987. Host specificity of trichodinid ectoparasites of freshwater fish. Parasitol. Today 3:88-90.

4. Woo, P.T.K. 1987. Immune response of fish to parasitic protozoa. Parasitol. Today 3:186-188.

SUGGESTED READINGS

Bykhovskaya-Pavlovskaya, I.E. 1962. Key to the parasites of freshwater fish. Academy of Science, Moscow. English translation, Israel Program for Scientific Translations, Jerusalem (1964). (An oustanding reference to ciliate parasites.)

Corliss, J.O. 1979. The ciliated protozoa. Characterization, classification and guide to the literature. Pergamon Press, Oxford. (Advanced treatise but essential to serious students of ciliates.)

Hoffman, G.L. 1967. Parasites of North American freshwater fishes. University of California Press, Berkeley. (Ciliates of North American fish are listed in this useful reference work.)

Levine, N.D. 1973. Protozoan parasites of domestic animals and of man, ed. 2. Burgess Publishing Co., Minneapolis.

CHAPTER 12

PHYLUM MESOZOA: PIONEERS OR DEGENERATES?

Mesozoa are tiny, ciliated animals that parasitize marine invertebrates. Their affinities with other phyla are obscure, chiefly because of the simplicity of their structure and their unusual biology. Digestive, circulatory, nervous, and excretory systems are lacking. Basically, a mesozoan's body is made of two layers of cells, but these are not homologous with the endoderm and ectoderm of diploblastic animals.

Two distinct groups traditionally are placed in the phylum Mesozoa: the classes Rhombozoa and Orthonectida. However, these two groups are so different in morphology and life cycles that they probably should be placed in separate phyla.[1] Since no one has thus far formally proposed the separation, we will include both within the phylum Mesozoa, although recognizing the artificiality of the scheme.

CLASS RHOMBOZOA

Rhombozoans are parasites of the renal organs of cephalopods, either lying free in the kidney sac or attached to the renal appendages of the vena cava. Partial life cycles are known for a few species, but certain details are lacking in all cases. Interesting histories of the group are presented by Stunkard.[7,8]

Order Dicyemida

The most prominent developmental stages in the cephalopod are the **nematogens** and the **rhombogens** (Fig. 12-1 and 12-2). Their bodies are composed of a **polar cap,** or **calotte,** and a **trunk.** The calotte is made up of two tiers of cells, usually with four or five cells in each. The anterior tier is called the **propolar;** the posterior is called the **metapolar.** The cells in the two tiers may be arranged opposite or alternate to each other, depending on the genus. The trunk comprises relatively large axial cells surrounded by a single layer of ciliated, somatic cells. The axial cells give rise to new individuals, as in the following description.

The earliest known stage in the cephalopod is a cil-

iated larva, the **larval stem nematogen.** The axial cells of the larval stem nematogen each contain a **vegetative nucleus** and a **germinative nucleus.** The germinative nucleus becomes an **agamete.** The stem nematogen grows larger while agametes continue to divide, becoming aggregates of cells, in a process much like the asexual, internal reproduction (germ balls) found in miracidia, sporocysts, and rediae of digenetic trematodes (Chapter 17). The animal is now called an **adult stem nematogen.**

Within the axial cell, agametes develop into vermiform embryos (Fig. 12-3) of **primary nematogens** that escape the body of the stem nematogen and attach to the kidney tissues of the host. Agametes within the axial cell of the primary nematogen produce many generations of identical vermiform embryos that develop into primary nematogens, building up a massive infection in the cephalopod. When the host becomes sexually mature, the production of primary nematogens ceases. Instead, the vermiform embryos form stages that become primary **rhombogens,** similar to nematogens in cell number and distribution but with a different method of reproduction and with lipoprotein- and glycogen-filled somatic cells (Fig. 12-2). These cells may become so engorged that they swell out, and the animal appears lumpy. Some primary nematogens metamorphose directly into rhombogens, and rhombogens derived by this route are referred to as **secondary rhombogens.**

Both primary and secondary rhombogens produce agametes in the axial cell that divide to become nonciliated **infusorigens.** An infusorigen is a mass of reproductive cells that represents either a hermaphroditic sexual stage or a hermaphroditic gonad.[5] It remains within the axial cell and produces male and female gametes, which fuse in fertilization. The zygotes detach from the infusorigen, and each then divides to become a hollow, ciliated ovoid stage called an **infusoriform larva,** which is the most complex stage in the life cycle.[6] This microscopic larva consists of a fixed

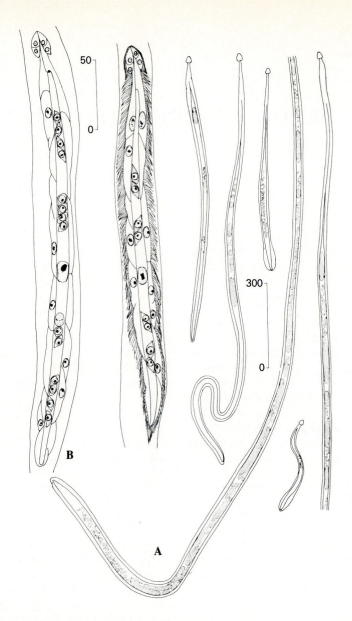

FIG. 12-1

Dicyemennea antarcticensis. **A,** Entire nematogens. **B,** Vermiform embryos within axial cells of nematogens. (Scales in micrometers.)

From Short, R.B., and F.G. Hochberg, Jr. 1970. J. Parasitol. 56:517-522.

number of cells of several different types (Fig. 12-2). The infusoriform larva escapes from the axial cell and parent rhombogen and leaves the host. It is the only stage known that can survive in seawater. Subsequent to leaving the host, the fate of the larva is unknown because attempts to infect new hosts with it have failed. It is possible than an alternate or intermediate host exists in the life cycle.

Order Heterocyemida

Although they are also parasites of cephalopods, the heterocyemids differ in morphology from the dicyemids. The nematogens of heterocyemids have no cilia or calotte and are covered by a syncytial external layer. Rhombogens are much like nematogens, and they produce infusorigens and infusoriform larvae, as in the dicyemids.

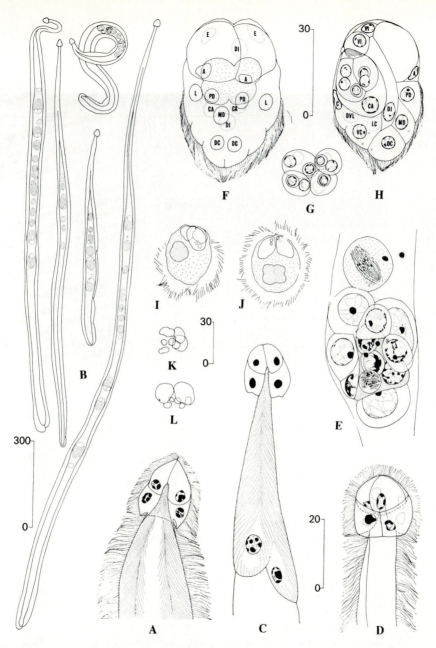

FIG. 12-2

Dicyemennea antarcticensis life stages (continued from Fig. 12-1). (Scale between **C** and **D** also applies to **A**. Scale between **F** and **H** also applied to **E** and **G**. Scale to right of **K** also applies to **I**, **J**, and **L**.) **A**, Nematogen, anterior end. **B**, Entire rhombogens. **C** and **D**, Rhombogens, anterior ends. **E**, Infusorigen. **F** to **J**, Infusoriform larvae. Abbrevations denoting cells (in **F**, only nuclei of cells are shown): *A*, Apical; *CA*, capsule; *C*, couvercle; *DC*, dorsal caudal; *DI*, dorsal interior; *E*, enveloping; *L*, lateral; *LC*, lateral caudal; *MD*, median dorsal; *PD*, paired dorsal; *VI*, ventral internal; *VI*, first ventral. **F**, Dorsal view, position of urn cells stippled. **G**, Urn cells. **H**, Side view, optical section. **I** and **J**, Views to show relative sizes of refringent bodies and urn cells. **K** and **L**, Refringent bodies.

From Short, R.B., and F.G. Hochberg, Jr. 1970. J. Parasitol. 56:517-522.

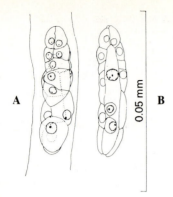

FIG. 12-3

Primary nematogens developing within adult stem nematogens of *Dicyema typoides,* mature. **A,** Somatic cell outlines and nuclei. **B,** Optical section.

From Short, R.B. 1964. J. Parasitol. 50:646-651.

CLASS ORTHONECTIDA

The Orthonectida are quite different from the Rhombozoa in their biology and morphology. The 17 known species parasitize marine invertebrates, including brittle stars, nemerteans, annelids, turbellarians, and molluscs. Complete life cycles are known for some.

Morphology and biology

The best known orthonectid is *Rhopalura ophiocomae,* a parasite of brittle stars along the coast of Europe (Figs. 12-4 and 12-5). Both sexual and asexual stages exist in the life cycle.

A **plasmodium stage** lives in the tissues and spaces of the gonads and genitorespiratory bursae of the ophiuroid *Amphipholis squamata* and may spread into the aboral side of the central disc, around the digestive system, and into the arms. Developing host ova degenerate, with ultimate castration, but male gonads usually are unaffected.[2] The multinucleate plasmodia are usually male or female but are sometimes hermaphroditic. Some of the nuclei are vegetative, whereas others are agametes that divide to form balls of cells called **morulas.** Each morula differentiates into an adult male or female, with a ciliated somatoderm of **jacket cells** and numerous internal cells that become gametes. Monoecious plasmodia that produce both male and female offspring may represent the fusion of two separate, younger plasmodia. Male ciliated forms are elongate and 90 to 130 μm long. Constrictions around the body divide it into a conical cap, a middle, and a terminal portion. A genital pore,

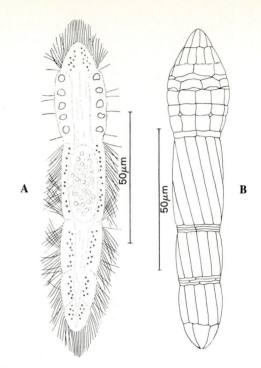

FIG. 12-4

Rhopalura ophiocomae, representing adult stages of an orthonectid mesozoan. Male: **A,** Living individual, as seen in optical section, showing distribution of cilia, lipid inclusions, crystal-like inclusions of the second superficial division of the body, and testis. **B,** Boundaries of jacket cells, at the surface; silver nitrate impregnation.

From Kozloff, E.N. 1969. J. Parasitol. 55:171-195.

through which sperm escape, is located in one of the constrictions. Jacket cells are arranged in rings around the body; the number of rings and their arrangement is of taxonomic importance.

There are two types of females in this species. One type is elongated, 235 to 260 μm long and 65 to 80 μm wide, whereas the other is ovoid, 125 to 140 μm long and 65 to 70 μm wide. Otherwise, the two forms are similar to each other and differ from the male in lacking constrictions that divide the body into zones. The female genital pore is located at about midbody. The oocytes are tightly packed in the center of the body.

Males and females emerge from the plasmodia and escape from the ophiuroid into the sea. There, tailed sperm somehow transfer into females, where they fertilize the ova. Within 24 hours after fertilization, the zygote has developed into a multicellular, ciliated larva that is born through the genital pore of its

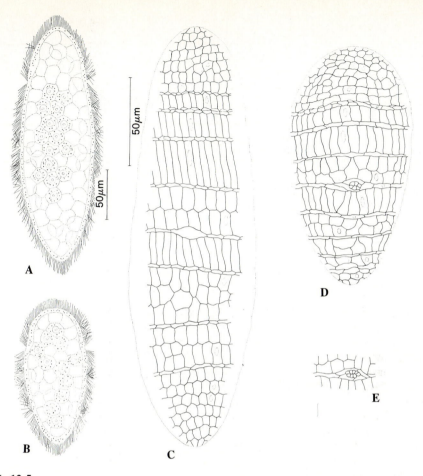

FIG. 12-5

Adult stages of *Rhopolura ophiocomae* (continued from Fig. 12-4). Female: **A,** Living specimen of elongate type, as seen in optical section. **B,** Living specimen of ovoid type, as seen in optical section. **C,** Boundaries of jacket cells of elongated type; silver nitrate impregnation. (The cells surrounding the genital pore have been omitted because they were not distinct; approximately proportions of nuclei of representative cells are based on specimens impregnated with Protargol.) **D,** Cell boundaries of ovoid type; silver nitrate impregnation. **E,** Genital pore of ovoid type; silver nitrate impregnation.

From Kozloff, E.N. 1969. J. Parasitol. 55:171-195.

mother and enters the genital opening of a new host.

It is not known whether a plasmodium is derived from an entire ciliated larva or from certain of its cells or whether one larva can propagate more than one plasmodium.

PHYLOGENETIC POSITION

The phylogenetic position of the Mesozoa is most obscure. Early taxonomists placed them between protozoa and sponges because of their cilia, small size, and simple cellularity. Certainly their structure and life cycles are no more complex than those of some Protozoa. A good argument has been made for considering rhombozoans to be primitive or degenerate Platyhelminthes. The ciliated larva is similar to a miracidium (p. 241) in some ways, and the internal reproduction by agametes in nematogens and rhombogens parallels similar processes in germinal sacs of digenetic trematodes.

Another possibility is that Mesozoa represent one or two independent lines of evolution that have pro-

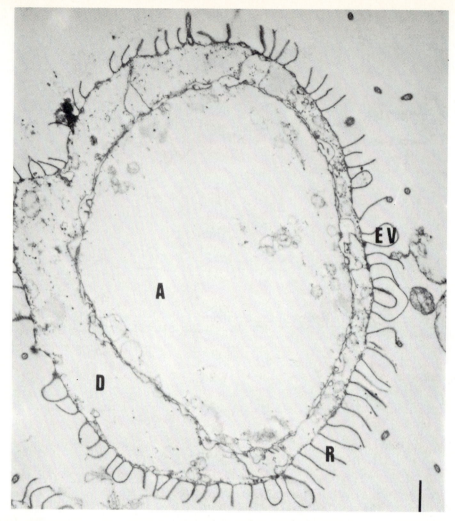

FIG. 12-6

Nematogen of *Dicyema aegira,* transverse section through somatic cells and axial cells. Note scarcity of organelles, imparting hyaline appearance to cells. Ruffles on somatic cells fused distally at several locations around periphery, forming large endocytotic vesicles. (× 8500.) *A,* Axial cell; *D,* somatic cell; *EV,* endocytotic vesicle; *R,* ruffle membrane.

From Ridley, R.K. 1968. J. Parasitol. 54:975-998.

ceeded no further, so they cannot be aligned with higher forms. The question cannot be answered with conviction at this time.

PHYSIOLOGY AND HOST-PARASITE RELATIONSHIPS

What little is known of the physiology of the Mesozoa was reviewed by McConnaughey.[3] Most of what is known concerns the rhombozoans and is based on the early observations of Nouvel.[4] Good ultrastructural studies of both rhombozoans and orthonectids are available.[2,5,6]

Most rhombozoans attach themselves loosely to the lining of the cephalopod kidney by their anterior cilia. They are easily dislodged and can swim about freely in their host's urine. The relationship appears to be entirely commensalistic; no pathogenic consequences of the infection can be discerned. However, a few species have morphological adaptations for gripping the renal cell surface more firmly, and, on dislodgement, the renal tissue shows an eroded appearance.

The ruffle membrane surface of the nematogens and rhombogens (Fig 12-6) evidently is an elaboration to facilitate uptake of nutrients. Ridley[5] showed that the membranes could fuse at various points and form endocytotic vesicles, and "transmembranosis" was suggested by uptake of ferritin. The peripheral cells of infusoriform larvae do not have ruffle membranes but do have microvilli.[6] Clearly the nutritive substances of nematogens and rhombogens must be derived largely or entirely from the host's urine, whereas the infusoriform must live for a period on stored food molecules. Oxygen is very low or absent in the cephalopod's urine, and the nematogens and rhombogens apparently are obligate anaerobes. The organisms live longer in vitro with nitrogen, or even in the presence of cyanide, than those maintained in urine under air or in the absence of cyanide. The infusoriform can live anaerobically only until its glycogen supply is consumed. Adult orthonectids, on the other hand, require aerobic conditions.

A plethora of questions remains. Especially in light of modern knowledge of the terminal reactions of glycolysis in parasitic Platyhelminthes and nematodes, it would be fascinating to compare the analogous metabolic pathways of Mesozoa. Numerous other intriguing questions have been enumerated by McConnaughey,[3] including the following:

1. What are the actual nutritional requirements of the Mesozoa, and to what extent are these met by the host's urine?
2. What are the factors responsible for differentiation of the nematogens from rhombogens, and do these include host hormones?
3. What is responsible for the host specificity, and how do the invasive stages find and invade young specimens of the correct host species?
4. Why is cephalopod urine usually sterile except for the Mesozoa, in spite of the fact that it is a good culture medium?

CLASSIFICATION OF PHYLUM MESOZOA

Class Rhombozoa

Order Dicyemida

FAMILY DICYEMIDAE

Genera
Dicyema, Pseudicyema, Pleodicyema, Dicyemennea

Order Heterocyemida

FAMILY CONOCYEMIDAE

Genera
Conocyema, Microcyema

Class Orthonectida

Order Orthonectida

FAMILY RHOPALURIDAE

Genera
Rhopalura, Stoecharthrum

FAMILY PELMATOSPHAERIDAE

Genus
Pelmatosphaera

REFERENCES

1. Dodson, E.O. 1956. A note on the systematic position of the Mesozoa. Syst. Zool. 5:37-40.
2. Kozloff, E.N. 1969. Morphology of the orthonectid *Rhopalura ophiocomae*. J. Parasitol. 55:171-195.
3. McConnaughey, B.H. 1968. The Mesozoa. In Florkin, M., and B.T. Scheer, editors. Chemical zoology, vol. 2. Porifera, Coelenterata, and Platyhelminthes, Academic Press, Inc., New York, pp. 537-570.
4. Nouvel, H. 1933. Recherches sur la cytologie, la physiologie et la biologie des dicyemides. Ann. Inst. Oceanogr. 13:163-255.
5. Ridley, R.K. 1968. Electron microscopic studies on dicyemid Mesozoa. I. Vermiform stages. J. Parasitol. 54:975-998.
6. Ridley, R.K. 1969. Electron microscopic studies on dicyemid Mesozoa. II. Infusorigen and infusoriform stages. J. Parasitol. 55:779-793.
7. Stunkard, H.W. 1954. The life history and systematic relations of the Mesozoa. Q. Rev. Biol. 29:230-244.
8. Stunkard, H.W. 1972. Clarification of taxonomy in the Mesozoa. Syst. Zool. 21:210-214.

SUGGESTED READINGS

Grassé, P.P., and M. Caullery. 1961. Embranchement des mésozoaires. In Grassé, P., editor. Traité de zoologie: anatomie, systématique, biologie, vol. 4. Platheminthes, Mésozoaires, Acanthocéphales, Némertiens. Masson & Cie, Paris, pp. 693-729. (A modern summary of the group.)

McConnaughey, B.H. 1968. The Mesozoa. In Florkin, M., and B.T. Scheer, editors. Chemical zoology, vol. 2. Porifera, Coelenterata, and Platyhelminthes. Academic Press, Inc., New York, pp. 537-570. (An excellent review of the physiology of Mesozoa.)

Stunkard, H.W. 1982. Mesozoa. In Parker, S.P., editor. Synopsis and classification of living organisms, vol. 1. McGraw-Hill, Inc., New York, pp. 853-855.

CHAPTER 13

INTRODUCTION TO PHYLUM PLATYHELMINTHES AND CLASS TURBELLARIA

The Platyhelminthes display several phylogenetic advances over what may be considered more primitive phyla, such as Porifera and Coelenterata. They are bilaterally symmetrical and have a definite "head end," with associated sensory and motor nerve elements. This increase in nervous function enabled them to invade a wide variety of ecological niches, including the bodies of other kinds of animals. In fact most platyhelminths are parasitic. A peculiarity of their physiology is their apparent inability to synthesize fatty acids and sterols *de novo,*[5] which may explain why flatworms are most often symbiotic with other organisms, either as commensals or parasites. The free-living acoel turbellarians, generally considered the most primitive worms in the phylum, also seem to lack this ability, indicating that the parasites may not have lost it secondarily as a response to parasitism.

Flatworms are so called because most are dorsoventrally flattened. They are usually leaf shaped or oval, but some are extremely elongate, such as the tapeworms. They range in size from nearly microscopic to over 200 feet in length. A coelom has not evolved in this phylum.

The **tegument** varies in structure among classes. Generally speaking, the Turbellaria and some free-living stages of Cestoidea and Trematoda have a ciliated epithelium, which is their primary mode of locomotion. This epithelium is very thin, being formed of a single layer of cells, and contains many glandular cells and ducts from subepithelial glands. Sensory nerve endings are abundant in the epithelium. The Trematoda and Cestoidea have lost the cilia except in protonephridia and certain larval stages. Instead, the tegument is a syncytial layer, the nuclei of which are in cell bodies (**cytons**) located beneath a superficial muscle layer. Embedded in the tegument in most free-living turbellarians and in the trematode *Rhabdiopoeus* are numerous rodlike bodies called **rhabdites.** Their function is not clear, but various authors have attributed lubrication, adhesion, and predator repellancy to them; they are generally absent in symbiotic Turbellaria.

Most of the body of a flatworm is made up of **parenchyma,** a loosely arranged mass of fibers and cells of several types. Some of these cells are secretory, others store food or waste products, and still others have huge mitochondria and function in regeneration. The internal organs are so intimately embedded in the parencyhma that dissecting them out is nearly impossible. The bulk of the parenchyma probably is composed of myocytons.

Muscle fibers course through the parenchyma. Contractile portions of the muscle fibers are rarely striated and are usually arranged in one or two longitudinal layers near the body surface. Circular and dorsoventral fibers also occur.

The **nervous system** in the primitive turbellarians consists of a simple nerve plexus under the epithelium with a slight concentration of cell bodies near the anterior end. In the more advanced turbellarians and in the trematodes and cestodes, the nerve system is a "ladder type," with a complex ganglion near the anterior end and with longitudinal nerve trunks extending from it to near the posterior end of the body (see Fig. 21-14). The number of trunks varies, but most are lateral and are connected by transverse commissures. Sensory elements are abundant, especially in the Turbellaria. Tactile cells, chemoreceptors, eyespots, and, rarely, statocysts have been found.

A **digestive system** is completely absent in cestodes. Primitive turbellarians and a few trematodes (*Anenterotrema, Austromicrophallus*) have only a mouth but no permanent gut; food is digested by individual cells of the parenchyma. Most flatworms have a mouth near the anterior end, and many turbellarians and most trematodes have a muscular **pharynx,** behind the mouth, with which they suck in food. The gut varies from a simple sac to a highly branched tube, but only rarely does the flatworm have an anus. Digestion is primarily extracellular, with phagocytosis by intestinal epithelium. Undigested wastes are eliminated through the mouth.

The functional unit of the **excretory system** is the **flame cell,** or **protonephridium** (see Fig. 21-17). This is a single cell with a tuft of cilia that extends into a delicate tubule. Excess water, which may contain soluble nitrogenous wastes, is forced into the tubule, which joins with other tubules, eventually to be eliminated through one or more excretory pores. Some species have an excretory bladder just inside the pore. Because the excreta are mainly excess water, this is often referred to as an **osmoreguatory system,** with excretion of other wastes considered a secondary function.

The **reproductive systems** follow a common pattern in all Platyhelminthes. However, extreme variations of the common pattern are found among groups. Most species are monoecious, but a few are dioecious. Because the reproductive organs are so important in identification of parasites and therefore are considered in great detail for each group, we will not discuss them here. Most hermaphrodites can fertilize their own eggs, but cross-fertilization is also known for many. Some turbellarians and cestodes practice **hypodermic impregnation,** which is sperm transfer by piercing the body wall with a male organ, the **cirrus,** and injecting sperm into the parenchyma of the recipient. How the sperm find their way into the female system is not known. Most worms, however, desposit sperm directly into the female tract. The young are usually born within egg membranes, but a few species are viviparous or ovoviviparous. Asexual reproduction is also common in trematodes and a few cestodes.

CLASSIFICATION OF PHYLUM PLATYHELMINTHES*

Class Turbellaria
Mostly free-living worms in terrestrial, freshwater, and marine environments; some are commensals or parasites of invertebrates, especially of echinoderms and molluscs.

Class Monogenea
All are parasitic, mainly on the skin or gills of fish; although most are ectoparasites, a few live within the stomodaeum, proctodaeum, or their diverticula.

*There are several different schemes of classification for the phylum Platyhelminthes because agreement has not been reached on which morphological and biological characters are most important in reflecting natural relationships. The classification that we propose here is a middle-of-the-road scheme, which we have found to be practical and universally understandable. We accept four classes in the phylum.

Class Trematoda
All are parasitic, mainly in the digestive tract, of all classes of vertebrates; there are three subclasses.

Subclass Digenea
At least two hosts in life cycle, the first almost always a mollusc; perhaps most diversification in bony marine fish, although many species in all other groups of vertebrates.

Subclass Aspidogastrea
Most have only one host, a mollusc; a few mature in turtles or fishes and have a mollusc or lobster intermediate host.

Subclass Didymozoidea
Tissue-dwelling parasites of fish; no complete life cycle is known, but an intermediate host may not be required.

Class Cestoidea
All are parasites, being common in all classes of vertebrates except agnathan classes; an intermediate host is required for almost all species.

CLASS TURBELLARIA

Most turbellarians are free-living predators, but many of the 12 orders contain species that maintain varying degrees and types of symbiosis. Of these, most are symbionts of echinoderms, but others are found on or in sipunculids, arthropods, annelids, molluscs, coelenterates, other turbellarians, and fish. At least 27 families have symbiotic species. A considerable degree of host specificity is manifested by these worms. Most symbionts are commensals; few are true parasites. Several degrees of these relationships are known within the class. Although it is tempting to array these in a series of ectocommensals, endocommensals, ectoparasites, and so on, to postulate how parasitism evolved in this phylum, it is clear that most individual cases are the end results of their particular situation and have not given rise to succeedingly complex associations. This is a predominately parasitic phylum, however, and a brief study of the commensal and parasitic turbellarians might indicate trends toward parasitism as it is found in the other classes.

Order Acoela

The acoels are entirely marine and are from 1 to several millimeters long. They possess several primitive characteristics, including the absence of an excretory system, pharynx, and permanent gut, and many have no rhabdites. Most are free living, feeding on algae, protozoa, bacteria, and various other microscopic organisms. A temporary gut with a syncytial lining appears whenever food is ingested, and digestion occurs in vacuoles within it. After digestion is completed the gut disappears.

Few species have adopted a symbiotic existence,

and it is difficult to decide which, if any, are true parasites. *Ectocotyla paguri* is the only ectocommensal known. It lives on hermit crabs, but nothing is known of its biology or feeding habits. Several species of acoels live in the intestines of Echinoidea and Holothuroidea. It is not known if any are parasites, but because no apparent harm comes to the hosts, they are usually considered endocommensals.

Order Neorhabdocoela

Most symbiotic turbellarians belong to this order. Again, most seem to be commensals, but a few are definitely parasitic. Neorhabdocoels are small, like acoels, but they have a permanent, straight gut and a complex, bulbous pharynx. Most are predators of small invertebrates. Of the four suborders in the order, three have symbiotic species. Within the suborders Dalyellioida and Temnocephalida there are species of considerable interest. The suborder Typhloplanoida has a species that is ectoparasitic on a polychaete worm but will not be considered here.

■ Suborder Dalyellioida

Family Fecampiidae. *Fecampia erythrocephala* lives in the hemocoel of decapod crustaceans. During their development in the host, the young worms lose their eyes, mouth, and pharynx; and they absorb their nutrients from the host's blood. When sexually mature, they mate and leave the host. After cementing itself to a substrate the flatworm shrinks until all internal tissues vanish, leaving only a bottle-shaped cocoon made of the degenerated epidermis. Each cocoon contains two eggs and several vitelline cells that produce two ciliated, motile juveniles. These swim about until contacting a crustacean.[1] Their mode of entry into a host is not known.

The host is not killed by the parasite but does suffer adverse effects of the hepatopancreas and ovaries. Because of its fertility, this parasite presents a high risk for culture of prawns in Atlantic and Mediterranean marine areas. *Fecampia* may illustrate a hypothetical stage in the origin of the Digenea.

Kronborgia amphipodicola is very unusual among the Turbellaria because it is dioecious.[2] Furthermore, there is pronounced sexual dimorphism: The males are 4 to 5 mm long, whereas the females are 20 to 30 mm long and can stretch to 45 mm. Both sexes lack eyes and digestive systems at all stages of their life cycles. They mature in the hemocoel of the tube-dwelling amphipod *Amphiscela macrocephala,* with the male near the anterior end and the female filling the rest of the available space. On reaching sexual maturity, the worms burrow out of the posterior end of the host, which becomes paralyzed and quickly dies. As if to add insult to injury, before the host is killed, it is castrated. After emergence from the amphipod, the female worm quickly secretes a cocoon around herself and attaches the cocoon to the wall of the burrow, protruding from it 2 to 3 cm. The male enters the cocoon, crawls down to the female, and inseminates her. He then leaves the cocoon and dies. The female produces thousands of capsules, each with two eggs and some vitelline cells, and then also dies. A ciliated larva hatches from each egg and eventually encysts on the cuticle of another amphipod. While in the cyst, the larva bores a hole through the host's body wall and enters the hemocoel to begin its parasitic existence.

The ultrastructure of *K. amphipodicola* has been studied.[4] The lateral membranes of the epidermal cells break down, and the epidermis thus becomes syncytial. Although short microvilli are not unusual on the outer surface of epithelial cells of free-living Turbellaria, the microvilli of *K. amphipodicola* are quite long and constitute an adaptation for increasing surface area to absorb nutrients. Subepidermal gland cells with long processes extending to the surface are thought to function in the escape of the worm from its host and in construction of the cocoon.

Family Umagillidae. Umagillids live in the digestive tract or coelom of Holothuroidea and Echinoidea. Crinoidea and Sipunculida also are infected. Traditionally considered harmless commensals, some species are now known to consume host intestinal cells, as well as commensal ciliated protozoa.[3] For example, *Syndesmis franciscanus* and *Syndesmis dendrastrorum* ingest host intestinal tissue along with intestinal contents, whereas *Syndesmis echinorum* subsists entirely on host intestinal tissue.[7]

Syndesmis spp. (Fig. 13-1) and *Syndisyrinx* spp. are found in the intestine of sea urchins and therefore are available to nearly any college laboratory with preserved or living sea urchins in its stock. Very little is known of their biology, but they appear to be excellent subjects for study. About 50 species have been described in this family.

Syndesmis franciscanus inhabits sea urchins of the genus *Strongylocentrotus* along the northwest coast of North America. It produces an egg capsule about every 1½ days; they are released one at a time into the intestine of the host and pass to the outside with feces. Each capsule contains two to eight oocytes and several hundred vitelline cells. Embryogenesis requires about 2 months. The worms hatch when eaten by a suitable host and mature with no further migration.[8]

■ Suborder Temnocephalida

These turbellarians are the only freshwater symbionts in the class. Most are ectocommensals on crus-

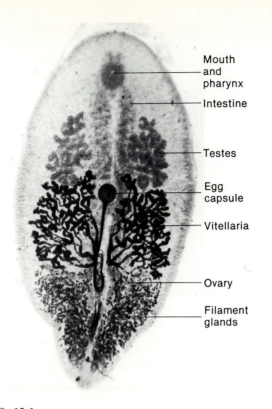

Mouth
and
pharynx

Intestine

Testes

Egg
capsule

Vitellaria

Ovary

Filament
glands

FIG. 13-1

Syndesmis sp., a rhabdocoel turbellarian from the intestine of a sea urchin. It is about 2.5 to 3 mm long.

Photograph by Warren Buss.

taceans (Fig. 13-2) in South and Central America, Australia, New Zealand (Fig. 13- 3), Madagascar, Sri Lanka, and India; a few are known from Europe. A few species occur on turtles, molluscs, and freshwater hydromedusae. Probably they are much more widespread but have gone undiscovered or unrecognized as the result of the paucity of trained specialists.

Temnocephalids are small and flattened, with tentacles at the anterior end and a weak, adhesive sucker at the posterior end (Fig. 13-4). They have leech-like movements, alternately attaching with the tentacles and posterior sucker. The tegument is syncytial with no or very few cilia and with a structure like that of trematodes, although adequate electron microscope studies have not been done. Rhabdites are located only at the anterior end, and mucous glands are mainly around the posterior sucker.

The biology of temnocephalids is simple, as far as it is known. Eggs are laid in capsules and attached to the exoskeleton of the host. Each hatches as an immature adult and matures with no further ado. What happens to those which are lost at ecdysis of the host is unknown, and, for that matter, the fate of the adults at that time is also unknown. It is possible that a free-living stage is present in the life cycle of these worms but has yet to be found.

The pattern of nutrition apparently does not differ from that of free-living rhabdocoels, with protozoa, bacteria, rotifers, nematodes, and other microscopic creatures serving as food. Cannibalism has been estab-

FIG. 13-2

Chela of New Zealand crayfish with several temnocephalids crawling on it.

Photograph by Wallaceville Animal Research Centre, New Zealand; courtesy William B. Nutting.

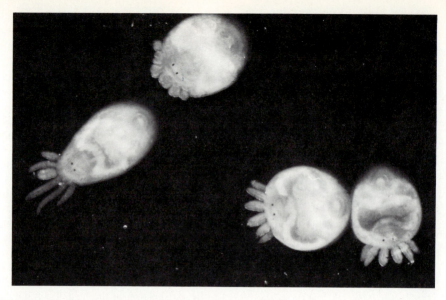

FIG. 13-3

Live temnocephalid turbellarians from a New Zealand crayfish.

Photograph by Wallaceville Animal Research Centre, New Zealand; courtesy William B. Nutting.

lished. The host serves only as a substrate for attachment.

Order Alloecoela

Alloeocoels are basically intermediate between Acoela and Tricladida and have an irregular gut. Most are marine, but a few inhabit brackish or fresh water, and a few are terrestrial. Several are commensal on snails, clams, and crustaceans, but *Ichthyophaga subcutanea* is clearly a parasite of marine teleost fish. It lives in cysts under the skin in the branchial and anal regions of its host and apparently ingests blood. Morphologically it has nonparasite features, such as eyes and a ciliated epithelium.[9]

Monocelis sp. lives within the valves of intertidal barnacles and snails during low tide but returns to the open water when the tide is in. This may illustrate a case of incipient endosymbiosis.

Order Tricladida

Tricladida are large worms, up to 50 cm in length, that occupy marine, freshwater, and terrestrial habitats. They are easily recognized by their tripartite intestine. Nearly all are free-living predators, feeding on small invertebrates and sucking the contents out of larger ones by means of their eversible pharynges.

Three genera, *Bdelloura, Syncoelidium,* and *Ectoplana,* live on the book gills of horseshoe crabs, *Limulus polyphemus.* Of these, *Bdelloura candida* (Fig. 13-5) is the most common. It has a large adhesive disc at its posterior end and well-developed eyespots. Apparently it feeds on particles of food torn apart by the gnathobases of its host and washed back to the gill area. No evidence of harm to its host has been detected. It lays its eggs in capsules on the book gill lamellae. The triclads may migrate from one horseshoe crab to another during copulation of their hosts, a sort of marine, verminous veneral disease! The biology and physiology of these worms would surely prove to be a rewarding area of research.

Order Polycladida

The polyclads have a complex gut with many radiating branches. Except for one freshwater species, they are all marine. No parasites are known in this group, and the few reported "commensals" are suspect of even that degree of symbiosis. Although some species are found together with hermit crabs, they are also found in empty shells. Others, such as the "oyster leech," *Stylochus frontalis,* live between the valves of oysters and are predators on the original owner, devouring large pieces of it at a time.[6]

The truly parasitic turbellarians show structural changes expected with their specialized way of life: losses of ciliated epidermis, eyes, mucous glands, and rhabdites. The various commensals, however, show few or no specializations over their free-living brethren. The prevalence of rhabdocoels in echinoderms

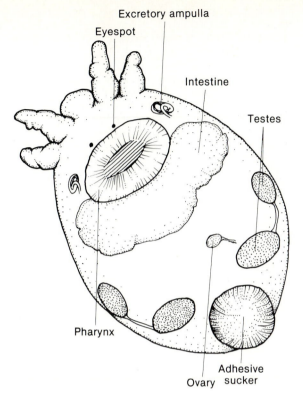

Eyespot

Excretory ampulla

Intestine

Testes

Pharynx

Ovary

Adhesive sucker

FIG. 13-4

Temnocephala tasmanica, a turbellarian from an Australian crayfish.

Drawing by Ian Grant.

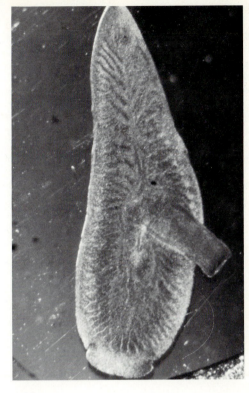

FIG. 13-5

Bdelloura candida, a triclad turbellarian from the gills of a horseshoe crab. Note the eyespots and the huge midventral pharynx. Overall length may reach 20 mm.

Photograph by Warren Buss.

may simply be a result of the rich fauna of ciliated, protozoan commensals in the latter, which offer rich pickings for the former. The origin of trematodes and cestodes from acoel ancestors, which became adapted to endocommensalism within molluscs and crustaceans, is not difficult to visualize.

REFERENCES

1. Bellon-Humbert, C. 1983. *Fecampia erythrocephala* Giard (Turbellaria Neorhabdocoela), a parasite of the prawn *Palaemon serratus* Pannart: the adult phase. Aquaculture 31:117-140.
2. Christiansen, A.P., and B. Kanneworff. 1965. Life history and biology of *Kronborgia amphipodicola* Christiansen and Kanneworff (Turbellaria, Neorhabdocoela). Ophelia 2:237-251.
3. Jennings, J.B., and D.F. Mettrick. 1968. Observations on the ecology, morphology and nutrition of the rhabdocoel turbellarian *Syndesmis franciscana* (Lehman, 1946) in Jamaica. Caribb. J. Sci. 8:57-69.
4. Lee, D.L. 1972. The structure of the helminth cuticle. In Dawes, B., editor. Advances in parasitology, vol. 10. Academic Press, Inc., New York, pp. 347-379.
5. Meyer, F., and H. Meyer. 1972. Loss of fatty acid biosynthesis in flatworms. In Van den Bossche, H., editor. Comparative biochemistry of parasites. Academic Press, Inc., New York, pp. 383-393.
6. Pearse, A.S., and G.W. Wharton. 1938. The oyster "leech" *Stylochus inimicus* Palomi, associated with oysters on the coasts of Florida. Ecol. Monogr. 8:605-655.
7. Shinn, G.L. 1981. The diet of three species of umagillid neorhabdocoel turbellarians inhabiting the intestine of echinoids. Hydrobiologia 84:155-162.
8. Shinn, G.L. 1983. The life history of *Syndisyrinx franciscanus,* a symbiotic turbellarian from the intestine of echinoids, with observations on the mechanism of hatching. Ophelia 22:57-79.
9. Syriamiatnikova, I.P. 1949. A new turbellarian of fish, *Ichthyophagia subcutanea* n. g. n. sp. C.R. Akad. Nauk 68:805-808.

SUGGESTED READINGS

Baer, J.F. 1961. Class des Temnocéphales. In Grassé, P., editor. Traité de zoologie: anatomie, systématique, biologie, vol. 4. Plathelminthes, Mésozoaires, Acanthocéphales, Némertiens. Masson & Cie, Parix, pp. 213-241. (A fairly up-to-date account of this group.)

Brooks, D.R., R.T. O'Grady, and D.R. Glen. 1985. The phylogeny of the Cercomeria. Brooks, 1982 (Platyhelminthes). Proc. Helminthol. Soc. Wash. 52:1-20.

Crezée, M. 1982. Turbellaria. In Parker, S., editor. Synopsis and classification of living organisms. Vol. 1. McGraw-Hill, Inc., New York, pp. 718-740. (A thorough and modern synopsis of the Turbellaria.)

Jennings, J.B. 1971. Parasitism and commensalism in the Turbellaria. In Dawes, B., editor. Advances in parasitology, vol. 9. Academic Press, Inc., New York, pp. 1-32. (The most readable account of the subject. Recommended for all parasitologists.)

Kozloff, E.N., and C.A. Westervelt, Jr. 1987. Redescription of *Syndesmis echinorum*. François, 1886 (Turbellaria: Neorhabdocoela: Umagillidae), with comments on distinctions between *Syndesmis* and *Syndisyrinx*. J. Parasitol. 73:184-193.

Nollen, P.M. 1983. Patterns of sexual reproduction among parasitic platyhelminths. Symposia of the British Society for Parasitology, vol. 20. Parasitology 86:99-120.

Whitfield, P.J., and N.A. Evans. 1983. Parthenogenesis and asexual multiplication among parasitic platyhelminths. Symposia of the British Society for Parasitology, vol. 20. Parasitology 86:121-160.

CHAPTER 14

CLASS MONOGENEA

The Monogenea are hermaphroditic flatworms that mainly are external parasites of vertebrates, particularly fish. Some species, however, are found internally in diverticula of the stomodeum or proctodeum and also in the ureters of fish and the bladders of turtles and frogs. A single species is known from mammals: *Oculotrema hippopotami* from the eye of the hippopotamus.[33] Nevertheless, monogeneans are primarily fish parasites, particularly of the gills and external surfaces. Although a few fish deaths have been attributed to monogeneans in nature, the worms are not usually regarded as hazardous to wild populations. However, like copepods and numerous other fish pathogens, monogeneans become a serious threat when fish are crowded together, as in fish farming.

The group has been somewhat neglected by parasitologists, despite their remarkable morphology and life cycles and their considerable economic importance. Probably fewer than half of the existing species have been described.[32]

These worms were aligned loosely with digenetic trematodes by early parasitologists. The first comprehensive overview of the group was by Braun[2] in 1889 and 1893. Next, Fuhrmann,[6] in 1928, helped establish the Monogenea as a category separate from the Digenea, although closely allied with it. Bychowsky,[3] in 1937, was apparently the first to propose Monogenea as a separate class, apart from and equal to Digenea. This viewpoint was not adopted widely until recently, and even today it is not universally accepted.[30] We believe that the separation of these worms into two classes in fully justified on morphological and biological grounds.

Monogeneans are often very particular about both the species of host and the site where they live on that host, restricting themselves to extremely narrow niches in many cases. Thus one species may live only at the base of a gill filament, whereas another is found only at its tip.[28] Furthermore, many species are found on certain gill arches but not on others within the same fish. It is possible that such niche specificity is influenced by the physical attachment abilities of the highly specialized, posterior attachment organ, the opisthaptor. A similar phenomenon is known in the case of the scolex of tetraphyllidean cestodes of elasmobranchs. Some monogeneans remain fixed to the original site of attachment and cannot relocate later. Others, especially those on the skin, move about actively, leechlike, relocating at will. Certain species are found only on young fish, whereas others occur only on mature fish. Nutritional requirements of the parasites may play a role in the determination of such host specificity, but in some cases the free-swimming larvae are particularly attracted by mucus produced by the epidermis of their host species.[13]

The life span of monogeneans varies from a few days to several years. Many are incapable of living more than a short time after the death of the host. Monogeneans can seem to be absent from a fish population, when actually the worms had dropped off after the fish were caught. For this reason dead fish should not be transported to the laboratory in water, it is far better to risk them drying so that the worms will remain on the gills. A recommended procedure is to place the fish on ice until examination or to remove the gills and drop them into a 10% formalin solution while still in the field.

FORM AND FUNCTION
Body form

Monogeneans are basically bilaterally symmetrical, with partial asymmetry superimposed on a few species, particularly involving the opisthaptor. The body can be subdivided roughly into the following regions: **cephalic region** (anterior to pharynx), **trunk** (body proper), **peduncle** (portion of body tapered posteriorly), and **opisthaptor** (Fig. 14-1).

Most monogeneans are quite small, but a few are large, ranging from 0.03 to 20 mm long. Marine forms are usually larger than those from fresh water. All are capable of stretching and compressing their bodies, so one must take care that the worms are properly relaxed before fixing, or they may be contracted

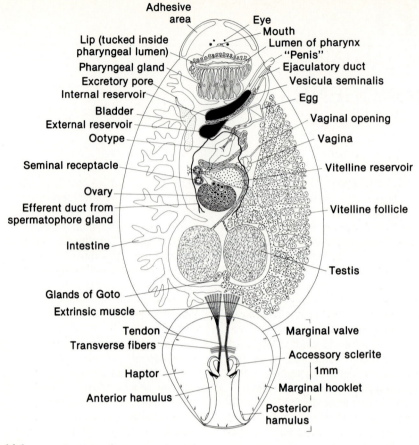

FIG. 14-1

Anatomy of an adult specimen of *Entobdella soleae* (ventral view).

From Kearn, G.C. 1971. In Fallis, M., editor. Ecology and physiology of parasites. University of Toronto Press, Toronto.

considerably. The dorsal side of the body is usually convex, while the ventral side is concave. The body is usually colorless or gray, but eggs, internal organs, or ingested food may cause it to be red, pink, brown, yellow, or black.

The anterior end of the body bears various adhesive and feeding organs, collectively called the **prohaptor**, which sometimes is associated with compound sense organs.[26] There are two main types of prohaptor: those which are not connected with the mouth funnel and those which are. The first (Fig. 14-2) is found on the more primitive types, in which the head end usually is truncated, lobated, or broadly rounded. This group usually bears **cephalic,** or **head glands,** which are unicellular organs that release sticky substances through individual or groups of ducts. The utility of the substances produced by the head glands for adhe-

sion is clear to anyone who has watched a monogenean with no anterior sucker progress in an inchworm-like manner, alternately attaching and releasing the anterior and posterior ends, down a fish gill filament. In one species of *Gyrodactylus* at least three different head gland types have been recognized. Two to eight clusters of such ducts, called **head organs,** are usual. These areas usually bear dense, long microvilli on the tegument, in contrast to the short, scattered microvilli on the remainder of the body. These microvilli may function to spread and mix the secretions of the different types of head glands. Some species in this group have shallow, muscular **bothria,** which serve as suckers, in conjunction with the head gland secretions. Most species have two bothria, but some species have four.

The second and more advanced, type of prohaptor

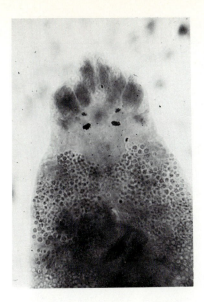
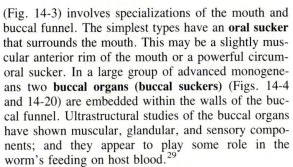

FIG. 14-2

Primitive prohaptor, not connected to mouth funnel.

Photograph by Warren Buss.

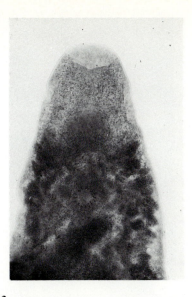

FIG. 14-3

Advanced prohaptor, with an oral sucker.

Photograph by Warren Buss.

(Fig. 14-3) involves specializations of the mouth and buccal funnel. The simplest types have an **oral sucker** that surrounds the mouth. This may be a slightly muscular anterior rim of the mouth or a powerful circumoral sucker. In a large group of advanced monogeneans two **buccal organs (buccal suckers)** (Figs. 14-4 and 14-20) are embedded within the walls of the buccal funnel. Ultrastructural studies of the buccal organs have shown muscular, glandular, and sensory components; and they appear to play some role in the worm's feeding on host blood.[29]

The posterior end of all monogeneans also bears a highly characteristic organ, the opisthaptor (Fig. 14-5). It is clear that, in a group whose primary habitat is the surface and gills of fish, great adaptive value will accrue from an efficient attachment organ that prevents dislodgment by strong water currents, particularly one that will allow the mouth end to "hang downstream" and graze at will. In the Monogenea the opisthaptor is such an organ, and it has been, unsurprisingly, subjected to immense adaptive radiation in this group.

The opisthaptor may extend for a considerable distance anteriorly along the trunk of the worm or may be confined to the posterior extremity. It may be sharply delineated from the body by a peduncle or may be merely a broad continuation of it. Opisthaptors develop into one or two basic types during ontog-

eny. The larva that hatches from the egg always has a tiny opisthaptor armed with sclerotized hooks or spines. This is retained in the adults of most species and either expands into the definitive opisthaptor or remains juvenile, while the adult organ develops from other sources near or surrounding it. In this first basic type the muscles expand into a large disc that often has shallow **loculi** or well-developed **suckers,** as well as large hooks called **anchors,** referred to by some workers as **hamuli.** The tiny hooklets of the larva usually can be found on the margins or ventral surface of the opisthaptor (Fig. 14-6).

The second type loses the larval opisthaptor entirely or retains it as a tiny organ on or in the adult opisthaptor or attached at the end of a muscular appendage (**lappet,** or **appendix**). In this large group the opisthaptor is the most highly specialized, being profoundly subdivided into individual adhesive organs and/or equipped with complex sclerotized clamps, or muscular valves, commonly with attendant hooks.

Because it has undergone such adaptive radiation and varies considerably among species, while remaining fairly constant within a species population, the opisthaptor is an important taxonomic character. Furthermore, the sclerotized hooks, bars, clamps, and so on, which are easily studied, are heavily relied on by specialists studying this group. It is possible that more attention to the anatomy of other organs would lead to

197

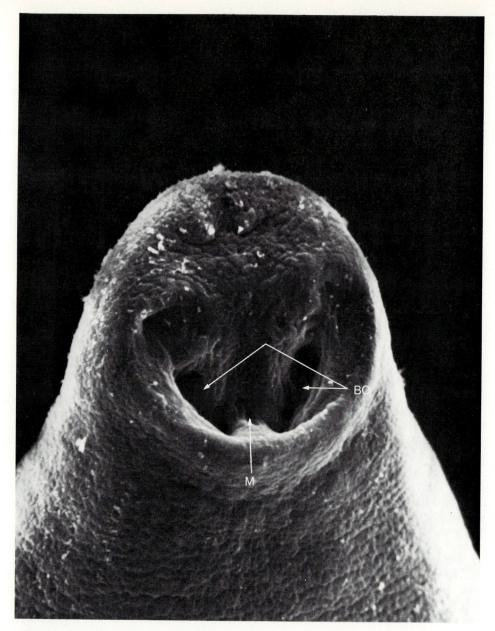

FIG. 14-4

Scanning electron micrograph of anterior end of *Lintaxine cokeri*. Openings of buccal organs *(BO)* and mouth *(M)* are indicated.

Courtesy John C. Mergo, Jr.

a more natural classification of the monogeneans. Histochemical and x-ray analysis show promise of clarifying classification.[12]

Hooks are characterized as being "marginal" or "central." **Marginal hooks,** which are not always strictly marginal, are usually the very tiny hooklets of the larval opisthaptor, some of which may be missing.

It takes careful and skillful microscopy to find these. **Central hooks** (Fig. 14-7) are the larger anchors, or hamuli, and occur in one to three pairs, usually in the center of the opisthaptor, although they may be displaced to the side or posterior margin of the disc. Central hooks often have **connecting bars,** or **accessory sclerites,** supporting them (Fig. 14-8). The homolo-

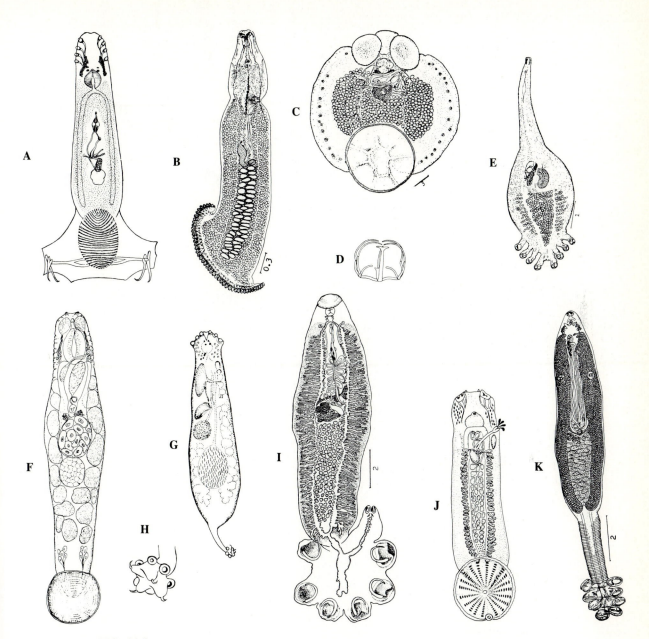

FIG. 14-5

A variety of monogeneans, showing variations of opisthaptors. **A,** *Diplectanum aculeatum.* **B,** *Neoaxine constricta.* **C,** *Capsala pricei.* **D,** Sclerites in clamp of *Neoaxine constricta.* **E,** *Diclidophora merlangi.* **F,** *Udonella caligorum.* **G,** *Aviella baikalensis.* **H,** Opisthaptor of *Aviella baikalensis.* **I,** *Erpocotyle borealis.* **J,** *Acanthocotyle lobianchi.* **K,** *Chimaericola leptogaster.* (Scales in millimeters, where available.)

After various authors, from Yamaguti, S. 1963. Systema helminthum, vol. 4. Monogenea and Aspidocotylea. Copyright © by Interscience Publishers, New York. Reprinted by permission of John Wiley & Son, Inc.

FIG. 14-6

Marginal opisthaptor hooklets from representative families of diclidophoridean Monogenea.

With permission from Llewellyn, J. 1963. In Dawes, B., editor. Advances in parasitology, vol. 1. Copyright Academic Press, Inc. (London) Ltd.

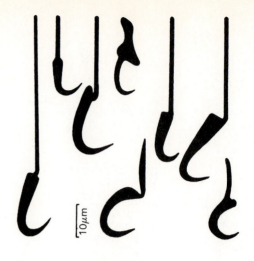

FIG. 14-7

Typical central hooks from opisthaptors of monogeneans.

From Bychowsky, B.E. 1961. Monogenetic trematodes, their systematics and phylogeny. American Institute of Biological Sciences, Washington, D.C.

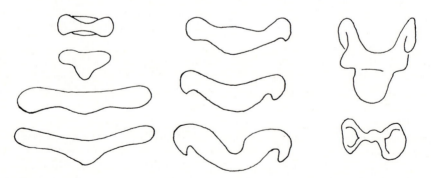

FIG. 14-8

Representative connecting bars, or accessory sclerites, of monogenean opisthaptors.

From Bychowsky, B.E. 1961. Monogenetic trematodes, their systematics and phylogeny. American Institute of Biological Sciences, Washington, D.C.

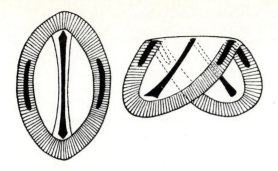

FIG. 14-9

Diagram of the primitive attaching clamp. On the left it is fully open; on the right it is partially closed. The sclerotized parts are black; the musculature is cross-hatched.

From Bychowsky, B.E. 1961. Monogenetic trematodes, their systematic and phylogeny. American Institute of Biological Sciences, Washington, D.C.

gies of central hooks and their supporting bars are not always clear. The protein in the anchors and marginal hooks appears to be keratin.

Rarely, **supplementary discs,** or **compensating discs,** are developed near the base of the opisthaptor (family Diplectanidae). These are accessory to the opisthaptor and are not technically a part of the larval or adult opisthaptor. They consist of a series of sclerotized lamellae or spines. **Suckers** are found on the ventral surface of the opisthaptors of many species. They range in number from two to eight.

Complex **clamps** are found on many species of highly evolved monogeneans. On some species the clamp is muscular, whereas on others it is mainly sclerotized. It functions as a pinching mechanism, aiding in adherence to the host. Although many variations of structure occur, all are based on a single, primitive type of clamp (Fig. 14-9). The identity of the material of which the clamps are constructed is enigmatic; it is not keratin, chitin, quinone-tanned protein, or collagen.[20] The number of clamps varies from eight to several hundred, distributed symmetrically in some species and asymmetrically in others. Finally, the combinations of hooks, suckers, and clamps vary among several families.

Tegument

As in the digeneans and cestodes, the tegument of monogeneans traditionally has been referred to as a cuticle because light microscopists could discern little structure within it. However, by use of the electron microscope, the "cuticle" has now been recognized as a living tissue, the **tegument.** Its fundamental structure appears to be similar to that of digeneans and ces-

todes, with some noteworthy differences. The surface layer of the tegument is, as in cestodes and digeneans, a syncytial stratum, laden with vesicles of various types and mitochondria, bounded externally by a plasma membrane and glycocalyx (fine filamentous layer on surface) and internally by a membrane and basal lamina. This stratum is the **distal cytoplasm,** and it is connected by trabeculae (**internuncial processes**) to the "cell bodies," or **cytons** (perikarya), located internal to a superficial muscle layer. Such is the case in all species studied so far except those of *Gyrodactylus,* in which trabeculae and cytons could not be observed. *Gyrodactylus* spp. are very peculiar in other respects and will be discussed further. Often the outer surface of the tegument is supplied with short, scattered microvilli; in some species these are absent, and shallow pits are observed.

A curious condition has been reported in certain species. Some areas of the body are without a tegument, and large pieces of the tegument are only loosely connected to the surface. In these areas the basal lamina constitutes the external covering. Rohde[27] contends that the condition is not artifactual but that pieces of the tegument are being secreted into the environment and that these cases might offer a clue to the adaptive value of the tegumental arrangement of the monogeneans, digeneans, and cestodes, that is, syncytial distal cytoplasm with internal perikarya. He has suggested that the transfer of an original superifical epithelium into the interior of the body may be a way to prevent permanent damage by the hosts. Since the tegument may be subjected to such damaging influences as host secretions, it then can easily be replaced from the cell machinery that is still intact and can be protected further below.

Muscular and nervous systems

The main musculature, other than that in the opisthaptor, appears to be the **superficial muscles,** immediately below the distal cytoplasm of the tegument, arranged in circular, diagonal, and longitudinal layers. The muscles of the opisthaptor in the suckers or inserted on the hooks and accessory sclerites are clearly important in adhesion. The mechanics of their operation have been explained in several species, an example of which is *Entobdella soleae*[13] (Fig. 14-10). This species lives on the skin of the sole, and its opisthaptor is well adapted to anchor the animal firmly in its relatively smooth and exposed site on the host. The disc-shaped opisthaptor forms an effective suction cup. Prominent muscles in the peduncle are inserted on a tendon that passes down to near the ventral surface of the disc, up over a notch in the accessory sclerites, and then to the proximal end of the large anchors. Contraction of the muscles erects the accessory sclerites so that their distal ends are directed down against the fish's skin and their proximal ends serve as a prop toward which the proximal ends of the anchors are pulled. This action tends to lift the center area of the opisthaptor, thus reducing pressure and creating suction, at the same time that the distal, pointed ends of the anchors are pushed downward to penetrate the host's epidermis.

Studies on neural and neuromuscular conduction have not been carried out, but cholinesterase was demonstrated in the nervous system of *Diclidophora merlangi*[10]; therefore at least some fibers are probably cholinergic.

The general pattern of the nervous system is the ladder type with **cerebral ganglia** in the anterior and several nerve trunks coursing posteriorly from them. The nerve trunks are connected by the ladder commissures, and additional nerves emanate from the cerebral ganglia to connect with the pharyngeal commissure. As would be expected, the adhesive organs of the opisthaptor are well innervated.

Monogeneans have a fairly wide variety of sense organs. Most have pigmented eyes in the free-swimming larval stage. Oncomiracidia of Monopisthocotylea usually have four eyespots, which persist in the adult, perhaps somewhat reduced, whereas the two larval eyespots of Polyopisthocotylea are lost during maturation. These are rhabdomeric eyes similar to those found in Turbellaria and some larval Digenea. In addition, what appears to be a nonpigmented ciliary photoreceptor has been found in the larva of *Entobdella*, with counterparts in structures described in larval Digenea. Several different types of ciliary sense organs in the tegument have been described, including single receptors (one modified cilium in a single nerve ending) and compound receptors (consisting both of several associated nerve endings, each with a single cilium, and of one or a few nerves, each with many cilia.[22]

Finally, a very interesting, nonciliated sense organ occurs on the opisthaptor of *Entobdella*. The disc surface of the opisthaptor is covered with more than 800 small papillae (Fig. 14-11), and beneath the tegument of each papilla are packed nerve endings that are doubled over and piled on top of one another. The function of these peculiar organs is believed to be mechanoreception, perhaps to sense contact with the host or detect local tensions in the opisthaptor. It is not known whether similar organs occur in other monogeneans.[23]

Osmoregulatory system

The excretory system has not been used as a tool for systematics in this group as it has in the Digenea. Typical of the Platyhelminthes, the excretory unit is the **flame cell protonephridium.** A thin-walled capillary leads from this unit to fuse with a succession of ducts leading to two lateral **excretory pores** near the anterior end of the worm. The terminal ducts are often each equipped with a contractile bladder at their distal ends.

The fine structure of the excretory system has been studied only in two species of *Polystomoides*.[27] The

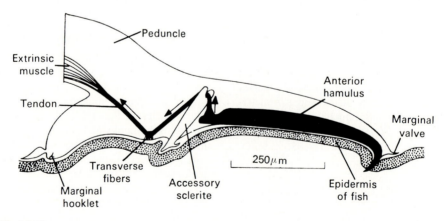

FIG. 14-10

A diagrammatic parasagittal section through the adhesive organ of *Entobdella soleae*. The arrows show the direction of movement of the tendon when the extrinsic muscle contracts. The posterior hamulus (anchor) has been omitted.

From Kearn, G.C. 1971. In Fallis, A.M., editor. Ecology and physiology of parasites. University of Toronto Press, Toronto.

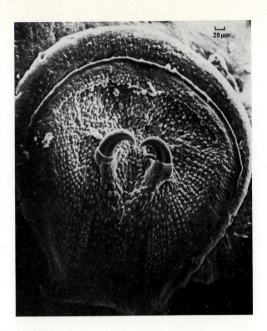

FIG. 14-11

Scanning electron micrograph of the opisthaptor of *Entobdella soleae,* showing the papillae that are thought to be sensory.

From Lyons, K.M. 1973. Z. Zellforsch. 137:471-480.

flame cell is generally similar to that of Digenea and Cestoda with minor differences. The internal surface area of the tubules is increased in a manner differing from that in either of the other groups, that is, by strongly reticulated walls (Fig. 14-12). Lateral or non-terminal flames are frequent.

Acquisition of nutrients

The mouth and buccal funnel often have associated suckers. Behind the buccal funnel a short **prepharynx** is followed by a muscular and glandular **pharynx.** This powerful sucking apparatus draws food into the system. In *Entobdella soleae* the pharynx can be everted and the pharyngeal lips closely applied to the host's skin.[13] The pharyngeal glands secrete a strong protease that erodes the host epidermis, and the lysed products are sucked up into the worm's gut by the pharyngeal muscles. Fortunately for the fish, its epidermis is capable of rapid migration and regeneration to close the wound left by the parasite's feeding.

Rohde[29] suggested that the buccal organs found in many polyopisthocotyleans may attach by muscular action to the host's gill filaments, the gill mucosa is breached by secreted enzymes, and the sensory elements detect whether blood has been found.

Posterior to the pharynx may be an **esophagus,** al-

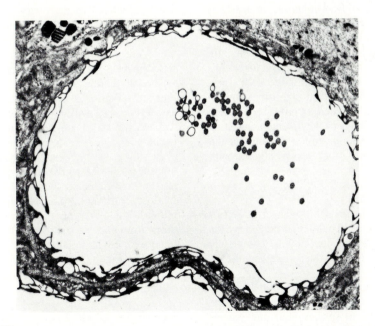

FIG 14-12

Cross section of an excretory tubule of *Polystomoides*. Note the reticulated walls and the median cilia. (×5400.)

From Rohde, K. 1975.Int. J. Parasitol. 3:331.

though it is absent in many species. The esophagus may be simple or have lateral branches and may have unicellular digestive glands opening into it.

In most monogeneans the **intestine** divides into two lateral **crura,** which are often highly branched and may even connect along their length. If the crura join near the posterior end of the body, it is common for a single tube to continue posteriad for some distance. There is no anus. Digestion and ultrastructure of the gut of several species have been studied.[7,11,27] Monopisthocotyleans seem to feed mostly on mucus and epithelial cells of their hosts, although feeding on blood has been demonstrated in *Dactylogyrus* spp. and some other Monopisthocotylea. In contrast, blood appears to be the dominant component of the diet of Polyopisthocotylea. It was believed formerly that the unusable breakdown product of hemoglobin digestion, hematin, was eliminated in the gut of polyopisthocotyleans by sloughing off gut cells containing it so that parts of the cecal wall were denuded. However, ultrastructural studies on *Polystomoides* and *Diclidophora* have shown that the cecal epithelium is not discontinuous but that the hematin-containing cells are interspersed with a different kind of cell called the "connecting cell."[9] In *Diclidophora* both the hematin cell and the connecting cell have their luminal surface increased by long, thin lamellae, but only the hematin cell has such lamellae in *Polystomoides.* It appears that the digestion of hemoglobin in *Diclidophora,* at least, is mostly or entirely intracellular: the protein is taken into the cell by pinocytosis and digested within an extensive, intracellular reticular space, and the hematin is subsequently extruded by temporary connections between the reticular system and the gut lumen. Indigestible particles are eliminated through the mouth in all monogeneans. Finally, *Diclidophora* can absorb neutral amino acids through its tegument, suggesting the possibility that direct absorption of low-molecular weight organic compounds from seawater could supplement its blood diet.[8]

Male reproductive system (Fig. 14-13)

Monogeneans are hermaphroditic with cross-fertilization usually taking place. This is epitomized by the genus *Diplozoon,* in which individuals completely fuse into pairs with their genital ducts together, the ultimate in "oneness."

Testes usually are rounded or ovoid, but they may be lobated. Most species have only one testis, but the number varies according to species, and one species has more than 200 per individual. Each testis has a **vas efferens,** which expands or fuses into an ejaculatory duct. There is no trace of a cirrus pouch or eversible cirrus, in the sense of those in cestodes or trem-

atodes. In some cases the ejaculatory duct is simple and terminates within a shallow, sometimes suckerlike, **genital atrium,** which propels sperm into the female system at copulation. A higher degree of development is found in many species in which the tissues surrounding the terminal ejaculatory duct are thickened and muscular, forming a papilla-like **penis.** Hooks of consistent size and form for each species commonly arm the distal end of the penis. In many the lining of the distal ejaculatory duct is sclerotized, sometimes for a considerable portion of its length. (We will use the term **sclerotized,** although the chemical nature of the stabilized protein is unknown.) A simple, saclike seminal vesicle is present in some species. Unicellular **prostatic glands** are usually present.

Still another type of copulatory organ exists in several families, in which the ejaculatory duct joins with a complex **sclerotized copulatory apparatus.** These vary widely among species but are similar within a species and, therefore, are important taxonomic characters. The structures are contained in a membranous sac and are controlled by muscles.

Female reproductive system (Fig. 14-13)

The single **ovary** of all species of Monogenea is usually anterior to the testes. Between species it varies in shape from round or oval to elongate or lobated. The **oviduct** leaves the ovary and courses toward the ootype, receiving the vitelline, vaginal, and genitointestinal ducts along the way. More specifically, the oviduct extends from the ovary to the confluence with the vitelline duct; the remainder is often referred to as the **female sex duct.** A seminal receptacle is present, either as a simple swelling of the oviduct or as a special sac with a separate duct to the oviduct.

The **vitellaria** are abundant, usually extending throughout the parenchyma and often even into the opisthaptor. Despite their many ramifications, the vitellaria consist basically of left and right groups. Each has an efferent duct; they fuse midventrally near the oviduct, forming a small **vitelline reservoir.** Each vitelline follicle consists of a few cells surrounded by a thin, muscular membrane. The vitelline ducts are lined with ciliated epithelium.

The vagina may be present or not or, when present, may be doubled. Vaginal openings are dorsal, ventral, or lateral. The terminal portion is sclerotized in some species; in others the vaginal pore is multiple or surrounded by spines. In some species, such as *Entobdella* (Fig. 14-1), the vagina may be much smaller than the penis, and sperm transfer is achieved by deposition of a spermatophore adjacent to the vagina of the mating partner, rather than by direct copulation.

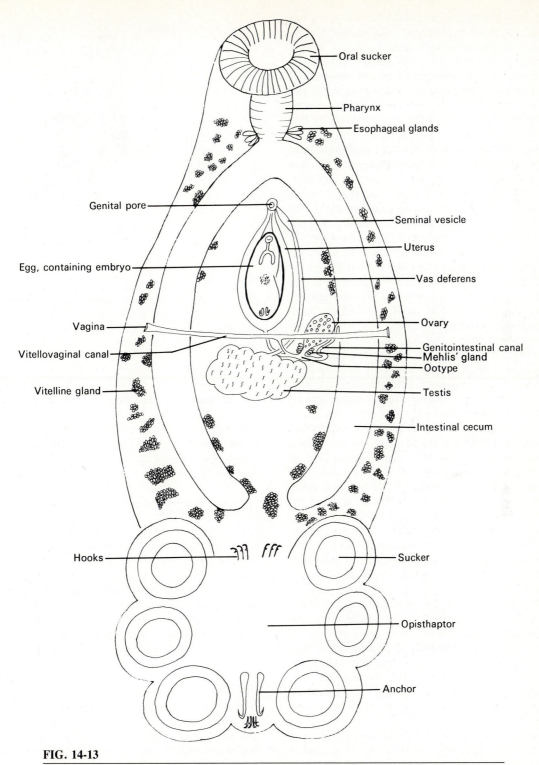

FIG. 14-13

Polystomoidella oblongum, illustrating the male and female reproductive systems.

Redrawn from Cable, R.M. 1958. An illustrated laboratory manual for parasitology. Burgess Publishing Co., Minneapolis.

Diclidophora merlangi, which does not have a vagina, practices a kind of hypodermic impregnation.[24] The sucker-like penis of an individual attaches at a ventrolateral position posterior to the genital openings of its partner, draws up a papilla of tegument into the penis, and breaches the tegument with spines in the penis. Sperm enter and make their way between cells of the partner to the seminal receptacle, a distance of 1 to 2 mm.

A very curious structure called the **genitointestinal canal** is present in most Polyopisthocotylea. It is a connection between the oviduct and a leg of the intestine or one of its branches. The function, if any, of the duct is unknown. Sometimes yolk granules and sperm are observed in the gut, presumably having arrived there through the genitointestinal canal. One hypothesis is that the canal represents a vestige of a mechanism by which eggs are passed into the intestine to be expelled through the mouth; another hypothesis is that "surplus" reproductive materials are digested and reabsorbed in the gut.[27] The canal is found in many tubellarians, especially polyclads, and its function there is obscure also.

After being fertilized in the oviduct or ovary itself, the zygote and attendant vitelline cells pass into the ootype, a muscular expansion of the female duct (Fig. 14-14). In the species studied **Mehlis' gland** around the ootype comprises two cell types, mucous and serous. (The ootype epithelium may also be secretory.) The function of Mehlis' gland is not known. It was formerly thought to contribute shell material, but in the Monogenea, as in the Digenea and Cestoda, the shell material seems to come from the vitelline cells. The shape of the egg is apparently determined by the walls of the ootype. In *Entobdella* the tetrahedral egg shape is imparted by four pads in the ootype walls.[13] The eggs of many monogeneans have a filament at one or both ends, also characteristic of a given species. The filament may have an adhesive property that serves to attach the egg to the host or substrate on which it falls after release into the open water. Further observations on egg development of *Entobdella soleae* are given by Kearn.[15]

It is generally believed that the protein in the eggshell is stabiized by a process of quinone tanning to form sclerotin, but some work indicates the stabilization may not be by quinone tanning but by means of dityrosine and disulfide links as in resilin and keratin.[25]

Although many eggs may be produced (*Polystoma* produces one to three eggs every 10 to 15 seconds), they are passed out of the worm fairly rapidly; therefore not many may be found within the parent at one time. Some species may store a few eggs in the

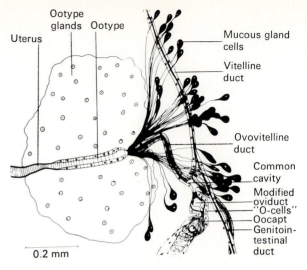

FIG. 14-14

Polystomoides malayi: ventral view of the female reproductive tract in the region of the ootype.

From Rohde, K., and A. Ebrahimzadeh. 1969. Z. Parasitenkd. 33:113.

ootype and then pass them to the outside directly through a pore, but, in most species, the eggs pass from the ootype into a uterus, which courses anteriad to open into the genital atrium, together with the ejaculatory duct. Hence the uterus, at least in most cases, does not function as a vagina as in digeneans.

DEVELOPMENT

The life cycles of a few species of Monogenea have been well studied, but little or nothing is known about most. With the exception of the viviparous Gyrodactylidae, monogeneans usually have a simple, direct life cycle involving an egg, oncomiracidium, and adult. Some evidence suggests that two species of gastrocotylids that parasitize predatory fish do not infect their definitive hosts directly but undergo a period of development on fish preyed on by the parasite's definitive hosts.[18]

Oncomiracidium

The oncomiracidium (Fig. 14-15) hatches from the egg and rather resembles a ciliate protozoan in size and shape. It is elongate and bears three zones of cilia, one in the middle and one at each end. The zones of ciliated epidermal cells are separated by an interciliary, nonnucleate syncytium. It has been shown in *Entobdella* that the nuclei of the interciliary regions are

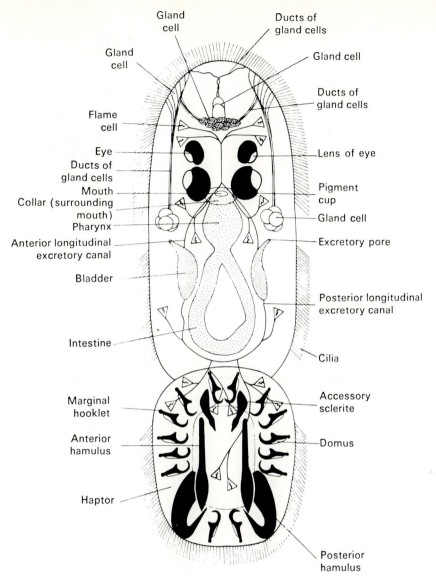

FIG. 14-15

The oncomiracidium of *Entobdella soleae* (ventral view).

From Kearn, G.C. 1971. In Fallis, A.M., editor. Ecology and physiology of parasites. University of Toronto Press, Toronto.

actually extruded during embryogenesis. Subsequently, the cytons of the "presumptive adult" tegument, which are located within the superficial muscle layer, extend processes out to underlay the ciliated cells and join the syncytial interciliary regions. The animal is thus ready for rapid shedding of the ciliated cells on attachment to the host; the stimulus for this shedding in *Entobdella* is mucus from the host epidermis, and the shedding takes only 30 seconds.

The oncomiracidium has cephalic glands with efferent ducts opening on the anterior margin and, as previously noted, has one or two pairs of eyes. The digestive tract is well differentiated, and the excretory pores are already formed. The posterior end always is developed into an attachment organ that bears hook sclerites, and these sclerites are retained in the adult. The larvae swim about until they contact a host; then they attach, lose their ciliated cells, and develop into adults. Oncomiracidia of *Entobdella soleae* swim alternately toward and away from light, apparently performing a search pattern for their host, a bottom-dwelling flatfish.[14] Rates of development into adults are largely unknown.

Inasmuch as the oncomiracidium is free swimming and the primary hosts of Monogenea are fish, with some in other vertebrates that are aquatic or amphibious, it would seem that infection of new hosts would not be a problem. However, the free-swimming life of the oncomiracidium is short; its potential hosts are widely dispersed most of the time and, in any case, can swim much faster than the larva. In addition, potential hosts may not even be present in the aquatic habitat except during breeding season. Thus it is of great selective value for the worm's egg production to be closely related to its host's reproduction, for instance, to coincide with a time when the host will be concentrated in spawning areas. Other features of the host's habits will also enhance chances for infection. Such correlation has been shown in several species, and similar adaptations are probably more widely prevalent.[19] Some of these adaptations will be illustrated in the discussion of life cycles that follow. Of course, some hosts are quite lethargic or are available during long periods in circumscribed areas; in these cases there may be little or no correlation of monogenean reproduction with host habits.

- ### *Dactylogyrus* species (Fig. 14-16)

A large number of species in this genus have been described, and some of them, such as *D. vastator, D. anchoratus,* and *D. extensus,* are of great economic importance as pathogens of hatchery fish. *Dactylogyrus* has large anchors on its opisthaptor and lives on the gill filaments of its host. Heavy infections cause loss of blood, erosion of epithelium, and access for secondary bacterial or fungal infections. Irritation to the gills stimulates increased mucus production, which often smothers the fish. Heavy infections may kill the host; massive die-offs are common in the crowded situations of fish culture ponds.

The life cycles and factors influencing the economically important species are reasonably well known.[4] The main features of the life cycle correspond to the preceding general outline. *Dactylogyrus vastator* on carp shows marked seasonal fluctuations correlated with temperature. Each worm deposits 4 to 10 eggs per 24 hours during the summer, and this rate increases with increasing temperature. The eggs require 4 to 5 days with temperatures between 20° and 28° C for embryonation, but this rate slows with temperatures, down to 4° C, at which point development is completely suppressed. The adult worms are adversely affected by lower temperatures so that the number of

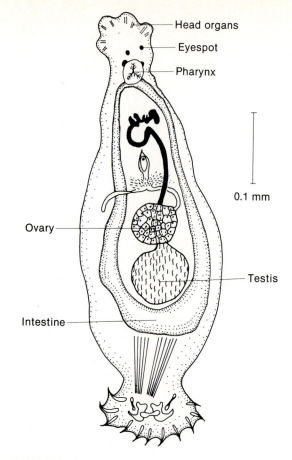

FIG. 14-16

Dactylogyrus vastator.

Redrawn from Bychowsky, B.E. 1933. In Bychowsky, B.E. 1961. Monogenetic trematodes, their systematics and phylogeny. American Institute of Biological Sciences, Washington, D.C.

parasites on the fish decreases greatly during the winter. The net effect is that the parasite population builds up over the summer, but the eggs deposited toward the end of the season winter over and result in a mass emergence in the spring to infest the young-of-the-year fish.

- ### *Gyrodactylus* species (Fig. 14-17)

Despite the (unfortunate) similarity in name and pathologic effects with *Dactylogyrus, Gyrodactylus* is a very different organism. It is also significant economically as an important pest particularly of trout, bluegills, and goldfish in fish ponds.[5] The family Gyrodactylidae is unusual among the Monogenea in that it is viviparous. The young are retained in the uterus

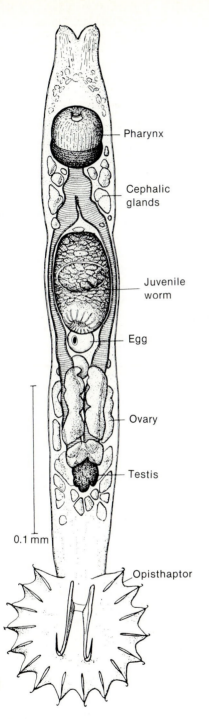

FIG. 14-17

Gyrodactylus cylindriformis, ventral view.

Pharynx

Cephalic glands

Juvenile worm

Egg

Ovary

Testis

0.1 mm

Opisthaptor

From Mueller, J.F., and H.J. Van Cleave. 1932. Roosevelt Wildl. Ann. 3:73-154.

until they develop into functional subadults. Inside such a developing juvenile, one can often see a second juvenile developing, with a third juvenile inside of it and a fourth inside the third! The exact mechanism of this unique embryogenesis is unknown, but it may be considered a type of sequential polyembryony; as many as four individuals usually result from one zygote. After birth the young worm begins feeding on its host and gives birth to the juvenile "remaining" inside. Only then can an egg from its own ovary be fertilized and repeat the sequence. Since only a day or so is required for a worm to mature after birth and give birth to another worm already developing within it, massive infection can build up quickly.

Certain other peculiarities of *Gyrodactylus* have been described that may be correlated with their unusual embryogenesis. One of these is the possession of an **ovovitellarium,** a fused mass of ova and vitelline cells. Another is the fact that the "distal cytoplasm" of the tegument apparently is not connected to cytons in the parenchyma. The epidermis of the developing embryo has nuclei that are not present in the adult tegument. The tegument of *Gyrodactylus* may be embryologically equivalent to the ciliated epidermal cells on the oncomiracidium of other monogeneans.[21]

Not having an oncomiracidium, *Gyrodactylus* must depend on transmission of the adult or subadult from one host to another. Since these forms appear unable to swim, it is clear that the prospective host must be quite close to the worm's current host for the transferral to take place, but it is not known how the worm detects the proximity of a new potential host. Random leaps are unlikely; infection can spread through schooled fish in a pond.[19]

■ Polystomatidae

Polystoma integerrimum (Fig. 14-18) is a parasite of Old World frogs. It is of particular interest because it is known that the worm's reproductive cycle is synchronized with that of its host by means of host hormones, a mechanism to provide a ready supply of hosts to the hatching oncomiracidia. Furthermore, two different types of adults develop: normal and neotenic.

Adult worms live in the urinary bladder of their host. They are dormant during the winter, while the frogs hibernate, but become active in the spring along with their hosts. When the frog's gonads begin to swell and produce gametes, the worms begin to copulate and produce eggs that are released into the surrounding urine. In the laboratory, maturation and stimulation of worm gamete production can be elicited by injecting the frogs with pituitary extract, although it is not known whether the effect is direct by way of

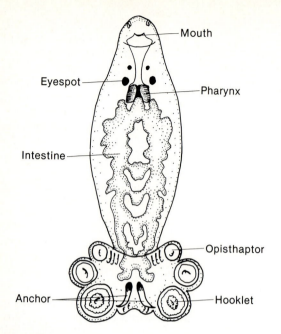

Mouth

Eyespot

Pharynx

Intestine

Opisthaptor

Anchor

Hooklet

FIG. 14-18

Polystoma integerrimum, a parasite of Old World frogs.
Redrawn from Zeller, E. 1872. Z. Wiss. Zool. 22:1-28.

the gonadotropins or stimulated by the injection via the gonadal hormones of the host.[31] In nature the eggs are voided into the water in the frog's spawning area. Depending on the temperature, the oncomiracidia hatch in 20 to 50 days. By then the frog's eggs have developed into tadpoles that will be the next host generation. Tadpoles first have external, then internal, gills and breathe by sucking water into the mouth, over the gills, and out the side of the pharynx through slits, much like fish. An oncomiracidium contacting a gill attaches, metamorphoses, and begins producing eggs within 20 to 25 days.

The gill form of *P. integerrimum* is considerably different in morphology from the bladder form. Its body is narrower, and the opisthaptor is not sharply set off from the body. The intestine has fewer lateral branches, and the ovary is a different shape. Furthermore, there is no uterus or genitointestinal canal. Some authors consider the gill stage to be neotenic.[1]

Eggs from these worms hatch in 15 to 20 days, when the water is somewhat warmer. These larvae also attach to the gills of tadpoles, but by now the tadpoles are older, and worm maturation is delayed. When the tadpoles begin their metamorphosis, including resorption of the gills, the worms migrate to the bladder. The migration occurs over the ventral skin of the tadpole at night and takes only about a minute.[18]

Oddly, even when the metamorphosing tadpole is exposed to newly hatched larvae, the worms go first to the gills and then to the bladder, and this has been taken as evidence that the migration is controlled endogenously and not by stimuli from the host. It takes much longer, 4 to 5 years, for the bladder form to mature and begin egg production.

A similar species, *P. nearcticum,* occurs in tree frogs in the United States. It also has a neotenic gill form but no slowly developing and then migrating immature gill form. Oncomiracidia enter the cloaca directly when they contact metamorphosing tadpoles. Larvae of *Protopolystoma xenopi* also enter the cloaca of its host, *Xenopus laevis,* directly. Interestingly, *Xenopus* remains in water all year around, and reproduction of *Protopolystoma* continues correspondingly.

At the other extreme are *Pseudodiplorchis americanus* and *Neodiplorchis scaphiopodis,* parasites in the urinary bladder of spadefoot toads (*Scaphiopus* spp.) in Arizona. The toads live in one of the hottest and driest areas in the United States and spend about 10 months per year beneath the soil in a state of torpor. They breed during only one to three nights per year in temporary pools formed by the brief desert rains. These conditions present an extraordinary challenge to a parasite that depends on an aqueous environment for transmission, and transmission must be exquisitely coordinated with the activities of the hosts. During the time that the toads are in hibernation, the encapsulated oncomiracidia become fully developed in the uterus of the adult worms.[34] When the toad enters water, the larvae are deposited, pass out with the urine, hatch within seconds, and are fully infective for the next host. The oncomiracidia are much larger than most other oncomiracidia (up to 600 μm), have a much longer free-swimming life (up to 48 hours), and are more resistant to drying (up to 1 hour). Thus, if they do not reach another host during the first night when the toads are spawning, they have a good chance of succeeding the next night. When the oncomiracidia contact another spadefoot toad as it is floating with the end of its nose above water, they crawl on the toad's skin up *out* of the water and enter its nares. Then they migrate to the lungs, undergo a period of development, and finally migrate to the urinary bladder by passing through the stomach and intestine. An unidentified host factor protects them as they pass through this lethal environment.[35]

■ *Diplozoon paradoxum* (Fig. 14-19)

Diplozoon paradoxum is a common parasite of the gills of species of European cyprinid fish. Like *Dactylogyrus, Diplozoon* exhibits a strong seasonal variation in its reproductive activity. Virtually no gametes

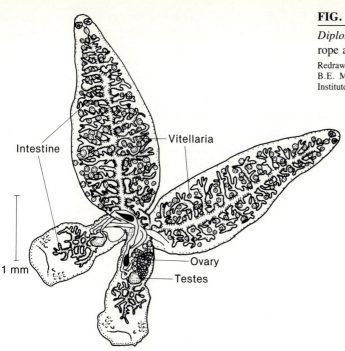

FIG. 14-19

Diplozoon paradoxum, a parasite of freshwater fishes in Europe and Asia.

Redrawn from Bychowsky, B.E., and L.F. Nagibina. 1961. In Bychowsky, B.E. Monogenetic trematodes, their systematics and phylogeny. American Institute of Biological Sciences, Washington, D.C.

are produced during the winter, but gonads begin to function during the spring, reaching a peak during May to June and continuing through the summer. The eggs, which have a long, coiled filament at their ends, can hatch about 10 days after they are deposited, and light intensity and turbulence of the water, as might be caused by host feeding or spawning activity, stimulate hatching. The oncomiracidium bears two clamps on its opisthaptor with which it attaches to a gill filament; it loses its cilia almost immediately. The worm feeds and begins to grow, adding another pair of clamps to the opisthaptor. A small sucker also appears on the ventral surface, and a tiny papilla on the dorsal surface, slightly more posterior than the sucker. When this stage (Fig. 14-20) was first discovered, it was thought to represent a new genus, and it was named *Diporpa.* When *Diporpa* was recognized to be a juvenile stage of *Diplozoon,* the stage was referred to as a **diporpa.** Curiously, in contrast to *Dactylogyrus* spp., *Diplozoon,* even diporpae, rarely infect young-of-the-year fish.

A diporpa juvenile can live for several months, but it cannot develop further until encountering another diporpa, and unless this happens, the diporpa usually perishes by winter. When one diporpa finds another, each attaches its sucker to the dorsal papilla of the other. Thus begins one of the most intimate associations of two individuals in the animal kingdom. The two worms fuse completely, with no trace of parti-

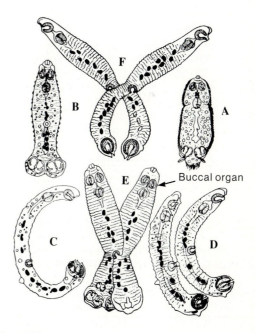

FIG. 14-20

Development of *Diplozoon paradoxum.* **A,** Freshly hatched, free-swimming juvenile. **B,** Diporpa juvenile. **C** to **F,** Diporpa juveniles attaching themselves to one another.

From Baer, J.G. 1952. Ecology of animal parasites. University of Illinois Press, Urbana, Ill.

tions separating them. The fusion stimulates maturation. Gonads appear; the male genital duct of one terminates near the female genital duct of the other, permitting cross-fertilization. Two more pairs of clamps develop in the opisthaptor of each. Adults apparently can live in this state for several years.

METABOLISM

Although ultrastructure, behavior, and some aspects of their physiology have received attention in recent years, knowledge of the metabolism of monogeneans is very meager. Ectoparasitic species store little glycogen, but endoparasitic forms store about the same amounts as do digeneans.[31] In light of the fact that these ectoparasites live in habitats with medium to high oxygen concentrations, it would not be surprising if they depended heavily on aerobic energy metabolism. Indeed, certain observations imply that this may be true. *Entobdella soleae,* whose habitat is the lower surface of a bottom-dwelling flatfish, has characteristic, rhythmic, undulating body movements that appear to be clearly ventilatory in function. The frequency and amplitude of the undulations are increased be lower oxygen concentration and increased temperature. Furthermore, when ambient oxygen concentration is high, the parasite is somewhat contracted and wrinkled. At lower oxygen concentrations the worm stretches out, increasing its surface area and decreasing the distance through which oxygen must diffuse to the interior. Experiments with metabolic inhibitors on *Diclidophora merlangi* suggest that this species has a functional tricarboxylic acid cycle and a classical cytochrome system.[31] Research in monogenean metabolism will surely be rewarding for those who enter the field.

PHYLOGENY

Although some modern authors, for example, Schell,[30] persist in allying the Monogenea with the Digenea and Aspidogastrea as an order or subclass within the Trematoda, others have argued that the Monogenea are more closely related to the Cestoidea.[4,16] There is certainly an undeniable similarity in the structure of the hooks found in oncomiracidia of monogenes to those in the newly hatched larvae of Cestodaria and Eucestoda. Furthermore, there is a chemical similarity in all of these; that is, they are composed of a protein stabilized by disulphide cross-linkages.[20] No such structures are found in larval or adult Digenea or Aspidogastrea.

Llewellyn[17] proposed that the Digenea and Aspidogastrea arose separately from Monogenea, both lines

from rhabdocoel turbellarian ancestors. He suggested that the monogene ancestor began feeding on the surface of bottom-dwelling, slow-moving, early vertebrates.[16] It had a posterior adhesive pad. As the descendant vertebrates became more efficient and rapid swimmers, there was a selective advantage for the development of hooks on the adhesive pad, and the adaptive radiation of the opisthaptor followed. He suggested further that some lines could have become endoparasitic in the digestive tracts of their hosts and eventually could have given rise to the Cestodaria and the Eucestoda.[17] Strobilation (see p. 313) arose not as an adaptation to produce more eggs but to increase the efficiency of egg production by a replication of ootypes; that is, a bottleneck in egg production is eliminated by a multiplicity of ootypes because only one egg can be formed at a time in an ootype.

CLASSIFICATION OF CLASS MONOGENEA

Class Monogenea
Hermaphroditic, dorsoventrally flattened, elongate or oval worms with a syncytial tegument; conspicuous posterior adhesive organ present (opisthaptor) that is muscular, sometimes divided into loculi, usually with sclerotized anchors, hooks, and/or clamps, often subdivided into individual suckers or clamps without sclerites; an anterior adhesive organ (prohaptor) usually present, consisting of one or two suckers, grooves, glands, or expanded ducts from deeper glands; eyes, when present, usually of two pairs; mouth near anterior, pharynx usually present; gut usually with two simple or branched stems often anastomosing posteriorly, rarely a single median tube or sac; male genital pore usually in atrium common with female pore; genitointestinal duct present or absent; reproduction oviparous or viviparous; vitelline follicles extensive, usually lateral; two lateral osmoregulatory canals present, each with expanded vesicle opening dorsally near anterior end; parasites on or in aquatic vertebrates, especially fishes, or rarely on aquatic invertebrates; cosmopolitan.

Subclass Monopisthocotylea
Opisthaptor a single unit that may be subdivided into shallow loculi, usually developed directly from the larval haptor; one to three pairs of large anchors usually present, commonly with tiny marginal hooks; prohaptor glandular or with paired suckers or pseudosuckers; oral sucker absent; eyes often present; genitointestinal canal absent; seminal receptable is enlargement of vagina.

Order Acanthocotyloidea
Opisthaptor developed separately from larval haptor, which remains as tiny structure with 14 marginal and 2 central hooklets; functional haptor muscular, without anchors, but radially arranged spines or septa may be present; parasites of marine fishes: cosmopolitan.

FAMILY
Acanthocotylidae

Order Capsaloidea
Opisthaptor large, circular, muscular, often divided by septa into shallow loculi; anchors, when present, lacking connecting bars;

marginal hooklets present or absent; prohaptor, when present, with two glandular areas, two lateral suckers, or single pseudosucker; cirrus always lacking accessory piece; parasites of fishes; cosmopolitan.

FAMILIES
Bothitrematidae, Calceostomatidae, Capsalidae, Dactylogyridae, Dioncidae, Diplectanidae, Loimoidae, Microbothriidae, Monocotylidae, Protogyrodactylidae, Tetraoncidae, Tetraoncoididae

Order Gyrodactyloidea
Opisthaptor rounded or bilobed, with none, one, or two pairs of anchors supported by one to three bars; marginal larval hooks present; prohaptor with groups of cephalic glands with ducts opening on margin of anterior end; parasites of fishes, amphibians, cephalopods, and crustaceans; cosmopolitan.

FAMILY
Gyrodactylidae

Order Udonelloidea
Opisthaptor muscular, lacking armature or septa; prohaptor poorly developed, but with lateral head organs or pseudosuckers; parasites of marine fishes or of copepods parasitic on marine fishes.

FAMILY
Udonellidae

Subclass Polyopisthocotylea
Opisthaptor complex, with suckers, clamps, or anchor complexes, commonly subdivided; larval haptor absent or reduced to pad-supporting terminal anchors; marginal hooklets usually absent; mouth surrounded by sucker, striated fringe, or with paired buccal organs inside buccal cavity; prohaptor usually without adhesive glands; eyes usually absent; genitointestinal canal usually present; gut usually with two crura, sometimes joined posteriorly; testes usually numerous; seminal receptacle present or absent.

Order Avielloidea
Opisthaptor at end of long stalk, with six suckers on margin and four large anchors in middle; intestine a simple unbranched sac; prohaptor with two or three pairs of glandular organs; four eyes present; single testis; eggs lacking polar filaments; parasites of freshwater teleosts; Lake Baikal, Russia.

FAMILY
Aviellidae

Order Chimaericoloidea
Posterior portion of body elongate and slender, with small opisthaptor present near larval haptor; opisthaptor with two rows of simple clamps; prohaptor a simple, weak, circumoral sucker; eyes and intraoral suckers absent; reproductive organs in anterior portion of body; parasites of Holocephali; Atlantic, Mediterranean, Pacific.

FAMILY
Chimaericolidae

Order Diclidophoroidea
Opisthaptor commonly subdivided into two lateral rows of four each, suckers or clamps; prohaptor usually simple; buccal organs present in oral cavity; intestine bifurcate; eyes absent; parasites of fishes or crustaceans parasitic on fishes; cosmopolitan.

FAMILIES
Dactylocotylidae, Diclidophoridae, Discocotylidae, Gastrocotylidae, Hexostomatidae, Macrovalvitrematidae, Mazocraeidae, Octolabeidae, Plectanocotylidae, Protomicrocotylidae, Pterinotrematidae

Order Diclybothrioidea
Opisthaptor with three pairs of clamps or suckers, each surrounding a large anchor; also, posterior appendix on opisthaptor that bears three pairs of large and one pair of very small hooks and in some species also a rudimentary pair of suckers on posterior margin; prohaptor with two lateral, sucker-shaped depressions; two pairs of eyes present; intestine bifurcate, branched, anastomosing near posterior end of body; parasites of Acipenseriformes, Selachii, and Polyodontidae; cosmopolitan.

FAMILIES
Diclybothriidae, Hexabothriidae

Order Diplozooidea
Adults permanently fused in pairs forming an X shape; opisthaptor rectangular or bilobed, with four pairs, or numerous, clamps; one pair of posterior anchors also present in some species; intestine a single tube with many branches; genital pore in posterior half of body; parasites of the gills of freshwater fishes; cosmopolitan.

FAMILY
Diplozoidae

Order Megaloncoidea
Opisthaptor complex, with three pairs of anchor complexes, each consisting of several sclerotized elements; marginal hooklets and eyes absent; two terminal appendices, each with two pairs of hooks, present on posterior margin of opisthaptor of some species; prohaptor with pair of buccal organs; parasites of marine teleosts; Japan, China.

FAMILIES
Anchorophoridae, Megaloncidae

Order Microcotyloidea
Opisthaptor symmetrical or asymmetrical, with numerous clamps arranged symmetrically or asymmetrically; anchors may also be present at posterior end; prohaptor with pair of buccal organs; intestine bifurcate, not anastomosed posteriorly; eyes absent; parasites of marine fishes; cosmopolitan.

FAMILIES
Allopyragraphoridae, Axinidae, Cemocotylidae, Heteromicrocotylidae, Microcotylidae, Pyragraphoridae

Order Polystomatoidea
Opisthaptor with two or six well-developed suckers, with or without anchors, with larval hooklets present; mouth surrounded by oral sucker; intestine bifurcate, rejoined or not; eyes usually absent; parasites of fishes, amphibians, reptiles, and on eyes of hippopotami; cosmopolitan.

FAMILIES
Polystomatidae, Sphyranuridae

REFERENCES

1. Baer, J.G., and L. Euzet. Classe des monogènes. Monogenoidea Bychowsky. In Grassé, P., editor. Traité de zoologie: anatomie, systématique, biologie, vol. 4. Masson & Cie, Paris, pp. 243-325.

2. Braun, M. 1889-1893. Trematodes Rudolphi 1808. In Bronn's Klassen und Ordnugen des Tierreichs 4:306-925.

3. Bychowsky, B.E., 1937. Ontogenese und phylogenetische Beziehungen der parasitischen Platyhelminthes. Izvest. Akad. Nauk SSSR Seria Biol. 4:1353-1383.

4. Bychowsky, B.E., 1957. Monogenetic trematodes: their systematics and phylogeny. Akademii Nauk SSSR, Moscow. (English translation, 1961: Hargis, W.J., editor: American Institute of Biological Sciences, Washington, D.C.)

5. Cone, D.K., and P.H. Odense. 1984. Pathology of five species of *Gyrodactylus* Nordmann, 1832 (Monogenea). Can. J. Zool. 62:1084-1088.

6. Fuhrmann, O. 1928. Trematoda. Zwite Klasse der Cladus Platyhelminthes. In Kükenthal's and Krumbach's Handbuch der Zoologie 2:1-140.

7. Halton, D.W. 1974. Hemoglobin absorption in the gut of a monogenetic trematode, *Diclidophora merlangi*. J. Parasitol. 60:59-66.

8. Halton, D.W. 1978. Trans-tegumental absorption of L-alanine and L-leucine by a monogenean, *Diclidophora merlangi*. Parasitology 76:29-37.

9. Halton, D.W., E. Dermott, and G.P. Morris. 1968. Electron microscope studies on *Diclidophora merlangi* (Monogenea: Polyopisthocotylea). I. Ultrastructure of the cecal epithelium. J. Parasitol. 54:909-916.

10. Halton, D.W., and G.P. Morris. 1969. Occurrence of cholinesterase and ciliated sensory structures in fish-gill fluke, *Diclidophora merlangi* (Trematoda: Monogenea). Z. Parasitenkd. 33:21-30.

11. Jennings, J.B. 1968. Nutrition and digestion. In Florkin, M., and B.T. Scheer, editors. Chemical zoology, vol. 2, sect. III. Platyhelminthes, Mesozoa. Academic Press, Inc., New York, pp. 303-326.

12. Kayton, R.J. 1983. Histochemical and x-ray elemental analysis of the sclerites of *Gyrodactylus* spp. (Platyhelminthes: Monogenoidea) from the Utah chub, *Gila atraria* (Girard). J. Parasitol. 69:862-865.

13. Kearn, G.C. 1971. The physiology and behaviour of the monogenean skin parasite *Entobdella soleae* in relation to its host *(Solea solea)*. In Fallis, A.M., editor. Ecology and physiology of parasites. University of Toronto Press, Toronto, pp. 161-187.

14. Kearn, G.C. 1980. Light and gravity responses of the oncomiracidium of *Entobdella soleae* and their role in host location. Parasitology 81:71-89.

15. Kearn, G.C. 1985. Observations on egg production in the monogenean *Entobdella soleae*. Int. J. Parasitol. 15:187-194.

16. Llewellyn, J. 1963. Larvae and larval development of monogeneans. In Dawes, B., editor. Advances in parasitology, vol. I. Acadmic Press, Inc., New York, pp. 287-326.

17. Llewellyn, J. 1965. The evolution of parasitic platyhelminths. In Taylor, A.E.R., editor. Evolution of parasites. Third Symposium of the British Society for Parasitology. Blackwell Scientific Publications Ltd. Oxford, Eng., pp.47-78.

18. Llewellyn, J. 1968. Larvae and larval development of monogeneans. In Dawes, B., editor. Advances in parasitology, vol. 6, Academic Press, Inc., New York, pp. 373-383.

19. Llewellyn, J. 1972. Behavior of monogeneans. In Canning, E.U., and C.A. Wright, editors. Behavioural aspects of parasite transmission. Linnaean Society of London. Academic Press, Inc. (London) Ltd., pp. 19-30.

20. Lyons, K.M. 1966. The chemical nature and evolutionary significance of monogenean attachment sclerites. Parasitology 56:63-100.

21. Lyons, K.M. 1970. Fine structure of the outer epidermis of the viviparous monogenean *Gyrodactylus* sp. from the skin of *Gasterosteus aculeatus*. J. Parasitol. 56:1110-1117.

22. Lyons, K.M. 1972. Sense organs of monogeneans. In Canning, E.U., and C.A. Wright, editors. Behavioural aspects of parasite transmission. Linnaean Society of London. Academic Press, Inc. (London) Ltd., pp. 181-199.

23. Lyons, K.M. 1973. The epidermis and sense organs of the Monogenea and some related groups. In Dawes, B., editor. Advances in parasitology, vol. 11. Academic Press, Inc., New York, pp.193-232.

24. Macdonald, S., and J. Caley. 1975. Sexual reproduction in the monogenean *Diclidophora merlangi*: tissue penetration by sperms. Z. Parasitenkd. 45:323-334.

25. Ramalingam, K. 1973. Chemical nature of the egg shell in helminths. II. Mode of stabilization of egg shells of monogenetic trematodes. Exp. Parasitol. 34:115-122.

26. Rees, J.A., and G.C. Kearn. 1984. The anterior adhesive apparatus and an associated compound sense organ in the skin-parasitic monogenean *Acanthocotyle lobianchi*. Z. Parasitenkd. 70:609-625.

27. Rohde, K. 1975. Fine structure of the Monogenea, especially *Polystomoides* Ward. In Dawes, B., editor. Advances in parasitology, vol. 13, Academic Press, Inc., New York, pp. 1-33.

28. Rohde, K. 1977. Habitat partitioning in Monogenea of marine fishes. Z. Parasitenkd. 53:171-182.

29. Rohde, K. 1979. The buccal organ of some Monogenea Polyopisthocotylea. Zoologica Scripta 8:161-170.

30. Schell, S.C. 1982. Trematoda. In Parker, S.P., editor. Synopsis and classification of living organisms, vol. 1. McGraw-Hill, Inc., New York, pp. 740-807.

31. Smyth, J.D., and D.W. Halton. 1983. The physiology of trematodes, ed. 2. Cambridge University Press, Cambridge, England.

32. Sproston, N.G. 1946. A synopsis of the monogenetic trematodes. Trans. Zool. Soc. Lond. 23:183-600.

33. Thurston, J.P., and R.M. Laws. 1965. *Oculotrema hippopotami* (Trematoda: Monogenea) in Uganda. Nature 205:1127.

34. Tinsley, R.C., and C.M. Earle. 1983. Invasion of vertebrate lungs by the polystomatid monogeneans *Pseudodiplorchis americanus* and *Neodiplorchis scaphiopodis*. Parasitology 85:501-517.

35. Tinsley, R.C., and H.C. Jackson. 1986. Intestinal migration in the life cycle of *Pseudodiplorchis americanus* (Monogenea). Parasitology 93:451-469.

SUGGESTED READINGS

Hargis, W.J., Jr. 1957. The host specificity of monogenetic trematodes. Exp. Parasitol. 6:620-625.

Hargis, W.J., Jr., A.R. Lawler, R. Morales-Alamo, and D.E. Zwerner. 1969. Bibliography of the monogenetic trematode literature of the world. Virginia Insitute of Marine Science, Special Scientific Report, Gloucester Point, Va. 55:1-95.

Llewellyn, J. 1963. Larvae and larval development of monogeneans. In Dawes, B., editor. Advances in parasitology, vol. 1. Academic Press, Inc., New York, pp. 287-326. (An excellent review on the subject, in which the author makes important phylogenetic hypotheses.)

Sproston, N.G. 1946. A synopsis of the monogenetic trematodes. Trans. Zool. Lond. 25:185-600. (An extremely important monograph on the systematics of the group.)

Yamaguti, S. 1963. Systema Helminthum, vol. 4, Interscience Publishers, New York. (An easy-to-use key to all genera of Monogenea, with many illustrations.)

CHAPTER **15**

CLASS TREMATODA: SUBCLASS ASPIDOGASTREA

The Aspidogastrea constitute a small group of Digenea-like worms that seem poorly adapted to parasitism. Although a discrete group, they are much more similar to the Digenea than to the Monogenea and so are given status as a subclass of the Trematoda. They have established a loosely parasitic relationship with molluscs, in most species, but some are facultative or obligate parasites of fishes or turtles.[2]

Two other names have often been used for this group: Aspidocotylea and Aspidobothria. Although the latter undoubtedly has priority, most of the literature has accumulated under the name Aspidogastrea, which is the title we shall use here.

By any name, this group of organisms has attracted little attention because they are of no medical or known economic importance. Nevertheless these innocuous little worms are of considerable biological interest: they seem to represent a step between free-living and parasitic organisms. A comprehensive review of Aspidogastrea has been presented by Rohde.[5]

FORM AND FUNCTION
Body form

Externally, aspidogastreans exhibit three basic types of anatomy, corresponding to the three families that have been established for them. The Aspidogastridae (Fig. 15-1) have a huge **ventral sucker,** extending most of the length of the body. This sucker (also known as an **opisthaptor** or **Baer's disc**) has muscular **septa** in longitudinal and transverse rows, dividing it into shallow depressions called **alveoli** or **loculi** (Fig. 15-2). The number, shape, and arrangement of these loculi are of considerable taxonomic importance. Hooks or other sclerotized structures are never present. Between the marginal loculi are usually **marginal bodies,** which are secretory organs, or short tentacles, also presumably secretory in nature. Exceptionally, both are absent.

In the Stichocotylidae (see Fig. 15-10) a longitudinal series of individual suckers occurs instead of a single complex of loculi, whereas in Rugogastridae the ventral holdfast is made up of transverse **rugae** (Fig. 15-3).

The **marginal bodies** are round to oval organs and are connected to each other by fine ducts. They consist of gland cells, storage chambers, and secretory ducts. Although a sensory function has been suggested for the marginal bodies, no indication exists that their function is other than secretory. The tentacles of *Lophotaspis* (Fig. 15-4) are probably modified marginal organs.

The **longitudinal septum** is a peculiar morphological characteristic of the Aspidogastrea. It is a horizontal layer of connective tissue and muscle in the anterior part of the body, projecting like a shelf and dividing the body into dorsal and ventral compartments. The function of the septum is not known, but it might be correlated with pressures exerted by contraction of the giant ventral sucker.

Tegument

Although only one species *(Multicotyle purvisi)* has been studied adequately, the tegument of this species seems to be basically similar to that of other groups of parasitic flatworms. It is syncytial and has an outer stratum of **distal cytoplasm,** containing numerous vesicles of various types, and mitochondria. The tegumental nuclei are in **cytons** internal to the superficial muscle layer and connected to the distal cytoplasm by internuncial processes. The cytons are rich in Golgi complexes. A mucoid layer of variable thickness is found on the outer surface membrane, and in some areas the surface membrane has riblike elevations to support the thick mucoid layer.

Digestive system

The digestive tract is simple. The **mouth** is funnel-like in some species, whereas in others it is surrounded by a muscular sucker or several muscular lobes. At the base of the mouth funnel is a spheroid **pharynx,** a powerful muscular pump. The **intestine,** or **cecum,** is a single, simple sac that usually extends to near the posterior end of the body. Its epithelial cells

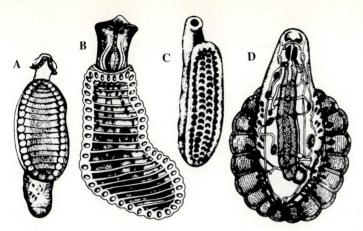

FIG. 15-1

Examples of the family Aspidogastridae. **A,** *Lobatostomum ringens*. **B,** *Cotylogaster michaelis*. **C,** *Lophotaspis vallei*. **D,** *Cotylaspis insignis*.

FIG. 15-2

Scanning electron micrograph, ventral view of *Cotylogaster occidentalis*. The cup-shaped buccal funnel is at left, and the neck is semiretracted into the body by a fold. Note the smooth tegument and prominent alveoli in the ventral haptor. (×92).

From Ip, H.S., S.S. Desser, and I. Weller. 1982. Trans. Am. Microsc. Soc. 101:253-261.

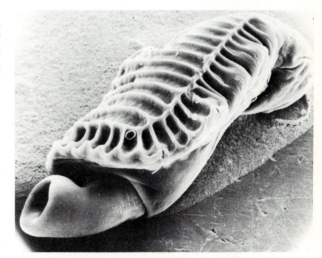

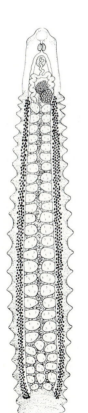

FIG. 15-3

Rugogaster hydrolagi from the rectal gland of a ratfish.

From Schell, S.C. 1973. J. Parasitol. 59:803-805.

FIG. 15-4

Lophotaspis interiora From an alligator snapping turtle.

From Ward, H.B., and S.H. Hopkins. 1932. J. Parasitol. 18:69-78.

bear a complex reticulum of lamellae on their luminal surface, presumably vastly increasing the absorptive surface. A layer of muscles, usually of both circular and longitudinal fibers, surrounds the cecum.

Osmoregulatory system

This system consists of numerous **flame cell protonephridia** connected to capillaries feeding into larger excretory ducts and eventually into an **excretory bladder** near the posterior end of the body. The flame cells are peculiar in that their ciliary membranes continue beyond the tips of the cilia and anchor apically in the cytoplasm of the flame cell. Lateral or nonterminal ciliary flames have been reported in a number of species. The small capillaries have numerous microvilli projecting into their lumina, and the larger capillaries and excretory ducts are abundantly provided with lamellar projections of their surface membranes, thus suggesting secretory-absorptive function. The **excretory pore** is dorsosubterminal or terminal and usually single.

Nervous system

The nervous system of aspidogastreans is very complex for a parasitic flatworm, reminiscent of a condition more typical of free-living forms. As in the Turbellaria, there is a complex set of anterior nerves called the **cerebral commissure** and a modified ladder type of peripheral system. A wide variety of sensory receptors has been observed, mostly around the mouth and on the margins of the ventral disc. In a specimen of *Multicotyle purvisi* 6.1 mm long, Rohde[5] counted 360 dorsal and 260 ventral receptors in the prepharyngeal region and 140 in the oral cavity, not counting free nerve endings below the tegument. Three types of "ciliated sense organs" (sensilla) have been described on the body of *Cotylogaster occidentalis*.[3]

A complex system of connectives and commisures occurs in the ventral disc and walls of the alveoli, indicating a high degree of neuromuscular coordination. The septum, intestine, pharynx, prepharynx, cirrus pouch, uterus, and genital and excretory openings are all innervated by plexuses. Some cells in the nervous system are positive for paraldehyde-fuchsin stain, indicating possible neurosecretory function.

Reproductive systems

The male reproductive system of Aspidogastrea is similar to that of the Digenea (see Fig. 17-16). One, two, or many **testes** are present, located posterior to the ovary (Fig. 15-5). The **vas deferens** expands to form an **external seminal vesicle** before it enters the **cirrus pouch** to become the **ejaculatory duct.** A cirrus pouch is absent in some species. The **cirrus** is unarmed and opens through the genital pore into a common genital atrium, located on the midventral surface just anterior to the leading margin of the ventral disc. The axonemes in the spermatozoon filament have the 9 + 1 structure, as is the case in other platyhelminth sperm (Fig. 15-6).

The female reproductive system consists of an ovary, vitelline cells, uterus, and associated ducts. The **ovary** (Fig. 15-5) is lobated or smooth and empties its products into an **oviduct.** The oviduct is peculiar among the Platyhelminthes in that its lumen is divided into many tiny chambers by septa, and the lining along much of its length is ciliated (Fig. 15-7). Each septum has a small hole in it through which the eggs pass. The oviduct empties onto the **ootype,** which is surrounded by Mehlis' gland cells (Fig. 15-8). A short tube, leading from the ootype and ending blindly in the parenchyma or, in a few cases, connecting with the excretory canal, is called **Laurer's canal** and probably represents a vestigial vagina.

Vitelline follicles occur in two lateral fields, each of which has a main **vitelline duct** that fuses with that from the other field to form a small **vitelline reservoir,** which in turn opens into the ootype. Finally, a **uterus** extends from the ootype to course toward the genital

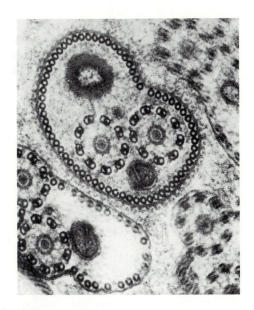
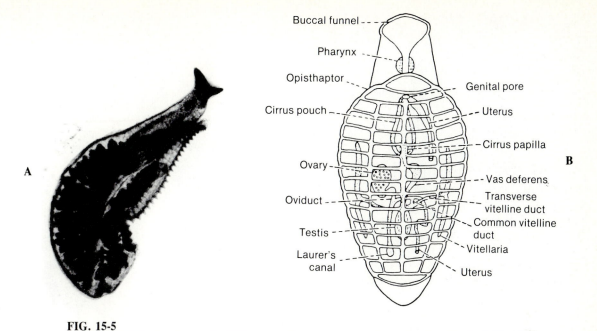

FIG. 15-5

Aspidogaster conchicola, a common parasite of freshwater clams. **A,** Lateral view, showing general body form. **B,** Ventral view. The gut and the portion of the uterus between the descending limb and the terminal portion have been omitted in **B.**

A photograph by Warren Buss; **B** from Olsen, O.W. 1974. Animal parasites, their life cycles and ecology, ed. 3. University Park Press, Baltimore.

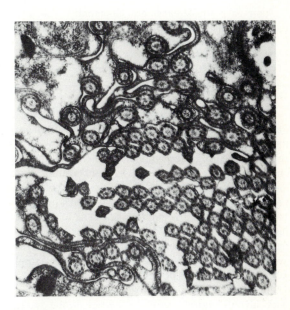

FIG. 15-6

Cross section of sperm filament of *Aspidogaster conchicola.*
Photograph by Ronald P. Hathaway.

FIG. 15-7

Section of proximal oviduct of *Aspidogaster conchicola,* Showing the cilia that line much of its length.
Photograph by Ronald P. Hathaway.

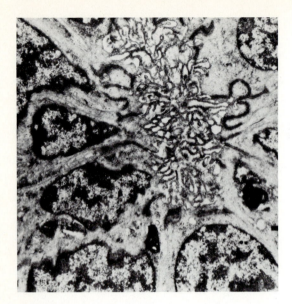

FIG. 15-8

Ootype of *Aspidogaster conchicola* surrounded by Mehlis' gland cells.

Photograph by Ronald P. Hathaway.

atrium, usually with a posterior loop and anterior, distal stem. The distal end of the uterus has powerful muscles in its walls and is called the **metraterm.** This propels the eggs out of the system.

Some aspidogastreans are apparently self-fertilizing, with the cirrus depositing sperm in the terminal end of the uterus, which serves as a vagina. Self-fertilization does not occur in some species.[6]

DEVELOPMENT

As in other platyhelminths with separate vitellaria, the eggs of aspidogastreans are ectolecithal; that is, most of the yolk supply of the embryo is derived from separate cells packaged with the zygote inside the eggshell. Some species are completely embryonated when they pass from the parent and hatch within a matter of hours, whereas others require 3 to 4 weeks of embryonation in the external environment. The larvae (**cotylocidia**) (Fig. 15-9) hatching from the egg in most species have a number of ciliary tufts that are effective in swimming. They possess a mouth, pharynx, simple gut, and a prominent posterior-ventral disc without alveoli; there are no hooks. As the worm develops in its host, alveoli begin to form, tier by tier, in the anterior part of the ventral disc. The original cup of the disc remains apparent for some time behind the new ventral sucker, then disappears forever. The larval ultra-

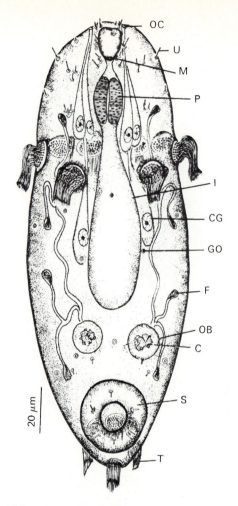

FIG. 15-9

Composite drawing of cotylocidium of *Cotylogaster occidentalis*. *C*, Concretion; *CG*, cephalic gland; *GO*, opening of goblet-like gland cells; *F*, flame cell; *I*, intestine; *M*, mouth; *OB*, osmoregulatory bladder; *OC*, opening of cephalic gland; *P*, pharynx; *S*, sucker; *T*, tuft of cilia; *U*, uniciliated sensory structure.

From Fredericksen, D.W. 1978. J. Parasitol. 64:961-976.

structure has been studied in *M. purvisi* and in *C. occidentalis*.[1,5] The tegument pattern is similar to that of the adult, with the distal cytoplasmic, syncytial layer at the surface and with internal cytons. Between the ciliary tufts and covering most of the body, the tegument surface in *M. purvisi* bears unique filiform structures called **microfila.** These have one central filament and about 9 to 12 peripheral filaments, differing from microvilli in that they do not have a cytoplasmic core. Their function is unknown, but it has been suggested that they help the larva to float.[5] In contrast, the tegument of *C. occidentalis* bears short microvilli with an external glycocalyx coat.[1]

Most aspidogastreans have a direct life cycle, requiring no intermediate host. Those parasitic in vertebrates appear to require an intermediate host; no case is known in which the free larva is directly infective to vertebrates. Individuals can be removed from their definitive hosts and are capable of surviving for several days in water or saline, suggesting that they are rather generalized physiologically and not highly specialized for parasitism. Furthermore, if they are eaten by a fish or turtle, they can live for a considerable length of time in this new host. Therefore it is not uncommon to find an aspidogastrean in the intestine of a fish, although it normally parasitizes a mollusc. Some species have so little host specificity that they can mature both in clams and in fish, although those in fish may be larger and produce more progeny.[1] Others apparently will not mature in a mollusc and need a fish final host.[2] *Lobatostoma manteri* preadults develop in any of several species of snails but must reach the intestine of a snail-eating fish *(Trachinotus blochi)* to mature.[6,7]

The following life cycles illustrate the biology of two families in the subclass.

Aspidogaster conchicola (see Fig. 15-5)

This common representative of the Aspidogastridae is most often found in the pericardial cavity of freshwater clams in Europe, Africa, and North America, although it is known from other molluscs, fish, and turtles. The adult is 2.5 to 3 mm long by 1 mm wide; it is oval, with a long, mobile "neck" with a buccal funnel at its end. The loculi on the ventral sucker are arrayed in four longitudinal rows, totaling 64 to 66.

When the eggs hatch within the host mollusc, the young can develop without further migration. If the egg or cotylocidium leaves the mollusc and is drawn into the incurrent siphon of the same or another clam, it can reach the nephridiopore and migrate through the kidney into the pericardium.

The cotylocidium is 13 to 17 μm long at hatching, lacks external cilia, and bears a simple posterior sucker without lobuli. Growth and metamorphosis are rapid.

Lophotaspis vallei, also in Aspidogastridae, may use a marine snail as intermediate host. Mature forms have been found in marine turtles, but it is possible that they normally mature in molluscs.

Stichocotyle nephropsis (Fig. 15-10)

This parasite lives in the bile ducts of rays in the Atlantic Ocean. It has been found in lobsters and other

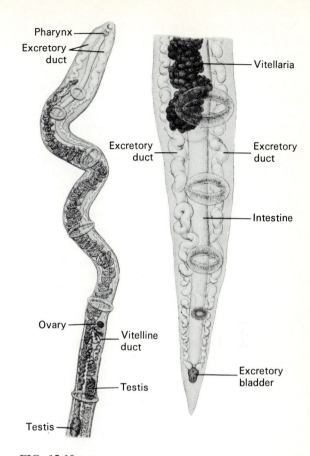

FIG. 15-10

Stichocotyle nephropsis from the bile ducts of rays.
From Odhner, T. 1910. Kungl. Svenska Vetern. Handl. 45:3-16

crustaceans and is thought to employ them as intermediate hosts. The adult is slender, 115 mm long, and has 24 to 30 separate suckers along its ventral surface. This is the only species in Stichocotylidae and the only aspidogastrean found in crustaceans. It is possible that the crustacean is not a normal host in the life cycle of this parasite and that the occurrence of *S. nephropsis* in these animals is accidental. It also is probable that this worm does not belong in Aspidogastrea.

PHYLOGENETIC CONSIDERATIONS

Superficially, the Aspidogastrea appear to be a link between the Monogenea and Digenea. Their anatomy is rather digenean, whereas their biology is suggestive of the Monogenea. This tiny group of worms displays sufficient individuality that it becomes apparent that the Aspidogastrea are distinct and separate from both groups.

Aspidogastreans differ morphologically from both Digenea and Monogenea in that the ventral sucker develops as a new structure, unrelated to any homologue in larvae or adults of the other two groups. The frontal septum of aspidogastreans is not found in either other group; neither is the septate, ciliated oviduct. However, the predominance of mollusc hosts and the presence of Laurer's canal and a highly developed nervous system are suggestive of Digenea. The simple, sac-like cecum and undemanding physiological requirements are primitive, more in keeping with the Turbellaria. Rohde[6] believes that an organism such as *Lobatostoma manteri,* with its mollusc intermediate host and fish definitive host, probably lies close to the ancestral protodigenean stock.

Clearly, then, the Aspidogastrea are a small group of animals, found mainly in molluscs, that appear not to be highly specialized for parasitism. Aspidogastreans retain sufficient identity to justify placement in a group of their own, closest to Digenea, but undoubtedly as a subclass of the Trematoda.

CLASSIFICATION OF SUBCLASS ASPIDOGASTREA

Subclass Aspidogastrea
Trematodes with single, large, ventral sucker subdivided by septa into numerous, shallow loculi or with one ventral row of individual suckers; no sclerotized armature on any species; mouth with or without sucker, sometimes lobated; pharynx well developed; intestine with a single or double median sac; testis single, double, or numerous; cirrus pouch present or absent; genital pores median, in front of sucker; ovary single, pretesticular; vagina absent, Laurer's canal sometimes present; vitellaria follicular, usually lateral but occasionally otherwise; eggs lacking polar prolongations; excretory pores on or near posterior end; development direct, without metamorphosis; parasites of molluscs, fishes, and turtles.

Order Aspidogastrida

FAMILY ASPIDOGASTRIDAE
Body oval or elongate; ventral sucker with numerous shallow loculi; one or two testes present; vitellaria follicular, lateral; parasites of molluscs, fishes, or turtles; cosmopolitan.

Subfamilies
Aspidogasterinae, Cotylaspidinae, Rohdellinae

Order Stichocotylida

FAMILY STICHOCOTYLIDAE
Body elongate, slender; ventral surface with longitudinal row of separate suckers; two testes present; vitellaria tubular, unpaired; parasites of Batoidea.

FAMILY RUGOGASTRIDAE
Body elongate; most of ventral and lateral body surface has transverse rugae; musculature of buccal funnel weakly developed; pharynx, prepharynx, and esophagus present; two ceca; testes multiple; ovary pretesticular; Laurer's canal present, seminal receptacle absent; vitellaria distributed along ceca, uterus ventral to testes; eggs operculate; parasites of Holocephali.

FAMILY MULTICALYCIDAE
Body elongate. Holdfast composed of fused suckers, otherwise similar to Rugogastridae.

REFERENCES

1. Fredericksen, D.W. 1978. The fine structure and phylogenetic position of the cotylocidium larva of *Cotylogaster accidentalis* Nickerson 1902 (Trematoda: Aspidogastridae). J. Parasitol. 64:961-976.
2. Hendrix, S.S., and R.M. Overstreet. 1977. Marine aspidogastrids (Trematoda) from fishes in the northern Gulf of Mexico. J. Parasitol. 63:810-817.
3. Ip, H.S., S.S. Desser, and I. Weller. 1982. *Cotylogaster occidentalis* (Trematoda: Aspidogastrea): scanning electron microscopic observations of sense organs and associated surface structures. Trans. Am. Microsc. Soc. 100:253-261.
4. Ip, H.S., and S.S. Desser. 1984. Transmission electron microscopy of the tegumentary sense organs of *Cotylogaster occidentalis* (Trematoda: Aspidogastrea). J. Parasitol. 70:563-575.
5. Rohde, K. 1972. The Aspidogastrea, especially *Multicotyle purvisi* Dawes, 1941. In Dawes, B., editor. Advances in parasitology, vol. 10. Academic Press, Inc., New York, pp. 77-151.
6. Rohde, K. 1973. Structure and development of *Lobatostoma manteri* sp. nov. (Trematoda: Aspidogastrea) from the Great Barrier Reef, Australia. Parasitology 66:63-83.
7. Rohde, K. 1975. Early development and pathogenesis of *Lobatostoma manteri* Rohde (Trematoda: Aspidogastrea). Int. J. Parasitol. 5:597-607.

SUGGESTED READINGS

Baer, J.G., and C. Joyeux. 1961. Classe des trématodes (Trematoda Rudolphi). In Grassé, P., editor. Traité de zoologie; anatomie, systématique, biologie, vol. 4. Masson & Cie, Paris, pp. 561-570. (This brief discussion gives all the salient knowledge of the Aspidogastrea.)

Dollfus, R.P. 1958. Trématodes. Sous-classe Aspidogastea. Ann. Parasitol. 33:305-395. (A detailed summary of knowledge of this group to 1958.)

Gibson, D.I., and S. Chinabut. 1984. *Rohdella siamensis* gen. et sp. nov. (Aspidogastridae: Rohdellinae subfam. nov.) from freshwater fishes in Thailand, with a reorganization of the classification of the subclass Aspidogastrea. Parasitology 88:383-393.

Yamaguti, S. 1963. Systema Helminthum, vol. 4. Interscience Publishers, New York. (The most useful taxonomic treatment of the group.)

CHAPTER 16

SUBCLASS DIDYMOZOIDEA

This small group of trematodes shows many affinities with the Digenea and in fact traditionally has been placed as a family within that subclass. However, because of certain morphological and biological peculiarities of these worms, it seems best to remove them from the Digenea.

All known species are tissue-dwelling parasites of fishes, especially those in marine environments. Most live within gills, often in pairs, but other organs, such as skin, kidney, buccal and intestinal epithelium, and ovary, are also infected. A review of the group is given by Baer and Joyeux.[1]

MORPHOLOGY

The body of didymozoids is elongate, usually flattened but occasionally globular or lobated. Most are only a few millimeters long, but *Nematobothrium* sp. reaches a length of 2.5 m, and a worm found in an ocean sunfish (Fig. 16-1) may reach a length of 40 feet![4]

In several genera, such as *Didymozoon* and *Wedlia*, the anterior end of the body is long and thin and inserts at an angle somewhere along the length of the stout or globular posterior body (Fig. 16-2). The body is long and threadlike in *Nematobothrium, Metanematobothrium, Atalostrophion*, and others. In *Diplotrema* and *Phacelotrema* two or even three individuals become completely fused at their posterior ends.

Many species are dioecious and exhibit striking sexual dimorphism. Most live in encysted pairs.

The mouth is at or near the anterior end of all species. An oral sucker is absent in some but present in varying degrees of development in others. A ventral sucker is apparently absent in most species of didymozoids or is present in young individuals only. A true pharynx is absent or feebly developed. The intestine has two limbs that may reach to near the posterior end of the body or undergo varying degrees of atrophy. The intestine is completely absent in a few species.

The excretory system is of the flame bulb proto-nephridia type. A voluminous excretory bladder extends to near the pharynx in some species; in some it has no external opening.[8]

The gonads of the Didymozoidea are generally tubular and filiform. One or two threadlike testes occur, but a cirrus pouch is absent. The vas deferens may be expanded into a terminal seminal vesicle. The single ovary is also threadlike and leads to a short oviduct that is surrounded for most of its length by voluminous Mehlis glands. An ootype and Laurer's canal are absent. The uterus extends to near the anterior end of the body, then descends to near the posterior end, and then winds to near the mouth, where it opens with the male pore. The eggs are very small, operculate, and numerous.

BIOLOGY

A complete life cycle is known for no species of Didymozoidea. The fragmentary knowledge available suggests that development is direct, without the intervention of an intermediate host, although larvae shed from a barnacle were thought to be didymozoids.[2]

Some species are undoubtedly hermaphroditic, probably fertilizing their own eggs.[7] Others are dioecious with sexual dimorphism, but vestigial organs of the opposite sex may be present.[5] Sex determination has not been demonstrated experimentally, but available evidence indicates that the sexuality of one individual may be determined by the presence of another individual. Ishii[3] showed that the first worm in a location develops into a female, whereas any later arrival has its femininity inhibited and thus becomes a male. A similar phenomenon is known among some molluscs, echiurids, and parasitic crustaceans and is suspected in dioecious tapeworms.

The egg contains a larva that appears to be directly infective to the definitive host. On hatching, the unciliated larva has a short gut with an oral sucker that is surrounded by two or more circles of spines. The mode of infection to the definitive host is unknown; it

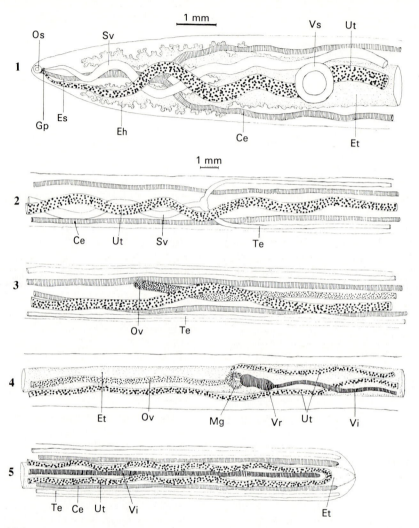

FIG. 16-1

Schematic drawings of *Nematobibothriodes histoidii* from *Mola mola*. **1,** Anterior end of worm. Note minute oral sucker, relatively large ventral sucker, absence of cirrus apparatus, and diverticulated excretory horns. **2,** Union of testes to form seminal vesicle. The excretory tube is omitted to make other organs stand out more clearly. **3,** Beginning of the ovary. All organs are filamentous. Excretory tube omitted for clarity. **4,** Ovary entering female genital complex at Mehlis' gland. Testes and ceca omitted for clarity. **5,** Posterior end of body. *Ce,* Cecum; *Eh,* excretory horn; *Es,* esophagus; *Et,* excretory tube; *Gp,* genital pore; *Mg,* Mehlis' gland; *Os,* oral sucker; *Ov,* ovary; *Sv,* seminal vesicle; *Te,* testis; *Ut,* uterus; *Vi,* vitellarium; *Vr,* vitelline reservoir; *Vs,* ventral sucker.

From Noble, G.A. 1975. J. Parasitol. 62:224- 227.

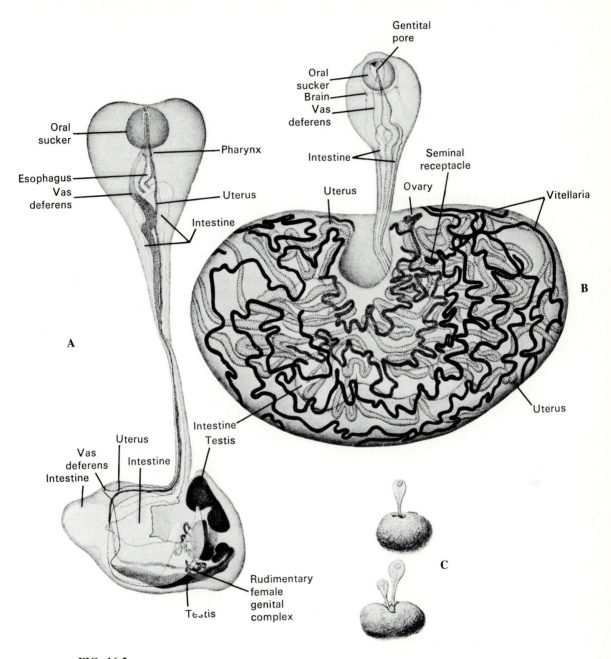

FIG. 16-2

Koellikeria bipartia, a common parasite of the intestinal mucosa of various tunas. **A,** Male. **B,** Female. **C,** Female *(top)* and male inside cavity of female *(bottom).*

Modified from Odhner, T. 1910. Zur Anatomie der Didymozoen: ein Getrenntgeschlechtlicher Trematode mit rudimentärem Hermaphroditismus. Almqvist & Wiksells, Uppsala, Sweden.

may be by direct penetration to the site of maturation or by migration to that site through the bloodstream.

Attempts to infect snails and crustaceans with *Ovarionematobothrium texomensis,* a parasite of the ovaries of buffalo fish in North America, have been unsuccessful.[5] This worm is only found in mature ovaries. Immature worms have never been reported. Presumably the worms are stimulated to mature by the host's hormones, then die, disintegrate, and pass out of the fish at spawning. Many fish do not spawn, in which case the ovaries and worms are resorbed. It has been shown that less than 7% of the parasite's eggs are viable in nonspawning fish.[6] How fish become infected is unknown.

CLASSIFICATION OF FAMILY DIDYMOZOIDEA

FAMILY DIDYMOZOIDAE

With previously discussed characters of subclass.

Subfamilies

Didymozoinae, Adenodidymocystiinae, Annulocystiinae, Colocyntotrematinae, Didymocodiinae, Gonopodasmiinae, Koellikeriinae, Metadidymozoinae, Nemathobothriinae, Neodidymozoinae, Neodiplotrematinae, Nephrodidymotrematinae, Opepherocystiinae, Opepherotrematinae, Osteodidymocodinae, Patellokoellikeriinae, Phacelotrematinae, Philopinninae, Pseudocolocyntotrematinae, Reniforminae, Sicuotrematinae, Skrjabinozoinae

REFERENCES

1. Baer, J.G., and C. Joyeux. 1961. Sous-classe des Didymozoides. Didymozoidea subcl. nov. In Grassé, P.P., editor. Traité de zoologie: anatomie, systématique, biologie, vol. 4, part I. Plathelminthes, Mésozaires, Acanthocéphales, Némertiens. Masson & Cie, Paris, pp. 678-685.
2. Cable, R.M., and F.M. Nahhas. 1962. *Lepas* sp., second intermediate host of didymozoid trematodes. J. Parasitol. 48:34.
3. Ishii, N. 1935. Studies on the family Didymozoonidae (Monticelli, 1888). Jpn. J. Zool. 6:279-335.
4. Noble, G.A. 1975. Description of *Nematobibothrioides histoidii* (Noble, 1974) (Trematoda: Didymozoidae) and comparison with other genera. J. Parasitol. 61:224-227.
5. Self, J.T., L.E. Peters, and C.E. Davis, 1963. The egg, miracidium, and adult of *Nematobothrium texomensis* (Trematoda: Digenea). J. Parasitol. 49:731-736.
6. Whittaker, F.H. 1973. Application of histochemistry in studies on the life cycle of *Nematobothrium texomensis* (Trematoda: Didymozoidae). Helminthologia 11:217-220 (dated 1970).
7. Williams H.H. 1959. The anatomy of *Köllikeria filicolis* (Rudolphi, 1819), Cobbold, 1860 (Trematoda: Digenea) showing that the sexes are not entirely separate as hitherto believed. Parasitology 49:39-53.
8. Yamaguti, S. 1951. Studies on the helminth fauna of Japan, part 48. Trematodes of fishes. X. Arb. Med. Fak. Okayama 7:315-334.

SUGGESTED READINGS

Self, J.T., L.E. Peters, and C.E. Davis. 1961. The biology of *Nematobothrium texomensis* McIntosh et Self, 1955 (Didymozoidae), in the buffalo fishes of Lake Texoma. J. Parasitol. 47:42-43.

Yamaguti, S. 1971. Synopsis of digenetic trematodes of vertebrates, vol. 1. Keigaku Publishing Co., Tokyo, pp. 248-276. (Diagnostic keys and descriptions of all subfamilies and genera, by the foremost authority on the group.)

CHAPTER 17

SUBCLASS DIGENEA: FORM, FUNCTION, BIOLOGY AND CLASSIFICATION

The digenetic trematodes, or flukes, are among the most common and abundant of parasitic worms, second only to nematodes in their distribution. They are parasites of all classes of vertebrates, especially marine fish, and some species, as adults or juveniles, inhabit nearly every organ of the vertebrate body. Their development occurs in at least two hosts, the first a mollusc or, very rarely, an annelid. Many species include a second and even a third intermediate host in their life cycles. Several species cause economic losses to society through infections of domestic animals, and others are medically important parasites of humans. Because of their importance, the Digenea have stimulated vast amounts of research, and the literature on the group is immense. We will summarize the morphology and biology of the group, illustrating it with some of the more important species.

Trematode development will be considerd in detail later (p. 240), but it is necessary to outline a "typical" life cycle briefly here. A ciliated, free-swimming larva, the **miracidium,** emerges from the egg and penetrates the first intermediate host, usually a snail. At the time of penetration or soon after, the ciliated epithelium is discarded, and the miracidium metamorphoses into a rather simple, saclike form, the **sporocyst.** Within the sporocyst, a number of embryos develop asexually to become **rediae.** The redia is somewhat more differentiated than the sporocyst, possessing, for example, a pharynx and a gut, neither of which was present in the miracidium or the sporocyst. Additional embryos develop within the redia, and these become **cercariae.** The cercaria emerges from the snail and usually has a tail to aid in swimming. Although many species require further development as **metacercariae** before they are infective to the definitive host, the cercariae are properly considered juveniles; they have organs that will develop into the adult digestive tract and suckers, and genital primordia are often present. The fully developed, encysted metacercaria is infective to the definitive host and develops there into the adult trematode.

FORM AND FUNCTION
Body form

Flukes exhibit a great variety of shapes and sizes, as well as variations in internal anatomy. They range in size from the tiny *Levinseniella minuta,* only 0.16 mm long, to the giant *Fascioloides magna,* which reaches 5.7 cm in length and 2.5 cm in width.

Most flukes are dorsoventrally flattened and oval in shape, but some are as thick as they are wide; some species are filiform, round or even wider than they are long. Flukes usually possess a powerful oral sucker that surrounds the mouth, and most also have a midventral acetabulum or ventral sucker. The words **distome, monostome,** and **amphistome** are sometimes used as descriptive terms, although they formerly had taxonomic significance, and, of course, they refer to suckers, not mouths. If a worm has only an oral sucker, it is called a monostome (Fig. 17-1); with an oral sucker and an acetabulum at the posterior end of the body, it is an amphistome (Fig. 17-2); and if the acetabulum is elsewhere on the ventral surface, the worm is referred to as a distome (Fig. 17-3). The oral sucker may have muscular lappets, as in *Bunodera* (Fig. 17-4), or there may be an anterior adhesive organ with tentacles, as in *Bucephalus* (Fig. 17-5). *Rhopalias* spp., parasites of American opposums, have a spiny, retractable proboscis on each side of the oral sucker (Fig. 17-6). In species of Hemiuridae the posterior part of the body telescopes into the anterior portion.

Tegument

The tegument of trematodes, like cestodes, traditionally has been considered a nonliving, secreted cuticle; and, likewise, study with the electron microscope has revealed that the body covering of trema

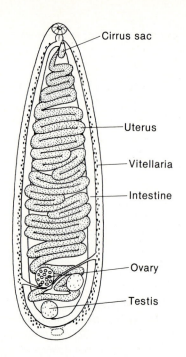

FIG. 17-1

Cyclocoelum lanceolatum, a common monostome fluke from the air sacs of shore birds.

Drawing by William C. Ober.

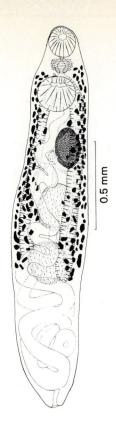

FIG. 17-3

Alloglossidium hirudicola, a distome trematode from leeches.

From Schmidt, G.D., and K. Chaloupka. 1969. J. Parasitol. 55:1185-1186.

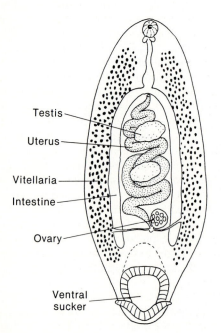

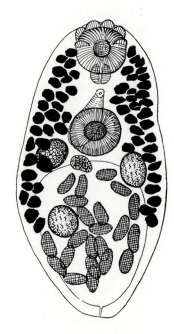

FIG. 17-2

Zygocotyle lunata, an amphistome fluke from ducks.

Drawing by William C. Ober.

FIG. 17-4

Bunodera sacculata from yellow perch. Note the muscular lappets on the oral sucker.

From Van Cleave, H.J., and J.F. Mueller, 1932. Roosevelt Wildl. Ann. 3:9-71.

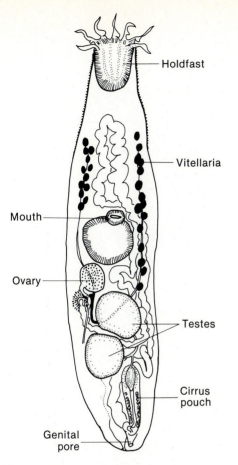

FIG. 17-5

Bucephalus polymorphus from European and Asian fishes.

Redrawn from Yamaguti, S. 1971. Synopsis of digenetic trematodes of vertebrates. Keigaku Publishing C., Tokyo.

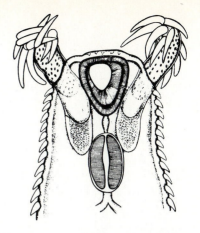

FIG. 17-6

Rhopalias coronatus, a parasite of American opossums. Retractable proboscides are located on each side of the mouth.

From Yamaguti S. 1971. Synopsis of digenetic trematodes of vertebrates, Keigaku Publishing Co., Tokyo.

todes is a living, complex tissue. In contrast to the cestodes, trematodes have a gut, and it has been assumed that the digestive tract functions in the absorption of nutrients. Perhaps for this reason, the tegument of trematodes has excited somewhat less investigation than that of cestodes; nevertheless, substantial knowledge of tegument structure in several species has accumulated.[73] In addition, the tegument may be important in absorption of some nutrients, even though a gut is present.[85]

In common with the Monogenea and the Cestoidea, digenetic trematodes have a "sunken" epidermis; that is, there is a distal, anucleate layer. The cell bodies containing the nuclei (cytons) lie beneath a superficial layer of muscles, connected to the distal cytoplasm by way of internuncial processes (Fig. 17-7). Because the distal cytoplasm is continuous, with no intervening cell membranes, the tegument is syncytial. Although

this is the same general organization found in the cestodes, trematode tegument differs in many details, and striking differences in structure may occur in the same individual from one region of the body to another. Ornamentation such as spines is often present in certain areas of the trematode's body and may be discernible with the light microscope, but the scanning electron microscope is the preferred instrument for such studies. The oral and ventral suckers of *Schistosoma mansoni* are densely beset with spines, and much of the male's dorsum bears bosses with 50 to 250 spines.[78] Papillae, many with crater-like sensory openings, are interspersed (Fig. 17-8). Bosses are absent from the male's gynecophoral canal (see p. 266) and from the female, but the females have many *anteriorly directed* spines on their posterior ends (Fig 17-9). The spines consist of crystalline actin[23]; their bases lie above the basement membrane of the distal cytoplasm, and their apices project above the surface, although generally they are covered by the outer plasma membrane.

The distal cytoplasm usually contains vesicular inclusions, more or less dense, and sometimes several recognizable types in the tegument of the same worm. The function of the vesicles is unclear, although in some cases they contribute to the outer surface. The surface membrane of *S. mansoni* is continuously renewed by multilaminar vesicles moving outward through the distal cytoplasm, perhaps to replace membrane damaged by host antibodies.[54] Kemp et al.[59] have shown that the outer layers with host antibody adsorbed to them are indeed shed by the worm. In *Megalodiscus* the contents of some vesicles seem to be emptied to the outside.[7]

229

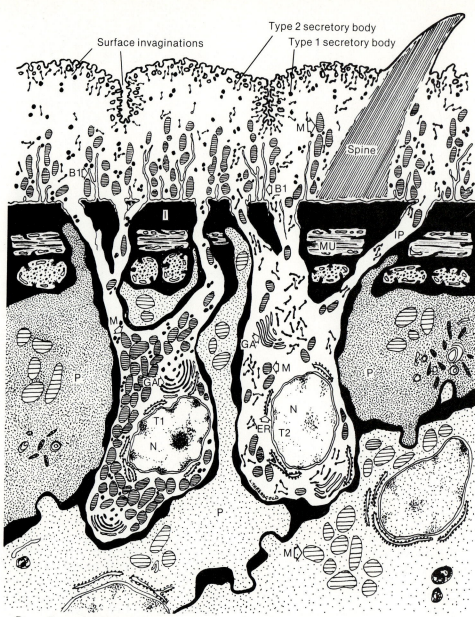

P = Parenchymal cell
T1 = Type 1 tegumentary cyton
T2 = Type 2 tegumentary cyton
GA = Golgi complex
I = Interstitial material (connective tissue)
IP = Internuncial process

MU = Muscle
BI = Basal invagination
N = Nucleus
ER = Granular endoplasmic reticulum
M = Mitochondria

FIG. 17-7

Diagrammatic drawing of the structure of the tegument of *Fasciola hepatica*.

Drawing by L.T. Threadgold.

FIG. 17-8

Tegument of *Schistosoma mansoni:* dorsal region in distal third of male, showing spines in interbossal spaces and papillae with openings.

From Miller, F.H., G.S. Tulloch, and R.E. Kuntz. 1972. J Parasitol. 58:693-698.

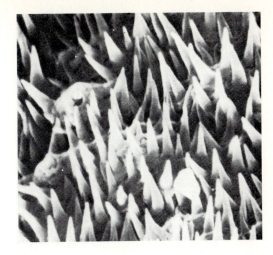

FIG. 17-9

Schistosoma mansoni: dense covering of forward-projecting spines at posterior tip of female.

From Miller, F.H., G.S. Tulloch, and R.E. Kuntz. 1972. J. Parasitol. 58:693-698.

The vesicles of the distal cytoplasm are produced in Golgi bodies found in the cytons and passed outward through the internuncial processes, although Golgi bodies occasionally occur in the distal cytoplasm as well.

Mitochondria are found in the distal cytoplasm in most species examined, although not in *Megalodiscus* and some paramphistomes.[7,35]

The outer surface of adult trematodes is not modified for absorption by the elaboration of microvillus-like microtriches, as in the cestodes, but some structural features that increase surface area occur. The tegument of some trematodes is penetrated by many deep pits (Fig. 17-10) and channels.[68,103] The lack of mitochondria in the distal cytoplasm of some species has been associated with limited absorptive capabilities of their tegument.[7,35]

Miracidia of *Fasciola* and *Schistosoma* (at least) are covered by ciliated epithelial cells with nuclei as is typical for such cells.[119] The epithelial cells are interrupted by "intercellular ridges," extensions of cells whose perikarya lie beneath the superficial muscle layer and that bear no cilia, although some microvilli may be present (Fig. 17-11). On loss of the ciliated epithelium, metamorphosis to the sporocyst involves a spreading of the distal cytoplasm over the worm's surface, although whether this comes from the intercellular ridges is unclear. Well-developed microvilli are present on the surface of both the sporocyst and redia.

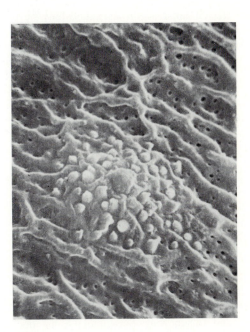

FIG. 17-10

A tubercle and spines on the tegument of a male *Schistosoma mansoni.* Note the many pits in ths surface.

With permission from Hockley, D.J. 1973. In Dawes, B., editor. Advances in parasitology, vol. 11. Copyright Academic Press, Inc. (London) Ltd.

SUBCLASS DIGENEA

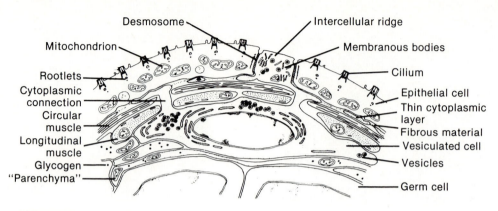

FIG. 17-11

Miracidium of *Fasciola hepatica*. Line drawing reconstruction of transverse segment of body wall in region of the germ cell cavity.

From Wilson, R.A 1969. J. Parasitol. 55:124-133.

The *luminal surface* of the tegumental cells in the redia may be thrown into a large number of flattened sheets that extend to other cells in the body wall and to the cercarial embryos contained in the lumen. Nutritive molecules, such as glucose, and molecules up to the size of horseradish peroxidase may be passed through the tegument to the developing cercariae.[7]

The early embryos of the cercariae are covered with a primary epidermis below which a definitive epithelium forms. The nuclei of this secondary epithelium sink into the parenchyma, and the final form of the cercarial tegument results in an organization similar to that of the adult. Cystogenic cells in the parenchyma begin to secrete cyst material, which is passed into the distal cytoplasm of the tegument. The metacercarial cyst is formed when the cercarial tegument is sloughed off, and the cyst material it contains undergoes chemical and/or physical changes to envelop the worm in its cyst. The cystogenic cells in the parenchyma then flow toward the surface, their nuclei being retained beneath the superficial muscles, and a thin layer of cytoplasm spreads over the organism to become the definitive adult tegument. A contrasting mode of cercarial tegument formation has been reported for *S. mansoni*.[52] The definitive tegument, including its nuclei, is formed beneath the primitive tegument and is syncytial. Subsequently, the nuclei of the tegument become pyknotic and are lost, as processes from subtegumental cells grow outward and join the distal cytoplasmic layer (Fig. 17-12). The end result is the same.

The tegument is variously interrupted by cytoplasmic projections of gland cells, by openings of excretory pores, and by nerve endings. Both miracidia and cercariae may have penetration glands that open at the anterior, and the adults of some species have prominent glandular organs opening to the exterior.

Muscular and nervous systems

The muscles that occur most consistently throughout the Digenea are the superficial muscle layers (formerly called "subcuticular"); and these usually comprise circular, longitudinal, and diagonal layers enveloping the rest of the body like a sheath below the tegument. The degree of muscularization varies considerably in the group; some species have a rather feeble musculature, some are very robust and strong, and other examples fit all conditions in between. The deep musculature found in cestodes is generally absent in trematodes. Muscles are often more prominent in the anterior parts of the body, and strands connecting the dorsal to the ventral superficial muscles are usually found in the lateral areas. The fibers are smooth, and the nuclei occur in cytons called "myoblasts," connected to the fiber bundles and located in various sites around the body, often in syncytial clusters. It is likely that the "parenchymal" cells are in fact cytons of muscle cells, as has been found in cestodes.[74] The suckers and pharynx are supplied with radial muscle fibers, often very strongly developed. A network of fibers may surround the intestinal ceca, helping to fill and empty these structures.

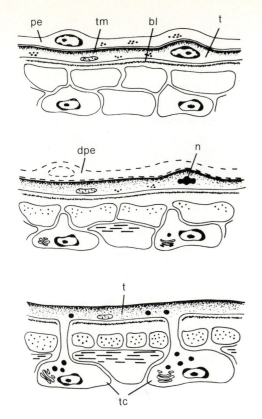

FIG. 17-12

Diagram summarizing three stages in the formation of the tegument during cercarial development. *Top,* Germ ball covered with a primitive epithelium *(pe)* and the tegument *(t),* which has a thickened outer membrane *(tm)* and an underlying basal lamina *(bl). Center,* Young cercaria with a degenerating primitive epithelium *(dpe)* and a pyknotic tegumental nucleus *(n). Bottom,* Cercaria nearly ready to emerge from the sporocyst; the primitive epithelium has been lost, and the tegument *(t)* is connected to nucleated tegumental cytons *(tc).*

With permission from Hockley, D.J. 1973. In Dawes, B., editor. Advances in parasitology, vol. 11. Copyright Academic Press, Inc. (London) Ltd.

The organization of the nervous system is the typical platyhelminth ladder type. A pair of cerebral ganglia is connected by a supraesophageal commissure. From this, several nerves issue anteriorly, and three main pairs of trunks—dorsal, lateral, and ventral—supply the posterior parts of the body. The ventral nerves are usually best developed, and a variable number of commissures link these and the other longitudinal nerves. Branches provide motor and sensory endings to muscles and tegument. The anterior end,

especially the oral sucker, is well supplied with sensory endings.

Sensory endings in the Digenea are an interesting array of types, particularly in the miracidia and cercariae. The adults require no orientation to such stimuli as light and gravity, and in the few forms in which the ultrastructure has been studied, only one type of sensory ending has been described.[53] This is a bulbous nerve ending in the tegument that has a short, modified cilium projecting from it, and the cilium is enclosed throughout its length by a thin layer of tegument. The general structure is similar to sense organs described in cestodes (see Fig. 21-15). These structures generally have been regarded as tangoreceptors in trematodes.

Cercariae and miracidia show more variety in sense organs, doubtlessly related to the adaptive value of finding a host quickly.[9] Uniciliated bulbous endings are found on the anterior portion of the cercariae of *S. mansoni,* similar to but smaller than those on the adult. The tegumentary sheath opens at the ciliary apex. In addition, a bulbous type with a long (7 μm) unsheathed cilium is widely distributed over the body of the cercaria, and its lateral areas bear small, flask-shaped endings containing five or six cilia and opening to the outside through a 0.2 μm pore (Fig. 17-13). This latter type is thought to be a chemosensory ending. Most trematode miracidia bear a pair of conspicuous lateral papillae between the first and second series of plates of ciliated epithelium.

In a number of species, these papillae consist of a large, bulbous nerve ending (or two endings) covered with an evagination of the intercellular ridge. The function of these endings has been rather puzzling, since there is no obvious organelle for the transduction of stimuli. A uniciliated bulbous ending, as described, is always located just anterior to each papilla, and assuming some flexibility of the papilla and that its contents are slightly different in specific gravity from the rest of the miracidium, it could impinge on the ciliated ending and give the organism information retarding its orientation with respect to gravity.[9] *Gigantocotyle explanatum* miracidia have a curious "conical organ," thought to provide information on spatial orientation or angular acceleration.[34]

Eyespots are present in many species of miracidia and in some cercariae. Although also present in some adult trematodes, eyespots do not appear to function in them. The structure of those in several different miracidia has been investigated and is generally similar to such organs found in Turbellaria and some Annelida. The eyespots consist of one or two cup-shaped pigment cells surrounding the parallel rhabdomeric

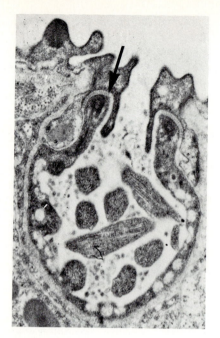

FIG. 17-13

Multiciliated pit in anterior body tegument of cercaria of *Schistosoma mansoni*. Arrow indicates septate desmosome (× 36,000.)

From Morris, G.P. 1971. Z. Parasitenkd. 36:20.

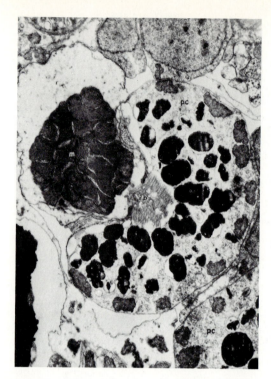

FIG. 17-14

Ultrastructure of an eyespot in a miracidium of *Diplostomum spathaceum*. Only one retinular cell is visible in this section. Note the closely packed mitochondria and the rhabdomeric microvilli. *m*, Mitochondrion; *pc*, pigment cell; *rb*, rhabdomere (× 15,500.)

From Brooker, B.E. 1972. In Canning, E.U., and C.A. Wright, editors. Behavioral aspects of the parasite transmission. Academic Press, Inc., New York. Copyright by the Linnean Society of London.

microvilli of one or more retinular cells (Fig. 17-14). The mitochondria of the retinular cells are packed in a mass near the rhabdomere. Because the rhabdomeres are the photoreceptors, the cup shape of the pigment cells allows the organism to distinguish light direction. Interestingly, some miracidia do not have eyespots yet can orient with respect to light. Some cells in the miracidia of *Diplostomum spathaceum* and *S. mansoni* have large vacuoles; into these vacuoles project a number of cilia, each of which has a conspicuous membrane evagination. These membranes, which are stacked in a lamellar fashion, might be photoreceptors, thus providing, in the case of *S. mansoni*, a means of light sensitivity for a miracidium without eyespots.[9]

Another apparent chemoreceptor described in the miracidium of *D. spathaceum* consists of two dorsal papillae between the first series of ciliated plates. Each papilla consists of a nerve ending and has radiating from it a number of modified cilia, which are parallel to the surface of the miracidium. These sensory endings are strikingly similar to the olfactory receptors of the vertebrate nasal epithelium!

Acetylcholine (ACh) appears to be an important neurotransmitter in trematodes, and 5-hydroxytryptamine (5-HT) is also of some importance.[11] ACh and its synthetic and degradative enzymes, acetylcholinesterase and choline acetylase, have been found in *S. mansoni* and *F. hepatica*. Motor activity of *S. mansoni* is stimulated by ACh blocking agents; therefore it is concluded that the cholinergic endings inhibit muscle activity. The 5-HT is the corresponding excitatory neurotransmitter in *S. mansoni* because incubation in 5-HT or in compounds that cause release or inhibit reuptake of 5-HT by its storage sites causes greatly increased motility. Some properties of the cholinergic endings have been investigated, and several differences between the worm's and the host's receptors have been shown. They respond differently to certain drugs, and the kinetic properties and effect of inhibitors of the worm's acetylcholinesterase and choline acetylase differ from the host's.

Active transport systems in the tegument of *S. mansoni* are important in ion regulation within the worm and directly or indirectly affect muscle function. Interference with the transport of Na^+ and K^+ leads to a depolarization of the tegumental membrane potential and a large, nonreversible increase in muscle tension.[37] An external source of the amino acid L-glutamine, along with a physiological concentration of phosphate, is necessary to maintain the membrane potential.[66] Apparently, in the absence of L-glutamine and phosphate, intracellular pH homeostasis cannot be maintained.

Osmoregulatory system

The osmoregulatory system of Digenea is based on the flame bulb **protonephridium,** so called because it is closed at the inner end and opens to the exterior by way of a pore. The **flame bulb,** or cell, is flask shaped and contains a tuft of fused cilia to provide the motive force for the fluid in the system. The number of flame bulbs, from a few to many, depends on the species and its size or the extent of its parenchyma. Since the parenchyma is substantially solid, the more massive the worm, the more extensive the flame bulb system required to drain it. Some forms have developed accessory "circulatory" systems. Each flame cell extends processes into the surrounding parenchyma in a "starlike" fashion.[18] The ductules of the flame cells join collecting ducts, those on each side eventually feeding into a bladder in the adult that opens to the outside with a single pore. The pore is almost always located near the posterior end of the worm. The bladder is usually referred to as the **excretory bladder,** and we will use the adjectives osmoregulatory and excretory interchangeably in reference to this system, although there is little evidence to indicate whether either or both are most appropriate. In some trematodes the walls of the collecting ducts are supplied with microvilli, indicating that some transfer of substances, absorption or secretion, is probably occurring.[99] That the system is osmoregulatory may be inferred from the fact that among free-living Platyhelminthes, freshwater Turbellaria have much better developed protonephridial systems than do marine planarians. Trematodes normally have two free-swimming stages in which an efficient water-pumping system is found to be necessary in those which occur in fresh water.

The embryogenesis of the excretory bladder has been used as a basis for the distinction between the superorders Epitheliocystidia and Anepitheliocystidia. However, doubts about the validity of this distinction have been cast by studies with the electron microscope that show that the excretory bladders of all digeneans are syncytial.[44] The shape of the bladder, whether it is Y-, V-, or I-shaped, may have diagnostic value at lower taxonomic levels.

A supplementary "lymphatic system" is found in several families. This is an independent system of irregular, fluid-filled, contractile tubules of uncertain function. Its presence is sometimes used as a taxonomic character.

The primary nitrogenous excretory product of trematodes appears to be ammonia, although excretion of uric acid and urea has been reported. Whether the excretion takes place through the tegument, ceca, or excretory system is not known.

Acquisition of nutrients and digestion

Halton[49] studied feeding and digestion in a variety of trematodes by histological and histochemical methods. Eight species from six distinct habitats within the vertebrate host were studied, and substantial diversity in digestive mechanisms and morphology was observed. Two lung flukes of frogs, *Haematoloechus medioplexus* and *Haplometra cylindracea,* feed predominately on blood from the capillaries. Both species draw a plug of tissue into their oral sucker and then erode its surface by a pumping action of the strong, muscular pharynx. Other trematode species characteristically found in the intestine, urinary bladder, rectum, and bile ducts feed more or less by the same mechanism, although their food may consist of less blood and more mucus and tissue from the wall of their habitat, and it may even include gut contents (as does that of *Diplodiscus subclavatus* in the rectum of frogs). In species without a pharynx that feed by this mechanism, the walls of the esophagus are quite muscular, and this apparently serves the function of the pharynx. In contrast, *S. mansoni,* living in the blood vessels of the hepatoportal system and immersed in its semifluid blood food, has no necessity to breach host tissues, and interestingly enough, this species has neither pharynx nor muscular esophagus. Digestion in most species studied is predominately extracellular in the ceca, but in *Fasciola hepatica* it occurs by a combination of intracellular and extracellular processes. One of the frog lung flukes, *Haematoloechus cylindracea,* has pear-shaped gland cells in its anterior end, and a nonspecific esterase is secreted from these cells through the tegument of the oral sucker, beginning the digestive process even before the food is drawn into the ceca. Various degrees of adaptation exist in the blood-feeding flukes, concerning their abilities to eliminate the unwanted iron component of the hemoglobin molecule. In *F. hepatica,* in which the final digestion of hemoglobin is intracellular, the iron is excreted through the extretory system and tegument. The fate of the iron in *H. cylindracea* is unclear, but apparently it is stored, within the worm, tightly bound to protein. The extracellular digestion in *H. medioplexus* and *S. mansoni* produces insoluble end products within the cecal lumen, and these are periodically regurgitated.

In *S. mansoni* the end products are a heterogeneous population of molecules, but the worms digest and incorporate some of both the globin and heme moieties of hemoglobin.[38] All species investigated have microvilli on the gastrodermal cells, although these vary from short (1 to 15 μm) and irregular to long (10 to 20 μm) and are organized into a definite brush border, according to species. The ceca of trematodes apparently do not bear any gland cells, but the gastrodermal cells themselves may secrete some digestive enzymes in certain species: proteases, a dipeptidase, an aminopeptidase, lipases, acid and alkaline phosphatases, and esterases have been detected.[95,102] A protease from *S. mansoni* has marked substrate specificity for hemoglobin and an optimal pH of 3.9 to 4.5.[102] Interestingly, female *S. mansoni* ingest far more blood than do males, and their protease is 4.8 times more active on a weight basis.[102] This is attributed to the requirement for egg production.

The fine structure of the gut cells has been studied in a variety of species.[30,31,41,42] It is clear that the structures interpreted as microvilli in several species at the light microscope level are in fact flattened plate-like or lamelloid processes projecting into the lumen (Fig. 17-15). However, digitiform processes more like microvilli are borne by the gut cells of some species. *Fasciola hepatica* and *Echinostoma hortense* have distal and/or marginal filamentous extensions on the lamellae. In all cases, the absorptive surface area is vastly greater than if the cell surface were flat. Within the gut cells of both *Gorgodera amplicava* and *Haematoloechus medioplexus* are abundant rough endoplasmic reticulum, many mitochondria, and frequent Golgi bodies and membrane-bound, vesicular inclusions. High activity of acid phosphatase is found in the vesicles of *Paragonimus kellicotti* and *H. medioplexus,* and after incubation in ferritin the material is found within them. No evidence of "transmembranosis" has been found, but the vesicles may be lysosomes that would function in degradation of nutritive materials after phagocytosis.

It is not surprising that trematodes can absorb small molecules through the tegument. The uptake of nutrients through the surface of sporocysts and rediae has already been mentioned. Absorption of nutrients through the tegument of adults has been studied by radioautography, by ligature of the anterior end, and by incubation in radioactive substrate. In the few species examined, glucose was absorbed through the tegument and not by way of the gut, although it has not always been clear whether the worms might not have been able to absorb this hexose by the intestinal route had they been feeding normally.[86] In the case of *Philophthalmus megalurus,* it must be assumed that the trematodes were "feeding" in vitro, since they absorbed the amino acids tyrosine and leucine only through the gut, whereas glucose was absorbed mostly through the tegument.[81] Thymidine was absorbed by *P. megalurus* by *both* routes. *Gorgoderina* can absorb tyrosine, thymidine, adenosine, and glucose through its tegument, whereas *Haematoloechus* can absorb glucose via its tegument but arginine only by its gut.[83,84,86] Also, is it known that *S. mansoni* takes in glucose only through its tegument,[90] and male *S. mansoni* somehow pass glucose to the female lying in the gynecophoral canal.[26] *Fasciola* and *Fascioloides* spp. absorb several amino acids by way of their tegument.[56]

Megalodiscus temperatus cannot absorb glucose or galactose across its tegument,[100] and this species, as well as several other paramphistomes, has no mitochondria in the tegumental cytoplasm.[35] For that reason it is believed that the tegument of these trematodes has little or no absorptive capacity.

Reproductive systems

Most trematodes are hermaphroditic (important exceptions are the schistosomes), and some are capable of self-fertilization. Others, however, require cross-fertilization to produce viable progeny. Some species inseminate themselves readily; others will do so if there is only one worm present, but they always seem to cross-inseminate when there are two or more in the host.[82] A few instances are known in which adult trematodes can reproduce parthenogenetically.[117]

■ Male reproductive system

(Fig. 17-16). The male reproductive system usually includes two testes, although some species have from one testis to several dozen. Their shape varies from round to highly dendritic, according to species. Each testis has a vas efferens that connects with the other to form a vas deferens. This courses toward the genital pore, which is usually found within a shallow genital atrium. The genital atrium is most often on the midventral surface, anterior to the acetabulum, but it can be found nearly anywhere, including at the posterior end, beside the mouth, or even dorsal to the mouth in

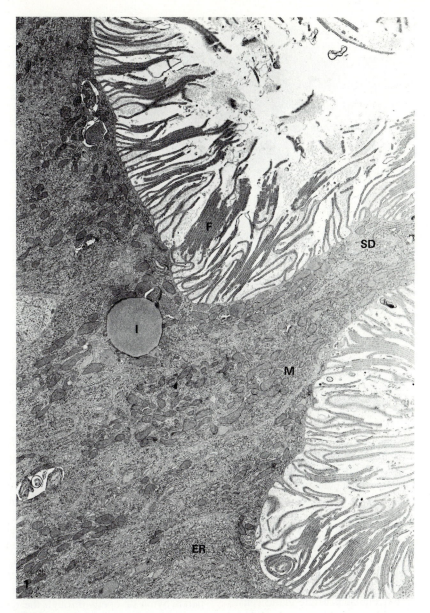

FIG. 17-15

Apical portion of the cecal epithelium of *Paragonimus kellicotti*. The apical surface has numerous folds *(F)* extending into the lumen. The cecal epithelial cells are joined by septate desmosomes *(SD)*. The cytoplasm contains a well-developed granular endoplasmic reticulum *(ER)* and numerous mitochondria *(M)*. An inclusion *(I)* is indicated.

From Dike, S.C. 1969. J. Parasitol. 55:113.

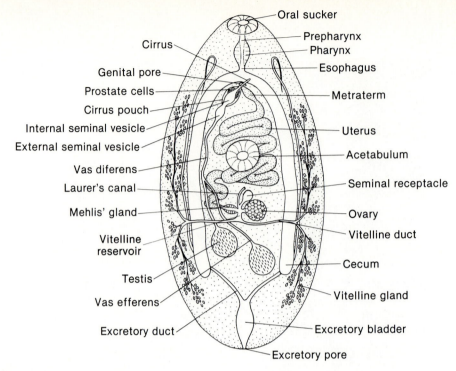

Cirrus
Genital pore
Prostate cells
Cirrus pouch
Internal seminal vesicle
External seminal vesicle
Vas diferens
Laurer's canal
Mehlis' gland
Vitelline reservoir
Testis
Vas efferens
Excretory duct

Oral sucker
Prepharynx
Pharynx
Esophagus
Metraterm
Uterus
Acetabulum
Seminal receptacle
Ovary
Vitelline duct
Cecum
Vitelline gland
Excretory bladder
Excretory pore

FIG. 17-16

Diagrammatic representation of digenetic fluke, showing male and female reproductive systems.
Drawing by William C. Ober.

some species. Before reaching the genital pore, the vas deferens usually enters a muscular cirrus pouch where it may expand into an **internal seminal vesicle** for sperm storage. Constricting again, the duct forms a thin ejaculatory duct, which extends the rest of the length of the cirrus pouch and forms, at its distal end, a muscular cirrus. The cirrus is the male copulatory organ. It can be invaginated into the cirrus pouch and evaginated for transfer of sperm to the female system. The cirrus may be naked or covered with spines of different sizes. The ejaculatory duct is usually surrounded by numerous unicellular **prostate gland cells.** At this point a muscular dilation may form a **pars prostatica.**

Much variation in these terminal organs occurs among families, genera, and species. The cirrus pouch and prostate gland may be absent, with the vas deferens expanded into a powerful seminal vesicle that opens through the genital pore, as in *Clonorchis.* The vas deferens may expand into an **external seminal vesicle** before continuing into the cirrus pouch. Other, more specialized modifications are described and illustrated by Yamaguti.[121]

■ **Female reproductive system** (Fig. 17-16)

The single ovary in the female reproductive tract is usually round or oval, but it may be lobated or even branched. The short oviduct is provided with a proximal sphincter, the **ovicapt,** that controls the passage of ova. The oviduct and most of the rest of the female ducts are ciliated. A seminal receptacle forms as an outpocketing of the wall of the oviduct. It may be large or small, but it is almost always present. At the base of the seminal receptacle there often arises a slender tube, Laurer's canal, which ends blindly in the parenchyma or opens through the tegument. Laurer's canal is probably a vestigial vagina that no longer functions as such (with a few possible exceptions), but it may serve to store sperm in some species.

Unlike other animals, but in common with the cestodes and some Turbellaria, the yolk is not stored in the ovum but is contributed by separate cells called vitelline cells. The vitelline cells are produced in follicular vitelline glands, usually arranged in two lateral fields and connected by ductules to the main right and left vitelline ducts. These ducts carry the vitelline cells to a single, median vitelline reservoir, from which ex-

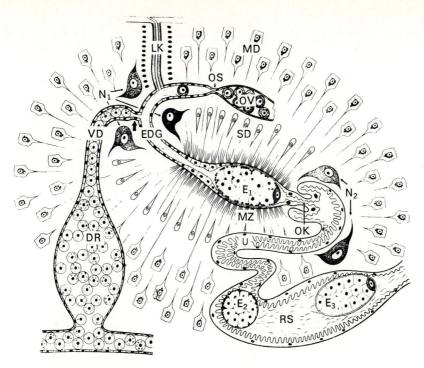

FIG. 17-17

Schematic representation of the oogenotop of *Fasciola hepatica*. *DR*, Vitelline reservoir; E_1, egg in ootype, with beginning shell formation; E_2, egg in the lower uterus, shell granules almost coalesced; E_3, egg in the uterus, shell formation almost completed; *EDG*, ovovitelline duct; *LK*, Laurer's canal; *MD*, mucus cells of Mehlis' gland; *MZ*, area where Mehlis' glands open into the ootype; N_1, nerve plexus I; N_2, nerve plexus II, *OK*, ootype sphincter; *OS*, oviduct sphincter; *OV*, oviduct; *RS*, seminal receptacle; *SD*, serous cells of Mehlis' gland; *U*, uterus; *VD*, vitelline duct.

From Gönnert, R. 1962. Z. Parasitenkd. 21:477.

tends the common vitelline duct joining the oviduct. The distribution of vitelline glands tends to be constant within a species and so is an important taxonomic character. After the junction with the common vitelline duct the oviduct expands slightly to form the ootype. Numerous unicellular Mehlis' glands surround the ootype and deposit their products into it by means of tiny ducts.

The structural complex just described (Fig. 17-17), as well as the upper uterus, is called the "egg-forming apparatus," or **oogenotop**.[46] Beyond the ootype, the female duct expands to form the uterus, which extends to the female genital pore. The uterus may be short and fairly straight, or it may be long and coiled or folded. The distal end of the uterus is often quite muscular and is called the **metraterm.** The metraterm functions as ovijector and as a vagina. The female genital pore opens near the male pore, usually together with it in the genital atrium. In some species,

such as in the Heterophyidae, the genital atrium is surrounded by a muscular sucker called a **gonotyl.**

At the time the ova leave the ovary, they may not have completed meiosis and thus are not, strictly speaking, ova at all but oocytes. Meiosis is completed after sperm penetration. The first meiotic division may reach pachytene or diplotene, at which point meiotic activity is arrested, and the chromosomes may return to a diffuse state. After sperm penetration, the chromosomes quickly reappear as bivalents and proceed from the first meiotic metaphase. The two meiotic divisions occur, with extrusion of polar bodies, and the male and female pronuclei fuse.[16,60]

As the oocyte leaves the ovary and proceeds down the oviduct, it becomes associated with several vitelline cells and a sperm emerging from the seminal receptacle. These all come together in the area of the ootype, and there are contributions from the cells of Mehlis' gland as well. It was long thought that Mehlis' gland contributed the shell material, and the organ was often referred to as the "shell gland" in older texts. However, it is now clear that the bulk of the shell material is contributed by the vitelline cells, and the function of the Mehlis' gland has been in considerable doubt. In at least some species two distinct types of secretions are released—"mucoid dense bodies" and "membranous bodies."[8] The mucoid dense bodies may serve as an adhesive mediating coalescence of vitelline globules to form the shell, or they may serve as a lubricant for the various components in the ootype. The membranous bodies aggregate to surround the oocyte, two or three vitelline cells, and some spermatozoa.[8] Globules released from the vitelline cells coalesce against the membranous aggregate; therefore, the aggregate forms a kind of template for the shell material before stabilization.

"Stabilization" of structural proteins (e.g., sclerotin, keratin, and resilin) to impart qualities of physical strength and intertness occurs by cross-linkage to amino acid moieties in adjacent protein chains. Most trematode eggshells appear to be stabilized primarily by the quinone-tanning process of sclerotization.[25,116] (see Fig. 34-2). In some trematodes, keratin or elastin may be the major structural proteins.

DEVELOPMENT

At least two hosts serve in the life cycle of a typical digenetic fluke. One is a vertebrate (with a few exceptions) in which sexual reproduction occurs, and the other is usually a mollusc in which one or more generations are produced by an unusual type of asexual reproduction that will be discussed further. A few species have asexual generations that develop in annelids.

This alternation of sexual and asexual generations in different hosts is one of the most striking biological phenomena. The variability and complexity of life cycles and ontogeny have stimulated the imaginations of zoologists for more than 100 years, creating a huge amount of literature on the subject. Even so, many mysteries remain, and research on questions of trematode life cycles remains active.

As many as six recognizably different body forms may develop during the life cycle of a single species of trematode (see p. 249 for summary). In a given species certain stages may be repeated during ontogeny, whereas stages found in other species may be absent. So many variations occur that few generalizations are possible. Therefore we will first examine each form separately and then illustrate the subject with a few examples.

Embryogenesis

Apart from the fact that the embryo produced by the sexual adult begins with a fertilized egg, the early embryogenesis of progeny produced asexually and sexually is basically similar. The first cleavage produces a **somatic cell** and a **propagatory cell,** which are cytologically distinguishable. Daughter cells of the somatic cell will contribute to the body tissues of the embryo, whether miracidium, sporocyst, redia, or cercaria. Further divisions of the propagatory cell may each produce another somatic cell and another propagatory cell, but at some point, propagatory cell divisions produce only more propagatory cells. Each of these will become an additional embryo in the miracidium, sporocyst, or redia. In the developing cercaria the propagatory cells become the gonad primordia. Thus the propagatory cells are the germinal cells in the asexually reproducing forms, and they give rise to the germ cells in the sexual adult. As noted previously, the miracidium metamorphoses into the sporocyst; however, if a sporocyst stage is absent in a particular species, redial embryos develop in the miracidium to be released after penetration of the intermediate host. The youngest embryos developing in a given stage are usually seen in the posterior portions of its body and are often referred to as **germ balls.**

The nature of this asexual reproduction has long been controversial; it has at various times been thought to represent **budding, polyembryony,** or **parthenogenesis.** The view of early zoologists—that it is an example of metagenesis (strictly speaking, an alternation of generations in which the asexual generation reproduces by budding)—was discarded when it was realized that the specific reproductive cells (the propagatory cells) are kept segregated in the germinal sacs. The most widely held opinion has been that the process is one of sequential polyembryony,[27,28] that is, production of multiple embryos from the same zygote with no intervening gamete production as, for example, in monozygotic twins in humans. Whitfield and Evans[117] reviewed the evidence for parthenogenesis and found it insubstantial. They felt that the asexual reproduction in Digenea "most probably represents a budding process in which the development of the buds is initiated by the division of diploid totipotent (propagatory) cells."

Larval and juvenile development
■ Egg

The structure referred to as an "egg" of trematodes is not an ovum but the developing (or developed) embryo enclosed by its shell, or capsule. The egg capsule of most flukes has an **operculum** at one end, through which the larva eventually will escape. It is not clear how the operculum is formed, but it appears that the embryo presses pseudopodium-like processes against the inner surface of the shell while it is being formed, thereby forming a circular groove. An operculum is absent in the eggshell of blood flukes. Considerable variation exists in the shape and size of fluke eggs, as well as in the thickness and coloration of the capsules.

In many species the egg contains a fully developed miracidium by the time it leaves the parent; in others development has advanced to only a few cell divisions by that time. In some species *(Cyclocoelum and Heronimus)* the miracidium hatches while still in the uterus. For eggs that embryonate in the external environment, certain factors are known to influence embryonation. Water is necessary, since the eggs desiccate rapidly in dry conditions. Development is stimulated by high oxygen tension, although eggs can remain viable for long periods under conditions of low oxygen. Eggs of *F. hepatica* will not develop outside a pH range of 4.2 to 9.[92] Temperature is critical, as would be expected. Thus *F. hepatica* requires 23 weeks to develop at 10° C, whereas it takes only 8 days at 30° C. However, above 30° C development again slows and completely stops at 37° C. Eggs are killed rapidly at freezing. Light may be a factor influencing development in some species, but this has not been thoroughly investigated.

Eggs of many species will hatch freely in water, whereas others hatch only when eaten by a suitable intermediate host. Factors stimulating hatching have been investigated for several species. Light and osmotic pressure are important in species that hatch in water, and osmotic pressure, carbon dioxide tension, and, probably, host enzymes initiate hatching in those which must be eaten before they will hatch. The eggs of *Transversotrema patialense* hatch spontaneously and show what appears to be a circadian rhythm, hatching at the same time of day whether kept under constant light or a 12 hour/12 hour, light/dark cycle.[14] The time of hatching is correlated with the time the snail intermediate host is nearby.

The mechanism of hatching has been studied most in *F. hepatica*.[118] The miracidium is surrounded by a thin **vitelline membrane,** which also encloses a pad-like **viscous cushion** between the anterior end of the miracidium and the operculum. Light stimulates hatching activity. Some factor apparently is released by the miracidium that alters the permeability of the membrane enclosing the viscous cushion. The latter structure contains a mucopolysaccharide that becomes hydrated and greatly expands the volume of the viscous cushion. The considerable increase in pressure

within the egg causes the operculum to pop open, remaining attached at one point, and the miracidium rapidly escapes, propelled by its cilia. The nonoperculated eggs of *Schistosoma* are fully embryonated when passed from the host, and they hatch spontaneously in fresh water. They release substantial quantities of leucine aminopeptidase, and this enzyme may help digest the capsule from the inside.[120] Unlike leucine aminopeptidases from other sources, the enzyme produced by schistosome miracidia is inhibited by NaCl, which would prevent hatching while in the body of the host.

Miracidium

The typical miracidium (Fig. 17-18) is a tiny, ciliated organism that could easily be mistaken for a protozoan by the casual observer. It probably is quite similar to the acoeloid ancestor of the Platyhelminthes. It is piriform, with a retractable **apical papilla** at the anterior end. The apical papilla has no cilia but bears five pairs of duct openings from glands and two pairs of sensory nerve endings (Fig. 17-19). The gland

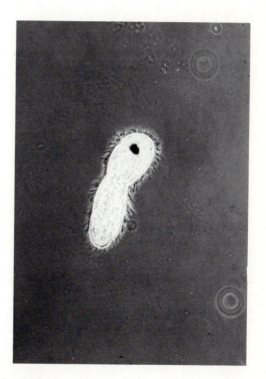

FIG. 17-18

Miracidium of *Alaria* sp.

Photograph by Jay Georgi.

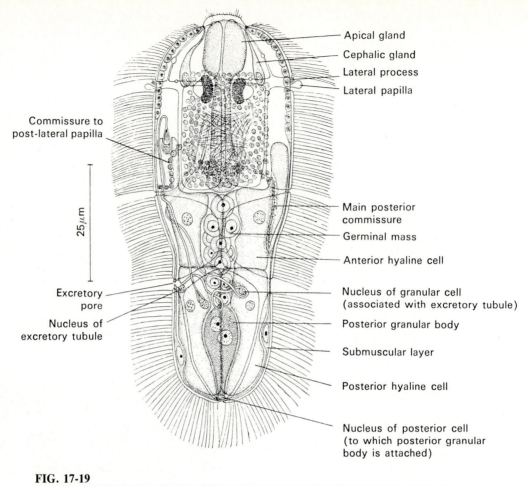

Apical gland
Cephalic gland
Lateral process
Lateral papilla

Commissure to
post-lateral papilla

25 μm

Main posterior
commissure
Germinal mass
Anterior hyaline cell

Excretory
pore

Nucleus of
excretory tubule

Nucleus of granular cell
(associated with excretory tubule)

Posterior granular body

Submuscular layer

Posterior hyaline cell

Nucleus of posterior cell
(to which posterior granular
body is attached)

FIG. 17-19

Miracidium of *Neodiplostomum intermedium*, dorsal view.

From Pearson, J.C. 1961. Parasitology 51:133-172.

ducts connect with **penetration glands** inside the body. A prominent **apical gland** can be seen in the anterior third of the body. This probably also secretes histolytic enzymes. An apical stylet is present on some species, and spines are found on others. The sensory nerve endings connect with nerve cell bodies that in turn communicate with a large ganglion. Miracidia have a variety of sensory organs and endings, including adaptations for photoreception, chemoreception, tangoreception, and statoreception.

The outer surface of a miracidium is covered by flat, ciliated epidermal cells, the number and shape of which are constant for a species. Underlying this are longitudinal and circular muscle fibers. Cilia are restricted to protruding **ciliated bars** in the genus *Leucochloridiomorpha* (Brachylaimidae) and family Bucephalidae, and they are absent altogether in the fam-

ilies Azygiidae and Hemiuridae. One or two pairs of protonephridia are connected to a pair of posterolateral excretory pores.

In the posterior half of the miracidium are found propagatory cells, or germ balls (embryos), which will be carried into the sporocyst stage to initiate further generations.

Free-swimming miracidia are very active, swimming at a rate of about 2 mm per second, and they must find a suitable molluscan host rapidly, since they can survive as free-living organisms for only a few hours. Snail-finding behavior of miracidia has been reviewed.[20] In many cases the mucus produced by the mollusc is a powerful attractant for miracidia.

On contacting a proper mollusc, the miracidium attaches to it with the apical papilla, which actively contracts and extends, undergoing an auger-like motion.

Cytolysis of snail tissues can be seen as the miracidium embeds itself deeper and deeper. As penetration proceeds, the miracidium loses its ciliated epithelium, although this may be delayed until penetration is complete. The miracidium takes about 30 minutes to complete penetration and begin the next phase of its life cycle as the sporocyst.

■ Sporocyst

Metamorphosis of the miracidium into the sporocyst involves extensive changes. In addition to the loss of the ciliated epithelial cells, the new tegument with its microvilli forms as described previously (p. 231). The subtegumental muscle layer of the miracidium is retained, as are the protonephridia, but all other miracidial structures generally disappear. The sporocyst has no mouth or digestive system; it absorbs nutrients from the host tissue, with which it is in intimate contact, and the entire structure serves only to nurture the developing embryos. The sporocyst (or other stage with embryos developing within it, that is, the miracidium or redia) may be referred to as a **germinal sac.** Often sporocysts grow near the site of penetration, such as foot, antenna, or gill, but they may be found in any tissue, depending on the species; and sometimes they may become very slender and extended, branched, or ramified.

The embryos in the sporocyst may develop into another sporocyst generation **(daughter sporocysts);** into a different form of germinal sac, the redia; or directly into cercariae (Fig. 17-20).

An interesting sporocyst adaptation is found in the genus *Leucochloridium.* The highly branched daughter sporocysts contain encysted cercariae, and they extend as swollen, brightly colored brood sacs into the tentacles of their snail host. There they pulsate rapidly: 70 times a minute at summer temperatures.[115] Their color and movement serve to attract the attention (and appetite) of their bird definitive hosts.

■ Redia

Rediae (Fig. 17-21) burst their way out of the sporocyst or leave through a terminal birth pore and usually migrate to the hepatopancreas or gonad of the molluscan host. They are commonly elongate and blunt at the posterior end and may have one or more stumpy appendages called **procrusculi.** More active than most sporocysts, they crawl about within their host. They have a rudimentary, but functional, digestive system, consisting of a mouth, muscular pharynx, and short, unbranched gut. Rediae pump food into their gut by means of the pharyngeal muscles, as previously described in adults. They not only feed on

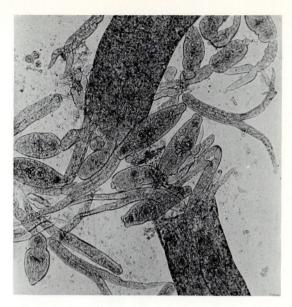

FIG. 17-20

Ruptured sporocyst releasing furcocercous cercariae.

Photograph by James Jensen.

host tissue but also they can prey on sporocysts of their own or other species.[70] The luminal surface of their gut is greatly amplified by flattened, lamelloid or ribbon-like processes.[64] The gut cells are apparently capable of phagocytosis. The outer surface of the tegument also functions in absorption of food, and it is provided with microvilli or lamelloid processes.

The embryos in the redia develop into daughter rediae or into the next stage, the cercaria, which emerges through a birth pore near the pharynx. The epithelial lining of the birth pore in *Cryptocotyle lingua* (and probably other species) is highly folded so that it can withstand the extreme distortion produced by the exit of a cercaria.[55] It appears that rediae must reach a certain population density before they stop producing more rediae and begin producing cercariae: young rediae have been transplanted from one snail to another through more than 40 generations without cercariae being developed.[33] This is an interesting parallel to certain free-living invertebrates that reproduce parthenogenetically only as long as certain environmental conditions are maintained.[17]

■ Cercaria

The cercaria represents the juvenile stage of the vertebrate-inhabiting adult. There are many varieties of cercariae, and most have specializations that enable

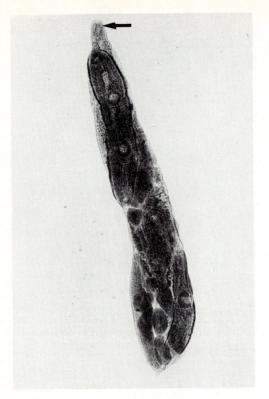

FIG. 17-21

A redia. Note the large, muscular pharynx *(arrow)* just inside the mouth.

Photograph by Warren Buss.

them to survive a brief free-living existence and make themselves available to their definitive or second intermediate hosts (Fig. 17-22). Most have tails that aid them in swimming, but many have rudimentary tails or none at all; these cercariae can only creep about, or they may remain within the sporocyst or redia that produced them until they are eaten by the next host.

The structure of a cercaria is easily studied, and cercarial morphology often has been considered a more reliable indication of phylogenetic relationships among families than is the morphology of adults. Cercariae are widely distributed, abundant, and easily found; hence they have attracted much attention from zoologists. The name *Cercaria* can be used properly in a generic sense for a species in which the adult form is unknown, as is done with the term *Microfilaria* among some nematodes.

Most cercariae have a mouth near the anterior end, although it is midventral in the Bucephalidae. The mouth is usually surrounded by an oral sucker, and a prepharynx, muscular pharynx, and a forked intestine are normally present. Each branch of the intestine is simple, even those which are ramified in the adult. Many cercariae have various glands opening near the anterior margin, often called "penetration glands" because of their assumed function. The schistosome cercaria, which has been best studied in this regard, has no less than four distinguishable types of such glands[105]:

1. **Escape glands.** So called because their contents are expelled during emergence of the cercaria from the snail, but their function is not known.
2. **Head gland.** Secretion is emitted into the matrix of the tegument and is thought to function in the postpenetration adjustment of the schistosomule.
3. **Postacetabular glands.** Produce mucus, help cercariae adhere to surfaces, and have other possible functions.

FIG. 17-22

A few of the many types of cercariae. **A,** Amphistome cercaria. **B,** Monostome cercaria. **C,** Gymnocephalous cercaria. **D,** Gymnocephalous cercaria of pleurolophocercous type. **E,** Cystophorous cercaria. **F,** Trichocercous cercaria. **G,** Echinostome cercaria. **H,** Microcercous cercaria. **I,** Xiphidiocercaria. **J,** Ophthalmoxiphidiocercaria. **K** to **O,** Furocercous types of cercariae. **K,** Gasterostome cercaria. **L,** Lophocercous cercaria. **M,** Apharyngeate furcocercous cercaria. **N,** Pharyngeate furcocercous cercaria. **O,** Apharyngeate monostome fucocercous cercaria without oral sucker. **P,** Cotylocercous cercaria. **Q,** Rhopalocercous cercaria. **R,** Cercariaea. **S,** Rattenkönig, or rat-king, cercariae.

From Olsen, O.W. 1974. Animal parasites, their life cycles and ecology. © 1974 University Park Press, Baltimore.

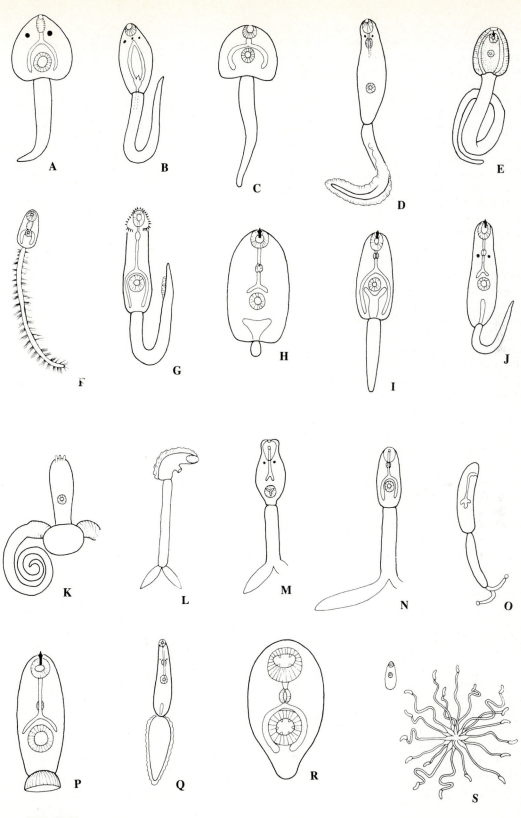

FIG. 17-22

For legend see opposite page.

4. **Preacetabular glands.** Secretion contains calcium and a variety of enzymes including a protease. The function of these glands seems most important in actual penetration of host skin.

Secretory **cystogenic cells** are particularly prominent in cercariae that will encyst on vegetation or other objects.

Many morphological variations exist in cercariae that are constant within a species (or larger taxon); thus certain descriptive terms are of value in categorizing the different varieties. Some of the more commonly used terms follow:

1. **Xiphidiocercaria.** Cercaria with stylet in anterior margin of oral sucker
2. **Ophthalmocercaria.** Cercaria with eyespots
3. **Cercariaeum.** Cercaria without a tail
4. **Microcercous cercaria.** Those with small, knoblike tail
5. **Furcocercous cercaria.** Those with forked tail
6. **Cercariae ornatae.** Cercaria tail with a fin

The excretory system is well developed in the cercaria. In some cercariae the excretory vesicle empties through one or two pores in the tail. Because the number and arrangement of the protonephridia are constant for a species, these are important taxonomic characters. Each flame cell has a tiny **capillary duct** that joins with others to form the **anterior** or **posterior collecting tubules,** one each of which joins to form an **accessory tubule.** Accessory tubules join to form a **common collecting tubule** on each side (Fig. 17-23). When the common collecting tubules extend to the region of the midbody and then fuse with the excretory vesicle, the cercaria is called **mesostomate.** If the tubules extend to near the anterior end and then pass posteriad to join the vesicle, the cercaria is known as **stenostomate.** The number and arrangement of flame cells can be expressed conveniently by the *flame cell formula.* For example, $2[(3 + 3) (3 + 3)]$ means that both sides of the cercaria, 2, have three flame cells on each of the two accessory tubules, $(3 + 3)$, on the anterior collecting tubule, plus the same arrangment on the posterior collecting tubule.

Mature cercariae emerge from the mollusc and begin to seek their next host. Many remarkable adaptations can be found among cercariae that enable them to do this. Most are active swimmers, of course, and rely on chance to place them in contact with an appropriate organism. Some species are photopositive, dispersing themselves as they swim toward the surface of the water, but then become photonegative and return to the bottom where the next host is. Some opisthorchioid cercariae remain quiescent on the bottom until a fish swims over them; the resulting shadow activates

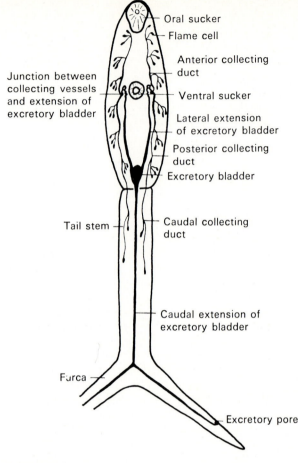

FIG. 17-23

Diagrammatic representation of the excretory system of a fork-tailed cercaria. The caudal flame cells are absent from the tail in nonfurcate forms.

From Erasmus, D.A. 1972. The biology of trematodes. Crane, Russak & Co., New York.

them to swim upward. Some plagiochioid cercariae cease swimming when in a current; hence, when drawn over the gills of a crustacean host, they can attach and penetrate rather than swim on. Large, pigmented azygiids and bivesiculids are enticing to fish, which eat them and become infected. Some cercariae float; some unite in clusters; some creep at the bottom. Cercariae of *Schistosoma mansoni,* which directly penetrate the warm-blooded definitive host, concentrate in a thermal gradient near a heat source (34° C).[22] In the case of certain cystophorous hemiurid cercariae, the body is withdrawn into the tail, which becomes a complex injection device.[76] The second intermediate host of the trematode is a copepod crustacean, which attempts to eat the caudal cyst bearing the cercaria. When the narrow end of the cyst is broken by the mandibles of the copepod, a delivery tube is rapidly everted into the mouth of the copepod, piercing its midgut, and then the cercaria slips through the delivery tube into the hemocoel of the crustacean![76] These and many more adaptations help ensure that the trematode will reach its next host.

■ Mesocercaria

Species of the strigeoid genus *Alaria* have a unique larval form, the **mesocercaria,** which is intermediate between the cercaria and metacercaria (see Fig. 18-3).

■ Metacercaria

Between the cercaria and the adult is a quiescent stage, the metacercaria, although this stage is absent in the families of blood flukes. The metacercaria is usually encysted, but this stage in *Brachycoelium, Halipegus, Panopistus,* and others is not. Most metacercariae are found in or on an intermediate host, but some (Fasciolidae, Notocotylidae, and Paramphistomidae) encyst on aquatic vegetation, sticks, and rocks or even free in the water.

The cercaria's first step in encysting is to cast off its tail. Cyst formation is most elaborate in metacercariae encysting on inanimate objects or vegetation. The cystogenic cells of *F. hepatica* are of four types, each with the precursors of a different cyst layer. Metacercariae encysting in intermediate hosts have thinner and simpler cyst walls, with some components contributed by the host.

Development that occurs in the metacercaria varies widely according to species, from those in which a metacercaria is absent *(Schistosoma)* to those in which the gonads mature and viable eggs are produced *(Proterometra)*. Often, some amount of development is necessary in the metacercaria before the organism is infective for the definitive host. We can arrange metacercariae in three broad groups on this basis[32]:

1. Species whose metacercariae encyst in the open, on vegetation and inanimate objects, for example, *Fasciola* spp. Members of this group can infect the definitive host almost immediately after encystment, in some cases within only a few hours, with no growth occurring.
2. Species that do not grow in the intermediate host but that require at least several days of physiological development to infect the definitive host, for example, Echinostomatidae.
3. Species whose metacercariae undergo growth and metamorphosis before they enter their rest-

ing stage in the second intermediate host and that usually require a period of weeks for this development, for example, Diplostomatidae.

These developmental groups are correlated with the longevity of the metacercariae: those in group 1 must live on stored food and can survive the shortest time before reaching the definitive host, whereas those in groups 2 and 3 obtain some nutrient from their intermediate hosts and so can remain viable for the longest periods—in one case up to 7 years. After the required development the metacercaria goes into the quiescent stage and remains in readiness to excyst on reaching the definitive host. *Zoogonus lasius,* a typical example of group 2, has a high rate of metabolism for the first few days after infecting its second intermediate host, a nereid polychaete, and then drops to a low level, only to return to a high rate on excystation.[112] Metacercariae of *Bucephalus haimeanus* remain active in the liver of their fish host and increase threefold in size. They take up nutrients from degenerating liver cells, including large molecules, by pinocytosis.[50] Metacercariae of *Clinostomum marginatum* take up glucose both by facilitated diffusion and by active transport.[110]

The metacercarial stage has a high selective value for many species of trematodes. It may provide a means for transmission to a definitive host that does not feed on the first intermediate host or is not in the environment of the mollusc, or it may permit survival over unfavorable periods, such as the absence of the definitive host during a particular season.

Development in the definitive host

Once the cercaria or metacercaria has reached its definitive host, it matures in a variety of ways: either by penetration (if a cercaria) or by excystation (if a metacercaria) and then by migration, growth, and morphogenesis to gamete production. If the species does not have a metacercaria and the cercaria penetrates the definitive host directly, as in the schistosomes, the most extensive growth, differentiation, and migration will be necessary. At the other extreme the adult form may already have been attained as a metacercaria, the gonads may almost be mature, or some eggs may even be present in the uterus; and little more than excystation is needed before the production of progeny *(Bucephalopsis, Coitocaecum, Transversotrema)*. A very few species *(Proctoeces maculatus*[1] and *Proterometra dickermani)* reach sexual reproduction in the mollusc and apparently do not have a vertebrate definitive host; for example, several species of Macroderoididae develop to sexual maturity in leeches (Fig. 16-3),[111] and *Allocorrigia filiformis* matures in the antennal gland of a crayfish.[109] These are some-times considered examples of neoteny.

Normally, development in the definitive host begins with excystation of the metacercaria, and species with the heaviest, most complex cysts appear to require the most complex stimuli for excystation, such as those with cysts on vegetation, for example, *F. hepatica*. The outer cyst of *F. hepatica* is largely removed by digestive enzymes, but escape from the inner cyst requires the presence of a temperature of about 39° C, a low oxidation-reduction potential, carbon dioxide, and bile. Comparable conditions produce maximal encyst-ment of *Himasthla quiessetensis*.[61] This combination of conditions is not likely to be present anywhere but in the intestine of a homoiothermic vertebrate, and, like the conditions required for the hatching or exsheathment of some nematodes, the requirements constitute an adaptation that avoids premature escape from the protective coverings. Such an adaptation is less important to metacercariae that are not subjected to the widely varying physical conditions in the external environment, such as those encysted within a second intermediate host. These have thinner cysts and excyst on treatment with digestive enzymes. A number of species require the presence of a bile salt(s) or excyst more rapidly in its presence.[6,36,40,104]

After excystation in the intestine, a more or less extensive migration is necesssary if the final site is in some other tissue. The main sites of the "tissue" parasites are the liver, lungs, and circulatory system. Probably the most common way to reach the liver is by way of the bile duct *(Dicrocoelium dendriticum),* but *F. hepatica* burrows through the gut wall into the peritoneal cavity and finally, wandering through the tissues, reaches the liver. *Clonorchis sinensis* usually penetrates the gut wall and is carried to the liver by the hepatoportal system. *Paragonimus westermani* penetrates the gut wall, undergoes a developmental phase of about a week in the abdominal wall, and then reenters the abdominal cavity and makes its way through the diaphragm to the lungs.

A remarkable physiological aspect of trematode life cycles is the sequence of totally different habitiats in which the various stages must survive, with physiological adjustments that must often be made extremely rapidly. As the egg passes from the vertebrate, it must be able to withstand the rigors of the external environment in fresh water or seawater, if only for a period of hours, before it can reach haven in the mollusc. There, conditions are quite different from both the water and the vertebrate. The trematode's physiological capacities must again be readjusted on escape from the intermediate host and again on reaching the second intermediate or definitive host. Environmental change may be somewhat less dramatic if the second intermediate host is a vertebrate, but often it is an invertebrate. Although the adjustments must be extensive, the nature of these physiological adjustments made by trematodes during their life cycles has been little investigated, the most studied in this respect being *Schistosoma*.[21] *Transversotrema patialense* has been considered ectoparasitic because it lives under the scales of its fish host. However, it must be osmotically sheltered there because it loses its ability to survive in tap water within 5 minutes after the cercariae reach the host.[79]

Penetration of the definitive host is a hazardous phase of the life cycle of schistosomes, and it requires an enormous amount of energy. The hazards include a combination of the dramatic changes in the physical environment, in the physical and chemical nature of the host skin through which it must penetrate, and in host defense mechanisms. Depending on the host species, losses at this barrier may be as high as 50%, and the glycogen content of the newly penetrated schistosomula (**schistosomule** is the name given the young developing worm) is only 6% of that found in cercariae. Among the most severe physical conditions the organism must survive is the sequence of changes in ambient osmotic pressure. The osmotic pressure of fresh water is considerably below that in the snail, and that in the vertebrate is twice as great as in the mollusc. Assuming that the osmotic pressure of the cercarial tissues approximates that in the snail, the trematode must avoid taking up water after it leaves the snail and avoid a serious water loss after it penetrates the vertebrate. Aside from the possible role of the osmoregulatory organs (protonephridia and others), there appear to be major changes in the character, and probably permeability, of the cercarial surface. The cercarial surface is coated with a fibrillar layer, or glycocalyx, which is lost on penetration of the vertebrate, and with it is lost the ability to survive in fresh water; 90% of schistosomula recovered from mouse skin 30 minutes after penetration die rapidly if returned to fresh water. That chemical changes have occurred is indicated by the fact that the schistosomule surface is much less easily dissolved by a number of chemical reagents, including 8 M urea, than that of the cercaria. The antigenic determinants of the schistosomule as compared with the cercaria are changed also. When cercariae are incubated in immune serum, a thick envelope forms around them called the CHR *(cercarien-hüllenreaktion),* but schistosomula do not give this reaction.

Before penetration, the schistosome cercaria must be attracted to the host skin and remain there, exploring for entry sites. The cercariae are apparently attracted to the host skin by the amino acid arginine,[47] whereas the most important stimulus for actual penetration seems to be the skin lipid film, specifically essential fatty acids, such as linoleic and linolenic acids, and certain nonessential fatty acids.[93] Human skin surface lipid applied to the walls of their glass container will cause cercariae to attempt to penetrate it, lose their tails, evacuate their preacetabular glands, and become intolerant to water. The presence of the penetration-stimulating substances causes loss of osmotic protection and a reduction of the CHR, even in cercariae free in the water.[48a] Successful penetration and transformaton has been correlated with cercarial production of eicosanoids, such as leukotrienes and prostaglandins[43,94] (fatty acid derivatives with potent pharmacological activity).

After penetration, the tegument of the developing schistosomule undergoes a remarkable morphogenesis. Within 30 minutes, numerous subtegumental cells have connected with the distal cytoplasm and are passing abundant "laminated bodies" into it. These bodies have *two* trilaminar limiting membranes (therefore, heptalaminar). These bodies move to the surface of the tegument to become the new tegumental outer membrane; the old cercarial outer membrane, along with its remaining glycocalyx, is cast off. The schistosomule outer membrane is almost entirely heptalaminar 3 hours after penetration. There is some evidence that these tegumental changes may be hormonally mediated.[107] During the next 2 weeks the main changes in the tegument are a considerable increase in thickness and the development of many invaginations and deep pits. These pits increase the surface area fourfold between 7 and 14 days after penetration. It may be assumed that this represents an adaptation for nutrient absorption through the tegument.

Summary of life cycle

Following is the basic pattern of a digenetic trematode life cycle:

egg → miracidium → sporocyst →
$\qquad$ redia → cercaria → metacercaria → adult

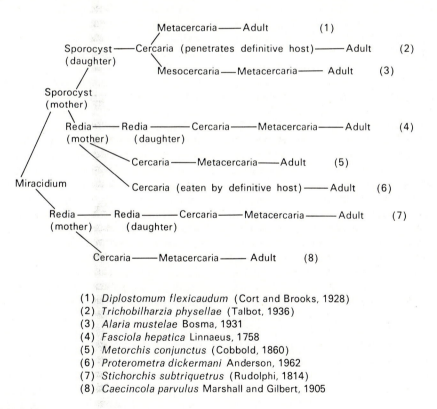

(1) *Diplostomum flexicaudum* (Cort and Brooks, 1928)
(2) *Trichobilharzia physellae* (Talbot, 1936)
(3) *Alaria mustelae* Bosma, 1931
(4) *Fasciola hepatica* Linnaeus, 1758
(5) *Metorchis conjunctus* (Cobbold, 1860)
(6) *Proterometra dickermani* Anderson, 1962
(7) *Stichorchis subtriquetrus* (Rudolphi, 1814)
(8) *Caecincola parvulus* Marshall and Gilbert, 1905

FIG. 17-24

Some life cycles of digenetic trematodes.

From Schell, S.C. 1970. How to know the trematodes. William C. Brown Co., Publishers, Dubuque, Iowa.

The student should learn this pattern well, since it is the theme on which to base the variations, the most common of which are (1) more than one generation of sporocysts or rediae, (2) deletion of either sporocyst or redial generations, and (3) deletion of metacercaria. Much less common are cases in which miracidia are produced by sporocysts and forms with adult morphology in the mollusc producing cercariae (these in turn lose their tails and produce another generation of cercariae).[57] Fig. 17-24 shows some possible life cycles.

METABOLISM
Energy metabolism

Investigators have accorded the metabolism of trematodes considerable attention.* Compared with that of certain vertebrates, however, trematode metabolism is meagerly known, and the metabolism of larval stages has received scant attention. Furthermore, because of size, availability, and/or medical importance, the metabolism of *Schistosoma* and *Fasciola* has been much more thoroughly investigated than that of other species.

In terms of energy derivation from nutrient molecules, adult cestodes and trematodes have much in common. Their main sources of energy are from the degradation of carbohydrate from glycogen and glucose. They are facultative anaerobes, and even in the presence of oxygen, they depend greatly on glycolysis and excrete large amounts of short-chain acid end products[51] (Fig. 17-25). In other words the energy potential in the glucose molecule is far from completely realized. The worm has what is, for all practical purposes, an inexhaustible food supply. Interestingly, some workers have found that the glucose in bile was insufficient to account for the needs of *F. hepatica*; the main sources of its energy in vivo remain unclear.[24] Nonetheless, these and other trematodes may be subjected to very low oxygen concentrations part or all of the time, and their metabolic end products can be further catabolized by their host. Even in the case of the schistosomes, which live in the blood and presumably have an abundant oxygen supply, glycolysis is the main energetic pathway. The worms can survive 5 to 6 days under nitrogen or in 1 mM cyanide, but they consume oxygen when it is available. Inhibitions of oxygen uptake by cyanide, comparable with those to be expected in normal aerobic tissues, have been shown in several other trematodes. Cheah and Prichard[19] concluded that *F. hepatica* has a branched electron transport system: one being the

classical mammalian type, with cytochrome a_3, and the other with cytochrome o as terminal electron acceptors. Propionate production by *F. hepatica* was reduced by 30% under aerobic as opposed to anaerobic conditions. Cheah and Prichard believe that the branched chain is an adaptation of large parasites to low environmental oxygen. However, NADH–cytochrome c oxidoreductase, succinate–cytochrome c oxidoreductase, NADH oxidase, and cytochrome c–oxygen oxidoreductase were all present in the trematode. Barrett[2] concluded that parasitic helminths generally are capable of oxidative phosphorylation but that its contribution to overall energy balance is difficult to assess.

The ability to catabolize certain of the citric acid cycle intermediates and the presence of various enzymes in that cycle have been shown in several trematode species. In fact all enzymes necessary for a functional citric acid cycle are present in *F. hepatica*,[71] but the levels of aconitase and isocitrate dehydrogenase activities are so low that the cycle is considered of minor importance at most. Evidence exists for a functional citric acid cycle in schistosomes,[23] but its significance is still not clear. The primary role of some of the enzymes may lie in metabolic paths other than the Krebs cycle (succinic dehydrogenase and malic dehydrogenase) (Fig. 17-25). Some enzymes may be vestiges of previous phylogenetic or ontogenetic stages, or the cycle may simply function at a lower level than usual in strict aerobes.

Although little is known of the lipid metabolism, we have no evidence that lipids are used as energy sources or energy storage compounds. However, trematodes may contain considerable lipid, and sizable quantities may be excreted. *Fasciola hepatica* excretes about 2% of its net weight per day as polar and neutral lipids (including cholesterol and its esters), and this is mainly by way of its excretory system.[15]

Consequently, it is not surprising that digeneans contain large amounts of stored glycogen: 9% to 30% of dry weight, according to species. Amounts in female *S. mansoni* are unusually low: only about 3.5% of dry weight. Although the glycogen content of cestodes may range higher than 30%, it is still surprising that trematodes, even tissue-dwelling species, store so much, since the availability of their food should not be subject to the vagaries of their host's feeding schedule, as it is with cestodes. In cases in which measurements have been performed, a large proportion of the trematode's glycogen is consumed under starvation conditions in vitro. In fact the maintenance of a high glycogen concentration in the worms may be

*References 3, 23, 71, 102, 108, 113.

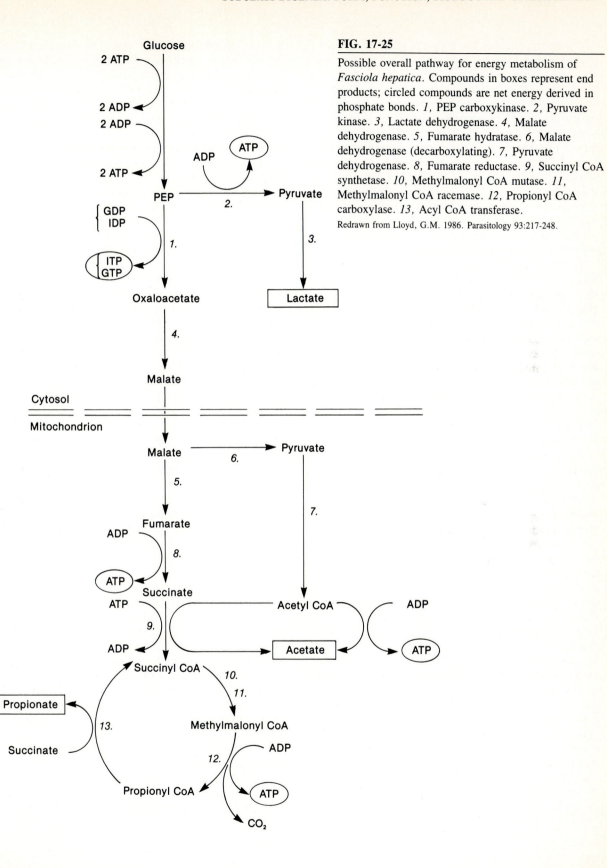

FIG. 17-25

Possible overall pathway for energy metabolism of *Fasciola hepatica*. Compounds in boxes represent end products; circled compounds are net energy derived in phosphate bonds. *1*, PEP carboxykinase. *2*, Pyruvate kinase. *3*, Lactate dehydrogenase. *4*, Malate dehydrogenase. *5*, Fumarate hydratase. *6*, Malate dehydrogenase (decarboxylating). *7*, Pyruvate dehydrogenase. *8*, Fumarate reductase. *9*, Succinyl CoA synthetase. *10*, Methylmalonyl CoA mutase. *11*, Methylmalonyl CoA racemase. *12*, Propionyl CoA carboxylase. *13*, Acyl CoA transferase.

Redrawn from Lloyd, G.M. 1986. Parasitology 93:217-248.

of critical importance. The action of niridazole, an antischistosomal drug, has been attributed to the fact that it causes glycogen depletion in the schistosome, and the mechanism of action is very interesting. The glucose moieties in glycogen are mobilized for glycolysis by the action of glycogen phosphorylase, as in other systems, and the extent of the mobilization is controlled by how much of the enzyme is in the physiologically active *a* form. Niridazole inhibits the conversion of phosphorylase *a* to the inactive *b* form; thus the phosphorolysis of glycogen is uncontrolled, the glycogen stores of the worm are depleted, and it is finally killed if the niridazole concentration is maintained.[13] As with any good chemotherapeutic agent, the corresponding host enzyme is not affected.

The mode of action of organic trivalent antimonials, traditional antischistosomal drugs, has been attributed to their inhibition of a critical enzyme in glycolysis, phosphofructokinase (PFK). The PFK of the schistosomes is much more sensitive to the antimonials than is the corresponding host enzyme.[12] However, evidence has accumulated that the action on PFK does not fully account for the effect of these drugs.[5,23] Antimonials have severe side effects on the host and have now been replaced by other compounds.

As in cestodes, the terminal reactions in the glycolytic sequence may be quite different from those to which we are accustomed in mammals. An exception to this statement, however, is found in the schistosomes. These worms are referred to as **homolactic producers;** *S. mansoni* consumes glucose equivalent to up to 20% of its dry weight per hour, and more than 80% of this is accounted for as lactate.[10] *Dicrocoelium dendriticum* and *F. hepatica* also consume considerable amounts of glucose (though at a much lower rate than does *S. mansoni*) but with different glycolytic end products. *Dicrocoelium dendriticum* produces about 40% lactate, 30% acetate, 30% propionate, and 3% succinate, whereas *F. hepatica* excretes 8% lactate, 24% acetate, 68% propionate, and only traces of succinate.[63] The small amount of lactate produced by *F. hepatica* is presumably explained by the low activity of its lactic dehydrogenase.[88] All four of the short-chain acids seem to be produced through carbon dioxide fixation by phosphoenol pyruvate carboxykinase (PEPCK), since the activity of pyruvate kinase (PK) is very low in *Dicrocoelium* and *Fasciola* spp. (Fig. 17-25), and pyruvate is apparently produced by decarboxylation of malate by malate dehydrogenase (decarboxylating). Activity of PK in *Schistosoma* is much higher than that of PEPCK. Acetate in *F. hepatica* is formed by an oxidative decarboxylation to acetyl-CoA and then cleavage of acetyl-CoA.[4]

Köhler, Bryant, and Behm[62] found that succinate was decarboxylated by succinyl- and propionyl-CoA to form propionate and that a net, substrate-level phosphorylation of ADP to ATP occurred (Fig. 17-25). They proposed a reaction sequence and pointed out the advantage to the worm of propionate rather than succinate excretion, that is, the production of two more moles of ATP for each mole of glucose oxidized. The reactions were strongly inhibited by the anthelmintic mebendazole.

The phosphofructokinase reaction is rate limiting in *Fasciola,* as in *Schistosoma,* and 5-hydroxytryptamine (5-HT), which is known to be an excitatory neurotransmitter in *Schistosoma,* has a very interesting effect on PFK in Fasciola. The amine occurs in and is synthesized by *F. hepatica,* and the activity of PFK in homogenates of the fluke is increased by 5-HT. Cyclic 3′,5′-AMP duplicates the effect of the 5-HT, and it appears that the activation of the PFK is a result of 5-HT stimulation of cyclic 3′,5′-AMP synthesis. Epinephrine, a related catecholamine, which increases formation of cyclic 3′,5′-AMP in particulate fractions of the mammalian liver, has no effect on the same fractions from the fluke.[75]

A pentose phosphate pathway may function in schistosomes, but critical enzymes for a glyoxylate cycle have not been found. In contrast, the pentose cycle in *Fasciola* appears to be minimal, but enzymes necessary for the glyoxylate path are all present.[71]

The astonishing ability of trematodes to survive the radical changes in environment requires important adjustments in their energy metabolism. Clearly the ability to derive every possible ATP from every glucose unit would be of great selective value to the free-swimming miracidium or cercaria that does not feed. In all species investigated so far the miracidia and cercariae are obligate aerobes, killed by short exposures to anaerobiosis. Pyruvate is used rapidly by cercariae of *S. mansoni;* carbon dioxide is produced from all three of the pyruvate carbons. The miracidium may have a functional citric acid cycle within the egg. Study of the sporocysts is difficult because of the problem of separating host from parasite tissue; however, the same drugs that kill sporocysts within the snail affect the adults, and sporocysts (or the cercariae within) produce lactic acid under aerobic conditions. Therefore, the metabolism of the sporocysts is thought to be like that of the adults. Use of inhibitors suggests that a functional citric acid cycle is present in cercariae; they exhibit a Pasteur effect, and cytochromes *a*/*a*³, *b,* and *c* are all present. Immediately after penetration, schistosome energy metabolism seems to undergo a major adjustment. The ability to use pyruvate drops dramatically, and the schisotosomula again produce lactate aerobically.[23] They continue to use oxygen when available, as do the adults, and the oxygen consumption of both cercariae and schistosomula is inhibited by more than 80% in the presence of 0.2 mM cyanide. Such developmental studies on the metabolism of other trematodes would be very interesting, but few have been reported. The oxygen consumption of adult *Gynaecotyla adunca,* an intestinal parasite of fish and birds, dropped sharply 24 hours after excystation, then even more after 48 and 72 hours. Juvenile *F. hepatica,* living in the liver parenchyma, have a cyanide-sensitive respiration but are facultative anaerobes, thus they seem to be in transition from the aerobic cercariae to the anaerobic adults.[106]

Transamination ability appears limited, but the α-ketoglutarate-glutamate transaminase reaction is active.[29,114]

Ammonia and urea are both important end products in degradation of nitrogenous compounds in *Fasciola* and *Schistosoma* spp., and both worms excrete several amino acids as well. A full complement of the enzymes necessary for the ornithine-urea cycle is not present in *Fasciola;* the urea produced must be by other pathways.[58]

Synthetic metabolism

Stimulated by the search for chemotherapeutic agents, researchers have studied purine metabolism in *Schistosoma. Schistosoma mansoni* cannot synthesize nucleotides de novo.[98] Neither glycine nor glucose is incorporated into purine nucleotide bases, but adenine is taken up from the medium and synthesized into the nucleotides. Kurelec[65] showed that *F. hepatica* and *Paramphistomum cervi* could not synthesize carbamyl phosphate and concluded that they depended on their hosts for both pyrimidines and arginine. In light of the high arginine requirement of *S. mansoni,* it would seem probable that the situation is the same in that species.

The requirement of schistosomes for arginine is so high, in fact, that they reduce the level of serum arginine to almost zero in mice with severe infections. The worms more rapidly take up arginine than histidine, tryptophan, or methionine. Proline from the host is also rapidly consumed by the male schistosome through both the gut and the tegument, but only a little is absorbed by the tegument of the female.[97] Interestingly, the proline consumed is concentrated in the ventral arms of the gynecophoral canal, the region of contact with the female. Glycogen concentrations in male and female schistosomes fluctuate in a parallel

manner.[69] This and the foregoing have suggested that the embrace of the male has a nutritive as well as a sexual function.

It has already been pointed out that trematodes excrete lipids; this seems a rather profligate custom, since it appears that they cannot synthesize their own complex lipids. Meyer, Meyer, and Bueding[77] compared the synthetic capacity of *S. mansoni* with that of the free-living planarian *Dugesia dorotocephala* and found that neither flatworm could synthesize fatty acids or sterols de novo. Both can synthesize their complex lipids provided that they are supplied with a source of long-chain fatty acids. Meyer and coworkers noted that of the Platyhelminthes studied so far, including two species of cestodes, all lack the biosynthetic pathways for all three classes of lipids: sterols, saturated fatty acids, and unsaturated fatty acids. If this deficiency is common to the entire phylum, it may be a factor predisposing the phylum's members to a symbiotic way of life.

Biochemistry of the tegument

Recognition that the tegument of schistosomes represents their barrier of defense against the host has led in recent years to much investigation of its structure and chemistry. This research has shown that the tegumental surface is active and complex. The heptalaminate structure of the tegument in schistosomes has already been mentioned (p. 249). The vesicles and granules in the distal cytoplasm appear to replace the outer membranes continuously, and turnover is quite rapid.[23] A variety of carbohydrates, including mannose, glucose, galactose, *N*-acetyl glucosamine, *N*-acetyl galactosamine, and sialic acid, are exposed on the surface. There are receptors for both host antigens (e.g., blood group antigens) and antibodies, including IgG, IgA, and IgM. At least some of the antibodies that bind to the schistosome surface are not anti-schistosome antibodies; therefore these antibodies are apparently binding to Fc receptors. A number of enzymes have been detected, including alkaline phosphatase, ATPase, alkaline phosphodiesterase, glycerol triphosphatase, glucose-6-phosphatase, and a protease that can cleave the IgG bound there.

It appears that the aggregate of host molecules bound on the tegument, plus the rapid turnover of tegumental membrane, effectively shields the schistosome from the host's immune defenses; and chemicals that could disrupt the integrity of the membrane could allow host immune effectors to recognize the worm. Such an action appears to be important in the effect of praziquantel, a drug that is highly effective against many flatworms. In addition, praziquantel affects permeability to calcium ions, allowing a rapid influx and resulting in a muscular tetany. Some evidence suggests that both the effects on Ca^{++} metabolism and on tegumental structure are necessary for the lethal effect of the drug.[5] Praziquantel is not effective against *Fasciola hepatica,* possibly because the tegument of *Fasciola* is much thicker than that of *Schistosoma.* Also, tetanic contraction of *Fasciola* in vitro requires 100 times greater concentration of praziquantel than that required for *Schistosoma.*[5]

PHYLOGENY OF DIGENETIC TREMATODES

Numerous schemes have been suggested for the origin of the Digenea. Various authors have derived the ancestral form from Monogenea, Aspidogastrea, and even insects.[87,101] However, most authorities today believe that digenetic trematodes evolved from free-living turbellarians, probably rhabdocoels or rhabdocoel-like ancestors.[44] Whatever the ancestral digenean, any system of their phylogeny must rationalize the evolution of their complex life cycles in terms of natural selection, a most perplexing task.

Digeneans display much more host specificity to their molluscan hosts than to their vertebrate hosts. This may imply that they established themselves as parasites of molluscs first and then added a vertebrate host as a later adaptation. It is not difficult to imagine a small, rhabdocoel-like worm, which feeds on soft bodied invertebrates, invading the mantle cavity of a mollusc and feeding on its tissues. In fact the main hosts of known endocommensal rhabdocoels are molluscs and echinoderms. The stage of the protodigenean in the mollusc was probably a developmental one, with a free-living, sexually reproducing adult. There are several reasons for believing this. First, a free reproductive stage would be of selective value in dispersion and transferral to new hosts. Precisely this "life-style" is shown by *Fecampia,* a rhabdocoel symbiont of various marine crustaceans. Second, the possession of a cercarial stage is surprisingly ubiquitous among digeneans, and most of these are adapted for swimming. Those without tails show evidence that the structure has been secondarily lost.

If one grants that the present adult represents the ancestral adult, which was free living, it is clear that additional (asexual) multiplication in the mollusc would have been advantageous also, and the alternation of the two reproductive generations could have been established. It is likely that such free-living adults would often be eaten by fish, and individuals in the population that could survive and maintain themselves in the fish's digestive tract for a period of time

would have selective advantage in extending their reproductive life. *Fecampia,* for example dies after depositing its eggs.

Further evidence that the parasite was originally free living as an adult is demonstrated by the fluke still having to leave the snail to infect the next host. With few exceptions, the fluke is incapable of infecting a definitive host while still in its first intermediate host, even if eaten. In most cases when a life cycle requires that the fluke be eaten within a mollusc, it has left its first host and penetrated a second to become infective. Some workers, however, feel that the protodigenean adult was a parasite of molluscs, as in modern aspidogastreans.[44,91]

It is not unlikely that the miracidium represents the larval form of the fluke's ancestor; all digeneans still have them, even though they are not all now free swimming.

Once the basic two-host cycle with two reproducing generations was established in the protodigenean, it became less difficult to visualize how further elaborations of the life cycle could have been selected for. The ecological value of the metacercaria, already mentioned, may have had some protective function as well. The mesocercaria of *Alaria* has a clear value in that the definitive host normally does not feed on the second intermediate host. Many other examples could be cited.

It may be assumed that the digenean adaptation to vertebrate hosts has occurred relatively recently. Digenetic trematodes are very common in members of all classes of vertebrates except Chondrichthyes; extremely few species of digeneans are found in sharks and rays. The urea in the tissue of most elasmobranchs, which plays such an important role in their osmoregulation, is quite toxic to the flukes on which it has been tested. It is supposed that elasmobranchs did not have digenean parasites when that particular osmoregulatory adaptation was evolved and that the urea has since proved a barrier to invasion of the elasmobranch habitat by flukes. The situation is quite the opposite with the cestodes; sharks and rays have a rich tapeworm fauna, and their cestodes either tolerate the urea or degrade it.[89]

CLASSIFICATION OF SUBCLASS DIGENEA

The most widely accepted system of classification of the higher taxa has been based on that of LaRue[67], using characteristics of the excretory bladder (superorders Anepitheliocystidia and Epitheliocystidia). Due to doubts cast upon this system based on ultrastructural studies,[44] we will not here divide the Digenea at the superorder level. The lists of families are fairly complete, but not all specialists will agree with their placement.

Subclass Digenea

Hermaphroditic, dorsoventrally flattened to cylindrical, elongate, or oval worms with a syncytial tegument; ventral adhesive organ variously placed posterior to mouth, sometimes lacking; oral sucker usually present; sclerotized structures lacking in ventral or oral sucker; eyes usually lacking in adults; gut usually with two simple or branched stems, occasionally rejoining posteriorly; vitelline follicles usually lateral; osmoregulatory canals joining a common excretory bladder; bladder opening through single pore at posterior end; sexually reproducing adults alternating with asexually reproducing larval forms; adults endoparasitic in vertebrates, rarely ectoparasitic, rarely in invertebrates; larval forms in invertebrates, usually gastropod molluscs.

Order Strigeata

Cercariae with forked tail, usually with two suckers, miracidium usually with two pairs of protonephridia.

Superfamily Strigeoidea

Cercaria with thin tail bearing two long rami, pharynx present; adult body usually divided by constriction into anterior and posterior portions; accessory suckers and/or penetration glands often present on anterior portion of body; genital pore usually terminal; parasites of reptiles, birds, and mammals.

FAMILIES

Bolbocephalidae, Brauninidae, Cyathocotylidae, Diplostomatidae, Proterodiplostomatidae, Strigeidae

Superfamily Clinostomatoidea

Cercaria with short rami on tail; oral sucker replaced with protractile penetration organ, ventral sucker rudimentary; pigmented eyespots present; parasites of reptiles, birds, and mammals.

FAMILY

Clinostomatidae

Superfamily Schistosomatoidea

Cercaria with short caudal rami; pharynx absent, oral sucker replaced with protractile penetration organ; eyespots pigmented or not; adults in vascular system of definitive host; no second intermediate host in life cycle; parasites of fishes, reptiles, birds, and mammals.

FAMILIES

Aporocotylidae, Sanguinicolidae, Schistosomatidae, Spirorchiidae

Superfamily Azygioidea

Cercaria furcocystocercous; oral sucker present, ventral sucker sometimes absent; adults often very large; parasites of fishes.

FAMILIES

Aphanhysteridae, Azygiidae, Bivesiculidae

Superfamily Transversotrematoidea

Cercaria with short rami; base of tail with two appendages; pharynx absent; adult transversely elongate, intestine fused at distal ends; parasites of the dermis of fishes.

FAMILY

Transversotrematidae

Superfamily Cyclocoeloidea

Cercaria with short, bilobed tail, or tail absent; adults elongate, flat; acetabulum absent or rudimentary; oral sucker present or absent; in- testine united at posterior end of body; parasites of the respiratory system of birds.

FAMILLY

Cyclocoelidae

Superfamily Brachylaemoidea

Cercaria with or without tail, usually developing in terrestrial snails; genital pore postequatorial; parasites of amphibians, birds, and mammals.

FAMILIES

Brachylaemidae, Harmotrematidae, Leucochloridiidae, Liolopidae, Ovariopteridae, Thapariellidae

Superfamily Fellodistomatoidea

Cercaria furcocerous, developing in marine bivalves; adults plump; ceca sometimes united posteriorly; parasites of marine fishes.

FAMILIES

Fellodistomatidae, Maseniidae, Monodhelminthidae

Superfamily Bucephaloidea

Mouth midventral in both cercaria and adult; tail of cercaria very short, with two very long rami; acetabulum absent; parasites of fishes and amphibians.

FAMILIES

Bucephalidae, Sinicovothylacidae

Order Echinostomata

Miracidium with single pair of protonephridia; cercaria with simple tail.

Superfamily Echinostomatoidea

Cercaria and adult often with circumoral collar usually armed with spines; parasites of reptiles, birds, and mammals.

FAMILIES

Balfouridae, Campulidae, Cathemaisiidae, Echinostomatidae, Fasciolidae, Haplosplanchnidae, Philophthalmidae, Psilotrematidae, Rhopaliasidae, Rhytidodidae

Superfamily Paramphistomoidea

Cercaria without penetration glands; monostomes or amphistomes; adults always amphistomes; no second intermediate host in life cycle; adults parasites of fishes, amphibians, reptiles, birds, and mammals.

FAMILIES

Angiodictyidae, Gastrodiscidae, Gastrothylacidae, Heronimidae, Paramphistomidae, Mesometridae

Superfamily Notocotyloidea

Cercaria monostomous, without pharynx; adults monostomous, occasionally with tegumental glands on ventral surface; parasites of birds and mammals.

FAMILIES

Notocotylidae, Pronocephalidae, Rhabdiopoeidae

Order Plagiorchiata

Superfamily Plagiorchioidea

Cercaria typical distomes, with oral stylet; metacercaria usually in invertebrates; parasites of fishes, amphibians, reptiles, birds, and mammals.

FAMILIES

Anchitrematidae, Batrachotrematidae, Brachycoelidae, Cephalogonimidae, Dicrocoeliidae, Dolochoperidae, Echinoporidae, Eucotylidae, Haematoloechidae, Haplometridae, Lecithodendriidae, Lissorchiidae, Macroderidae, Macroderoididae, Mesotretidae, Microphallidae, Ochetosomatidae, Omphalometridae, Pachypsolidae, Plagiorchiidae, Plectognathrematidae, Prosthogonimidae, Stomylotrematidae, Urotrematidae

Superfamily Allocreadioidea

Cercariae of many varied types, usually with eyespots; adults also variable, with or without eyespots; oral sucker usually simple, but occasionally with appendages; acetabulum in anterior half of body; testes always in hindbody; ovary almost always pretesticular; parasites of fishes, amphibians, reptiles, and mammals.

FAMILIES

Acanthocolpidae, Allocreadiidae, Apocreadiidae, Collyriclidae, Gekkonotrematidae, Gorgocephalidae, Homalometridae, Lepocreadiidae, Megaperidae, Monorchiidae, Octotestidae, Opecoelidae, Opistholebetidae, Gorgoderidae, Glyauchenidae, Schistorchiidae, Tetracladiidae, Troglotrematidae, Zoogonidae

Order Opisthorchiata

Cercaria with excretory vessels in tail; oral stylet never present.

Superfamily Isoparorchioidea

Adults parasitic in swim bladder of fish; body large, usually plump; testes postacetabular; vitellaria posterior.

FAMILIES

Aerobiotrematidae, Albulatrematidae, Cylindrorchiidae, Dictysarcidae, Isopharochiidae, Pelorohelminthidae, Tetrasteridae

Superfamily Opisthorchioidea

Cercaria with well-developed penetration glands; oral sucker protractile, ventral sucker rudimentary; tail variable; parasites of fishes, amphibians, reptiles, birds, and mammals.

FAMILIES

Acanthostomidae, Cryptogonimidae, Heterophyidae, Opisthorchiidae

Superfamily Hemiuroidea

Adults: tegument usually unspined; tail appendage sometimes present; testes usually anterior to ovary; vitellaria usually postovarian; parasites of fishes.

FAMILIES

Accacoeliidae, Bathycotylidae, Botulidae, Dinuridae, Halipegidae, Hemiceridae, Hirudinellidae, Lampritrematidae, Lechithasteridae, Lechithochiriidae, Mabiaramidae, Ptychogonimidae, Sclerodistomidae, Syncoeliidae

FAMILIES OF UNCERTAIN RELATIONSHIP

The following families, although justifiably distinct, have unknown life cycles and ontogeny and thus cannot be aligned with major groups at this time:

Acanthocollaritrematidae, Achillurbainiidae, Atractotrematidae, Botulisaccidae, Braunotrematidae, Callodistomatidae, Cortrematidae, Diplangidae, Eumegacetidae, Haploporidae, Jubilariidae, Laterotrematidae, Lobatovetelliovariidae, Mesotretidae, Meristocotylidae, Moreauiidae, Nasitrematidae, Ommatobrephidae, Pholeteridae, Prostogonotrematidae, Sigmaperidae, Treptodemidae, Waretrematidae

REFERENCES

1. Aitken-Ander, P., and N.L. Levin. 1985. Occurrence of adult and developmental stages of *Protoeces maculatus* (Trematoda: Digenea) in the gastropod *Crepidula convexa*. Trans. Amer. Microsc. Soc. 104:250-260.

2. Barrett, J. 1976. Bioenergetics in helminths. In Van den Bossche, H., editor. Biochemistry of parasites and host-parasite relationships. Elsevier/North Holland Biomedical Press, Amsterdam.

3. Barrett, J. 1981. Biochemistry of parasitic helminths. University Park Press, Baltimore, 308 pp.

4. Barrett, J., G.C. Coles, and K.G. Simpkin. 1978. Pathways of acetate and propionate production in adult *Fasciola hepatica*. Int. J. Parasitol. 8:117-123.

5. Bennett, J.L., and D.P. Thompson. 1986. Mode of action of antitrematodal agents. In Campbell, W.C., and R.S. Rew (eds.). Chemotherapy of parasitic diseases. Plenum Press, New York. pp. 427-443.

6. Bock, D. 1986. *In vitro* excystment of the metacercaria of *Plagiorchis species 1* (Trematoda, Plagiorchiidae). Int. J. Parasit. 16:641-645.

7. Bogitsh, B.J. 1968. Cytochemical and ultrastructural observation on the tegument of the trematode *Megalodiscus temperatus*. Trans. Am. Microsc. Soc. 87:477-486.

8. Bogitsh, B.J. 1987. Further observations on eggshell formation in *Haematoloechus medioplexus* (Trematoda: Digenea). Trans. Am. Microsc. Soc. 106:373-378.

9. Brooker, B.E. 1972. The sense organs of trematode miracidia. In Canning, E.U., and C.A. Wright, editors. Behavioural aspects of parasite transmission. Linnean Society of London. Academic Press Inc. (London) Ltd., pp. 171-180.

10. Bueding, E. 1950. Carbohydrate metabolism of *Schistosoma mansoni*. J. Gen. Physiol. 33:475-495.

11. Bueding, E., and J. Bennett. 1972. Neurotransmitters in trematodes. In Van den Bossche, H., editor. Comparative biochemistry of parasites. Academic Press, Inc., New York, pp. 95-99.

12. Bueding, E., and J. Fisher. 1966. Factors affecting the inhibition of phosphofructokinase activity of *Schistosoma mansoni* by trivalent antimonials. Biochem. Pharmacol. 15:1197-1211.

13. Bueding, E., and J. Fisher. 1970. Biochemical effects of niridazole on *Schistosoma mansoni*. Mol. Pharmacol. 6:532-539.

14. Bundy, D.A.P. 1981. Periodicity in the hatching of digenean eggs: a possible circadian rhythm in the life-cycle of *Transversotrema patialense*. Parasitology 83:13-22.

15. Burren, C.H., I. Ehrlich, and P. Johnson. 1967. Excretion of lipids by the liver fluke (*Fasciola hepatica* L). Lipids 2:353-356.

16. Burton, P.R. 1960. Gametogenesis and fertilization in the frog lung fluke, *Haematoloechus medioplexus* Stafford (Trematoda: Plagiorchiidae). J. Morphol. 107:92-122.

17. Cable, R.M. 1971. Parthenogenesis in parasitic helminths. Am. Zool. 11:267-272.

18. Cardell, R.R., Jr. 1962. Observations on the ultrastructure of the body of the cercaria of *Himasthla quissetensis* (Miller and Northup, 1926). Trans. Am. Microsc. Soc. 81:124-131.

19. Cheah, K.S., and R.K. Prichard. 1975. The electron transport systems of *Fasciola hepatica* mitochondria. Int. J. Parasitol. 5:183-186.

20. Christensen, N.O. 1980. A review of the influence of host-

and parasite-related factors and environmental conditions on the host-finding capacity of the trematode miracidium. Acta Trop. 37:303-318.

21. Clegg, J.A. 1972. The schistosome surface in relation to parasitism. In Taylor, A.E.R., and R. Muller, editors. Functional aspects of parasite surfaces, vol. 10. Blackwell Scientific Publications Ltd., Oxford, Eng., pp. 23-40.

22. Cohen, L.M., H. Neimark, and L.K. Eveland. 1980. *Schistosoma mansoni:* response of cercariae to a thermal gradient. J. Parasitol. 66:362-364.

23. Coles, G.C. 1984. Recent advances in schistosome biochemistry. Parasitology 89:603-637.

24. Coles, G.C., K.G. Simpkin, and J. Barrett. 1980. *Fasciola hepatica:* energy sources and metabolism. Exp. Parasitol. 49:122-127.

25. Cordingley, J.S. 1987. Trematode eggshells: novel protein biopolymers. Parasitol. Today 3:341-344.

26. Cornford, E.M., and M.E. Huot. 1981. Glucose transfer from male to female schistosomes. Science 213:1269-1271.

27. Cort, W.W. 1944. The germ cell cycle in the digenetic trematodes. Q. Rev. Biol. 19:275-284.

28. Cort, W.W., D.J. Ameel, and A. Van der Woude. 1954. Parasitological reviews—germinal development in the sporocysts and rediae of the digenetic trematodes. Exp. Parasitol. 3:185-225.

29. Daugherty, J.W. 1952. Intermediary protein metabolism in helminths. I. Transaminase reactions in *Fasciola hepatica.* Exp. Parasitol. 1:331-338.

30. Dike, S.C. 1967. Ultrastructure of the ceca of the digenetic trematodes *Gorgodera amplicava* and *Haematoloechus medioplexus.* J. Parasitol. 53:1173-1185.

31. Dike, S.C. 1969. Acid phosphatase activity and ferritin incorporation in the ceca of digenetic trematodes. J. Parasitol. 55:111-123.

32. Dönges, J. 1969. Entwicklungs-und Lebensdauer von Metacercarien. Z. Parasitenkd. 31:340-366.

33. Dönges, J. 1970. Transplantation of rediae—a device for solving special problems in trematodology. J. Parasitol. 54:82-83.

34. Dunn, T.S., R.E.B. Hanna, and W.A. Nizami. 1987. Sensory receptors of the miracidium of *Gigantocotyle explanatum* (Trematoda: Paramphistomidae). Int. J. Parasit. 17:1131-1140.

35. Dunn, T.S., R.E.B. Hanna, and W.A. Nizami. 1987. Ultrastructural and cytochemical observations on the tegument of three species of paramphistomes (Platyhelminthes: Digenea) from the Indian water buffalo, *Bubalus bubalis.* Int. J. Parasit. 17:1153-1161.

36. Fashuyi, S.A. 1986. Excystment of the metacercaria of the trematode *Mesocoelium monodi.* Int. J. Parasit. 16:237-239.

37. Fetterer, R.H., R.A. Pax, and J.L. Bennett. 1981. Na^+-K^+ transport, motility and tegumental membrane potential in adult male *Schistosoma mansoni.* Parasitology 82:97-109.

38. Foster, L.A., and B.J. Bogitsh. 1986. Utilization of the heme moiety of hemoglobin by *Schistosoma mansoni* schistosomules *in vitro.* J. Parasit. 72:669-676.

39. Fried, B. 1986. Chemical communication in hermaphroditic digenetic trematodes. J. Chem. Ecol. 12:1659-1677.

40. Fried, B., and G.B. Ramundo. 1987. Excystation and cultiva-

tion *in vitro* and *in vivo* of *Cyathocotyle bushiensis* (Trematoda) metacercariae. J. Parasit. 73:541-545.

41. Fujino, T., and Y. Ishii. 1978. Comparative ultrastructural topography of the gut epithelia of the lung fluke *Paragonimus* (Trematoda: Troglotrematidae). Int. J. Parasitol. 8:139-148.

42. Fujino, T., and Y. Ishii. 1979. Comparative ultrastructural topography of the gut epithelia of some trematodes. Int. J. Parasitol. 9:435-448.

43. Fusco, A.C., B. Salafsky, and K. Delbrook. 1986. *Schistosoma mansoni:* production of cercarial eicosanoids as correlates of penetration and transformation. J. Parasit. 72:397-404.

44. Gibson, D.I. 1987. Questions on digenean systematics and evolution. Parasitology 95:429-460.

45. Granzer, M., and W. Haas. 1986. The chemical stimuli of human skin surface for the attachment response of *Schistosoma mansoni* cercariae. Int. J. Parasit. 16:575-579.

46. Gönnert, R. 1962. Histologische Untersuchungen über den Feinbau der Eibildungsstatte (Oogenotop) von *Fasciola hepatica.* Z. Parasitenkd. 21:475-492.

47. Granzer, M., and W. Haas. 1986. The chemical stimuli of human skin surface for the attachment response of *Schistosoma mansoni* cercariae. Int. J. Parasit. 6:575-579.

48. Haas, W., and R. Schmitt. 1982. Characterization of chemical stimuli for the penetration of *Schistosoma mansoni* cercariae. I. Effective substances, host specificity. Z. Parasitenkd. 66:293-307.

48a. Haas, W., and R. Schmitt. 1982. Characterization of chemical stimuli for the penetration of *Schistosoma mansoni* cercariae. II. Conditions and mode of action. Zeitschr. Parasitenkd. 66:309-319.

49. Halton, D.W. 1967. Observations on the nutrition of digenetic trematodes. Parasitology 57:639-660.

50. Higgins, J.C. 1979. The role of the tegument of the metacercarial stage of *Bucephalus haimeanus* (Lacaze-Duthiers, 1854) in the absorption of particulate material and small molecules in solution. Parasitology 78:99-106.

51. Hochachka, P.W., and T. Mustafa. 1972. Invertebrate facultative anaerobiosis. Science 178:1056-1060.

52. Hockley, D.J. 1972. *Schistosoma mansoni:* the development of the cercarial tegument. Parasitology 64:245-252.

53. Hockley, D.J. 1973. Ultrastructure of the tegument of *Schistosoma.* In Dawes, B., editor. Advances in parasitology, vol. 2. Academic Press, Inc., New York, pp. 233-305.

54. Hockley, D.J., and D.J. McLaren. 1973. *Schistosoma mansoni:* changes in the outer membrane of the tegument during development from cercaria to adult worm. Int. J. Parasitol. 3:13-25.

55. Irwin, S.W.B., L.T. Threadgold, and N.M. Howard. 1978. *Cryptocotyle lingua* (Creplin) (Digenea: Heterophyidae): observations on the morphology of the redia, with special reference to the birth papilla and release of cercariae. Parasitology 76:193-199.

56. Isseroff, H., and C.P. Read. 1969. Studies on membrane transport. VI. Absorption of amino acids by fascioliid trematodes. Comp. Biochem. Physiol. 30:1153-1159.

57. James, B.L. 1964. The life cycle of *Parvatrema homoeotecnum* sp. nov. (Trematoda; Digenea) and a review of the family Gymnophallidae Morozov, 1955. Parasitology 54:1-41.

58. Janssens, P.A., and C. Bryant. 1969. The ornithine-urea cycle in some parasitic helminths. Comp. Biochem. Physiol. 30:261-272.

59. Kemp, W.M., P.R. Brown, S.C. Merritt, and R.E. Miller. 1980. Tegument-associated antigen modulation by adult male *Schistosoma mansoni*. J. Immunol. 124:806-811.

60. Khalil, G.M., and R.M. Cable. 1968. Germinal development in *Philophthalmus megalurus* (Cort, 1914) (Trematoda: Digenea). Z. Parasitenkd. 31:211-231.

61. Kirschner, K., and W.J. Bacha, Jr. 1980. Excystment of *Himasthla quissetensis* (Trematoda: Echinostomatidae) metacercariae in vitro. J. Parasit. 66:263-267.

62. Köhler, P., C. Bryant, and C.A. Behm. 1978. ATP synthesis in a succinate decarboxylase system from *Fasciola hepatica* mitochondria. Int. J. Parasitol. 8:399-404.

63. Köhler, P., and D.F. Stahel. 1972. Metabolic end products of anaerobic carbohydrate metabolism of *Dicrocoelium dendriticum* (Trematoda). Comp. Biochem. Physiol. 43B:733-741.

64. Krupa, P.L., A.K. Bal, and G.H. Cousineau. 1967. Ultrastructure of the redia of *Cryptocotyle lingua*. J. Parasitol. 53:725-734.

65. Kurelec, B. 1972. Lack of carbamyl phosphate synthesis in some parasitic platyhelminths. Comp. Biochem. Physiol. 43B:769-780.

66. Lane, C.A., R.A. Pax, and J.L. Bennett. 1987. L-Glutamine: an amino acid required for maintenance of the tegumental membrane potential of *Schistosoma mansoni*. Parasitology 94:233-242.

67. LaRue, G.R. 1957. The classification of digenetic Trematoda: a review and a new system. Exp. Parasitol. 6:306-349.

68. Leitch, B., A.J. Probert, and N.W. Runham. 1984. The ultrastructure of the tegument of adult *Schistosoma haematobium*. Parasitology 89:71-78.

69. Lennox, R.W., and E.L. Schiller. 1972. Changes in dry weight and glycogen content as criteria for measuring the postcercarial growth and development of *Schistosoma mansoni*. J. Parasitol. 58:489-494.

70. Lie, K.J., D. Heyneman, and N. Kostanian. 1975. Failure of *Echinostoma lindoense* to reinfect snails already harboring that species. Int. J. Parasitol. 5:483-486.

71. Lloyd, G.M. 1986. Energy metabolism and its regulation in the adult liver fluke *Fasciola hepatica*. Parasitology 93:217-248.

72. Llewellyn, J. 1965. The evolution of parasitic platyhelminths. In Taylor, A., editor. Evolution of parasites. Blackwell Scientific Publications Ltd., Oxford, Eng., pp. 47-78.

73. Lumsden, R.D. 1975. Surface ultrastructure and cytochemistry of parasitic helminths. Exp. Parasitol. 37:267-339.

74. Lumsden, R.D., and R. Specian. 1980. The morphology, histology, and fine structure of the adult stage of the cyclophyllidean tapeworm *Hymenolepis diminuta*. In Arai, H.P., editor. Biology of the tapeworm *Hymenolepis diminuta*. Academic Press, Inc., New York, pp. 157-280.

75. Mansour, T.E. 1967. Effect of hormones on carbohydrate metabolism of invertebrates. Fed. Proc. 26:1179-1185.

76. Matthews, B.F. 1981. *Cercaria vaullegeardi* Pelseneer, 1906 (Digenea: Hemiuridae); the infection mechanism. Parasitology 83:587-593.

77. Meyer, F., H. Meyer, and E. Bueding. 1970. Lipid metabolism in the parasitic and free-living flatworms, *Schistosoma mansoni* and *Dugesia dorotocephala*. Biochem. Biophys. Acta 210:257-266.

78. Miller, F.H., Jr, G.S. Tulloch, and R.E. Kuntz. 1972. Scanning electron microscopy of integumental surface of *Schistosoma mansoni*. J. Parasitol. 58:693-698.

79. Mills, C.A. 1979. The influence of differing ionic environments on the cercarial, post-cercarial and adult stages of the ectoparasitic digenean *Transversotrema patialense*. Int. J. Parasitol. 9:603-608.

80. Nollen, P.M. 1968. Autoradiographic studies on reproduction in *Philophthalmus megalurus* (Cort, 1914) (Trematoda). J. Parasitol. 54:43-48.

81. Nollen, P.M. 1968. Uptake and incorporation of glucose, tyrosine, leucine, and thymidine by adult *Philophthalmus megalurus* (Cort, 1914) (Trematoda), as determined by autoradiography. J. Parasitol. 54:295-304.

82. Nollen, P.M. 1983. Patterns of sexual reproduction among parasitic platyhelminths. Symposia of the British Society for Parasitology, vol. 20. Parasitology 86(4):99-120

83. Nollen, P.M., A.L. Restaino, and R.A. Alberico. 1973. *Gorgoderina attenuata:* uptake and incorporaton of tyrosine, thymidine, and adenosine. Exp. Parasitol. 33:468-476.

84. Pappas, P.W. 1971. *Haematoloechus medioplexus:* uptake, localization, and fate of tritiated arginine. Exp. Parasitol. 30:102-119.

85. Pappas, P.W., and C.P. Read. 1975. Membrane transport in helminth parasites: a review. Exp. Parasitol. 37:469-530.

86. Parkening, T.A., and A.D. Johnson. 1969. Glucose uptake in *Haematoloechus medioplexus* and *Gorgoderina* trematodes. Exp. Parasitol. 25:358-367.

87. Pigulevskii, S.V. 1958. On the question of the phylogeny of flatworms. Rabot. Gelmintol. 80 Let. Skrjabin, 265-270.

88. Prichard, R.K., and P.J. Schofield. 1968. The glycolytic pathway in adult liver fluke, *Fasciola hepatica*. Comp. Biochem. Physiol. 24:697-710.

89. Read, C.P., L.T. Douglas, and J.E. Simmons, Jr. 1959. Urea and osmotic properties of tapeworms from elasmobranchs. Exp. Parasitol. 8:58-75.

90. Rogers, S.H., and E. Bueding. 1975. Anatomical localization of glucose uptake by *Schistosoma mansoni* adults. Int. J. Parasitol. 5:369-371.

91. Rohde, K. 1971. Phylogenetic origin of trematodes. Parasitol. Schrift. 21:17-27.

92. Rowcliffe, S.A., and C.B. Ollerenshaw. 1960. Observations on the bionomics of the egg of *Fasciola hepatica*. Ann. Trop. Med. Parasitol. 54:172-181.

93. Salafsky, B., Y.-S. Wang, A.C. Fusco, and J. Antonacci. 1984. The role of essential fatty acids and prostaglandins in cercarial penetration *(Schistosoma mansoni)*. J. Parasit. 70:656-660.

94. Salafsky, B., Y.-S. Wang, M.B. Kevin, H. Hill, and A.C. Fusco. 1984. The role of prostaglandins in cercarial *(Schistosoma mansoni)* response to free fatty acids. J. Parasit. 70:584-591.

95. Sauer, M.C.V., and A.W. Senft. 1972. Properties of a proteolytic enzyme from *Schistosoma mansoni*. Comp. Biochem. Physiol. 42B:205-220.

96. Senft, A.W. 1963. Observations on amino acid metabolism of

Schistosoma mansoni in a chemically defined medium. Ann. N.Y. Acad. Sci. 113:272-288.

97. Senft, A.W. 1968. Studies in proline metabolism by *Schistosoma mansoni*. I. Radioautography following in vitro exposure to radioproline C[14]. Comp. Biochem. Physiol. 27:251-261.

98. Senft, A.W., R.P. Miech, P.R. Brown, and D.G. Senft. 1972. Purine metabolism in *Schistosoma mansoni*. Int. J. Parasitol. 2:249-260.

99. Senft, A.W., D.E. Philpott, and A.H. Pelofsky. 1961. Electron microscope observations of the integument, flame cells and gut of *Schistosoma mansoni*. J. Parasitol. 47:217-229.

100. Shannon, W., Jr., and B.J. Bogitsch. 1971. *Megalodiscus temperatus:* comparative radioautography of glucose-^{3}H and galactose-^{3}H incorporation. Exptl. Parasitol. 29:309-319.

101. Sinitsin, D. 1931. Studien über die Phylogenie der Trematoden. IV. The life histories of *Plagioporus silicus* and *Plagioporus virens,* with special reference to the origin of Digenea. Z. Wiss. Zool. 138:409-456.

102. Smyth, J.D., and D.W. Halton. 1983. The physiology of trematodes, ed. 2. Cambridge University Press, Cambridge. 446 p.

103. Sobhon, P., E.S. Upatham, and D.J. McLaren. 1984. Topography and ultrastructure of the tegument of adult *Schistosoma mekongi*. Parasitology 89:511-521.

104. Spellman, S.J., and A.D. Johnson. 1987. *In vitro* excystment of the black spot trematode *Uvulifer ambloplitis* (Trematoda: Diplostomatidae). Int. J. Parasit. 17:897-902.

105. Stirewalt, M.A. 1974. *Schistosoma mansoni:* cercaria to schistosomule. In Dawes, B., editor. Advances in parasitology, vol. 12. Academic Press, Inc., New York, pp. 115-182.

106. Tielens, A.G.M., P. van der Meer, and S.G. van den Bergh. 1981. The aerobic energy metabolism of the juvenile *Fasciola hepatica*. Mol. Biochem. Parasitol. 3:205-214.

107. Torpier, G., M. Hirn, P. Nirde, M. de Reggi, and A. Capron. 1982. Detection of ecdysteroids in the human trematode, *Schistosoma mansoni*. Parasitology 84:123-130.

108. Trager, W. 1986. Living together. The biology of animal parasitism. Plenum Press, New York. 467 pp.

109. Turner, H.M. 1984. Orientation and pathology of *Allocorrigia filiformis* (Trematoda: Dicroeliidae) from the antennal glands of the crayfish *Procambarus clarkii*. Trans. Amer. Microsc. Soc. 103:434-437.

110. Uglem, G.L., and O.R. Larson. 1987. Facilitated diffusion and active transport systems for glucose in metacercariae of *Clinostomum marginatum* (Digenea). Int. J. Parasit. 17:847-850.

111. Vande Vusse, F.J., T.D. Fish, and M.P. Neumann. 1981. Adult Digenea from upper midwest hirudinid leeches. J. Parasitol. 67:717-720.

112. Vernberg, W.B., and F.J. Vernberg. 1971. Respiratory metabolism of a trematode metacercaria and its host. In Cheng, T.C., editor. Aspects of the biology of symbiosis. University Park Press, Baltimore, pp. 91-102.

113. Ward, P.F.V. 1982. Aspects of helminth metabolism. Parasitology 84:177-194.

114. Watts, S.D.M. 1970. Transamination in homogenates of rediae of *Cryptocotyle lingua* and of sporocysts of *Cercaria emasculans* Pelseneer, 1900. Parasitology 61:499-504.

115. Wesenberg-Lund, C. 1931. Biology of *Leucochloridium*. Kongl. Danske Vidensk. Selsk. Skr. Nat. Math. Afd., ser. 9, 4:89-142.

116. Wharton, D.A. 1983. The production and function morphology of helminth egg-shells. Symposia of the British Society for Parasitology, vol. 20. Parasitology 86(4):85-97.

117. Whitfield, P.J. and N.A. Evans. 1983. Parthenogenesis and asexual multiplication among parasitic platyhelminths. Parasitology 86:121-160.

118. Wilson, R.A. 1968. The hatching mechanism of the egg of *Fasciola hepatica* L. Parasitology 58:79-89.

119. Wilson, R.A. 1969. Fine structure of the tegument of the miracidium of *Fasciola hepatica* L. J. Parasitol. 55:124-133.

120. Xu, Y., and M.H. Dresden. 1986. Leucine aminopeptidase and hatching of *Schistosoma mansoni* eggs. J. Parasit. 72:507-511.

121. Yamaguti, S. 1971. Synopsis of digenetic trematodes of vertebrates, vol. 1. Keigaku Publishing Co., Tokyo.

SUGGESTED READINGS

Baer, J.G. and C. Joyeux. 1961. Classe des Trématodes (Trématoda Rudolphi). In Grassé, P.P., editor. Traité de zoologie: anatomie, systématique, biologie, vol. 4, part I. Plathelminthes, Mésozoaires, Acanthocéphales, Némertiens. Masson & Cie, Paris, pp. 561-692. (A well-illustrated overview of trematodes.)

Barrett, J. 1981. Biochemistry of parasitic helminths. University Park Press, Baltimore, 308 pp.

Buttner, A. 1951. La Progénèse chez les trematodes digentiques, sa signification, ses manifestations. Contribution à l'étude de son determinism. Ann. Parasitol. 25:376-434.

Cable, R.M. 1972. Behaviour of digenetic trematodes. Zool. J. Linn. Soc. 51(suppl. 1):1-18.

Dawes, B. 1946. The Trematoda, with special reference to British and other European forms. Cambridge University Press, Cambridge, Eng. (A classic reference work, of value to all interested in trematodes.)

Gibson, E.I. 1987. Questions on digenean systematics and evolution. Parasitology 95:429-460.

Hyman, L.H. 1951. The invertebrates, vol. 2. Platyhelminthes and Rhynchocoela. The acoelomate Bilateria. McGraw-Hill Book Co., New York. (A standard reference to all aspects of Trematoda.)

Lloyd, G.M. 1986. Energy metabolism and its regulation in the adult liver fluke *Fasciola hepatica*. Parasitology 93:217-248.

Schell, S.C. 1985. Handbook of trematodes of North America north of Mexico. University Press of Idaho, Moscow, Idaho.

Smyth, J.D., and D.W. Halton. 1983. The physiology of trematodes, ed. 2. Cambridge University Press, Cambridge. 446 p.

Trager, W. 1986. Living together. The biology of animal parasitism. Plenum Press, New York. 467 pp.

CHAPTER 18

ORDER STRIGEATA

Of the several superfamilies in this order, only two, Strigeoidea and Schistosomatoidea, are of much economic or medical significance. The latter, however, contains some of the most important disease agents of humans.

SUPERFAMILY STRIGEOIDEA

Strigeoidea are bizzare in appearance, with their bodies divided into two portions (Fig. 18-1). The anterior portion usually is spoon or cup shaped, with accessory **pseudosuckers** on each side of the oral sucker. Behind the acetabulum is a spongy, padlike organ referred to as the **adhesive** or **tribocytic organ.** This structure secretes proteolytic enzymes that digest host mucosa, probably functioning both as an accessory holdfast and as a digestive-absorptive organ. The hindbody contains most of the reproductive organs, although vitelline follicles often extend into the forebody. The genital pore is located at the posterior end.

Most strigeoids are quite small and are found commonly in the digestive tracts of fish-eating vertebrates. Their cercariae are easily recognized by the fact that they have both a pharynx and a forked tail. No adult strigeoids are known to parasitize humans, but they are so ubiquitous and their biology so interesting that we will briefly consider a few species.

Family Diplostomatidae

■ *Alaria americana*

The genus *Alaria* contains several very similar species, all of which mature in the small intestines of carnivorous mammals. *Alaria americana* is found in various species of Canidae in northern North America. They are about 2.5 to 4 mm long, with the forebody longer than the hindbody. The forebody has a pair of ventral flaps that are narrowest at the anterior end (Fig. 18-2 *A*). A pointed process flanks each side of the oral sucker. The tribocytic organ is relatively large and elongate and has a ventral depression in its center.

The life cycles of *Alaria* spp. are remarkable in that the worms may require four hosts before they can develop to maturity (Fig. 18-2). The eggs are unembryonated when laid, and they hatch in about 2 weeks. The miracidium swims actively and will attack and penetrate any of several species of planorbid snails.[27] Mother sporocysts develop in the renal veins and produce daughter sporocysts in about 2 weeks. Daughter sporocysts migrate to the digestive gland and need about a year to mature and begin producing cercariae. The furcocercous cercaria leaves the snail during daylight hours and swims to the surface, where it hangs upside down. Occasionally it sinks a short distance and then returns to the surface. If a tadpole swims by, the resulting water currents stimulate the cercaria to swim after it. If it contacts the tadpole, the cercaria will quickly attack, drop its tail, penetrate the skin, and begin wandering within the amphibian. In about 2 weeks the cercaria has transformed into a **mesocercaria,** an unencysted form between a cercaria and a metacercaria. It is then infective to the next host, which may be the definitive host or a paratenic host. If a canid eats an infected tadpole or adult frog, the mesocercariae are freed by digestion, penetrate into the coelom, and then move to the diaphragm and lungs. After about 5 weeks in the lungs, the mesocercaria has transformed into a **diplostomulum metacercaria** (Fig. 18-2, *H*). Diplostomula migrate up the trachea and then to the intestine, where they mature in about a month.

Tadpoles, however, are not always available to terrestrial canids and, furthermore, are distasteful to all but the hungriest carnivores. This ecological barrier is overcome when a water snake eats the infected tadpole or frog and thereby becomes a paratenic host. A snake (or other animal) can accumulate large numbers of mesocercariae in its tissues, rendering a heavy infection to the definitive host when the animal is eaten. The mesocercariae then migrate, develop into diplostomula, and mature in the intestine, as do those from

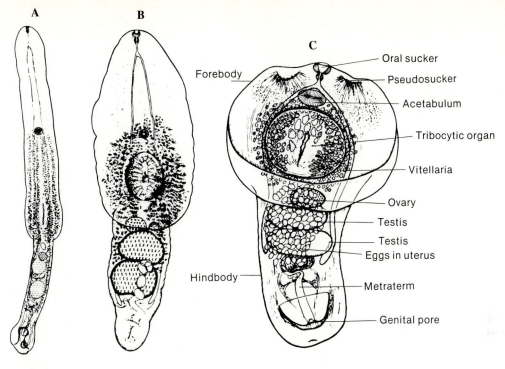

FIG. 18-1

Typical strigeoid trematodes, illustrating body forms. **A,** *Mesodiplostomum gladiolum* Dubois, 1936. **B,** *Pseudoneodiplostomum thomasi* (Dollfus, 1935). **C,** *Proalarioides serpentis* Yamaguti, 1933.

From Yamaguti, S. 1971. Synopsis of digenetic trematodes of vertebrates, vol. 2. Keigaku Publishing Co., Tokyo.

tadpoles. Life cycles of other species of *Alaria* are similar.

Mature *Alaria* spp. are quite pathogenic, causing severe enteritis that often kills the definitive hosts in severe infections. Also, the mesocercaria is pathogenic, especially when accumulated in large numbers. Fig. 18-3 is from a fatal infection of mesocercariae in a human.

Shoop and Corkum[32] demonstrated that a related species, *Alaria marcianae,* can be transmitted to a juvenile definitive host through the milk of its mother. In this species the parasite matures normally in the intestine of the adult male and nonlactating females but remains as diplostomula disseminated throughout the tissues of lactating females. In an experimental infection, a single cat infected 21 of her offspring via milk over the course of five litters and still harbored infective juveniles after 3 years.

■ *Uvulifer ambloplitis*

Several species of strigeoid trematodes cause black spots in the skin of fish, one of which is *Uvulifer am-* bloplitis. This is a parasite of the kingfisher, a fish-eating bird that is widely distributed across the United States. The spoon-shaped forebody of the parasite is separated from the longer hindbody by a slender constriction. Adults are 1.8 to 2.3 mm long.

The eggs, which are unembryonated when laid, hatch in about 3 weeks. The miracidium will penetrate snails of the genus *Helisoma* and will tranform into mother sporocysts that retain the eyespots of the first larva. Daughter sporocysts invade the digestive gland and produce cercariae in about 6 weeks. The cercariae escape from the tissues of the snail and rise to the surface of the water, where they are sensitive to the passing of fish. If they contact a fish, especially a centrarchid or percid, they drop their tails and penetrate the skin. Once inside the dermis, the flukes metamorphose into **neascus metacercariae** and secrete a delicate, hyaline cyst wall around themselves. A neascus is similar to a diplostomulum except that the forebody is spoon shaped, without anterolateral "points." The fish host responds to the neascus by laying down layers of melanin granules. The result is a conspicuous

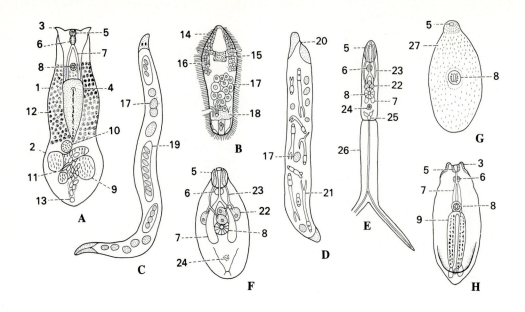

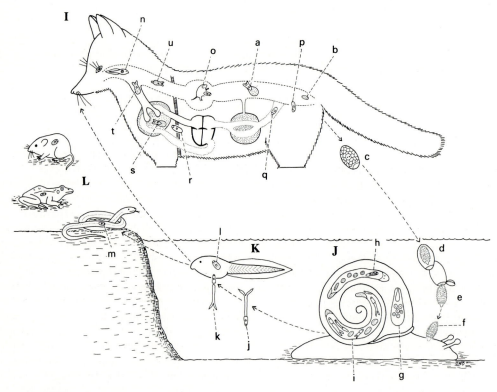

FIG. 18-2

For legend see opposite page.

FIG 18-2

Life cycle of *Alaria americana*. **A,** Ventral view of adult fluke. **B,** Miracidium. **C,** Mother sporocyst. **D,** Daughter sporocyst. **E,** Fork-tailed cercaria. **F,** Mesocercaria showing some internal organs. **G,** External view of mesocercaria. **H,** Diplostomulum metacercaria. **I,** Fox definitive host. **J,** Planorbid snail *(Helisoma)*, first intermediate host. **K,** Tadpole *(Rana, Bufo)*, second intermediate host. **L,** Paratenic hosts (snakes, frogs, mice). *1,* Forebody; *2,* hindbody; *3,* lappet or pseudosucker; *4,* holdfast (tribocytic) organ; *5,* oral sucker; *6,* pharynx; *7,* cecum; *8,* ventral sucker; *9,* testes; *10,* ovary; *11,* eggs in uterus; *12,* vitelline glands; *13,* common genital pore; *14,* cilia; *15,* eyespot; *16,* flame cell; *17,* germ cells; *18,* excretory opening; *19,* daughter sporocyst with germ balls or developing cercariae; *20,* birth pore; *21,* cercaria; *22,* penetration glands; *23,* duct of penetration glands; *24,* genital primordium; *25,* excretory bladder; *26,* forked tail; *27,* body spines; *a,* adult fluke in small intestine; *b,* egg passing out of body in feces; *c,* unembryonated egg; *d,* embryonated egg; *e,* egg hatching; *f,* miracidium penetrating *Helisoma* snail; *g,* young mother sporocyst;, *h,* mature mother sporocyst; *i,* daughter sporocyst; *j,* cercaria free in water in characteristic resting position; *k,* cercaria penetrating tadpole, casting tail as it enters; *l,* mesocercaria; *m,* mesocercaria in snake, frog, and mouse paratenic hosts; *n,* infection of definitive host by swallowing tadpole, second intermediate host; *o,* infection of definitive host by swallowing infected paratenic host; *p,* mesocercariae migrate through gut wall into coelom; *q,* mesocercariae enter hepatoportal vein, but it has not been shown that they reach the lungs by way of the blood; *r,* mesocercariae pass through the diaphragm and penetrate the lungs; *s,* in the lungs the mesocercariae transform to a diplostomulum stage; *t,* diplostomulae migrate up trachea; *u,* diplostomulae are swallowed, go to small intestine, and develop to maturity *(a)* in 5 to 6 weeks.

From Olsen, O.W. 1974. Animal parasites, their life cycles and ecology. © 1974 University Park Press, Baltimore.

"black spot" indicating the presence of a metacercaria (Fig. 18-4). When such fish are heavily infected, they are often discarded as diseased by fishermen. Kingfishers become infected when they eat such a fish. The flukes mature in 27 to 30 days.

Other, related flukes also cause "black spot" in a wide variety of fish and have similar life cycles. When a neascus larva is encountered for which the adult genus is unknown, it is proper to refer it to the genus *Neascus*. This is also true for *Diplostomulum, Tetracotyle,* and *Cercaria*. In fact new species can be named in these genera; of course, when the adult form becomes known, the species reverts to the proper genus.

Family Strigeidae

■ *Cotylurus flabelliformis*

This is a common parasite of wild and domestic ducks in North America. Adult flukes are 0.55 to 1 mm long. The forebody is cup shaped, with the acetablum and tribocytic organ located at its depths. The hindbody is short and stout and is curved dorsad.

Adult worms live in the small intestines of ducks and lay unembryonated eggs. These hatch in about 3 weeks, and the miracidia attack snails of the family Lymnaeidae *(Lymnaea, Stagnicola)*. The mother sporocyst produces daughter sporocysts that migrate to the digestive gland and grow into slender, wormlike bodies. After about 6 weeks they begin to release furcocercous cercariae, which work their way free into the water. The cercariae are very active, and if they contact a snail of the same family Lymnaeidae, they penetrate, migrate to the ovotestis, and transform into **tetracotyle metacercariae.** The tetracotyle is similar to a diplostomulum except that it has an extensive system of excretory canals that often are filled with excretory products. The canals are called the **reserve bladder system.** When the snail is eaten by ducks, the flukes are released by digestion and mature in about 1 week.

On the other hand, if the cercaria enters a snail of the families Planorbidae or Physidae, it will attack sporocysts or rediae of other species of flukes already present and will develop into a tetracotyle metacer-

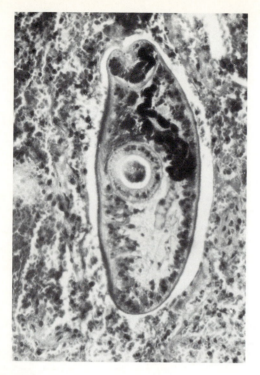

FIG. 18-3

Mesocercaria of *Alaria americana* in human lung biopsy. The case proved to be fatal, with nearly every organ of the body infected, presumably as a result of eating undercooked frog's legs.

From Freeman, R., et al. 1976. Am J. Trop. Med. Hyg. 25:803-807.

caria within them. This is an example of hyperparasitism among larval parasites. These too, will mature in about a week, if eaten by a duck.

• • •

Strigeoid trematodes, then, exhibit complex life cycles that involve several unrelated hosts. Their adaptability is amazing, when one considers the differences in environments provided by snails, pond water, fish, amphibians, reptiles, and birds or mammals. It is also remarkable that a parasite that may require more than a year to complete its larval development can become sexually mature in its definitive host in a week or less and die a few days later. Such is the pattern in the life cycles of the strigeoids.[28]

SUPERFAMILY SCHISTOSOMATOIDEA

Flukes of the superfamily Schistosomatoidea are peculiar in that they have no second intermediate host in their life cycles and also in that they mature in the blood vascular system of their definitive hosts. Most species are dioecious. The tantalizing puzzle of what advantage there is in being dioecious in a class of worms that is overwhelmingly monoecious is discussed by Popiel.[30] They are parasites of fishes, turtles, birds, and mammals throughout the world. Several species are parasites of humans, causing misery and death wherever they are distributed.[36] The families Aporocotylidae, Sanguinicolidae, and Spirorchidae parasitize fish and turtles and are of little economic importance. The family Schistosomatidae, however, includes species that are among the most

FIG. 18-4

A minnow, *Pimephales* sp., infected with neascus-type metacercariae, "black spot."

Photography by John S. Mackiewicz.

dreaded parasites of humans. To date, 10 species and 2 varieties of schistosomes have been reported in Africa. The taxonomy of this group and the validity of some species of parasites have been subjects of controversy for years; the final decision on species recognition awaits in-depth studies on the basic biology of schistosomes.

Family Schistosomatidae

■ *Schistosoma* species and schistosomiasis

Three species of schistosomes are of vast medical significance: *Schistosoma haematobium*, *S. mansoni*, and *S. japonicum*—all parasites of humans since antiquity. Bloody urine was a well-recognized disease symptom in northern Africa in ancient times. At least 50 references to this condition have been found in surviving Egyptian papyri, and calcified eggs of *S. haematobium* have been found in Egyptian mummies dating from about 1200 BC. Hulse[11] has presented a well-reasoned hypothesis that the curse that Joshua placed on Jericho can be explained by the introduction of *S. haematobium* into the communal well by the invaders. The removal of the curse occurred after the abandonment of Jericho and subsequent droughts had eliminated the snail host, *Bulinus truncatus*. Today Jericho (Ariha, Jordan) is well known for its fertile lands and healthy, well-nourished people.

The first Europeans to record contact with *S. haematobium* were surgeons with Napoleon's army in Egypt (1799-1801). They reported that **hematuria** (bloody urine) was prevalent among the troops, although the cause, of course, was unknown. Nothing further was learned about **schistosomiasis haematobia** for more than 50 years, until a young German parasitologist, Theodor Bilharz, discovered the worm that caused it. He announced his discovery in letters to his former teacher, Von Siebold, naming the parasite *Distomum haematobium*.[6] Tragically, Bilharz died of typhus at the age of 37. During the next few years, 30% to 40% of the population in Egypt were discovered to be infected with *S. haematobium,* and it even was found in an ape dying in London. The peculiar morphology of the worm made it apparent that it could not be included in the genus *Distomum,* so in 1858 Weinland proposed the name *Schistosoma*. Three months latter Cobbold named it *Bilharzia*, after its discoverer. This latter name became widely accepted throughout the world and was even given the slang name "Bill Harris" by British soldiers serving in Europe during World War I. Today, however, the strict rules of zoological nomenclature decree that *Schistosoma* has priority and is thus the current name for the parasite. Even so, health officers in many parts of the world erect signs next to ponds and streams that warn prospective bathers of the dangers of "bilharzia." Nonetheless, *Schistosoma* is an apt name, referring to the "split body" (gynecophoral canal) of the male.

While information was accumulating on the biology of *S. haematobium,* doubts were being raised as to whether it was a single species or whether two or more species were being confused. The problem was confounded by the occurrence of terminally spined eggs in both urine and feces. Whenever eggs with lateral spines were noticed, they were ignored as being "abnormal." Sir Patrick Manson, in 1905, decided that intestinal and vesicular (urinary bladder) schistosomiasis usually are distinct diseases, caused by distinct species of worms. He reached this conclusion when he examined a man from the West Indies who had never been to Africa and who passed laterally spined eggs in his feces but none at all in his urine.[22] Sambon argued in favor of the two-species concept in 1907, and he named the parasites producing laterally spined eggs *Schistosoma mansoni*. (Japanese zoologists had already detected still another species by this time, but their reports were generally unknown to Europeans.) However, the eminent German parasitologist Looss disagreed and brought the full sway of his reputation and dialectic against the notion and even stated that he had seen a female worm with both kinds of eggs in its uterus. Sambon, undaunted, replied that, until Professor Looss could "show me an actual specimen, I am bound to place the worm capable of producing the two kinds of eggs with the phoenix, the chimaera and other mythical monsters."[31]

The question was finally resolved by Leiper[18] in 1915. He first visited Japan to acquaint himself with the work of Miyairi and Suzuki on *S. japonicum*. Then, working in Egypt, he discovered that cercariae emerging from the snail *Bulinus* could infect the vesicular veins of various mammals, and they always produced eggs with terminal spines. Those emerging from a different snail, *Biomphalaria*, infected the intestinal veins and produced laterally spined eggs. It was soon determined that *S. mansoni* had a broad distribution in the world, having been widely scattered by the slave trade. It is now widespread in Africa and the Middle East and is the only blood fluke of humans in the New World, with the possible exception of a small focus of *S. haematobium* in Surinam.[20] The original endemic area of *S. mansoni* is thought to have been the Great Lakes region of central Africa.

While Cobbold, Weinland, Bancroft, Sambon, and others were wrestling with the problem of *S. haematobium* and *S. mansoni,* Japanese researchers were investigating a similar disease in their country. For

TABLE 18-1

Comparative morphology of the three primary species of human schistosomes

Characteristic	*S. haematobium*	*S. mansoni*	*S. japonicum*
Tegumental papillae	Small tubercles	Large papillae with spines	Smooth
Size			
Male			
Length	10 to 15 mm	10 to 15 mm	12 to 20 mm
Width	0.8 to 1 mm	0.8 to 1 mm	0.5 to 0.55 mm
Female			
Length	ca 20 mm	ca 20 mm	ca 26 mm
Width	ca 0.25 mm	ca 0.25 mm	ca 0.3 mm
Number of testes	4 to 5	6 to 9	7
Position of ovary	Near midbody	In anterior half	Posterior to midbody
Uterus	With 20 to 100 eggs at one time; average 50	Short; few eggs at one time	Long; may contain up to 300 eggs; average 50
Vitellaria	Few follicles, posterior to ovary	Few follicles, posterior to ovary	In lateral fields, posterior quarter of body
Egg	Elliptical, with sharp terminal spine; 112 to 170 μm × 40 to 70 μm	Elliptical, with sharp lateral spine; 114 to 175 μm × 45 to 70 μm	Oval to almost spherical; rudimentary lateral spine; 70 to 100 μm × 50 to 70 μm

years physicians in the provinces of Hiroshima, Saga, and Yamanachi had recognized an endemic disease characterized by an enlarged liver and spleen, ascites, and diarrhea. At autopsy they noted eggs of an unknown helminth in various organs, especially in the liver. In 1904 Professor Katsurada of Okayama recognized that the larvae in these eggs resembled those of *S. haematobium*. Because he was unable to make a postmortem examination of an infected person, he began examining local dogs and cats, in hopes that they were reservoirs for the parasite. He soon found adult worms containing eggs identical to those from humans and named them *Schistosoma japonicum*. The experimental elucidation of the life cycle by various Japanese researchers was a milestone in the history of parasitology and formed the basis for Leiper's work on blood flukes in Egypt. The distribution of this parasite is limited to Japan, China, Taiwan, the Philippines, Celebes, and Thailand.

Morphology. Although the three species are generally similar structurally, several differences in detail are listed in Table 18-1. Considerable sexual dimorphism exists in the genus *Schistosoma*, the males being shorter and stouter than the females (Fig. 18-5). The males have a ventral, longitudinal groove, the **gynecophoral canal,** where the female normally resides. The mouth is surrounded by a strong oral sucker, and

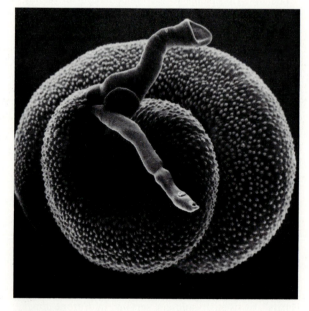

FIG. 18-5

Scanning electron micrograph of male and female *Schistosoma mansoni*. The female is lying in the gynecophoral groove in the ventral surface of the male.

Photograph by Sue Carlisle Ernst.

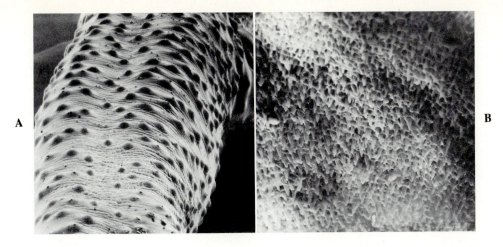

FIG. 18-6

Tegumental tubercles of *Schistosoma haematobium.* **A,** Dorsal surface. **B,** Toothlike tubercles in oral sucker.

Photographs by Robert E. Kuntz.

the acetabulum is near the anterior end. There is no pharynx. The paired intestinal ceca converge and fuse at about the midpoint of the worm and then continue as a single gut to the posterior end. The male possesses five to nine testes, according to species, each of which has a delicate vas efferens, and these combine to form the vas deferens. The latter dilates to become the seminal vesicle, which opens ventrally through the genital pore immediately behind the ventral sucker. Cirrus pouch, cirrus, and prostate cells are absent.

The suckers of the females are smaller and not so muscular as those of the males, and the tegumental tubercles (Fig. 18-6), if any, are confined to the ends of the female. The ovary is anterior or posterior to or at the middle of the body, and the uterus is correspondingly short or long, depending on the species. On the basis of differences between the species, La Roux proposed that the genera *Afrobilharzia* and *Sinobilharzia* be erected for *S.mansoni* and *S. japonicum,* respectively, but his suggestion has not been generally accepted.

Biology. Adult worms live in the veins that drain certain organs of their host's abdomen, and the three species have distinct preferences: *S. haematobium* lives principally in veins of the urinary bladder plexus; *S. mansoni* prefers the portal veins draining the large intestine; and *S. japonicum* is more concentrated in the veins of the small intestine. The female worm is often found in the gynecophoral canal of the male worm, where copulation takes place; there may be other physiological reasons for this habitus as well.

The worms work their way "upstream" into smaller veins, and the female may leave the gynecophoral canal to reach still smaller venules to deposit eggs. The eggs (Fig. 18-7) must then traverse the wall of the venule, some intervening tissue, and the gut or bladder mucosa before they are in a position to be expelled from the host. The mechanism by which this "escape" is achieved is not at all clear and has been the subject of much speculation. The spines on the eggs are often credited with contributing to the expulsion, but one must remember that the feat is accomplished by *S. japonicum,* which has only the most rudimentary spine. One likely explanation has been provided by Kuba,[17] who observed that, when eggs are lodged adjacent to the wall of the venule, a small blood thrombus (clot) forms around each of them. As the thrombus is slowly infiltrated with fibroblasts (organized) and overgrown with endothelial cells, the eggs are isolated from the blood flow, and the venule wall thins. The eggs then pass through as the host repair processes continue. How the egg gets from there to the lumen of the gut or bladder is less clear. Damian (see suggested readings) postulated that the extravasated egg stimulates a granuloma to form around it. The granuloma, consisting of motile cells, then moves to the intestinal lumen, carrying the egg with it. Once in the lumen the granuloma disintegrates. In any case about two thirds of the eggs do not make it, and large numbers build up in the gut or bladder wall, particularly in chronic cases in which the wall is toughened by a great amount of connective (scar) tissue. Of

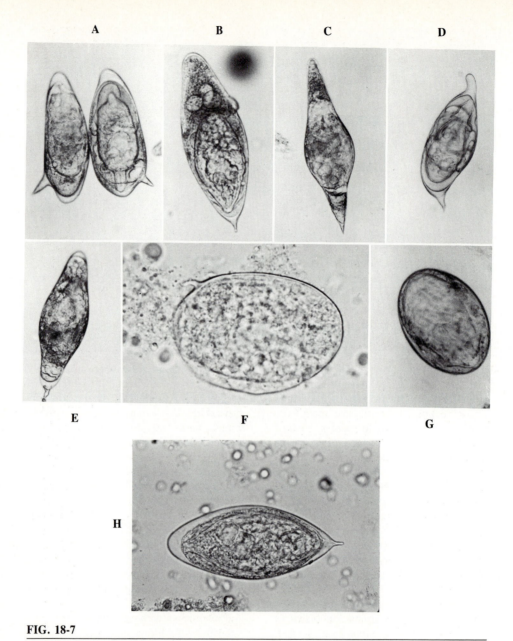

FIG. 18-7

Eggs of schistosome flukes, **A,** *Schistosoma mansoni.* **B,** *Schistosoma intercalatum.* **C,** *Schistosoma bovis.* **D,** *Schistosoma rhodhaini.* **E,** *Schistosoma mattheei.* **F,** *Schistosoma japonicum.* **G,** *Schistosomatium douthitti.* **H,** *Schistosoma haematobium.*

Photogragphs by Robert E. Kuntz and Jerry A. Moore.

course, many eggs are never expelled from the venules but are swept away by the blood, eventually to lodge in the liver or capillary beds of other organs. By the time the eggs reach the outside by way of the urine or feces, they are completely embryonated and hatch when exposed to the lower osmolarity of fresh water.

The mechanism of hatching is poorly understood. The first indication of hatching is activation of the cilia on the miracidium. This increases until the miracidium is a veritable spinning ball. Then, suddenly, an osmotically induced vent opens on the side of the egg, and the miracidium emerges (Fig. 18-8). The miracidium usually contracts a few times to completely clear

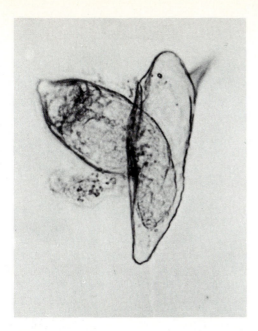

FIG. 18-8

Miracidium of *Schistosoma mansoni* escaping from its eggshell.

Photograph by Robert E. Kuntz.

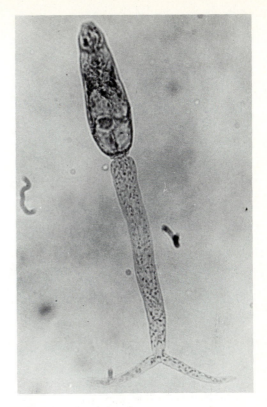

FIG. 18-9

Cercaria of *Schistosoma mansoni*.

Photograph by Warren Buss.

the shell, after which it rapidly swims away. However, some eggs will not hatch, no matter how active the miracidium becomes, and others will hatch before the larva becomes activated.

The miracidium is typical and swims ceaselessly during its short life. If hatching from an old egg, it will live only 1 to 2 hours; in optimal conditions it will survive for 5 to 6 hours. Although the miracidia of schistosomes do not have eyespots, they apparently have photoreceptors, and they are positively phototropic.[3] When miracidia enter the vicinity of a snail host, they are stimulated to swim more rapidly and change direction much more frequently, thus increasing their chances of encountering the host. Following are the most important snails:

1. For *S. haematobium,* several species of *Bulinus* and *Physopsis,* possibly also *Planorbarius*
2. For *S. mansoni, Biomphalaria alexandrina* in northern Africa, Saudia Arabia, and Yemen; *B. sudanica, B. rupellii, B. pfeifferi,* and others in the genus in other parts of Africa; *B. glabrata* in the Western Hemisphere; and *Tropicorbis centrimetralis* in Brazil
3. For *S. japonicum,* several species of *Oncomelania*

After penetration of the snail the miracidium sheds its epithelium and begins development into a mother sporocyst, usually near its point of entrance. After about 2 weeks the mother sporocyst, which has four protonephridia, gives birth to daughter sporocysts, which usually migrate to other organs of the snail, if there is room. The mother sporocyst continues producing daughter sporocysts for up to 6 to 7 weeks.[24] There is no redial generation.

The furcocercous cercariae (Fig. 18-9) start to emerge from the daughter sporocysts and the snail host about 4 weeks after initial penetration by the miracidium. The cercaria has a body 175 to 240 μm long by 55 to 100 μm wide and a tail 175 to 250 μm long by 35 to 50 μm wide, bearing a pair of furci 60 to 100 μm long. The oral sucker is absent, being replaced by a head organ composed of penetration glands, and the ventral sucker is small and covered with minute spines. Four types of glands open through bundles of ducts at the anterior margin of the head organ (p. 244).

There is no second intermediate host in the life cycle. The cercariae alternately swim to the surface of the water and slowly sink toward the bottom, continuing to live this way for 1 to 3 days. If they come into

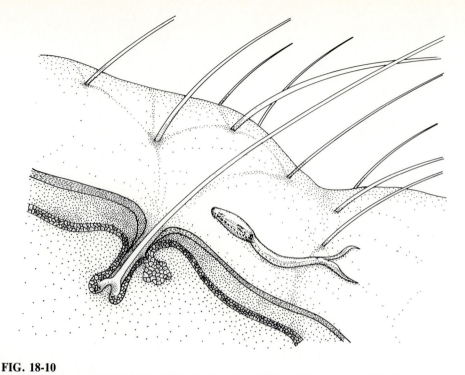

FIG. 18-10

Diagram of a cercaria of *Schistosoma* sp. in exploring position on skin.

Drawing by Ian Grant.

contact with the skin of a prospective host, such as a human, they attach and creep about for a time as if seeking a suitable place to penetrate (Fig. 18-10). They are attracted to secretions of the skin. Experiments have shown they are not attracted to sugars. They exhibit a weak response to a mixture of electrolytes, urea, and lactate, but a strong positive response to arginine. Upon attraction to arginine, the cercaria begin to produce arginine themselves from postacetabular glands. This may attract other cercariae in the neighborhood.[7] They require only half an hour or less to completely penetrate the epidermis, and they can disappear through the surface in 10 to 30 seconds. Penetration is accompanied by a vigorous wiggling, together with secretion of the products of the head organ. The tail is cast off in the process. The worms are somewhat smaller now that the penetration glands have emptied their contents. Within 24 hours the **schistosomula,** as they are now called, enter the peripheral circulation and are swept off to the heart. Some of the schistosomula may migrate through the lymphatics to the thoracic duct and thence to the subclavian veins and heart. Leaving the right side of the heart, the small worms wriggle their way through the pulmonary capillaries to gain access to the left heart and systemic circulation. It appears that only the schistosomula that enter the mesenteric arteries, traverse the intestinal capillary bed, and reach the liver by the hepatoportal system can continue to grow. After undergoing a period of about 3 weeks of development in the liver sinusoids, the young worms migrate to the walls of the gut or bladder (according to species), copulate, and begin producing eggs. The entire prepatent period is about 5 to 8 weeks. Adults may live 20 to 30 years.[15]

Unpaired female worms do not become sexually mature and have the appearance of starving. Their esophageal musculature is weak and thin, they produce little of at least some digestive enzymes, and they ingest about one fourth as many erythrocytes as paired females. It is believed that a growth-stimulating function results from the muscular action of the clasping male, which helps the immature female to pump blood into her intestine.[8]

Epidemiology. Human waste in water containing intermediate hosts of the *Schistosoma* worm is the sin-

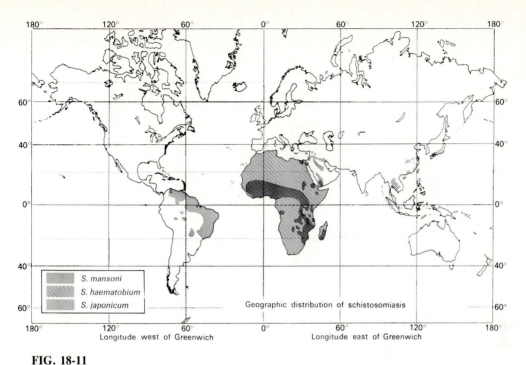

S. mansoni
S. haematobium
S. japonicum

Geographic distribution of schistosomiasis

Longitude west of Greenwich

Longitude east of Greenwich

FIG. 18-11

Geographical distribution of schistosomiasis.

AFIP neg. no. 68-4866-3.

gle most important epidemiological factor in schistosomiasis, and the availability of suitable species of snail host will determine the endemicity of the particular species of *Schistosoma*. The latter is well illustrated by the fact that although both *S. mansoni* and *S. haematobium* are widespread in Africa, only *S. mansoni* became established in the New World by the slave trade, almost certainly because snails suitable for only that species were present there (Fig. 18-11). Survival of these parasites depends on human insistence on polluting water with their organic wastes. Adequate sewage treatment is sufficient to eliminate schistosomiasis as a disease of humans (Fig. 18-12). However, really adequate sewage treatment has not yet been realized in the most advanced civilizations, and any treatment at all is lacking in many areas where the flukes are prevalent. Tradition, at once the salvation and the bane of culture, prompts people to use the local waterway for sewage disposal instead of foul-smelling outhouses (Fig. 18-13). A bridge across a small stream becomes a convenient toilet; a grove of mango trees over a rivulet is a haven for children who bombard the area with their feces (Fig. 18-14). Especially vulnerable to infection are farmers who wade in their irrigation water, fishermen who wade in their

lakes and streams, children who play in any contaminated body of water, and women who wash clothes in streams. A focus of infection in Brazil was a series of ditches in which watercress was grown for food. In some Moslem countries the religious requirement of ablution, that is, washing the anal or urethral orifices after urination or defecation, is an important factor in transmission. Not only is the convenient water source to perform ablution likely to be a contaminated river or canal, the deposition of additional feces and urine in its vicinity is ensured.

Clearly the economic and education level of the population will influence the transmission of the disease, and age and sex are important factors as well. Males usually show the highest rates of infection and the most intense infections, and the most hazardous age is the second decade of life. This appears to reflect occupational and recreational differences, rather than sex or age resistance to infection. In Surinam, where both sexes work in the fields, the highest (and equal) prevalence is found in adults of both sexes. Certain other factors, such as immunity and the cessation of egg release in chronic infections, must be considered when a survey of a population is carried out and transmission is studied. The fact that buildup of

271

FIG. 18-12

Government-encouraged pit latrines in the Philippines serve as a means of prevention of schistosomiasis.

Photograph by Robert E. Kuntz.

FIG. 18-13

Typical native dwelling in a schistosomiasis area of the Philippines. *Oncomelania,* the snail host of *S. japonicum,* is found in the stream near the house, which is also likely to be contaminated by human feces.

Photograph by Robert E. Kuntz.

granulation tissue in the gut or bladder wall prevents release of eggs into the feces or urine will mask infection unless immunodiagnostic or biopsy methods are used. Furthermore, even though such persons may be gravely disabled, they are removed from the cycle of transmission.

During the course of the infection some protective immunity to superinfection is elicited, either by repeated exposures to the cercariae or by the presence of the adults, although the adult worms themselves are not affected by the immune response. This amount of protective immunity may prevent the disease from becoming an even worse scourge than it is. For an excellent review of immunity in human schistosomiasis see Butterworth and Hagan.[5]

It is of utmost importance to recognize that agriculatural projects intended to increase food production in underdeveloped countries have, in many cases, created more misery than they have alleviated, by extending snail habitats.[37] A $10 million irrigation project in southern Rhodesia had to be abandoned 10 years after it was started because of schistosomiasis.[26] The Aswan High Dam in Egypt, much acclaimed in its inception, may have its benefits canceled by the increase in disease it has caused. Retraint of the wide fluctuations in the water level of the Nile, although making possible four crops per year by perennial irrigation, has also created conditions vastly more congenial to snails.[34] Before the dam construction, perennial irrigation was already practiced in the Nile delta region, and the prevalence of schistosomiasis was about 60%; in the 500 miles of river valley between Cairo and Aswan, where the river was subject to annual floods, the prevalence was only about 5%. Four

FIG. 18-14

Slow-running streams and protecting tropical vegetation provide ideal habitats for *Biomphalaria,* a snail host for *S. mansoni* in Puerto Rico.

Photograph by Robert E. Kuntz.

years after the dam was completed the prevalence of *S. haematobium* ranged from 19% to 75%, with an average of 35%, between Cairo and Aswan, or an average sevenfold increase! In the area above the dam, prevalence was very low before its construction; in 1972, 76% of the fishermen examined in the impounded area were infected. A 1982 study showed a continued increase in prevalence in six villages of upper Egypt.[16]

The cost, in terms of productivity, of *S. haematobium* to a native may well equal the worker's per capita income. Even oil production in some countries of the Middle East may be slightly, but directly, affected by the presence of schistosomes in oil refinery employees. Recently, blood fluke was found to be reintroduced into the oil fields of Jordan.

Currently, only the small island of Vieques, off the coast of Puerto Rico has been completely eradicated of schistosomiasis. After 80 years of effort in the Caribbean, abatement of transmission has been accomplished only in Puerto Rico, Saint Kitts, Saint Martin, and parts of Saint Lucia. Promising results have been obtained in Surinam and Venezuela. In all other centers of endemicity, schistosomiasis has remained at the same level or has increased.[4]

The role of reservoir hosts and of strains of the parasite have some importance as epidemiological factors, depending on the species. Members of no less than seven mammalian orders have been successfully infected experimentally with *S. mansoni;* however, certain monkeys and a variety of rodents are probably important natural reservoir hosts in Africa and tropical America. *Schistosoma haematobium* is more host specific than is *S. mansoni,* and it is thought that no natural reservoir hosts exist for it. The opposite is true of *S. japonicum,* which seems to be the least host specific. It can develop in dogs, cats, horses, swine, cattle, caribou, rodents, and deer; but there seems to be more than one race of this worm, and the susceptibility of a given host varies. For example, *S. japonicum* is widely prevalent in rats in Taiwan, but it is rare in humans there.

Pathogenesis. The pathogenesis of schistosomiasis differs somewhat according to the species of the fluke involved, mainly because of the preferred site of each. Progression of the disease caused by all three species is commonly divided into three phases: (1) the initial phase, 4 to 10 weeks after infection, characterized by fever and toxic or allergic phenomena; (2) the intermediate stage, 2.5 months to several years after infection, with pathological changes in intestinal or urinary tracts and eggs in excreta; and (3) the final phase, with complications involving the gastrointestinal, renal, and other systems and often with no eggs being passed. The initial phase is similar for all species: intermittent fever, frequently a skin rash, abdominal pain, bronchitis, enlargement of liver and spleen, and diarrhea. The most serious damage is done by the eggs in all three species. With *S. mansoni,* the large intestine is most notably affected, especially the sigmoid colon and rectum. The eggs lodged in the venules and submucosa (Fig. 18-15) act as foreign bodies and cause inflammatory reactions with leukocytic and then fibroblastic infiltration. These finally become small fibrous nodules called granulomata, or **pseudotubercles,** so called because of their resemblance to the localized nodules of tissue reaction (tubercles) in tuberculosis. Small abscesses occur, and the occlusion of small vessels leads to necrosis and ulceration. Often a high eosinophilia (high leukocyte count in blood, predominately of eosinophils) is fol-

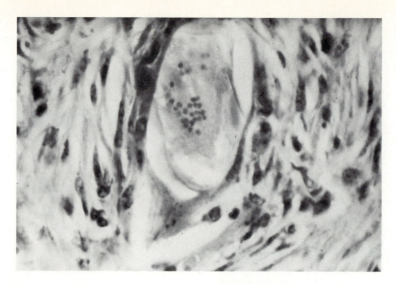

FIG. 18-15

Egg of *Schistosoma mansoni* embedded in intestinal wall.

Photograph by David F. Oetinger.

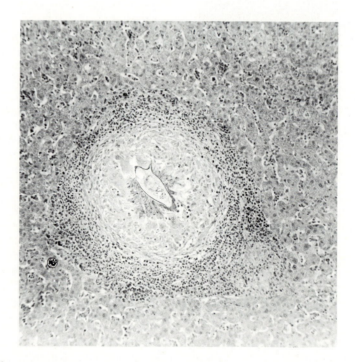

FIG. 18-16

Egg of *Schistosome mansoni* in liver, surrounded by granulomatous cells. Note leukocytic infiltration around the granuloma.

AFIP neg. no. 64-6532.

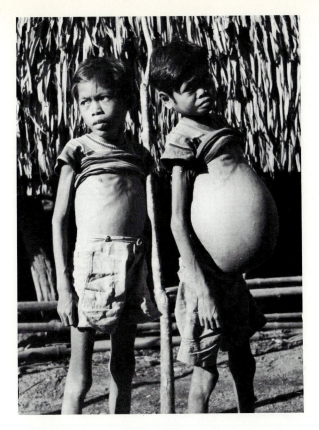

FIG. 18-17

Ascites in advanced schistosomiasis japonica, Leyte, Philippines *(right)*. This is an example of dwarfing caused by schistosomiasis. The male on the left is 13 years old; the one on the right is 24 years old.

Photograph by Robert E. Kuntz.

lowed by leukopenia (lowered white cell count). The clinical symptoms include abdominal pain and diarrhea with blood, mucus, and pus. Many eggs are swept back into the liver, where they lodge in the hepatic capillary bed (Fig. 18-16) and cause a similar foreign body reaction with pseudotubercle formation. As the eggs accumulate and the fibrotic reactions in the liver continue, a periportal cirrhosis and portal hypertension ensue. A marked enlargement of the spleen (splenomegaly) occurs, partly because of eggs lodged in it and partly because of the chronic passive congestion of the liver. Ascites (accumulation of fluid in the abdominal cavity) is common at this stage (Fig. 18-17). Some eggs pass the liver, lodging in the lungs, nervous system, or other organs and produce pseudotubercles there. An interesting review of the immunopathology of granuloma formation and fibrosis in

schistosomiasis is presented by Phillips and Lammie.[29]

Schistosoma japonicum causes pathological changes in the intestine and liver similar to those of *S. mansoni,* except that the small intestine is more extensively involved and the large intestine less so. Frequently fibrous nodules containing nests of eggs are found on the serosal and peritoneal surfaces. Eggs of *S. japonicum* seem to reach the brain more often than do those of the other species, and neurological disease, including coma and paralysis, occurs in about 9% of cases. **Schistosomiasis japonica** is the most grave of the three, and the prognosis is unfavorable in heavy infections without early treatment.

Infection with *S. haematobium* is considered the least serious. Since the adults of this species live in the venules of the urinary bladder, the chief symptoms are associated with the urinary system: cystitis (inflammation of the bladder), hematuria, and pain on urination. The onset of hematuria is usually gradual and becomes marked as the disease develops and the bladder wall becomes more ulcerated. Pain is most intense at the end of urination.[12] The changes in the bladder wall (Fig. 18-18) are associated with the foreign body reactions around the eggs, that is, pseudotubercles, fibrous infiltration, thickening of the muscularis layer, and ulceration. Malignant changes sometimes occur. Chronic cases lead to genital, ureteral, and kidney involvement and to lesions in other parts of the body, as with other species.

Diagnosis and treatment. As in diagnosing many other helminths, the demonstration of eggs in the excreta is the most straightforward mode of diagnosis. However, the number of eggs produced per female schistosome, even for *S. japonicum,* is far smaller than most other helminth parasites of humans; therefore, direct smears must be augmented by concentration techniques and other diagnostic methods, such as biopsy and immunodiagnosis. In the case of *S. mansoni,* which produces the smallest number of eggs, only 47% of patients could be diagnosed after three direct smears.[21] With concentration techniques, such as gravity or centrifugal sedimentation, more than 90% of the coprologically demonstrable cases could be diagnosed. Nevertheless, it must be remembered, particularly in chronic cases, that few or no eggs may be passed. In such cases, rectal, liver, or bladder biopsies may be of great value, but these require the services of specialists and the availability of appropriate surgical facilities. Hence a substantial research effort has been directed at finding sensitive, accurate, and reliable immunodiagnostic methods. Of these, the simple intradermal test is valuable. There is little im-

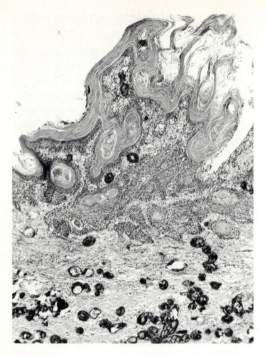

FIG. 18-18

Schistosomiasis of the urinary bladder. In this case many eggs of *S. haematobium* can be seen in all the layers of the bladder. Many of the eggs are calcified. The epithelium has undergone squamous metaplasia. Note also leukocytic infiltration, granulosis, and ulceration.

AFIP neg. no. 65-6779.

mune protection against the adult worms; the phrase "immune serum" is used in the immunological sense of serum that contains antibodies.

The intradermal test becomes positive from 4 to 8 weeks after infection and remains so for years, even after the patient may be cured. The reaction is a histamine or immediate hypersensitivity type (Chapter 3). A small amount of antigen is injected intradermally, and the area of the wheal produced is measured exactly 15 minutes later and compared with the area of control injection (antigen vehicle alone). With good antigen and careful administration, this test may have an efficacy up to 95%.

Much activity is being devoted to develop more sensitive immunodiagnostic tests. Hillyer et al.[9] compared enzyme-linked immunosorbent assay (ELISA), radioimmunoassay (RIA) and circumoval precipitation (COP) tests using antigens from eggs of *S. mansoni*. They concluded that specificity was highest with the COP test, which correctly identified 95% of infected individuals. The COP reaction depends on the forma-

tion of a precipitate around lyophilized eggs incubated in immune serum. Because this test becomes negative about 8 months after the eggs have died, it can be used to confirm cure.

On the other hand, a World Health Organization collaborative study on antigens for immunodiagnosis of schistosomiasis concluded that although egg antigens yielded a higher sensitivity than antigens from adult worms, no particular method for detecting antibodies was superior.[23]

Difficulty in treating **schistosomiasis** is a major factor contributing to the disease as a world health problem. Until recently the most effective drugs were the organic trivalent antimonials, but these are quite toxic to humans and must be given carefully—in small doses over a period from 2 to 6 weeks, depending on the drug. Problems inherent in treating large numbers of people over wide areas in underdeveloped countries with such drugs are obvious. Some other drugs (lucanthone hydrochloride and niridazole) are less toxic but also less effective. They may inhibit egg production and cause the worms to move back into the liver for a time, but evidence exists that the worms can recover and begin producing eggs again. The drug praziquantel has supplanted the more dangerous antimonials. Other drugs are undergoing field trials and show promise of effectiveness (hycanthone, metriphonate, oxamniquine).[10]

By the stage in the disease when liver damage is extensive, all chemotherapeutic treatment is contraindicated, and surgical intervention may be necessary. At the late stages, prognosis is poor, and the treatment can only be supportive.

Control. Control of schistosomiasis is exceedingly difficult, depending ultimately on the almost intractable task of persuading masses of uneducated, poor people to change their customs and traditions. Although draining snail habitats is of value, it was pointed out before that many more good habitats are now being created by efforts to increase agricultural production. Use of chemical molluscicides has met with some success but problems involved include determination and application of the proper quantity in a given body of water, dilution, effects on other organisms in the environment, and errors in estimating the physical and chemical characteristics of the water. Molluscicidal control of *S. japonicum* is virtually ineffective because *Oncomelania* is amphibious and only visits water to lay its eggs. In Puerto Rico some success in control of *Biomphalaria* has been achieved with predatory snails (*Marisa cornuarietis*, *Tarebia granifera maniensis*). In one experiment on flowing streams in Puerto Rico, release of 20,000 *M. cornua-*

rietis resulted in control of *Biomphalaria* (and their schistosome infections) at a cost of only 5% to 10% of the expenditure necessary for equivalent control with chemical molluscicides.[14]

Between 1966 and 1981 a model schistosomiasis control program was conducted by Saint Lucia's Ministry of Health and the Rockefeller Foundation. The techniques employed and results obtained are summarized in a book by Jordan.[15] One interesting approach involved naturally occurring freshwater shrimp, *Macrobrachium* spp., which avidly feed on *Biomphalaria* spp., even in preference to other snails.[15] One specimen of *M. carcinus* was observed to eat 33 snails in 14 days. The potential for snail control using these crustaceans is doubtful, however, because they are large, luscious, and are highly regarded as food by people in most localities.

Snail-eating fish have been cultured and released in infected waters with some success.

The development of an effective vaccine would have great potential value in the control of schistosomiasis,[10] and this area of research is being actively pursued. Some protection can be conferred by vaccination with irradiated cercariae and/or schistosomula,[1] but for practical use, a long-acting, killed antigen would be more desirable. It is possible that such a vaccine can be developed.[19]

It has been found that irradiated schistosome cercariae could stimulate high levels of resistance to *S. japonicum* and *S. mansoni* in rhesus monkeys.[13,33] Trial vaccines against *S. bovis* in cattle, using schistosomulae derived from irradiated cercariae, showed great promise in Sudan.

■ Other schistosomes of lesser medical importance

Schistosoma mattheei-intercalatum "complex". In Africa hundreds of cases are known in which terminal-spined eggs are recovered from stools only. The worms have been named *Schistosoma intercalatum*, but they are morphologically indistinguishable from *S. mattheei*, a natural parasite of African ruminants and primates. If only one species is involved, *S. mattheei* is the correct name. To further confuse the issue, cases are known in which eggs of this species are passed in both stool and urine by human patients.[2]

Schistosoma mekongi. In isolated regions of Kampuchea and neighboring Laos there exists a species of *Schistosoma* long thought to be *S. japonicum*. Now it is known to be a separate species, *S. mekongi*,[35] although it shares several characteristics with *S. japonicum*: (1) both have eggs of similar shape and small size, with similar spines; (2) both species use prosobranch snails, whereas all other species that infect mammals parasitize pulmonate snails; (3) adults occupy the same sites in the definitive host; and (4) the prepatent period in both is longer than in the other species in mammals.

Differences between *S. mekongi* and *S. japonicum* follow: (1) they have different snail hosts, (2) there are slightly smaller eggs in *S. mekongi,* and (3) the prepatent period of *S. mekongi* is 7 or 8 days longer than that of *S. japonicum.*[35]

Schistosoma mekongi uses the operculate snail *Tricula aperta* (syn. *Lithoglyphopsis aperta*) as its molluscan host. The known endemic areas of infection are in the area of the Kampuchean provincial capital of Kratie on the Mekong River, including surrounding villages, and the Khong Island in the Mekong River in Laos. Carnivores, including the domestic dog, serve as reservoir hosts.

Schistosome cercarial dermatitis ("swimmer's itch"). Several species in the genus *Schistosoma* are known to cause a severe rash when their cercariae penetrate the skin of an unsuitable host. Hence *S. spindale,* a parasite of ruminants in India, Malaya, Africa, and Sumatra, and *S. bovis* from ruminants, equines, and primates in Europe, Africa, and the Middle East are agents of dermatitis in humans throughout their range.

More important, several species of bird schistosomes are distributed throughout the world and cause "swimmer's itch" when their cercariae attack anyone on whose skin the organisms land. The genera *Trichobilharzia, Gigantobilharzia, Ornithobilharzia, Microbilharzia,* and *Heterobilharzia* are the guilty parties.

For the most part the skin reaction is a product of sensitization, with repeated infections causing increasingly severe reactions.[25] When the cercaria penetrates the skin and is unable to complete its migration, the host's immune responses rapidly kill it. At the same time the cercariae release allergenic substances that cause inflammation and, typically, a pus-filled pimple (Fig. 18-19). The reaction may also be general, with an itching rash produced over much of the body. The condition is not a serious threat to health but is a terrific annoyance, much like poison ivy, which interrupts a summer vacation, for instance, or decreases the income of someone who rents lakefront cottages to the summer crowd. In the United States the problem is most serious in the Great Lakes area, but it has been reported from nearly all states.

Control depends mainly on molluscicides, but their usefulness is limited because they threaten sport fishing by poisoning the fish and, of course, because of the other problems previously mentioned. Ocean

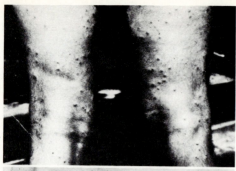

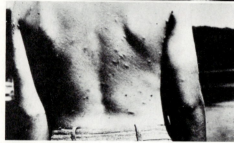

FIG. 18-19

Cercarial dermatitis, or "swimmer's itch," caused by cercariae of avian blood flukes.

AFIP neg. no. 77203.

beaches are occasionally infested with avian schistosome cercariae, for which no control has yet been devised.

REFERENCES

1. Bickle, Q.D., M.G. Taylor, M.J. Doenhoff, and G.S. Nelson. 1979. Immunization of mice with gamma-irradiated intramuscularly injected schistosomula of *Schistosoma mansoni.* Parasitology 79:209-222.
2. Biale, D.M. 1966. The occurrence of terminal spined eggs, other than those of *Schistosoma haematobium,* in human beings in Rhodesia. Cent. Afr. J. Med. 12:103-109.
3. Brooker, B.E. 1972. The sense organs of trematode miracidia. In Canning, E.U., and C.A. Wright, editors. Behavioural aspects of parasite transmission. Linnean Society of London. Academic Press Inc. (London) Ltd., pp. 171-180.
4. Bundy, D.A.P. 1984. Caribbean schistosomiasis. Parasitology 89:377-406.
5. Butterworth, A.E., and P. Hagan. 1987. Immunity in human schistosomiasis. Parasitol. Today 3:11-16.
6. Foster, W.D. 1965. A history of parasitology. E. & S. Livingstone, Edinburgh.
7. Granzer, M., and W. Haas. 1986. The chemical stimuli of human skin surface for the attachment response of *Schistosoma mansoni* cercariae. Int. J. Parasitol. 16:575-579.
8. Gupta, B.C., and P.F. Basch. 1987. The role of *Schistosoma* mansoni males in feeding and development of female worms. J. Parasitol. 73:481-486.
9. Hillyer, G.V., E.R. Tiben, W.B. Knight, I.G. de Rios, and R.P. Pellay. 1979. Immunodiagnosis of infection with *Schistosoma mansoni:* comparison of ELISA, radioimmunoassay, and precipitation tests performed with antigens from eggs. Am. J. Trop. Med. Hyg. 28:661-669.
10. Hoffman, D.B., et al. 1979. Control of schistosomiasis. Report of a workshop. Am. J. Trop. Med. Hyg. 28:249-259.
11. Hulse, E.V. 1971. Joshua's curse and the abandonment of ancient Jericho: schistosomiasis as a possible medical explanation. Med. Hist. 15:376-386.
12. Hunter, G.W., III, J.C. Swartzwelder, and D.F. Clyde, 1976. The schistosomes. In Tropical medicine, ed. 5. W.B. Saunders Co., Philadelphia.
13. James, S.L., and A. Sher. 1986. Prospects for a nonliving vaccine against schistosomiasis. Parasitol. Today 2:134-137.
14. Jobin, W.R., and A. Laracuente. 1979. Biological control of schistosome transmission in flowing water habitats. Am. J. Trop. Med. Hyg. 28:916-917.
15. Jordan, P. 1985. Schistosomiasis—the St. Lucia Project. Cambridge Unviersity Press, Cambridge.
16. King, C.L., F.D. Miller, M. Hussein, R. Rarkat, and A.S. Monto. 1982. Prevalence and intensity of *Schistosoma haematobium* infection in six villages in Upper Egypt. Am. J. Trop. Med. Hyg. 31:320-327.
17. Kuba, N. 1963. Histopathological study on mechanism of extrusion of schistosome ova from blood vessels. Jpn. J. Vet. Sci. 25:289-297.
18. Leiper, R.T. 1915. Observations on the mode of spread and prevention of vesicle and intestinal bilharziosis in Egypt, with additions to August, 1916. Proc. R. Soc. Med. 9:145-172.
19. Maddison, S.E., S.B. Slemenda, F.W. Chandler, and I.G. Kagan. 1978. Studies on putative adult worm-derived vaccines and adjuvants for protection against *Schistosoma mansoni* infection in mice. J. Parasitol. 64:986-993.
20. Maldonado, J.F. 1967. Schistosomiasis in America. Editorial Cientifico-Medica, Barcelona.
21. Maldonado, J.F., J. Acosta-Matienzo, and F. Velez-Herrera. 1954. Comparative value of fecal examination procedures in the diagnosis of helminth infections. Exp. Parasitol. 3:403-416.
22. Manson, P. 1905. Lectures on tropical diseases. Constable, London, p. 54.
23. Mott, K.E., and H. Dixon. 1982. Collaborative study on antigens for immunodiagnosis of schistosomiasis. Bull. WHO 60:729-753.
24. Okabe, K. 1964. Biology and epidemiology of *Schistosoma japonicum* and schistosomiasis. In Morishita, K., Y. Komiya, and H. Matsubayashi, editors. Progress of medical parasitology in Japan, vol. 1. Meguro Parasitological Museum, Tokyo, pp. 185-218.
25. Olivier, L.J. 1949. Schistosome dermatitis, a sensitization reaction. Am J. Hyg. 49:209-301.
26. Osmundsen, J.A. 1965. Science: battle is on against a dread crippler. New York Times, Aug. 22, 1965, p. 8E.
27. Pearson, J.C. 1956. Studies on the life cycles and morphology of the larval stages of *Alaria orisaemoides* Augustine and Uribe, 1927 and *Alaria canis* La Rue and Fallis. Can. J. Zool. 34:295-387.

28. Pearson, J.C. 1959. Observations on the morphology and life cycle of *Strigea elegans* Chandler and Rausch, 1947 (Trematoda: Strigidae). J. Parasitol. 45:155-170, 171-174.

29. Phillips, S.M., and P.J. Lammie. 1986. Immunopathology of granuloma formation and fibrosis in schistosomiasis. Parasitol. Today 2:296-302.

30. Popiel, I. 1986. The reproductive biology of schistosomes. Parasitol. Today 2:10-15.

31. Sambon, L.W. 1909. What is *Schistosoma mansoni* Sambon, 1907? J. Trop. Med. 12:1-11.

32. Shoop, W.L., and K.C. Corkum. 1987. Maternal transmission by *Alaria marcianae* (Trematoda) and the concept of amphiparatenesis. J. Parasitol. 73:110-113.

33. Taylor, M.G., and Q.D. Bickle. 1986. Irradiated schistosome vaccines. Parasitol. Today 2:132-134.

34. Van der Schalie, H. 1974. Aswan Dam revisited. Environment 16(9):18-20, 25-26.

35. Voge, M., D. Bruckner, and J.I. Bruce, 1978. *Schistosoma mekongi* sp. n. from man and animals, compared with four geographic strains of *Schistosoma japonicum*. J. Parasitol. 64:577-584.

36. Wright, W.H. 1968. Schistosomiasis as a world problem. Bull. N.Y. Acad. Med 44:301-302.

37. Wright, W.H. 1972. A consideration of the economic import of schistosomiasis. Bull. WHO 197:559-566.

SUGGESTED READINGS

Ansari, N., editor. 1973. Epidemiology and control of schistosomiasis (bilharziasis). S. Karger AG, Basel. (An official publication of the World Health Organization, outlining advances in the area.)

Berrie, A.D. 1970. Snail problems in African schistosomiasis. In Dawes, B., editor. Advances in parasitology, vol. 8. Academic Press, Inc., New York. pp. 43-96.

Bruce, J.I., and S. Sornmani, editors. 1980. The Mekong schistosome. Malacological Reviews, Suppl. 2. Whitmore Lake, Mich.

Damian, R.T. 1987. Presidential Address—The exploitation of host immune responses by parasites. J. Parasitol. 73:1-13.

Loker, E.S. 1983. A comparative study on the life-histories of mammalian schistosomes. Parasitology 87:343-369.

McCully, R.M., C.N. Barron, and A.W. Cheever. 1976. Schistosomiasis (bilharziasis). In Binford, C.H., and D.H. Connor, editors. Pathology of tropical and extraordinary diseases, vol. 2, sect. 10. Armed Forces Institute of Pathology, Washington, D.C.

Smithers, S.R., and R.J. Terry. 1969. The immunology of schistosomiasis. In Dawes, B., editor. Advances in parasitology, vol. 7. Academic Press, Inc., New York, pp. 41-93.

CHAPTER 19

ORDER ECHINOSTOMATA

Members of the order Echinostomata often show little resemblance to one another in the adult stages, but studies on their embryology have shown common ancestries through developmental similarities. Often, though by no means always, the tegument bears well-developed scales or spines, particularly near the anterior end. The acetabulum is near the oral sucker. In many cases a second intermediate host is absent, and the metacercaria encysts on underwater vegetation or debris. Most species are parasitic in wild animals, but a few are important as agents of disease in humans and/or their domestic animals.

SUPERFAMILY ECHINOSTOMATOIDEA

Parasites of the superfamily Echinostomatoidea infect all classes of vertebrates and are found in marine, freshwater, and terrestrial environments. Some are among the most common parasites encountered, and a few cause devastating losses to agriculture.

Family Echinostomatidae

Echinostomes are easily recognized by their circumoral collar of peglike spines, hence their name (Fig. 19-1). The spines are arranged either in a single, simple circle or in two circles, one slightly lower than, and alternating with, the other. The collar is interrupted ventrally and at each end has a group of "corner spines." The size, number, and arrangement of these spines are of considerable taxonomic importance in both cercariae and adults. Echinostomes typically are slender worms with large preequatorial acetabula, pretesticular ovaries, and tandem testes, although exceptions occur. The vitellaria are voluminous and mainly postacetabular. These worms are parasites of the intestine or bile duct of reptiles, birds, and mammals, particularly those frequenting aquatic environments.

■ *Echinostoma revolutum* (Fig. 19-2)

Echinostoma revolutum is a cosmopolitan parasite that is one of the most common and abundant of all trematodes of warm-blooded, semiaquatic vertebrates. It shows little host specificity and appears to be at home in any kind of bird or mammal that eats the metacercaria. It is especially common in ducks, geese, muskrats, beavers, and shore birds and has been found in many terrestrial birds. Experimentally it develops well in rabbits, rats, dogs, and guinea pigs. It is a fairly common parasite of humans in the orient, particularly in Taiwan and Indonesia.[2] Routine inspections of snails of the genera *Physa, Lymnaea, Helisoma, Paludina,* and *Segmentina*—all common genera—usually reveal infections with *E. revolutum.* Metacercariae encyst in molluscs, planaria, fish, and tadpoles. Infection of the definitive host is accomplished when the definitive host eats one of these. Humans are usually infected by eating raw mussels or snails.

Morphologically, the worm is easily identified. Its circumoral collar bears 37 spines in a double circle, of which five are corner spines at each end. The operculate eggs are large, 90 to 126 μm by 54 to 71 μm, and few of them occur in the uterus at any one time. The genital pore is median and preacetabular, and the cirrus pouch is large, passing dorsal to the voluminous acetabulum. The short uterus has an ascending limb only. Overall size varies greatly.

■ *Echinostoma ilocanum*

The eggs of *E. ilocanum* were first found in the stool of a prisoner in Manila in 1907. The organism has since been found commonly throughout the East Indies and China. Tubangui[14] found that the Norway rat was an important reservoir of infection.

The morphology of *E. ilocanum* is similar to that of *E. revolutum,* but it differs in collar spine number and arrangement. *E. ilocanum* has 49 to 51 spines, with 5

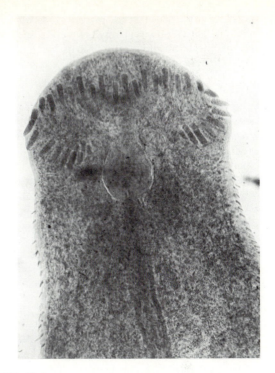

FIG. 19-1

Anterior end of *Echinostoma* sp., showing the double crown of peg-like spines on the circumoral collar.

Photograph by Warren Buss.

or 6 corner spines at each end. The double row of spines is continous dorsally, and the testes are deeply lobate.

The biology of *E. ilocanum* is similar to that of *revolutum,* with the metacercaria encysting in a freshwater mollusc. Infected snails are eaten raw by native peoples, who thereby become infected. The worms cause inflammation at their sites of attachment within the small intestine. Intestinal pain and diarrhea may develop in severe cases.

■ Other echinostomatid species reported from humans

Several species of echinostomes in different genera, which normally parasitize wild animals, have been reported from humans. These include *E. lindoense* in Celebes and possibly Brazil; *E. malayanum* from India (Fig. 19-3), southeast Asia, and the East Indies; *E. cinetorchis* from Japan, Taiwan, and Java; *E. melis,* which is circumboreal; and *Hypoderaeum conoidum* in Thailand. Others are *Himasthla muehlensi* in New York; *Paryphostomum surfrartyfex* in Asia Minor; and *Echinochasmus perfoliatus* from eastern Europe and

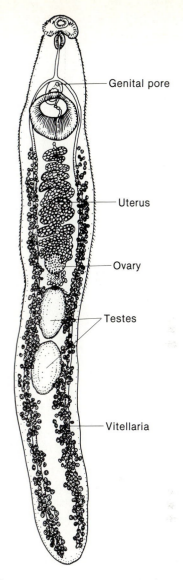

FIG. 19-2

Echinostoma revolutum, a common intestinal parasite of aquatic birds and mammals.

Drawing by Ian Grant.

from Asia. These are only some of the species of echinostomes of wild animals that have found their ways into humans, but considering their lack of host specificity, others probably do so fairly often and remain undetected. This points to the practical importance of systematic surveys of animal parasites, since when zoonotic infections are found, it immediately becomes

FIG. 19-3

Echinostoma malayanum, a parasite of humans in southern Asia.

From Odhner, T. 1913. Zool. Anz. 41:577- 582.

necessary to identify the pathogen and determine how the person became infected. Faunal surveys are the only way to fulfill this need.

Family Fasciolidae

Members of the family Fasciolidae are large, leaf-shaped parasites of mammals, mainly of planteaters. They have a tegument covered with scalelike spines, and the acetabulum is close to the oral sucker. The testes and ovary are dendritic, and the vitellaria are extensive, filling most of the postacetabular space. There is no second intermediate host in the life cycle; the metacercariae encyst on submerged objects or free in the water. One important species lives in the intestinal lumen, but most parasitize the liver of mammals.

■ *Fasciola hepatica* (Fig. 19-4)

Although *Fasciola hepatica* is rare in humans in most countries, it has been known as an important parasite of sheep and cattle for hundreds of years. Because of its size and economic importance, it has been

the subject of many scientific investigations and is probably the best known of any trematode species. The first published record of it is that of Jean de Brie in 1379. He was well acquainted with a disease of sheep called "liver rot," in which the liver of a diseased animal is infected with large, flat worms. In 1668 the great pragmatist Francisco Redi was the first to illustrate this fluke, thereby stimulating others to investigate its biology. Leeuwenhoek was interested in the organism but apparently was distracted by all of the other beings he found with his microscope.

The cercaria and redia of *F. hepatica* were described in 1737 by Jan Swammerdam, a man with a remarkable ability to see and understand microscopic objects with the use of a primitive microscope. Linnaeus gave the worm its name in 1758 but considered it to be a leech. Pallas, in 1760, was the first to find it in a human. Professor C.L. Nitzsch, in 1816, was the first to recognize the similarity of cercariae and adult liver flukes. Thus the history of *F. hepatica* parallels the history of trematodology itself. In 1844 Johannes Steenstrup published a landmark book, *Alternation of Generations,* in which he postulated that trematodes have two generations, one adult and one not.

By the mid-1800s, circumstantial evidence indicated that molluscs were involved in the transmission of *F. hepatica*. In 1880 George Rolleston, professor of anatomy and physiology at Oxford, was convinced that a common slug was the intermediate host of *F. hepatica*. Although he was wrong in this assumption, he recommended that A.P. Thomas undertake an investigation to determine the life cycle of this parasite. Thomas was a 23-year-old demonstrator at the time but took on this formidable task with zeal. He soon found the snail *Lymnaea truncatula* to be infected with rediae and cercariae that were similar in many regards to *Fasciola*. Then he successfully infected this snail with miracidia and followed its development through the sporocyst, redia, and cercarial stages.

At the same time as this "lowly" Oxford demonstrator was investigating liver rot, fascioliasis was also engrossing the mind of the greatest parasitologist then living, Rudolph Leuckart. After a series of false starts Leuckart traced the development of this parasite through the same species of snail and, as a final irony, published his results 10 days before Thomas published his. Credit is given to both men equally, but one can scarcely refrain from lending sympathy to the young Englishman who elucidated the first trematode life cycle in a truly scientific manner, without the advantages of a large budget and long experience.

Neither Thomas nor Leuckart determined the mode of infection of the definitive host. This was done by

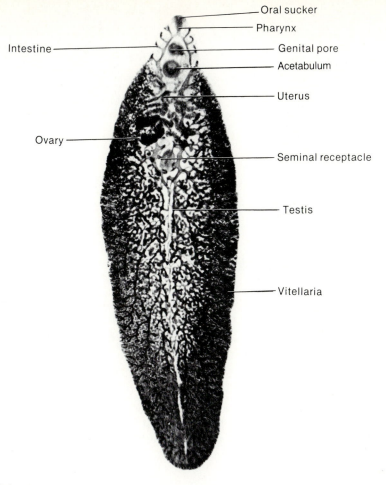

Oral sucker
Pharynx
Intestine
Genital pore
Acetabulum
Uterus
Ovary
Seminal receptacle
Testis
Vitellaria

FIG. 19-4

Fasciola hepatica, the sheep liver fluke.

Courtesy Turtox/Cambosco.

Adolph Lutz, a Brazilian working in Hawaii, who demonstrated in 1892 to 1893 that ruminants become infected by eating juveniles encysted on vegetation. That Lutz was actually working with a different species, *F. gigantica,* is immaterial, since the biology of both species is the same.

Morphology. *Fasciola hepatica* is one of the largest flukes of the world, reaching a length of 30 mm and a width of 13 mm. It is rather leaf shaped, pointed posteriorly and wide anteriorly, although the shape varies somewhat. The oral sucker is small but powerful and is located at the end of a cone-shaped projection at the anterior end. The marked widening of the body at the base of the so-called oral cone gives the worm the appearance of having shoulders. The combi-

nation of an oral cone and "shoulders" is an immediate means of identification. The acetabulum is somewhat larger than the oral sucker and is quite anterior, at about the level of the shoulders. The tegument is covered with large, scalelike spines, reminding one of echinostomes, to which they are closely related. The intestinal ceca are highly dendritic and extend to near the posterior end of the body.

The testes are large and greatly branched, arranged in tandem behind the ovary. The smaller, dendritic ovary lies on the right side, shortly behind the acetabulum, and the uterus is short, coiling between the ovary and the preacetabular cirrus pouch. The vitelline follicles are extensive, filling most of the lateral body and becoming confluent behind the testes. The opercu-

FIG. 19-5

Life cycle of *Fasciola hepatica*. **A,** Adult worm in bile duct of sheep or other mammal. **B,** Egg. **C,** Miracidium. **D,** Mother sporocyst. **E,** Mother sporocyst with developing rediae. **F,** Redia with developing cercariae. **G,** Free-swimming cercaria. **H,** Metacercaria, encysted on aquatic vegetation.

Drawing by Carol Eppinger.

late eggs are 130 to 150 μm by 63 to 90 μm.

Biology (Fig. 19-5). Adult *F. hepatica* live in the bile passages of the liver of many kinds of mammals, especially of ruminants. Humans are occasionally infected. In fact, fascioliasis is one of the major causes of hypereosinophilia in France.[5] The flukes feed on the lining of the biliary ducts. Their eggs are passed out of the liver with the bile and into the intestine to be voided with the feces. If they fall into water, the eggs will complete their development into miracidia and hatch in 9 to 10 days, during warm weather. Colder water retards their development. On hatching, the miracidium has about 24 hours in which to find a suitable snail host, which in the United States is *Fossaria modicella* or *Stagnicola bulimoides*. In other parts of the world different but related snails are the important first intermediate hosts. Mother sporocysts produce first-generation rediae, which in turn produce daughter rediae that develop in the snail's digestive gland. Cercariae begin emerging 5 to 7 weeks after infection. If the water in which the snails live dries up, the snails burrow into the mud and survive, still infected, for months at a time. When water is again present, the snails emerge and rapidly shed many cercariae.

The cercaria has a simple, club-shaped tail about twice its body length. Once in the water, the cercaria quickly attaches to any available object, drops its tail, and produces a thick, transparent cyst around itself. If it does not encounter an object within a short time, it will drop its tail and encyst free in the water. When a mammal eats metacercariae encysted on vegetation or in water, juvenile flukes excyst in the small intestine. They immediately penetrate the intestinal wall, enter the coelom, and creep over the viscera until contacting the capsule of the liver. Then they burrow into the liver parenchyma and wander about for almost 2 months, feeding and growing and finally entering bile ducts.[7] The worms become sexually mature in another month and begin producing eggs. Adult flukes are known to live as long as 11 years. The behavior of juvenile stages within the definitive host has been studied by Sukhdeo and Mettrick.[13]

Epidemiology. Infection begins when metacercaria-infected aquatic vegetation is eaten or when water containing metacercariae is drunk. Humans can be infected by eating watercress, a common green plant. Human infection is common in parts of Europe, northern Africa, Cuba, South America, and other locales. Surprisingly, few cases are known in humans in the United States, although the worm is fairly common in parts of the South and West. Sheep, cattle, and rabbits are the most common reservoirs of infection.

Pathology. Little damage is done by larvae penetrating the intestinal wall and the capsule surrounding the liver (Glisson's capsule), but much necrosis results from the migration of flukes through the parenchyma of the liver. During this time, they feed on liver cells and blood.[6] Anemia sometimes results from heavy infections. There is evidence that this anemia is not caused by hematophagia but probably by a chemical released, perhaps proline.[11] Further, it is known that deposition of bile duct collagen is induced by proline from the worm.[15]

Worms in the bile ducts cause inflammation and edema, which in turn stimulate the production of fibrous tissue in the walls of the ducts (pipestem fibrosis) (see Fig. 20-27). Thus thickened, the ducts can handle less bile and are less responsive to the needs of the liver. Back pressure causes atrophy of the liver parenchyma, with concomitant cirrhosis and possibly icterus. In heavy infections the gallbladder is damaged, and the walls of the bile ducts are eroded completely through, with the worms then reentering the parenchyma, causing the large abscesses. Migrating juveniles frequently produce ulcers in ectopic locations, such as the eye, brain, skin, and lungs.

Ingestion of raw sheep or goat liver in the Middle East may result in adult worms establishing in the nasopharynx. The resulting respiratory blockage is called **halzoun** in Arabic. Recent research casts doubt on this disease entity, placing the full blame on pentastomid larvae and leeches. At this writing, *F. hepatica* remains suspect.

Diagnosis and treatment. Whenever liver blockage coincides with a history of watercress consumption, fascioliasis should be suspected. Specific diagnosis depends on finding the eggs (Fig. 19-6) in the stool. A false record can result when the patient has eaten infected liver and *Fasciola* eggs pass through with the feces. Daily examination during a liver-free

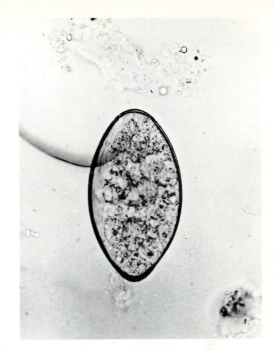

FIG. 19-6

Egg of *Fasciola hepatica*. Eggs of this species are 130 to 150 μm by 63 to 90μ m.

Photograph by Jay Georgi.

diet will unmask this false diagnosis. An enzyme-linked immunosorbent assay (ELISA) test also is available. Early diagnosis is important before irreparable damage to the liver occurs. Prevention in humans depends on eschewing raw watercress. In domestic animals the problem is much more difficult to avoid. Snail control is always a possibility, although this is almost impossible in areas of high precipitation, where nearly every cow track is filled with water and snails. Reservoir hosts, particularly rabbits, can maintain infestation of a pasture when pasture rotation is attempted as a control measure. *Fasciola hepatica* is one of the most important disease agents of domestic stock throughout the world and shows promise of remaining so for years to come. Losses are enormous because of mortality, condemned livers, reduction of milk and meat production, secondary bacterial infection, and expensive anthelmintic treatment.

Several drugs are effective in chemotherapy of fascioliasis, both in humans and in domestic animals. One of these, rafoxanide, apparently acts by uncoupling oxidative phosphorylation in the fluke.[9]

Some work indicates that the microsporidean

FIG. 19-7

Fasciola gigantica, a liver fluke. It may be as large as 75 mm long.

Photograph by Warren Buss.

Nosema eurytremae could be a potential biological control of *F. hepatica.*[3]

■ Other fasciolid trematodes

Fasciola gigantica (Fig. 19-7), a species that is longer and more slender but is otherwise very similar to *F. hepatica,* is found in Africa, Asia, and Hawaii, being relatively common in herbivorous mammals, especially cattle, in these areas. The morphology, biology, and pathology are nearly identical to those of *F. hepatica,* although different snail hosts are necessary.

Fasciola jacksoni causes anemia, loss of weight, hypoproteinemia, abdominal and submandibular edema, and sometimes death in Asian elephants.[4]

Fascioloides magna is a giant in a family of large flukes, reaching nearly 3 inches in length and 1 inch in width. Formerly strictly an American species, it

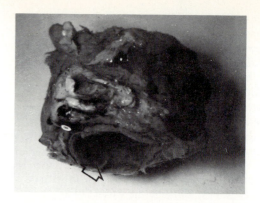

FIG. 19-8

Large calcareous cyst from the liver of a steer (*arrow* points to its opening). This cyst contained two *Fascioloides magna* and is about 3 inches wide.

Photograph by Warren Buss.

was first discovered in Italy, in an American elk in a zoo. Now the fluke has become established in Europe, mainly in game reserves. It is easily distinguished from *Fasciola* spp. by its large size and the absence of a cephalic cone and "shoulders." The life cycle is similar to that of *F. hepatica,* except that the adults live in the liver parenchyma rather than in the bile ducts. Because of their large size, they cause extensive damage. Often, but not always, they become encased in a calcareous cyst of host origin (Fig. 19-8). Their excretory system produces great amounts of melanin, which fills their excretory canals and also the cyst containing them. The normal hosts are probably elk and other Cervidae, but cattle are commonly infected in endemic areas. Human infections have not been found.

Fasciolopsis buski (Fig. 19-9) is a common parasite of humans and pigs in the Orient. Stoll[12] estimated 10 million human infections in 1947. The number may be greater today. Although basically a typical fasciolid, it is peculiar because it lives in the small intestine of its definitive host rather than in the liver. It is elongate-oval, reaching a length of 20 to 75 mm and a width of up to 20 mm. There is no cephalic cone or "shoulders." The acetabulum is larger than the oral sucker and is located close to it. Another difference from "typical" fasciolids is the presence of unbranched ceca. The dendritic testes are tandem in the posterior half of the worm. The ovary is also branched and lies in the midline anterior to the testes. Vitelline follicles are extensive, filling most of the lateral parenchyma all the way to the caudal end. The uterus is short, with

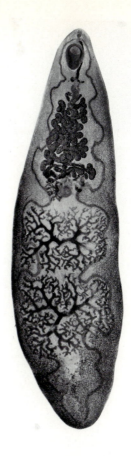

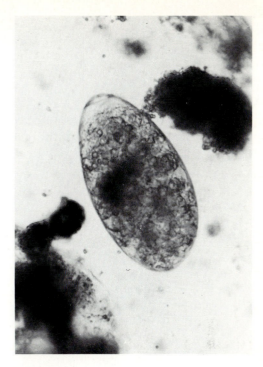

FIG. 19-10

Egg of *Fasciolopsis buski* in a human stool in Taiwan. Eggs of this species are 130 to 140 μm by 80 to 85 μm.

Photograph by Robert E. Kuntz.

FIG. 19-9

Fasciolopsis buski, an intestinal fluke in the Fasciolidae. It may reach 75 mm long.

Photograph by Robert E. Kuntz.

an ascending limb only. The eggs are almost identical to those of *F. hepatica.*

The life cycle of *F. buski* parallels that of *F. hepatica.* Each worm produces about 25,000 eggs (Fig. 19-10) per day, which take up to 7 weeks to mature and hatch at 27° to 32° C. Several species of snails of the genera *Segmentina* and *Hippeutis* (Planorbidae) serve as intermediate hosts. Cercariae encyst on underwater vegetation, including cultivated water chestnut, water caltrop, lotus, bamboo, and other edible plants (Fig. 19-11). Metacercariae are swallowed when these plants are eaten raw or when they are peeled or cracked with the teeth before eating. The worms excyst in the small intestine, grow, and mature in about 3 months without further migration. Infection, then, depends on human or pig feces being introduced directly or indirectly into bodies of water in which edible plants grow.[10]

Pathological conditions resulting from *F. buski* are toxic, obstructive, and traumatic. Inflammation at the site of attachment provokes excess mucous secretion, which is a typical symptom of infection. Heavy infections block the passage of food and interfere with normal digestive juice secretions. Ulceration, hemorrhage, and abscess of the intestinal wall result from long-standing infections. Chronic diarrhea is symptomatic. Another aspect of disease is a typical profound toxemia caused by absorption of the worm's metabolites. This **verminous intoxication** results from sensitization phenomena and may eventually cause the death of the patient. Treatment is usually effective in early or lightly infected cases. Late cases do not fare so well. Prevention is easy. Immersion of vegetables in boiling water for a few seconds will kill the metacercariae. Snail control should be attempted whenever it is impractical to prevent the use of night soil as a fertilizer.

FIG. 19-11

Woman harvesting water caltrop, which serves as a medium for transport of metacercariae of *Fasciolopsis buski* to humans in Taiwan.

Photograph by Robert E. Kuntz.

SUPERFAMILY PARAMPHISTOMOIDEA

The superfamily Paramphistomoidea contains the "amphistomes," flukes in which the acetabulum is located at or near the posterior end. Usually they are thick, fleshy worms with the genital pore preequatorial and the ovary usually posttesticular. Species are found in fish, amphibians, reptiles, birds, and mammals. Of several families in this group, we will consider three.

Family Paramphistomidae

Members of the family Paramphistomidae parasitize mammals, especially herbivores. Several species in different genera of this family parasitize sheep, goats, cattle, cervids, water buffalo, elephants, and other important animals. One species was found once in humans.

■ *Paramphistomum cervi*

The species *Paramphistomum cervi* lives in the rumen of domestic animals throughout most of the world. Adults are almost conical in shape and are pink when living. The testes are slightly lobated.

The life cycle is similar to that of *F. hepatica* and, in North America at least, they develop in the same snail hosts. The cercariae are large and pigmented and have eyespots. There is no second intermediate host; the metacercariae encyst on aquatic vegetation. When eaten, the worms excyst in the duodenum, penetrate the mucosa, and migrate anteriad through the tissues. On reaching the abomasum, or true stomach, they return to the lumen and creep farther forward to the rumen. There they attach among the villi and mature in 2 to 4 months.

Paramphistomum cervi is a particularly pathogenic species. Migrating juveniles cause severe enteritis and hemorrhage, often killing the host. Secondary bacterial infection often complicates the problem. No adequate prevention or treatment is known.

Stichorchis subtriquetrus (Fig. 19-12) is a parasite of beavers, occurring throughout their range. Like *P. cervi,* the metacercariae encyst on underwater objects, including sticks that beavers embed in the bottom of their ponds and streams. When they later eat the bark off these sticks, they swallow any attached metacercariae. A peculiarity of this worm's early embryogenesis is the development of a mother redia within the miracidium. This is released in the snail immediately after penetration, and the remainder of the miracidium

FIG. 19-12

Stichorchis subtriquetrus, a stomach parasite of the American beaver. These parasites are about 10 mm long.

Photograph by Warren Buss.

then disintegrates. Adult worms live in the stomach and reportedly have been causing mortality in beavers in the Soviet Union.

Family Diplodiscidae

Flukes in the family Diplodiscidae have a pair of posterior diverticula in the oral sucker.

■ *Megalodiscus temperatus*

Megalodiscus temperatus and other genera and species of amphistomes are common parasites of the rectum and urinary bladder of frogs. They measure up to 6 mm long and 2.25 mm wide at the posterior end. The posterior sucker is equal to about the greatest width of the body.

The life cycle of this species is similar to that of other amphistomes in that no second intermediate host is required. Miracidia hatch soon after the eggs reach water and penetrate snails of the genus *Helisoma*. Cercariae have eyespots and swim toward lighted areas. If they contact a frog, they will encyst almost immediately on its skin, especially on its dark spots. Frogs molt the outer layers of their skin regularly and not infrequently will eat the sloughed skin. Metacercariae excyst in the rectum and mature in 1 to 4 months. If a tadpole eats a cercaria, the worm will encyst in the stomach and excyst when it reaches the rectum. At metamorphosis, when the amphibian's intestine shortens considerably, the flukes migrate anteriad as far as the stomach and then to the rectum. These parasites are of no economic importance, but they are an easily obtained amphistome for general studies.

Family Gastrodiscidae

The morphology of the family Gastrodiscidae is essentially similar to that of Paramphistomidae and perhaps should not be separate from it. One of its species is a common parasite of humans in restricted areas of the world.

■ *Gastrodiscoides hominis*

This typical amphistome is cone shaped, fleshy, and pink. It is an important parasite of humans in Assam, India, southeastern Asia, and the Philippines, inhabiting the lower small intestine and the upper colon. Rodents and primates are reservoirs.

Adult worms are 5 to 8 mm long by 5 to 14 mm wide at the ventral disc, which occupies about two thirds of the ventral surface. There is a conspicuous posterior notch in the rim of the ventral sucker.

The complete life cycle of *G. hominis* is unknown, but the planorbid snail *Helicorbus coenosus* serves as an experimental host in India.[8] Presumably, humans are infected by eating uncooked aquatic plants. An adult worm draws a mass of mucosal tissue into the ventral sucker and remains attached for some time, causing a nipple-like projection on the intestinal lining.[1] The most common symptom is mucoid diarrhea. Treatment and prevention have not been well studied.

REFERENCES

1. Ahluwalia, S.S. 1960. *Gastrodiscoides hominis* (Lewis and McConnell) Leiper 1913. Indian J. Med. Res. 48:315-325.
2. Bonne, C., G. Bras, and Lie Kian Joe. 1948. Five human echinostomes in the Malayan Archipelago. Med. Monandbl. 23:456-465.
3. Canning, E.U., G.C. Higby, and J.P. Nicholas. 1979. An experimental study of the effects of *Nosema eurytremae* (Microsporida: Nosematidae) on the liver fluke *Fasciola hepatica.* Parasitology 79:381-392.
4. Caple, I.W., M.R. Jainudeen, T.D. Buick, and C.Y. Song. 1979. Some clinico-pathologic findings in elephants (*Elephas maximus*) infected with *Fasciola jacksoni.* J. Wildl. Dis. 14:110-115.
5. Danis, M., J.-P. Nozais, and J. Chandenier. 1985. La fasciolose humaine en France. L'Action Veterinaire 907:1-3.
6. Dawes, B. 1961. On the early stages of *Fasciola hepatica* penetrating into the liver of an experimental host, the mouse: a histological picture. J. Helminthol., R.T. Leiper Supplement, pp. 41-52.
7. Dawes, B., and D.C. Hughes. 1970. Fascioliasis: the invasion stages in mammals. In Dawes, B., editor. Advances in parasitology, vol. 8. Academic Press, Inc., New York, pp. 259-274.
8. Dutt, S.C., and H.D. Srivastava. 1966. The intermediate host and the cercaria of *Gastrodiscoides hominis* (Trematoda: Gastrodiscidae). J. Helminthol. 40:45-52.
9. Prichard, R.K. 1978. The metabolic profile of adult *Fasciola*

hepatica obtained from rafoxanide-treated sheep. Parasitology 76:277-288.

10. Sadun, E.H., and C. Maiphoom. 1953. Studies on the epidemiology of the human intestinal fluke, *Fasciolopsis buski* (Lankester) in central Thailand. Am. J. Trop. Med. Hyg. 2:1070-1084.

11. Spengler, R.N., and H. Isseroff. 1981. Fascioliasis: is the anemia caused by hematophagia? J. Parasitol. 67:886-892.

12. Stoll, N.R. 1947. This wormy world. J. Parasitol. 33:1-18.

13. Sukhdeo, M.V.K., and D.F. Mettrick. 1986. The behavior of juvenile *Fasciola hepatica*. J. Parasitol. 72:492-497.

14. Tubangui, M.A. 1931. Worm parasites of the brown rat, *Rattus norvegicus,* in the Philippine Islands, with special reference to those that may be transmitted to human beings. Phil. J. Sci. 46:537-591.

15. Wolf-Spengler, M.L., and H. Isseroff. 1983. Fascioliasis: bile duct collagen induced by proline from the worm. J. Parasitol. 69:290-294.

SUGGESTED READINGS

Boray, J.C. 1969. Experimental fascioliasis in Australia. In Dawes, B., editor. Advances in parasitology, vol. 7. Academic Press, Inc., New York, pp. 96-210. (A very interesting general account of fascioliasis, together with many experimental approaches. Accent is on special problems of Australia.)

Connor, D.H., and R.C. Neafie. 1976. Fasciolopsiasis. In Binford, C.H., and D.H. Connor, editors. Pathology of tropical and extraordinary diseases, vol. 2, sect. 10. Armed Forces Institute of Pathology, Washington, D.C.

Dawes, B., and D.L. Hughes. 1964. Fascioliasis: the invasive stages of *Fasciola hepatica* in mammalian hosts. In Dawes, B., editor. Advances in parasitology, vol. 2. Academic Press, Inc., New York, pp. 97-168.

Horak, I.G. 1971. Paramphistomiasis of domestic ruminants. In Dawes, B., editor. Advances in parasitology, vol. 9. Academic Press, Inc., New York, pp. 33-72. (A thorough discussion of paramphistomiasis, mainly outside North America.)

Kendall, S.B. 1965. Relationships between the species of *Fasciola* and their molluscan hosts. In Dawes, B., editor. Advances in parasitology, vol. 3. Academic Press, Inc., New York, pp. 59-98.

Kendall, S.B. 1970. Relationships between the species of *Fasciola* and their molluscan hosts. In Dawes, B., editor. Advances in parasitology, vol. 8. Academic Press, Inc., New York, pp. 251-258. (This short article brings the earlier one by this author up to date).

Leuckart, R. 1882. The developmental history of the liver fluke, second part. (English translation: Hoogewey, J.H.) Ill. Vet. 2:8-10.

Meyers, W.M., and R.C. Neafie. 1976. Fascioliasis. In Binford, C.H., and D.H. Connor, editors. Pathology of tropical and extraordinary diseases, vol. 2, sect. 10. Armed Forces Institute of Pathology, Washington, D.C.

Pantelouris, E.M. 1965. The common liver fluke, *Fasciola hepatica* L. Pergamon Press, Oxford, Eng. (A general reference to *F. hepatica.*)

Reinhard, E.G. 1957. Landmarks of parasitology. I. The discovery of the life cycle of the liver fluke. Exp. Parasitol. 6:208-232.

CHAPTER **20**

ORDERS PLAGIORCHIATA AND OPISTHORCHIATA

ORDER PLAGIORCHIATA

Even more than in the superorder Anepitheliocystidia, adults of the order Plagiorchiata often show little resemblance to each other. However, they have many larval and juvenile similarities. The wall of the excretory bladder is epithelial and of mesodermal origin (although some workers are beginning to doubt this). The cercaria has a simple tail, and an oral stylet is common.

SUPERFAMILY PLAGIORCHIOIDEA

Members of the superfamily Plagiorchioidea parasitize fishes, amphibians, reptiles, birds, and mammals and thereby are among the most commonly encountered flukes. They inhabit hosts in marine, freshwater, and terrestrial environments. Most are unimportant parasites of wild animals, but a few are important disease agents of humans and domestic animals. Cercariae possess an oral stylet, and the metacercaria usually encysts in an invertebrate intermediate host. Of the many families in the superfamily, we will discuss four.

Family Dicrocoeliidae

This is one of the three major families of liver flukes that will be considered in this book (see also Fasciolidae, Opisthorchiidae). Some species, however, parasitize the gallbladder, pancreas, or intestine. All are parasites of terrestrial or semiterrestrial vertebrates and use land snails as first intermediate hosts. All dicrocoeliids are medium sized and flattened, with a subterminal oral sucker and a powerful acetabulum in the anterior half of the body. The body is usually pointed at both ends. The ceca are simple. Testes are preequatorial, and the ovary is posttesticular. The voluminous uterus has a descending and ascending limb, commonly filling most of the medullary parenchyma.

Most dicrocoeliids parasitize amphibians, reptiles, birds, and wild mammals and as such are of little economic importance. One cosmopolitan species is an important parasite of domestic mammals and, occasionally, of humans.

■ *Dicrocoelium dendriticum* (Fig. 20-1)

Dicrocoelium dendriticum is common in the bile ducts of sheep, cattle, goats, pigs, and cervids and rarely is found in humans. It is common throughout most of Europe and Asia and has foci in North America and Australia, where it was recently introduced. It is commonly known as the **lancet fluke** because of its bladelike shape.

Morphology. *Dicrocoelium dendriticum* is 6 to 10 mm long by 1.5 to 2.5 mm at its greatest width, near the middle. Both ends of the body are pointed. The ventral sucker is larger than the oral sucker and is located near it. The large, lobate testes lie almost in tandem directly behind the acetabulum, and the small ovary lies immediately behind them. Loops of the uterus fill most of the body behind the ovary. The vitellaria are lateral and restricted to the middle third of the body. The operculate eggs are 36 to 45 μm by 22 to 30 μm.

Biology. *Dicrocoelium dendriticum* is an interesting example of a trematode that has dispensed with the aquatic environment at all stages of its life cycle. Adult *D. dendriticum* live in the bile ducts within the liver, much like *F. hepatica*. When laid, the eggs contain miracidia and must be eaten by land snails before they will hatch (Fig. 20-2). *Cionella lubrica* appears to be the most important snail host in the United States, whereas other species serve in other lands. On hatching in the snail's intestine, the miracidium penetrates the gut wall and transforms into a mother sporocyst in the digestive gland. Mother sporocysts produce daughter sporocysts, which in turn produce xiphidiocercariae (stylet-bearing cercariae). The fact that these cercariae possess well-developed tails probably indicates a recent aquatic origin. About 3 months after infection, cercariae may accumulate in the "lung" (mantle cavity) of the snail or on its body surface. The

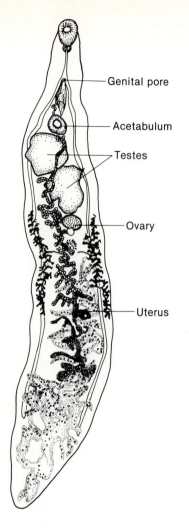

FIG. 20-1

Dicrocoelium dentriticum, a liver fluke of mammals.

Drawing by Ian Grant.

Labels on figure: Genital pore, Acetabulum, Testes, Ovary, Uterus

snail surrounds this irritant with thick mucus and eventually deposits these cercaria-containing **slime balls** as it crawls along. The slime ball may be expelled from the snail's pneumostome with some force. Slime balls are most abundantly produced during a wet period immediately following a drought. Individual slime balls may contain up to 500 cercariae each. Drying of the slime ball surface retards desiccation of the interior and thereby prolongs the lives of the cercariae within.

Continued development of the fluke depends on its ingestion by an ant, which becomes the second intermediate host. The common brown ant *Formica fusca*

is the arthropod host in North America. On eating the delectable slime balls or feeding them to their larvae, the ants become host to metacercariae, most of which encyst in the hemocoel and are then infective to the definitive host. Over 100 metacercariae may occur in a single ant. One or two, however, migrate to the subesophageal ganglion and encyst there.[13] These will not become infective, but they alter the ant's behavior in a most remarkable way. When the temperature drops in the evening, ants thus infected climb to the tops of grasses and other plants and grasp them firmly in their mandibles, whereas the uninfected nest-mates return to warmer digs. They remain attached until later the next day when they warm up and seemingly resume normal behavior.[3] This behavioral pattern keeps the infected ants near the tops of vegetation during the active periods of grazing by ruminants during the evening and morning hours but allows them to retreat to cooler places during the hot hours of the day. The parasite thus influences its intermediate host to behave in a manner encouraging passage to the definitive host. A similar phenomenon is found when the metacercariae of the dicrocoeliid *Brachylecithum mosquensis,* a parasite of American robins, encyst near the supraesophageal ganglion of carpenter ants, *Campanotus* spp. Instead of retreating from brightly lighted areas, as is normal for these ants, infected individuals actually seek such places and wander aimlessly or in circles on exposed surfaces. This makes them much more attractive to the bird definitive host.[5]

On being eaten by a definitive host, *D. dendriticum* excysts in the duodenum. It apparently is attracted by bile and quickly migrates upstream to the common bile duct and thence in the liver. The flukes mature in sheep in 6 or 7 weeks and begin producing eggs about a month later. Up to 50,000 *D. dendriticum* have been found in a single sheep.

Pathological conditions of dicrocoeliiasis are basically the same as that for fascioliasis, except that there is no trauma to the gut wall or liver parenchyma resulting from migrating larvae. General biliary dysfunction, with its several symptoms, is typical.

Numerous cases of *D. dendriticum* in humans have been reported. Most of these were false infections. That is, the eggs that were detected in the stool were actually part of a liver repast that the person had enjoyed a few hours earlier. A few genuine infections in humans have been diagnosed, however, mainly in Russia, Europe, Asia, and Africa. A human case was recently reported in New Jersey.[7] Many cases of human infection with a related species, *D. hospes,* have been reported from Africa.[16]

Adequate drug treatment is available for both hu-

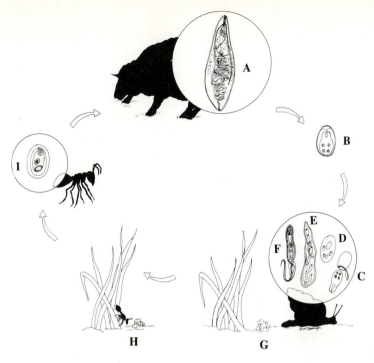

FIG. 20-2

Life cycle of *Dicrocoelium dendriticum*. **A,** Adult, in bile duct of sheep or other plant-eating mammal. **B,** Egg. **C,** Miracidium hatching from egg after being eaten by snail. **D,** Mother sporocyst. **E,** Daughter sporocyst. **F,** Cercaria. **G,** Slime balls containing cercariae. **H,** Ant, eating slime balls. **I,** Metacercaria encysted in ant; the definitive host becomes infected when it accidentally eats the ant.

Diagram by Carol Eppinger.

mans and domestic animals, but control promises to be difficult in the foreseeable future because of the ubiquity of land snails and ants.

Family Haematoloechidae

The flukes of the family Haematoloechidae are parasitic in the lungs of frogs and toads. They are of no economic or medical importance to humans, but because of their large size and easy availability, they are often the first live parasites seen by beginning students of biology. Their transparent beauty, enigmatic location, and fascinating biology have led more than one novitiate into a career in parasitology.

Haematoloecus medioplexus is a typical species among the more than 40 known species that are found in all parts of the world where amphibians occur (Fig. 20-3). This is a flat nonmuscular worm up to 8 mm long and 1.2 mm wide. The acetabulum is small and inconspicuous in this and related species because of their undisturbed site in the lung. The uterus is voluminous, with a descending limb reaching near the posterior end and then ascending with wide loops to the genital pore near the oral sucker. So many eggs fill the uterus that most internal organs are obscured. However, when living worms are placed in tap water, they will expel most of the eggs and thereby become transparent enough to study.

Adult flukes lay prodigious numbers of eggs, which are carried out of the respiratory tract by ciliary action and thence through the gut to the outside. When swallowed by a scavenging *Planorbula armigera* snail, the miracidium hatches and migrates to the hepatic gland, where it develops into a sporocyst. Cercariae escape the snail by night and live the free life for up to 30 hours. When sucked into the anal "lung" of a dragonfly nymph, the cercaria penetrates the thin cuticle and encysts in nearby tissues. When the insect metamorphoses into an adult, the metacercariae remain in the posterior end of the abdomen, to be eaten, along with the rest of the luckless dragonfly, by a frog or toad.

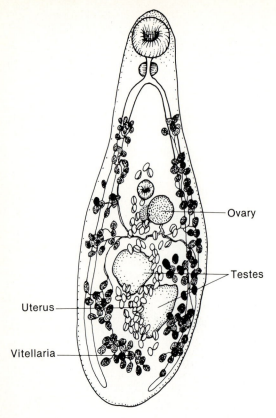

FIG. 20-3

Haematoloechus medioplexus, common in the lungs of frogs.

Drawing by Ian Grant.

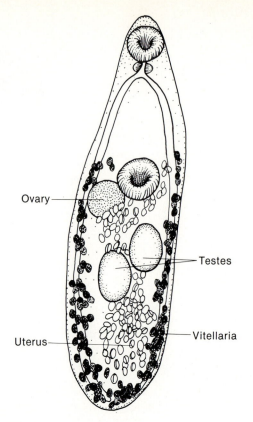

FIG. 20-4

Plagiorchis maculosus, a common parasite of swallows. Birds are infected when they eat mayflies containing metacercariae.

Drawing by Ian Grant.

Excystation occurs in the stomach. The little flukes creep through the stomach, up the esophagus, through the glottis, and into the respiratory tree. As many as 75 worms have been found in a single lung, although two or three are average.

The known life cycles of other haematoloechids are quite similar.

Family Plagiorchiidae

Another family of flukes often encountered in wild animals is Plagiorchiidae. They parasitize vertebrate classes from fishes through birds and mammals and are the "typical" flukes in the animal world. All species use aquatic snails as first intermediate hosts, and insects, such as mayflies and dragonflies, usually are the second intermediate hosts. The literature on plagiorchiids is replete with variations of life cycles and descriptions of the many species of worms encoun-

tered. Apparently this is a plastic group, adaptable to different situations. Representative species are *Plagiorchis maculosus,* cosmopolitan in swallows (Fig. 20-4); *P. nobeli* in American blackbirds; *Ochetosoma* in the mouth and esophagus of snakes; *Astiotrema* in fishes; and *Neoglyphe* in shrews. Several species of *Plagiorchis* have been reported from humans in Japan, Java, and the Philippines, but these no doubt were zoonotic infections.[9]

Family Prosthogonimidae

Most prosthogonimids are parasites living in the oviduct, bursa of Fabricius, or gut of birds. They are remarkably transparent, stain well, and make good examples for classroom studies of trematode morphology.

Prosthogonimus macrorchis (Fig. 20-5) is the oviduct fluke of domestic fowl and various wild birds in

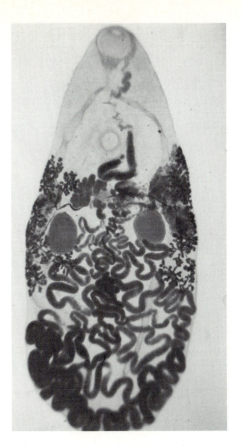

FIG. 20-5

Prosthogonimus macrorchis, an oviduct fluke of birds.

Photograph by Warren Buss.

North America. It causes considerable damage to the oviduct and can decrease or even prevent egg laying. Many have been found within eggs after being trapped in the membranes formed by the oviduct, presumably giving the cook an unexpected surprise.

The life cycle of *P. macrorchis* is similar to that of *Haematoloechus* spp. When the embryonated eggs are passed into water, they sink to the bottom. They do not hatch until eaten by a snail (*Amnicola*), where they burrow into the digestive gland, become sporocysts, and produce short-tailed xiphidiocercariae. When these are sucked into the rectal branchial chamber of a dragonfly nymph, they attach and penetrate into the hemocoel and encyst in the muscles of the body wall, remaining infective after the insect metamorphoses. When eaten by a bird, they excyst in the intestine, migrate downstream to the cloaca and into the bursa of Fabricius or the oviduct, and mature in

about a week. In male birds the infection is lost when the bursa atrophies. More than 30 species of *Prosthogonimus* are known from various areas around the world.

SUPERFAMILY ALLOCREADIOIDEA

Allocreadiids are mainly parasites of fishes, although species are found in amphibians, reptiles, and rarely mammals. Most are of no known importance to humans, but one family has species of great medical significance. The cercariae usually have eyespots; metacercariae encyst in arthropods or fishes.

Family Troglotrematidae

The Troglotrematidae are oval, thick flukes with a spiny tegument and dense vitellaria. They are parasites of the lungs, intestine, nasal passages, cranial cavities, and varous ectopic locations of birds and mammals in many parts of the world. We will illustrate the biology of this interesting group with discussions of two species.

■ *Paragonimus westermani*

Paragonimus westermani was first described from two Bengal tigers that had died in zoos in Europe in 1878. During the next 2 years, infections by this worm in humans were found in Formosa. It was very quickly found in the lungs, brain, and viscera of humans in Japan, Korea, and the Philippines. The life cycle was worked out by Kobayashi[17] and Yokagawa.[33] The major focus of infection today remains in the Orient, including India and the Philippines. It also appears to be endemic throughout the East Indies, New Guinea, the Solomon Islands, Samoa, western Africa, Peru, Colombia, and Venezuela. The taxonomy of the genus is difficult, and some of these reports may be of other, closely related species.[2] Paragonimiasis is an excellent example of a zoonosis.

Morphology. Adult worms (Fig. 20-6) are 7.5 to 12 mm long and 4 to 6 mm at their greatest width. They are very thick, however, measuring 3.5 to 5 mm in the dorsoventral axis. In life they are reddish brown, lending the worms the overall size, shape, and color of coffee beans. The tegument is densely covered with scalelike spines. The oral and ventral suckers are about equal in size, with the latter placed slightly preequatorially. The excretory bladder extends from the posterior end to near the pharynx. The lobated testes are at the same level, located at the junction of the posterior fourth of the body. A cirrus and cirrus pouch are absent. The genital pore is postacetabular.

The ovary is also lobated and is found to the left of

FIG. 20-6

Adult *Paragonimus westermani*

Photograph by Robert E. Kuntz and Jerry A. Moore.

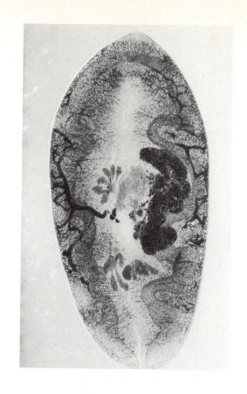

FIG. 20-7

Life cycle of *Paragonimus westermani*. **A,** Sexually mature lung fluke. **B,** The egg is coughed up, swallowed, and passed out with the feces. **C,** Miracidium. **D,** Sporocyst. **E,** Mother redia. **F,** Daughter redia. **G,** Cercaria. **H,** Metacercaria in the second intermediate host.

Diagram by Carol Eppinger.

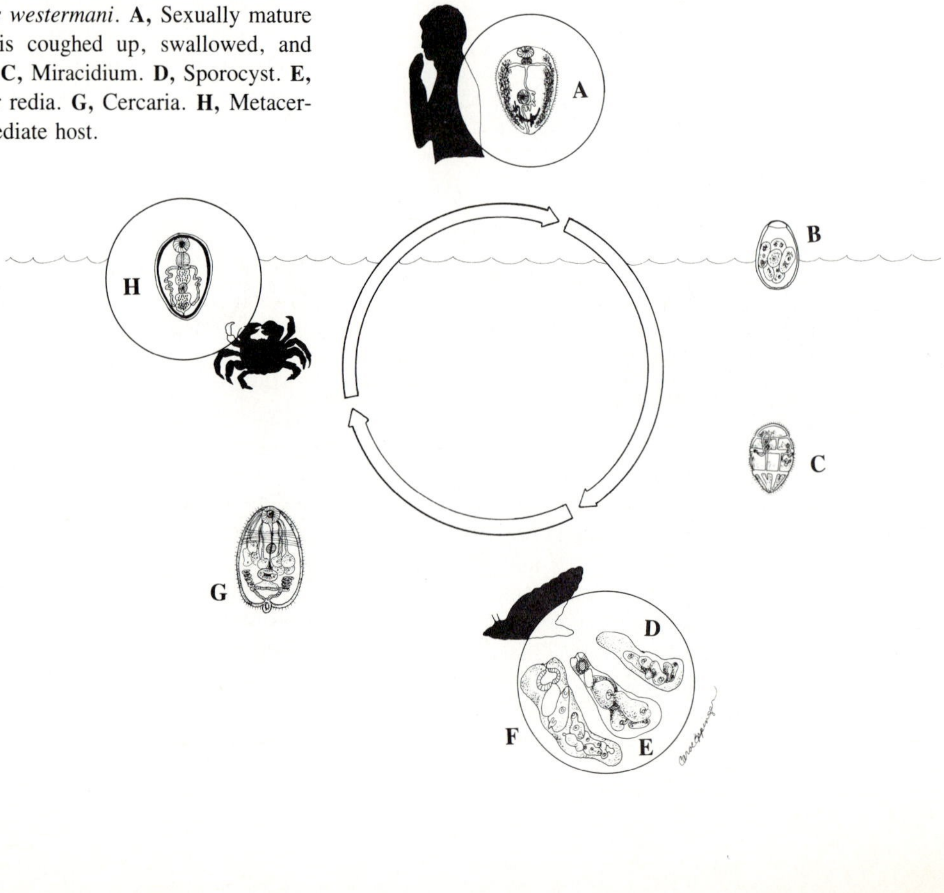

midline, slightly postacetabular. The uterus is tightly coiled into a "rosette" at the right of the acetabulum and opens into the common genital atrium with the vas deferens. Vitelline follicles are extensive in the lateral fields, from the level of the pharynx to the posterior end. The eggs are ovoid and have a rather flattened operculum set into a rim. They measure 80 to 118 μm by 48 to 60μm.

Identification of the 30 or so species of *Paragonimus* is difficult, with much emphasis being placed on the characters of the metacercaria and the shape of the tegumental spines.[20] Several nominal species probably will be synonymized after they have been properly studied.

Biology (Fig. 20-7). Adult *P. westermani* usually live in the lungs, encapsulated in pairs by layers of connective tissues (Figs. 20-8 and 20-9). They have been found in many other organs of the body, however (Fig. 20-10). Cross-fertilization usually occurs. The eggs (Fig. 20-11) are often trapped in surrounding tissues and cannot leave the lungs, but those which escape into the air passages are moved up and out by the ciliary epithelium. Most eggs escape before encapsulation is complete. Arriving at the pharynx, they are swallowed and passed through the alimentary canal to be voided with the feces. The larvae require from 16 days to several weeks in water before development of the miracidium is complete. Hatching is spontaneous, and the miracidium must encounter a snail in the family Thieridae if it is to survive. Since these snails usually live in swift-flowing streams, the chances of survival of any miracidium are slight. This is offset by the numbers of eggs produced by the adult. On entering a snail, the miracidium forms a sporocyst that produces rediae, which in turn develop many cercariae. These cercariae (Fig. 20-12) are microcercous, with spined, knoblike tails and minute oral stylets.

After escaping from the snail, cercariae become quite active, creeping over rocks in inchworm fashion, and attack crabs and crayfish of at least 11 species, encysting in the viscera and muscles. A common second intermediate host in Taiwan is *Eriocheir japonicus* (Fig. 20-13). Some evidence suggests that crustaceans may become infected by eating infected snails.[22] The metacercariae (Fig. 20-14) are pearly white in life and, through microscopic examination, can be identified to species by an expert. When the crustacean is eaten by a proper definitive host, the worms excyst in the duodenum, pierce its wall, and embed themselves in the abdominal wall. Several days later they reenter the coelom, penetrate the diaphragm (Fig. 20-15) and pleura, and enter the bronchioles of

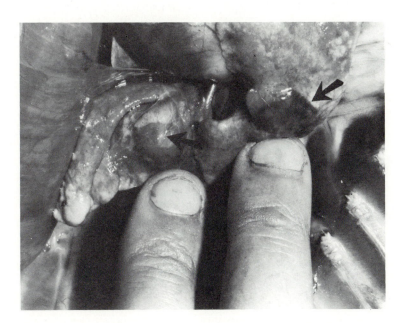

FIG. 20-8

Lung of a cat with two cysts containing adult *Paragonimus westermani (arrows).*

Photograph by Robert E. Kuntz.

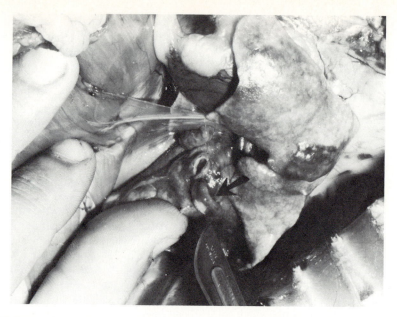

FIG. 20-9

Same specimen as in Fig. 20-8, with a worm dissected out of its cyst *(arrow)*.

Photograph by Robert E. Kuntz.

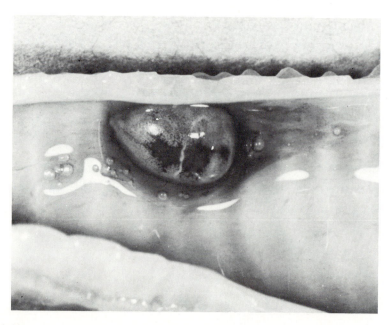

FIG. 20-10

Adult *Paragonimus westermani* in the trachea of an experimentally infected cat.

Photograph by Robert E. Kuntz.

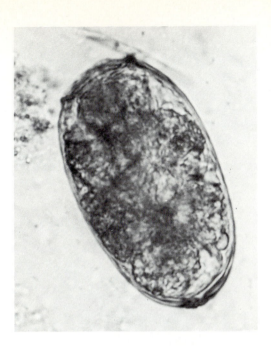

FIG. 20-11

Egg of *Paragonimus westermani* from the feces of a cat. Eggs average 87 by 50 μm.

Photograph by Robert E. Kuntz.

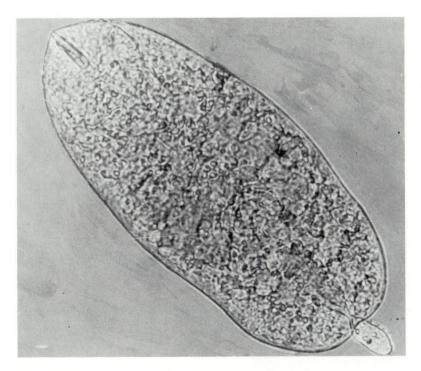

FIG. 20-12

Microcercous cercaria of *Paragonimus westermani*. It is about 500 μm long.

Photograph by Robert E. Kuntz.

FIG. 20-13

Eriocheir japonicus, second intermediate host for *Paragonimus westermani* in Taiwan.

Photograph by Robert E. Kuntz.

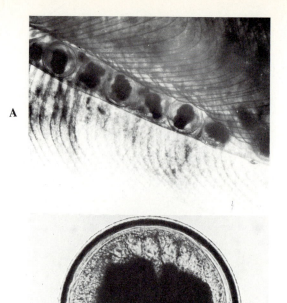

FIG. 20-14

Metacercariae of *Paragonimus westermani.* **A,** Several metacercariae in a gill filament of a crab. **B,** A single metacercaria. The opaque mass is characteristic for this genus. The size is 340 to 480 μm.

Photograph by Robert E. Kuntz.

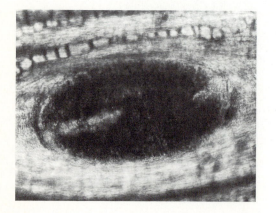

FIG. 20-15

Juvenile *Paragonimus westermani* penetrating the diaphragm of a civet (a small carnivorous mammal).

Photograph by Robert E. Kuntz.

the lungs. They mature in 8 to 12 weeks. Wandering juveniles may locate in ectopic locations, such as the brain, mesentery, pleura, or skin.

Epidemiology. The natural, definitive, and, therefore, reservoir, hosts of *P. westermani* are several species of carnivores, including felids, canids, viverids, and mustelids, as well as some rodents and pigs. Humans are probably a lesser source of infective eggs than are other mammals, but like the others, humans become infected when they eat raw or insufficiently cooked crustaceans. Crab collectors in some countries distribute their catch miles from their source, effectively propagating paragonimiasis (Fig. 20-16). Completely raw crab or crayfish is not as commonly eaten in the Orient as that prepared by marination in brine, vinegar, or wine, which coagulates the protein in the muscles, giving it a cooked appearance and taste but not affecting the metacercariae. Exposure commonly is effected by contamination of fingers or cooking utensils during food preparation (Fig. 20-17).[30] It is even possible that persons accidentally become infected when they smash rice-eating crabs in the paddies, splashing themselves with juices that contain metacercariae. Another factor of possible epidemiological significance in some ethnic groups is the medicinal use of juices strained from crushed crabs or crayfish.

Pathology. The early, invasive stages of paragonimiasis cause few or no symptomatic pathological conditions. Once in a lung or an ectopic site, the worm stimulates connective tissue proliferation that eventually will enshroud it in a brownish or bluish capsule. Such capsules often ulcerate and heal slowly. Eggs in surrounding tissues will themselves become centers of pseudotubercles. Worms in the spinal cord are known to cause paralysis, which sometimes is total. Fatal cases of *Paragonimus* in the heart have been recorded. Cerebral cases have the same results as those of cerebral cysticercosis (see p. 357).[19] Pulmonary cases usually cause chest symptoms, with breathing difficulties, chronic cough, and sputum containing blood or brownish streaks (fluke eggs). Fatal cases are common.

Diagnosis and treatment. The only sure diagnosis, aside from surgical discovery of the adult worm, is by finding the highly characteristic eggs in sputum, aspi-

FIG. 20-16

Crab collectors stringing crabs for a trip to market in Taiwan. This practice distributes *Paragonimus* far from its source.

Photograph by Robert E. Kuntz.

FIG. 20-17

Children "cooking" fresh-caught crabs on an open fire. Such practices contribute to the widespread incidence of infection in the Orient.

Photograph by Robert E. Kuntz.

rated pleural fluid, feces, or matter from a *Paragonimus*-caused ulcer. The pulmonary type is easily mistaken for tuberculosis, pneumonia, spirochaetosis, and other such illnesses; and x-ray examination may be incorrectly interpreted. Cerebral involvement requires differentiation from tumors, cysticercosis, hydatids, encephalitis, and others. Seroimmunological diagnosis is useful and particularly valuable in detecting ectopic infection. The intradermal test is practiced for surveys but must be followed by a complement fixation test on persons testing positive, because the dermal reaction persists for long periods after recovery from the disease.

Several drugs show promise in treatment of paragonimiasis especially praziquantel.[21] Clinical symptoms decrease after 5 to 6 years of infection, but worms are known to live for 20 years or more. Infection can be avoided by cooking crustaceans before eating them and by avoiding contamination with their juices.

Paragonimus kellicotti closely resembles *P. westermani*. It has been found in a wide variety of mammals (cat, dog, raccoon, opossum, skunk, mink, muskrat, bobcat, pig, goat, red fox, coyote, weasel) in North America east of the Rocky Mountains, and many details of its life history and pathogenesis are known.[27,28] The first intermediate host is *Pomatiopsis lapidaria*. Crayfish of the common genus *Cambarus* serve as second intermediate host, with the metacercariae usually encysting on the heart. Like *P. westermani*, *P. kellicotti* in the definitive hosts is usually found in cysts occupied by pairs of worms in the lung. How the migrating worms find each other is still unknown, but encounter with another worm may be necessary for them both to mature.[26] One case of infection in a human has been reported.

Other species of *Paragonimus* have been identified from a variety of wild animals.

- **■ *Nanophyetus salmincola* (Fig. 20-18)**

After 1814 it was known that dogs that ate raw salmon were prone to a disease so severe that scarcely 1 in 10 survived.[12] It was not until 1926 that this disease was shown to be associated with a minute fluke, whose metacercariae were common in the flesh and viscera of salmon.[6] It is now known that the disease itself is caused by the rickettsia *Neorickettsia helminthoeca*, which is transmitted to the dogs by the flukes. Infected dogs can be treated effectively with sulfanilamides and antibiotics.

The identity of the fluke was almost as elusive as that of the rickettsia it carried. When first described, it was placed in the family Heterophyidae, whose members it superficially resembles. Witenberg, in 1932, restudied the original specimens and found a cirrus pouch but no genital sucker or seminal receptacle. He concluded that it was actually a species of *Troglotrema* and transferred it to the Troglotrematidae. In 1935 Wallace returned it to *Nanophyetus* and established the subfamily Nanophyetinae within the Troglotrematidae, where the species resides today.

Morphology. Adult worms are 0.8 to 2.5 mm long and 0.3 to 0.5 mm wide. The oral sucker is slightly larger than the midventral acetabulum. The testes are side by side in the posterior third of the body. A cirrus pouch is present, but there is no cirrus. The small ovary is lateral to the acetabulum, and the uterus is short, containing only a few eggs at a time.

Biology. Adult *N. salmincola* live deeply embedded in crypts in the wall of the small intestine of at least 32 species of mammals, including humans, as well as in fish-eating birds. They produce unembryonated eggs that hatch in water after 87 to 200 days. The snail host in northwestern United States is *Oxytrema silicula*, an inhabitant of fast-moving streams. Experimental infection of snails in the laboratory has not been accomplished. Sporocysts have not been found,

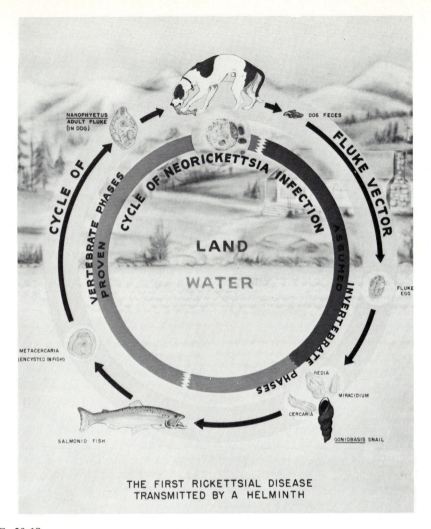

THE FIRST RICKETTSIAL DISEASE
TRANSMITTED BY A HELMINTH

FIG. 20-18

Life cycle of *Nanophyetus salmincola* and of the neorickettsia it harbors.

From Philip, C.B. 1959. Arch. Inst. Pasteur Tunis 36:595-603.

but rediae are well known, occurring in nearly all tissues of the snail. The xiphidiocercaria is microcercous. It penetrates and encysts in at least 34 species of fish, but salmonid fish are more susceptible than are fish of other families. Metacercariae can be found in nearly any tissue of the fish, but they are most numerous in the kidneys, muscles, and fins. Young fish have a high rate of mortality in heavy infection.[10] A variety of mammals and even two bird species (heron and merganser) can be infected with the trematode, but the raccoon and spotted skunk are clearly the main definitive hosts in nature.[25] The worm has been reported from humans in North America on at least 10 occa-

sions and is known to infest up to 98% of the people in some villages in Siberia.[8]

Pathology. Adult flukes themselves cause surprisingly little disease. Philip noted that the inflammatory changes in the intestine of dogs carrying hundreds of *Nanophyetus* were no more extensive than in animals infected with salmon poisoning disease by injection with lymph node suspensions. Salmon poisoning disease is restricted to dogs, coyotes, and other canids and does not affect humans. Its course in dogs is rapid and severe. After an incubation period of 6 to 10 days, the dog's temperature rises to 40° C to 42° C, often with edematous swelling of the face and discharge of pus from the eyes. The dog exhibits depression, loss of appetite, and increased thirst, then vomiting and diarrhea by 4 to 7 days after onset of symptoms. The fever usually lasts from 4 to 7 days, and the dog can be expected to die about 10 days to 2 weeks after onset; however, those that recover are immune for the rest of their lives.

Much remains to be learned about the biology of this fluke and the rickettsia it harbors. Dogs are extremely susceptible, and when untreated, the mortality is about 90%. The disease can be transmitted experimentally by injection of lymph node preparations from other infected dogs or by injecting eggs (evidence of transovarial transmission in the fluke), metacercariae, or adult flukes and digestive glands from infected snails.[23] The geographical range of the disease coincides with the distribution of the snail host of the fluke, and the proportion of salmonids within this range that are infected is extremely high. In light of these facts and the mortality in dogs, it is assumed that there must be some reservoir of the rickettsia, but the identity of that reservoir is not at all clear. Raccoons do not seem to be susceptible; after fluke infection or injection with infected lymph nodes, they have a transitory, low-grade fever, but attempts to transmit the disease from them to dogs by way of lymph node preparations were unsuccessful.[24]

ORDER OPISTHORCHIATA

The order Opisthorchiata contains three superfamilies. The Hemiuroidea and Isoparorchioidea parasitize fishes and a few amphibians in the adult stage, and although they are interesting, there is not room in this book to discuss them. The remaining superfamily is of more importance to humans and domestic animals. It will serve, then, to illustrate the biology of this order of parasites.

SUPERFAMILY OPISTHORCHIOIDEA

These are medium to small flukes, often spinose and with poorly developed musculature. The testes are at or near the posterior end, and a cirrus pouch is absent. A seminal receptacle is present, and the metraterm and ejaculatory ducts unite to form a common genital duct. Eggs are embryonated when passed, but hatching occurs only after ingestion by a suitable snail. Adults live in the intestine or biliary system of fishes, reptiles, birds, and mammals. Metacercariae are in fishes.

Family Opisthorchiidae

Opisthorchiids are delicate, leaf-shaped flukes with weakly developed suckers. Most are exceptionally transparent when prepared for study and so are popular subjects for parasitology classes. Adults are in the biliary system of reptiles, birds, and mammals. Two species are of substantial consequence to humans.

■ *Clonorchis sinensis*

Clonorchis sinensis was first discovered in the bile passages of a Chinese carpenter in Calcutta in 1875. Other infections were quickly discovered in Hong Kong and Japan. Today it is known that the "Chinese liver fluke" is widely distributed in Japan, Korea, China, Taiwan, and Vietnam, where it causes untold suffering and economic loss. Stoll estimated that 19 million persons were infected in eastern Asia in 1947. The number is probably higher today. Reports of this parasite outside the Orient involve infections acquired while visiting there or by eating frozen, dried, or pickled fish imported from endemic areas. Prevalence of infection among 150 New York City immigrant Chinese was 26%. Four cases were described by Sun.[29]

Morphology. Adults (Fig. 20-19) measure 8 to 25 mm long by 1.5 to 5 mm wide. The tegument lacks spines, and the musculature is weak. The oral sucker is slightly larger than the acetabulum, which is about a fourth of the way from the anterior end.

The male reproductive system consists of two large, branched testes in tandem near the posterior end and a large, serpentine seminal vesicle leading to the genital pore. A cirrus and cirrus pouch are absent. The pretesticular ovary is relatively small and has three lobes. The seminal receptacle is large and transverse and is located just behind the ovary. The uterus ascends in broad, tightly packed loops and joins the ejaculatory duct to form a short, common genital duct. The genital pore is median, just anterior to the acetabulum. Vitelline follicles are small and dense and are confined to the level of the uterus. Laurer's canal is conspicuous.

Biology. Chinese liver flukes mature in the bile ducts and produce up to 4000 eggs per day for at least 6 months. The mature egg (Fig. 20-20) is yellow-brown, 26 to 30 μm long and 15 to 17 μm wide. The operculum is large and fits into a broad rim of the eggshell. There is usually a small knob or curved spine on the abopercular end that helps distinguish the eggs of this species. When passed, the egg contains a well-developed miracidium that is rather asymmetrical in its internal organization.

Hatching of the miracidium will occur only after the egg is eaten by a suitable snail, of which *Parafossarulus manchouricus* is the most common and, therefore, most important first intermediate host throughout the Orient. The miracidium transforms into a sporocyst in the wall of the intestine or in other organs within 4

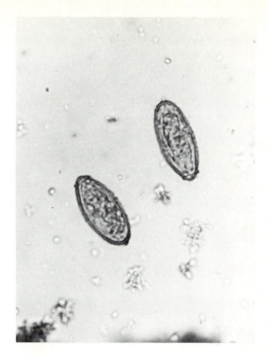

FIG. 20-19

Chinese liver fluke, *Clonorchis sinensis*. Adults measure 8 to 25 mm long by 1.5 to 5 mm wide.

Photograph by Robert E. Kuntz.

FIG. 20-20

Eggs of *Clonorchis sinensis* from a human stool. They are 26 to 30 μm long. Note the small knob on the abopercular end.

Photograph by Robert E. Kuntz and Jerry A. Moore.

hours after infection. These produce rediae within 17 days. Each redia produces from 5 to 50 cercariae. The cercaria (Fig. 20-21) has a pair of eyespots and is beset with delicate bristles and tiny spines. The entire cercaria is brownish. The tail has dorsal and ventral fins (**pleurolophocercous cercaria).**

The cercaria hangs upside down in the water and slowly sinks to the bottom. When contacting any object, it rapidly swims upward toward the surface and again begins to sink. Even a slight current of water will also cause this reaction. Thus, when a fish swims by, the cercaria is stimulated to react in a way favoring its contact with its next host. On touching the epithelium of the fish, the cercaria attaches with its suckers, casts off its tail, and bores through the skin, coming to rest and encysting under a scale or in a muscle (Fig. 20-22). Nearly a hundred species of fishes, mostly in Cyprinidae (Fig. 20-23), have been found naturally infected with metacercariae of *C. sinensis,* although some species are more susceptible than others. Thousands of metacercariae may accumu-

late in a single fish, but the number usually is much smaller. Metacercariae will also develop in the crustaceans *Caridina, Macrobrachium,* and *Palaemonetes;* and such metacercariae have been shown to be infective to guinea pigs.[31] The definitive host is infected when it eats raw or undercooked fish or crustaceans.

Mammals other than humans that have been found infected with adult *C. sinensis* are pigs, dogs, cats, rats, and camels.[18] Experimentally, rabbits and guinea pigs are highly susceptible. Perhaps any fish-eating mammal can become infected. Dogs and cats undoubtedly are important reservoir hosts. Birds may possibly be infected.

The young flukes excyst in the duodenum. The route of migration to the liver is not clear; conflicting reports have been published. It seems probable to us that the juveniles migrate up to the common bile duct to the liver. Young flukes have been found in the liver 10 to 40 hours after infection of experimental animals. The worms mature and begin producing eggs in about a month. The entire life cycle can be completed in 3 months under ideal conditions. Adult worms can live at least 8 years in humans.

FIG. 20-21

Cercaria of *Clonorchis sinensis*.

From Gibson, J.B., and T. Sun, 1971. In Marcial-Rojas, R.A., editor. Pathology of protozoal and helminthic diseases, with clinical correlation. © 1971 The Williams & Wilkins Co., Baltimore.

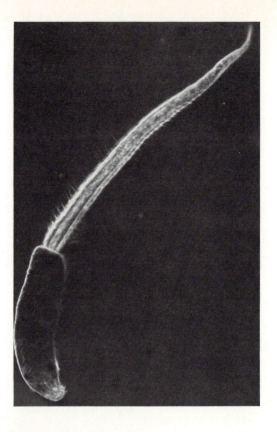

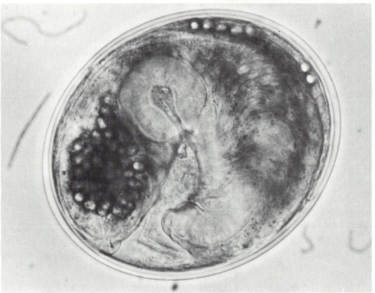

FIG. 20-22

Encysted metacercaria of *Clonorchis sinensis* from fish muscle. The oral and ventral suckers are clearly seen; the round bodies are excretory corpuscles.

From Gibson, J.B., and T. Sun, 1971. In Marcial-Rojas, R.A., editor. Pathology of protozoal and helminthic diseases, with clinical correlation. © 1971 The Williams & Wilkins Co., Baltimore.

FIG. 20-23

Grass carp, *Ctenopharyngodon idellus,* a common second intermediate host of *Clonorchis sinensis.* This fish is widely cultivated in the Orient.

From Gibson, J.B., and T. Sun. 1971. In Marcial-Rojas, R.A., editor. Pathology of protozoal and helminthic diseases, with clinical correlation. © 1971 The Williams & Wilkins Co., Baltimore.

Epidemiology. It is easy to see why clonorchiasis is common in countries in which raw fish is considered a delicacy (Fig. 20-24). In some areas the most heavily infected people are wealthy epicures who can afford beautifully cut and arranged slices of raw fish. On the other hand, the poor are also afflicted, since fish is often their only source of animal protein. The prevalence may range from an average of 14% in cities such as Hong Kong to 80% in some endemic rural areas. Although complete protection is achieved simply by cooking fish, it would be a futile exercise to try to get millions of people to change centuries-old eating habits. Even so, to educate these people to cook their fish would not change matters, since fuel is commonly a luxury that many cannot afford. Kim and Kuntz[15] discuss the epidemiology of clonorchiasis in Taiwan.

Fish farming is a mainstay of protein production throughout the Orient, in Europe, and, increasingly, in the United States. More protein in the form of fish can be harvested from an acre of pond than in the form of beef, beans, or corn from an acre of the finest farmland. The fastest growing fish are primary consumers of algae and other plants. Such ponds typically are fertilized with human feces (Fig. 20-25), which increases the growth rate of water plants and thereby that of the fish. Of course, this abets the life cycle of *C. sinensis.* Where fish farming is not so important, dogs and cats serve as reservoirs of infection, contaminating streams and ponds with their feces.

Metacercariae will withstand certain types of preparation of fish, such as salting, pickling, drying, and smoking. Because of this, people can become infected

FIG. 20-24

"Yue-shan chuk," thin slices of raw carp with rice soup, vegetable garnishing, and soy sauce—a Cantonese delicacy.

From Gibson, J.B., and T. Sun. 1971. In Marcial-Rojas, R.A., editor. Pathology of protozoal and helminthic diseases, with clinical correlation. © 1971 The Williams & Wilkins Co., Baltimore.

thousands of miles from an endemic area when they eat such imported fish.[4]

Pathology. The basic pathogenesis of *Clonorchis* infection is erosion of the epithelium lining the bile ducts (Fig. 20-26). The ultimate effect depends mainly on the intensity and duration of infection, and, fortunately, worm burdens are usually small. The mean intensity of infection in most endemic areas is 20 to 200 flukes, but as many as 21,000 have been removed at a single autopsy. Chronic defoliation of the biliary epithelium leads to gradual thickening and oc-

FIG. 20-25

Privy over a fish-culture pond in Hong Kong. The Chinese characters on the structure advertise a
worm medicine.

From Gibson, J.B., and T. Sun. 1971. In Marcial-Rojas, R.A., editor. Pathology of protozoal and helminthic diseases, with
clinical correlation. © 1971 The Williams & Wilkins Co., Baltimore.

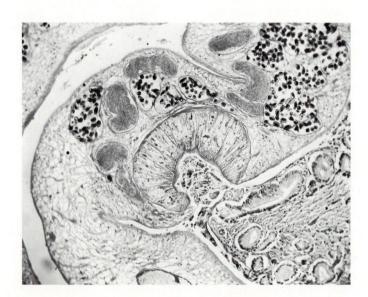

FIG. 20-26

Adult *Clonorchis sinensis* attached by its ventral sucker to biliary epithelium in a human.

From Gibson, J.B., and T. Sun. 1971. In Marcial-Rojas, R.A., editor. Pathology of protozoal and helminthic diseases, with
clinical correlation.© 1971 The Williams & Wilkins Co., Baltimore.

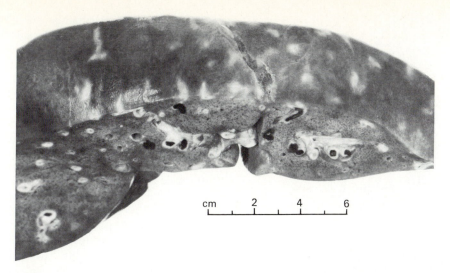

FIG. 20-27

Severe clonorchiasis with "pipestem fibrosis" in a human. The dilated, thick-walled bile ducts are full of flukes.

From Gibson, J.B., and T. Sun. 1971. In Marcial-Rojas, R.A., editor. Pathology of protozoal and helminthic diseases, with clinical correlation. © 1971 The Williams & Wilkins Co., Baltimore.

clusion of the ducts (Fig. 20-27). Pockets form in the walls of bile ducts, and complete perforation into surrounding parenchyma may result. Infiltrating eggs become surrounded by connective tissues, thereby interfering with liver function.

Ascites nearly always occurs in fatal cases, but its relationship to *Clonorchis* infection is uncertain. Jaundice is found in a small percentage of cases and is probably caused by bile retention when ducts are obstructed. Eggs, and sometimes entire worms, often become nuclei of gallstones. Cancer of the liver is more prevalent in Japan than elsewhere, and its relationship to clonorchiasis should be investigated.

Diagnosis and treatment. Diagnosis is based on the recovery of the characteristic eggs in the feces. Liver abnormalities just described should suggest clonorchiasis in endemic areas, but care must be taken to exclude cancer, hydatid disease, beriberi, amebic abscess, and other types of hepatic disease. Intradermal tests seem promising in making a diagnosis. A variety of drugs has been tried against *Clonorchis* infections; at present, praziquantel is the drug of choice.

■ *Opisthorchis felineus*

The cat liver fluke is very similar to *C. sinensis* but has a more European distribution. Originally de-

scribed from a domestic cat in Russia, it is common throughout southern, central, and eastern Europe, Turkey, southern Russia, Vietnam, India, and Japan. Besides cats and other carnivores, it parasitizes humans, probably infecting more than a million persons within its range.

The most obvious difference in morphology from *C. sinensis* is the shape of its testes, which are slightly lobed in *O. felineus* but greatly branched in the Chinese liver fluke. Their life cycles also are nearly identical. The only known snail first intermediate host is *Bithynia leachi*. The pleurolophocercous cercaria encysts within the muscles of several species of cyprinid fish. Its origin and pathogenesis parallel those of *C. sinensis*.

Other, related species of opisthorchiids are *O. viverrini* in wild and domestic carnivores (and possibly 3 million cases in humans) in southeast Asia and *Amphimerus pseudofelineus* in cats and wild carnivores in North America.

Family Heterophyidae

Heterophyids are tiny, teardrop-shaped flukes, usually maturing in the small intestine of fisheating birds and mammals. The suckers are usually feeble, with the acetabulum enclosed inside a sucker-like genital sinus, or **gonotyl,** which is greatly modified in differ-

309

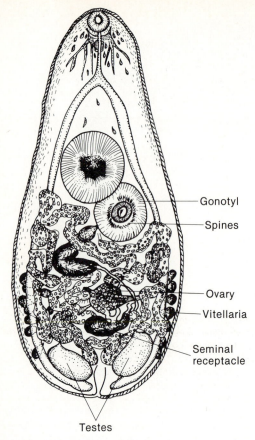

Gonotyl

Spines

Ovary

Vitellaria

Seminal
receptacle

Testes

FIG. 20-28

Heterophyes heterophyes. Its size is 1 to 1.7 mm long.
Drawing by Ian Grant.

ent species. A cirrus pouch is absent. The tegument is scaly, especially anteriorly. This is a large family with several subfamilies. Several species are important parasites of humans.

■ *Heterophyes heterophyes* (Fig. 20-28)

Heterophyes heterophyes is a minute fluke that was first discovered in an Egyptian in Cairo in 1851. It is common in northern Africa, Asia Minor, and the Far East, including Korea, China, Japan, Taiwan, and the Philippines. Because the eggs cannot be differentiated from other, related species, an accurate estimate of human infections cannot be made.

Morphology. The adults are 1 to 1.7 mm long and 0.3 to 0.4 mm at their greatest width. The entire body is covered with slender scales, most numerous near the anterior end. The oral sucker is only about 90 μm in diameter, whereas the acetabulum is around 230

μm wide and is located at the end of the first third of the body. The two oval testes lie side by side near the posterior end of the body. The vas deferens expands to form a sinuous seminal vesicle, which constricts again, becoming a short ejaculatory duct. The ovary is small, medioanterior to the testes, at the beginning of the last fourth of the body. A seminal receptacle and Laurer's canal are present. The uterus coils between the ceca and constricts before joining the ejaculatory duct to form a short common genital duct, which then opens into the genital sinus. The gonotyl is about 150 μm wide and has 60 to 90 toothed spines on its margin. Lateral vitelline follicles are few in number and are confined to the posterior third of the worm. The eggs are 28 to 30 μm by 15 to 17 μm.

Biology. Adult worms live in the small intestine, burrowed between the villi. The egg contains a fully developed miracidium when laid but hatches only when eaten by an appropriate freshwater or brackish-water snail (*Pironella conica* in Egypt, *Cerithidia cingula* in Japan). After penetrating the gut of the snail, the miracidium transforms into a sporocyst that produces rediae. A second-generation redia gives birth to cercariae with eyespots and finned tails (**ophthalmolophocercous cercariae**), which emerge from the snail. Like the cercaria of *Clonorchis sinensis,* that of *H. heterophyes* swims toward the surface of the water and slowly drifts downward. On contacting a fish, it penetrates the epithelium, creeps beneath a scale, and encysts in muscle tissue. Metacercariae are most abundant in various species of mullet, which are exposed when they enter estuaries or brackish-water shorelines. Several thousand metacercariae have been found in a single, small fish. The definitive host becomes infected when it eats raw or undercooked fish.

Epidemiology. For eggs to be available to estuarine and brackish-water snails, pollution must occur in these waters. Therefore boatmen, fishermen, and others who live by or on the water are often the main reservoirs of infection. Infected fish are distributed widely in fish markets. Other fish-eating mammals, such as cats, foxes, and dogs, serve as other reservoirs of infection.

Pathology. Each worm elicits a mild inflammatory reaction at the site of contact with the intestine. Heavy infections, which are common, cause damage to the mucosa and produce intestinal pain and mucous diarrhea. Perforation of the mucosa and submucosa sometimes occurs and allows eggs to enter the blood and lymph vascular systems and to be carried to ectopic sites in the body.[1] The heart is particularly affected, with tissue reactions in the valves and myocardium leading to heart failure. Kean and Breslau[14] reported that 14.6% of cardiac failure in the Philippines resulted from heterophyid myocarditis. Eggs in the brain or spinal cord lead to neurological disorders that are sometimes fatal. Two bizarre cases are known where adult *H. heterophyes* were found in the brains of humans, and in another case an adult worm was found in the myocardium.[1] Such infections are probably more common than previously thought, for experimental infections in laboratory animals often lead to ectopic flukes.[11] Immature flukes were found inside lymphoid follicles and Peyer's patches. Young flukes migrated from the sinuses in Peyer's patches via the lymphatics to the mesenteric lymph glands, which become enlarged and hyperplastic and contained mature worms.

Diagnosis is difficult when adult worms are not available. The eggs closely resemble those of several other heterophyids and are not very different from those of *C. sinensis*. Tetrachloroethylene has been found effective in treatment, but praziquantel is the drug of choice.

■ Other heterophyid parasites of humans

Heterophyes katsuradai is very similar to *H. heterophyes*. It has been found in humans near Kobe, Japan. Infection is acquired by eating raw mullet. *Metagonimus yokagawai* is a very common heterophyid in the Far East, the Soviet Union, and the Balkan region, where it infects humans. It superficially resembles *H. heterophyes,* but its acetabulum is displaced to the left, where it is fused with the gonotyl. The biology of *M. yokagawai* is identical to that of *H. heterophyes,* except that a different snail host (*Semisulcospira* spp.) is required and the second intermediate hosts are freshwater fish of several species. The definitive host becomes infected when it eats uncooked fish. Various fish-eating mammals are natural reservoirs, and even pelicans have been incriminated in this regard. Pathogenesis, diagnosis, and treatment are as for *H. heterophyes*.

Until proved otherwise, all species of Heterophyidae should be considered potential parasites of humans. More than a dozen species have been found infective to date.

REFERENCES

1. Africa, C.M., W. de Leon, and E.Y. Garcia. 1937. Heterophyidiasis. VI. Two more cases of heart failure associated with the presence of eggs in sclerosed veins. J. Philippine Isl. Med. Assoc. 17:605-609.
2. Alarcón de Noya, B., O. Noya G., J. Torres, and C. Botto. 1985. A field study of paragonimiasis in Venezuela. Am. J. Trop. Med. Hyg. 34:766-769.
3. Anokhin, I.A. 1966. Daily rhythm in ants infected with metacercariae of *Dicrocoelium lanceatum*. Dokl. Akad. Nauk SSSR 166:757-759.
4. Binford, C.H. 1934. Clonorchiasis in Hawaii. Report of cases in natives of Hawaii. Public Health Rep. 49:602-604.
5. Carney, W.P. 1969. Behavioral and morphological changes in carpenter ants harboring dicrocoeliid metacercariae. Am. Midland Natur. 82:605-611.
6. Donham, C.R., B.T. Simms, and F.W. Miller. 1926. So-called salmon poisoning in dogs (progress report). J. Am. Vet. Med. Assoc. 68:701-715.
7. Drabick, J.J., J.E. Egan, S.L. Brown, R.G. Vicr, B.M. Sandman, and R.C. Neafie. 1988. Dicroceliasis (lancet fluke disease) in an HIV seropositive man. JAMA 259:567-568.
8. Eastburn, R.L., T.R. Fritsche, and C.A. Terhune, Jr. 1987. Human intestinal infection with *Nanophyetus salmincola* from salmonid fishes. Am. J. Trop. Med. Hyg. 36:586-591.
9. Faust, E.C., P.F. Russell, and R.C. Jung. 1970. Craig and Faust's Clinical parasitology, ed. 8. Lea & Febiger, Philadelphia.
10. Gebhardt, G.A., R.E. Millemann, S.E. Knapp, and P.A. Nyberg. 1966. "Salmon poisoning" disease. II. Second intermediate host susceptibility studies. J. Parasitol. 52:54-59.
11. Hamdy, E.L., and E. Nicola. 1981. On the histopathology of the small intestine in animals experimentally infected with *H. heterophyes*. J. Egypt. Med. Assoc. 63:179-184.
12. Henry's Astoria Journal. 1814. In the Oregon country under the Union Jack. A reference book of historical documents for Scholars and Historians. 1962. Rayette Radio Ltd., Montreal.
13. Hohorst, W. 1964. Die Rolle der Ameisen in Entwicklungsgan des Lanzettegels *(Dicrocoelium dendriticum)*. Z. Parasitenkd. 22:105-106.
14. Kean, B.H., and R.C. Breslau. 1964. Parasites of the human heart. Grune & Stratton, Inc., New York, pp. 95-103.
15. Kim, D.C., and R.E. Kuntz. 1964. Epidemiology of helminth diseases: *Clonorchis sinensis* (Cobbold, 1875) Looss, 1907 on Taiwan (Formosa). Chinese Med. J. 11:29-47.
16. King, E.V.J. 1971. Human infection with *Dicrocoelium hospes* in Sierra Leone. J. Parasitol. 57:989.
17. Kobayashi, H. 1918. Studies on the lung-fluke in Korea. Mitteil. Med. Hochschule Keijo. 2:95-113.
18. Komiya, Y., and N. Suzuki. 1964. Biology of *Clonorchis sinensis*. In Morishita, K., et al., editors. Progress of medical parasitology in Japan, vol. 1. Meguro Parasitological Museum, Tokyo, pp. 551-645.
19. Madrigal, R.B., B. Rodriquez-Otiz, G.V. Solano, E.M.O. Oband, and P.J.R. Sotela. 1982. Cerebral hemorrhagic lesions produced by *Paragonimus mexicanus*. Am. J. Trop. Med. Hyg. 31:522-526.
20. Miyazaki, I. 1965. Recent studies on *Paragonimus* in Japan with special reference to *P. ohirai* Miyazaki, 1939, *P. iloktsuenensis* Chen, 1940 and *P. miyazakii* Kamo, Nishida, Hatsushika et Tomimura, 1961. In Morishita K., et al., editors. Progress of medical parasitology in Japan, vol. 2. Meguro Parasitological Museum, Tokyo, pp. 349-354.
21. Monson, M.H., J.W. Koenig, and R. Sachs. 1983. Successful treatment with praziquantel of six patients infected with the African lung fluke, *Paragonimus uterobilateralis*. Am. J. Trop. Med. Hyg. 32:371-375.
22. Noble, G.A. 1963. Experimental infection of crabs with *Paragonimus*. J. Parasitol. 44:352.
23. Nyberg, P.A., S.E. Knapp, and R.E. Millemann. 1967.

"Salmon poisoning" disease. IV. Transmission of the disease to dogs by *Nanophyetus salmincola* eggs. J. Parasitol. 53:694-699.

24. Philip, C.B. 1955. There's always something new under the "parasitological" sun (the unique story of the helminthborne salmon poisoning disease). J. Parasitol. 41:125-148.

25. Schlegel, M.W., S.E. Knapp, and R.E. Millemann. 1968. "Salmon poisoning" disease. V. Definitive hosts of the trematode vector, *Nanophyetus salmincola*. J. Parasitol. 54:770-774.

26. Sogandares-Bernal, F. 1966. Studies on American paragonimiasis. IV. Observations on pairing of adult worms in laboratory infections of cats. J. Parasitol. 52:701-703.

27. Sogandares-Bernal, F., and J.R. Seed. 1973. American paragonimiasis. Curr. Top. Comp. Pathobiol. 2:1-56.

28. Stromberg, P.C., and J.P. Dubey. 1978. The life cycle of *Paragonimus kellicotti* in cats. J. Parasitol. 64:998-1002.

29. Sun, T. 1980. Clonorchiasis: a report of four cases and discussion of unusual manifestations. Am. J. Trop. Med. Hyg. 29:1223-1227.

30. Suzuki, Z. 1958. Epidemiological studies on paragonimiasis in South Izu District, Shizuoka Prefecture, Japan Kiseichugaku Zasshi 7:560-572.

31. Tang, C.C., et al. 1963. Clonorchiasis in South Fukien with special reference to the discovery of crayfishes as second intermediate hosts. Chinese Med. J. 82:545-618.

32. Yokagawa, M. 1969. *Paragonimus* and paragonimiasis. In Dawes, B., editor. Advances in parasitology, vol. 7. Academic Press, Inc., New York. pp. 375-387.

33. Yokagawa, S. 1919. A study of the lung distoma. Third report, Formosan Endoparasitic Disease Research.

SUGGESTED READINGS

Dooley, J.R., and R.C. Neafie. 1976. Clonorchiasis and opisthorchiasis, In Binford, C.H., and D.H. Connor, editors. Pathology of tropical and extraordinary diseases, vol. 2, sect. 10. Armed Forces Institute of Pathology, Washington, D.C.

Holmes, J.C., and W.M. Bethel. 1972. Modification of intermediate host behaviour by parasites. In Canning, E.U., and C.A. Wright, editors. Behavioral aspects of parasite transmission. Linnaean Society of London. Academic Press Inc. (London) Ltd., pp. 123-149. (An outstanding summary of the subject. Should be required reading for all students of parasitology.)

Komiya, Y. 1966. *Clonorchis* and clonorchiasis. In Dawes, B., editor. Advances in parasitology, vol. 4. Academic Press, Inc., New York, pp. 53-106.

Meyers, W.M., and R.C. Neafie. 1976. Paragonimiasis. In Binford, C.H., and D.H. Connor, editors. Pathology of tropical and extraordinary diseases, vol. 2, sect. 10. Armed Forces Institute of Pathology, Washington, D.C.

Millemann, R.E., and S.E. Knapp. 1970. Biology of *Nanophyetus salmincola* and "salmon poisoning" disease. In Dawes, B., editor. Advances in parasitology, vol. 8. Academic Press, Inc., New York, pp. 1-41.

Miyata, I. 1965. The development of *Eurytrema pancreaticum* and *Eurytrema coelomaticum* in the intermediate host snails. In Morishita, K., et al., editors. Progress of medical parasitology in Japan, vol. 2. Meguro Parasitological Museum, Tokyo, pp. 348-357.

Travassos, L. 1944. Revisão da familia Dicrocoeliidae Odhner, 1910. Inst. Oswaldo Cruz Monogr. 2:1-357. (The definitive monograph on Dicrocoeliidae.)

Yokagawa, M. 1965. *Paragonimus* and paragonimiasis. In Dawes, B., editor. Advances in parasitology, vol. 3. Academic Press, Inc., New York, pp. 99-158. (A complete summary of paragonimiasis. Required reading by all who are interested in the subject.)

CHAPTER 21

CLASS CESTOIDEA: FORM, FUNCTION, AND CLASSIFICATION OF THE TAPEWORMS

Fear and superstition still abound among laypersons, who generally view tapeworms as the lowliest and most degenerate of creatures (Fig. 21-1). Most of the repugnance with which most people regard these animals derives from the fact that the tapeworms live in the intestine and are only seen when they are passed with the feces of the host. Furthermore, tapeworms seem to be generated spontaneously, and mystery is nearly always accompanied by fear. Finally, in a few instances their presence initiates disease conditions that traditionally have been difficult to cure. Be that as it may, a scientific approach to cestodology has increased understanding of tapeworms and shown that they are one of the most fascinating groups of organisms in the animal kingdom. Their complex life cycles and intricate host-parasite relationships are rivaled by few known organisms. Observations of cestodes no doubt began in earliest times and extend today into sophisticated laboratories throughout the world.

Historically, Hippocrates, Aristotle, and Galen appreciated the animal nature of tapeworms.[24] The Arabs suggested that the segments passed with the feces were a separate species of parasite from tapeworms; they called these segments the cucurbitini, after their similarity to cucumber seeds.[30] Andry, in 1718, was the first to illustrate the scolex of a tapeworm from a human (Fig. 21-2). Three common species in humans, *Taeniarhynchus saginatus, Taenia solium,* and *Diphyllobothrium latum,* were confused by all scientists until the brilliant efforts of Küchenmeister, Leuckart, Mehlis, Siebold, and others in the nineteenth century determined both the external and internal anatomy of these and other common species. These researchers also proved conclusively that bladder worms, hydatids, and coenuri were juvenile tapeworms and not separate species or degenerate forms in improper hosts. Although these organisms have been removed from the realms of ignorance and superstition within the past 150 years, much misconception persists.

Sexually mature tapeworms live in the intestine or

its diverticula (rarely in the coelom) of all classes of vertebrates. Two forms are known that mature in invertebrates: *Archigetes* spp. (order Caryophyllidea) in the coelom of a freshwater oligochaete and *Cyathocephalus truncatus* (order Spathebothriidea) in the hemocoel of an amphipod.[1]

FORM AND FUNCTION
Strobila

The **strobila** (Fig. 21-3) of cestodes is a structure unique among the Metazoa. Typically it consists of a linear series of sets of reproductive organs of both sexes; each set is referred to as a **proglottid** or **proglottis.** Cestodes with multiple proglottids are described as **polyzoic,** but members of the order Caryophyllidea and of the subclass Cestodaria have only one set of reproductive organs (**monozoic**) (Fig. 21-4). Some workers advise avoiding the use of the terms *polyzoic* and *monozoic* because such usage implies that polyzoic tapeworms are chains of zoids (individuals), which may not be the case.[48] Nonetheless, in the absence of a better term, we shall retain *polyzoic* to describe tapeworms with multiple sets of genitalia. Polyzoic cestodes may have only a few proglottids, but in others there may be thousands. Usually there are constrictions between proglottids, and the worms are said to be segmented. However, tissues such as tegument and muscle are continuous between proglottids, and no membranes separate them; thus "segment" may also be a misnomer.[48] Some polyzoic cestodes lack such constrictions between proglottids (order Spathebothriidea). Further, "segments" of tapeworms should not be confused with segments or metameres of metameric animals such as annelids and arthropods (p. 530). Yet, zoologists have no trouble calling zoids of coelenterates, such as *Obelia,* and many bryozoans as colonies of individuals, so a philosophical argument for a polyzoic organization of most tapeworms is not difficult to accept.

In many polyzoic species new proglottids are con-

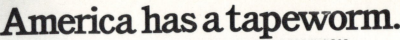

FIG. 21-1

Advertisements of this kind illustrate the low regard most people have for cestodes.

Courtesy SANE, Washington, D.C.

tinuously differentiated near the anterior end, a process called **strobilation.** Each proglottid moves toward the posterior end as a new one takes its place and, during the process, becomes sexually mature. By the time they approach the posterior end of the strobila, the proglottids will have copulated and produced eggs. A given proglottid can copulate with itself, with others in the strobila, or with those in other worms, depending on the species. After the proglottid contains fully developed eggs or shelled embryos, it is said to be **gravid.** When it reaches the end of the strobila, it often detaches and passes intact out of the host with the feces, as *Taenia* does, or disintegrates en route, releasing the eggs, as *Hymenolepis* does. This process is called **apolysis.** In some species the eggs are released from the gravid proglottid through a uterine pore, such as in *Diphyllobothrium* spp., or through tears or slits in the proglottid (as in Trypanorhyncha);

FIG. 21-2

Taeniid tapeworm, probably *T. saginatus,* illustrated for the first time by Andry in 1718.

the proglottid only detaches when it is senile or exhausted **(pseudapolysis** or **anapolysis).** In some forms the proglottids may be shed while immature and lead an independent existence in the gut while developing to maturity **(hyperapolysis),** as in some Tetraphyllidea. If the posterior margin of a proglottid overlaps the anterior of the following one, the strobila is said to be **craspedote;** if not, it is called **acraspedote** (Fig. 21-5).

Scolex

Most tapeworms bear a "head," or **scolex** (plural: **scolices**), at the anterior end that may be equipped with a variety of holdfast organs to maintain the position of the animal in the gut (Fig. 21-6). The scolex may be provided with suckers, grooves, hooks, spines, glands, or combinations of these (Fig. 21-7). However, the scolex can be simple or absent altogether. In some forms the holdfast function of the scolex is lost early in life, and a holdfast organ is formed by a distortion of the anterior end of the strobila called a **pseudoscolex** (Fig. 21-8). Some species penetrate the gut wall of the host to a considerable distance, with the scolex and a portion of the strobila then encapsulated by reacting host tissues.

315

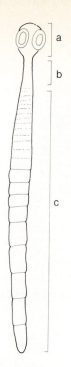
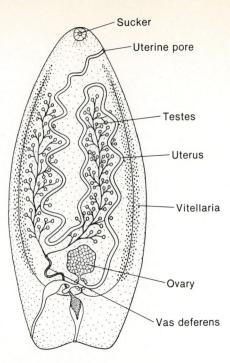

FIG. 21-3

Generalized diagram of a tapeworm, showing scolex *(a),* neck *(b),* and strobila *(c).*

From Schmidt, G.D. 1970. How to know the tapeworms. William C. Brown Co., Publishers, Dubuque, Iowa.

FIG. 21-4

Example of monozoic tapeworm. *Amphilina foliacea* (Amphilinidea). (See also Fig. 22-36).

Redrawn from Wardle, R.A., and J.A. McLeod. 1952. The zoology of tapeworms. Hafner Publishing Co., New York.

FIG. 21-5

A, Scolex and proglottids of *Paranoplocephala mamillana,* a craspedote cestode. **B,** A proglottid of *Dipylidium caninum,* an acraspedote species.

Photographs by Jay Georgi.

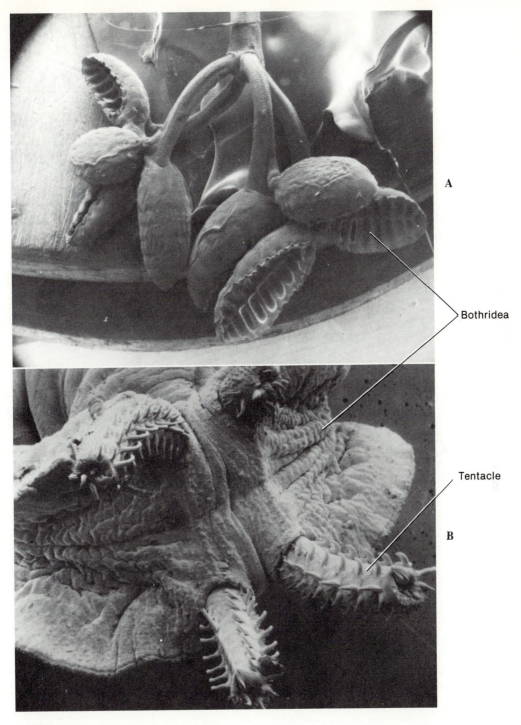

FIG. 21-6

Two types of holdfast organs. **A,** Bothridea of a tetraphyllidean. **B,** Spiny tentacles of a trypano-rhynchan.

Photographs by Frederick H. Whittaker.

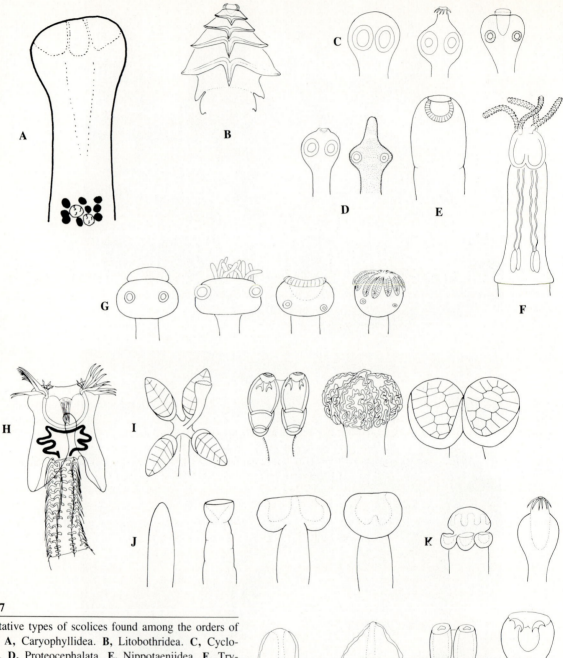

FIG. 21-7

Representative types of scolices found among the orders of cestodes. **A,** Caryophyllidea. **B,** Litobothridea. **C,** Cyclophyllidea. **D,** Proteocephalata. **E,** Nippotaeniidea. **F,** Trypanorhyncha. **G,** Lecanicephalidea. **H,** Diphyllidea. **I,** Tetraphyllidea. **J,** Spathebothriidea. **K,** Aporidea. **L,** Pseudophyllidea.

From Schmidt, G.D. 1970. How to know the tapeworms. William C. Brown Co., Publishers, Dubuque, Iowa.

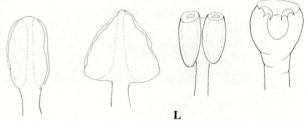

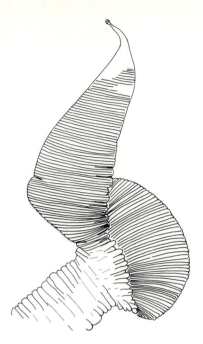

FIG. 21-8

Fimbriaria fasciolaris, a tapeworm with a pseudoscolex in addition to a tiny true scolex.

Modified from Mönnig, H.O. 1934. Veterinary helminthology and lutomology. William Wood, London.

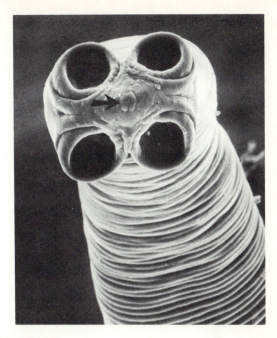

FIG. 21-9

Scolex of *Hymenolepis diminuta.* Note the apical organ *(arrow).* (×200.)

From Ubelaker, J.E., V.F. Allison, and R.D. Specian. 1973. J. Parasitol. 59:667-671.

Sucker-like organs on scolices of tapeworms can be divided into three types: **acetabula, bothria,** and **bothridia.** The acetabulum is more or less cup shaped, circular or oval in outline, and with a heavy muscular wall. There are normally four acetabula on a scolex, spaced equally around it. Bothridia usually are in groups of four; are quite muscular, projecting sharply from the scolex; and have highly mobile, leaflike margins. Bothria are usually two in number, although as many as six may occur, and take the form of shallow pits or longer grooves. Accessory suckers sometimes occur, and most cestodes have a variety of keratinaceous hooks for anchoring the scolex to the host gut. In acetabulate worms the hooks often are arranged in one or more circles anterior to the suckers, borne on a protrusible, dome-shaped area on the apex of the scolex called the **rostellum.** Both the presence or absence and the shape and arrangements of the hooks are of great taxonomic value. If the rostellum is armed with hooks,it is supplied internally with a heavy muscular pad, which becomes flat and disc-shaped when the hooks attach to the host's gut wall. Retraction of the central area of the pad allows withdrawal of the hooks.

Various kinds of gland cells have been reported in the scolices of a variety of tapeworms, but their function(s) remain(s) enigmatic.[38,39] In some Caryophyllidea and Pseudophyllidea, secretions of the glands may aid in adhesion of the scolex to the host gut mucosa.[29,63] The contents of one type of gland in the pseudophyllidean *Diphyllobothrium dendriticum* are expelled within 3 days after infection of the definitive host; another gland type remains active and is associated with the nervous system in the scolex.[27] *Hymenolepis diminuta,* a cyclophyllidean with an unarmed rostellum, has an invagination of the apical tegument termed an apical organ or anterior canal (Fig. 21-9).[75] Cells identified as modified tegumentary cytons (see p. 320) secrete material into the apical organ and through the surrounding rostellar tegument.[76] It is possible that these materials play some regulatory role in the development of the worms; there is circumstantial evidence that they are antigenic.[19,31] Apical organs are found in many other cestodes, but they may not be homologous or even structurally similar to that of *H. diminuta.* Apparently similar and homologous structures are found in the order Proteocephalata, where, at least in certain cases, their secretion has proteolytic activity and probably functions in penetration (p. 325).[17]

The scolex contains the chief neural ganglia of the worm (as will be discussed further), and it bears numerous sensory endings on its anterior surface, probably detecting both physical and chemical stimuli. Such sensory input may allow optimal placement of the scolex and entire strobila with respect to the gut surface and physicochemical gradients within the intestinal milieu.

Commonly, between the scolex and the strobila lies a relatively undifferentiated zone called the **neck,** which may be long or short. It contains stem cells that apparently are responsible for giving rise to new proglottids. In the absence of a neck, similar cells may be present in the posterior portion of the scolex. Praziquantel, a chemotherapeutic agent highly effective against cestodes, preferentially damages the tegument of the neck region and leaves the tegument of proglottids farther down the strobila unaffected.[7]

Tegument

Cestodes lack any trace of a digestive tract and therefore must absorb all required substances through their external covering. Because of this fact, the structure and function of the body covering have been of great interest to parasitologists, who have used electron microscopy and radioactive tracers to contribute much to this area of cestodology. Before 1960 the body covering of cestodes and trematodes was commonly referred to as a "cuticle," but it is now known that it is a living tissue with high metabolic activity, and most parasitologists prefer the term **tegument.**

Tegumental structure is generally similar in all cestodes studied, differing in details according to species. The tegument is covered by minute projections called **microtriches** (singular: microthrix) that are underlaid by the tegumental distal cytoplasm (Fig. 21-10). The distal cytoplasm is connected to cytons by internuncial processes that run through the superficial muscle layer (Fig. 21-11). The microtriches are similar in some respects to the microvilli found on gut mucosal cells and other vertebrate and invertebrate transport epithelia, and they completely cover the worm's surface, including the suckers. They have a dense distal portion set off from the base by a multilaminar plate (Fig. 21-12). The cytoplasm of the base is continuous with that of the rest of the tegument, and the entire structure is covered by a plasma membrane. A layer of carbohydrate-containing macromolecules, the **glycocalyx,** is found on the membrane. The microtriches serve to increase the absorptive area of the tegument, but they also may help the worm maintain its position in the host gut.[73] A number of phenomena, apparently depending on interaction of certain molecules with the glycocalyx, have been reported: enhancement of host amylase activity; inhibition of host trypsin, chymotrypsin, and pancreatic lipase; absorption of cations; and absorption of bile salts. Several of these seem to depend on absorption of the molecules to the glycocalyx, but present evidence suggests that this is not the

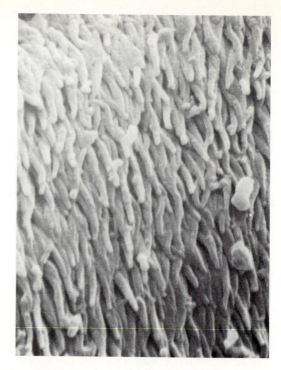

FIG. 21-10

Posteriorly directed microtriches on the surface of a proglottid of *Hymenolepis diminuta.* (×44,925.)

From Ubelaker, J.E., V.F. Allison, and R.D. Specian. 1973. J. Parasitol. 59:667-671.

case with trypsin inhibition.[70] When incubated in the presence of *H. diminuta,* trypsin seems to undergo a subtle conformational change that decreases its proteolytic activity. The functional value of such phenomena to the worm is uncertain, but interaction with nutrient absorption, protection against digestion by host enzymes, and maintenance of the integrity of the worm's surface membrane may be involved.

The distal cytoplasm beneath the microtriches contains abundant vesicles and electron-dense bodies, as well as numerous mitochondria. The tegumental nuclei are not found in this layer but lie in the cytons. The vesicles are secreted in the cytons, passed to the distal cytoplasm through the internuncial processes, and at least some of them contribute to microthrix and hook formation.[41,51,63] Although each cyton contains but one nucleus, the distal cytoplasm is continuous, with no intervening cell membranes; therefore the tegument of cestodes is a syncytium.

An organ of unknown function called a **tumulus** has been described in the tegument of adult pseudophyllideans (Fig. 21-13).[12,79] An evagination is connected via ducts to cytons lying beneath the superficial muscle, and materials produced in the cytons are trans-

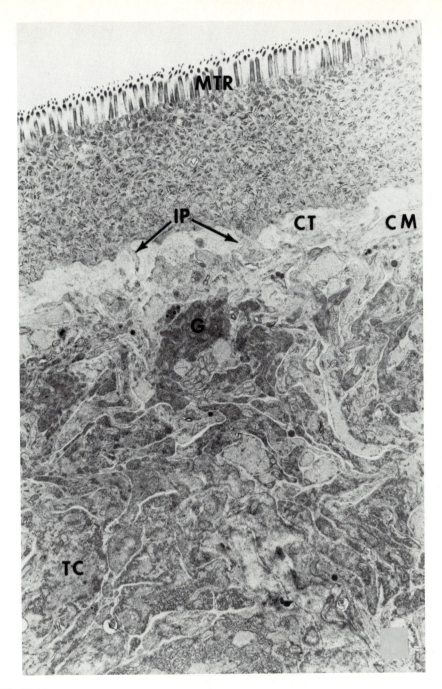

FIG. 21-11

Longitudinal section through immature proglottid of *Hymenolepis diminuta* showing nature of tegumentary cortical region. Basal tegumentary cytons (perikarya) *(TC)* are surrounded by glycogen-filled processes *(G)* of cortical myocytons. Internuncial processes (trabeculae) *(IP)* from tegumentary cytons extend through longitudinal and circular *(CM)* muscles, as well as connective tissue *(CT)* layer (basement lamina), before joining syncytial distal cytoplasm. Microtriches *(MTR)* line free surface of syncytial layer, and discoidal vesicles occupy distal cytoplasm. (× 5900.)

From Lumsden, R.D., and R.D. Specian. 1980. In Arai, H.P., editor. Biology of the rat tapeworm, *Hymenolepis diminuta*. Academic Press, Inc., New York.

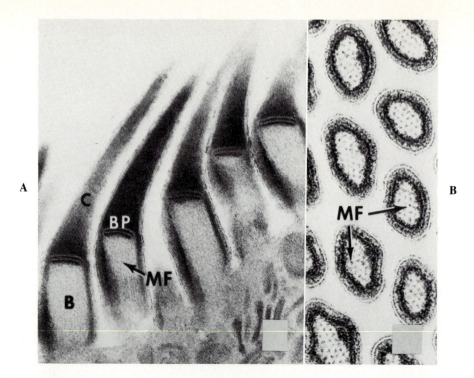

FIG. 21-12

A, Sagittal section of tegumental microtriches. Electron-opaque cap *(C)* separated from the base *(B)* by multilaminar baseplate *(BP)*. Microfilaments *(MF)* regularly arranged within base. Tegumental plasmalemma extends over entire length of each microthrix. (× 71,000.) **B,** Cross section through bases of tegumental microtriches revealing orderly array of microfilaments *(MF)* surrounded by accumulation of electron-dense material. (× 120,000.)

From Lumsden, R.D., and R.D. Specian. 1980. In Arai, H.P., editor. Biology of the rat tapeworm, *Hymenolepis diminuta*. Academic Press, Inc., New York.

ported to the tumulus, there to be secreted by an eccrine mechanism to the outside.[79]

Calcareous corpuscles

The tissues of most cestodes contain curious structures termed calcareous corpuscles, also found in the excretory canals of some trematodes.[13] They are secreted in the cytoplasm of differentiated calcareous corpuscle cells, which are themselves destroyed in the process. The corpuscles are from 12 to 32 μm in diameter, depending on the species, and consist of inorganic components, principally compounds of calcium, magnesium, phosphorus, and carbon dioxide embedded in an organic matrix. The organic matrix is organized into concentric rings and a double outer envelope; it contains protein, lipid, glycogen, mucopolysaccharides, alkaline phosphatase, RNA, and DNA. The bodies always contain a series of minor inorganic elements, and these, as well as the amount of

phosphate, are affected by the diet of the host. The possible function of the calcareous corpuscles has been the subject of much speculation. For example, mobilization of the inorganic compounds might buffer the tissues of the worm against the large amounts of organic acids produced in its energy metabolism (p. 337). Another suggestion has been that they might provide depots of ions or carbon dioxide for use when such substances are present in insufficient quantity in the environment, such as on initial establishment in the host gut. Confirmation of these proposals or discovery of the true function of the corpuscles will require the application of a creative scientific mind.

Muscular system

It has been shown that the muscle cells of *Hymenolepis diminuta* consist of two portions: the contractile myofibril and the noncontractile myocyton.[42] The contractile portion contains actin and myosin fibrils,

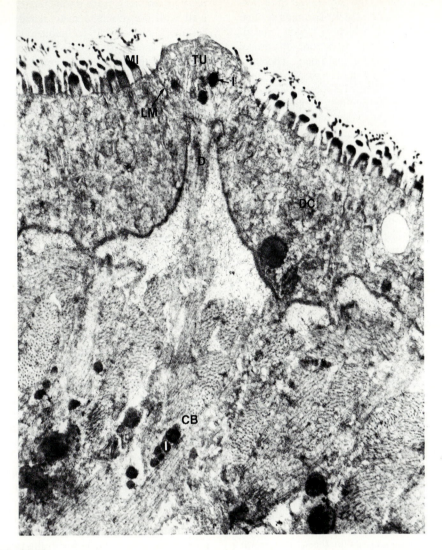

FIG. 21-13

Tegument of *Eubothrium salvelini*, showing tumulus. The tegumental basal lamina *(B)* is evaginated to form a duct *(D)* leading to the tumulus *(TU)*. Note inclusions *(I)* within the distal portion of a cytoplasmic bridge *(CB)* and under the surface of the tumulus. Also shown are the tegumental distal cytoplasm *(DC)*, muscle fibers *(MU)*, and microtriches *(MI)*. A possible lateral membrane *(LM)* is indicated. (× 14,500.)

From Tedesco, J.L., and J.R. Coggins. 1980. Int. J. Parasitol. 10:275-280.

and like muscles of other platyhelminths, it is nonstriated and lacks transverse sarcolemmal tubules (T tubules),[40] as might be expected of muscles with slow contraction. The myocyton comprises the bulk of the parenchyma of the worm and has traditionally been referred to as the "parenchymal" cell. It contains the nucleus, rough endoplasmic reticulum, free ribosomes, a vesicular Golgi apparatus, few mitochondria, and abundant glycogen. Lipid is stored here as well. Although these cytological details are best known for *H. diminuta*, it is highly likely that they pertain to all other cestodes and even trematodes.

The contractile portions of the muscle cells are arranged in discrete bundles in specific regions of the worms. Just internal to the distal cytoplasm are bundles of longitudinal and circular fibers. More powerful musculature lies below the superficial muscles. The longitudinal bundles are usually arranged around a central parenchymal area, which is largely free of contractile elements. There may be a zone of cortical parenchyma, also free of longitudinal fibers. There are numbers of dorsoventral and transverse fibers, and sometimes radial fibers as well. The pattern and relative development of muscle bundles are highly variable in the Cestoidea but constant within a species; therefore they are often valuable taxonomic characters.

Internal musculature of the scolex is complex, making the scolex extraordinarily mobile. Three distinct muscle types have been found in scolices of trypanorhynchs (p. 371): peripheral myofibers similar to those previously described, tentacle retractor muscles, and tentacle bulb muscles. The bulb muscles are obliquely striated and have numerous motor end plates, whereas motor innervation of the peripheral muscles and retractor muscles has not been found.[87]

Nervous system

The main nerve center of the cestode is in its scolex, and the complexity of ganglia, commissures, and motor and sensory innervation there depends on the number and complexity of other structures on the scolex. Among the simplest are the bothriate cestodes such as *Bothriocephalus,* which have only a pair of lateral cerebral ganglia united by a single ring and a transverse commissure. Arising from the cerebral ganglia is a pair of anterior nerves, supplying the apical region of the scolex; four short posterior nerves; and a pair of lateral nerves that continue posteriorly through the strobila. The bothria are innervated by small branches from the lateral nerves. In contrast, worms with bothridia or acetabula and hooks, rostellum, and so on may have a substantially more complex system of commissures and connectives in the scolex, with three to five longitudinal nerves running posteriorly from the cerebral ganglia through the strobila (Fig. 21-14). In addition to the motor innervation of the scolex, there may be many sensory endings, particularly at the apex of the tegument. Stretch receptors have been described.[62]

As the longitudinal nerves proceed posteriad, they are connected by intraproglottidal commissures in a ladder-like fashion. Smaller nerves emanate from them to supply the general body musculature and sensory endings. The cirrus and vagina are richly innervated, and sensory endings around the genital pore are more abundant than in other areas of the strobilar tegument.[90] Such an arrangement has obvious value.

Study of the neuroanatomy of cestodes is difficult because the nerves are unmyelinated and do not stain well with conventional histological stains. However, certain histochemical techniques that show sites of acetylcholinesterase activity delineate the nervous systems of cestodes beautifully, even the sensory endings in the tegument. This fact, as well as studies on the actions of cholinesterase inhibitors, lends support to the conclusion that neural transmission in cestodes is essentially cholinergic.[73] There is evidence to indicate, however, that there is no functional motor innervation to the somatic muscles in at least some cestodes.[86] Ward, Allen, and McKerr[86] found that 5-hydroxytryptamine excited spontaneous contractions by depolarizing somatic muscles of a trypanorhynch, but acetylcholine inhibited contractions by hyperpolarizing the muscle.

Sensory function probably includes both tactoreception and chemoreception, and cestodes possess at least two, and perhaps more, morphologically distinct types of sensory endings in their tegument.[9] One of these has a modified cilium projecting as a terminal process (Fig. 21-15), as is common in such cells in invertebrates.

Osmoregulatory system

In many families of cestodes the main "osmoregulatory" canals run the length of the strobila from the scolex to the posterior end. These are usually in two pairs, one ventrolateral and the other dorsolateral on each side (Fig. 21-16). Most often the dorsal pair is smaller in diameter than is the ventral pair, a useful criterion for determining the dorsal and ventral sides of a tapeworm. The canals may branch and rejoin throughout the strobila or may be independent. Usually a transverse canal joins the ventral canals at the posterior margin of each proglottid. The dorsal and ventral canals unite in the scolex, often with some degree of branching. Posteriorly, the two pairs of canals merge into an excretory bladder with a single pore to the outside. When the terminal proglottid of a polyzoic species is detached, the canals empty independently at the end of the strobila. Rarely the major canals also empty through short, lateral ducts. In some orders, such as Caryophyllidea and Pseudophyllidea, the canals form a network that lacks major dorsal and ventral ducts.

Embedded throughout the parenchyma are flame cell protonephridia (Fig. 21-17), whose ductules feed into the main canals. The cilia of the flame cell provide motive force to the fluid in the system.

Although the system is often called "osmoregulatory," it is also commonly referred to as "excretory," and there is some evidence that the latter is more descriptive. An analysis of fluid from the excretory ca-

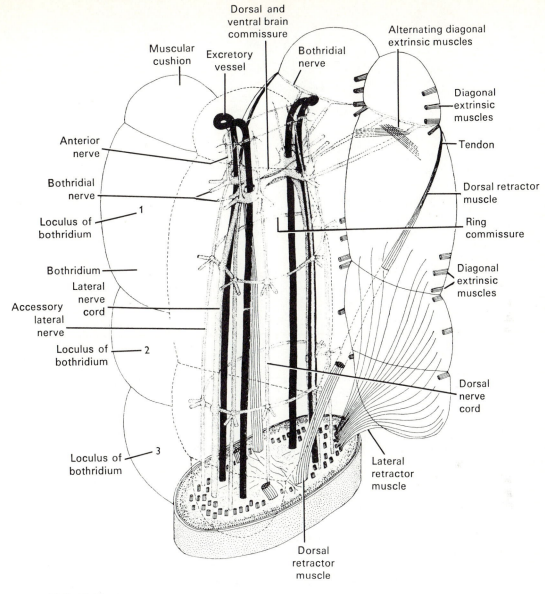

FIG. 21-14

Acanthobothrium coronatum. Reconstruction of nervous system of scolex. Excretory vessels and some muscles included.

From Rees, G., and H.H. Williams. 1965. Parasitology 55:617-651.

nals of *H. diminuta* demonstrated glucose, soluble proteins, lactic acid, urea, and ammonia; lipid was absent.[89]

In some cases, at least, the excretory ducts are lined with microvilli (Fig. 21-18), thus suggesting that the duct linings serve a transport function. Therefore possible functions of the system might include active transport of excretory wastes and resorption of substances such as ions from the excretory fluid.

Osmoregulation, on the other hand, may be another function of the tegumental surface. Although cestodes have been regarded as osmoconformers, with little ability to regulate their body volume in media of differing osmotic concentrations, there is some evidence

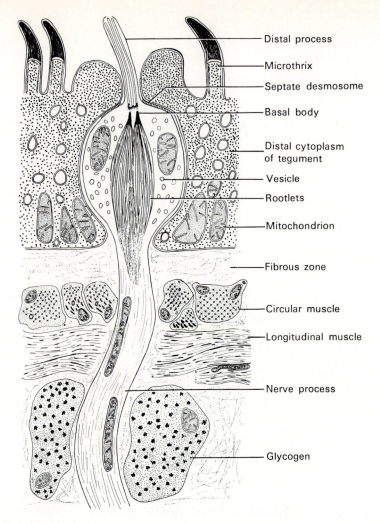

Distal process

Microthrix

Septate desmosome

Basal body

Distal cytoplasm of tegument

Vesicle

Rootlets

Mitochondrion

Fibrous zone

Circular muscle

Longitudinal muscle

Nerve process

Glycogen

FIG. 21-15

Schematic drawing of a longitudinal section through a sensory ending in the tegument of *Echinococcus granulosus*.

From Morseth, D.J. 1967. J. Parasitol. 53:492-500.

that they can regulate within the range of salinities they are likely to encounter in the host gut.[44] Lussier, Podesta, and Mettrick[44] found an increased flux of amino acids with increasing osmolarity, and they suggested that an active driving force alters amino acid fluxes to counteract the osmotic transfer of water.

Reproductive systems

Tapeworms are monoecious with the exception of a few rare species from birds and two from a stingray that are dioecious. Usually each proglottid has one complete set of both male and female systems, but some genera have two sets of each system (Fig. 21-5, *B*), and a few species in birds have one male and two female systems, in each proglottid.

As a proglottid moves toward the posterior end of the strobila, the reproductive systems mature, sperm are transferred, and ova are fertilized. Usually the male organs mature first and produce sperm that are stored until maturation of the ovary; this is called **protandry** or **androgyny.** In a few species the ovary matures first; this is called **protogyny** or **gynandry.** This may be an adaptation that avoids self-fertilization of the same proglottid. Many variations occur in struc

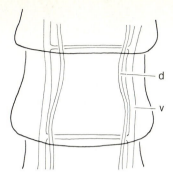

FIG. 21-16

Diagram showing the typical arrangement of dorsal *(d)* and ventral *(v)* osmoregulatory canals.

From Schmidt, G.D. 1970. How to know the tapeworms. William C. Brown Co., Publishers, Dubuque, Iowa.

ture, arrangement, and distribution of reproductive organs in tapeworms. These variations are useful at all levels of taxonomy.

■ Male reproductive system

The male reproductive system (Fig. 21-19) consists of one to many testes, each of which has a fine vas efferens. The vasa efferentia unite into a common vas deferens that channels the sperm toward the genital pore. The vas deferens may be a simple duct, or it may have sperm storage capacity in convolutions or in a spheroid external seminal vesicle. As the vas deferens leads into the cirrus pouch, which is a muscular sheath containing the terminal organs of the male system, it may form a convoluted ejaculatory duct or dilate into an internal seminal vesicle. The male copulatory organ is the muscular cirrus, which may or may not bear spines. It can invaginate into the cirrus pouch and evaginate through the cirrus pore. Commonly the reproductive pores of both sexes open into a common sunken chamber, the genital atrium, which may be simple or equipped with spines, stylets, glands, or accessory pockets. The cirrus pore may open on the margin or somewhere on the flat surface of the proglottid. If two male systems are present, they open on margins opposite from one another.

■ Female reproductive system

The female reproductive system (Fig. 21-20) consists of an ovary and associated structures, which are variable in size, shape, and location, depending on the genus. The entire complex is called the **oogenotop**. Vitelline cells, which contribute yolk and eggshell material to the embryo, may be arranged into a single, compact vitellarium, or they may be scattered as folli-

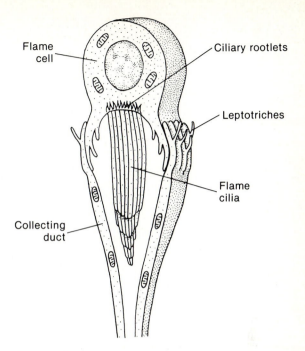

FIG. 21-17

Diagram of terminal organ of flame cell *(FC)* protonephridium in *Hymenolepis diminuta*. *F*, Flame composed of approximately 50 cilia; *R*, ciliary rootlets; *LT*, leptotriches; *DC*, cytoplasm of the collecting duct.

Redrawn from M.B. Hildreth; from Lumsden, R.D., and R.D. Specian. 1980. In Arai, H.P., editor. Biology of the rat tapeworm, *Hymenolepis diminuta*. Academic Press, Inc., New York. Drawing by William C. Ober.

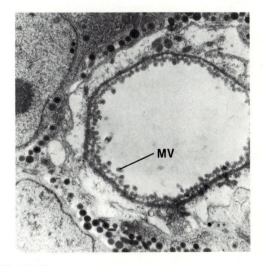

FIG. 21-18

Low-magnification electron micrograph of excretory duct of *Hymenolepis diminuta* showing bead-like microvilli *(MV)*.

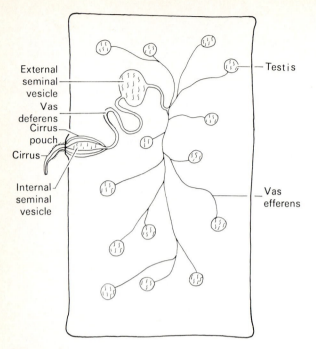

FIG. 21-19

Diagrammatic representation of a tapeworm male reproductive system.

From Schmidt, G.D. 1970. How to know the tapeworms. William C. Brown Co., Publishers, Dubuque, Iowa.

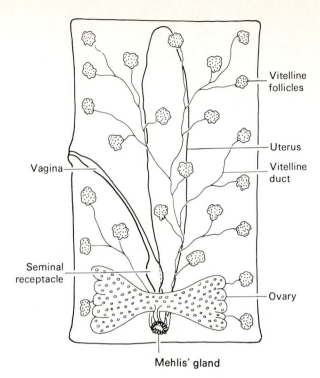

FIG. 21-20

Diagram of a tapeworm female reproductive system.

From Schmidt, G.D. 1970. How to know the tapeworms. William C. Brown Co., Publishers, Dubuque, Iowa.

cles in various patterns. Ova are called **ectolecithal** because they do not produce their own yolk. As oocytes mature, they leave the ovary through a single oviduct, which usually has a controlling sphincter, the **oocapt.** Fertilization occurs in the proximal oviduct. One or more cells from the vitelline glands pass through a common vitelline duct, sometimes equipped with a small vitelline reservoir, and join with the zygote. Together they pass into an area of the oviduct known as the ootype. This zone is surrounded by unicellular glands, called Mehlis' glands, which appear to secrete a thin membrane around the zygote and its associated vitelline cells. Eggshell formation is then completed from within by the vitelline cells and, in some cases, cells of the embryo. Eggs of Pseudophyllidea and Caryophyllidea are covered by a thick **capsule** of sclerotin. These capsules are apparently homologous with trematode eggshells and are formed in a similar manner (p. 239). Some of these embryonate in water after passing from the host and usually hatch to release a free-swimming larval stage that is eaten by the aquatic intermediate host. Shell formation in other cestodes is complicated by a variety of layers contributed by

embryonic cells.[18,83] These layers include the **coat, embryophore,** and **oncospheral membrane;** the capsule is thin or lacking. Three different types are recognized (Fig. 21-21): (1) *Dipylidium* type with a thin capsule and an embryophore (as in the cyclophyllidean genera *Dipylidium, Moniezia,* and *Hymenolepis* and in Tetraphyllidea and Proteocephalata); (2) *Taenia* type with a very thin capsule but with a thick embryophore (as in *Taenia* and *Echinococcus* spp.); and (3) *Stilesia* type, formed by species with no distinct vitellaria, with cellular covering apparently laid down by the uterine wall.[73] In contrast with the pseudophyllidean types, only one or a few vitelline cells associate with the zygote. During early embryogenesis some cells become segregated from the rest of the embryo, fuse, surround the embryo, and form the **outer envelope (OE)** (Fig. 21-22). Other cells become an **inner envelope (IE).** The vitelline cell contributes to the OE. The coat forms within the OE and adds to or replaces the capsule. The embryophore and oncospheral membrane are formed by the IE.

As the incipient egg and vitelline cells pass through the ootype, the secretions of Mehlis' glands are

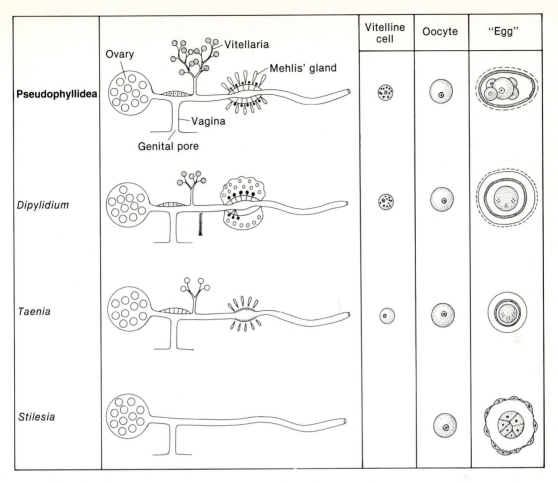

FIG. 21-21

Types of egg-forming systems in cestodes. **A,** Pseudophyllidean type, found also in the Caryophyllidea. A relatively thick capsule of sclerotin is formed from material from the vitelline cells. **B,** *Dipylidium* type, found in some Cyclophyllidea, Tetraphyllidea, and Proteocephalata. A thin coat from the outer envelope is added to the thin capsule. In tetraphyllideans 2 to 3 vitelline cells contribute to the capsule. **C,** *Taenia* type with thick embryophore and very thin capsule. **D,** *Stilesia* type, found in cestodes with no distinct vitellaria, in which cellular covering is apparently laid down by the uterine wall.

Modified from Löser, E. 1965. Z. Parasitenkd. 25:556-580. Drawing by William C. Ober.

added. The secretions may cause exocytosis of shell material from the vitelline cells and form the structural support component for the capsule.[83] Leaving the ootype, the developing larva passes into the uterus where embryonation is completed.

The form of the uterus varies considerably between groups. It may be reticulated, lobulated, or circular; it may be a simple sac or a simple or convoluted tube; or it may be replaced by other structures. In some tapeworms the uterus disappears, and the eggs, either singly or in groups, are enclosed within the hyaline **egg capsules** embedded within the parenchyma. In some species one or more fibromuscular structures called **paruterine organs** form, attached to the uterus. In this case the eggs pass from the uterus into the paruterine organ, which assumes the functions of a uterus. The uterus then usually disintegrates.

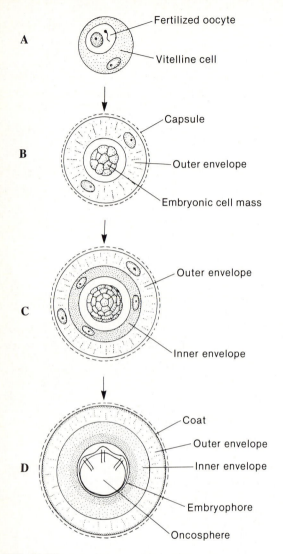

A — Fertilized oocyte

Vitelline cell

B — Capsule

Outer envelope

Embryonic cell mass

C — Outer envelope

Inner envelope

D — Coat

Outer envelope

Inner envelope

Embryophore

Oncosphere

FIG. 21-22

Diagram showing formation of embryonic envelopes in Cyclophyllidea. The organization of the envelopes is similar in other cestode orders. **A,** Fertilized oocyte surrounded by vitelline cell and capsule. **B,** Early phase of development showing formation of outer envelope from vitelline cell and embryonic blastomeres. **C,** Later phase showing formation of inner envelope from other blastomeres. **D,** Mature oncosphere with fully developed embryonic envelopes: coat; embryophore of inner envelope; oncospheral membrane (the oncospheral membrane cannot be seen by light microscopy).

Redrawn from Rybicka, K. 1965. Acta Parasitol. Pol. 13:25-34. Drawing by William C. Ober.

DEVELOPMENT

Nearly every life cycle known for tapeworms requires two hosts for its completion. One notable exception is *Vampirolepis nana,* a cyclophyllidean parasite of mice and humans, which can complete its juvenile stages within the definitive host (p. 363). Complete life cycles are known for only a comparatively few species of tapeworms. In fact there are several orders in which not a single life cycle has been determined. Among the life cycles that are known, much variety exists in juvenile forms and patterns of development.

Sexually mature tapeworms live in the intestine or its diverticula or, rarely, in the coelom of all classes of vertebrates. As stated earlier, two genera are known that can mature in invertebrates. A mature tapeworm may live for a few days or up to many years, depending on the species. During its reproductive life a single worm produces from a few to millions of eggs, each with the potential of developing into the adult form. Because of the great hazards obstructing the course of transmission and development of each worm, mortality is high.

Most tapeworms are hermaphroditic and are capable of fertilizing their own eggs. Sperm transfer is usually from the cirrus to the vagina of the same segment or between adjacent strobilae, if the opportunity affords. A few species are known in which a vagina is absent; **hypodermic impregnation** has been observed in some of these. In such cases the cirrus is forced through the body wall, and the sperm are deposited within the parenchyma (Fig. 21-23). It is not known how they find their way into the seminal receptacle.

A few species of tapeworms are known to be dioecious. In these cases it is not clear what determines the gender of a given strobila, since it appears that each has the potential of maturing as either male or female. Interaction between two or more strobilae is important in sex determination of dioecious forms. For example, in *Shipleya* (Cyclophyllidea, Dioecocestidae), if a single strobila is present in a host, it is usually female; if two are present, one is nearly always a male. Similar phenomena are known in certain flukes and nematodes. Much opportunity for research exists in this aspect of cestode biology.

Both invertebrates and vertebrates serve as intermediate hosts of tapeworms. Nearly every group of invertebrates has been discovered harboring juvenile cestodes, but the most common are crustaceans, insects, molluscs, mites, and annelids. As a general rule, when a tapeworm occurs in an aquatic definitive host, the juvenile forms are found in aquatic intermediate hosts. A similar assumption can be made for terrestrial hosts.

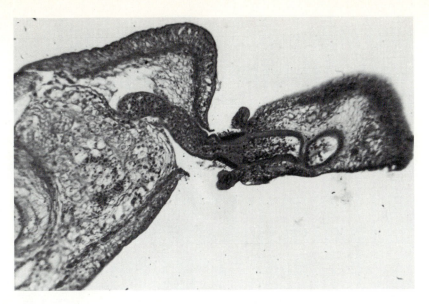

FIG. 21-23

Hypodermic impregnation of *Dioecotaenia cancellatum.* A cross-section view: the smaller male proglottid *(right)* has pierced the female proglottid *(left)* with its cirrus.

Photograph by Gerald D. Schmidt.

Vertebrate intermediate and paratenic hosts are found among fishes, amphibians, reptiles, and mammals. Tapeworms found in these hosts normally mature within predators whose diets include the intermediary.

Larval and juvenile development

Among the life histories that are known, much variety exists in the juvenile forms and details of development, but there seems to be a single basic theme[26]: (1) embryogenesis within the egg to result in a larva, the **oncosphere;** (2) hatching of the oncosphere after or before being eaten by the next host, where it penetrates to a parenteral (extraintestinal) site; (3) metamorphosis of the larva in the parenteral site into a juvenile **(metacestode)** usually with a scolex; and (4) development of the adult from the metacestode in the intestine of the same or another host. The oncospheres of all Eucestoda have three pairs of hooks (Fig. 21-24) and are thus also referred to as **hexacanths.** The free-swimming oncospheres hatching from the egg of some pseudophyllideans and a few trypanorhynchans have a ciliated inner envelope and are called **coracidia**[83] (Fig. 21-25). The larvae of cestodarians have 10 hooks (hence are **decacanths**), are also ciliated, and are called **lycophoras.**

In cestodes with free-swimming larvae the coracidium must be eaten by an intermediate host, usually an

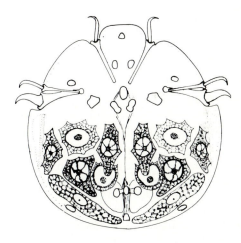

FIG. 21-24

Diagram of oncosphere of *Hymenolepis diminuta,* dorsal view.

From Ogren, R.E. 1967. Trans. Am. Microsc. Soc. 86:250-260.

arthropod, within a short time. There the coracidium sheds its ciliated IE and actively uses its six hooks to penetrate the gut of its host. In the hemocoel it metamorphoses into a **procercoid.** During this reorganization the oncospheral hooks are relegated to the posterior end in a structure known as the **cercomer.** The procercoid is defined as the stage in which the larval

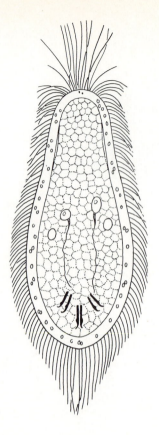

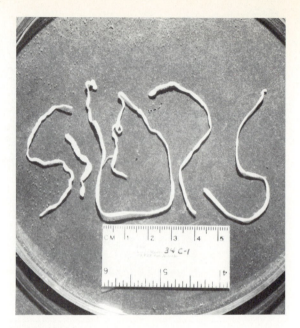

FIG. 21-26

Plerocercoids from the musculature of a vervet, an African primate.

Photograph by Robert E. Kuntz.

FIG. 21-25

Coracidium of *Diphyllobothrium erinacei*.

Redrawn from Neveu Lemaire, M. 1936. Traité d'helminthologie médicale et veterinaire. Paris. Drawing by William C. Ober.

hooks are still present but the definitive holdfast has not developed. It is regarded by some authors as a differentiating metacestode.[26] When the first intermediate host is consumed by the second intermediate host— often a fish—the procercoid penetrates the host gut into the peritoneal cavity and mesenteries and then commonly into the skeletal muscles. Development of the scolex characterizes the **plerocercoid** (Fig. 21-26), and there is commonly strobila formation at this stage, with or without concomitant proglottid formation. In the pseudophyllideans *Ligula* and *Schistocephalus,* development as plerocercoids proceeds so far that little growth occurs when these worms reach the definitive host, and the gonads mature within 72 hours and start producing eggs within 36 hours thereafter.[46,53] The Proteocephalata develop a first-stage plerocercoid in the arthropod intermediate host, with no intervening procercoid, and a second-stage plerocercoid in a parenteral site in the second intermediate host. (There

is some difference of opinion on this point, but we elect to follow conventional thought at this time.) In other species of this order, metacestode development (plerocercoid II) may be completed in the gut of the definitive host, or the metacestodes may develop through a sequence of sites: parenterally in an intermediate host, then parenterally in the definitive host, and finally enterally in the definitive host.[22,23,91] Coracidia, procercoids, and plerocercoids of pseudophyllideans and plerocercoids of proteocephalans are all plentifully supplied with penetration glands to aid in penetration of, and migration in, host tissues.[17,38]

The life cycles of cyclophyllideans, which are of most concern in this discussion, contrast in that neither a procercoid nor a plerocercoid is formed. In these the eggs are fully embryonated and infective when they pass from the definitive host, but they do not hatch until eaten by an intermediate host. The oncosphere penetrates the gut of the intermediate host to reach a parenteral site and metamorphoses to a **cysticercoid** or to a **cysticercus** type of metacestode. The cysticercoid (Figs. 21-27 and 21-28) is a solid-bodied organism with a fully developed scolex invaginated into its body. It is surrounded by cystic layers, and the cercomer, which contains the larval hooks, is outside the cyst. If not displaced mechanically, the cercomer

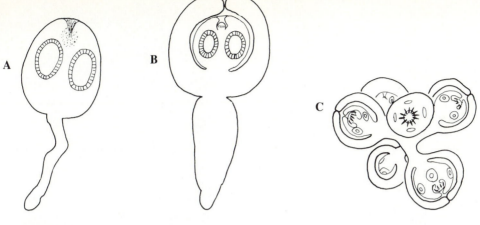

FIG. 21-27

Some types of cysticercoids. **A,** Simple, with no enclosing tissues. **B,** With scolex enclosed. **C,** Multiple.

From Schmidt, G.D. 1970. How to know the tapeworms. William C. Brown Co., Publishers, Dubuque, Iowa.

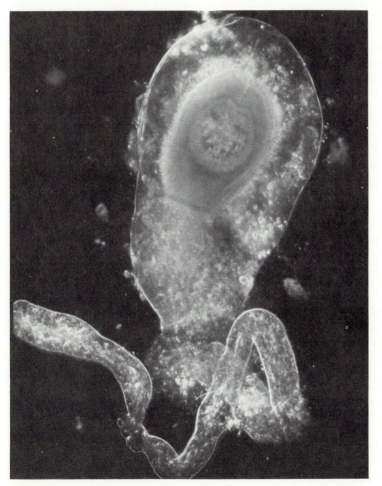

FIG. 21-28

Fully developed cysticercoid of *Hymenolepis diminuta*.

From Voge, M. 1969. In Schmidt, G.D., editor. Problems in systematics of parasites. © 1969 University Park Press, Baltimore.

will be digested away, along with parts of the cyst, in the gut of the definitive host. A few cysticercoids have been described that undergo asexual reproduction by budding.

Members of the cyclophyllidean family Taeniidae form a cysticercus metacestode (Figs. 22-11, 22-16), which differs from the cysticercoid in that the scolex is *introverted* as well as invaginated, and the scolex forms on a germinative membrane enclosing a fluid-filled bladder. Several variations from the simple cysticercus in the Taeniidae undergo asexual reproduction by budding (as will be discussed further). They are of considerable medical and veterinary importance.

Numerous other kinds of metacestodes can be distinguished from the typical forms described previously, but they are, for the most part, simply modifications of the following types:

1. **Sparganum**—a term originally proposed to be applied to any pseudophyllidean plerocercoid of unknown species but now usually used for some plerocercoids of the genus *Diphyllobothrium* (formerly *Spirometra*).

2. **Plerocercus**—a modified plerocercoid found in some Trypanorhyncha, in which the posterior forms a bladder, the **blastocyst,** into which the rest of the body can withdraw (as in *Gilquinia* spp.).

3. **Strobilocercoid**—a cysticercoid that undergoes some strobilation, found only in *Schistotaenia* spp.

4. **Tetrathyridium**—a fairly large, solid-bodied juvenile that can be regarded as a modified cysticercoid, developing in vertebrates that have ingested the cysticercoid encysted in the invertebrate host. It is known only in the atypical cyclophyllidean *Mesocestoides*.

5. Variations on cysticercus.
    a. **Strobilocercus** (Fig. 21-29)—a simple cysticercus in which some strobilation occurs within the cyst (for example, *Taenia taeniae-formis*).
    b. **Coenurus** (Figs. 21-30 and 22-19)—budding of a few to many scolices (called **proto-scolices**) from the germinative membrane of the cyst, each on a simple stalk invaginated into the common bladder (as in *Taenia multiceps*).
    c. **Unilocular hydatid** (Fig. 22-21)—up to several million protoscolices present; occasional sterile specimens. Usually there is an inner, or **endogenous, budding** of **brood cysts,** each with many protoscolices inside. **Exogenous budding** rarely occurs, resulting in two more hydatids called **daughter cysts.** This form may grow very large, sometimes containing

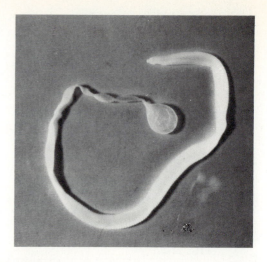

FIG. 21-29

Strobilocercus from the liver of a rat. Note the small bladder at the posterior end.

Photograph by James Jensen.

several quarts of fluid. Occasionally many protoscolices break free and sink to the bottom of the cyst, forming **hydatid sand** (Fig. 22-23), but this is probably rare in the living, normal cyst. This metacestode form is known only for the cyclophyllidean genus *Echinococcus*.

d. **Multilocular** or **alveolar hydatid** (Fig. 22-25)—known only for *Echinococcus multilocularis,* exhibiting extensive exogenous budding, when in an abnormal host such as humans, resulting in an infiltration of host tissues by numerous cysts. It forms a single mass with many little pockets that contain protoscolices when in a normal host.

Development in the definitive host

As with many other areas of parasitology, generalizations regarding this phase of development may be ill advised because detailed studies of relatively few species are available. However, substantial data have accumulated for the species that have been examined.

When the juvenile tapeworm reaches the small intestine of its definitive host, certain stimuli cause it to excyst, evaginate, or both and begin growth and sexual maturation. In encysted forms action of digestive enzymes in the host's gut may be necessary to at least partially free the organism from its cyst. In *H. diminuta* most of the cyst wall may be removed by treatment with pepsin and then with trypsin, but few worms will evaginate and emerge from the cyst unless bile salts are present.[69]

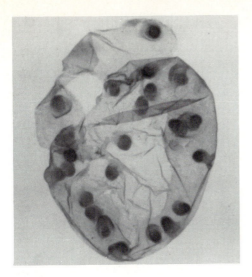

FIG. 21-30

Coenurus. Each round body in the bladder is an independent protoscolex.

Photograph by Warren Buss.

Smyth[73] emphasized the possible role of bile salts in determining host specificity. He found that the protoscolices of *Echinococcus granulosus* rapidly lysed in solutions of deoxycholic acid, and bile from rabbits and sheep, rich in deoxycholic acid, had the same effect. Bile from normal definitive hosts of *E. granulosus*, such as foxes and dogs, has relatively little deoxycholic acid and does not lyse the protoscolices. Without doubt, a combination of factors interacts to determine whether a tapeworm can develop in a given species of host. Among these factors are conditions that stimulate strobilation and maturation. The identity of such conditions is largely unknown, although in some instances they may be quite simple. In some pseudophyllideans with a well-developed strobila in the plerocercoid (for example, *Ligula* and *Schistocephalus*), an increase in temperature to that of their definitive host is all that is required for them to mature.[3] The temperature "activation" of such plerocercoids is accompanied by a great increase in the rate of carbohydrate catabolism, excretion of organic acids, and levels of tricarboxylic acid cycle intermediates.[8,37] A burst of neurosecretory activity has been correlated with activation of *Diphyllobothrium dendriticum* plerocercoids.[28] In *Echinococcus* it has been shown that contact of the rostellum with a suitable protein substrate is necessary to induce strobilar growth.[74]

As strobilar development begins, subsequent events are influenced by a variety of conditions, including size of the infecting juvenile, species of the worm and host, size and diet of the host, presence of other worms, and the immune and/or inflammatory state of the host intestine.[33] Under optimal conditions certain species have a burst of growth that must surely rival growth rates found anywhere in the animal kingdom. *Hymenolepis diminuta* can increase its weight by up to 1.8×10^6 times within 15 to 16 days.[64] Such rapid growth, accompanied by strictly organized differentiation, makes this worm a fascinating system for the study of development, particularly since the course of the growth may be altered experimentally. The growth of the worm is especially sensitive to the composition of the host diet with respect to carbohydrate. The situation is best known for *H. diminuta*, but the findings can be extended to other tapeworms, to some extent at least. *Hymenolepis diminuta* apparently has a high carbohydrate requirement, but it can only absorb glucose and, to a lesser degree, galactose across its tegument. This is true for other cestodes tested, although some can absorb a limited number of other monosaccharides and disaccharides.[14] For optimal growth the carbohydrate must be supplied in the host diet in the form of starch so that the glucose will be released as the digestion proceeds in the host gut. If glucose per se—or a disaccharide containing glucose, such as sucrose—is furnished in the host diet, the worm is placed at a competitive disadvantage for glucose with respect to the gut mucosa, physiological conditions in the gut are altered, or both, such that the worm's growth is substantially restrained.

Another important condition affecting worm growth is the presence of other tapeworms in the gut, the so-called **crowding effect.** This may be viewed as an interesting adaptation by which the parasite biomass is adjusted to the carrying capacity of the host. Again, although best known in *H. diminuta*, evidence exists that the crowding effect occurs in at least several other species.[61] Within certain limits, the weight of the individual worms in a given infection is, on the average, inversely proportional to the number of worms present. In consequence the total worm biomass and the number of eggs produced is the same and is maximal for that host, regardless of the number of worms present. The operational mechanism of the crowding effect is of considerable biological interest as a mode of developmental control. The most widely held view has been that the individual worms compete for available host dietary carbohydrate. However, the means by which the competition might be translated into lower rates of cell division and cell growth have not been elucidated, and the worms apparently secrete "crowding factors" that influence the development of other worms in the population.[34,67,92]

As the worm approaches maximal size, growth rate decreases, and production of new proglottids is only sufficient to replace those lost by apolysis. Although some species, such as *V. nana,* characteristically become senescent and pass out of the host after a period, others may be limited only by the length of their host's life. *Taeniarhynchus saginatus* has been known to live in a human for more than 30 years, and *H. diminuta* lives as long as the rat it inhabits. In fact Read[60] reported an "immortal" worm that he kept alive for 14 years by periodically removing it from its host, severing the strobila in the region of the germinative area, and then surgically reimplanting the scolex in another rat.

Finally, it should be noted that some tapeworms manage a surprising degree of mobility within their host's intestine. It has been known for some time that cestodes may establish initially in one part of the gut and then move to another as they grow. In fact, *D. dendriticum* in rats passes all the way to the large intestine within a few hours of infection, but less than 24 hours later it moves back to the duodenum to start growth.[2] More recently it has been discovered that *H. diminuta* actually undergoes a diurnal migration in the rat's gut (Fig. 21-31). This migration is correlated with the nocturnal feeding habits of rats and can be reversed by giving food to the rat only in the daytime. In fact migration of the worms is apparently mediated by vagal nerve stimulation of gastrointestinal function rather than the presence of food itself.[49]

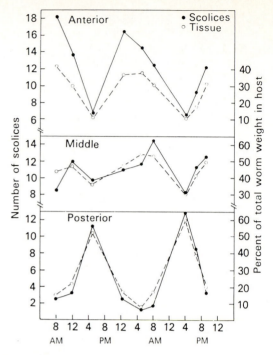

FIG. 21-31

Distribution of scolices and of wet tissue of *H. diminuta* in the host intestine at various times of the day. "Anterior" refers to the first 10 inches, "middle" to the second 10 inches, and "posterior" to the remainder of the small intestine. Each point is the mean of determinations from four host animals, representing 110 to 120 worms.

From Read, C.P. 1970. J. Parasitol. 56:643-652.

METABOLISM
Acquisition of nutrients

All nutrient molecules must be absorbed across the tegument; Pappas and Read[54] reviewed this subject in 1975. The mechanisms of absorption include active transport, mediated diffusion, and simple diffusion. Whether pinocytosis is possible at the cestode surface has been the subject of some dispute,[43] but the plerocercoids of *Schistocephalus* and *Ligula* are capable of this process.[32,82] Cysticerci of *Taenia crassiceps* are capable of pinocytosis, and the process is stimulated by the presence of glucose, yeast extract, or bovine serum albumin in the medium.[80,81]

Glucose seems to be the most important nutrient molecule to fuel energy processes in tapeworms. As noted before, the only carbohydrates that most cestodes can absorb are glucose and galactose, and although some tapeworms can absorb other monosaccharides and disaccharides, only glucose and galactose are known actually to be metabolized. The primary fate of galactose seems to be incorporation into membranes or other structural components, such as glycocalyx.[52] Galactose can be incorporated into glycogen but does not support net glycogen synthesis.[36] Both glucose and galactose are actively transported and accumulated in the worm against a concentration gradient. Of the two sugars, glucose has been studied more extensively. Glucose influx in a number of species is coupled to a sodium pump mechanism, that is, the maintenance of a sodium concentration difference across the membrane. The accumulation of glucose, in *H. diminuta* at least, is also sodium dependent. At least two transport sites for glucose are kinetically distinct in the tegument of *H. diminuta,* and the relative proportion of these sites changes during development.[65,78]

Amino acids are also actively transported and accumulated, although less is known about them than about glucose. However, efflux of amino acids from the worm is stimulated by the presence of other amino acids in the ambient medium; therefore the worm pool of amino acids rapidly comes to equilibrium with the amino acids in the intestinal milieu.

Purines and pyrimidines are absorbed by facilitated diffusion, and the transport locus is distinct from the amino acid and glucose loci.[45]

The actual mechanism of lipid absorption has not been investigated, but it is likely to be a form of diffusion. Fatty acids, monoglycerides, and sterols are absorbed at a considerably greater rate when they are in a micellar solution with bile salts.[5]

Only two cases of the requirements for external supplies of the various vitamins are substantiated. Investigations of vitamin requirements are difficult, as they are in some other parasites, because of limitations in in vitro cultivation techniques, because the worm may be less sensitive than its host to a vitamin-deficient diet, or both. In any case the pathogenesis of vitamin deficiency in the host may have indirect effects on the worm. The necessity for an external supply of a vitamin has been demonstrated unequivocally in only one case, that of pyridoxine and *H. diminuta*.[57,68] By inference we can assume that *Diphyllobothrium latum* has a requirement for vitamin B$_{12}$, since the worm accumulates unusually large amounts of it.[11] In some cases *D. latum* can compete so successfully with its host for the vitamin that the worm can cause pernicious anemia in persons genetically susceptible to its effects (Chapter 22).

Energy metabolism

Carbohydrate metabolism and electron transport systems in cestodes have been reviewed.[16,66] As do a number of other endoparasites that have been investigated, cestodes apparently derive their energy predominantly by anaerobic processes, even in the presence of oxygen. They take up oxygen when it is available, but it is not clear that the function of oxygen is as a terminal electron acceptor in an energy-producing series of reactions (for example, oxidative phosphorylation via the "classical" cytochrome system). Although earlier research indicated that some cytochromes (*b, c,* and cytochrome oxidase) might be present in some cestodes, later research failed to confirm that a cytochrome system was operating, and the function of such cytochromes as were present was a mystery. Use of more sensitive techniques now has provided evidence that a classical mammalian type of electron transport system is present in at least some cestodes but that the classical chain is probably of minor importance (Fig. 21-32) and that the major cytochrome system is a so-called *o*-type, similar to that reported in many bacteria. The significance of this system is that it seems to be an adaptation to facultative anaerobiosis. The terminal oxidase can transfer electrons to either fumarate or oxygen, depending on

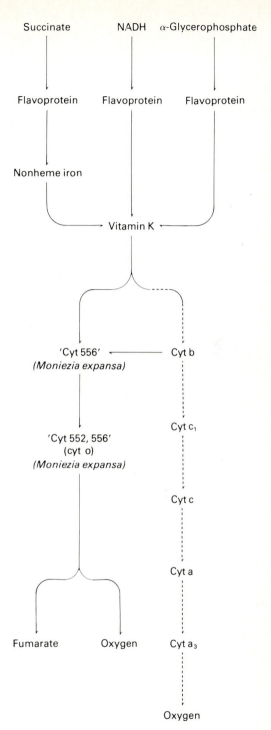

FIG. 21-32

Branched chain electron transport system with cytochrome *o*, facultatively transporting electrons to fumarate or oxygen. Evidence exists that a similar system operates in *Moniezia, Taenia,* and probably other cestodes. *Solid lines,* Major pathway; *dotted lines,* minor pathway.

Modified from Bryant, C. 1970. In Dawes, B., editor. Advances in parasitology, vol. 8. Academic Press, Inc., New York.

whether conditions are aerobic or anaerobic, and the products are either succinate or hydrogen peroxide, respectively. A peroxidase destroys the hydrogen peroxide before it reaches toxic levels. Under anaerobic conditions, the succinate formed in this pathway would be excreted; another mechanism for succinate formation and excretion, even under aerobic conditions, will be described later.

The Krebs tricarboxylic acid cycle is of little or no importance in the adult cestodes investigated so far, but a substantial amount of glucose carbon may flow through the Krebs cycle in certain metacestodes. As much as 40% of the carbohydrate utilized by protoscolices of *Echinococcus multilocularis* and the sheep strain of *E. granulosus* may be channeled into the Krebs cycle, and only 22% of the glycogen catabolized by plerocercoids of *Schistocephalus solidus* is accounted for by excreted acids.[37,47] Activity of the cycle increases in *S. solidus* when the plerocercoids are activated by an increase in the ambient temperature.[8]

The phosphogluconate pathway probably exists in cestodes and may be biologically important but quantitatively insignificant when compared to glycolysis.[66]

The most important series of reactions to produce energy in adult cestodes is the Embden-Meyerhof sequence, or glycolysis, and certain mitochondrial reactions that follow.[66] As noted above, some glucose carbon may enter the Krebs cycle to be degraded to carbon dioxide, but most is excreted as relatively reduced products, such as lactate, acetate, succinate, propionate, and alanine. Because the total energy derived from degradation of 1 mole of glucose in these reactions is that required to form 4 moles, or possibly 6, of adenosine triphosphate (ATP) from adenosine diphosphate (ADP) (Fig. 21-33), and because much more energy could be derived from a mole of glucose if it were completely oxidized, the energy metabolism of cestodes seems very inefficient. The biological significance of this style of energy metabolism is the subject of ongoing debate and will be discussed further. Nevertheless, because cestodes have very limited ability to degrade fatty and amino acids, their processes of carbohydrate storage and catabolism assume critical importance for energy production.

Indeed, juvenile and adult cestodes characteristically store enormous amounts of glycogen, ranging from about 20% to more than 50% of dry weight. Whereas the tissue-dwelling juveniles will be exposed to a reasonably constant glucose concentration maintained by the homeostatic mechanisms of the host, the adults must survive between host feeding periods. The large amount of stored glycogen serves at these times as an effective cushion. *Hymenolepis diminuta* con-

sumes 60% of its glycogen during 24 hours of host starvation and another 20% during the next 24 hours. When glucose is again available, the glycogen stores are rapidly replenished.[20]

Glucose from glycogen or absorbed directly from the host intestine is degraded by classical glycolysis as far as phosphoenolpyruvate (PEP), but at this point there is a branch in the pathway (Fig. 21-33). Either lactate is produced by dephosphorylation of PEP and reduction of pyruvate, or malate is produced by fixation of carbon dioxide to form oxaloacetate, which is then reduced to malate.[66] Both branches thus far are functionally equal because each generates a high-energy phosphate bond and reoxidizes the NADH formed in glycolysis; therefore the cytoplasmic redox balance is preserved. However, additional energy is obtained when the malate enters the mitochondria, where it undergoes a dismutation reaction. Part of the malate is oxidized and decarboxylated to pyruvate, which is then either decarboxylated and excreted as acetate or is transaminated and excreted as alanine. The other half of the malate is dehydrated to fumarate and then reduced to succinate. Reducing equivalents for the reduction of fumarate are provided by the oxidation of malate (Fig. 21-33). However, the oxidative decarboxylation of malate is NADP dependent in *H. diminuta* and *H. microstoma*, whereas the fumarate reduction is NAD dependent. Therefore a hydride ion must be transferred from NADPH to NAD, and this is accomplished by a NADPH:NAD transhydrogenase.[21] The reduction of the fumarate is accompanied by an electron transport-associated, net generation of ATP, and the catabolism of pyruvate to acetate probably also generates another ATP. Thus excretion of succinate and acetate produces two more ATPs than if the glucose carbon were excreted solely as lactate. In some cestodes propionate is formed by decarboxylation of succinate, generating additional ATP.[56] Alanine apparently also is an excreted end product and is formed by transamination from glutamate to pyruvate.[66,85] An advantage of alanine excretion would be that it is less acidic than lactate.

It is clear that the foregoing reactions derive more energy from carbohydrate when succinate and/or propionate are excreted rather than lactate, and it has been generally assumed that these processes are adaptations to obtain maximal energy in the hypoxic conditions prevailing in the host gut. Some workers have argued that succinate excretion is not related to hypoxia but rather is an adaptation to avoid tissue acidification and to dispose of H^+[58]; however, this hypothesis is weakened by the discovery of a strain of *H. diminuta* that excretes little succinate.[15] It has been observed that secretion of the organic acids by *H.*

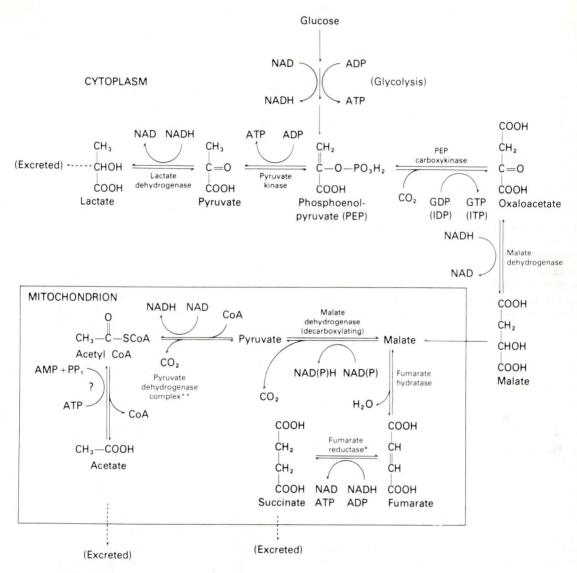

FIG. 21-33

Reactions forming the major end products of energy metabolism from phosphoenolpyruvate in *Hymenolepis diminuta* (adapted and proposed from various sources). These reactions yield additional ATP above that from classical glycolysis, with a balanced cytoplasmic oxidation-reduction and a balanced mitochondrial oxidation-reduction (ratio of succinate to acetate excreted, approximately 2:1). *This enzyme (fumarate reductase) usually is referred to as succinate dehydrogenase, but it acts in an opposite direction from mammalian systems, that is, as a fumarate reductase. In *Hymenolepis* the malate dehydrogenase (decarboxylating) reaction is NADP dependent, and the hydride ion is transferred to NAD for the fumarate reductase reaction by a transhydrogenase, thus maintaining the redox balance in the mitochondrion.[21] (**Watts and Fairbairn[88])

diminuta has another effect: regulation of the ambient pH to about 5. Inasmuch as trypsin activity at this pH is minimal, an important function and result of the acid excretion may be protection of the worms against digestion by host enzymes. This may be in addition to the elaboration of molecules with anti-trypsin activity, previously mentioned.[70]

Tapeworms probably do not derive any energy from degradation of lipids or proteins. *H. diminuta* has only a modest capacity for carrying out transaminations and can degrade only four amino acids.[85] The function served by much of the lipid in cestodes remains a mystery, since no one has been able to show that lipids are depleted at all during starvation, even though they may comprise up to 20% of worm dry weight or more than 30% in the parenchyma of gravid proglottids. *Schistocephalus solidus* has all the enzymes necessary for the β-oxidation sequence of lipids; nevertheless, it appears unable to catabolize them.[6] Lipids in cestodes may represent metabolic end products, since they are relatively nontoxic to store, and the parenchyma of gravid proglottids is discarded during apolysis.

Nitrogenous end products excreted by *H. diminuta* include considerable quantities of ammonia, α-amino nitrogen, and urea.[20]

Synthetic metabolism

Little need be said of the protein and nucleic acid synthetic abilities of cestodes. It is clear that they can absorb amino acids, purines, pyrimidines, and nucleosides from the intestinal milieu and synthesize their own proteins and nucleic acids (see, for example, Bolla and Roberts[10]). *Moniezia expansa* cannot synthesize carbamyl phosphate; therefore it depends on its host for both pyrimidine and arginine.

In contrast, capacity for synthesis of lipids appears minimal. The worm can neither synthesize fatty acids de novo from acetyl-CoA nor introduce double bonds into the fatty acids it absorbs.[35] *Hymenolepis diminuta* rapidly hydrolyzes monoglycerides after absorption, and it can then resynthesize triglycerides. It can lengthen the chain of fatty acids provided that the acid already contains 16 or more carbons. Similar observations have been reported for *Diphyllobothrium mansonoides*.[50] There is some evidence that *Hymenolepis microstoma* might be able to synthesize fatty acids *de novo*.[59]

Finally, *H. diminuta* cannot synthesize cholesterol, a biosynthesis that is known to require molecular oxygen in other systems. The normal precursors of cholesterol (acetate, hydroxymethylglutarate, and meval-onate) are converted by *H. diminuta* into 2-*cis*, 6-*trans* farnesol and lesser amounts of 2-*trans*, 6-*trans* farnesol, the latter being a normal precursor of cholesterol in mammals. These compounds are normally present in this cestode, a fact that is of interest because both isomers are mimics of juvenile hormone in insects.[25]

Hormonal effects of metabolites

Certain instances are known in which substances produced by cestodes have effects on their hosts that mimic the host's own hormones. Fish infected with the plerocercoids of *Ligula* are unable to reproduce; their gonads do not develop, and there is an apparent suppression of the presumed gonadotropin-producing cells in their pituitary glands. On the possibility that a sex steroid was produced by the worms and that the steroid interfered with gonadotrophin production and hence gonad development, the presence of such compounds was investigated.[4] None was found, and the mechanism of the effect on the host remains enigmatic.

A more surprising case is that of a substance produced by the plerocercoids of *Diphyllobothrium (=Spirometra) mansonoides*. Referred to as **plerocercoid growth factor** (PGF), the compound closely resembles human growth factor (hGF) and acts as a growth hormone in several mammalian species (Fig. 21-34).[55] It is definitely not the same compound, although the pituitary recognizes it as such and decreases its own production of growth hormone. Under normal circumstances administration of hGF can have antiinsulin and diabetogenic activities; these properties are not shared with PGF. Growth hormones from other vertebrates (except primates) are not active in humans because of the strict binding specificity of the receptor molecules in humans. However, PGF has the same binding specificity as hGF, and monoclonal antibodies raised against the unique epitope of hGF cross-react with PGF. This and other evidence has suggested the following hypothesis: the gene for PGF is a *human gene* (for hGF) that *has been sequestered by the tapeworm* during its evolution.[55] The possible value of PGF to the tapeworm has been a mystery, but there is some evidence of interaction with the immune system of the host. Perhaps suppression of one or more pathways by PGF might indirectly suppress the immune response, thus allowing the worms to evade host defenses while becoming established.[71] Nevertheless, there is significant potential for use of PGF in human endocrinology because of its lack of antiinsulin and diabetogenic activities.

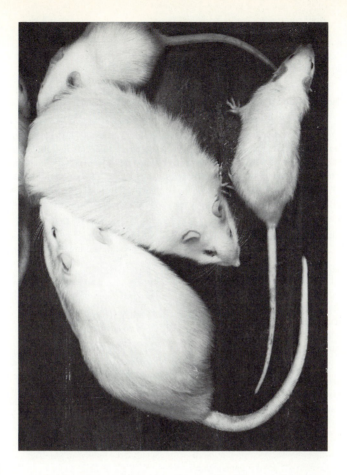

FIG. 21-34

Illustration of growth hormone-like action of *Diphyllobothrium mansonoides* plerocercoids. All rats were hypophysectomized when they weighed 90 g, but the two larger rats received 20 juvenile scolices of *D. mansonoides* approximately 1 month after the operation. The photograph was taken 6 months later, and the experimental animals outweigh the controls by three or four times.

From Mueller, J.F. 1974. J. Parasitol. 60:3-14.

CLASSIFICATION OF CLASS CESTOIDEA

Bilaterally symmetrical, elongate, dorsoventrally flattened worms; body usually consists of anterior holdfast (scolex), neck, and a chain of proglottids; digestive system absent; tegument syncytial, often spined; protonephridia drain into two pairs (usually) of longitudinal, dorsal, and ventral osmoregulatory canals, which themselves drain directly through the posterior end of the worm after the earliest proglottid detaches; proglottids monoecious, rarely dioecious; some species have more than one set of male and female systems per proglottid; adults obligate endoparasites in the intestine or bile duct of vertebrates, rarely in invertebrates; juvenile forms in invertebrates or vertebrates.

Subclass Cestodaria

Monozoic, with single set of reproductive organs; no scolex present, but there may be a small, proboscis-like organ at the ante-rior end; posterior end single, rounded, sometimes forming a crenulated rosette or a long slender cylinder; genital pores near posterior end; testes in two lateral, preovarian fields; ovary posterior; vitelline glands follicular, lateral; uterus N shaped or looped; uterine pore near anterior end.

FAMILIES

Amphilinidae, Austramphilinidae, Gyrocotylidae

Subclass Eucestoda

Polyzoic (except orders Caryophyllidea and Spathebothriidea), with one or more sets of reproductive systems per proglottid; scolex usually present; shelled embryo with six hooks; parasites of fishes, amphibians, reptiles, birds, and mammals.

Order Caryophyllidea

Scolex unspecialized or with shallow grooves or loculi or shallow

341

bothria; monozoic; genital pores midventral; testes numerous; ovary posterior; vitellaria follicular, scattered or lateral; uterus a coiled median tube, opening, often together with vagina, near male pore; parasites of teleost fishes and aquatic annelids.

FAMILIES
Caryophyllaeidae, Balanotaeniidae, Lytocestidae, Capingentidae

Order Spathebothriidea

Scolex feebly developed, undifferentiated or with funnel-shaped apical organ or one or two hollow, cup-like organs; constrictions between proglottids absent, proglottids distinguished internally; genital pores and uterine pore ventral or alternating dorsal and ventral; testes in two lateral bands; ovary dendritic; vitellaria follicular, lateral or scattered; uterus coiled; parasites of teleost fishes.

FAMILIES
Cyathocephalidae, Spathebothriidae, Bothrimonidae

Order Trypanorhyncha

Scolex elongate, with two or four bothridia and four eversible (rarely atrophied) tentacles armed with hooks; each tentacle invaginates into internal sheath provided with muscular bulb; neck present or absent; strobila apolytic, anapolytic, or hyperapolytic; genital pores lateral, rarely ventral; testes numerous; ovary posterior; vitellaria follicular, cortical, and encircling other reproductive organs; uterine pore present or absent; parasites of elasmobranchs.

FAMILIES
Dasyrhynchidae, Eutetrarhynchidae, Gilquiniidae, Gymnorhynchidae, Hepatoxylidae, Hornelliellidae, Lacistorhynchidae, Mustelicolidae, Otobothriidae, Paranybeliniidae, Pterobothriidae, Sphyriocephalidae, Tentaculariidae, Mixodigmatidae, Rhinoptericolidae

Order Pseudophyllidea

Scolex with two bothria, with or without hooks, neck present or absent; strobila variable; proglottids anapolytic; genital pores lateral, dorsal, or ventral; testes numerous; ovary posterior; vitellaria follicular, as in Trypanorhyncha; occasionally in lateral fields but not interrupted by interproglottidal boundaries; uterine pore present, dorsal or ventral; egg usually operculate, containing a coracidium; parasites of fishes, amphibians, reptiles, birds, and mammals.

FAMILIES
Amphicotylidae, Bothriocephalidae, Cephalochlamydidae, Diphyllobothriidae, Echinophallidae, Haplobothriidae, Parabothriocephalidae, Ptychobothriidae, Triaenophoridae

Order Lecanicephalidea

Scolex divided into anterior and posterior regions by transverse groove; anterior portion cushion-like or with unarmed tentacles, capable of being withdrawn into posterior portion, forming a large sucker-like organ; posterior portion usually with four suckers; neck present or absent; testes numerous; ovary posterior; vitellaria follicular, lateral or encircling proglottid; uterine pore usually present; parasites of elasmobranchs.

FAMILIES
Adelobothriidae, Balanobothriidae, Disculicepitidae, Lecanicephalidae

Order Aporidea

Scolex with simple suckers or grooves and armed rostellum; constrictions between proglottids absent, proglottids distinguished inter-

nally or separate proglottids not evident; genital ducts and pores, cirrus, ootype, and Mehlis' gland absent; hermaphroditic, rarely dioecious; vitelline cells mixed with ovarian cells; parasites of Anseriformes.

FAMILY
Nematoparataeniidae

Order Tetraphyllidea

Scolex with highly variable bothridia, sometimes also with hooks, spines, or suckers; myzorhynchus present or absent; proglottids commonly hyperapolytic; hermaphroditic, rarely dioecious; genital pores lateral, rarely posterior; testes numerous; ovary posterior; vitellaria follicular, usually medullary in lateral fields; uterine pore present or not; vagina crosses vas deferens; parasites of elasmobranchs.

FAMILIES
Onchobothriidae, Phyllobothriidae, Trilocubriidae

Order Dioecotaeniidea

Totally dioecious. Females much larger than males. Scolex with loculated bothridia. Parasites of elasmobranchs (rays).

FAMILY
Dioecotaeniidae

Order Diphyllidea

Scolex with armed or unarmed peduncle; two spoon-shaped bothridia present, lined with minute spines, sometimes divided by median, longitudinal ridge; apex of scolex with insignificant apical organ or with large rostellum bearing dorsal and ventral groups of T-shaped hooks; strobila cylindrical, acraspedote; genital pores posterior, midventral; testes numerous, anterior; ovary posterior; vitellaria follicular, lateral, or surrounding other organs; uterine pore absent; uterus tubular or saccular; parasites of elasmobranchs.

FAMILIES
Ditrachybothridiidae, Echinobothriidae

Order Litobothridea

Scolex a single, well-developed apical sucker; anterior proglottids modified, cruciform in cross section; neck absent; strobila dorsoventrally flattened, with numerous proglottids, each with single set of medullary reproductive organs; proglottids laciniated and craspedote, apolytic or anapolytic; testes numerous, preovarian; genital pores lateral; ovary two or four lobed, posterior; vitellaria follicular, encircling medullary parenchyma; parasites of elasmobranchs.

FAMILY
Litobothridae

Order Proteocephalata

Scolex with four suckers, often with prominent apical organ, occasionally with armed rostellum; neck usually present; genital pores lateral; testes numerous; ovary posterior; vitelline glands follicular, usually lateral, either cortical or medullary; uterine pore present or absent; parasites of fishes, amphibians, and reptiles.

FAMILIES
Proteocephalidae, Monticellidae

Order Cyclophyllidea

Scolex usually with four suckers; rostellum present or not, armed or not; neck present or absent; strobila usually with distinct segmentation, monoecious or rarely dioecious; genital pores lateral (ventral in Mesocestoididae); vitelline gland compact, single (double in Mesocestoididae), posterior to ovary (anterior or beneath ovary in Tetrabothriidae); uterine pore absent; parasites of amphibians, reptiles, birds, and mammals.

FAMILIES

Amabiliidae, Anoplocephalidae, Catenotaeniidae, Davaineidae, Dilepididae, Dioecocestidae, Diploposthidae, Hymenolepididae, Mesocestoididae, Nematotaeniidae, Progynotaeniidae, Taeniidae, Tetrabothriidae, Triplotaeniidae

Order Nippotaeniidea

Scolex with single sucker at apex, otherwise simple; neck short or absent; strobila small; proglottids each with single set of reproductive organs; genital pores lateral; testes anterior, ovary posterior; vitelline gland compact, single, between testes and ovary; osmoregulatory canals reticular; parasites of teleost fishes.

FAMILY

Nippotaeniidae

REFERENCES

1. Amin, O.M. 1978. On the crustacean hosts of larval acanthocephalan and cestode parasites in southwestern Lake Michigan. J. Parasitol. 64:842-845.

2. Archer, D.M., and C.A. Hopkins. 1958. Studies on cestode metabolism. III. Growth pattern of *Diphyllobothrium* sp. in a definitive host. Exp. Parasitol. 7:125-144.

3. Arme, C. 1966. Histochemical and biochemical studies on some enzymes of *Ligula intestinalis* (Cestoda: Pseudophyllidea). J. Parasitol. 52:63-68.

4. Arme, C., D.V. Griffiths, and J.P. Sumpter. 1982. Evidence against the hypothesis that the plerocercoid larva of *Ligula intestinalis* (Cestoda: Pseudophyllidea) produces a sex steroid that interferes with host reproduction. J. Parasitol. 68:169-171.

5. Bailey, H.H., and D. Fairbairn. 1968. Lipid metabolism in helminth parasites. V. Absorption of fatty acids and monoglycerides from micellar solution by *Hymenolepis diminuta* (Cestoda). Comp. Biochem. Physiol. 26:819-836.

6. Barrett, J., and W. Körting. 1977. Lipid catabolism in the plerocercoids of *Schistocephalus solidus* (Cestoda: Pseudophyllidea). Int. J. Parasitol. 7:419-422.

7. Becker, B., H. Mehlhorn, P. Andrews, and H. Thomas. 1981. Ultrastructural investigations on the effect of praziquantel on the tegument of five species of cestodes. Z. Parasitenkd. 64:257-269.

8. Beis, I., and J. Barrett. 1979. The content of adenine nucleotides and glycolytic and tricarboxylic acid cycle intermediates in activated and non-activated plerocercoids of *Schistocephalus solidus* (Cestoda: Pseudophyllidea). Int. J. Parasitol. 9:465-468.

9. Blitz, N.M., and J.D. Smyth. 1973. Tegumental ultrastructure of *Raillietina cesticillus* during the larval-adult transformation, with emphasis on the rostellum. Int. J. Parasitol. 3:561-570.

10. Bolla, R.I., and L.S. Roberts. 1971. Developmental physiology of cestodes. X. The effect of crowding on carbohydrate levels and on RNA, DNA, and protein synthesis in *Hymenolepis diminuta*. Comp. Biochem. Physiol. 40A:777-787.

11. von Bonsdorff, B. 1956. *Diphyllobothrium latum* as a cause of pernicious anemia. Exp. Parasitol. 5:207-230.

12. Boyce, N.P. 1976. A new organ in cestode surface ultrastructure. Can. J. Zool. 54:610-613.

13. von Brand, T. 1973. Biochemistry of parasites, ed. 2. Academic Press, Inc. New York.

14. von Brand, T., P. McMahon, E. Gibbs, and H. Higgins. 1964. Aerobic and anaerobic metabolism of larval and adult *Taenia taeniaeformis*. II. Hexose leakage and absorption; tissue glucose and polysaccharides. Exp. Parasitol. 15:410-429.

15. Bryant, C. 1983. Australian Society for Parasitology. Presidential Address. Intraspecies variations of energy metabolism in parasitic helminths. Int. J. Parasitol. 13:327-332.

16. Cheah, K.S. 1983. Electron-transport systems. In Arme, C., and P.W. Pappas, editors. The biology of the Eucestoda, vol. 2. Academic Press Ltd., London, pp. 421-440.

17. Coggins, J.R. 1980. Tegument and apical end organ fine structure in the metacestode and adult *Proteocephalus ambloplitis*. Int. J. Parasitol. 10:409-418.

18. Davis, R.E., and L.S. Roberts. 1983. Platyhelminthes—Eucestoda. In Adiyodi, K.G., and R.G. Adiyodi, editors. Reproductive biology of invertebrates, vol. I. Oogenesis, oviposition, and oosorption. John Wiley & Sons, Chicester, England, pp. 709-733.

19. Elowni, E.E. 1982. *Hymenolepis diminuta*: the origin of protective antigens. Exp. Parasitol. 53:157-163.

20. Fairbairn, D., G. Wertheim, R.P. Harpur, and E.L. Schiller. 1961. Biochemistry of normal and irradiated strains of *Hymenolepis diminuta*. Exp. Parasitol. 11:248-263.

21. Fioravanti, C.F. 1982. Mitochondrial malate dehydrogenase, decarboxylating ("malic" enzyme) and transhydrogenase activities of adult *Hymenolepis microstoma* (Cestoda). J. Parasitol. 68:213-220.

22. Fischer, H. 1968. The life cycle of *Proteocephalus fluviatilis* Bangham (Cestoda) from the smallmouth bass, *Micropterus dolomieu* Lacepede. Can. J. Zool. 46:569-579.

23. Fischer, H., and R.S. Freeman. 1973. The role of plerocercoids in the biology of *Proteocephalus ambloplitis* (Cestoda) maturing in smallmouth bass. Can. J. Zool. 51:133-141.

24. Foster, W.D. 1965. A history of parasitology. E. & S. Livingston, Edinburgh.

25. Frayha, G.J., and D. Fairbairn. 1969. Lipid metabolism in helminth parasites. VI. Synthesis of 2-*cis*, 6-*trans* farnesol by *Hymenolepis diminuta* (Cestoda). Comp. Biochem. Physiol. 28:1115-1124.

26. Freeman, R. 1973. Ontogeny of cestodes and its bearing on their phylogeny and systematics. In Dawes, B., editor. Advances in parasitology, vol. 11. Academic Press, Inc., New York, pp. 481-557.

27. Gustafsson, M.K.S., and B. Vaihela. 1981. Two types of frontal glands in *Diphyllobothrium dendriticum* (Cestoda, Pseudophyllidea) and their fate during the maturation of the worm. Z. Parasitenkd. 66:145-154.

28. Gustafsson, M.K.S., and M.C. Wikgren. 1981. Activation of the peptidergic neurosecretory system in *Diphyllobothrium dendriticum* (Cestoda: Pseudophyllidea). Parasitology 83:243-247.

29. Hayunga, E.G. 1979. The structure and function of the glands of three caryophyllid tapeworms. Proc. Helm. Soc. Wash. 46:171-179.

30. Hoeppli. R.J.C. 1959. Parasites and parasitic infections in early medicine and science. University of Malaya Press, Singapore.

31. Hopkins, C.A., and I.F. Barr. 1982. The source of antigen in an adult tapeworm. Int. J. Parasitol. 12:327-333.

32. Hopkins, C.A., L.M. Law, and L.T. Threadgold. 1978. *Schistocephalus solidus:* Pinocytosis by the plerocercoid tegument. Exp. Parasitol. 44:161-172.

33. Howard, R.J., et al. 1978. The effect of concurrent infection with *Trichinella spiralis* on *Hymenolepis microstoma* in mice. Parasitology 77:273-279.

34. Insler, G.D., and L.S. Roberts. 1980. Developmental physiology of cestodes. XV. A system for testing possible crowding factors in vitro. J. Exp. Zool. 211:45-54.

35. Jacobsen, N.S., and D. Fairbairn. 1967. Lipid metabolism in helminth parasites. III. Biosynthesis and interconversion of fatty acids by *Hymenolepis diminuta* (Cestoda). J. Parasitol. 53:355-361.

36. Komuniecki, R., and L.S. Roberts. 1977. Galactose utilization by the rat tapeworm, *Hymenolepis diminuta*. Comp. Biochem. Physiol. 57B:329-333.

37. Körting, W., and J. Barrett. 1977. Carbohydrate catabolism in the plerocercoids of *Schistocephalus solidus* (Cestoda: Pseudophyllidea). Int. J. Parasitol. 7:411-417.

38. Kuperman, B.I., and V.G. Davydov. 1982. The fine structure of glands in oncospheres, procercoids and plerocercoids of Pseudophyllidea (Cestoda). Int. J. Parasitol. 12:135-144.

39. Kuperman, B.I., and V.G. Davydov. 1982. The fine structure of frontal glands in adult cestodes. Int. J. Parasitol. 12:285-293.

40. Lumsden, R.D., and J. Bryam III. 1967. The ultrastructure of cestode muscle. J. Parasitol. 53:326-342.

41. Lumsden, R.D., J.A. Oaks, and J.F. Mueller. 1974. Brush border development in the tegument of the tapeworm, *Spirometra mansonoides*. J. Parasitol. 60:209-226.

42. Lumsden, R.D., and R.D. Specian. 1980. The morphology, histology, and fine structure of the adult stage of the cyclophyllidean tapeworm *Hymenolepis diminuta*. In Arai, H.P., editor. Biology of the rat tapeworm, *Hymenolepis diminuta*. Academic Press, Inc., New York, pp. 157-280.

43. Lumsden, R.D., L.T. Threadgold, J.A. Oaks, and C. Arme. 1970. On the permeability of cestodes to colloids: an evaluation of the transmembranosis hypothesis. Parasitology 60:185-193.

44. Lussier, P.E., R.B. Podesta, and D.F. Mettrick. 1978. *Hymenolepis diminuta:* amino acid transport and osmoregulation. J. Parasitol. 64:1140-1141.

45. MacInnis, A.J., F.M. Fisher, Jr., and C.P. Read. 1965. Membrane transport of purines and pyrimidines in a cestode. J. Parasitol. 51:260-267.

46. McCaig, M.L.O., and C.A. Hopkins. 1963. Studies on *Schistocephalus solidus*. II. Establishment and longevity in the definitive host. Exp. Parasitol. 13:273-283.

47. McManus, D.P., and J.D. Smyth. 1982. Intermediary carbohydrate metabolism in protoscoleces of *Echinococcus granulosus* (horse and sheep strains) and *E. multilocularis*. Parasitology 84:351-366.

48. Mehlhorn, H., B. Becker, P. Andrews, and H. Thomas. 1981. On the nature of the proglottids of cestodes: a light and electron microscopic study on *Taenia, Hymenolepis,* and *Echinococcus*. Z. Parasitenkd. 65:243-259.

49. Mettrick, D.F., and C.H. Cho. 1981. Effect of electrical vagal stimulation on migration of *Hymenolepis diminuta*. J. Parasitol. 67:386-390.

50. Meyer, F., S. Kimura, and J.F. Mueller. 1966. Lipid metabolism in the larval and adult forms of the tapeworm *Spirometra mansonoides*. J. Biol. Chem. 241:4224-4232.

51. Mount, P.M. 1970. Histogenesis of the rostellar hooks of *Taenia crassiceps* (Zeder, 1800) (Cestoda). J. Parasitol. 56:947-961.

52. Oaks. J.A., and R.D. Lumsden. 1971. Cytological studies on the absorptive surfaces of cestodes. V. Incorporation of carbohydrate-containing macromolecules into tegument membranes. J. Parasitol. 57:1256-1268.

53. Orr, T.S.C., and C.A. Hopkins. 1969. Maintenance of *Schistocephalus solidus* in the laboratory with observations on rate of growth of, and proglottid formation in, the plerocercoid. J. Fish. Res. Bd. Can. 26:741-752.

54. Pappas, P.W., and C.P. Read. 1975. Membrane transport in helminth parasites: a review. Exp. Parasitol. 37:469-530.

55. Phares, C.K. 1987. Plerocercoid growth factor: a homologue of human growth hormone. Parasitol. Today 3:346-349.

56. Pietrzak, S.M., and H.J. Saz. 1981. Succinate decarboxylation to propionate and the associated phosphorylation in *Fasciola hepatica* and *Spirometra mansonoides*. Mol. Biochem. Parasitol. 3:61-70.

57. Platzer, E.G., and L.S. Roberts. 1969. Developmental physiology of cestodes. V. Effects of vitamin deficient diets and host coprophagy prevention on development of *Hymenolepis diminuta*. J. Parasitol. 55:1143-1152.

58. Podesta, R.B., et al. 1976. Anaerobes in an aerobic environment: role of CO_2 in energy metabolism of *Hymenolepis diminuta*. In Van den Bossche, H., editor. Biochemistry of parasites and host-parasite relationships. Elsevier/North Holland Biomedical Press, Amsterdam.

59. Rath, E.A., and M. Walkey. 1987. Fatty acid and cholesterol synthesis in mice infected with the tapeworm *Hymenolepis microstoma*. Parasitology 95:79-92.

60. Read, C.P. 1967. Longevity of the tapeworm, *Hymenolepis diminuta*. J. Parasitol. 53:1055-1056.

61. Read, C.P., and J.E. Simmons, Jr. 1963. Biochemistry and physiology of tapeworms. Physiol. Rev. 43:263-305.

62. Rees, G. 1966. Nerve cells in *Acanthobothrium coronatum* (Rud.) (Cestoda: Tetraphyllidea). Parasitology 56:45-54.

63. Richards, K.S., and C. Arme. 1981. Observations on the microtriches and stages in their development and emergence in *Caryophyllaeus laticeps* (Caryophyllidea: Cestoda). Int. J. Parasitol. 11:369-375.

64. Roberts, L.S. 1961. The influence of population density on patterns and physiology of growth in *Hymenolepis diminuta* (Cestoda: Cyclophyllidea) in the definitive host. Exp. Parasitol. 11:332-371.

65. Roberts, L.S. 1980. Development of *Hymenolepis diminuta* in its definitive host. In Arai, H.P., editor. Biology of the rat tapeworm, *Hymenolepis diminuta*. Academic Press, Inc., New York, pp. 357-423.

66. Roberts, L.S. 1983. Carbohydrate metabolism. In Arme, C., and P.W. Pappas, editors. The biology of the Eucestoda, vol. 2. Academic Press Ltd., London, pp. 343-390.

67. Roberts, L.S., and G.D. Insler. 1982. Developmental physiology of cestodes. XVII. Some biological properties of putative "crowding factors" in *Hymenolepis diminuta*. J. Parasitol. 68:263-269.

68. Roberts, L.S., and F.N. Mong. 1973. Developmental physiology of cestodes. XIII. Vitamin B$_6$ requirement of *Hymenolepis diminuta* during in vitro cultivation. J. Parasitol. 59:101-104.

69. Rothman, A.H. 1959. Studies on the excystment of tapeworms. Exp. Parasitol. 8:336-364.

70. Schroeder, L.L., P.W. Pappas, and G.E. Means. 1981. Trypsin inactivation by intact *Hymenolepis diminuta* (Cestoda): some characteristics of the inactivated enzyme. J. Parasitol. 67:378-385.

71. Sharp, S.E., C.K. Phares, and M.L. Heidrick. 1982. Immunological aspects associated with suppression of hormone levels in rats infected with plerocercoids of *Spirometra mansonoides* (Cestoda). J. Parasitol. 68:993-998.

72. Smyth, J.D. 1962. Lysis of *Echinococcus granulosus* by surface-active agents in bile and the role of this phenomenon in determining host specificity in helminths. Proc. R. Soc. B 156:553-572.

73. Smyth, J.D. 1969. The physiology of cestodes. W.H. Freeman & Co., Publishers, San Francisco.

74. Smyth, J.D., H.J. Miller, and A.B. Howkins. 1967. Further analysis of the factors controlling strobilization, differentiation and maturation of *Echinococcus granulosus* in vitro. Exp. Parasitol. 21:31-41.

75. Specian, R.D., and R.D. Lumsden. 1980. The microanatomy and fine structure of the rostellum of *Hymenolepis diminuta*. Z. Parasitenkd. 63:71-88.

76. Specian, R.D., and R.D. Lumsden. 1981. Histochemical, cytochemical and autoradiographic studies on the rostellum of *Hymenolepis diminuta*. Z. Parasitenkd. 64:335-345.

77. Specian, R.D., R.D. Lumsden, J.E. Ubelaker, and V.F. Allison. 1979. A unicellular endocrine gland in cestodes. J. Parasitol. 65:569-578.

78. Starling, J.A. 1975. Tegumental carbohydrate transport in intestinal helminths: correlation between mechanisms of membrane transport and the biochemical environment of absorptive surfaces. Trans. Am. Microsc. Soc. 94:508-523.

79. Tedesco, J.L., and J.R. Coggins. 1980. Electron microscopy of the tumulus and origin of associated structures within the tegument of *Eubothrium salvelini* Schrank, 1790 (Cestoidea: Pseudophyllidea). Int. J. Parasitol. 10:275-280.

80. Threadgold, L.T., and J. Dunn. 1983. *Taenia crassiceps*: regional variations in ultrastructure and evidence of endocytosis in the cysticerus' tegument. Exp. Parasitol. 55:121-131.

81. Threadgold, L.T., and J. Dunn. 1984. *Taenia crassiceps*: basic mechanisms of endocytosis in the cysticercus. Exp. Parasitol. 58:263-269.

82. Threadgold, L.T., and C.A. Hopkins. 1981. *Schistocephalus solidus* and *Ligula intestinalis:* pinocytosis by the tegument. Exp. Parasitol. 51:444-456.

83. Ubelaker, J.E. 1983. The morphology, development and evolution of tapeworm larvae. In Arme, C., and P.W. Pappas, editors. Biology of the Eucestoda, vol. 1. Academic Press, London.

84. Uglem, G.L., and J.J. Just. 1983. Trypsin inhibition by tapeworms: antienzyme secretion or pH adjustment? Science 220:79- 81.

85. Wack, M., R. Komuniecki, and L.S. Roberts. 1983. Amino acid metabolism in the rat tapeworm, *Hymenolepis diminuta*. Comp. Biochem. Physiol. 74B:399-402.

86. Ward, S.M., J.M. Allen, and G. McKerr. 1986. Neuromuscular physiology of *Grillotia erinaceus* metacestodes (Cestoda: Trypanorhyncha) *in vitro*. Parasitology 93:121-132.

87. Ward, S.M., G. McKerr, and J.M. Allen. 1986. Structure and ultrastructure of muscle systems within *Grillotia erinaceus* metacestodes (Cestoda: Trypanorhyncha). Parasitology 93:587-597.

88. Watts, S.D.M., and D. Fairbairn. 1974. Anaerobic excretion of fermentation acids by *Hymenolepis diminuta* during development in the definitive host. J. Parasitol. 60:621-625.

89. Webster, L.A., and R.A. Wilson. 1970. The chemical composition of protonephridial canal fluid from the cestode *Hymenolepis diminuta*. Comp. Biochem. Physiol. 35:201-209.

90. Wilson, V.C.L.C., and E.L. Schiller. 1969. The neuroanatomy of *Hymenolepis diminuta* and *H. nana*. J. Parasitol. 55:261-270.

91. Wooten, R. 1974. Studies on the life history and development of *Proteocephalus percae* (Müller) (Cestoda: Proteocephalida). J. Helminthol. 48:269-281.

92. Zavras, E.T., and L.S. Roberts. 1985. Developmental physiology of cestodes: cyclic nucleotides and the identity of putative crowding factors in *Hymenolepis diminuta*. J. Parasitol. 71:96-105.

SUGGESTED READINGS

Arme, C., and P.W. Pappas, editors. 1983. The biology of the Eucestoda, 2 vols. Academic Press Ltd., London. (Up-to-date summary of cestodology; covers evolution and systematics, ecology, morphology and fine structure, development, biochemistry and physiology, pathology, immunology, and chemotherapy.)

Barrett, J. 1981. Biochemistry of parasitic helminths. University Park Press, Baltimore.

Fairbairn, D. 1970. Biochemical adaptation and loss of genetic capacity in helminth parasites. Biol. Rev. 45:29-72.

Hyman, L.H. 1951. The invertebrates, vol. 2. McGraw-Hill Book Co., New York. (A complete summary of knowledge of cestodes up to 1951.)

Read, C.P. 1959. The role of carbohydrates in the biology of cestodes. VIII. Exp. Parasitol. 8:365-382.

Schmidt, G.D. 1986. Handbook of tapeworm identification. CRC Press, Boca Raton, Fla.

Wardle, R.A., and J.A. McLeod. 1952. The zoology of tapeworms. Hafner Publishing Co., New York. (This monograph is the classic in its field. No student of tapeworms should be without it.)

Yamaguti, S. 1959. Systema helminthum, vol. 2. The cestodes of vertebrates. Interscience, New York.

CHAPTER 22

TAPEWORMS

Although most species of cestodes are parasites of wild animals, a few are of particular interest because they infect humans or domestic animals. All tapeworms of humans are in the two largest orders, Pseudophyllidea and Cyclophyllidea. We will consider these first.

Most orders of tapeworms are of little or no medical or economic importance. Still, they are interesting in their own right and deserve at least an introduction. Their diversity of morphology is astonishing, and the study of their many varieties of life cycles is a science in itself. Many opportunities are available for research on these worms. For example, entire orders of cestodes exist for which not a single life cycle is known. Following the discussion of Pseudophyllidea and Cyclophyllidea, we will give brief descriptions of the remaining orders.

ORDER PSEUDOPHYLLIDEA

Pseudophyllidean cestodes typically have a scolex with dorsal and ventral longitudinal grooves called **bothria.** These may be deep or shallow, smooth or fimbriated, and in some cases they are fused along all or part of their length, forming longitudinal tubes. Sclerotized hooks accompany the bothria in some species. The genital pores may be lateral or medial, depending on the species. The vitellaria are always follicular and scattered throughout the segment. The testes are numerous. Some species are fairly small, but the largest tapeworms known are in the Pseudophyllidea. For example, *Hexagonoporus* from the sperm whale measures more than 30 meters long. In addition, each segment has 4 to 14 complete sets of reproductive organs. It has up to 45,000 segments. The reproductive capacity of such an animal is staggering. Generally, the life cycles of pseudophyllideans involve crustacean first intermediate hosts and fish second intermediate hosts.

Family Diphyllobothriidae
■ *Diphyllobothrium latum*

Usually called the broad fish tapeworm, this cestode is common in fish-eating carnivores, particularly in northern Europe. It appears to exhibit a striking lack of host specificity, occurring in many canines and felines, mustelids, pinnipeds, bears, and humans. Most of these records, however, are misidentifications. In northeastern North America humans may also become infected with *D. ursi,* a closely related species. Humans seem to be quite suitable as hosts; *D. latum* is so common in some small areas of the world that nearly 100% of the human population are infected.[13] It is most abundant in Scandinavia, the Baltic states, and Russia and is present in the Arctic and Great Lakes areas of North America. It has also been found in Africa, Japan, South America, Ireland, and Israel, although some of these records are probably erroneous.

Morphology. The adult worm (Fig. 22-1) may attain a length of 30 feet and shed up to a million eggs a day. The species is anapolytic and characteristically releases long chains of spent proglottids, usually the first indication that the infected person has a secret guest.

The scolex (Fig. 22-2) is finger shaped and has dorsal and ventral bothria. Proglottids (Fig. 22-3) are usually wider than long. There are numerous testes and vitelline follicles scattered throughout the proglottid, except for a narrow zone in the center. The male and female genital pores open midventrally. The bilobed ovary is near the rear of the segment. The uterus consists of short loops and extends from the ovary to a midventral uterine pore.

Biology. (Fig. 22-4). The ovoid eggs measure about 60 by 40 μm and have a lidlike operculum at one end and a small knob on the other (Fig. 22-5). When released through the uterine pore, the shelled embryo is at an early stage of development, and it must be deposited in water for development to continue. Completion of development to coracidium takes from 8 days to several weeks, depending on the temperature. Emerging through the operculum (Fig. 22-6), the ciliated coracidium swims randomly about, where it may attract the attention of predaceous copepods of the genus *Diaptomus*. Soon after being eaten, the coracidium loses its ciliated epithelium and immediately begins to attack the wall of the midgut

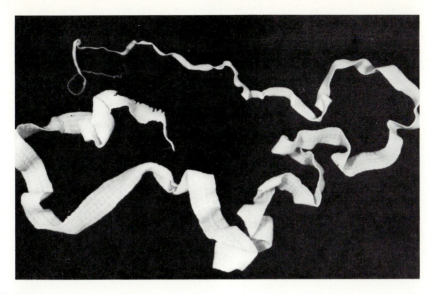

FIG. 22-1

Diphyllobothrium latum. The scolex is at the tip of the threadlike end at upper left.

Photograph by Warren Buss.

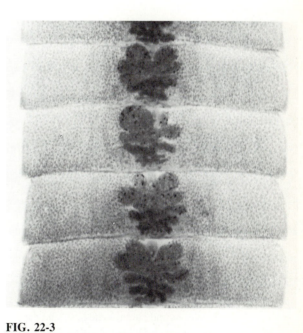

FIG. 22-2

Scolex of *Diphyllobothrium latum*. It is about 1 mm long. Note the dorsal bothrium at right.

Photograph by Warren Buss.

FIG. 22-3

Gravid proglottids of *Diphyllobothrium latum*, showing the characteristic rosette-shaped uterus.

Photograph by Larry Jensen.

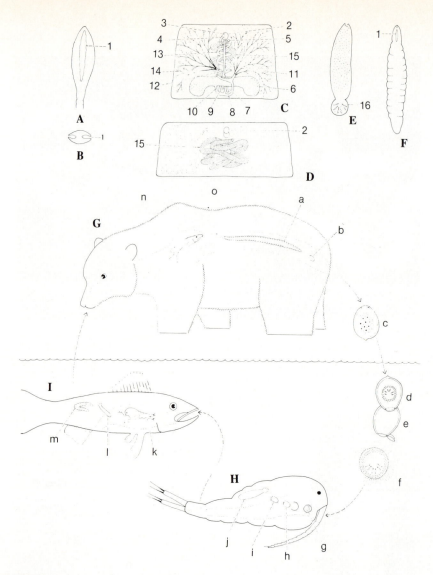

FIG. 22-4

A, Scolex of adult worm. **B,** Cross section of scolex. **C,** Mature proglottid. **D,** Ripe proglottid. **E,** Procercoid. **F,** Plerocercoid. **G,** Definitive host. **H,** Copepod first intermediate host. **I,** Walleyed pike *(Stizostedion vitreum)* second intermediate host. *1,* Bothrium; *2,* common genital atrium; *3,* male genital pore; *4,* female genital pore; *5,* uterine pore; *6,* bilobed ovary; *7,* Mehlis' gland surrounding ootype; *8,* vitelline duct; *9,* proximal portion of vagina; *10,* oviduct; *11,* vitelline glands; *12,* vagina; *13,* vas deferens; *14,* testes; *15,* uterus; *16,* cercomer with oncospheral hooks; *a,* adult in intestine of definitive host; *b,* egg passing out of intestine with feces; *c,* unembryonated egg; *d,* embryonated egg; *e,* hatched egg; *f,* ciliated six-hooked coracidium free in water; *g,* coracidium eaten by crustacean first intermediate host; *h,* coracidium sheds ciliated covering; *i,* coracidium migrates through intestinal wall into hemocoel; *j,* procercoid; *k,* infected crustacean is swallowed by fish second intermediate host, and the procercoid is liberated by the digestive enzymes; *l,* procercoid passes through intestinal wall into coelom and finally into muscles of fish where development continues to infective stage; *m,* plerocercoid in muscles; *n,* infection of definitive host occurs when infected fish are eaten; *o,* plerocercoids are liberated and develop to adults.

From Olsen, O.W. 1974. Animal parasites, their life cycles and ecology, ed. 3. University Park Press, Baltimore.

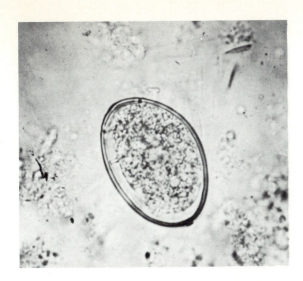

FIG. 22-5

Egg of *Diphyllobothrium latum* in a human stool; note operculum at upper end and small knob at opposite end. It is 40 to 60 μm long.

Photograph by David Oetinger.

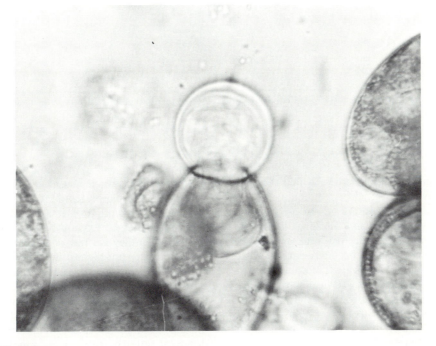

FIG. 22-6

Diphyllobothriid coracidium emerging from its eggshell. The cilia are moving so fast that they appear as a blur.

Photograph by Justus F. Mueller.

with its six tiny hooks. Once through the intestine and into the crustacean's hemocoel, it becomes parasitic, absorbing nourishment from the surrounding blood. In about 3 weeks it increases its length to around 500 μm, becoming an elongate, undifferentiated mass of parenchyma with a cercomer at the posterior end. It is now a procercoid (Fig. 22-7), incapable of further de-velopment until eaten by a suitable second intermediate host—any of several species of freshwater fish, especially pike and related fish, or any of the salmon family. The cercomer may be lost while still in the copepod or soon after the procercoid enters a fish. Large, predaceous fish eat comparatively few microcrustaceans but can still become infected by eating

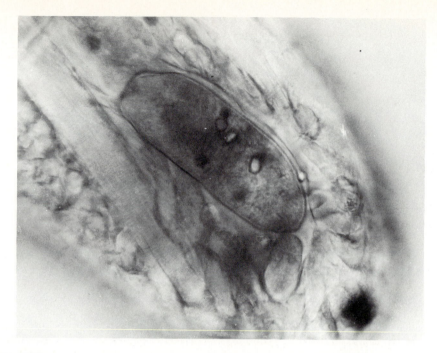

FIG. 22-7

Procercoid in the hemocoel of a copepod. Note the anterior pit, the posterior cercomer, and the internal calcareous granules.

Photograph by Justus F. Mueller.

smaller fish containing plerocercoids, which then migrate into the new host.

When the infected copepod is eaten by a fish, the procercoid is released in the fish's intestine. From here it bores its way through the intestinal wall and into the body muscles. Here it again becomes parasitic, absorbing nutrients and growing rapidly into a plerocercoid. Mature plerocercoids vary in length from a few millimeters to several centimeters. They are still mainly undifferentiated, but there may be evidence of shallow bothria at the anterior end. Usually plerocercoids are found unencysted and coiled up in the musculature, although they may be encysted in the viscera. They are easily seen as white masses in uncooked fish (Fig. 22-8), but when the flesh is cooked, the worms are seldom noticed. Plerocercoids of other pseudophyllideans, as well as those of proteocephalans and trypanorhynchans, are also found in fish and are often mistaken for those of *D. latum*. When the plerocercoid is ingested by a suitable host, it survives the digestive fate of its late host and begins a close relationship with a new one. The worms grow rapidly and may begin egg production by 7 to 14 days. Little of the growth may be attributed to the production of new proglottids but is caused by growth in primordia already in the plerocercoid. As much as 70% of the strobila may mature on the same day.[2]

Epidemiology. Obviously, persons become infected when they eat raw or undercooked fish. Hence infection rates are highest in countries where raw fish are eaten as a matter of course. Communities that dispose of sewage by draining it into lakes or rivers without proper treatment create an opportunity for a massive buildup of *D. latum* in local fish. These fish may be harvested for local consumption or shipped thousands of miles by refrigerated freight to distant markets. There an unsuspecting customer may gain infection in a restaurant or at home by tasting such dishes as gefilte fish during preparation.

Pathogenesis. Many cases of **diphyllobothriasis** are apparently asymptomatic or have poorly defined symptoms associated with other tapeworms, such as vague abdominal discomfort, diarrhea, nausea, and weakness. However, in a small number of cases the worm causes a serious megaloblastic anemia; virtually all of these cases are in Finnish people. It has been estimated that almost a fourth of the population of Finland may be infected with *D. latum*, and about 1000

FIG. 22-8

Two plerocercoids in the flesh of a perch.

From Vik, R. 1971. In Marcial-Rojas, R.A., editor. Pathology of protozoal and helminthic diseases, with clinical correlation. ©1971 The Williams & Wilkins Co., Baltimore.

of these will have pernicious anemia.[9] It was thought originally that toxic products of the worm produced the anemia, but it is now known that the large amount of vitamin B_{12} absorbed by the cestode, in conjunction with some degree of impairment of the patient's normal absorptive mechanism for vitamin B_{12}, is responsible for the disease. Nyberg[47] reported that an average of 44% of a single oral dose of vitamin B_{12} labeled with cobalt 60 was absorbed by *D. latum* in otherwise healthy patients, but in patients with tapeworm pernicious anemia 80% to 100% of the dose was absorbed by the cestode. The clinical symptoms of tapeworm pernicious anemia are similar in many respects to "classical" pernicious anemia (caused by a failure in intestinal absorption of vitamin B_{12}), except that expulsion of the worm generally brings a rapid remission of the anemia.

Diagnosis and treatment. Demonstration of the characteristic eggs or proglottids passed with the stool gives positive diagnosis. In the past a variety of drugs has been used against *D. latum* and other tapeworms; aspidium oleoresin (extract of male fern), mepacrine, dichlorophen, and even extracts of fresh pumpkin seeds (*Cucurbita* spp.) have anticestodal properties.[17]

However, the drug of choice for the past 20 years has been niclosamide (Yomesan). Its mode of action seems to be an inhibition of an inorganic phosphate–ATP exchange reaction associated with the worm's anaerobic electron transport system. Praziquantel is as effective as niclosamide and may supplant it.

■ Other pseudophyllideans found in humans

Several other species of *Diphyllobothrium* have been reported from humans in different parts of the world. These include *D. chordatum* and *D. pacificum,* parasites of pinnipeds in the northern and southern hemispheres, respectively, and *D. ursi* of bears. Other species of *Diphyllobothrium* reported from humans are probably synonyms of *D. latum*. An expert can clearly differentiate the species in animals from *D. latum*. The current fad in the United States of eating raw salmon as sushimi has led to infections.[53]

Diphyllobothrium erinacei, Digramma brauni, and *Ligula intestinalis* have also been reported from humans, but such occurrences must be considered rare. *Diplogonoporous grandis (D. balaenopterae)* has been reported numerous times from humans in Japan.[33] A parasite of whales, its plerocercoid occurs in marine fish, the mainstay of the Japanese protein diet. The proglottid is easily recognized, since it has two sets of male and female reproductive organs in each segment.

Sparganosis. With the exception of the forms with scolex armature, it is impossible to distinguish the species of plerocercoids found in humans by examining their morphology. When procercoids of some species are ingested accidentally, usually by swallowing an infected copepod in drinking water, they can migrate from the gut and develop into plerocercoids, sometimes reaching a length of 14 inches. The infection is called **sparganosis** and may cause severe pathological consequences. Cases have been reported from most countries of the world but are most common in the Orient. Yamane, Okada, and Takihara[63] reported a living sparganum that had infected a woman's breast for at least 30 years.

Other means of infection are by ingestion of insufficiently cooked amphibians, reptiles, birds, or even mammals such as pigs.[15] Plerocercoids present in these animals may then infect the person indulging in such delicacies. Many Chinese are infected in this way because of their tradition of eating raw snake to cure a panoply of ills.[36]

A third method of infection results from the oriental treatment of skin ulcers, inflamed vagina, or inflamed eye (Fig. 22-9), by poulticing the area with a split frog or flesh of a vertebrate that may be incidentally infected with spargana. The active worm then crawls

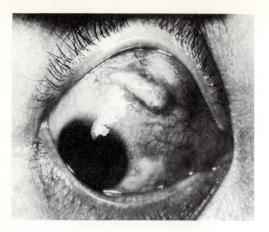

FIG. 22-9

Right eye of patient with sparganosis. Note the protruding mass in the upper conjunctiva.

From Wang, L.T., and J.H. Cross. 1974. J. Formosan Med. Assoc. 73:173-177.

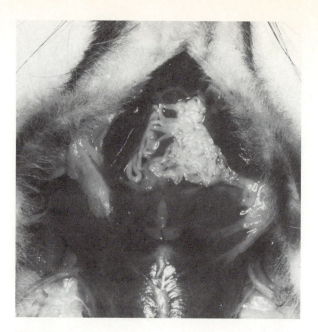

FIG. 22-10

Spargana in subcutaneous connective tissues of a wild rat in Taiwan.

Photograph by Robert E. Kuntz.

into the orbit, vagina, or ulcer and establishes itself. Most cases of sparganosis in the Orient are probably caused by *D. erinacei,* a parasite of carnivores.

In North America most spargana are probably *D. mansonoides,* a parasite of cats.[45] It usually does not proliferate, except by occasionally breaking transversely, and may live up to 10 years in a human.[60] The current public awareness of the symptoms of cancer has led to an increase in reported cases of sparganosis in this country. Subdermal lumps are no longer ignored by the average person, and more than one physician has been shocked to find a gleaming, white worm in a lanced nodule. Wild vertebrates are commonly infected with spargana (Fig. 22-10).

Rarely a sparganum will be proliferative, splitting longitudinally and budding profusely. Such cases are very serious, since many thousands of worms can result, with the infected organs becoming honeycombed.

Treatment of sparganosis is usually by surgery, but some success has been obtained by chemotherapy.[35]

ORDER CYCLOPHYLLIDEA

The most characteristic morphological features of cyclophyllideans are a single compact vitelline gland and a scolex with four suckers. A rostellum, which usually is armed with hooks, is commonly present. The genital pores are lateral in all except the family Mesocestoididae, in which they are midventral. The single vitellarium is usually postovarian but may be preovarian. The number of testes varies from one to several hundred, depending on the species. Most spe-

cies are rather small, although some are giants of more than 30 feet in length. Most tapeworms of birds and mammals belong to this order.

Family Taeniidae

The largest cyclophyllideans are in the family Taeniidae, as are the most medically important tapeworms of humans. A remarkable morphological similarity occurs among most species in the family; a striking exception is *Echinococcus,* which is much smaller than cestodes of the other genera. An armed rostellum is present on most species and, when present, is not retractable. The testes are numerous, and the ovary is a bilobed mass near the posterior margin of the proglottid. The metacestodes are various types of bladderworms (Fig. 22-11), and mammals serve as their intermediate hosts.

■ *Taeniarhynchus saginatus*

Taeniarhynchus saginatus is by far the most common taeniid of humans, occurring in nearly all countries where beef is eaten. The beef tapeworm, as it is usually known, lacks a rostellum or any scolex armature (Fig. 22-12). Individuals of this exceptionally large species may attain a length of over 75 feet, but 10 to 15 feet is much more common. Even the smaller

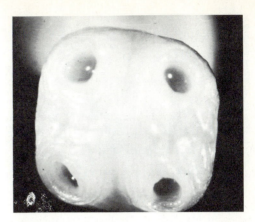

FIG. 22-12

En face view of the scolex of *Taeniarhynchus saginatus*. Note the absence of a rostellum or armature.

AFIP neg. no. 65-12073-2.

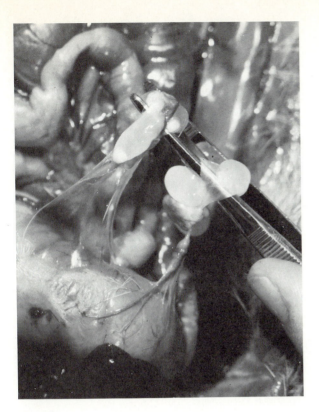

FIG. 22-11

Cysticerci of *Taenia pisiformis* in the mesenteries of a rabbit.

Photograph by John Mackiewicz.

specimens may consist of as many as 2000 proglottids.

Morphology. The scolex, with its four powerful suckers, is followed by a long, slender neck. Mature proglottids are slightly wider than long, whereas gravid ones are much longer than wide. Usually the gravid proglottid passed in the feces is first noticed and taken to a physician for diagnosis. Because the eggs of this species cannot be differentiated from those of *Taenia solium,* the next most common taeniid of humans, accurate diagnosis depends on a critical examination of a gravid uterus. Of course, if the entire worm is passed, the combination of unarmed scolex and taeniid type of proglottid (Fig. 22-13) leads to an unmistakable diagnosis.

The spherical eggs are characteristic of the Taeniidae (Fig. 22-14). A thin, hyaline, outer membrane is usually lost by the time the egg is voided with the feces. The embryophore is very thick and riddled with numerous tiny pores, giving it a striated appearance in optical section. Unfortunately, the egg sizes of several

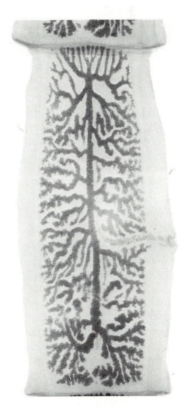

FIG. 22-13

Gravid proglottid of *Taenia* sp. The uterus is characteristic for this genus, consisting of a median stem with lateral branches.

Photograph by Warren Buss.

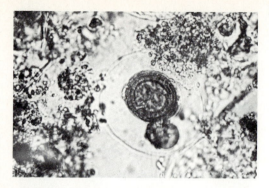

FIG. 22-14

Taenia egg in human feces. The thin outer membrane is often lost at this stage.

Photograph by David Oetinger.

FIG. 22-15

Scolex of *Taenia solium*. Note the large rostellum with two circles of hooks.

Photograph by David Oetinger.

taeniids in humans overlap, making diagnosis of species impossible on this character alone.

Biology. When gravid, the proglottids detach and either pass out with the feces or migrate out of the anus. Each segment behaves like an individual worm, crawling actively about, as if searching for something. The segments are easily mistaken for trematodes or even nematodes at this stage. As a segment begins to dry up, a rupture occurs along the midventral body wall, allowing eggs to escape. The eggs are fully embryonated and infective to the intermediate host at this time; they remain viable for many weeks. Cattle are the usual intermediate host, although cysticerci have also been reported from llamas, goats, sheep, giraffes, and even reindeer (perhaps incorrectly).

When eaten by a suitable intermediate host, the egg hatches in the duodenum under the influence of gastric and intestinal secretions. The released hexacanth quickly penetrates the mucosa and enters an intestinal venule, to be carried throughout the body. Typically it leaves a capillary between muscle cells and enters a muscle fiber, developing into an infective cysticercus in about 2 months. This metacestode is white, pearly, and about 10 mm at its greatest diameter and contains a single, invaginated scolex. Humans are probably an unsuitable intermediate host, and the few records of *T. saginatus* cysticerci in humans are most likely misidentifications. Before the beef cysticercus was known to be a juvenile form of *T. saginatus,* it was placed in a separate genus under the name of *Cysticercus bovis.* The disease produced in cattle is thus known as **cysticercosis bovis,** and flesh riddled with the juveniles is called **measly beef.**

A person who eats infected beef, cooked insufficiently to kill the juveniles, becomes infected. The invaginated scolex and neck of the cysticercus evaginate in response to bile salts. The bladder is digested by the host or absorbed by the scolex, and budding begins. Within 2 to 12 weeks the worm will begin shedding gravid proglottids.

Epidemiology. Human infection is highest in areas of the world where beef is a major food and sanitation is of little concern. Thus, in several developing nations of Africa and South America, for instance, ample opportunity exists for cattle to eat tapeworm eggs and for people to eat infected flesh. Many persons are content to eat a chunk of meat that is cooked in a campfire, charred on the outside and raw on the inside. Local custom may have profound effect on infection rates. Hence in India there may be a high rate of infection in a Moslem population, whereas Hindus, who do not eat beef, are unaffected. In the United States, federal meat inspection laws and a high degree of sanitation combine to keep the incidence of infection low. However, only 80% of the cattle slaughtered in the United States are federally inspected, and studies have shown that standard inspection procedures fail to detect a fourth of infected cattle.[18] One wonders if backyard cookery and the popularity of steak tartare might not contribute to an increase in taeniiasis.

Despite the high level of sanitation in any country, it still is possible for cattle to be exposed to the eggs of this parasite. One infected person who defecates in a pasture or cattle-feeding area can quickly infect an entire herd. The use of human feces as fertilizer can have the same effect. The eggs are known to remain viable in liquid manure for 71 days, in untreated sewage for 16 days, and on grass for 159 days.[31] Cattle are coprophagous and often will eat human dung, wherever they find it. In India, where cattle roam at will, it is common for a cow to follow a person into the woods, in hopes of obtaining a fecal meal.[13]

Prevention of human infection is easy; when meat is cooked until it is no longer pink in the center, it is safe to eat, since cysticerci are killed at 56° C. Furthermore, meat is also rendered safe by freezing at $-5°$ C for at least a week.

Pathogenesis. Disease characteristics of *T. saginatus* infection are similar to those of infection by any large tapeworm, except that the avitaminosis B_{12} found in association with *D. latum* is unknown. Verminous intoxication, caused by absorption of the worm's excretory products, is common, with the characteristic symptoms of dizziness, abdominal pain, headache, localized sensitivity to touch, and nausea. Delirium is rare but does occur. Diarrhea and intestinal obstruction are common. Hunger pains, universally accepted by lay people as a symptom of tapeworm infection, are not common, but *loss* of appetite is frequent. In addition, it is difficult to estimate the psychological effects on an infected person of observing continued migration of proglottids out of the anus.

Diagnosis and treatment. Identification of taeniid eggs according to species is impossible. Therefore accurate diagnosis depends on examination of a scolex or a gravid proglottid. The latter is characterized by having 15 to 20 lateral branches on each side. Because these branches tend to fuse in old segments, freshly passed specimens must be obtained for reliable results.

Numerous taeniicides have been used in the past. Today niclosamide and praziquantel are the drugs of choice.

■ *Taenia solium*

The most potentially dangerous adult tapeworm of humans is the pork tapeworm, *Taenia solium,* because of the possibilities of self-infection with cysticerci. Furthermore, it is possible to infect others in the same household with juveniles of this parasite, often with grave results.

Morphology. The scolex of the adult (Fig. 22-15) bears a typical, nonretractable taeniid rostellum, armed with two circles of 22 to 32 hooks measuring 130 to 180 μm long. Whereas the scolex of *T. saginatus* is cuboidal and up to 2 mm in diameter, that of *T. solium* is spheroid and only half as large. The strobila has been reported as being as long as 30 feet, but 6 to 10 feet is much more common. Mature proglottids are wider than long and are nearly identical to those of *T. saginatus,* differing in number of testes (150 to 200 in *T. solium,* 300 to 400 in *T. saginatus*). Gravid proglottids are longer than wide and have the typical taeniid uterus, a medial stem with 7 to 13 lateral branches.

Biology. The life cycle of *T. solium* (Fig. 22-16) is in most regards like that of *T. saginatus,* except that the intermediate hosts are pigs instead of cattle. Gravid proglottids passed in the feces are laden with eggs infective to swine. When eaten, the oncospheres develop into cysticerci (*Cysticercus cellulosae*) in the muscles and other organs. A person easily becomes infected when a bladderworm is eaten along with insufficiently cooked pork. Evaginating by the same process as in *T. saginatus,* the worm attaches to the mucosa of the small intestine and matures in 5 to 12 weeks. Specimens of *T. solium* have been known to live for as long as 25 years. Pathogenesis caused by the adult worm is similar to that in taeniiasis saginatus.

■ Cysticercosis

Unlike those of most other species of *Taenia,* the cysticerci of *T. solium* develop readily in humans. Infection occurs when embryonated eggs pass through the stomach and hatch in the intestine. Persons who are infected by adult worms may contaminate their households or food with eggs that are accidentally eaten by themselves or others. Possibly, a gravid proglottid may migrate from the lower intestine to the stomach or duodenum, or it may be carried there by reverse peristalsis. Subsequent release and hatching of many eggs at the same time results in a massive infection by cysticerci. Fortunately, this parasite is not common anywhere in the world.

Virtually every organ and tissue of the body may harbor cysticerci. Most commonly they are found in the subcutaneous connective tissues. The second most common site is the eye, followed by the brain (Fig. 22-17), muscles, heart, liver, lungs, and coelom. A fibrous capsule of host origin surrounds the metacestode, except when it develops in the chambers of the eye. The effect of any cysticercus on its host depends on where it is located. In skeletal muscle, skin, or liver, little noticeable pathogenesis usually results, except in massive infection. Ocular cysticercosis may cause irreparable damage to the retina, iris, uvea, or choroid. A developing cysticercus in the retina may be

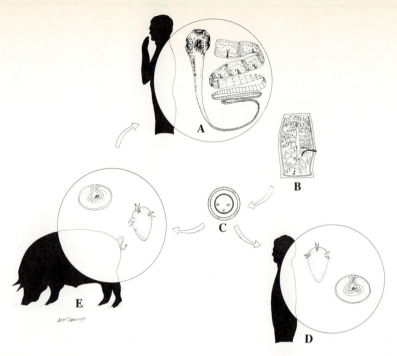

FIG. 22-16

Life cycle of *Taenia solium*. **A,** Adult tapeworm in the small intestine of a human. **B,** Gravid proglottids detach from the strobila and migrate out of the anus or pass with feces. **C,** Egg. **D,** If eaten by a human, the oncosphere hatches, migrates to some site in the body, and develops into a cysticercus. **E,** Cysticerci will also develop if the eggs are eaten by a pig. The life cycle is completed when a person eats pork containing live cysticerci.

Diagram by Carol Eppinger.

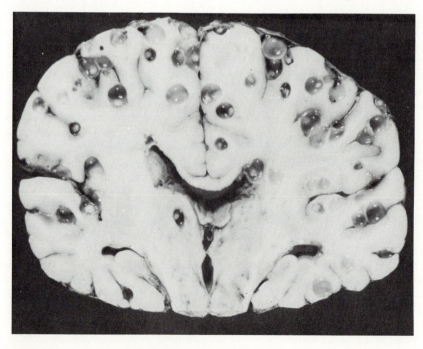

FIG. 22-17

Human brain containing numerous cysticerci of *Taenia solium*.

From Flisser, A. 1988. Parasitol. Today 4:131-137.

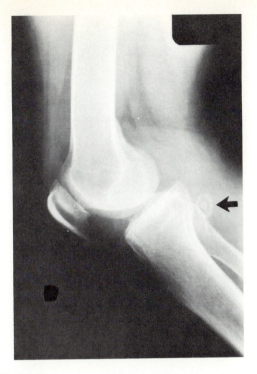

FIG. 22-18

Cysticercus cellulosae: partially calcified cyst *(arrow)* found in a routine x-ray examination of a human leg.

From Roudabush, R.L., and G.A. Ide. 1975. J. Parasitol. 61:512.

mistaken for a malignant tumor, resulting in the unnecessary surgical removal of the eye. Removal of the cysticercus by fairly simple surgery is usually successful.

Cysticerci occur rarely in the spinal cord but commonly in the brain.[7] Symptoms of infection are vague and rarely diagnosed except at autopsy. Radiological diagnosis of neurocysticercosis is practical.[12] Pressure necrosis may cause severe central nervous system malfunction, blindness, paralysis, disequilibrium, obstructive hydrocephalus, or disorientation. Perhaps the most common symptom is epilepsy of sudden onset. When this occurs in an adult with no family or childhood history of epilepsy, cysticercosis should be suspected.[20]

When a cysticercus dies, it elicits a rather severe inflammatory response. Many of them may rapidly prove to be fatal to the host, particularly if the worms are located in the brain. This was observed frequently in former British soldiers of whom a high proportion who had served in India became infected. Other types of cellular reaction also occur, usually resulting in eventual calcification of the parasite (Fig. 22-18). If

this occurs in the eye, there is little chance of corrective surgery.

A cysticercus will rarely become proliferative, developing branching extensions that destroy even more of the host's tissues. The increased danger of such an organism is obvious; furthermore, there is less chance of successful surgery in such cases. Praziquantel has been used successfully to treat cerebral cysticercosis.[42]

Prevention of cysticercosis depends on early detection and elimination of the adult tapeworm and a high level of personal hygiene. Fecal contamination of food and water must be avoided, and the use of untreated sewage on vegetable gardens eschewed. The majority of cases apparently originate from such sources, including contamination by infected food handlers.[23]

Curiously, swine cysticercosis is fairly common, even in countries where the adult is rare. It is evident that the multitudes of eggs produced by even one adult worm overcome great odds to infect pigs and thereby continue the species.

■ **Other taeniid species of medical importance**

Taeniarhynchus confusus is very similar to *T. saginatus.* The gravid segments are about twice as long as those of *T. saginatus,* whereas the uterus has branched arms that are fewer in number. It has been reported from humans in Africa and the United States; cattle are satisfactory intermediate hosts.

Taenia multiceps, T. glomeratus, T. brauni, and *T. serialis* are all characterized by developing a coenurus type of bladderworm (Fig. 22-19). This is similar to a cysticercus but has many rather than a single protoscolex. Such coenuri occasionally occur in humans, particularly in the brain, eye, muscles, or subcutaneous connective tissue, where they often grow to be longer than 40 mm. The resulting pathogenesis is similar to that of cysticercosis. The adults are parasites of carnivores, particularly dogs, with herbivorous mammals serving as intermediate hosts. Accidental infection of humans occurs when the eggs are ingested. Coenuriasis of sheep, caused by *T. multiceps,* causes a characteristic vertigo called "gid," or "staggers."

■ *Echinococcus granulosus*

The genus *Echinococcus* contains the smallest tapeworms in the Taeniidae. However, their juvenile forms are often huge and are capable of infecting humans, resulting in **hydatidosis,** a very serious disease in many parts of the world. An interesting review of the ecology and distribution of this and other species of *Echinococcus* was given by Rausch.[49]

Echinococcus granulosus uses carnivores, particularly dogs and other canines, as definitive hosts. Many

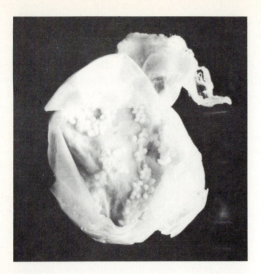

FIG. 22-19

Coenurus metacestode of *Taenia serialis,* from the muscle of a rabbit, that has been opened to show the numerous protoscolices arising from the germinal epithelium. The cyst is about 4 inches wide.

Photograph by James Jensen.

FIG. 22-20

Adult *Echinococcus granulosus* from the intestine of a dog.
Courtesy Ann Arbor Biological Center.

mammals may serve as intermediate hosts, but herbivorous species are most likely to become infected by eating the eggs on contaminated herbage.

The adult (Fig. 22-20) lives in the small intestine of the definitive host. It measures 3 to 6 mm long when mature and consists of a typically taeniid scolex, a short neck, and usually only three proglottids. The nonretractable rostellum bears a double crown of 28 to 50 (usually 30 to 36) hooks. The anteriormost proglottid is immature; the middle one is usually mature; and the terminal one is gravid. The gravid uterus is an irregular longitudinal sac. The eggs cannot be differentiated from those of *T. solium* and *T. saginatus.* The ripe segment detaches and develops a rupture in its wall, releasing the eggs, which are fully capable of infecting an intermediate host.

Hatching and migration of the oncosphere are the same as previously described for *T. saginatus,* except that the liver and lungs are the usual sites of development. By a very slow process of growth, the oncosphere metamorphoses into a type of bladderworm called a **unilocular hydatid** (Fig. 22-21). In about 5 months the hydatid has developed a thick outer, laminated, noncellular layer and an inner, thin, nucleated germinal layer. The inner layer eventually produces the protoscolices that are infective to the definitive host. Protoscolices (Fig. 22-22) are usually produced singly into the lumen of the bladder as in a coenurus

and also within **brood capsules.** The latter are small cysts, containing 10 to 30 protoscolices, which usually are attached to the germinal layer by a slender stalk; they may break free and float within the hydatid fluid. Similarly, individual scolices and brood capsules may break free and sink to the bottom of the bladder, where they are known as **hydatid sand** (Fig. 22-23) (although this may happen only in dead cysts). Rarely germinal cells penetrate the laminated layer and form **daughter capsules.** When the hydatid is eaten by a carnivore, the cyst wall is digested away, freeing the protoscolices, which evaginate and attach among the villi of the small intestine. A small percentage of hydatids lack protoscolices and are sterile, being unable to infect a definitive host. The worm matures in about 56 days and may live for 5 to 20 months.

Epidemiology. The life cycle of *E. granulosus* in wild animals may involve a wolf-moose, wolf-reindeer, dingo-wallaby, or other carnivore-herbivore relationship, which is known as **sylvatic echinococcosis.** Humans are seldom involved as accidental intermediate hosts in these cases. However, ample opportunities exist for human infection in situations in

FIG. 22-21

Several unilocular hydatids in the lung of a sheep. Each hydatid contains many protoscolices.

Photograph by James Jensen.

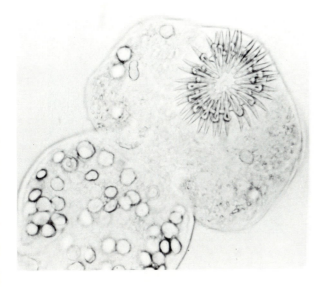

FIG. 22-22

Protoscolex of *Echinococcus granulosus*, removed from a hydatid cyst.

Photograph by Sharon File.

which domestic herbivores are raised in association with dogs. For example, hydatid disease is a very serious problem in sheep-raising areas of Australia, New Zealand, North and South America, Europe, Asia, and Africa. Similarly, goats, camels, reindeer, and pigs, together with dogs, maintain the cycle in various parts of the world. Dogs are infected when they feed on the offal of butchered animals, and herbivores are infected when they eat herbage contaminated with dog dung. Humans are infected with hydatids when they

accidentally ingest *Echinococcus* eggs, usually as a result of fondling dogs.

Local traditions may contribute to massive infections. Some primitive tribes of Kenya, for instance, are said to relish dog intestine roasted on a stick over a campfire. Because cleaning of the intestine may involve nothing more than squeezing out its contents, and cooking may entail nothing more than external scorching, these people probably have the highest rate of infection with hydatids in the world. Nelson and Rausch[46] discussed how some infants in Kenya become infected when their parents encourage the family dog to clean up the mess when the child vomits or defecates, by licking the child's face and anal area. A further complication lies in the lack of burial of the dead by the Turkana people of Kenya. When the corpses are eaten by carnivores, humans become true intermediate hosts of *E. granulosus*.[41]

A different set of circumstances leads to infection in tanners in Lebanon, where dog feces are used as an ingredient of the tanning solution. Scats picked off the street are added to the vats, and any eggs present may infect their handler by contamination.[56] Sheep-

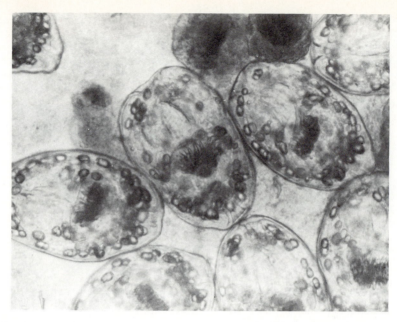

FIG. 22-23

"Hydatid sand," consisting of unattached protoscolices of *Echinococcus granulosus*.
Photograph by Robert E. Kuntz.

herders in the United States and elsewhere risk infection by living closely with their dogs. Surveys of cattle, hogs, and sheep in abattoirs reveal that *E. granulosus* is distributed throughout most of the United States, with greatest concentrations in the deep South and far West. Recent outbreaks have been diagnosed in California and Utah. This disease can be eliminated from an endemic area only by interrupting the life cycle by denying access by dogs to offal, by destroying stray dogs, and by a general education program.[26,43]

Pathogenesis. The effects of a hydatid may not become apparent for many years after infection because of its usual slow growth. Up to 20 years may elapse between infection and overt pathogenesis. If infection occurs early in life, the parasite may be almost as old as its host.[4]

The type and extent of pathological conditions depend on the location of the cyst in the host. As the size of the hydatid increases, it crowds adjacent host tissues and interferes with their normal functions (Fig. 22-24). The results may be very serious. Clinical effects may be manifested relatively early in the infection, before much growth occurs, if the parasite is lodged in the nervous system. When bone marrow is affected, the growth of the hydatid is restricted by lack of space. Chronic internal pressure caused by the parasite usually causes necrosis of the bone, which be-

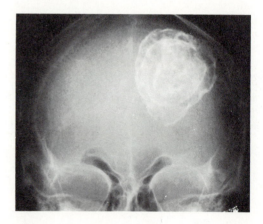

FIG. 22-24

Partially calcified hydatid cyst in the brain.
AFIP neg. no. 68-2740.

comes thin and fragile; characteristically, the first sign of such an infection is a spontaneous fracture of an arm or leg. When the hydatid grows in an unrestricted location, it may become enormous, containing more than 15 quarts of fluid and millions of protoscolices. Even if it does not occlude a vital organ, it can still cause sudden death if it ruptures. Hydatid fluid is quite proteinaceous, inducing an adverse host reaction called anaphylactic shock. Unconsciousness and death are nearly instantaneous in such instances.

Diagnosis and treatment. When hydatids are found, it is often during routine medical x-ray examinations or exploratory surgery. An intradermal immunological test (Casoni's test) is available for use in suspected cases. The antigen, which is manufactured from the proteins in hydatid fluid, is inoculated into the skin; if a hydatid is or has been in the patient, a characteristic wheal develops at the site of injection. If all signs of the inoculation disappear almost immediately, the results are negative for hydatidosis. Other tests have been developed, employing complement fixation, hemagglutination, fluorescent antibody, latex slide agglutination, bentonite flocculation, and precipitin. None is 100% accurate, but they are of sufficient accuracy to make tests most useful on a probability basis. The development of monoclonal antibodies for immunodiagnosis of hydatidosis in humans shows much promise.[19]

Surgery remains the only routine method of treatment and then only when the hydatid is located in an unrestricted location; however, in preliminary studies mebendazole caused regression of *E. granulosus* and *E. multilocularis* hydatids in human patients.[6,61] A high rate of surgical success is obtained on ocular hydatidosis. The typical procedure involves incising the surrounding adventitia until the capsule is encountered and aspirating the hydatid fluid with a large syringe. Considerable delicacy is required at this point, since fluid spilled into a body cavity can quickly cause fatal anaphylactic shock. After aspiration of the cyst contents, 10% formalin is injected into the hydatid to kill the germinal layer. This fluid is withdrawn after 5 minutes, and the entire cyst is then excised.

■ *Echinococcus multilocularis*

Echinococcus multilocularis is primarily boreal in its distribution. It is known from Europe, Asia, and North America, having been discovered as far south as Wyoming and Iowa.[38] Human cases were known previously in the United States only from Alaska,[62] but a human infection has been reported from Minnesota.[25] Cases have also been reported from South America and New Zealand. The adult is mainly a parasite of foxes, but dogs, cats, and coyotes may also serve as definitive hosts. The hydatid develops in several species of small rodents such as voles, lemmings, and mice.

The adult is very similar to *E. granulosus,* differing from it in the following characteristics: (1) *E. granulosus* is 3 to 6 mm long, whereas *E. multilocularis* is only 1.2 to 3.7 mm long; (2) the genital pore of *E. granulosus* is about equatorial, but it is preequatorial

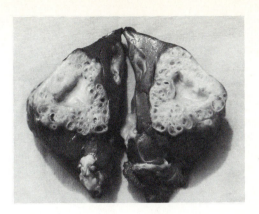

FIG. 22-25

Alveolar hydatid cyst on the liver of an experimentally infected rhesus monkey.

Photograph by Robert Rausch.

in *E. multilocularis;* (3) *E. granulosus* has 45 to 65 testes with a few located anterior to the cirrus pouch; *E. multilocularis* has 15 to 30 testes, all located posterior to the cirrus pouch.

The juvenile form (Fig. 22-25) differs in several respects from that of *E. granulosus.* Instead of developing a thick, laminated layer and growing into large, single cysts, this parasite has a thin outer wall that grows and infiltrates processes into the surrounding host tissues like a cancer. Each process may have several small, fluid-filled pockets containing several protoscolices. In humans and other unnatural hosts the pockets typically lack protoscolices. In natural intermediate hosts the cyst is more regular. In humans, pieces of the cyst sometimes break off and metastasize to other parts of the body. Because of its type of construction, this metacestode form is called an **alveolar** or **multilocular hydatid.** Some authorities, especially in the Soviet Union, place this species in a separate genus, *Alveococcus,* because of its unique form.

Human infection with alveolar hydatid is rare because the normal life cycle is sylvatic rather than urban and because humans do not seem to be very good hosts. Although protoscolices may not develop in human hosts, the germinal membrane is still viable.[51] Anyone handling wild foxes may be exposed to infection. Thus this disease is most common among professional trappers and among handlers of sled dogs, where the dogs catch and eat wild mice as a regular part of their diet.

Diagnosis of alveolar hydatid is difficult, particularly because the protoscolices may not be found. Even at necropsy it may be mistaken for a malignant tumor. As a result of the difficulties of liver surgery, excision is usually practical only when the hydatid is localized near the tip of a lobe of the liver; infections of the hilar area are inoperable. The infiltrative nature of the cyst and its slow rate of growth may advance the disease to an inoperable state without its presence being detected.

Alveolar hydatidosis can be prevented only by avoiding dogs and their feces in endemic regions, by carefully washing all strawberries, cranberries, and the like that may be contaminated by dung, and by regularly worming dogs that may be liable to infection. Because it is a sylvatic disease it is much more difficult to eradicate than *E. granulosus*.[26]

■ *Echinococcus vogeli*

This parasite of canids in Central and South America is rarely known to cause hydatidosis in humans. Previously attributed to *E. oligarthrus*, a tapeworm of felids, all known cases are now identified as *E. vogeli*.[16] The most important source of infection for humans is the domestic dog. The suitability of humans as hosts for *E. vogeli* seems to be intermediate between that of *E. granulosus* and *E. multilocularis;* although polycystic in humans, *E. vogeli* produces relatively large, fluid-filled vesicles with numerous protoscolices. Its natural intermediate host is a rodent called a paca.[50]

OTHER TAPEWORMS OF HUMANS
Family Hymenolepididae

The huge family Hymenolepididae consists of numerous genera with species only in birds and mammals. Excepting *Vampirolepis nana* and *Hymenolepis diminuta,* the remaining several hundred species are of little or no economic or health importance. The family offers considerable taxonomic difficulties because of the large number of species and the immense and far-flung literature that has accumulated. However, their morphology is relatively simple, compared with, for example, the Pseudophyllidea, and most species are small, transparent, and easy to study.

The most obvious morphological feature of the group is the small number of testes, usually one to four. The combination of few testes, usually unilateral genital pores, and large external seminal vesicle allows easy recognition of the family. All (except *V. nana*) require arthropod intermediate hosts.

■ *Vampirolepis nana* (Fig. 22-26)

Commonly called the dwarf tapeworm, *Vampirolepis nana* (previously known as *Hymenolepis nana*) is a cosmopolitan species that is the most common cestode of humans in the world, especially among children. Rates of infection run from 1% in the southern United

---

FIG. 22-26

Life cycle of *Vampirolepis nana,* the dwarf tapeworm. **A,** Scolex. **B,** Rostellar hook. **C,** Mature proglottid. **D,** Gravid proglottid. **E,** Embryonated egg. **F,** Cysticercoid. **G,** Cysticercoid in villus of intestine. **H,** Mouse definitive host. **I,** Larval intermediate beetle host *(Tenebrio molitor, Tribolium confusum).* **J,** Adult beetle. *1,* Scolex; *2,* sucker; *3,* armed rostellum; *4,* handle of hook; *5,* guard; *6,* blade; *7,* common genital pore; *8,* cirrus pouch; *9,* testis; *10,* vagina; *11,* seminal receptacle; *12,* ovary; *13,* oviduct; *14,* vitelline gland; *15,* longitudinal excretory canal; *16,* gravid uterus; *17,* shell, or outer membrane, of egg; *18,* thick inner membrane with terminal filaments; *19,* oncosphere; *20,* tail of cystercoid; *21,* oncospheral hooks; *22,* intestinal villus; *23,* cysticercoid. *a,* adult worm in small intestine; *b,* gravid proglottid detached from strobila; *c,* egg in feces passing out of intestine; *d,* egg free; *e,* egg swallowed and returned to small intestine; *f,* egg hatches in small intestine; *g,* oncosphere burrows into intestinal wall; *h,* cysticercoid develops in villus; *i,* cysticercoid breaks out of intestinal wall and evaginates; *j* to *l* and *a,* cysticercoid attaches to intestinal wall and grows to sexually mature cestode; *m,* some infective eggs do not leave the body but hatch in the intestine, initiating internal autoinfection in which development proceeds as in *e* to *l; n,* eggs passed in feces are infective to arthropod intermediate hosts; *o,* eggs swallowed by larva of beetle or flea hatch in intestine; *p,* oncospheres enter hemocoel and develop into cysticercoids; *q,* eggs swallowed by adult beetles hatch in intestine; *r,* oncospheres migrate from intestine into hemocoel; *s,* tailed cysticercoids develop; *t,* infected beetles swallowed by definitive hosts; *u,* cysticercoids released from beetle; *v,* cysticercoids evaginate and shed tail; *j* to *l* and *a,* cysticercoids develop into adult worms.

From Olsen, O.W. 1974. Animal parasites, their life cycles and ecology, ed. 3. © 1974 University Park Press, Baltimore.

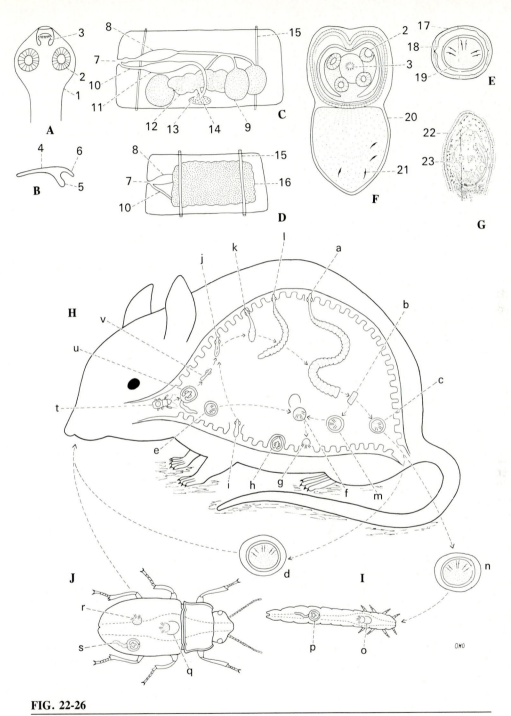

FIG. 22-26

For legend see opposite page.

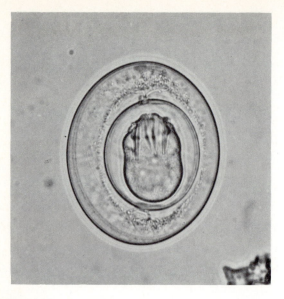

FIG. 22-27

Egg of *Vampirolepis nana.* Note the polar filaments on the inner membrane and the well-developed oncosphere. Its size is 30 to 47 μm.

Photograph by Jay Georgi.

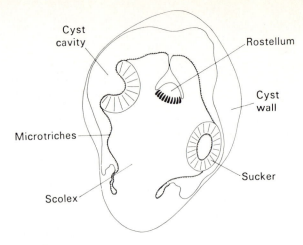

FIG. 22-28

Vampirolepis nana: diagrammatic representation of a longitudinal section through a cysticercoid from a mouse villus.

From Caley, J. 1975. Z. Parasitenkd. 47:218-228.

States to 9% in Argentina, to 97.3% in Moscow.[34]

As its name implies, this is a small species, seldom exceeding 40 mm long and 1 mm wide. The scolex bears a retractable rostellum armed with a single circle of 20 to 30 hooks. The neck is long and slender, and the proglottids are wider than long. The genital pores are unilateral, and each mature segment contains three testes. After apolysis the gravid segments disintegrate, releasing the eggs, which measure 30 to 47 μm in diameter. The oncosphere (Fig. 22-27) is covered with a thin, hyaline, outer membrane and an inner, thick membrane with polar thickenings that bear several filaments. The heavy embryophores that give taeniid eggs their characteristic striated appearance are lacking in this and the other families of tapeworms infecting humans.

The life cycle of *V. nana* is unique in that an intermediate host is optional (Fig. 22-26). When eaten by a person or a rodent, the eggs hatch in the duodenum, releasing the oncospheres, which penetrate the mucosa and come to lie in the lymph channels of the villi. Here each develops into a cysticercoid (Fig. 22-28). In 5 to 6 days the cysticercoid emerges into the lumen of the small intestine, where it attaches and matures.

This direct life cycle is doubtless a recent modification of the ancestral two-host cycle, found in species of *Hymenolepis,* since the cysticercoid of *V. nana* can still develop normally within larval fleas and beetles (Fig. 22-29). One reason for the facultative nature of the life cycle is that *V. nana* cysticercoids can develop at higher temperatures than can those of other hymenolepidids. Direct contaminative infection by eggs is probably the most common route in human cases, but accidental ingestion of an infected grain beetle or flea cannot be ruled out.

Besides humans, domestic mice and rats also serve as suitable hosts for *V. nana.*[24] Some authors contend that two subspecies exist: *V. nana nana* in humans and *V. nana fraterna* in murine rodents. Differences do seem to exist in the physiological host-parasite relationships of these two subspecies, since higher rates of infection result from eggs obtained from the same host species than from the other.[48] This is a good example of **allopatric speciation** in action.

Pathological results of infection by *V. nana* are rare and usually occur only in massive infections. Heavy infections can occur through autoinfection,[28] and the symptoms are similar to those already described for *Taeniarhynchus saginatus* intoxication. Treatment with niclosamide is efficacious but may have to be repeated in a month to remove the worms that were developing in villi at the time of treatment. Praziquantel acts very rapidly against *V. nana* and *H. diminuta.*[1,5] In vitro it produces vacuolization and disruption of the tegument in the neck of the worms but not in more posterior portions of the strobila.

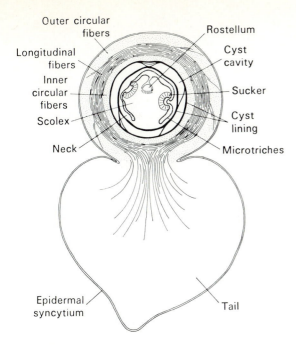

FIG. 22-29

Diagrammatic representation of a longitudinal section through a cysticercoid of *Vampirolepis nana* from the insect host.

From Caley, J. 1975. Z. Parasitenkd. 47:218-228.

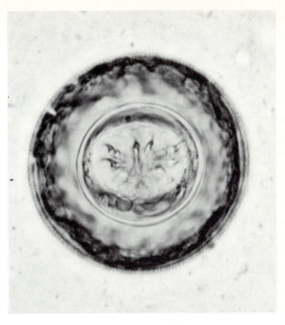

FIG. 22-30

Egg of *Hymenolepis diminuta*. It is 40 to 50 μm wide.

Photograph by Jay Georgi.

■ *Hymenolepis diminuta*

Hymenolepis diminuta is a cosmopolitan worm that is primarily a parasite of domestic rats, but many cases of human infection have been reported. It is a much larger species than *V. nana* (up to 90 cm) and differs from it in lacking hooks on its rostellum. Typical of the genus, it has unilateral genital pores and three testes per proglottid. The eggs (Fig. 22-30) are easily differentiated from those of *V. nana*, since they are larger and have no polar filaments. It has been demonstrated experimentally that more than 90 species of arthropods can serve as suitable intermediate hosts. Stored-grain beetles (*Tribolium* spp.) are probably most commonly involved in infections of both rats and humans. A household shared with rats is also likely to have its cereal foods infested with beetles.

The ease with which this parasite is maintained in laboratory rats and beetles makes it an ideal model for many types of experimental studies; its physiology has been more thoroughly examined than that of any other tapeworm.

Treatment is as recommended for *V. nana*.[32] An important book on this species was edited by Arai (see Suggested Readings). Its early embryology was beautifully demonstrated by Coil.[14]

Family Davaineidae

■ *Raillietina* species

The following species of *Raillietina* have been reported from humans: *R. siriraji*, *R. asiatica*, *R. garrisoni*, *R. celebensis*, and *R. demarariensis*. All normally parasitize domestic rats and possibly represent no more than two actual species. The genus is easily recognized by its large rostellum with hundreds of tiny, hammer-shaped hooks and by its spiny suckers (Fig. 22-31). The life cycle is known for none of them, but because several other species of *Raillietina* are known to use ants or other insects as intermediate hosts, it seems probable that the epidemiology of this infection is similar to that of *H. diminuta*. The clinical pathological conditions of raillietiniasis are unknown.

Raillietina cesticillus is one of the most common poultry cestodes in North America, and a wide variety of grain, dung, and ground beetles serve as intermediate hosts. The genus is very large, with species in many birds and mammals. The closely related *Davainea* is identical to *Raillietina*, except that its strobila is very short, consisting of only a few proglottids.

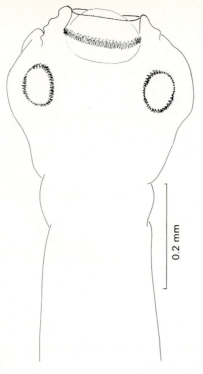

FIG 22-31

Scolex of *Raillietina*. The suckers are weak and have a double circle of spines, and the massive rostellum has many hammer-shaped hooks.

Drawing by Thomas Deardorff.

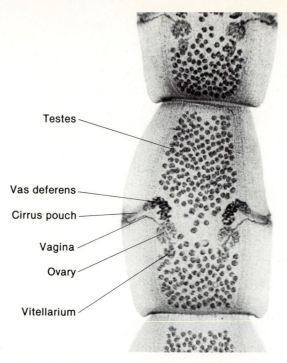

FIG. 22-32

Mature segment of *Dipylidium caninum*, the "double-pored tapeworm" of dogs and cats. The two vitelline glands are directly behind the larger ovaries. The smaller spheres are testes.

Courtesy Ann Arbor Biological Center.

Davainea spp. are found in galliform birds and use terrestrial molluscs as intermediate hosts. Other genera infect a wide variety of hosts, from passeriform birds to scaly anteaters. The family is divided into three subfamilies depending on characteristics of the uterus: a persistent, saccular uterus (Ophryocotylinae); a paruterine organ present (Idiogeninae); and a uterus replaced by egg capsules (Davaineinae).[55]

Family Dilepididae

■ *Dipylidium caninum*

A cosmopolitan, common parasite of domestic dogs and cats, *Dipylidium caninum* has been found many times in children.[44] It is easily recognized because each segment has two sets of male and female reproductive systems and a genital pore on each side (Fig. 22-32). The scolex has a retractable, rather pointed rostellum with several circles of rosethorn-shaped hooks. The uterus disappears early in its development and is replaced by hyaline, noncellular egg capsules, each containing 8 to 15 eggs. Gravid proglottids de-

tach and either wander out of the anus or are passed with feces. They are very active at this stage and are the approximate size and shape of cucumber seeds. As the detached segments begin to desiccate, the egg capsules are released. Fleas are the usual intermediate hosts, although chewing lice have also been implicated. Unlike the adult, a larval flea has simple, chewing mouthparts and feeds on organic matter, which may include *Dipylidium* egg capsules. The resulting cysticercoids survive their host's metamorphosis into the parasitic adult stage, when the flea may be nipped or licked out of the fur of the dog or cat, thereby completing the life cycle. This, by the way, is an example of **hyperparasitism,** since the flea is itself a parasite.

Nearly every reported case of infection of humans has involved a child. This may reflect adult resistance or may simply be a result of the familiarity of a child with a dog, with increased chances of a flea being accidentally swallowed. The symptoms and treatment are the same as for *Vampirolepis nana*.[32]

Basically, the only feature separating this family

from Hymenolepididae is an increased number of testes, usually more than 12. This family, too, consists of hundreds of species that parasitize birds and mammals. Taxonomic difficulties also attend this family.

Like the Davaineidae, the Dilepididae have three subfamilies: uterus reticular, ring shaped, or saccular (Dilepidinae); uterus replaced with egg capsules (Dipylidinae); and uterus replaced with one or more paruterine organs (Paruterininae). This last subfamily is raised to the level of an independent order by some workers.

Family Anoplocephalidae
■ *Bertiella studeri*

Normally a parasite of Old World primates, *Bertiella studeri* has been reported many times from humans, especially in southern Asia, the East Indies, and the Philippines. The scolex is unarmed, and the proglottids are much wider than they are long, with the ovary located between the middle of the segment and the cirrus pouch. The egg is characteristic: 45 to 50 μm in diameter, with a bicornuate **pyriform apparatus** on the inner shell.

Ripe segments are shed in chains of about a dozen at a time. The intermediate hosts are various species of oribatid mites.[59] These free-living animals feed on organic detritus and readily ingest eggs of *B. studeri*. Accidental ingestion of mites infected with cysticercoids completes the life cycle within primates. No disease has been ascribed to infection by *B. studeri*. Treatment is as for *Vampirolepis nana*.

Bertiella mucronata is similar to *B. studeri* and has also been reported from humans. It appears to be a parasite of New World monkeys, and children may become infected when living with a pet monkey and the ubiquitous oribatid mite. Distinguishing this species from *B. studeri* normally requires a specialist.

■ *Inermicapsifer madagascariensis*

Inermicapsifer madagascariensis is normally parasitic in African rodents, but it has been reported repeatedly in humans in several parts of the world, including South America and Cuba. Baer[3] concluded that humans are the only definitive host outside Africa.

The scolex is unarmed. The strobila is up to 42 cm long. Mature proglottids are somewhat wider than long and bear a centrally located ovary. The uterus is replaced by egg capsules in ripe segments, each capsule containing six to ten eggs, which do not possess a pyriform apparatus.

The life cycle of this parasite is unknown but undoubtedly involves an arthropod intermediate host. Clinical pathological conditions have not been stud-

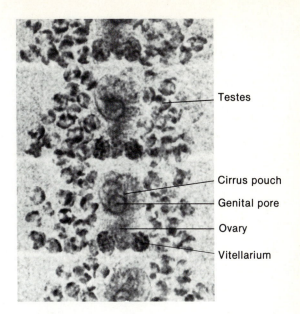

FIG. 22-33

Mesocestoides sp., a cyclophyllidean cestode with a midventral genital pore and a bilobed vitellarium.

Photograph by Larry Shults.

ied, and treatment is similar to that described for other species.

Other reports of anoplocephalids in humans are rare, such as that of *Mathevotaenia* sp. in an infant in Thailand.[37]

Family Mesocestoididae
■ *Mesocestoides* species

Unidentified specimens of the genus *Mesocestoides,* whose definitive hosts are normally various birds and mammals, have occasionally been reported from humans in Denmark, Africa, United States, Japan, and Korea.[30] The ventromedial location of the genital pores is clearly diagnostic of the genus (Fig. 22-33). The complete life cycle is not known for any species in this difficult family, but many have a rodent or reptile intermediate host, in which a cysticercoid type of larva known as a **tetrathyridium** (Fig. 22-34) develops. Neither mammals nor reptiles can be infected directly by eggs, so a first host must be involved. As yet, such a host has not been identified (Fig. 22-35). Pathological conditions and treatment of humans have not been studied. *Mesocestoides* is very curious in that it may undergo asexual multiplication in the definitive host (Fig. 22-35)—not by budding, as in coenuri and

FIG. 22-34

Tetrathyridial metacestodes of *Mesocestoides* sp. in the mesenteries of a baboon, *Papio cyanocephalus*.

Photograph by Robert E. Kuntz.

hydatids, but by longitudinal fission of the scolex! An inwardly directed protuberance of the tegument between the suckers, the "apical massif," has morphocytogenetic power.[27]

The scolex has four simple suckers and no rostellum. Each proglottid has a single set of male and female reproductive systems; the genital pores are median and ventral. Otherwise the morphology is typically cyclophyllidean, with a paruterine organ replacing the uterus in most species.

Mesocestoides spp. are widespread in carnivores throughout most of the world. No complete life cycle is known, since a first intermediate host has never been discovered (Fig. 22-35). Rodents and reptiles are the most common second intermediate hosts, in which the parasite develops into a tetrathyridium (Fig. 22-34). When eaten by the predator, the tetrathyridium develops into an adult (Fig. 22-33).

Some specialists consider this family to represent a distinct order of tapeworms.

Family Triplotaeniidae

This bizarre group contains the single genus *Triplotaenia*, species of which have two strobilae attached to each neck. The scolex bears four simple suckers. External segmentation is lacking; each strobila is twisted and fringed. These are parasites of Australian marsupials. Beveridge[8] considers this group of three species to represent a subfamily of Anoplocephalidae.

Family Dioecocestidae

Except for *Dioecotaenia* representing an order in rays, the only dioecious tapeworms are found in this family. All are parasites of shorebirds, grebes, or herons. Some species are completely dioecious, whereas others are regionally so. There are wide variations in scolex types, although all have four suckers. In some species, such as *Shipleya* in dowitchers, both sexes have secondary sex organs of the opposite sex. Hence the male has a uterus, and the female a cirrus and cirrus pouch, but each has only an ovary or testes. For a review of the genera in this family see Schmidt.[54,55]

Family Progynotaeniidae

This group is given separate family status because the female reproductive system develops well before the male system, a rarity among cestodes. The scolex bears an armed rostellum. The strobila is small and weak. The proglottids are hermaphroditic except for *Gynandrotaenia* and *Thomasitaenia*, which alternate male and female proglottids. Vaginas are absent. No complete life cycles are known in this family.

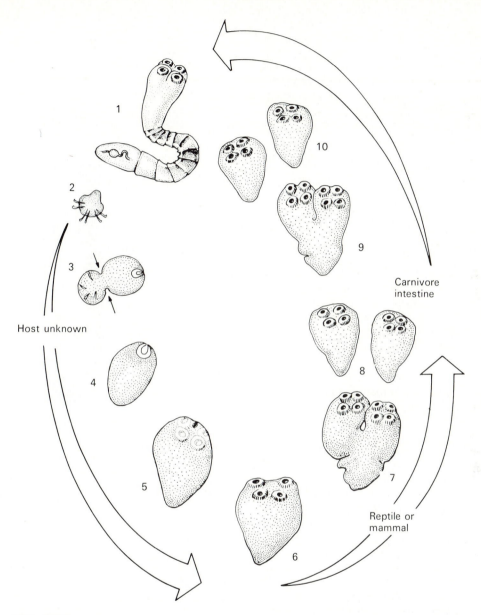

FIG. 22-35

Developmental sequence of *Mesocestoides corti,* illustrating early developmental stages (*2* to *5*), tetrathyridium and asexual multiplication in second intermediate host (*6* to *8*), and asexual multiplication with subsequent formation of adult worms in intestine of definitive host (*9* to *10*). Not illustrated here is the potential reinvasion of tissues from intestinal lumen of carnivore with continuing asexual multiplication of tetrathyridial stage.

From Voge, M. 1969. In Schmidt, G.D. Problems in systematics of parasites. © 1969 University Park Press, Baltimore.

ORDER CARYOPHYLLIDEA

Caryophyllideans are intestinal parasites of freshwater fishes, except for a few that mature in the coelom of freshwater oligochaete annelids. All are monozoic, showing no trace of internal or external proglottidization (Fig. 22-36). The scolex is never armed and usually is quite simple, bearing shallow depressions (loculi); is frilled; or is entirely smooth (Fig. 22-37). Some species seem to lack a scolex altogether. The anterior end of the worm is very motile, however, and functions well as a holdfast. Some species induce a pocket in the wall of the host's intestine in which one or more worms remain.

Each worm has a single set of male and female reproductive organs. In most the ovary is near the posterior end. Testes fill the median field of the body, and vitelline follicles are mainly lateral. Male and female genital pores open near each other on the midventral surface.

Catfishes, true minnows (Cyprinidae), and suckers are the most common hosts of the Caryophyllidea. Intermediate hosts are aquatic annelids. After the oligochaete eats the egg, the oncosphere hatches and penetrates into the coelom. There it grows into a procercoid with a prominent cercomer, similar to that of *Diphyllobothrium*. When eaten by a fish, the procercoid loses its cercomer and grows directly into an adult.

The biology and morphology of Caryophyllidea and Pseudophyllidea are similar, the main differences being the absence, in the Caryophyllidea, of a plerocercoid and a strobilated adult. Also, caryophyllideans use annelids as intermediate hosts, whereas pseudophyllideans employ crustaceans.

It has been suggested that segmented adults once existed but became extinct with their hosts, probably

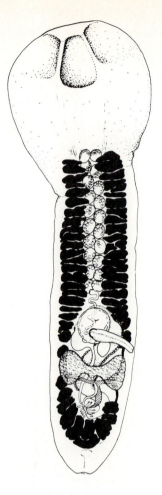

FIG. 22-36

Penarchigetes oklensis, a typical caryophyllaeid cestode, from a spotted sucker.

From Mackiewicz, J.S. 1969. Proc. Helm. Soc. Wash. 36:119-126.

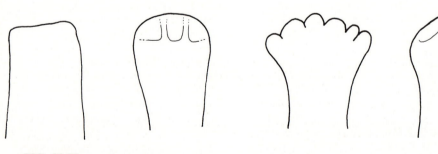

FIG. 22-37

Typical scolices of Caryophyllidea.

From Schmidt, G.D. 1970. How to know the tapeworms. William C. Brown Co., Publishers, Dubuque, Iowa.

aquatic reptiles. However, this did not happen before the plerocercoid developed neoteny in the fish second intermediate host. If this hypothesis is true, extant caryophyllidean species actually are **neotenic plerocercoids.** Support is lent to this idea by the existence of several species of *Archigetes,* which become sexually mature *while in the annelid.* The reproductive adult retains the cercomer and infects no additional host, although it can live for some time if eaten by a fish. *Archigetes,* then, appears to be a **neotenic procercoid.** This order was thoroughly reviewed by Mackiewicz.[39,40]

ORDER SPATHEBOTHRIIDEA

These are peculiar parasites of marine and freshwater teleost fishes. Their most striking characteristic is a complete absence of external constrictions, while internally possessing a typically linear series of reproductive systems. The scolex always lacks armature. It may be totally undifferentiated, as in *Spathebothrium;* it may be a shallow funnel-shaped organ, as in *Cyathocephalus;* or it may consist of one or two powerful cup-like organs (Fig. 21-7, *J*). The genital pores are ventral, the testes are in two lateral bands, the ovary is dendritic, and the vitellaria are follicular and lateral or scattered. The uterus is rosette-like and opens ventrally, usually near the vaginal pore.

This order is circumboreal in distribution. It appears to be an ancient group; some species are found both in the North Atlantic and North Pacific. No life cycles are known. Although these worms are of no known economic importance, they remain an interesting zoological group that should be studied further. *Bothrimonus,* a common genus in North America, has been investigated more fully.[10]

ORDER TRYPANORHYNCHA

Trypanorhynchans are all parasites of the spiral intestine of sharks and rays. No complete life cycle has been experimentally developed, but infective metacestodes are common in marine molluscs,[11] crustaceans, and fishes. The fishes undoubtedly are paratenic hosts, as may be some of the molluscs and crustaceans.

The scolex (Fig. 21-7, *F*) of trypanorhynchan is an extraordinary organ. It usually is elongate with two or four shallow, muscular holdfast organs called bothridia, which may be covered with minute spines. Four eversible tentacles (atrophied in *Apororhynchus*) emerge from the apex of the scolex. The tentacles are armed with an astonishing array of hooks and spines (Fig. 22-38), shaped and arranged differently in each species. Interpretation of the hook arrangement is dif-

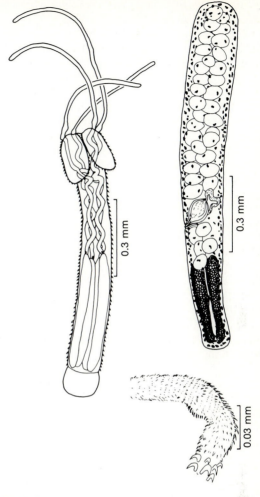

FIG. 22-38

Basic morphology of *Eutetrarhynchus,* a typical genus of Trypanorhyncha.

From Kovacs, K.J., and G.D. Schmidt, 1980. Proc. Helm. Soc. Wash. 47:10-14.

ficult but must be accomplished before species identification is possible. Each tentacle invaginates into an internal tentacle sheath, provided at its base with a muscular bulb. A retractor muscle originates at the base or front end of the bulb, courses through the tentacle sheath, and inserts inside the tip of the tentacle. When the retractor muscle contracts, it invaginates the tentacle, detaching it from host tissues. When the bulb contracts, it hydraulically evaginates the tentacle, driving it deep into the host's intestinal wall. This process is very similar to that which manipulates the proboscis of an acanthocephalan (Chapter 32.)

A neck is present or absent; the strobila varies from hyperapolytic to anapolytic. Proglottids are nearly identical to those of the orders Lecanicephalidea and Tetraphyllidea, discussed next. The single ovary is basically bilobed and posterior. Vitellaria are follicular, cortical, and lateral or circummedullary. The uterus is a simple sac, usually in the anterior two thirds of the gravid proglottid. Testes are few to many and medullary, and cirrus pouch and cirrus often are huge relative to the proglottid. All genital pores are lateral.

The infective stage, called a plerocercus, may or may not bear a posterior sac, the **blastocyst,** into which the scolex is inverted. Plerocerci may be so plentiful in the flesh of certain fish or shrimps as to make them unpalatable, and thereby unsalable. This is the only known economic importance of the Trypano-

rhyncha. They have never been reported from humans. However, they remain among the most enigmatic and challenging invertebrates for the taxonomist. Classical reviews of the order are presented by Dollfus.[21,22] A key to families and genera is provided by Schmidt.[54,55]

ORDER LECANICEPHALIDEA

The main characteristic separating this order from the closely related Tetraphyllidea is the scolex, which is separated into a posterior portion that usually bears four simple suckers, and an anterior portion that is either a simple pad (invaginable, forming a sucker) or

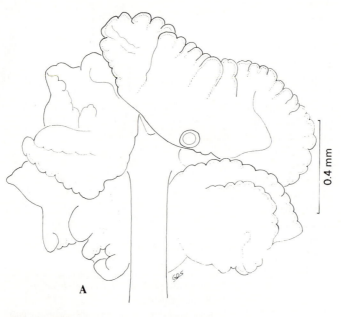

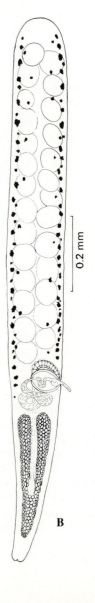

FIG. 22-39

Phyllobothrium kingae, **A,** showing unarmed bothridia with accessory suckers, and **B,** mature proglottid.

From Schmidt, G.D. 1978. Proc. Helm. Soc. Wash. 45:132-134.

subdivided into simple or feathered tentacles. The scolex lacks hooks or other sclerotized armature except in the single genus *Balanobothrium*. The proglottids are similar to those described for the Trypanorhyncha. Typical scolices are illustrated in Fig. 21-7, *G*.

All known lecanicephalids are uncommon parasites of elasmobranch fishes, including many genera of rays. Sharks are not known to be infected. No complete life cycle is known, but penaeid shrimps serve as intermediate hosts for some. Probably the best reference to this order is the very old monograph by Southwell.[58]

ORDER TETRAPHYLLIDEA

Tetraphyllideans are notable for their astonishing variety of scolex forms (Figs. 21-6, *A* and 21-7, *I*). Basically there are four bothridia, which may be stalked or sessile, smooth or crenate, or subdivided into loculi or major units. Often there are accessory suckers (Fig. 22-39) and/or hooks (Fig. 22-40) or spines. An apical, stalked, sucker-like organ, the **my-zorhynchus,** is present on some. A neck is present or absent. The strobila and proglottids are essentially identical to those of the Lecanicephalidea and Trypanorhyncha, and, like members of these orders, adult tetraphyllideans are all parasites of the spiral intestine of elasmobranchs.

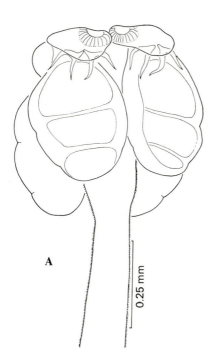

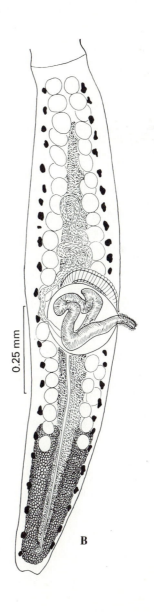

FIG. 22-40

Acanthobothrium urolophi, an armed tetraphyllidean. **A,** Scolex. **B,** Mature proglottid.

From Schmidt, G.D. 1973. Proc. Helm. Soc. Wash. 40:91-93.

As far as is known, the life cycles are also similar. No complete cycle has been discovered, but infective plerocercoids are common in molluscs, crustaceans, and fishes. Again, an old but useful monograph on this order is that of Southwell.[57]

ORDER DIPHYLLIDEA

Only two genera are known in this order, both from elasmobranchs. *Echinobothrium* has a powerful rostellum armed with rows of large hooks (Fig. 21-7 *H*). The hooks are longest in the middle of each row, decreasing in size toward the ends. Two simple bothridia are present, covered with very small, hairlike spines. Posterior to the bothridia is a long peduncle armed with longitudinal rows of straight spines, each with a transverse root. The proglottids are similar to those of Tetraphyllidea.

Ditrachybothridium lacks armature and has a small rostellum, short peduncle, and spoon-shaped bothridia. It is known from a single species from the coast of Scotland.

ORDER LITOBOTHRIDEA

This unusual order consists of five species from thresher sharks from the Pacific. The scolex is a single, well-developed apical sucker. The anterior proglottids are modified with dorsal and ventral spurs, the whole serving as a holdfast organ (Fig. 21-7, *B*). The strobila has numerous proglottids, each with a single set of medullary reproductive organs (Fig. 22-41). The testes are numerous and postovarian. Genital pores are lateral. The vitellaria are follicular, encircling the medullary parenchyma. Nothing is known of the life cycles of these rare worms, which appear to be closely related to the Tetraphyllidea.

ORDER APORIDEA

This small order consists of two genera that are so different they probably should not be placed in the same order (Fig. 21-7, *K*). Both are aberrant, even by cestode standards, and both are parasites of anseriform birds.

The scolex of *Nematoparataenia* has four large, forwardly directed suckers and a massive, glandular rostellum armed with an undulating row of about 1000 very small hooks. The strobila is cylindrical with a longitudinal groove on one side. External and internal segmentation is lacking. The testes are follicular, filling most of the medulla of the mature regions. The ovary also is follicular, mainly surrounding the testes. The follicular vitelline cells are mixed with the ovarian cells. The only known species are found in the

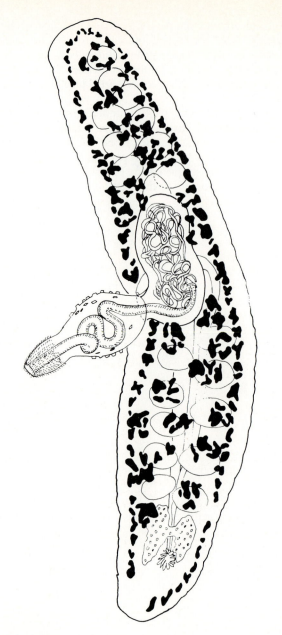

FIG. 22-41

Mature proglottid of *Litobothrium alopias*.

From Dailey, M.D. 1969. Proc. Helm. Soc. Wash. 36:218-224.

small intestine of swans in Australia, New Zealand, and Sweden.

By contrast, the scolex of *Apora* has four narrow grooves but no suckers, and the slender rostellum is armed with 10 hooks in a simple circle. Separate proglottids cannot be recognized. The testes are follicular and are arranged in a semicircle in cross section. The vasa efferentia join together without opening to the

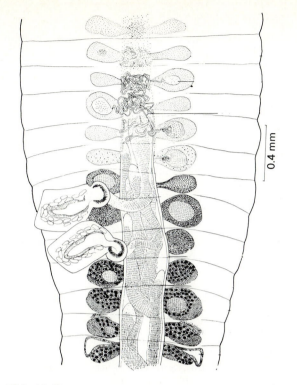

FIG. 22-42

Dioecotaenia cancellatum, a dioecious tetraphyllidean from stingrays. Note the two male proglottids with their cirri injected into the much larger female.

From Schmidt, G.D. 1969. J. Parasitol. 55:271-275.

surface. The ovary is follicular, surrounding the testes. Vitellaria are also follicular and are mixed with the ovarian cells. The worms are found under the gizzard lining of ducks in Russia and North America.

ORDER DIOECOTAENIIDEA

This order, consisting of two species, is of interest mainly in being completely dioecious and in having internal morphology entirely different from any other cestodes (Fig. 22-42.) They are parasites of rays in the Atlantic.

ORDER NIPPOTAENIIDEA

Nippotaeniids are parasites of freshwater gobiid fishes in Japan, Russia, and New Zealand. The scolex (Fig. 21-7, *E*) has a single apical sucker; otherwise it is simple. The strobila is small, with a few proglottids. The genital pores are lateral; the ovary, posterior; and the testes, anterior. A single, compact vitellarium lies between the testes and ovary. The order was reviewed by Hine.[29]

ORDER PROTEOCEPHALIDEA

The members of this large order of many genera are all parasites of freshwater fishes, amphibians, or reptiles and so are of little or no economic importance. The scolices (Fig. 21-7, *D*) are much like those occurring in the Cyclophyllidea, bearing four simple suckers and occasionally armature or a rostellum. The proglottids, however, are much more like those of the Tetraphyllidea. Genital pores are lateral. The ovary is posterior, and the numerous testes fill most of the region anterior to it. The vitellaria are follicular and are restricted to the lateral margins of the proglottid.

Complete life cycles are known for several species. All involve a cyclopoid crustacean intermediate host, in which the worm develops into a procercoid (the plerocercoid I, according to some authors; see p. 332). This metacestode has a well-developed scolex and a cercomer at the posterior end. In some species the procercoid is directly infective to the definitive host; in others it burrows into the viscera for a time before reemerging and maturing in the lumen of the gut. Paratenic hosts are common in proteocephalid life cycles. The boring action of the plerocercoid (as it is now called, since it loses the cercomer as it penetrates the intestinal wall) may be highly pathogenic to the host. For example, *Proteocephalus ambloplites* in bass in North America sometimes is known to castrate its fish host.

Subclass Cestodaria

All tapeworms discussed so far belong to the subclass Eucestoda. Another, small group of monozoic cestodes, however, is so different that it is placed in its own subclass.

ORDER AMPHILINIDEA

The body of these worms is dorsoventrally flattened, with an indistinct, proboscis-like holdfast at the anterior end (Fig. 21-4). The genital pores are near the posterior end; the uterine pore is near the anterior end. The ovary is posterior, the vitelline follicles are bilateral, and the testes are preovarian. The uterus is N shaped or looped. The oncosphere (**lycophora**) is provided with 10 hooks.

Amphilineans are parasites of the body cavity or intestine of fishes and turtles in Asia, Japan, Europe, North America, Sri Lanka, Brazil, Africa, East Indies, and Australia. They are of no medical or economic importance. An excellent study on the structure and life history of *Austramphilina* was presented by Rohde and Georgi.[52]

ORDER GYROCOTYLIDEA

The anterior end of these worms is provided with a small, inversible holdfast organ. The posterior end is a frilled, rosette-like organ, and the lateral margins may be frilled (*Gyrocotyle*, Fig. 22-43), or it is a long, simple cylinder, and the lateral margins are smooth (*Gyrocotyloides*). The ovary is posterior; the uterus has extensive lateral loops, terminating in a midventral pore in the anterior half. Testes are anterior. The genital pores are near the anterior end. Gyrocotylideans also have a decacanth lycophora larva. They are parasites of the spiral intestine of Holocephali.

Some authorities consider the Gyrocotylidea to be more closely aligned with the Monogenea, but because of obvious similarities to the Amphilinidea and because of the absence of a digestive system, we prefer to refer them to the Cestodaria.

REFERENCES

1. Andrews, P., and H. Thomas. 1979. The effect of praziquantel on *Hymenolepis diminuta* in vitro. Tropenmed. Parasitol. 30:391-400.
2. Archer, D.M., and C.A. Hopkins. 1958. Studies on cestode metabolism, III. Growth pattern of *Diphyllobothrium* sp. in a definitive host. Exp. Parasitol. 7:125-144.
3. Baer, J.G. 1956. The taxonomic position of *Taenia madagascariensis* Davaine, 1870, a tapeworm parasite of man and rodents. Ann. Trop. Med. Parasitol. 50:152-156.
4. Barnett, L. 1939. Hydatid disease: errors in teaching and practice. Br. Med. J. 2:593-599.
5. Becker, B., H. Mehlhorn, P. Andrews, and H. Thomas. 1980. Scanning and transmission electron microscope studies on the efficacy of praziquantel on *Hymenolepis nana* (Cestoda) in vitro. Z. Parasitenkd. 61:121-133.
6. Bekhti, A., et al. 1977. Treatment of hepatic hydatid disease with mebendazole: preliminary results in four cases. Br. Med. J. 2:1047-1051.
7. Berman, J.D., P.C. Beaver, A.W. Cheever, and E.A. Quindlen. 1981. Cysticercus of 60-milliliter volume in human brain. Am. J. Trop. Med. Hyg. 30:616-619.
8. Beveridge, I. 1976. A taxonomic revision of the Anoplocephalidae (Cestoda: Cyclophyllidea) of Australian marsupials. Aust. J. Zool. Suppl. Ser. 44:1-110.
9. von Bonsdorff, B. 1956. *Diphyllobothrium latum* as a cause of pernicious anemia. Exp. parasitol. 5:207-230.
10. Burt, M.D.B., and I.M. Sandeman. 1969. Biology of *Bothrimonus* (=*Diplocotyle*) (Pseudophyllidea: Cestoda). I. History, description, synonymy, and systematics. J. Fish. Res. Bd. Can. 26:975-996.
11. Cake, E.W., Jr. 1976. A key to larval cestodes of shallow-water, benthic mollusks of the northern Gulf of Mexico. Proc. Helm. Soc. Wash. 43:160-171.
12. Camargo, C.A., and W.H. Marshall. 1987. Radiological diagnosis of neurocysticercosis. Parasitol. Today 3:30-31.
13. Chandler, A.C., and C.P. Read. 1961. Introduction to parasitology, ed. 10, John Wiley & Sons, Inc., New York.

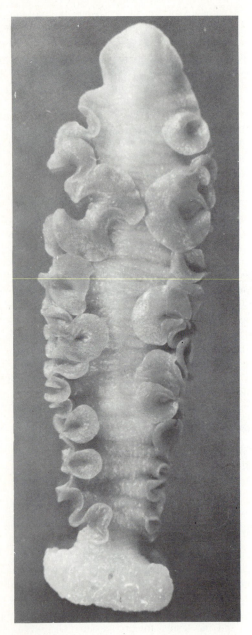

FIG. 22-43

Gyrocotyle parvispinosa from the ratfish, *Hydrolagus colliei*.

Photograph by Warren Buss.

14. Coil, W.H. 1986. The early embryology of *Hymenolepis diminuta* (Cestoda). Proc. Helm. Soc. Wash. 53:38-47.

15. Corkum, K.C. 1966. Sparganosis in some vertebrates of Louisiana and observations of human infection. J. Parasitol. 52:444-448.

16. D'Alessandro, A., R.L. Rausch, C. Cuello, and N. Aristizabal. 1979. *Echinococcus vogeli* in man, with a review of polycystic hydatid disease in Columbia and neighboring countries. Am. J. Trop. Med. Hyg. 28:303-317.

17. Davis, A. 1973. Drug treatment in intestinal helminthiases. World Health Organization, Geneva.

18. Dewhirst, L.W., J.D. Cramer, and J.J. Sheldon. 1967. An analysis of current inspection procedures for detecting bovine cysticercosis. J. Am. Vet. Med. Assoc. 150:412-417.

19. DiFelice, G., and A. Siracusano. 1987. Monoclonal antibodies for immunodiagnosis of human hydatidosis. Parasitol. Today 3:25-26.

20. Dixon, H.B.F., and D.W. Smithers. 1934. Epilepsy in cysticercosis *(Taenia solium)*. A study of seventy-one cases. Q. J. Med. 3:603-616.

21. Dollfus, R.P. 1942. Etudes critiques sur les Tétrarhynques du Museum de Paris. Arch. Mus. nat. Hist. nat. Ser. 6, 19:1-466.

22. Dollfus, R.P. 1946. Notes diverses sur des Tétrarhynques. Mem. Mus. nat. Hist. nat. n. S. 22:179-220.

23. Dooley, J.F. 1980. Health precautions in Mexico. Letter to the editor. J.A.M.A. 243:1524.

24. Ferretti, G., F. Gabriele, and C. Palmas. 1981. Development of human and mouse strain of *Hymenolepis nana* in mice. Int. J. Parasitol. 11:424-430.

25. Gamble, W.G., M. Segal, P.M. Schantz, and R.L. Rausch. 1979. Alveolar hydatid disease in Minnesota. First human case acquired in the contiguous United States. J.A.M.A. 241:904-907.

26. Gemmell, M.A., J.R. Lawson, and M.G. Roberts. 1987. Towards global control of cystic and alveolar hydatid diseases. Parasitol. Today 3:144-151.

27. Hess, E. 1980. Ultrastructural study of the tetrathyridium of *Mesocestoides corti* Hoeppli, 1925: tegument and parenchyma. Z. Parasitenkd. 61:135-159.

28. Heyneman, D. 1962. Studies of helminth immunity: I. Comparison between lumenal and tissue phases of infection in the white mouse by *Hymenolepis nana* (Cestoda: Hymenolepididae). Am. J. Trop. Med. Hyg. 11:46-63.

29. Hine, P.M. 1977. New Species of *Nippotaenia* and *Amurotaenia* (Cestoda: Nippotaeniidae) from New Zealand freshwater fishes. J.R. Soc. N.Z. 7:143-155.

30. Hutchison, W.F., and J.B. Martin. 1980. *Mesocestoides* (Cestoda) in a child in Mississippi treated with paromomycin sulfate (Humantin). Am. J. Trop. Med. Hyg. 29:478-479.

31. Jepsen, A., and H. Roth. 1952. Epizootiology of *Cysticercus bovis*-resistance of the eggs of *Taenia saginata*. Report 14. Int. Vet. Cong. 22:43-50.

32. Jones, W.E. 1979. Niclosamide as a treatment for *Hymenolepis diminuta* and *Dipylidium caninum* infection in man. Am. J. Trop. Med. Hyg. 28:300-302.

33. Kamegai, S., A. Ichihara, H. Nonobe, M. Machida, and T. Hara. 1968. A case of human infection with the immature worm of *Diplogonoporus grandis*. Res. Bull. Meguro Parasitol. Mus. 2:1-8.

34. Karnaukov, V.K., and A.I. Laskovenko. 1984. Clinical picture and treatment of rare human helminthiases (*Hymenolepis diminuta* and *Dipylidium caninum*). Meditsinskaya Parazitol. i Parazitornye Bolezni no. 4, 77-79.

35. Keller, M. 1937. Sur une nouvelle méthode de traitment du la sparganose oculaire. Ann. l'Ecole Super. Med. Pharm. Indochine. 1:77-89.

36. Kuntz, R.E. 1963. Snakes of Taiwan. Q. J. Taiwan Mus. 16:1-79.

37. Lamom, C., and G.J. Greer. 1986. Human infection with an anoplocephalid tapeworm of the genus *Mathevotaenia*. Am. J. Trop. Med. Hyg. 35:824-826.

38. Leiby, P.D., W.P. Carney, and C.E. Woods. 1970. Studies on sylvatic echinococcosis, III. Host occurrence and geographic distribution of *Echinococcus multilocularis* in the north central United States. J. Parasitol. 56:1141-1150.

39. Mackiewicz, J.S. 1972. Caryophyllidea (Cestoidea): a review. Exp. Parasitol. 31:417-512.

40. Mackiewicz, J.S. 1982. Caryophyllidea (Cestoidea): perspectives. Parasitology 84: 397-417.

41. MacPherson, C.L. 1983. An active intermediate host role for man in the life cycle of *Echinococcus granulosus* in Turkana, Kenya. Am. J. Trop. Med. Hyg. 32:397-404.

42. Markwalder, K., K. Hess, A. Valovanis, and F. Witassek. 1984. Cerebral cysticercosis: treatment with praziquantel. Am. J. Trop. Med. Hyg. 33:273-280.

43. McManus, D.P., and J.D. Smyth. 1986. Hydatidosis: changing concepts in epidemiology and speciation. Parasitol. Today 2:163-167.

44. Moore, D.V. 1962. A review of human infections with the common dog tapeworm, *Dipylidium caninum,* in the United States. Southwestern Vet. 15:283-288.

45. Mueller, J.F. 1974. The biology of *Spirometra*. J. Parasitol. 60:3-14.

46. Nelson, G.S., and R.L. Rausch. 1963. *Echinococcus* infections in man and animals in Kenya. Ann. Trop. Med. Parasitol. 57:136-149.

47. Nyberg, W. 1958. Absorption and excretion of vitamin B_{12} in subjects infected with *Diphyllobothrium latum* and in noninfected subjects following oral administration of radioactive B_{12}. Acta Haematol. 19:90-98.

48. Pampiglione, S. 1962. Indagine sulla diffusion dell' imenolepiasi nella Sicilia occidentale. Parassitologia 4:49-58.

49. Rausch, R.L. 1967. On the ecology and distribution of *Echinococcus* spp. (Cestoda: Taeniidae), and characteristics of their development in the intermediate host. Ann. Parasitol. 42:19-63.

50. Rausch, R.L., A. D'Allessandro, and V.R. Rausch. 1981. Characteristics of the larval *Echinococcus vogeli* Rausch and Bernstein, 1972 in the natural intermediate host, the paca, *Cuniculus paca* L. (Rodentia: Dasyproctidae). Am. J. Trop. Med. Hyg. 30:1043-1052.

51. Rausch, R.L., and J.F. Wilson. 1973. Rearing of the adult *Echinococcus multilocularis* Leuckart, 1863, from sterile larvae from man. Am. J. Trop. Med. Hyg. 22:357-360.

52. Rohde, K., and M. Georgi. 1983. Structure and development of *Austramphilina elongata* Johnston, 1931 (Cestodaria: Amphilinidea). Int. J. Parasitol. 13:273-287.

53. Ruttenber, A.J., et al. 1984. Diphyllobothriasis associated with salmon consumption in Pacific coast states. Am. J. Trop. Med. Hyg. 33:455-459.

54. Schmidt, G.D. 1970. How to know the tapeworms. William C. Brown Co., Publishers, Dubuque, Iowa. 266 p.

55. Schmidt, G.D. 1986. Handbook of tapeworm identification. CRC Press, Boca Raton, Florida, 675 p.

56. Schwabe, C.W., and K.A. Daoud. 1961. Epidemiology of echinococcosis in the Middle East. I. Human infection in Lebanon, 1949-1959. Am. J. Trop. Med. Hyg. 10:374-381.

57. Southwell, T. 1925. A monograph on the Tetraphyllidea. Liverpool University Press, Liverpool School of Tropical Medicine Memoir n.s., no. 2, pp. 1-368.

58. Southwell, T. 1930. The fauna of British India, including Ceylon and Burma. Cestoda, vol. 1. Taylor & Francis, London.

59. Stunkard, H. 1940. The morphology and life history of the cestode *Bertiella studeri*. Am. J. Trop. Med. 20:305-332.

60. Swartzwelder, J.C., P.C. Beaver, and M.W. Hood. 1964. Sparganosis in southern United States. Am. J. Trop. Med. Hyg. 13:43-48.

61. Wilson, J.F., M. Davidson, and R.L. Rausch. 1978. A clinical trial of mebendazole in the treatment of alveolar hydatid disease. Ann. Rev. Resp. Dis. 118:747-757.

62. Wilson, J.F., and R.L. Rausch. 1980. Alveolar hydatid disease. A review of clinical features of 33 indigenous cases of *Echinococcus multilocularis* infection in Alaskan Eskimos. Am. J. Trop. Med. Hyg. 29:1340-1355.

63. Yamane, Y., N. Okada, and M. Takihara. 1975. On a case of long term migration of *Spirometra erinacei* larva in the breast of a woman. Yonago Acta Medica. 19:207-213.

SUGGESTED READINGS

Arai, H.P., editor. 1980. Biology of the tapeworm *Hymenolepis diminuta*. Academic Press, Inc., New York.

Arme, C., and P.W. Pappas, editors. 1983. Biology of the Eucestoda. Academic Press, London. 2 Vols.

Binford, C.H., and D.H. Connor, editors. 1976. Pathology of tropical and extraordinary diseases, sect. 11. Disease caused by cestodes. Armed Forces Institute of Pathology, Washington, D.C.

Hoffman, G.L. 1967. Parasites of North American freshwater fishes. University of California Press, Berkeley. (Keys to all genera of tapeworms of North American fishes with lists of species and hosts.)

Schmidt, G.D. 1985. Handbook of tapeworm identification. CRC Press, Boca Raton, Fla. 675 p.

Soulsby, E.J.L. 1965. Textbook of veterinary clinical parasitology, vol. 1. F. A. Davis Co., Philadelphia. (A useful reference to tapeworms of veterinary significance.)

Spasskaya, L.P. 1966. Cestodes of birds of SSSR. Akademii Nauk SSSR, Moscow. (A useful, illustrated survey of tapeworms of northern birds.)

Stunkard, H.W. 1962. The organization, ontogeny and orientation of the cestodes. Q. Rev. Biol. 37:23-34.

Thompson, R.C.A., editor. 1986. The biology of *Echinococcus* and hydatid disease. George Allen and Unwin, London.

CHAPTER 23

PHYLUM NEMATODA: FORM, FUNCTION, AND CLASSIFICATION

Nematodes are among the most abundant animals on earth. Of course, more species of insects have been described, but when one realizes that nearly every kind of insect examined harbors at least one species of parasitic nematode and when one further calculates the number of kinds of nematodes parasitic in the rest of the animal kingdom, there is no contest. There are also many species of nematodes that parasitize plants. Finally, the species of free-living marine, freshwater, and soil-dwelling nematodes probably far outnumber those that are parasitic. Numbers of individuals are often extremely high: 90,000 were found in a single rotting apple; 1074 individuals, representing 36 species, were counted in 6 to 7 ml of mud; and 3 to 9 billion per acre may be found in good farmland in the United States.[71]

Obviously, these animals have a lot going for them. We hope that this chapter will give some insight into their enormous success.

Most nematodes are small, inconspicuous, and apparently unimportant to humans and therefore attract the attention only of specialists. A few, however, cause diseases of extreme importance to humans and domestic and wild plants and animals. These are the roundworms that attract the attention of parasitologists.

HISTORICAL ASPECTS

Ancient people were probably familiar with the larger nematodes, which they encountered when they slew game or disemboweled an opponent at war. The earliest records mention the worms or contain recognizable allusions to them. Aristotle discussed the worm we now call *Ascaris lumbricoides,* and the Ebers Papyrus of 1550 BC Egypt described clinical hookworm disease, as did Hippocrates, Lucretius, and the ancient Chinese. Moses wrote of a scourge that probably was caused by the guinea worm. The eggs of *Ascaris lumbricoides* and *Trichuris trichiura* were found in the intestine of a 2300-year-old body found preserved in a peat bog in the Orkney Islands.[19]

The Arabians Avicenna and Avenzoar, who kept parasitology alive during the Dark Ages in Europe, studied elephantiasis, differentiating it from leprosy.[40]

Linnaeus, in 1758, placed the roundworms in his class Vermes, along with all other worms and wormlike animals. Goeze, Zeder, and Rudolphi made great advances in recognition of various nematodes, although they still believed the worms arose by spontaneous generation. Further work by Gegenbauer, Huxley, Hatschek, Leuckart, Beneden, Diesing, Linstow, Looss, Railliet, Stossich, and many others established the nematodes as a distinct and important group of animals.

The name Nematoda is a modification of Rudolphi's Nematoidea and was placed in Nemathelminthes, itself first considered a class in phylum Vermes, by Gegenbaur in 1859 and later was elevated to phylum status. Although still employed by some authors today, the name Nemathelminthes is little used. The name Aschelminthes was proposed by Grobben in 1910 as a superphylum to contain several divergent groups of wormlike animals that had in common a pseudocoelomic body cavity. Hyman resurrected the name Aschelminthes for use as a phylum containing the classes Nematoda, Rotifera, Priapulida, Gastrotricha, Nematomorpha, and Kinorhyncha.[50] Today most authorities consider each of these groups as separate phyla, sometimes placing them in the superphylum Aschelminthes, recognizing their rather distant phyletic relationship. Attempts by a few students of free-living nematodes to change Nematoda to Nema or Nemata have not been widely accepted.

Curiously, studies of nematodes have developed along two separate lines, with parasitologists claiming the parasites of vertebrates and nematologists accounting for free-living and plant- and invertebrate-parasitic roundworms. This doubtless is because of parasitologists' historical concern for parasites of medical and

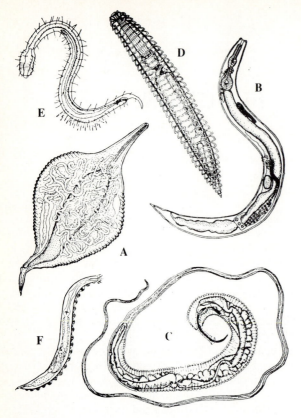

FIG. 23-1

Variety of form in the nematodes. **A,** *Tetrameres*. **B,** *Rhabditis*. **C,** *Trichuris*. **D,** *Criconema*. **E,** *Chaetosoma*. **F,** *Bunonema*.

From Crofton, H.D. 1966. Nematodes. Hutchinson University Library, London.

veterinary importance. Exceptions occur, of course; a few individuals, such as Chitwood, Bird, Crites, Inglis, Mawson, and Schuurmanns-Steckhoven, have made significant contributions to both areas. Each discipline publishes in its own journals, uses unique terminology and taxonomic formulas, and to a large extent employs different techniques for handling and preparing specimens for study. Recent trends suggest that the two disciplines are beginning to merge. If so, the science of nematology will assuredly benefit, since each school of thought will profit from the special knowledge of the other, and the waste engendered by lack of communication will be minimized.

FORM AND FUNCTION

Typical nematodes are elongate, are tapered at both ends, are bilaterally symmetrical, and possess a pseudocoel, that is, a body cavity derived from the embryonic blastocoel. There are varieties of this basic shape, however (Fig. 23-1). The digestive system is complete, with a mouth at the extreme anterior end and an anus near the posterior tip. The lumen of the pharynx is characteristically triradiate. The body is covered with a noncellular cuticle that is secreted by an underlying hypodermis and is shed four times during ontogeny. The muscles of the body wall are only one layer thick and are distinguished by all being *longitudinally arranged* with no separate circular layer. The excretory system consists of lateral canals, ventral glands, or both, which open near the anterior end through a ventral excretory pore. Except for some sensory endings of modified cilia, neither cilia nor flagella are present, even in the male gamete. Most nematodes are dioecious and show considerable sexual dimorphism: the females are usually larger, and the tail of the male is more curled. Some species are hermaphroditic, and others are parthenogenetic. The female reproductive system opens through a ventral genital pore; the male system opens into a cloaca, together with the digestive system. Adult nematodes vary in size from less than 1 mm, as in the genus *Caenorhabditis*, to more than a meter, as in *Dracunculus*.

A considerable body of knowledge has accumulated on the function and structure (both at the light and electron microscope levels) of nematodes, far beyond our ability to review within the confines of this chapter. Many reviews and literature references are available.*

Body wall

The body wall of nematodes comprises the cuticle, hypodermis, and body wall musculature. The outermost covering is the **cuticle,** a complex structure of great functional significance to the animals. The cuticle also lines the stomodeum, proctodeum, excretory pore, and vagina. It is basically of three regions—the cortical, the middle (also called homogeneous or matrix), and the inner fibrous layers—which are commonly subdivided. The cuticle in *Ascaris* spp. seems to be fairly typical, and a total of nine regions can be distinguished (Fig. 23-2). The cortex is covered by what appears to be a thin layer of lipid less than 0.1 μm thick, although some evidence suggests that this may be a trilaminar membrane.[13] The main body of the cortex is divided into an **external** and an **internal cortical layer.** The external layer is a stabilized protein related to keratin, recognized because disulfide cross-links have been detected (Chapter 34). In some species quinone tanning may play a role in stabilizing

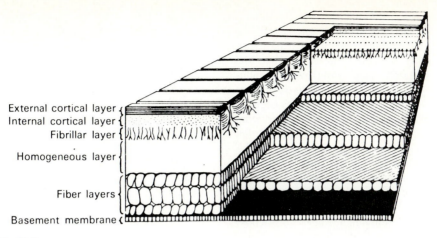

FIG. 23-2

Diagram showing transverse, longitudinal and tangential sections of the cuticle of *Ascaris*. The strands of each of the three fiber layers run at an angle of about 75 degrees to the longitudinal axis of the worm, and the strands of the middle layer run about 135 degrees from those of the inner and outer layers.

From Bird, A.F., and K. Deutsch. 1957. Parasitology 47:319-329.

the protein in this layer. The inner cortical layer, as well as the other layers in the cuticle, is primarily of collagen, a protein type also abundant in vertebrate connective tissue. Cuticlin is another stabilized protein reported from *Ascaris* cuticle; it has dityrosine cross-links (p. 532).[42] The external layer is amorphous and electron dense and is interrupted by a series of fine, transverse grooves. These are punctuated by a series of apparent pores leading into canals extending through the inner cortical layer to the **fibrillar layer.** The pore canals may function as means to transport substances to the surface of the cuticle, as skeletal supports, or both. Even under the electron microscope one can distinguish little structure in the **middle layer,** although some researchers have reported fine striations. Three **fiber layers,** each of parallel strands of collagen-like protein, run at an angle of about 75 degrees to the longitudinal axis of the worm. Strands of the middle fiber layer run at an angle of about 135 degrees to those of the inner and outer layers, which are parallel to each other, thus forming a lattice-like arrangement. The fibrous layers form an important component of the hydrostatic skeleton; the strands themselves are not extensible but allow for longitudinal expansion and contraction of the overlying cuticle by changes in the angles between the layers. The innermost layer of the cuticle is the **basal lamella,** a layer of fine fibrils that merges with the underlying hypodermis. Transverse annulations occur in the cuticle, aiding flexibility of the animal. These striations are more prominent in some species than in others. The

FIG. 23-3

Scanning electron micrograph of *Toxocara cati,* illustrating cervical alae.

Photograph by John Ubelaker.

cuticle of the parasitic juveniles of mermithids (described later) is quite different in structure.[8]

Cuticular markings and ornamentations of many types occur in various kinds of nematodes. These include shallow **punctations,** deeper **pores,** and **spines** of varying complexity.[51,129] Lateral or sublateral cuticular thickenings called **alae** are present in many species. Cervical alae (Fig. 23-3) are found on the anterior part of the body, caudal alae are on the tail ends of some males; and longitudinal alae, when present, extend the entire length of both sexes. The lateral alae may be of value to the animal when it is swimming or lend greater stability on solid substrate, when the nematode is crawling on its side by dorsoventral undulations, as in the juvenile of *Nippostrongylus brasiliensis*[60] (Fig. 23-4). Longitudinal ridges occur in many adult trichostrongylids. Cuticular ultrastructure has been studied in *N. brasiliensis;* Lee reported that the ridges are supported by a series of struts or skeletal rods in the middle layer of the cuticle[58] (Fig. 23-5). The struts are held erect by collagenous fibers inserted in the cortical and fibrous layers, but the middle layer itself is fluid filled and contains hemoglobin. The function of the longitudinal ridges in trichostrongylids is both to aid in locomotion, as the worm moves between villi with a corkscrew type of motion (Fig. 23-6), and to abrade the microvillar surface, thus helping to obtain food in the absence of biting or piercing mouthparts.[60] The arrangement of these rods is called the **synlophe.**

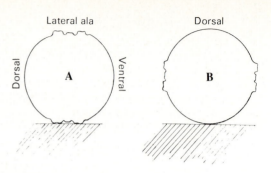

FIG 23-4

Outline of a transverse section through a third-stage juvenile to show the more stable position when it lies on its side **(A),** when moving by two-dimensional undulatory propulsion, and the less stable position, when it lies on its ventral surface **(B).**

From Lee, D.L. 1964. In Taylor, A.E.R., editor. *Nippostrongylus* and *Toxomplasma.* Blackwell Scientific Publications Ltd., Oxford, Eng.

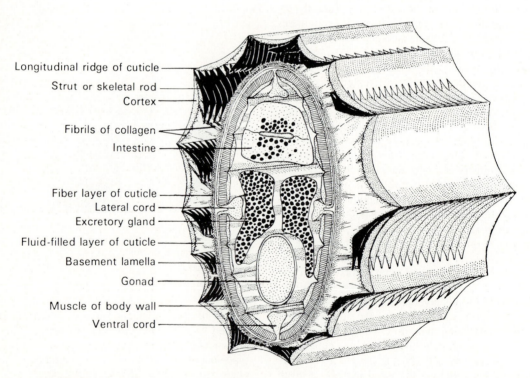

Longitudinal ridge of cuticle
Strut or skeletal rod
Cortex
Fibrils of collagen
Intestine
Fiber layer of cuticle
Lateral cord
Excretory gland
Fluid-filled layer of cuticle
Basement lamella
Gonad
Muscle of body wall
Ventral cord

FIG. 23-5

Stereogram of a thick section taken from the middle region of an adult *Nippostrongylus brasiliensis* to show the arrangement of the various layers of the cuticle and other internal organs.

From Lee, D.L. 1965. Parasitology 55:174.

The **hypodermis** lies just beneath the basal lamella of the cuticle. It is usually syncytial in adult worms, and the nuclei lie in four thickened portions that project into the pseudocoel, the **hypodermal cords.** The hypodermal cords run longitudinally and divide the somatic musculature into four quadrants. On the large nematodes, these may be discernible with the naked eye as pale lines. The dorsal and ventral cords contain longitudinal nerve trunks, whereas the lateral cords contain the lateral canals of the excretory system in most species. Especially in the regions of the cords, the hypodermis contains mitochondria and endoplasmic reticulum. An important function of the hypodermis is secretion of the cuticle, described in the section on development. In at least two genera of trichuroids, *Trichuris* and *Capillaria,* specialized areas of the hypodermis, the **bacillary bands,** occur. They open through pores lateral to the esophagus in *Trichuris* and extend the length of the body in *Capillaria.* The function of the bacillary bands is unknown, but they contain both apparently glandular and nonglandular cells.[128]

Musculature and pseudocoel

The somatic musculature is technically a part of the body wall, but it is convenient to consider its function along with that of the pseudocoel, and indeed they,

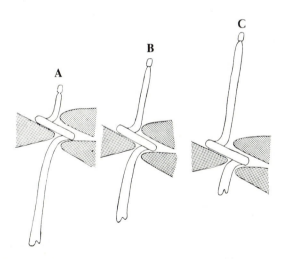

FIG. 23-6

Locomotion of an adult *Nippostrongylus* removed from the intestine of the host and placed among moist sand grains. It is probable that similar movements are performed by the nematode among the villi of the host intestine.

From Lee, D.L. 1965. The physiology of nematodes. Oliver & Boyd Ltd., Edinburgh.

along with the cuticle, function together as a **hydrostatic skeleton,** which will be discussed further.

Certain terms to describe forms and arrangement of body wall muscles were originally coined because of their supposed taxonomic value; they are still in use because they are convenient to describe forms and arrangements of muscle cells. If there are no rows or only two rows of cells between adjacent cords, the arrangement is **holomyarian.** If there are a few rows (two to five) per quadrant, the arrangement is described as **meromyarian,** and if there are many, it is called **polymyarian.** The **platymyarian** muscle cell is rather ovoid in cross section; contains its contractile fibrils at one side, adjacent to the hypodermis; and has a noncontractile portion of about the same width bulging into the pseudocoel (Fig. 23-7, *B*). The noncontractile portion contains the nuclei, large mitochondria with numerous cristae, ribosomes, endoplasmic reticulum, glycogen, and lipid. The **coelomyarian** cell is more spindle shaped, with the contractile portion at the distal end in the shape of a narrow U (Fig. 23-8). The distal end of the U is placed against the hypodermis, the contractile fibrils extend up along its sides, and the space in the middle is tightly packed with mitochondria. In some cases the elongate contractile portion does not sandwich the mitochondria, but these organelles are concentrated in the distal portion of the cell body close to the contractile fibrils.[126] The "cell body," or myocyton, bulges medially into the pseudocoel. It contains the nucleus, some mitochondria, endoplasmic reticulum, a Golgi body, and a large amount of glycogen. One important function of the coelomyarian cyton seems to be as a glycogen storage depot. In the **circomyarian** cell the contractile fibrils at the periphery entirely encircle the noncontractile portion of the cell.

The myofilaments seem to be essentially similar in all muscle types and are of two sizes: thick filaments of about 23 nm in diameter and thin filaments of about 8 nm. The thick filaments are made up of subunits of about 5 nm. It is believed that contraction occurs in a manner similar to the Hanson-Huxley model for vertebrate striated muscle, the thick filaments containing myosin and the thin filaments of actin. The actin filaments slide past the myosin filaments in contraction. The A, H, and I band typical of striated muscle can be distinguished but Z lines are absent. Thus the structure of nematode muscle is similar to vertebrate striated muscle and insect flight muscle and is referred to as "obliquely striated"[90-92] (Fig. 23-9).

Nematode muscles are unusual: processes run from the cytons to the nerves, rather than processes from the nerve cytons running to the muscles (Fig. 23-10). Another unusual feature of nematode muscle cells is

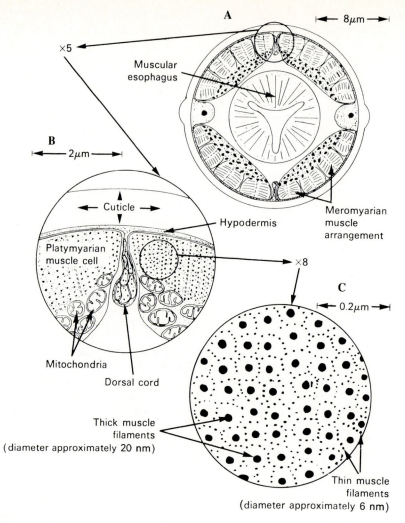

FIG. 23-7

Diagrams depicting a typical meromyarian-platymyarian muscle type at different magnifications. **A,** Whole transverse section. **B,** Portion of muscle cells on either side of dorsal nerve cord. **C,** Two types of muscle filaments as seen at high resolution with the aid of the electron microscope.

From Bird, A.F. 1971. The structure of nematodes. Academic Press, Inc., New York.

their frequent muscle-muscle connectives, at least in coelomyarian types.[127] These occur most often in the anterior regions of the worms and between the innervation processes of the muscle cells, although they may be between cytons. There is sometimes cytoplasmic continuity between the cells (cytoplasmic bridges) and sometimes continuity of the external layers (sarcolemmal bridges). A higher degree of muscular coordination is thought to result from transmission of nerve impulses between muscle cells that are so connected.

The somatic musculature and the rest of the body wall enclose a fluid-filled cavity, the **pseudocoelom,** or **pseudocoel** (Fig. 23-11). The pseudocoel differs from the true coelom in that it is derived from a persistent embryonic blastocoel, rather than being a cavity within the endomesoderm, and, therefore it has no peritoneal (mesodermal) lining. The nematode coelom functions as a hydrostatic skeleton. Hydrostatic skeletons are widespread in invertebrates. Their function depends on the enclosure of a volume of fluid, the

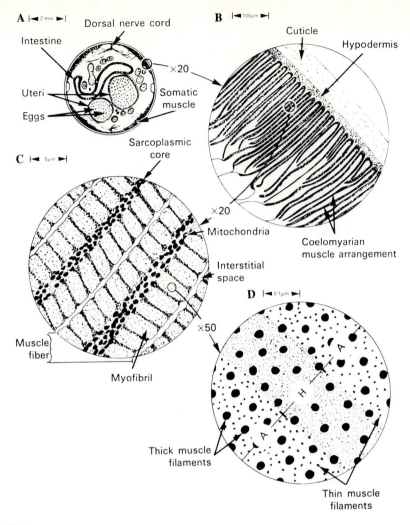

FIG. 23-8

Diagrams depicting the typical polymyarian-coelomyarian muscle type of *Ascaris lumbricoides* over a wide range of magnifications. **A,** Whole transverse section. **B,** Part of the muscle quadrant between the dorsal nerve and lateral hypodermal cord. **C,** Fibers of two muscle cells. **D,** An *H* and two *A* bands and the two types of muscle filaments at high resolution.

From Bird, A.F. 1971. The structure of nematodes. Academic Press, Inc., New York.

ability of muscle contraction to apply pressure to that fluid and the transmission of the pressure in all directions in the fluid as the result of its incompressibility. Thus, in a simple case, simultaneous contraction of circular muscles and relaxation of longitudinal muscles will cause an animal to become thinner and longer, whereas relaxation of circular muscles and a contraction of longitudinal muscles make an animal shorter and thicker. However, in nematodes the somatic musculature is entirely composed of longitudi-

nal fibers, and the muscles act not against other antagonistic muscles but against forces exerted by the internal pressure on the cuticle.[46] An increase in efficiency and strength of this system in locomotion can only be achieved by an increase in the pressure of the pseudocoelomic fluid. Measurements on the hydrostatic pressure (turgor) of the fluid of *Ascaris* have shown that the pressure can average from 70 to 120 mm Hg and vary up to 210 mm Hg.[44,46] This is an order of magnitude higher than the pressure in the

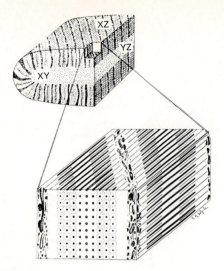

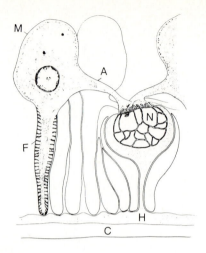

FIG. 23-9

Diagram of a muscle fiber showing the pattern of striation in three planes. In the XZ plane the myofilaments are staggered with the result that the striations are oblique rather than transverse. A second consequence of the stagger is that the adjacent rows of myofilaments do not reach the XY plane in phase, resulting in the appearance of striation in this plane also. The YZ plane shows cross-striation.

From Rosenbluth, J. Reproduced from *The Journal of Cell Biology,* 1965, vol. 25, p. 510, by copyright permission of The Rockefeller University Press.

FIG. 23-10

Diagram of muscle cells and myoneural junctions in transverse section. The myocyton *(M),* containing the nucleus of the muscle cell, is continuous with the core of the striated fiber *(F)* and with the elongate arm *(A).* The arm subdivides as it approaches the nerve cord *(N).* The individual axons comprising the nerve cord are embedded in a trough-like extension of the hypodermis *(H),* which underlies the animal's cuticle *(C).*

From Rosenbluth, J. Reproduced from *The Journal of Cell Biology,* 1965, vol. 26, p. 580, by copyright permission of The Rockefeller University Press.

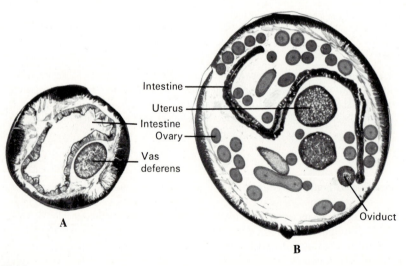

FIG. 23-11

Cross sections of male and female *Ascaris suum.* Space between organs is pseudocoel. **A,** Male. **B,** Female.

Photograph by Warren Buss.

body fluids of animals with hydrostatic skeletons in other phyla. Thus the limitations imposed by this high internal pressure determine many features of the nematode body plan.

The mechanism of body movement can be summarized as follows: The longitudinal muscles in an area on one side of the body contract; those on the other side relax. Force of the contraction is transmitted through the pseudocoel by the enclosed fluid, stretching the cuticle on the side opposite the contraction and forming a curve in the body. The elastic cuticle, being compressed on the concave side and stretched on the convex side of the curve, antagonizes the contracted muscles to bring the body back to its resting state on relaxation. Thus the worm moves in its characteristic undulating fashion.

The pseudocoelomic fluid is known as **hemolymph.** In *Ascaris,* and probably in other nematode parasites of animals, it is a clear, pink, almost cell-free, complex solution. Aside from its structural significance, it almost certainly is important in transport of solutes from one tissue to another. These solutes include a variety of electrolytes, proteins, fats, and carbohydrates. The proteins include albumins, globulins, hemoglobin, and several enzymes. Curiously, the fluid has far less chloride than would be required to balance the cations present, and the anion deficiency is made up mostly of volatile and nonvolatile organic acids.[35]

A peculiar and unique cell type found in the pseudocoel is the **coelomocyte.** Usually two, four, or six such cells, ovoid or many-branched, lie in the pseudocoel, attached to surrounding tissues. Although often small, in some species they are enormous; in *Ascaris* the coelomocytes are 5 mm by 3 mm by up to 1 mm thick. Their function is still obscure, although they may have a role in the accumulation and storage of vitamin B_{12} and in protein synthetic and secretory function.[15]

Nervous system

The nervous system of nematodes is relatively simple. Because the mechanical coordination of movement in different parts of the body is accomplished by local changes in hemolymph volume, local reflex networks are not needed.

There are two main concentrations of nerve elements in nematodes, one in the esophageal region and one in the anal area, connected by longitudinal nerve trunks. The most prominent feature of the anterior concentration is the **nerve ring,** or **circumesophageal commissure.** In *Ascaris* the nerve ring comprises eight cells, four of which are nerve cells and four of which are supporting, or **glial,** cells. The ring lies close to

the outer wall of the esophagus and is fairly easily seen in most species. Because its location is constant within a species, it is a good taxonomic character. The ring serves as a commissure for the **ventral, lateral,** and **dorsal** cephalic **ganglia** (Fig. 23-12, *A*), which are usually paired. The ventral ganglia are largest; the dorsal ganglia are smallest. Emanating from each ganglion posteriorly are the **longitudinal nerve trunks,** which become embedded in the hypodermal cords, and, again, the ventral nerve is largest. Proceeding anteriorly from the lateral ganglia are two **amphidial nerves.** Six **papillary nerves,** which are derived directly from the nerve ring, innervate the cephalic sensory papillae surrounding the mouth.

The ventral nerve trunk runs posteriorly as a chain of ganglia, the last of which is the **preanal ganglion.** The preanal ganglion gives rise to two branches that proceed dorsally into the pseudocoel to encircle the rectum, thus forming the **rectal commissure,** or **posterior nerve ring.** Other posterior nerves and ganglia are depicted in Fig. 23-12, *B*. The peripheral nervous system consists of a latticework of nerves that interconnect with fine commissures and supply nerves to sensory endings within the cuticle.

The main sense organs are the cephalic and caudal papillae, the amphids, the phasmids, and, in certain free-living species, the ocelli. The pattern of sensory papillae on the head of a nematode is a very important taxonomic character. The primitive pattern of lips surrounding the mouth of the ancestral nematodes is thought to have been two lateral, two dorsolateral, and two ventrolateral, each of which was supplied with sensory papillae (Fig. 23-13). In addition to the papillae forming the **inner** and **outer labial circles,** there were four **cephalic papillae,** one located behind the lips in each of the dorsolateral and ventrolateral quadrants. Most parasitic nematodes are modified from this basic form. Labial papillae are often lost or fused together, and cephalic papillae usually are quite reduced in size. However, some papillae are found on all species, and careful study will reveal all 16 nerve endings on most species, even those which have lost all semblances of lips. The pattern of lips and papillae on nematodes is studied by slicing the anterior tip from the worm with a sharp blade and orienting the end for an en face view on a microscope slide. Studies with the electron microscope have shown that the sensory endings of the papillae are modified cilia.[12] The papillae are probably tactile receptors.

The **amphids** are a pair of somewhat more complex sensory organs that open on each side of the head at about the same level as the cephalic circle of papillae. They are most conspicuous in marine, free-living forms and usually are reduced in animal parasites. The

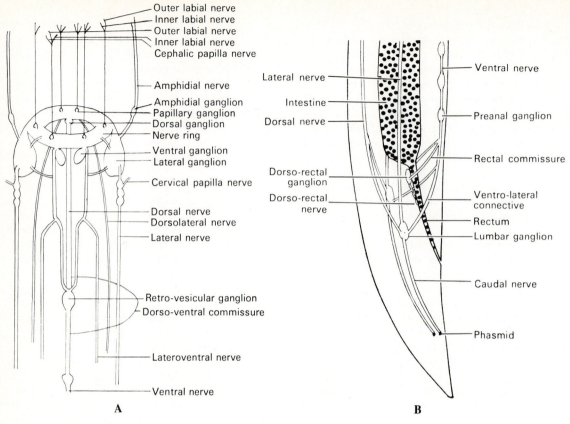

FIG. 23-12

Diagrammatic representation of the nervous system of a nematode. **A,** Anterior end. **B,** Posterior end.

From Crofton, H.D. 1966. Nematodes. Hutchinson University Library, London.

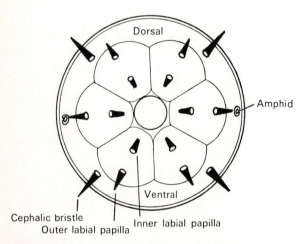

FIG. 23-13

Diagram of the anterior end of a hypothetical primitive nematode to show the arrangement of the sense organs.

Modified from de Connick. 1951. Ann. Soc. R. Zool. Belg. 81:26; from Crofton, H.D. 1966. Nematodes. Hutchinson University Library, London.

amphidial opening, which ordinarily is at the tip of a papilla, leads into a deep, cuticular pit, at the base of which is a nerve bulb with several nerve processes (Fig. 23-14). The sensory endings are modified cilia, up to 23 in one amphid, in contrast to the one to three per papilla. Until modified cilia were discovered in the sense organs of nematodes, it was thought that these worms had no cilia. Of course, their structure is rather different from ordinary kinetic cilia. They have no kinetosomes, and the microtubules usually diverge from the normal 9 + 2 pattern (p. 36), for example, to 9 + 4, 8 + 4, or 1 + 11 + 4. The amphids are considered chemoreceptors but may have a secretory function in some species; extracts of hookworm amphids inhibit clotting of vertebrate blood.[112]

Most parasitic nematodes have a pair of cuticular papillae, the **deirids,** or **cervical papillae,** at about the level of the nerve ring, and other sensory papillae are found at different levels along the body of many species. **Caudal papillae** (Fig. 23-15) are more elabo-

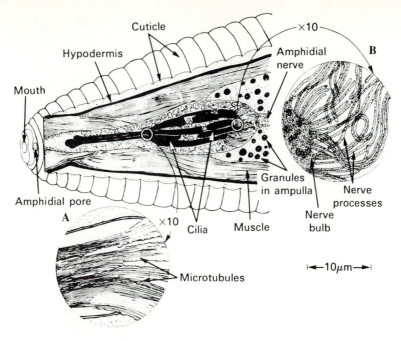

FIG. 23-14

Scale diagram of part of the tip of the head of *Meloidogyne* sp. cut open to reveal one of the two amphids.

From Bird, A.F. 1971. The structure of nematodes. Academic Press, Inc., New York.

FIG. 23-15

Ventral view of female *Toxascaris* sp., showing caudal papillae.

Photograph by Jay Georgi.

rately developed in the male, aiding in copulation. The pattern of distribution is an important taxonomic character. These papillae reach maximal development in the order Strongylata, where they help to form a complex copulatory bursa (Chapter 26). Also near the posterior end of most parasitic nematodes is a bilateral pair of cuticle-lined organs, the **phasmids.** The phasmids are similar in structure to the amphids and are of two basic types: one glandular, with an excretory function, and one sensory, involved in chemoreception.[77] These enigmatic structures are found neither in most free-living nematodes nor in the parasitic Dioctophymata and Trichurata. The presence or absence of phasmids is one criterion to separate the classes Adenophorea (= Aphasmidia, without phasmids) and Secernentea (= Phasmidea, with phasmids). Although difficult to see in some species, in most the phasmids are easily recognized by their cuticle-lined ducts (Fig. 23-16) that open at the apices of papillae near the tip of the tail.

Neural and neuromuscular transmission in nematodes is predominantly cholinergic[64] because the neurotransmitter at the synapses is acetylcholine. In *Ascaris* the noncontractile part of the muscle cell has a membrane resting potential of 20 to 30 mV, and the contractile part, a resting potential of 40 to 60 mV. The muscle cell undergoes spontaneous depolarization in the innervation arm and then generation of action potential in a repeated or oscillatory manner.[26] This spontaneous rhythmic spike production is contrasted with vertebrate skeletal muscle, in which an action potential (spike) is initiated by transmission of a nerve impulse across the neuromuscular junction. It is similar to cardiac muscle, in which there is a myogenic, rhythmic spike production. The rate of spontaneous firing is increased with lowered resting potential and decreased with higher resting potential. The role of the nerve fibers is primarily one of modulation, and there are both excitatory and inhibitory fibers. Stimulation of excitatory fibers releases acetylcholine at the neuromuscular junction, depolarizes the muscle membrane, and increases the rate of spikes. The inhibitory fibers release γ-aminobutyric acid (GABA), hyperpolarize the muscle, and decrease the rate of action potentials. Rhythmic spikes disappear altogether at resting potentials greater than 40 mV.

Some effective drugs against nematode parasites of vertebrates interfere with neuromuscular transmission, effectively paralyzing the worms, which then pass out of the host. Piperazine hyperpolarizes the muscle membrane to about 45 mV.[26] Ivermectin stimulates release of the inhibitory transmitter GABA from nerve endings and enhances binding of GABA to its receptors on the postsynaptic membrane.[25, 116] The benzim-

FIG. 23-16

Ventral view of female *Toxascaris* sp., showing ducts *(arrows)* leading to phasmids.

Photograph by Jay Georgi.

idazoles (mebendazole, parbendazole, fenbendazole) appear to have two modes of action.[95] They inhibit mitochondrial electron transport, especially the fumarate reductase system (see p. 406), thus inhibiting energy metabolism; and they also bind with tubulin, which interferes with microtubule-dependent processes such as acetylcholinesterase secretion, and paralyze the worms.

Digestive system and acquisition of nutrients

The digestion system is complete in most nematodes, with mouth, gut, and anus, although in mermithids and a few filariids the anus is atrophied. The stomodeum (buccal cavity and esophagus) and proctodeum (rectum) are lined with cuticle, and the cuticular lining is shed with the molting of the exterior cuticle.

The mouth is usually a circular opening surrounded by a maximum of six lips. Few parasitic nematodes possess as many as six lips; in some they have fused in pairs to form three. In many species the lips are absent altogether, whereas in others two lateral lips develop as new structures derived from the inner margin of the mouth. Regardless of the mor-

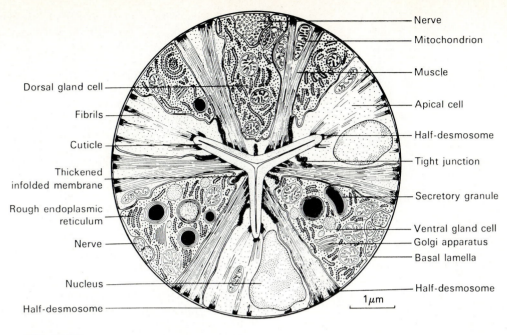

Dorsal gland cell

Fibrils

Cuticle

Thickened
infolded membrane

Rough endoplasmic
reticulum

Nerve

Nucleus

Half-desmosome

Nerve

Mitochondrion

Muscle

Apical cell

Half-desmosome

Tight junction

Secretory granule

Ventral gland cell

Golgi apparatus

Basal lamella

Half-desmosome

1 μm

FIG. 23-17

Diagram of a transverse section through the posterior part of the esophagus of *Nippostrongylus brasiliensis* to show the arrangement of various cells, cell membranes, and cellular organelles. Reconstructed from several electron micrographs.

From Lee, D.L. 1968. J. Zool. 154:9-18.

phology of a given species, it is a variation of the primitive, six-lipped form.

A buccal cavity lies between the mouth and esophagus of most nematodes. The size and shape of this area vary among species and are important taxonomic characters. In some species the cuticular lining is quite thick, forming a rigid structure known as a **buccal capsule**; in others the lining is thin. The cavity may be elongate, reduced or absent altogether, with a mouth that opens almost immediately into the lumen of the esophagus. Buccal armament is often present in parasitic and predaceous nematodes. The elements arise from the cavity wall or as anterior projections of the esophagus. Some nematodes have teeth of both types.

Food ingested by a nematode moves into a muscular region of the digestive tract known as the esophagus or pharynx. This is a pumping organ that sucks food into the alimentary canal and forces it into the intestine. It appears to be necessary because of the high turgor of the coelom. The esophagus assumes a variety of shapes, depending on the order and species of nematode, and for this reason is an important taxonomic character. It is highly muscular and cylindrical and often has one or more enlargements (**bulbs**). The

lumen of the esophagus is lined with cuticle and is triradiate in cross section, with one radius directed ventrad and the other two pointed laterodorsad (Fig. 23-17). Radial muscles insert on the cuticular lining in the interradii and run the length of the esophagus. Interspersed among the muscles are three esophageal glands, one in each of the interradial zones. The dorsal gland is usually more extensive than are the ventrolaterals. Each gland is usually uninucleate and opens independently into the lumen of the esophagus, although the dorsal one commonly opens farther anteriad. In some species the dorsal gland opens into the buccal cavity or even on the margin of the mouth. The secretions produced by these glands are digestive: amylase, proteases, pectinases, chitinases, and cellulases have been detected in them. In hookworms the secretions have anticoagulant properties.[112] In some species the glands fuse together near the posterior end of the esophagus, and in some nematodes, such as the Spiruroidea and many filariids, the posterior portion of the esophagus is mostly glandular. In some species the glands, especially the dorsal gland, are so extensive that much of their mass lies outside the esophagus proper (Fig. 23-18). In the specialized esophagus of

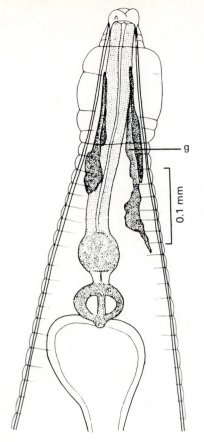

FIG. 23-18

Syphacia, a rodent pinworm with enlarged esophageal glands *(g).*

From Schmidt, G.D., and R.E. Kuntz. 1968. Parasitology 58:845-854.

FIG. 23-19

Diagrams to show the structure and function of the *Rhabditis* type of pharynx during feeding. Food particles, small enough to pass through the buccal cavity, are drawn into the lumen of the metacorpus by sudden dilation of the procorpus and metacorpus (**A**). Closure of the lumen of the pharynx in these regions expels excess water (**B**), and the mass of food particles is passed backward along the isthmus (**B** and **C**). Food is drawn between the bulb flaps of the posterior bulb by dilation of the haustrulum, which inverts the bulb flaps (**A**), and is passed to the intestine by closure of haustrulum and by dilation, followed by closure of the pharyngeal-intestinal valve (**B**). The bulb flaps contribute to the closure of the valve in the posterior bulb and, when they invert (**A**), also crush food particles.

From Lee, D.L. 1965. The physiology of nematodes. Oliver & Boyd Ltd., Edinburgh.

the Trichuroidea, the anterior portion is a thin-walled, muscular tube, whereas the posterior portion is a very thin tube surrounded by a column of single cells, the **stichocytes,** the entire structure being referred to as the **stichosome.** The ultrastructure of the stichocytes suggests that they are secretory, and they communicate with the esophageal lumen by small ducts.[106]

Rapid contraction of the buccal muscles and anterior esophageal muscles opens the mouth and dilates the anterior end of the esophagus, sucking food in (Fig. 23-19). Internal hydrostatic pressure closes the mouth and esophageal lumen when the muscles relax. The food is passed down the esophagus by the posteriorly progressing wave of muscle contraction opening the lumen for it until it reaches the intestine. The posterior bulb of many species appears to function as a one-way, nonregurgitation valve for food in the intestine. Thus the mechanism is a kind of peristalsis in which the force moving the food is not the contraction of circular muscles but the closure of the esophageal lumen by hydrostatic pressure behind the food. The frequency of pumping has been recorded as two to 24 per second.[29]

In a few ascaroids (*Contracaecum, Multicaecum, Polycaecum,* and others) one to five posteriorly directed esophageal ceca originate from a short, glandular **ventriculus** between the body of the esophagus and the intestine (Fig. 23-20).

The intestine is a simple, tubelike structure, extending from the esophagus to the proctodeum, and is constructed of a single layer of intestinal cells.

In females a short terminal, cuticle-lined rectum runs between the anus and intestine. In the male the rectum is further specialized in its terminal portion to

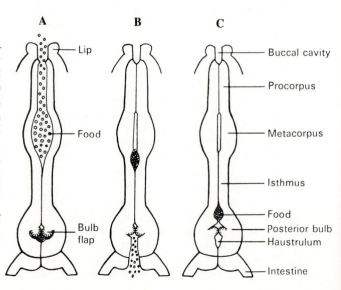

receive the products of the reproductive system and, therefore, is a cloaca. The dorsal wall of the cloaca is usually invaginated into two pouches, the spicule sheaths, which contain the copulatory spicules to be described along with the reproductive system. The vas deferens opens into the ventral wall of the cloaca.

The intestine is nonmuscular. Its contents are forced posteriad by the action of the esophagus as it adds more food to the front end of the system and perhaps by locomotor activity of the worm. Internal pressure of the coelom causes the intestine to be bilaterally flattened when empty. Between the dorsal wall of the cloaca and the body wall is a powerful muscle bundle called the **depressor ani.** This is a misnomer because, when it contracts, the anus is opened; it is therefore a dilator rather than a depressor. Defecation is also caused by hydrostatic pressure surrounding the intestine when the anus is opened. The force of the hydrostatic pressure is demonstrated by the fact that *Ascaris* can project its feces nearly 2 feet, when lifted from the saline solution maintaining it.[29]

The wall of the intestine consists of tall, simple columnar cells with prominent brush borders of microvilli.[107] Each microvillus is about 0.1 μm wide and consists of several core filaments surrounded by a filamentous coat of mucoprotein (Fig. 23-21). Although several digestive enzymes have been identified in the intestinal lumen, intestinal digestion is probably of minor importance in most forms because of the rapid rate of food movement through the intestine.

The number of intestinal cells varies from about 30 in some free-living species to more than a million in the larger parasitic forms. These cells rest on a basement membrane, which is attached to random extensions of the body wall musculature. It is probable that the intestine serves as the primary means of excretion of nitrogenous waste products, in addition to its function in nutrient absorption. Crofton[29] states that the intestine of *Ascaris lumbricoides* is emptied by defecation every 3 minutes under experimental conditions. Such a rapid turnover of materials must surely limit the amount of enzymatic action possible in the intestinal lumen but would favor the excretion of water-soluble waste products.

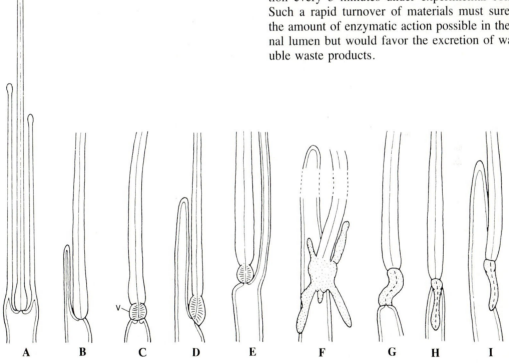

FIG. 23-20

Variations in esophagi in some ascaroid nematodes: **A,** *Crossophorus.* **B,** *Angusticaecum.* **C,** *Toxocara. v,* ventriculus. **D,** *Porrocaecum.* **E,** *Paradujardinia.* **F,** *Multicaecum.* **G,** *Anisakis.* **H,** *Raphidascaris.* **I,** *Contracaecum.*

Redrawn from Hartwich, G. 1974. CIH keys to the nematode parasites of vertebrates, no. 2. Commonwealth Agricultural Bureaux, Farnham Royal, Bucks, Eng.

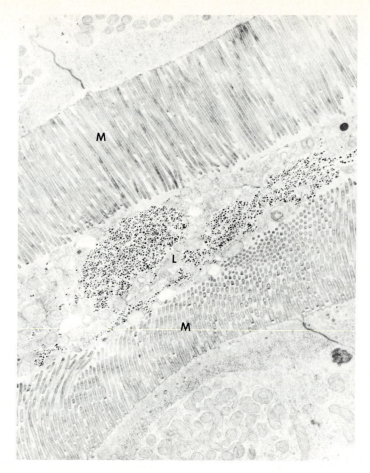

FIG. 23-21

Cross section of intestine showing microvilli *(M)* of dorsal and ventral sides. Cellular debris fills the lumen *(L)*. (× 10,800.)

From Sheffield, H.G. 1964. J. Parasitol. 50:365-379.

The food of parasitic nematodes, especially those in the intestine of their hosts, is high in amino acids and sugars. Nematodes feed extravagantly and wastefully, and the thin-walled intestine with its brush border is an efficient absorptive mechanism.

Members of the family Mermithidae are unusual in that the adults are free living, but the juveniles are parasitic in invertebrates, primarily insects. The adults do not feed, and at no stage is there a functional gut. The body wall in the adults and the first-stage juveniles has a structure typical of other nematodes, described previously, but the body wall in the parasitic juveniles is greatly modified for absorption of nutrients.[8] The cuticle is very thin, and the hypodermis is thick and metabolically active, with microvilli underlying the cuticle. It is connected by cytoplasmic bridges to a food storage organ, the **trophosome.**[9] During the sometimes long, nonfeeding adult life, the worm apparently lives on nutrients stored in the trophosome. Because mermithids almost always kill their host, they have received some attention as biological control agents of insect pests, for example, *Romanomermis culicivorax,* a parasite of mosquitoes.[80,83]

Excretion and osmoregulation

An excretory system has been observed in all parasitic nematodes except the aphasmidians Trichurata and Dioctophymata. Although the several types of excretory systems obviously are derived from one or two basic organs and all have an external opening called the excretory pore, it has not been completely proved that the systems are excretory in function. Strong evi-

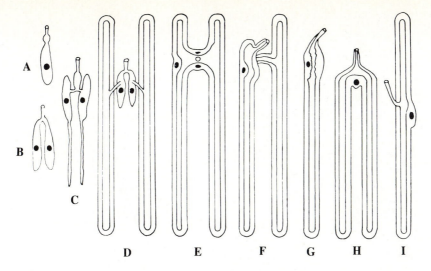

FIG. 23-22

Excretory systems. **A,** Single renette in a dorylaimid. **B,** Two-celled renette in *Rhabdias*. **C,** Larval *Ancylostoma*. **D,** Rhabditoid type. **E,** Oxyuroid type. **F,** *Ascaris*. **G,** *Anisakis*. **H,** *Cephalobus*. **I,** *Tylenchus*.

From Crofton, H.D. 1966. Nematodes. Hutchinson University Library, London.

dence exists that most excretion occurs through the intestine.[96] As in the evolution of excretion in other animal groups, the excretory system of nematodes probably originated as an osmoregulatory system, any excretion of metabolic wastes by means of this system being secondarily acquired. Following tradition, we shall refer to it as an excretory system.

The presence of an excretory system is apparently primitive and probably evolved first in fresh-water forms. There are no flame cells or nephridia; in fact the nematode excretory system seems to be unique in the animal kingdom. The two basic types are **glandular** and **tubular.** The glandular type is typical of the free-living Aphasmidea and may be involved in secretion of enzymes, proteins, or mucoproteins; it will not be considered here.

Several varieties of tubular excretory systems occur in parasitic forms (Fig. 23-22). Basically two long canals in the lateral hypodermis connect to each other by a transverse canal near the anterior end. This transverse canal opens to the exterior by means of a median, ventral duct and pore, the excretory pore. This pore is conspicuous in most species; its location is fairly constant within a species and, therefore, is a useful taxonomic character.

Several variations of this basic, H-shaped system are common. The arms anterior to the transverse canal may be absent, forming a U-shaped system. In some species one entire lateral half is missing, resulting in an asymmetrical system. The posterior crura may be short, resulting in an inverted-U system. Many parasitic nematodes have a pair of large, granular, subventral gland cells associated with the transverse ducts. These are probably secretory in function.

The ability to osmoregulate varies greatly among nematodes and is correlated generally with the requirements of their habitats. The body fluids of species parasitic in animals may be somewhat different in osmotic pressure from the tissues they inhabit but not dramatically so. For example, *Ascaris* hemolymph is about 320 to 350 mosmol, whereas pig intestinal contents are around 400 mosmol.[45] *Ascaris* clearly can control its electrolyte concentrations to some degree: chloride ion concentration of host intestinal contents varies between 34 and 102 mM, but *Ascaris* hemolymph is fairly constant at around 52 mM. Adults of most parasitic species cannot tolerate media much different in osmotic pressure from their hemolymph; when placed in tap water, they will burst, sometimes within minutes, from addition of the imbibed water to the already high internal pressure. Of course, freshwater and terrestrial nematodes, including juveniles of many parasitic species, must withstand (and regulate in) extremely hypotonic conditions.

Details of water and ion excretion are poorly known. Contractions of excretory canals and the ampulla near the excretory pore have been observed in several species. Contractions of the ampulla in free-living, third-stage juveniles of *Ancylostoma* and *Nippostrongylus* are inversely proportional to the salt con-

395

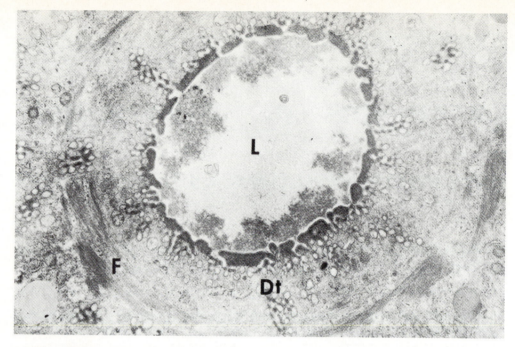

FIG. 23-23

Anisakis excretory gland. Transverse section through main excretory canal, showing round canal with interrupted dense material lining the lumen *(L)*, ramifying drainage tubules *(Dt)*, and congregated vesicles surrounding main canal and drainage tubules. Filaments *(F)* appear in circular arrangement around main canal. (× 10,000.)

From Lee, H. 1973. J. Parasitol. 59:289-298.

centration of the solution in which they are maintained. Some evidence suggests that osmoregulation by the ampulla is part of a homeostatic mechanism to maintain constant volume so that locomotor activity is not impaired. Nelson and Riddle[74] ablated different portions of the excretory system of a free-living nematode *(Caenorhabditis elegans)* with a laser microbeam. Destruction of the pore cell, duct cell, or excretory cell led to accumulation of water and death of the worm, but ablation of the gland cell had no effect.

Information on ultrastructure of the excretory system strongly suggests that the system functions in osmoregulation and perhaps in excretion of waste products and in secretion as well.[66] The surface area of the peripheral cell membrane may be greatly increased by numerous bulbular invaginations, and on the interior the lumen is perforated by drainage ductules, or canaliculi (Fig. 23-23). Filaments that are presumably contractile may surround the lumen of the duct. The hydrostatic pressure in the pseudocoel is thought to provide filtration pressure to excrete substances through the canals embedded in the hypodermal cords.

The ultrastructure of the gland cells clearly suggests secretory function. Enzymes responsible for exsheathment (shedding the old cuticle at ecdysis) are produced there by various strongyle juveniles. A variety of nematodes excrete substances antigenic for their hosts through the excretory pores. Lee[60,61] suggested that digestive enzymes were secreted by adult *Nippostrongylus* to act in conjunction with the abrading action of the cuticle.

The major nitrogenous waste product of nematodes is ammonia. In normal saline *Ascaris* excretes 69% of the total nitrogen excreted as ammonia and 7% as urea. Under conditions of osmotic stress, these proportions can be changed to 27% ammonia and 52% urea. Amino acids, peptides, and amines may be excreted by nematodes. Other excretory products include carbon dioxide and a variety of fatty acids. The fatty acids are end products of energy metabolism and will be considered further. The role of the excretory system in the elimination of the foregoing substances is not well established. Juvenile *Nippostrongylus* excrete several primary aliphatic amines through their excretory pore. It has been shown that a large proportion of nitrogenous waste products can be excreted via the intestine and anus by *Ascaris*,[96] and it would seem that the cuticle must play a major role in ammonia excretion in most nematodes.

Reproduction

Most nematodes are dioecious, although a few monoecious species are known. Parthenogenesis also exists in some. Sexual dimorphism usually attends dioecious forms, with females growing larger than the males. Furthermore, males have a more coiled tail and often have associated external features, such as bursae, alae, and papillae. Such dimorphism achieves the ultimate in the Tetrameridae and the plant-parasitic Heteroderidae, where the males have typical nematode anatomy, but the females are little more than swollen bags of uteri.

The gonads of nematodes are solid cords of cells that are continuous with the ducts that lead to the external environment. This allows the reproductive systems to function in spite of the high turgor of the pseudocoelom.

■ Male reproductive system

Usually there is a single testis in nematodes, although two have been found in a few species. This organ may be relatively short and uncoiled, but in the larger animal parasites it appears as a long, threadlike structure that is coiled around the intestine and itself at various levels of the body. Two zones usually can be distinguished: the **germinal zone,** incorporating the blind end and in which spermatogonial divisions take place, and the **growth zone.** The end of the growth zone merges with a more tubular structure, the **seminal vesicle,** which is a sperm storage organ. The seminal vesicle merges into the vas deferens, which is usually divided into an anterior, glandular region and a posterior, muscular region, the ejaculatory duct. The ejaculatory duct opens into the cloaca. Some species have a pair of cement glands near the ejaculatory duct that secrete a hard, brown material to plug the vulva after copulation.

Nearly all nematodes have a pair of sclerotized, acellular, copulatory spicules (Fig. 23-24). They originate within dorsal outpocketings of the cloacal wall and are controlled by proximal muscles. Each spicule is surrounded by a fibrous spicule sheath. The spicule structure varies between species but is fairly constant among individuals within a species, making the size and morphology of the spicules two of the most important taxonomic characters. A dorsal sclerotization of the cloacal wall, the **gubernaculum,** occurs in many species. It guides the exsertion of the spicules from the cloaca at copulation. In several strongyloid

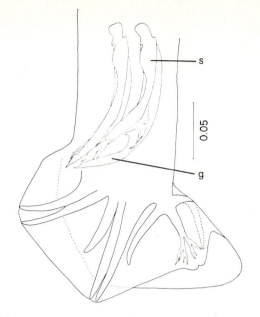

FIG. 23-24

Bursate nematode *Molineus,* showing complex spicules *(s)* and a gubernaculum *(g).*

From Schmidt, G.D. 1965. J. Parasitol. 51:164-168.

genera an additional ventral sclerotization of the cloaca, the **telamon,** has the same general function as that of the gubernaculum. Both structures are important taxonomic characters. The spicules are inserted into the vulva at copulation. They are not true intromittent organs, since they do not conduct the sperm, but are another adaptation to cope with the high internal hydrostatic pressure. The spicules must hold the vulva open while the ejaculatory muscles overcome the hydrostatic pressure in the female and rapidly inject sperm into her reproductive tract.

Nematode spermatozoa are unusual among those studied in the animal kingdom in that they lack a flagellum and acrosome. Furthermore, internal organization of organelles differs markedly from that of all other sperm previously described. The sperm are rather diverse between species in cytological characteristics; Foor[39] recognized at least four types among those so far described. As mature sperm in the seminal vesicle of the male, the types range from small, rounded structures to ameboid cells with distinct anterior and posterior cytoplasm to elongate, tadpole-shaped structures, with "head" and "tail." In all types the nucleus is not bounded by a nuclear membrane. Some types undergo further morphological development after insemination of the female; for example, the "tadpole" becomes ameboid, and it is thought that

397

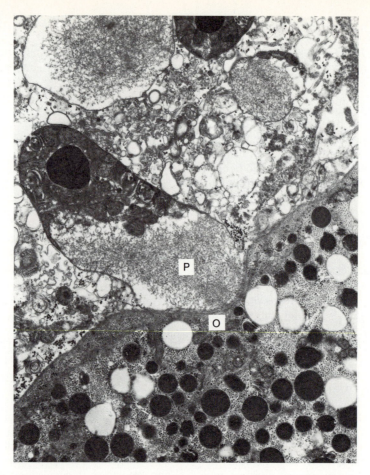

FIG. 23-25

Angiostrongylus cantonensis sperm in contact with an oocyte *(O)* in the uterus of the female worm. Note large pseudopod *(P)* and continuity between the membrane specialization and the sperm plasma membrane. (× 14,500.)

From Foor, W.E. 1970. Biol. Reprod. 2(suppl.):177-202.

the sperm may not be able to fertilize an ovum until these changes have occurred.[110] Motility has not been easy to discern, although clearly the gametes must travel up the female tract. The "tail" of the tadpole-shaped sperm apparently is nonmotile, but the "head" can put out pseudopodia.[63] Good ultrastructural evidence for pseudopodial movement is available[39] (Fig. 23-25). Ugwunna[114] suggested that the considerable amount of smooth endoplasmic reticulum, characteristic of nematode spermatozoa, functions in pseudopod formation. Activation of nematode sperm does not seem to be associated with increased levels of cyclic AMP, as it is with flagellated sperm.[104]

■ Female reproductive system

Most female nematodes have two ovaries, although some have from one to more than six. The general pattern of structure of the reproductive system in the female is similar to that in the male, except the gonopore is independent of the digestive system: a linear series of structures, with the gonad at the internal or proximal end, followed by developmental, storage, and ejective areas. The number of tracts per female and their disposition relative to each other are given descriptive terms. If a species has only one ovary and uterus, it is called **monodelphic.** Monodelphic species usually have the vulva near the anus. The more com-

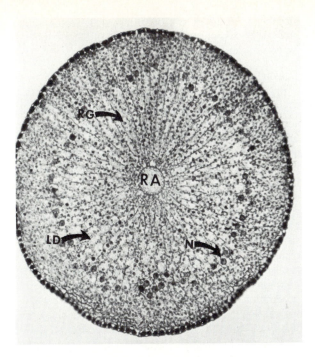

FIG. 23-26

Ascaris lumbricoides: transverse section through growth zone of ovary. *LD,* Lipid droplet; *N,* nucleus; *RA,* rachis; *RG,* refringent granule. (× 440.)

From Foor, W.E. 1967. J. Parasitol. 53:1245-1261.

mon situation, in which there are two ovaries and uteri, is termed **didelphic.** More than two uteri and associated structures is a condition called **polydelphic.** If the two uteri converge from opposite directions at their junction with the vagina, they are called **amphidelphic.** When they are parallel and converge from an anterior direction, they are **prodelphic;** if they converge from a posterior direction, they are **opisthodelphic.**

The ovaries are solid cords of cells that produce gametes and move them distally into the terminal portion of the system. The proximal end of the ovary is the **germinal zone,** which produces oogonia; the oogonia become oocytes and move into the **growth zone** of the ovary, toward the oviduct. In the large ascarids the oocytes are attached to a central supporting structure, the **rachis.** In *Ascaris* the germinal zone is very short, and most of the 200 to 250 cm length of the ovary comprises oocytes attached in a radial manner by cytoplasmic bridges to the rachis[38] (Fig. 23-26). The oocytes increase in size as they move down the rachis, and about 3 to 5 cm from the oviduct, they become detached from it. In some nematodes the rachis ends at the beginning of the growth zone, and the oocytes

pass single file down the growth zone, increasing greatly in volume.[68]

The proximal end of the oviduct in most nematodes is a distinct **spermatheca,** or sperm storage area. As the oocytes enter the oviduct (spermathecal area), they are penetrated by sperm; only then do they undergo their meiotic divisions. A polar body is extruded at the first division, and the second division is followed by expulsion of another polar body. Concurrent with these events, shell formation is occurring, which will be described in the section on development.

The wall of the uterus has well-developed circular and oblique muscle fibers, and these move the developing embryos ("eggs") distally by peristaltic action. The shape of the eggs may be molded by the uterus, and uterine secretory cells may contribute additional material to the eggshells. The distal end of the uterus is usually quite muscular and is known as the **ovijector.** The ovijectors of the uteri fuse to form a short vagina that opens through a ventral, transverse slit in the body wall, the **vulva.** The vulva may be located anywhere from near the mouth to immediately in front of the anus, depending on the species. The vulva never opens posterior to the anus and only very rarely into the rectum to form a cloaca. When the lips of the vulva protrude, they are said to be **salient.** The muscles of the vulva act as dilators, and constriction of the circular muscles of the ovijectors both expel the eggs and restrain more proximal, undeveloped eggs from being expelled because of hydrostatic pressure.

Mating behavior

Clearly, adult worms of opposite sexes must find each other and copulate for reproduction to occur. Both chemotactic and thigmotactic mechanisms operate in these processes.

Pheromone sex attractants have now been shown for about 40 species of nematodes,[67] usually by means of an in vitro assay. For example, male *N. brasiliensis* migrate toward a source of medium in which females have been incubated or which contains an aqueous extract of females.[16] Studies on the chemical composition of nematode pheromones suggest that there may be a "medley" of attractants, more complex than hitherto realized.[67] The use of pheromones as biological control agents has some potential.

Having found each other, copulation is facilitated by thigmotactic responses mediated by the papillae. The female in some species seeks the coiled posterior end of the male, which she enters. The caudal papillae of the male detect the vulva; this excites a probing response of the spicules, leading to sperm transfer. Females of some species have vulvar papillae. Curi-

FIG. 23-27

Female *Ascaris lumbricoides* strangled by a shoe eyelet. This illustrates the tropism of female nematodes to seek the coiled tail of males.

From Beaver, P.C. 1964. Am J. Trop. Med. Hyg. 13:295-296.

ously, if no males are present within a host, females of some species tend to wander, seeking a constriction to squeeze through. This may result in dire consequences to the host if a bile duct, for example, is selected for exploration. Other unexpected results of this behavior have been recorded (Fig. 23-27). Serial copulations of a female with several males have been observed.[2]

DEVELOPMENT

Historically, studies on the development of nematodes have led to fundamental discoveries in zoology. For example, van Beneden,[11] in 1883, was the first to elucidate the meiotic process and realize that equal amounts of nuclear material were contributed by sperm and egg after fertilization. Boveri[17] (1899) first demonstrated the genetic continuity of chromosomes and determinate cleavage, that is, embryogenesis in which the fate of the blastomeres is determined very early. Both men based their insights on studies of nematode material.

Not surprisingly, in such a successful group as nematodes, details of development and life history differ greatly among the various groups. However, the general pattern is remarkably similar for all species known. The four juvenile stages and the adult are each separated from the preceding one by an ecdysis, or molting of cuticle. The juvenile stages are referred to conventionally as "larvae." The first-stage juvenile is quite similar in body form to the adult. No real metamorphosis occurs during ontogeny, and, with certain possible exceptions, all somatic cells of the adult may be present in the embryo.[29]

Eggshell formation

Penetration of the ovum by the sperm initiates the process by which protective layers are produced around the zygote and developing embryo. The fully formed shell in most nematodes consists of three layers: (1) an outer **vitelline layer,** often not detectable by light microscopy; (2) a **chitinous layer;** and (3) a **lipid layer,** innermost and called the "vitelline membrane" in older literature.[12] A fourth, **proteinaceous** layer is contributed by uterine cell secretions in some nematodes *(Ascaris, Thelastoma, Meloidogyne)* and consists of an acid mucopolysaccharide–tanned protein complex. Formation of the shell layers has been best studied in *Ascaris,* but it seems likely that the process is similar in other nematodes. Immediately after sperm penetration a new plasma membrane forms beneath the original; the old plasma membrane becomes the vitelline layer and separates from the peripheral cytoplasm, and the cytoplasm shrinks back, leaving an electron-lucid space within which the chitinous layer forms[38,65] (Fig. 23-28). Refringent bodies, previously dispersed throughout the cytoplasm, migrate to the periphery and extrude their contents, the fusion of which forms the lipid layer. The so-called chitinous layer is probably supportive or structural in function and also contains protein; the proportion of chitin present varies among groups from great (ascaroids, oxyuroids) to very small (strongyloids). Resistance to desiccation and to penetration of polar substances is conferred by the lipid layer. At least in ascarids, this layer is composed of 25% protein and 75% **ascarosides.** Ascarosides are very interesting and unique glycosides (compounds with a sugar and an alcohol joined by a glycosidic bond). In ascarosides the sugar is **ascarylose** (3,6-dideoxy-L-arabinohexose), and the alcohols are a series of secondary monols and diols containing 22 to 37 carbon atoms.[52] The ascarosides render the eggshell virtually impermeable to substances other than gases and lipid solvents; further discussion of the resistance of *Ascaris* eggs is in Chapter 27. Whether ascarosides are present in the lipid layers of nematode eggs other than ascaroid is not known. Some water can pass across the lipid layer of *Ascaris* and at least some other nematodes, but the embryos can continue to develop despite water stress.[109,124]

Eggshell formation is similar in oxyurids, but there are two uterine layers, and the lipid layer in some *(Syphacia* spp., for example) is thin and of doubtful protective value.[122,123] Oxyurids and some other nematodes also have an operculum, which is a specialized

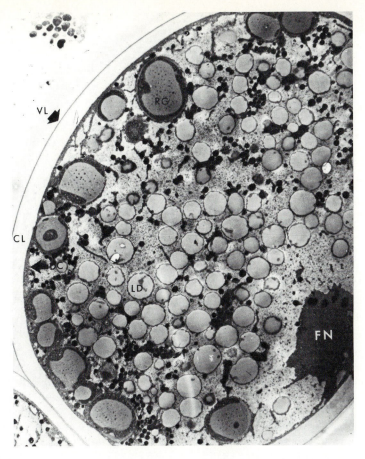

FIG. 23-28

Low magnification electron micrograph of newly fertilized egg, showing vitelline layer *(VL);* incipient chitinous layer *(CL);* dense, particulate, cortical cytoplasm *(DC);* and numerous lipid droplets *(LD).* Female nucleus *(FN)* lies near surface, and refringent granules *(RG)* have migrated to position just beneath cortical cytoplasm. After extrusion the contents of the refringent granules will become the lipid layer (× 3800.)

From Foor, W.E. 1967. J. Parasitol. 53:1245-1261.

area on the egg to facilitate hatching of the juvenile.[124] Trichurid eggs have an opercular plug at each end.

Embryogenesis

Molecular biology of early development has been studied in *Ascaris,* and some data indicate similar phenomena in *Parascaris.*[37] During oogenesis in other well-studied systems (such as amphibians and echinoderms), there is considerable synthesis of ribosomal and informational RNA (rRNA, mRNA). These are conserved in the unfertilized egg, and little or no RNA synthesis follows fertilization and early cleavage. In contrast, there is almost no RNA in mature *Ascaris*

oocytes, and fertilization is followed by a burst of rRNA synthesis in the *male* pronucleus. This may reflect an adaptation to the fact that the female nuclei are otherwise occupied at this time, undergoing their maturation divisions, and the burden of RNA production in preparation for protein synthesis falls on the male pronucleus.[37]

The determinate cleavage of the nematode embryo is the clearest and best documented example of germinal lineage in the animal kingdom.[29] Because of the early determination of the fate of each cell (blastomere) in the cleaving embryo, names or letter designations can be given to each blastomere, and the tissues that will develop from each are known (Fig. 23-

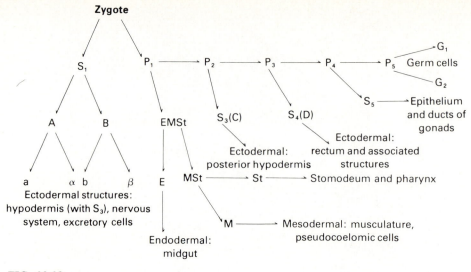

FIG. 23-29

Cell lineage of nematodes. The two cells produced at the first cleavage of the zygote are P_1 and S_1. The diagram indicates the progeny of these cells and the tissues to which they give rise.

29). At the first cleavage the zygote produces one cell that will give rise to somatic tissues and one cell whose progeny will comprise more somatic cells and the germinal cells. The early cleavages of nematodes are marked by a very curious phenomenon called **chromatin diminution.** The chromosomes fragment, and only the middle portions are retained, the ends being extruded to the cytoplasm to degenerate. Oddly, during the diminution there is an approximate doubling of the histone/DNA ratio. The explanation for this observation is unclear; the distance between the nucleosomes does not decrease, nor is there a detectable free histone pool.[34]

Since chromatin diminution only occurs in the somatic cells, the germ line can be recognized by its full chromatin complement. Finally, the only cells left with complete chromosomes are G_1 and G_2, which will give rise to the gonads. Interestingly, further differentiation of the nuclei in the various tissues seems to go in both directions with respect to chromatin content. Some, such as muscle and ganglia nuclei, further diminish until DNA can no longer be detected, whereas others, particularly those with protein synthetic activity as excretory and pharyngeal glands and uterine cells, exhibit polyploidy with respect to DNA content.[111] Clearly there must be a great redundancy of the genes left after chromatin diminution.

Rather typical morula and blastula stages are formed. Gastrulation is by invagination and also by epiboly (movement of the micromeres down over the macromeres).

In the fully formed embryo the nuclei other than the germinal cells cease to divide; thus all cells of the adult are present at this time. The phenomenon is known as **cell** or **nuclear constancy,** or **eutely,** and it is characteristic of several aschelminth phyla. There are some exceptions: cells of the intestine and hypodermis of the large nematodes divide further, but in most species growth after embryogenesis is a matter of cell enlargement rather than cell division. The number of cells per individual is fairly constant within a species and varies among species.

The timing, site, and physical requirements for embryogenesis vary greatly among species. In some the cleavage will not begin until the egg reaches the external environment and oxygen is available. Others begin (or even complete) embryogenesis before the egg passes from the host, whereas in some the juveniles complete development and hatch within the female nematode (ovoviviparity).

Studies on embryonation of *Ascaris* eggs have revealed a most fascinating sequence of biochemical epigenetic adaptation: adaptive appearance and disappearance of biochemical pathways through ontogeny, based on repression and derepression of genes.[37] The energy metabolism of adult *Ascaris* is anaerobic, but that of the embryonating egg is obligately aerobic. Dependence on pathways such as glycolysis would not only be wasteful of the limited stored nutrient in the embryos but also a toxic concentration of acidic end products would soon build up as the result of impermeability of the eggshell. Eggs survive temporary anaerobiosis, but they do not develop unless oxygen is present. They are completely embryonated and infective after 20 days at 30° C, and throughout this time a tricarboxylic acid cycle and cytochrome *c*-cytochrome oxidase electron transport system are present. The infective stage is the second-stage juvenile, having undergone one molt in the egg. (A second molt before hatching has been reported, which would make the infective stage the third, as in most other nematodes.[69]) The egg hatches in the host intestine and the juvenile goes through a tissue migration. It breaks out into the lung alveoli, travels up the trachea, and then is swallowed to gain access to the intestine, where it becomes an adult. Cytochrome oxidase is still present in the juveniles recovered from the lungs, and they require oxygen for motility. Oxidase activity disappears from the fourth-stage juvenile in the intestine and is essentially repressed through adult life. A similar phenomenon has been observed with regard to the enzymes of the glyoxylate cycle.[7] It was shown some years ago that embryonating *Ascaris* eggs consume both lipid and carbohydrate reserves during the first 10 days of embryonation and then *resynthesize carbohydrate* (glycogen and trehalose) from fat[78] (Fig. 23-30). Derivation of energy from lipids normally requires degradation to two-carbon fragments in the form of acetyl-CoA. Then the acetyl-CoA is oxidized in the tricarboxylic acid cycle. Most higher animals cannot synthesize carbohydrates from acetyl-CoA, but many plants and microorganisms can accomplish this feat because they have two essential enzymes, **isocitrate lyase** and **malate synthase,** to perform the glyoxylate cycle. Other enzymes necessary for the complete cycle are normally associated with the tricarboxylic acid cycle (Figs. 4-12 and 23-35). Isocitrate lyase and malate synthase have been demonstrated in several other nematodes and in the trematode *Fasciola hepatica.* However, *Ascaris* is the only metazoan in which the glyoxylate cycle and its role in the conversion of fat to carbohydrate has been established.[7] Finally, all activity of the two critical enzymes seems to be repressed in the adult muscle. It may be supposed that the synthesized trehalose may play a role in egg hatching, and the glycogen accumulation may be in anticipation of the juveniles' needs during their tissue migration.

Egg hatching

Hatching mechanisms of nematodes were reviewed by Perry and Clarke.[79] Hatching of nematodes whose juveniles are free living before becoming parasitic occurs spontaneously. This probably is a result of synthesis of lipid-hydrolyzing enzymes in the subventral esophageal glands, these structures only becoming active in the terminal phases of embryogenesis.[12] A number of species, however, will hatch only after being swallowed by a prospective host. On reaching the infective stage, such eggs remain dormant until the proper stimulus is applied, and this requirement has the obvious adaptive value of preventing premature hatching. Ascarid eggs require a combination of conditions: temperature about 37° C, a moderately low oxidation-reduction potential (the presence of an oxidizing agent reversibly inhibits hatching),[48] a high carbon dioxide concentration, and a pH of about 7. These conditions are present in the gut of many warm-blooded vertebrates, and, indeed, *Ascaris* will hatch in a wide variety of mammals and even in some birds, but all four conditions are unlikely to be present simultaneously in the external environment. The first change detectable on application of the stimulus is a rapid change in permeability; trehalose from the perivitelline fluid leaks from the eggs. The lipid layer is now permeable to chitinase secreted by the juvenile. Esterases and proteinases also are secreted, and these enzymes attack the hard shell, digesting it sufficiently for the worm to force a hole in it and escape.[35] First-stage juveniles of some nematodes, such as *Trichuris,* possess a stylet on their anterior end, and when the juveniles are acti-

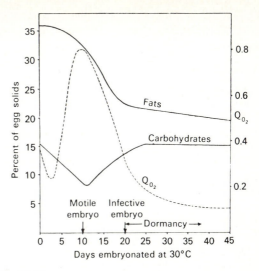

FIG. 23-30

Changes in oxygen consumption, fat, and carbohydrates during embryonation of *Ascaris* eggs. Fats are connected to carbohydrates by the glyoxylate cycle, which requires the enzymes isocitrate lyase and malate synthase. These enzymes are present in the embryo but apparently are absent in the adult.

From Fairbairn, D. 1960. In Sasser, J.N., and W.R. Jenkins, editors. Nematology. University of North Carolina Press, Chapel Hill.

vated by the hatching stimulus, they penetrate the operculum (polar plug) with the stylet and emerge from the eggshell.[76]

Growth and ecdysis

Unlike most arthropods, there is growth in body dimensions of nematodes between molts of their cuticle (Fig. 23-31). After the fourth molt in large nematodes such as *Ascaris,* there is considerable increase in size, and the cuticle itself continues to grow after the last ecdysis. The molting process has been studied in several species. First the hypodermis detaches from the basal lamella of the old cuticle and starts to secrete a new one, beginning with the cortical layers. By the time the new cuticle is secreted, it may be substantially folded under the old cuticle, to be stretched out later after ecdysis. In some cases the old cuticle up to the cortical layer is dissolved and resorbed through the new cuticle. This is particularly important when conservation of materials and space is a consideration, such as in the first molt of *Ascaris* and less so when there is plenty of food and the old cuticle is very complex in structure, as in the fourth molt of *Nippostrongylus.*[61] Escape from the old cuticle seems to be facilitated by several enzymes. A collagenase-like enzyme attacks the old cuticle.[87] A lipase may also be secreted, but its role, if any, is unclear. Present evidence indicates that leucine aminopeptidase, formerly thought to be the exsheathing enzyme, is inactive against the cuticular material.[87] It may mediate the release of the other enzymes from the excretory cells.

Growth and ecdysis are controlled to an as yet undetermined degree by neurosecretory mechanisms. **Neurosecretion** is a process in which substances with endocrine function are secreted by nerve cells, or modified nerve cells, and its action is well documented in vertebrates and some arthropods. On the basis of staining reactions, neurosecretory activity has been suggested in several nematodes, but Davey and Kan[33] have shown that *Terranova decipiens* fails to undergo ecdysis in the absence of a neurosecretory cycle. Furthermore, extracts of isolated neurosecretory cells activate leucine aminopeptidase in isolated excretory cells. Insect juvenile hormone activates the excretory cells in the intact worms and leads to ecdysis, apparently by changing the permeability of the cells and allowing entry of water into them.[32]

A common adaptation in many parasites is a resting stage at some point in their development, enabling them to survive adverse conditions while awaiting access to a new host. Such **developmental arrest** or **hypobiosis**[102] is of particular interest in nematodes, not only because of the variety of stages and situations

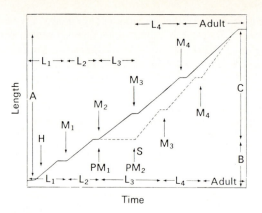

FIG. 23-31

Idealized form of the basic life cycle of nematodes. The life cycle of a free-living nematode is represented by a solid line. Hatching *(H)* is "spontaneous," and there are four molts (M_1 to M_4). The broken line represents a life cycle in which a change in environment is necessary to stimulate *(S)* the completion of the second molt (PM_2). A to C are different environments. L_1 to L_4 are the conventional designations of the "larval" stages.

From Rogers, W.P., and R.I. Sommerville. With permission from Advances in parasitology, vol. 1, edited by B. Dawes. Copyright Academic Press, Inc. (London) Ltd.

in which it takes place but also because fundamentally similar processes are demonstrated in some free-living species.[88] An example is *Rhabditis dubia,* which lives in cow dung. It may go through an indefinite number of generations, developing normally, but when unsuitable conditions occur, special third-stage juveniles called **dauer juveniles** are produced. The dauer juveniles develop no further but await access to psychodid flies, to which they attach. When the fly obligingly transports them to a new pile of cow dung, they detach and proceed with development. Another species is *R. coarctata,* which uses dung beetles for transport and in which dauer juveniles are produced every generation. Several other examples could be cited, and a particularly interesting one was reported by Hominick and Aston.[47] The dauer juveniles of *Pelodera strongyloides* attach not to invertebrates but to mice, where they enter the hair follicles of the abdominal skin and molt to fourth-stage juveniles. They will develop no further at the body temperature of the mouse, and the mouse may accumulate hundreds, or even thousands, of nematode juveniles during its life. When the mouse dies and its body cools, the nematodes rapidly emerge and, in the presence of a food source, molt to the adult stage. The mouse seems little inconvenienced by its passengers.

Typically, dauer juveniles do not feed but have stored reserves in their intestinal cells and are more resistant to desiccation than are other stages. One reason for their resistance seems to be an incomplete second ecdysis. When the third-stage cuticle is secreted, the old second-stage cuticle is retained in place as a sheath. A wide variety of parasitic nematodes produce infective third-stage juveniles that are quite comparable to dauer juveniles. They develop no further until a new host is available, remaining ensheathed in the second-stage cuticle. They live on stored food reserves and usually exhibit behavior patterns that enhance the likelihood of reaching a new host. For example, third-stage juveniles of *Haemonchus* and *Trichostrongylus* migrate out of the fecal mass and onto vegetation that is eaten by the host. Third-stage juveniles of species that penetrate the host skin, such as hookworms and *Nippostrongylus,* migrate onto small objects (sand grains, leaves, and others) and move their anterior ends freely back and forth, in the same manner as do some dauer juveniles (Fig. 23-32). In both dauer juveniles and infective juveniles, a more or less specific stimulus is required for development and completion of the ecdysis of the second-stage cuticle. Those that penetrate skin usually exsheath in the processes of penetration, but the stimulus for exsheathment of swallowed juveniles (*Haemonchus, Trichostrongylus,* and others) is very similar to that required for hatching of *Ascaris* eggs, including carbon dioxide, temperature, redox potential, and pH. In fact Rogers and Sommerville[88] considered infective eggs fundamentally the same as infective juveniles and dauer juveniles. For most nematodes tested, carbon dioxide seems to be the most important stimulus for hatching or exsheathing.[81,82]

Nematodes with intermediate hosts normally undergo hypobiosis at the third stage and remain dormant until they reach the definitive host. Some species are astonishingly plastic in their capacities to sustain more than one developmental arrest in their ontogenies. For example, if some species of hookworms and ascarids infect an unsuitable host, they enter another developmental arrest and lie dormant in the host tissues until they receive another stimulus to migrate.[72] In several of these, the older animal is an unsuitable host, and the worms lie dormant until they are stimulated by the hormones of host pregnancy. They then migrate to the uterus or mammary glands and infect the infant by way of the placenta in utero or the milk after birth.[23] Some species, for example, *Strongyloides ratti,* may not undergo a second developmental arrest at this stage, but if the lactating female is infected, the juveniles are somehow diverted from completing their migration to the adult's intestine and mi-

FIG. 23-32

Third-stage infective juveniles of *Nippostrongylus brasiliensis,* illustrating the typical behavior of crawling up on pebbles, blades of grass, or the like and waving their anterior ends to and fro. In this photograph of living worms, they have mounted granules of charcoal and even each other.
Photograph by L.S. Roberts.

grate instead to the mammary glands and infect the suckling young.[125] In some species adverse environmental conditions, such as chilling or the onset of the dry season, can predispose infective juveniles to undergo hypobiosis when they reach a definitive host.[1,103] Thus maturation and production of eggs at a time when progeny cannot survive is avoided.[4] More examples of adaptational arrests in development will be found in the nematode life cycles described in the chapters to follow.

METABOLISM
Energy metabolism

Probably more is known about the nematodes than about any other group of parasitic helminths.* *Ascaris* was one of the first organisms in which cytochrome was demonstrated.[54] Nevertheless, numerous questions await resolution.

Parasitic nematodes are a very diverse group, occupying a number of different habitats, and, not surprisingly, their energy metabolisms, although basically similar, exhibit a wide variety of minor variations. As

*References 6, 18, 20-22, 37, 41, 64, 86, 97, 108, 115, 119.

TABLE 23-1

Examples of end products of energy metabolism excreted by some nematodes

Substance excreted	Ascaris	Trichinella juveniles	Heterakis	Litomosoides	Dracunculus	Dirofilaria	Caenorhabditis†	Ancylostoma	Trichuris
Lactate	T*	T	T	+	+	+	T	−	+
Propionate	+	+	+	−	−	−	T	+	+
Acetate	+	+	+	+	−	−	T	+	−
Pyruvate	−	−	+	−	−	−	T	−	−
Succinate	−	−	+	−	−	−	T	−	−
α- Methylbutyrate	+	−	−	−	−	−	T	+	−
n-Valeric acid	+	+	−	−	−	−	T	−	+
Isocaproic acid	+	−	−	−	−	−	T	−	−
n-Caproic acid	+	+	−	−	−	−	T	−	−
Acetylmethyl carbinol	+	−	−	+	−	−	T	−	−
Isobutyric acid	−	−	−	−	−	−	T	+	−
n-Butyric acid	+	+	−	−	−	−	T	−	T
Tiglic acid	+	−	−	−	−	−	T	−	−
C_6 acids (unidentified)	−	+	−	−	−	−	T	T	−

Modified from Lee, D.L. 1965. The physiology of nematodes. Oliver & Boyd Ltd., Edinburgh.
*T, Trace; +, present; −, absent or not investigated.
†Free-living nematode.

indicated in Table 23-1, a number of different compounds appear to be end products of anerobic carbohydrate metabolism. However, these compounds apparently are all derived from the products of a similar pathway (Fig. 23-33). The scheme is best documented for the pig roundworm, *Ascaris suum*. Glucose is converted to phosphoenolpyruvate (PEP) through classical Embden-Meyerhof glycolysis. Pyruvate kinase activity (see Fig. 21-33) is low, and the PEP is converted to oxaloacetate by the carbon dioxide-fixing enzyme, PEP carboxykinase, rather than to pyruvate. Since cytoplasmic malate dehydrogenase activity is high, the oxaloacetate is rapidly converted to malate. In addition, this reaction oxidizes the NADH, formed previously in glycolysis, to NAD and maintains the oxidation-reduction balance of the system, a role normally played by lactate dehydrogenase.

Cytoplasmic malate enters the mitochondria and is used by a dismutation reaction, that is, the reduction of one metabolite by another, which requires that NAD shuttle between the surfaces of the two substrate-specific dehydrogenases. Half of the malate is oxidized to pyruvate and carbon dioxide by the action of malic enzyme and generates intramitochondrial reducing power in the form of NADH. The NADH shuttles to fumarate reductase, which reduces the remaining malate, via fumarate, to succinate. This last reaction results in a site I, electron transport-associated phosphorylation of ADP to ATP. The importance of

the ATP generated and the maintenance of redox balance in these reactions is indicated by the observation that some antinematodal agents (tetramisole and thiabendazole) are effective because they block the fumarate reductase.

A wide array of other end products are excreted besides succinate, of which the most abundant quantitatively in *Ascaris* spp. are α-methylbutyrate and α-methylvalerate. These branched-chain acids are formed by condensations of propionate units or a propionate and acetate (Fig. 23-34), the propionate and acetate arising from decarboxylations of succinate and pyruvate, respectively.[100,101] The exact significance of such reactions and their products is incompletely known, but succinate decarboxylation to propionate generates an ATP.[99] Other reactions in the sequence may generate additional energy or help regulate redox potential or both.[56]

Other electron transport reactions are present in *Ascaris* spp., but their importance and sequence are still not known with certainty. Mitochondria contain *a*-, *b*-, and *c*-type cytochromes and a very low concentration of functional cytochrome a_3,[28] but a classical electron transport system is unlikely to be of physiological importance.[55] The worms have an alternative oxidase pathway that is insensitive to cyanide. Succinate and malate are oxidized in mitochondrial preparations with hydrogen peroxide as an end product. The toxicity of hydrogen peroxide and the fact that it is

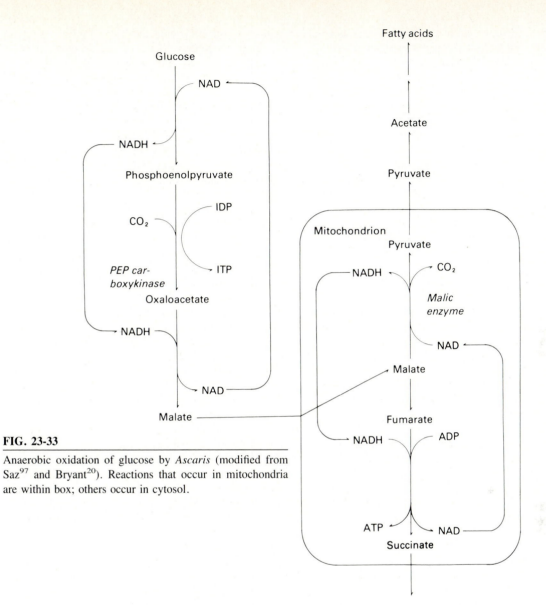

FIG. 23-33

Anaerobic oxidation of glucose by *Ascaris* (modified from Saz[97] and Bryant[20]). Reactions that occur in mitochondria are within box; others occur in cytosol.

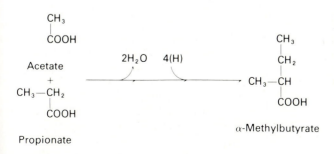

FIG. 23-34

Mode of α-methylbutyrate formation in *Ascaris* muscle; α-methylvalerate is formed in a similar manner except that the α-carbon of one propionate unit condenses with the carboxyl carbon of another propionate unit.[100,101]

produced in terminal electron transport by this worm may account for the observation that oxygen at atmospheric concentrations drastically reduces survival time of *Ascaris* in vitro.[43] Although oxygen is unnecessary in the energy metabolism of these "anaerobic" nematodes and may be toxic in moderate concentrations, small amounts may be necessary in certain biosynthetic reactions.

In addition to the foregoing, other nematodes have been studied that survive and metabolize carbohydrates in the absence of oxygen for extended periods; examples are *Heterakis gallinarum* and *Trichuris vulpis*. These organisms serve as particularly good examples of how some parasites have solved the metabolic problem of reoxidizing NADH (Chapter 4) in the absence of oxygen as a terminal electron acceptor. Some adults *(Haemonchus contortus)* apparently have a tricarboxylic acid cycle that operates when oxygen is present and an *Ascaris* type of system operative in the absence of oxygen.[120,121] *Rhabdias bufonis,* a parasite in the lungs of frogs, also apparently has alternative systems,[3] in spite of the unlikelihood of its being subjected to anaerobiosis.

However, some other adult nematodes seem to be obligate aerobes with respect to their energy metabolism, requiring the presence of at least low concentrations of oxygen for survival and motility. Nonetheless, even among these, glucose is not oxidized completely to carbon dioxide and water, and substantial quantities of various reduced end products are excreted. Some species apparently have a classical cytochrome system, or, alternatively, the electrons may be transported by a flavoprotein and terminal flavin oxidase to oxygen, producing hydrogen peroxide.

Oxygen requirements of various nematodes have been shown by examining several different parameters. *Nippostrongylid brasiliensis,* a trichostrongyle parasite in the intestine of rats, and *Litomosoides carinii* and *Brugia pahangi,* both filarial worms found in the body cavities of rodents, exhibit a **Pasteur effect.** That is, they consume more glucose in the absence of oxygen than in its presence. Thus they probably derive at least some energy from reaction sequences with oxygen as the terminal electron acceptor. *N. brasiliensis* and *L. carinii* can survive short periods of anaerobiosis but are killed by longer periods (a few hours).[85,98] *Nippostrongylus* has a fluid-filled layer in its cuticle (see Fig. 23-5) that contains hemoglobin, and the hemoglobin loads and unloads oxygen in the living animal.[105] Thus the worm can exploit areas in the intestine that are quite hypoxic. *B. pahangi* produces three times more lactate in the absence of oxygen than in its presence (when glucose is present).[6]

The character of the terminal electron transport re-actions is still unclear in these nematodes; *L. carinii* has both cytochrome *c* and cytochrome oxidase, whereas *N. brasiliensis* has cytochrome oxidase but apparently no cytochrome *c*. Nevertheless, oxygen consumption of *L. carinii* is inhibited by cyanide and by drugs called **cyanine dyes.** The cyanine dyes have no inhibitory effect on mammalian cytochrome systems, and they have chemotherapeutic activity against *L. carinii* in vivo. Unfortunately, these drugs are ineffective against the filarial parasites of humans, probably because these species are facultatively anaerobic. Other filarial worms *(Brugia pahangi, Dipetalonema viteae)* are facultative anaerobes, and the end product of their energy metabolism is lactate; lactate is the most important glycolytic end product in the dog heartworm, *Dirofilaria immitis.*[49,73] The tricarboxylic acid cycle is of little or no importance in the energy metabolism of any of these filarial worms. Interestingly, the lactate production in *D. immitis* glycolysis seems not to be derived from pyruvate via the pyruvate kinase route but rather from malate by way of a PEP carboxylase route, with carbon dioxide fixation[70] (see Fig. 21-33). The lactate dehydrogenase activity is high in this worm, and the reaction of pyruvate to lactate appears essential to reoxidize the NADH generated earlier in glycolysis. Suramin, a drug used against another filarial worm, *Onchocerca volvulus,* is a potent inhibitor of *D. immitis* lactate dehydrogenase.

Some evidence suggests that the extent of dependence on aerobic pathways in nematodes is correlated with body diameter, that is, the larger nematodes are more anaerobic, whereas the smaller ones can get to the oxygen near the mucosa and are more aerobic.[41] Fry and Jenkins view the parasitic nematodes as "metabolic opportunists, combining the versatility of an anaerobic and aerobic energy metabolism."[41]

All of this discussion has been in reference to adult nematodes. Different stages in the life cycles may show dramatic biochemical adaptations in energy metabolism, such as the aerobic embryonating eggs and the anaerobic adult of *Ascaris.* *Strongyloides* is another interesting example of biochemical epigenetic adaptation.[57] *Strongyloides* has a complex life cycle with free-living adults (males and females) and parasitic adults (parthenogenetic females only). The first three juvenile stages of both types are free living, but those destined to become parasitic undergo developmental arrest at the third stage, until penetration of the host (Chapter 25). All free-living stages are subjected to a selective pressure common to other free-living animals—to use as completely as possible the energetic value in their nutrient molecules—and they have a complete tricarboxylic acid cycle and probably a cytochrome system. In contrast, the parasitic females have neither a complete tricarboxylic acid cycle nor cytochrome system, a situation similar to many other intestinal helminths. Juveniles of several other parasitic species have apparently functional tricarboxylic acid cycles,[14,118] although in some species the significance of the cycle may lie in regulation of four-carbon intermediates rather than energy production. Oddly, it has been reported that the first and second stage juveniles of *Ancylostoma tubaeforme* and *Haemonchus contortus* are anaerobic, and the infective third stage is aerobic.[75]

The normal pathway of fatty acid oxidation is referred to as β-oxidation, so called because the β-carbon of the fatty acetyl-CoA is oxidized, and the two-carbon fragment, acetyl-CoA, is cleaved off to enter the tricarboxylic acid cycle. It would be expected, therefore, that the presence of β-oxidation enzymes would be correlated with a functional tricarboxylic acid cycle (though not necessarily so), and in the few cases investigated this is the case. β-oxidation of fatty acids has been found in embryonating *Ascaris* eggs and in free-living *Strongyloides* juveniles and adults.[57,117] It has been observed that tissue lipids gradually disappear (are consumed?) by infective eggs or juveniles of several species. Interestingly, β-oxidation could be demonstrated in neither *Ascaris* muscle nor parasitic females of *Strongyloides*.

Synthetic metabolism

Synthetic metabolism of nematodes has not been as intensively studied as has energy metabolism, probably because, in contrast to prokaryotes, energy pathways of helminths usually offer the better sites for chemotherapy. However, there are several points of interest. In light of the enormous number of progeny produced by an organism such as *Ascaris,* protein and nucleic acid synthetic ability must be correspondingly great. In this connection the RNA metabolism of fertilized *Ascaris* eggs deserves further comment (see earlier discussion of embryogenesis). The young oocytes have nucleoli and large amounts of cytoplasmic RNA, and these presumably are responsible for the very large amount of yolk protein synthesized in the developing oocyte. By the time the oocyte matures, the nucleoli and most of the cytoplasmic RNA have disappeared.[53] At the same time the sperm contains little or no RNA. Immediately after fertilization, there is a massive ribosomal RNA synthesis in the male pronucleus, along with a smaller amount of informational RNA, while the female pronucleus is going through its maturation divisions. Kaulenas and Fairbairn suggested that the female genome, therefore, is responsible for the high rate of oocyte production

and yolk synthesis, whereas ribosomes provided by the male genome largely support shell formation and cleavage. Although the sperm brings with it no ribosomes, it carries a protein of uncertain function called ascaridine.[35] Ascaridine contains no phosphorus, sulfur, or purines, and two amino acids (aspartic acid and tryptophane) account for 35% and 15% of the total nitrogen, respectively. The protein is contained in refringent granules that coalesce during sperm formation to become the **refringent body.** It is clear that the protein in the refringent body is directly related to the ribosome formation just after fertilization, and the ascaridine may well be precursor material for the ribosomes.[39]

As mentioned, the developing oocytes in the ovary are sites of much protein synthesis. Presumably, much of the amino acid supply is furnished by the nearby intestinal absorption, but evidence exists that some amino acids are synthesized in the ovaries as well. The ovaries contain active transaminases, which form amino acids from the corresponding α-keto acids derived from carbohydrate metabolism. In addition, the ovaries can condense pyruvate with ammonia to form the amino acid alanine. In this connection it is interesting that some free-living nematodes can synthesize a wide variety of amino acids from a simple substrate, such as acetate. When incubated in a medium containing glycine, glucose, and acetate, *Caenorhabditis briggsae* synthesizes an array of "nonessential" and "essential" amino acids. This was the first metazoan known that could synthesize "essential" amino acids, and it was later found that *C. briggsae* could synthesize glycine by a transamination of glyoxylate, the glyoxylate having been produced by the action of isocitrate lyase[94] (Fig. 23-35).

At least some nematodes can synthesize polyunsaturated fatty acids de novo but apparently are unable to synthesize sterols de novo.[93] *Ascaris* incorporates acetate into long-chain fatty acids, probably by the malonyl-CoA pathway as found in vertebrates.[10] The nonsugar parts of the ascarosides (the alcohols) in *Ascaris* spp. are synthesized from long-chain fatty acids. This involves a condensation in which the carboxyl carbon of one fatty acid condenses with carbon number 2 of another, with the elimination or a molecule of carbon dioxide.[36] The ascarylose is freely synthesized by *Ascaris* ovaries from glucose or glucose-1-phosphate, and the end product of the synthesis is probably ascarylose-dinucleotidephosphate, which then condenses with the nonsugar moiety to give the ascaroside. *Dirofilaria immitis* can synthesize all classes of complex lipids, including cholesterol.[113]

As noted in the discussion of the body wall, much collagen is found in the cuticle of *Ascaris*. These sta-

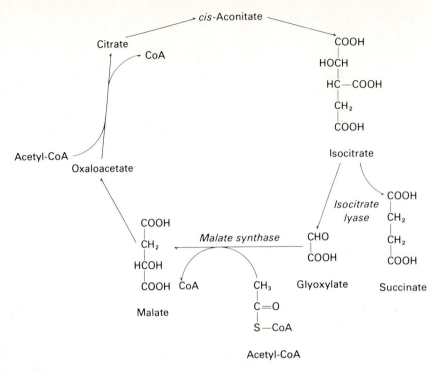

FIG. 23-35

Glyoxylate cycle: a pathway found in plants, microorganisms, and nematodes that can convert fatty acids or acetate to carbohydrates. Each turn of the cycle requires 2 molecules of acetyl-CoA and yields 1 molecule of succinate. Enzymes required other than isocitrate lyase and malate synthase are normally found in the tricarboxylic acid cycle.

bilized proteins are important factors in the resistance and strength of the nematode's cuticle. Collagens are stabilized by bonds between lysine residues in the subunits, and they are unusual in that they contain around 12% proline and 9% hydroxyproline; hydroxyproline is an amino acid rarely found in other proteins. In addition to cuticle, collagens exist in muscle, intestine, and reproductive organs of *Ascaris* and these collagens differ in their hydroxyproline content. Collagen synthesis in *Ascaris* has been studied by several authors.[24] In all systems the normal collagen precursor is a polypeptide called protocollagen, and the proline in protocollagen is hydroxylated to hydroxyproline by the enzyme **protocollagen proline hydroxylase (PPH** or **proline monoxygenase)**. Among other cosubstrates, this enzyme requires molecular oxygen to carry out the hydroxylation of proline. This is an example of a biosynthetic reaction that requires oxygen in an organism that is anaerobic with respect to its energy metabolism. PPH activity appears in the embryonating *Ascaris* egg, rises to a peak at the time of the larval molt, decreases, and then rises to another peak coinciding with infectivity. In comparing PPH from adult muscle with that from embryonating eggs, Cain and Fairbairn[24] found that oxygen concentration greater than 5% inhibited the enzyme from muscle but not the PPH from the embryos.

CLASSIFICATION OF PHYLUM NEMATODA
Parasites of vertebrates

Class Adenophorea (=Aphasmidea)
Amphids generally well developed (except in parasitic forms), well behind the lips, often with complex pores; caudal and hypodermal glands common; phasmids absent; excretory system lacking lateral canals, formed of a single, ventral, glandular cell or entirely absent; deirids always absent; mostly free living, some parasitic on plants or in invertebrates or vertebrates.

Order Trichurata
Anterior end more slender than posterior end; lips and buccal capsule absent or much reduced; esophagus a very slender, capillary-like tube, embedded within one or more rows of large, glandular cells (stichocytes) along its posterior portion; bacillary band present; both sexes with a single gonad; males with one spicule or none; eggs with polar plugs (opercula); parasites of nearly all organs of all classes of vertebrates.

FAMILIES

Anatrichosomatidae, Capillariidae, Cystoopsidae, Trichinellidae, Trichosomoididae, Trichuridae

Order Dioctophymata

Stout worms, often very large. Anterior cuticle spinose in some species; esophageal glands highly developed, multinucleate; lips and buccal capsule reduced, replaced by muscular oral sucker in Soboliphymatidae; esophagus cylindrical; nerve ring far anterior; anus at posterior end in both sexes; male with bell-shaped muscular copulatory bursa without rays; both sexes with a single gonad; male with single spicule; eggs deeply sculptured or pitted; parasites of birds and mammals.

FAMILIES

Dioctophymatidae, Eustrongylidae, Soboliphymatidae

Class Secernentia (=Phasmidea)

Amphids generally poorly developed, with small, simple pores near or on the lips; caudal and hypodermal glands absent; phasmids present; excretory system with one or two lateral canals, with or without associated glandular cells; deirids commonly present; free living or parasitic in plants or invertebrates or vertebrates.

Order Rhabditata

Tiny to small worms, commonly with six small lips; esophagus muscular, divided into anterior corpus, median isthmus, and posterior bulb; pseudobulb often present between corpus and isthmus; bulb usually absent in parasitic stages; buccal capsule small or absent; tail conical in both sexes, spicules equal, gubernaculum usually present; parasitic generations parthenogenetic (males usually unknown) or hermaphroditic, alternating with gonochoristic, free-living generations; parasites of lungs of amphibians and reptiles or of intestines of amphibians, reptiles, birds, and mammals.

FAMILIES

Rhabdiasidae, Strongyloididae

Order Strongylata

Commonly long, slender worms; esophagus usually swollen posteriorly but lacking definite bulb; male with well-developed copulatory bursa supported by sensory rays; usually oviparous; eggs thin shelled, rarely developed beyond morula when laid; parasites of all classes of vertebrates (rare in fishes).

FAMILIES

Amidostomatidae, Ancylostomatidae, Angiostrongylidae, Cloacinidae, Cyathostomidae, Deletrocephalidae, Diaphanocephalidae, Dictyocaulidae, Filaroididae, Heligmosomatidae, Ichthyostrongylidae, Metastrongylidae, Oesophagostomatidae, Ollulanidae, Pharyngostrongylidae, Protostrongylidae, Pseudaliidae, Stephanuridae, Strongylacanthidae, Strongylidae, Syngamidae, Trichostrongylidae

Order Ascaridata

Commonly large, stout worms; usually three lips present, less often two or none; esophagus simple, muscular, lacking specializations, occasionally with posterior ventriculus, with or without appendix; intestine occasionally with one or more appendices at esophago-intestinal junction; preanal sucker present in a few species; eggs usually with sculptured uterine layer, unembryonated when laid, juveniles infective to host in second (perhaps third) stage; parasites of all classes of vertebrates.

FAMILIES

Acanthocheilidae, Angusticaecidae, Anisakidae, Ascaridae, Ascarididae, Crossophoridae, Goeziidae, Heterocheilidae, Inglisonematidae, Oxyascarididae, Toxocaridae

Order Oxyurata

Medium-sized to small worms, commonly with sharply pointed tails; lips, when present, usually three in number; often reduced or absent; esophagus with posterior bulb; caudal alae often well developed; preanal sucker present in some species; eggs thin shelled, usually fully embryonated when laid, and life cycle usually direct; parasites of arthropods and all classes of vertebrates.

Superfamily Oxyuroidea

Tails of both sexes usually sharply pointed; spicules small: one or both absent in several species; eggs often with operculum; no intermediate host in life cycle.

FAMILIES

Heteroxynematidae, Oxyuridae, Ozolaimidae, Pharyngodonidae, Syphaciidae

Superfamily Atractoidea

Head usually with complex ornamentation; intestinal cecum present in *Cruzia;* one ovary present (except in *Probstmayria);* viviparous.

FAMILIES

Atractidae, Crossocephalidae, Cruziidae, Hoplodontophoridae, Labiduridae, Schrankianidae, Travnematidae

Superfamily Cosmocercoidea

Three or six lips present; males with two equal spicules, caudal papillae numerous.

FAMILIES

Cosmocercidae, Gyrinicolidae, Lauroiidae

Superfamily Heterakoidea

Three well-defined lips; preanal sucker present on males; two spicules; eggs not embryonated when laid.

FAMILIES

Aspidoderidae, Heterakidae, Spinicaudidae, Strongyluridae

Superfamily Kathlanioidea

Mouth complex, with three or six lips; buccal capsule well developed, sometimes with teeth; preanal sucker or powerful preanal musculature present.

FAMILY

Kathlaniidae

Superfamily Subuluroidea

Lips absent or quite reduced; anterior end of esophagus commonly with teeth; preanal sucker present.

FAMILIES

Maupasinidae, Parasubuluridae, Subuluridae

Order Spirurata

Mouth surrounded by six small lips, or by a cuticular ring, or with two lateral pseudolabia; buccal capsule present; cephalic ornamentation common; esophagus usually divided into an anterior, muscular portion and a posterior, glandular portion, never with posterior bulb; spicules usually unequal in size and shape.

FAMILIES

Acuariidae, Ascaropsidae, Cobboldinidae, Crassicaudidae, Desmidocercidae, Gnathostomatidae, Gongylonematidae, Habronematidae, Haplonematidae, Hedruridae, Physalopteridae, Pneumospiruridae, Rhabdochonidae, Rictulariidae, Salobrellidae, Schistorophidae, Seuratidae, Spinitectidae, Spirocercidae, Spiruridae, Streptocaridae, Tetrameridae, Thelaziidae

Order Camallanata

Lips absent; buccal capsule present or absent or replaced with large bilateral, sclerotized valves; esophagus long, distinctly divided into anterior muscular and posterior glandular portions; spicules unequal and dissimilar or equal and similar. Ovoviviparous; anus and vulva may be atrophied in females; parasites of tissue, coelom, air bladder, circulatory system or digestive system of aquatic and terrestrial vertebrates, including humans.

FAMILIES

Anguillicolidae, Camallanidae, Dracunculidae, Oceanicucullanidae, Philometridae, Phlyctainophoridae, Skrjabillanidae, Tetanonematidae

Order Filariata

Mouth simple, lacking lips; buccal capsule absent in most species; esophagus usually divided into anterior muscular and posterior glandular portions; spicules usually unequal and dissimilar; oviparous or ovoviviparous. Parasites of tissues or respiratory system of terrestrial vertebrates.

FAMILIES

Aproctidae, Desmidocercidae, Diplotriaenidae, Filariidae, Onchocercidae, Setariidae

REFERENCES

1. Altaif, K.I., and W.H. Issa. 1983. Seasonal fluctuations and hypobiosis of gastro-intestinal nematodes of Awassi lambs in Iraq. Parasitology 86:301-310.
2. Anderson, R.V., and H.M. Darling. 1964. Embryology and reproduction of *Ditylenchus destructor* Thorne, with emphasis on gonad development. Proc. Helm. Soc. Wash. 31:240- 256.
3. Anya, A.O., and G.M. Umezurike. 1978. Respiration and carbohydrate energy metabolism of the lung-dwelling parasite *Rhabdias bufonis* (Nematoda: Rhabdiasoidea). Parasitology 76:21-27.
4. Armour, J., and M. Duncan. 1987. Arrested larval development in cattle nematodes. Parasitol. Today 3:171-176.
5. Barrett, J. 1983. Biochemistry of filarial worms. Helminthol. Abstracts A 42:1-18.
6. Barrett, J., A.H.W. Mendis, and P.E. Butterworth. 1986. Carbohydrate metabolism in *Brugia pahangi* (Nematoda: Filaroidea). Int. J. Parasit. 16:465-469.
7. Barrett, J., C.W. Ward, and D. Fairbairn. 1970. The glyoxylate cycle and the conversion of triglycerides to carbohydrates in developing eggs of *Ascaris lumbricoides*. Comp. Biochem. Physiol. 35:577-586.
8. Batson, B.S. 1979. Body wall of juvenile and adult *Gastromermis boophthorae* (Nematoda: Mermithidae): ultrastructure and nutritional role. Int. J. Parasitol. 9:495-503.
9. Batson, B.S. 1979. Ultrastructure of the trophosome, a food-storage organ in *Gastromermis boophthorae* (Nematoda: Mermithidae). Int. J. Parasitol. 9:505-514.
10. Beames, C.G., Jr., B.G. Harris, and F.A. Hopper, Jr. 1967. The synthesis of fatty acids from acetate by intact tissue and muscle extract of *Ascaris lumbricoides suum*. Comp. Biochem. Physiol. 20:509-521.
11. van Beneden, E. 1883. Recherches sur la maturation de l'oeuf et la fécondation *(Ascaris megalocephala)*. Arch. Biol. 4:265-641.
12. Bird, A.F. 1971. The structure of nematodes. Academic Press, Inc., New York.
13. Bird, A.F., and J. Bird, 1969. Skeletal structures and integument of Acanthocephala and Nematoda. In Florkin, M., and B.T. Scheer, editors. Chemical zoology, vol. 3. Academic Press, Inc., New York, pp. 253-288.
14. Boczoń, K., and J.W. Michejda. 1978. Electron transport in mitochondria of *Trichinella spiralis* larvae. Int. J. Parasitol. 8:507-513.
15. Bolla, R.I., P.P. Weinstein, and G.D. Cain. 1972. Fine structure of the coelomocyte of adult *Ascaris suum*. J. Parasitol. 58:1025-1036.
16. Bone, L.W., L.K. Gaston, and S.K. Reed. 1980. Production and activity of the Kav 0.64 pheromone fraction of *Nippostrongylus brasiliensis*. J. Parasitol. 66:268-273.
17. Boveri, T. 1899. Die Entwicklung von *Ascaris megalocephala* mit besonderer Rücksicht auf die Kernverhältnisse. Festschrift für C. Von Kupffer. Jena, Germany.
18. von Brand, T. 1979. Biochemistry and physiology of endoparasites. Elsevier/North Holland Biomedical Press, Amsterdam.
19. Brothwell, D. 1987. The bog man and the archeology of people. Harvard University Press, Cambridge, Massachusetts.
20. Bryant, C. 1975. Carbon dioxide utilization and the regulation of respiratory metabolic pathways in parasitic helminths. In Dawes, B., editor. Advances in parasitology, vol. 13. Academic Press, Inc., New York, pp. 36-69.
21. Bryant, C. 1978. The regulation of respiratory metabolism in parasitic helminths. In Lumdsen, W.H.R., R. Muller, and J.R. Baker, editors. Advances in parasitology, vol. 16. Academic Press, Inc., New York.
22. Bryant, C. 1982. Biochemistry. In Cox, F.E.G., editor. Modern parasitology. Blackwell Scientific Publications Ltd., Oxford, Eng., pp. 84-115.
23. Burke, T.M., and E.L. Roberson. 1985. Prenatal and lactational transmission of *Toxocara canis* and *Ancylostoma caninum:* experimental infection of the bitch before pregnancy. Int. J. Parasitol. 15:71-75.
24. Cain, G.D., and D. Fairbairn. 1971. Protocollagen proline hydroxylase and collagen synthesis in developing eggs of *Ascaris lumbricoides*. Comp. Biochem. Physiol. 40B:165-179.
25. Campbell, W.C. 1985. Ivermectin: an update. Parasitol. Today 1:10-16.
26. del Castillo, J. 1969. Pharmacology of Nematoda. In Florkin, M., and B.T. Scheer, editors. Chemical zoology, vol. 3. Academic Press, Inc., New York, pp. 521-554.
27. Chappell, L.H. 1979. Physiology of parasites. John Wiley & Sons, Inc., New York.
28. Cheah, K.S., and B. Chance. 1970. The oxidase systems of *Ascaris*-muscle mitochondria. Biochem. Biophys. Acta 223:55-60.
29. Crofton, H.D. 1966. Nematodes. Hutchinson University Library, London.

30. Croll, N.A., editor. 1976. The organization of nematodes. Academic Press, New York.

31. Croll, N.A., and B.E. Matthews. 1977. Biology of nematodes. John Wiley & Sons, Inc., New York.

32. Davey, K.G. 1979. Molting in a parasitic nematode, *Phocanema decipiens;* the role of water uptake. Int. J. Parasitol. 9:121-125.

33. Davey, K.G., and S.P. Kan. 1968. Molting in a parasitic nematode, *Phocanema decipiens.* IV. Ecdysis and its control. Can. J. Zool. 46:893-898.

34. Davis, A.H., and C.E. Carter. 1980. Chromosome diminution in *Ascaris suum.* Exp. Cell Res. 128:59-62.

35. Fairbairn, D. 1960. The physiology and biochemistry of nematodes. In Sasser, J.N., and W.R. Jenkins, editors. Nematology. University of North Carolina Press, Chapel Hill, pp. 267-296.

36. Fairbairn, D. 1969. Lipid components and metabolism of Acanthocephala and Nematoda. In Florkin, M., and B.T. Scheer, editors. Chemical zoology, vol. 3. Academic Press, Inc., New York, pp. 361-378.

37. Fairbairn, D. 1970. Biochemical adaptation and loss of genetic capacity in helminth parasites. Biol. Rev. 45:29-72.

38. Foor, W.E. 1967. Ultrastructural aspects of oocyte development and shell formation in *Ascaris lumbricoides.* J. Parasitol. 53:1245-1261.

39. Foor, W.E. 1970. Spermatozoan morphology and zygote formation in nematodes. Biol. Reprod. 2 (suppl.):177-202.

40. Foster, W.D. 1965. A history of parasitology. E. & S. Livingstone, Edinburgh.

41. Fry, M., and D.C. Jenkins. 1984. *Nematoda:* aerobic respiratory pathways of adult parasitic species. Exp. Parasitol. 57:86-92.

42. Fujimoto, D. 1975. Occurrence of dityrosine in cuticlin, a structural protein from *Ascaris* cuticle. Comp. Biochem. Physiol. 51B:205-207.

43. Harpur, R.P. 1962. Maintenance of *Ascaris lumbricoides* in vitro: a biochemical and statistical approach. Can. J. Zool. 40:991-1011.

44. Harpur, R.P. 1964. Maintenance of *Ascaris lumbricoides* in vitro. III. Changes in the hydrostatic skeleton. Comp. Biochem. Physiol. 13:71-85.

45. Harpur, R.P., and J.S. Popkin, 1965. Osmolality of blood and intestinal contents in the pig, guinea pig, and *Ascaris lumbricoides.* Can. J. Biochem. 43:1157-1169.

46. Harris, J.E., and H.D. Crofton. 1957. Structure and function in the nematodes: internal pressure and cuticular structure in *Ascaris.* J. Exp. Biol. 34:116-130.

47. Hominick, W.M., and A.J. Aston. 1981. Association between *Pelodera strongyloides* (Nematoda: Rhabditidae) and wood mice, *Apodemus sylvaticus.* Parasitology 83:67-75.

48. Hurley, L.C., and R.I. Sommerville. 1982. Reversible inhibition of hatching of infective eggs of *Ascaris suum* (Nematoda). Int. J. Parasitol. 12:463-465.

49. Hutchison, W.F., and A.C. Turner. 1979. Glycolytic end products of the adult dog heartworm, *Dirofilaria immitis.* Comp. Biochem. Physiol. 62B:71-73.

50. Hyman, L.H. 1951. The invertebrates: Acanthocephala, Aschelminthes, and Entoprocta, the pseudocoelomate Bilateria, vol. 3. McGraw-Hill Book Co., New York.

51. Inglis, W.G. 1964. The structure of the nematode cuticle. Proc. Zool. Soc. Lond. 143:465-502.

52. Jezyk, P.F., and D. Fairbairn. 1967. Metabolism of ascarosides in the ovaries of *Ascaris lumbricoides* (Nematoda). Comp. Biochem. Physiol. 23:707-719.

53. Kaulenas, M.S., and D. Fairbairn. 1968. RNA metabolism of fertilized *Ascaris lumbricoides* eggs during uterine development. Exp. Cell Res. 52:233-251.

54. Keilin, D. 1925. On cytochrome, a respiratory pigment common to animals, yeasts and higher plants. Proc. R. Soc. Ser. B, 98:312-339.

55. Köhler, P., and R. Bachmann. 1980. Mechanisms of respiration and phosphorylation in *Ascaris* muscle mitochondria. Molec. Biochem. Parasitol. 1:75-90.

56. Komuniecki, R., P.R. Komuniecki, and H.J. Saz. 1981. Relationships between pyruvate decarboxylation and branched-chain volatile acid synthesis in *Ascaris* mitochondria. J. Parasitol. 67:601-608.

57. Körting, W., and D. Fairbairn. 1971. Changes in beta-oxidation and related enzymes during the life cycle of *Strongyloides ratti* (Nematoda). J. Parasitol. 57:1153-1158.

58. Lee, D.L. 1965. The cuticle of adult *Nippostrongylus brasiliensis.* Parasitology 55:173-181.

59. Lee, D.L. 1966. The structure and composition of the helminth cuticle. In Dawes, B. editor. Advances in parasitology, vol. 4. Academic Press, Inc., New York, pp. 187-254.

60. Lee, D.L. 1969. *Nippostrongylus brasiliensis:* some aspects of the fine structure and biology of the infective larva and the adult. In Taylor, A.E.R., editor. *Nippostrongylus* and *Toxoplasma.* Symposia of the British Society of Parasitology, vol. 7. Blackwell Scientific Publications Ltd., Oxford, Eng., pp. 3-16.

61. Lee, D.L. 1970. Moulting in nematodes: the formation of the adult cuticle during the final moult of *Nippostrongylus brasiliensis.* Tissue Cell 2:139-153.

62. Lee, D.L. 1972. The structure of the helminth cuticle. In Dawes, B., editor. Advances in parasitology, vol. 10. Academic Press, Inc., New York, pp. 347-379.

63. Lee, D.L., and A.O. Anya. 1967. The structure and development of the spermatozoon of *Aspiculuris tetraptera* (Nematoda). J. Cell Sci. 2:537-544.

64. Lee, D.L., and H.J. Atkinson. 1977. Physiology of nematodes, ed. 2. Columbia University Press, New York.

65. Lee, D.L., and P. Lestan. 1971. Oogenesis and eggshell formation in *Heterakis gallinarum* (Nematoda). J. Zool. 164:189-196.

66. Lee, H., I. Chen, and R. Lin. 1973. Ultrastructure of the excretory system of *Anisakis* larva (Nematoda: Anisakidae). J. Parasitol. 59:289-298.

67. MacKinnon, B.M. 1987. Sex attractants in nematodes. Parasitol. Today 3:156-158.

68. MacKinnon, B.M. 1987. An ultrastructural and histochemical study of oogenesis in the trichostrongylid nematode *Heligmosoides polygyrus.* J. Parasitol. 73:390-399.

69. Maung, M. 1978. The occurrence of the second moult of *Ascaris lumbricoides* and *Ascaris suum.* Int. J. Parasitol. 8:371-378.

70. McNeill, K.M., and W.F. Hutchison. 1972. Carbohydrate metabolism of *Dirofilaria immitis.* In Bradley, R.E., and G.

Pacheco, editors. Canine heartworm disease: the current knowledge. Department of Veterinary Science, University of Florida, Gainesville.

71. Meglitsch, P.A. 1972. Invertebrate zoology, ed. 2. Oxford University Press, New York.

72. Michel, J.F. 1974. Arrested development of nematodes and some related phenomena. In Dawes, B., editor. Advances in parasitology, vol. 12. Academic Press, Inc., New York, pp. 280-366.

73. Middleton, K.R., and H.J. Saz. 1979. Comparative utilization of pyruvate by *Brugia pahangi, Dipetalonema viteae,* and *Litomosoides carinii.* J. Parasitol. 65:1-7.

74. Nelson, F.K., and D.L. Riddle. 1984. Functional study of the *Caenorhabditis elegans* secretory-excretory system using laser microsurgery. J. Exp. Zool. 231:45-56.

75. Onwuliri, C.O.E. 1985. Energy metabolism in the developing larval stages of *Ancylostoma tubaeforme* and *Haemonchus contortus:* glycolytic and tricarboxylic acid cycle enzymes. Parasitology 90:169-177.

76. Panesar, T.S., and N.A. Croll. 1981. The hatching process in *Trichuris muris* (Nematoda: Trichuroidea). Can. J. Zool. 59:621-628.

77. Paramonov, A.A. 1954. On the structure and function of the phasmids. Trudy Gelmint. Lab Akad. Nauk SSSR 7:19-49. (Translation available from G.D. Schmidt.)

78. Passey, R.F., and D. Fairbairn. 1957. The conversion of fat to carbohydrate during embryonation of *Ascaris* eggs. Can. J. Biochem. Physiol. 35:511-525.

79. Perry, R.N., and A.J. Clarke. 1981. Hatching mechanisms of nematodes. Parasitology 83:435-449.

80. Petersen, J.J. 1982. Current status of nematodes for the biological control of insects. In Anderson, R.M. and E.U. Canning. Parasites as biological control agents. Symposia of the British Society for Parasitology, vol. 19. Parasitology 84(4):177-204.

81. Petronijevic, T., W.P. Rogers, and R.I. Sommerville. 1985. Carbonic acid as the host signal for the development of parasitic stages of nematodes. Int. J. Parasitol. 15:661-667.

82. Petronijevic, T., W.P. Rogers, and R.I. Sommerville. 1986. Organic and inorganic acids as the stimulus for exsheathment of infective juveniles of nematodes. Int. J. Parasitol. 16:163-168.

83. Platzer, E.G. 1980. Nematodes as biological control agents. California Agric. 34(3):27.

84. Poinar, G.O., Jr. 1983. The natural history of nematodes. Prentice-Hall, Inc., Englewood Cliffs, N.J.

85. Roberts, L.S., and D. Fairbairn. 1965. Metabolic studies on adult *Nippostrongylus brasiliensis* (Nematoda: Trichostrongyloidea). J. Parasitol. 51:129-138.

86. Rogers, W.P. 1969. Nitrogenous components and their metabolism: Acanthocephala and Nematoda. In Florkin, M., and B.T. Scheer, editors. Chemical zoology, vol. 3. Academic Press, Inc., New York, pp. 379-428.

87. Rogers, W.P. 1982. Enzymes in the exsheathing fluid of nematodes and their biological significance. Int. J. Parasitol. 12:495-502.

88. Rogers, W.P., and R.I. Sommerville. 1963. The infective stage of nematode parasites and its significance in parasitism. In Dawes, B., editor. Advances in parasitology, vol. 1. Academic Press, Inc., New York, pp. 109-177.

89. Rogers, W.P., and R.I. Sommerville. 1968. The infectious process and its relation to the development of early parasitic stages of nematodes. In Dawes, B., editor. Advances in parasitology, vol. 6. Academic Press, Inc., New York, pp. 327-348.

90. Rosenbluth, J. 1965. Ultrastructural organization of obliquely striated muscle fibers in *Ascaris lumbricoides.* J. Cell Biol. 25:495-515.

91. Rosenbluth, J. 1965. Ultrastructure of somatic cells in *Ascaris lumbricoides.* II. Intermuscular junctions, neuromuscular junctions, and glycogen stores. J. Cell Biol. 26:579-591.

92. Rosenbluth, J. 1967. Obliquely striated muscle. III. Concentration mechanism of *Ascaris* body muscle. J. Cell Biol. 34:15-33.

93. Rothstein, M. 1970. Nematode biochemistry. XI. Biosynthesis of fatty acids by *Caenorhabditis briggsae* and *Panagrellus redivivus.* Int. J. Biochem. 1:422-428.

94. Rothstein, M., and H. Mayoh. 1964. Glycine synthesis and isocitrate lyase in the nematode, *Caenorhabditis briggsae.* Biochem. Biophys. Res. Comm. 14:43-47.

95. Sangster, N.C., R.K. Prichard, and E. Lacey. 1985. Tubulin and benzimidazole-resistance in *Trichostrongylus colubriformis* (Nematoda). J. Parasitol. 71:645-651.

96. Savel, J. 1955. Études sur la constitution et le métabolism protéiques d'*Ascaris lumbricoides* Linné, 1758. Rev. Path. Comp. Hyg. Gen. Comp. 55:52-121.

97. Saz, H.J. 1972. Comparative biochemistry of carbohydrates in nematodes and cestodes. In Van den Bossche, H., editor. Comparative biochemistry of parasites. Academic Press, Inc., New York, pp. 33-47.

98. Saz, D.K., T.P. Bonner, M. Karlin, and H.J. Saz. 1971. Biochemical observations on adult *Nippostrongylus brasiliensis.* J Parasitol. 57:1159-1162.

99. Saz, H.J., and S.M. Pietrzak. 1980. Phosphorylation associated with succinate decarboxylation to propionate in *Ascaris* mitochondria. Arch. Biochem. Biophys. 202:388-395.

100. Saz, H.J., and A. Weil. 1960. The mechanism of the formation of α-methylbutyrate from carbohydrate by *Ascaris lumbricoides* muscle. J. Biol. Chem. 235:914-918.

101. Saz, H.J., and A. Weil. 1962. Pathway of formation of α-methylvalerate by *Ascaris lumbricoides.* J. Biol. Chem. 237:2053-2056.

102. Schad, G.A. 1977. The role of arrested development in the regulation of nematode populations. In Esch, G.W., editor. Regulation of parasite populations. Academic Press, Inc., New York, pp. 111-167.

103. Schad, G.A. 1983. Arrested development of *Ancylostoma caninum* in dogs: influence of photoperiod and temperature on induction of a potential to arrest. In Meerovitch, E., editor. Aspects of parasitology. Institute of Parasitology, McGill University, Montreal, pp. 361-391.

104. Sepsenwol, S., M. Nguyen, and T. Braun. 1986. Adenylate cyclase activity is absent in inactive and motile sperm in the nematode parasite, *Ascaris suum.* J. Parasitol. 72:962-964.

105. Sharpe, M.J., and D.L. Lee. 1981. Observations on the structure and function of the haemoglobin from the cuticle of *Nippostrongylus brasiliensis* (Nematoda). Parasitology 83:411-424.

106. Sheffield, H.G. 1963. Electron microscopy of the bacillary band and stichosome of *Trichuris muris* and *T. vulpis.* J. Parasitol. 49:998-1009.

107. Sheffield, H.G. 1964. Electron microscope studies on the intestinal epithelium of *Ascaris suum*. J. Parasitol. 50:365-379.

108. Slutzky, G.M., editor. 1981. The biochemistry of parasites. Pergamon Press, Oxford.

109. Smales, L.R. 1984. The egg-shell of *Labiostrongylus eugenii* (Nematoda, Strongyloidea): structure and function. Int. J. Parasitol. 14:231-239.

110. Sommerville, R.I., and P.P. Weinstein. 1964. Reproductive behavior of *Nematospiroides dubius* in vivo and in vitro. J. Parasitol. 50:401-409.

111. Swartz, F.J., M. Henry, and A. Floyd. 1967. Observations on nuclear differentiation in *Ascaris*. J. Exp. Zool. 164:297-307.

112. Thorson, R.E. 1956. The effect of extracts of the amphidial glands, excretory glands, and esophagus of adults of *Ancylostoma caninum* on the coagulation of dog's blood. J. Parasitol. 42:26-30.

113. Turner, A.C., and W.F. Hutchison. 1979. Lipid synthesis in the adult dog heartworm, *Dirofilaria immitis*. Comp. Biochem. Physiol. 64B:403-405.

114. Ugwunna, S.C. 1986. The origin and some of the functions of smooth endoplasmic reticulum in *Ancylostoma caninum* sperm cells. Int. J. Parasitol. 16:289-296.

115. Van den Bossche, H., editor. 1976. Biochemistry of parasites and host-parasite relationships. Elsevier North Holland Biomedical Press, Amsterdam.

116. Wann, K.T. 1987. The electrophysiology of the somatic muscle cells of *Ascaris suum* and *Ascaridia galli*. Parasitology 94:555-566.

117. Ward, C.W., and D. Fairbairn. 1970. Enzymes of β-oxidation and their function during development of *Ascaris lumbricoides* eggs. Dev. Biol. 22:366-387.

118. Ward, C.W., and P.J. Schofield. 1967. Comparative activity and intracellular distribution of tricarboxylic acid cycle enzymes in *Haemonchus contortus* larvae and rat liver. comp. Biochem. Physiol. 23:335–359

119. Ward, P.F.V. 1982. Aspects of helminth metabolism. Parasitology 84:177-194.

120. Ward, P.F.V., and N.S. Huskisson. 1978. The energy metabolism of adult *Haemonchus contortus*, in vitro. Parasitology 77:255-271.

121. Ward, P.F.V., and N.S. Huskisson. 1980. The role of carbon dioxide in the metabolism of adult *Haemonchus contortus*, in vitro. Parasitology 80:73-82.

122. Wharton, D.A. 1979. The structure and formation of the egg-shell of *Hammerschmidtiella diesingi* (Hammerschmidt) (Nematoda: Oxyuroidea). Parasitology 79:1-12.

123. Wharton, D.A. 1979. The structure and formation of the egg-shell of *Syphacia obvelata* Rudolphi (Nematoda: Oxyurida). Parasitology 79:13-28.

124. Wharton, D.A. 1980. Nematode egg-shells. Parasitology 81:447-463.

125. Wilson, P.A.G., M. Cameron, and D.S. Scott. 1978. *Strongyloides ratti* in virgin female rats: studies of oestrous cycle effects and general variability. Parasitology 76:221-227.

126. Wright, K.A. 1964. The fine structure of the somatic muscle cells of the nematode *Capillaria hepatica* (Bancroft, 1893). Can. J. Zool. 42:483-490.

127. Wright, K.A. 1966. Cytoplasmic bridges and muscle systems in some polymyarian nematodes. Can. J. Zool. 44:329-340.

128. Wright, K.A. 1968. Structure of the bacillary band of *Trichuris myocastoris*. J. Parasitol. 54:1106-1110.

129. Wright, K.A., and W.D. Hope. 1968. Elaborations of the cuticle of *Acanthonchus duplicatus* Wieser, 1959 (Nematoda: Cyatholaimidae) as revealed by light and electron microscopy. Can. J. Zool. 46:1005-1011.

SUGGESTED READINGS

Anderson, R.C., A.G. Chabaud, and S. Willmott. 1976-1985. CIH keys to the nematode parasites of vertebrates. Commonwealth Agricultural Bureaux, Farnham Royal, Bucks, England. (A ten-part up-to-date series of keys for the identification of nematode parasites.)

Barrett, J. 1976., Bioenergetics in helminths. In Van den Bossche, H., editor. Biochemistry of parasites and host-parasite relationships. Elsevier/North Holland Biomedical Press, Amsterdam.

Bird, A.F. 1971. The structure of nematodes. Academic Press, Inc., New York. (This is a very good summary of nematode morphology. Indispensible for students of nematodology.)

Bryant, C. 1978. The regulation of respiratory metabolism in parasitic helminths. In Lumsden, W.H.R., R. Muller, and J.R. Baker, editors. Advances in parasitology, vol. 16. Academic Press, Inc., New York.

Crites, J.L. 1969. Problems in systematics of parasitic nematodes. In Schmidt, G.D., editor. Problems in systematics of parasites. University Park Press, Baltimore, pp. 77-87. (A philosophical discussion of nematode systematics.)

Crofton, H.D. 1966. Nematodes. Hutchinson University Library, London. (A very useful summary of nematode characteristics.)

Croll, N.A., editor. 1976. The organization of nematodes. Academic Press, Inc., New York.

Croll, N.A., and B.E. Matthews. 1977. Biology of nematodes. John Wiley & Sons, Inc., New York.

Grassé, P.P. 1965. Traité de zoologie: anatomie, systématique, biologie, vol. 4, parts 2 and 3. Némathelminthes (Nématodes-Gordiacés), rotifères-gastrotriches, kinorinques. Masson & Cie, Paris. (An indispensable reference for serious students of nematodes.)

Lee, D.L., and H.J. Atkinson. 1977. Physiology of nematodes, ed. 2. Columbia University Press, New York.

Levine, N.D. 1980. Nematode parasites of domestic animals and of man, ed. 2. Burgess Publishing Co., Minneapolis. (An excellent general reference. The introductory chapter is useful for anatomy; Levine's proposed standard endings for taxa are used throughout.)

Skrjabin, K.I., et al. 1949-1952. Key to parasitic nematodes, vols. 1-3. Akademii Nauk SSSR, Moscow. (Excellent taxonomic presentation of several groups of parasitic nematodes. Volumes 1 and 3 have been translated into English.)

Skrjabin, K.I., et al. 1953-1972. Essentials of nematodology, vols. 1-22. Akademii Nauk SSSR, Moscow. (The most complete taxonomic compilation of information on parasitic nematodes in the world. Several volumes have been translated into English.)

CHAPTER 24

ORDERS TRICHURATA AND DIOCTOPHYMATA: APHASMIDIAN PARASITES

Some of the most dreaded, disfiguring, and debilitating diseases of humans are caused by nematodes. In addition, agriculture suffers mightily from attacks by these animals. Nematodes normally parasitic in wild animals can occasionally infect humans and domestic animals, causing mystifying diseases. Furthermore, nonparasitic nematodes may accidentally find their way into a vertebrate and become a short-lived, but pathogenic, parasite. Many thousands of nematodes are known to parasitize vertebrates; many still are unknown. A few examples are presented here and in the following chapters, selected for their interest as parasites of humans and as illustrations of parasitism as exemplified by nematodes.

ORDER TRICHURATA

This order of nematodes contains, among others, three genera of medical importance: *Trichuris, Capillaria,* and *Trichinella*. They have morphological and biological peculiarities that place them in the class Aphasmidia, a group of mostly nonparasitic worms.

Family Trichuridae

Whipworms, members of the family Trichuridae, are so called because they are threadlike along most of their body, and then they abruptly become thick at the posterior end, reminiscent of a whip with a handle (Fig. 24-1). The name *Trichocephalus* ("thread-head"), in widespread use in some countries, was coined when it was realized that the "thread" was the anterior end rather than the tail, but the term *Trichuris* has priority. There are many species in a wide variety of mammalian hosts, and one is a very important parasite of humans.

■ *Trichuris trichiura*

Morphology. *Trichuris trichiura* measures from 30 to 50 mm long, with males being somewhat smaller than females. The mouth is a simple opening, lacking lips. The buccal cavity is tiny and is provided with a minute spear. The esophagus is very long, occupying about two thirds of the body length and consists of a thin-walled tube surrounded by large, unicellular glands, the **stichocytes.** The entire structure often is referred to as the **stichosome.** The anterior end of the esophagus is somewhat muscular and lacks stichocytes. Both sexes have a single gonad, and the anus is near the tip of the tail. Males have a single spicule that is surrounded by a spiny spicule sheath. The ejaculatory duct joins the intestine anterior to the cloaca. In the female the vulva is near the junction of the esophagus and the intestine. The uterus contains many unembryonated, lemon-shaped eggs, each with a prominent opercular plug at each end (Fig. 24-2).

The excretory system is absent. The ventral surface of the esophageal regions bears a wide band of minute pores, leading to underlying glandular and nonglandular cells.[34,41] This **bacillary band** is typical of the order. Although the function of the cells in the bacillary band is unknown, their ultrastructure suggests that the gland cells may have a role in osmotic or ion regulation, and the nongland cells may function in cuticle formation and food storage.

Biology. Estimates of egg production range from 1000 to 7000 per day. Embryonation is completed in about 21 days in soil, which must be moist and shady. When swallowed, the infective juvenile hatches in the small intestine and enters the crypts of Lieberkühn. After a short period of development, it reenters the intestinal lumen and migrates to the ileocecal area, where it matures in about 3 months. Adults live for several years, so large numbers may accumulate in a person, even in areas in which the rate of new infection is low.

Epidemiology. The two requirements for *T. trichiura* to become a serious health problem are poor standards of sanitation in which human feces are deposited on the soil and the combination of physical conditions that allows the worm's survival and development: a warm climate, high rainfall and humidity, moisture-

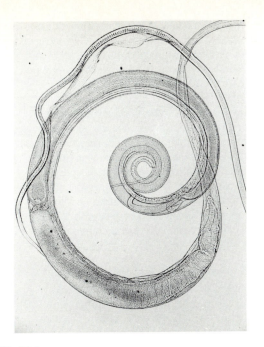

FIG. 24-1

Male *Trichuris*. Note the slender anterior end and the stout posterior end with a single, terminal spicule

Photograph by Jay Georgi.

FIG. 24-2

Egg of *Trichuris trichiura*. It measures 50 to 54 μm by 22 to 23 μm.

Photograph by Robert E. Kuntz.

retaining soil, and dense shade. Although generally coextensive in distribution with *Ascaris, T. trichiura* is more sensitive to the effects of desiccation and direct sunlight. Approximate physical conditions exist in much of the world, including parts of the southeastern United States, where the prevalence of infection may reach 20% to 25%, mainly in small children. Stoll[39] calculated the world prevalence at 355 million. It is doubtful that the figure is lower today, considering the population increase in tropical areas of the world; *T. trichiura* is perhaps the most common nematode of humans after *Ascaris* and *Enterobius*. Small children are most commonly infected, either by drinking contaminated water or by placing egg-contaminated fingers in their mouths.

Pathology. Fewer than 100 worms rarely cause clinical symptoms, and the majority of infections are symptomless. A heavier burden may result in a variety of conditions, occasionally terminating in death. The anterior ends of the worms burrow in the mucosa (Fig. 24-3), where the worms consume blood cells, although blood loss by this mechanism is negligible. Trauma to the intestinal epithelium and underlying submucosa, however, can cause a chronic hemorrhage that may result in anemia. Secondary bacterial infec-

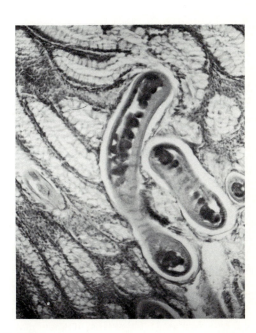

FIG. 24-3

Section of large intestine. Note the sections of *Trichuris trichiura* embedded in the mucosa.

Photograph by Robert E. Kuntz.

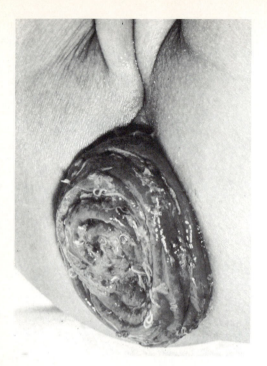

FIG. 24-4

Prolapse of the rectum caused by whipworm infection

Courtesy University of Miami School of Medicine, From Beck, J.W., and
J.E. Davies. 1981. Medical parasitology, ed. 3. The C.V. Mosby Co., St.
Louis.

tion, possibly coupled with allergic responses, results
in colitis, proctitis, and in extreme cases, prolapse of
the rectum (Fig. 24-4). Inflammation of the appendix
incriminates whipworm as a possible cause of appendicitis, although the mere presence of the worms in an
infected appendix is not proof of a causal relationship.
Amebic dysentery may be complicated by *T. trichiura*. Among other symptoms of infection are insomnia, nervousness, loss of appetite, vomiting, urticaria,
prolonged diarrhea, constipation, flatulence, and verminous intoxication. It is believed, however, that the
more severe "toxic" effects are not solely the result of
the worm infection but involve other factors, such as
malnutrition.[38]

Diagnosis and treatment. Specific diagnosis depends on demonstrating a worm or egg in the stool.
The eggs, with distinctive bipolar plugs, are 50 to 54
μm by 22 to 23 μm and have smooth outer shells.
Their structure and formation have been reported by
Preston and Jenkins.[31] Clinical symptoms may be
confused with those of hookworm, amebiasis, or acute
appendicitis.

Because of their location in the cecum, appendix,
or lower ileum, whipworms are difficult to reach with

oral drugs or medicated enemas. Mebendazole is effective and is the drug of choice.[18] Training of children and adults in sanitary disposal of feces and washing of hands is necessary to prevent reinfection.

■ Other *Trichuris* species

Some 60 to 70 other species of *Trichuris* have
been described from a wide variety of mammals. Several species occur in wild and domestic ruminants, of
which *T. ovis* is the most important in domestic sheep
and cattle. *Trichuris suis*, which is indistinguishable
from *T. trichiura*, is found in swine. *T. vulpis* is
found in the cecum of dogs, foxes, and coyotes and is
common in the United States except in the drier areas.
This species occasionally infects humans.[20]

Family Capillariidae

Members of the genus *Capillaria* look very much
like *Trichuris* spp., except that the transition between
the anterior, filiform portion and the posterior, stout
portion is gradual, rather than sudden. Other morphological features are similar. A large genus, *Capillaria*
includes species that are parasitic in nearly all organs
and tissues of all classes of vertebrates.

■ *Capillaria hepatica*

Biology. *Capillaria hepatica* is a parasite of the
liver, mainly of rodents, but it has been found in a
wide variety of mammals, including humans. The female deposits eggs in the liver parenchyma, where
they have no means of egress until eaten by a predator
or until the liver decomposes after death. The eggs
cannot embryonate while in the liver, so a new host
cannot be infected when it eats an egg-laden liver.
The eggs merely pass through the digestive tract of the
predator with feces. Embryonation occurs on the soil,
and new infection is by contamination. After hatching
in the small intestine, the juveniles migrate to the
liver, where they mature.

Epidemiology. As with *T. trichiura*, infection occurs when contaminated objects, food, or water is ingested. Unlike whipworm, however, human feces are
not the source of contamination; more likely, the feces
of carnivores or flesh-eating rodents are involved.
Eggs of *C. hepatica* have been found in several species of earthworms. These transport hosts may facilitate infections in normal definitive hosts.[33]

Pathology. Wandering of adult *C. hepatica* through
the host liver causes loss of liver cells and thereby loss
of normal function. Large areas of parenchyma may
be replaced by masses of eggs (Fig. 24-5). Rarely,
eggs will be carried to the lungs or other organs by the
bloodstream.

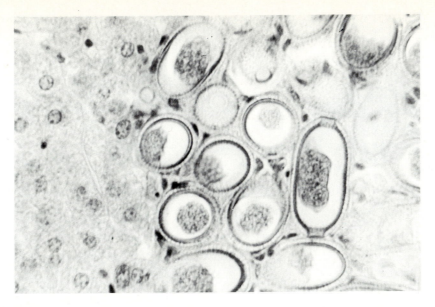

FIG. 24-5

Eggs of *Capillaria hepatica* in liver. Note the extensive damage to hepatic parenchyma.

Photograph by Warren Buss.

Diagnosis and treatment. Verified cases of this parasite in humans are rare, partly because of difficulties of diagnosis. Most cases have been determined after death, but liver biopsy has uncovered others. Untreated cases may be fatal.[3] Clinical symptoms resemble numerous liver disorders, especially hepatitis with eosinophilia. Specific diagnosis depends on demonstrating the eggs, which closely resemble those of *Trichuris* except that they measure 51 to 67 μm by 30 to 35 μm and have deep pits in the shells. Treatment for this disease is poorly known. A patient was successfully treated with a combination of prednisone, disophenol, and pyrantel tartrate, whereas albendazole shows promise as the drug of choice.[26,30]

Discovery of *C. hepatica* eggs in human feces may indicate the presence of a spurious infection caused by eating an infected liver.

■ *Capillaria philippinensis*

Capillaria philippinensis was discovered in 1963 as a parasite of humans in the Philippines. In contrast to *C. hepatica*, *C. philippinensis* is an intestinal parasite. Its appearance as a human pathogen was sudden and unexpected. One or two isolated cases were followed by an epidemic in Luzon in 1967 that killed several dozen persons.[11] It also is known from Thailand[7] and Iran. Probably a zoonotic disease, it possibly is transmitted between humans by fecal contamination. The original animal host remains unknown, but circum-stantial evidence points to fish as the source of human infection. *Capillaria philippinensis* has been transmitted to mammals by experimentally infected fish.[7]

Morphology. This parasite is very small; males measure 2.3 to 3.17 mm, and females measure 2.5 to 4.3 mm long. The male has small caudal alae and a spineless spicule sheath. The esophagus of the female is about half as long as the body. The female produces typical *Capillaria*-type eggs that lack pits. Unlike most parasitic nematodes; many eggs hatch while still in the intestine, building a heavy infection.

Epidemiology. Humans remain the only known host of this parasite, although probably it is shared by some species of wild animals. Intensive surveys of the Philippine fauna have so far failed to identify any reservoir host. Once established in a village, human transmission is readily maintained, particularly during the rainy season when soil contaminated with feces abounds in the form of mud.

Pathology. The worms repeatedly penetrate the mucosa of the small intestine and reenter the lumen, especially in the jejunum, leading to progressive degeneration of the epithelium and submucosa. The two primary results are pathological malabsorption of nutrients and violent diarrhea. Abdominal distention and pain usually occur. Death appears to be caused by emaciation and ion imbalance, leading to fatal shock.[9]

Diagnosis and treatment. Both adults and eggs, as well as juveniles, are abundant in feces of heavily in-

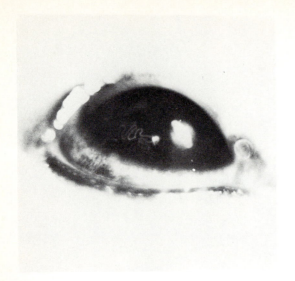

FIG. 24-6

Anatrichosoma ocularis in the eye of a tree shrew, *Tupaia glis.*

From File, S.K. 1974. J. Parasitol. 60:985-988.

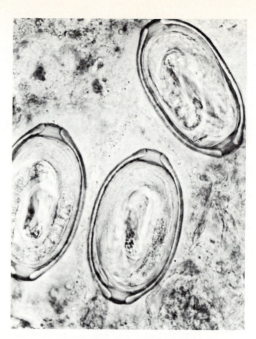

FIG. 24-7

Eggs of *Anatrichosoma ocularis* from eye secretions.

Courtesy Sharon K. File.

fected persons, and at least one of them is necessary for specific diagnosis. An intradermal test also is available but may not prove useful.

Mebendazole is very effective in curing this disease.[36] Control is unsatisfactory because the source of infection in nature is unknown. Epidemics can be prevented by a high standard of hygiene.

■ Other *Capillaria* species

Several species of *Capillaria* are important parasites of domestic animals. *Capillaria aerophila* is a lung parasite of cats, dogs, and other carnivores and has been reported several times from humans in the Soviet Union. It is probably the most destructive parasite of commercial fox farms.

Capillaria annulata and *C. caudinflata* infect the esophagus and crop of chickens, turkeys, and several other species of birds. Unlike most species in the genus, an intermediate host (earthworm) is required in the life cycle.

Few species of terrestrial vertebrates, wild or domestic, are free from at least one species of *Capillaria.*

■ *Anatrichosoma* species

Species of *Anatrichosoma* are very similar to *Capillaria,* except that they lack a spicule and spicule sheath. They have been reported from the tissues of a wide variety of Asian and African monkeys and ger-

bils and from the North American opossum. *Anatrichosoma ocularis* (Fig. 24-6) lives in the corneal epithelium of tree shrews, *Tupaia glis.* The eggs of this genus (Fig. 24-7) have the polar plugs characteristic of the order. No species is known to parasitize humans, but the species in monkeys, at least, should be considered as potential zoonoses.

Family Trichinellidae

■ *Trichinella spiralis*

Curiously, the smallest nematode parasite of humans, which exhibits the most unusual life cycle, is one of the most widespread and clinically important parasites in the world. *Trichinella spiralis* actually is a complex of at least four sibling species, subspecies, or strains, according to various authors. Taxonomic problems arise from the fact that although adults show no morphological differences, there are variations in metabolism, clinical symptoms, reactions to drugs, and host reactions.[16] No doubt evolutionary processes are underway in *Trichinella,* with eventual complete separateness of species in the future. Until then we can only describe what we know about different populations, whether or not the four share their gene pools when sympatric. *Trichinella spiralis* occurs in temperate zones, *T. nelsoni* is a tropical form, and *T. nativa* is arctic in distribution. *Trichinella pseudospiralis* is Nearctic, found first in raccoons. These sibling species are morphologically indistinguishable by SEM studies.[24] Dick and Chadee[13] reported that gene flow can occur among populations of *T. spiralis*, *T. nativa*, and *T. pseudospiralis*.

These parasites are responsible for the disease variously known as trichinosis, trichiniasis, or trichinelliasis. It is common in carnivorous mammals, including rodents and humans, primarily on the circumboreal continents. *Trichinella spiralis* is less common in tropical regions but is well known in Mexico, parts of South America, Africa, southern Asia, and the Middle East. Incidence of infection is always higher than suspected because of the vagueness of symptoms, which usually suggests other conditions; more than 50 different diseases have been diagnosed incorrectly as trichinosis.

Morphology. The males (Fig. 24-8) measure 1.4 to 1.6 mm long and are more slender at the anterior than the posterior end. The anus is nearly terminal and has a large papilla on each side of it. A copulatory spicule is absent. Like other members of the order Trichurata, stichocytes are arranged in a row following a short muscular esophagus. Females are about twice the size of males, also tapering toward the anterior end. The anus is nearly terminal. The vulva is located near the middle of the esophagus, which is about a third the length of the body. The single uterus is filled with developing eggs in its posterior portion, whereas the anterior portion contains fully developed, hatching juveniles.

Biology. The biology of this organism is unusual in that the same animal serves as both definitive and intermediate host, with the juveniles and the adults located in different organs.

When the infective juveniles are swallowed and reach the small intestine of the host, they rapidly molt four times and enter the intestinal mucosa.[21] Copulation and insemination occur, apparently within the mucosal epithelium, from 30 to 32 hours after infection.[17] Interestingly, the worms are in an intracellular location, lying directly in the cytoplasm and threading through a serial row of intestinal cells.[42] During her sojourn in the intestinal epithelium, the female gives birth to about 1500 juveniles over a period of 4 to 16 weeks. Eventually the spent female dies and is absorbed by the host. Males die shortly after copulation.

Most juveniles are carried away by the hepatoportal system through the liver, then to the heart, lungs, and the arterial system, which distributes them throughout the body (Fig. 24-9). During this migration, they have

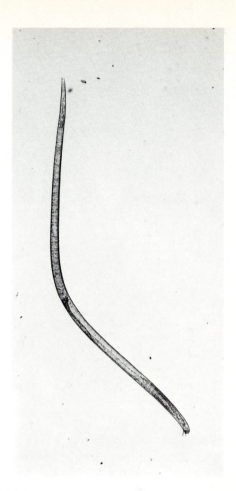

FIG. 24-8

Male *Trichinella spiralis* (1.4 to 1.6 mm in length) from the intestine of a rat.

Photograph by Jay Georgi.

been found in literally every kind of tissue and space in the body. When they reach skeletal muscle, they penetrate individual fibers and begin to grow, eventually forming a spiral and becoming encysted by infiltrating leukocytes (Fig. 24-10). Although one worm per cyst is most common, up to seven have been observed in a single cyst within a single muscle cell.

Some muscles are much more heavily invaded than are others, but the reasons for this are not understood. Most susceptible are muscles of the eye and tongue and masticatory muscles, then diaphragm and intercostals, and finally the heavy muscles of the arms and legs. Juveniles of *T. spiralis* invade only slow twitch fibers, whereas those of *T. pseudospiralis* invade both fast twitch and slow twitch fibers.[4] The juveniles absorb their nutrients from the enclosing muscle cell and increase in length to about 1 mm in about 4 to 8

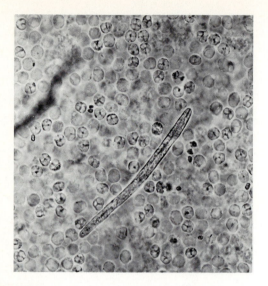

FIG. 24-9

Live *Trichinella spiralis* juvenile migrating in the blood of a rat.

Photograph by Jay Georgi.

weeks, at which time they are infective to their next host.

The cyst wall becomes gradually thicker during this time, and finally achieves a length of 0.25 to 0.5 mm. The enclosing muscle cell degenerates, as the result of movements of the juvenile, loss of nutrients, and reaction to the parasite's metabolic wastes. It now is called a **nurse cell.** The juveniles enter a developmental arrest and can live for months while encysted. Gradually, after about 10 months, the host reactions begin to calcify the cyst walls and, eventually, the worms themselves. However, some are believed to be viable up to 30 years.[18]

It was the presence of such calcified cysts in human cadavers that led to the discovery of this species in 1835 by James Paget, a medical student in London. Noticing that his subject had gritty particles in its muscles that tended to dull his scalpels, he studied the particles and demonstrated their wormlike nature to his fellow students. He then showed them to the eminent anatomist Richard Owen, who reported on them further and gave them their scientific name. It was another 25 years before it was determined that these minute animals cause disease.

As in other nematodes, there are four juvenile stages, but various authors have disagreed on whether the infective juvenile is the first, second, third, or fourth stage.[23] Kozek[21] observed four molts in the intestinal phase; therefore the encysted, infective juvenile would correspond to the first stage. However,

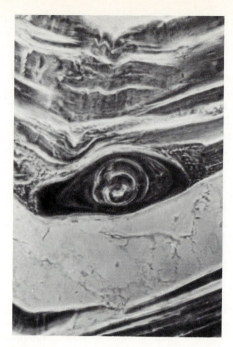

FIG. 24-10

Juvenile of *Trichinella spiralis* encysted in a muscle fiber.

there is considerable development while the worm is encysted, and the degree of maturity attained by the time the juvenile becomes infective is substantially greater than that in first stage juveniles of other nematodes. After ingestion, the enclosing muscle cells and cyst wall are digested off, freeing the worm, which rapidly matures and enters the intestinal mucosa, as noted above. Some juveniles, particularly in hosts with some protective immunity acquired from a previous infection, may not enter the mucosa and may pass out in the feces. These can then infect another animal that eats the feces.[29]

Most mammals are susceptible to infection. **Sylvatic trichinosis** occurs between wild carnivores and their prey or carrion. Hence bears, badgers, foxes, and even walruses are commonly heavily infected. **Urban trichinosis** occurs primarily as an unfortunate triangle between humans, rats, and pigs.

Trichinosis may best be considered a zoonotic disease, since humans can scarcely be important in the life cycle of the parasite. Unless an infected person is eaten by a carnivorous predator or becomes the occupant of a cannibal's pot, both unlikely eventualities these days, the parasites are at a dead-end course.

Epidemiology. Because sylvatic trichinosis involves wild mammals, humans are infected only when they interject themselves into the sylvatic food chain.

Eskimos, Indians, and others who rely on wild carnivores for food and urban dwellers who return home with the spoils of the hunt are all subject to infection with *Trichinella*. Fatal cases of trichinosis are common among those who eat undercooked or underfrozen bear, wild pig, cat, dog, or walrus meat. Theoretically, any wild mammals may be a source of infection, but, of course, most rarely find their way to the dinner table. In Alaska, polar bears, black bears, and walrus are common sources of infection.[10,27,32] Arctic explorers have been killed by *Trichinella* acquired from uncooked polar bear meat. The cause of death of the three members of the ill-fated André polar expedition of 1897 was determined *50 years later* by finding *Trichinella* juveniles in museum specimens of the polar bear meat that the men had been eating before they died.[40] A strain of *Trichinella* isolated from a polar bear in Canada is known to have survived freezing at $-15°$ C for 12 months, although it lost its ability to survive freezing after passage in mice.[12] An outbreak at Barrow, Alaska, in 1980 was found due to meat of grizzly bear that had been stored frozen.[10]

Another form, *T. pseudospiralis*, does not form cysts in muscle, can infect birds, and has a lower infectivity for laboratory animals.[6] The epidemiological importance of *T. pseudospiralis* has yet to be assessed.

Urban trichinosis is epidemiologically more important to humans because of the close relationship among rats, pigs, and people. Infected pork is our most common source of infection. Pigs become infected by eating offal or trichinous meat in garbage or by eating rats, which are ubiquitous in pig farms. It is usually concluded that garbage containing raw pork scraps is the usual source of infection for pigs, but it should not be overlooked that pigs will greedily devour dead or even live rats when they can catch them. The rats probably maintain their infections by cannibalism, although it has been demonstrated that rat and mouse feces can contain juveniles capable of infecting rats, pigs, or humans.[23,29] Juveniles in the muscles of mice are known to alter the behavior of the rodent, making them more vulnerable to predation.[43] Infection can spread from pig to pig when they nip off and eat each other's tails, a common practice in crowded piggeries.[37] It has been suggested that prenatal infection is possible.[8] Piggish cannibalism also is involved. The worm can survive and remain infective after the anaerobic digestion of sewage sludge.[15]

The importance of cooking pork thoroughly before it is eaten cannot be overstated. A roast or other piece of solid meat is safe when all traces of pink have disappeared. Many persons are careful about this but become careless when cooking sausage, which is equally dangerous. Raw sausage is a delicacy among many

peoples of the world, particularly in the areas where trichinosis is a chronic health problem. Even a casual taste to determine proper seasoning can be fatal: Heavily infected pork may contain more than 100,000 juveniles per ounce. If half of these are females and each produces 1500 offspring, a single bite of meat can theoretically contain 1.5 million juveniles. Five juveniles per gram of body weight are usually fatal for human beings. Particularly important in transmission is meat processed by "backyard butchers" (individuals who slaughter their own stock) or by very small packing houses. Sausage from such sources is unlikely to be diluted with uninfected meat. In December 1975, 30 symptomatic cases (5 confirmed) around Springfield, Massachusetts, were attributed to eating sausage prepared by a "backyard butcher." Between 1966 and 1979 the incidence of trichinosis in the United States was between 125 and 150 cases per year; of these, 69% were single-source outbreaks.

Nevertheless, a taste for rare pork can still be satisfied, since freezing at $-15°$ C for 20 days destroys all parasites, in the temperate zone strain, at least.

Survivors of trichinosis have varying degrees of immunity to further infection. Duckett, Denham and Nelson[14] demonstrated that this immunity in mice can be passed by a mother to her young by suckling. Suckling rats rapidly expel *Trichinella* if the mother is immune. Her serum antibodies are passed through the milk to the pups.[2]

Pathogenesis. The pathogenesis of *Trichinella* infection can be considered in three successive stages: penetration of adult females into the mucosa, migration of juveniles, and penetration and encystment in muscle cells.

First symptoms may appear between 12 hours and 2 days after ingestion of infected meat. Commonly this phase is clinically inapparent because of low-grade infection or is misdiagnosed because of the vagueness of symptoms. When the gravid females penetrate the intestinal epithelium, they cause traumatic damage to the host tissues; the host begins to react to their waste products, and enteric bacteria are introduced into the wounds they cause. These wounds result in intestinal inflammation and pain, with symptoms of food poisoning, such as nausea, vomiting, sweating, and diarrhea. Respiratory difficulties may occur, and red blotches erupt on the skin in some cases. This period usually terminates with facial edema and fever 5 to 7 days after the first symptoms.

During migration the newborn juveniles damage blood vessels, resulting in localized edema, particularly in the face and hands. Wandering juveniles may also cause pneumonia, pleurisy, encephalitis, meningitis, nephritis, deafness, peritonitis, brain or eye damage, and subconjunctival or sublingual hemorrhage. Death resulting from myocarditis (inflammation of the heart muscle) may occur at this stage. Although the juveniles do not stay in the heart, they migrate through its muscle, causing local areas of necrosis and infiltration of leukocytes.

By the tenth day after the first symptoms appear, the juveniles begin penetration of muscle fibers. Attendant symptoms are again varied and vague: intense muscular pain, difficulty in breathing or swallowing, swelling of masseter muscles (occasionally leading to a misdiagnosis of mumps), weakening of pulse and blood pressure, heart damage, and various nervous disorders, including hallucination. Extreme eosinophilia is common but may not be present even in severe cases. Death is usually caused by heart failure, respiratory complications, toxemia, or kidney malfunction.

Diagnosis and treatment. Most cases of trichinosis, particularly subclinical cases, go undetected. Routine examinations rarely detect juveniles in feces, blood, milk, or other secretions. Although muscle biopsy is seldom employed, it remains an accurate diagnostic if trichinosis is suspected. Pressing the tissue between glass plates and examining it by low-power microscopy is useful, although digestion of the muscle in artificial gastric enzymes for several hours provides a much more reliable diagnostic technique. **Xenodiagnosis,** feeding suspected biopsies to laboratory rats, may be employed. Several immunodiagnostic techniques have been developed, none of which is 100% effective but which are useful nonetheless. An ELISA test shows promise.[1]

No really satisfactory treatment for trichinosis is known. Treatment is basically that of relieving the symptoms by use of analgesics and corticosteroids. Purges during the initial symptoms may dislodge the females that have not yet begun penetrating the intestinal epithelium. Thiabendazole has been shown effective in experimental animals, but results in clinical cases have been variable. Prolonged, oral high-dose mebendazole therapy has been reported effective in one case that was unresponsive to steroid treatment.[22]

Despite immense research, trichinosis remains an important disease of humans, one that has the potential of striking anyone, anywhere. One hopeful note: for unknown reasons the incidence of infection has slowly but steadily declined throughout the world.

ORDER DIOCTOPHYMATA

The few members of the order Dioctophymata are parasites of aquatic birds and terrestrial mammals. Most are of no economic or medical importance to humans. Of the three families in the order, Soboliphymi-

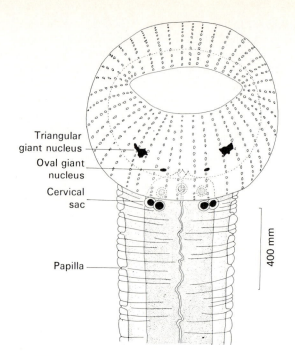

Triangular giant nucleus

Oval giant nucleus

Cervical sac

Papilla

400 mm

FIG. 24-11

Soboliphyme baturini from a short-tailed weasel. Note the swollen mouth capsule typical of this genus.

From Schmidt, G.D., and J.M. Kinsella. 1965. Trans. Am. Microsc. Soc. 84:413-415.

dae has the single genus *Soboliphyme*, parasites of shrews and mustelid carnivores (Fig. 24-11), and Eustrongylidae has the genera *Eustrongylides* and *Hystrichis,* both parasites of wild birds, although infection of humans by juvenile *Eustrongylides* has been reported.[35] The other family is of more concern and will be considered more fully.

Family Dioctophymatidae

The family Dioctophymatidae has three genera, *Dioctowittius, Mirandonema,* and *Dioctophyme,* each with a single species. The first is known only from a snake and the second from a Brazilian carnivore, whereas the third has been reported from a wide variety of mammals, including humans, in many parts of the world.

■ *Dioctophyme renale*

Morphology. *Dioctophyme renale* is truly a giant among nematodes, with the males up to 20 cm long and 6 mm wide and the females up to 100 cm long and 12 mm wide. They are blood red and rather blunt at the ends. The male has a conspicuous, bell-shaped copulatory bursa that lacks any supporting rays or papillae (Fig. 24-12). A single, simple spicule is 5 to 6

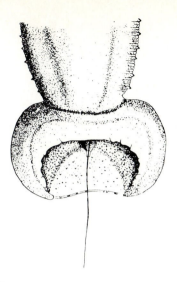

FIG. 24-12

Posterior end of a male *Dioctophyme renale*. Note the single spicule and the powerful copulatory bursa.

From Stefanski, W. 1928. Ann. Parasitol. Hum. Comp. 6:93-100.

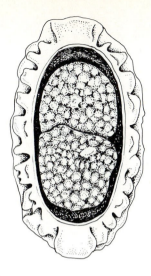

FIGS. 24-13

Eggs of *Dioctophyme renale*, showing the corrugated shell and the two-cell embryo. This stage is released by the female worm.

From Mace T.F., and R.C. Anderson. 1975. Can. J. Zool. 53:1552-1568.

mm long. The vulva is near the anterior end. Eggs (Fig. 24-13) are lemon shaped, with deep pits in the shells, except at the poles.

Biology. The life cycle of this parasite has been described.[19,25] The thick-shelled eggs require about 0.5 to 3 months in water to embryonate, depending on the temperature. The first-stage juvenile, which bears an oral spear, hatches when eaten by the aquatic oligochaete annelid *Lumbriculus variegatus*. There it penetrates into the ventral blood vessel, where it develops into the third-stage juvenile. When the small annelid is swallowed, the juvenile migrates to a kidney of the new host, where it matures. If the annelid is eaten by any of several species of fish and frogs, the juvenile will encyst in the muscle or viscera, using the fish or frog as a paratenic host.[28] When swallowed by a definitive host, the juvenile penetrates the stomach wall to the submucosa. After about 5 days it migrates to the liver and remains in the liver parenchyma for about 50 days; then it migrates to the kidney. Usually the right kidney is invaded, perhaps because of the proximity of the stomach to the right lobes of the liver, from which the worms can migrate directly to the kidney.[25] The worms mature in the kidney, and the eggs are voided from the host in its urine. In rare human infections the worms usually have been found in a kidney, but one was a third-stage juvenile in a subcutaneous nodule.[5]

Epidemiology. Probably any species of large mammal can serve as definitive host. Because of their fish

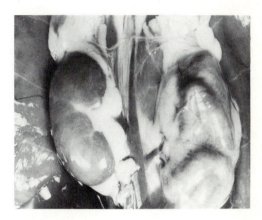

FIG. 24-14

Ferret dissection: the kidney on the left is normal, whereas that on the right is distended by an adult *Dioctophyme renale*. (Anterior of animal is toward the bottom.)

Photograph by Arthur E. Woodhead. Courtesy Ann Arbor Biological Center.

diets, mustelids, canids, and bears are particularly susceptible, as are humans. However, even such non-fish-eating mammals as cows, horses, and pigs can become infected, when they accidentally ingest an infected annelid. Thorough cooking of fish and drinking of only pure water will prevent infection in people.

Pathology. Pressure necrosis caused by the growing worm, together with its feeding activities, reduces

425

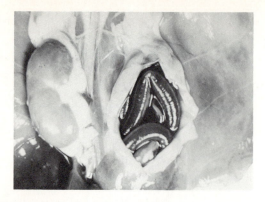

FIG. 24-15

Specimen in Fig. 24-14 with infected kidney opened to reveal the worm. The organ is reduced to a hollow shell.

Photograph by Arthur E. Woodhead. Courtesy Ann Arbor Biological Center.

the infected kidney to a thin-walled, ineffective organ (Figs. 24-14 and 24-15). Loss of kidney function is compounded by uremic poisoning. The worms will sometimes penetrate the renal capsule and wander in the coelomic cavity.

Diagnosis and treatment. The rarity of this parasite makes physicians unlikely to suspect its presence. Demonstration of the characteristic eggs in the urine is the only positive means of diagnosis, aside from surgical discovery of the worm itself. Surgical removal of the worm is the only treatment known.

REFERENCES

1. Anonymous. 1986. Scared of trichinosis? Check with ELISA. Science News 130:73.
2. Appleton, J.A., and D.D. McGregor. 1984. Rapid expulsion of *Trichinella spiralis* in suckling rats. Science 226:68-72.
3. Attah, E.B., S. Nagarajan, E.N. Obineche, and S.C. Gera. 1983. Hepatic capillariasis. Am. J. Clin. Pathol. 79:127-130.
4. Bagheri, A., J.E. Ubelaker, G.L. Stewart, and B. Wood. 1986. Muscle fiber selectivity of *Trichinella spiralis* and *Trichinella pseudospiralis*. J. Parasitol. 72:277-282.
5. Beaver, P.C., and J.H. Theis. 1979. Dioctophymatid larval nematode in a subcutaneous nodule from man in California. Am. J. Trop. Med. Hyg. 28:206-212.
6. Belosevic, M., and T.A. Dick. 1980. Chemical attraction in the genus *Trichinella*. J. Parasitol. 66:88-93.
7. Bhaibulaya, M., S. Indra-ngarm, and M. Ananthapruti. 1979. Freshwater fishes of Thailand as experimental intermediate hosts for *Capillaria philippinensis*. Int. J. Parasitol. 9:105-108.
8. Bourns, T.K.R. 1952. The discovery of trichina cysts in the diaphragm of a six-week old child. J. Parasitol. 38:367.
9. Canlas, B.D., Jr., B.O. Cabrera, and U. Diaz. 1967. Human intestinal capillariasis. II. Pathological features. Acta Med. Philippina 4:84-91.
10. Centers for Disease Control. 1979. Trichinosis associated with meat from a grizzly bear-Alaska. Morbid. Mortal. Weekly Rep. 30:115-116, 121.
11. Diaz, U., B.O. Cabrera, and B.D. Canlas, Jr. 1967. Human intestinal capillariasis. I. Clinical features. Acta Med. Philippina 4:72-83.
12. Dick, T.A., and M. Belosevic. 1978. Observations on a *Trichinella spiralis* isolate from a polar bear. J. Parasitol. 64:1143-1145.
13. Dick, T.A., and K. Chadee. 1983. Interbreeding and gene flow in the genus *Trichinella*. J. Parasitol. 69:176-180.
14. Duckett, M.G., D.A. Denham, and G.S. Nelson. 1972. Immunity to *Trichinella spiralis*. V. Transfer of immunity against the intestinal phase from mother to baby mice. J. Parasitol. 58:550-554.
15. Fitzgerald, P.R., and T.B.S. Prakasam. 1978. Survival of *Trichinella spiralis* larvae in sewage sludge anaerobic digesters. J. Parasitol. 64:445-447.
16. Flokhart, H.A. 1986. *Trichinella* speciation. Parasitol. Today 2:1-3.
17. Gardiner, C.H. 1976. Habitat and reproductive behavior of *Trichinella spiralis*. J. Parasitol. 62:865-870.
18. Hunter, G.W., III, J.C. Swartzwelder, and D.F. Clyde. 1976. Intestinal nematodes. In Tropical medicine, ed. 5. W.B. Saunders Co., Philadelphia.
19. Karmanova, E.M. 1968. Essentials of nematodology, vol. 20. Dioctophymidea of animals and man and the diseases caused by them. Akademii Nauk SSSR, Moscow.
20. Kenney, M., and V. Yermakov. 1980. Infection of man with *Trichuris vulpis*, the whipworm of dogs. Am. J. Trop. Med. Hyg. 29:1205-1208.
21. Kozek, W.J. 1971. The molting pattern in *Trichinella spiralis*. I. A light microscope study. J. Parasitol. 57:1015-1028.
22. Levin, M.L. 1983. Treatment of trichinosis with mebendazole. Am. J. Trop. Med. Hyg. 32:980- 983.
23. Levine, N.D. 1980. Nematode parasites of domestic animals and of man, ed. 2. Burgess Publishing Co., Minneapolis.
24. Lichtenfels, J.R., K.D. Murrell, and P.A. Pilitt. 1983. Comparison of three subspecies of *Trichinella spiralis* by scanning electron microscopy. J. Parasitol. 69:1131-1140.
25. Mace, T.F., and R.C. Anderson. 1975. Development of the giant kidney worm, *Dioctophyma renale* (Goeze, 1782) (Nematoda: Dioctophymatoidea). Can. J. Zool. 53:1552-1568.
26. Markus, M.B., and R.F. Cheetham. 1985. Chemotherapy for *Capillaria hepatica* infection. J. Antimicrob. Chemother. 15:790-791.
27. Maynard, J.E., and F.P. Pauls. 1962. Trichinosis in Alaska. A review and report of two outbreaks due to bear meat, with observations of serodiagnosis and skin testing. Am. J. Hyg. 76:252-261.
28. Measures, L.N., and R.C. Anderson. 1985. Centrarchid fish as paratenic hosts of the giant kidney worm, *Dioctophyme renale* (Goeze, 1782), in Ontario, Canada. J. Wildl. Dis. 21:11-19.
29. Olsen, O.W. and H.A. Robinson. 1958. Role of rats and mice in transmitting *Trichinella spiralis* through their feces. J. Parasitol. 44(sect. 2):35.
30. Pereira, V.G., and L.C. Mattosinho Franca. 1983. Successful treatment of *Capillaria hepatica* infection in an acutely ill adult. Am. J. Trop. Med. Hyg. 32:1272-1274.
31. Preston, C.M., and T. Jenkins. 1984. *Trichuris muris:* structure and formation of the egg-shell. Parasitology 89:263-273.

32. Rausch, R. 1953. Animal-borne diseases. Publ. Health Rep. 68:533.

33. Romashov, B.V. 1983. Details of the life cycle of *Hepaticola hepatica* (Nematoda, Capillariidae). Parazitologicheskie issledovaniya v Zapovednikakh, Moscow, pp. 49-58.

34. Sheffield, H.G. 1963. Electron microscopy of the bacillary band and stichosome of *Trichuris muris* and *T. vulpes*. J. Parasitol. 49:998-1009.

35. Shirazian, D., E.L. Schiller, C.A. Glaser, and S.L. Vonderfect. 1984. Pathology of larval *Eustrongylides* in the rabbit. J. Parasitol. 70:803-806.

36. Singston, C.N., T.C. Banzon, and J.H. Cross. 1975. Mebendazole in the treatment of intestinal capillariasis. Am. J. Trop. Med. Hyg. 24:932-934.

37. Smith, H.J. 1975. Trichinae in tail musculature of swine. Can. J. Comp. Med. 39:362-363.

38. Spencer, H. 1973. Nematode diseases. I. In Spencer, H., editor. Tropical pathology. Springer-Verlag, Berlin, pp. 457-509.

39. Stoll, N.R. 1947. This wormy world. J. Parasitol. 33:1-18.

40. Sundman, P.O. 1970. The flight of the Eagle. Random House, Inc., New York.

41. Wright, K.A. 1968. Structure of the bacillary band of *Trichuris myocastoris*. J. Parasitol. 54:1106-1110.

42. Wright, K.A. 1979. *Trichinella spiralis:* an intracellular parasite in the intestinal phase. J. Parasitol. 65:441-445.

43. Zohar, A.S., and M.R. Rau. 1986. The role of muscle larvae of *Trichinella spiralis* in the behavioral alterations of the mouse host. J. Parasitol. 72:464-466.

SUGGESTED READINGS

Campbell, W.C., editor. 1983. *Trichinella* and trichinellosis. Plenum Press, New York.

Neafie, R.C., and D.H. Connor. 1976. Trichuriasis. In Binford, C.H., and D.H. Connor, editors. Pathology of tropical and extraordinary diseases, vol. 2, sect. 9. Armed Forces Institute of Pathology, Washington, D.C.

Neafie, R.C., D.H. Connor, and J.H. Cross. 1976. Capillariasis. In Binford, C.H., and D.H. Connor, editors. Pathology of tropical and extraordinary diseases, vol. 2, sect. 9. Armed Forces Institute of Pathology, Washington, D.C.

Stehr-Green, J.K., P.M. Schantz, and E.M. Chisolm. 1986. Trichinosis surveillance, 1984. Morbid. Mortal. Weekly Rep. (CDC Surveillance Summaries) 35:1155-1156.

CHAPTER 25

ORDER RHABDITATA: PIONEERING PARASITES

The tiny worms in the order Rhabditata appear to bridge the gap between free-living and parasitic modes of life, since several species alternate between free-living and parasitic generations. Most species inhabit decaying organic matter and are common in soil, foul water, decaying fruit, and so on. For this reason they often find their way into the bodies of larger animals; the digestive, reproductive, respiratory, and excretory tracts are particularly susceptible, as are open wounds. Once in such locations, they may become facultatively parasitic for a time or may simply pass through the body. Rarely, they can cause serious disease.[5]

Since most species are similar to each other, it usually requires the services of a specialist to differentiate the pathogenic species from nonpathogenic ones. Species are more difficult to identify during free-living phases than during parasitic stages.

Some of the behavioral and developmental adaptations of free-living species *(Rhabditis, Pelodera)* in this order were mentioned in Chapter 23 (dauer juveniles). It is not difficult to visualize how analogous adaptations for transfer to new food supplies in the ancestors of the frankly parasitic species could have been preadaptations to parasitism. The diversity in life cycles of parasites in Rhabditata further illustrates this evolutionary opportunism. Thus, within the order, one can observe completely free-living species, species with preadaptations for parasitism, facultative parasites, obligate parasites, and even species that seem to produce obligate free-living or obligate parasitic forms, depending on conditions. Families that have obligate parasites or parasites interspersed with free-living adults are Rhabdiasidae and Strongyloididae. Although the latter is of far more medical and veterinary importance, Rhabdiasidae is of interest in illustrating the diversity of life cycles in the order. Its members are often encountered in frog dissections.

FAMILY RHABDIASIDAE

Rhabdias bufonis and *R. ranae,* common parasites in the lungs of toads and frogs, have a very curious life cycle. The parasitic adult is a **protandrous hermaphrodite,** that is, an individual that is a functional male before it becomes a female. Sperm (chromosome number [N] = 5 or 6)[16] are produced in an early male phase and stored in a seminal receptacle. Then the gonad produces functional ova (N = 6) that are fertilized by the stored sperm. The resulting shelled zygotes (2N = 11 or 12) are passed up the trachea of the host, and then swallowed, embryonating along the way. The juveniles hatch in the intestine of the frog, and the first-stage juveniles accumulate in the cloaca to be voided with the feces. The first-stage juveniles often are referred to as **rhabditiform** because the posterior end of their esophagus has a prominent **bulb** that is separated from the anterior portion **(corpus)** by a narrower region **(isthmus)** (Fig. 25-1). These juveniles undergo four molts to produce a generation of free-living males (2N = 11) and females (2N = 12), a dioecious generation. This nonparasitic generation feeds on bacteria and other inhabitants of humus soil. The progeny of this generation hatch in utero and proceed to consume the internal organs of their mother, destroying her. "How sharper than a serpent's tooth!" They escape from the female's body and, by now, have become third-stage infective juveniles. They are referred to as filariform at this stage because the esophagus has no terminal bulb or isthmus. The filariform juveniles undergo developmental arrest unless they penetrate the skin of a toad or frog. After penetration, they lodge in various tissues; those which reach the lungs mature into hermaphroditic adults, whereas the rest apparently expire. This type of life cycle is **indirect,** or **heterogonic;** that is, a free-living generation is interspersed between parasitic generations.

Rhabdias fuscovenosa is a common parasite in the lungs of some kinds of aquatic snakes (Natricinae). The life cycle differs somewhat from that of *R. bufonis* and *R. ranae* in that most of the eggs from the parasitic forms yield filariform larvae. Few free-living adults are found[4]; therefore the life cycle in this species is predominately **direct,** or **homogonic,** and the worm is an obligate parasite.

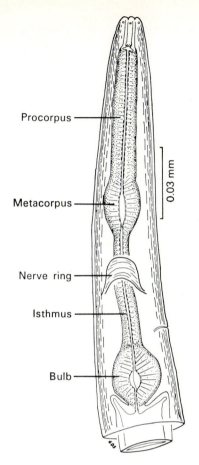

FIG. 25-1

Typical rhabditiform esophagus.

From Schmidt, G.D., and Robert E. Kuntz. 1972. Proc. Helminth. Soc. Wash. 39:189-191.

Labels on figure: Procorpus, Metacorpus, Nerve ring, Isthmus, Bulb, 0.03 mm

FAMILY STRONGYLOIDIDAE
Strongyloides spp.

Some species of this family are among the smallest nematode parasites of humans, the males being even smaller than *Trichinella*. *Strongyloides stercoralis* is the most common, widespread species; *S. fuelleborni* occurs commonly in parts of Africa.[14] *Strongyloides stercoralis* also infects other primates, as well as dogs, cats, and some other mammals, and races from various geographical areas vary in infectivity for different hosts.[12] *Strongyloides fuelleborni* has been reported from a variety of primates. Other species are parasites of other mammals, for example, *S. ratti* in rats, *S. ransomi* in swine, and *S. papillosus* in sheep. They also are common in birds, amphibians, and reptiles. The species of *Strongyloides* are remarkable in

their ability, at least in some cases, to maintain homogonic, parasitic life cycles or repeat free-living generations indefinitely, depending on conditions. The parasitic generation apparently consists only of parthenogenetic females, since sperm have not been found in the seminal receptacle of the females.[3] Parasitic males, though rare, have been reported. The free-living generation consists of both males and females. The following description pertains to *S. stercoralis* (Fig. 25-2).

■ Morphology

Parthenogenetic females reach a length of 2.0 to 2.5 mm, whereas parasitic males (if published reports are accurate) are about 0.7 mm long and appear identical to free-living males. The buccal capsule of both sexes is small, and they possess a long, cylindrical esophagus that lacks a posterior bulb. The vulva is in the posterior third of the body; the uteri are divergent and contain only a few eggs at a time. The free-living adults both have a rhabditiform esophagus. The male is up to 0.9 mm long and 40 to 50 μm wide. The male has two simple spicules and a gubernaculum; its pointed tail is curved ventrad. The female is stout and has a vulva that is about equatorial; the uteri generally contain more eggs than do those of the parasitic female.

■ Biology (Fig. 25-3)

The parasitic females anchor themselves with their mouths to the mucosa of the small intestine or burrow their anterior ends into the submucosa. They are found occasionally in the respiratory, biliary, or pancreatic system. They produce several dozen, thin-shelled, partially embryonated eggs a day and release them into the gut lumen or submucosa. The eggs measure 50 to 58 μm by 30 to 34 μm. They hatch during passage through the gut or within the submucosa, and the juveniles escape to the lumen. These first-stage juveniles are 300 to 380 μm long, and they usually are passed with the feces. The juveniles go on either to develop into free-living adults or to become infective, filariform juveniles with a developmental arrest at the third stage; they are now 490 to 630 μm long. The filariform juveniles develop no further unless they gain access to a new host by skin penetration or ingestion. If by skin penetration, they are carried by the blood to the lungs, where they exit into the alveoli, travel up the trachea, are swallowed, and mature in the small intestine. If they are ingested, a lung migration appears unnecessary. The free-living adults, in turn, can produce successive generations of free-living adults. Both parasitic and free-living females can produce ju-

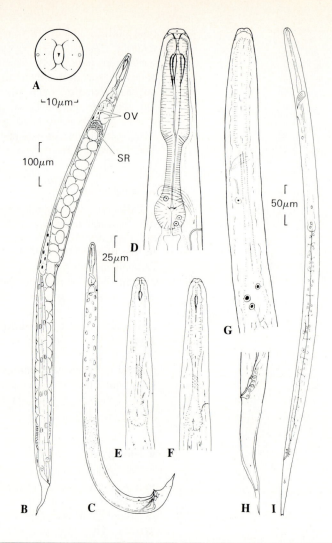

FIG. 25- 2

Strongyloides stercoralis. **A,** Free-living female, en face view. **B,** Free-living female, lateral view (*OV,* ovary; *SR,* seminal receptacle containing sperm). **C,** Free-living male. **D,** Anterior end of free-living female, showing details of esophagus. **E,** Newly hatched, first-stage juvenile obtained by duodenal aspiration from human. **F,** First-stage juvenile from freshly passed feces of same patient as was juvenile in **E. G,** Second-stage juvenile developing to the filariform stage; cuticle is separating at anterior end. **H,** Tail of same juvenile as shown in **G;** notched tail is developing, and cuticle is separating at tip and in rectum. **I,** Filariform juvenile.

From Little, M.D. 1966. J. Parasitol. 52:69-84.

veniles that will become filariform, infective juveniles and juveniles that will mature into free-living adults. In other words, the homogonic and heterogonic life cycles seem to be mixed in a random fashion.

The mechanism that determines whether a given embryo will become a free-living male or female or a parasitic female is still unclear. Some evidence suggests that the parasitic females are haploid in *S. ratti.*[3] If this is so, then all haploid embryos will produce parasitic females, and diploid embryos will produce free-living adults. Parasitic females of both *S. ransomi* and *S. papillosus* are apparently diploid, and even the free-living females reproduce parthenogenetically, although males may be present.[18] Environmental conditions that appear to encourage production of one or the other type of juvenile may *in fact* strongly favor the *survival* of either the free-living adults or filariform larvae.[7] Some important differences in the en-

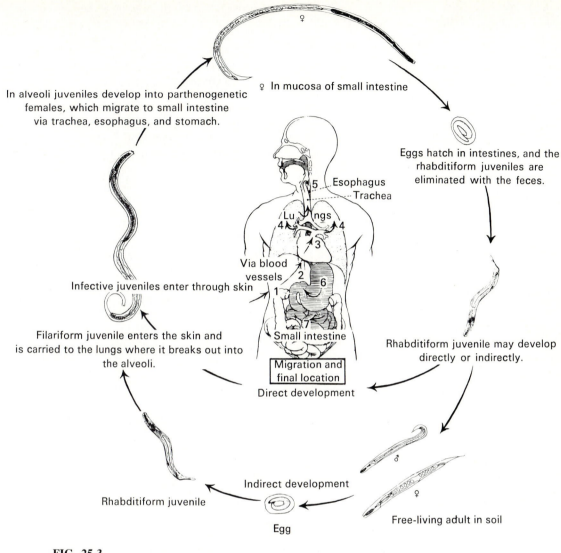

FIG. 25-3

Life cycle of *Strongyloides stercoralis* in humans.

Modified from Medical protozoology and helminthology. 1959. U.S. Naval Medical School, Bethesda, Md.

ergy metabolism of free-living and parasitic *Strongy-loides* have already been mentioned (Chapter 23), and such factors may have bearing on the question.

Still other permutations of the life cycle are possible. If the juveniles have time to molt twice during their transit down the digestive tract, they may penetrate the lower gut mucosa or perianal skin, go through their migration, and mature. This process is called **autoinfection.** Some researchers have maintained, furthermore, that filariform juveniles could repenetrate the small gut mucosa in a process of **hyper-**

infection, but others have questioned whether this can occur. Nevertheless, whether by maintenance of the original adult worms, by autoinfection, or otherwise, cases are known in which patients have had *Strongyloides* infections for up to 40 years. In 1984 Pelletier[15] reported on 142 American ex-prisoners of war who had worked on the Burma-Thailand Railroad in World War II. Of these, 52 had symptomatic, previously unrecognized strongyloidiasis. Such extreme longevity has also been reported in British and Australian ex-prisoners of war.

■ Epidemiology

People typically become infected with strongyloids by contacting the juveniles in contaminated soil or water. These primarily are parasites of tropical regions but extend well into temperate zones in several continents. Like most filthborne diseases, strongyloidiasis is most prevalent under conditions of low sanitation standards. Although traditionally a disease of poor, uneducated persons in depressed areas of the world, it is often found wherever conditions of filth prevail, such as some mental institutions. It could also become important among the affluent with "vacation hideways," who rely on inadequate sewage disposal facilities, such as in mountain environments.

■ Pathology

The effects of strongyloidiasis may be described in three stages: **invasion, pulmonary,** and **intestinal.**

Penetration of the skin by invasive juveniles results in slight hemorrhage and swelling, with intense itching at the site of entry. If pathogenic bacteria are introduced with the juveniles, inflammation may result as well.

During migration through the lungs, damage to lung tissues results in massive host-cell reactions, which often delay or prevent further migration. When this happens, the worms may establish themselves in the pulmonary tissues and begin reproducing as if they were in the intestine, again demonstrating their extraordinary adaptability. A burning sensation in the chest, a nonproductive cough, and other symptoms of bronchial pneumonia may accompany this phase. The pulmonary phase has been suspected of reactivating quiescent pulmonary tuberculosis.[13]

After being swallowed, the juvenile females enter the crypts of the intestinal mucosa, where they rapidly mature and invade the tissues. They rarely penetrate deeper than the muscularis mucosae, but some cases of deeper penetration have been reported. The worms migrate randomly through the mucosa, depositing eggs. An intense, localized burning sensation or aching pain in the abdomen usually is felt at this time. Destruction of tissues by adult worms and juveniles results in sloughing of patches of mucosa, with fibrotic changes in chronic cases, sometimes with death resulting from septicemia (bacterial infection of the blood) following ulceration of the intestine.

Infection with *S. stercoralis* is most commonly asymptomatic, but cases are known in which the host was an asymptomatic carrier for years and then developed serious disease.

In immunocompromised patients, the host-parasite relationship is altered in favor of the parasite, and fulminant, fatal hyperinfection can occur. Mortality has been observed in cases treated with high-dose steroid therapy and in patients with AIDS.[10,11,19] The course of infection in dogs and monkeys treated with immunosuppresive drugs has been studied.[6,8,17] Though less grave, chronic strongyloidiasis can lead to relapsing colitis.[1]

■ Diagnosis and treatment

Demonstration of rhabditiform (or occasionally filariform) juveniles in freshly passed stools is a sure means of diagnosis. A direct fecal smear is often effective in cases of massive infections, and various concentration techniques, such as Baermann isolation or zinc flotation with centrifugation, will increase chances in cases in which fewer worms are present. Rarely, after purgation or in severe diarrhea, embryonating eggs may be seen in the stool. These resemble hookworm eggs but are more rounded.

Difficulties arise, however, because of day-to-day variability in numbers of juveniles in the feces. Furthermore, once autoinfection or hyperinfection becomes established, the number of juveniles passed in the feces may decrease. Duodenal aspiration is a very accurate technique, but only applies to duodenal infections: Juveniles further down the intestine cannot be obtained.

Once juveniles are obtained, difficulties of identification can occur. First-stage juveniles are similar to rhabditiform hookworm juveniles, which may be present if the stool was constipated or had remained at room temperature long enough for hookworm eggs to hatch. Two morphological features can be useful to separate the two: *Strongyloides* has a short buccal cavity and a large genital primordium, whereas hookworm juveniles have a long buccal cavity and a tiny genital primordium (p. 442).

If the stool has been exposed to soil or water, species of *Rhabditis* or related genera may invade it, compounding the confusion. Furthermore, filariform juveniles appear in cases of constipation or autoinfection. These, however, are easily recognized by their notched tails. Although no immunological tests are currently available, there is promise that such tests will become available in the near future.

Several drugs are effective in treatment of strongyloidiasis, but most have undesirable side effects. Thiabendazole currently is preferred; it has a high percentage of cure and minimal side effects.[9] Cambendazole has been reported as even more effective than thiabendazole in this infection, however.[2]

REFERENCES

1. Berry, A.J., et al. 1983. Chronic relapsing colitis due to Strongyloides stercoralis. Am. J. Trop. Med. Hyg. 32:1289-1293.
2. Bicalho, S.A., O.J. Leao, and Q. Pena, Jr. 1983. Cambendazole in the treatment of human strongyloidiasis. Am. J. Trop. Med. Hyg. 32:1181-1183.
3. Bolla, R.I., and L.S. Roberts. 1968. Gametogenesis and chromosomal complement in *Strongyloides ratti* (Nematoda: Rhabdiasoidea). J. Parasitol. 54:849-855.
4. Chu, T. 1936. Studies on the life cycle of *Rhabdias fuscovenosa* var. *catanensis* (Rizzo, 1902). J. Parasitol. 22:140-160.
5. Gardiner, C.H., D.S. Koh, and T.A. Cardella. 1981. *Micronema* in man: third fatal infection. Am. J. Trop. Med. Hyg. 30:586-589.
6. Grove, D.I., P.J. Heenan, and C. Northern. 1983. Persistent and disseminated infections with *Strongyloides stercoralis* in immunosuppressed dogs. Int. J. Parasitol. 13:483-490.
7. Hansen, E.L., B.J. Buecher, and W.S. Cryan. 1969. *Strongyloides fülleborni:* environmental factors and free-living generations. Exp. Parasitol. 26:336-343.
8. Harper, J.W., III, et al. 1984. Experimental disseminated strongyloidiasis in *Erythrocebus patas*. I. Pathology. Am. J. Trop. Med. Hyg. 33:431-443.
9. Hunter, G.W., III, J.C. Swartzwelder, and D.F. Clyde. 1976. Intestinal nematodes. In Tropical medicine, ed. 5. W.B. Saunders Co., Philadelphia.
10. Kalb, R.E., and M.E. Grossman. 1986. Periumbilical purpura in disseminated strongyloidiasis. JAMA 256:1170-1171.
11. Katner, H. 1988. Personal communication.
12. Levine, N.D. 1980. Nematode parasites of domestic animals and of man, ed. 2. Burgess Publishing Co., Minneapolis.
13. Palmer, E.D. 1944. A consideration of certain problems presented by a case of strongyloidiasis. Am. J. Trop. Med. 24:249-254.
14. Pampiglioni, S., and M.L. Ricciardi. 1971. The presence of *Strongyloides fülleborni* von Linstow, 1905, in man in central and east Africa. Parassitologia 13:257-269.
15. Pelletier, L.L., Jr. 1984. Chronic strongyloidiasis in World War II Far East ex-prisoners of war. Am. J. Trop. Med. Hyg. 33:55-61.
16. Runey, W.M., G.L. Runey, and F.H. Lauter. 1978. Gametogenesis and fertilization in *Rhabdias ranae* Walton 1929. I. The parasitic hermaphrodite. J. Parasitol. 64:1008-1014.
17. Schad, G.A., M.E. Hellman, and D.W. Muncey. 1984. *Strongyloides stercoralis:* hyperinfection in immunosuppressed dogs. Exp. Parasitol. 57:289-296.
18. Triantaphyllou, A.C., and D.J. Moncol. 1977. Cytology, reproduction, and sex determination of *Strongyloides ransomi* and *S. papillosus*. J. Parasitol. 63:961-973.
19. Vieyra-Herrera, G., et al. 1988. *Strongyloides stercoralis* hyperinfection in a patient with acquired immune deficiency syndrome. Acta Cytol. 32:277-278.

SUGGESTED READINGS

Blunden, A.S., L.F. Khalil, and P.M. Webbon. 1987. *Halicephalobus deletrix* infection in a horse. Equine Vet. J. 19:255-260. (A good historical account of *Micronema* in humans and horses.)
Cable, R.M. 1971. Parthenogenesis in parasitic helminths. Am. Zool. 11:267-272. (Contains a short discussion of parthenogenesis in nematodes, with emphasis on *Strongyloides.)*
Little, M.D. 1966. Comparative morphology of six species of *Strongyloides* (Nematoda) and redefinition of the genus. J. Parasitol. 52:69-84.
Meyers, W.M., D.H. Connor, and R.C. Neafie. 1976. Strongyloidiasis. In Binford, C.H., and D.H. Connor, editors. Pathology of tropical and extraordinary diseases, vol. 2, sect. 9. Armed Forces Institute of Pathology, Washington, D.C.

CHAPTER **26**

ORDER STRONGYLATA: BURSATE PHASMIDIANS

The large order Strongylata is of great economic and medical importance. It seems to have evolved directly from rhabditoid-type ancestors. One feature in common with nearly all species is a broad copulatory bursa on the posterior end of the males. Most, but not all, species are parasites of the intestine of vertebrates and have a direct life cycle, requiring no intermediate hosts. Of the numerous families in the order, the Ancylostomidae, Trichostrongylidae, and superfamily Metastrongyloidea are the most important to humans. Examples from these groups illustrate the parasitological significance of the order.

FAMILY ANCYLOSTOMIDAE

Members of the family Ancylostomidae are commonly known as **hookworms.** They live in the intestine of their hosts, attaching to the mucosa and feeding on blood and tissue fluids sucked from it.

Morphology

Much similarity of morphology and biology exists among the numerous species in this family, so we will first give them a general consideration. Most species are rather stout, and the anterior end is curved dorsad, giving the worm a hooklike appearance. The buccal capsule is large and heavily sclerotized and usually is armed with cutting plates, teeth, lancets, or a dorsal cone (Fig. 26-1). A **dorsal gutter** extends along the middorsal wall of the buccal capsule, emptying the dorsal esophageal gland into it. Lips are reduced or absent. In one subfamily (Arthrostominae), the buccal capsule is subdivided into several articulated plates.

The esophagus is stout, with a swollen posterior end, giving it a club shape. It is mainly muscular, corresponding to its action as a powerful pump. The esophageal glands are extremely large and are mainly outside of the esophagus, extending posteriad into the body cavity. Cervical papillae are present near the rear level of the nerve ring.

Males have a conspicuous copulatory bursa, consisting of two broad **lateral lobes** and a smaller **dorsal lobe,** all supported by fleshy rays (Fig. 26-2). These rays follow a common pattern in all species, varying only in relative size and point of origin; consequently, they are important taxonomic characters. The spicules are simple, needlelike, and similar. A gubernaculum is present.

Females have a simple, conical tail. The vulva is postequatorial, and two ovaries are present. About 5% of the daily output of eggs is found in the uteri at any one time; the total production is several thousand per day for as long as 9 years.

Biology (Fig. 26-3)

Hookworms mature and mate in the small intestine of their host. The embryos develop into the two-, four-, or several-cell stage by the time they are passed with the feces (Fig. 26-4). The species infecting humans cannot be diagnosed reliably by the egg alone. The eggs require warmth, shade, and moisture for continued development. Coprophagous insects may mix the feces with soil and air, thus hastening embryonation, which may occur within 24 to 48 hours in ideal conditions (Fig. 26-5). Urine is fatal to developing embryos. Newly hatched first-stage juveniles have a rhabditiform esophagus with its characteristic constriction at the level of the nerve ring, such as that occurring in the rhabditiform juveniles of *Strongyloides* spp. In fact differentiation of hookworm juveniles from those of *Strongyloides* is difficult for the beginner.

The juveniles live in the feces, feeding on fecal matter, and molt their cuticle in 2 to 3 days. The second-stage juvenile, which also has a rhabditiform esophagus, continues to feed and grow and, after about 5 days, molts to the third stage, which is infective to the host. The second-stage cuticle may be retained as a loose-fitting sheath until penetration of a new host, or it may be lost earlier. The third-stage filariform juvenile has a **strongyliform esophagus;** that is, it has a posterior bulb, but the bulb is not separated from the corpus by an isthmus. The intestine is filled with food granules that sustain the worm through the nonfeeding, third stage. It is similar to the filariform juvenile of *Strongyloides,* but can be distinguished

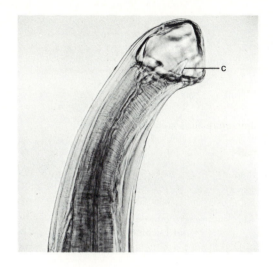

FIG. 26-1

Lateral view of the anterior of *Bunostomum* sp., a hook-worm of ruminants. Note the large buccal capsule typical of hookworms, and the dorsal cone *(c)*. The dorsal flexure of the head is also typical of hookworms.

Photograph by Jay Georgi.

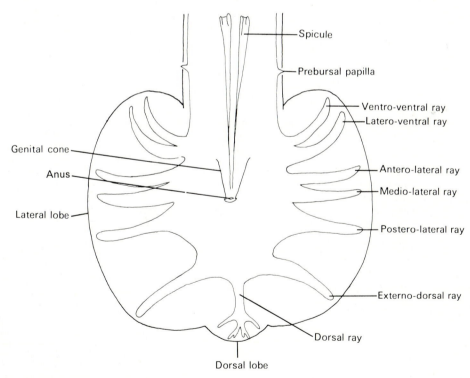

FIG. 26-2

Ventral view of a typical strongyloid copulatory bursa. The basic pattern is found in all of the Strongylata.

Drawing by G.D. Schmidt.

435

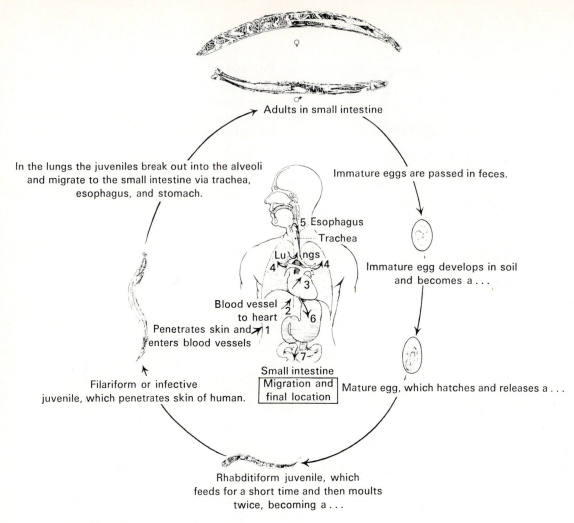

Adults in small intestine

In the lungs the juveniles break out into the alveoli and migrate to the small intestine via trachea, esophagus, and stomach.

Immature eggs are passed in feces.

5 Esophagus
Trachea
Lungs
4 4
3
Blood vessel
to heart 2 6
Penetrates skin and 1 enters blood vessels
7
Small intestine

Immature egg develops in soil and becomes a ...

Mature egg, which hatches and releases a ...

Migration and final location

Filariform or infective juvenile, which penetrates skin of human.

Rhabditiform juvenile, which feeds for a short time and then moults twice, becoming a ...

FIG. 26-3

The life cycle of hookworms.

From Medical protozoology and helminthology. 1959. U.S. Naval Medical School, Bethesda, Md.

FIG. 26-4

Hookworm egg from the feces of a cat.

Photograph by Jay Georgi.

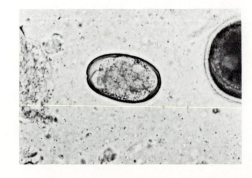

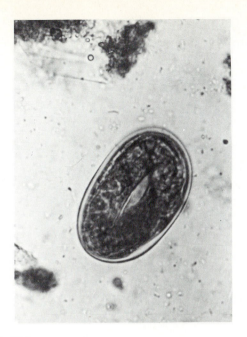

FIG. 26-5

Fully embryonated hookworm egg.

Photograph by R.E. Kuntz and J. Moore.

from it by the tail, which is pointed in hookworms and notched in *Strongyloides* (p. 432).

Third-stage juveniles live in the upper few millimeters of soil, remaining in the capillary layer of water surrounding soil particles. They are killed quickly by freezing or desiccation. There is a short, vertical migration in the soil, depending on the weather or time of day; when the surface of the ground is dry, they migrate a short distance into the soil, following the retreating water. When the surface of the ground is wet, after rain or morning dew, they move to the surface and extend themselves like a snake, waving back and forth in a manner that allows maximal opportunity for contact with a host. Thousands of juveniles often group together, crawling over each other and waving rhythmically in unison (see Fig. 23-32). They can live for several weeks under ideal conditions.

Infection usually occurs when third-stage juveniles contact the skin and burrow into it, although transplacental transmission and transmission in mother's milk occur in some species. Any epidermis can be penetrated by the juveniles, although the parts most often in contact with the soil, such as hands, feet, and buttocks, are most often attacked. If juveniles are introduced into the mouth, they usually pierce the oral mucosa, although some species can survive if they are swallowed.

Juveniles that wander through subcuticular tissues

soon die and are absorbed, but those which enter vessels are carried to the heart and thereby to the lungs. After breaking into the air spaces, they are carried by ciliary action up the respiratory tree to the glottis and are swallowed. Arriving in the small intestine, they attach to the mucosa, grow, and molt to the fourth stage, which has an enlarged buccal capsule. After further growth and a final molt, the worm becomes sexually mature. At least 5 weeks are required from the time of infection to egg production.

Several genera and many species of hookworm plague humans and domestic and wild mammals. The following two species infect an estimated 900 million people.

■ *Necator americanus*

Necator americanus, the "American killer," was first discovered in Brazil, then Texas, but was later found indigenous in Africa, India, southeast Asia, China, and the southwest Pacific islands. It probably was introduced into the New World with the slave trade. The worm has had an important impact on the development of the southern United States, as well as other regions of the world in which it occurs. In 1947 Stoll[27] estimated 384.3 million infected persons in the world, with 1.8 million cases in North America. The numbers are probably higher today, considering the increase in world population since 1947. Surveys in the southeast United States show average prevalence of about 4%, with up to 15% in certain areas.[32]

Necator americanus has a pair of dorsal and a pair of ventral cutting plates surrounding the anterior margin of the buccal capsule (Fig. 26-6). In addition, a pair of subdorsal and a pair of subventral teeth are near the rear of the buccal capsule.

Males are 7 to 9 mm long and have a bursa diagnostic for the genus (Fig. 26-7).[16] The needlelike spicules have minute barbs at their tips and are fused distally. Females are 9 to 11 mm long and have the vulva located in about the middle of the body. The life cycle is generally as described previously, although infection apparently cannot occur by swallowing juveniles.[33] Dönges and Madecki[7] could find no evidence of transmission through mother's milk in Nigeria. Female worms produce about 9000 eggs per day.

Primarily a tropical parasite, *N. americanus* is the dominant species in humans in most of the world. About 95% of the hookworms in the southern United States are this species. Adults survive in the gut up to 15 years.[5]

■ *Ancylostoma duodenale*

Ancylostoma duodenale is abundant in southern Europe, northern Africa, India, China and southeastern

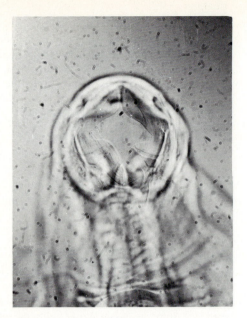

FIG. 26-6

En face view of the mouth of *Necator americanus*. Note the two broad cutting plates in the ventrolateral margins *(top)*.

Photograph by L.S. Roberts.

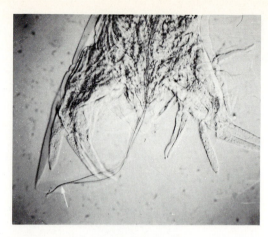

FIG. 26-7

Copulatory bursa and spicules of *Necator americanus*. The spicules are fused at their distal ends *(arrow)* and form a characteristic hook.

Photograph by L.S. Roberts.

Asia, as well as in other scattered locales, including small areas of the United States, Caribbean Islands, and South America. It is known in mines as far north as England and Belgium; since Lucretius, in the first century, it was known to cause a serious anemia in miners. Generally speaking, it is never as abundant as *N. americanus*.

The anterior margin of the buccal capsule has two ventral plates, each with two large teeth that are fused at their bases (Fig. 26-8). A pair of small teeth is found in the depths of the capsule.

Adult males are 8 to 11 mm long and have a bursa characteristic for the species (Fig. 26-9). The needle-like spicules have simple tips and are never fused distally. Females are 10 to 13 mm long, with the vulva located about a third of the body length from the posterior end. A single female can lay from 25,000 to 30,000 eggs per day. Adults may live up to 5 years.[5]

This is the first hookworm for which a life cycle was elucidated. In 1896 Arthur Looss, working in Egypt, was dropping cultures of *Ancylostoma* larvae into the mouths of guinea pigs when he spilled some of the culture onto his hand. He noticed that it produced an itching and redness and wondered if infection would occur this way. He began examining his feces at intervals and, after a few weeks, found that he was passing hookworm eggs. He next placed some ju-

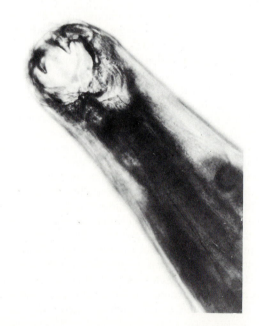

FIG. 26-8

Ancylostoma duodenale: dorsal view. Notice the powerful ventral teeth.

AFIP neg. no. N-41730-2.

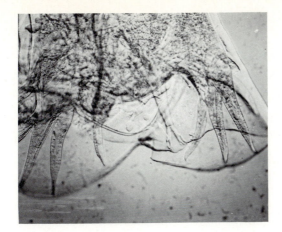

FIG. 26-9

Copulatory bursa and spicules of *Ancylostoma duodenale*. The tips of the spicules are not fused into a hook, as in *Necator americanus*.

Photograph by L.S. Roberts.

FIG. 26-10

Ancylostoma caninum, male.

Courtesy Ann Arbor Biological Center.

veniles on the leg of an Egyptian boy who was to have his leg amputated within an hour. Subsequent microscopic sections showed juveniles penetrating the skin. Looss' monograph on the morphology and life cycle of *A. duodenale* remains one of most elegant of all works on helminthology.[18]

It is possible for swallowed juveniles to develop normally without a migration through the lungs, but this is probably a fairly rare means of infection.

■ Other hookworms reported from humans

Ancylostoma ceylanicum is normally a parasite of carnivores in Sri Lanka, southeast Asia, and the East Indies, but it has been reported from humans in the Philippines.[29] A very similar species, *A. braziliense,* is found in domestic and wild carnivores in most of the tropics. Although it has been reported from humans in Brazil, Africa, India, Sri Lanka, Indonesia, and the Philippines, these infections probably were *A. ceylanicum. Ancylostoma braziliense* probably is the most common cause of creeping eruption (p. 442). *Ancylostoma malayanum,* a parasite of bears in Malaysia and India, has been reported once in a human.

Ancylostoma caninum (Fig. 26-10) is the most common hookworm of domestic dogs and cats, especially in the northern hemisphere. It has been found in humans on at least five occasions, and the worm also is a common cause of creeping eruption. The species has proved to be a useful tool in studying hookworm biology because of the ease of maintaining it in the laboratory. The anthelminthic closantel is effective against both adults and juveniles in arrested development in tissues.[8]

■ Hookworm disease

An important distinction is to be made between hookworm infection and hookworm disease. Far more people are infected with the worm than exhibit symptoms of the disease. Whether disease is manifested depends strongly on two factors: the number of worms present and the nutritional condition of the infected person. In general fewer than 25 *N. americanus* in a person will cause no symptoms, 25 to 100 worms lead to light symptoms, 100 to 500 produce considerable damage and moderate symptoms, 500 to 1000 result in severe symptoms and grave damage, and with more than 1000 worms, very grave damage may be accompanied by drastic and often fatal consequences. Because *Ancylostoma* spp. suck more blood, fewer worms cause greater disease; for example, 100 worms may cause severe symptoms. However, the clinical disease is intensified by the degree of malnutrition, corresponding impairment of the host's immune response, and other considerations. Little protective immunity is acquired by infection.[5]

Epidemiology. From the discussion on the biology of hookworms, it is obvious that the combination of poor sanitation and appropriate environmental conditions is necessary for high endemicity.

Environmental conditions conducive to transmission of the disease have been well studied, and they are the conditions that favor the development and survival of the juveniles. The disease is restricted to warmer parts of the world (and to specialized habitats such as mines in more severe climates) because the juveniles will not develop to maturity at less than 17° C, with 23° to 30° C being optimal. Frost kills eggs and juveniles. Oxygen is necessary for hatching of eggs and juvenile development because their metabolism is aerobic. This means, among other things, that the juveniles will not develop in undiluted feces or waterlogged soil. Therefore a loose, humusy soil that has reasonable drainage and aeration is favorable. Both heavy clay and coarse sandy soils are unfavorable for the parasite, the latter because the juveniles are also sensitive to desiccation. Alternate drying and moistening are particularly damaging to the juveniles; hence very sandy soils become noninfective after brief periods of frequent rainfall. However, the juveniles live in the film of water surrounding soil particles, and even apparently dry soil may have enough moisture to enable survival, particularly below the surface. The juveniles are quite sensitive to direct sunlight and survive best in shady loca-

tions, such as coffee, banana, or sugarcane plantations. Workers and other adults in such situations often have preferred defecation sites, not out in the open where the juveniles would be killed by the sun, of course, but in shady, cool, secluded spots beneficial for juvenile development. Repeated return of people to the defecation site exposes them to continual reinfection. Furthermore, the use of preferred defecation sites makes it possible for hookworms to be endemic in otherwise quite arid areas. Another physical condition conducive to survival of juveniles is pH; they develop best near neutrality, and acid or alkaline soils inhibit development, as does the acid pH of undiluted feces (pH 4.8 to 5). Chemical factors have an influence. Urine mixed with the feces is fatal to the eggs, and several strong chemicals that may be added to feces as disinfectants of fertilizers are lethal to the free-living stages. Salt in the water or soil inhibits hatching and is fatal to juveniles.

The longevity of the worms is important in transmission to new hosts, continuity of infection in a locality, and introduction to new areas. The juveniles can survive in reasonably good environmental conditions for about 3 weeks, except in protected sites like mines, where they can last for a year. There is some dispute as to the life span of the adult, but a good estimate is 5 to 15 years. If a person is removed from an endemic area, the infection is lost in about that time.

The degree of soil contamination is an important factor in transmission. Obviously, a higher average number of worms per individual will seed the soil with more eggs. Promiscuous defecation, associated with poverty and ignorance, keeps soil contamination high. The use of night soil as fertilizer for crops is an especially important factor in the Orient.

Because the worm penetrates the skin, habits of going barefoot in tropical countries make an elemental contribution to transmission.

Race is another important epidemiological factor: white people are about 10 times more susceptible to hookworm than are black persons. Prevalence and numbers of worms per individual tend to be higher among whites than blacks. The image of the lazy, apathetic poor whites ("poor white trash") in the Old South, as contrasted with the more industrious blacks on comparable socioeconomic levels, is generally attributed to the differential effects of hookworm.

A new dimension in the epidemiology of hookworm disease was the discovery by Schad et al.[24] that juveniles will survive in the muscles of paratenic hosts. Thus, *A. duodenale*, at least, can be transmitted by eating undercooked meat.

Pathogenesis. Hookworm disease manifests three main phases of pathogenesis: the cutaneous or invasion period, the migration or pulmonary phase, and the intestinal phase. Another pathogenic condition, caused when a juvenile enters an unsuitable host, will be discussed separately.

The cutaneous phase begins when juveniles penetrate the skin. They do little damage to the superficial layers, since they seem to slip through tiny cracks between skin scales or to penetrate hair follicles. Once in the dermis, however, their attack on blood vessels initiates a tissue reaction that may isolate and kill the worms. If, as usually happens, pyogenic bacteria are introduced into the skin with the invading juvenile, a urticarial reaction will result, causing a condition known as **ground itch.**

The pulmonary phase occurs when the juveniles break out of the lung capillary bed into the alveoli and progress up the bronchi to the throat. Each site hemorrhages slightly, with serious consequences in massive infections; however, this is rare. The phase is usually asymptomatic, although there may be some dry coughing and sore throat. A pneumonitis may result in severe infections.

The intestinal phase is the most important period of pathogenesis. On reaching the small intestine, the young worm attaches to the mucosa with its strong buccal capsule and teeth, and it begins to feed on blood (Fig. 26-11). In heavy infections worms are found from the pyloric stomach to the ascending colon, but usually they are restricted to the anterior third of the small intestine. The worms move from place to place, but bleeding quickly ceases at a lesion that a worm has left, contrary to prevalent belief.[26] The worms pass substantially more blood through their digestive tracts than would appear necessary for their nutrition alone, but the reason for this is unknown. Recent estimates of blood loss per worm are about 0.03 ml per day for *Necator* and around 0.26 ml per day for *Ancylostoma*. Up to 200 ml of blood may be lost by patients with heavy infections, but around 40% or so of the iron may be reabsorbed by the patient before it leaves the intestine.[14] Nevertheless, a moderate hookworm infection will gradually produce an iron deficiency anemia as the body reserves of iron are used up. The severity of the anemia depends on the worm load and the dietary iron intake of the patient. Slight, intermittent abdominal pain, loss of normal appetite, and desire to eat soil (geophagy) are common manifestations of moderate hookworm disease. (Certain areas in the southern United States became locally famous for the quality of their clay soil, and people traveled for miles to eat it. In fact in the early 1920s an enterprising person began a mail-order business, shipping clay to hookworm sufferers throughout the country!)

In very heavy infections, patients suffer severe protein deficiency, with dry skin and hair, edema, and

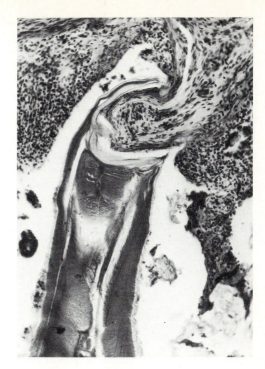

FIG. 26-11

Hookworm attached to intestinal mucosa. Notice how the ventral tooth in the depth of the buccal capsule lacerates the host tissue.

AFIP neg. no. N-33818.

potbelly in children and with delayed puberty, mental dullness, heart failure, and death. Intestinal malabsorption is not a marked feature of infection with hookworms, but hookworm disease is usually manifested in the presence of malnutrition and is often complicated by *Ascaris, Trichuris,* and/or malaria infections. The drain of protein and iron is catastrophic to one who is subsisting on a minimal diet. Such chronic malnutrition, particularly in the young, often causes irreversible damage, resulting in stunted growth and below-average intelligence. Impairment in ability to produce gamma globulin results in lowered antibody response to the hookworm as well as to other infectious agents. No living organism could be expected to live up to its potential under such conditions, and it is small wonder that development has been so difficult for many tropical countries. An interesting approach to the control of polycythemia (an overabundance of red blood cells) was reported by Walterspiel et al.[31] Adult *Ancylostoma duodenale* were transferred from a dog to a child to induce chronic blood loss. Results were inconclusive, but the technique shows promise as a therapeutic process.

Diagnosis, treatment, and control. Demonstration of hookworm eggs or the worms themselves in the feces is, as usual, the only definitive diagnosis of the disease. Demonstration of eggs in direct smears may be difficult, however, even in clinical cases, and one of the several concentration techniques should be used. If estimation of worm burden is necessary, techniques are available that give reliable data on egg counts (Beaver and Stoll techniques).

It is not necessary or possible to distinguish *Necator* eggs from those of *Ancylostoma,* but care should be taken to differentiate *Strongyloides* infections. This is not a problem unless some hours pass between time of defecation and time of examination of feces or unless the feces were obtained from a constipated patient. Then the hookworm eggs may have hatched, and the juveniles must be distinguished from those of *Strongyloides*. It is desirable, nevertheless, to distinguish *Necator* and *Ancylostoma* spp. in studies on the efficacy of various drugs or chemotherapeutic regimens because the two species are not equally sensitive to particular drugs. Differentiation can be accomplished by recovery of adults after anthelmintic treatment or by culturing the juveniles from the feces. Hookworm antigens and the potential for vaccination are discussed by Hotez, LeTrang, and Cerami.[11] Mebendazole is the drug of choice for treatment, as it removes both species of hookworm and also any concurrent infection with *Ascaris lumbricoides*. Single-dose therapy is inexpensive, convenient, and effective.[1]

Treatment for hookworm disease should always include dietary supplementation. In many cases provision of an adequate diet alleviates the symptoms of the disease without worm removal, but treatment for the infection should be instituted, if only for public health reasons.

Control of hookworm disease depends on lowering worm burdens in a population to the extent that remaining worms, if any, can be sustained within the nutritional limitations of the people, without causing symptoms. Mass treatment campaigns do not eradicate the worms but certainly lower the "seeding" capacity of their hosts. Education and persuasion of the population in the sanitary disposal of feces are also vital. The economic dependence on night soil in family gardens remains one of the most persistent of all problems in parasitology.

Recognizing these factors, the Sanitary Commission of the Rockefeller Foundation initiated a hookworm campaign in 1913. Beginning in Puerto Rico, where it was estimated that a third of all deaths resulted from hookworm disease, then extending throughout the southeastern United States, the commission would first survey an area. Residents of the area were examined for infection, and then treated with anthelmintics. Thousands of latrines were provided, together with instructions on how to use and maintain them. It is a study in human nature to note that many persons refused to use latrines and could be persuaded only with great difficulty. As the result of the efforts of this and other similar hygiene commissions, hookworm prevalence has been reduced greatly from the earlier levels in the southern United States, most of the Caribbean, and a few other areas of the world. Nevertheless, hookworm remains one of the most important parasites of humans in the world today.

■ Creeping eruption

Also known as **cutaneous larva migrans,** creeping eruption is caused by invasive juvenile hookworms of species or strains normally maturing in animals other than humans. The juveniles manage to penetrate the skin of humans but are incapable of successfully completing migration to the intestine. However, before they are overcome by immune responses, they result in distressing and occasionally serious complications of the skin (Fig. 26-12). Possibly any species of hookworm can cause this condition, but those of cats, dogs, and other domestic animals are most likely to come into contact with people. *Ancylostoma braziliense* appears to be the most common agent, followed by *A. caninum*. Both are common parasites of domestic dogs and cats, at least in some localities.

After entering the top layers of epithelium, the juveniles are usually incapable of penetrating the basal layer (stratum germinativum), so they begin an aimless wandering. As they tunnel through the skin, they leave a red, itchy wound that usually becomes infected by pyogenic bacteria. The worms may live for weeks or months. It is known that some can enter muscle fibers and become dormant.[17] The juveniles can attack any part of the skin, but because people's feet and hands are more in contact with the ground, they are most often affected. Thiabendazole has revolutionized treatment of creeping eruption, and topical application of a thiabendazole ointment has supplanted all other forms of treatment.

FAMILY TRICHOSTRONGYLIDAE

Many genera and an immense number of species comprise the family Trichostrongylidae. They are parasites of the small intestine of all classes of vertebrates, causing great economic losses in domestic animals, especially ruminants, and in a few cases causing disease in humans.

Trichostrongylids (Fig. 26-13) are small, very slen-

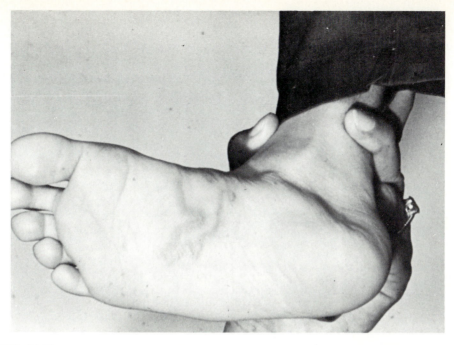

FIG. 26-12

Creeping eruption caused by infection with *Ancylostoma* sp. juvenile.

Photograph courtesy F. Battistine, with permission of H. Zaiman, editor. A pictorial presentation of parasites.

der worms with a rudimentary buccal cavity in most cases. Lips are reduced or absent, and teeth rarely are present. The cuticle of the head may be inflated. Males have a well-developed bursa, and the spicules vary from simple to complex, depending on the species. Females are considerably larger than males. The vulva is located anywhere from preequatorial to near the anus, according to the species. Thin-shelled eggs are laid while in the morula stage.

Life cycles are similar in all species. No intermediate host is required; the eggs hatch in soil or water and develop directly to infective third-stage juveniles. Some infections may occur through the skin, but as a rule the juveniles must be swallowed with contaminated food or water. Enormous numbers of juveniles may accumulate on heavily grazed pastures, causing serious or even fatal infections in ruminants and other grazers. Since their life cycles are similar, a given host usually is infected with several species, and the severe pathogenesis results from the cumulative effects of all the worms.

■ *Haemonchus contortus*

Haemonchus contortus lives in the "fourth stomach," or abomasum, of sheep, cattle, goats, and wild ruminants of many species. The species has been reported in humans in Brazil and Australia. It is one of the most important nematodes of domestic animals, causing a severe anemia in heavy infections.

The small buccal cavity contains a single well-developed tooth, which pierces the mucosa of the host. Blood is sucked from the wound, giving the transparent worms a reddish color. The large females have the white ovaries wrapped around the red intestine, lending it a characteristic red and white appearance, leading to its common names—"twisted stomach worm" and "barber-pole worm." Prominent cervical papillae are found near the anterior end. The male bursa is powerfully developed, with an asymmetrical dorsal ray (Fig. 26-14). The spicules are 450 to 500 μm long, each with a terminal barb. The vulva has a conspicuous anterior flap in many individuals but not in all. The frequency of occurrence of the vulvar flap seems to vary according to strain.

Infection occurs when the third-stage juvenile, still wearing the loosely fitting second-stage cuticle, is eaten with forage. Exsheathment takes place in the forestomachs. Arriving in the abomasum or upper duodenum, the worm molts within 48 hours, becoming a fourth-stage juvenile with a small buccal cap-

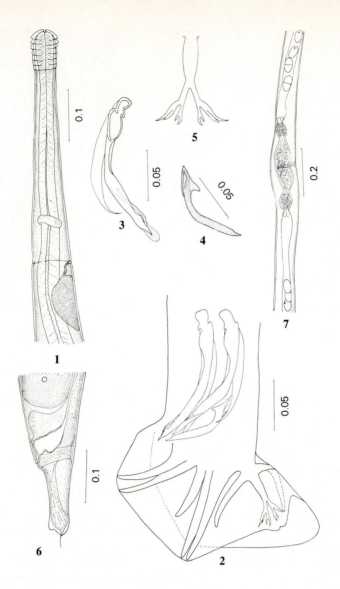

FIG. 26-13

Molineus mustelae, showing the characters typical of the Trichostrongylidae. **1,** Anterior end, lateral view. **2,** Posterior end of male. **3,** Complex spicules, lateral view. **4,** Gubernaculum, lateral view. **5,** Dorsal ray. **6,** Posterior end of female, lateral view. **7,** Midregion of female, showing ovijectors. All scales are in millimeters.

From Schmidt, G.D. 1965. J. Parasitol. 51:164-168.

sule. It feeds on blood, which forms a clot around the anterior end of the worm. The worm molts for a final time in 3 days and begins egg production about 15 days later.

The anemia, emaciation, edema, and intestinal disturbances caused by these parasites result principally from loss of blood and injection of hemolytic proteins into the host's system. The host often dies with heavy infections, but those which survive usually develop an immunity and effect a self-cure.

■ *Ostertagia* **species**

Ostertagia spp. are similar to *H. contortus* in host and location but differ in color, being a dirty brown— hence their common name "brown stomach worm." The buccal capsule is rudimentary and lacks a tooth. Cervical papillae are present. The male bursa is symmetrical. The vulva has a large anterior flap, and the tip of the female's tail bears several cuticular rings.

The life cycle is similar to that of *Haemonchus* except that the third-stage juvenile burrows into the abo-

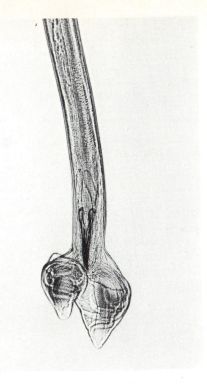

FIG. 26-14

Haemonchus contortus: ventral view of male, showing asymmetrical copulatory bursa.

Photograph by Jay Georgi.

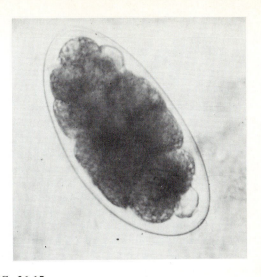

FIG. 26-15

Trichostrongylus egg found in a human stool.

Photograph by David Oetinger.

masal mucosa, where it molts. Returning to the lumen, it feeds, molts, and begins producing eggs about 17 days after infection. *Ostertagia* spp. suck blood but not as much as does *Haemonchus*.

Some common species of *Ostertagia* are *O. circumcincta* in sheep, *O. ostertagi* in cattle and sheep, and *O. trifurcata* in sheep and goats. *Ostertagia ostertagi* and *O. circumcincta* have been reported from humans in Russia. It is possible that the infection followed eating insufficiently cooked abomasum of sheep, cattle, or goats. It is estimated that *O. ostertagi* costs $600 million in the United States alone, due to economic losses to the cattle industry.[25]

■ *Trichostrongylus* species

Trichostrongylus spp. are the smallest members of the family, seldom exceeding 7 mm in length. Many species exist parasitizing the small intestine of ruminants, rodents, pigs, horses, birds, and humans.

They are colorless, lack cervical papillae, and have a rudimentary, unarmed buccal cavity. The male bursa is symmetrical, with a poorly developed dorsal lobe. Spicules are brown and distinctive in size and shape in each species. The vulva lacks an anterior flap.

The life cycle is similar to that of *Ostertagia* except that the worms pass through the abomasum and burrow into the mucosa of the duodenum, where they molt. After returning to the lumen, they bury their heads in the mucosa and feed, grow, and molt for the last time. Egg production begins about 17 days after infection.

Common species of *Trichostrongylus* are *T. colubriformis* in sheep; *T. tenuis* in chickens and turkeys; *T. capricola, T. falcatus,* and *T. rugatus* in ruminants; *T. retortaeformis* and *T. calcaratus* in rabbits; and *T. axei* in a wide variety of mammals.

Eight species of *Trichostrongylus* have been reported in humans, with records from nearly every country of the world: six species in Armenia alone. Rate of infection varies from very low to as high as 69% in southwest Iran[23] and 70% in a village in Egypt.[13] Stoll[27] estimated 5.5 million cases of human infection in the world.

Pathological conditions are identical in humans and other infected animals. Traumatic damage to the intestinal epithelium may be produced by burrowing juveniles and feeding adults. Systemic poisoning by metabolic wastes of the parasites and possible thyroid deficiency, hemorrhage, emaciation, and mild anemia may develop in severe infections.

Diagnosis can be made by finding the characteristic eggs (Fig. 26-15) in the feces or by culturing the juveniles in powdered charcoal. The juveniles are very similar to those of hookworms and *Strongyloides* spp., and careful differential diagnosis is required.

445

Treatment with thiabendazole or with bephenium hydroxynaphthoate has proved effective. Cooking vegetables adequately will prevent many infections in humans.

■ Other trichostrongylids

In addition to the species from ruminants already mentioned, *Cooperia curticei*, *Nematodirus spathiger*, and *N. filicollis* should be included in the group of trichostrongyles that often occur in the same host and cause so much damage. *Hyostrongylus rubidus* is a serious pathogen of swine and can cause death when present in large numbers. *Amidostomum anseris* burrows under the horny lining of the gizzard in ducks and other waterfowl, and it seems to be one of the main causes of mortality in overwintering Canada geese. *Nematospiroides dubius* in mice and *Nippostrongylus brasiliensis* in rats are of no direct medical or veterinary significance, but because they are easily kept in the laboratory, they serve as important tools for investigation of trichostrongyle nematodes.

Superfamily Metastrongyloidea

Metastrongyloidea are the bursate nematode lungworms of mammals. Most species for which the life cycles are known require an invertebrate intermediate host, and some also employ a vertebrate or invertebrate transport host. Most species mature in terrestrial mammals, although several species in numerous genera are important parasites of marine mammals. Taxonomy of the group is unsettled, with several schemes in use. The metastrongyles are fairly homogeneous morphologically, with buccal cavities reduced or absent and bursal lobes and rays reduced. Most are parasitic in the bronchioles, but some inhabit the pulmonary arteries, heart, muscles, and frontal sinuses.

Dictyocaulidae
■ *Dictyocaulus filaria*

This important parasite of sheep and goats shows close relationship to the Trichostrongylidae; for example, the life cycle does not involve an intermediate host. Adults live in the bronchi and bronchioles, where the females produce embryonated eggs. The eggs hatch while being carried out of the respiratory tree by ciliary action. First-stage juveniles appear in the feces and develop to the third stage in contaminated soil, without feeding. The cuticles of both the first and second stages are retained by the third stage until the worm is eaten by a definitive host, then cuticles of all these stages are shed together. Fourth-stage juveniles penetrate the mucosa of the small intestine, enter the lymphatic system, mix with the blood, and

are carried to the lungs, where they enter alveoli and migrate to the bronchioles. They commonly cause death to their host.

These worms are slender and quite long, males reaching 80 mm and females 100 mm. The bursa is small and symmetrical; the spicules are short and boot shaped in lateral view. The uterus is near the middle of the body.

Dictyocaulus arnfieldi in horses and *D. viviparus* in cattle are similar to *D. filaria* in morphology and biology.

Angiostrongylidae
■ *Angiostrongylus cantonensis*

Angiostrongylus cantonensis was first discovered in the pulmonary arteries and heart of domestic rats in China in 1935. Later the worm was found in many species of rats and bandicoots, and it may mature in other mammals throughout southeast Asia, the East Indies, Madagascar, and Oceanica, with infection rates as high as 88%. As a parasite of rats, it attracted little attention, but 10 years after its initial discovery, it was found in the spinal fluid of a 15-year-old boy in Taiwan. It has been discovered since in humans in Hawaii, Tahiti, the Marshall Islands, New Caledonia, Thailand, New Hebrides, the Loyalty Islands, and other places in the eastern hemisphere (Fig. 26-16). It is now known to exist in Louisiana. This illustrates the value of basic research in parasitology to medicine, because when the medical importance of the parasite was realized, the reservoir of infection in rats already was known. Surveys of parasites endemic to the wild fauna of the world remain the first step in understanding the epidemiology of zoonotic diseases.

Angiostrongylus cantonensis is a delicate, slender worm with a simple mouth and no lips or buccal cavity. Males are 15.9 to 19 mm long, whereas females attain 21 to 25 mm. The bursa is small and lacks a dorsal lobe. Spicules are long, slender, and about equal in length and form. An inconspicuous gubernaculum is present. In the female the intertwining of the intestine and uterine tubules gives the worm a conspicuous barber-pole appearance. The vulva is about 0.2 mm in front of the anus. The eggs are thin shelled and unembryonated when laid.

Biology. The eggs are laid in the pulmonary arteries, are carried to the capillaries, and break into the air spaces, where they hatch. The juveniles migrate up the trachea, are swallowed, and are expelled with the feces.

A number of types of molluscs serve as satisfactory intermediate hosts, including slugs and aquatic and terrestrial snails. Terrestrial planarians, freshwater

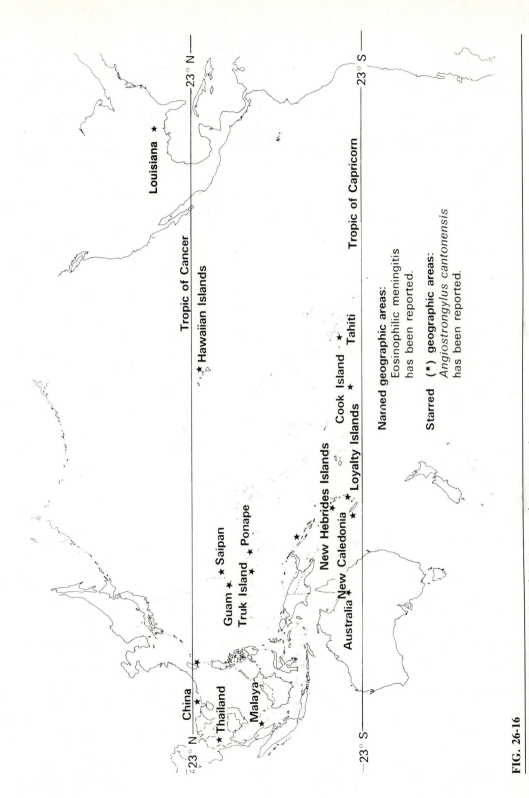

FIG. 26-16

The known geographical distribution of *Angiostrongylus cantonensis*.
AFIP neg. no. 68-5861.

shrimp, land crabs, and coconut crabs serve as paratenic hosts. Frogs have been found naturally infected with infective juveniles.[3] Experimentally, Cheng[6] infected American oysters and clams, and Wallace and Rosen[30] succeeded in infecting crabs. All juveniles thus produced were infective to rats.

When eaten by a definitive host, the third-stage juveniles undergo an obligatory migration to the brain, which they leave 4 weeks after ingestion. Maturation occurs in the pulmonary arteries in about 6 weeks. Many wander in the body and mature in other locations, primarily in the central nervous system, meninges, and eyes.

Epidemiology. Humans or other mammals become infected when third-stage juveniles are ingested. There may be several avenues of human infection, depending on the food habits of particular peoples. In Tahiti, where freshwater shrimp are known to be infected, it is a common practice to catch and eat them raw or to make sauce out of their raw juices. It is possible to eat slugs or snails accidentally with raw vegetables or fruit, so this route of infection should be considered. In Thailand and Taiwan, raw snails are often considered a delicacy. Heyneman and Lim[9] showed that infective juveniles escape from slugs and are left behind in their mucous trail. These investigators also found juveniles on lettuce sold in a public market in Malaya. Thus, although the epidemiology of angiostrongyliasis is not completely known, ample opportunities for infection exist.

Pathology. For many years a disease of unknown cause was recognized in tropical Pacific islands and was named **eosinophilic meningoencephalitis.** Patients with this condition have high eosinophil counts in peripheral blood and spinal fluid in about 75% of the cases and increased lymphocytes in cerebrospinal fluid. Nervous disorders commonly accompany these symptoms, occasionally followed by death. It is now known that *A. cantonensis* is at least one cause of this condition.

The presence of worms in blood vessels of the brain and meninges, as well as free-wandering worms in the brain tissue itself, results in serious damage. Some effects of such infection are severe headache, fever in some cases, paralysis of the fifth cranial nerve, stiff neck, coma, and death. Destruction of brain and spinal cord cells by trauma and immune responses evoked by dead worms results in vague symptoms for which the cause is most difficult to diagnose.

Diagnosis and treatment. When the symptoms described above appear in a patient in areas of the world where *A. cantonensis* exists, angiostrongyliasis should be suspected. It should be kept in mind that many of these symptoms can be produced by hydatids, cysticerci, flukes, *Strongyloides, Trichinella,* various juvenile ascarids, and possibly other lungworms. Alicata[2] and Ash[4] differentiate the juveniles of several species of metastrongylids that could be confused with *A. cantonensis.*

Thiabendazole shows promise in treating early, invasive stages, but little is known of treating the adults. Dead worms in blood vessels and the central nervous system may be more dangerous than live ones. A spinal tap to relieve headache may be recommended.

Angiostrongylus costaricensis parasitizes the mesenteric arteries of rats in Central and South America, southern North America, and Cuba.[20] Morera and Céspedes[21] found more than 70 human cases of infection with this parasite. The worms matured in the mesenteric arteries and their intramural branches. Most damage was to the wall of the intestine, especially the cecum and appendix, which had become thickened and necrotic, with massive eosinophilic infiltration.[19] Abdominal pain and high fever were the most evident symptoms. Evidently, intestinal disorders were caused by blockage of the arterioles supplying the area by eggs and juveniles of the parasites. No symptoms of meningoencephalitis, typical of *A. cantonensis,* were noted. However, eosinophilic meningoencephalitis has been reported in patients infected with *A. cantonensis* in Cuba.[22] It is possible that *A. costaricensis* was involved.

Ubelaker[28] convincingly argued that both *A. cantonensis* and *A. costaricensis* belong in the genus *Parastrongylus.* We leave them in *Angiostrongylus* for now because of the immense amount of literature that has accumulated under that name.

Other species of *Angiostrongylus* are potential parasites of humans, perhaps actually causing a disease that has yet to be diagnosed. North American species are *A. michiganensis* and *A. blarina* in shrews, *A. schmidti* from rice rats, and *A. gubernaculatus* in mustelid carnivores. A key to species of the world is given by Kinsella.[12]

■ Other metastrongylids

Protostrongylus rufescens parasitizes the bronchioles of ruminants in many parts of the world. Its intermediate hosts are terrestrial snails, in which it develops to the third stage, and the definitive host is infected when the snail is eaten along with forage. Mountain sheep in America are seriously threatened by this and related species, which take a high toll of lambs every spring. Hibler, Lange, and Metzger[10] demonstrated transplacental transmission of *Protostrongylus* spp. in bighorn sheep.

Muellerius capillaris lives in nodules in the parenchyma of the lungs of sheep and goats in most areas of the world. The life cycle is similar to that of *Protostrongylus rufescens,* involving a snail intermediate host.

Metastrongylus apri mainly infects swine, but sheep, cattle, and three cases in humans have been reported. Adults live in the bronchioles, and the eggs may hatch as in *Dictyocaulus,* or they may appear in the feces before hatching. An earthworm intermediate host is required for development to the infective third stage. This lungworm is also known to serve as a vector for the virus that causes **swine influenza.**[15] The nematodes serve as reservoirs for the disease, since they may live up to 3 years while carrying the virus within their bodies.

REFERENCES

1. Abadi, K. 1985. Single dose mebendazole therapy for soil-transmitted nematodes. Am. J. Trop. Med. Hyg. 34:129-133.
2. Alicata, J.E. 1963. Morphological and biological differences between the infective larvae of *Angiostrongylus cantonensis* and those of *Anafilaroides rostratus.* Can. J. Zool. 41:1179-1183.
3. Ash, L.R. 1968. The occurrence of *Angiostrongylus cantonensis* in frogs of New Caledonia with observations on paratenic hosts of metastrongyles. J. Parasitol. 54:432-436.
4. Ash, L.R. 1970. Diagnostic morphology of the third-stage larvae of *Angiostrongylus cantonensis, Angiostrongylus vasorum, Aelurostrongylus abstrusus,* and *Anafilaroides rostratus* (Nematoda: Metastrongyloidea). J. Parasitol. 56:249-253.
5. Behnke, J.M. 1987. Do hookworms elicit protective immunity in man? Parasitol. Today 3:200-206.
6. Cheng, T.C. 1965. The American oyster and clam as experimental intermediate hosts of *Angiostrongylus cantonensis.* J. Parasitol. 51:296.
7. Dönges, J., and O. Madecki. 1968. The possibility of hookworm infection through breast milk. German Med. Monthly 13:391-392.
8. Guerrero, J., M.R. Page, and G.A. Schad. 1982. Anthelminthic activity of closantel against *Ancylostoma caninum* in dogs. J. Parasitol. 68:616-619.
9. Heyneman, D., and B.L. Lim. 1967. *Angiostrongylus cantonensis:* proof of direct transmission with its epidemiological implications. Science 158:1057-1058.
10. Hibler, C.P., R.E. Lange, and C.J. Metzger. 1972. Transplacental transmission of *Protostrongylus* spp. in bighorn sheep. J. Wildl. Dis. 8:389.
11. Hotez, P.J., N. LeTrang, and A. Cerami. 1987. Hookworm antigens: the potential for vaccination. Parasitol. Today 3:247-249.
12. Kinsella, J.M. 1971. *Angiostrongylus schmidti* sp. n. (Nematoda: Metastrongyloidea), from the rice rat. *Oryzomys palustris,* in Florida, with a key to the species of *Angiostrongylus* Kamensky, 1905. J. Parasitol. 57:494-497.
13. Lawless, D.K., R.E. Kuntz, and C.P.A. Strome. 1956. Intestinal parasites in an Egyptian village of the Nile Valley with emphasis on the protozoa. Am. J. Trop. Med. Hyg. 5:1010-1014.
14. Layrisse, M., A. Paz, N. Blumenfeld, and M. Roche. 1961. Hookworm anemia: iron metabolism and erythrokinetics. Blood 18:61-72.
15. Lee, D.L. 1971. Helminths as vectors of micoorganisms. In Fallis, A.M., editor. Ecology and physiology of parasites. University of Toronto Press, Toronto, pp. 104-122.
16. Levine, N.D. 1980. Nematode parasites of domestic animals and of man, ed. 2. Burgess Publishing Co., Minneapolis.
17. Little, M.D., N.A. Halsey, B.L. Cline, and S.P. Katz. 1983. *Ancylostoma* larva in a muscle fiber of man following cutaneous larva migrans. Am. J. Trop. Med. Hyg. 32:1285-1288.
18. Looss, A. 1898. Zur Lebensgeschichte des *Ankylostoma duodenale.* Cbt. Bakt. 24:441-449, 483-488.
19. Loría-Cortés, R., and J.F. Lobo-Sanahuja. 1980. Clinical abdominal angiostrongylosis. A study of 116 children with intestinal eosinophilic granuloma caused by *Angiostrongylus costaricensis.* Am. J. Trop. Med. Hyg. 29:538-544.
20. Morera, P. 1985. Abdominal angiostrongyliasis: a problem of public health. Parasitol. Today 1:173-175.
21. Morera, P., and R. Céspedes. 1971. *Angiostrongylus costaricensis* n. sp. (Nematoda: Metastrongyloidea), a new lungworm occurring in man in Costa Rica. Rev. Biol. Trop. 18:173-185.
22. Pascual, J.E., R.P. Bouli, and H. Aguiar. 1981. Eosinophilic meningoencephalitis in Cuba, caused by *Angiostrongylus cantonensis.* Am. J. Trop. Med. Hyg. 30:960-962.
23. Sabha, G.H., F. Arfaa, and H. Bijan. 1967. Intestinal helminthiasis in the rural area of Khuzestan, south-west Iran. Ann. Trop. Med. Parasitol. 61:352-357.
24. Schad, G.A., et al. 1984. Paratenesis in *Ancylostoma duodenale* suggests possible meat-borne human infection. Trans. R. Soc. Trop. Med. Hyg. 78:203-204.
25. Smith, G., and B.T. Granfell. 1985. The population biology of *Ostertagia ostertagi.* Parasitol. Today 1:76-81.
26. Spencer, H. 1973. Nematode diseases. I. In Spencer, H., editor. Tropical pathology. Springer-Verlag, Berlin, pp. 457-509.
27. Stoll, N.R. 1947. This wormy world. J. Parasitol. 33:1-18.
28. Ubelaker, J.C. 1986. Systematics of species referred to the genus *Angiostrongylus.* J. Parasitol. 72:237-244.
29. Velasquez, C., and B.C. Cabrera. 1968. *Ancylostoma ceylanicum* (Looss), in a Filipino woman. J. Parasitol. 54:430-431.
30. Wallace, G.D., and L. Rosen. 1966. Studies on eosinophilic meningitis. 2. Experimental infection of shrimp and crabs with *Angiostrongylus cantonensis.* Am. J. Epidemiol. 84:120-141.
31. Walterspiel, J.N., G.A. Schad, and G.R. Buchanan. 1984. Direct transfer of adult hookworms *(Ancylostoma duodenale)* from dog to child for therapeutic purposes. J. Parasitol. 70:217-219.
32. Warren, K.S. 1974. Helminthic diseases endemic in the United States. Am. J. Trop. Med. Hyg. 23:723-730.
33. Yoshida, Y., K. Okamoto, A. Higo, and K. Imai. 1960. Studies on the development of *Necator americanus* in young dogs. Jpn. J. Parasitol. 9:735-743.

SUGGESTED READINGS

Alicata, J.E. 1988. *Angiostrongylus cantonensis* (eosinophilic meningitis): historical events in its recognition as a new parasitic disease of man. J. Wash. Acad. Sci. 78:38-46.

Cross, J.H., editor. 1979. Studies on angiostrongyliasis in eastern Asia and Australia. U.S. Naval Medical Research Unit No. 2 (Special Publication), 164 pp. (Available from NAMRU-2, Box 14, APO San Francisco, CA 96236.)

Dooley, J.R., and R.C. Neafie. 1976. Angiostrongyliasis: *Angiostrongylus cantonensis* infections. In Binford, C.H., and D.H. Connor, editors. Pathology of tropical and extraordinary diseases, vol. 2, sect. 9. Armed Forces Institute of Pathology, Washington, D.C.

Frenkel, J.K. 1976. Angiostrongyliasis: *Angiostrongylus costaricensis* infections. In Binford, C.H., and D.H. Connor, editors. Pathology of tropical and extraordinary diseases, vol. 2, sect. 9. Armed Forces Institute of Pathology, Washington, D.C.

Looss, A. 1898. Zur Lebensgeschichte des *Ankylostoma duodenale*. Cbt. Bakt. 24:441-449, 483-488. (For the parasite historian: the first account of a hookworm life cycle.)

Meyers, W.M., and R.C. Neafie. 1976. Creeping eruption. In Binford, C.H., and D.H. Connor, editors. Pathology of tropical and extraordinary diseases, vol. 2, sect. 9. Armed Forces Institute of Pathology, Washington, D.C.

Meyers, W.M., R.C. Neafie, and D.H. Connor. 1976. Ancylostomiasis. In Binford, C.H., and D.H. Connor, editors. Pathology of tropical and extraordinary disease, vol. 2, sect. 9. Armed Forces Institute of Pathology, Washington, D.C.

Stoll, N.R. 1972. The osmosis of research: example of the Cort hookworm investigations. Bull. N.Y. Acad. Med. 48:1321-1329.

Travassos, L. 1937. Revisão da familia Trichostrongylidae Leiper, 1912. Monogr. Inst. Oswaldo Cruz. I. (Out of date but still a valuable reference, especially the 295 plates.)

CHAPTER 27

ORDER ASCARIDATA: LARGE INTESTINAL ROUNDWORMS

The ascaroid worms are typically large, stout, intestinal parasites with three large lips. However, there are minute species with small or no lips, large species with two lips, and small species with well-defined lips. A preanal sucker is found on males of some. In one large group the esophagointestinal junction is highly specialized, with muscular or glandular appendages. Usually, however, the esophagus is simple and muscular. The life cycle is usually simple, lacking an intermediate host, although such a host is required in a few species.

Of the several families in this order, we will emphasize the Ascaridae and Toxocaridae, which have the most medical importance. We will discuss other families briefly.

FAMILY ASCARIDAE

The ascarids are among the largest of nematodes, some species achieving a length of 18 inches or more. Cervical, lateral, and caudal alae are absent, as are any esophageal ceca or ventriculi. Three large rounded or trapezoidal lips are present; interlabia are absent. Spicules are simple and equal. This family contains one of the oldest associates of people— *Ascaris,* the large intestinal roundworm.

Ascaris lumbricoides and *Ascaris suum*

Because of their great size, abundance, and cosmopolitan distribution, these nematodes may well have been the first parasites known to humans. Certainly the ancient Greeks and the Romans were familiar with them, and they were mentioned in the Ebers Papyrus. It is probable that *A. lumbricoides* was originally a parasite of pigs that adapted to humans when swine were domesticated and began to live in close association with humans—or perhaps it was a human parasite that we gave to pigs. This is not surprising, since the physiologies of people and swine are remarkably similar, as on occasion, are their eating and social habits. Today two populations of this parasite exist,

one in humans and one in pigs. They show a strong host specificity, but the two forms are so close morphologically that they were long considered the same species. Sprent[32] pointed out slight differences in the tiny denticles on the dentigerous ridges along the inner edge of the lips. This difference seems consistent and is much clearer when the structures are viewed with the scanning electron microscope,[37] therefore we can now consider the two as separate species, *A. suum* from pigs and *A. lumbricoides* from humans. This seems to be a good example of evolution in action.[24] Each species may diverge even further with time, now that they have been reproductively isolated in many parts of the world where pigs no longer enjoy the homes of their masters. Although it generally is considered that *A. suum* cannot infect humans, the topic is still controversial. Considerable evidence indicates that of least some cases of human infection are due to *A. suum.*[36]

Aside from the host specificity and the characteristics of the denticles, there are few, if any, other differences in the two species, and the following remarks on morphology and biology apply to both equally.

■ Morphology

In addition to their great size (Fig. 27-1), these species are characterized by having three prominent lips, each with a dentigerous ridge (Fig. 27-2), and no interlabia or alae. Lateral lines are visible grossly.

Males are 15 to 31 cm long and 2 to 4 mm at greatest width. The posterior end is curved ventrad, and the tail is bluntly pointed. Spicules are simple, nearly equal, and measure 2 to 3.5 mm long. No gubernaculum is present.

Females are 20 to 49 cm long and 3 to 6 mm wide. The vulva is about one third the body length from the anterior end. The ovaries are extensive, and the uteri may contain up to 27 million eggs at a time, with 200,000 being laid per day. Fertilized eggs (Fig. 27-3) are oval to round, 45 to 75 μm long by 35 to 50 μm

FIG. 27-1

Ascaris lumbricoides, males *(right)* and females *(left).* Females are up to 18 inches long.

Courtesy Ann Arbor Biological Center.

wide, with a thick, lumpy outer shell (**mammillated, uterine, or proteinaceous layer**) that is contributed by the uterine wall. When the eggs are passed in the feces, the mammillated layer is bile stained to a golden brown. The eggs are usually uncleaved when laid and passed in the feces. An unfertilized female or one in early stages of oviposition commonly deposits unfertilized eggs (Fig. 27-4) that are longer and narrower, measuring 88 to 94 μm long by 44 μm wide. Because formation of vitelline, chitinous, and lipid layers of the egg ensues only after sperm penetration (p. 400), only the proteinaceous layer can be clearly distinguished in unfertilized eggs.

■ Biology

A period of 9 to 13 days is the minimal time required for the embryo to develop into an active first-stage juvenile. Although extremely resistant to low temperature, desiccation, and strong chemicals, embryonation is retarded by such factors. Sunlight and high temperatures are lethal in a short time. The juvenile molts to the second stage before hatching through an indistinct operculum (Fig. 27-5). Contrary to the

pattern of most parasitic nematodes, the apparent second-stage juvenile is infective to the host (see discussion of embryonation in Chapter 23). However, a second molt within the egg has been reported in both *A. suum* and *A. lumbricoides;* thus the infective juvenile may in fact be the third stage, as is common among other nematodes.[22]

Infection occurs when embryonated eggs are swallowed with contaminated food and water. They hatch in the duodenum, where the juveniles penetrate the mucosa and submucosa and enter lymphatics or venules. After passing through the right heart, they enter the pulmonary circulation and break out of capillaries into the air spaces. Many worms get lost during this migration and accumulate in almost every organ of the body, causing acute tissue reactions.

While in the lungs, the juveniles molt twice (or complete the second molt and then molt again) during a period of about 10 days, to a length of 1.4 to 1.8 mm. They then move up the respiratory tree to the pharynx, where they are swallowed. Many juveniles make this last step of their migration before molting to the fourth stage, but these cannot survive the gastric juices in the stomach. Fourth-stage juveniles are resistant to such a hostile environment and pass through the stomach to the small intestine, where they mature. Within 60 to 65 days after being swallowed, they begin producing eggs. It seems curious that the worms embark on such a hazardous migration only to end up where they began. One theory to account for it suggests that the migration simulates an intermediate host, which normally would be required for the juvenile of the ancestral form to develop to the third stage. Another possibility is that the ancestor was a skin penetrator for which the migration was a developmental necessity.

■ Epidemiology

The dynamics of *Ascaris* infection are essentially the same as for *Trichuris.* Indiscriminate defecation, particularly near habitations, "seeds" the soil with eggs that remain viable for many months or even years. The resistance of *Ascaris* eggs to chemicals is almost legendary. They can embryonate successfully in 2% formalin, in potassium dichromate, and in 50% solutions of hydrochloric, nitric, acetic, and sulfuric acid, among other similar inhospitable substances.[30] This extraordinary chemical resistance is the result of the lipid layer of eggshell, which contains the ascarosides.

The longevity of *Ascaris* eggs also contributes to the success of the parasite. Brudastov et al.[5] infected themselves with eggs kept for 10 years in soil at Samarkand, Russia. Of these eggs, 30.7% to 52.7%

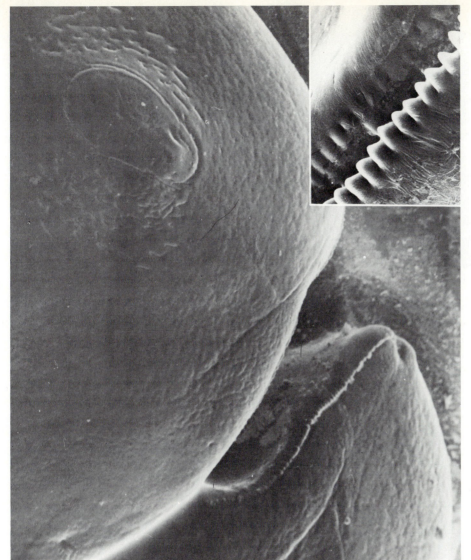

A

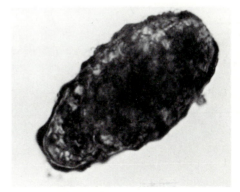

B

FIG. 27-2

A, Lips of *Ascaris lumbricoides*. Note the large double papilla on the upper lip and the dentigerous ridge on the lower one. **B,** Enlarged view of the denticles of *Ascaris suum*.

Photographs by John Ubelaker.

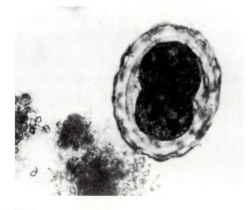

FIG. 27-3

Fertilized egg of *Ascaris lumbricoides* from a human stool. Eggs of this species are 45 to 75 μm long.

Photograph by Robert E. Kuntz.

FIG. 27-4

Unfertilized egg of *Ascaris lumbricoides* from a human stool. Such eggs are 88 to 94 μm long.

Photograph by Robert E. Kuntz.

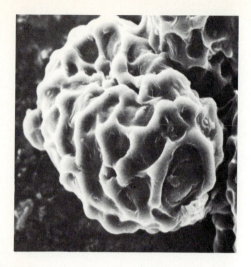

FIG. 27-5

Scanning electron micrograph showing egg of *Ascaris lumbricoides*. An operculum is visible at one end.

Photograph by John Ubelaker.

were found to be still infective. Because of this longevity, it is impossible to prevent reinfection when houseyards have been liberally seeded with eggs, even when proper sanitation habits are initiated later.

Contamination, then, is the typical means of infection. Children are the most likely to become infected (or reinfected) by eating dirt or placing soiled fingers and toys in their mouths. In regions in which night soil is used as fertilizer, principally the Orient, Germany, and certain Mediterranean countries, uncooked vegetables become important vectors of *Ascaris* eggs. Experimental support for this came from Mueller,[23] who seeded a strawberry plot with eggs; he and volunteers ate unwashed strawberries from this plot every year for 6 years and became infected each year. Cockroaches have been found to carry and disseminate *Ascaris* eggs.[6]

Even windborne dust can carry *Ascaris* eggs, when conditions permit. Bogojawlenski and Demidowa[4] found *Ascaris* eggs in nasal mucus of 3.2% of schoolchildren examined in the Soviet Union. From the nasal mucosa to the small intestine is a short trip in children. Dold and Themme[12] found *Ascaris* eggs on 20 German banknotes in actual circulation.

There can be little doubt that ascariasis is present in epidemic proportions in the southeastern United States.[11] Surveys in various states between 1956 and 1970 showed prevalences of 20% to 60% in the childhood population. Of 26,489 stool samples submitted to the South Carolina State Laboratory in 1969, 15% contained *A. lumbricoides*.[11] Because no active cam-

paign of eradication has been mounted since then, there is no reason to believe that the figure is any lower now. Worldwide about 1 billion persons, almost one quarter of the world population, are infected.[9]

■ Pathogenesis

Little damage is caused by the penetration of intestinal mucosa by newly hatched worms. Juveniles that become lost and wander and die in anomalous locations, such as the spleen, liver, lymph nodes, or brain, often elicit an inflammatory response that may cause vague symptoms that are difficult to diagnose and may be confused with other diseases. This is apparently the fate of most *A. suum* juveniles in humans. Transplacental migration into a developing fetus is also known. Allergy and immunopathology of ascariasis was reviewed by Coles.[7]

When the juveniles break out of lung capillaries into the respiratory system, they cause a small hemorrhage at each site. Heavy infections will cause small pools of blood to accumulate, which then initiate edema with resultant clogging of air spaces. Accumulations of white blood cells and dead epithelium add to the congestion, which is known as **Ascaris pneumonitis (Loeffler's pneumonia).** Large areas of lung can become diseased, and when bacterial infections become superimposed, death can result. One instance is known in which a perhaps unbalanced parasitology graduate student vented his ire on his roommates by "seeding" their breakfast with embryonated *Ascaris* eggs. They almost died before their malady was diagnosed.[1]

Pathogenesis of adult worms can be discussed conveniently in two categories: normal worm activities in the small intestine and wandering worms.

Although it is probable that *Ascaris* occasionally sucks blood from the intestinal wall, its main food is liquid contents of the intestinal lumen. In moderate and heavy infections the resulting theft of nourishment can cause malnutrition and underdevelopment in small children.[11,41] Abdominal pains and sensitization phenomena—including rashes, eye pain, asthma, insomnia, and restlessness—often result as allergic responses to metabolites produced by the worms.

A massive infection can cause fatal intestinal blockage[2] (Frontispiece and Fig. 27-6). Why, in one case, do large numbers of worms cause no apparent problem, whereas in another, the worms knot together to form a mass that completely blocks the intestine? It is known that certain drugs, such as tetrachloroethylene, if used to treat hookworm, can aggravate *Ascaris* to knot up, but other factors are still unknown. Penetration of the intestine or appendix is not uncommon.

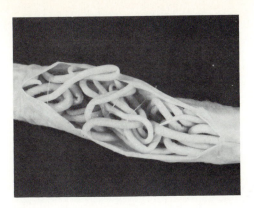

FIG. 27-6

Intestine of a pig, nearly completely blocked by *Ascaris suum* (threads were inserted to hold worms in place). Such heavy infections are also fairly common with *A. lumbricoides* in humans.

Photograph by L.S. Roberts.

The resulting peritonitis is usually quickly fatal. According to Louw,[21] 35.5% of all deaths in acute abdominal emergencies of children in Capetown were caused by *Ascaris*.

Wandering adult worms cause various conditions, some serious, some bizarre, all unpleasant. The tropism of a female to squirm through the coiled tail of a male causes her to wander if no males are present. A similar restlessness can be observed in even higher forms of animals. Overcrowding may also lead to wandering. A downstream wandering leads to the appendix, which can be clogged or penetrated, or to the anus, with an attendant surprise for the unsuspecting host. Upstream wandering leads to the pancreatic and bile ducts, possibly occluding them with grave results. Multiple liver abscesses have resulted from such invasion.[28] Worms reaching the stomach are aggravated by the acidity and writhe about, often causing nausea. The psychological trauma induced in one who vomits an 18-inch ascarid is difficult to quantify. Aspiration of the vomited worm can result in death.[11] Worms that reach the esophagus, usually while the host is asleep, may crawl into the trachea, causing suffocation or lung damage, or into the eustachian tubes and middle ears, causing extensive damage, or may simply exit through the nose or mouth, causing a predictable consternation.

■ **Diagnosis and treatment**

Accurate diagnosis of migrating juveniles is impossible at this time. Demonstration of juveniles in sputum is definitive, provided the technician can identify them.

Most diagnoses are made by identifying the characteristic, mammillated eggs in the stool or by an appearance of the worm itself. So many eggs are laid each day by one worm that one or two direct fecal smears are usually sufficient to demonstrate at least one. Otherwise, diagnosis is difficult if not impossible. *Ascaris* should be suspected when any of the previously listed pathogenic conditions are noted. Most light infections are asymptomatic, and presence of worms may be determined only by spontaneous elimination of spent individuals from the anus.

Mebendazole is the drug of choice, with pyrantel pamoate as an alternative. No efficient treatment of migrating juveniles has been discovered.

Parascaris equorum

This large nematode is the only ascarid found in horses. *Parascaris equorum* is a cosmopolitan species that also infects the mule, ass, and zebra. It is very similar in gross appearance to *A. lumbricoides* but is easily differentiated by its huge lips, which give it the appearance of having a large, round head.

The life cycle is similar to that of *A. lumbricoides*, involving a lung migration. Resulting pathogenesis is especially important in young animals, with pneumonia, bronchial hemorrhage, colic, and intestinal disturbances resulting in unthriftiness and morbidity. Intestinal perforation or obstruction is common. Older horses are usually immune to infection. Prenatal infection is not known to occur. Piperazine is the preferred drug and is usually administered with a purgative to prevent intestinal obstruction.

Baylisascaris procyonis

This is a very common intestinal parasite of raccoons in North America. Other, similar species are found in bears, skunks, badgers, and other carnivores. When embryonated eggs are swallowed by a raccoon, they will hatch in the small intestine and mature. However, *Baylisascaris* also can become paratenic when the eggs are swallowed by a different vertebrate. Most commonly the paratenic host is a rodent, bird, or lagomorph.[18] In these animals the parasite juveniles wander, often invading the central nervous system. Resulting debilitation, sometimes confused with rabies, makes them vulnerable to attack by raccoons, in which the juveniles are freed by digestion to grow to maturity. Unfortunately, the juveniles affect humans in the same way. Several proven and suspected cases of infection with *B. procyonis* have been reported, with fatal cases in children.[15] Retinal involvement is common.

The epidemiology of infection requires close contact between humans and raccoons. The scavenging raccoon is a bold animal that is comfortable near, and sometimes in, human dwellings and outbuildings. They have preferred defecation sites, which are dangerous sources of infection to humans and other animals. The eggs can remain infective for years under ideal conditions, so once an area is contaminated it is nearly impossible to decontaminate it.[17] Raccoons are popular pets, especially when young. Because they can become infected while very young, by ingesting eggs from their mother's skin or fur, wild-caught cubs may well be infected. This is another way this dangerous parasite can be brought into contact with humans.

Other species of *Baylisascaris* may have similar pathogenicity, but most hosts are not as likely to come in close contact with humans. Pet skunks infected with *B. columnaris* are potential hazards, however.

FAMILY TOXOCARIDAE
Toxocara canis and *Toxocara cati*

These two species are cosmopolitan parasites of domestic dogs and cats and their relatives, and they have been found as adults in humans on several occasions. Except for the cervical alae, the biology and morphology of *T. canis* and *T. cati* are similar. The alae of *T. canis* are long and narrow, whereas those of *T. cati* are short and broad. The following remarks apply generally to both.

It is not uncommon for 100% of puppies and kittens to be infected in enzootic areas. As the result of prenatal infections (to be discussed further), one may expect even puppies in well-cared-for kennels to be infected at birth, and they are treated accordingly. The casual owner of a new puppy or kitten is likely to be startled by the vomiting by the pet of a number of large, active worms. Older dogs and cats seem to develop strong immunity to further infection, and they harbor adult worms less often. The reported age resis-

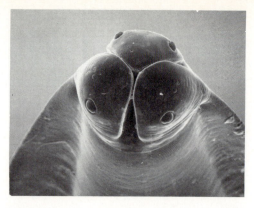

FIG. 27-7

Scanning electron micrograph of *Toxocara cati:* en face view illustrating the three lips with sensory papillae.
Photograph by John Ubelaker.

tance of dogs not previously exposed to *T. canis* is partly related to size of infective dose; a smaller number of eggs administered is more likely to lead to patent infection.[13]

Adults look basically like *Ascaris*, only they are much smaller. Three lips are present (Fig. 27-7). Unlike *Ascaris*, however, *Toxocara* has prominent cervical alae in both sexes. Males are 4 to 6 cm, and females are 6.5 to more than 15 cm long. The brownish eggs are almost spherical, with surficial pits, and are unembryonated when laid. *Toxocara canis* and *T. cati* have slightly different eggs.

■ Biology (Fig. 27-8)

Adult worms live in the small intestine of their host (Fig. 27-9), producing prodigious numbers of eggs, which are passed with the host's feces. Development of the second-stage juvenile takes 5 to 6 days under

FIG. 27-8

Life cycle of *Toxocara canis.* **A,** En face view of adult. **B,** Dorsal view of anterior end of adult. **C,** Lateral view of anterior end of adult. **D,** Lateral view of posterior end of adult male. **E,** Ventral view of tail of adult male. **F,** Lateral view of tail of adult female. **G,** Embryonated egg. **H,** Dog definitive host. **I,** Pup definitive host. **J,** Mouse host. **K,** Human in which visceral larva migrans occurs, as in mice. *1,* Dorsal lip; *2,* lateroventral lip; *3,* labial papillae; *4,* mouth; *5,* cervical alae; *6,* preanal papillae; *7,* anus; *8,* spicules; *9,* postanal papillae; *10,* esophagus; *11,* ventriculus; *12,* intestine; *13,* anus; *14,* excretory pore; *15,* excretory gland; *16,* nerve ring; *a,* adult worms relatively rare in intestine of adult dogs but common in young ones (*i'*); *b,* eggs voided with feces; *c,* eggs undeveloped when passed in feces; *d,* egg in morula stage; *e,* first stage juvenile in egg; *f,* second stage infective juvenile in egg; *g,* infective egg being swallowed by adult dog; *h,* eggs hatch in stomach; *i,* juveniles (stippled) go into small intestine; *j,* juveniles penetrate intestinal wall and enter hepatic portal vein; *k,* juveniles leave capillaries in liver; *l,* some juveniles

Continued on facing page.

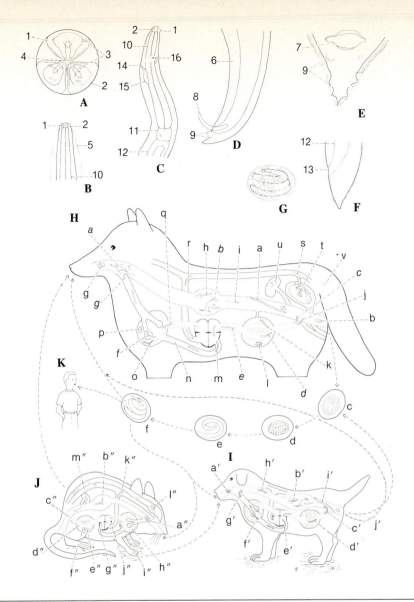

accumulate in liver parenchyma; *m,* some juveniles continue in blood stream through right side of heart; *n,* juveniles enter lungs; *o,* juveniles enter lung parenchyma; *p,* juveniles in lung parenchyma; *q,* some juveniles pass through lungs to left side of heart; *r,* juveniles enter general circulation; *s,* juveniles enter umbilical artery and eventually go into fetus; *t,* juveniles in liver, heart, and lungs of fetus; *u,* juveniles in kidneys; *v,* juveniles in skeletal muscles; *a̲,* mouse with juveniles in tissues being swallowed; *b̲,* juveniles (unstippled) freed from mouse by digestive processes in stomach; *c̲,* juveniles enter hepatic portal vein; *d̲* and *e̲,* juveniles migrate to lungs via heart; *f̲,* juveniles enter bronchioles from blood vessels; *g̲,* juveniles migrate up trachea, are swallowed, and develop to adults, but infrequently in grown dogs; *a′,* embryonated eggs swallowed by puppy; *b′,* eggs hatch in stomach; *c′* to *e′,* juveniles (stippled) enter hepatic portal vein and migrate through liver, heart, and up into lungs; *f′,* juveniles enter alveoli and migrate up trachea; *g′,* juveniles in pharynx; *h′,* juveniles in esophagus are swallowed; *i′,* juveniles develop readily to adults and lay eggs (when infected mice are eaten, the juveniles [unstippled] migrate and develop in the same manner as those from eggs); *j′,* infective juvenile passed in feces; *a″,* embryonated eggs swallowed by mice and other rodents; *b″,* eggs hatch in intestine; *c″,* juveniles enter hepatic portal vein; *d″,* some juveniles enter liver parenchyma; *e″,* juveniles remain in liver; *f″* and *g″,* some juveniles migrate through liver and heart; *h″,* some juveniles enter lung parenchyma; *i″,* juveniles remain in lung parenchyma; *j″* and *k″,* other juveniles pass through heart and into arterial circulation; *l″* and *m″,* juveniles enter central nervous system and kidneys where they remain inactive and without further development until released in stomach of dogs.

Courtesy Olsen, O.W. 1974. Animal parasites, their life cycles and ecology, ed. 3. University Park Press, Baltimore.

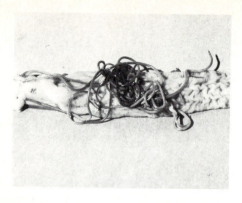

FIG. 27-9

Intestine of a domestic cat, opened to show numerous *Toxocara cati*.

Photograph by Robert E. Kuntz.

optimal conditions. The fate of ingested juveniles depends on the age and immunity of the host. If the puppy is young and has had no prior infection, the worms hatch and migrate through the portal system and lungs and back to the intestine, as in *A. lumbricoides*.

If the host is an older dog, particularly with some immunity acquired from past infection, the juveniles do not complete the lung migration. They wander through the body, eventually becoming inactive but remaining alive for a long period. If a bitch becomes pregnant, the dormant juveniles apparently are activated by host hormones and reenter the circulatory system, where they are carried to the placentas. There they can penetrate through to the fetal bloodstream, where they complete a lung migration en route to the intestine. Thus a puppy can be born with an infection of *Toxocara*, even though the dam has shown no sign of infection.

A third option in the life cycle of *T. canis* is offered when a rodent eats embryonated eggs. In this unfavorable host the juvenile begins to migrate but becomes lost and dormant. If the rodent is eaten by a dog that is not immune, the worms promptly migrate through the lungs to the intestine. Although this adaptability favors survival of the parasite, it bodes ill for certain accidental hosts, such as humans.

According to Sprent,[33] the life cycle of *T. cati* varies from that of *T. canis* in an apparent absence of prenatal infection. Some development occurs in the intestinal or stomach wall, and the lung migration may be omitted. Also, transport hosts may play a greater role.

Visceral larva migrans

When eggs of *T. canis,* and probably also of *T. cati*, are eaten by an improper host, the juveniles hatch and begin the typical liver-lung-intestine migration. However, they do not complete the normal migration but undergo developmental arrest and begin an extended, random wandering through the body. The resulting disease entity is known as **visceral larva migrans,** in contrast to cutaneous larva migrans (p. 000). In its widest sense, visceral larva migrans can be caused by a variety of spirurid, filariid, strongylid, and other nematodes in addition to *Toxocara*. Thus Sprent[34] listed 36 species known to occur in humans in Australia and southeast Asia. Even hookworm juveniles that cause cutaneous larva migrans occasionally enter the deep tissues of the body, thereby initiating a visceral disease. In some of these, humans are true intermediate hosts, since some development of the worm occurs.

In a narrower sense, visceral larva migrans occurs only when the juvenile maintains an extended migration but *does not undergo further development itself.* In these cases the infected animal is a paratenic host, and the infection reflects a normal element in the life cycle of the parasite. The behavior of the juveniles in humans is essentially the same as in a natural paratenic host.[3]

Therefore, although several genera of nematodes can cause visceral larva migrans, we will emphasize *Toxocara,* which is by far the most common cause of the condition in human beings, with *T. canis* seeming to be the most important species.

▪ Epidemiology

Only a few years ago it was generally assumed that dog and cat worms could not infect humans or were not dangerous to them. We have come to realize since the early 1950s that the assumption is not true, particularly in the case of *T. canis*. Actually, very few cases of visceral larva migrans have been reported worldwide since then, but "most observers believe that this is the very small tip of a very large iceberg."[20] About 20% of adult dogs and 98% of puppies in the United States are infected with *T. canis;* therefore the risk of exposure is very high. Most cases are either unrecognized or unreported. One of us knows of two local cases, one of which terminated fatally, the other with loss of an eye, but neither case found its way into the literature.

Dogs and cats defecating on the ground seed the area with eggs, which embryonate and become infective to any mammal eating them, including children. Considering the crawling-walking age of small children as the time when virtually every available object goes into the mouth for a taste, it is not surprising that the disease is most common in children between 1 and 3 years old. That favorite outdoor playground, the children's sandbox, unfortunately also constitutes an ideal cat toilet and *Toxocara* embryonation site. In the urban setting the dog owner looks on the city park as the perfect place to "walk" the dog, while the parent at the same time brings young children to play on the "seeded" grass. An especially unhappy fact in the epidemiology of larva migrans is the high risk to children by exposure to the environment of puppies and kittens.[29] Finally, a factor to contemplate in light of the foregoing is the durability and longevity of *Toxocara* eggs, which are comparable to those of *Ascaris* (discussed before).

■ Pathogenesis

Characteristic symptoms of visceral larva migrans include fever, pulmonary symptoms, hepatomegaly, and eosinophilia. Extent of damage usually is related to the number of juveniles present and their ultimate homestead in the body. Deaths have occurred when juveniles were especially abundant in the brain; however, relatively few juveniles can be life threatening in the presence of a severe allergic reaction.[3] There seems little doubt that most cases result in rather minor, transient symptoms, which are undiagnosed or misdiagnosed. The most common site of larval invasion is the liver (Fig. 27-10), but no organ is exempt. Eventually each juvenile is surrounded by a granulomatous host reaction that blocks further migration. Beaver[3] suggested that this reaction might actually be advantageous to the parasite, since nourishment and protection against further defenses of the host are gained.

Juveniles in the eye cause chronic inflammation of the inner chambers or retina or provoke dangerous granulomas of the retina. These reactions can lead to blindness in the affected eye. Ocular involvement has been reported in 245 patients with an average age of 7.5 years.[40] Other lesions destroy lung, liver, kidney, muscle, and nervous tissues.

■ Diagnosis and treatment

Diagnosis is difficult. A liver biopsy may demonstrate the characteristic granuloma surrounding the juvenile, but to obtain a biopsy containing a larva may be a matter of luck. A high eosinophilia is suggestive, especially if the possibility of other parasitic infections can be eliminated. Various serological and cutaneous tests are being perfected for diagnostic purposes.[10,29]

No dependably effective treatment is known at this time, although several drugs have been used.[20] Mebendazole has been reported as successful. Control consists of periodic worming of household pets, especially young animals, and proper disposal of the animal's feces. Dogs and cats should be restrained, if possible, from eating available transport hosts. Children's sandboxes should be covered when not in use.

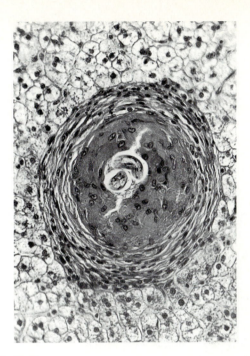

FIG. 27-10

Toxocara canis juvenile in liver of a monkey at 9 months' infection. The juvenile rests in a matrix of epithelioid cells surrounded by a fibrous capsule lacking intense inflammatory reaction.

From Beaver, P.C. 1955. J. Parasitol. 55:3-12.

Toxocara vitulorum (= *Neoascaris vitulorum*)

The only ascarid that occurs in cattle is *T. vitulorum,* which is almost cosmopolitan in calves. It has also been reported in zebu, water buffalo, and sheep.

Adults appear much like *Ascaris,* except that there is a small, glandular ventriculus at the posterior end of the esophagus, typical of *Toxocara* spp. Lateral alae are absent. Males are 15 to 26 cm long, whereas females reach a length of 22 to 30 cm. Like those of *T. canis,* the eggs have a pitted outer layer.

The life cycle is direct. Transplacental infection has been suggested, but infection by ingestion of juveniles in mother's milk has now been proved.[39] Adult hosts are refractory to intestinal infection, but the second-stage juveniles invade their tissues. At parturition, the juveniles migrate to the mammary glands and are swallowed by the nursing calf. Calves can pass eggs as early as 23 days after birth.[16]

Young calves may succumb to verminous pneumonia during the migratory stages of the parasites. Diarrhea or colic in later stages results in unthriftiness and subsequent economic loss to the owner. Piperazine is effective in diagnosed cases, but prevention of infection is difficult.

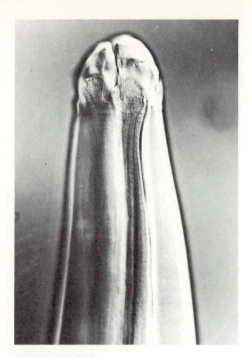

FIG. 27-11

Anterior end of *Toxascaris leonina*, an intestinal parasite of dogs and cats. Note the narrow cervical alae as compared with the broad alae of *Toxocara cati*.

Photograph by Jay Georgi.

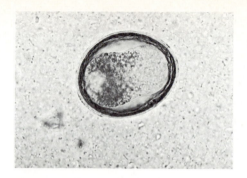

FIG. 27-12

Egg of *Toxascaris leonina*. The size is 75 to 85 by 60 to 75 μm.

Photograph by Jay Georgi.

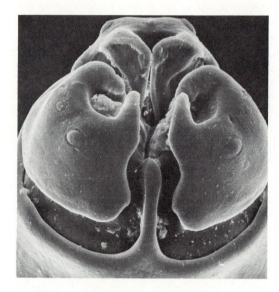

FIG. 27-13

Lagochilascaris turgida. Note the prominent cleft in the tip of each lip, typical of the genus (lagochil means harelip).

Photograph by John Sprent.

Toxascaris leonina

Toxascaris leonina is a cosmopolitan parasite of dogs and cats and related canids and felids. It is similar in appearance to *Toxocara* spp., being recognized in the following ways: (1) the body tends to flex dorsally in *Toxascaris*, ventrally in *Toxocara*; (2) alae of *T. cati* are short and wide, whereas they are long and narrow in *T. canis* and *T. leonina* (Fig. 27-11); (3) the surface of the egg of *T. leonina* is smooth (Fig. 27-12) but pitted in *Toxocara* spp.; and (4) the tail of male *Toxocara* is constricted abruptly behind the anus, whereas it gradually tapers in *Toxascaris*.

The life cycle of *T. leonina* is simple. Ingested eggs hatch in the small intestine, where the juveniles penetrate the mucosa. After a period of growth, they molt and return directly to the intestinal lumen, where they mature.

Although they are mildly pathogenic, their main importance is a diagnostic one: it may be useful to separate *T. leonina* from *Toxocara* spp. in identification procedures because *T. leonina* is considered of little importance as a source of visceral larva migrans.

Lagochilascaris minor

Lagochilascaris minor is evidently closely related to *Toxocara* and *Toxascaris*, differing only in minor morphological details (Fig. 27-13). Little is known about its biology in nature. It has been found in the stomach, pharynx, and trachea of various wild cats in South America and the Caribbean, with three related species in other felids and opossums in the same areas and in Africa and North America.

Lagochilascaris minor has been reported in humans at least 28 times, usually in the tonsils, nose, or

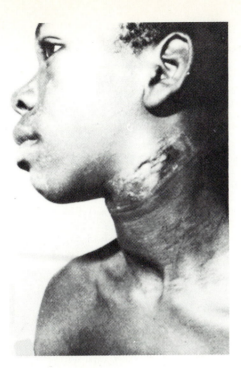

FIG. 27-14

Abscess in the neck of a 15-year-old native of Surinam. It contained numerous adults, juveniles, and eggs of *Lagochilascaris* sp. After treatment with thiabendazole, the fistula closed, and the abscess healed, leaving only a small scar.

From Oostburg, B.F., 1971. Am. J. Trop. Med. Hyg. 20:580-583.

neck.[35,38] A fatal brain infection has been reported.[27] When present, the worms cause abscesses that may contain from one to more than 900 individuals (Fig. 27-14). They can mature in these locations, and they produce pitted eggs, much like those of *Toxocara*. It is common for human infections to last many years or to rapidly kill the infected person. *Lagochilascaris minor* epitomizes the zoonotic infection. How humans become infected is unknown. However, Smith, Bowman, and Little[31] discovered that a related species, *L. sprenti* in American opossums, requires a mammal intermediate host. Some worms in experimentally infected mice became sexually mature in ectopic lesions. Because all human cases have also been ectopic infections, it seems probable that humans are unnatural, accidental hosts. The definitive host is not known. One of us examined several dozen marsupials, reptiles, and carnivores for *Lagochilascaris* in an endemic area of Trinidad and Tobago without finding it.

FAMILY ANISAKIDAE
Anisakis species

The many species in the genus *Anisakis* are parasites of the stomachs of marine fishes, birds, and mammals. Although the life cycles generally are unknown, fish of many species serve as paratenic hosts. Humans enter the picture when juvenile worms are eaten. Two aspects of the situation are important to humans: esthetics and public health. The first relates to the disgust experienced by persons who find large, stout worms in the flesh of the meal they are preparing or eating. Many a finnan haddie has ended up in the garbage pail when *Anisakis* juveniles were discovered, a rather common occurrence at that!

More important, *Anisakis* juveniles can produce a variety of pathological conditions in humans, when eaten in raw, salted, or pickled fish. Intestinal obstruction, colic, abscesses, and peritonitis commonly result from infections with these worms. The stomach commonly is afflicted with a peculiar, tumorlike growth at the site of attachment. Most cases have been reported from Japan, Europe, and Scandinavia, where raw fish is relished. Fatalities have been recorded.[19,26]

The taxonomy of this group of worms is confused and contradictory. For the sake of convenience, most parasitologists refer to the juvenile stages as "anisakis type" and let it go at that. The extreme abundance of such worms in marine fishes the world over suggests that fish dishes must always be well cooked or deep frozen before they are eaten.[25]

FAMILY ASCARIDIIDAE

The family Ascaridiidae is of particular interest because it appears to represent an evolutionary link between the Ascaridata and the Oxyurata. The esophagus is typically ascaridoid, whereas the male posterior end is typically heterakoid (p. 468). The type genus *Ascaridia* has numerous species in birds, including the large nematode of chickens, *A. galli*.

Ascaridia galli is a cosmopolitan parasite of the small intestine of domestic fowl. Males reach a length of 77 mm and females, 115 mm.

Typical of ascarids, the second-stage juvenile hatches from the egg after it is ingested with contaminated food or water. No extensive migration is involved in the life cycle. Instead, 8 or 9 days after infection, the juveniles molt to the third stage and begin to burrow into the mucosa, where they generally remain with their tails still in the intestinal lumen. After molting to the fourth stage at about 18 days, they return to the lumen, where they undergo their final molt and mature. Probably the majority of worms complete their three molts and attain maturity without ever leaving the lumen. Those that attack the mucosa, however, cause extensive damage, which may result in unthriftiness or even death. Adult chickens seem to be refractory to infection.

REFERENCES

1. Anonymous. 1970. LIer sought in roommates' poisoning. Newsday, Feb. 27.
2. Baird, J.K., et al. 1986. Fatal human ascariasis following secondary massive infection. Am. J. Trop. Med. Hyg. 35:314-318.
3. Beaver, P.C. 1969. The nature of visceral larva migrans. J. Parasitol. 55:3-12.
4. Bogojawlenski, N.A., and A. Demidowa. 1928. Ueber den Nachweis von Parasiteneiern auf der menschlichen Nasenschleimhaut. Russian J. Trop. Med. 6:153-156.
5. Brudastov, A.N., V.R. Lemelev, S.K. Kholnukhanedov, and L.N. Krasnos. 1971. The clinical picture of the migration phase of ascariasis in self-infection. Medskaya Parazitol. 40:165-168.
6. Burgess, N.R.H. 1984. Hospital design and cockroach control. Trans. R. Soc. Trop. Med. Hyg. 78:293-294.
7. Coles, G.C. 1985. Allergy and immunopathology of ascariasis. In Crompton, D.W.T., M.C. Nesheim, and Z.S. Pawlowski, editor. Ascariasis and its public health importance. Taylor and Francis, London.
8. Crompton, D.W.T. 1985. Chronic ascariasis and malnutrition. Parasitol. Today 1:47-52.
9. Crompton, D.W.T., and J.J. Tulley. 1987. How much ascariasis is there in Africa? Parasitol. Today 3:123-127.
10. Cypess, R.H., M.H. Karol, J.L. Zidian, L.T. Glickman, and D. Gitlin. 1977. Larva-specific antibody in patients with visceral larva migrans. J. Infect. Dis. 135:633-640.
11. Darby, C.P., and M. Westphal. 1972. The morbidity of human ascariasis. J.S.C. Med. Assoc. 68:104-108.
12. Dold, H., and H. Themme. 1949. Ueber die Möglichkeit der uebertragung der Askaridiasis durch Papiergeld. Dtsch. Med. Wochenschr. 74:409.
13. Dubey, J.P. 1978. Patent Toxocara canis infection in ascarid-naive dogs. J. Parasitol. 64:1021-1023.
14. Fleming, W.J., and J.W. Caslick. 1977. Rabies and cerebrospinal nematodiasis in woodchucks (Marmota monax) from New York. Cornell Vet. 68:391-395.
15. Fox, A.S., et al. 1985. Fatal eosinophilic meningoencephalitis and visceral larva migrans caused by the raccoon ascarid Baylisascaris procyonis. N. Engl. J. Med. 312:1619-1623.
16. Herlich, H., and D.A. Porter. 1954. Experimental attempts to infect calves with Neoascaris vitulorum. Proc. Helm. Soc. Wash. 21:75-77.
17. Kazacos, K.R. 1982. Contaminative ability of Baylisascaris procyonis infected raccoons in an outbreak of cerebrospinal nematodiasis. Proc. Helm. Soc. Wash. 49:155-157.
18. Kazacos, K.R. 1986. Raccoon ascarids as a cause of larva migrans. Parasitol. Today 2:253-255.
19. Kliks, M.M. 1983. Anisakiasis in the western United States: four new case reports from California. Am. J. Trop. Med. Hyg. 32:526-532.
20. Levine, N.D. 1980. Nematode parasites of domestic animals and of man, ed. 2. Burgess Publishing Co., Minneapolis.
21. Louw, J.H. 1966: Abdominal complications of Ascaris lumbricoides infestation in children. Br. J. Surg. 53:510-521.
22. Maung, M. 1978. The occurrence of the second moult of Ascaris lumbricoides and Ascaris suum. Int. J. Parasitol. 8:371-378.
23. Mueller, G. 1953. Untersuchungen ueber die Lebensdauer von Askarideiern in Gartenerde. Zentralbl. Bakt. I. Orig. 159:377-379.
24. Nadler, S.A. 1987. Biochemical and immunological systematics of some ascaridoid nematodes: genetic divergence between congeners. J. Parasitol. 73:811-816.
25. Oshima, T. 1987. Anisakiasis—is the sushi bar guilty? Parasitol. Today 3:44-48.
26. Overstreet, R.M., and G.W. Meyer. 1981. Hemorrhagic lesions in the stomach of a rhesus monkey caused by a piscine ascaridoid nematode. J. Parasitol. 67:226-235.
27. Rosenberg, S., et al. 1986. Fatal encephalopathy due to Lagochilascaris minor infection. Am. J. Trop. Med. Hyg. 35:575-578.
28. Rossi, M.A., and F.W. Bisson. 1983. Fatal case of multiple liver abscesses caused by adult Ascaris lumbricoides. Am. J. Trop. Med. Hyg. 32:523-525.
29. Schantz, P.M., D. Meyer, and L.T. Glickman. 1979. Clinical, serologic, and epidemiologic characteristics of ocular toxocariasis. Am. Trop. Med. Hyg. 28:24-28.
30. Schwartz, B. 1960. Evolution of knowledge concerning the roundworm Ascaris lumbricoides. Smithsonian Report for 1959. The Smithsonian Institution, Washington, D.C., pp. 465-481.
31. Smith, J.L., D.D. Bowman, and M.D. Little. 1983. Life cycle and development of Lagochilascaris sprenti (Nematoda: Ascarididae) from opossums (Marsupialia: Didelphidae) in Louisiana. J. Parasitol. 69:736-745.
32. Sprent, J.F.A. 1952. Anatomical distinction between human and pig strains of Ascaris. Nature 170:627-628.
33. Sprent, J.F.A. 1956. The life history and development of Toxocara cati (Schrank, 1788) in the domestic cat. Parasitology 46:54-78.
34. Sprent, J.F.A. 1969. Nematode larva migrans. N.Z. Vet. J. 17:39-48.
35. Sprent, J.F.A. 1971. Speciation and development in the genus Lagochilascaris. Parasitology 62:71-112.
36. Taffs, L.F. 1985. Ascaris in man: a reply to Dr. Denham. Trans. R. Soc. Trop. Med. Hyg. 79:732.
37. Ubelaker, J.E., and V.F. Allison. 1972. Scanning electron microscopy of the denticles and eggs of Ascaris lumbricoides and Ascaris suum. In Arceneaux, C.J., editor. Thirtieth Annual Proceedings of the Electron Microscopy Society of America. Claitor's Publishing Division, Baton Rouge.
38. Volcan, G., F.R. Ochoa, C.E. Medrano, and Y. de Valera. 1982. Lagochilascaris minor infection in Venezuela. Am. J. Trop. Med. Hyg. 31:1111-1113.
39. Warren, E.G. 1971. Observations of the migration and development of Toxocara vitulorum in natural and experimental hosts. Int. J. Parasitol. 1:85-99.
40. Warren, K.S. 1974. Helminthic diseases endemic in the United States. Am. J. Trop. Med. Hyg. 23:723-730.
41. Willett, W.C., W.L. Kilama, and C.M. Kihamia. 1979. Ascaris and growth rates: a randomized trial of treatment. Am. J. Public Health 69:987-991.

SUGGESTED READINGS

Chabaud, A.G. 1974. Keys to subclasses, orders and superfamilies. In Anderson, R.C., A.G. Chabaud, and S. Willmott, editors. CIH keys to the nematode parasites of vertebrates. Commonwealth Agricultural Bureaux, Farnham Royal, Bucks, Eng.

Crompton, D.W.T., M.C. Nesheim, and Z.S. Pawlowski, editors. 1985. Ascariasis and its public health significance. Taylor and Francis, London.

Douvres, F.W., F.G. Tromba, and G.M. Malakatis. 1969. Morphogenesis and migration of *Ascaris suum* larvae developing to fourth stage in swine. J. Parasitol. 55:689-712.

Hartwich, G. 1974. Keys to genera of the Ascaridoidea. In Anderson, R.C., A.G. Chabaud, and S. Willmott, editors. CIH keys to the nematode parasites of vertebrates. Commonwealth Agricultural Bureaux, Farnham Royal, Bucks, Eng.

Neafie, R.C., and D.H. Connor. 1976. Visceral larva migrans; ascariasis. In Binford, C.H., and D.H. Connor. Pathology of tropical and extraordinary diseases, vol. 2, sect. 9. Armed Forces Institute of Pathology, Washington, D.C.

Sprent, J.F.A. 1983. Observations on the systematics of ascaridoid nematodes. In Stone, A.R., H.M. Platt, and L.F. Khalil, editors. Concepts in nematode systematics. Academic Press Ltd., London.

Willmott, S. 1974. General introduction, glossary of terms. In Anderson, R.C., A.G. Chabaud, and S. Willmott, editors. CIH keys to the nematode parasites of vertebrates. Commonwealth Agricultural Bureaux, Farnham Royal, Bucks, Eng.

CHAPTER 28

ORDER OXYURATA: PINWORMS

Members of the Oxyurata are called "pinworms" because they typically have slender, sharp-pointed tails, especially the females. All pinworms have one feature in common: a conspicuous muscular bulb on the posterior end of the esophagus (Fig. 28-1). Three lips are present around the mouth of the more primitive species (Fig. 28-2), but the lips are reduced or absent in more advanced forms. Caudal and cervical alae are common, and males of many species have a preanal sucker. Life cycles typically are direct, with no intermediate host required. Pinworms are common in mammals, birds, reptiles, and amphibians but are rare in fish. Most domestic birds and mammals harbor pinworms, but, curiously, they are absent in dogs and cats. Terrestrial arthropods, especially insects and millipedes, commonly are infected. Two species, *Enterobius vermicularis* and *E. gregorii*, are among the most common nematode parasites of humans. Because pinworms usually inhabit the large intestine and apparently feed only on bacteria and other intestinal contents, it has been suggested that they are not parasitic. The following discussion shows that the pinworms of humans, at least, qualify for the status of parasites, as defined in the introductory chapter of this book.

The oxyurids will be illustrated by examples from humans, rodents, and domestic fowl.

FAMILY OXYURIDAE
Enterobius vermicularis and *E. gregorii*

In some ways pinworms are rather paradoxical among the nematode parasites of humans. For one thing they are not tropical in their distribution, thriving best in the temperate zones of the world. Furthermore, pinworms often are found in families at high socioeconomic levels, where after introduction into the premises by one member, they rapidly become a "family affair." It is fair to say, however, that the greatest pinworm problems are among institutionalized persons, such as in orphanages and mental hospitals, where conditions facilitate transmission and reinfection.

The fact that these worms inhabit at least 500 million persons is perhaps less surprising than the fact that practically nothing is being done to eliminate it. At least part of the reason is simple and practical: pinworms cause no obvious debilitating or disfiguring effects. Their presence is an embarrassment and an irritation, like acne or dandruff. Resources of democracies, kingdoms, and dictatorships could scarcely be expected to mobilize to combat such an innocuous foe, particularly when they seem to have so much trouble mounting efforts against more disabling infectious agents.

And yet, is enterobiasis so unimportant after all? Certainly it is important to the millions of persons who suffer the discomforts of infection. Furthermore, a great deal of money is spent in efforts to be rid of pinworms. The frantic efforts by persons to rid their households of the tiny worms often lead to what has been called a "pinworm neurosis." The mental stress and embarrassment suffered by families who know they harbor parasites are unmeasurable but very real consequences of infection, especially when multiplied by the vast number of persons involved. Finally, the pathogenesis of these worms may be greatly underrated.[2]

■ Morphology

Both sexes have three lips surrounding the mouth, followed by a cuticular inflation of the head (Fig. 28-1). Females of *E. vermicularis* and *E. gregorii* are nearly identical. Males, however, are easily differentiated by the single spicule which is 100 to 141 μm long in *E. vermicularis* (Fig. 28-3) and 68 to 80 μm long in *E. gregorii*. Other, more subtle differences, are described by Hugot and Tourte-Schaffer.[3] Males of both species are 1 to 4 mm long and have the posterior ends strongly curved ventrad. The conspicuous caudal alae are supported by papillae.

Females measure 8 to 13 mm long and have the posterior end extended into a long, slender point (Fig. 28-4), giving pinworms their name. The vulva opens between the first and second thirds of the body. When gravid, the two uteri contain thousands of eggs, which

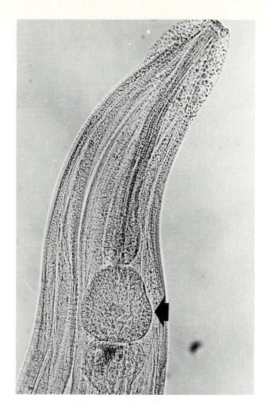

FIG. 28-1

Anterior end of the pinworm *Enterobius vermicularis*. Note the large esophageal bulb *(arrow)* and the swollen cuticle at the head end, typical of this genus.

Photograph by Warren Buss.

FIG. 28-2

Pharyngodon, a pinworm of reptiles, showing the primitive, three-lipped condition. Each lip has a lateral notch.

Photograph by John Ubelaker.

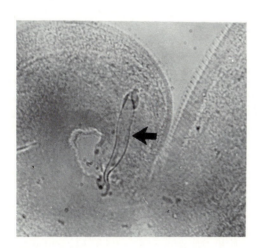

FIG. 28-3

Posterior end of a male *Enterobius vermicularis,* illustrating the single spicule *(arrow).*

Photograph by Warren Buss.

are elongate-oval and flattened on one side (Fig. 28-5), measuring 50 to 60 μm by 20 to 30 μm.

■ Biology

Adult worms congregate mainly in the ileocecal region of the intestine, but they commonly wander throughout the gastrointestinal tract from the stomach to the anus. They attach themselves to the mucosa where they presumably feed on epithelial cells and bacteria. Gravid females begin migrating within the lumen of the intestine, commonly passing out of the anus onto the perianal skin. As they crawl about, both within the bowel and on the outer skin, they leave a trail of eggs. One worm may deposit from 4,600 to 16,000 eggs. Females die soon after oviposition, whereas males die soon after copulation. Consequently, it is usual to find many more females than males within a host.

When laid, each egg contains a partially developed juvenile, which can develop to infectivity within 6 hours at body temperature.[4] Ovic juveniles are resistant to putrefaction and disinfectants but succumb to dehydration in dry air within a day.

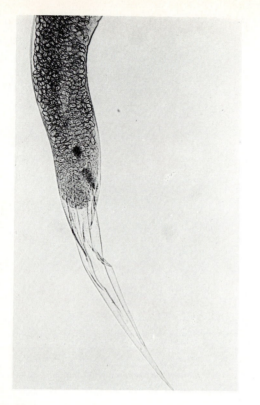

FIG. 28-4

Posterior end of a gravid female *Enterobius vermicularis*. The long pointed tail lends this species the name "pinworm."

Photograph by Warren Buss.

Reinfection occurs by two routes. Most often the eggs, containing third-stage juveniles, are swallowed and hatch in the duodenum. They slowly move down the small intestine, molting twice to become adults by the time they arrive at the ileocecal junction. Total time from ingestion of eggs to sexual maturity of the worms is 15 to 43 days.

If the perianal folds are unclean for long periods, the attached eggs may hatch and the juveniles wander into the anus and hence to the intestine, a process known as **retrofection.** Hatching of the eggs while still inside the intestine apparently does not occur, except, perhaps, during constipation.

■ **Epidemiology**

Clothing and bedding rapidly become seeded with eggs when an infection occurs. Even curtains, walls, and carpets become sources of subsequent infection (or reinfections). The microscopic eggs are very light and are wafted about by the slightest air currents, by which they are deposited throughout the building. The

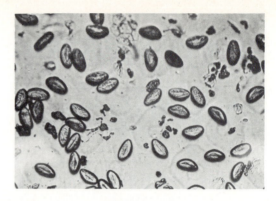

FIG. 28-5

Eggs of *Enterobius vermicularis*. They are 50 to 60 μm long.

Photograph by David Oetinger.

eggs remain viable in cool, moist conditions for up to a week.

The most common means of infection is by placing soiled fingers or other objects into the mouth, as well as by use of contaminated bedding, towels, and so on. Obviously, it becomes next to impossible to avoid contamination when eggs are abundant. Furthermore, it remains impossible to avoid reinfection when retrofection occurs.

Airborne eggs may be inhaled and subsequently swallowed, or they may remain in the nose until they hatch. This, together with nose-picking, accounts for the occasional case of pinworm in the nose. Contrary to popular belief, pinworms cannot be transmitted by dogs and cats because these animals are free of pinworms.

White people seem more susceptible to pinworms than are black people.

■ **Pathogenesis**

About one third of infections are completely asymptomatic, and in many more, clinical symptoms are negligible. Nevertheless, very large numbers of worms may be present and lead to more serious consequences. Pathogenesis has two aspects: damage caused by worms within the intestine and damage resulting from egg deposition around the anus. Minute ulcerations of the intestinal mucosa from attachment of adults may lead to mild inflammation and bacterial infection.[7] Very rarely, pinworms will penetrate into the submucosa with fatal results. The movements of the females out the anus to deposit eggs, especially when the patient is asleep, lead to a tickling sensation of the perianus, causing the patient to scratch. The subsequent vicious circle of bleeding, bacterial infection, and intensified itching can lead to a nightmare of discomfort.

It is common for worms to wander into the vulva where they remain for several days, causing a mild irritation. Cases have been reported where pinworms have wandered up the vagina, uterus, and oviducts into the coelom, to become encysted in the peritoneum. They have been known to become encapsulated within an ovarian follicle.[1]

Children with heavy pinworm infection are often nervous, restless, and irritable and may suffer from loss of appetite, nightmares, insomnia, weight loss, and perianal pain.

■ Diagnosis and treatment

Positive diagnosis can be made only by finding eggs or worms on or in the patient. Ordinary fecal examinations are usually unproductive because few eggs are deposited within the intestine and passed in the feces. Heavy infections can be discovered by examining the perianus closely under bright light, during the night or early morning. Wandering worms glisten and can be seen easily. When adults cannot be found, eggs often can be, as they are left behind in the perianal folds. A short piece of cellophane tape, held against a flat, wooden applicator or similar instrument, sticky side out, is pressed against the junction of the anal canal and the perianus. The tape is then reversed and stuck to a microscope slide for observation. If a drop of xylene or toluene is placed on the slide before the tape, it will dissolve the glue on the tape and clear away bubbles, simplifying the search for the characteristic, flat-sided eggs (Fig. 28-5). It is desirable for the physician to teach the parent how to prepare the slide, since it should be done just after awakening in the morning, certainly before bathing the child for a trip to the doctor's office.

Numerous home remedies and over-the-counter medications have been in use for many years, with results ranging from poor to completely ineffective. Today the preferred drugs are pyrantel pamoate and mebendazole (Vermox). Both are highly effective, inexpensive, and safe. Treatment should be repeated after about 10 days to kill worms acquired after the first dose, and sanitation procedures should be instituted concurrently. All members of the household should be treated simultaneously, regardless of whether the infection has been diagnosed in all.

Although diagnosis and cure of enterobiasis are easy, preventing reinfection is more difficult. Personal hygiene is most important. Completely sterilizing the household is a gratifyingly difficult activity but of limited usefulness. Nevertheless, at time of treatment, all bed linens, towels, and the like should be washed in

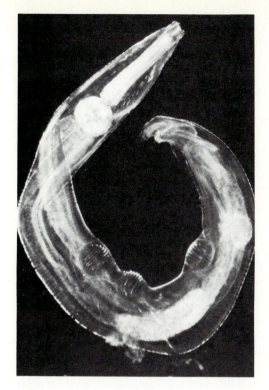

FIG. 28-6

Syphacia, a pinworm of rodents. Note the three corrugated mamelons on the ventral surface of the male.
Photograph by Warren Buss.

hot water, and the household should be cleaned as well as possible to lower the prevalence of infective eggs in the environment. If all persons are undergoing chemotherapy while reasonable care is taken to avoid reinfection, the family infection can be eradicated—until the next time a child brings it home from school.

FAMILY SYPHACIIDAE
Syphacia species

The tiny parasites of the family Syphaciidae (*Syphacia* spp.) are rarely found in humans but are commonly encountered in their natural hosts, wild and domestic rodents. Laboratory rats and mice are frequently infected by *S. muris* and by *S. obvelata,* which lives in the cecum and has been reported from humans.[6]

Male *Syphacia* are easily recognized by their **mamelons,** two or three ventral, serrated projections (Fig. 28-6). Females are typical pinworms, with long, pointed tails. The eggs are operculated. The life cycle is direct: the worms mature in the cecum or large intestine.[5] No migration within the host is known.

FAMILY HETERAKIDAE
Heterakis gallinarum

These large pinworms are cosmopolitan in domestic chickens and related birds. Probably, they were brought to the U.S. in imported ringnecked pheasants. They live in the cecum, where they feed on its contents.

Three large lips and a bulbar esophageal swelling are found in this genus, as are lateral alae. Males are as long as 13 mm and possess wide caudal alae supported usually by 12 pairs of papillae (Fig. 28-7). The tail is sharply pointed, and there is a prominent preanal sucker. The spicules are strong and dissimilar, and a gubernaculum is not present.

Females are typical pinworms, with the vulva near the middle of the body and with a long, pointed tail.

Many species of *Heterakis* are known in birds, particularly in ground feeders, and one species, *H. spumosa,* is cosmopolitan in rodents.

■ Biology

The eggs of *H. gallinarum* are in the zygote stage when laid. They develop to the infective stage in 12 to 14 days at 22° C and can remain infective for 4 years in soil. Infection is contaminative: when embryonated eggs are eaten, the second-stage juveniles hatch in the gizzard or duodenum and pass down to the ceca. Most complete their development in the lumen, but some penetrate the mucosa, where they remain for 2 to 5 days without further development. Returning to the lumen they mature, about 14 days after infection.

If eaten by an earthworm, the juvenile may hatch and become dormant in the worm's tissues, remaining infective to chickens for at least a year. Since the nematodes do not develop further until eaten by a bird, the earthworm is a paratenic host.

■ Epidemiology

As a result of the longevity of the eggs, it is difficult to eliminate *Heterakis* from a domestic flock. Thus, although adult chickens may effect a self-cure, infective eggs are still available the following spring, when new chicks are hatched. Furthermore, as earthworms feed in contaminated soil, they accumulate large numbers of juveniles, which in turn causes massive infections in the unlucky birds that eat them.

■ Pathogenesis

In heavy infections the cecal mucosa may thicken and bleed slightly. Generally speaking, *Heterakis* is not highly pathogenic in itself.

However, a flagellate protozoan, *Histomonas meleagridis,* is transmitted between birds within eggs of

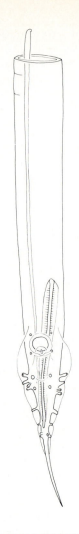

FIG. 28-7

Posterior end of *Heterakis variabilis,* a parasite of pheasants that is similar to *H. gallinarum*. Note the conspicuous preanal sucker.

From Inglis, W.G., G.D. Schmidt, and R.E. Kuntz. 1971. Rec. S. Aust. Mus. 16:1-14.

Heterakis gallinarum. This protozoan is the etiological agent of **histomoniasis,** a particularly serious disease in turkeys. The protozoan is eaten by the nematode and multiplies in the worm's intestinal cells, in the ovaries, and, finally, in the embryo within the egg (p. 92). Hatching of the worm within a new host releases *Histomonas*. Hence we encounter the curious phenomenon of one parasite acting as a true intermediate host and vector of another.

■ Diagnosis and treatment

Heterakis gallinarum can be diagnosed by finding the eggs in the feces of its host. The worms are effectively eliminated with mebendazole. Usually a flock of birds is routinely fed this or other drugs in its feed or water. This treatment, together with rearing the birds on hardware cloth, will eliminate the parasite from the flock. Birds that are allowed to roam the barnyard usually are infected.

REFERENCES

1. Beckman, E.N., and J.B. Holland. 1981. Ovarian enterobiasis—a proposed pathogenesis. Am. J. Trop. Med. Hyg. 30:74-76.
2. Bijlmer, E. 1946. Exceptional case of oxyuriasis of the intestinal wall. J. Parasitol. 32:359-366.
3. Hugot, J.P., and C. Tourte-Schaffer. 1985. Etude morphologique des oxyures parasites de l'homme: *Enterobius vermicularis* et *E. gregorii*. Ann. Parasitol. Hum. Comp. 60:57-64.
4. Hulinská, D. 1968. The development of the female *Enterobius vermicularis* and the morphogenesis of its sexual organ. Folia Parasitol. 15:15-27.
5. Prince, M.J.R. 1950. Studies on the life cycle of *Syphacia obvelata,* a common nematode parasite of rats. Science 111:66-67.
6. Riley, W.A. 1920. A mouse oxyurid, *Syphacia obvelata,* as a parasite of man. J. Parasitol. 6:89-92.
7. Shubenko-Gabuzova, I.N. 1965. Appendicitis in enterobiasis. Med. Parasitol. Dis. 34:563-566.

SUGGESTED READINGS

Beaver, P.C., J.J. Kriz, and T.J. Lau. 1973. Pulmonary nodule caused by *Enterobius vermicularis*. Am. J. Trop. Med. Hyg. 22:711-713.

Little, M.D., C.J. Cuello, and A. D'Alessandro. 1973. Granuloma of the liver due to *Enterobius vermicularis*. Report of a case. Am. J. Trop. Med. Hyg. 22:567-569.

Skrjabin, K.I., N.P. Schikhobolova, and E.A. Lagodovskaya. 1960-1967. Essentials of nematodology, vols. 8, 10, 13, 15, and 18. Oxyurata. Akademii Nauk SSSR, Moscow. (Indispensable reference works for the oxyurid taxonomist.)

Skrjabin, K.I., N.P. Schikhobolova, and A.A. Mosgovoi. 1951. Key to parasitic nematodes, vol. 2. *Oxyurata* and *Ascaridata*. Akademii Nauk SSSR, Moscow. (A useful key to genera, with lists of species.)

CHAPTER 29

ORDER SPIRURATA: A POTPOURRI OF NEMATODES

Spirurids are parasitic in all classes of vertebrates and employ an intermediate host in their development, usually an arthropod. They are a very large, heterogeneous group, with many species. The many variations of morphology in this order make generalization difficult, but most have two lateral lips, called **pseudolabia,** and an esophagus that is divided into anterior muscular and posterior glandular portions. The lips do not represent the fusion of primitive lips but are evolutionarily new structures that originate in anterior shifting of tissues from within the buccal walls. Although the esophagus usually has both muscular and glandular portions, there are species whose esophagus is primarily muscular and others in which it is mainly glandular. Some of these, however, are of uncertain taxonomic position. Spirurid spicules are usually dissimilar in size and shape.

Spirurids seldom parasitize humans; when they do, they are only as zoonoses. Several, however, are important in domestic livestock; the rest are parasites of wild animals. When a nematode is found in a wild animal, the chances are 50% that it will be a spirurid, particularly if insects are a preferred part of the host's diet.

Of the many families in this order, a few that demonstrate the diversity in the group will be examined briefly.

FAMILIES ACUARIIDAE AND SCHISTOROPHIDAE

Nematodes of these two families, all parasites of birds, exhibit very peculiar morphological structures at their head ends. Acuariids have four grooves or ridges, called **cordons,** which begin two dorsally and two ventrally at the junctions of the lateral lips and proceed posteriad for varying distances (Fig. 29-1). The cordons may be straight, sinuous, recurving, or even anastomosing in pairs.

The schistorophids, which are very closely related to the acuariids, do not have cordons but instead possess four extravagant cuticular projections, sometimes simple, sometimes serrated, or even feathered (Fig. 29-2).

Both specializations, cordons and cuticular projections of the head, seem to correlate with the parasite's location within the host, the stomach. Most mature under the koilon, or gizzard lining, where they cause considerable damage to the underlying epithelium. How these anterior modifications aid the parasite is not known.

Common genera of acuariids are *Acuaria* and *Cheilospirura* in terrestrial birds and *Echinuria, Skrjabinoclava, Chevreuxia,* and *Cosmocephalus* in aquatic birds.

Some genera of schistorophids are *Torquatella, Viquiera,* and *Serticeps* in terrestrial birds and *Schistorophus, Ancyracanthopsis,* and *Sciadiocara* in aquatic birds.

With the exception of *Echinuria* spp. in ducks, geese, and swans, these parasites are of little or no economic importance. However, they represent an interesting example of adaptive radiation.

FAMILY GNATHOSTOMATIDAE

Family Gnathostomatidae contains the genera *Tanqua* from reptiles, *Echinocephalus* in elasmobranchs, and *Gnathostoma* in the stomachs of carnivorous mammals (Fig. 29-3). These distinctive nematodes have two powerful, lateral lips, followed by a swollen "head," which is separated from the rest of the body by a constriction. Internally, four peculiar, glandular cervical sacs, reminiscent of acanthocephalan lemnisci, hang into the coelom from their attachments near the anterior end of the esophagus. The head bulb is divided internally into four hollow areas called **ballonets.** Each cervical sac has a central canal, which is continuous with a ballonet. The functions of these organs are unknown.

Gnathostoma spp. are particularly interesting because of their widespread distribution and their peculiar biology and because they often cause disease in humans in some areas of the world. In the United States *G. procyonis* is common in the stomachs of rac-

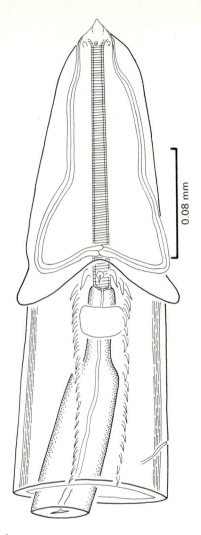

FIG. 29-1

Cordonema venusta, from the stomach of an aquatic bird (dipper). Note the helmetlike inflation of the cuticle, which bears two cordons on each side. The cordons of each side join at their posterior ends in this genus.

From Schmidt, G.D., and R.E. Kuntz. 1972. Parasitology 64:235-244.

coons and opossums, and *G. spinigerum* has been reported from a wide variety of carnivores in the Orient. *Gnathostoma doloresi* is common in pigs in the Orient (Fig. 29-4). Of the 20 or so species that have been described, *G. spinigerum* has most consistently been demonstrated as a cause of disease in humans, so we will examine this parasite in more detail.

Gnathostoma spinigerum

In 1836 Richard Owen discovered *G. spinigerum* in the stomach wall of a tiger that had died in the Lon-

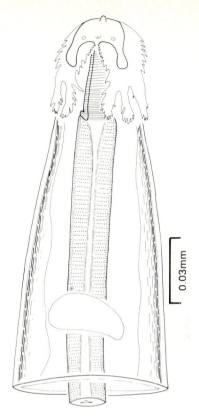

FIG. 29-2

Sobolevicephalus chalcyonis from the stomach of a kingfisher. The head cuticle has four feathered projections.

From Schmidt, G.D., and R.E. Kuntz, 1972. Parasitology 64:264-278.

don Zoo. Since then, it has been found in many kinds of mammals in several countries, although it is most common in southeast Asia.

■ Morphology

The body is stout and pink in life. The swollen head bulb is covered with four circles of stout spines. The anterior half of the body is covered with transverse rows of flat, toothed spines, followed by a bare portion. The posterior tip of the body has numerous tiny cuticular spines.

Males are 11 to 31 mm long and have a bluntly rounded posterior end. The anus is surrounded by four pairs of stumpy papillae. The spicules are 1.1 mm and 0.4 mm long and are simple with blunt tips.

Females are 11 to 54 mm long and also have a blunt posterior end. The vulva is slightly postequatorial in position. The eggs are unembryonated when laid, 65 to 70 μm by 38 to 40 μm in size, and have a polar cap at only one end. The outer shell is pitted.

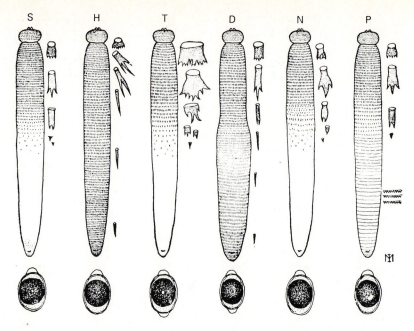

FIG. 29-3

Morphological comparison among six species of female *Gnathostoma; S, G. spinigerum; H, G. hispidum; T, G. turgidum; D, G. doloresi; N, G. nipponicum; P, G. procyonis.* This figure indicates the arrangement and shape of the cuticular spines and fresh fertilized uterine eggs, which at times may show various developmental stages when preserved.

From Miyazaki, I. 1966. In Morishita, K., Y. Komiga, and H. Matsubayashi, editors. *Progress of medical parasitology in Japan*, vol. 3. Meguro Parasitological Museum, Tokyo.

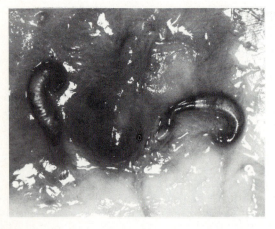

FIG. 29-4

Gnathostoma doloresi attached to the stomach mucosa of a pig.

Photograph by Robert E. Kuntz.

■ **Biology** (Fig. 29-5)

The eggs complete embryonation and hatch in about a week at 27° to 31° C.[7] The actively swimming juvenile is eaten by a cyclopoid copepod, where it penetrates into the hemocoel and develops further into a second-stage juvenile in 7 to 10 days. The second-stage juvenile already has a swollen head bulb covered with four transverse rows of spines.

When the infected crustacean is eaten by a vertebrate second intermediate host, the second-stage juvenile penetrates the intestine of its new host and migrates to muscle or connective tissue, where it molts to the third stage. The third stage is infective to a definitive host. However, if it is eaten by the wrong host, it may wander in that animal's tissues without further development. More than 35 species of paratenic hosts are known, among which are crustaceans, freshwater fishes, amphibians, reptiles, birds, and mammals, including humans. The biology of this parasite in Japan was reviewed by Miyazaki.[8]

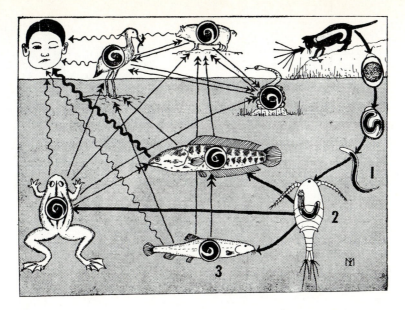

FIG. 29-5

Life history of *G. spinigerum*. *1*, First-stage juvenile (averaging 0.27 mm long) swimming in the water; *2*, second-stage juvenile (averaging 0.5 mm long) parasitic in cyclops; *3*, encysted third-stage juvenile (averaging 3 to 4 mm long) in the muscle of the second intermediate host. Double arrow in the figure indicates "secondary infection." Humans are susceptible to the infection from a variety of second intermediate hosts.

From Miyazaki, I. 1966. In Morishita, K., Y. Komiya, and H. Matsubayashi, editors. Progress of medical parasitology in Japan, vol. 3. Meguro Parasitological Museum, Tokyo.

Adult worms are found embedded in tumorlike growths in the stomach wall of the definitive host. They begin producing eggs about 100 days after infection.

■ Epidemiology

Human infection results from eating a raw or undercooked intermediate or paratenic host containing third-stage juveniles. In Japan this is most often a fish, whereas in Thailand, domestic duck and chicken are probably the most important vectors.[5] However, any amphibian, reptile, or bird may harbor juveniles and thereby contribute an infection if eaten raw.

■ Pathology

In humans the third-stage juveniles usually migrate to the superficial layers of the skin, causing **gnathostomiasis externa.** They may become dormant in abscessed pockets in the skin, or they may wander, leaving swollen red trails in the skin behind them. This creeping eruption resembles larva migrans caused by hookworms or fly larvae.

If the worms remain in the skin with little wandering, they cause relatively little disease. Often they will erupt out of the skin spontaneously. However, erratic migration may take them into an eye, the brain, or the spinal cord, with serious results that even may cause death.

■ Diagnosis and treatment

Diagnosis depends on recovery and accurate identification of the worm. An intradermal test, using an antigen prepared from *G. spinigerum,* has been employed with success in Japan. Gnathostomiasis should be suspected in an endemic area when a localized edema is accompanied by leukocytosis with a high percentage of eosinophils.

Chemotherapy has not been effective against this zoonosis. The only effective treatment is surgical removal of the worm.

Prevention is the most realistic means of controlling this disease. Cooling by any means will kill the worms. In regions where ritualistic consumption of raw fish is an important tradition, the fish should be well frozen before preparation, or marine fish should be used. Consumption of raw, previously unfrozen fish in any area of the world is dangerous for a variety of parasitological reasons.

FAMILY PHYSALOPTERIDAE

Members of the family Physalopteridae are mostly rather large, stout worms that live in the stomachs or intestines of all classes of vertebrates. All have two large, lateral pseudolabia, usually armed with teeth. The head papillae are on the pseudolabia. The cuticle at the base of the lips is swollen into a "collar" in some genera. Caudal alae are well developed on males. Spicules are equal or unequal, and a gubernaculum is absent. This family has a tendency toward **polydelphy,** or many ovaries and uteri. Of the several genera in this family, we will briefly consider *Physaloptera.*

Physaloptera species

In the genus *Physaloptera* the triangular pseudolabia are armed with varying numbers of teeth, and a conspicuous cephalic collar is present. The male has numerous pedunculated caudal papillae and caudal alae that join anterior to the anus. In a few species the cuticle of the posterior end is inflated into a prepuce-like sheath, which encloses the tail. Three species are found in Amphibia, around 45 species in reptiles, 24 in birds, and nearly 90 in mammals.

Physaloptera praeputialis (Fig. 29-6) lives in the stomachs of domestic and wild dogs and cats throughout the world except Europe. It is common in dogs, cats, coyotes, and foxes in the United States. A flap of cuticle covers the posterior ends of both sexes. Its life cycle is incompletely known, but development has been experimentally obtained in cockroaches.

Physaloptera rara is the most common physalopterid of carnivores in North America, to which it is apparently restricted. It is similar to *P. praeputialis* but lacks the posterior cuticular flap. The life cycle involves an insect intermediate host, usually a field cricket, in which it develops to the third stage. A paratenic host, such as a snake, is commonly necessary in the life cycle because of the feeding habits of the definitive hosts.

Physaloptera caucasica is the only species recorded from humans.[10] It is normally parasitic in African monkeys. Most recorded cases in humans were from Africa, although several records, some based only on eggs found in patients' feces, have been reported from South and Central America, India, and the Middle East. It is possible that some of these were misidentified.

The life cycle is unknown but most likely involves insect intermediate hosts and a vertebrate paratenic host. Humans may become infected by eating either of these types of host.

Symptoms include vomiting, stomach pains, and

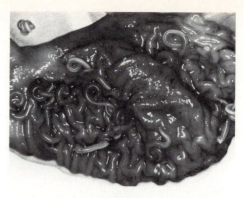

FIG. 29-6

Physaloptera praeputialis in the stomach of a domestic cat. Photograph by Robert E. Kuntz.

eosinophilia. Tentative diagnosis can be made by demonstrating the eggs in a fecal sample or by obtaining an adult specimen for accurate identification.

FAMILY TETRAMERIDAE

Members of the family Tetrameridae are bizarre in their degree of sexual dimorphism. Whereas males exhibit typical nematoid shape and appearance, the females are greatly swollen and often colored bright red. The three genera in this family are all parasitic in the stomachs of birds. The genus *Geopetitia* is represented by five rare species, which live in cysts on the outside of the proventriculus or gizzard of birds, where they communicate with the enteric lumen through a tiny pore. The posterior end of the female is distorted and swollen; the anterior portion is normal. Both sexes are colorless.

The other two genera in the family are very common parasites that live in the branched secretory glands of the proventriculus, although males can be found wandering throughout the organ.

Tetrameres (Fig. 29-7) is a large genus of about 50 species, mainly parasites of aquatic birds. A well-developed, sclerotized buccal capsule is present in both sexes. Males are typically nematoid in form, lacking caudal alae and possessing spicules that are vastly dissimilar in size. Lateral, longitudinal rows of spines are present on many species. Females, however, are greatly swollen, with only the front and back ends retaining the appearance of a nematode. In addition, females are blood-red with a black, saclike intestine. They are easily seen as reddish spots in the wall of the proventriculus, where they mature with the tail end near the lumen of that organ. The vulva is near the anus and thus is available to males who find it. The

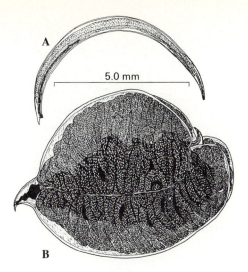

FIG. 29-7

Tetrameres strigiphila from owls. **A,** Male. **B,** Female.

From Pence, D.B. 1975. Parasitology 61:494-498.

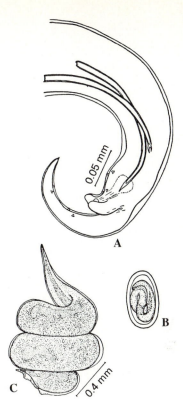

FIG. 29-8

Microtetrameres aguila, a parasite of the proventriculus of eagles. **A,** Posterior end of male, lateral view. **B,** Embryonated egg. **C,** Female.

From Schell, S.C. 1953. Trans. Am. Microsc. Soc. 72:227-236.

eggs are embryonated when laid. The intermediate hosts are crustaceans or insects. Definitive hosts are water birds, chickens, owls, and hawks.

The terrestrial counterpart of *Tetrameres* is *Microtetrameres* (Fig. 29-8). About 40 species have been described in this genus, all of which live in the proventricular glands of insectivorous birds. Females of this genus are also swollen and, in addition, are twisted into a spiral. Morphological and biological characteristics are otherwise similar to those of *Tetrameres,* with terrestrial crustaceans and insects serving as intermediate hosts.

The quaint little worms in this family are familiar to all who survey the parasites of birds, since they are common in many species of hosts. Because of the difficulties of taxonomy of the group, only a small percentage of actual species have been described. The economic importance of these nematodes is slight. Even though a hundred or more females are commonly embedded in the proventriculus of a single duck, for example, they seem to have little effect on the overall health of their host.

FAMILY GONGYLONEMATIDAE

This family contains the single genus *Gongylonema,* which has several species that are found in the upper digestive tracts of birds and mammals. Morphologically, they resemble several spiruriids, except that the cuticle of the anterior end is covered with large bosses, or irregular scutes, arranged in eight longitudinal rows (Fig. 29-9). Cervical alae are present,

as are cervical papillae. The posterior end of the male bears wide caudal alae, which are supported by numerous pedunculated papillae.

Of the 25 or so species in this genus, *G. pulchrum* is probably the best known. Primarily a parasite of ruminants and swine, the worm has also been reported from monkeys, hedgehogs, bears, and humans.[6,9] It has been demonstrated experimentally that the life cycle involves an insect intermediate host, either a dung beetle or a cockroach. Although these would appear rather unpalatable fare for people, they have been ingested often enough: numerous cases of human gongylonemiasis have been reported. In normal hosts the worms invade the esophageal epithelium, where they burrow stitchlike in shallow tunnels. In an abnormal host, such as humans, they behave similarly but do not mature and seem to wander further, being found often in the epithelium of the tongue, gums, or buccal cavity. Their active movements, together with resulting irritation and bleeding, soon make their pres-

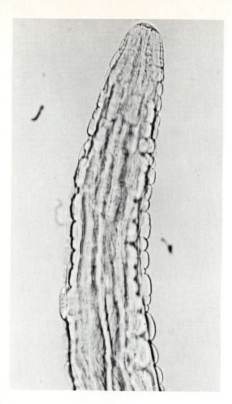

FIG. 29-9

Anterior end of *Gongylonema,* demonstrating the cuticular bosses typical of the genus.

Photograph by Warren Buss.

ence known. Treatment is surgical removal of worms that can be seen. Chemotherapy is seldom employed.

Gongylonema neoplasticum and *G. orientale* in domestic rats are thought to induce neoplastic tumors. They are not known to infect humans.

FAMILY SPIROCERCIDAE

The family Spirocercidae is closely related to the Spiruridae and the genus *Cyathospirura;* all are parasites of mammals. Of these, *Spirocerca lupi* is the most interesting because of its complex life cycle and its relationship to esophageal cancer in dogs.[1,2]

Spirocerca lupi

This stout worm is bright pink to red when alive. The mouth is surrounded by six rudimentary lips, and the buccal capsule is well developed, with thick walls. A short muscular portion of the esophagus is followed by a longer glandular portion. Males are 30 to 54 mm long, with a left spicule 2.45 to 2.8 mm long and a right spicule 475 to 750 μm long. Females are 50 to 80 mm long, with the vulva 2 to 4 mm from the anterior end. The eggs are cylindrical and embryonated when laid.

■ Biology

Adults normally are found in clusters, entwined within the wall of the upper digestive tract of dogs, although they have been reported in other organs, mostly the dorsal aorta, and in a wide variety of carnivorous hosts. Hounds seem to be the breeds most frequently infected in the United States. This is probably related to their opportunities for exposure rather than breed susceptibility.

Embryonated eggs pass out of the host with its feces. Any of several species of scarabaeid dung beetles can serve as intermediate host. A wide variety of paratenic hosts is known, including birds, reptiles, and other mammals. Dogs can become infected by eating dung beetles or infected paratenic hosts. Domestic dogs are probably most often infected by eating the offal of chickens that have third-stage juveniles encysted in their crops.

Once in the stomach of a definitive host, the juveniles penetrate its wall and enter the wall of the gastric artery, migrating up to the dorsal aorta and forward to the area between the diaphragm and the aortic arch. They remain in the wall of the aorta for 2½ to 3 months, after which they emerge and migrate to the nearby esophagus, which they penetrate. After establishing a passage into the lumen of the esophagus, they move back into the submucosa or muscularis where they complete their development about 5 to 6 months after infection. Eggs pass into the esophageal lumen through the tiny passage formed by the worm.

Many worms get lost during migration and may be found in abnormal locations, including lung, mediastinum, subcutaneous tissue, trachea, urinary bladder, and kidney.

■ Epidemiology

Spirocerca lupi is most common in warm climates but has been found in Manchuria and northern regions of the Soviet Union. Many questions are still unanswered, such as why is the parasite distributed so sporadically throughout the United States and the world, and what factors influence the change in prevalence of the infection in a given area? The attractiveness of dog feces to susceptible beetle species is a factor, as is the application of pesticides in an endemic area. Certainly the successful transfer of third-stage junveniles from one paratenic host to another increases the parasite's chances for survival.

FIG. 29-10

Hound with severe hypertrophic pulmonary osteoarthropathy associated with esophageal sarcoma.
From Bailey, W.S. 1963. Ann. N.Y. Acad. Sci. 108:890-923.

■ Pathology

When the third-stage juvenile penetrates the mucosa of the stomach, it causes a small hemorrhage in the area. This irritation commonly causes the dog to vomit. The lesions in the aorta caused by the migrating worms are often severe, with hemorrhage accounting for death in some dogs with heavy infections. Destruction of tissues in the wall of the aorta with subsequent scarring is typical of this disease and may lead to numerous aneurysms.

Worms that leave the aorta and migrate upward come in contact with the tissues surrounding the vertebral column, where they frequently cause, by a mechanism not yet elucidated, a condition known as **spondylosis.** This deformation may be so severe as to cause adjacent vertebrae to fuse. Hypertrophic pulmonary osteoarthropathy, with inflamed and swollen joints, is a common sequel to this disease (Fig. 29-10).

The most striking lesion associated with spirocercosis is in the wall of the esophagus, where the worms mature. Here their presence stimulates the formation of a **reactive granuloma,** made up of fibroblasts. The granulomas are rather more loosely organized than is usually the case in granulomatous reactions, and the cell structure is characteristic of incipient **neoplasia** (cancer). Some of the granulomas change to **sarcomas,** true cancerous growths (Fig. 29-11). The worms may continue to live inside these tumors for some time, or they may be extruded or compressed and killed by the rapidly growing tissue. The precise oncogenic factor responsible for stimulation of neoplasia is still unknown. Although several other helminths have been thought to be associated with malignancy, in no other instance is there as strong evidence for a cause-effect relationship as in canine spirocercosis.

The only known case of spirocercosis in humans was that of a fatal, prenatal infection reported in Italy.[4] The baby was born prematurely and died 12 days later. Mature worms were found in the wall of the terminal ileum. The mother may have become infected by eating a coprophagous beetle or undercooked chicken.

■ Diagnosis and treatment

Diagnosis in dogs usually is performed by demonstrating the characteristic eggs, which measure 40 μm by 12 μm and have nearly parallel sides, in a fecal examination. At necropsy, aortic scarring and aneurysms and esophageal granuloma or sarcoma are considered diagnostic, even if worms are no longer present.

Disophenol is effective against *S. lupi,* but if exten-

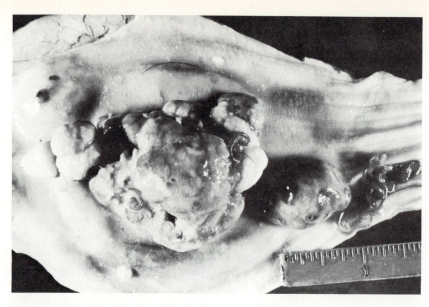

FIG. 29-11

Esophageal sarcoma associated with *Spirocerca lupi* infection. Pedunculated masses protrude into the lumen; adult *S. lupi* are partially embedded in the neoplasm.

From Bailey, W.S. 1972. J. Parasitol. 58:3-22. Photograph courtesy Department of Pathology and Parasitology, Auburn University School of Veterinary Medicine, Auburn, Ala.

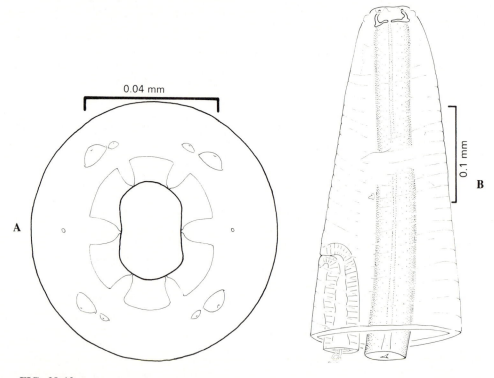

FIG. 29-12

Thelazia digiticauda from under the nictitating membrane of the eye of a kingfisher. **A,** En face view. **B,** Female, lateral view of anterior end.

From Schmidt, G.D., and R.E. Kuntz. 1971. Parasitology 63:91-99.

sive aneurysms or a sarcoma have already developed, the treatment will not affect these conditions.

FAMILY THELAZIIDAE

Members of the family Thelaziidae live on the surface of the eye in birds and mammals, usually remaining in the lacrimal ducts or conjunctival sacs or under the nictitating membrane. Most are parasites of wild animals, but two species of *Thelazia* have been reported from humans.

These worms lack lips but show evidence of the primitive condition in having a hexagonal mouth (Fig. 29-12, *A*). The buccal capsule is well developed, with thick walls. Alae and cuticular ornamentations are absent, except for conspicuous transverse striations near the anterior end (Fig. 29-12, *B*). These are deep, and their overlapping edges ostensibly aid movements across the smooth surface of the cornea.

Thelazia callipaeda is a parasite of dogs and other mammals in southeast Asia, China, and Korea, and *T. californiensis* parasitizes deer and other mammals in western North America. Both species have been reported from humans several times.[3]

Little is known of the biology of these worms except that filth flies *(Musca* and *Fannia)* are capable of serving as intermediate hosts. Probably, when an infected fly is swallowed, the third-stage juveniles migrate up the esophagus to the pharynx and then up the lacrimal ducts to the orbits.

REFERENCES

1. Bailey, W.S. 1963. Parasites and cancer: sarcoma in dogs associated with *Spirocerca lupi*. Ann. N.Y. Acad. Sci. 108:890-923.
2. Bailey, W.S. 1971. *Spirocerca lupi:* a continuing enquiry. J. Parasitol. 58:3-22.
3. Bhaibulaya, M., S. Prasertsilpa, and S. Vajrasthira. 1970. *Thelazia callipaeda* Railliet and Henry, 1910, in man and dog in Thailand. Am. J. Trop. Med. Hyg. 19:476-479.
4. Biocca, E. 1959. Infestazione umana prenatale da *Spirocerca lupi* (Rud. 1809). Parassitologia 1:137-142.
5. Daengsvang, S., P. Thienprasitthi, and P. Chomcherngpat. 1966. Further investigations on natural and experimental hosts of larvae of *Gnathostoma spinigerum* in Thailand. Am. J. Trop. Med. Hyg. 15:727-729.
6. Feng, L.C., M.S. Tung, and S.C. Su. 1955. Two Chinese cases of *Gongylonema* infection. A morphological study of the parasite and clinical study of the case. Chin. Med. J. 73:149-162.
7. Miyazaki, I. 1954. Studies on *Gnathostoma* occurring in Japan (Nematoda: Gnathostomidae). II. Life history of *Gnathostoma* and morphological comparison of its larval forms. Kyushu Mem. Med. Sci. 5:123-140.
8. Miyazaki, I. 1966. Gnathostoma and gnathostomiasis in Japan. In Morishita, K., Y. Komiya, and H. Matsubayashi, editors. Progress of medical parasitology in Japan, vol. 3. Meguro Parasitological Museum, Tokyo, pp. 529-586.
9. Thomas, L.J. 1952. *Gongylonema pulchrum*, a spirurid nematode infecting man in Illinois, U.S.A. Proc. Helm. Soc. Wash. 19:124-126.
10. Vandepitte, G., J. Michaux, J.L. Fain, and F. Gatti. 1964. Premieres observations congolaises de physaloptérose humaine. Ann. Soc. Belg. Med. Trop. 44:1067-1076.

SUGGESTED READINGS

Chabaud, A.G. 1954. Valeur des charactèrs biologiques pour la systématique des nématodes spirurides. Vie Milieu 5:299-309.

Chabaud, A.G. 1975. Keys to the order Spirurida, part 2. In Anderson, R.C., A.G. Chabaud, and S. Willmott, editors. CIH keys to the nematode parasites of vertebrates. Commonwealth Agricultural Bureaux, Farnham Royal, Bucks, Eng.

Chitwood, B.G., and E.E. Wehr. 1934. The value of cephalic structure as characters in nematode classification, with special reference to the superfamily Spiruroidea. Z. Parasitenkd. 7:273-335.

Skrjabin, K.I. 1949. Key to parasitic nematodes, vol. 1. Spirurata and Filariata. Akademii Nauk SSSR, Moscow. (English translation, 1968.) (Useful keys to genera with lists of species.)

Skrjabin, K.I., A.A. Sobolev, and V.M. Ivaskin. 1963-1967. Essentials of nematodology, vols. 11, 12, 14, 16, and 19. Spirurata of animals and man and the diseases they cause. Akademii Nauk SSSR, Moscow. (The most complete monographs on the subject.)

CHAPTER 30

ORDER CAMALLANATA: GUINEA WORMS AND OTHERS

The order Camallanata appears to be transitional between the Spirurata, which are primarily parasites of the digestive tract, and the Filariata, mainly parasites of tissues. The morphology of several families, such as the Dracunculidae, is strikingly filariid, whereas that of others, such as Camallanidae and Cucullanidae, is quite spirurid. All but Cucullanidae are ovoviviparous, like most filariids, but the arthropod intermediate host must be eaten to complete transmission, as in the spirurids. Most species are uncommon parasites of economically unimportant vertebrate hosts and, therefore, will not be discussed here. Two families, Camallanidae and Philometridae, are commonly encountered in fishes, and a third, Dracunculidae, has a species of great medical importance to humans. These three families will serve to illustrate the order.

FAMILY CAMALLANIDAE

Included in the family Camallanidae are several similar genera that inhabit the intestines of fishes, amphibians, and reptiles. Their most conspicuous character is the head, in which the buccal capsule has been replaced with a pair of large, bilateral sclerotized valves (Fig. 30-1). The complex ornamentation of these valves (Fig. 30-2) is a useful taxonomic character.

The genus *Camallanus* is common in freshwater fishes and turtles in the United States. *Camallanus oxycephalus* is often seen as a bright red worm extending from the anus of a crappie *(Pomoxis)* or other warm-water panfish. The life cycles of all species that have been investigated involve a cyclopoid copepod crustacean as intermediate host. Development proceeds to maturity in the intestine of the vertebrate with no tissue migration.

FAMILY PHILOMETRIDAE

Two common genera in the family Philometridae

are *Philometra*, with a smooth cuticle, and *Philometroides*, with a cuticle covered with bosses. Each is a tissue parasite of fishes. The mouth is small, there is no sclerotized buccal capsule, and the esophagus is short. Males of many species are unknown. Gravid females live under the skin, in the swim bladder, or in the coelom of fishes, where they release first-stage juveniles. After reaching the external environment, the juveniles develop further if they are eaten by a cyclopoid crustacean. The microcrustacean, containing third-stage juveniles, must be eaten by the definitive host if the worm is to survive. Development to the adult is not well known. Males and females mate in the deep tissues of the body, and the males die soon after. The females then migrate to their definitive site, where the young are released. *Philometra oncorhynchi*, a parasite of salmon in the western United States and Canada, apparently passes out with the fish's eggs when it spawns, bursts in the fresh water, and, thus, releases its juveniles.[14]

Philometroides, under the skin of the head and fins of suckers (Catostomidae), are familiar sights to those who work with these fish in the United States (Fig. 30-3).

FAMILY DRACUNCULIDAE

Members of the family Dracunculidae are tissue-dwellers of reptiles, birds, and mammals. All have life cycles involving aquatic intermediate hosts. Morphological characteristics of the several genera and species are remarkably similar, with small differences between those in reptile hosts, for example, and those in mammals.

Several species of *Dracunculus* are known from snakes, and one is common in snapping turtles in the United States. The genus *Micropleura*, is found in crocodilians and turtles in South America and India, whereas *Avioserpens* has species in aquatic birds.

The genus *Dracunculus* is also known from mam-

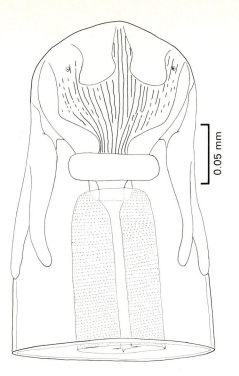

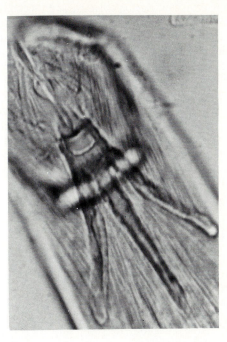

FIG. 30-1

Head of *Camallanus marinus*, lateral view. In this genus the buccal cavity is replaced by large, sclerotized valves with various markings.

From Schmidt, G.D., and R.E. Kuntz. 1969. Parasitology 59:389-396.

FIG. 30-2

Dorsal view of the head of *Camallanus marinus*, showing the large, sclerotized trident characteristic of this genus.

Photograph by G.D. Schmidt.

FIG. 30-3

Philometroides sp. in the skin of a fin of a white sucker, *Catastomus commersoni*.

Photograph by John S. Mackiewicz.

mals. In the Americas a species known as *D. insignis* is common in muskrats, opossums, and raccoons and other carnivores, especially those occupying semiaquatic environments. *Dracunculus medinensis* is prevalent in circumscribed areas of Africa, India, and the Middle East. It has been reported from humans in the United States several times, but these cases may have been caused by *D. insignis*. In fact, *D. insignis* may well be *D. medinensis,* perhaps in a form attenuated in its pathogenicity to humans. This might explain the scarcity of reports of it in humans in this country. It has been shown to be infective to rhesus monkeys.[2]

However, *D. medinensis* is not scarce in humans in all countries of the world, so we will examine it in greater detail.

Dracunculus medinensis

Dracunculus medinensis has been known since antiquity, particularly in the Middle East and Africa, where it causes great suffering even today. For example, 2.5 million cases per year occur in Nigeria alone.[7,13] Because of its large size and the conspicuous effects of infection, it is not surprising that the parasite was mentioned by classical authors. The Greek, Agatharchidas of Cnidus, who was tutor to one of the sons of Ptolemy VII in the second century BC, gave a lucid description of the disease: ". . . the people taken ill on the Red Sea suffered many strange and unheard of attacks, amongst other worms, little snakes, which came out upon them, gnawed away their legs and arms, and when touched retracted, coiled themselves in the muscles, and there gave rise to the most unsupportable pains."[4] The Greek and Roman writers Paulus Aegineta, Soranus, Aetius, Actuarus, Pliny, and Galen all described the disease, although most of them probably never saw an actual case. The Spanish and Arabian scholars Avicenna, Avenzoar, Rhazes, and Albucasis also discussed this parasite, probably from firsthand observations. In 1674 Velschius described winding the worm out on a stick as a cure. European parasitologists remained ignorant of this worm until about the beginning of the nineteenth century, when British army medical officers began serving in India. Information about *D. medinensis* slowly accumulated, but it remained for a young Russian traveler and scientist, Aleksej Fedchenko, to give the first detailed account of the morphology and life cycle of the worm in 1869 to 1870.[5] His discovery that humans become infected by swallowing infected *Cyclops* pointed the way to a means of prevention of dracunculiasis. However, the disease is still common in some areas today, and certain details of the worm's biology are still unknown. An ex-

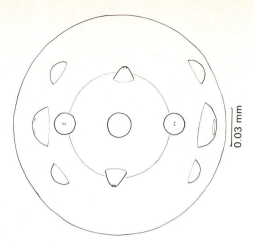

FIG. 30-4

En face view of *Dracunculus,* showing the arrangement of papillae and amphidial pores.

Drawing by G.D. Schmidt.

cellent review of this parasite is given by Muller.[11]

■ Morphology

Dracunculus medinensis is one of the largest nematodes known. Adult females have been recorded up to 800 mm long, although the few males known do not exceed 40 mm. The mouth is small and triangular and is surrounded by a quadrangular, sclerotized plate. Lips are absent. Cephalic papillae are arranged in an outer circle of four double papillae at about the same level as the amphids and an inner circle of two double papillae, which are peculiar in that they are dorsal and ventral (Fig. 30-4). The esophagus has a large glandular portion that protrudes and lies alongside the thin muscular portion.

In the female the vulva is about equatorial in young worms; it is atrophied and nonfunctional in adults. The gravid uterus has an anterior and a posterior branch, each of which is filled with hundreds of thousands of embryos. The intestine becomes squashed and nonfunctional as a result of the pressure of the uterus.

A major difficulty in the taxonomy of dracunculids is the sparsity of discovered males. The few specimens known range from 12 to 40 mm long; the spicules are unequal and 490 to 730 μm long. The gubernaculum ranges from 115 to 130 μm long. Genital papillae vary considerably in published descriptions. In fact, in monkeys, at least, males taken from a single animal have varying numbers of papillae. It is possible that more than one species is responsible for dracunculiasis, or there may be a complex of subspecies. Because of the technical difficulties of obtaining many

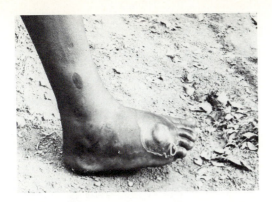

FIG. 30-5

Blister, caused by a female *Dracunculus medinensis,* in the process of bursting. There has been an unusually severe tissue reaction resulting in a very large blister. A loop of the worm can be seen protruding through the skin.

With permission from Muller, R. 1971. In Dawes, B., editor. Advances in parasitology, vol. 9. Copyright Academic Press, Inc. (London) Ltd.

FIG. 30-6

Living nauplius of *Cyclops vernalis* with a juvenile of *Dracunculus medinensis* in its hemocoel.

Photograph by Ralph Muller.

specimens, it remains for an experimental approach to solve the taxonomy of the species.

■ Biology

Dracunculus medinensis is ovoviviparous. When the parasite is gravid, the thousands of embryos in the uteri cause a high internal pressure. At this stage the female has migrated to the skin of the host. Usually the legs and feet are infected, but nearly any portion of the body is susceptible. Internal pressure and progressive senility cause the body wall and uterus of the parasite to burst, forcing a loop of the uterus through, freeing many juveniles. The juveniles cause a violent allergic reaction that causes a blister in the skin of the host (Fig. 30-5). This eventually ruptures, forming an exit for the young worms, which trickle out onto the surface of the skin. Sometimes, instead of the body wall rupturing, the uterus forces itself out of the mouth of the worm. Muscular contractions of the body wall force juveniles out in periodic spurts, with more than half a million ejected at a single time. These contractions are instigated by cool water, which causes the worm and its uterus to protrude through the wound. As portions of the uterus empty, they disintegrate, and adjacent portions move into the ulcer. Eventually all of the worms will be "used up," and the wounds will heal.

The first-stage juvenile must enter directly into water, after leaving its mother and host, to survive. It can live for 4 to 7 days but is able to infect an intermediate host for only 3 days. To develop further, it must be eaten by a cyclopoid crustacean. Once in the intestine of their new host, the juveniles penetrate into the hemocoel, especially dorsad to the gut, where they develop to the infective third stage in 12 to 14 days at 25° C (Fig. 30-6).

Infection of the definitive host is effected when infected copepods are swallowed with drinking water. The released juveniles penetrate the duodenum, cross the abdominal mesenteries, pierce the abdominal muscles, and enter the subcutaneous connective tissues, where they migrate to the axillary and inguinal regions. The third molt occurs about 20 days after infection, and the final one at about 43 days. Females are fertilized by the third month after infection. Males die between the third and seventh months, become encysted, and degenerate. Gravid females migrate to the skin of the extremities between the eighth and tenth months, by which time the embryos are fully formed. Between 10 and 14 months after initial infection, the female causes a blister in the skin.

Little is known about the physiology of this parasite, but the gut is often filled with a dark-brown material, suggesting that the worms feed on blood. Glycogen is stored in several tissues of the mature female. Glucose utilization and the rate of formation of lactic acid are not affected by the presence or absence of oxygen.[3] The blister formation in the definitive host is an immunological response to parasite antigens.

■ Epidemiology

To become infected, a person must swallow a copepod that had been exposed to juveniles previously re-

FIG. 30-7

Pond in the Mabauu area of Sudan, in the Sahel savannah zone. Conditions such as these favor the transmission of the guinea worm.

Photograph by J. Bloss; with permission from Muller, R. 1971. In Dawes, B., editor. Advances in parasitology, vol. 9. Copyright Academic Press, Inc. (London) Ltd.

FIG. 30-8

Step well at Kantarvos, near Kherwara, India, infected with *Dracunculus medinensis*.

Photograph by A. Banks; with permission from Muller, R. 1971. In Dawes, B., editor. Advances in parasitology, vol. 9. Copyright Academic Press, Inc. (London) Ltd.

leased from the skin of a definitive host. Thus three conditions must be met before the parasite's life cycle can be completed: The skin of an infected individual must come in contact with water, the water must contain the appropriate species of microcrustaceans, and the water must be used for drinking. There is circumstantial evidence that infection can be acquired by eating a fish paratenic host.[8]

It is curious that a parasite life cycle that is so dependent on water is most successfully completed under conditions of drought.

In some areas of Africa, for instance, people depend on rivers for their water. During periods of normal river flow, few or no new cases of dracunculiasis occur. During the dry season, however, rivers are reduced to mere trickles with occasional deep pools, which are sometimes enlarged and deepened by those who depend on them as a water source (Fig. 30-7). Planktonic organisms flourish in this warm, semistag-

nant water, and a cyclopean population explosion occurs. At the same time any bathing, washing, and water drawing bring infected persons in contact with water, into which juveniles are shed. When such water is drunk, many infected copepods may be downed at a quaff.

In areas of India the step well (Fig. 30-8) is a time-honored method of exposing ground water. These wells, often centuries old, have steps leading into the water, on which water bearers enter the well to fill their jars and, incidentally, release juveniles into the water at the same time.

In many desert areas the populace depends on deep wells, which are crustacean free, during the dry season. Most villages also have one or more ponds that fill during the rainy season and become a source of infection with *Dracunculus*. Most villagers prefer the pond water because they have to pay for well water, and, moreover, the well water is usually saline.

With these examples in mind, it is no wonder that a parasite with an aquatic life cycle should thrive in a desert environment, since all animals, humans and beasts alike, depend on isolated waterholes for their existence. So does *D. medinensis*. On the other hand, this dependence on isolated waterholes exposes a weakness in the parasite's life cycle. Guinea worm is the only helminthic disease transmitted solely by drinking water. If safe water can be provided in endemic areas, this parasite is sure to be eradicated.[6] For example, in Ivory Coast, Africa, active surveillance reduced the disease from 4971 cases in 1976 to 592 cases in 1985.[13]

Pathogenesis

Dracunculiasis may result in three major disease conditions: the emergence of adult worms, secondary bacterial infection, and nonemergent worms.

At the onset of migration to the skin, the female worm elicits an allergic reaction caused by the release of metabolic wastes into the host's system. The reaction may produce a rash, nausea, diarrhea, dizziness, and localized edema. The worms remain just under the skin for about a month before a reddish papule develops. This rapidly becomes a blister. The feet and legs are most often affected, although the blister may appear nearly anywhere on the surface of the body. On rupture of the blister, the allergic reactions usually subside. The site of the blister becomes abscessed, but this heals rapidly if serious secondary complications do not occur. A tiny hole remains, through which the worm protrudes. When the worm is removed or is expelled, healing is completed. Infection, however, does not confer immunity and a person may be reinfected many times.

Serious complications can result from the introduction of bacteria under the skin by the retreating worm. In parts of Africa this is the third most common mode of entry of tetanus spores.[9] Other complications are abscesses, synovitis, arthritis, bubo, and other infections.

Worms that fail to reach the skin often cause complications in deeper tissues of the body, although many die and are absorbed or calcified, with no apparent effect on the host. Chronic arthritis, with a calcified worm in or alongside the joint, is common. More serious symptoms, such as paraplegia, result from a worm in the central nervous system. Adult worms have also been found in the heart and urogenital system.

Commonly, when worms do not emerge, they eventually begin to degenerate and release powerful antigenic substances. These cause aseptic abscesses, which also can lead to arthritis. These abscesses can

FIG. 30-9

Ancient woodcut showing removal of guinea worm by winding it on a stick.
Velschius, 1674.

be large, with up to half a liter of fluid containing leukocytes and, frequently, numerous embryos. Usually, however, the worms become calcified.

Diagnosis and treatment

The appearance of an itchy, red papule that rapidly transforms into a blister is the first strong symptom of dracunculiasis. On a few occasions, the patient can feel or see the worm in the skin before papule formation. After the blister ruptures, juveniles can be obtained by placing cold water on the wound; when mounted on a slide, they can be seen actively moving about under a low-power microscope.

When a part of the worm emerges, diagnosis is fairly evident, although the drying, disintegrating worm does not show the typical morphology of a nematode. An occasional sparganum may be diagnosed as dracunculiasis. Immunological tests show promise but are not yet perfected.

Pulling out guinea worms by winding them on a stick is a treatment used successfully since antiquity (Fig. 30-9).

FIG. 30-10

Seal of the American Medical Association and the double-serpent caduceus of the military medical profession. Might the serpent on a staff originally have depicted the removal of guinea worm?

And they journeyed from Mount Hor by way of the Red Sea, to compass the land of Edom . . .

And the Lord sent fiery serpents among the people, and they bit the people; and much people of Israel died . . . And the Lord said unto Moses, "Make thee a fiery serpent and set it upon a pole; and it shall come to pass that everyone that is bitten, when he looketh upon it, shall live."

NUMBERS 21:6

This excerpt from the Old Testament is a pretty fair account of dracunculiasis and its treatment. A serpent on a pole and a worm on a stick are not that different, after all. Moses and his people were, at the time, near the Gulf of Akaba, where *Dracunculus* is still endemic. Also, the Israelites had for some time been in a drought area, existing on water where they could find it. This is consistent with the epidemiology of dracunculiasis.

The staff with serpent carried by Aesculapius, the Roman god of medicine, adopted today as the official symbol of medicine (and the double-serpent caduceus of the military), may well depict the removal of *Dracunculus* (Fig. 30-10). This form of cure is still widely used (Fig. 30-11). If cold water is applied to the worm, she will expel enough juveniles to allow about 5 cm of her body to be pulled out. The procedure is repeated once a day, complete removal requiring about 3 weeks. In some areas of India the worms are said to be sucked out by native doctors using a crude aspirator!

An alternate method is removal of the complete worm by surgery. This is often successful when the entire worm is near the skin and also in the case of deep abscesses containing worms that failed to reach the skin. However, if the worm is threaded through a

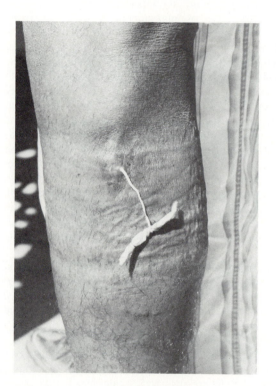

FIG. 30-11

Uncomplicated case of dracunculiasis. The worm is being pulled out through a small hole left after the ulcer is mostly healed.

With permission from Muller, R. 1971. In Dawes, B., editor. Advances in parasitology, vol. 9. Copyright Academic Press, Inc. (London) Ltd.

tendon or deep fascia or is broken into several pieces, it may be impossible to remove completely.

Several chemotherapeutic agents show promise against this disease organism, particularly niridazole, thiabendazole, and metronidazole.[1] After a short time the worms are spontaneously ejected, or they may be pulled out with minimal effort. Muller[12] found that worms treated with these drugs in vivo showed no histological changes that would indicate that the drugs killed them. Since these drugs have an antiinflammatory effect, he suggested that this may be the sole factor in expelling the worms. He supported this idea by applying 2.5% hydrocortisone cream, an antiinflammatory agent, to new blisters, and the worms were pulled out easily after 5 days, leaving little tissue reaction behind. Juveniles in the uterus were still infective to copepods.

Prevention and control lie in interrupting the life cycle. The most logical means of doing this is by chemical treatment of ponds and wells to eliminate copepods. On a cost/effectiveness basis, DDT would probably be the most effective compound. Those who fear DDT will have to pay a higher price, either by using more expensive compounds or by living with *D. medinensis*. One compound that shows promise, among others, is Abate (O,O,O′, O′-tetramethyl O,O′-thiodiphenylene phosphorothioate).[10]

REFERENCES

1. Antani, J., H.V. Srinivas, K.R. Krishnamurthy, and B.R. Jahagirdar. 1970. Metronidazole in dracunculiasis. Am. J. Trop. Med. Hyg. 19:821-822.

2. Beverly-Burton, M., and V.F.J. Crichton. 1973. Identification of guinea-worm species. Trans. R. Soc. Trop. Med. Hyg. 67:152.

3. Bueding, E., and J. Oliver-Gonzalez. 1950. Aerobic and anaerobic production of lactic acid by the filarial worm *Dracunculus insignis*. Br. J. Pharmacol. Chemother. 5:62-64.

4. Cobbold, T.S. 1864. Entozoa. Groombridge, London.

5. Fedchenko, A.P. 1870. Concerning the structure and reproduction of the guinea worm *(Filaria medinensis)*. Proc. Imp. Soc. Friends Nat. Sci. Anthropol. Ethnograph. 8:columns 71-81. (Translation in Am. J. Trop. Med. Hyg. 20:511-523.)

6. Hopkins, D.R. 1987. Dracunculiasis eradication: a mid-decade status report. Am. J. Trop. Med. Hyg. 37:115-118.

7. Ilegbodu, V.A., et al. 1986. Impact of guineaworm disease on children in Nigeria. Am. J. Trop. Med. Hyg. 35:962-964.

8. Kobayashi, A., et al. 1986. Human case of dracunculiasis in Japan. Am. J. Trop. Med. Hyg. 35:159-161.

9. Lauckner, T.R., A.M. Rankin, and F.C. Adi. 1961. Analysis of medical admissions to University College Hospital, Ibadan. W. Afr. Med. J. 10:3.

10. Muller, R. 1970. Laboratory experiments on the control of *Cyclops* transmitting guinea worm. Bull. WHO 42:563-567.

11. Muller, R. 1971. *Dracunculus* and dracunculiasis. In Dawes, B., editor. Advances in parasitology, vol. 9. Academic Press, Inc., New York, pp. 73-151.

12. Muller, R. 1971. The possible mode of action of some chemotherapeutic agents in guinea worm disease. Trans. R. Soc. Trop. Med. Hyg. 65:843-844.

13. Muller, R. 1985. Guineaworm eradication—the end of another disease? Parasitol. Today 1:39, 58.

14. Platzer, E.G., and J.R. Adams. 1967. The life history of a dracunculoid *Philonema onchorhynchi*, in *Onchorhynchus nerka*. Can. J. Zool. 45:31-43.

SUGGESTED READINGS

Chabaud, A.G. 1975. Keys to genera of the order Spirurida, part 1. In Anderson, R.C., A.G. Chabaud, and S. Willmott, editors. CIH keys to the nematode parasites of vertebrates. Commonwealth Agricultural Bureaux, Farnham Royal, Bucks, Eng.

Ivashkin, V.N., A.A. Sobolev, and L.A. Hromova. 1971. Essentials of nematodology, vol. 22. Camallanata of animals and man and the diseases they cause. Akademii Nauk SSSR, Moscow. (A most valuable reference to all species in this order.)

Neafie, R.C., D.H. Connor, and W.M. Meyers. 1976. Dracunculiasis. In Binford, C.H., and D.H. Connor, editors. Pathology of tropical and extraordinary diseases, vol. 2, sect. 9. Armed Forces Institute of Pathology, Washington, D.C.

CHAPTER 31

ORDER FILARIATA: FILARIAL WORMS

The filariids are tissue-dwelling parasites, with the exception of the Diplotriaenidae that live in the air sacs of birds. They are among the most highly evolved of the parasitic nematodes, appearing to have arisen from the Camallanata by way of the Spirurata. All species employ arthropods as intermediate hosts, most of which deposit third-stage juveniles on the skin with their bite (Fig. 31-1). They are parasitic in all classes of vertebrates except fish. Generally speaking, filariids are slender worms with reduced lips and buccal capsule. Most are parasites of wild animals, particularly of birds, but several are very important disease organisms of humans and domestic animals. The majority of these belong to the large family Onchocercidae.

FAMILY ONCHOCERCIDAE

Members of the family Onchocercidae live in the tissues of amphibians, reptiles, birds, and mammals. Most are of no known medical or economic importance, but a few cause some of the most tragic, horrifying, and debilitating diseases in the world today. Of these, species of *Wuchereria, Brugia, Onchocerca,* and *Loa* will be considered in detail. Short mention will be made of others.

Wuchereria bancrofti

Perhaps the most striking disease of humans is the clinical entity known as **elephantiasis** (Fig. 31-2). The horribly swollen parts of the body afflicted with this condition have been known since antiquity. The ancient Greek and Roman writers likened the thickened and fissured skin of infected persons to that of the elephant, although they also confused leprosy with this condition. Actually elephantiasis is a nonsense word, since literally translated it means "a condition caused by elephants." The word is so deeply entrenched, however, that it is not likely ever to be abandoned. Classic elephantiasis is a rather rare consequence of infection by *Wuchereria bancrofti* and by at least two other species of filariids.

Infection by *W. bancrofti* and other filariids is best referred to as filariasis. **Bancroftian filariasis** *(W. bancrofti)* is the most widespread of the filariases of humans, extending throughout central Africa, the Nile Delta, Turkey, India and southeast Asia, the East Indies, the Philippine and Oceanic Islands, Australia, and parts of South America (in short, across a broad equatorial belt). It was probably brought to the New World by the slave trade. A nidus of infection remained in the vicinity of Charleston, South Carolina, until it died out spontaneously in the 1920s.[6]

Filariasis was a cause of great psychological concern to American armed forces in the Pacific theater in World War II. Although thousands of cases of filariasis were in fact contracted by American servicemen, no single case of classical elephantiasis resulted. Some persons experienced symptoms for as long as 16 years.[28]

The evolution of knowledge of this disease and its cause remains one of the classics of medical history.[10] Two species are known in the genus.

■ Morphology

Adult worms are long and slender with a smooth cuticle and bluntly rounded ends. The head is slightly swollen and bears two circles of well-defined papillae. The mouth is small; a buccal capsule is lacking.

The male is about 40 mm long and 100 μm wide. Its tail is fingerlike. The female is 6 to 10 cm long and 300 μm wide. The vulva is near the level of the middle of the esophagus.

■ Biology

Adult *Wuchereria* live in the major lymphatic ducts of humans, tightly coiled into nodular masses. They are normally found in the afferent lymph channels near the major lymph glands in the lower half of the body. Rarely they invade a vein. The females are ovoviviparous, producing thousands of juveniles known as **microfilariae** (Fig. 31-3). Microfilariae are not as differentiated as are normal first-stage juveniles and are sometimes considered advanced embryos. The microfilariae of *W. bancrofti* retain the egg membrane as a "sheath" (not to be confused with the sheath of some

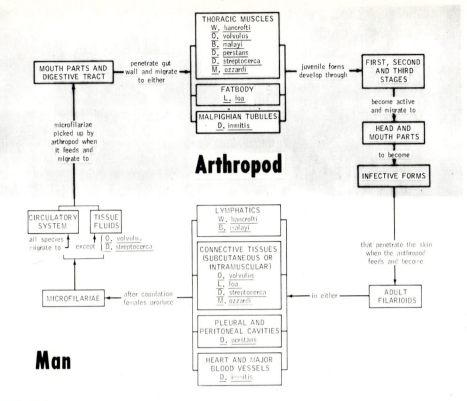

FIG. 31-1

Filarioid diseases. Relationships of life cycles of filarial parasites to mode of transmission.

AFIP neg. no. 67-19037-1.

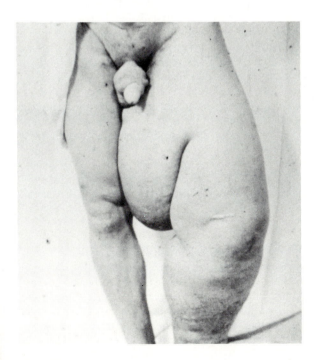

FIG. 31-2

Elephantiasis caused by infection with *Wuchereria bancrofti* in Venezuela.

Photo by R.E. Kuntz. Courtesy H. Zaiman, editor. A pictorial presentation of parasites.

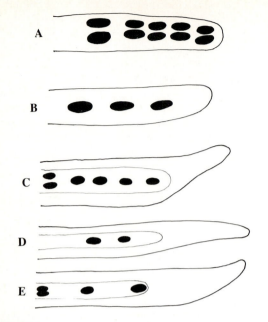

FIG. 31-3

Presence or absence of a sheath and the arrangement of nuclei in the tail are useful criteria in identifying microfilariae. **A,** *Mansonella perstans.* **B,** *Mansonella ozzardi.* **C,** *Loa loa.* **D,** *Wuchereria bancrofti.* **E,** *Brugia malayi.*

third-stage strongyle juveniles, which is the second-stage cuticle). The sheath is rather delicate and close fitting but can be detected where it projects at the anterior and posterior ends of the microfilaria. When stained, several internal nuclei and primordia of organs can be seen in the microfilariae. The location of these and the presence or absence of a sheath are used to identify the several species of microfilariae found in humans.

The microfilariae are released into the surrounding lymph by the female. Some may wander into the adjacent tissues, but most are swept into the blood through the thoracic duct. Throughout much of the geographical distribution of this parasite, there is a marked **periodicity** of microfilariae in the peripheral blood; that is, they can be demonstrated at certain times of the day, whereas at other times, they virtually disappear from the peripheral circulation. The maximal number usually can be found between 10 PM and 2 AM. For this reason, night-feeding mosquitoes are the primary vectors of *Wuchereria* in areas where microfilarial periodicity occurs. During the day the microfilariae are concentrated in blood vessels of the deep tissues of the body, predominately in the pulmonary vessels proximal to the pulmonary arterioles.[25] Causes of the periodicity remain obscure, but they apparently do not in-

volve daily release of a new generation of progeny by the adult female. Stimuli, such as arterial oxygen tension and body temperature, probably are involved. Administration of pure oxygen to a patient during peak microfilaremia can cause the microfilariae to localize in the deep tissues. Reversal of the patient's sleep schedule causes reversal of periodicity so that microfilaremia becomes diurnal. The adaptive value of the periodicity is difficult to explain. Although it is clearly advantageous for the microfilariae to be present in the peripheral blood when the vector is likely to be feeding, what value is there in being absent when the vector is not feeding? Some strains of *Wuchereria* have a diurnal periodicity. The periodicity is unimportant clinically, but it has significant diagnostic and epidemiological implications.

In certain areas of the South Pacific, including Fiji, Samoa, the Philippines, and Tahiti, a strain of *Wuchereria* is common that shows a diurnal periodicity, referred to as **subperiodic.** The morphology of the adults is identical to that of *Wuchereria* producing periodic microfilariae; most investigators believe that only one species is involved, although some designate the subperiodic type as a separate species named *W. pacifica.* Daytime-feeding mosquitoes are the major vectors of the subperiodic strain.

Microfilariae are ingested by the mosquito along with its blood meal. They lose their sheath in the first 2 to 6 hours in the insect's stomach, after which they penetrate the gut of the host and reach the thoracic muscles. The first cuticular molt occurs about 2 days later. The second-stage juvenile is a short, sausage-shaped worm (**sausage stage**) (see Fig. 31-7) in which most of the organ systems are present. Within 2 weeks the second molt takes place. The juvenile is now an elongate, slender **filariform** third stage (see Fig. 31-8) and development ceases. The filariform juveniles are 1.4 to 2 mm long and are infective to the definitive host. They migrate throughout the hemocoel, eventually reaching the labium, or proboscis sheath, from which they escape when the mosquito is feeding. They enter the skin through the wound made by the mosquito. After migrating though the peripheral lymphatics, the worms settle in the larger lymph vessels, where they mature.

■ Epidemiology

Many mosquito vectors of *Wuchereria* have a preference for human blood and often breed near human habitation. At least 77 species and subspecies of mosquitoes in the genera *Anopheles, Aedes, Culex,* and *Mansonia* are known intermediate hosts for *Wuchereria.* In areas in which the periodic strain of *Wuchereria* is found the mosquito vectors are primarily night feeders. The species of mosquito serving as vector in a particular area seems to depend more on coincidence (which species feeds when the juveniles within the definitive host are available) than on physiological determinants of host specificity. Nevertheless, the periodicity has practical epidemiological significance because this fact determines which mosquito species must be controlled and, consequently, what control measures must be applied.

Suitable breeding sites for mosquitoes abound in tropical areas. Some sites are difficult or impossible to control, such as tree holes and hollows at the bases of palm fronds, whereas others can be controlled with a degree of effort. Hollow coconuts, killed while still green by rats gnawing holes in them, fall, fill with rainwater, and become havens for developing mosquito larvae. These can be collected and burned. Even dugout canoes that are unused for a few days can partially fill with rainwater and become mosquito nurseries. Conditions for transmission of *Wuchereria* vary from locality to locality and country to country. The epidemiologist must consider each case independently within the framework of the biology of the vector and host, putting the economic and technical resources that are available to best advantage.

■ Pathogenesis

Pathogenesis in filariasis depends heavily on inflammatory and immune responses, and these are predominately responses to the adult worms, particularly the female; little or no disease is caused by the microfilariae. There are three clinical phases: the **incubation** stage, the **acute** or **inflammatory** stage, and the **obstructive** phase or stage of complications caused by the chronic lymphedema. The incubation phase is the time between infection and the appearance of microfilariae in the blood. It is largely symptomless, but there may be transient lymphatic inflammation with mild fever and malaise.

The acute inflammatory stage follows when the females reach maturity and start releasing microfilariae. Intense lymphatic inflammation occurs, usually in the lower half of the body, with chills, fever, and toxemia. The area of the affected lymphatic is swollen and painful, and the overlying skin may be reddened. The attack usually subsides after a few days, but this and the other manifestations described further often recur at frequent intervals.

Additional common symptoms in the acute stage of filariasis include **inguinal lymphadenitis** (inflammation of the lymph nodes in the inguinal region), **orchitis** (inflammation of the testes, usually with sudden enlargement and considerable pain), **hydrocele** (forcing of lymph into the tunica vaginalis of the testis or spermatic cord), and **epididymitis** (inflammation of the spermatic cord). Acute febrile episodes called **elephantoid fever** recur frequently. These are marked by sudden onset, rigors and sweating, and fever to 104° F and endure from a few hours to several days. On the histological level, extensive proliferation of the lining cells occurs in the lymphatics, with much inflammatory cell infiltration, especially of polymorphonuclear leukocytes and eosinophils, around the lymphatics and adjacent veins. The most prominent cells in the infiltration become lymphocytes, plasma cells, and eosinophils, as the most acute phase subsides. Abscesses around dead worms may exist, with accompanying bacterial infection. Microfilariae may disappear from the peripheral blood during and after the acute phase, presumably because the lymphatic vessel containing the female becomes blocked.

The obstructive phase is marked by **lymph varices, lymph scrotum, hydrocele, chyluria,** and **elephantiasis.** Lymph varices are "varicose" lymph ducts, caused when lymph return is obstructed and the lymph "piles up," greatly dilating the affected duct. This causes chyluria, or lymph in the urine, a common symptom of filariasis. The chyle gives the urine a milky appearance, and some blood is often present. A feature of the chronic obstructive phase is progressive infiltration of the affected areas with fibrous connective tissue, or "scar" formation, after inflammatory episodes. However, dead worms are sometimes calcified instead of absorbed, usually causing little further difficulty.

In a certain proportion of cases, thought to be associated with repeated attacks of acute lymphatic inflammation, the condition known as elephantiasis gradually develops. This is a chronic lymphedema with much fibrous infiltration and thickening of the skin. In men the organs most commonly afflicted with elephantiasis are scrotum, legs, and arms; in women the legs and arms are usually afflicted, with vulva and breasts being affected more rarely. Elephantoid organs are composed mainly of fibrous connective tissues, granulomatous tissue, and fat. The skin becomes thickened and cracked, and invasive bacteria and fungi further complicate the matter. Microfilariae usually are not present.

Elephantiasis is thus seen to be a result of complex immune responses of long duration. After worms die and are absorbed, the symptoms gradually disappear. Repeated superinfections over many years are usually necessary to cause elephantiasis. Casual visitors into endemic areas may well become infected with the parasite, however, and suffer from localized edema and painful inflammation of the lymphatic system, but they may have no microfilariae in their peripheral blood. This condition may persist for many years, subsiding and recurring from time to time.[28]

One of the most perplexing problems of the disease is that of why microfilariae are so rarely found in the peripheral blood if a person is first infected as an adult. In World War II, 10,431 United States naval personnel were infected with *W. bancrofti,* yet only 20 showed a microfilaremia.[3] However, among adults indigenous to an endemic area, there is a high incidence of microfilaremia. It is possible that this phenomenon may be explained by a condition of immune tolerance. Transplacental infection has been demonstrated in some filarial worms, and if such occurred with *W. bancrofti,* invasion of the human embryo might result in later failure of the individual's immune system to recognize the microfilariae as foreign. It must be, however, that the tolerance is not complete because the person would soon be overwhelmed by the juveniles if a certain proportion were not destroyed. It seems likely that the clinical disease is determined by individual reactions to continual and sometimes massive antigenic stimulation. Also, it appears that associated bacterial infections play a role, but the nature of this role is not understood.

■ Diagnosis and treatment

Demonstration of microfilariae in the blood is a simple and fairly accurate diagnostic technique,[14] provided that thick blood smears are made during the period when the juveniles are in the peripheral blood. The technician must be able to distinguish this species from others that could be present. X-ray examinations can detect dead, calcified worms. Because microfilariae often cannot be demonstrated, especially in newcomers to an endemic area, the intradermal skin test is valuable. A preparation of powdered *Dirofilaria immitis* (the heart filarial worm of dogs) in saline gives nearly 100% accuracy in diagnosis, although false positives result when the patient is infected with other species of filariids.[5] Filariasis should always be suspected if clinical symptoms occur about 3 months or more after arrival in an endemic area. Diagnosis and treatment have been discussed by Partono.[22]

The drug of choice at this time is diethylcarbamazine (Hetrazan), which eliminates microfilariae from the blood and, with careful administration, usually kills the adults.[15] Metronidazole may be effective where diethylcarbamazine fails.

Swollen limbs are sometimes successfully treated by applying pressure bandages, which force the lymph out of the swollen area. This may gradually reduce the size of the member, which returns to nearly normal. Any connective tissue proliferation that might have developed will not be affected, however. Surgical removal of elephantoid tissue is often possible.

Prevention primarily remains protection against the bite of mosquitoes when in endemic areas. Insect repellent, mosquito netting, and other preventive measures should be rigorously used by persons temporarily visiting such places. Long-term protection requires mosquito control and mass chemotherapy of indigenous people to eliminate microfilariae from the circulating blood, where they are available to mosquitoes.

Brugia malayi

It was first noticed in 1927 that a microfilaria, different from that of *W. bancrofti,* occurred in the blood of natives of Celebes. It was not until 1940 that the adult form was found in India, and a year later it was discovered in Indonesia. *Brugia malayi* is now known to parasitize humans in China, Korea, Japan, southeast Asia, India, Sri Lanka, the East Indies, and the Philippines. Much of its distribution overlaps that of *Wuchereria.*

The morphology of this parasite is very similar to that of *W. bancrofti,* although the male is only about half as large. The number of anal papillae of males differs slightly between the two species, and the left spicule of *B. malayi* is a little more complex than that of *W. bancrofti.* These are feeble differences on which to separate two genera, but, because a large literature is accumulating on Malayan filariasis under the name of *Brugia,* we reluctantly follow common usage.

■ Morphology

Males are 13.5 to 20.5 mm long and 70 to 80 μm wide. The tail is curved ventrad and bears three or four pairs of adanal and three or four pairs of postanal papillae. The spicules are unequal and dissimilar, and a small gubernaculum is present.

Females are 80 to 100 mm long by 240 to 300 μm wide. The finger-like tail is covered with minute cuticular bosses. The vulva is near the level of the middle of the esophagus.

■ Biology and pathology

The life cycle of *B. malayi* is nearly identical to that of *W. bancrofti.* Mosquitoes of the genera *Mansonia, Aedes,* and *Culex* are intermediate hosts. Adults live in the lymphatics and cause the same disease symptoms as *W. bancrofti,* although elephantiasis, when it occurs, is more restricted to the legs; the genitalia are rarely affected.[15]

The microfilaria is somewhat similar to that of *Wuchereria* but can be differentiated from it by the presence of nuclei in the tail tip. There are both periodic and subperiodic strains.

Diagnosis and treatment are as for *Wuchereria.*

Control is also primarily by mosquito eradication. Because *Mansonia* is the major vector in many areas, herbicides can be put to good advantage in eliminating the aquatic plants that the mosquito larvae depend on as their source of oxygen. Larvae of this genus pierce the stems of aquatic vegetation to tap air, obviating the need for the wiggler to reach the surface regularly.

Another species of *Brugia*, *B. timori*, was first known from its distinctive microfilariae, and since then adults have been described.[23] It has been found only from the Lesser Sunda Islands of southeast Indonesia and can cause severe disease in affected populations. It shows nocturnal periodicity and is transmitted by *Anopheles barbirostris;* there is no known animal reservoir.[24] Another species, *W. lewisi,* is known from its microfilariae in Brazil. It has been common in the history of filaria research that the microfilaria was discovered years before the adult was found. It would not be surprising if more unknown species exist in humans in various parts of the world. *B. beaveri* exists in raccoons and bobcats in the United States. Several infections in humans with *Brugia* may have been this species.[2,13]

Onchocerca volvulus

River blindness is a disease caused by this large filariid worm in areas of Africa (where 30 million are infected), Arabia, Guatemala, Mexico, Venezuela, and Colombia. It probably has an even greater distribution than is presently known, since it is a cryptic disease that does not always manifest overt symptoms. When it does, the symptoms often are overlooked by health authorities, most of whom are busy with more pressing and immediate problems. An estimated 2000 infected persons lived as immigrants in London in 1975.[29]

Onchocerciasis, as the condition is also known, is not a fatal disease. However, it does cause disfigurement and blindness in many cases; in some small communities in Africa and Central America, most of the people of middle age and over are blind. Eradication of this disease from the earth would not result in the "parasitologist's dilemma," since it would not increase the birthrate or increase the chances for infant survival. It would, instead, free hundreds of thousands of persons from a debilitating disease and thereby remove this economic burden from developing nations. Onchocerciasis was extensively reviewed by Nelson.[20]

■ Morphology

The morphology of *Onchocerca volvulus* is not very different from that of *W. bancrofti*. The worms characteristically are knotted together in pairs or groups in the subcutaneous tissues (Fig. 31-4). They are slender and blunt at both ends. Lips and a buccal capsule are absent, and two circles of four papillae each surround

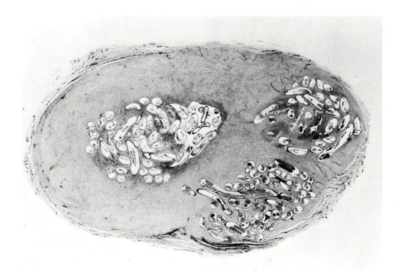

FIG. 31-4

Cross section of a fibrous nodule (**onchocercoma**) removed from the chest of an African. It contained several worms bound together in a mass.

From Connor, D.H., et al. 1970. Hum. Pathol. 1:553-579. AFIP neg. no. 69-3625.

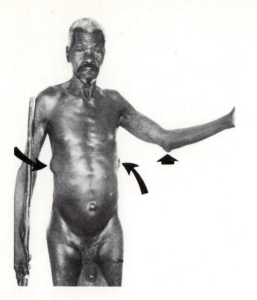

FIG. 31-5

Several nodules *(arrows)* filled with *Onchocerca volvulus* are found in the skin of this man. Note also the elephantoid scrotum and the depigmentation and wrinkling of the skin of the upper arms, also symptoms of onchocerciasis.

From Connor, D.H., et al. 1970. Hum. Pathol. 1:553-579. AFIP neg. no. 68-10071-3.

FIG. 31-6

A black fly, *Simulium damnosum,* biting the arm of a human. This insect is a major vector of *Onchocerca volvulus* in Africa.

From Connor, D.H., et al. 1970. Hum. Pathol. 1:553-579. AFIP neg. no. 68-2763-1.

the mouth. The esophagus is not conspicuously divided.

Males are 19 to 42 cm long by 130 to 210 μm wide; females are 33.5 to 50 cm long by 270 to 400 μm wide, with the vulva just behind the posterior end of the esophagus. The tail of the male is curled ventrad and lacks alae; it bears four pairs of adanal and six or eight pairs of postanal papillae. The microfilariae are unsheathed.

■ Biology

Adult worms locate under the skin, where they become encapsulated by host reactions. If this is over a bone, such as at a joint or over the skull, a prominent nodule appears (Fig. 31-5). The location of these nodules is correlated with the geographical area. In Africa most infections are below the waist, whereas in Central America they are usually above the waist. This is probably an adaptation to the biting preferences of the insect vectors, since the microfilariae are concentrated in the areas where the insects prefer to bite. Perhaps the cause is simply that "bush country" Americans are more inclined to wear trousers than are their African counterparts.

The unsheathed microfilariae remain in the skin, where they can be ingested by the black fly intermediate hosts *Simulium* spp. (Fig. 31-6). These ubiquitous pests become infected when they take a blood meal. Their mouthparts are not adapted for deep piercing, so much of their food consists of tissue juices, which contain numerous microfilariae in infected persons. The first-stage juvenile migrates from the intestinal tract of the fly to its thoracic muscles. There it molts to the sausage stage (Fig. 31-7) and then molts again to the infective, filariform stage (Fig. 31-8). The filariform juvenile moves to the labium of the fly and can infect a new host when the insect next feeds. Mature worms appear in the skin in less than a year.

Onchocerca volvulus was probably introduced to the Americas with African slaves. It became established in Central America and has since mutated sufficiently to cause different clinical symptoms in its definitive host and to differ in its infectivity to various vectors and laboratory animals. That the species has done this within about 400 years is an indication of the mutability of dioecious parasites with high reproductive capacity. Evidently, the parasite is expanding its distribution northward through Mexico.[8] Humans appear to be the only natural definitive host for *O. volvulus*. The physiology of this parasite has not been studied.

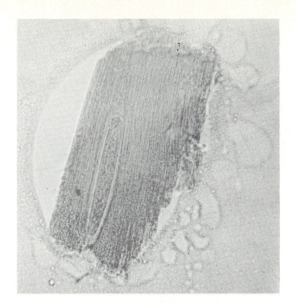

FIG. 31-7

Piece of thoracic muscle from *Simulium damnosum*. Note the second, or "sausage-stage," juvenile of *Onchocerca volvulus*.

Photograph by John Davies.

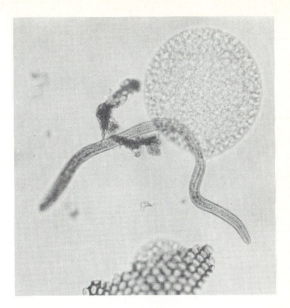

FIG. 31-8

Third-stage, or filariform, juvenile of *Onchocerca volvulus*, dissected from the head of a *Simulium damnosum*.

Photograph by John Davies.

Epidemiology

Generally speaking, onchocerciasis is a model system for the landscape epidemiologist. *Simulium* spp. live their larval stages only in clear, fast-running streams. The adult flies survive only where there is high humidity and plenty of streamside vegetation. It was long known by certain African natives that the disease was associated with rivers (and even with black flies, although it was not officially "discovered" until 1926), and they gave it the name river blindness. Anyone who intrudes into such an area is viciously attacked by these insects. Wild-caught black flies are often infected with a variety of species of filariids, most of which are still unidentified, but in areas endemic with *O. volvulus* the juveniles can often be recognized as that species.

Surprisingly, foci of onchocerciasis occur in the arid savannah of west Africa and the desert areas along the Nile near the Egypt-Sudan border. The epidemiology in these areas has not been thoroughly studied, but it is certain to depend on adaptations for survival of the black fly vectors.

Pathogenesis

Two different elements contribute to the pathogenesis of onchocerciasis: the adult worms and the microfilariae. Of these, the adult is the least pathogenic, of-ten causing no symptoms whatever and, at the worst, stimulating the growth of palpable subcutaneous nodules called **onchocercomas** (Fig. 31-5), especially over bony prominences. In the African strain these nodules are most frequent in the pelvic area, with a few along the spine, chest, and knees. The Venezuelan form is much like that in Africa, but in Central America the nodules are mostly above the waist, especially on the neck and head. These nodules are relatively benign, causing some disfigurement but no pain or ill health. The number of nodules may vary from one to well over a hundred. They consist mainly of collagen fibers surrounding one to several adult worms.[12] Rarely the nodule will degenerate to form an abscess, or the worm will become calcified.

True elephantiasis is sometimes caused by this worm (Fig. 31-9), and another condition, known as "hanging groin," is common in some areas of Africa. A loss of elasticity of the skin causes a sagging of the groin into pendulous sacs, often containing lymph nodes. The testes and scrotum are not affected, and hydrocele does not accompany the condition. Females are similarly affected (Fig. 31-10). Leonine face is a rare complication in Central America and Africa. Onchocerciasis frequently causes hernias, especially femoral hernia, in Africa.

The presence of microfilariae in the skin often re-

495

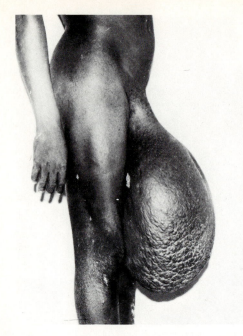

FIG. 31-9

Severe elephantoid scrotum on a native of Ubangi territory. It was removed surgically and a good cosmetic result was obtained. The scrotum weighed 20 kg and, when viewed microscopically, was an edematous mass of interlacing collagen and smooth muscle fibers.

From Connor, D.H., et al. 1970. Hum. Pathol. 1:553-579. AFIP neg. no. 68-8582-9.

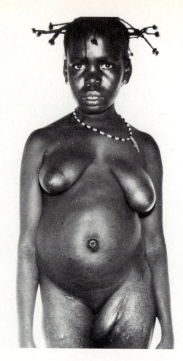

FIG. 31-10

"Hanging groin," or adenolymphocele. The tissue was excised and contained a group of lymph nodes embedded in subcutaneous tissue. The nodes contained many microfilariae of *Onchocerca volvulus*.

From Connor, D.H., et al. 1970. Hum. Pathol. 1:553-579. AFIP neg. no. 68-10066-1.

sults in a severe **dermatitis** caused either by allergic responses or toxic effects after the death of the juveniles. The first symptom is an intense itching, which may lead to secondary bacterial infection, often accompanied by dyspigmentation of the skin in small or extensive areas. This is followed by a thickening, discoloration, and cracking of the skin. These symptoms parallel those of avitaminosis A, and it has been suggested that they result from the parasite's competition for or interference with vitamin A metabolism. The last stage of the skin lesion is characterized by loss of elasticity, which gives the patient a look of premature aging. **Depigmentation** is accentuated and may extend over large areas, especially of the legs (Fig. 31-11). Patients at this stage are often misdiagnosed as leprous. The distressing effects on the life-style of these persons can only be imagined.

Microfilariae in advanced cases often are located in the deeper part of the dermis and are not detected by skin-snip biopsy.

By far the most dreadful complications of onchocer-

ciasis are those of the eyes. It has been calculated that the number of blind persons per 100,000 is more than 1500 in areas endemic for onchocerciasis, compared with 250 per 100,000 in a random sample in Europe.[20] The rate of impaired vision may reach 30% in some communities of Africa, where blindness exceeds 10% of the adult population. In these areas, and in similar areas of Guatemala, it is not unusual to see a child with good vision leading a string of blind adults to the local market.

Ocular complications are less common in the rain forest areas of Africa but are frequent in the savannah. The reason for this remains one of the enigmas of parasitology. It is possible that different strains of worms have different tropisms for the cornea.

Lesions of the eye take many years to develop; most affected persons are over 40 years old. However, in Central America, with more worms concentrated on the head, young adults also show symptoms.

The earliest and most common complications begin when microfilariae invade the cornea. This causes in-

FIG. 31-11

An 11-year-old boy with severe dermatitis characterized by depigmentation, wrinkling, and thickening of the skin. He also has elephantoid changes of the penis and scrotum and onchocercomas over the knees.

From Connor, D.H., et al. 1970. Hum. Pathol. 1:553-579. AFIP neg. no. 68-7912-1.

FIG. 31-12

Skin snip from a patient with onchocerciasis. Note the emerging microfilariae.

Photograph by Warren Buss.

flammation of sclera, or white of the eye, followed by an invasion of fibrous tissue, leading to extensive vascularization of the cornea, which, in turn, severely impairs vision. Subsequent fibrosis may lead to complete blindness.

In some cases there is damage to the retina, with complications of the optic nerve. The exact cause of this condition is incompletely known, but it probably results from immune responses to dead microfilariae, perhaps in conjunction with toxic metabolites of the parasite. Microfilariae in the chambers of the eye are easily demonstrated in onchocerciasis. In fact often the ophthalmologist first diagnoses the disease during routine ocular inspection. Many aspects of ocular complications of river blindness parallel those of avitaminosis A.

A bizarre manifestation of onchocerciasis in Uganda is **dwarfism.** At first thought to be pygmies, these cases are now known to derive from normal parents and are always infected with *O. volvulus.* The symptoms are pituitary deficiency, and there is little doubt that the pituitary has been damaged, either directly or indirectly, by microfilariae.

Adult worms may live as long as 16 years.

■ Diagnosis and treatment

The best method of diagnosis is the demonstration of microfilariae in bloodless skin snips. These are made by raising a small bit of skin with a needle and slicing it off with a razor or scissors. The bit of skin is then placed in saline on a slide and observed with a microscope for emerging microfilariae (Fig. 31-12). These must be differentiated from other species that might be present. Nodules may be aspirated, but no microfilariae will be found if only males or dead worms are present. Skin-snip biopsies can be taken anywhere, but if only a single snip is available, it should be taken from the buttock. If the snip is so deep as to draw blood, it might be contaminated with other species of filariid. Also, in old cases the microfilariae may be so deep as to elude the snip. Each case may demonstrate other, overt symptoms that obviate the need for demonstrating the microfilariae.

In parts of Africa there may be some periodicity of microfilariae in the skin, but this has not been adequately studied. The most specific, sensitive, and practically useful immunodiagnostic test so far is the indirect hemagglutination test.[16]

Treatment of onchocerciasis is by two methods, surgical and chemotherapeutic. Excision of nodules, especially those around the head, may be effective in lowering both the rate of eye damage and the number of new infections within a population. It is a simple operation that can be performed under rather primitive conditions.

Suramin is an effective drug; it kills the adults and causes a slow disappearance of the microfilariae.[15]

Nodules should be removed surgically where possible because abscess formation may follow death of the adults. Some ocular damage is irreversible, but improvement in vision may follow suramin therapy. Diethylcarbamazine kills microfilariae very rapidly but does not affect adult worms. In fact it kills the microfilariae too rapidly, since tissue reactions to dead larvae are usually violent, resulting in massive eruptions of the skin, extreme prostration, and sometimes death caused by anaphylactic shock. Under hospital conditions, these drugs can be administered, along with antihistamines and/or corticosteroids, to diminish the side effects, with good results. Ivermectin currently is the drug of choice.[1] In an act of humanitarian generosity, the manufacturer of ivermectin has announced that it will give the drug free of charge to countries that need it.[17a]

Prevention may best be accomplished by eliminating black flies from inhabited areas. It has been demonstrated clearly that DDT applications to swift-running streams will destroy all simuliids.[11] Similar results have been obtained by spraying DDT along river banks from airplanes or helicopters. The widespread use of DDT has caused international uneasiness about its possible, and as yet largely unknown, effects on humans and other organisms. This should be cautiously weighed against the ultimate goal of eradicating one of humankind's most devastating diseases. Other, more biodegradable insecticides are being developed for use in place of DDT.

An ongoing program of black fly control in west Africa promises to free 10 million people otherwise condemned to an arduous battle for economic and social survival.[9]

Loa loa

Loa loa is the "eye worm" of Africa, which produces **loaiasis** or **fugitive** or **Calabar swellings.** It is distributed in the rain forest areas of west Africa and equatorial Sudan. Although it was established for a short time in the West Indies, where it was first discovered during slavery, it no longer exists there.

The morphology of *L. loa* is typical of the family: a simple head with no lips and eight cephalic papillae; a long, slender body; and a blunt tail. The cuticle is covered with irregular, small bosses, except at the head and tail. Males are 20 to 34 mm long by 350 to 430 μm wide. The three pairs of preanal and five pairs of postanal papillae are often asymmetrical. The spicules are uneven and dissimilar, 123 and 88 μm long. Females are 20 to 70 mm long and about 425 μm wide. The vulva is about 2.5 mm from the anterior end, and the tail is about 265 to 300 μm long.

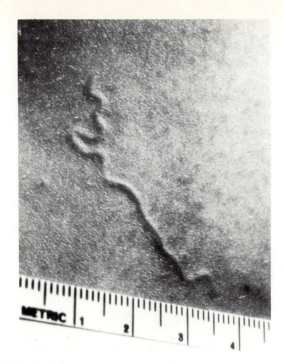

FIG. 31-13

Adult female *Loa loa* visible under the skin of a patient.

From Price, D.L., and H.C. Hopps. 1971. In Marcial-Rojas, R.A., editor. Pathology of protozoal and helminthic diseases, with clinical correlation. © 1971. The Williams & Wilkins Co., Baltimore. AFIP neg. no. 67-5366.

■ Biology

Adults live in subcutaneous tissues (Fig. 31-13), including back, chest, axilla, groin, penis, scalp, and eyes in humans and several other primates. Infections of deep tissues are also known, including fatal encephalitis.[19] The microfilariae (Fig. 31-3) are periodic, appearing in the peripheral blood in maximal numbers during daylight hours and concentrating in the lungs at night. The intermediate host is any of several species of deer fly, genus *Chrysops*, which feeds by slicing the skin and imbibing the blood as it wells into the wound. The worms develop to the third-stage, filariform juveniles in the fat body of the fly, after which they migrate to the mouthparts. The prepatent period in humans is about a year, and adult worms may live at least 15 years.

■ Pathogenesis

These worms have a tendency to wander through the subcutaneous connective tissues, provoking inflammatory responses as they go. When they remain in one spot for a short time, the host reaction results in

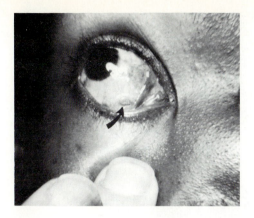

FIG. 31-14

Adult female *Loa loa* coiled under the conjunctival epithelium *(arrow)* of the eye of an African from the Congo.

From Price, D.L., and H.C. Hopps. 1971. In Marcial-Rojas, R.A., editor. Pathology of protozoal and helminthic diseases, with clinical correlation. © 1971. The Williams & Wilkins Co., Baltimore. AFIP neg. no. 67-5368-1.

localized "Calabar swellings," which disappear when the worm moves on. Localized inflammation of the area is most evident in white hosts. Adult worms also have an annoying habit of migrating through the conjunctiva and cornea (Fig. 31-14), with swelling of the orbit and psychosomatic results to the host. The overall disease is rather benign compared with those of most other filariids of humans.

■ **Diagnosis and treatment**

Demonstration of typical microfilariae in the blood (Fig. 31-3) is ample proof of loaiasis. The visual observation of a worm in the cornea or over the bridge of the nose is also indicative of this species. Finally, transient swellings of the skin are suspect, although sparganosis or onchocerciasis may be confused with loaiasis before the parasite is excised and examined. Surgical removal is simple and effective, providing the worm is properly located, but most of the worms are inapparent. Chemotherapy is as in filariasis bancrofti. Control of deer flies, which breed in swampy areas of the forest, is extremely difficult.

Other filariids found in humans

Mansonella ozzardi is a filariid parasite of the New World, with distribution known to encompass northern Argentina, the Amazon drainage, the northern coast of South America, Central America, and several islands of the West Indies. It has never been found in the Old World. Adults live in the body cavity, threaded among the mesenteries and peritoneum, and

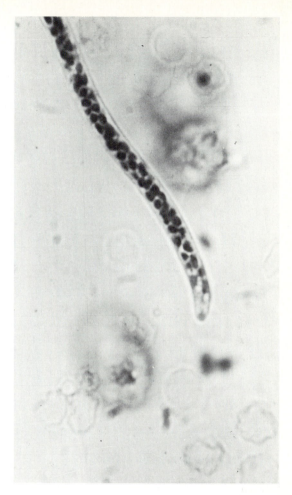

FIG. 31-15

Tail end of microfilaria of *Mansonella perstans*. The infection was acquired in Nigeria.

Photograph by M.G. Schultz. Courtesy H. Zaiman, editor. A pictorial presentation of parasites.

in subcutaneous tissues. Its manifestations can mimic bancroftian filariasis, with polylymphadenitis, lymphedema, elephantiasis, and hepatomegaly.[17]

The species was redescribed by Orihel and Eberhard.[21] The intermediate hosts are species of *Culicoides* and *Simulium*.[27]

Mansonella perstans (formerly *Dipetalonema perstans*) exists in people in tropical Africa and South America. Several primates have been incriminated as reservoir hosts. Adult worms live in the coelom and produce unsheathed microfilariae (Fig. 31-15). Intermediate hosts are species of biting midges of the genus *Culicoides*. They appear to cause little pathological effect.

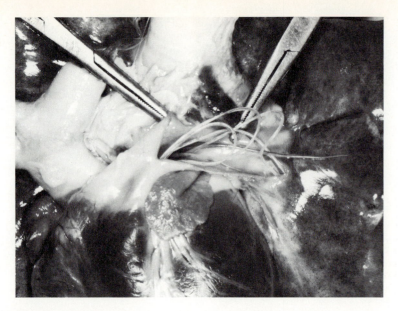

FIG. 31-16

Dirofilaria immitis protruding from the opened pulmonary artery of a German shepherd dog.

Photograph by Robert E. Kuntz.

FIG. 31-17

The life cycle of *Dirofilaria immitis*. **A,** Anterior end of adult of female worm. **B,** En face view of head of adult worm. **C,** Caudal end of adult male worm. **D,** Ventral view of caudal end of male. **E,** Sinistral view of caudal end of adult female. **F,** Microfilaria from blood (other stages of juveniles similar to those of species infecting man). **G,** Dog definitive host. **H,** Mosquito (*Aedes* spp.) intermediate host in feeding position on dog. *1,* Inner circle of cephalic papillae; *2,* outer circle of cephalic papillae; *3,* amphid; *4,* mouth without lips; *5,* muscular portion of esophagus; *6,* glandular portion of esophagus; *7,* intestine; *8,* anus; *9,* nerve ring; *10,* uterus containing unsheathed microfilariae; *11,* preanal papillae; *12,* adanal papillae; *13,* postanal papillae; *14,* long spicule; *15,* short spicule; *16,* nerve ring of microfilaria; *17,* excretory cell; *18,* G_1 cell; *19,* anal space; *20,* tail cells; *a,* adult worms in heart and pulmonary artery; *b* to *d,* microfilariae born in and circulating through bloodstream; *e,* microfilariae in peripheral blood and available to feeding mosquitoes; *f,* section of skin of dog; *g,* microfilaria being sucked up from peripheral blood vessel; *h,* microfilaria in stomach of mosquito; *i,* microfilaria entering malpighian tubules; *j,* microfilaria changes to sausage stage and prepares to undergo first molt; *k,* second-stage juvenile; *l,* sausage stage elongating and preparing for second molt; *m,* third-stage filariform juvenile in malpighian tubule; *n,* infective third-stage juvenile having escaped from malpighian tubule; *o,* juvenile migrating through thorax; *p* and *q,* juvenile entering and migrating down labium; *r,* juvenile escaping from labellum onto the skin; *s,* juvenile entering skin; *t,* juveniles in subcutaneous tissue (also muscle and adipose tissues); *u,* juvenile entering peripheral vessels from tissues 85 to 129 days after infection; *v,* juvenile entering general circulation; *w,* juvenile entering heart.

Courtesy Olsen, O.W. 1974. Animal parasites. Their life cycles and ecology, ed. 3. © 1974. University Park Press, Baltimore.

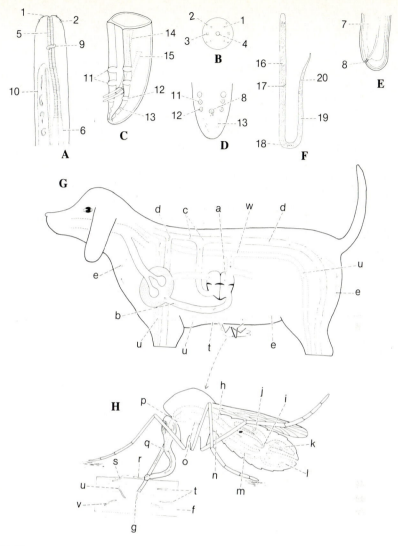

FIG. 31-17

For legend see opposite page.

Mansonella streptocerca (formerly *Dipetalonema streptocerca*) is a common parasite in the skin of humans in many of the rain forests of Africa.[18] It probably is a normal parasite of chimpanzees.

Dipetalonema sp. has been reported three times from the eyes of patients in Oregon.[4]

Dirofilaria immitis (Fig. 31-16) is parasitic in the right side of the heart and pulmonary artery of dogs and other mammals throughout most of the world. It has been found in humans several times, including cases in the United States. The worm is transmitted by several species of mosquito (Fig. 31-17). It often results in right-sided heart failure and pulmonary complications in dogs, and it constitutes a dangerous pathogen for dogs in many areas of the United States. In humans the symptoms are vague and unpredictable. Human pulmonary dirofilariasis was reviewed by Ciferri,[7] who discussed 60 cases. Infection of the heart and inferior vena cava has been described in detail.[26]

The range of this parasite in the United States appears to be increasing rapidly. Several drugs are effective in killing adult worms, and microfilaricides also are available. In some cases surgical removal of adult worms is warranted.

Dirofilaria conjunctivae has been found in subcutaneous lesions of humans in several European and American countries. It probably is not an actual species but rather is a zoonotic expression of *D. repens,* a parasite of dogs in Europe, Asia, and South America, and *D. tenuis,* a raccoon parasite in North America.

REFERENCES

1. Aziz, M.A., 1986. Ivermectin vs. onchocerciasis. Parasitol. Today 2:233-235.
2. Baird, J.K., et al. 1986. North American brugian filariasis: report of nine infections of humans. Am. J. Trop. Med. Hyg. 35:1205-1209.
3. Beaver, P.C. 1970. Filariasis without microfilaremia. Am. J. Trop. Med. Hyg. 19:181-189.
4. Beaver, P.C., E.A. Meyer, E.L. Jarroll, and R.C. Rosenquist. 1980. *Dipetalonema* from the eye of a man in Oregon, U.S.A. Am. J. Trop. Med. Hyg. 29:369-372.
5. Bozicevich, J., and A.M. Hutter. 1944. Intradermal and serological tests with *Dirofilaria immitis* antigen. Publ. Health Rep. 53:2130-2138.
6. Chernin, E. 1987. The disappearance of bancroftian filariasis from Charleston, South Carolina. Am. J. Trop. Med. Hyg. 37:111-114.
7. Ciferri, F. 1982. Human pulmonary dirofilariasis in the United States: a critical review. Am. J. Trop. Med. Hyg. 31:302-308.
8. Editorial Board. 1986. Is New World onchocerciasis spreading? Parasitol. Today 2:131.
9. Fatoyinbo, A. 1975. Initial success in battle against West African "river blindness." Trop. Med. Hyg. News 24:6-13.
10. Foster, W.D. 1965. A history of parasitology. E. & S. Livingstone, Edinburgh.
11. Garnham, P.C.G., and J.P. McMahon. 1947. The eradication of *Simulium neavei* Raubaud, from an onchocerciasis area in Kenya Colony. Bull. Ent. Res. 37:619-628.
12. George, G.H., J.R. Palmieri, and D.H. Connor. 1985. The onchocercal nodule: interrelationship of adult worms and blood vessels. Am. J. Trop. Med. Hyg. 34:1144-1148.
13. Gutierrez, Y., and R.E. Petras. 1982. *Brugia* infection in Northern Ohio. Am. J. Trop. Med. Hyg. 31:1128-1130.
14. Hoegaerden, M. van, and B. Ivanoff. 1986. A rapid, simple method for isolation of viable microfilariae. Am. J. Trop. Med. Hyg. 35:148-151.
15. Hunter, G.W., III, J.C. Swartzwelder, and D.F. Clyde. 1976. Tissue-inhabiting nematodes: the Filaroidea. In Tropical medicine, ed. 5. W.B. Saunders Co., Philadelphia.
16. Ikeda, T., et al. 1979. A sero-epidemiological study of onchocerciasis with the indirect hemagglutination test. J. Parasitol. 65:855-861.
17. Jörg, M.E. 1983. Filariasis por *Mansonella ozzardi* (Manson 1897) Faust 1929 en la Argentina, con descripción de un caso grave. Prensa Medica Argentina 70:181-192.
17a. Lindley, D. 1987. Merck's new drug free to WHO for river blindness programme. Nature 329:752.
18. Meyers, W.M., et al. 1972. Human streptocerciasis. A clinicopathologic study of 40 Africans (Zairians) including identification of the adult filaria. Am. J. Trop. Med. Hyg. 21:528-545.
19. Negesse, et al. 1985. Loiasis: "Calabar swellings" and involvement of deep organs. Am. J. Trop. Med. Hyg. 34:537-546.
20. Nelson, G.S. 1970. Onchocerciasis. In Dawes, B., editor. Advances in parasitology, vol. 8. Academic Press, Inc., New York, pp. 173-224.
21. Orihel, T.C., and M.L. Eberhard. 1982. *Mansonella ozzardi:* a redescription with comments on its taxonomic relationships. Am. J. Trop. Med. Hyg. 31:1142-1147.
22. Partono, F. 1985. Diagnosis and treatment of lymphatic filariasis. Parasitol. Today 1:52-57.
23. Partono, F., et al. 1977. *Brugia timori* sp. n. (Nematoda: Filaroidea) from Flores Island, Indonesia. J. Parasitol. 63:540-546.
24. Purnomo, D, T. Dennis, and F. Partono, 1977. The microfilaria of *Brugia timori* Partono et al. 1977 (= Timor microfilaria, David and Edeson, 1964): morphologic description with comparison to *Brugia malayi* of Indonesia. J. Parasitol. 63:1001-1006.
25. Spencer, H. 1973. Nematode diseases. II. Filarial diseases. In Spencer, H., editor. Tropical pathology. Springer-Verlag, Berlin, pp. 511-559.
26. Takeuchi, T., K. Asami, S. Kobayashi, M. Masuda, M. Tanabe, S. Miura, M. Asakawa, and T. Murai. 1981. *Dirofilaria immitis* infection in man: report of a case of the infection in heart and inferior vena cava from Japan. Am. J. Trop. Med. Hyg. 30:966-969.
27. Tidwell, M.A., and M.A. Tidwell. 1982. Development of *Mansonella ozzardi* in *Simulium amazonicum, S. argenticutum,* and *Culicoides insinuatus* from Amazonas, Colombia. Am. J. Trop. Med. Hyg. 31:1137-1141.
28. Trent, S. 1963. Reevaluation of World War II veterans with filariasis acquired in the South Pacific. Am. J. Trop. Med. Hyg. 12:877-887.
29. Woodhouse, D.F. 1975. Tropical eye diseases in Britain. Practitioner 214:646-653.

SUGGESTED READINGS

Chabaud, A., and R.C. Anderson. 1959. Nouvel essai de classification des filaries (Superfamille des Filarioidea) II, 1959. Ann. Parasitol. 34:64-87. (An accurate, easy-to-use key to the genera of Filariata.)
Chernin, E. 1983. Sir Patrick Manson's studies on the transmission and biology of filariasis. Rev. Infect. Dis. 5:148-166.
Chernin, E. 1983. Sir Patrick Manson: an annotated bibliography and a note on a collected set of his writings. Rev. Infect. Dis. 5:353-386.
Connor, D.H., et al. 1970. Onchocerciasis, onchocercal dermatitis, lymphadenitis, and elephantiasis in the Ubangi Territory. Human Pathol. 1:553-579.
Duke, B.O.L. 1971. The ecology of onchocerciasis in man and animals. In Fallis, A.M., editor. Ecology and physiology of parasites. University of Toronto Press, Toronto, pp. 213-222.
Duke, B.O.L. 1971. Onchocerciasis. Br. Med. Bull. 28:66-71.

Goodwin, L.G., E.A. Ottesen, and B.A. Southgate. 1984. Recent advances in research on filariasis. Tr. Roy. Soc. Trop. Med. Hyg. 78 (Suppl.):1-28.

Khanna, N.N., and G.K. Joshi. 1971. Elephantiasis of female genitalia. A case report. Plast. Reconstr. Surg. 48:374-381.

Meyers, W.M., et al. 1976. Diseases caused by filarial nematodes. In Binford, C.H., and D.H. Connor. Pathology of tropical and extraordinary disease, vol. 2, sect. 8. Armed Forces Institute of Pathology, Washington, D.C.

Sasa, M. 1976. Human filariasis. A global survey of epidemiology and control. University Park Press, Baltimore.

Sonin, M.D. 1966, 1968. Essentials of nematodology, vol. 17 and 21. Filariata of animals and man. Akademii Nauk SSSR, Moscow. (Comprehensive treatments of the Aproctoidea and Diplotriaenoidea.)

CHAPTER 32

PHYLUM ACANTHOCEPHALA: THORNY-HEADED WORMS

Few zoologists and still fewer veterinarians and physicians ever encounter a thorny-headed worm. Compared with parasitic platyhelminths or nematodes, they are fairly rare. Still, representatives are to be found inhabiting the intestines of fishes, amphibians, reptiles (rarely), birds, and mammals, where they have established a parasitic relationship with their host and, occasionally, cause serious disease.

The first recognizable description of an acanthocephalan in the literature is that of Redi, who, in 1684, reported white worms with hooked, retractable proboscides in the intestines of eels. From the time of Linnaeus to the end of the nineteenth century, all species were placed in the collective genus *Echinorhynchus* Zoega in Mueller, 1776, although Koelreuther is credited with naming the genus *Acanthocephalus* in 1771. Hamann[19] divided *Echinorhynchus,* which by then had become large and unwieldy, into *Gigantorhynchus, Neorhynchus,* and *Echinorhynchus,* thereby beginning the modern classification of the Acanthocephala.

Lankester[27] proposed elevating the order Acanthocephala, proposed by Rudolphi in 1808, to the level of phylum. This suggestion was not widely accepted until Van Cleave[50,51] convincingly argued in its favor. Today the Acanthocephala is widely accepted as a separate phylum.

FORM AND FUNCTION

The morphology of the acanthocephalans reflects an extensive adaptation to their parasitic mode of life and enteric habitat. There appears to have been an evolutionary reduction in muscular, nervous, circulatory, and excretory systems and a complete loss of a digestive system. The remaining animal seems little more than a pseudocoelomate bag of reproductive organs with a spiny holdfast at one end. The worms range in size from the tiny *Octospiniferoides chandleri,* only 0.92 to 2.4 mm long, to *Oligacanthorhynchus*

longissimus, exceeding a meter in length. Extensive reviews of acanthocephalan physiology have been published.[8,37]

General body structure

Superficially the acanthocephalan body is seen to consist of an anterior proboscis, a neck, and a trunk (Fig. 32-1).

The proboscis is variable in shape, from spherical to cylindrical, depending on the species (Fig. 32-2). It is covered by a tegument and has a thin, muscular wall within which are embedded the roots of recurved, sclerotized hooks. The sizes, shapes, and numbers of these hooks are among the most useful characters in the taxonomy of the worms. The proboscis is hollow and fluid filled. Attached to its inner apex is a pair of muscles, called **proboscis inverter muscles,** which extend the length of the proboscis and neck and insert in the wall of a muscular sac called the **proboscis receptacle.** The proboscis receptacle itself is attached to the inner wall of the proboscis. Its morphology varies somewhat depending on the family, but, generally speaking, it consists of one or two layers of muscle fibers. When the proboscis inverter muscles contract, the proboscis invaginates into the proboscis receptacle, with the hooks completely inside. When the proboscis receptacle contracts, it forces the proboscis to evaginate by a hydraulic system.[20] A nerve ganglion called the **brain** or **cerebral ganglion** is located within the receptacle. The proboscis and its receptacle are sometimes referred to as the **presoma.**

The neck is a smooth, unspined zone between the most posterior hooks of the proboscis and an infolding of the body wall. **Neck retractor muscles** attach this infolding of the body wall to the inner surface of the trunk. In some species other muscles, called **protrusers,** attach to the proboscis receptacle. When the proboscis retractor and the neck retractor muscles contract, the entire anterior end is withdrawn into the trunk. Some species have a sensory pit on each side of

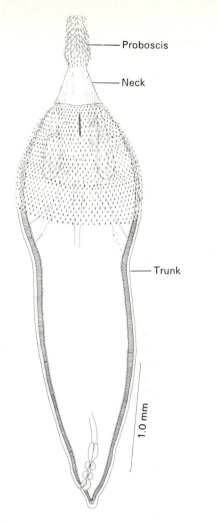

FIG. 32-1

Polymorphus swartzi, a parasite of ducks, showing the main body divisions.

From Schmidt, G.D. 1965. J. Parasitol. 51:809-813.

the neck, and two similar pits are found on the tip of the proboscis of many species.

The rest of the body, posterior to the neck, is called the **trunk,** or **metasoma.** Like the proboscis and neck, it is covered by a tegument and has muscular internal layers. Many species have simple, sclerotized spines embedded in the trunk wall that maintain close contact with the mucosa of the host's intestine. The trunk contains the reproductive system (Fig. 32-3) and also functions in absorbing and distributing nutrients from the host's intestinal contents. In the living worm the trunk is bilaterally flattened, usually with numerous transverse wrinkles, but when the worm is placed in a

hypotonic solution, such as tap water, it swells and becomes turgid. This is desirable for ease of study of the specimen, since it places the internal organs in constant relationship with each other, and it usually forces the introverted proboscis to evaginate.

Body wall

The body wall is a complex syncytium containing nuclear elements and a series of internal, interconnecting canals called the **lacunar system.** In some species the nuclei are gigantic (Fig. 32-3) but few in number. In others the nuclei fragment during larval development and are widely distributed throughout the trunk wall. When entire nuclei are present, their number is constant for each species, demonstrating the principle of **eutely,** or nuclear constancy. Development of the wall was described by Butterworth.[4]

■ **Tegument**

The tegument has no true layers, but several regions differ in their construction. These are, beginning with the outermost, the (1) surface coat, (2) striped zone, (3) vesicular zone, (4) felt zone, (5) radial fiber zone, and (6) basement lamina (Fig. 32-4). Inside the tegument is a layer of irregular connective tissue, followed by circular and longitudinal muscle layers. Like that of the trematodes and cestodes, the tegument is syncytial, but unlike those groups, the nuclei are in the basal region of the tegument, not in cytons separated from the distal cytoplasm. The **surface coat,** or **glycocalyx,** a filamentous material, was formerly known as the epicuticle. It is, for instance, about 0.5 μm thick on *Moniliformis moniliformis,* an acanthocephalan of rats and the most commonly investigated species in the laboratory. The surface coat is composed of acid mucopolysaccharides and neutral polysaccharides and/or glycoproteins.[58] The surface coat fits the definition of a glycocalyx, a carbohydrate-rich coat found on a variety of eukaryotic and prokaryotic cells. The stabilized system of polyelectrolytic filaments in the surface coat constitutes an extensive surface for molecular interactions, including those involved in transport functions and enzyme-substrate interactions.

Immediately beneath the surface coat and limited by the trilaminar outer membrane is the **striped zone.** This zone is 4 to 6 μm thick and is punctuated by a large number of crypts about 2 to 4 μm deep that open to the surface by pores[5] (Fig. 32-4). These crypts give this zone a striped appearance *(Streifenzone)* under the light microscope. The crypts increase the surface area of the worm by 44 times that of a smooth surface. A filamentous molecular sieve is seen in the necks of the crypts, but particles of less than about 8.5 nm can gain access to the crypts and undergo pinocytosis by

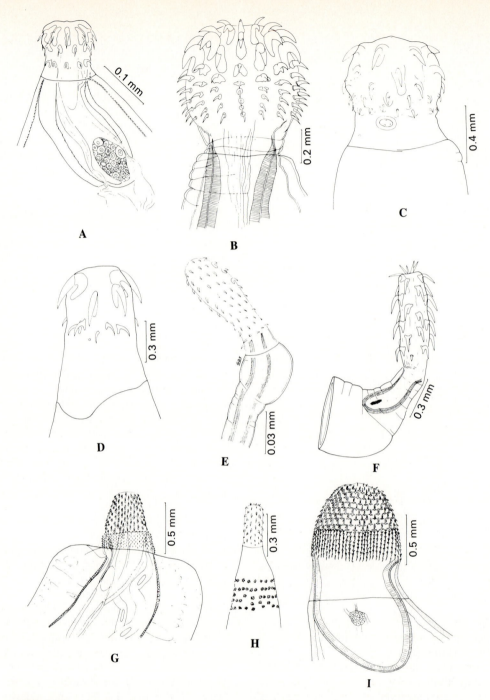

FIG. 32-2

Examples of different types of acanthocephalan proboscides. **A,** *Octospiniferoides australis*. **B,** *Sphaerechinorhynchus serpenticola*. **C,** *Oncicola spirula*. **D,** *Acanthosentis acanthuri*. **E,** *Pomphorhynchus yamagutii*. **F,** *Paracanthocephalus rauschi*. **G,** *Mediorhynchus wardae*. **H,** *Palliolisentis polyonca*. **I,** *Owilfordia olseni*.

A, E, and H from Schmidt, G.D., and E.J. Hugghins. 1973. J. Parasitol. 59:829-838; B from Schmidt, G.D., and R.E. Kuntz. 1966. J. Parasitol. 52:913-916; C from Schmidt, G.D. 1972. In Fiennes, R.N.T.W., editor. Pathology of simian primates. S. Karger, Basel, pp. 144-156; D from Schmidt, G.D. 1975. J. Parasitol. 61:865-867; F from Schmidt, G.D. 1969. Reproduced by permission of the National Research Council of Canada from the Canadian Journal of Zoology, vol. 17, 1969, pp. 383-385; G from Schmidt, G.D., and A.G. Canaris. 1967. J. Parasitol. 53:634-637; I from Schmidt, G.D., and R.E. Kuntz. 1967. J. Parasitol. 53:130-141.

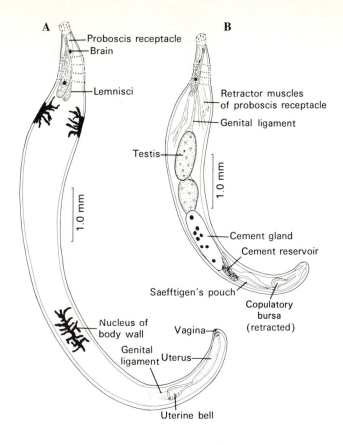

FIG. 32-3

Quadrigyrus nickoli, illustrating basic acanthocephalan morphology. **A**, Female. **B**, Male.

From Schmidt, G.D., and E.J. Hugghins. 1973. J. Parasitol. 59:829-835.

the crypt membrane.[5] The importance of pinocytosis in the acquisition of nutrients by the worms is unknown. In the deeper aspects of the striped zone, numerous lipid droplets, mitochondria, Golgi complexes, and lysosomes are found.

The striped zone grades into a region of numerous, closely packed, randomly arranged fibrils known as the **felt-fiber zone.** Mitochondria, numerous glycogen particles, vesicles, and, occasionally, lipid droplets and lysosomes also are found in the felt-fiber zone. The **radial fiber zone** is just within the felt-fiber zone and makes up about 80% of the thickness of the body wall. It contains large bundles of filaments that course radially through the cytoplasm, large lipid droplets, and nuclei of the body wall. Here, too, are many glycogen particles, mitochondria, Golgi complexes, and lysosomes. Rough endoplasmic reticulum is found in the perinuclear cytoplasm. The nuclei have numerous nucleoli. The lacunar canals course through the radial fiber zone.

The structure of the proboscis wall is similar to that of the trunk, except with fewer crypts, a thinner radial zone, and the absence of a felt zone.

■ Lacunar system and muscles

The system of fluid-filled channels in the body wall called the lacunar system has been long known, but its function has been enigmatic. A fascinating picture of the relationship of this curious system to the functioning of the body wall muscles has emerged, largely as a result of the efforts of Miller and Dunagan and their co-workers.[29-31,57]

The lacunar system is present in two parts, apparently unconnected to each other: that in the proboscis and neck and that in the trunk. The presomal lacunar system has channels that run into two structures called **lemnisci** that grow from the base of the neck into the pseudocoelom. Each lemniscus has a central canal that is continuous with the presomal lacunar system. The function of the lemnisci is unknown, although it may

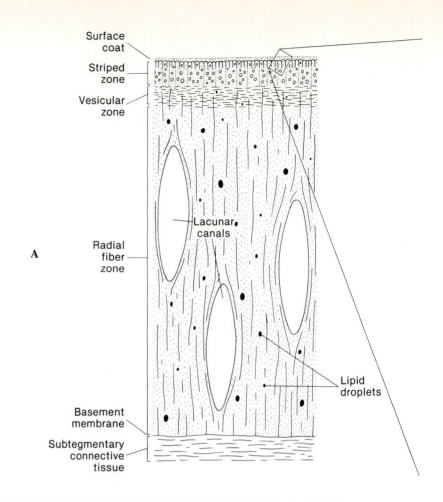

FIG. 32-4

Tegument of *Moniliformis moniliformis*. **A,** Diagram of transverse section to show layers. The vesicular zone is transitional between the striped zone and the felt-fiber zone; it contains many vesicles and mitochondria with poorly developed cristae. Lacunar canals are in the radial fiber zone. **B,** Electron micrograph showing the major features of the striped zone. The worm is coated with a finely filamentous surface coat *(SC)*. Numerous surface crypts *(C)* appear as large scattered vesicular structures with elements occasionally appearing to course to the surface of the helminth. The crypts are separated by patches of moderately electron-opaque material (*), giving the zone its striped appearance under the light microscope. Mitochondria *(M),* glycogen particles, microtubules, and other cytoplasmic details are evident in the inner portion of the striped zone. Bundles of fine cytoplasmic filaments *(f)* extend between this region and the deeper cytoplasm of the body wall. (×42,000.)

A drawing by William C. Ober; **B** from Byram, J.E., and F.M. Fisher. 1974. Tissue Cell 5:559.

contribute to the hydraulics of the proboscis mechanism. The metasomal lacunar system consists of a complicated network of interconnecting canals. In most species there are two main longitudinal canals, either dorsal and ventral or lateral. These are connected by numerous irregular or regular, transverse canals. The location and arrangement of the lacuni are used as taxonomic characters. In addition, at least in some species, there is a pair of medial longitudinal channels, each connected periodically by short radial canals to circular canals coursing between the dorsal and ventral longitudinal channels[29] (Fig. 32-5). The medial longitudinal channels lie on the pseudocoel side of the body-wall muscles, and the radial canals pierce the muscle layers to intercept the ring canals. Some of the ring canals give rise to branches that run throughout the radial fiber zone of the tegument (Fig. 32-4).

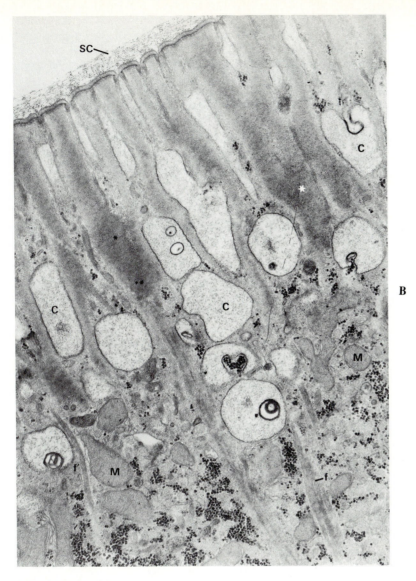

FIG. 32-4, Cont'd.

For legend see opposite page.

The body-wall muscles are composed of a longitudinal layer surrounded by a circular muscle layer (Fig. 32-6). These muscles have a very curious structure. They are hollow, with tubelike cores and numerous, anastomosing interconnectives.[30] It has been found that the lumina of the muscles are continuous with the lacunar system; therefore circulation of the lacunar fluid may well bring nutrients to and remove wastes from the muscles. Although there is no heart or other circulatory organ, contraction of the circular muscles would force fluid into the longitudinal components and vice versa. Thus, the lacunar system seems to function as an effective fluid transport system and possibly a hydrostatic skeleton.[33]

Acanthocephalan muscles are peculiar in other respects. They are electrically inexcitable, have low membrane potentials, and are slow conductors.[21] They are characterized by rhythmic, spontaneous depolarizations. Although the muscles appear to be stimulated by acetylcholine, nervous control of contraction is at present unclear. It is believed that nerves initiate contractions via the **rete system,** which is a highly

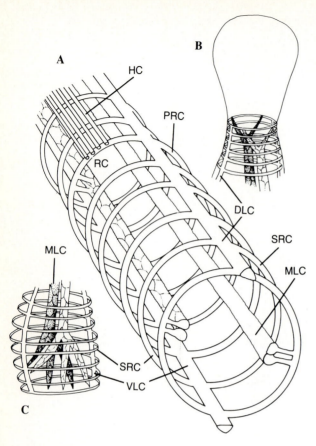

FIG. 32-5

Organization of lacunar system in *Macracanthorhynchus hirudinaceus*. **A,** Midmetasomal region. **B,** Region near neck, presomal lacunar system not indicated. **C,** Near posterior end of metasoma. *DLC,* Dorsal longitudinal channel; *HC,* hypodermal canal (in radial fiber zone); *MLC,* medial longitudinal channel; *PRC,* primary ring canal; *RC,* radial canal; *SRC,* secondary ring canal; *VLC,* ventral longitudinal channel.

From Miller, D.M., and T.T. Dunagan. 1976. Proc. Helm. Soc. Wash. 43:99-106.

branched, anastomosing network of thin-walled tubules lying on the medial surface of the longitudinal muscles or between the longitudinal and circular muscle layers.[57] The rete system itself seems to be modified muscle cells.

Reproductive system

Acanthocephalans are dioecious and usually demonstrate some degree of sexual dimorphism in size, with the female being larger (Fig. 32-3). In both sexes one or two thin **ligament sacs** are attached to the posterior end of the proboscis receptacle and extend to near the distal genital pore. Within these sacs are the gonads and some accessory organs of the reproductive systems. In some species the ligament sacs are permanent; in others they break down as the worm matures.

■ Male reproductive system

Two testes normally occur in all species, and their location and size are somewhat constant for each species. Spermiogenesis has been described.[56] Each testis has a vas efferens through which mature spermatozoa, which appear as slender, headless threads, travel to a common vas deferens and/or to a small penis. Several accessory organs also are present, the most obvious of which are the **cement glands.** These syncytial organs, numbering from one to eight, contain one or more giant nuclei or several nuclear fragments. In many species they are joined in places by slender bridges. They secrete a **copulatory cement** of tanned protein, which is stored in a **cement reservoir,** in some species, until copulation occurs. At that time the cement plugs the vagina after sperm transfer and rapidly hardens to form a **copulatory cap.** This remains attached to the posterior end of the female during subsequent development of the embryos within her body but eventually disintegrates.

Another male accessory sex organ is the **copulatory bursa** (Fig. 32-7), a bell-shaped specialization of the distal body wall that is invaginated into the posterior end of the body cavity except during copulation. A muscular sac, **Saefftigen's pouch,** is attached to the base of the bursa. When it contracts, fluid is forced into the lacunar system of the bursa, and, by hydrostatic pressure, it is everted. Many sensory papillae line the bursa; when it contacts the posterior end of a female, it clasps the female by muscular contraction, and sperm transfer is effected with a small penis.

■ Female reproductive system

The ovary of the female acanthocephalan is peculiar in that it fragments into **ovarian balls** early in life, often while the worm is still a juvenile in the intermediate host. These balls of oogonia float freely within the ligament sac, increasing slightly in size before insemination occurs. The posterior end of the ligament sac is attached to a muscular **uterine bell** (Fig. 32-8). This organ allows mature eggs to pass through into the uterus and vagina and out the genital pore, while returning immature eggs to the ligament sac. Its mechanism has been described.[54]

After copulation, the spermatozoa migrate from the vagina, through the uterus and uterine bell, and into the ligament sac. There they begin fertilizing the oocytes of the ovarian balls. After the first few cleav-

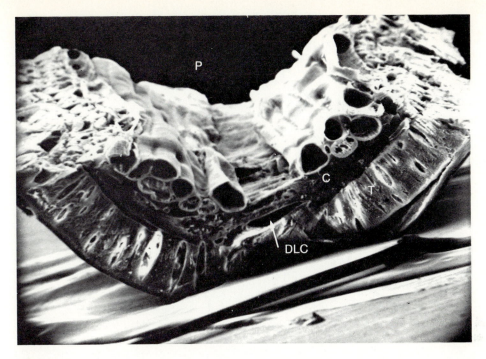

FIG. 32-6

Scanning electron micrograph of body wall, *Oligacanthorhynchus tortuosa*. *C*, Circular muscle; *P*, pseudocoel; *T*, tegument, showing hypodermal lacunar canals; *DLC*, dorsal longitudinal channel.

Courtesy D.M. Miller and T.T. Dunagan.

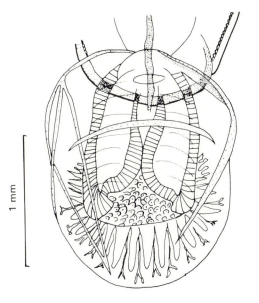

FIG. 32-7

Extended copulatory bursa of *Owilfordia olseni*. Note the numerous sensory papillae.

From Schmidt, G.D., and R.E. Kuntz. 1967. J. Parasitol. 53:130-141.

ages the embryos detach from the ovarian ball and float freely in the pseudocoelomic fluid. This exposes underlying oocytes for fertilization. Thus several stages of early embryogenesis may be found in a single female. Eventually, from this one copulation, many thousands or even millions of embryonated eggs are produced and released by each female. These, when sufficiently mature, pass from the host in its feces, where they may become available to the proper intermediate host.

As the shelled embryos are pushed into the uterine bell by peristaltic action, two possible routes are available. They may pass back into the pseudocoelom through slits in the bell or on into the uterus. Fully developed embryos are slightly longer than immature ones and, therefore, cannot pass through the bell slits[54]; hence they are passed on into the uterus, whereas immature eggs are retained for further maturation. The efficiency of the sorting is quite high, and apparently no immature forms are passed into the uterus.

Excretory system

Excretion in most species appears to be effected by diffusion through the body wall. However, members

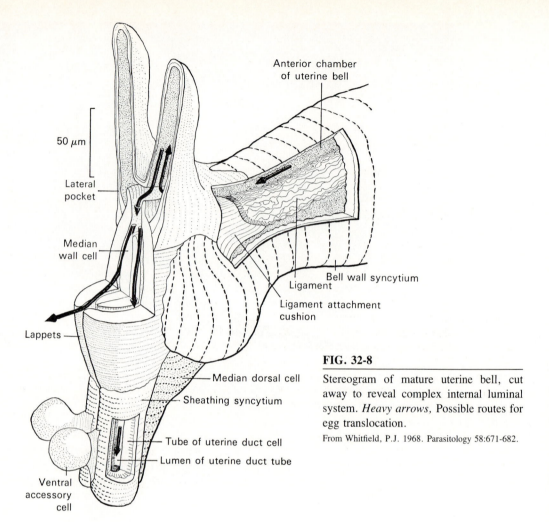

Anterior chamber
of uterine bell

50 μm

Lateral
pocket

Median
wall cell

Bell wall syncytium

Ligament

Ligament attachment
cushion

Lappets

Median dorsal cell

Sheathing syncytium

Tube of uterine duct cell

Lumen of uterine duct tube

Ventral
accessory
cell

FIG. 32-8

Stereogram of mature uterine bell, cut away to reveal complex internal luminal system. *Heavy arrows,* Possible routes for egg translocation.

From Whitfield, P.J. 1968. Parasitology 58:671-682.

of one family in the class Archiacanthocephala, the Oligacanthorhynchidae, are unique in possessing two **protonephridial excretory organs.** Each comprises many anucleate flame bulbs with tufts of cilia and may be encapsulated or not, depending on the species.[15,26] In the male these organs are attached to the vas deferens and empty through it; in the female they are attached to the uterine bell and empty into the uterus.

Acanthocephalans show little ability to osmoregulate, swelling in hypotonic, balanced saline or sucrose solutions and becoming flaccid in hypertonic solutions. The osmotic pressure of their pseudocoelomic fluid is close to or somewhat above that of the intestinal contents. They take up sodium and potassium, swelling in hypertonic solutions of sodium chloride or potassium chloride at 37° C. In balanced saline they lose sodium and accumulate potassium against a con-

centration gradient. Their hexose transport mechanism is not sodium coupled.

Nervous system

The nervous system of acanthocephalans is simple. The cerebral ganglion consists of only 54 to 88 cells in the species studied; it lies in the proboscis receptacle.[14,34] Relatively few nerves issue from the ganglion, the largest of which are the anterior proboscis nerve and the lateral posterior nerves.[13] Nerves supply the two lateral sense organs and the apical sense organ, if present. A large multinucleate cell referred to as a "support cell" is located ventral and slightly anterior to the cerebral ganglion.[32] Processes from the support cell lead to the lateral and apical sensory organs, but these processes are not nerves. Their function is unknown, but they may be secretory and help explain the inflammatory reaction of the host to the worm's proboscis.

DEVELOPMENT AND LIFE CYCLES

Each species of Acanthocephala uses at least two hosts in its life cycle. The first is an insect, crustacean, or myriapod, and the arthropod must eat an egg that was voided with the feces of a definitive host. Development proceeds through a series of stages to that which is infective to a definitive host. Many species, when eaten by a vertebrate that is an unsuitable definitive host, can penetrate the gut and encyst in some location where they survive without further development. This unsuitable vertebrate becomes a paratenic host, since if it is eaten by the proper definitive host, the parasite excysts, attaches to the intestinal mucosa, and matures. Such adaptability has survival value. For example, ecological gaps exist in the food chain between a microcrustacean and a large predaceous fish or between a grasshopper and an eagle. The paratenic host is one member in a food chain that bridges such a gap and, incidentally, ensures the survival of the parasite. It has been demonstrated that the behavior of the intermediate host may be altered to favor the probability of its being eaten by the definitive host.[35,36]

The manner of early embryogenesis is an unusual characteristic of the group. Early cleavage is spiral, although this pattern is somewhat distorted by the spindle shape of the eggshell. At about the 4- to 34-cell stage, the cell boundaries begin to disappear, and the entire organism becomes syncytial. Gastrulation occurs by migration of nuclei to the interior of the embryo.[42] They continue to divide but become smaller, until they form a dense core of tiny nuclei, the **inner nuclear mass.** These nuclei give rise to all internal organ systems of the worm. In some species the uncondensed nuclei remaining in the peripheral area give rise to the tegument; in some the tegument is derived from a nucleus that separates from the inner mass, whereas in others there are contributions from both.

The fully embryonated larva that is infective to the arthropod intermediate host is called the **acanthor.** The acanthor is an elongate organism that is usually armed at its anterior end with six or eight bladelike hooks. The hooks may be replaced by smaller spines in some species. The hooks or spines with their muscles are called the **aclid organ** or **rostellum.** The hooks aid in penetration of the gut of the intermediate host.[55] The acanthor is a resting, resistant stage and will undergo no further development until it reaches the intermediate host. Under normal environmental conditions, the acanthors may remain viable for months or longer. Acanthors of *Macracanthorhynchus hirudinaceus* can withstand subzero temperatures and desiccation, and they can remain viable for up to 3½ years in the soil. The acanthors of some species completely penetrate the gut, coming to lie in the host's hemocoel, whereas others stop just under the serosa. In both cases the worm then becomes parasitic on the arthropod, absorbing nutrients and enlarging, thus initiating the developmental stage known as the **acanthella.** The end of the acanthor that bears the aclid organ apparently becomes the anterior end of the adult in some species, whereas others exhibit a curious 90-degree change in polarity, in which the anterior end of the adult develops from the side of the acanthor. During the acanthella stage, the organ systems develop from the central nuclear mass and the hypodermal nuclei of the acanthor.

At termination of this development, the juvenile is an infective stage called a **cystacanth.** In most species the anterior and posterior ends invaginate, and the entire cystacanth becomes encased in a hyaline envelope. The parasite then must be eaten by the definitive host before it can fulfill its potential. Obviously mortality is very high, since only a tiny fraction of the immense number of eggs produced may survive the numerous hazards involved in completion of the life cycle.

Complete life cycles are known for only about 20 species in the phylum, although we have partial information on several more. The following examples illustrate the pattern followed in the life histories of the three major groups.

Class Eoacanthocephala

Neoechinorhynchus saginatus is an eoacanthocephalan parasite of various species of suckers and of creek chubs, *Semotilus atromaculatus,* fish distributed from Maine to Montana. Its life cycle and embryology were described by Uglem and Larson[47] (Fig. 32-9). When the eggs are eaten by the common ostracod crustacean *Cypridopsis vidua,* they hatch within an hour and begin penetrating the gut within 36 hours. After penetration, the unattached larva begins to enlarge and rearrange its nuclei, initiating the formation of internal organs. By 16 days after infection, the acanthella has developed into an infective cystacanth. Time required for maturation within the fish has not been determined.

Other eoacanthocephalan life cycles are similar, although paratenic hosts are known for *N. cylindratus*[53] and *N. emydis.*[22]

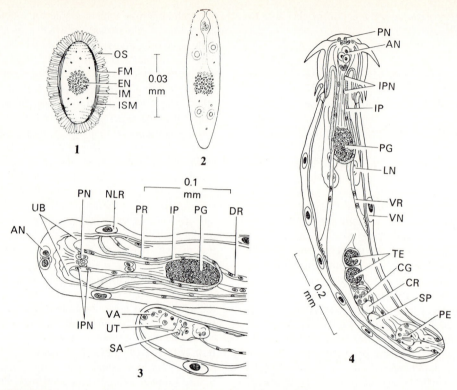

FIG. 32-9

Stages in development of *Neoechinorhynchus saginatus*. **1,** Shelled acanthor from body cavity of adult female. **2,** Acanthor from gut of ostracod, 1 hour after feeding. **3,** Female acanthella, age 12 days (one lemniscal nucleus omitted for clarity). **4,** Late male acanthella, age 14 days (neck retractors omitted). *AN,* Apical nuclei; *CG,* cement gland; *CR,* cement reservoir; *DR,* dorsal retractor of proboscis receptacle; *EN,* condensed nuclear mass; *FM,* fertilization membrane; *IM,* inner membrane; *IP,* proboscis inverter; *IPN,* proboscis inverter nuclei; *ISM,* inner shell membrane; *LN,* lemniscal nucleus; *NLR,* lemniscal ring nuclei; *OS,* outer shell; *PE,* penis; *PG,* brain anlage; *PN,* proboscis nuclear ring; *PR,* proboscis receptacle muscle sheath; *SA,* selector apparatus; *SP,* Saefftigen's pouch; *TE,* testes; *UB,* uncinogenous bands; *UT,* uterus; *VA,* vagina; *VN,* giant nucleus of ventral trunk wall; *VR,* ventral retractor of proboscis receptacle.

From Uglem, G.L., and O.R. Larson. 1969. J. Parasitol. 55:1212-1217.

Class Palaeacanthocephala

Plagiorhynchus cylindraceus is a palaeacanthocephalan that is common in robins and other passerine birds in North America. Its life cycle and embryology were described by Schmidt and Olsen[43] (Fig. 32-10). When the eggs are eaten by the terrestrial isopod crustacean *Armadillidium vulgare,* they hatch in the midgut within 15 minutes to 2 hours. Active entrance of the acanthor into the gut wall occurs within 1 to 12 hours, and the acanthor lies within the tissues of the gut wall. After 15 to 25 days of apparent dormancy, it migrates to the outside of the gut, where it clings loosely to the serosa. Progressive changes follow in which the overall size increases, and the organs of the mature worm are delineated. The cystacanth appears fully developed in 30 to 40 days but is not infective to the definitive host until 60 to 65 days. On ingestion of an infected isopod by the definitive host, the proboscis of the cystacanth evaginates, pierces the cyst, and attaches to the gut wall, where it develops to maturity.

Nickol and Oetinger[38] found encapsulated *P. cylindraceus* in the mesenteries of a shrew. This illustrates how the interjection of a paratenic host into a life cycle may doom a parasite rather than serve it, since it is unlikely, although possible, that a robin would eat a

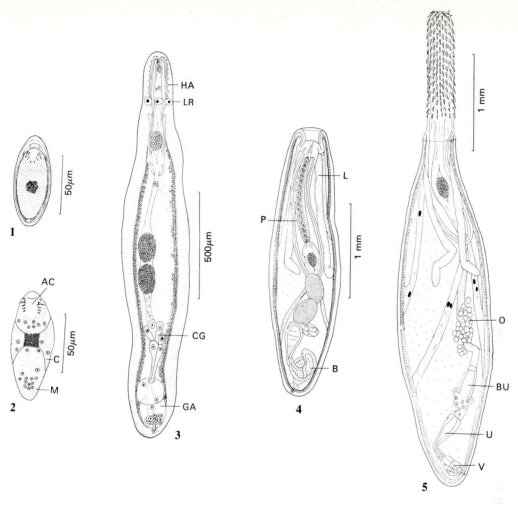

FIG. 32-10

Plagiorhynchus cylindraceus, anatomy and development. **1,** Egg, containing mature acanthor. **2,** Acanthor after escape from egg (*AC*, aclid organ; *C*, cortex; *M*, medulla). **3,** Acanthella, 30 days (*CG*, cement gland; *GA*, anlagen of copulatory apparatus; *HA*, hook anlage; *LR*, nucleus of lemniscal ring). **4,** Cystacanth, 37 days (*B*, bursa; *L*, lemniscus; *P*, proboscis). **5,** Cystacanth, 60 days (*BU*, uterine bell; *O*, ovarian balls; *U*, uterus; *V*, vagina).

From Schmidt, G.D., and O.W. Olsen. 1964. J. Parasitol. 50:721-730.

shrew. Chances of the worm invading a hawk or owl would be improved, however, and if the parasite were preadapted to survive in such a host, its host range would thus be extended.

Class Archiacanthocephala

Macracanthorhynchus hirudinaceus is a cosmopolitan archiacanthocephalan parasite of pigs. Its life cycle has been known since 1868 and was more recently reported in detail by Kates.[23] When the eggs are eaten by white grubs (larvae of the beetle family Scarabaei-

dae), they hatch in the midgut within an hour and penetrate its lining. Within 5 to 20 days after infection, the developing acanthellas are found free in the hemocoel or attached to the outer surface of the serosa. By 60 to 90 days after infection, the cystacanth is infective to the definitive host. Pigs are infected by eating the grubs or the adult beetles, which have metamorphosed with their parasites intact.

Most archiacanthocephalans are parasites of predaceous birds and mammals, so paratenic hosts often are involved in life cycles within this class.

EFFECTS OF THE PARASITE ON ITS HOST

Because most acanthocephalans are parasitic in wild animals, their host-parasite relations have been little studied. Surveys of effects of these parasites on wild mammals and captive primates have been discussed.[39,41]

The nature of damage to intestinal mucosa is primarily traumatic, by penetration of the proboscis, and is compounded by the tendency of the worm to release its hold occasionally and reattach at another place. Complete perforation of the gut sometimes occurs, and in mammals, at least, the results are often rapidly fatal (Fig. 32-11). Great pain accompanies this phase: infected monkeys show evident distress, and Grassi and Calandruccio[18] recorded the symptoms of pain and delirium experienced by Calandruccio after he voluntarily infected himself with cystacanths of *Moniliformis moniliformis,* a common parasite of domestic rats.

It is suspected that secondary bacterial infection is responsible for localized and generalized peritonitis, hemorrhage, pericarditis, myocarditis, arteritis, cholangiolitis, and other complications.

In view of the invasive nature of the parasites, it is surprising that they elicit so little inflammatory response in many cases. The reaction seems mainly a result of the traumatic damage, with granulomatous infiltration and sometimes collagenous encapsulation around the proboscis. Some species show evidence that antigens are released from the proboscis (as in *M. hirudinaceus*), and the inflammatory response is intense. It is clear that the pathogenesis caused by acanthocephalans can be severe, but little consideration usually is given to the effects of this group of parasites as a controlling factor of wildlife populations.

Little chemotherapy has been developed for acanthocephalans. Various authors have proposed chenopodium and castor oil, calomel and santonin, carbon tetrachloride, and tetrachloroethylene for primates and pigs, with varying results. Oleoresin of aspidium has been used successfully in human cases but is not recommended for children. Mebendazole was used successfully in a 12-month-old child.[16] Control of intermediate hosts is helpful in preventing infection of domestic or captive animals.

ACANTHOCEPHALA IN HUMANS

Records of Acanthocephala in humans are few, no doubt because of the nature of the intermediate and paratenic hosts involved in the life cycles of the parasites. Few people of the world eat such animals as insects, microcrustaceans, toads, or lizards, at least without cooking them first. However, human infec-

FIG. 32-11

Complete perforation of the large intestine of a squirrel monkey by *Prosthenorchis elegans.*

From Schmidt, G.D. 1972. In Fiennes, R.N.T.W., editor. Pathology of simian primates, vol. 2. S. Karger AG, Basel.

tions with five different species have been reported.[40] *Macracanthorhynchus hirudinaceus* has been recognized occasionally as a parasite of humans from 1859 to the present.[11] Nine *M. hirudinaceus* were recovered from a 1-year-old child in Austin, Texas, in 1983.[1] *Moniliformis moniliformis* has been found repeatedly in people. *Acanthocephalus rauschi* is known only from specimens taken from the peritoneum of an Alaskan Eskimo, an obvious case of accidental parasitism, since the proper host is undoubtedly a fish. The zest of Eskimos for raw fish probably contributes to such zoonotic infections rather commonly. *Corynosoma strumosum,* a common seal parasite, also has been found in humans. More puzzling is a case of *Acanthocephalus bufonis,* a toad parasite, in an Indonesian. In this instance it is probable that the man ate a raw paratenic host.[40]

Thus it seems that the Acanthocephala do not pose much of a threat to human health. They are much more important as parasites of wild and captive animals, where sudden epizootics have been known to kill a great number of individuals in a short time.

PHYLOGENETIC RELATIONSHIPS

Although the ancestors of most parasitic groups of animals seem more clear, the thorny-headed worms are closely related to no known form. Their affinities with the nematodes and Nematomorpha seem indicated by the presence of a pseudocoel and exterior "cuticle." However, the tegument is quite different from the cuticle of nematodes, and the body wall structure (including lacunar system and tubular muscles), the eversible, spined proboscis with its accompanying mechanisms, the nervous system, and the complete lack of a digestive system set these worms well apart from other phyla in the superphylum Aschelminthes. They seem to represent a small relict of a larger population, most members of which have become extinct. This has left phylogenetic gaps between the Acanthocephala and other phyla and, indeed, between several groups within the phylum as well. A further paradox is the fact that the Archiacanthocephala, parasites of the most highly evolved vertebrates, are in many ways more primitive than are the Eoacanthocephala, which are parasites of fishes. Golvan[17] discussed the subject in detail and proposed a hypothetical ancestor, which he named *Protacanthocephala*.

Very interesting fossils were found in the mid-Cambrian Burgess Shale of British Columbia in 1911.[52] The worms, named *Ottoia prolifica*, show many similarities to existing acanthocephalans (Fig. 32-12). Vertebrates had not yet evolved in the mid-Cambrian; so *Ottoia*, if a parasite, could only have parasitized invertebrates, such as trilobites, which were abundant then. More likely it was a predator or scavenger.

METABOLISM

Because of the availability of a good laboratory subject (*M. moniliformis* in rats), investigators have been able to accumulate some knowledge of acanthocephalan metabolism. However, the problem of assessing the general applicability of observations on *M. moniliformis* and the few others reported is acute. (*Moniliformis dubius* is a junior synonym of *M. moniliformis*, and much literature has accumulated on the physiology of the organism under that name.)

Acquisition of nutrients

Since the Cestoda and the Acanthocephala are both groups that must obtain all nutrient molecules through their body surfaces, comparisons between the two are quite interesting, particularly in light of their structural differences.[28] Some areas of divergence will be pointed out in the brief account to follow.

Acanthocephalans can absorb at least some triglycerides, amino acids, nucleotides, and sugars. Amino acids are absorbed, at least partially, by stereospecific membrane transport systems in *M. moniliformis* and *M. hirudinaceus*.[49] The surface of *M. moniliformis*

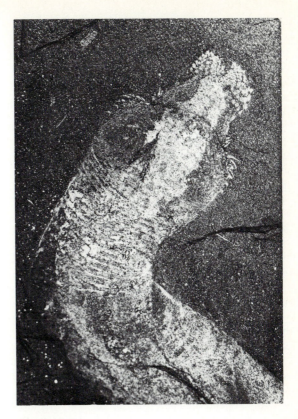

FIG. 32-12

Ottoia prolifica: a close-up of presoma, showing a partially inverted proboscis. Note the conspicuous trunk spines.

Photograph by H.B. Whittington. Geological Survey of Canada no. 40972; from Crompton, D.W.T. 1975. In Symbiosis. Symposia of the Society for Experimental Biology. Cambridge University Press, Cambridge, Eng.

contains peptidases, which can cleave several dipeptides, and the amino acid products are then absorbed by the worm.[48] Absorbed thymidine is incorporated into DNA in the perilacunar regions and into the nuclei of the ovarian balls and testes. Nuclei in the body wall are not labeled by radioactive thymidine; therefore it is assumed that the DNA synthesized there is mitochondrial.

Like the tapeworm *Hymenolepis diminuta*, *M. moniliformis* has an absolute dependence on host dietary carbohydrate for growth and energy metabolism as an adult.[44,45] The worm can absorb glucose, mannose, fructose, and galactose, as well as several glucose analogs. In contrast to *H. diminuta*, *M. moniliformis* can grow and mature in the host fed a diet containing fructose as the sole carbohydrate source.[9,10,24]

Absorption of glucose is through a single transport locus, whereas transport of mannose, fructose, and galactose is mediated both by the glucose locus and another site, referred to by Starling and Fisher[45] as the "fructose site." Maltose and glucose-6-phosphate (G6P) are absorbed also, but first they are hydrolyzed to glucose by enzymes in or on the tegumental surface. Acid phosphatase is localized in the crypts of the tegument[6]; kinetic evidence suggests that the hydrolysis of the G6P and part of the glucose absorption occurs there, while some of the glucose diffuses back out of the crypts and is absorbed by the outer surface.[45] A most interesting observation, and one in sharp contrast to tapeworm and other glucose transport systems, is that glucose absorption by *M. moniliformis* is not coupled to cotransport of sodium. Therefore an alternative explanation for the mediated transport of glucose is required. Part of the explanation may lie in the fact that glucose is rapidly phosphorylated, and this removal of free glucose from the vicinity of the tegumental transport loci would form a metabolic sink for the flow of additional hexose down its concentration gradient. However, substantial amounts of free glucose are found in the body wall.[46] Evidence suggests that the free glucose pool is not derived directly from absorbed glucose but is first "shuttled" through the nonreducing disaccharide trehalose. The glucose would then be deposited, perhaps by intervention of a membrane-bound trehalase, in an internal membranous compartment that by some means can resist the efflux of the glucose it contains. The scheme offers an interesting possible metabolic role for trehalose, perhaps similar to that in insects.

Energy metabolism

Moniliformis moniliformis can ferment the hexoses it absorbs: glucose, fructose, galactose, and mannose. As in trematodes and cestodes, glycolysis is an important degradative pathway, and some of the regulatory enzymes have been investigated.[7,12] The tricarboxylic acid cycle apparently does not operate in *M. moniliformis* or *M. hirudinaceus,* although there is evidence for it in *Echinorhynchus gadi,* a parasite of cod.

As in the other helminth parasites, the energy metabolism is adapted for facultative anaerobiosis. The terminal reactions of glycolysis derive more energy from glucose than would classical glycolysis alone and also reoxidize reduced coenzymes and produce less toxic end products. *Moniliformis* fixes carbon dioxide, and the principal enzyme of carbon dioxide fixation is phosphoenolpyruvate carboxykinase (see Fig. 21-34). It is likely that some or all of the reactions shown in Fig. 21-34 occur in acanthocephalans, although species vary. Lactate and succinate are the main end products of glucose degradation in *Polymorphus minutus*. Interestingly, the main end products of

glycolysis in *M. moniliformis* are ethanol and carbon dioxide with a small amount of lactate and only traces of succinate, acetate, and butyrate. The presence of the enzymes necessary to produce pyruvate from PEP (pyruvate kinase), lactate from pyruvate (lactate dehydrogenase), and ethanol (and carbon dioxide) from pyruvate (pyruvate decarboxylase and alcohol oxidoreductase) in *M. moniliformis* has been confirmed.[25] Even though PEP carboxykinase activity is high, it must be regulated in such a way that the major end products are ethanol and lactate, rather than succinate. One possible explanation may lie in the fact that fumarate hydratase (Fig. 21-34) is very low or absent. Thus the terminal reactions of glycolysis in *M. moniliformis* resemble those of yeast rather than those of other intestinal helminths. However, the alcohol oxidoreductase of the acanthocephalan requires NADP rather than being NAD dependent, as is the yeast enzyme.

Lipids apparently are not used as energy sources. Körting and Fairbairn[25] found that endogenous lipids were not metabolized during in vitro incubation of *M. dubius*. This was correlated with the fact that enzymes necessary for the β-oxidation of lipids were low in activity, and one of them seemed to be completely absent.

Electron transport in acanthocephalans has been studied very little. Oxidation of both succinate and NADH lead to reduction of cytochrome b.[3] Two pathways for reoxidation of this compound have been postulated, the major one independent of cytochrome c and cytochrome oxidase. This is somewhat similar to the branched-chain electron transport postulated for the cestode *Moniezia expansa* (see Fig. 21-32). It could provide an additional means for reoxidation of NADH and generation of ATP by anaerobic mechanisms.

CLASSIFICATION OF PHYLUM ACANTHOCEPHALA

Class Archiacanthocephala

Main longitudinal lacunar canals dorsal and ventral or just dorsal; hypodermal nuclei few; giant nuclei present in lemnisci and cement glands; two ligament sacs persist in females; protonephridia present in one family; cement glands separate, pyriform; eggs oval, usually thick shelled; parasites of birds and mammals; intermediate hosts are insects or myriapods.

Order Moniliformida

Trunk usually pseudosegmented; proboscis cylindrical, with long, approximately straight rows of hooks, sensory papillae present; proboscis receptacle double walled, outer wall with muscle fibers usually arranged spirally; proboscis retractor muscles pierce posterior end of receptacle or somewhat ventral; brain near posterior end or near middle of receptacle; protonephridial organs absent.

FAMILY
Moniliformidae

Order Gigantorhynchida

Trunk occasionally pseudosegmented; proboscis a truncate cone, with approximately longitudinal rows of rooted hooks on the anterior portion and rootless spines on the basal portion; sensory pits present on apex of proboscis and each side of neck; proboscis receptacle single walled with numerous accessory muscles, complex, thickest dorsally; proboscis retractor muscles pierce ventral wall of receptacle; brain near ventral, middle surface of receptacle; protonephridial organs absent.

FAMILY
Gigantorhynchidae

Order Oligacanthorhynchida

Trunk may be wrinkled but not pseudosegmented; proboscis subspherical, with short, approximately longitudinal rows of few hooks each; sensory papillae present on apex of proboscis and each side of neck; proboscis receptacle single walled, complex, thickest dorsally; proboscis retractor muscle pierces dorsal wall of receptacle; brain near ventral, middle surface of receptacle; protonephridial organs present.

FAMILY
Oligacanthorhynchidae

Order Apororhynchida

Trunk short, conical, may be curved ventrally; proboscis large, globular, with tiny spinelike hooks (which may not pierce the surface of the proboscis) arranged in several spiral rows; proboscis not retractable; neck absent or reduced; protonephridial organs absent.

FAMILY
Apororhynchidae

Class Palaecanthocephala

Main longitudinal lacunar canals lateral; hypodermal nuclei fragmented, numerous, occasionally restricted to anterior half of trunk; nuclei of lemnisci and cement glands fragmented; spines present on trunk of some species; single ligament sac of female not persistent throughout life; protonephridia absent; cement glands separate, tubular to spheroid; eggs oval to elongate, sometimes with polar thickenings of second membrane; parasites of fishes, amphibians, reptiles, birds, and mammals.

Order Echinorhynchida

Trunk never pseudosegmented; proboscis cylindrical to spheroid, with longitudinal, regularly alternating rows of hooks, sensory papillae present or absent; proboscis receptacle double walled; proboscis retractor muscles pierce posterior end of receptacle; brain near middle or posterior end of receptacle; parasites of fishes and amphibians.

FAMILIES
Diplosentidae, Echinorhynchidae, Fessisentidae, Heteracanthocephalidae, Heterosentidae, Hypoechinorhynchidae, Illiosentidae, Pomporhynchidae, Rhadinorhynchidae

Order Polymorphida

Proboscis spheroid to cylindrical, armed with numerous hooks in alternating longitudinal rows; proboscis receptacle double walled, with brain near center; parasites of reptiles, birds, and mammals.

FAMILIES
Centrorhynchidae, Plagiorhynchidae, Polymorphidae

Class Eocanthocephala

Main longitudinal lacunar canals dorsal and ventral, often no larger in diameter than irregular transverse commissures; hypodermal nuclei few, giant, sometimes ameboid; proboscis receptacle single walled; proboscis retractor muscle pierces posterior end of receptacle; brain near anterior or middle of receptacle; nuclei of lemnisci few, giant; two persistent ligament sacs in female; protonephridia absent; cement gland single, syncytial, with several nuclei, with cement reservoir appended; eggs variously shaped; parasites of fish, amphibians, and reptiles.

Order Gyracanthocephalida

Trunk small or medium-sized, spined; proboscis small, spheroid, with a few spiral rows of hooks.

FAMILY
Quadrigyridae

Order Neoechinorhynchida

Trunk small to large, unarmed; proboscis spheroid to elongate, with hooks arranged variously.

FAMILIES
Neoechinorhynchidae, Tenuisentidae

Class Polyacanthocephala

Trunk spinose. Hypodermic nuclei many and small. Main longitudinal lacunar canals dorsal and ventral. Many hooks, in longitudinal rows. Proboscis receptacle single-walled. Cement glands elongate, with giant nuclei. Protonephridia absent. Parasites of fishes and (?) crocodilians.

Order Polyacanthorhynchida

With characters of the class.

FAMILY
Polyacanthorhynchidae

REFERENCES

1. Associated Press. 1983. Austin tot suffers parasite usually found only in pigs. Lubbock Avalanche Journal, July 6, 1983.
2. Bryant, C. 1970. Electron transport in parasitic helminths and protozoa. In Dawes, B., editor. Advances in parasitology, vol. 8. Academic Press, Inc., New York, pp. 139-172.
3. Bryant, C., and W.L. Nicholas. 1966. Studies on the oxidative metabolism of *Moniliformis dubius* (Acanthocephala). Comp. Biochem. Physiol. 17:825-840.
4. Butterworth, P.E. 1969. The development of the body wall of *Polymorphus minutus* (Acanthocephala) in the intermediate host, *Gammarus pulex*. Parasitology 59:373-388.
5. Byram, J.E., and F.M. Fisher, Jr. 1973. The absorptive surface of *Moniliformis dubius* (Acanthocephala). I. Fine structure. Tissue Cell 5:553-579.
6. Byram, J.E., and F.M. Fisher, Jr. 1974. The absorptive surface of *Moniliformis dubius* (Acanthocephala). II. Functional aspects. Tissue Cell 6:21-42.
7. Cornish, R.A., J. Wilkes, and D.F. Mettrick. 1981. The levels of some metabolites in *Moniliformis dubius* (Acanthocephala). J. Parasitol. 67:754-756.
8. Crompton, D.W.T. 1970. An ecological approach to acantho-

cephalan physiology. Cambridge University Press, Cambridge, Eng.

9. Crompton, D.W.T., A. Keymer, A. Singhvi, and M.C. Nesheim. 1983. Rat-dietary fructose and the intestinal distribution and growth of *Moniliformis* (Acanthocephala). Parasitology 86:57-71.

10. Crompton, D.W.T., A. Singhvi, and A. Keymer. 1982. Effects of host dietary fructose on experimentally stunted *Moniliformis* (Acanthocephala). Int. J. Parasitol. 12:117-121.

11. Dingley, D., and P.C. Beaver. 1985. *Macracanthorhynchus ingens* from a child in Texas. Am. J. Trop. Med. Hyg. 34:918-920.

12. Donahue, M.J., N.J. Yacoub, M.R. Kaeini, S. Tu, R.A. Hodzi, and B.G. Harris. 1981. Studies on potential carbohydrate regulatory enzymes and metabolite levels in *Macracanthorhynchus hirudinaceus* (Acanthocephala). J. Parasitol. 67:756-758.

13. Dunagan, T.T., and D.M. Miller. 1970. Major nerves in the anterior nervous system of *Macracanthorhynchus hirudinaceus* (Acanthocephala). Comp. Biochem. Physiol. 37:235-242.

14. Dunagan, T.T., and D.M. Miller. 1975. Anatomy of the cerebral ganglion of the male acanthocephalan, *Moniliformis dubius*. J. Comp. Neurol. 164:483-494.

15. Dunagan, T.T., and D.M. Miller. 1986. A review of protonephridial excretory systems in Acanthocephala. J. Parasitol. 72:621-632.

16. Goldsmid, J.M., M.E. Smith, and F. Fleming. 1974. Human infections with *Moniliformis* sp. in Rhodesia. Ann. Trop. Med. Parasitol. 68:363-364.

17. Golvan, Y.J. 1958. Le phylum Acanthocephala. Premiére note. Sa place dans l'echelle zoologique. Ann. Parasitol. 33:539-602.

18. Grassi, B., and S. Calandruccio. 1888. Ueber einen *Echinorhynchus*, welcher auch in Menschen parasitiert und dessen Zwischenwirt ein *Blaps* ist. Zentralbl. Bakteriol. Parasitenkd. Orig. 3:521-525.

19. Hamann, O. 1892. Das system der Acanthocephalen. Zool. Anz. 15:195-197.

20. Hammond, R.A. 1966. The proboscis mechanism of *Acanthocephalus ranae*. J. Exp. Biol. 45:203-213.

21. Hightower, K., D.M. Miller, and T.T. Dunagan. 1975. Physiology of the body wall muscles in an acanthocephalan. Proc. Helm. Soc. Wash. 42:71-80.

22. Hopp, W.B. 1954. Studies on the morphology and life cycle of *Neoechinorhynchus emydis* (Leidy), an acanthocephalan parasite of the map turtle, *Graptemys geographica* (La Sueur). J. Parasitol. 40:284-299.

23. Kates, K.C. 1943. Development of the swine thorn-headed worm, *Macracanthorhynchus hirudinaceus*, in its intermediate host. Am. J. Vet. Res. 4:173-181.

24. Keymer, A., D.W.T. Crompton, and D.E. Walters. 1983. Parasite population biology and host nutrition: dietary fructose and *Moniliformis* (Acanthocephala). Parasitology 87:265-278.

25. Körting, W., and D. Fairbairn. 1972. Anaerobic energy metabolism in *Moniliformis dubius* (Acanthocephala). J. Parasitol. 58:45-50.

26. Krapf, K., and T.T. Dunagan. 1987. Structural features of the protonephridia in female *Macracanthorhynchus hirudinaceus* (Acanthocephala). J. Parasitol. 73:1176-1181.

27. Lankester, R. 1900. A treatise on zoology. Adam & Charles Black, London.

28. Lumsden, R.D. 1975. Surface ultrastructure and cytochemistry of parasitic helminths. Exp. Parasitol. 37:267-339.

29. Miller, D.M., and T.T. Dunagan. 1976. Body wall organization of the acanthocephalan, *Macracanthorhynchus hirudinaceus:* a reexamination of the lacunar system. Proc. Helm. Soc. Wash. 43:99-106.

30. Miller, D.M., and T.T. Dunagan. 1977. The lacunar system and tubular muscles in Acanthocephala. Proc. Helm. Soc. Wash. 44:201-205.

31. Miller, D.M., and T.T. Dunagan. 1978. Organization of the lacunar system in the acanthocephalan, *Oligacanthorhynchus tortuosa*. J. Parasitol. 64:436-439.

32. Miller, D.M., and T.T. Dunagan. 1983. A support cell to the apical and lateral sensory organs in *Macracanthorhynchus hirudinaceus* (Acanthocephala). J. Parasitol. 69:534-538.

33. Miller, D.M., and T.T. Dunagan. 1985. New aspects of acanthocephalan lacunar system as revealed in anatomical modeling by corrosion cast method. Proc. Helm. Soc. Wash. 52:221-226.

34. Miller, D.M., T.T. Dunagan, and J. Richardson. 1973. Anatomy of the cerebral ganglion of the female acanthocephalan, *Macracanthorhynchus hirudinaceus*. J. Comp. Neurol. 152:403-415.

35. Moore, J. 1983. Responses of an avian predator and its isopod prey to an acanthocephalan parasite. Ecology 64:1000-1015.

36. Moore, J. 1984. Altered behavioral responses in intermediate hosts—an acanthocephalan parasite strategy. Am. Natural. 123:572-577.

37. Nicholas, W.L. 1973. The biology of the Acanthocephala. In Dawes, B., editor. Advances in parasitology, vol. 11. Academic Press, Inc., New York, pp. 671-706.

38. Nickol, B.B., and D.F. Oetinger. 1968. *Prosthorhynchus formosus* from the short-tailed shrew *(Blarina brevicauda)* in New York State. J. Parasitol. 54:456.

39. Schmidt, G.D. 1969. Acanthocephala as agents of disease in wild mammals. Wildl. Dis. 53:1-10.

40. Schmidt, G.D. 1971. Acanthocephalan infections of man, with two new records. J. Parasitol. 57:582-584.

41. Schmidt, G.D. 1972. Acanthocephala of captive primates. In Fiennes, R.N.T.W., editor. Pathology of simian primates, vol. 2. S. Karger AG, Basel.

42. Schmidt, G.D. 1973. Early embryology of the acanthocephalan *Mediorhynchus grandis* Van Cleave, 1916. Trans. Am. Microsc. Soc. 92:512-516.

43. Schmidt, G.D., and O.W. Olsen. 1964. Life cycle and development of *Prosthorhynchus formosus* (Van Cleave 1918) Travassos, 1926, an acanthocephalan parasite of birds. J. Parasitol. 50:721-730.

44. Starling, J.A. 1975. Tegumental carbohydrate transport in intestinal helminths: correlation between mechanisms of membrane transport and the biochemical environment of absorptive surfaces. Trans. Am. Microsc. Soc. 94:508-523.

45. Starling, J.A., and F.M. Fisher, Jr. 1975. Carbohydrate transport in *Moniliformis dubius* (Acanthocephala). I. The kinetics and specificity of hexose absorption. J. Parasitol. 61:977-990.

46. Starling, J.A., and F.M. Fisher, Jr. 1979. Carbohydrate transport in *Moniliformis dubius* (Acanthocephala). III. Post-absorp-

tive fate of fructose, mannose, and galactose. J. Parasitol. 65:8-13.

47. Uglem, G.L., and O.R. Larson. 1969. The life history and larval development of *Neochinorhynchus saginatus* Van Cleave and Bangham, 1949 (Acanthocephala: Neoechinorhynchidae). J. Parasitol. 55:1212-1217.

48. Uglem, G.L., P.W. Pappas, and C.P. Read. 1973. Surface amino peptidase in *Moniliformis dubius* and its relation to amino acid uptake. Parasitology 67:185-195.

49. Uglem, G.L., and C.P. Read. 1973. *Moniliformis dubius:* uptake of leucine and alanine by adults. Exp. Parasitol. 34:148-153.

50. Van Cleave, H.J. 1941. Relationships of the Acanthocephala. Am. Natur. 75:31-47.

51. Van Cleave, H.J. 1948. Expanding horizons in the recognition of a phylum. J. Parasitol. 34:1-20.

52. Wallcott, C.D. 1911. Middle Cambrian annelids. Cambrian geology and paleontology II. Smithsonian Misc. Coll. 57:109-144.

53. Ward, H.L. 1940. Studies on the life-history of *Neoechinorhynchus cylindratus* (Van Cleave 1913) (Acanthocephala). Trans. Am. Microsc. Soc. 59:327-347.

54. Whitfield, P.J. 1970. The egg sorting function of the uterine bell of *Polymorphus minutus* (Acanthocephala). Parasitology 61:111-126.

55. Whitfield, P.J. 1971. The locomotion of the acanthor of *Moniliformis dubius* (Archiacanthocephala). Parasitology 62:35-47.

56. Whitfield, P.J. 1971. Spermiogenesis and spermatozoan ultrastructure in *Polymorphus minutus* (Acanthocephala). Parasitology 62:415-430.

57. Wong, B.S., D.M. Miller, and T.T. Dunagan. 1979. Electrophysiology of acanthocephalan body wall muscles. J. Exp. Biol. 82:273-280.

58. Wright, R.D., and R.D. Lumsden. 1968. Ultrastructural and histochemical properties of the acanthocephalan epicuticle. J. Parasitol. 54:1111-1123.

SUGGESTED READINGS

Amin, O.M. 1987. Key to the families and subfamilies of Acanthocephala, with the erection of a new class (Polyacanthocephala) and a new order (Polyacanthorhynchida). J. Parasitol. 73:1216-1219.

Bullock, W.L. 1969. Morphological features as tools and as pitfalls in acanthocephalan systematics. In Schmidt, G.D., editor. Problems in systematics of parasites. University Park Press, Baltimore, pp. 9-43. (A useful, philosophical discussion of the subject, with recommended techniques for study.)

Crompton, D.W.T. 1970. An ecological approach to acanthocephalan physiology. Cambridge University Press, Cambridge, Eng. (An outstanding summation of the subject.)

Crompton, D.W.T. 1975. Relationships between Acanthocephala and their hosts. Symposium of the Society for Experimental Biology, vol. 29. Symbiosis. Cambridge University Press, Cambridge, Eng., pp. 467-504.

Crompton, D.W.T., and B.B. Nickol, eds. 1985. Biology of the Acanthocephala. Cambridge University Press, Cambridge, Eng.

Golvan, Y.J. 1969. Systematiques des Acanthocéphales (Acanthocéphala Rudolphi 1801). L'order des Palaeacanthocephala Meyer 1931. La super-famille des Echinorhynchoidea (Cobbold 1876) Golvan et Houin 1963. Mem. Mus. Nat. Hist. Nat. 47:1-373. (A very up-to-date account of this important superfamily. Besides descriptions of each species, it contains a key to genera and a host list.)

Pappas, P.W., and C.P. Read. 1975. Membrane transport in helminth parasites: a review. Exp. Parasitol. 37:469-530.

Petrochenko, V.I. 1956, 1958. Acanthocephala of domestic and wild animals, vols. 1 and 2. Akademii Nauk SSSR, Moscow. (English translations: Israel Program for Scientific Translations, 1971.) (An indispensable resource for students of the phylum. Descriptions are given for nearly every species known at the time of writing.)

Schmidt, G.D. 1972. Revision of the class Archiacanthocephala Meyer, 1931 (Phylum Acanthocephala), with emphasis on Oligacanthorhynchidae Southwell and MacFie, 1925. J. Parasitol. 58:290-297. (A modern classification of this difficult class.)

Yamaguti, S. 1963. Systema Helminthum, vol. 5. Acanthocephala. Interscience, New York. (In most regards a practical key to all genera of Acanthocephala known to 1963. Lists of all species and their hosts are included.)

CHAPTER 33

PHYLUM PENTASTOMIDA: TONGUE WORMS

The pentastomids, or tongue worms, are wormlike parasites of the respiratory systems of vertebrates. About 100 species are known. As adults, most live in the respiratory system of reptiles, especially snakes, lizards, amphibians, and crocodilians, but one species lives in the air sacs of sea birds, and another inhabits the nasopharynx of canines and felines. The latter species is occasionally found as transient nymphs in the nasopharynx of humans, whereas other species, in their nymphal stages, also parasitize humans. Thus pentastomids are certainly of zoological interest and also are of some medical importance.[22]

The evolutionary relationships of pentastomids are obscure. Certain similarities with the Annelida have been pointed out, but most modern taxonomists align them with the Arthropoda.[15,19] It is possible that the group reached its zenith in the Mesozoic age of reptiles and that today's few species are relicts derived from those ancestors. Wingstrand[24] proposed that the Pentastomida be regarded as an order of the crustacean class Branchiura (Chapter 35). His conclusion is based on a demonstration that the spermatozoa of the two groups are almost identical with regard to structure and development and that this type of spermatozoon represents a type of its own, not encountered in other animals. His argument is strengthened by the fact that each major crustacean group is characterized by its own type of spermatozoa, and if the Pentastomida and the Branchiura were unrelated, their sperm structure and development would represent a most extraordinary example of convergence in detail. Certainly, the dissimilarity of the adult morphology is great, but this in itself would not preclude Wingstrand's thesis. The adult thoracican barnacle is equally as dissimilar to an adult *Sacculina* (Chapter 35), but they both belong to the class Cirripedia based on developmental evidence. Riley and co-workers,[16] on the basis of embryogenesis, tegumental structure, and gametogenesis, concluded that pentastomids should be regarded as a subclass of Crustacea, closely allied to the Branchiura. For the present, however, we will adopt the traditional view and retain the Pentastomida in a phylum of its own.

MORPHOLOGY

The body of a pentastomid (Fig. 33-1) is elongate, usually tapering toward the posterior end, and often showing distinct segmentation, forming numerous **annuli.** It is indistinctly divided into an anterior **forebody** and a posterior **hindbody,** which is bifurcated at its tip in some species.

The exoskeleton contains chitin,[23] which is sclerotized around the mouth opening and accessory genitalia. A striking characteristic of all adult pentastomids is the presence of two pairs of sclerotized hooks in the mouth region (Fig. 33-2). These may be located at the ends of stumpy stalks or may be nearly flush with the surface of the cephalothorax; in either case they can be withdrawn into cuticular pockets. The hooks are single in some species and double in others. The apparently double hooks are actually single, with an accessory hooklike protrusion of the cuticle. In some species the hook articulates against a basal fulcrum. The hooks are manipulated by powerful muscles and are used to tear and embed the mouth region into host tissues.

The body cuticle in some species also has circular rows of simple spines; the annuli may overlap enough to make the abdomen look serrated. There usually are transverse rows of cuticular glands, with conspicuous pores, which apparently function in regulation of the hydromineral balance in the hemolymph.[2]

The chitinous cuticle is similar to that of the arthropods, although it is thin and weak.[23] The muscles, too, are arthropodan in nature, being striated and segmentally arranged. The only sensory structures so far recognized are papillae, especially on the exoskeleton of the cephalothorax. The digestive system is simple and complete, with the anus opening at the posterior end of the abdomen. The mouth is permanently held open by its sclerotized lining, the **cadre,** which may

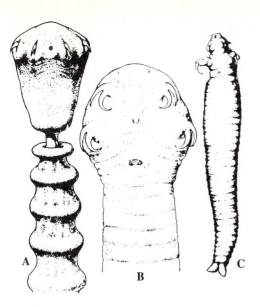

FIG. 33-1

Examples of pentastome body types. **A,** Anterior end of *Armillifer annulatus*. **B,** Head of *Leiperia gracilis*. **C,** Entire specimen of *Raillietiella mabuiae*.

Modified from Heymons, R. 1935. Pentastomida. In Bronn's Klass. Ord. Tier. 5[4], book 1; from Baer, J.G. 1952. Ecology of animal parasites. The University of Illinois Press, Urbana.

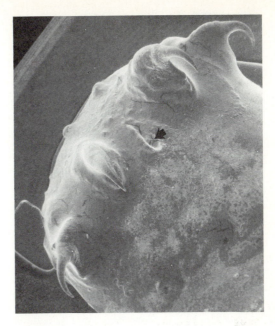

FIG. 33-2

Anterior end of a pentastome. Note both the mouth *(arrow)* between the middle hooks and the apical sensory papillae.

Photograph by John Ubelaker.

be circular, oval, or **U** shaped, and is an important taxonomic character. The nervous system is arthropodan in nature and has been described by Doucet.[4]

Pentastomids are dioecious and show sexual dimorphism in that males are usually smaller than females. The male has a single, tubular testis (two in *Linguatula*), which occupies one third to one half of the body cavity (Fig. 33-3, A). It is continuous with a seminal vesicle, which in turn connects to a pair of ejaculatory organs. These each have a duct that extends to a terminal penis that fits into a **dilator organ.** The dilator organ, which is usually sclerotized, serves as an intromittent organ in some species and as a dilator and guide for the cirrus in others. The male genital pore is midventral on the anterior abdominal segment, near the mouth.

In the female a single ovary extends nearly the length of the body cavity (Fig. 33-3, B). It may bifurcate at its distal end to become two oviducts. These unite to form the uterus. The oviducts and uterus usually are extensively coiled within the body. One or more diverticulae of the uterus serve as seminal receptacles. The uterus terminates as a short vagina that opens through the female gonopore, at the anterior end of the abdomen in the order Cephalobaenida or at

the posterior end in the order Porocephalida.[7] The female mates once; the male may be polygamous.

BIOLOGY (Fig. 33-4)

Adult pentastomids feed on tissue fluids and blood cells of their host. They appear to stimulate a strong host immune response, but because they are long lived, they must evade its consequences.[17] Their frontal and subparietal glands elaborate a lamellate secretion, which is poured over the entire surface of the cuticle, and this may protect vital areas of the parasite from antibody action.[17]

The female, depending on its size, may produce several million fully embryonated eggs, which pass up the trachea of its host, are swallowed, and then pass out with the feces. The intact egg appears to be surrounded by two shell membranes—an outer, thin membrane and an inner, thick one.[6] The inner layer, however, consists of three distinct layers. A characteristic of the pentastomid egg is the **facette,** a permanent, funnel-shaped opening through the inner membrane complex, with an inner opening extending toward the larva. In the embryo a gland called the **dorsal organ** (to be described further) secretes a mucoid

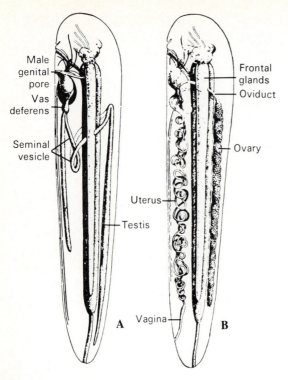

FIG. 33-3

Reproductive systems of the pentastome *Waddycephalus teretiuscules*. **A,** Male. **B,** Female.

Modified from Spomer. In Heymons, R. 1935. Pentastomida. In Bronn's Klass. Ord. Tier. 5(4), book 1; from Baer, J.G. 1952. Ecology of animal parasites. The University of Illinois Press, Urbana.

substance that pours through the facette and ruptures the original outer membrane, which is lost. The mucoid material then flows over the inner membrane to form a new outer membrane, which is sticky when wet.[11] The viscid eggs cling together, sometimes resulting in massive infections in the intermediate host. The eggs can withstand drying for at least 2 weeks, in the case of *Porocephalus crotali,* and they can remain viable in water at refrigerator temperatures for about 6 months.[6]

The larva that hatches from the egg is an oval, tailed creature with four stumpy legs, each with one or two retractable claws. The claws are manipulated by a combination of muscle fibers and an inner hydraulic mechanism. A **penetration organ** is located at the anterior end of the body. This is composed of a median spear and two lateral, pointed forks; together with the clawed legs, these structures can tear through the tissues of the intermediate host. A pair of ducts open on either side of the median spear. Their origin and the nature of their selection are unknown. Accessory spinelets are present around the penetration organ of many species.

Between the anterior legs is a simple mouth, surrounded by a U-shaped sclerotized cadre. An esophagus extends into the dorsal part of the body and expands into a blind sac; a thin hindgut is present in some species. Within the body cavity are a number of irregular giant cells, some with neutrophilic and others with eosinophilic granules. Their function is unknown. A consistent feature of the pentastomid embryo is the dorsal organ, referred to before. It consists of a number of gland cells surrounding a central hollow vesicle. The vesicle opens through the cuticle by a dorsal pore.

Complete life cycles are known for few species, but partial information is available for several. With the exceptions of *Reighardia sternae* in birds and a few species of *Linguatula* in mammals, all pentastomids mature in reptiles. The intermediate hosts are various fishes, amphibians, reptiles, insects, or rarely, mammals. Typically, after ingestion by a poikilothermous vertebrate, the larva hatches and penetrates the intestine and migrates randomly in the body, finally becoming quiescent and metamorphosing into a nymph (Fig. 33-5). The nymph is infective to the definitive host; when eaten by the latter, it penetrates the host's intestine and bores into the lung, where it matures (Fig. 33-6). Each developmental stage undergoes one to several molts of the cuticle. The nymphal instars are difficult to differentiate. Some species even become sexually mature before completing the final ecdysis. A definitive host that eats the egg can also serve as intermediate host, similar to the case of *Trichinella*. The parasites, however, probably cannot migrate to the lung and mature. Whereas vertebrates are the intermediate hosts for the Porocephalida, cockroaches are used by some species of *Raillietiella*,[1,10] and *Reighardia sternae* in gulls has a direct life cycle. *Subtriquetra subtriquetra,* a parasite of the nasopharynx of South American crocodilians, differs in having a free-living larva, which somehow finds its fish intermediate host. Pentastomid reproductive biology was reviewed by Riley.[14]

The following two life cycles illustrate the biology of the group.

Porocephalus crotali

The life cycle of *P. crotali* of crotalid snakes was experimentally demonstrated by Esslinger[5] (Fig. 33-4), using white mice as intermediate hosts; further elaboration was provided by Riley.[13] On hatching, the larva penetrates the duodenal mucosa and works its way to the abdominal cavity. Complete penetration can be accomplished within an hour after the egg is

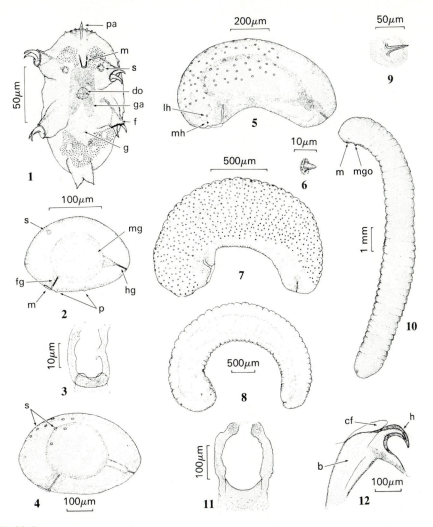

FIG. 33-4

Developmental stages of *Porocephalus crotali* in experimental intermediate hosts (camera lucida drawings made from living specimens). **1,** Primary larva (ventral view) after release from egg. **2,** First nymphal stage (nymph I) in left lateral view (all succeeding nymphs identically oriented). **3,** Mouth ring of nymph I (en face view with anterior margin uppermost). **4,** Nymph II. **5,** Nymph III. **6,** Lateral mouth hook of nymph III. **7,** Nymph IV. **8,** Nymph V (individual stigmata not shown). **9,** Lateral mouth hook of nymph V. **10,** Nymph VI (infective stage), male, removed from enveloping cuticle of nymph V. **11,** Mouth ring of nymph VI. **12,** Lateral mouth hook of nymph VI. *b,* Base of mouth hook; *cf,* cuticular fold or auxiliary hook; *do,* dorsal organ; *f,* foot or leg; *fg,* foregut; *g,* gut; *ga,* ganglion; *h,* external claw-like portion of mouth hook; *hg,* hindgut; *lh,* lateral mouth hook; *m,* mouth ring; *mg,* midgut; *mgo,* male genital opening; *mh,* medial mouth hook; *p,* papilla; *pa,* penetrating apparatus; *s,* stigma.

From Esslinger, J.H. 1962. J. Parasitol. 48:452-456.

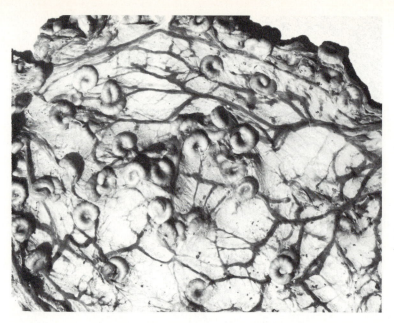

FIG. 33-5

Nymphs of *Porocephalus* sp. in the mesenteries of a vervet, *Cercopithicus aethops*.

Photograph by Robert E. Kuntz; from Self, J.T. 1972. Trans. Am. Microsc. Soc 91:2-8.

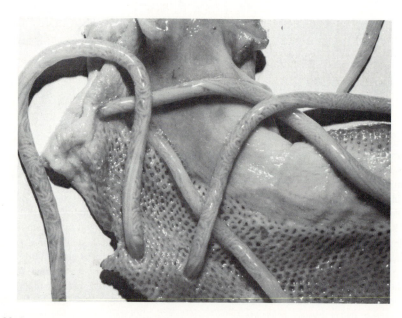

FIG. 33-6

Kiricephalus pattoni in the lung of an Oriental rat snake, *Ptyas mucosus*.

Photograph by Robert E. Kuntz.

swallowed. After wandering about for 7 or 8 days the larva molts and becomes lightly encapsulated in host tissue. The subsequent nymphal stages are devoid of the larval characteristics, having lost the legs, penetration apparatus, and tail. During the next 80 days or so, *P. crotali* molts five more times, gradually increasing in size and becoming segmented. The mouth hooks appear during the fourth nymphal instar and increase in size through subsequent ecdyses. The sexes can be differentiated after the fifth molt. After the sixth molt, the nymph becomes heavily encapsulated and dormant. When eaten by a snake, the nymph is activated; it quickly penetrates the snake's intestine and usually passes directly to the lung, since the lung and intestine are adjacent to each other. It buries its forebody into lung tissues, feeds on blood and tissue fluids, and matures.

Linguatula serrata

Linguatula serrata is unusual among the Pentastomida in that the adults live in the nasopharyngeal region of mammals. Cats, dogs, foxes, and other carnivores are the normal hosts of this cosmopolitan parasite. Apparently, almost any mammal is a potential intermediate host.

Adult *L. serrata* embed their forebody into the nasopharyngeal mucosa, feeding on blood and fluids. Females live at least 2 years and produce millions of eggs.[8] The eggs are about 90 by 70 μm, with an outer shell that wrinkles when dry. Eggs exit the host in nasal secretions or, if swallowed, with the feces. When swallowed by an intermediate host, the four-legged larva hatches in the small intestine, penetrates the intestinal wall, and lodges in tissues, particularly in lungs, liver, and lymph nodes. There the nymphal instars develop, with the infective stage becoming surrounded by host tissues. When eaten by a definitive host, the infective nymph either attaches in the upper digestive tract or quickly travels there from the stomach, eventually reaching the nasopharynx. Females begin egg production in about 6 months.

PATHOGENESIS

There are two aspects of **pentastomiasis** in humans. **Visceral pentastomiasis** results when eggs are eaten and nymphs develop in various internal organs, and **nasopharyngeal pentastomiasis** results when nymphs that are eaten locate in the nasopharynx. Both types are rather common in some parts of the world.

Visceral pentastomiasis

Several species of pentastomids have been found encysted in humans. Probably the most commonly in-

volved species is *Armillifer armillatus*, which has been reported from the liver, spleen, lungs, eyes, and mesenteries of people in, among other places, Africa, Malaysia, the Philippines, Java, and China.[3,21,22] Other reported species are *A. moniliformis, Pentastoma najae, L. serrata,* and *Porocephalus* sp.

Most infections cause few if any symptoms and, therefore, go undetected. In fact most recorded cases were found at autopsy, after death from other causes. However, infection of the spleen, liver, or other organs causes some tissue destruction. Ocular involvement may cause vision damage.[12] Prior visceral infection may sensitize a person, resulting in an allergy to subsequent infection.[9,20] The host response to nymphs is often highly inflammatory, although little pathological response is elicited in definitive reptilian hosts. Dead nymphs are often calcified and are sometimes detected in x-ray films. Others begin a slow deterioration, causing a mononuclear cell response, with a subsequent abscess and granuloma formation. Experimentally produced heavy infections in rodents may kill them, indicating that visceral pentastomiasis possibly may be more important in human medicine than usually is thought.[20]

Nasopharyngeal pentastomiasis

When nymphs of *L. serrata* invade the nasopharyngeal spaces of humans, they cause a condition usually called **halzoun,** also known as **marrara** or **nasopharyngeal linguatulosis.** According to Schacher et al.[18]

Halzoun has been a clinically well recognized but aetiologically obscure disease in the Levant since its original description by Khouri (1905); in the Sudan it is known as the marrara syndrome. In Lebanon the disease is linked in the popular mind with the eating of raw or undercooked sheep or goat liver or lymph nodes; in the Sudan it is linked with the ingestion of various raw visceral organs of sheep, goats, cattle or camels. A few minutes to half an hour or more after eating, there is discomfort, and a prickling sensation deep in the throat; pain may later extend to the ears. Oedematous congestion of the fauces, tonsils, larynx, eustachian tubes, nasal passages, conjunctiva and lips is sometimes marked. Nasal and lachrymal discharges, episodic sneezing and coughing, dyspnoea, dysphagia, dysphonia and frontal headache are common. Complications may include abscesses in the auditory canals, facial swelling or paralysis and sometimes asphyxiation and death.

At various times, this condition was suspected to be caused by the trematodes *Fasciola hepatica, Clinostomum complanatum,* or *Dicrocoelium dendriticum* or by leeches. However, the recovery of *L. serrata* nymphs from the nasal passages and throats of patients in India, Turkey, Greece, Morocco, and Lebanon indicate that this species is at least the main cause of the condition in these areas. It is possible that the parasites can become mature if not removed or lost initially.

The epidemiology of this condition depends on cultural food patterns, in which nymphs are ingested when visceral organs, primarily liver or mesenteric lymph nodes of domestic herbivores, are consumed raw or undercooked.[18]

CLASSIFICATION OF PHYLUM PENTASTOMIDA
(After Riley[15])

Order Cephalobaenida
Mouth anterior to hooks; hooks lacking fulcrum; vulva at anterior end of abdomen.

FAMILY CEPHALOBAENIDAE
Parasites of snakes, lizards and amphibians.

Genera
Cephalobaena, Raillietiella

FAMILY REIGHARDIIDAE
Parasites of marine birds.

Genus
Reighardia

Order Porocephalida
Mouth between or below level of anterior hooks; hooks with fulcrum; vulva near posterior end of body.

FAMILY SEBEKIDAE
Parasites of crocodilians, chelonians.

Genera
Sebekia, Alofia, Leiperia

FAMILY SUBTRIQUETRIDAE
Parasites of crocodilians.

Genus
Subtriquetra

FAMILY SAMBONIDAE
Parasites of monitor lizards, snakes.

Genera
Sambonia, Elenia, Waddycephalus, Parasambonia

FAMILY DIESINGIDAE
Parasites of chelonians.

Genus
Diesingia

FAMILY POROCEPHALIDAE
Parasites of snakes.

Genera
Porocephalus, Kiricephalus

FAMILY ARMILLIFERIDAE
Parasites of snakes.

Genera
Armillifer, Cubirea, Gigliolella

FAMILY LINGUATULIDAE
Parasites of mammals.

Genus
Linguatula

REFERENCES

1. Ali, J.H., and J. Riley. 1983. Experimental life-cycle studies of *Raillietiella gehyrae* Bovien, 1927 and *Raillietiella frenatus* Ali, Riley and Self, 1981: pentastomid parasites of geckos utilizing insects as intermediate hosts. Parasitology 86:147-160.

2. Banaja, A.A., J.L. James, and J. Riley. 1977. Observations on the osmoregulatory system of pentastomids: the tegumental chloride cells. Int. J. Parasitol. 7:27-40.

3. Dönges, J. 1966. Parasitäre abdominalcysten bei Nigeriarern. Z. Trop. Parasitol. 17:252-256.

4. Doucet, J. 1965. Contribution á l'étude anatomique, histologique et histochimique des pentastomes (Pentastomida). Mem. Office Rech. Sci. Tech. Outre-Mer. Paris 14:1-150.

5. Esslinger, J.H. 1962. Development of *Porocephalus crotali* (Humboldt, 1808) (Pentastomida) in experimental intermediate hosts. J. Parasitol. 48:452-456.

6. Esslinger, J.H. 1962. Morphology of the egg and larva of *Porocephalus crotali* (Pentastomida). J. Parasitol. 48:457-462.

7. Fain, A. 1961. Les pentastomides d'Afrique central. Mus. R. Afr. Cent. Ann. 92:1-115.

8. Hobmeier, A. and M. Hobmeier. 1940. On the life cycle of *Linguatula rhinaria*. Am. J. Trop. Med. 20:199-210.

9. Khalil, G.M., and J.F. Schacher. 1965. *Linguatula serrata* in relation to halzoun and the marrara syndrome. Am. J. Trop. Med. Hyg. 14:736-746.

10. Lavoippierre, M.M.J., and M. Lavoippierre. 1966. An arthropod intermediate host of a pentastomid. Nature 210:845-846.

11. Osche, G. 1963. Die systematische Stellung und Phylogenie der Pentastomids. Embryologische und vergleichendanatomische Studien an *Reighardia sternae*. Z. Morphol. Ökol. Tiere 52:487-596.

12. Rendtdorff, R.C., M.W. Deiwesse, and W. Murrah. 1962. The occurrence of *Linguatula serrata*, a pentastomid, within the human eye. Am. J. Trop. Med. Hyg. 11:762-764.

13. Riley, J. 1981. An experimental investigation of the development of *Porocephalus crotali* (Pentastomida: Porocephalida) in the western diamondback rattlesnake *(Crotalus atrox)*. Int. J. Parasitol. 11:127-132.

14. Riley, J. 1983. Recent advances in our understanding of pentastomid reproductive biology. Parasitology 86:59-83.

15. Riley, J. 1986. The biology of pentastomids. In Baker, J.R., and R. Muller, editors. Advances in parasitology, vol. 25. Academic Press, New York. pp. 45-128.

16. Riley, J., A.A. Banaja, and J.L. James. 1978. The phylogenetic relationships of the Pentastomida: the case for their inclusion within the Crustacea. Int. J. Parasitol. 8:245-254.

17. Riley, J., J.L. James, and A.A. Banaja. 1979. The possible role of the frontal and sub-parietal gland systems of the pentastomid *Reighardia sternae* (Diesing, 1864) in the evasion of the host immune response. Parasitology 78:53-66.

18. Schacher, J.F., S. Saab, R. Germanos, and N. Boustany. 1969. The aetiology of halzoun in Lebanon: recovery of *Linguatula serrata* nymphs from two patients. Trans. R. Soc. Trop. Med. Hyg. 63:854-858.

19. Self, J.T. 1969. Biological relationships of the Pentastomida: a bibliography of the Pentastomida. Exp. Parasitol. 24:63-119.
20. Self, J.T. 1972. Pentastomiasis: host responses to larval and nymphal infections. Trans. Am. Microsc. Soc. 91:2-8.
21. Self, J.T., H.C. Hopps, and A.O. Williams. 1972. Porocephaliasis in man and experimental mice. Exp. Parasitol. 32:117-126.
22. Self, J.T., H.C. Hopps, and A.O. Williams. 1975. Pentastomiasis in Africans, a review. Trop. Geogr. Pathol. 27:1-13.
23. Trainer, J.E., Jr., J.T. Self, and K.H. Richter. 1975. Ultrastructure of *Porocephalus crotali* (Pentastomida) cuticle with phylogenetic implications. J. Parasitol. 61:753-758.
24. Wingstrand, K.G. 1972. Comparative spermatology of a pentastomid, *Raillietiella hemidactyli,* and a branchiuran crustacean, *Argulus foliaceus,* with a discussion of pentastomid relationships. Kong. Danske Vidensk. Selsk. Biol. Skrift. 19:1-72.

CHAPTER 34

PHYLUM ARTHROPODA: FORM, FUNCTION, AND CLASSIFICATION

The arthropods are an enormous assemblage of animals, far outnumbering in species all other animals put together. Insects are the most numerous of the arthropods: 700,000 to 800,000 species have been described. The next largest groups are the arachnids, with between 50,000 and 60,000 species, and crustaceans, with close to 30,000 species. The numbers of species in the remaining arthropod groups are much smaller and include few, if any, parasites. Both the Crustacea and the arachnid order Acari contain many important and interesting parasites; later chapters will be devoted to each. Parasitic insects are extremely important ecologically, medically, and economically. In fact, if we define "parasite" broadly—to include intermittent parasites (such as mosquitoes and biting flies), parasitoids that feed on but then kill their hosts, as well as more conventionally "acceptable" parasites—then some 15% of all insects are parasites. Thus, of all animal species on earth, 1 in 10 is a parasitic insect.[5]

Traditionally the arthropods have comprised a single phylum of metameric, coelomate animals, probably descended from an annelid or annelid-like ancestor. Features that the arthropods share with the Annelida include their being triploblastic (three primary germ layers in the embryo), eucoelomate, bilaterally symmetrical, and metameric and possessing a tubular gut from mouth to anus. In addition, the basic plan of the nervous system of annelids and arthropods is similar, with a chain of segmental ganglia running ventral to the gut and a single pair of supraesophageal ganglia (brain) anterior and dorsal to the gut. However, several important features contrast with the annelids. The evolutionary development of the constellation of characteristics associated with the exoskeleton may be referred to as **arthropodization.** The arthropods have a cuticle containing chitin; in almost all of them this cuticle forms a firm exoskeleton. The more rigid sections of the exoskeleton are articulated with each other by thinner, flexible cuticular joints. The appendages are generally composed of several such articulated sections, hence "arthropod" (from the Greek, *arthron,* "joint," + *podos,* "foot"). Other features of arthropods include the fact that they grow through a series of molts, or ecdyses; that their main body cavity is a space (**hemocoel**) filled with blood (**hemolymph**); and that the heart has ostia. Their coelom has become vestigial and is seen only as transient embryonic structures or as portions of certain excretory organs. The features unique to the arthropods are those associated with their cuticular exoskeleton; even the hemocoel and heart with ostia are correlated with the decline in importance of the hydrostatic function served by the coelom in their soft-bodied ancestors.

In recent years some investigators[3,20,21] have maintained that arthropodization occurred more than once and that arthropods as a group do not constitute a natural phylum; that is, they are polyphyletic. Others believe that the preponderance of evidence supports monophyly of the phylum Arthropoda.[7,24] Recent classifications of Crustacea have considered the higher taxa in that group at the class rank, thus elevating Crustacea to at least subphylum status.[8,22] Therefore, without taking a position on the possible polyphyly of arthropods, we will consider the Arthropoda a single phylum with three extant subphyla: Crustacea, Uniramia, and Chelicerata. The largest groups of extant chelicerates are the spiders (order Araneae) and the mites and ticks (order Acari), both in the class Arachnida. The Uniramia includes the Insecta and the myriapod classes (millipedes, centipedes, and some smaller groups).

GENERAL FORM AND FUNCTION

As indicated by the numbers of species we cited before, the arthropod body plan has been enormously successful evolutionarily, allowing its posessors to exploit almost every type of niche capable of supporting metazoan life. The insects and arachnids have radiated

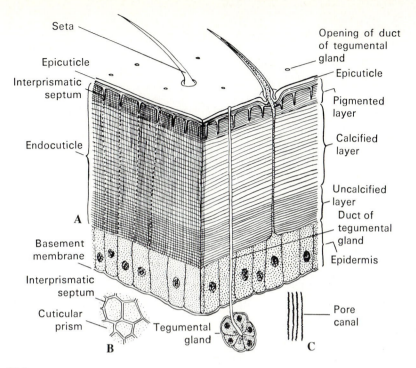

FIG. 34-1

A, Diagram showing structure of crustacean cuticle. All layers are secreted by the hypodermis (epidermis). The thin epicuticle is of sclerotized protein, and the procuticle (endocuticle) contains protein, chitin, and mineral salts. Protein in the uncalcified layer is unsclerotized. The procuticle of insects and arachnids is divided into the highly sclerotized exocuticle and the less sclerotized endocuticle. **B,** Horizontal section through pigmented layer of endocuticle. **C,** Pore canals as they appear in vertical sections.

From Dennell, R. 1960. In Waterman, T.H., editor. The physiology of Crustacea, vol. 1. Academic Press, Inc., New York.

immensely in the terrestrial environment, and the crustaceans exhibit analogous radiation in marine habitats. Both insects and crustaceans are widely prevalent in freshwater niches. The major reason for their success is the unique combination of arthropodization and cuticular exoskeleton.

Cuticle

The **cuticle** is made up of several layers, all secreted by the **hypodermis** (Fig. 34-1), which sometimes is called the epidermis. The substance of the cuticle is nonliving, but the molting process of the animal involves an active flux of resorption and deposition. The cuticle contains protein, lipid, and polysaccharide components. Much of the protein is stabilized (rendered relatively inert chemically) by processes of **sclerotization,** which is thought to involve cross-linking of the amino acid chains in adjacent protein molecules. The cross-linkages of sclerotization in the arthropods appear to be achieved mainly

by two different means, both of which are preceded by the production of N-acetyldopamine from the amino acid tyrosine[2,15] (Fig. 34-2, A). In the first of these the N-acetyldopamine is first oxidized by phenoloxidase to a quinone, which then spontaneously links (without enzymatic catalysis) to amino groups in adjacent amino acid chains of proteins (Fig. 34-2, A). This process is known as **quinone tanning.** Alternatively, the β-carbon of the N-acetyldopamine is activated by an enzyme that leads to the formation of covalent bonds between the β-carbon and adjacent protein molecules, a process called β-**sclerotization** (Fig. 34-2, A). So far, β-sclerotization is known only in insects and has not been found in the few chelicerates and crustaceans investigated.[25] In addition, protein stabilization in cuticle by the formation of dityrosine and trityrosine cross-links has been reported, as in the protein resilin, and disulfide crosslinks, as in keratin, may be possible[2] (Fig. 34-2, B and C). Some workers believe that sclerotization does not occur by covalent

531

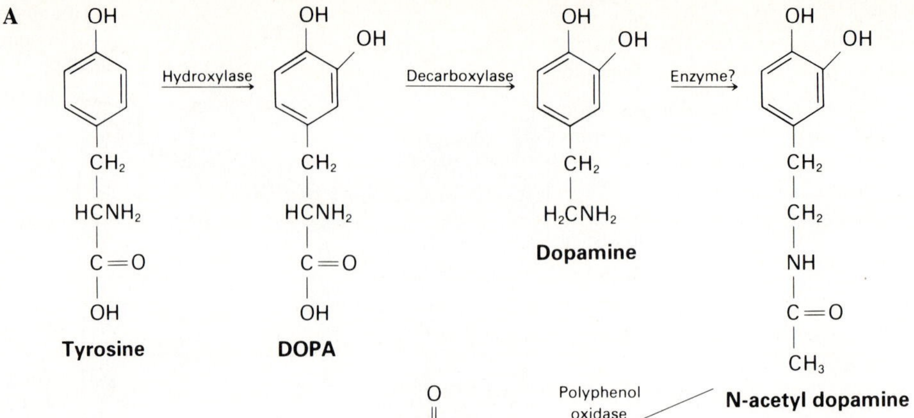

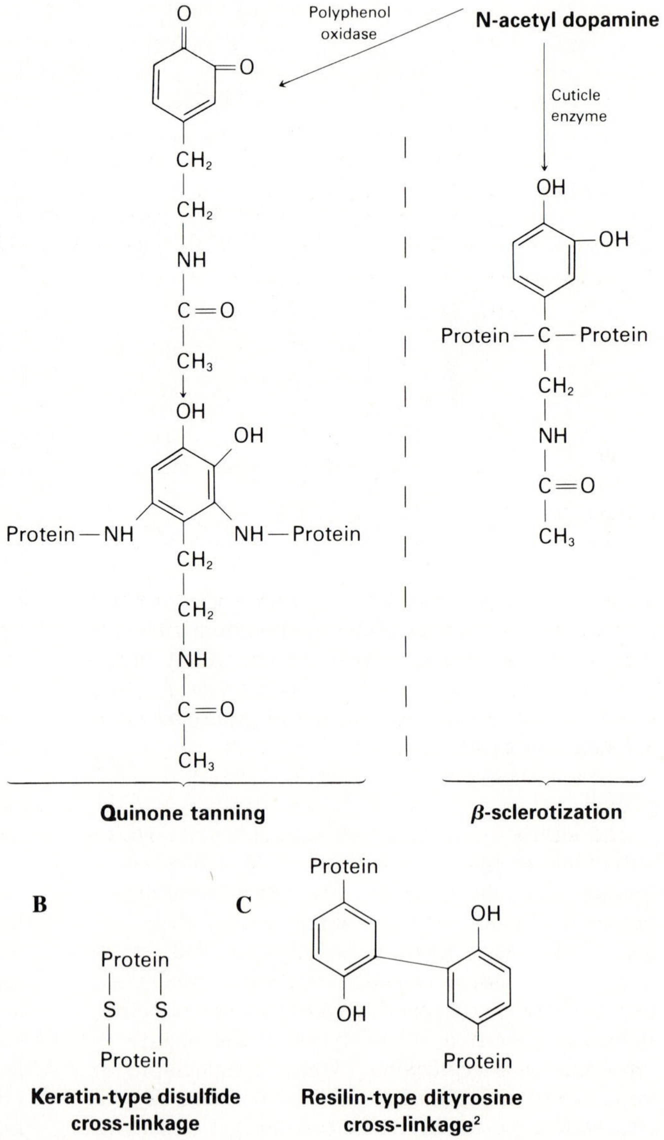

FIG. 34-2

A, Proposed reaction sequence for stabilization of sclerotins by quinone tanning in insects and other arthropods and by β-sclerotization in insects (based on Richards[25] and on Riddiford and Truman[27]). **B,** Disulfide cross-linkages of adjacent amino acid chains, as in keratin; probably also occurs in arthropod cuticle.[2] **C,** The highly elastic protein, resilin, stabilized by dityrosine and trityrosine cross-links, is sometimes found in insect cuticle, either mixed with other proteins or in almost pure form.[2]

cross-linking of the protein chains but by controlled dehydration driven by quinones and other chemicals secreted into the cuticle.[33] In any case, when the protein is thus stabilized, it is virtually insoluble except by vigorous chemical treatment.

The main polysaccharide component of cuticle is chitin. Chitin is a polymer of N-acetylglucosamine linked by 1,4- α-glycosidic bonds in long, unbranched molecules of high molecular weight. In contrast to the popular impression of the phrase "chitinous exoskeleton," chitin is flexible and contributes little to the rigidity of the cuticle. The hardness of the exoskeleton is conferred by the sclerotized protein and by deposition of inorganic salts. However, evidence exists that the chitin is bonded to the protein components, although the structural significance of this linkage is unclear.[25]

The outermost layer of the cuticle, the **epicuticle,** is very thin and contains stabilized protein (probably tanned), sometimes called **cuticulin,** but no chitin. Covering the cuticulin, or perhaps interspersed in its structure, insects and arachnids usually have a lipoidal layer that is of great value in preventing water loss through the cuticle. Over the lipid, they have a "varnish" layer that protects the wax from abrasion. The waxy and varnish layers are apparently absent in Crustacea. Beneath the epicuticle lies the thicker **procuticle,** often called **endocuticle** in crustaceans. This portion lends strength and weight to the exoskeleton. The procuticle contains protein, chitin, and, in crustaceans, substantial deposits of calcium carbonate, along with some calcium phosphate and other inorganic salts. In crustaceans the hardened layers containing inorganic salts and sclerotized proteins (**sclerotins**) are the **pigmented** and **unpigmented calcified layers.** The uncalcified layer also contains chitin and protein, but here the protein is unsclerotized, and the layer is membranous and flexible. The procuticle of insects and arachnids is subdivided into **exocuticle** and **endocuticle,** the endocuticle being much less sclerotized than the exocuticle. (Note the confusing terminology used in reference to the different groups: the endocuticle layer of insects and arachnids does not correspond to the layer often called endocuticle in crustaceans, which in that group refers to the entire procuticle.)

Obviously, if all areas of the cuticle were hard and massive, the animal would be encased in an immovable box. The joints are thinner areas with little calcification or sclerotization and may be quite flexible. A sclerotized area limited by a suture line or flexible membrane is called a **sclerite.** The larger the amount of sclerotization and calcification in a sclerite, the greater its strength and mass. The entire thickness of the sclerites of many adult insects may be sclerotized; hence such sclerites do not contain any endocuticle.[25] The main dorsal sclerite on a given somite is termed the **tergite,** the ventral sclerite is the **sternite,** and lateral sclerites are **pleurites.** Muscles are inserted on the inner surface of the cuticle, extending through the hypodermis, and a portion of cuticle for a particular muscle insertion may project inward as a spine (**apodeme**) or as a ridge (**apophysis**). The bearing surfaces between articulated sclerites are **condyles** (Fig. 34-3).

The protective function of the arthropodan exoskeleton is obvious, but it does not explain the success of its possessors. Metamerism conferred compartmentalization of their soft-bodied ancestors, thus allowing the use of a "separate" hydrostatic skeleton compartment in each metamere, with localized movement of its appendages, greater efficiency, and development of greater complexity in muscle arrangement and nervous system. Nevertheless, in comparison with the arthropodan system of levers, the movements of worms are limited and inefficient. Such a system, in genetically and evolutionarily plastic groups, led to the enormous radiation we see today. Insertion of the muscles in the unyielding exoskeleton, coupled with fulcra provided by the condyles (Fig. 34-3) in the flexible joints, made very fine control of movements mechanically possible. Increase in complexity of movements meant corresponding evolution of nervous elements to coordinate the movements. Furthermore, small changes in a given sclerite could substantially increase efficiency of a particular body part for a given function, and arthropods have many sclerites. Thus the arthropodan cuticle gave natural selection raw material with which to work.

Appendages have been modified to fulfill many functions in a variety of ways. In addition to their primitive function of locomotion, appendages display a wide array of specialization for food gathering, defense, reproduction, reception of stimuli, feeding, prehension, and respiration. Because a study of most arthropods requires a familiarity with the basic arrangement of appendages in the group, we will introduce appendages later.

Molting

Although the arthropod cuticle conferred many advantages of evolutionary potential, it conferred evolutionary problems as well, most important of which was growth of an animal enclosed in a nonexpansible covering. The solution is a series of **molts,** or **ecdyses,** through which all arthropods go during their ontogeny. Much of the physiological activity of any arthropod is related to the molting cycle, and, in the groups in which the processes have been best investigated (in-

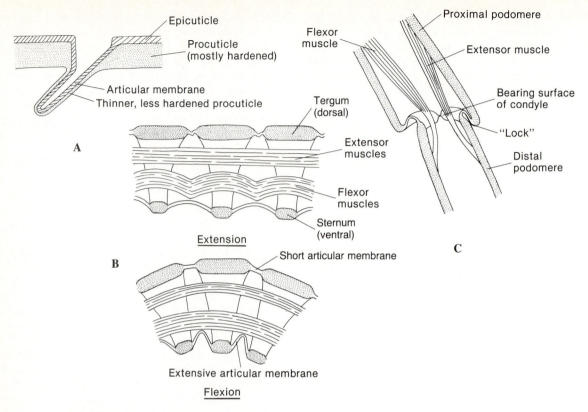

FIG. 34-3

Diagram of articulation and musculature of arthropod joints. **A,** Flexibility is provided by thinner cuticle between sclerites. **B,** Greater flexion in one direction such as in the abdomen of a crayfish is made possible by the larger areas of thin cuticle on that side. **C,** Movement of a joint is by contraction of muscles inserted on opposite sides of the articulation, and the bearing surfaces of the joints are the condyles.

Drawing by William C. Ober.

sects and malacostracan crustaceans), the physiological activity is controlled by hormones.

Growth in tissue mass occurs during an interecdysial period, and dimensional increase occurs immediately after molting, while the new cuticle is still soft. The stages of the animal between each molt are referred to as **instars.** In the Malacostraca a **molt-inhibiting hormone (MIH)** is produced in the **X-organ,** which is composed of neurosecretory cells in the eyestalk, and **molting hormone (MH)** is produced in the **Y-organs,** a pair of glands near the mandibular adductor muscles (in decapods). Structures comparable to X-organs are found within the head of sessile-eyed Crustacea.[1] In insects at least three hormones are directly involved in the molting process: **prothoracicotropic hormone (PTTH,** brain hormone), **ecdysone,** and **bursicon.** PTTH is secreted by neurosecretory

cells in the brain and released by neurohemal organs called **corpora cardiaca** (Fig. 34-4). The hormone is carried in the hemolymph to the **prothoracic glands,** which are stimulated to produce ecdysone. Ecdysone stimulates molting and is analogous, perhaps homologous, to MH of crustaceans. Bursicon is secreted from the perivisceral neurohemal organs associated with ventral nerve ganglia. It regulates postecdysial hardening of the cuticle.

As the level of MIH decreases and that of MH increases in the crustacean or the level of ecdysone increases in the insect, the organism undergoes a series of changes preparatory for a molt, the **preecdysial** period. DNA synthesis in the hypodermal cells is stimulated, then RNA and protein synthesis. The next effect of the ecdysial hormones is to cause the hypodermis to detach from the old procuticle **(apolysis)** and start se-

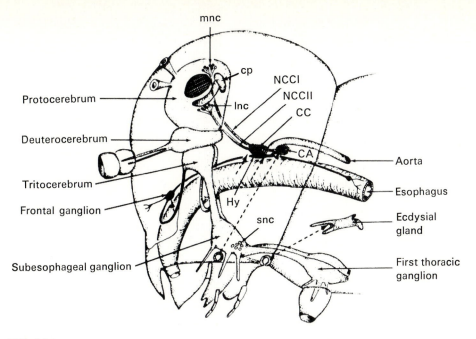

FIG. 34-4

Generalized central nervous system of insect, showing sources of hormones in the head and pro-thorax. Neurosecretory hormones are produced by the median *(mnc)*, lateral *(lnc)*, and subesoph-ageal *(snc)* neurosecretory cells and perhaps also by the corpora pedunculata *(cp)*. Hormones from the neurosecretory centers *(NCCI and NCCII)* in the protocerebrum pass in two paired nerves to be stored in the corpora cardiaca *(CC)*. Hormones are also secreted by the corpora allata *(CA)* and the prothoracic or ecdysial glands. The dashed lines indicate the original embryonic origin of the organs indicated and their subsequent migration route during development. *(Hy)*, Hypocerebral ganglion.

From Jenkin, P.M. 1962. Animal hormones: a comparative survey, part 1. Pergamon Press Ltd., Oxford, Eng.

creting a new epicuticle (Fig. 34-5). At the same time beneath the old procuticle, enzymes (including chiti-nases and proteinases) begin to dissolve it. As the so-lution proceeds, the products of the reactions are re-sorbed into the animal's body, including amino acids, *N*-acetylglucosamine, and calcium and other ions. These materials are thus salvaged and later are incor-porated into the new cuticle.

Almost immediately after apolysis, the new epicuti-cle becomes limited in permeability, thus protecting the new procuticle from the enzymes dissolving the old cuticle above.[18,32,36] Solution proceeds, and the products may be resorbed through the body surface or swallowed to be absorbed in the midgut.[36] The old cu-ticle is not completely dissolved; the epicuticle and sclerotized exocuticle remain in insects, and in crusta-ceans the epicuticle and calcified regions remain, al-though some decalcification of these layers occurs. At the time of ecdysis, the old cuticle splits, normally along particular lines of weakness, or dehiscence, and

the organism climbs out of its old clothes. The animal must expand to split the old cuticle and again before the new cuticle hardens. Insects inhale air; crustaceans expand by rapid inhibition of water, a process aided by the fact that the osmotic pressure in the tissues and blood has been increased before the molt by the cal-cium ions mobilized from the cuticle.[28] The increase in blood and tissue volume causes the small wrinkles in the still soft cuticle to smooth out, increasing the body dimensions, and the cuticle begins to harden again. In this postecdysial period, sclerotization of the protein and redeposition of calcium salts in the procu-ticle occur, and more procuticle is secreted. The ani-mal is highly vulnerable to attack by predators while its cuticle is soft.

The length of the subsequent **intermolt** phase de-pends on the species involved, its age and stage of de-velopment, the season or annual cycle, and any inter-ceding diapause (to be discussed later).

Decapod crustaceans that molt on an annual cycle

535

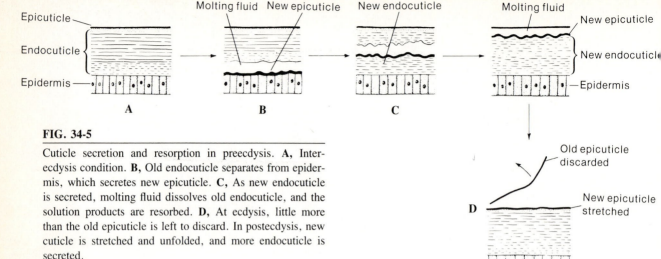

FIG. 34-5

Cuticle secretion and resorption in preecdysis. **A,** Interecdysis condition. **B,** Old endocuticle separates from epidermis, which secretes new epicuticle. **C,** As new endocuticle is secreted, molting fluid dissolves old endocuticle, and the solution products are resorbed. **D,** At ecdysis, little more than the old epicuticle is left to discard. In postecdysis, new cuticle is stretched and unfolded, and more endocuticle is secreted.

Drawing by William C. Ober.

and have a long intermolt period are said to be **anecdysic.**[16] Species in which one ecdysial cycle grades rapidly into another are **diecdysic.** Some crabs reach maximal size and stop molting, undergoing "terminal anecdysis." These phases are controlled by MH and MIH produced by the Y- and X-organs described previously. In crustaceans other than malacostracans, little is known of the hormonal control of molting. Barnacles, at least, seem to be in a permanent diecdysis.[6] Furthermore, many copepod parasites of fish cease molting when they reach the adult stage, although they continue to grow actively. For example, a female *Lernaeocera* is about 2 mm long after her last molt, but she may attain an ultimate size of up to 60 mm *without molting*. What changes in the cuticle when the copepod reaches sexual maturity, and what causes the change? We do not know, but it is clear that the change permits continuous growth. However, more is known about the mechanisms allowing such expansion in some parasitic insects and ticks. The fourth-stage nymph of the bug *Rhodnius* takes a large blood meal that necessitates stretching its abdominal cuticle threefold.[35] Before this blood meal, its cuticle is stiff and inextensible, but this condition changes as the bug feeds. The change is mediated by neurosecretory axons running to the hypodermis, apparently stimulating an enzyme discharge, which affects the substance of the cuticle. After feeding, additional cuticle material is deposited, probably to provide protection and a template for the cuticle of the next instar.[13] The female cattle tick, *Boophilus microplus*, ingests 150 times its body weight in blood after molting to the adult, increasing in length from 2.5 mm to 11 mm.

Filshie[12] reported that, before the last molt, the epicuticle is laid down as a highly folded layer. The subsequent expansion is accommodated by unfolding of the inexpansible epicuticle and stretching and growth of the underlying procuticle.

Development
■ Embryogenesis

Embryogenesis will not be discussed in detail here; an excellent treatment is given by Anderson.[3] Almost all arthropods have highly yolky eggs. In most members of the Crustacea and Insecta, the nuclei undergo several divisions within the yolk mass (**intralecithal** cleavage) and then migrate to the periphery to become the **blastoderm.** Yolk is concentrated in the interior of the embryo (**centrolecithal**), and differentiation proceeds in the superficial areas. A blastoderm is also formed in the chelicerates, but the initial cleavages are sometimes **holoblastic** (complete).

■ Postembryonic development

The typical larva that hatches from the crustacean egg is called the **nauplius** (Fig. 34-6). The nauplius has only three pairs of appendages: antennules, antennae, and mandibles. These are different in form from the adult appendages and have locomotive function. The nauplius undergoes several ecdyses, usually adding somites and appendages at each molt. Nauplii typically have several instars; and later ones may be referred to as **metanauplii.**

Change in body form from larval to adult is referred to as **metamorphosis,** whether the change is gradual over several instars, or more abrupt, from one instar

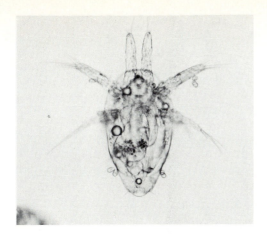

FIG. 34-6

Copepod nauplius. The anteriormost, uniramous appendages are the antennules, followed by the biramous antennae and mandibles. Note the internal parasitic nematode and the external phoretic protozoa.

Photograph by Ralph Muller.

to the next. Such development is described as **indirect.** In **direct** development, all larval instars are suppressed, and a juvenile, rather than a larva, hatches with segmentation and appendages complete.

The adaptibility and wide variation of the Crustacea are demonstrated by the variety of developmental patterns in the various groups. Thus, even within the same class, such as the Copepoda or the Malacostraca, some forms may be slightly metamorphic, and some strongly metamorphic. Specific examples will be cited in Chapter 35.

Of the Insecta, only certain wingless orders (Collembola, Thysanura, and Diplura) have direct development. The Pterygota, which are of concern in parasitology, are all metamorphic. In all cases the young that hatches from the egg is quite different from the adult. In several orders (Dermaptera, Dictyoptera, Mallophaga, Anoplura, and Hemiptera) the juvenile instars are called **nymphs,** and they become gradually more like the adult **(imago)** with each ecdysis **(gradual metamorphosis,** or **hemimetabolous** development) (Fig. 34-7). The wing buds develop externally **(exopterygote)** and can be observed readily in the nymphal instars. In other orders (Neuroptera, Coleoptera, Strepsiptera, Siphonaptera, Diptera, Lepidoptera, and Hymenoptera) the juvenile instars are called **larvae,** and they abruptly metamorphose into the imago after a relatively quiescent instar called the **pupa** (Fig. 34-8). This type of development is referred to as **complete metamorphosis (holometabolous** development), and the wing buds in the larvae are internal **(endopterygote)** (Fig. 34-9). The nymphs of hemimetabolous insects have well-developed appendages, and their habits are generally similar to those of the adult. They have compound eyes, and rudiments of the external genitalia and wings are present beginning in an early instar, increasing in size and complexity in later instars. The larvae of holometabolous insects are extremely diverse, usually differing radically in form and habits from the adult. Consequently, they usually occupy a different niche from the adult and avoid competition with it. Eyes, when present, are simple ocellus-like stemmata; the cuticle is usually less sclerotized than that of the adult; and cephalic appendages and legs are commonly reduced or absent. Several different forms of holometabolous larvae are recognized. The **protopod** type occurs in some parasitic Hymenoptera and Diptera and is little more than a precociously hatched embryo (Fig. 34-10). The **polypod** type occurs in Lepidoptera and some Hymenoptera; it has several abdominal processes **(prolegs),** which are used in locomotion. **Oligopod** larvae are common in Coleoptera and Neuroptera. They have a well-developed head and thoracic legs. **Campodeiform** oligopod larvae are usually active predators with a relatively highly sclerotized cuticle and powerful mouthparts. **Scarabaeiform** oligopod larvae, such as are found in May beetles, are C shaped, have less powerful mouthparts, and have a lightly sclerotized body cuticle. The **apodous** larvae of most Diptera and Hymenoptera and some Coleoptera have no legs, and the head is often much reduced.

Postembryonic development of chelicerates is characteristically direct, with a tiny facsimile of the adult (nymph) hatching from the egg. Some arachnids undergo one or two "larval" molts while still within the

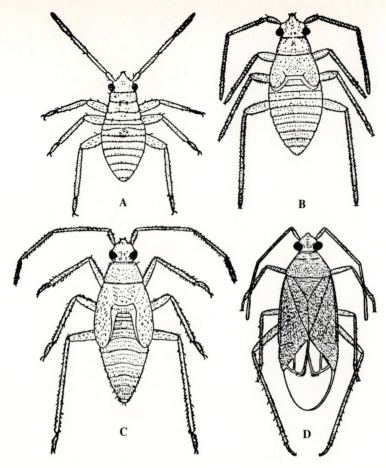

FIG. 34-7

Hemimetabolous development in a hemipteran, *Lygocoris pabulinus*. **A,** Second instar nymph. **B,** Fourth instar. **C,** Fifth instar. **D,** Imago.

After Petherbridge and Thorpe; from Richards, O.W., and R.G. Davies. 1978. Imms' outlines of entomology. Chapman & Hall Ltd., London.

FIG. 34-8

Immature stages of the black blister beetle, *Epicauta pennsylvanica*. **A,** Unfed first instar. **B,** Newly molted fifth instar. **C,** Pupa. (**A,** ×17; **B,** ×9; **C,** ×5.)

After Horsfall; reprinted, by permission, from A Textbook of Entomology, Third Edition, by Herbert H. Ross, copyright © 1965 by John Wiley & Sons, Inc.

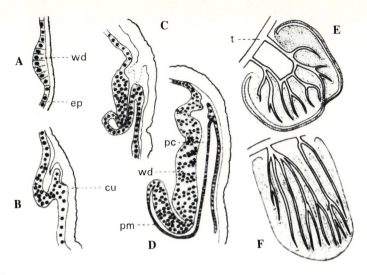

FIG. 34-9

Developing wing buds (imaginal discs) in a moth, *Ephestia kuehniella*. **A** to **D,** Sections through successively older wing discs *(wd).* **E** and **F,** Surface view of younger and older wing discs showing tracheal supply *(t). ep,* Epidermis; *pc,* peripodial cavity; *cu,* cuticle; *pm,* peripodial membrane.

After Köhler; from Richards, O.W., and R.G. Davies. 1978. Imms' outlines of entomology. Chapman & Hall Ltd., London.

egg. The number of nymphal instars varies depending on the type of arachnid. A six-legged larva hatches from the egg in members of the order Acari and becomes an eight-legged nymph at the first molt. In most mites there are three nymphal instars: the **protonymph,** the **deutonymph,** and the **tritonymph.** However, the hard ticks (Ixodidae) have only one nymphal stage, and the soft ticks (Argasidae) may have as many as eight.

How development is controlled is best known for the insects, and the mechanism is essentially similar in both holometabolous and hemimetabolous forms. A **juvenile hormone (JH)** is produced by the **corpora allata** (Fig. 34-4). Although three different chemical forms of the hormone are present in varying proportions in different insects, all three forms produce similar effects.[27] The best known and documented of such actions is the so-called status quo effect; that is, the JH acts on the tissues of the insect to maintain larval or nymphal characteristics (impedance of maturation). The level (titer) of JH in the blood of the insect decreases as it develops through its juvenile instars; consequently, the tissues and organs become progressively more adult-like. Finally, the titer drops to an undetectable level at about the beginning of the last nymphal instar in hemimetabolous insects, and the adult emerges at the next ecdysis. Disappearance of JH from the blood of holometabolous forms usually occurs about midway through the last larval instar, and the next molt produces the pupa. The status quo action of JH is believed to result from its directing the kinds of RNA produced after ecdysone stimulation, but the mechanism by which it does this is still unknown.[27] If a high titer of JH is present, the RNA produced will lead to synthesis of larval proteins; if little or no JH is in the blood, then the RNA for synthesis of proteins with adult characteristics will be produced. Interestingly, although a shutdown of JH production by the corpora allata is necessary for maturation to the adult form, the corpora allata are reactivated in the adult of many insects and again secrete JH. This hormone is necessary for egg maturation in females and proper development of the sex accessory glands in males. Furthermore, the adult of almost all insects does not molt again, but ecdysone is again secreted and also plays a role in reproductive function.

■ Diapause

Without doubt, another factor contributing greatly to the evolutionary success of arthropods has been their ability to withstand adverse environmental conditions. Under circumstances in which normal physiological function would be impossible, for example, subfreezing temperatures or extremely dry conditions,

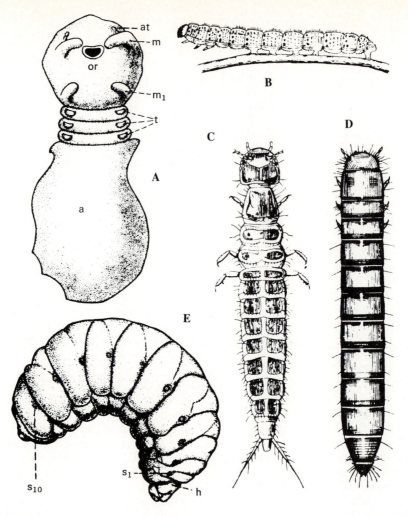

FIG. 34-10

Types of endopterygote larvae. **A,** Protopod type, found in some parasitic Diptera and Hymenoptera, has poorly defined segmentation, rudimentary appendages, and incompletely differentiated internal organs. **B,** Polypod, typical of Lepidoptera (butterflies and moths). **C** and **D,** Oligopod, as found in beetles (Coleoptera). **E,** Apodous type, found in Hymenoptera, Diptera, some Coleoptera, usually lives among abundant food. *a,* Abdomen; *at,* antenna; *m,* mandible; *m₁,* maxilla; *h,* head; *s₁* and *s₁₀,* spiracles; *t,* thoracic limbs; *or,* mouth.

A after Kulagin; **E** after Nelson; **A** to **E** from Richards, O.W., and R.G. Davies. 1978. Imms' outlines of entomology. Chapman & Hall Ltd., London.

many arthropods can enter a period of developmental arrest known as **diapause.** Again, although diapause occurs in many crustaceans and arachnids, much more is known about it from insects. Knowledge of the role of diapause is very important in understanding the biology of many arthropod vectors of parasitic diseases.

Diapause in insects may occur in the egg, larva, pupa, or adult, depending on the species. In the immature stages it is marked by a cessation of development and prolongation of that stage; in the adult, reproduction is inhibited. In addition, other physiological processes more or less cease, and the animal is quiescent. In all insects that have been studied, the mechanism of diapause initiation and termination is hormonal, but different hormones are involved, depending on the stage in the life cycle.[27]

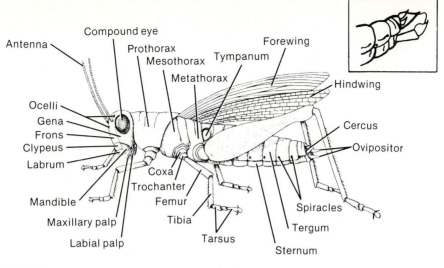

Antenna
Compound eye
Prothorax
Mesothorax
Tympanum
Forewing
Metathorax
Hindwing
Ocelli
Gena
Frons
Clypeus
Labrum
Cercus
Ovipositor
Coxa
Trochanter
Femur
Mandible
Tibia
Spiracles
Maxillary palp
Tarsus
Tergum
Labial palp
Sternum

FIG. 34-11

External features of a relatively generalized insect, the grasshopper *Romalea*. The terminal segment of a male with external genitalia is shown in inset.

From Hickman, C.P., Jr., L.S. Roberts, and F.M. Hickman. 1988. Integrated principles of zoology, ed. 8. The C.V. Mosby Co., St. Louis.

External morphology

The detailed morphology and related physiology of a group as enormous and diverse as the arthropods are far beyond the scope of this book. However, the parasitology student needs to know a modicum of structure to identify parasitic arthropods and to understand host-parasite relationships. A brief coverage will be given here, and more specialized details will be included in appropriate chapters that follow.

The evolution of arthropodization was accompanied or followed by a great deal of specialization of the ancestral metameres. This included fusion of certain body segments to form a head, with incorporation of the appendages they bore to become feeding and sensory appendages; regionalization of certain other body segments bearing locomotory appendages; loss of appendages from segments in other regions; and concentration of some internal organs in specific regions of the body. The specialization of metameres into recognizable, distinct body regions is called **tagmatization.** The different body regions are called **tagmata** (singular: **tagma**), and differences in tagmatization are characteristic of the major groups of arthropods.

■ Form of the pterygote insects

In all members of the class Insecta the tagmata comprise the **head, thorax,** and **abdomen.** The head is made up of six fused metameres (metameres usually are called **somites** in the arthropods). It is not known whether all of the head somites bore appendages in the insect ancestor, but only four pairs remain in modern insects. The bases of the freely movable, sensory **antennae** are above or between the eyes (Fig. 34-11). The **mandibles** are usually the primary feeding appendages and are borne ventrally, lateral to the mouth. Immediately posterior to the mandibles are the **maxillae** and, following these, the **labium,** whose parts are illustrated in Fig. 34-12. The labium is composed of fused second maxillae. Anteriorly, the mouth is covered by an upper lip, or **labrum.** A tonguelike lobe, the **hypopharynx,** arises from the floor of the mouth; the hypopharynx and labrum are not considered appendages, although they may serve important feeding functions in various insects. The generalized mouthparts shown in Fig. 34-12 are highly modified for the specialized feeding habits found in many insects, as will be apparent in following chapters. In addition to the appendages, photoreceptor organs are found on the head of most insects: a pair of **compound eyes** and one or more simple eyes (**ocelli**).

The thorax of hexapods comprises three somites, the **prothorax, mesothorax,** and **metathorax,** each of which bears a pair of legs. Each leg usually is divided into five segments, or **podomeres** (Fig. 34-11). The

541

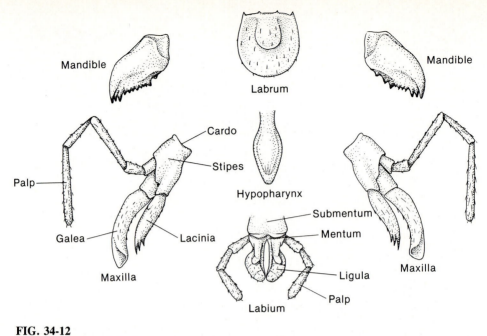

FIG. 34-12

Anterior view of the mouthparts of a grasshopper.

Drawing by William C. Ober.

basal segment of each leg, or **coxa,** articulates with the **trochanter,** which is fixed to the following podomere, the **femur.** The femur is often the largest of the podomeres and articulates distally with the more slender **tibia.** Distal to the tibia is the **tarsus,** which is subdivided into two to five segments. The **pretarsus** consists of claws or other structures attached to the terminal tarsal segment.

Adult pterygote insects characteristically have wings, although some, such as fleas, lice, and worker ants and termites, have lost their wings during the course of evolution. Both the mesothorax and metathorax bear a pair of wings. In the order Diptera the metathoracic wings are reduced to balancing organs called **halteres.** In male Strepsiptera the mesothoracic wings are reduced to halteres, whereas the females are highly modified parasites with no wings at all. Apart from whether the wing buds develop internally or externally, growth of the wings is essentially similar in endopterygote and exopterygote insects. The wings develop from evaginations of thoracic epidermis and thus consist of a double layer of epidermis. The layers are penetrated by canals called **lacunae,** and the lacunae contain nerves, tracheae (respiratory tubules), and blood. The epidermal cells atrophy as the wing approaches full development; thus the wing consists of two thin layers of cuticle secreted by the epidermis, and it is supported by the more heavily sclerotized

veins, which are the remains of the lacunae. The pattern of wing venation is constant within a species and, therefore, often is of value in taxonomy. To help describe wing venation in particular species and to facilitate use of this characteristic in keys, a special nomenclature for wing venation is used by entomologists (Fig. 34-13).

The abdomen of adult insects consists primitively of 11 somites plus a terminal **telson,** although all segments usually can be discerned only in the embryo in the more specialized pterygotes. Appendages also can be seen on the abdominal segments of the embryo, but the abdomen of the adult pterygote bears appendages only on the genital segments (**external genitalia**) and sometimes on the eleventh somite (the **cerci**). The cerci may be absent in many higher insects, they may remain as vestigial appendages, or they may be filiform sensory organs. The ninth genital segment bears the external genitalia of males, which include an intromittent organ: the **penis,** or **aedeagus.** The female gonopore opens behind the eighth sternum in most insects, and the paired appendages of the eighth and ninth abdominal somites form the egg-laying organ, or **ovipositor.**

In addition, most insects bear openings into the respiratory system called **spiracles** (Fig. 34-11). The more generalized insects commonly have spiracles on the mesothoracic and metathoracic pleura, as well as

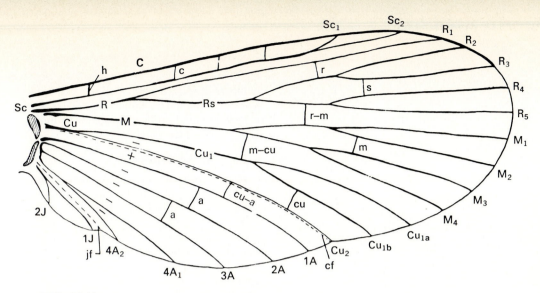

FIG. 34-13

Diagram of typical venation of wing in modern insects, showing standardized abbreviations for the veins:

Costa (C) usually forms the thickened anterior margin of the wing. It is unbranched.

Subcosta (Sc) runs immediately below the costa, always in the bottom of a trough between costa and radius. Typically the subcosta is divided into two branches.

Radius (R) is the next main vein. It is a stout one and connects at the base with an axillary sclerite. It is divided into two main branches, R_1 and radial sector Rs. Radial sector is frequently divided into four main branches.

Media (M) is one of three veins articulating with some of the small median axillary sclerites. The base is usually in a depression. Typically it is divided into four branches: M_1, M_2, M_3, and M_4.

Cubitus (Cu) also articulates with the median axillary sclerites and has two main branches. Its basal portion and Cu_2 are in a depression, but Cu_1 runs along a ridge and is usually branched.

Cubital furrow (cf) is a definite crease along which the wing folds. It is not a vein but is one of the most important landmarks for identifying the cubital and anal veins, which it separates.

Anal veins (1A, 2A, 3A, etc.) form a set and are united or close together at the base and closely associated with the third axillary sclerites.

Jugal furrow (jf) is a crease separating the anal region or fold from the jugal fold, which is the small area at the basal posterior corner of the wing. This fold also is one of the most stable wing landmarks.

Jugal veins (1J, 2J) are short veins in the jugal fold.

Cross-veins. Definite names are given to the kinds of cross-veins, based on the veins they connect. These cross-veins have standard abbreviations, which are never written in capital letters; these are outlined in the following table. Numbers are used to denote individual cross-veins in a series, for example, fourth costal cross-vein and third radiomedial cross-vein. There is one notable exception: the cross-vein between costa and subcosta at the base of the wing is called the humeral cross-vein and indicated by *h*. In orders such as the Trichoptera and Lepidoptera in which the cross-veins are greatly reduced, the cross-vein between R_1 and R_2 is called the radial *r*, and that between R_3 and R_4 is called the sectorial cross-vein *s*.

Veins connected	Name of cross-veins	Abbreviation
Costa to subcosta or R_1	Costal	*c*
Branches of radius	Radial	*r*
Radius to media	Radiomedial	*r-m*
Branches of media	Medial	*m*
Media to cubitus	Mediocubital	*m-cu*
Branches of cubitus	Cubital	*cu*
Cubitus to anal	Cubitoanal	*cu-a*
Various anals	Anal	*a*

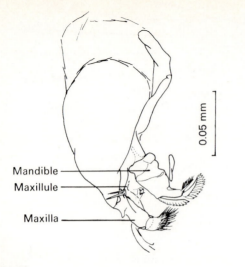

FIG. 34-14

Mouthparts (one side drawn) of *Ergasilus cerastes*, a parasite of catfish (*Ictalurus* spp.).

From Roberts, L.S. 1969. J. Parasitol. 55:1268.

eight spiracles along each side of the abdomen. The numbers of spiracles are often reduced in the more advanced forms. The spiracles usually have closing mechanisms, of value in respiratory function and in helping to reduce water loss.

■ Form of the Crustacea

In the Crustacea the head is usually not clearly set off from the trunk: one or more thoracic somites are commonly fused with the head, whereas some thoracic segments are distinct (see Fig. 34-15). Thus the normal tagmatization of crustaceans is a **cephalothorax,** a free **thorax,** and an **abdomen,** although the degree of prominence and of fusion of the tagmata varies greatly from group to group. As in the insects, the head proper seems to have been formed by fusion of six primitive somites. The cephalothorax, and sometimes even the entire body, may be covered by a **carapace,** which arises as a fold from the posterior margin of the head.

The anteriormost appendages of the head are the **antennules** (first antennae), followed by the **antennae** (second antennae). Crustaceans are the only arthropods with two pairs of antennae. The antennules and antennae are usually sensory, but in some forms they

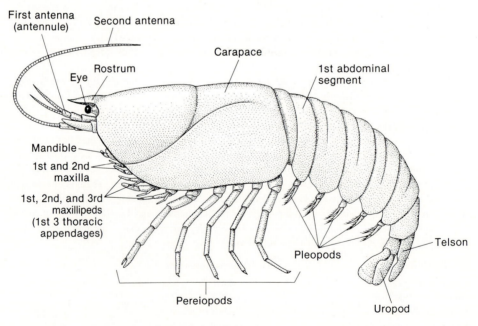

FIG. 34-15

Lateral view of a generalized malacostracan crustacean.

Drawing by William C. Ober.

may be adapted for locomotion or prehension. The two types of eyes found in the crustaceans are **median eyes** and **compound eyes.** The median eye, also called the **nauplius** eye, sometimes persists into the adult and is the only kind of eye found in copepods. It consists of three or four pigment-cup ocelli. Adults of most species have a pair of compound eyes. They may be sessile, or they may be mounted on stalks and be very convex, with an angle of vision of 180 degrees or more. Feeding appendages on the head are the **mandibles, maxillules** (first maxillae), and **maxillae** (second maxillae) (Fig. 34-14). One or more pairs of appendages borne by the thoracic segments of the cephalothorax may be incorporated into the mouthparts and are then called **maxillipeds.** They are followed by the other thoracic appendages, the **pereiopods,** and finally by the abdominal appendages, or **pleopods.** The pereiopods and pleopods may be variously modified for walking, swimming, or copulation. The number of pereiopods and pleopods varies from group to group, and in some groups pleopods are absent. The abdomen terminates in a **telson,** which may be flanked by the posteriormost pleopods, called **uropods.**

The appendages of Crustacea were primitively **biramous** (having two branches) (Figs. 34-15 and 34-16), and this condition prevails in at least some appendages of all living species during their lives. The terminology applied by various workers to crustacean appendages has not been blessed with uniformity. At least two systems are currently in wide use, and we have given the alternative term for each structure in parentheses. The lateral branch is the **exopod** (exopodite), and the medial one is the **endopod** (endopodite), each of which may contain several segments, varying by appendage and according to species. The endopod and exopod are borne on a **basis** (basipodite); the basis, in turn, is attached to the **coxa** (coxopodite), together being referred to as the **protopod.** Processes from the protopod are termed **endites** and **exites,** and the exites may be called **epipods** (epipodites). Several of these terms usually are not applied to the appendages of the subclass Branchiopoda, but because no symbiotic branchiopods are known, description of their limbs is not necessary. The two branches of the legs may not be homologous through all crustacean classes.[29]

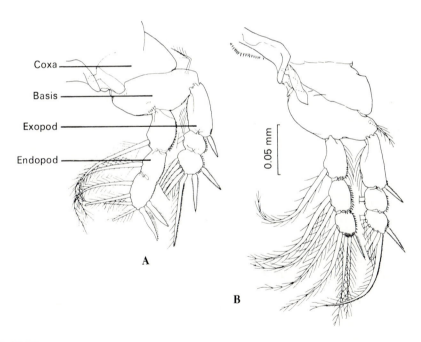

FIG. 34-16

First **(A)** and second **(B)** thoracic appendages of *Ergasilus megaceros* (Copepoda), a parasite of the sucker *Catastomus commersoni.* The terminal segments of the first endopod are fused, the ancestral condition being indicated by the presence of vestigial condyles. The medial side of the coxa may be modified for food handling in some Crustacea and is called a gnathobase.

From Roberts, L.S. 1970. Trans. Am. Microsc. Soc. 89:144.

■ Form of the Acari

The primary tagmata in the class Arachnida are a cephalothorax (**prosoma**) and an abdomen (**opisthosoma**). The somites of these tagmata are fused to a greater or lesser degree, depending on the order. In the spiders (order Araneae) the fusion is complete in almost all species, but the prosoma and the opisthosoma are distinct. In the other large order of arachnids, the Acari, which is of concern to parasitology, even these tagmata are fused, and the opisthosoma is defined rather arbitrarily as the region posterior to the legs. This situation has given rise to a special nomenclature for the body regions applied only to the Acari, given by Savory[30] (Fig. 34-17) as follows:

PROTEROSOMA
- GNATHOSOMA (segments of mouth and its appendages)
- PROPODOSOMA (segments of first and second legs)

HYSTEROSOMA
- METAPODOSOMA (segments of third and fourth legs)
- OPISTHOSOMA (segments posterior to legs)

PODOSOMA

IDIOSOMA

The **proterosoma** can usually be distinguished from the **hysterosoma** by a boundary between the second and third pairs of legs. Dorsally, the **idiosoma** is often covered by a single, sclerotized plate, the **carapace**. The **gnathosoma**, or **capitulum**, is usually sharply set off from the idiosoma, and it carries the feeding appendages. These appendages are the **chelicerae**, usually with three podomeres, and the **pedipalps**, whose free segments may vary from one to five in different groups (Fig. 34-18). The chelicerae may be **chelate** (pincer-like) in scavenging and predatory mites, whereas those of parasitic mites are usually modified to form stylets or bear teeth for piercing. The bases of the pedipalps are lateral to and just posterior to the bases of the chelicerae. The pedipalps may be leglike or chelate, or they may be reduced in size and serve as sense organs. Ventrally the fused coxae of the pedipalps extend forward to form the **hypostome**, which, together with a labrum, forms the **buccal cone** (Fig. 34-19). The chelicerae are located over the buccal cone; in the hard ticks the chelicerae are in sheaths, from which they can be protracted (Fig. 34-19, *B*). The dorsal part of the capitulum projects forward over the chelicerae as a **rostrum**, or **tectum**.

Mites typically have four pairs of legs, as in other arachnids, but only one to three pairs may be present.

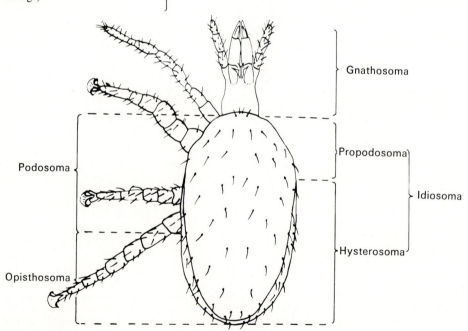

FIG. 34-17

Divisions of the body of a mite.

The podomeres of the legs may vary from two to seven, but six is the usual number: **coxa, trochanter, femur, patella, tibia,** and **tarsus.** The tarsi of most mites each bears a pair of claws. Spiracles may or may not be present, and their position and existence are important criteria for distinguishing the suborders. The anus is near the posterior end of the body, but the location of the gonopore is more variable, being found as far anteriad as the first legs in some forms. Some male mites have an intromittent organ, or **aedeagus.** The gonopore commonly opens through a more heavily sclerotized area, the **genital plate.** Other plates or shields are found on the idiosoma; their location and form are of taxonomic value.

The body and legs of most mites are well supplied with sensory setae, which may be simple and hairlike, plumose, or leaflike. They are mostly tactile, and movement of a seta stimulates nerve cells at its base.

One or two pairs of simple eyes are found laterally on the propodosoma in members of most suborders. Some mites have paired **Claparedé organs,** or **urstigmata,** between the coxae of the first and second legs. The urstigmata are believed to be humidity receptors. Ticks have a depression in the first tarsi called **Haller's organ,** which bears four different kinds of sensory setae.[4] Haller's organ is a humidity and olfactory receptor and is of considerable value in finding hosts.[17,31]

Internal structure
■ Hemocoel and circulatory system

The coelom arises embryologically as a cavity in the endomesoderm in animals that possess it. The spacious, fluid-filled coelom was of great adaptive value as a hydrostatic skeleton in the early coelomate animals, which were covered with a soft, flexible

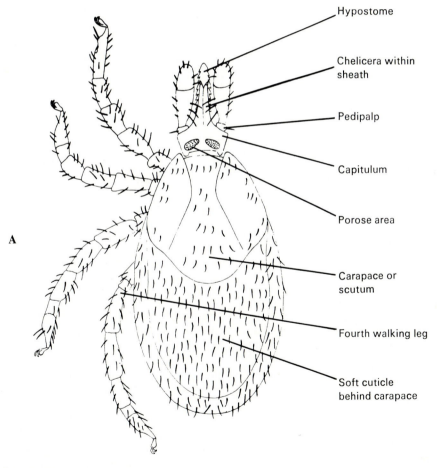

Hypostome

Chelicera within sheath

Pedipalp

Capitulum

Porose area

Carapace or scutum

Fourth walking leg

Soft cuticle behind carapace

A

FIG. 34-18

Female ixodid (hard) tick. **A,** Dorsal view.

From Snow, K.R. 1970. The arachnids: an introduction. Columbia University Press, New York.

Continued.

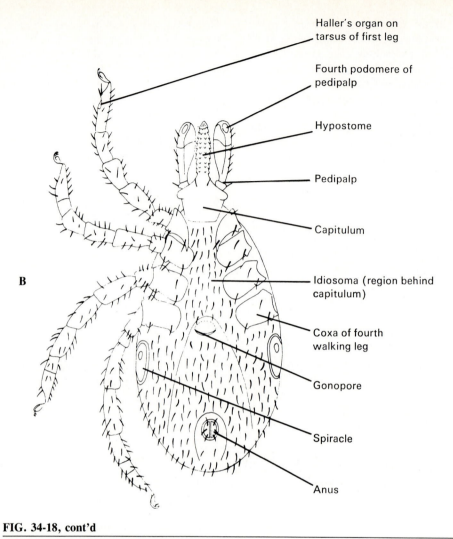

Haller's organ on
tarsus of first leg

Fourth podomere of
pedipalp

Hypostome

Pedipalp

Capitulum

Idiosoma (region behind
capitulum)

Coxa of fourth
walking leg

Gonopore

Spiracle

Anus

B

FIG. 34-18, cont'd

Female ixodid (hard) tick. **B,** Ventral view.

integument.[10] With the evolutionary development of a cuticular exoskeleton, this adaptive value of a coelom disappeared, as did the need for a separate, closed circulatory system. Hence the main body cavity of all arthropods is a blood-filled hemocoel. The hemocoel arises embryologically when the incipient coelomic spaces in the embryo open to join the space around the yolk, sometimes considered the remnant blastocoel. The internal organs, therefore, are neither covered by mesodermal peritoneum nor suspended by mesodermal mesenteries. However, the lack of need for a closed circulatory system did not obviate the necessity for circulation of blood (hemolymph). This is facilitated in all larger arthropods by a dorsal, tubular heart (Fig. 34-20). The heart receives hemolymph from the

surrounding pericardial sinus through pairs of lateral openings, the **ostia.** The ostia are one-way valves, so that when the heart contracts, the ostia close, and the hemolymph cannot flow back into the pericardial sinus. It flows anteriorly toward the head and, in many arthropods, into a system of arteries that distribute it through the body. From the arteries the hemolymph exits into a system of spaces, or sinuses, that constitute the hemocoel, where it bathes the tissues. Its route back to the heart may be directed by a system of partitions in the sinuses to ensure circulation of the hemolymph through the body; circulation also is aided by body movements. Formed elements in the blood are mostly **amebocytes.**

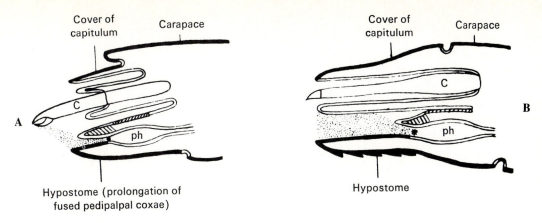

FIG. 34-19

Diagrammatic longitudinal section through the capitulum of acarines. **A,** Mite, **B,** Hard tick. The hypostome and the labrum *(crosshatched)* form the buccal cone, the anterior of which surrounds the preoral food canal *(shaded)*. The mouth, designated by an asterisk, leads into the muscular pharynx *(ph)*. The chelicerae *(C)* of the tick lie in a sheath and can be protracted and retracted.

From Snow, K.R. 1970. The arachnids: an introduction. Columbia University Press, New York.

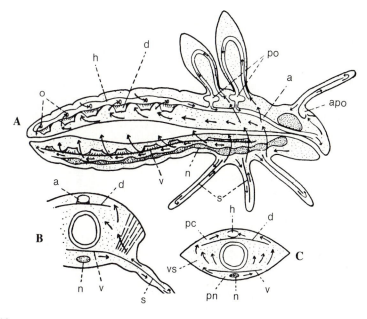

FIG. 34-20

Insect circulatory system showing route of blood circulation. Although the blood flows from the arteries into the open hemocoel, its circulation through the body is assured by partitions. **A,** Schematic of insect with fully developed circulatory system. **B,** Transverse section of **A. C,** Transverse section of abdomen. *Arrows,* Course of circulation; *a,* aorta; *apo,* accessory pulsatile organ of antenna; *d,* dorsal diaphragm with aliform muscles; *h,* heart; *n,* nerve cord; *o,* ostia; *pc,* pericardial sinus; *pn,* perineural sinus; *po,* mesothoracic and metathoracic pulsatile organs; *s,* septa dividing appendages; *v,* ventral diaphragm; *vs,* visceral sinus.

From Wigglesworth, V.B. 1972. Principles of insect physiology, ed. 7. Methuen & Co. Ltd., London.

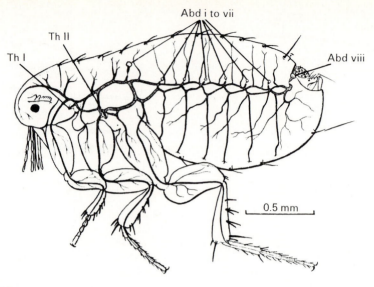

FIG. 34-21

Half of the tracheal system of the flea *Xenopsylla*. The main tracheae and locations of the spiracles are shown.

From Wigglesworth, V.B. 1972. Principles of insect physiology, ed. 7. Methuen & Co. Ltd., London.

FIG. 34-22

Diagram of trachea of an insect. The tracheae are virtually impermeable to liquids, but the finely branching tracheoles, leading into the tissues, are freely permeable, and their tips normally contain fluid. Oxygen primarily diffuses through the tracheolar walls, and elimination of carbon dioxide takes place more generally through the tracheal walls and body surface. Taenidia are chitinous bands that strengthen the tracheae.

From Hickman, C.P., Jr., L.S. Roberts, and F.M. Hickman. 1988. Integrated principles of zoology, ed. 8. The C.V. Mosby Co., St. Louis.

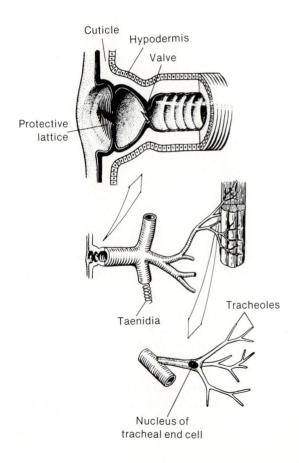

■ **Respiratory system**

Gas exchange often takes place directly through the body wall in very small arthropods, in which specialized respiratory organs, and even a heart, are absent. Larger Crustacea have **gills,** which are areas with greatly increased surface area covered by a thin cuticle, through which the hemolymph circulates. Most insects, as well as many Acari, have **tracheal systems** for breathing air (Fig. 34-21). This network of tubes lined with cuticle opens to the surface of the animal by a relatively small number of spiracles but ramifies throughout the body into a large number of very fine **tracheoles** (Fig. 34- 22). The cuticle of the tracheae, but not that of the tracheoles, is shed at ecdysis. Tracheal systems apparently evolved independently in the

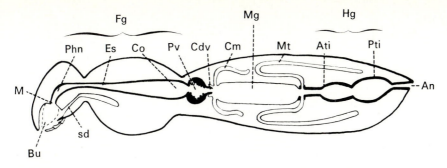

FIG. 34-23

Diagram of the digestive system of an insect. *An,* Anus; *Ati,* anterior intestine; *Bu,* buccal cavity; *Cdv,* cardia valve; *Cm,* cecum; *Co,* crop; *Es,* esophagus; *Fg,* foregut; *Hg,* hindgut; *M,* mouth; *Mg,* midgut; *Mt,* malpighian tubules; *Phn,* pharynx; *Pti,* posterior intestine; *Pv,* proventriculus; *sd,* salivary duct.

From Fox, R.M., and J.W. Fox. 1966. Introduction to comparative entomology. Reinhold Publishing Corp., New York.

insects and arachnids. Arachnid tracheal systems are thought to have been derived from another type of organ for breathing air, the book lungs. Book lungs occur in several arachnid orders but not in Acari. Ventilation of the tracheal system is accomplished by pressure of body muscles on the walls of elastic tracheae, on tracheal air sacs, or both.

■ **Nervous system**

The nervous system of all arthropods was derived from an ancestral condition similar to the nervous system possessed by annelids. This included a dorsal ganglionic mass **(brain),** with nerves to supply cephalic sense organs, connected to a **sub-esophageal ganglion** by commissures around the esophagus. The subesophageal ganglion and subsequent ganglia in each metamere were connected to each other by a double, ventral nerve cord. Later evolution involved considerable shortening of the nerve cord with fusion of the segmental ganglia to result in fewer, larger ganglia. The primitive first segmental ganglion (originally subesophageal) has moved around the esophagus and forms part of the brain **(tritocerebrum)** in modern arthropods (Fig. 34-4). It innervates the antennae of crustaceans and the chelicerae of chelicerates, as well as the labium and parts of the digestive tract. The remaining segmental ganglia display increased fusion correlated with increasing specialization of the arthropod groups. This tendency is shown at its extreme by the ticks and mites, which have only a single ganglionic mass, the brain, that surrounds the esophagus (see Fig. 34-24). In these animals the digestive tract, legs, genitalia, and musculature are innervated by nerves from the subesophageal part of the brain, which apparently represents all of the fused segmental ganglia.

■ **Digestive system**

The functional morphology of the digestive tract and its associated organs is diverse among the arthropods, but it is important to an understanding of host-parasite relationships.

A **foregut, midgut,** and **hindgut** can be recognized in most crustaceans. Part of the foregut may be expanded into an enlarged, **triturating stomach,** bearing calcareous ossicles, chitinous ridges, or denticles on its walls. The midgut is often enlarged to form a stomach, and it usually bears one or more pairs of **ceca.** One pair of ceca may be modified to form a **digestive gland,** or **hepatopancreas,** which produces digestive enzymes. Absorption is confined to the midgut and tubules of the digestive gland.

The digestive tract of insects also can be divided into a foregut, midgut, and hindgut. Distinct regions of the foregut are the **esophagus, crop,** and **proventriculus** (Fig. 34-23). The esophagus takes the form of a muscular **pharynx** in insects that suck fluid meals from their hosts. The crop functions as a storage chamber, and the form of the proventriculus is correlated with the type of food of the insect. It functions as a gizzard in insects that eat solid food, but in sucking insects the proventriculus is only a valve that regulates passage of food into the midgut. A pair of salivary glands usually lies beneath the midgut; the glands open into the buccal cavity by a common duct. Their secretions contain digestive enzymes and a variety of other substances in various insects, including anticoagulants in bloodsucking species. The midgut is the

551

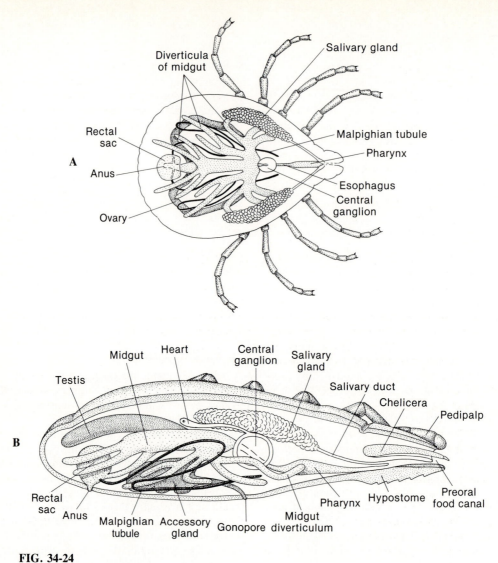

FIG. 34-24

Internal anatomy of hard tick. **A,** Dorsal view of female. **B,** Lateral view of male.

Drawings by William C. Ober.

principal site of digestive and absorptive function, as is usual in arthropods. The midgut of many insects secretes a thin, chitinous layer, the **peritrophic membrane,** which invests the food mass. Permeable to enzymes and the products of digestion, the peritrophic membrane is believed to protect the delicate epithelial lining of the midgut. Insects living on a liquid diet usually do not secrete a peritrophic membrane.

Gastric ceca are found near the anterior end of the midgut of most insects; the ceca increase the absorptive area of the midgut. The hindgut, divided into **intestine** and **rectum,** functions not only in the elimina-

tion of wastes but also is important in the regulation of water and ions in the insect's body.

In the Acari the mouth leads into the muscular, sucking **pharynx,** which lies partly in the buccal cone (Fig. 34-24). A slender esophagus proceeds posteriad through the brain to the stomach, or **ventriculus.** A large pair of **salivary glands** above the ventriculus and esophagus opens by means of ducts into the **salivarium** in the buccal cone over the labrum. The secretion of these glands contains anticoagulant and histolytic components in bloodsucking forms. The ventriculus has up to five pairs of ceca, which contain secretory and absorptive cells. The hindgut may be a short tube leading from the midgut to the anus, or an enlarged portion, the **rectal sac,** may precede the anus. In some of the Prostigmata and Metastigmata the ventriculus has lost its connection with the midgut and ends blindly. In some of these forms the indigestible food residues are removed from the body by a remarkable process called **schizeckenosy.**[23] The residues are stored in gut cells that detach from the epithelial lining and move into the posterodorsal gut lobes. When one of the lobes fills with waste-laden gut cells, the lobe breaks free from the ventriculus and is extruded through a split in the posterodorsal cuticle.

■ Excretory system

The principal organs of excretion in the Crustacea are pairs of **antennal** and **maxillary glands,** opening to the outside on or near the bases of the antennae or maxillae, respectively. Both pairs are often present in larvae; adults normally retain only one or the other. The principal nitrogenous excretory products are ammonia with some amines and small amounts of urea and uric acid. Considerable excretion of ammonia also takes place across the gills.

Analogous excretory organs found in some mites and other arachnids are the **coxal glands,** which open to the outside at the bases of one or more pairs of appendages. Regardless of whether coxal glands are present, most mites have a pair of **Malpighian tubules** (Fig. 34-24). These thin-walled tubules are closed at their inner ends and open into the midgut near its junction with the hindgut. Waste from the hemocoel is picked up by the tubule walls and excreted into their lumen as guanine, the principal nitrogenous excretory product. In those Prostigmata and Metastigmata whose ventriculus does not connect with the hindgut, an anteriorly-directed excretory canal connects with the hindgut, and guanine is excreted by these organs through the "anus" (uropore).

Almost all insects have Malpighian tubules, ranging in number from 4 to more than 100 (Fig. 34-23). As in the arachnids, the Malpighian tubules of insects open into the intestine near the midgut-hindgut junction. Uric acid is excreted, usually as an ammonium, potassium, or sodium salt. Water in the urine is reabsorbed by the proximal Malpighian tubules or by the rectal wall, and the sodium and potassium are resorbed as bicarbonates, leaving virtually insoluble, free uric acid to precipitate.[26,34] Thus the water and cations are recycled, a mechanism of great value in conserving water. Bloodsucking forms, however, produce copious amounts of fluid urine after feeding, thus ridding themselves of excess water. Some insects excrete substantial quantities of other nitrogenous compounds in addition to uric acid.

■ Reproductive systems

Most Crustacea are dioecious, and the gonopores open on a sternite or at the base of a trunk appendage. The male may have a penis, or appendages may be modified for copulation. Many crustaceans have nonflagellate, nonmobile sperm. In some groups the male places a packet of sperm (**spermatophore**) in the seminal receptacle or on the body surface of the female. Many Crustacea retain the fertilized eggs during embryonation, either in a brood chamber, attached to certain appendages, or within a sac formed during extrusion of the eggs.

Acarines are dioecious and males of different groups have a single testis, a pair of testes, or multiple testes. The **vas deferens** leads to the **ejaculatory duct.** One or more **accessory glands,** which may function as a seminal vesicle, may be associated with the vas deferens. Females have single or paired **ovaries,** which lead to a **uterus** in some and, in others, directly to the **vagina.** The vagina opens medially on the ventral side of the idiosoma or toward the posterior end. Some mites have a **copulatory bursa,** separate from the vagina. The bursa opens into a **seminal receptacle,** which is connected to the ovaries. In acarines with an aedeagus, the sperm may be introduced directly into the vagina or into the bursa. In males without an aedeagus, the chelicerae may bear a special sperm transfer organ (**spermatodactyl**), which picks up sperm from the male's gonopore and deposits them in one or both of a pair of special copulatory receptacles between the third and fourth coxae of the female. Some mites transfer sperm in a spermatophore, placed at the female gonopore by the male. In others the spermatophore is placed on a stalk secreted by the male, and it is then picked up by the female genital apparatus when she moves over it.[19]

Insects have a pair of **testes.** The **vasa deferentia** from the testes usually lead to a common, median **ejaculatory duct,** which opens to the outside by the aedeagus (Fig. 34-25, A). **Accessory glands** join the ejaculatory duct and in many cases provide the material comprising the spermatophores. The paired **ovaries** of the females are subdivided into **ovarioles** (Fig. 34-25, B). Each ovary usually has four to eight ovarioles, but some insects' ovaries have more than 200, and only one ovariole graces the ovary of the viviparous Diptera. The upper end of the ovariole produces the oocytes and nurse cells. The developing oocytes become larger by the accumulation of yolk produced by the nurse cells and surrounding follicular cells. The **common oviduct** enlarges into a vagina, which opens

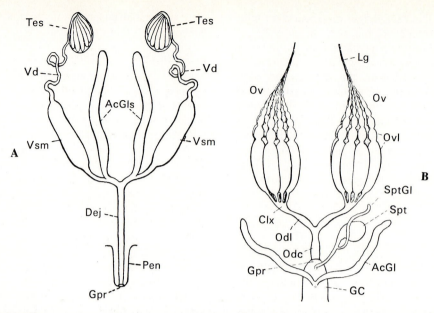

FIG. 34-25

General structure of insect reproductive organs. **A,** Male. *AcGls,* Accessory glands; *Dej,* ejaculatory duct; *Pen,* penis; *Gpr,* gonopore; *Tes,* testis; *Vd,* vasa deferentia; *Vsm,* seminal vesicle. **B,** Female. *Lg,* ovarian ligament; *Ov,* ovary; *Ovl,* ovariole; *Clx,* calyx; *Odl,* lateral oviduct; *Odc,* common oviduct; *Gpr,* gonopore; *GC,* genital chamber; *AcGl,* accessory gland; *Spt,* spermatheca; *SptGl,* spermathecal gland.

From Principles of insect morphology by R.E. Snodgrass, Copyright © 1935 by McGraw-Hill Book Co., Inc. Used with the permission of McGraw-Hill Book Co.

to the exterior behind the eighth or ninth abdominal sternite. The **seminal receptacle** connects to the oviduct or vagina by a slender **spermathecal duct.** Accessory glands (**colleterial glands**) also open into the common oviduct or vagina, and these may produce a substance that cements the eggs together or to the substrate when they are laid or produce material for an egg capsule (**ootheca**).

CLASSIFICATION OF ARTHROPODAN TAXA WITH SYMBIOTIC MEMBERS

This classification of the Crustacea relies heavily on Kabata,[14] Marcotte,[22] and Bowman and Abele,[8] except that, in accord with Boxshall and Lincoln,[9] we have not accepted Maxillopoda as a class. Classification of the Arachnida is according to Savory,[30] and diagnoses of the orders of pterygotes mainly follow Richards and Davies.[26]

Subphylum Crustacea
Head appendages consisting of two pairs of antennae, one pair of mandibles, and two pairs of maxillae; mostly aquatic; respiration usually with gills, sometimes through general body surface; head usually not clearly defined from trunk; cephalothorax usually with dorsal carapace; appendages, except first antennae (antennules), primitively biramous; sexes usually separate; development primitively with nauplius stage.

Class Ostracoda
Body entirely enclosed in bivalve carapace; body unsegmented or indistinctly segmented; no more than two pairs of trunk appendages. (Only a few species recorded as gill parasites of marine teleosts and elasmobranchs.)

Subclass Myodocopa

FAMILY
Cypridinidae

Class Copepoda
Typically with elongate, segmented body consisting of head, thorax, and abdomen; thorax with seven somites, of which first and sometimes second fused with head to form cephalothorax; thoracic appendages biramous except maxillipeds and often fifth swimming legs uniramous; no appendages on abdomen except pair of rami on telson; no carapace; compound eyes absent, but median nauplius eye often present; gonopores on "genital segment," usually considered last somite of thorax. (Parasitic forms may not fit some or much of foregoing diagnosis and may be highly modified as adults and sometimes as juveniles.)

Order Calanoida
(No symbiotic members, but included because of ecological importance.) Antennules very long with 16 to 26 articles; buccal cavity open; antennae, mandibles, and maxillules biramous; mandibles gnathostomous; maxillae and maxilliped uniramous; first thoracic legs biramous, multiarticular, with plumose setae for swimming;

last thoracic leg uniramous, modified, or missing; heart present in many; large order of important marine and freshwater planktonic organisms, never symbiotic.

Order Monstrilloida

Nauplii free swimming, then becoming sac-like endoparasites of marine polychaetes, prosobranch gastropods, and, occasionally, echinoderms; adults planktonic, without antennae, mouthparts, or functional gut; adult thoracic legs biramous for swimming.

Order Siphonostomatoida

Adult segmentation often reduced or lost; antennules reduced or elongate and multiarticulate; antennules may end in single massive claw for attachment to host; labrum and labium prolonged into siphon or tube, sometimes with some fusion; mandibles enclosed in buccal siphon, uniramous; maxillules ancestrally biramous, modified or reduced in derived forms; maxillae subchelate or brachiform (like human arm) for attachment to host; maxilliped subchelate or absent, sometimes absent in female only; adult thoracic limbs may be normal swimming appendages in some, in majority variously modified and reduced; adults ectoparasitic or endoparasitic on freshwater and marine fish and on various invertebrates.

FAMILIES (representative)
Caligidae, Cecropidae, Dichelesthiidae, Lernaeopodidae, Pandaridae, Pennellidae, Sphyriidae, Xenocoelomatidae

Order Cyclopoida

Antennules short with 10 to 16 articles; buccal cavity open; antennae uniramous; mandibles and maxillules usually biramous; mandibles gnathostomous; free-living planktonic and benthic; commensal and ectoparasitic.

FAMILIES
Ascidicolidae, Enterocolidae, Lernaeidae, Notodelphyidae

Order Poecilostomatoida

Adult segmentation often lost with copepodid metamorphosis; antennules often insignificant in size; buccal cavity slit-like; antennae often end in many small claws for attaching to host; mandibles with falcate (falcatus = sickle-shaped) gnathobase, rami missing; maxillules much reduced; maxillae reduced with denticulate inward pointing claw or slender, armed grasping claws; maxillipeds subchelate in males, often missing in females; adult thoracic limbs variously modified and reduced; adults parasitic on mostly marine invertebrates and fishes.

FAMILIES (representative)
Bomolochidae, Chondracanthidae, Clausiidae, Ergasilidae, Lichomolgidae, Philichthyidae, Sarcotacidae, Tuccidae

Order Harpacticoida

Antennules short with fewer than 10 articles; buccal cavity open; antennae and mandibles biramous; mandibles gnathostomous; maxillules usually biramous; various degrees of fusion, reduction, and loss of rami in cephalic and thoracic appendages; heart absent; mostly free living, benthic, epibenthic, planktonic.

Class Tantulocarida

No recognizable cephalic appendages; solid median cephalic stylet; six free thoracic somites, each with pair of appendages, anterior five biramous; six abdominal somites; anterior five thoracic appendages with well-developed protopod and large endite arising from base of protopod; recently described[9] class of minute, copepod-like ectopar-

asites of other deep-sea benthic crustaceans. Examples: *Basipodella, Deoterthron.*

Class Branchiura

Body with head, thorax, and abdomen; head with flattened, bilobed, cephalic fold incompletely fused to first thoracic somite; thorax with four pairs of appendages, biramous, and with proximal extension of exopod of first and second legs; abdomen without appendages, unsegmented, bilobed; eyes compound; both pairs of antennae reduced; claws on antennules; maxillules often forming pair of suctorial discs; maxillae uniramous; gonopore at base of fourth leg; ectoparasites of marine and freshwater fishes, occasionally of amphibians.

Order Argulidea

FAMILIES
Argulidae, Dipteropeltidae

Class Cirripedia

Sessile or parasitic as adults, head reduced and abdomen rudimentary; paired, compound eyes absent; body segmentation indistinct; usually hermaphroditic; in nonsymbiotic and epizoic forms, carapace becomes mantle, which secretes calcareous plates; antennules become organs of attachment; antennae disappear; young hatches as nauplius and develops to bivalved cypris larva; all marine.

Order Thoracica

With six pairs thoracic appendages, alimentary canal; usually nonsymbiotic, although some epizoic and commensal on whales, fishs, sea turtles, crabs. Examples: *Chelonibia, Conchoderma, Coronula, Xenobalanus.*

Order Acrothoracica

Bores into mollusc shells or coral; females usually with four pairs thoracic appendages, gut present, no abdomen; dioecious; males very small, without gut and appendages except antennules; parasitic on outside of mantle of female.

Order Ascothoracica

With segmented or unsegmented abdomen; usually six pairs thoracic appendages; gut present; parasitic on echinoderms and soft corals. Example: *Trypetesa.*

Order Rhizocephala

Adults with no segmentation, gut, or appendages; with root-like absorptive processes through tissue of host; common parasites of decapod crustaceans.

FAMILIES
Lernaeodiscidae, Peltogastridae, Sacculinidae

Class Malacostraca

Distinctly segmented bodies, typically with eight somites in the thorax and six somites plus the telson in the abdomen (except seven in Nebaliacea); all segments with appendages; antennules often biramous; first one to three thoracic appendages often maxillipeds; primitively with carapace covering head and part of all of thorax, but lost in some orders; gills usually thoracic epipods; female gonopores on sixth thoracic segment; male gonopores on eighth thoracic segment; largest subclass, marine, freshwater, few terrestrial; many free living, but parasitic members relatively few, found in only 3 of the 10 to 12 extant orders commonly recognized.

SUPERORDER PERACARIDA
Without carapace or with carapace leaving at least four free thoracic somites; first thoracic somite fused with head; brood pouch in fe-

male (typically formed from modified thoracic epipods, the oostegites); several small, marine orders, and the two large orders with parasitic members.

Order Amphipoda

No carapace; ventral brood pouch of oostegites; antennules often biramous; eyes usually sessile; gills on thoracic coxae, first thoracic limbs maxillipeds, second and third pairs usually prehensile (gnathopods); usually bilaterally compressed body form; marine, freshwater, and terrestrial; free living and symbiotic.

Suborder Hyperiidea

Head and eyes very large; only one thoracic somite fused with head; pelagic or symbiotic in medusae or tunicates.

FAMILIES

Hyperiidae, Phronimidae

Suborder Caprellidea

Two thoracic somites fused with head; abdomen much reduced, with vestigial appendages; so-called skeleton shrimp and whale lice.

FAMILIES

Caprellidae, Cyamidae

Order Isopoda

No carapace; ventral brood pouch of oostegites; antennules usually uniramous, sometimes vestigial; eyes sessile; gills on abdominal appendages; second and third appendages usually not prehensile; body usually dorsoventrally flattened.

Suborder Gnathiidea

Thorax much wider than abdomen; first and seventh thoracic somites reduced, seventh without appendages; larvae parasitic on marine fishes.

FAMILY

Gnathiidae

Suborder Flabellifera

Flattened body, with ventral coxal plates sometimes joined to body; telson fused with next abdominal somite, and other abdominal somites may be fused; uropods flattened, forming tail fan; marine, free living, and ectoparasitic on fishes.

FAMILIES WITH PARASITIC MEMBERS

Aegidae, Crallanidae, Cymothoidae

Suborder Epicaridea

Females greatly modified for parasitism; somites and appendages fused, reduced, or absent; mouthparts modified for sucking, mandible for piercing, and maxillae reduced or absent; males small but less modified; marine parasites of Crustacea.

FAMILIES

Bopyridae, Cryptoniscidae, Dajidae, Entoniscidae, Phryxidae

SUPERORDER EUCARIDA

All thoracic segments fused with and covered by carapace; no oostegites or brood pouch; eyes on stalks; usually with zoea larval stage.

Order Decapoda

First three pairs thoracic appendages modified to maxillipeds (therefore appendages on remaining five thoracic somites equal ten); includes crabs, lobsters, shrimp.

Suborder Pleocyemata

Eggs carried by female and brooded on pleopods, hatch as zoeae.

INFRAORDER BRACHYURA

Carapace broad; abdomen reduced and tightly flexed beneath cephalothorax; first legs in form of heavy chelipeds; typical crabs.

FAMILIES WITH SYMBIOTIC MEMBERS

Parthenopidae, Pinnotheridae

Subphylum Uniramia

All appendages uniramous; head appendages consisting of one pair of antennae, one pair of mandibles, and one or two pairs of maxillae.

Class Insecta

Body with distinct head, thorax, and abdomen; one pair of antennae; thorax of three somites; abdomen with variable number, usually 11 somites; thorax usually with two pairs of wings (sometimes one pair or none) and three pairs of jointed legs; separate sexes; usually oviparous; gradual or abrupt metamorphosis, few with direct development.

Subclass Apterygota

Primitive wingless insects; development direct or slight metamorphosis.

Orders

Collembola, Diplura, Protura, Thysanura

Subclass Pterygota

Insects with wings (some secondarily wingless); all metamorphic; includes 97% of all insects. (Although members of all orders serve as hosts, only those with some medical or veterinary importance follow, in addition to orders that have appreciable numbers of symbiotic members.)

Order Dermaptera

Forewings represented by small tegmina; hindwings large, membranous, and complexly folded; mouthparts for biting, ligula bilobed; body terminated by forceps; earwigs; few ectoparasites of mammals *(Arixenia, Hemimerus);* some intermediate hosts of nematodes.

Order Dictyoptera

Antennae nearly always filiform with many segments; mouthparts for biting; legs similar to each other or forelegs raptorial, tarsi with five segments; forewings more or less thickened into tegmina with marginal costal vein; cerci many segmented, ovipositor reduced and concealed; eggs contained in an ootheca; cockroaches, mantids; none symbiotic, but some implicated in mechanical transmission of human pathogens; some are intermediate hosts of Acanthocephala. Examples: *Blatta, Blatella, Periplaneta, Supella.*

Order Mallophaga

Wingless; mouthparts modified biting type; prothorax free, mesothorax and metathorax often imperfectly separated; tarsi of one or two segments, with one or two claws; cerci absent; metamorphosis slight; ectoparasitic in all stages on birds, less frequently mammals; biting lice. Examples: *Menacanthus, Menopon, Piagetiella, Trichodectes.*

Order Anoplura

Wingless; mouthparts modified for sucking and piercing; retracted when not in use; thoracic segments fused; tarsi unisegmented, claws single; cerci absent; metamorphosis slight; ectoparasitic in all stages on mammals; sucking lice. Examples: *Haematopinus, Pediculus, Phthirus.*

Order Hemiptera
Wings variably developed with reduced or greatly reduced venation, forewings often more or less corneous, wingless forms frequent; mouthparts for piercing and sucking with mandibles and maxillae stylet-like and lying in the projecting grooved labium, palps never evident; metamorphosis gradual with an incipient pupal instar sometimes present; true bugs, aphids, scale insects, etc.; many free living, with some ectoparasites of birds and mammals. Examples: *Cimex, Leptocimex, Rhodnius, Triatoma.*

Order Neuroptera
Small to large, soft-bodied insects with two pairs of membranous wings without anal lobes, venation generally with many accessory branches and numerous costal veinlets; mouthparts for biting; antennae well developed; cerci absent; complete metamorphosis; campodeiform larvae with biting or suctorial mouthparts; alder flies, lacewings, ant lions, etc.; few parasites of freshwater sponges and of spiders' egg cocoons.

FAMILIES
Mantispidae, Sisyridae *(Climacia, Sisyra)*

Order Coleoptera
Minute to large insects whose forewings are modified to form elytra and abut down line of dorsum; hindwings membranous, folded beneath elytra, or absent; prothorax large; mouthparts for biting; metamorphosis complete, larvae of diverse types but never typical polypod; beetles; largest order of animals (more than 330,000 species), but 1.5% of which are protelean parasites of insects, few ectosymbionts of mammals.

FAMILIES
Leptinidae, Meloidae (some), Platypsyllidae, Rhipiphoridae, Staphylinidae (some)

Order Strepsiptera
Minute; males with branched antennae and degenerate biting mouthparts; forewings modified into small clublike processes; hindwings very large, plicately folded; females almost always extensively modified as internal parasites of other insects; larviform and devoid of wings, legs, eyes, and antennae; all protelean parasites of insects. Examples: *Corioxenos, Elenchus, Eoxenos, Stylops.*

Order Siphonaptera
Very small; wingless; laterally compressed body; mouthparts for piercing and sucking; complete metamorphosis with vermiform larvae; pupation in silk cocoons; adults all parasitic on warm-blooded animals; fleas. Examples: *Pulex, Ctenocephalides, Xenopsylla, Tunga.*

Order Diptera
Moderate sized to very small; single pair of membranous wings (forewings), hindwings modified into halteres; mouthparts for sucking or for piercing as well and usually forming a proboscis; complete metamorphosis with vermiform larvae; flies and mosquitoes; many species of invertebrates and vertebrate protelean parasites, vertebrate and insect ectoparasites. Examples: *Aedes, Anopheles, Bombylius, Chrysops, Conops, Culex, Glossina, Hippobosca, Melophagus, Phlebotomus, Simulium, Stomoxys, Stylogaster, Tabanus.*

Order Lepidoptera
Small to very large insects clothed with scales; mouthparts with galeae usually modified into a spirally coiled suctorial proboscis, mandibles rarely present; complete metamorphosis with larvae phytophagous, polypodous; butterflies and moths; large order with mostly free-living members, few insect protelean parasites and mammal ectoparasites. Examples: *Bradypodicola, Calpe, Cyclotorna, Fulgoraecia.*

Order Hymenoptera
Minute to moderate sized; membranous wings, hindwings smaller and connected with forewings by hooklets, venation specialized by reduction; mouthparts for biting and licking; abdomen with first segment fused with thorax; sawing or piercing ovipositor present; complete metamorphosis with usually polypodous or apodous larvae; sawflies, ants, bees, wasps, ichneumon flies, etc.; enormous insect order, about half of which are protelean parasites, mainly of other insects.

SUPERFAMILIES
Bethyloidea, Chalcidoidea (many), Cynipoidea (some), Evanioidea, Ichneumonoidea, Orussoidea, Proctorupoidea (Serphoidea), Trigonaloidea, Vespoidea (some)

Subphylum Chelicerata
Mostly terrestrial; respiration by gills, book lungs, tracheae, or through general body surface; first pair of appendages modified to form chelicerae; pair of pedipalps and usually four pairs of legs in adults; no antennae; tagmatization of prosoma (cephalothorax) and opisthosoma (abdomen), usually unsegmented.

Class Arachnida
Adult body fundamentally composed of 18 somites, divisible into 6-unit prosoma and 12-unit opisthosoma, but segmentation often obscured in either or both of these tagmata; eyes, if present, simple (ocelli), not more than 12; chelicerae of two or three podomeres, chelate or unchelate; pedipalps of six podomeres, may be chelate or leglike, often with gnathobases; respiration through general body surface or by book lungs or tracheae (or both); sexes separate, with orifices on lower side of second opisthosomatic somite.

Orders
Acari, Amblypygi, Araneae, Opiliones, Palpigradi, Pseudoscorpiones, Ricinulei, Schizomida, Scorpiones, Solifugae, Uropygi

Order Pseudoscorpiones
Prosoma undivided; opisthosoma with 12 distinguishable somites; chelicerae of two articles, chelate; pedipalps large, six articles, chelate; no pedicel; no telson; several pseudoscorpions symbiotic on mammals, prey on ectoparasites (lice, mites).[11] Examples: *Lasiochernes, Megachernes, Chiridiochernes.*

Order Acari
Highly specialized arachnids, in which modifications of segmentation divide body into proterosoma and hysterosoma, usually distinguishable as boundary between second and third pairs of legs; segments of mouth and its appendages borne on gnathosoma (capitulum), more or less sharply set off from rest of body (idiosoma); typically four pairs of legs but sometimes three, two, or one pair; podomeres of legs often six but varying from two to seven; position of respiratory and genital openings variable; includes also free-living suborders Notostigmata, Tetrastigmata.

Suborder Mesostigmata

Several sclerotized plates on dorsal and ventral surfaces, single pair of spiracular openings between second and fourth coxae; large group, many free living. Parasitic examples: *Dermanyssus, Ornithonyssus, Sternostoma.*

Suborder Metastigmata

Large acarines (ticks); hypostome with recurved teeth, used as holdfast organ; sensory organ (Haller's organ) on tarsus of first leg with olfactory and hygroreceptor setae; single pair of spiracular openings close to coxae of fourth legs except in larvae; all parasitic. Examples: *Amblyomma, Argas, Boophilus, Dermacentor, Ixodes, Ornithodoros, Otobius, Rhipicephalus.*

Suborder Prostigmata

Spiracular openings, when present, paired and located either between the chelicerae or on dorsum of anterior portion of hysterosoma; usually weakly sclerotized; chelicerae vary from strongly chelate to reduced; pedipalps simple, fang-like, or clawed; terrestrial and aquatic free-living, phytophagous, and parasitic forms. Examples: *Demodex, Trombicula.*

Suborder Cryptostigmata

Oribatid or beetle mites (so called because of superficial resemblance to beetles); spiracles absent, although some have trachea associated with paired dorsal pseudostigmata and with bases of first and third legs; free living, but some are vectors of tapeworms *(Galumna, Oppia).*

Suborder Astigmata

Mostly slow moving and weakly sclerotized; no spiracles, respire through body surface; free-living and parasitic forms. Examples: *Megninia, Otodectes, Sarcoptes.*

REFERENCES

1. Amar, R. 1948. Un organe endocrine chez *Idotea* (Crustacea Isopoda). C.R. Acad. Sci. Paris 227:301-303.
2. Andersen, S.O. 1976. Cuticular enzymes and sclerotization in insects. In Hepburn, H.R., editor. The insect integument. Elsevier Scientific Publishing Co., Amsterdam, pp. 121-144.
3. Anderson, D.T. 1973. Embryology and phylogeny in annelids and arthropods. Pergamon Press, Oxford, Eng.
4. Arthur, D.R. 1956. The morphology of the British Prostriata with particular reference to *Ixodes hexagonus* Leach. III. Parasitology 46:261-307.
5. Askew, R.R. 1971. Parasitic insects. American Elsevier Publishing Co., Inc., New York.
6. Barnes, H., and J.J. Gonor. 1958. Neurosecretory cells in the cirripede, *Pollicipes polymerus.* J. Mar. Res. 17:81-102.
7. Boudreaux, H.B. 1979. Arthropod phylogeny with special reference to insects. John Wiley & Sons, Inc., New York.
8. Bowman, T.E., and L.G. Abele. 1982. Classification of the recent Crustacea. In Abele, L.G., editor. The biology of Crustacea, vol. 1. Systematics, the fossil record, and biogeography. Academic Press, Inc., New York, pp. 1-27.
9. Boxshall, G.A., and R.J. Lincoln. 1983. Tantulocarida, a new class of Crustacea ectoparasitic on other crustaceans. J. Crust. Biol. 3:1-16.
10. Clark, R.B. 1964. Dynamics in metazoan evolution. The origin of the coelom and segments. Clarendon Press, Oxford, Eng.
11. Durden, L.A. 1987. Predator-prey interactions between ectoparasites. Parasitol. Today 3:306-308.
12. Filshie, B.K. 1976. The structure and deposition of the epicuticle of the adult female cattle tick *(Boophilus microplus)*. In Hepburn, H.R., editor. The insect integument, Elsevier Scientific Publishing Co., Amsterdam, pp. 193-206.
13. Hillerton, J.E. 1979. Changes in the mechanical properties of the extensible cuticle of *Rhodnius* through the fifth larval instar. J. Insect Physiol. 25:73-77.
14. Kabata, Z. 1979. Parasitic Copepoda of British fishes. Ray Society, London.
15. Karlson, P., and E.C. Sekeris. 1976. Control of tyrosine metabolism and cuticle sclerotization by ecdysone. In Hepburn, H.R., editor. The insect integument. Elsevier Scientific Publishing Co., Amsterdam, pp. 145-156.
16. Knowles, F.G.W., and D.B. Carlisle. 1956. Endocrine control in the Crustacea. Biol. Rev. 31:396-473.
17. Krantz, G.W. 1971. A manual of acarology. Oregon State University Bookstores, Inc., Corvallis.
18. Krishnan, G. 1951. Phenolic tanning and pigmentation of the cuticle in *Carcinus meanas*. Q. J. Microsc. Soc. 92:333-344.
19. Lipovsky, L.J., G.D. Byers, and E.H. Kardos. 1957. Spermatophores—the mode of insemination of chiggers (Acarina: Trombiculidae). J. Parasitol. 43:256-262.
20. Manton, S.M. 1973. Arthropod phylogeny—a modern synthesis. J. Zool. 171:111-130.
21. Manton, S.M. 1977. The Arthropoda: habits, functional morphology, and evolution. Clarendon Press, Oxford, Eng.
22. Marcotte, B.M. 1982. Evolution within the Crustacea, Part 2: Copepoda. In Abele, L.G., editor. The biology of Crustacea, vol. 1. Systematics, the fossil record, and biogeography. Academic Press, Inc., New York, pp. 185-197.
23. Mitchell, R.D., and M. Nadchatram. 1969. Schizeckenosy: the substitute for defectation in chigger mites. J. Nat. Hist. 3:121-124.
24. Paulus, H.F. 1979. Eye structure and the monophyly of the Arthropoda. In Gupta, A.P., editor. Arthropod phylogeny. Van Nostrand Reinhold Co., New York, pp. 299-383.
25. Richards, A.G. 1978. The chemistry of insect cuticle. In Rockstein, R., editor. Biochemistry of insects. Academic Press, Inc., New York, pp. 205-232.
26. Richards, O.W., and R.G. Davies. 1978. Imms' outlines of entomology, ed. 6, Chapman & Hall Ltd., London.
27. Riddiford, L.M., and J.W. Truman. 1978. Biochemistry of insect hormones and insect growth regulators. In Rockstein, R., editor. Biochemistry of insects. Academic Press, Inc., New York, pp. 307-357.
28. Robertson, J.D. 1960. Ionic regulation in the crab, *Carcinus meanas* (L.) in relation to the moulting cycle. Comp. Biochem. Physiol. 1:183-212.
29. Sanders, H.L. 1957. The Cephalocarida and crustacean phylogeny. Syst. Zool. 6:112-128, 148.
30. Savory, T. 1977. Arachnida, ed. 2. Academic Press, Inc., New York.
31. Snow, K.R. 1970. The arachnids: an introduction. Columbia University Press, New York.
32. Travis, D.R. 1955. The moulting cycle of the spiny lobster, *Panulirus argus*, Latreille. II. Preecdysial histological and histochemical changes in the hepatopancreas and integumental tissues. Biol. Bull. 108:88-112.
33. Vincent, J.F.V., and J.E. Hillerton. 1979. The tanning of in-

sect cuticle—a critical review and a revised mechanism. J. Insect Physiol. 25:653-658.

34. Wigglesworwth, V.B. 1972. Principles of insect physiology, ed. 7. Chapman & Hall Ltd., London.

35. Wigglesworth, V.B. 1976. The distribution of lipid in the cuticle of *Rhodnius*. In Hepburn, H.R., editor. The insect integument. Elsevier Scientific Publishing Co., Amsterdam, pp. 89-106.

36. Zacharuk, R.Y. 1976. Structural changes of the cuticle associated with moulting. In Hepburn, H.R., editor. The insect integument. Elsevier Scientific Publishing Co., Amsterdam, pp. 299-321.

SUGGESTED READINGS

Barnes, R.D. 1987. Invertebrate zoology, ed. 5. Saunders College Publishing, Philadelphia. (A good, general invertebrate text with chapters on each of the arthropod groups.)

Barrington, E.J.W. 1979. Invertebrate structure and function, ed. 2. John Wiley & Sons, Inc., New York. (An excellent treatment from the functional standpoint.)

Clarke, K.U. 1973. The biology of the Arthropoda. American Elsevier Publishing Co., Inc., New York.

Fox, R.M., and J.W. Fox. 1964. Introduction to comparative entomology. Reinhold Publishing Corp., New York. (Covers chelicerates and insects but not Crustacea.)

Harwood, R.F., and M.T. James. 1979. Entomology in human and animal health, ed. 7. Macmillan Publishing Co., Inc., New York. (Up to date. One of the best texts available in medical entomology.)

Marshall, A.J., and W.D. Williams, editors. 1972. Textbook of zoology. Invertebrates. American Elsevier Publishing Co., Inc., New York. (Especially good on structure and taxonomy, each chapter written by an expert on the group.)

Snodgrass, R.E. 1935. Principles of insect morphology. McGraw-Hill Book Co., New York. (This and the following are classics that remain valuable references on arthropod structure.)

Snodgrass, R.E. 1965. A textbook of arthropod anatomy. (Facsimile of the 1952 edition.) Hafner Publishing Co., Inc., New York.

Strickland, G.T. 1984. Hunter's tropical medicine, ed. 6. W.B. Saunders Co., Philadelphia.

U.S. Department of Health, Education, and Welfare. 1960 (1979 revision). Introduction to arthropods of public health importance. HEW Pub. No. (CDC) 79-8139. U.S. Government Printing Office, Washington, D.C. (A short, concise introduction to the subject, with a key to some common classes and orders of public health importance.)

CHAPTER 35

PARASITIC CRUSTACEANS

A most interesting array of morphological adaptations for symbiosis can be found among the crustaceans. In addition to the academic interest they hold, numerous parasitic crustaceans are of substantial economic importance. Despite this, they are often neglected in parasitology courses, as well as in courses in invertebrate zoology.

Some crustacean parasites have been known since antiquity, although the fact that they were crustaceans, or even arthropods, was not recognized until the early nineteenth century. Aristotle and Pliny recorded the affliction of tunny and swordfish by large parasites we would now recognize as lernaeocerid copepods. In 1554 Rondelet[54] figured a tunny with one of the copepods in place near the pectoral fin. In 1746 Linnaeus first established the genus *Lernaea,*[37] and, in his 1758 edition of *Systema Naturae,* he called the species (from European carp) *Lernaea cyprinacea.*[38] Various other highly modified copepods were described in the latter half of the eighteenth century and early nineteenth, but they had so few obvious arthropod features that they were variously classified as worms, gastropod molluscs, cephalopod molluscs, and annelids. Finally, Oken[43] (1815-1816) associated these animals with other parasitic copepods that could be recognized as such. Based on Surriray's important observation that their young, when hatched, resemble those of *Cyclops,* deBlainville[5] (1822) firmly established these animals as crustaceans. Copepoda is not the only crustacean class with members so modified for parasitism as to be superficially unrecognizable as arthropods, and some of these will be discussed as well.

The status of higher taxa in the Crustacea continues in a state of flux, even the status of the taxon "Crustacea" itself. In the latest, most authoritative classification of recent crustaceans, Bowman and Abele[7] begin their presentation with "Phylum, Subphylum, or Superclass Crustacea, Pennant 1777." Since 1981 two entirely new classes of crustaceans have been described. Dahl[13,14] proposed the subclass Maxillopoda to contain the Mystacocarida, Cirripedia, Copepoda, and Branchiura, which had themselves been traditionally considered subclasses. Bowman and Abele concurred in this view, using Maxillopoda as a class and the other groups as subclasses. However, other workers have strongly disagreed with the Maxillopoda concept.[8] It seems to us prudent not to adopt the Maxillopoda as a class; therefore Mystacocarida, Cirripedia, Copepoda, and Branchiura become classes of the subphylum Crustacea (or phylum or superclass, if you prefer).

CLASS COPEPODA

The copepods constitute one of the largest crustacean classes, second only to the Malacostraca, and they are extremely important both as free-living organisms and as parasites. Their evolutionary versatility is displayed by the fact that several groups of copepods have been able to exploit symbiotic niches and by the spectrum of adaptations to symbiosis that they show, ranging from the slight to the extreme. Although lernaeocerids are so bizarre that eighteenth century biologists did not recognize them as arthropods, many parasitic and commensal copepods are comparatively much less highly modified. In fact one can arrange examples of the various groups in an arbitrary series to demonstrate the progression from little to very high specialization.[30]

We shall cite but a few examples to illustrate the trends in adaptation to parasitism in copepods. Some of these trends are (1) reduction in locomotor appendages; (2) development of adaptations for adhesion, both by modification of appendages and by development of new structures; (3) increase in size and change in body proportions, caused by much greater growth of genital or reproductive regions; (4) fusion of body somites and loss of external evidence of segmentation; (5) reduction of sense organs; and (6) reduction in numbers of instars that are free living, both by passing more stages before hatching and by larval instars becoming parasitic. "Typical" or primitive copepod development can be regarded as gradual metamorphosis with a series of **copepodid** instars suc-

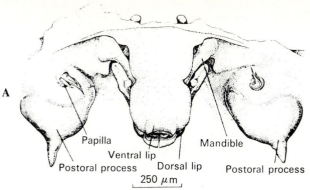

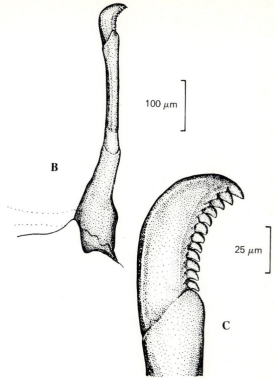

FIG. 35-1

Oral region of *Caligus curtus* (a siphonostome), which parasitizes a variety of marine fishes. **A,** The base of the mandible can be seen as it extends into the tube formed by the dorsal and ventral lips. **B** and **C,** The mandible is a long, flat blade with teeth at its distal tip.

From Parker, R.R., Z. Kabata, L. Margolis, and M.D. Dean. 1968. J. Fish. Res. Bd. Can. 25:1960.)

ceeding the nauplii. Copepodid larvae bear considerable similarity to the adults except in dramatically metamorphic families like Lernaeopodidae and Pennellidae.

The higher taxonomy of the Copepoda has also undergone recent revision. Kabata[31] resurrected and modernized an older concept, arranging the copepods on the basis of their mouth structure: gnathostome, poecilostome, and siphonostome. Gnathostomous mandibles are fairly short, broad, biting structures with teeth at their ends, and the buccal cavities are large and widely open. This is apparently the ancestral condition and is possessed by several copepod orders. Poecilostome mouths are rather similar, except that they are somewhat slitlike and have falcate (sickle-shaped) mandibles (Fig. 34-14). The siphonostome condition is characterized by a more or less elongate, conical, siphon-like mouth formed by the labrum and labium (Fig. 35-1, *A*), and the mandibles are stylet-like and enclosed within the siphon (Fig. 35-1, *B*). The possession of poecilostome and of siphonostome mouths forms the basis for recognition of the orders Poecilostomatoida and Siphonostomatoida, respectively, whereas copepods in several orders have gnathostomous mouths.

Order Cyclopoida

The order Cyclopoida is a large group of copepods, most species of which are free living. Free-living cyclopoids occupy important niches as primary consumers in many aquatic habitats, particularly freshwater. Several familes of cyclopoids are parasites on invertebrates, and the Lernaeidae is a highly specialized group of fish parasites.

■ Family Lernaeidae

Lernaeidae is a relatively small family that parasitizes freshwater teleosts. Often quite large and conspicuous, some species, especially *Lernaea cyprinacea,* are serious pests of economically important fishes. Therefore they are among the best known parasitic copepods. The genus *Lernaea* was established in 1746 by Linnaeus, and its ontogeny was clarified by Grabda[20] in 1963. *Lernaea cyprinacea* can infect a variety of fish hosts and even frog tadpoles.[57] The anterior of the parasite is embedded in the host's flesh and is anchored there by large processes that arise from the parasite's cephalothorax and thorax, hence the common name "anchor worm." They cause damage to the scales, skin, and underlying muscle tissue. There may be considerable inflammation, ulceration, and secondary bacterial and fungous infection. If the fish is small relative to the parasite, it can easily be killed by infection with several individuals. A fully developed *L. cyprinacea* may be more than 12 mm long. Epizootics of this pest occur in wild fish popula-

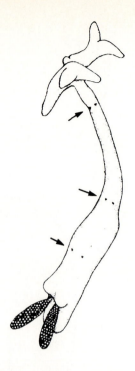

FIG. 35-2

Lernaea cyprinacea, the "anchor worm," is a serious pest of a variety of fishes, including several of economic importance. The anterior holdfast ("horns") is embedded in the host's flesh, and the posterior part of the body projects to the exterior. The swimming legs *(arrows)* do not participate in the rapid, final growth of the adult female (up to 16 mm long) and so remain proportionately very tiny.

From Kabata, Z. 1970. Diseases of fishes, book 1: Crustacea as enemies of fishes. (Snieszko, S.F., and H.R. Axelrod, editors) T.F.H. Publications, Inc., Neptune City, N.J.

tions, and it is a serious threat wherever fish are raised in hatcheries.[48]

Lernaeids are among the most highly specialized copepods. Once the sexually mature female is fertilized, she embeds her anterior end beneath a scale, near a fin base or in the buccal cavity. At that point the parasite is less than 1.5 mm long and is superficially quite similar to *Cyclops* or other unspecialized cyclopoids. The female begins to grow rapidly, reaching "normal" size in little more than a week. The largest specimen recorded was 15.9 mm (22 mm including the egg sacs).[20] Interestingly, the swimming legs and mouthparts do not take part in this growth so that they quickly become inconspicuous. At the same time the large anchoring processes, two ventral and two dorsal (Fig. 35-2), grow into the fish's muscle. The body segmentation becomes blurred, the location of the somites being recognized only by finding the tiny

legs. The result is an embryo-producing machine that bears practically no resemblance to an arthropod and that has its head permanently anchored in its food source. It is little wonder that the early taxonomists had such trouble correctly placing *Lernaea* in their system.

Nevertheless, as deBlainville[5] reported in 1882, the larvae can be clearly recognized as crustacean and are typical nauplii. The primitive series of naupliar instars has been shortened to three. When the nauplii hatch, they contain enough yolk material within their bodies to eliminate the need for feeding in any of the three naupliar stages. The third nauplius molts to give rise to the first copepodid, and this marks the end of the free-living life of a *Lernaea*. Thus the length of time spent as a free-living organism has been markedly shortened, compared with the primitive condition, and the free-living instars do not even feed.

The *L. cyprinacea* studied by Grabda[20] would develop to maturity only on crucians *(Carassius carassius);* if the copepodid were attached to some other species, it had to detach and find the proper host. Therefore it is curious that adult *L. cyprinacea* have been reported from such a variety of hosts belonging to remotely related families (Cyprinidae, Salmonidae, Centrarchidae, Catostomidae, and Ameiuridae) in several continents (Europe, North America, Asia and Africa). Some of these may represent mistaken identifications or strain or subspecies differences. Some 45 or so species of *Lernaea* have been described, but we consider only 28 valid[21]; these taxonomic problems of practical importance cannot be considered settled.

Order Poecilostomatoida

This order illustrates a progression from little specialized parasites (Ergasilidae) to some highly modified and bizarre forms (Philichthyidae, Sarcotacidae). Poecilostomes have been especially successful as symbionts of other invertebrates, particularly with cnidarian hosts. Of 1475 species of copepods known from invertebrates, 416 belong to this order, and 373 species of poecilostomes are associated with cnidarians.[23]

■ Family Ergasilidae

Ergasilids are among the most common copepods parasitic of fishes. They have been a "thorn in the flesh for many valuable fisheries in the Old World" for a long time[30] and often frequent the gills of a variety of fishes in North America.[52,59] *Ergasilus* spp. are primarily parasites of freshwater hosts but are common on several marine fishes, especially the more euryhaline ones such as sticklebacks, killifish, and mullets.

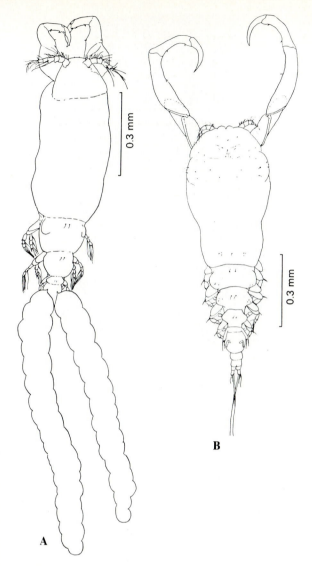

FIG. 35-3

Examples of *Ergasilus*, a common parasite of freshwater and some marine fishes. **A,** *Ergasilus celestis,* from eels *(Anguilla rostrata),* and burbot *(Lota lota),* bearing egg sacs. **B,** *Ergasilus arthrosis,* reported from several species of freshwater hosts, nonovigerous.

From Roberts, L.S. 1969. J. Fish. Res. Bd. Can. 26:1000, 1008.

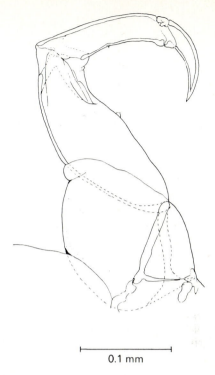

FIG. 35-4

Antenna of *E. centrarchidarum,* a common parasite of members of the sunfish family (Centrarchidae). The antennae of *Ergasilus* are usually modified into a powerful organ used to grasp their host's gill filament, with the third and fourth joints opposable with the second.

From Roberts, L.S. 1970. Trans. Am. Microsc. Soc. 89:27.

ber of segments in the antenna has been reduced to four, and the terminal segment is characteristically in the form of a sharp claw. The third and fourth segments are opposable with the second (subchelate) (Fig. 35-4). Rather than depending on muscle and heavy sclerotization of the antennae, the antennal tips may be fused or locked so that the gill filament is completely encircled *(E. amplectens, E. tenax)* (Fig. 35-5). When removed from their position on the gill, most *Ergasilus* can swim reasonably well; their pereiopods retain the primitive, flat copepod form, with setae and hairs well adapted for swimming. The first legs, however, show adaptation for their feeding habit. These appendages are supplied with heavy, blade-like spines; in some species the second and third endopodal segments are fused, presumably lending greater rigidity to the leg. Such modifications increase the ability of the animal to rasp off mucus and tissue from the gill to which it is clinging (Fig. 35-6). The

Ergasilidae show primitive morphological characteristics reminiscent of cyclopoids, with few, but effective adaptations for parasitism. The antennules are sensory, but the antennae have become modified into powerful organs of prehension (Fig. 35-3). *Ergasilus* females usually are found clinging by their antennae to one of the fish's gill filaments. The primitive num-

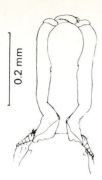

FIG. 35-5

Tips of the antennae of *Ergasilus tenax* "lock" together, completely encircling the host's gill filament.

From Roberts, L.S. 1965. J. Parasitol. 51:989.

first legs dislodge epithelial and underlying cells in this manner and sweep them forward to the mouth[18] (Fig. 35-7). It is easy to see that a heavy infestation with *Ergasilus* could severely damage gill tissue, interfere with respiration, open the way to secondary infection, and lead to death. Epizootics of *Ergasilus* on mullet *(Mugil)* were recorded in Israel; in one case up to 50% of the stock in some ponds was lost, and hundreds of dead mullet were found daily.[55] Rogers and Hawke[53] found large numbers of *Ergasilus* infesting skin lesions of shad *(Dorosoma)* in Tennessee; they believed the copepods were the primary cause of the moribund condition of the fish.

Ergasilus spp. hatch as typical nauplii; *E. sieboldi* has three naupliar and five copepodid stages, all free living.[60] The adult males are planktonic as well, and the female is fertilized before attaching to the fish host. Only the female has been found as a parasite. The males of very few species are known; in one of these, even the females are planktonic as adults *(E. chautauquaensis,* which may be the only nonparasitic species in the genus), although females of several other species are sometimes encountered in the plankton.[9]

■ **Family Lichomolgidae**

Lichomolgids are symbionts with a wide variety of marine invertebrates, including serpulid polychaetes, alcyonarian and madreporarian corals, ascidians, sea anemones, nudibranchs, holothurians, starfish, pelecypods, and sea urchins. It is evident that many species are involved, and many are yet to be discovered. The Lichomolgidae (family here broadly accepted) are divided by Humes and Stock[24,25] into five families, embracing 76 genera and 324 species.

Lichomolgids are generally cyclopoid in body form,

FIG. 35-6

Ergasilus labracis in situ on gills of striped bass, *Morone saxatilis* (two specimens are indicated by arrows.) The gill operculum has been removed. Note also that the fish is infected by an isopod, *Lironeca ovalis,* partly hidden under gill *(right).*

Photograph by L.S. Roberts.

with a retention of segmentation and swimming legs (Fig. 35-8). Segments of the antennae are reduced to three or four, and they often end in one to three terminal claws. The antennae are apparently adapted for prehension in much the same manner as those of the ergasilids. Higher specialization is shown in some species, in which one or more swimming legs may be reduced or vestigial. The copepodid larvae are often found parasitic on the same host as are the adults, and relatively little time is apparently spent in the free-living naupliar stages.

■ **Families Philichthyidae, Sarcotacidae**

Little is known of these families, but they deserve at least brief mention because of their great specialization for parasitism.

The general appearance of philichthyids is startling, to say the least, with unlikely looking processes emanating from their bodies (Fig. 35-9). This is a small group, completely endoparasitic in the subdermal canals of teleosts and elasmobranchs, that is, the frontal

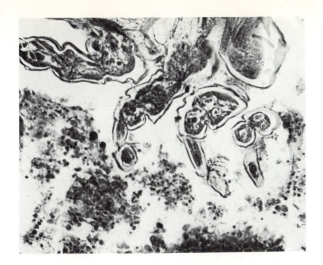

FIG. 35-7

Section of *Ergasilus sieboldi* in situ showing damage to gills inflicted by thoracic appendages. The tissue is rasped off, and the parasite feeds on detached epithelial, mucous, and blood cells. The first legs *(top left)* are particularly important in directing dislodged tissue anteriorly toward the mouth. (×200.)

From Einszporn, T. 1965. Acta. Parasitol. Polon. 13:380.

mucous passages and sinuses and the lateral line canal. Some species retain external evidence of segmentation, but in others it is less apparent. Because they have little use for organs of attachment, such appendages are reduced. The males are much smaller than the females and are less highly modified.

Sarcotacids are also endoparasitic copepods and are probably the most highly specialized of any copepod parasite of a vertebrate (Fig. 35-10). They live in cysts in the muscle or abdominal cavity of their fish hosts. Their appendages are vestigial, and they appear to feed on blood from the vascular wall of the cyst. The adult female is little more than a reproductive bag within the cyst and may reach several centimeters in size. Males are much smaller, and one lives in each cyst, mashed between the wall of the cyst and the huge body of its mate.

Nothing is known of the development and many other aspects of the biology of sarcotacids and philichthyids. Their sites on the hosts are so unobtrusive that these fascinating organisms probably occur much more widely than they have been reported; a diligent search would doubtless reveal a number of new species.

FIG. 35-8

Typical lichomolgid, *Ascidioxynus jamaicensis,* from the branchial sac of an ascidian, *Ascidia atra;* dorsal view of female.

From Humes, A.G., and J.H. Stock. 1973. Smithson. Contr. Zool., no. 127, p. 143.

Order Siphonostomatoida

The members of this large group are mostly parasites of fishes. Only 31 species have been recorded from cnidarians, but Humes[23] believes that this small number is probably due to neglect and is not real. Although even the most primitive siphonostomes show some adaptations to parasitism, like the poecilostomes, an array from generalized to extremely modified and bizarre can be demonstrated. The majority of siphonostomes are parasites of marine fishes, and since aquaculture of marine fishes has not yet been widely practiced, the actual or potential economic importance of the various siphonostomes is unknown. In western Japan, where culture of yellowtail *(Seriola)* is practiced intensively in small bays, *Caligus spinosus* has inflicted considerable damage.[26]

■ Family Caligidae

Adult caligids are evidently arthropods, although they have departed from the "typical" free-living copepod plan. They have, at least, some adaptations for prehension, tend to be larger than most free-living

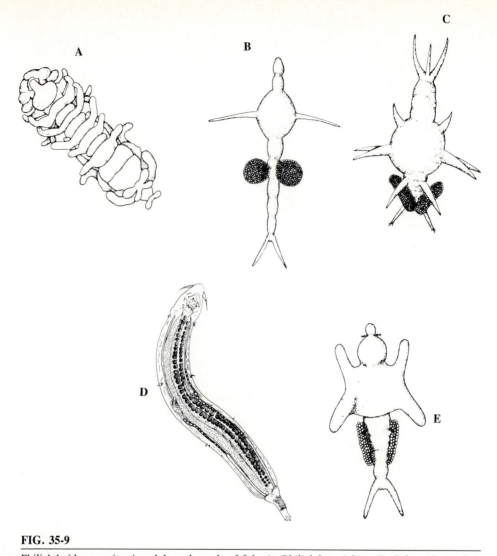

FIG. 35-9

Philichthyids, parasites in subdermal canals of fish. **A,** *Philichthys xiphiae*. **B,** *Sphaerifer leydigi*.
C, *Colobomatus sciaenae*. **D,** *Lerneascus nematoxys*. **E,** *Colobomatus muraenae*.

From Kabata, Z. 1970. Diseases of fishes; book 1: Crustacea as enemies of fishes. (Snieszko, S.F., and H.R. Axelrod, editors.) T.F.H. Publications, Inc., Neptune City, N.J.

groups, and have some dorsoventral flattening for closer adhesion to the host surface. Some tend to be more sedentary, being mostly confined to the fish's branchial chamber, but the adults of many species can move rapidly over the host's surface (fins, gills, and mouth). They can swim and change hosts.

The usual caligid body form shows a fusion of the ancestral body somites: a large, flat cephalothorax, followed by one to three free thoracic segments, a large genital segment, and a smaller unsegmented abdomen. The more primitive caligids, such as *Dissonus*

(Fig. 35-11), have three segments between the cephalothorax and genital somites, whereas the more advanced forms, like *Caligus* (Fig. 35-12), have only one.[45] *Caligus curtus'* principal appendages for prehension are the antennae and maxillipeds. It has two **lunules** on the anterior margin of the cephalothorax, which function as accessory organs of adhesion. The cephalothorax is roughly disc shaped and bears a flexible, membranous margin. The posterior portion of the disc is not formed by the cephalothorax itself but by the greatly enlarged, fused protopod of the third tho-

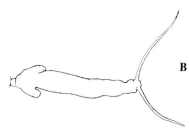

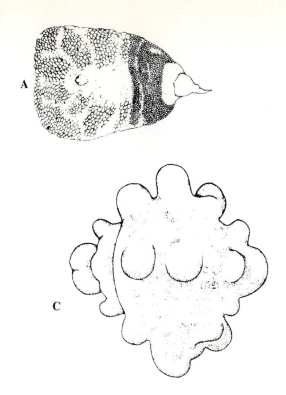

FIG. 35-10

Sarcotacids may be the most highly specialized copepod parasites of vertebrates. **A,** *Sarcotaces* sp., female. **B,** *Sarcotaces* sp., male. **C,** *Ichthyotaces pteroisicola.*

From Kabata, Z. 1970. Diseases of fishes, book 1: Crustacea as enemies of fishes. (Snieszko, S.F., and H.R. Axelrod, editors.) T.F.H. Publications, Inc., Neptune City, N.J.

racic legs (Fig. 35-13). Membranes on the margin of the protopods match those on the cephalothorax, and the arrangement forms an efficient suction disc, when the cephalothorax is applied to the fish's surface and is arched.

Caligus on stationary or sluggishly moving fish performs settling movements at intervals to increase effectiveness of its "suction cup."[33] Settling involves slight, rapid, rotational movements on the longitudinal axis of the body, mediated by the maxillae, while the first and second thoracic legs pump water from beneath the parasite. Settling movements are rarely observed when caligids are attached to vigorously swimming fish, and water pressure against the parasite, always facing the current, is assumed sufficient to keep it pressed closely to its host. The first and second legs are the swimming structures, whereas the uniramous fourth legs are used least. Interestingly, the third and fourth legs of the more primitive *Dissonus* are little modified; they are similar to the second legs, and all three pairs resemble the second legs of *C. curtus.*

The feeding apparatus of *C. curtus* is a good example of the tubular mouth type of the siphonostomes. The mouth tube is carried in a folded position parallel to the body axis, but it can be erected so that its tip can be applied directly to the host surface. The tip of the tube bears flexible membranes analogous to those

FIG. 35-11

Dissonus nudiventris, a more primitive caligid with three segments between the cephalothorax and genital somite.

From Kabata, Z. 1970. Diseases of fishes, book 1: Crustacea as enemies of fishes. (Snieszko, S.F., and H.R. Axelrod, editors.) T.F.H. Publications, Inc., Neptune City, N.J.

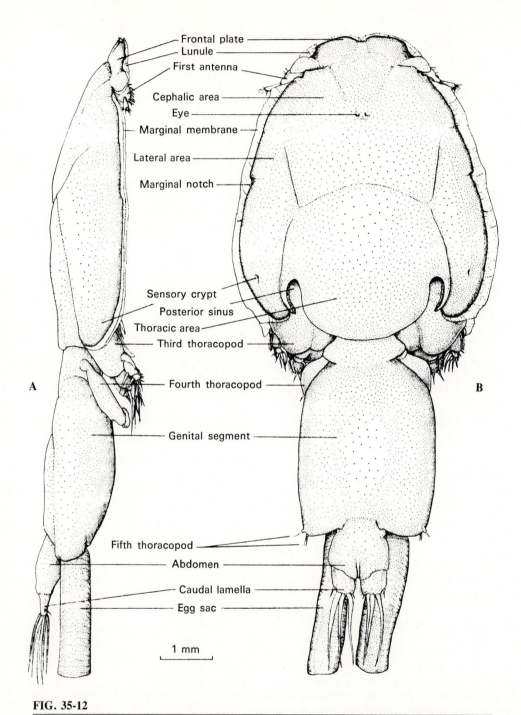

Frontal plate
Lunule
First antenna
Cephalic area
Eye
Marginal membrane
Lateral area
Marginal notch
Sensory crypt
Posterior sinus
Thoracic area
Third thoracopod
Fourth thoracopod
Genital segment
Fifth thoracopod
Abdomen
Caudal lamella
Egg sac

A

B

1 mm

FIG. 35-12

Caligus curtus, a more advanced caligid with only one segment between the cephalothorax and genital somite. **A** and **B,** Female, lateral and dorsal views. **C,** Female, ventral view. **D,** Male, lateral view.

From Parker, R.R., Z. Kabata, L. Margolis, and M.D. Dean. 1968. J. Fish. Res. Bd. Can. 25:1951-1952.

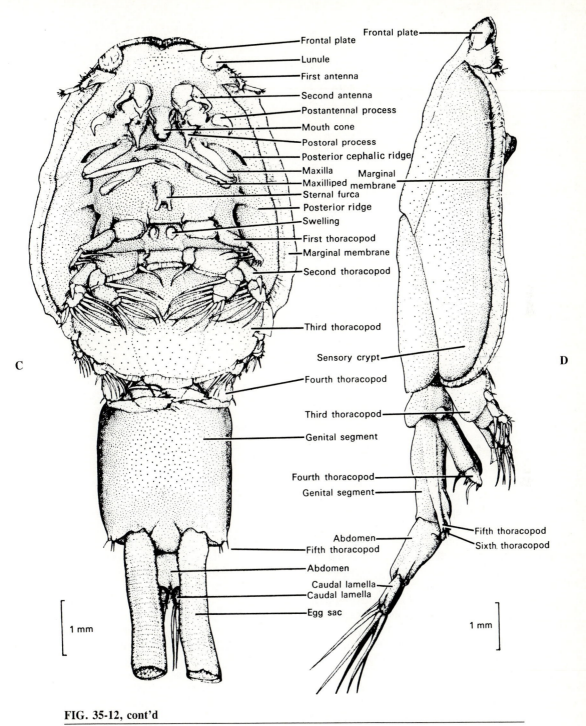

Frontal plate
Lunule
First antenna
Second antenna
Postantennal process
Mouth cone
Postoral process
Posterior cephalic ridge
Maxilla
Maxilliped
Sternal furca
Posterior ridge
Swelling
First thoracopod
Marginal membrane
Second thoracopod

Third thoracopod

Fourth thoracopod

Genital segment

Fourth thoracopod
Genital segment

Abdomen
Fifth thoracopod

Abdomen

Caudal lamella
Caudal lamella

Egg sac

Frontal plate

Frontal plate

Marginal membrane

Sensory crypt

Third thoracopod

Fourth thoracopod
Genital segment

Fifth thoracopod
Sixth thoracopod

Caudal lamella

C

D

1 mm

1 mm

FIG. 35-12, cont'd

For legend see p. 568.

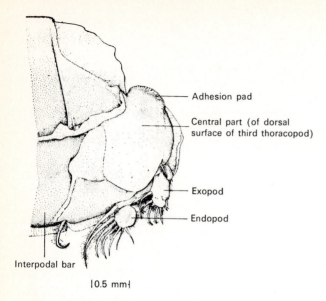

FIG. 35-13

Third thoracic leg of *Caligus curtus*. Note greatly enlarged, fused protopod with flexible marginal membrane.

From Parker, R.R., Z. Kabata, L. Margolis, and M.D. Dean. 1968. J. Fish Res. Bd. Can. 25:1966.

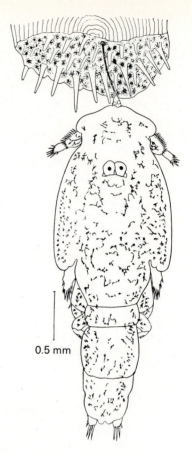

FIG. 35-14

Chalimus larva of *Caligus rapax*.

From Wilson, C.B. 1905. Proc. U.S. Nat. Mus. 28:549.

on the margin of the cephalothorax, again increasing the efficiency of the organ as a suction device. The bases of the mandibles are lateral to and outside the mouth tube. They enter the buccal cavity through longitudinal canals so that their tips lie within the opening of the cone. The mandibular tips bear a sharp cutting blade on one side and a row of teeth on the other. Thus the mandibles can work back and forth like little pistons in their canals, piercing and tearing off bits of host tissue to be sucked up by the muscular action of the mouth tube.

Although the developmental stages of caligids are mentioned in the older literature, reasonably complete descriptions for species of *Caligus* and *Lepeophtheirus* have more recently become available.[26,36] Both genera have only two naupliar stages, and these appear not to feed. The second nauplius molts to produce the first copepodid. The first copepodid must find a host or perish. If it finds its host, the copepodid clings to the fish with its prehensile antennae and molts to produce the specialized type of copepodid called the **chalimus** (Fig. 35-14). Three more chalimus instars follow, all of them attached to the host by the frontal filament. The actual attachment process of a caligid is unknown, but it is probably similar to that of the lernaeopodid. The chalimus backs off from its point of attachment, thus pulling more filament out of the fron-

tal organ, while stroking the filament with its maxillae.[36] The four chalimus instars are followed by two preadult stages in all caligids. The preadults are detached from the frontal filament, and they, as well as the adults, have the capacity for free movement over the host's body. Males are parasitic and not much smaller than females.

■ Family Lernaeopodidae

The lernaeopodids are common, widespread parasites of fish, which frequently occur on freshwater hosts. *Salmincola,* a parasite of salmonids, has caused great damage to hatchery stocks in North America.[30] Lernaeopodids are substantially more modified away from the ancestral copepod form than are the Caligidae. Virtually all external signs of segmentation have disappeared in the adult (Fig. 35-15), as is the case with the Lernaeidae and Pennellidae. Similarly, the adult females are permanently anchored in one place on the host. However, in contrast to these families, lernaeopodid females are attached almost completely outside the host; the anchor, or **bulla,** is nonliving and is formed from head and maxillary gland secretions. The maxillae themselves are fused to the bulla, and they are often huge. Occasionally, the maxillae are very short, as in *Clavella* (Fig. 35-16); however, in these cases a very long, mobile cephalothorax provides a "grazing range" similar in extent to that possible with longer maxillae.

The maxillipeds are usually modified to form powerful grasping structures, and although they were primitively posterior to the maxillae, in most species they are now located and function more anteriorly. The bases of the maxillae mark the approximate posterior limit of the cephalothorax, and the rest of the body is the trunk, or fused thoracic and genital segments. The abdomen and swimming legs are absent or vestigial. There is extreme sexual dimorphism. The males are pygmies and are free to move around in search of females after the last chalimus stage. Both the maxillae and maxillipeds of the males are used as powerful grasping organs. The males, however, do not use the bullae to anchor themselves.

The fascinating lernaeopodid development of *Salmincola californiensis* has been studied by Kabata and Cousens[32] (Fig. 35-17). As it hatches from the egg, the nauplius molts simultaneously to the copepodid. After the cuticle hardens, the copepodid must find a host within about 24 hours, or it dies. It attaches to the host with the prehensile hooks on the antennae and the powerful claws of the maxillae, and then it must find a suitable position on the fish for placement of the frontal filament. It wanders over the host's skin until it finds a solid structure, such as a bone or fin ray, close to the surface. The maxillipeds excavate a small cavity at that position and press the anterior end of the cephalothorax into the cavity. The terminal plug of the frontal filament detaches and is fixed to the underlying host structure by a rapidly hardening cement produced by the frontal gland. The copepodid moves backward, pulling the filament out of the frontal gland, and if the attachment site is favorable and the copepodid has not been too much damaged by detachment of the frontal filament, it soon molts to the first chalimus stage.

These hazards destroy many copepodids, but even after the copepodid is safely attached to a host, it must pass through four chalimus instars. Each chalimus molt involves a complicated series of maneuvers in which the frontal filament is detached by the maxillae, then reattached when the molt is completed (Fig. 35-18). The fourth chalimus finally breaks free of the frontal filament.

After this chalimus is free, it must find a suitable

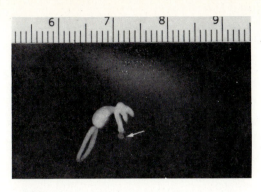

FIG. 35-15

Salmincola inermis, a lernaeopodid parasite of whitefish, *Coregonus* spp. The huge maxillae are fused to the bulla *(arrow),* which is embedded in the host's flesh, anchoring the female to that site. The powerful maxillipeds can be seen *anterior* to the maxillae, and the mouth is at the tip of the anteriormost cone-like projection.

Photograph by L.S. Roberts.

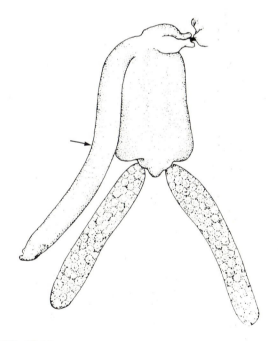

FIG. 35-16

The maxillae are very short in *Clavella,* but the cephalothorax *(arrow)* is long and mobile, providing an extended "grazing range."

From Kabata, Z. 1970. Diseases of fishes, book 1: Crustacea as enemies of fishes. (Snieszko, S.F., and H.R. Axelrod, editors) T.F.H. Publications, Inc., Neptune City, N.J.

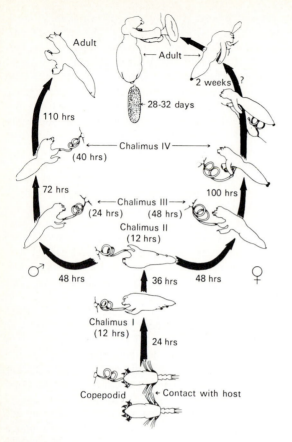

FIG. 35-17

Life cycle of *Salmincola californiensis*. The time periods in parentheses refer to the duration of that particular stage, whereas those without parentheses denote time from first contact with host.

From Kabata, Z., and B. Cousens. 1973. J. Fish. Res. Bd. Can. 30:901.

location for its permanent residence. With its antennae and mouth appendages, it rasps out a site for the bulla, now developing in the frontal organ. The maxillipeds cannot be used because they are the principal means of prehension. After molting, the bulla is everted, placed in the excavation, and detached from the anterior end. These processes are again dangerous for the parasite, which loses considerable body fluid, causing substantial mortality. Finally, the linked tips of the maxillae must find the opening in the implanted bulla, where they connect with small ducts and secrete cement from the maxillary glands. If and when this last maneuver is successful, the parasite is permanently attached to its host and can graze at will on the surface epithelium. It is no surprise that many copepods fail in this complicated series of developmental events; rather it is amazing that so many succeed.

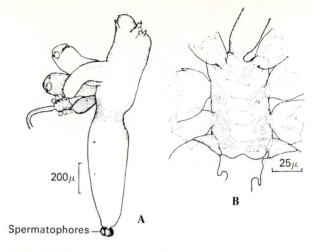

FIG. 35-18

Early fourth chalimus of female *Salmincola californiensis*, **A,** showing maxillae embedded in frontal filament. **B,** Enlarged end of frontal filament, showing tips of maxillae embedded in it (at bottom), along with the molted cuticle of maxillae tips from earlier chalimus stages.

From Kabata, Z., and B. Cousens. 1973. J. Fish. Res. Bd. Can. 30:888.

■ Family Pennellidae

The Pennellidae (formerly Lernaeoceridae) are widespread and conspicuous parasites of marine fish and mammals. They carry the evolutionary tendencies mentioned earlier to the extreme. Even the small ones are usually large by free-living standards, and the large ones are the mammoths of the copepod world. *Pennella balaenopterae* from whales may be more than a foot long! Their loss of external segmentation, obscuration of swimming appendages in the adult, and invasion of host tissue by their anterior ends are reminiscent of the cyclopoid family Lernaeidae. However, pennellids tend to be more invasive of the circulatory system, sense organs, and viscera than are lernaeids. Each species usually has a characteristic site into which the anterior end grows and feeds. Several species, including all *Lernaeocera* spp., invade particular parts of the circulatory system, normally a large blood vessel. (The large trunk, bearing the reproductive organs and ovisacs, is external to the fish surface.) Common sites are the heart, branchial vessels, and ventral aorta. On the Atlantic cod *Gadus morhua*, *L. branchialis* (Fig. 35-19) invades the bulbus arteriosus. The parasite generally attaches in the branchial area, and the cephalothorax may have to grow into and follow the ventral aorta for some distance. The associated pathogenesis is severe and is likely to have an impact on commercial fisheries. Two or more mature

FIG. 35-19

Lernaeocera branchialis from the Atlantic cod, *Gadus morhua*. The voluminous trunk of the organism, containing the reproductive organs, along with the coiled egg sacs, protrudes externally from the host in the region of the gills. The anterior end *(right)* extends into the flesh of the host, and the antlers are embedded in the wall of the bulbus arteriosus, which is severely damaged. The antlers rarely penetrate the lumen of the bulbus, since this would lead to thrombus formation and death of both the parasite and host.

Photograph by L.S. Roberts.

FIG. 35-20

Phrixocephalus longicollum, a lernaeocerid whose antlers proliferate into a luxuriant, intertwining growth.

From Kabata, Z. 1970. Diseases of fishes, book 1: Crustacea as enemies of fishes. (Snieszko, S.F., and H.R. Axelrod, editors.) T.F.H. Publications, Inc., Neptune City, N.J.

parasites on haddock *(Melanogrammus aeglefinus)* can cause the fish to be as much as 29% underweight, have less than half the normal amount of liver fat, and have a decrease of half the hemoglobin content of the blood.[29] At a 15% prevalence on haddock and 5% on cod, this parasite can be calculated to cause enormous losses.[28] Concurrent infections with a trypanosome can compound the damage.[34]

The form of adult females is grotesque. Anchoring processes, sometimes referred to as antlers, emanate from the anterior end. These are often more elaborate than those found in lernaeids. The greatest development of the antlers seem to be in *Phrixocephalus*, where many branches are found (Fig. 35-20). *Lernaeolophus* and *Pennella* have curious, branched outgrowths at the posterior part of the trunk, the function of which is unknown. As in the lernaeids, the appendages do not participate in the metamorphosis undergone by the rest of the female body; therefore they are so small compared with the rest of the body as to be hardly discernible.

The life cycles of pennellids are unique among the copepods in that they often require an intermediate host. Usually the intermediate host is another species of fish, but it may be an invertebrate. *Lernaeocera branchalis* apparently has only one nic nicliar stage, which leads a brief pelagic existence. The copepodid infects a flounder and undergoes several chalimus in-

stars. The female is fertilized as a late chalimus while on the intermediate host and then detaches from the frontal filament. She undergoes another pelagic phase to search out the definitive host, a species of gadid (cod family). The copepod attaches in the gill cavity; the anterior end burrows into the host tissue, aided by the strong antennae; and the dramatic metamorphosis begins. At the time she leaves the intermediate host, the female is only 2 to 3 mm long and is copepodan in appearance. In her metamorphosis she loses all semblance of external segmentation and grows to 40 mm or more. Adult females of *Cardiodectes medusaeus* are found on lanternfishes, with their anterior ends embedded in the bulbus arteriosus of the heart (Fig. 35-21).[47] The intermediate hosts are thecosomate gastropods. The 0.6 mm preadult female leaves the gastropod to find a definitive host, where she grows into an adult up to 15 mm in length. Both *Lernaeocera* and *Cardiodectes* feed on blood. *Cardiodectes* completely digests the hemoglobin and stores the waste iron as ferritin crystals; the manner in which *Lernaeocera* disposes of the excess iron is unknown.[47]

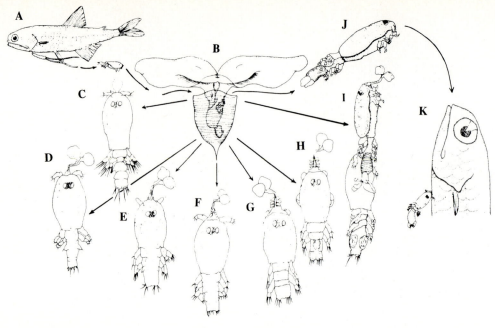

FIG. 35-21

Life cycle of *Cardiodectes medusaeus*. **A,** Lanternfish definitive host with adult copepod bearing egg sacs. **B,** Gastropod intermediate host. **C,** Copepodid. **D,** Chalimus I. **E,** Chalimus II, female. **F,** Chalimus II, male. **G,** Chalimus III, female. **H,** Chalimus III, male. **I,** Adult male and preadult female mating. **J,** Postmated, preadult female, carrying spermatophores leaving mantle cavity in search of definitive host. **K,** Preadult female grasping ventral body surface of definitive host. Not drawn to scale; preadult female is 0.6 mm in length; adult female is up to 15 mm.

From Perkins, P.S. 1983. J. Crust. Biol. 3:70-87.

Order Monstrilloida

The Monstrilloida are parasites of polychaetes and gastropods. They have the distinction of being parasitic only during their larval stages. The adult is the free-living dispersal agent. They have the further distinction among the Crustacea of having only one pair of antennae. Neither monstrilloid larvae nor adults have a mouth or functional gut. The nauplius penetrates its host, which is either a polychaete or a prosobranch gastropod, depending on the species. It molts to become a rather undifferentiated larva with one to three pairs of apparently absorptive appendages (Fig. 35-22). Progressive differentiation and copepodid stages ensue, and the adult finally breaks out of the host to reproduce. Thus only the adult and the nauplius are free living, and the larval stages absorb food in a manner analogous to that of a tapeworm. This strange lifestyle, in addition to marked sexual dimorphism, makes for some taxonomic problems. Of 29 reported species in the genus *Monstrilla,* 20 are known from only one sex, and the males and females of the same species have at times served as bases for descriptions of two separate species.[40]

Other strange copepods could be described. Ho, Katsumi, and Honma[22] found *Coelotrophus nudus* in the coelomic cavity of a sipunculan worm. Neither the males nor the females had a mouth; the females had no appendages, and the males had only one pair, which were used to grasp the female. The copepods were assigned to the family Antheacheridae, which Bowman and Abele[7] were not able to place in an order. It is very difficult to classify copepods that have completely lost so many of their appendages.

CLASS BRANCHIURA

The class Branchiura is relatively small in numbers of species but great in its destructive potential in fish culture. All species are ectoparasites of fishes, although some can use frogs and tadpoles as hosts, too. They are dorsoventrally flattened, reminiscent of caligid copepods with which they are sometimes con-

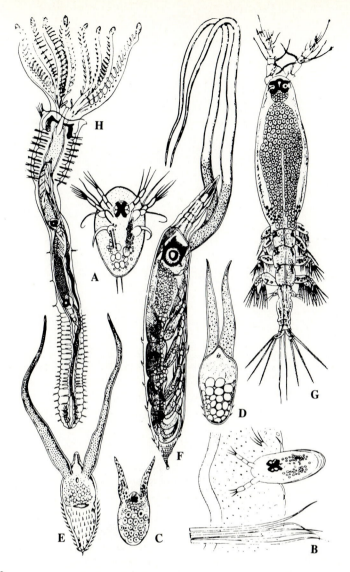

FIG. 35-22

Haemocera danae, a monstrillid parasite of polychaete annelids. **A,** Nauplius. **B,** Nauplius penetrating integument of host. **C** to **E,** Successive larval stages, showing development of absorptive appendages. **F,** Fully developed copepodid within spiny sheath. **G,** Adult female. **H,** Polychaete containing two copepodids in coelom.

From Baer, J.G. 1952. Ecology of animal parasites. University of Illinois Press, Urbana.

fused, and can adhere closely to the host's surface. Some species are moderately large, up to 12 mm or so. The most common, cosmopolitan genus is *Argulus* (Fig. 35-23). *Argulus* spp. can swim well as adults; the females must leave their hosts to deposit eggs on the substrate. Many *Argulus* spp. are not host specific and so have been recorded from a large number of fish species.

The branchurian carapace is expanded laterally to form respiratory alae. They have two pairs of antennae. Homologies of the remaining head appendages have been disputed, but the best evidence suggests that the only appendages in the suctorial proboscis, or mouth tube, are the mandibles.[39] The large, prominent sucking discs are modified maxillules. Immediately posterior to the maxillular discs are the large

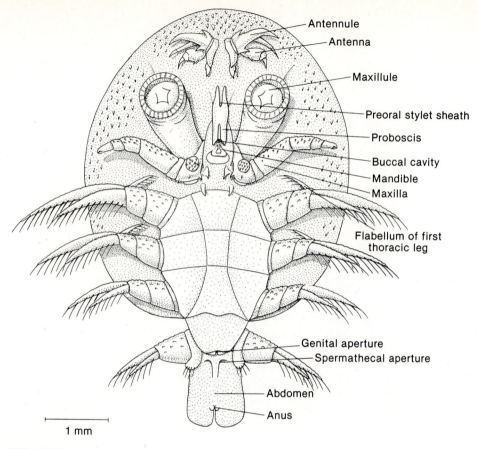

Antennule
Antenna
Maxillule
Preoral stylet sheath
Proboscis
Buccal cavity
Mandible
Maxilla
Flabellum of first thoracic leg
Genital aperture
Spermathecal aperture
Abdomen
Anus

1 mm

FIG. 35-23

Ventral view of *Argulus viridis,* female. Note suctorial proboscis, modification of maxillules into sucking discs, and lateral expansion of carapace into alae.

Drawing by William C. Ober.

maxillae, apparently used to maintain the animal's position on its host and to clean the other appendages. *Argulus* has four pairs of thoracic swimming legs of the typically crustacean biramous form. The exopods of the first two pairs often bear an odd, recurved process, the flabellum, thought by some to indicate affinities with the subclass Branchiopoda (Fig. 35-24). An unsegmented abdomen follows the four segments of the thorax.

The Branchiura were long associated taxonomically with the Copepoda, but present knowledge does not justify this. Among other characteristics that differ from the copepods, branchiurans have a carapace, compound eyes, and an unsegmented abdomen behind the genital apertures, and no thoracic segments are completely fused with the head. Another feature present in most branchiurans, but not in other Crusta-

cea, is the piercing stylet, or "sting." It is located on the midventral line, just posterior to the antennae (Fig. 35-25). The function of this curious organ is unknown. It has been observed to pierce the host's skin, and the gland cells with their ducts leading down the stylet are suggestive of toxic secretions. Many authors still believe that the stylet is used for feeding, and the fact that its tip is too small to allow passage of host erythrocytes has led to the probably erroneous conclusion that the animals cannot feed on blood. Causey[11] said that only "textbook species of *Argulus*" do not feed on blood. In fact Martin[39] found that the stylet bore no direct relation to the proboscis. Her figures showed no connection to the digestive tract, and she considered it "possible that its [the stylet's] function is not very important."

The development of *Argulus japonicus* and

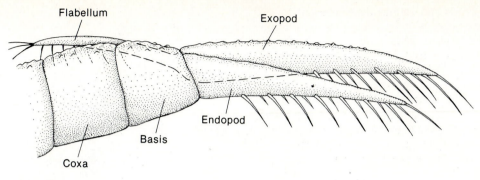

FIG. 35-24

First thoracic appendage of *Argulus viridis*, showing flabellum.

Drawing by William C. Ober.

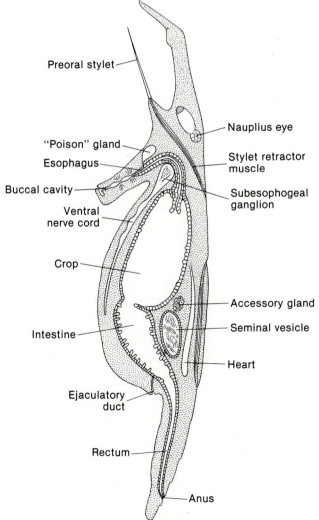

FIG. 35-25

Median longitudinal section of male *Argulus viridis*, semidiagrammatic, with preoral stylet extruded.

Drawing by William C. Ober.

Chonopeltis brevis and various developmental stages of other species have been described.[19,58] Here again is a difference between the Branchiura and Copepoda. Whereas the development of copepods is metamorphic, that of most branchiurans can best be described as direct. As noted, the eggs are laid on the substrate (no ovisacs, as in copepods), and the organism that hatches is not a larva, but a juvenile. In the first instar of *A. japonicus* even the sexes can be distinguished! *Argulus japonicus* has seven juvenile instars, but there may be fewer in other species.[41]

One of the most noteworthy developmental changes that occurs through the ecdysial series is that the primitive form of the maxillules is gradually lost, and the suckers develop in their place.

The development of *C. brevis* offers an intriguing puzzle.[19] The organism occurs on African fishes and seems to be one of the more modified branchiurans. Its thoracic legs have become reduced, and it has lost the ability to swim. However, in common with *Argulus* spp., the females must leave the host to deposit eggs, and the juveniles, which also cannot swim, are found on different host species from those of the adults! Their mode of dispersal remains unknown.

CLASS CIRRIPEDIA

The most familiar cirripedes belong to the order Thoracica, the barnacles. They are important members of the littoral and sublittoral benthic fauna and are economically important as fouling organisms. Some members of the Thoracica are commonly found growing on other animals. Interestingly, *Conchoderma virgatum* is often found on *Pennella,* a good example of hyperparasitism. (Numerous species of parasitic copepods frequently have epizooic suctorians, hydroids, algae, and so on growing on them, encouraged by the fact that the copepod is in terminal anecdysis; that is, it does not molt further.) However, other orders of cirripedes contain some fascinating organisms that are among the most highly specialized parasites known. These are parasites of other invertebrates, and space will permit consideration only of the most important order, the Rhizocephala.

Order Rhizocephala

Members of the order Rhizocephala are highly specialized parasites of decapod malacostracans. The decapods include the animals most of us know as crabs, crayfish, lobsters, and shrimp. The Sacculinidae are primarily parasites of a variety of brachyurans ("true" crabs), the Peltogastridae are found on hermit crabs (anomurans), and the Lernaeodiscidae prefer the anomuran families Galatheidae and Porcellanidae as hosts.

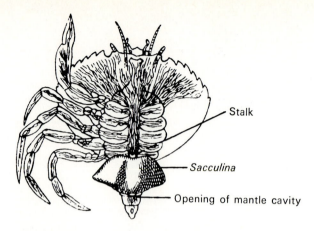

FIG. 35-26

Shore crab, *Carcinus,* infected with a mature rhizocephalan, *Sacculina.*

Modified from Calman, W.T. In Borradaile, L.A., F.A. Potts, L.E.S. Eastham, and J.T. Saunders, editors. 1956. The invertebrata. A manual for the use of students, ed. 2. Cambridge University Press, Cambridge, Eng.

As adults, the rhizocephalans resemble arthropods even less than do lernaeids and pennellids. They have no gut or appendages, not even reduced ones, but get nutrients by means of rootlike processes ramifying through the tissues of the crab host (Fig. 35-26). They start life much as do many other crustaceans, with a nauplius larva, but the nauplius has no mouth or gut. The nauplius undergoes four molts, and the fifth larval instar is referred to as a **cyprid** or **cypris** (Fig. 35-27) because of its resemblance to the free-living ostracod *Cypris.*

Thus far the rhizocephalan life cycle is not unlike that of a normal, thoracican barnacle. The cypris of a barnacle, however, would attach to a suitable spot on the substrate by its antennules and metamorphose to the adult form; the halves of the carapace become the mantle and secrete the calcareous covering plates. However, the female cypris of *Sacculina carcini,* which is the best-known rhizocephalan, attaches to a brachyuran with its antennules.

Next, most of the differentiated structures, including swimming legs and their muscles, are shed from between the two valves of the carapace. The remaining, largely undifferentiated cell mass forms a peculiar larva called the **kentrogon.** The kentrogon may be likened to a living hypodermic syringe. The cell mass within is actually injected into the body of the crab at the base of a seta or other vulnerable spot where the cuticle is thin. The mass of cells migrates to a site just ventral to the host intestine and begins to grow (Fig. 35-27). As the absorptive processes grow out into the crab's tissue, the central mass also begins to enlarge.

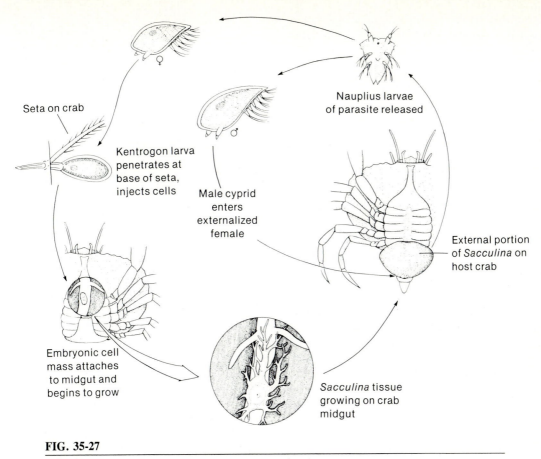

Seta on crab

Kentrogon larva
penetrates at
base of seta,
injects cells

Male cyprid
enters
externalized
female

Embryonic cell
mass attaches
to midgut and
begins to grow

Nauplius larvae
of parasite released

External portion
of *Sacculina* on
host crab

Sacculina tissue
growing on crab
midgut

FIG. 35-27

Life cycle of *Sacculina,* a rhizocephalan parasite of crabs.

From Hickman, C.P., Jr., L.S. Roberts, and F.M. Hickman. 1988. Integrated principles of zoology, ed. 8. The C.V. Mosby Co., St. Louis.

This mass contains the developing gonads of the female. As it grows larger, it appears to press against the host hypodermis in the ventral cephalothorax and thereby prevents cuticle secretion. Finally, the weakened cuticle overlying the parasite breaks open, and the gonadal mass of *Sacculina* becomes external.[16] After the parasite becomes externalized, the crab can no longer molt; therefore further development of the host essentially ceases. The externalized *Sacculina* attracts one or more male cyprids. These extrude a mass of cells that come to lie within male cell receptacles in the female and undergo spermatogenesis. Formerly, these cells were interpreted as testes, and *Sacculina* was thought to be hermaphroditic.

The life histories of peltogastrids and lernaeodiscids seem to be similar to those of sacculinids in many respects, although further ecdyses of the host are not hindered.[6,50]

In light of the invasiveness of rhizocephalans, it is not surprising that there is a range of pathogenic ef-fects of their hosts, including damage to the hepatic, blood, and connective tissues and to the thoracic nerve ganglion of infected crabs.[50] However, some of the most interesting effects are on the hormonal and reproductive processes of the host, the so-called parasitic castration. Crabs exhibit some degree of sexual dimorphism, and morphological differences between the sexes are especially pronounced in the Brachyura. In the normal sequence of ecdyses the secondary sexual characteristics of the respective sexes become increasingly apparent as the crab approaches maturity. When the young male crab is infected with *Sacculina,* various degrees of "feminization" are exhibited in the subsequent instars. The manifestations vary, depending on the species of host and its degree of development when infected, but they may include a broader, more completely segmented abdomen and alteration of the pleopods toward the female type. In female crabs the effects seem to be more complex, involving some aspects of both hyperfeminization and hypofeminization. Somewhat similar effects of parasitic castration have been reported in the hosts of the Peltogastridae and Lernaeodiscidae. The mechanism of host castration has yet to be explained satisfactorily. It may be a result of copious nutrient withdrawal by the large parasites, interference with the host endocrine system, or both.[4]

Whatever the mechanism of host castration, however, the combined results of the parasite structure and the castration lead to an astonishing diversion of host behavior to promote parasite survival. Ritchie and Høeg[51] have provided details in the case of *Lernaeodiscus porcellanae* infecting a porcellanid crab, *Petrolisthes cabrilloi*. The externa of the parasite is in the same position and is the same size as the egg mass of the crab, since it would be carried by the crab's abdomen. Reacting as though the parasite were in fact its egg mass, the crab protects, grooms, and ventilates the parasite. If the grooming legs of the crab are removed artificially, the externa of the parasite soon becomes fouled and necrotic. At the time the parasite begins to release its larvae, the crab performs spawning behavior. It comes from its normal hiding place, stands high on its legs, and waves its abdomen back and forth. Thus the nauplii of the parasite are released into the current created by the host!

CLASS MALACOSTRACA

Although the malacostracans constitute the largest class of Crustacea, with members widespread and abundant in marine and freshwater habitats, comparatively few are symbiotic. Those which are, by and large, are confined to the peracaridan orders Amphipoda and Isopoda. The isopods have been particularly successful in this regard, and some of them have become highly modified for parasitism. The eucaridan order Decapoda is the largest order of crustaceans, but few of its members are symbiotic.

Order Amphipoda

Free-living amphipods are widely prevalent aquatic organisms, often abundant along the seashore. Not many symbiotic species have been described, but some of the ones that have are quite common. Some Hyperiidae (Fig. 35-28), are frequent parasites of jellyfish (*Aurelia, Cyanea*) and Phronimidae are found in the tunic of planktonic ascidians (*Salpa*), apparently killing the tunicate itself and taking over its gelatinous case. *Laphystius sturionis* (suborder Gammaridea) is a relatively unmodified amphipod, parasitizing a variety of marine fishes.[30] The most interesting symbiotic amphipods are among the Caprellidae, and most unlikely looking amphipods they are. Caprellidae, or "skeleton

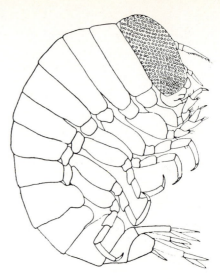

FIG. 35-28

Hyperia galba, an amphipod that is commonly found on jellyfish *(Aurelia)* in the North Atlantic. Hyperiidae typically have very large compound eyes that cover nearly the entire sides of the head.

From Kunkel, B.W. 1913. The Arthrostraca of Connecticut. State Geological and Natural History Survey.

shrimp," are predators that stalk around on hydroid or ectoproct colonies or algae and catch their prey with raptorial second legs, but the Cyamidae are curious ectoparasites of whales (Fig. 35-29). The abdomen is vestigial in both families. In contrast with most amphipods, cyamids are dorsoventrally flattened, a clearly adaptive characteristic in their ectoparasitic habitat. The second and the fifth through seventh legs are strongly modified adhesive organs.

Order Isopoda

Members of the order Isopoda have had more success in terrestrial environments than have other crustaceans, limited though that may be, and they are abundant in a variety of marine and freshwater habitats. Furthermore, they exploit the parasitic mode of existence more extensively than do other malacostracans.

The gnathiidean and flabelliferan families parasitic on marine fishes have relatively few modifications. *Gnathia* is a parasite only as a larva, the **praniza**, which was originally described as a separate genus before it true identity was recognized. The praniza stage (Fig. 35-30) attaches to a fish host and feeds on blood until its gut is hugely distended. It then leaves its host and molts to become an adult. The adults are benthic and do not feed. Because of considerable sexual dimorphism, the male was also described in a separate

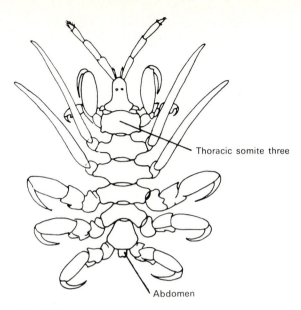

FIG. 35-29

Paracyamus, an amphipod parasite of whales. Cyamids are ectoparasites and are dorsoventrally flattened with several pairs of legs modified for clinging to their hosts.

Modified from Sars, G.O.; from Calman, W.T. 1909. Crustacea. In Lankester, R., editor. A treatise on zoology, vol. 8. Adam & Charles Black, London; from Meglitsch, P.A. 1972. Invertebrate zoology, ed. 2. Oxford University Press, Oxford, Eng.

genus, *Anceus*. Some of the Cymothoidae are of economic importance as fish parasites. The young of *Lironeca amurensis* and some other species burrow under a scale on their host. As the isopod grows, the underlying skin stretches to accommodate it; finally, the enveloped crustacean communicates to the exterior only by a small hole. *Lironeca ovalis* (Fig. 35-31) is a common parasite of a variety of teleosts in the Atlantic Ocean and has been reported along the United States coast from Mississippi to Massachusetts.[34a,56] The parasite is usually found beneath the gill operculum, where, on small host individuals, it causes a marked pressure atrophy of the adjacent gills (Fig. 35-32). Moser and Sakanari[42] showed that *L. vulgaris* juveniles slowed their swimming activity in the presence of fish mucus, and white color (either paper or fish skin) plus mucus induced a settling response.

The epicaridean isopods are highly specialized parasites of other Crustacea. The adult females of some species are comparable in loss of external segmentation and appendages to the most advanced copepods and the Rhizocephala. Portions of the appendages that are not lost, however, are the oostegites forming the brood pouch. These may become enormously devel-

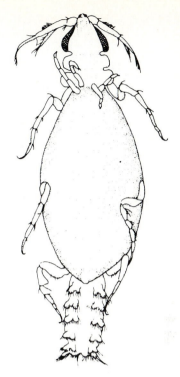

FIG. 35-30

Praniza larva of the isopod *Gnathia*. The praniza is the only parasitic stage of this isopod. The gut becomes greatly distended with blood from its fish host.

From Kabata, Z. 1970. Diseases of fishes, book 1: Crustacea as enemies of fishes. Snieszko, S.F., and H.R. Axelrod, editors. T.F.H. Publications, Inc., Neptune City, N.J.

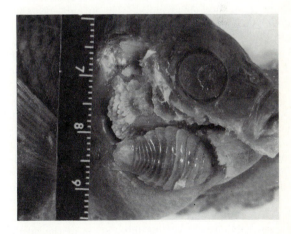

FIG. 35-31

Lironeca ovalis is a common parasite of fish along the Atlantic coast of the United States. Here it is found on a *Lepomis gibbosus* (operculum removed) from Chesapeake Bay.

Photograph by L.S. Roberts.

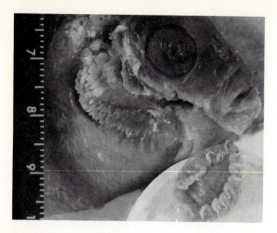

FIG. 35-32

Lironeca ovalis, same specimen as in Fig. 35-31, removed from its site on the gills. The gills show pressure atrophy and traumatic damage.

Photograph by L.S. Roberts.

oped, whereas most of the other appendages disappear or become vestigial.

Examples of the family Entoniscidae are *Pinnotherion vermiforme* (Fig. 35-33), a parasite of a brachyuran crab, *Pinnotheres pisum* (itself a parasite of the mussel *Mytilus edulis),* and *Portunion conformis,* a parasite of the shore crab *Hemigrapsus oregonensis.* The larval isopod (cryptoniscus, see below) enters the crab's hemocoel, apparently through the gills. In the gonads or alongside skeletal apodemes, it molts to an apodous juvenile.[35] The juvenile becomes invested with a cellular covering produced by the host. The female isopod becomes quite large and occupies the space normally filled by a host ovary, but the males remain small and live on the surface of the female. The female's oostegites are produced into extensive, thin lamellae that, together with the host-produced sheath, form a brood chamber. Extending from the sides of the abdomen are highly vascularized pleural lamellae, apparently of respiratory function, but the pleopods are vestigial. The short esophagus leads into a peculiar "cephalogaster," a contractile organ for sucking blood that apparently has absorptive function as well.[2] Upon hatching, the larvae are retained in the brood chamber until released through a pore formed between the sheath and the thin cuticle of the host's gill cavity.

In the Cryptoniscidae, morphological modification appears even more extreme. After infection of the definitive host, the female *Ancyroniscus bonnieri* feeds heavily, and the gorged isopod begins to produce eggs. During this process most of the internal organs,

including digestive and nervous system, disappear, and the animal becomes increasingly distended with eggs. Finally, all that is left is a large, pulsatile sac of eggs, which are freed by rupture of the sac.[10]

The life histories of epicarideans are of great interest. The larva that hatches is an **epicaridium,** quite isopod-like in appearance. The epicaridium has blood-sucking mouthparts and attaches to a free-swimming, calanoid copepod. There it feeds, grows rapidly,[1] and molts to become a **microniscus** larva in 3 to 4 days.[15] In a few more days the microniscus molts several times and develops into a **cryptoniscus,** which again is free swimming and must find a suitable definitive host. In the Bopyridae, Entoniscidae, and Dajidae, the first isopod to infect the definitive host becomes a female, and subsequent cryptoniscus larvae become small males, sometimes living as parasites within the female brood sac. In the bopyrid, *Stegophryxus hyptius,* the cryptoniscus actually has to enter the female's brood pouch to become a male; a masculinizing substance derived from feeding on the female may be responsible.[49] It seems clear that sex determination in a number of epicarideans is epigamic, depending on circumstances other than the chromosomal complement of the gametes. On the other hand, there is evidence that sex determination in some bopyrids can be genetic.[44]

Effects of epicarideans on their hosts are similar to those described for rhizocephalans, including parasitic castration. Secondary sexual characteristics of the male host are lost, the host becomes feminized, and the gonads of both males and females are suppressed or atrophied.[4] If the parasite dies (perhaps killed by host defense mechanisms), reproductive capacity of the crab may return.[35]

Feminization of brachyuran males by epicarideans is not as striking as that produced by rhizocephalans.[50] It is interesting that some rhizocephalans are themselves hyperparasitized by epicarideans, and these, in turn, induce castration of their rhizocephalan hosts!

Order Decapoda

The largest order of crustaceans, the decapods, contains a relatively small number of symbiotic species, although some of them are quite common. They are of interest because of the slight, but definite, modifications for parasitism that they illustrate. Pinnotherids are frequent commensals with polychaetes, in their tubes or burrows, or are parasites in the mantle cavity of pelecypods. *Pinnotheres pisum* has been mentioned, and *P. ostreum* is found in the commercially important oyster *Crassostrea virginica* (Fig. 35-34). *Pinnotheres ostreum* interferes with the feeding of its

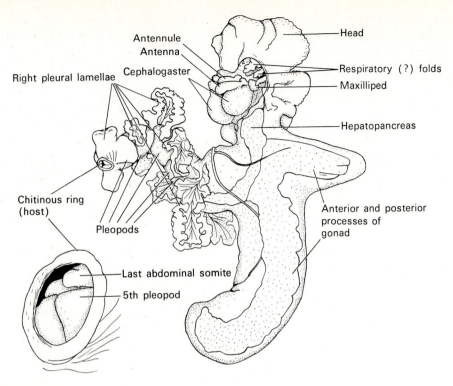

FIG. 35-33

Young female *Pinnotherion vermiforme,* an entoniscid isopod parasite of a crab, *Pinnotheres pisum*. The parasite develops in a closely investing, thin layer of host origin, which communicates with the branchial chamber by a small opening. The opening into the host's branchial chamber is surrounded by a somewhat thickened ring of cuticle (enlarged at left). The vascularized pleural lamellae extending from the abdomen are prominent. Note the vestigial nature of the appendages and presence of the peculiar contractile "cephalogaster."

Redrawn from Atkins, D. 1933. Proc. Zool. Soc. Lond., p. 331. Atkins noted that this specimen probably had not yet spawned; the respiratory (?) folds and pleural lamellae had not yet reached their final stage of complexity; and the cephalogaster was abnormal in that the two lobes were unequal in size.

FIG. 35-34

Pinnotheres ostreum damages the gills of its host, the commercial oyster *(Crassostrea virginica)*. The carapace is soft, and the eyes and chelae are reduced.

Photograph by L.S. Roberts.

host and damages its gills sufficiently to cause female oysters to become males.[3] The same sex change can be produced experimentally by starving the oysters.

Pinnotherids are modified relatively little from the typical, free-living brachyuran. The adult females tend to be white or cream colored with thin, soft cuticle in the carapace and with reduced eyes and chelae. Younger stages, including males, have hard carapaces and more well-developed eyes and chelae.[12] Pinnotherids appear to find their hosts by means of a chemosensory mechanism.[17]

A few species of symbiotic decapods belonging to other infraorders, such as Caridea and Anomura, are known. They live in such places as the mantle cavities of clams, the tubes of polychaetes, and on the stems of sea pens (Octocorallia) and are modified for symbiosis to about the same extent as are pinnotherids. The supposed rarity of these forms is probably a result of the failure of collectors to look for them.[27] The decapods involved in cleaning symbiosis have already been alluded to in Chapter 2.

REFERENCES

1. Anderson, C. 1975. Larval metabolism of the epicaridian isopod parasite *Probopyrus pandalicola* and metabolic effects of *P. pandalicola* on its copepod intermediate host *Acartia tonsa*. Comp. Biochem. Physiol. 50A:747-751.

2. Atkins, D. 1933. *Pinnotherion vermiforme* Giard and Bonnier, an entoniscid infecting *Pinnotheres pisum*. Proc. Zool. Soc. Lond., pp. 319-363.

3. Awaiti, P.R., and H.S. Rai, 1931. *Ostrea cucullata*. Indian Zool. Mem. 3:1-107.

4. Beck, J.T. 1980. The effects of an isopod castrator, *Probopyrus pandalicola*, on the sex characters of one of its caridean shrimp hosts, *Palaemonetes paludosus*. Biol. Bull. 158:1-15.

5. deBlainville, M.H.D. 1882. Mémoire sur les Lernées (Lernaea, Linn.). J. Physiol. (Paris) 95:372-380, 437-447.

6. Bower, S.M, and N.A. Sloan. 1985. Morphology of the externa of *Briarosaccus callosus* Boschma (Rhizocephala) and the relationship with its host *Lithodes aequispina* Benedict (Anomura). J. Parasitol. 71:455-463.

7. Bowman, T.E., and L.G. Abele. 1982. Classification of the recent Crustacea. In Abele, L.G., editor. The biology of Crustacea, vol. 1. Systematics, the fossil record, and biogeography. Academic Press, Inc., New York, pp. 1-27.

8. Boxshall, G.A., and R.J. Lincoln. 1983. Tantulocarida, a new class of Crustacea ectoparasitic on other crustaceans. J. Crust. Biol. 3:1-16.

9. Bricker, K.S., J.E. Gannon, L.S. Roberts, and B.G. Torke. 1978. Observations on the ecology and distribution of free-living Ergasilidae (Copepoda, Cyclopoida). Crustaceana 35:313-317.

10. Caullery, M., and F. Mesnil. 1920. *Ancyroniscus bonnieri* C. et M., epicaride parasite d'un sphéromide (*Dynamene bidentata* Mont.). Bull. Sci. Fr. Belg. 34:1-36.

11. Causey, D. 1959. "Ye crowlin' ferlie," or on the morphology of *Argulus*. Turtox News 37:214-217.

12. Christensen, A.M., and J.J. McDermott. 1958. Life-history and biology of the oyster crab, *Pinnotheres ostreum* Say. Biol. Bull. 114:146-179.

13. Dahl, E. 1956. Some crustacean relationships. In Wingstrand, K.G., editor. Bertil Hanström, Zoological papers in honour of his sixty-fifth birthday November 20th, 1956. Zoological Institute, Lund, pp. 138-147.

14. Dahl, E. 1963. Main evolutionary lines among recent Crustacea. In Whittington, H.B., and W.D.I. Rolfe, editors. Phylogeny and evolution of Crustacea. Special Publication, Museum of Comparative Zoology, Cambridge, Mass., pp. 1-15.

15. Dale, W.E., and G. Anderson. 1982. Comparison of morphologies of *Probopyrus bithynis*, *P. floridensis*, and *P. pandalicola* larvae reared in culture (Isopoda, Epicaridea). J. Crust. Biol. 2:392-409.

16. Day, J.H. 1935. The life history of *Sacculina*. Q. J. Microsc. Sci. 77:549-583.

17. Derby, C.D., and J. Atema. 1980. Induced host odor attraction in the pea crab *Pinnotheres maculatus*. Biol. Bull. 158:26-33.

18. Einszporn, T. 1965. Nutrition of Ergasilus *sieboldi* Nordmann. II. The uptake of food and the food material. Acta Parasitol. Polon. 13:373-380.

19. Fryer, G. 1961. Larval development in the genus *Chonopeltis* (Crustacea: Branchiura). Proc. Zool. Soc. Lond. 137:61-69.

20. Grabda, J. 1963. Life cycle and morphogenesis of *Lernaea cyprinacea* L. Acta. Parasitol. Polon. 11:169-199.

21. Harding, J.P. 1950. On some species of *Lernaea*. Bull. Mus. (Nat. Hist.) Zool. 1:1-27.

22. Ho, J., F. Katsumi, and Y. Honma. 1981. *Coelotrophus nudus* gen. et sp. nov., an endoparasitic copepod causing sterility in a sipunculan *Phascolosoma scolops* (Selenka and De Man) from Sado Island, Japan. Parasitology 82:481-488.

23. Humes, A.G. 1985. Cnidarians and copepods: a success story. Trans. Am. Microsc. Soc. 104:313-320.

24. Humes, A.G., and J.H. Stock. 1972. Preliminary notes on a revision of the *Lichomolgidae*, cyclopoid copepods mainly associated with marine invertebrates. Bull. Zool. Mus. Amsterdam 2:121-133.

25. Humes, A.G., and J.H. Stock. 1973. A revision of the Family Lichomolgidae Kossmann, 1877, cyclopoid copepods mainly associated with marine invertebrates. Smithsonian Contrib. Zool., no. 127.

26. Izawa, K. 1969. Life history of *Caligus spinosus* Yamaguti, 1939 obtained from cultured yellow tail, *Seriola quinqueradiata* T. and S. (Crustacea: Caligoida). Report of Faculty of Fisheries, Perfectural University of Mie 6:127-157.

27. Johnson, D.S. 1967. On some commensal decapod crustaceans from Singapore (Palaemonidae and Porcellanidae). J. Zool. 153:499-526.

28. Kabata, Z. 1955. The scientist, the fisherman, and the parasite. Scot. Fish. Bull. 4:13-14.

29. Kabata, Z. 1958. *Lernaeocera obtusa* n. sp.; its biology and its effects on the haddock. Mar. Res. Scot., pp. 1-26.

30. Kabata, Z. 1970. Diseases of fishes, book 1: Crustacea as enemies of fishes. (Snieszko, S.F., and H.R. Axelrod, editors.) T.F.H. Publications, Inc., Neptune City, N.J.

31. Kabata, Z. 1979. Parasitic Copepoda of British fishes. Ray Society, London.

32. Kabata, Z., and B. Cousens. 1973. Life cycle of *Salmincola*

californiensis (Dana, 1852) (Copepoda: Lernaeopodidae). J. Fish. Res. Bd. Can. 30:881-903.

33. Kabata, Z., and G.C. Hewitt. 1971. Locomotion mechanisms in Caligidae (Crustacea: Copepoda). J. Fish. Res. Bd. Can. 28:1143-1151.

34. Khan, R.A., and D. Lacey. 1986. Effect of concurrent infections of *Lernaeocera branchialis* (Copepoda) and *Trypanosoma murmanensis* (Protozoa) on Atlantic cod, *Gadus morhua*. J. Wildl. Dis. 22:201-208.

34a. Kunkel, B.W. 1918. The Arthrostraca of Connecticut. Conn. St. Geol. Nat. Hist. Surv. Bull., no. 26.

35. Kuris, A.M., G.O. Poinar, and R.T. Hess. 1980. Post-larval mortality of the endoparasitic isopod castrator *Portunion conformis* (Epicaridea: Entoniscidae) in the shore crab, *Hemigrapsus oregonensis*, with a description of the host response. Parasitology 80:211-232.

36. Lewis, A.G. 1963. Life history of the caligid copepod *Lepeophtheirus dissimulatus* Wilson, 1905 (Crustacea: Caligoida). Pacific Sci. 17:195-242.

37. Linnaeus, C. 1746. Fauna suecica sistems animalia suecica regni, ed. 1. Stockholm.

38. Linnaeus, C. 1758. Systema naturae, ed. 10. Stockholm.

39. Martin, M.F. 1932. On the morphology and classification of *Argulus* (Crustacea). Proc. Zool. Soc. Lond., pp. 771-806.

40. McAlice, B.J. 1985. On the male of *Monstrilla helgolandica* Claus (Copepoda, Monstrilloida). J. Crust. Biol. 5:627-634.

41. Meehean, O.L. 1940. A review of the parasitic Crustacea of the genus *Argulus* in the collection of the United States National Museum. Proc. U.S. Nat. Mus. 88:459-527.

42. Moser, M., and J. Sakanari. 1985. Aspects of host location in the juvenile isopod *Lironeca vulgaris* (Stimpson, 1857). J. Parasitol. 71:464-468.

43. Oken, L. 1816. Lehrbuch der Naturgeschichte, vols. 1 and 2. Dritter Theil, Zoologie, Jena.

44. Owens, L., and J.S. Glazebrook. 1985. Sex determination in the Bopyridae. J. Parasitol. 71:134-135.

45. Parker, R.R., Z. Kabata, L. Margolis, and M.D. Dean. 1968. A review and description of *Caligus curtus* Miller, 1785, type species of its genus. J. Fish. Res. Bd. Can. 25:1923-1969.

46. Perkins, P.S. 1983. The life history of *Cardiodectes medusaeus* (Wilson), a copepod parasite of lanternfishes (Myctophidae). J. Crust. Biol. 3:70-87.

47. Perkins, P.S. 1985. Iron crystals in the attachment organ of the erythrophagous copepod *Cardiodectes medusaeus* (Penneliidae). J. Crust. Biol. 5:581-605.

48. Putz, R.E., and J.T. Bowen. 1968. Parasites of freshwater fishes. IV. Miscellaneous, the anchor worm *(Lernaea cyprinacea)* and related species. U.S. Department of Interior, Bureau of Sport Fisheries and Wildlife, Division of Fisheries Research FDL-12.

49. Reinhard, E.G. 1949. Experiments on the determination and differentiation of sex in the bopyrid *Stegophryxus hyptius* Thompson. Biol. Bull. 96:17-31.

50. Reinhard, E.G. 1956. Parasitic castration of Crustacea. Exp. Parasitol. 5(1):79-107.

51. Ritchie, L.E., and J.T. Høeg. 1981. The life history of *Lernaeodiscus porcellanae* (Cirripedia: Rhizocephala) and coevolution with its porcellanid host. J. Crust. Biol. 1:334-347.

52. Roberts, L.S. 1970. *Ergasilus* (Copepoda: Cyclopoidia): revision and key to species in North America. Trans. Am. Microsc. Soc. 39:134-161.

53. Rogers, W.A., and J.P. Hawke. 1978. The parasitic copepod *Ergasilus* from the skin of the gizzard shad *Dorosoma cepedianum*. Trans. Am. Microsc. Soc. 97:244.

54. Rondelet, G. 1554. Libri de piscibus marinus, vol. 1. Lyon.

55. Sarig, S. 1971. Diseases of fishes, book 3: The prevention and treatment of diseases of warmwater fishes under subtropical conditions, with special emphasis on fish farming. (Snieszko, S.F., and H.R. Axelrod, editors.) T.F.H. Publications, Inc., Neptune City, N.J.

56. Sindermann, C.J. 1970. Principal diseases of marine fish and shellfish. Academic Press, Inc., New York.

57. Tidd, W.M. 1962. Experimental infestations of frog tadpoles by *Lernaea cyprinacea*. J. Parasitol. 48:870.

58. Tokioka, T. 1936. Larval development and metamorphosis of *Argulus japonicus*. Mem. Coll. Sci. Kyoto. Ser. B, 12:93-114.

59. Wilson, C.B. 1911. North American parasitic copepods belonging to the Family Ergasilidae. Proc. U.S. Nat. Mus. 39:263-400.

60. Zmerzlaya, E.I. 1972. *Ergasilus sieboldii* Nordmann, 1832, its development biology and epizootic significance. (English summary.) Izv. Gos-NIORKh 80:132-177.

CHAPTER 36

PARASITIC INSECTS: ORDERS MALLOPHAGA AND ANOPLURA, THE LICE

Riddle: What we caught we threw away; what we didn't we kept.

HOMER

Until recently in human history, lice and fleas were such common companions of *Homo sapiens* that they were considered one of life's inevitable nuisances for rich and poor, royalty and beggar alike. Not so long ago, the education of a princess included instructions that "it was bad manners to scratch when one did it by habit and not by necessity, and that it was improper to take lice or fleas or other vermin by the neck to kill them in company, except in the most intimate circles."[22] Although lice remain widely prevalent, mostly among the very poor and primitive people, we in modern, industrialized countries tend to think of them as pests of the past. Hence many upper- and middle-class parents in the United States today are astonished to receive notes from the school nurse that their offspring must seek treatment for head lice. Lice are part of our cultural heritage, giving rise to words and phrases in daily use, although most people are unconscious of the origins of the idioms. How many students, when complaining that their professor is a "lousy teacher," realize they are literally stating that the august personage is infested with lice? Even fewer persons appreciate the original meanings of the phrases "nit-picking" and "going over with a fine-toothed comb," which refer to removal of louse eggs (nits) from a companion's hair. "Getting down to the nitty-gritty" and "nitwit" may thus assume new dimensions for some readers.

Lice traditionally are assigned to two orders, the Mallophaga and the Anoplura, both of which were probably derived from a common ancestor with the free-living book lice (Psocoptera). Some workers consider the Mallophaga not to be a natural phyletic unit and so join the Anoplura and Mallophaga in a single order, Phthiraptera.[7] We adopt the conventional arrangement, as have other authors. The most critical difference between the two orders is the structure of the mouthparts, modified for chewing in the Mallophaga and for sucking in the Anoplura. Both orders are highly adapted for parasitism: they have no free-living stages, and they soon die when separated from their host.

These insects are wingless, are dorsoventrally flattened, and have reduced or no eyes; their tarsal claws are often enlarged, an adaptation for clinging to hair and feathers.

Development is hemimetabolous. The eggs are cemented to the feathers or hairs of their hosts (Fig. 36-1), and there are three nymphal instars.

ORDER MALLOPHAGA

Mallophagans commonly are referred to as the biting lice, but *chewing* is preferable, since anoplurans bite, too. About 3000 species parasitize various birds and mammals. None is of direct medical importance, but some species may become significant pests on domestic animals. They feed primarily on feathers and hair, but some eat sebaceous secretions, mucus, and sloughed epidermis. Most will eat blood, if available, such as that resulting from scratching by the host. *Menacanthus stramineus,* a louse of chickens and turkeys, chews into developing quills to feed on blood, and some species on small birds pierce the skin to feed on blood.

Morphology

Most lice are small, from 1 to a few millimeters in length. *Laemobothrion circi* is virtually a giant among lice at almost a centimeter long.[1] The head of malloph-

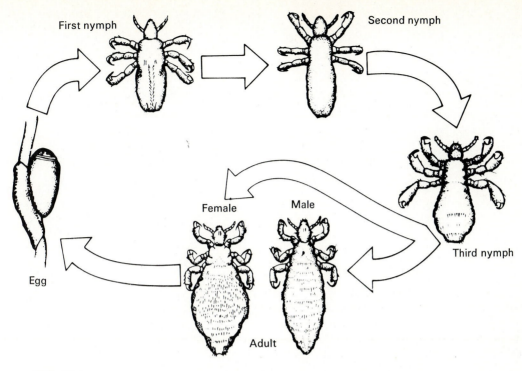

First nymph

Second nymph

Third nymph

Female

Male

Adult

Egg

FIG. 36-1

Life cycle of the head louse, *Pediculus humanus capitis*. The eggs (nits) are cemented to hairs and require 5 to 10 days to hatch. The life cycle requires about 21 days from egg to egg.

From Pratt, H.D., and K.S. Littig. 1973. Lice of public health importance and their control. Department of Health, Education, and Welfare Pub. No. (CDC) 77-8265. U.S. Government Printing Office, Washington, D.C.

agans is usually broader than the prothorax and without ocelli. The short antennae have three to five segments, and the tarsi have one or two segments. Among the parasitic insects, the mouthparts of the Mallophaga are the most similar to the primitive chewing apparatus of the free-living forms (Figs. 34-11 and 34-12). The mandibles are the most conspicuous of these appendages, whereas the maxillae and labium are reduced. The mandibles cut off pieces of feather or hair, and the labrum pushes them into the mouth.[1]

Three suborders of Mallophaga are recognized: Amblycera, Ischnocera, and Rhynchophthirina. The Amblycera are the most generalized and least host specific. They have maxillary palps, which have been lost in the other two suborders, and their antennae are carried in grooves in the head (Fig. 36-2). The filiform, easily seen antennae of the Ischnocera (Fig. 36-3) distinguish them from the Amblycera. The Ischnocera are more specialized, more host specific, and more limited in food preferences. That is, their food is more confined to hairs and feathers (keratin) than that

of others in the order. The Ischnocera of birds are so specialized that they are usually limited to a particular region of their host's body or to a certain part of a feather. The Rhynchophthirina is a much smaller suborder than the other two, comprising only two species: *Haematomyzus elephantis* on African and Indian elephants (Fig. 36-4) and *H. hopkinsi* on warthogs. Although of the chewing type, the mouthparts of these lice are carried at the end of a projecting structure, and the insects feed on blood.

Examples

Since the majority of Mallophaga are parasites of birds, it is not surprising that domestic fowls host a number of species. The most common amblycerans are *Menopon gallinae*, the shaft louse of fowl, and *Menacanthus stramineus*, the yellow body louse of chickens and turkeys. The shaft louse is about 2 mm in length and usually causes little economic loss. *Menacanthus stramineus*, however, is about 3 mm long and may occur in large numbers, up to 35,000 per

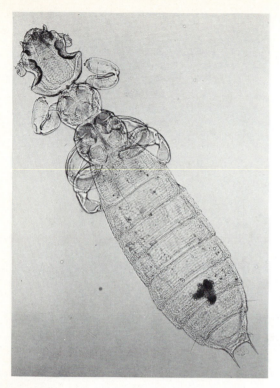

FIG. 36-2

Gliricola porcelli (Mallophaga, Amblycera), a chewing louse of guinea pigs. Antennae are normally held in the deep grooves on the sides of the head.

Photograph by Jay Georgi.

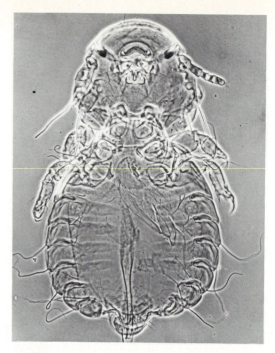

FIG. 36-3

Goniocotes gallinae (Mallophaga, Ischnocera), the fluff louse of fowl. The antennae are clearly visible and do not lie in grooves on the head.

Photograph by Jay Georgi.

bird.[9] It frequents lightly feathered areas such as the breast, the thigh, and around the anus, gnawing through the skin to reach the quills of the pinfeathers. This irritation can cause restlessness and disrupt the bird's feeding. The birds often are unthrifty, with reduced egg production and retarded development.

Important ischnoceran parasites of fowl include *Goniocotes gallinae,* called the fluff louse because it is found in the fluff at the base of the feather (Fig. 36-3); *Goniodes dissimilis,* the brown chicken louse; *Lipeurus caponis,* the wing louse; *Cuclotogaster heterographus,* the chicken head louse; *Chelopistes meleagridis,* the large turkey louse; *Oxylipeurus polytrapzius,* the slender turkey louse; *Columbicola columbae,* the slender pigeon louse (Fig. 36-5); and *Anaticola crassicornis* and *A. anseris,* duck lice.

Amblyceran parasites of mammals include *Gyropus ovalis* and *Gliricola porcelli* (Fig. 36-2) of guinea pigs and *Heterodoxus spiniger* on dogs. The guinea pig parasites are important because of the wide use of their hosts in laboratory experiments. *Heterodoxus spiniger* is common on dogs in warmer parts of the

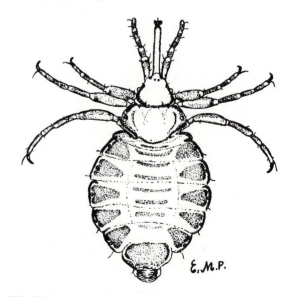

FIG. 36-4

Haematomyzus elephantis (Mallophaga, Rhynchophthirina), a parasite of Indian and African elephants. The chewing mouthparts are at the end of a long proboscis.

From Lapage, G. 1956. Veterinary parasitology. Oliver & Boyd Ltd., London.

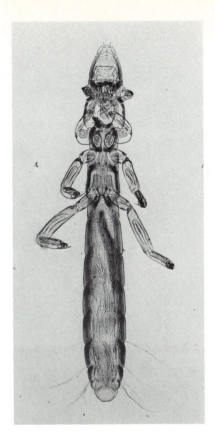

FIG. 36-5

Columbicola columbae (Mallophaga, Ischnocera), the slender pigeon louse.

Photograph by Jay Georgi.

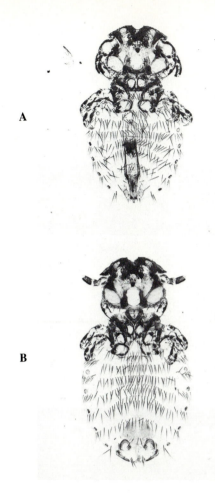

A

B

FIG. 36-6

Trichodectes canis (Mallophaga, Ischnocera), the chewing louse of dogs. **A,** Male. **B,** Female.

Photographs by Jay Georgi.

world. Several ischnocerans are pests of other domestic mammals, including *Bovicola bovis* on cattle; *B. equi* on horses, mules, and donkeys; *B. ovis* on sheep; *B. caprae* on goats; *Trichodectes canis* (Fig. 36-6) on dogs; and *Felicola subrostratus* on cats. The species of *Bovicola,* when abundant, cause considerable irritation to their hosts, although the lice are only 1.5 to 1.8 mm long. *Bovicola bovis* causes cattle to rub against solid objects and bite at their skin in an attempt to alleviate the irritation, with consequent abrasions and hair loss. The reddish brown color of *B. bovis* distinguishes it from the sucking lice commonly found on cattle. Irritation caused by *T. canis* can become severe, especially on puppies. *Trichodectes canis* is an important intermediate host, along with dog and cat fleas, of the tapeworm *Dipylidium caninum,* which also can develop in humans who accidentally ingest the insects while fondling their pets (p. 366). Chewing lice have been implicated as intermediate hosts for several other endoparasites.[9]

Elephants may suffer a severe dermatitis caused by the rhynchophthirinan *Haematomyzus elephantis*[17] (Fig. 36-4).

ORDER ANOPLURA

The sucking lice are a much smaller group than are the Mallophaga, with fewer than 500 species,[10] parasitizing only mammals. Morphologically they are more specialized than the chewing lice, but medically their importance and impact on human history are infinitely greater. Two species parasitize humans, *Pediculus humanus* (Fig. 36-7) and *Phthirus pubis* (Fig. 36-8), of which *P. humanus* is the more impor-

FIG. 36-7

Pediculus humanus (Anoplura), the head and body louse of humans.

Photograph by Warren Buss.

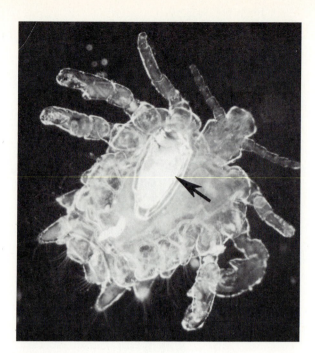

FIG. 36-8

Phthirus pubis (Anoplura), the pubic, or crab, louse of humans. *Arrow.* Developing egg.

Photograph by Warren Buss.

tant. The several species on domestic mammals are of considerable veterinary significance.

Morphology

The anoplurans superficially resemble the chewing lice, with their small, wingless, flattened bodies, but the anopluran head is narrower than the prothorax. The sucking mouthparts are retracted into the head when the animal is not feeding (Fig. 36-9). Each leg has a single tarsal segment with a large claw, an adaptation for clinging to the hairs of the host. The first legs, with their terminal claws, are often smaller than the other legs, and the third legs and their claws are usually largest. Eyes, if present, are small, and there are no ocelli. The antennae are short, clearly visible, and are composed of a scape, a pedicel, and a flagellum that is divided into three subsegments. All three subsegments of the flagellum bear tactile hairs, and subsegments two and three bear chemoreceptors.[19]

Mode of feeding

Lavoipierre[11] distinguished two distinct feeding methods used by bloodsucking arthropods. One of these he termed **solenophage** (from the Greek, "pipe" + "eating") for arthropods that introduce their mouthparts directly into a blood vessel to withdraw blood; the other he called **telmophage** (from the Greek, "pool" + "eating") for those whose mouthparts cut through the skin and vessels to produce and feed from a small pool of blood. Anoplurans are true solenophages.[12] Their proboscis is formed from the maxillae, hypopharynx, and labium, which are produced into long, thin **stylets** (Fig. 36-10). The maxillae are flattened and rolled transversely to form the food canal, and the salivary duct passes down the hypopharynx. The third member of the stylet bundle, or **fascicle**, is the labium, which bears three serrated lobes at its tip. Mandibles are absent in most adult anoplurans.

Feeding of *Haematopinus suis* was observed by Lavoipierre,[12] and the process in other anoplurans is likely to be quite similar. The anterior tip of the head is formed by the labrum, which bears in its interior several strong, recurved teeth (Figs. 36-9 and 36-11).

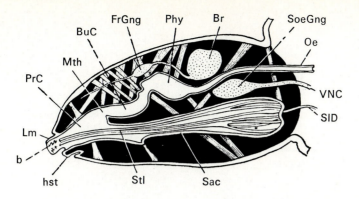

FIG. 36-9

Piercing and sucking apparatus of Anoplura. At rest, the buccal teeth *(b)* are within the labrum *(Lm)*, but when the louse bites its host, the labrum is everted, and the buccal teeth cut into the epidermis of the host. *PrC,* Preoral cavity, or "buccal funnel"; *Mth,* mouth; *BuC,* buccal cavity, first chamber of the sucking pump; *FrGng,* frontal ganglion; *Phy,* pharynx, second chamber of the sucking pump; *Br,* brain; *SoeGng,* subesophageal ganglion; *Oe,* esophagus; *VNC,* ventral nerve cord; *SID,* salivary duct; *Sac,* inverted sac holding the fascicle; *Stl,* stylet bundle, or fascicle; *hst,* hypostome.

From Principles in insect morphology by R.E. Snodgrass. Copyright © 1935 by McGraw-Hill Book Co. Used with permission of McGraw-Hill Book Co.

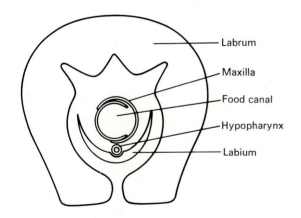

FIG. 36-10

Transverse section through the mouthparts of a sucking louse.

From Askew, R.R. 1971. Parasitic insects. American Elsevier Publishing Co., Inc., New York.

When the louse begins to feed, it places the tip of the labrum on the skin of its host and begins to evert the structure, including the buccal teeth. The teeth serve to cut through the horny outer layer of the skin and, when the labrum is fully everted, are oriented so that their cutting edges point away from the central axis of the labrum (Fig. 36-11). During feeding, the everted labrum and its teeth anchor the louse in place. The stylets evert and probe the tissues until they penetrate a blood vessel, usually a venule, whereupon the louse begins to suck blood. The louse sucks by means of a two-chambered pump in its head, the first chamber comprising the buccal cavity and the second,

the pharynx (Fig. 36-9). Contraction of muscles inserted on the walls of these structures and on the inner surface of the head cuticle serves to dilate the chambers.

How can the louse know, when it is probing the tissue with its fascicle, that it has penetrated a venule and can begin to suck blood? The stimulus for a number of hematophagous insects, presumably including lice, is the detection by chemoreceptors of adenine nucleotides, particularly ADP and ATP.[8] The nucleotides are released by platelets that aggregate in the bitten region as a result of the damage done to the blood vessel by the probing fascicle.

591

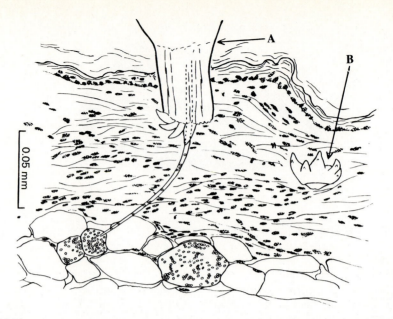

FIG. 36-11

Schematic diagram illustrating the feeding mechanism of *Haemotopinus suis,* drawn from a histologic section of the insect's mouthparts embedded in a mouse's skin. The labrum *(A)* is anchored in the dermis by the everted buccal teeth, and the stylets are inserted in a venule. *B,* Lateral view of everted buccal teeth.

From Lavoipierre, M.M.J. 1967. Exp. Parasitol. 20:303-311.

Pediculus humanus (Fig. 36-7)

Two distinct forms of *P. humanus* parasitize humans, the body louse *(P. humanus humanus)* and the head louse *(P. humanus capitis.)* The body louse also has been called *P. humanus corporis* and *P. humanus vestimenti,* although laypersons may know them by such common names as "cooties," "graybacks," "mechanized dandruff," or even more vulgar appellations. The two subspecies are difficult to distinguish morphologically, although they have slight differences. The subspecies will interbreed and are only slightly interfertile.[1] A very convincing argument for separate species status for the head louse (*Pediculus capitus*) and the body louse (*Pediculus humanus*) was given by Busvine.[6] It seems likely that the body louse descended from a head louse ancestor after humans began wearing clothes. Body lice are much more common in cooler parts of the world; in tropical areas persons who wear few clothes usually have only head lice.[16] This makes typhus (discussed further) a disease of the cooler climates because only body lice are vectors. Curiously, however, head lice can serve as hosts for the typhus organism and have a high potential for transmitting it.[14] Body lice are extremely unusual among the Anoplura in that they spend most of their time in their host's clothing, visiting the host's body only during feeding. They nevertheless stay close to the body and are most commonly found in areas where the clothing is in close contact.

The eggs (nits) of body lice are cemented to fibers in the clothes. They have a cap at one end to admit air and to facilitate hatching (Fig. 36-12). The eggs hatch in about a week, and the combined three nymphal stages usually require 8 to 9 days to mature when they are close to the host's body. Lower temperature lengthens the time of the complete cycle; for example, if the clothing is removed at night, the life cycle will require 2 to 4 weeks. If the clothing is not worn for several days, the lice will die. The female can lay 9 or 10 eggs per day, up to a total of about 300 eggs in her lifetime; therefore she has a high reproductive potential. Fortunately, the potential is usually not realized. It is typical to find no more than 10 lice per host, although as many as a thousand have been removed from the clothes of one person.[16]

Body lice normally do not leave their host voluntarily, but their temperature preferences are rather strict. They will depart when the host's body cools after death or if the person has a high fever. Nevertheless, they travel from one host to another fairly easily, and a person may acquire them by contact with infested people in such congested areas as crowds, buses, and

FIG. 36-12

Nit of *Phthirus pubis* cemented to a hair. The nits of *Pediculus* spp. are essentially similar. Note the operculum with pores.

Micrograph by John Ubelaker.

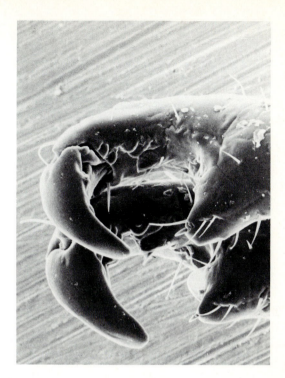

FIG. 36-13

Scanning electron micrograph of the tarsi of the second and third legs of *Phthirus pubis*, with the terminal claw, nicely illustrating the adaptation for grasping the hairs of its host.

Micrograph by John Ubelaker.

trains. Of course, they may be acquired easily by donning infested clothing or occupying bedding recently vacated by a person with lice. Potential for transmission is highest when people are in crowded, institutionalized conditions or in war or prison camps, where sanitation is bad and clothing cannot be changed often.

Head lice tend to be somewhat smaller than body lice: 1 to 1.5 mm for males and 1.8 to 2 mm for females, contrasted with 2 to 3 mm and 2 to 4 mm for male and female body lice, respectively.[16] The nits of both are about 0.8 mm by 0.3 mm. The nits of head lice are cemented to hairs. The lice are usually most prevalent on the back of the neck and behind the ears, and they do not infest the eyebrows and eyelashes. They are easily transmitted by physical contact and stray hairs, even under good sanitary conditions. Accordingly, they sometimes occur among schoolchildren. As in the case of body lice, however, the heaviest infestations are associated with crowded conditions and poor sanitation.

Infestation with lice (**pediculosis**) is not life threatening, unless the lice carry a disease organism, but it

can subject the host to considerable discomfort. The bites cause a red papule to develop, which may continue to exude lymph. The intense pruritis induces scratching, which frequently leads to dermatitis and secondary infection. Symptoms may persist for many days in sensitized persons. Years of infestation lead to a darkened, thickened skin, a condition called **vagabond's disease.** In untreated cases of head lice the hair becomes matted together from exudate, a fungus grows, and the mass develops a fetid odor. This condition is known as **plica polonica.** Large numbers of lice are found under the mat of hair.

Phthirus pubis (Fig. 36-8)

The origin of the common name of this insect, **crab louse** or, more popularly, **crabs,** is evident from its appearance. From 1.5 to 2 mm long and nearly as broad as long, the grasping tarsi on the two larger pairs of legs are reminiscent of a crab's pincer (Fig. 36-13). *Phthirus pubis* dwells primarily in the pubic region, but it may also be found in the armpits and, rarely, in the beard, mustache, eyebrows, and eyelashes. *Phthirus* is less active than is *Pediculus,* and it

FIG. 36-14

Haematopinus suis, the pig louse.

Photograph by Warren Buss.

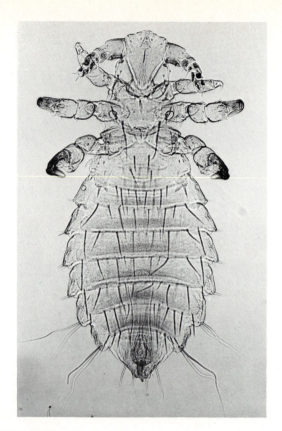

FIG. 36-15.

Polyplax spinulosa (Anoplura), parasitic on brown and black rats. This louse transmits murine typhus from rat to rat, although not from rats to humans. Another species of *Polyplax, P. serrata,* parasitizes mice.

Photograph by Jay Georgi.

may remain in the same position for some time with its mouthparts inserted in the skin. The bites can cause an intense pruritis but fortunately do not seem to transmit disease organisms.

The nits are cemented to hair, and the complete life cycle requires less than a month. The female deposits only about 30 eggs during her life.

Transmission is characteristically venereal, although infection by contact with bedding or other objects can also occur. The sexual revolution of recent years seems to have caused an increase in the frequency of crab lice. Not long ago, one of us received an envelope, addressed to the "Department of Microscopic Analysis, University of Massachusetts," along with the rather urgent instruction, "Please microscope these specimens immediately. They were taken from a living organism." Once could easily sense the distress and embarrassment of the sender.

Other anoplurans of note

Like many of the Mallophaga, the Anoplura tend to be host specific. Interestingly, *Pediculus humanus* can also live and breed on pigs,[1] and *Haematopinus suis* of swine will readily feed on humans when it is hungry.[3] The principal effects on the host are irritation, weight loss, and anemia in heavy infestations.

The USDA estimated that the combined effects of chewing and sucking lice amounted to a $47 million loss each in the cattle and sheep industries in 1965.[21] *Haematopinus suis* (Fig. 36-14) on swine is considered their most serious infection after hog cholera.[21] Other species of *Haematopinus* infest cattle: *H. eurysternus,* the short-nosed cattle louse; *H. quadripertusus,* the cattle tail louse; and *H. tuberculatus,* primarily a parasite of water buffalo. The most serious economic losses resulting from lice on cattle are caused by *H. eurysternus. Haematopinus asini* is a parasite of horses, mules, and donkeys. *Haematopinus* spp. are large, blind lice; female *H. suis* are as long as 6 mm. Species of the genus *Linognathus* parasitize cattle, sheep, goats, and dogs, and *Solenopotes capillatus* is found on cattle. *Pediculus mjobergi* is found on New World monkeys, sometimes becoming a problem in *zoos; it may be a subspecies of P. humanus.*[9] Like Pediculus, different members of the genus *Linognathus* may specialize on different regions of the body: *L. pedalis* is found on the legs of sheep, whereas *L. ovillus* predominates on the head.

Polyplax spinulosa (Fig. 36-15) of *Rattus* can transmit *Rickettsia typhi,* the causative agent of murine typhus, carried also by fleas (Chapter 38).

LICE AS VECTORS OF HUMAN DISEASE

Three important human diseases are transmitted by *Pediculus humanus humanus:* epidemic, or louseborne, typhus; trench fever; and relapsing fever.

Epidemic, or louse-borne, typhus

Typhus is caused by a rickettsial organism, *Rickettsia prowazekii.* Rickettsias are bacteria that usually are obligate intracellular parasites. Various species can inflict vertebrate and/or invertebrate hosts with effects ranging from symptomless to severe. Epidemic typhus has had an enormous impact on human history, detailed in Zinsser's classic book *Rats, Lice and History.*[22] Typhus epidemics tend to coincide with conditions favoring heavy and widely prevalent infestations of body lice, for example, during and after wars and under conditions of crowding, stress, poverty, and mass migration. Mortality rates during epidemics may approach 100%. It is not certain which or how many of the great epidemics of earlier human history were caused by typhus, but historical accounts of the decimation of the Christian and Moorish armies in Spain during 1489 and 1490 are clear. Typhus reduced the French army besieging Naples in 1528 from 25,000 to 4000, leading to its defeat, to the crowning of Charles V of Spain as Holy Roman Emperor, and to the dominance of Spain as a European power for more than a century. The Thirty Years' War can be divided epidemiologically into two periods: 1618 to 1630, when the chief scourge was typhus, and 1630 to 1648, when the major epidemic was plague. Zinsser contends that between 1917 and 1921, there "were no less and probably more than twenty-five million cases of typhus in the territories controlled by the Soviet Republic, with from two and one-half to three million deaths."[22] Numerous other examples could be cited. As recently as 1971, 10,272 cases and 106 deaths were reported throughout the world.[9]

The disease starts with a high fever (39.5° to 40.0° C), which continues for about 2 weeks, and backache, intense headache, and often bronchitis and bronchopneumonia. There is malaise, vertigo, and loss of appetite, and the face becomes flushed. A petechial rash appears by the fifth or sixth day, first in the axilla and on the flanks, then extending to the chest, abdomen, back, and extremities. The palms, soles, and face are rarely affected.[3] After about the second week, the fever drops, and profuse sweating begins. At this point, stupor ends with clearing consciousness, which is followed either by convalescence or by an increased involvement of the central nervous system and death. The rash often remains after death, and subdermal hemorrhagic areas frequently appear. The disease can be treated effectively by broad-spectrum antibiotics of the tetracycline group and chloramphenicol. Also, although prior vaccination with killed *R. prowazekii* does not result in complete protection, the severity of the disease is greatly ameliorated in persons who have been vaccinated.

Curiously, typhus is a fatal disease for lice. When the louse picks up the rickettsia along with the blood meal from the human host, the organism invades the louse's gut epithelial cells and multiplies so plentifully that the cells become distended and rupture. After about 10 days, so much damage has been done to the insect's gut that the louse dies. For several days before its demise, however, the louse's feces contain large numbers of the rickettsias. Scratching the louse bites, the human is inoculated with the typhus organism from the louse feces or when the offending creature is crushed. The louse's strong preference for normal body temperature causes it to leave the febrile patient and search for a new host, thus aiding in the rapid spread of the disease in epidemics. A person can also become infected with typhus by inhaling dried louse feces or getting them in the eye. *Rickettsia prowazekii* can remain viable in dried louse feces for as long as 60 days at room temperature.[9]

Because the infection is fatal to lice, transovarial transmission cannot occur, and humans are an important reservoir host. After surviving the acute phase of the disease, humans can be asymptomatic but capable of infecting lice for many years. The disease can recrudesce and produce a mild form known as **Brill-Zinsser disease.** Flying squirrels (*Glaucomys volans*) also can be a reservoir host, with the infection transmitted by lice (*Neohaematopinus sciuropteri*) and fleas (*Orchopeas howardii*).[2,20] Some recent cases in the United States are probably caused by contact with such animals.[13] The human and possibly the animal reservoirs could provide the source to begin a new epidemic in the event of a war, famine, or other disaster. As Harwood and James[9] point out, "Current standards of living in well-developed countries have largely eliminated the disease there, but its cause lies smoldering, ready to erupt quickly and violently under conditions favorable to it."

Tragically, both Ricketts and Prowazek, the pioneers of typhus research, became infected with typhus and died in the course of their work.

595

Trench fever

Trench fever is a nonfatal but very debilitating disease caused by another rickettsia, *Rochalimaea quintana,* transmitted by *Pediculus humanus humanus.* Epidemics occurred in Europe during World Wars I and II, and foci have since been discovered in Egypt, Algeria, Ethiopia, Burundi, Japan, China, Mexico, and Bolivia. *Rochalimaea quintana* is unusual for a pathogenic rickettsia in that it grows extracellularly and even can be cultivated on a cell-free, blood-agar plate. In the louse the rickettsia multiples in the lumen of the gut. Infection of humans occurs by contamination of abraded skin with louse feces or a crushed louse or by inhalation of louse feces. The organism is not pathogenic for the louse; thus, the vector remains infective for the duration of its life.

A latent infection period lasts about 10 to 30 days, toward the end of which the person may experience headache, body pain, and malaise. The temperature then rises rapidly to 39.5° to 40.0° C, accompanied by headache; pain in the back and legs, especially in the shins; dizziness; and postorbital pain in movement of the eyes. A typhus-like rash appears, usually early in the attack, on the chest, back, and abdomen but disappears within 24 hours. The fever continues for as long as a week and occasionally for several weeks. Convalescence is often slow, and the initial attack is followed in about half the cases by a regularly or irregularly relapsing fever curve. Tetracyclines are effective in treatment.

Humans are the primary reservoirs, and *R. quintana* has been recovered from the blood of convalescents as long as 8 years after the initial attack.[5]

Relapsing fever

The third important disease of humans transmitted by body lice is epidemic relapsing fever, which is caused by a spirochete, *Borrelia recurrentis.* Mortality is usually low, but the fatality rate can reach more than 50% in groups of undernourished people.[16] The louse picks up the bacterium along with the blood meal, and the spirochete penetrates the insect's gut to reach the hemocoel. It multiples in the hemolymph but does not invade the salivary glands, gonads, or Malpighian tubules. Therefore the spirochetes are not injected into the next host with the louse saliva; neither are they transmitted by the louse transovarially to its progeny or egested in its feces. Transmission is accomplished only when the louse is crushed by host scratching, which releases the spirochetes in the hemolymph. Hence the infectious organisms gain entrance through abraded skin, but evidence also indicates that they can penetrate unbroken skin.[4] Louse-borne relapsing fever apparently has disappeared from the United States, but scattered foci are in South America, Europe, Africa, and Asia. Ethiopia had 4700 cases and 29 deaths in 1971.[9] Frequent epidemics occurred in Europe during the eighteenth and nineteenth centuries, and major epidemics befell Russia, Central Europe, and North Africa during and after World Wars I and II. During the war in Vietnam an epidemic occurred in the Democratic People's Republic of Vietnam.[15]

Clinically, louse-borne relapsing fever is indistinguishable from the tick-borne relapsing fevers that are caused by other species of *Borrelia* (Chapter 41). After an incubation period of 2 to 10 days, the victim is struck rather suddenly by headache, dizziness, muscle pain, and a fever that develops rapidly to 40.0° to 40.5° C. Transitory rash is common, especially around the neck and shoulders, then extending to the chest and abdomen. The patient is severely ill for 4 to 5 days, when the temperature suddenly falls, accompanied by profuse sweating. Considerable improvement is seen for 3 to 10 days, and then another acute attack occurs. The cycle may be repeated several times in untreated cases. Antibiotic treatment is effective but complicated in this disease by serious systemic reactions to the drugs.

Humans are the only reservoirs, and epidemics are associated with the same kind of conditions as louse-borne typhus. The diseases often occur together.

CONTROL OF LICE

For complete information on control of lice on humans, the reader should consult Pratt and Littig.[16] A variety of commercial preparations containing insecticides effective against lice are available. No fewer than six brands recently were found on the shelves of a supermarket in Lubbock, Texas. Good personal hygiene with ordinary laundering of garments, including dry cleaning of woolens, will control body lice. Devices for large-scale treatment of civilian populations, troops, and prisoners of war, which blow insecticide dust into clothing, are effective and have controlled or prevented typhus epidemics.

Lice on pets and domestic animals can be controlled by insecticidal dusts and dips. Normal, healthy mammals and birds usually apply some natural louse control by grooming and preening themselves. Poorly nourished or sick animals that do not exhibit normal grooming behavior often are heavily infested with lice. Many species of passerine birds show an interesting behavior known as "anting." This may represent another natural method of louse control, although not one that many humans would want to imitate. The bird settles on the ground and fluffs up its feathers near a colony of ants, allowing the ants to crawl into its plumage or even picking up ants in its bill and applying them to the feathers. However, the bird uses only ant species whose workers exude or spray toxic substances in attack and defense but do not sting. Ants in two subfamilies of Formicidae are acceptable, and these either spray formic acid or exude droplets of a repugnatorial fluid from their anus.[18] The worker ants, angry at the bird intruder, liberally anoint its feathers with noxious fluids. Numbers of dead and dying lice have been found in the plumage of birds immediately after anting. Anting very likely evolved as an adaptation to combat ectoparasites, although this may not be its only function.

REFERENCES

1. Askew, R.R. 1971. Parasitic insects. American Elsevier Publishing Co., Inc., New York.
2. Bozeman, F.M., S.A. Masiello, M.S. Williams, and B.L. Elisberg, 1975. Epidemic typhus rickettsiae isolated from flying squirrels. Nature 255:545-547.
3. Burgdorfer, W. 1976. Epidemic (louse-borne) typhus. In Hunter, G.W., J.C. Swartzwelder, and D.F. Clyde. Tropical medicine, ed. 5. W.B. Saunders Co., Philadelphia.
4. Burgdorfer, W. 1976. The relapsing fevers. In Hunter, G.W., J.C. Swartzwelder, and D.F. Clyde. Tropical medicine, ed. 5. W.B. Saunders Co., Philadelphia.
5. Burgdorfer, W. 1976. Trench fever. In Hunter, G.W., J.C. Swartzwelder, and D.F. Clyde. Tropical medicine, ed. 5. W.B. Saunders Co., Philadelphia.
6. Busvine, J.R. 1978. Evidence from double infestations for the specific status of human head lice and body lice (Anoplura). Syst. Ent. 3:1-8.
7. Clay, T. 1970. The Amblycera (Phthiraptera: Insecta). Bull. Br. Mus. Nat. Hist. (Entomol.) 25:73-98.
8. Galun, R. 1977. The physiology of hematophagous insect/animal host relationships. In White, D., editor. Proceedings of the Fifteenth International Congress of Entomology. Entomological Society of America, College Park, Md., pp. 257-265.
9. Harwood, R.F., and M.T. James. 1979. Entomology in human and animal health, ed. 7. Macmillan Publishing Co., Inc., New York.
10. Kim, K.C., and H.W. Ludwig. 1978. The family classification of the Anoplura. Syst. Entomol. 3:249-284.
11. Lavoipierre, M.M.J. 1965. Feeding mechanisms of blood-sucking arthropods. Nature 208:302-303.
12. Lavoipierre, M.M.J. 1967. Feeding mechanism of *Haematopinus suis,* on the transilluminated mouse ear. Exp. Parasitol. 20:303-311.
13. McDade, J.E., C.C. Shepard, M.A. Redus, V.F. Newhouse, and J.D. Smith. 1980. Evidence of *Rickettsia prowazekii* infections in the United States. Am. J. Trop. Med. Hyg. 29:277-284.
14. Murray, E.S., and S.B. Torrey. 1975. Virulence of *Rickettsia prowazekii* for head lice. Ann. N.Y. Acad. Sci. 266:25-34.
15. Pan American Health Organization. 1973. Proceedings of the International Symposium on the Control of Lice and Louse-borne Diseases. PAHO Scientific Pub. No. 263, Washington, D.C.
16. Pratt, H.D., and K.S. Littig. 1973. Lice of public health importance and their control. U.S. Department of Health, Education, and Welfare Pub. No. (CDC) 77-8265. U.S. Government Printing Office, Washington, D.C.
17. Raghavan, R.S., K.R. Reddy, and G.A. Khan. 1968. Dermatitis in elephants caused by the louse *Haematomyzus elephantis* (Piaget, 1869). Indian Vet. J. 45:700-701.
18. Simmons, K.E.L. 1966. Anting and the problem of self-stimulation. J. Zool. 149:145-162.
19. Slifer, E.H., and S.S. Sekhon. 1980. Sense organs on the antennal flagellum of the human louse, *Pediculus humanus* (Anoplura). J. Morphol. 164:161-166.
20. Sonenshine, D.E., F.M. Bozeman, M.S. Williams, S.A. Masiello, D.P. Chadwick, N.I. Stocks, D.M. Lauer, and B.L. Elisberg. 1978. Epizootiology of epidemic typhus *(Rickettsia prowazekii)* in flying squirrels. Am. J. Trop. Med. Hyg. 27:339-349.
21. Steelman, C.D. 1976. Effects of external and internal arthropod parasites on domestic livestock production. Ann. Rev. Entomol. 21:155-178.
22. Zinsser, H. 1934. Rats, lice and history. Little, Brown & Co., Boston.

CHAPTER 37

PARASITIC INSECTS: ORDER HEMIPTERA, THE BUGS

Although the layperson often refers to all insects, and sometimes even bacteria and viruses, as "bugs," to the zoologist the only true bugs are members of the order Hemiptera. It is one of the larger insect orders, containing more than 55,000 species, but relatively few (about 100 species) are ectoparasites of mammals and birds.[1] Two suborders are commonly recognized, the Homoptera and the Heteroptera. The Homoptera (considered a separate order by some authors) include such insects as aphids, scale insects, leaf hoppers, and cicadas. Because the Homoptera feed on the juices of plants, they are not of public health or veterinary importance, but in many cases they are very important agricultural pests. Both the Homoptera and the Heteroptera have sucking mouthparts, which are folded back beneath the head and thorax when at rest (Fig. 37-1). The members of the suborders with wings are easily distinguished: both pairs of wings of homopterans are wholly membranous, whereas the forewings (**hemielytra**) of heteropterans are divided into a heavier, leathery proximal portion and a membranous distal portion.

Hemipterans are exopterygotes with hemimetabolous development; consequently, the nymphs have habitat and feeding preferences similar to those of the adults. Like the Homoptera, most heteropterans suck the juices of plants, but many are predatory, using their mouthparts to suck the body fluids of smaller arthropods. Although some normally plant-feeding forms suck blood in rare instances,[13] relatively few heteropterans obtain most or all of their nutrition from the blood of birds or mammals. Most of these are found in the families Cimicidae and Reduviidae.

MOUTHPARTS AND FEEDING

Regardless of whether they feed on plant or animal juices, the basic plan of homopteran and heteropteran mouthparts is similar.[4] The labrum is short and inconspicuous, but the labium is elongate and forms a tube containing the mandibles and maxillae (Fig. 37-2).

The maxillae enclose a food canal, through which the fluid food is drawn, and a salivary canal, through which the saliva is injected. The mandibles run alongside the maxillae in the labial tube and, together with the maxillae, comprise the **fascicle.** The tips of the mandibles and maxillae may be barbed or spined.

The operation of the mouthparts in the bedbug *Cimex* and in the reduviid *Rhodnius* has been studied.[2,6] Both are solenophages. *Rhodnius* applies the tip of its labial tube to the skin of the host, anchors itself with the barbed tips of the mandibles, and slides the maxillae against each other, piercing the dermal tissue (Fig. 37-3). The route of the maxillary penetration curves because the tip of one maxilla is hooked and the other is spiny. In *Cimex* the entire fascicle penetrates, and the labium folds at its joints (Fig. 37-4). In both cases, feeding begins when a blood vessel is pierced. Some, perhaps all, bloodsucking bugs have sensory receptors on the tips of the mandibles and maxillae,[10] which are undoubtedly essential to the success of the feeding process.

Like the lice, both the bloodsucking reduviids (Triatominae) and the Cimicidae depend for part of their nutrition on endosymbiotic bacteria. The bacteria are contained in the epithelial cells of the triatomine gut, and the cimicids bear theirs in two disc-shaped **mycetomes** in the abdomen beside the gonads (Fig. 37-5). The mutualistic bacteria are essential for the growth and maturation of the bugs; aposymbiotic triatomines, those which were artificially "cured" of their bacteria, can reach only the second, third, or fourth instars before death, depending on the species.[5]

FAMILY CIMICIDAE

The cimicids include a variety of small, wingless bugs that feed on the blood of warm-blooded animals, primarily birds and bats. *Cimex lectularius, C. hemipterus,* and *Leptocimex boueti* are known as bedbugs and attack humans. *Cimex lectularius* is cosmopolitan but is found primarily in the temperate zones,

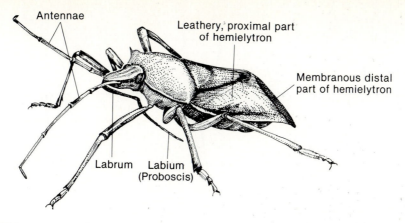

FIG. 37-1

Diagram of typical bug (Hemiptera). The labium forms a tube within which lies the fascicle (maxillae and mandibles.)

Drawing by Ian Grant.

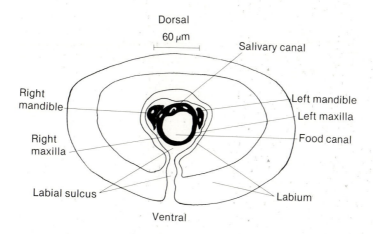

FIG. 37-2

Cross section of the proboscis of the nonfeeding *Rhodnius prolixus* close to the base of the labium.

From Lavoipierre, M.M.J., G. Dickerson, and R.M. Gordon. 1959. Ann. Trop. Med. Parasitol. 53:235-250.

whereas *C. hemipterus* tends to be more tropical, and *L. boueti* is confined to West Africa. Of the 22 genera of cimicids, 12 are parasites of bats. The bird feeders of the group most often attack birds that commonly nest in caves. Few other animals serve as hosts. It is believed that caves were probably the home of the ancestors of the Cimicidae,[1] and it is not unreasonable to assume that humans acquired bedbugs during their cave-dwelling period. All three species that parasitize humans will also feed on bats. In addition, *Cimex* spp. will feed on chickens, and *C. lectularius* readily attacks rodents and some other domestic animals. Although bedbugs are not known to carry any human

disease, they can be extremely annoying. Persons who are chronically infested by bedbugs suffer loss of sleep, sores from infected bites, iron and hemaglobin deficiencies, and rarely mechanical transmission of hepatitis B virus.[8] For an excellent treatment of the taxonomy, ecology, morphology, reproduction, and control of cimicids, the student should consult Usinger's monograph.[12]

Morphology

Cimicids are reddish brown bugs, up to about 8 mm long (Fig. 37-6). Their appendages are not particularly adapted for clinging to their hosts, and they can run

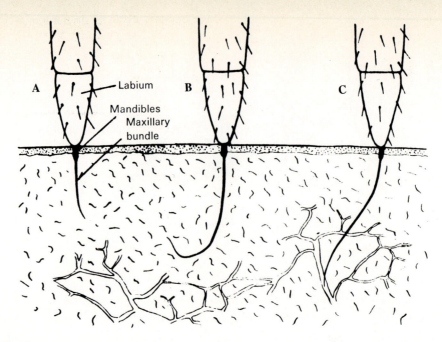

FIG. 37-3

Schematic diagram showing successive stages in the introduction of the fascicle of *Rhodnius prolixus* into the skin of a rodent. **A,** The maxillary bundle is being thrust into the tissues. **B,** Probing has commenced, and the flexible maxillary bundle is shown bending sharply. **C,** The tip of the maxillary bundle has entered the lumen of a vessel. The barbed mandibles act as anchors to the fascicle while the maxillae are projected deep into the tissues.

From Lavoipierre, M.M.J., G. Dickerson, and R.M. Gordon. 1959. Ann. Trop. Med. Parasitol. 53:235-250.

FIG. 37-4

Schematic diagram showing successive stages in the introduction of the fascicle of *Cimex lectularius* into the ear of a rodent. **A,** The fascicle (mandibles and maxillae) is being thrust into the tissues. **B,** Probing has commenced, and the flexible fascicle is shown bending in the tissues. **C,** The tip of the maxillary bundle has entered the lumen of a vessel, but the mandibles remain outside. Both the mandibles and the maxillae enter and probe the tissues of the host as a compact bundle (fascicle), and the labium is progressively bent to allow the fascicle to be projected deep into the tissues.

From Dickerson,G., and M.M.J. Lavoipierre. 1959. Ann. Trop. Med. Parasitol. 53:347-357.

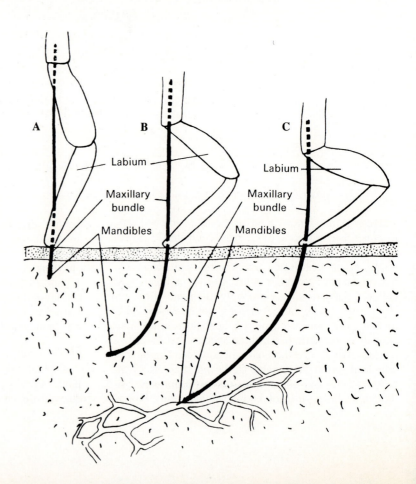

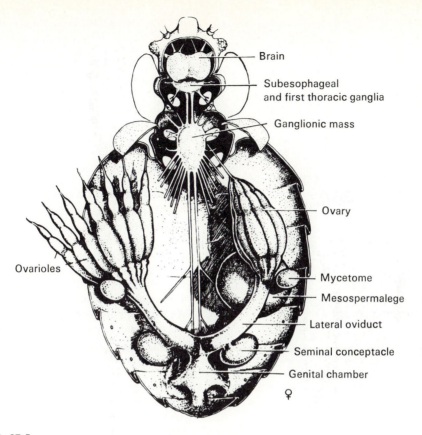

FIG. 37-5

Internal anatomy of *Cimex lectularius*. Female reproductive organs, mycetomes, and parts of nervous system.

Drawing by Catts; from Usinger, R.L. 1966. Monograph of Cimicidae (Hemiptera-Heteroptera). Entomological Society of America, College Park, Md.

rather rapidly, perhaps because they do not stay on their hosts for periods any longer than the 5 to 10 minutes of feeding. They are flattened dorsoventrally. The adults have no wings, although rudimentary wing pads form in the nymphs. Doubtless, this is an adaptation for inhabiting the narrow crevices in which they reside between feedings. Their antennae have four segments, and the distal two are much more slender than are the medial ones. Cimicids have conspicuous compound eyes but no ocelli, and the first thoracic somite forms a rim around the posterior portion of the head. The tergum of the first thoracic somite (**pronotum**) of *C. lectularius* is about two and a half times broader than long, the pronotum of *C. hemipterus* is just more than twice as broad as long, and the pronotum of *L. boueti* is only a bit wider than the head. Scent glands opening on the ventral side of the third thoracic somite produce an oily secretion and give bedbug-infested dwell-

ings a disagreeable odor. The secretion is probably a defense against predators.

Biology

Bedbugs are nocturnal, emerging from their daytime hiding places to feed on their resting hosts during the night. The peak activity of *C. lectularius* is just before dawn.[7] The bites cause little reaction in some people, whereas in others they cause considerable inflammation as a result of allergic reactions to the bug's saliva. The annoyance may disturb sleep, and persistent feeding may reduce the person's hemoglobin count significantly.[4]

Bedbugs can survive long periods of starvation. Adults commonly live without food for more than 4 months and have been recorded to survive up to 18 months.[1] Cannibalism is common.[3]

Cimicids practice a rather startling type of copula-

FIG. 37-6

Cimex lectularius.

Photograh by Warren Buss.

tion known as traumatic insemination, employed among other insects only by the related families Anthocoridae and Polyctenidae.[1] The male has a copulatory appendage, the **paramere,** which curves strongly to the left; the right paramere has been lost. The aedeagus is small and lies immediately above the base of the paramere. During copulation the paramere of the male *Cimex* stabs into a notch (paragenital sinus) near the right side of the posterior border of the female's fifth abdominal sternite[12] (Fig. 37-7). The sperm enter a pocket **(spermalege)** from which they emerge into the hemocoel and make their way to organs at the base of the oviducts, the **seminal conceptacles** (Figs. 37-5 and 37-7). The seminal conceptacles are analogous, but not homologous, to the spermathecae of other insects. From the conceptacles, the sperm travel by minute ducts in the walls of the oviducts to the ovarioles and ova. Male cimicids mate with females repeat-

edly, and homosexual behavior is common. In the laboratory they enthusiastically mate with females of other species, although viable young are not produced from these interspecific matings.

The female lays from 200 to 500 eggs in batches of 10 to 50. The eggs hatch in about 10 days, and the five nymphal instars must each have at least one blood meal. A blood meal is also necessary before the males will mate and the females oviposit. The time from egg to maturity is between 37 and 128 days, but this time is lengthened by periods of starvation.

Epidemiology and control

Their shape makes it possible for bedbugs to insinuate themselves into a variety of tight places. Their daytime sites can be seams of mattresses and box springs, wooden bedsteads, cracks in the wall, behind loose wallpaper, and so on. These sites may be some distance from their host, which they find by following a temperature, and perhaps a carbon dioxide, gradient. They can be transported from one dwelling to another in secondhand furniture, suitcases, bedding, laundry, and other items. Only a single female is required to create the nucleus of a new infestation. Transmission in public conveyances and gathering places occurs frequently.[4]

Control of bedbugs by application of residual insecticides to the areas of likely hiding places is usually effective, although resistance to some insecticides has been encountered. A high level of domestic cleanliness certainly helps in control.

FAMILY REDUVIIDAE

Most reduviids are predators on other insects and are commonly called "assassin bugs" for this reason. They often are valuable for their predation on pest species; *Reduvius personatus* may even enter houses and feed on bedbugs! Most of these can, but usually do not, bite humans, and the bite is quite painful. One of us has vivid memories of a college field trip on which an unidentified reduviid was picked up carelessly with the fingers—unidentified because the insect was abruptly released when it bit one of the careless fingers.

One subfamily of the Reduviidae, the Triatominae, is of great public health significance because its members are vectors for *Trypanosoma cruzi,* the causative agent of Chagas' disease (Chapter 5). The prevalence of Chagas' disease is currently estimated at more than 10 million cases in South and Central America.[4] The Triatominae characteristically feed on blood of various vertebrates. In contrast to the assassin bug, as might be expected of forms that must suck blood several

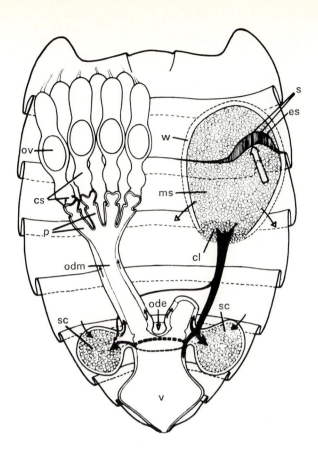

FIG. 37-7

Diagram of paragenital system and process of insemination in Cimicidae, based principally on *Cimex* type. Right ovary and nearly all of corresponding lateral oviduct omitted, and spermalege shown farther forward than is actually the case in *Cimex*. *Broad white arrow,* Course followed by paramere of male in reaching ectospermalege (hatched and crossed by three black bands representing scars of copulation). *Black arrows,* Normal routes of migration of spermatozoa from mesospermalege to bases of ovarioles. *Small arrows with white points,* Migratory routes never or rarely used in *Cimex* but seen in other Cimicidae. *cl,* Conductor lobe; *cs,* syncitial body; *es,* ectospermalege; *ms,* mesospermalege; *ode,* paired ectodermal oviduct; *odm,* paired mesodermal oviduct; *ov,* oocyte; *p,* pedicel; *s,* scars or traces of copulation; *sc,* seminal conceptacle; *v,* vagina; *w,* wall of mesospermalege.

From Usinger, R.L. 1966. Monograph of Cimicidae (Hemiptera-Heteroptera). Entomological Society of America, College Park, Md.

minutes unnoticed by the host, their bite is essentially painless. They are called "kissing bugs" because they often bite the lips of sleeping persons.

Morphology

Triatomines (Fig. 37-8) are relatively large bugs, ranging up to 34 mm in length. They usually have wings, which are held in a concavity on top of the abdomen. The head is narrow, and large eyes are located midway or far back on the sides of the head. Two ocelli may be present behind the eyes. The antennae are slender and in four segments. The apparently three-segmented labial tube folds backward at rest into a groove between the forelegs. The bugs can make a squeaking sound by rubbing the labium against ridges in the groove (stridulation).

Biology

The various reduviids characteristically frequent different sites; for example, some species are normally found on the ground, some in trees, and some in human dwellings. The eggs, numbering from a few dozen to a thousand, depending on species, are deposited in the normal habitat of the adult. There are usually five nymphal instars. Triatomines do not seem to be very choosy about their food sources; whatever vertebrate is available in their habitat is apparently acceptable. Triatomines that inhabit human dwellings feed on humans, dogs, cats, and rats; other species depend more on wild animals.

Epidemiology and control

The epidemiology of trypanosomiasis cruzi is discussed further in Chapter 5. Apparently all species of triatomines are suitable hosts for *T. cruzi*. The importance of a particular species depends on its domesticity (**synanthropism**). Although certain species may be more or less susceptible to local strains of *T. cruzi*, the most important vectors are *Panstrongylus megistus*, *Triatoma infestans*, *T. dimidiata*, and *Rhodnius prolixus*. The relative importance of each varies with locality (Fig. 37-9). The insects are nocturnal and hide by day in cracks, crevices, and roof thatching. Poorly constructed houses are thus a significant epidemiological factor.

Dogs, cats, and rats are important reservoir hosts

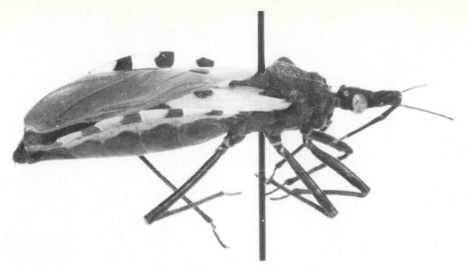

FIG. 37-8

Specimen of *Triatoma dimidiata* discovered feeding on a parasitologist.
Photograph by Warren Buss.

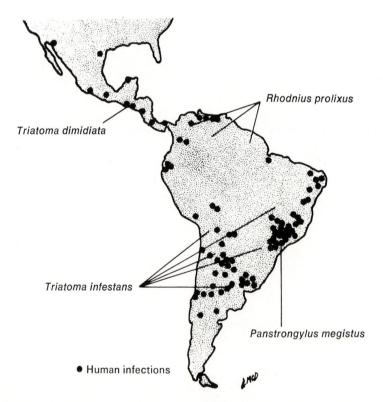

FIG. 37-9

Distribution of Chagas' disease in humans and of its four principal vectors.
AFIP neg. no. 65-5015.

around human habitations, and there is a wide variety of sylvatic reservoirs, the most important of which is the opossum, *Didelphis marsupialis*. The opossum is a common and successful marsupial occurring from the northern United States to Argentina. Other important reservoirs include armadillos, bats, squirrels, wild rats and mice, guinea pigs, and sloths. The number of triatomines in a house increases with the number of persons living in that house.[9]

Numbers of triatomines in a household can be reduced by reducing the number of hiding places for the bugs, that is, by improvements in construction. Replacement of thatched roofs with sheet metal is of great aid, and even whitewashing mud walls helps. Reducing other food sources for the bugs around the dwelling, such as dogs, birds, and rats, is of value. For example, removal of stacked firewood from near houses, and replacement of dirt floors with concrete, nearly eliminates *Triatoma dimidiata* infestation.[14] As with bedbugs, residual insecticides around the potential hiding places are effective in control. Precocene II, a natural product extracted from the plant *Ageratum* sp., shows promise as a fumigant against triatomines. It is cytotoxic to the corpora allata, preventing production of juvenile hormone. Precocene blocks oogenesis in adult females and causes immatures to molt precociously to sterile adults.[11]

Triatomine bugs are known in the United States from New England to California. Similarly, *T. cruzi* has been found from coast to coast in wild mammals, including wood rats, raccoons, opossums, and skunks. Recently, several cases of human infection have been diagnosed in Arizona.

REFERENCES

1. Askew, R.R. 1971. Parasitic insects. American Elsevier Publishing Co., Inc., New York.
2. Dickerson, G., and M.M.J. Lavoipierre. 1959. Studies on the methods of feeding of blood-sucking arthropods. II. The method of feeding adopted by the bed-bug *(Cimex lectularius)* when obtaining a blood-meal from the mammalian host. Ann. Trop. Med. Parasitol. 53:347-357.
3. Durden, L.A. 1987. Predator-prey interactions between ectoparasites. Parasitol. Today 3:306-308.
4. Harwood, R.F., and M.T. James. 1979. Entomology in human and animal health, ed. 7. Macmillan Publishing Co., Inc., New York.
5. Koch, A. 1967. Insects and their endosymbionts. In Henry, S.M., editor. Symbiosis, vol. 2. Academic Press, Inc., New York, pp. 1-106.
6. Lavoipierre, M.M.J., G. Dickerson, and R.M. Gordon. 1959. Studies on the methods of feeding of blood-sucking arthropods. I. The manner in which triatomine bugs obtain their blood-meal, as observed in the tissues of the living rodent, with some remarks on the effects of the bite on human volunteers. Ann. Trop. Med. Parasitol. 53:235-250.
7. Mellanby, K. 1939. The physiology and activity of the bedbug *(Cimex lectularius* L.) in a natural infestation. Parasitology 31:200-211.
8. Newberry, K., and E.J. Jansen. 1986. The common bedbug *Cimex lectularius* in African huts. Trans. R. Soc. Trop. Med. Hyg. 80:653-658.
9. Piesman, J., I.A. Sherlock, and H.A. Christensen. 1983. Host availability limits population density of *Panstrongylus megistus*. Am. J. Trop. Med. Hyg. 32:1445-1450.
10. Pinet, J.M., J. Bernard, and J. Boistel. 1969. Etude électrophysiologique des récepteurs des stylets chez une punaise hématophage: *Triatoma infestans*. C.R. Séances Soc. Biol. 163:1939-1946.
11. Tarrant, C., E.W. Cupp, and W.S. Bowers. 1982. The effects of precocene II on reproduction and development of triatomine bugs (Reduviidae: Triatominae). Am. J. Trop. Med. Hyg. 31:416-420.
12. Usinger, R.L. 1966. Monograph of Cimicidae (Hemiptera-Heteroptera). Entomological Society of America, College Park, Md.
13. Usinger, R.L., and J.G. Myers. 1929. Facultative bloodsucking in phytophagous Hemiptera. Parasitology 21:472-480.
14. Zeledón, R., and L.G. Vargas. 1984. The role of dirt floors and of firewood in rural dwellings in the epidemiology of Chagas' disease in Costa Rica. Am. J. Trop. Med. Hyg. 33:232-235.

CHAPTER **38**

PARASITIC INSECTS: ORDER SIPHONAPTERA, THE FLEAS

"The combined effects of Nero and Kubla Khan, of Napolean and Hitler, all the Popes, all the Pharoahs, and all the incumbents of the Ottoman throne are as a puff of smoke against the typhoon blast of fleas' ravages through the ages."[13] How can this be? Ravages of *fleas?* The apparent exaggeration disappears when one realizes that fleas transmit the dreaded plague, the killer of millions of people from the dawn of civilization through the beginning of the twentieth century. Of all humanity's major diseases and important insect pests, the combination of the flea and plague bacillus has had the greatest impact on human history. We will return to the subject of plague later in this chapter.

The 2000 species of fleas are small insects, from just less than a millimeter to a few millimeters long. Most are parasites of mammals, but approximately a hundred species regularly are found on birds. They are rather heavily sclerotized, bilaterally flattened, and secondarily wingless. Loss of wings is an advanced condition commonly found in parasitic insects. Some are tan or yellow, but they are commonly reddish brown to black. Their mouthparts are of the piercing-sucking type; the adults feed exclusively on blood. The larvae usually are not parasitic but feed on debris and materials associated with the nest or surroundings of the host, especially the feces of the adult fleas. Strong evidence suggests that fleas descended from a winged ancestor much like the present-day scorpion flies (Mecoptera). In fact several features of the jumping mechanism, which is well developed in most fleas, seem to be homologous with flight structures of flying insects.[17]

MORPHOLOGY

The head is broadly joined to the thorax and often bears a **genal ctenidium** (Fig. 38-1). Ctenidia are series of rather stout, peglike spines often found on the posterior margin of the first thoracic tergite (**pronotal ctenidium**) as well as on the head. Many species lack ctenidia, however. The ctenidia and the backwardly

directed setae on the body are adaptations that help the flea retain itself among the fur or feathers of its host. The width of the space between adjacent spine tips in the ctenidia of a particular flea species is correlated with the diameter of the hairs of its usual host, being slightly wider than the maximal diameter of its host's hairs.[7] Thus, in backward movement, the hairs tend to catch between the ctenidial spines. Because of the resistance to being dragged backward, removal of the flea by host grooming or preening is much more difficult. Obviously, these structures do not impede forward progress of the flea between the hairs; neither do the antennae, which fold back into grooves on the sides of the head. The antennae appear trisegmented, but the apparent terminal segment of the antenna actually consists of 9 or 10 segments. When present, the eyes are simple and have only a single, small lens. Ocelli are absent. Fleas bear a peculiar sensory organ near their posterior end called the **pygidium** that apparently functions to detect air currents.[2]

The legs are strong, and the hind legs are commonly much larger than are the other two pairs and are modified for jumping.

Jumping mechanism

Many fleas are champion jumpers. The **oriental rat flea,** *Xenopsylla cheopis,* can jump more than a hundred times its body length, and cat and human fleas are capable of a standing leap of 33 cm high.[20] In terms of proportionate body length, this would be the equivalent of a 6-foot human executing a standing high jump of almost 800 feet! It is not at all self-evident how fleas accomplish this remarkable feat. *Xenopsylla* reaches an acceleration of $140 \times g$ in a little more than a millisecond, yet the fastest single muscle twitches (as in the locust, for example) require 15 milliseconds to reach peak force. The answer lies in the use of the flight structures that the flea inherited from its ancestors. Fleas have **resilin** in the **pleural arch,** an area between the internal ridges of the metapleuron and metanotum (Fig. 38-1). The pleural arch is ho-

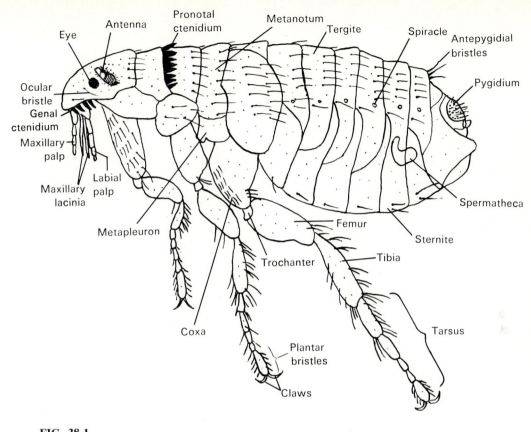

FIG. 38-1

Diagram of a flea.

mologous to the wing-hinge ligament in dragonflies, locusts, and scorpion flies.[20] Resilin (p. 531) is a stabilized protein with highly unusual elastic properties. It is a better "rubber" than rubber, releasing 97% of its stored energy on returning from a stretched position, compared with only 85% in most commercial rubber. When the flea prepares to jump, it rotates its hind femurs up so that they lie almost parallel to the coxae, and the flea is resting on the tarsi of its front two pairs of legs and on the trochanters and tarsi of its hind legs. The resilin pad is compressed and is maintained in that condition by catch structures on certain sclerites (two on each side of the flea). In effect, it has cocked itself. To take off, the flea must exert the relatively small muscular action to unhook the catches, allowing the resilin to expand. Then the flea rotates its femurs down toward the substrate and pushes off with its hind tarsi. By using resilin, the fleas have circumvented two major limitations of muscle: the relatively slow rate of contraction and relaxation and the poor performance at low temperature. Because resilin does not become deformed under prolonged strain, once the

jumping mechanism is cocked, little energy is required to retain this state. The flea can lie in waiting, ready to hop aboard a host in a fraction of an instant. Interestingly, fleas that do not need to jump to reach their preferred hosts tend to have reduced pleural arches, whereas those which prefer the largest hosts (for example, deer, sheep, cats, and humans) have the largest pleural arches and are the best jumpers.[20]

Mouthparts and mode of feeding

Like the Anoplura and Hemiptera, the Siphonaptera have piercing-sucking mouthparts, but the structure is different. The broad maxillae bear conspicuous, segmented palps (Fig. 38-2), as does the slender labium. The piercing fascicle comprises two elongate maxillary lobes (**laciniae**) and a median, unpaired **epipharynx.** In the past the maxillary laciniae often were considered mandibles and the epipharynx, the labrum.[6] The laciniae lie closely on each side of the epipharynx, and the fascicle is held in a channel formed by grooves in the inner side of the labial palps (Fig. 38-

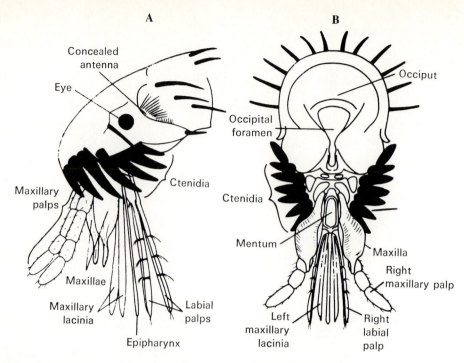

FIG. 38-2

Head and mouthparts of a flea. **A,** Side view. **B,** Ventral view.

Reprinted with permission of Macmillan Publishing Co., Inc., from Entomology in Human and Animal Health, 7th Edition (A Revision of Herms' Medical Entomology), by Robert F. Harwood and Maurice T. James. Copyright © 1979 by Macmillan Publishing Co., Inc.

3). The hypopharynx cannot be demonstrated, and the labium is rudimentary. Lavoipierre and Hamachi[12] reported that several species of fleas are solenophages, but Rothschild et al. believed that *Spilopsyllus cuniculi* is primarily a telmophage.[6,19] The laciniae are the cutting organs, and piercing is achieved by the back-and-forth cutting action of these structures. The tip of the epipharynx, but not the laciniae, enters a small blood vessel.[6] Saliva is ejected into the area near the puncture of the vessel by the laciniae, but it does not enter the vessel lumen. Very little damage is done by the penetrating fascicle, and, after withdrawal from the skin, hemorrhage is scant or absent.

DEVELOPMENT

Fleas undergo holometabolous development, and the larvae thus are quite different in form and habits from the adults. Although the adult female usually oviposits while on the host, the eggs are not sticky and so drop off the host's body. This often happens in the host's nest or lair, where there is a supply of detritus and flea feces on which the larvae feed. The eggs are relatively large (about 0.5 mm) (Fig. 38-4), providing many essential nutrients to the larvae. Under

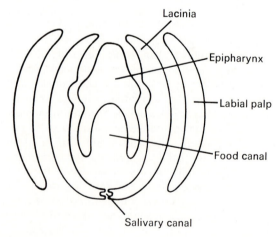

FIG. 38-3

Diagrammatic representation of transverse section through flea mouthparts.

From Askew, R.R. 1971. Parasitic insects. American Elsevier Publishing Co., Inc., New York.

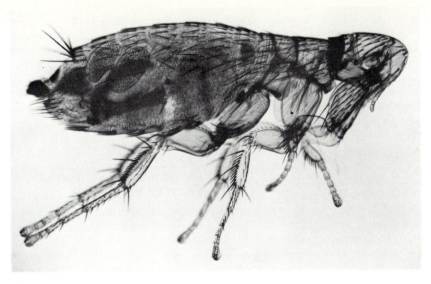

FIG. 38-4

Leptopsyllus segnis, the European mouse flea, which is also common in parts of the United States. Note the large size of the eggs visible within this cleared specimen.

Photograph by Jay Georgi.

favorable conditions, the larvae hatch within 2 to 21 days, the larval instars (usually three) require 9 to 15 days, and the pupa completes development in as short a time as a week. Low temperatures and those as high as the host's body temperature retard development. Low temperatures can extend the larval period to more than 200 days and the pupal stage to nearly a year. Because the larvae cannot close their spiracles, they are sensitive to low humidity. High humidity tends to favor egg laying in the adults and, of course, is apt to be the prevalent condition in nests and burrows. The larvae are white, legless, and eyeless, resembling the maggots of some Diptera (Fig. 38-5). They have chewing mouthparts and stout body hairs. The pupa spins a loose, silken cocoon from its salivary secretions, often picking up debris from the surroundings in the cocoon.

In common with other lair parasites such as bedbugs and in contrast with lice, fleas can survive long periods as adults without food, particularly under conditions of high humidity. Unfed *Pulex irritans* have survived 125 days at 7° to 10° C and *Xenopsylla cheopis* for 38 days, whereas periodically fed *P. irritans* may live up to 513 days and *X. cheopis* to 100 days.[6] Periodically fed *Ctenophthalmus wladimiri* have survived for more than 3 years at 7° to 10° C and 100% relative humidity. Such longevity has clear epidemiological importance because it allows flea-transmitted pathogens to survive long periods when vertebrate

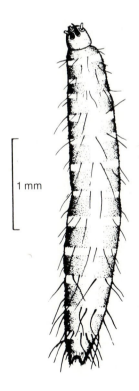

1 mm

FIG. 38-5

Third instar larva of *Spilopsyllus cuniculi*, the European rabbit flea, showing mostly the ventral surface.

From Askew, R.R. 1971. Parasitic insects. American Elsevier Publishing Co., Inc., New York.

hosts are absent. Cases are known in which long survival occurs even in the face of highly adverse conditions: *Glaciopsyllus* larvae, pupae, and some adults can withstand freezing in their host's nest, being covered with ice for 9 months out of the year.[17]

Two genera of the subfamily Spilopsyllinae, *Spilopsyllus* and *Cediopsylla,* are unusual in that their reproduction is closely controlled by their host's hormones. *Spilopsyllus cuniculi,* the **European rabbit flea,** is a relatively sedentary flea, commonly attaching itself to its host's ears for long periods. However, it does not breed on the adult rabbit. About 10 days before the pregnant doe gives birth, the fleas on the doe begin to mature sexually. This coincides with a rise in cortisol and corticosterone in the pregnant rabbit, which are the hormones that stimulate flea maturation. The hormone levels also are high in the newborn rabbit. By the time the rabbit's young are born, the flea's eggs are ripe, and the fleas detach from the doe's ear and move onto her face. As she tends her young, the fleas hop onto the newborn rabbits and feed voraciously, mate, and lay eggs. After about 12 days, the fleas leave the young and return to the doe. Good evidence indicates that the growth hormone somatotropin, present in young rabbits, constitutes the stimulus for fleas to mate and lay eggs. Reproductive control of *C. simplex* by host hormones is essentially similar to that of *S. cuniculi.*[18] This remarkable coordination of the flea's reproduction with that of the host assures that the flea's eggs will be ripe at just the right moment to be deposited into the host's nest and assures the larvae of a plentiful supply of food.[16]

EXAMPLES

In general, fleas are not very host specific, although they have preferred hosts. Most can transfer from one of their hosts to another or to a host of a different species. Their common names (for example, rat flea, chicken flea, and human flea) refer only to the preferred host and do not imply that they attack the host exclusively. In the United States 19 different species have been recorded as biting humans.[6]

Fleas can be grouped into four categories according to the degree of attachment to the host:
1. Some rodent fleas, such as *Conorhinopsylla* and *Megarthroglossus* spp., are seldom on the host but occur abundantly in its nest.
2. Most fleas spend most of their time on the host as adults but can transfer easily from one host individual to another.
3. Females of the **sticktight flea,** *Echidnophaga gallinacea,* attach permanently to the host skin by their mouthparts.

4. Females of the **chigoe,** *Tunga penetrans,* burrow beneath the host skin and become stationary, intracutaneous, and subcutaneous parasites.

In addition, a species is known (*Uropsylla tasmanica,* on Tasmanian devils) in which the larvae burrow beneath the skin and live as endoparasites. The larvae of *Hoplopsyllus* on the arctic hare live as ectoparasites in the fur of the host. Of the four general categories that precede, however, those which become permanently attached play little or no role in disease transmission.

FAMILIES CERATOPHYLLIDAE AND LEPTOPSYLLIDAE

The **northern rat flea,** *Nosopsyllus fasciatus,* is a common parasite of domestic rats and mice (*Rattus* and *Mus* spp.) throughout Europe and North America, and it has been recorded on many other hosts, including humans. Although it may be of some importance in transmission of plague from rat to rat, it is not regarded as an important plague vector because it usually does not bite humans, and it is widespread in temperate climates where plague is not commonly a problem.[15] The **ground squirrel flea,** *Diamanus montanus,* is found from Nebraska and Texas to the Pacific coast and may be of some importance in transmission of plague in wild rodents.

Ceratophyllus niger and *C. gallinae* are bird fleas, although both will bite humans. *Ceratophyllus niger* is the **western chicken flea** and can be distinguished easily from another common chicken flea, the sticktight (*E. gallinacea*), in that *C. niger* is larger and does not attach permanently. The **European chicken flea,** *C. gallinae,* commonly parasitizes a wide variety of other birds, especially passerines.

Leptopsyllus segnis is the **European mouse flea** (Fig. 38-4), but it is common throughout the Gulf states and in parts of California. It is more common on *Rattus* than on *Mus,* and although it can be infected with plague, it is not an important vector because it does not readily bite humans.

FAMILY PULICIDAE

Pulex irritans, the **human flea** (Figs. 38-6 and 38-7), and other species of medical and veterinary importance are members of this family.

Pulex irritans has been recorded from a variety of hosts, including pigs, dogs, coyotes, prairie dogs, ground squirrels, and burrowing owls, but some of the records may refer to another species, *P. simulans,* which occurs in the central and southwestern United States and in Central and South America.[22] Both species lack genal and pronotal ctenidia, and their meta-

FIG. 38-6

Male *Pulex irritans*, the human flea. The copulatory apparatus is visible in the abdomen. Rothschild[16] has called the genital organs of male fleas the most elaborate in the animal kingdom.

Photograph by Jay Georgi.

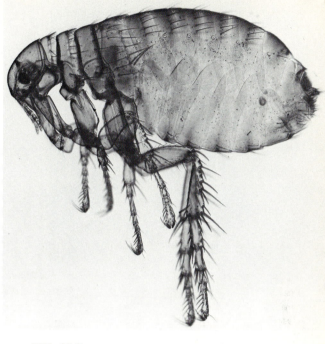

FIG. 38-7

Female *Pulex irritans*.

Photograph by Jay Georgi.

coxae have a row or patch of short spinelets on the inner side of that podomere. However, the maxillary laciniae of *P. irritans* extend only about half the length of the forecoxae, whereas those of *P. simulans* extend about three fourths the distance. *Pulex irritans* can transmit plague, and it has been implicated in the transmission from person to person in some epidemics. Kalkofen[10] found that more than 80% of fleas on dogs were *P. irritans* in a survey in Georgia and that dogs seemed to be the preferred hosts. Since dogs are susceptible to plague (see Kalkofen[10] for references), this has important public health implications.

Echidnophaga gallinacea is an important poultry pest, but it also attacks cats, dogs, horses, rabbits, humans, and other animals. It is called the **sticktight flea** because it buries its fascicle in the skin of the host and remains in place. The maxillary laciniae are broad and coarsely serrate (Fig. 38-8). This flea is widespread in tropical and subtropical regions and also occurs in the United States as far north as Kansas and Virginia.[4] Like those of *Tunga penetrans*, the thoracic segments are reduced, being shorter together than the head or the first abdominal segment; however, *E. gallinacea* has a patch of spinelets on the inner side of its metacoxa (absent in *Tunga*). It prefers to attach in areas with few feathers, such as the comb, wattles, around the eyes, and around the anus of the host. The infestation causes ulcers, into which the female deposits the eggs. The larvae hatch in the ulcers but then drop to the ground to develop off the host, as in most other fleas. Heavy infestations may kill the chickens.

Ctenocephalides canis (Fig. 38-9) and *C. felis* are the **dog** and **cat fleas,** and they can be distinguished from the other common fleas by the presence of a genal ctenidium with more than five teeth. In spite of the names, both species attack cats and dogs, as well as humans, other mammals, and occasionally chickens.[6] *Ctenocephalides felis* is more common than is *C. canis* on dogs in North America. They can be very annoying pests of humans, particularly when cats and dogs are kept on the premises. Oddly, they do not occur in the mid– to north–Rocky Mountain area.

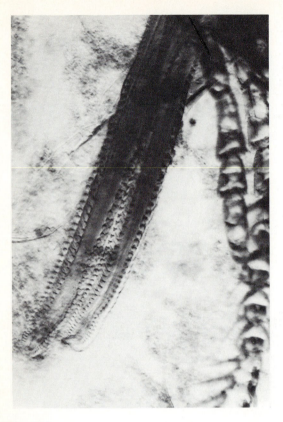

FIG. 38-8

Maxillary laciniae of *Echidnophaga gallinacea*, the stick-tight flea. The laciniae are broad and coarsely serrate, and the thoracic somites are much reduced compared with most other fleas. Maxillary palps are to the right in this photograph.

Photograph by Larry S. Roberts.

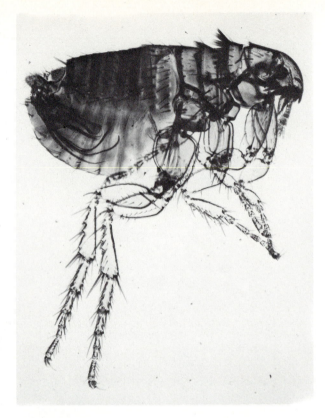

FIG. 38-9

Ctenocephalides canis, the dog flea. This species and *C. felis* bite humans frequently and are the source of much annoyance. They are the most common intermediate hosts of the tapeworm, *Dipylidium caninum*, of dogs and cats.

Photograph by Jay Georgi.

Xenopsylla cheopis (Fig. 38-10) is called the **oriental** or **tropical rat flea**, although it is almost cosmopolitan on *Rattus* spp., except in cold climates. In the United States it ranges as far north as New Hampshire, Minnesota, and Washington.[15] Like *Pulex irritans*, *X. cheopis* lacks both genal and pronotal ctenidia. It can be distinguished by the location of the ocular bristle, which originates in front of the eye in *X. cheopis* and beneath the eye in *P. irritans*. Females can be recognized easily by the dark-colored spermatheca, since this is the only species in the United States with a pigmented spermatheca (compare Figs. 38-7 and 38-10).[15] *Xenopsylla cheopis* has enormous public health significance because it is the most important vector of plague and of murine typhus. *Xenopsylla brasiliensis*, an African species that has become established in South America and India, appears to be

a more important plague vector in Kenya and Uganda. Some other species of *Xenopsylla* are implicated or suspected as vectors in various plague outbreaks.

FAMILY TUNGIDAE

Tunga penetrans is called the **chigoe, jigger, chigger, chique,** and **sand flea** (Fig. 38-11). Its name is said to result from the irritation it causes, prompting the host to "jig" about. This flea is apparently a native of Central and South America and the West Indies, from which it was introduced to Africa in the seventeenth century and again in the nineteenth century. It spread all over tropical Africa and then to India. *Tunga penetrans* attacks several other mammals besides humans, particularly swine.

The female chigoe penetrates the skin, most com-

FIG. 38-10

Xenopsylla cheopis, the oriental, or tropical, rat flea. This flea is the most common vector of plague and murine typhus. The spermatheca is darkly pigmented and clearly visible.

Photograph by Jay Georgi.

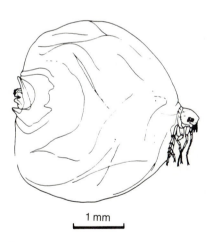

1 mm

FIG. 38-11

Engorged female *Tunga penetrans,* the chigoe, or jigger. This stage is found in a subcutaneous sinus that communicates with the outside by a small pore, through which the larvae escape. The legs are degenerating by the time the female is engorged.

From Askew, R.R. 1971. Parasitic insects. American Elsevier Publishing Co., Inc., New York.

monly around the nail bases of the hands and feet or between the toes. Only a small aperture through the skin is left to communicate with the outside world. The male does not penetrate host skin, but it copulates with the female after she has reached her final position. When she enters the skin, she is barely 1 mm long, but she gradually expands to about the size of a pea. Her body is enclosed in a sinus, into which she lays her eggs. After hatching, the larvae exit through the aperture and develop on the ground. The presence of the female causes extreme itching, pain, inflammation, and often secondary infection (Fig. 38-12). Tetanus and gangrene occasionally are complications. Autoamputation has been attributed to results of infection with this flea and its secondary infection in Angola.[4] Surgical removal with careful sterilization and dressing of the wound is the recommended remedy.

FLEAS AS VECTORS
Plague

Plague, also known as pest and black death, is caused by a bacterium, *Yersinia pestis* (formerly *Pasturella pestis*). The great pandemics of plague have probably had more profound effect on human history than have the effects of any other single infection. The pandemic of the fourteenth century alone, for exam-

613

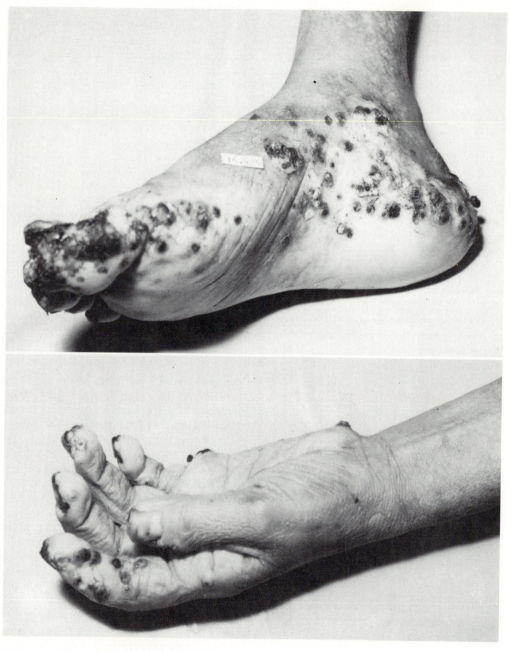

FIG. 38-12

Lesions caused by *Tunga penetrans*.

Photographs by Rodolfo Céspedes, Hospital San Juan de Dios, San José, Costa Rica; from Hunter, G.W., J.C. Swartzwelder, and D.F. Clyde. 1976. Tropical medicine, ed. 5. W.B. Saunders Co., Philadelphia.

ple, took 25 million lives, or a fourth of the population of Europe, and has been called the worst disaster that has ever befallen humanity.[11] Adequate treatment of this subject in relation to its importance is far beyond the scope of this book. Interested readers may avail themselves of numerous sources on plague, including a concise description of the disease itself by Blount,[3] and highly readable accounts of plague and fleas in history by Lehane[13] and Pollitzer.[14] A fascinating picture of human behavior faced by the terror of universal catastrophe is given by Langer.[11] The last pandemic began in the interior of China toward the end of the nineteenth century, reached Hong Kong and Canton by 1895 and Bombay and Calcutta in 1896, and then spread throughout the world, including numerous port cities in the United States. In the period 1898 to 1908, more than 548,000 people per year died from plague in India.[14] Between 1900 and 1972 there were 992 cases in the United States, 416 of them in Hawaii, of which 720 were fatal.[15] The disease has decreased in incidence and severity in recent years: between 1958 and 1972 there were 51 cases in the United States, of which only 9 were fatal. It is not clear if this improvement results entirely from better medical care. Furthermore, whereas the disease was formerly centered in seaports, from which it spread out in an epidemic, recent cases in the United States have been virtually all rural (campestral); that is, they were contracted after contacts with wild rodents in the countryside, rather than with *Rattus* in the cities. It is

problematic whether the urban plague spread into the wild rodents in the United States or whether it was there before 1900. The world distribution of plague as of 1969 is shown in Fig. 38-13.

Plague is essentially a disease of rodents, from which it is contracted by humans through the bites of fleas, particularly *Xenopsylla cheopis*. The bacteria are consumed by the fleas along with their blood meal, and the organisms multiply in the flea's gut, often to the extent that passage of food through the proventricular teeth is blocked. When the flea next feeds, the new blood meal cannot pass the obstruction, but is contaminated by the bacteria and then regurgitated back into the bite wound. The propensity of a particular flea species to have its gut blocked by growth of *Yersinia pestis* is an important determinant of its efficacy as a vector. *Xenopsylla cheopis* is a good vector because it becomes blocked easily, feeds readily both on infected rodents and humans, and is abundant near human habitations.[15] Although the black, or roof rat, *Rattus rattus*, is common and abundant around human habitations, it appears not to be a reservoir of infection.[21]

The three forms of plague are **bubonic, primary pneumonic,** and **primary septicemic.** Bubonic plague, which is the most common type in epidemics, demonstrates definite bubo formation in about 75% of cases. **Buboes** are swollen lymph nodes in the groins or armpits; they sometimes reach the size of a hen's egg (Fig. 38-14) and may rupture to the outside. They are

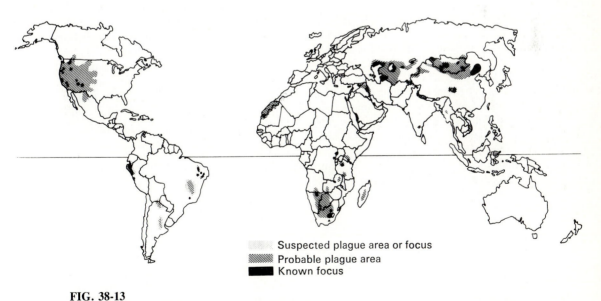

Suspected plague area or focus
Probable plague area
Known focus

FIG. 38-13

Known and probable foci and areas of plague, 1969.

From Hunter, G.W., J.C. Swartzwelder, and D.F. Clyde. 1976. Tropical medicine, ed. 5. W.B. Saunders Co., Philadelphia.

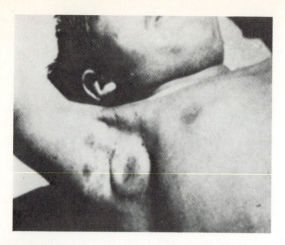

FIG. 38-14

Plague bubo in right axilla of human.

AFIP neg. no. ACC 219900-7-B.

hard, tender, and filled with bacteria. Pneumonic plague is a condition in which the lungs are most heavily involved, producing a pneumonia-like disease that is highly contagious to other persons. Primary septicemic plague is a generalized blood infection with little or no prior lymph node swelling, apparently because the blood is invaded too rapidly for the typical nodal inflammation to develop. However, secondary septicemic plague usually occurs in pneumonic and sometimes in bubonic plague.[3] Bubonic plague is fatal in about 25% to 50% of untreated cases, and pneumonic and septicemic plague are usually fatal.

The incubation period after the bite of the flea is usually 2 to 4 days, followed by a chill and rapidly rising temperature to 39.5° and 40° C. The lymph nodes draining the site of the infection swell, becoming hemorrhagic and often necrotic. Damage to vascular and lymphatic endothelium in many parts of the body lead to petechial and diffuse hemorrhages. At first there is mental dullness, followed by anxiety or excitement, and then delerium or lethargy and coma. If the patient is to recover, the fever begins to drop in 2 to 5 days. If untreated, the disease may cause death within 5 days. The patient usually dies within 3 days of the onset of primary septicemic plague. Treatment with antibiotic and antitoxin is usually effective. Persons traveling into area in which plague is known to occur with any regularity should be vaccinated.[3]

The rapid and serious effects of plague infection are primarily caused by a potent toxin released by *Yersinia pestis*. Actually, the bacteria release two toxins with identical effects.[9] Some animals, for example, rats and mice, are more sensitive to the toxins than are others (rabbit, dog, monkey, and chimpanzee). Evi-

dence indicates that the toxins act on the mitochondrial membranes of susceptible animals, inhibiting ion uptake and interfering with normal functioning of the respiratory chain.[9]

Conditions conducive to high rat and flea populations contribute to plague outbreaks. The disease may exist in rodent populations in acute, subacute, and chronic forms. Epidemics among humans usually closely follow epizootics, with high mortality among rats. When the rat dies, its fleas depart and seek greener pastures. Meteorological conditions are important, doubtless because of their effects on the flea population. Plague is usually seen in temperate climates during summer and autumn and in the tropics during the cool months. Extreme heat and dryness inhibit spread of plague. Epidemics of primary pneumonic plague occur only during conditions of low temperatures and high humidity, which are conditions that favor survival of the bacilli in sputum droplets.

Campestral plague (so called because it is associated with animals of open areas rather than wooded or sylvatic country) is widespread among wild rodents and rabbits in the United States west of the hundredth meridian. Cases among humans are reported sporadically, usually after the human has contacted the wild rodents and their fleas. New Mexico has had the highest case rate. One Californian contracted bubonic plague with secondary pneumonic plague after hunting ground squirrels and was the source of 13 cases of primary pneumonic plague in other persons, with 12 deaths.[15] A number of cases have been associated with skinning, cooking, and eating wild rabbits and hares; but the persons may have been bitten by the rabbit's fleas.[1] Campestral plague constitutes a reservoir of potentially great importance.

Throughout recorded history, plague has been cyclical, smoldering in endemic foci, and then giving rise to great outbreaks. The world seems to be in a remission phase at present, but the foci are still present.

Murine typhus

Murine typhus, also called endemic or flea-borne typhus, is caused by *Rickettsia mooseri* (=*R. typhi*) which is morphologically indistinguishable from *R. prowazekii*. It occurs in warmer climates throughout the world. It can infect a wide range of small mammals, including the opossum, *Didelphis marsupialis,* but the most important reservoir is *Rattus norvegicus,* in which it causes slight disease symptoms. Murine typhus can be transmitted from rat to rat by *Xenopsylla cheopis; Nosopsyllus fasciatus; Leptopsyllus segnis;* the rat louse, *Polyplax spinulosa;* and the tropical rat mite, *Ornithonyssus bacoti.* In humans the disease is a rather mild, febrile illness of about 14 days' duration, with chills, severe headaches, body pains, and rash. It tends to be more severe in elderly persons. *Xenopsylla cheopis* is considered the primary vector transmitting the disease to humans, either through the bite or by contamination of skin abrasions with flea feces by scratching. Ingestion of infected fleas and their feces also can produce infection in rats.[5] The rickettsias proliferate in the midgut cells of the flea but do not kill it. Rupture of the midgut cells releases the organisms into the gut of the flea. Before 1945 the incidence of murine typhus was high in the United States and reached a peak of 5401 cases in 1945. After the institution of a rat control program, use of DDT, and increasing use of antibiotics, the reported incidence dropped dramatically, ranging between 18 and 36 cases per year between 1969 and 1972.

Myxomatosis

The myxoma virus causes a disease in rabbits and is transmitted by several bloodsucking arthropods, including mosquitoes, fleas, and mites.[23] The principal vector in England is *Spilopsyllus cuniculi*, and the disease causes considerable losses in the domestic rabbit industry. The virus was apparently introduced from South America, where the rabbits are relatively resistant to myxomatosis.[16] It was intentionally introduced into Australia to control the abundant rabbits there. Unfortunately, the principal vectors in Australia were mosquitoes, which were not ideal vectors for rabbit control. The mosquitoes confer a selective advantage on an attenuated "field strain" of the virus and are most abundant during the warm months, when the rabbits have the best chance of surviving the disease.[23] Consequently, resistance to the virus in rabbit populations was unintentionally selected for. More recently, introduction of *Spilopsyllus cuniculi* has offered more hope of better rabbit control, along with reintroduction of virulent virus strains.[23]

Other parasites

Nosopsyllus fasciatus is a vector for the nonpathogenic *Trypanosoma lewisi* of rats (p. 69). *Ctenocephalides canis, C. felis,* and *Pulex irritans* serve as intermediate hosts of *Dipylidium caninum,* a common tapeworm of cats and dogs (p. 366). *Nosopsyllus fasciatus* and *Xenopsylla cheopis* can serve as vectors for the rat tapeworm, *Hymenolepis diminuta,* and the mouse tapeworm, *Vampirolepis nana,* can develop in *X. cheopis, C. felis, C. canis,* and *P. irritans* (pp. 362 and 365). All of these fleas acquire the tapeworms as larvae when they consume the eggs passed in the feces of the vertebrate host, retaining the cysticercoids in their hemocoel through metamorphosis to the adult. All three species can be transmitted to humans if the person inadvertently ingests an infected flea.

A filarial worm of dogs, *Dipetalonema reconditum,* which lives in the subcutaneous, connective, and perirenal tissues, is transmitted by *C. canis* and *C. felis.* The microfilariae are picked up by the fleas in their blood meal, develop to the infective stage in the flea's fat body in about 6 days, migrate to the head, then pass to the wound when the flea next feeds. *Dipetalonema reconditum* is of slight or no pathogenicity, but its microfilaria may be easily confused with those of the serious pathogen *Dirofilaria immitis* (p. 501). For techniques to distinguish the two, see Ivens, Mark, and Levine.[8]

CONTROL OF FLEAS

Many of us occasionally need to control fleas around our homes or on our pets. It is sometimes extremely important for public health reasons to control rat fleas and, more importantly, their hosts. For more complete instructions on and techniques on flea and rat control, the reader should consult Pratt and Stark.[15]

Within habitations one should keep debris that harbors larval fleas to a minimum, for example, under carpets, in floor crevices, and in pet bedding. Some persistent insecticides may be used indoors. A wider variety of insecticides is available for use outside on the ground in areas frequented by pets and other animals. It is important to keep premises where livestock are maintained as free from debris, manure, and other litter as possible. Various insecticidal flea powders for use on dogs and cats are available, but repeated application may be necessary because they easily pick up more fleas in outdoor areas not treated with insecticide. In recent years flea collars with slow-release vapors have proven very effective.

Personal protection may be achieved by use of insect repellants. In areas in which *Tunga penetrans* is found, it is important to wear shoes.

REFERENCES

1. Anonymous. 1984. 1983 sets high mark for plague. Lubbock Avalanche-Journal. March 24.
2. Askew, R.R. 1971. Parasitic insects. American Elsevier Publishing Co., Inc., New York.
3. Blount, R.E., Jr. 1976. Plague. In Hunter, G.W., J.C. Swartzwelder, and D.F. Clyde. Tropical medicine, ed. 5. W.B. Saunders Co., Philadelphia.
4. Grothaus, R.H., and D.E. Weidhaas. 1976. Class Insecta (Hexapoda). In Hunter, G.W., J.C. Swartzwelder, and D.F. Clyde. Tropical medicine, ed. 5. W.B. Saunders Co., Philadelphia.
5. Farhang Azad, A., and R. Traub. 1985. Transmission of mu-

rine typhus rickettsiae by *Xenopsylla cheopis,* with notes on experimental infection and effects of temperature. Am. J. Trop. Med. Hyg. 34:555-563.

6. Harwood, R.F., and M.T. James. 1979. Entomology in human and animal health, ed. 7. Macmillan Publishing Co., Inc., New York.

7. Humphries, D.A. 1967. The function of combs in fleas. Entomol. Mon. Mag. 102:232-236.

8. Ivens, V.R., D.L. Mark, and N.D. Levine. 1978. Principal parasites of domestic animals in the United States. Special Pub. No. 52. Colleges of Agriculture and Veterinary Medicine, University of Illinois, Urbana.

9. Kadis, S., T.C. Montie, and S.J. Ajl. 1969. Plague toxin. Sci. Am. 220(3):93-100.

10. Kalkofen, U.P. 1974. Public health implications of *Pulex irritans* infestations of dogs. J. Am. Vet. Med. Assoc. 165:903-905.

11. Langer, W.L. 1964. The black death. Sci. Am. 210(2):114-121.

12. Lavoipierre, M.M.J., and M. Hamachi. 1961. An apparatus for observations on the feeding mechanism of the flea. Nature 192:998-999.

13. Lehane, B. 1969. The compleat flea. The Viking Press, New York.

14. Pollitzer, R. 1954. Plague. World Health Organization, Geneva.

15. Pratt, H.D., and H.E. Stark. 1973. Fleas of public health significance and their control. U.S. Department of Health, Education, and Welfare Pub. No. (CDC) 75-8267. U.S. Government Printing Office, Washington, D.C.

16. Rothschild, M. 1965. Fleas. Sci. Am. 213(6):44-53.

17. Rothschild, M. 1975. Recent advances in our knowledge of the order Siphonaptera. Ann. Rev. Entomol. 20:241-259.

18. Rothschild, M., and B. Ford. 1972. Breeding cycle of the flea *Cediopsylla simplex* is controlled by breeding cycle of host. Science 178:625-626.

19. Rothschild, M., B. Ford, and M. Hughes. 1970. Maturation of the male rabbit flea *(Spilopsyllus cuniculi)* and the Oriental rat flea *(Xenopsylla cheopis):* some effect of mammalian hormones on development and impregnation. Trans. Zool. Soc. Lond. 32:105-188.

20. Rothschild, M., Y. Schlein, K. Parker, C. Neville, and S. Sternberg. 1973. The flying leap of the flea. Sci. Am. 229(5):92-100.

21. Schwan, T.G., D. Thompson, and B.C. Nelson. 1985. Fleas on roof rats in six areas of Los Angeles County, California: their potential role in the transmission of plague and murine typhus to humans. Am. J. Trop. Med. Hyg. 34:372-379.

22. Smit, F.G.A.M. 1958. A preliminary note on the occurrence of *Pulex irritans* L. and *Pulex simulans* Baker in North America. J. Parasitol. 44:523-526.

23. Sobey, W.R., and D. Conolly. 1971. Myxomatosis: the introduction of the European rabbit flea *Spilopsyllus cuniculi* (Dale) into wild rabbit populations in Australia. J. Hyg. 69:331-346.

CHAPTER 39

PARASITIC INSECTS: ORDER DIPTERA, THE FLIES

Time is fun and we're having flies.

KERMIT THE FROG

Among all the orders of insects, the Diptera stands out as by far the most medically important, directly or indirectly causing each year more than a million human deaths.[6] Moreover, various flies contribute to disfiguring, debilitating diseases of many kinds, either as vectors of pathogenic organisms or as active parasites in their own right. And who could not express the vexation caused by ravenous hoards of mosquitoes, black flies, midges, deer flies, and others? The order is so vast, with more than 80,000 species in 140 families, that we can do no more than introduce the subject in this text. Published information on mosquitoes alone would fill a small library.

The name "fly" is loosely used. Correctly, it applies to insects that have a pair of wings on the mesothorax and a reduced pair, the **halteres,** on the metathorax. Halteres are knoblike appendages that function as gyroscopic balance organs. Of course, some parasitic flies secondarily have lost all wings. Flies are holometabolous, with obtect, coarctate, or puparious pupae. Thus dragonflies, mayflies, and so on are not actually flies. Such common names are always written as one word, whereas true flies such as the house fly and horse fly are written as two. Because the order is large and its members vary considerably in feeding habits, both as larvae and adults, the structure of mouthparts is also diverse. We can divide them into five subtypes[10]: mosquito, horse fly, house fly, stable fly, and louse fly. In addition, the mouthparts of some larvae are of medical interest. This topic will be covered in the discussion of each appropriate group.

The order Diptera is divided for convenience into three suborders: Nematocera, Brachycera, and Cyclorrhapha. Nematocera are considered to be the most primitive of flies and the Cyclorrhapha the most advanced. Taxonomic characters most used in the Diptera are the mouthparts, head sutures, ocelli, antennae, wing venations, tarsi, and placement of bristles **(chaetotaxy).** Male genitalia offer useful taxonomic characters at the genus and species levels.

SUBORDER NEMATOCERA

The antennae of species in this group are many segmented and filamentous. They may be plumose, especially in the males, but basically they are simple and longer than the head. The wings have many veins, which is a primitive character. The larvae are active, with a well-developed head capsule, and the pupae often are free swimming. Most life cycles involve aquatic larval and pupal stages, although some develop in bogs or wet soil.

Family Simuliidae

Simuliids (Fig. 31-6) are commonly called **black flies,** although many species are gray or tan. They are small, 1 to 5 mm long. The prescutum of their mesonotum is reduced, giving them a humpbacked appearance, which explains their other common name, **buffalo gnat.** The wings are broad and iridescent, with strongly developed anterior veins. The antennae are filiform, usually with 11 segments. The eyes of the female are separated, whereas those of the male are contiguous above the antennae. There are no ocelli.

Mouthparts are of the horse fly subtype (p. 631), although delicate. Downes[2] indicated that serrated teeth on the edges of the mandibles are cutting structures, whereas recurved teeth on the maxillary lacinia serve to anchor the mouthparts during feeding.

Black flies are found worldwide but are most abundant in north temperate and subarctic zones. Females of most species feed on blood as well as nectar, but males feed only on plant juices. Mating occurs in flight, when females fly into swarms of hovering males.

Larval development can occur only in running, well-oxygenated water. Hence black flies are most numerous near rivers and streams, although they are known to travel several miles when aided by winds.

A female simuliid produces 200 to 800 eggs, laying them on the surface of the water, where they rapidly sink. In some species the female lands at the water's

619

edge, crawls down a rock or plant to deposit the eggs underwater, and then crawls back out of the water and flies away.

On hatching, the larva (Fig. 39-1), with its modified salivary glands, spins a silken mat on some underwater object. It attaches itself to this mat with a hooked sucker at the posterior end of its abdomen. Thus its head hangs downstream and, with fanlike projections around the mouth, filters protozoa, algae, and other small organisms and organic detritus from the passing water. Often the larval numbers are so great that they form a solid covering on a favorable location, such as a cement spillway or the downstream side of a rock or log. A larva is capable of changing locations rapidly by stretching out, spinning a new mat and clinging to it with its mandibles, and then releasing the old mat and hooking onto the new one. The six or seven larval instars require 7 to 12 days under ideal conditions of temperature and food availability, but this time may be greatly extended. Some species overwinter as larvae.

Before pupation, the larva spins a flimsy cocoon around itself. After molting, the pupa remains nearly immobile, respiring through long filamentous gills on its anterior end. The number and arrangement of these filaments are of taxonomic importance. The pupal stage lasts from a few days to 3 or 4 weeks. To emerge, the callow imago first cuts a T-shaped slit in the pupal thorax, through which it crawls. It quickly fills its air sacs with air extracted from the water, forming an internal balloon; releases its hold on the substrate; and shoots to the surface. One to six generations may mature per year, depending on the locality.

Classification and identification of simuliids are often difficult because of numerous complexes of sibling species. The most important genus is *Simulium*, with more than 800 species. Also important medically are *Prosimulium* and *Cnephia* in North America and *Austrosimulium* in Australia and New Zealand. Black flies are fairly host specific. Few species will bite humans, but those which do are extremely vexatious.

In North America *Prosimulium mixtum* can be very annoying, and *Cnephia pecuarum*, the southern buffalo gnat, has been known to ravage and kill entire herds of livestock. *Simulium vittatum* is widespread in the United States and is particularly irritating to livestock. *Simulium meridionale*, the turkey gnat, torments poultry, biting them on the combs and wattles.

All fishermen and campers in the northern United States and Canada are familiar with the attacks of *S. venustum*, which often occur in such numbers as to ruin a vacation. Vast numbers of *S. arcticum* killed more than a thousand cattle in western Canada annually from 1944 to 1948. *Simulium colombaschense*, of central and southern Europe, killed 16,000 cattle, horses, and mules in 1923 and 13,900 in 1934.[10]

Individuals react differently to the bites of black flies. Few people have little or no reaction; most develop local reactions in the form of reddened, itching wheals. **Black fly fever,** a combination of nausea, headache, fever, and swollen limbs, occurs in particularly sensitive persons.

Black flies are the vectors of *Onchocerca volvulus,* the cause of human onchocerciasis, discussed in detail in Chapter 31. The most common vector in Africa is *S. damnosum* (Fig. 31-6), although *S. neavei* also is important. In the New World *S. ochraceum, S. metallicum, S. callidum,* and *S. exiguum* are the most efficient vectors because of their preference for humans and their activity during the same hours as those of humans.

Onchocerca gutterosa commonly is transmitted to cattle by *S. ornatum* in Europe. In Australia *Simulium* and *Culicoides* spp. infect cattle with *Onchocerca gibsoni,* causing considerable loss to flesh and hides.

The malaria-like bird disease caused by *Leucocytozoon* spp. is transmitted by various species of *Simulium* (Chapter 9). *Leucocytozoon smithi* of turkeys is transmitted by *S. congreenarum, S. slossonae, S. nigroparvum,* and *S. occidentale. Leucocytozoon simondi,* a severe pathogen of ducks and other anatids, is transmitted by *S. rugglesi, S. anatinum,* and other species.

Family Psychodidae

Two subfamilies in this family are of medical importance: the Psychodinae contain the **moth flies,** which are of little serious importance; the Phlebotominae consist of the **sand flies,** of great importance in many parts of the world. Some authorities consider these two groups to represent separate families, but we follow recent workers on this point.[19,20]

■ Subfamily Psychodinae

The body and the ovoid wings are densely covered with hairs. This character, plus the rooflike position of the wings when at rest, suggest the appearance of a tiny moth, hence the common name "moth fly."

Psychodids breed in substrates rich in organic decomposition. *Psychoda alternata,* the **trickling filter fly,** is often found in incredible numbers in sewage disposal plants; it and other species are common in cesspools, washbasins, and drains where the larvae develop in the gelatinous linings of drain pipes. Emerging adults may be so numerous as to constitute a genuine annoyance to householders. *Psychoda alternata* is a cosmopolitan cause of such situations; in ad

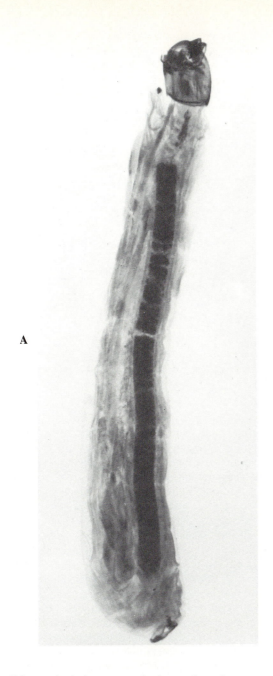

A

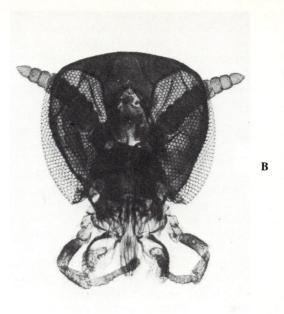

B

FIG. 39-1

Simulium sp. **A,** Larva. **B,** Head. Note the short, cutting-type mandibles.

Photographs by, **A,** Warren Buss and, **B,** Jay Georgi.

dition to its being a pest, its larvae have been reported in pseudomyiasis, and it can be involved in mechanical transmission of nematodes of livestock.[29]

■ Subfamily Phlebotominae

In contrast to the moth flies, sand flies are not so hairy and hold their wings at rest 60 degrees from the body, not rooflike. More important, the mouthparts are of the horse fly subtype, with cutting mandibles. Females feed on plant fluids and on blood by telmoph-

agy, whereas males feed mainly on plant juices and never on blood. The fascicle structures of mouthparts in this group have been shown to vary somewhat depending on the type of usual host, related to its skin structure.[19] Many species feed on reptiles or amphibians, whereas others feed on birds and mammals, including humans.

Adult sand flies (Fig. 5-15) usually feed at night or at twilight and early morning, although some are day feeders. They are weak flyers, capable of navigating only short distances, and so are inactive when any wind blows. When inactive, they hide in crevices, animal burrows, termite hills, or any other dark place with high humidity. Because of their soft, delicate exoskeleton, their survival depends on avoiding hot, dry places. However there are desert species that thrive by hiding in dry mud cracks, burrows, and so on, emerging only during the few humid night hours. Coincidentally, these are often the same hours of some human activities, such as water fetching.

There are two Old World genera of sand flies, *Phlebotomus* and *Sergentomyia,* and the American form, *Lutzomyia. Lutzomyia* spp. occur as far north as Canada, but medically important sand flies are primarily south of Texas. *Lutzomyia diabolica* and *L. shannoni* are the only known anthropophilic sandflies in the United States. Both experimentally have been shown to transmit *Leishmania mexicana*. The other North American species feed mainly on reptiles and rodents.

Phlebotomines do not breed in aquatic situations but require a combination of darkness, high humidity, and organic debris on which the larvae feed. These requirements are met in animal burrows, crevices, and hollow trees; under logs and dead leaves; and so on. Several eggs are laid at a time. The tiny, white larvae feed on such matter as animal feces, decaying vegetation, and fungi. They have simple, chewing mandibles. The four larval instars require 2 to 10 weeks before pupation. The pupa develops in about 10 days.

Sand flies are surprisingly good vectors of disease, considering their weakness and fragility. They are the vectors of the leishmaniases, bartonellosis, and some viral diseases.

The visceral and cutaneous leishmaniases of humans are discussed in Chapter 5. The many species of *Leishmania* in other vertebrates are beyond the scope of this book, but all are transmitted by phlebotomines. Recent reviews have been published.[16,18,21]

The bacterium *Bartonella bacilliformis* causes a disease known as **Carrión's disease,** with two clinical forms, **Oroya fever** and **verruga peruana.** It is found in Ecuador, southern Colombia, and the Andean region of Peru, being transmitted by *Lutzomyia verrucarum* and probably *L. colombiana.* Oroya fever is a sometimes fatal, visceral form of the disease, accompanied by bone, joint, and muscle pains; anemia; and jaundice. Verruga peruana is a mild, nonfatal cutaneous form. The disease is named after Daniel Carrión, who inoculated himself with organisms obtained from a verruga patient and subsequently developed Oroya fever. Before he died of it, he recognized that the two entities were actually expressions of the same disease.

Sand fly fever is transmitted by *Phlebotomus papatasi, P. sergenti,* and others in much of the Old World. Also known as **papatasi fever** and **three-day fever,** it occurs in the Mediterranean region, eastward to central Asia, southern China, and India. Sand fly fever is a nonfatal, febrile, viral disease of short duration but with a long convalescence period. Sand flies acquire the virus when they feed, but because males also have been shown to be infected, it seems probable that transovarial transmission occurs. The fly, then, also is a reservoir. Sporadic epidemics occur, such as in Yugoslavia in 1948 when three fourths of the population (1.2 million persons) acquired it.[10]

Family Ceratopogonidae

This family comprises the **biting midges,** also called **punkies, no-see-ums,** and "sand flies." They are very

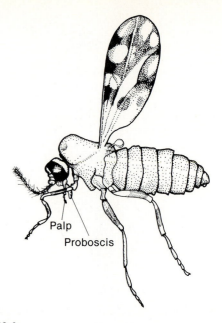

FIG. 39-2

Female *Culicoides* sp.
Drawing by Ian Grant.

small, usually less than 1 mm long, but what they lack in size they make up in ferocity. The majority are daytime feeders that cannot cope with blowing winds so are most pesky on hot, still days. Their small size enables them to crawl through ordinary window screening; some species, particularly in the tropics, enter houses freely.

Most of the 50 or more genera feed on insects. A few feed on poikilothermic vertebrates, and only four genera feed on mammals: *Culicoides* (Fig. 39-2), *Forcipomyia, Austroconops,* and *Leptoconops.* All four have species that happily feed on humans. Only females feed on blood. In addition to their small size, biting midges are recognized by their narrow wings, which have few veins, often are distinctly spotted, and are folded over the abdomen when at rest.

These flies breed in a wide variety of situations, a factor no doubt contributing to their worldwide range. Larvae are aquatic or subaquatic or develop in moist soil, tree holes, decaying vegetation, and cattle dung. *Leptoconops* larvae have been found as deep as 3 feet in the soil. Some species breed readily in salt or brackish water, notably mangrove swamps and salt marshes. The life cycle (Fig. 39-3) is completed in from 6 months to as long as 3 years.

Of the 800 or so species of *Culicoides,* several are known to bite humans. They may be so annoying as to

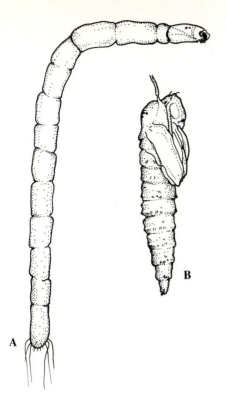

FIG. 39-3

Juvenile stages of *Culicoides* sp. **A,** Larva. **B,** Pupa.
Drawings by Ian Grant.

affect tourism in infested areas, such as beaches, mountain lodges, and Caribbean resorts. Farm workers often are intensely annoyed, and domestic livestock are plagued by these tiny flies. An estimated 10,000 *Culicoides* have been witnessed on a single cow.[24]

Culicoides furens is common and widespread from Massachusetts to Brazil, throughout the West Indies, and on the Pacific coast from Mexico to Ecuador. Species of *Leptoconops* abound in south temperate and tropical areas of the world. *Leptoconops torrens* and *L. carteri* are common pests in the western United States. The taxonomy of North American species of ceratopogonids has been treated by Wirth and Atchley.[32]

Three apparently nonpathogenic filariid nematodes are transmitted to humans by ceratopogonids: *Mansonella perstans* and *M. streptocerca* in Africa and *Mansonella ozzardi* in South and Central America. These are discussed briefly in Chapter 31. *Onchocerca cervicalis* of horses and *O. gibsoni* of cattle are transmitted by *Culicoides* spp., as are other filariids of domestic and wild animals.

Blood-dwelling protozoan parasites also use *Culicoides* spp. as vectors. *Hepatocystis* in monkeys and other arboreal mammals, some species of *Haemoproteus* in birds, and various species of *Leucocytozoon* are transmitted by ceratopogonids. These are dealt with in Chapter 8.

In addition, viruses are transmitted by the bites of these midges. *Orbivirus,* the etiological agent of **bluetongue,** is spread by *C. variipenis* (Fig. 39-2) in North America and by other species of *Culicoides* in Africa and Asia Minor. Bluetongue is a hemorrhagic disease of sheep, cattle, bison, deer, and other ruminants. It causes some mortality, but possibly of more importance, it is responsible for unthriftiness of infected animals, with loss of flesh, wool, and breeding. Transmission of encephalitis viruses, bovine ephemeral fever, and African horse sickness has also been implicated with the bites of these tiny midges.

Family Culicidae

Mosquitoes are the most important insect vectors of human disease and the most common of blood-sucking arthropods. They feed on amphibians, reptiles, birds, and mammals, some with considerable host specificity, others with catholic tastes. Mosquitoes have greatly affected the course of human events and continue to do so even today when we have an arsenal of insecticides at our disposal and a vast amount of knowledge about these insects and the diseases they carry. More than a million people die every year from malaria, and other mosquito-borne diseases cause incalculable misery, poverty, and debilitation. The annoyance of hordes of ravenous mosquitoes is in itself enough to affect real estate values, tourist industries, and outdoor activities. There is considerable concern as to whether mosquitoes can transmit the AIDS virus. At this writing transmission has not been demonstrated, but the potential should not be ignored.

Domestic and wild animals suffer mightily from mosquito attack. Recent figures show a loss of $25 million in cattle production in the United States, $10 million of which is in reduced milk production.[27] Nesting birds, migrating caribou, intrepid explorers, and domestic homeowners all serve the seemingly insatiable appetite of the female mosquito.

Mosquitoes are uncommonly successful insects. More than 3000 species have been described, at least 150 of these in North America (Fig. 39-4). Extensive treatises have been published on all stages of the mosquito life cycle; for a partial listing of such resources see Harwood and James.[10]

Morphologically, mosquitoes are fairly simple insects. They are readily differentiated from the superficially similar Dixidae, Chaoboridae, and Chironomi-

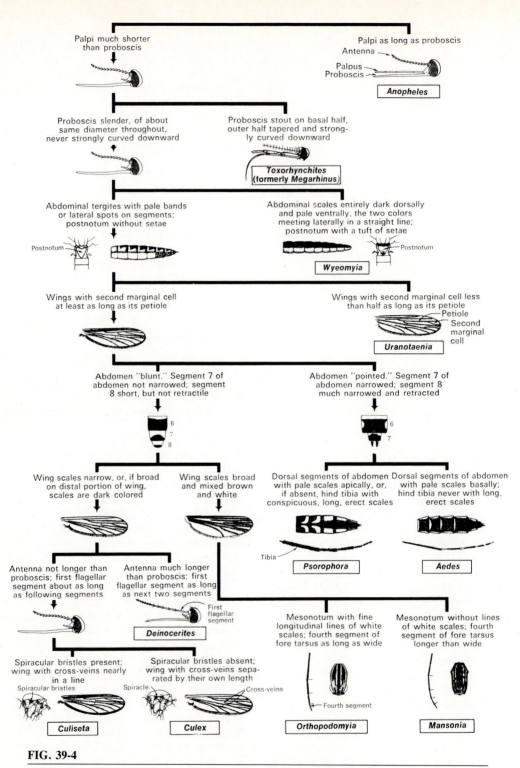

FIG. 39-4

Pictorial key to genera of female mosquitoes of the United States.

Courtesy Communicable Disease Center. 1953. U.S. Public Health Service, Washington, D.C.

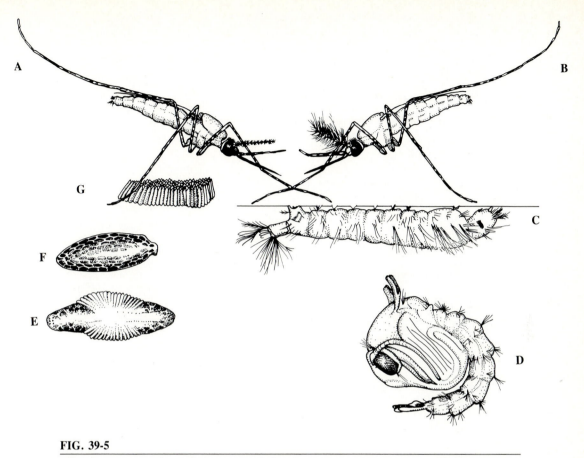

FIG. 39-5

Life history stages of *Anopheles*. **A,** Adult female. **B,** Adult male. **C,** Larva. **D,** Pupa. **E,** Egg.
F, Egg of *Aedes* sp. **G,** Egg raft of *Culex* sp. **F** and **G** are included for comparison.
Drawings by Ian Grant.

dae by the combination of slender wings with scales
on the veins and margins and elongate mouthparts that
form a proboscis. The fascicle consists of six stylets in
the mosquito subtype. The two mandibles, two maxillae, hypopharynx, and labrum-epipharynx are loosely
ensheathed in the elongate labium, which has a lobe-
like tip, the **labella.** The fascicle is inserted into a
blood vessel, and blood is pumped into the food channel formed by the labrum-epipharynx and hypopharynx. Males lack mandibular stylets and do not feed on
blood. Mosquito antennae are long and filamentous
with 14 or 15 segments. Whorls of hairs on the antennae are quite plumose in males of most species. Male
terminalia are complex and are taxonomic characters
useful to experts in differentiating species.

Although most species, particularly females, are
fairly easily identified, complexes of cryptic species
can be differentiated only by advanced techniques
such as cross-mating, electron microscopy of sensilla,
study of polytene chromosomes, or comparison of

electrophoretic patterns of enzyme systems.

Mosquitoes undergo complete metamorphosis, with
egg, larva, pupa, and adult stages (Fig. 39-5). Larval
and pupal stages can develop only in water. Eggs are
deposited singly on water or soil or in rafts of eggs on
water. They either hatch quickly or, in the case of
those on soil, after a period of drought followed by
flooding. Most floodwater mosquitoes hatch after the
first flooding, but some remain for subsequent flood-
ings, in some cases up to 4 years later.[33]

Mosquito larvae are called **wigglers** (Fig. 39-6).
Most hang suspended from the surface of the water by
a prominent breathing siphon, or **air tube.** The majority of wigglers are filter feeders or browse on microorganisms inhabiting solid substrates. Some larvae are
predaceous on other insects, including wigglers. Four
larval instars precede the pupa.

The pupa is called a **tumbler.** It is remarkably active, with a pair of trumpet-shaped breathing tubes on
the thorax, with which it breaks the surface of the wa-

625

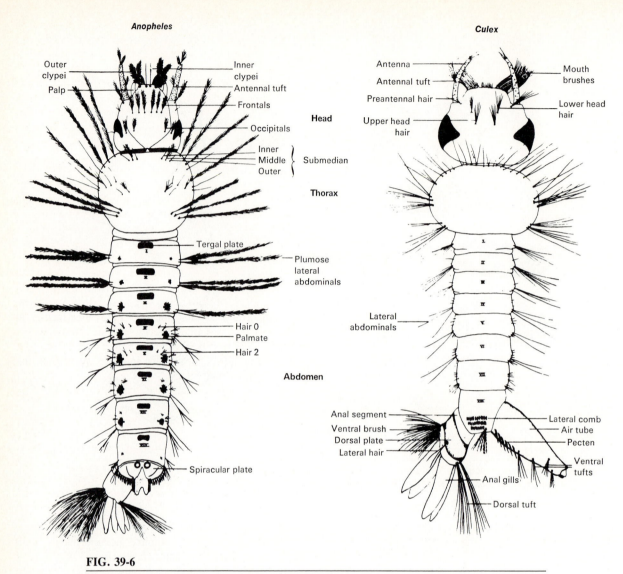

FIG. 39-6

Mosquito larvae, showing basic taxonomic characters.

Courtesy Communicable Disease Center. 1953. U.S. Public Health Service, Washington, D.C.

ter to respire. At the faintest disturbance it swims quickly to the bottom in a tumbling action. The tumbler's life span is short, usually 2 to 3 days. When fully developed, the skin on the thorax splits, and the adult quickly emerges to fly away. Adult females live for 4 to 5 months, especially if they undergo a period of hibernation. During hot summer months of greatest activity, females live only about 2 weeks. Males live about a week, but under optimal conditions of food and humidity, their life span may extend to more than a month. A key to genera of adults is provided in Fig. 39-4.

■ Subfamily Culicinae

Adult members of the Culicinae have a scutellum with a trilobed posterior margin, in dorsal view. The abdomen is densely covered with scales. Eggs are laid in rafts or singly on soil. Larvae have a prominent air tube and hang by it nearly perpendicular to the surface of the water. About 1700 species, in more than 20 genera, are placed in this subfamily. Most species are in the genera *Culex* and *Aedes*.

Genus *Culex*. *Culex* females (Fig. 39-7) are easily identified by the rounded tips of their abdomens. The palps of the female are less than half as long as the

FIG. 39-7

Culex sp., female.

Photograph by Warren Buss.

proboscis. They have no thoracic spiracular or postspiracular bristles. The larva has a long, slender, air tube bearing many hair tufts. Most *Culex* spp. are bird feeders but do not have a narrow host specificity. They overwinter as inseminated females. Several species are important vectors of bird malaria parasites and arboviruses.

Culex tarsalis, a robust, handsome mosquito, is widespread and common in the semiarid western United States and in the southern states as far northwest as Indiana. Its coloration is distinctive: nearly black with a white band on the *lower half* of each leg joint and a prominent white band in the middle of the proboscis. *Culex tarsalis* breeds in water in almost any sunny location. It is a bird feeder, most active at night when it is attracted to the high carbon dioxide concentration around trees and shrubs. There it finds roosting birds. *Culex tarsalis* is not reluctant to feed on humans and other mammals, however, and hence

is the main vector of **western equine encephalitis (WEE)** and also transmits **St. Louis encephalitis** virus. WEE is normally a bird disease, with no apparent symptoms, but it can be acquired by other hosts. Horses are particularly susceptible, with a high rate of mortality. Humans also can be infected; it is not as commonly fatal in humans as in horses but can be severe in children. In adults it results in fever and drowsiness; hence it is sometimes called **sleeping sickness.** Rarely, following a coma, a person may have reduced physical capabilities.

Culex pipiens, the **house mosquito,** is nearly worldwide in distribution. It is a plain, brown insect that breeds freely around human habitation, laying egg rafts in tin cans, tires, cisterns, clogged rain gutters, and any other receptacle of water. It enters houses readily and is a night feeder, causing consternation in many a bedroom. It actually is a complex of species with slight physiological differences, only some of which are understood. The members of this complex are important, not only for their annoyance factor, but because they are major vectors of the filarial worms *Wuchereria bancrofti* and *Dirofilaria immitis* (Chapter 31). They also transmit bird malaria, avian pox, and arbovirus encephalitides.

Culex tritaeniorhynchus is the most important vector of Japanese encephalitis virus in the Orient.

Genus *Aedes.* The following characteristics define the genus *Aedes* (Fig. 39-8): the posterior end of the female abdomen is rather pointed, postspiracular bristles are present on the thorax, the female claws are toothed, and pulvilli are absent or hairlike. Larvae have siphons bearing only one pair of posteroventral hair tufts. Because nearly half of North American mosquitoes are *Aedes* spp., and many of the rest are *Culex,* the pointed abdomen of the female usually is all one needs for field identification of this genus.

Species of *Aedes* are notable for their ferocity. Most are diurnal or crepuscular in their activities, as contrasted with the night-biting *Culex* spp. They lay their eggs singly on water, mud, or soil that is likely to be flooded. Mosquitoes of this genus are not only among the most obnoxious of bloodsucking insects but also are extremely important medically because of the diseases they transmit.

Two species, *Aedes dorsalis* and *A. vexans,* are scourge-mates in the western United States. Both are fierce daytime biters, and at a single swat one may kill half a dozen of each species. *Aedes dorsalis* has a wide range, including most of the Holarctic region, North Africa, and Taiwan. It breeds in salt marshes as well as fresh water. *Aedes vexans* overlaps the range of *A. dorsalis* and includes South Africa and the Pacific Islands.

FIG. 39-8

Aedes sp., female.

Photograph by Warren Buss.

Aedes dorsalis is easily recognized as a straw-colored, medium-sized mosquito of the utmost persistence. *Aedes vexans* is brown to black with white bands encompassing *both halves* of each leg joint. It is aptly named.

Among the many other species of *Aedes* must be mentioned the snow-water mosquitoes of the far north and western mountains. A difficult complex of species, they are all characterized by their immense numbers and ferocious appetites. Usually there is only one generation per year: the females lay eggs singly in low-lying areas destined to become flooded by melting snow water the following year. Although they transmit no known diseases to humans and domestic animals, the presence of snow-water mosquitoes in such numbers precludes carefree sport by those who venture into their domain.

Several species of *Aedes* are tree-hole breeders. Species that have adapted to breeding in small containers or leaf axils appear to have derived from tree-hole breeders, such as *A. aegypti. Aedes triseriatus* is a widespread tree-hole breeder east of the Rocky Mountains. It is similar in appearance to *A. vexans* but lacks white rings on the tarsi. It is an important vector of California (La Crosse) encephalitis virus. The most common tree-hole breeder in the United States is *A. hendersoni,* which is very similar to *A. triseriatus. Aedes sierrensis* is an American Pacific coast species. *Aedes albopictus,* the Asian tiger mosquito, was discovered in Houston, Texas in 1985. Apparently it arrived in this country in a shipload of used tires. It now is found in most areas east of the Rocky Mountains. Experimentally it is a good vector for dengue, equine encephalitis, yellow fever, and La Crosse virus.[11]

Many species of *Aedes* are vectors of a variety of virus diseases. The topic is too extensive to delve into here; for a summary see Harwood and James.[10] However, we will mention *A. aegypti,* the **yellow fever mosquito** (Fig. 39-9), because of its importance and wide distribution. It is found within a belt from 40° N to 40° S latitude, except in hot, dry locations. It is common in much of the southern United States. A beautiful mosquito, it is jet black or brown with silvery white or golden stripes on the abdomen and legs; the last tarsal segment is white. A lyre-shaped pattern covers the dorsal surface of the thorax. *Aedes aegypti* is a tree-breeding species in sylvatic situations, but when associated with human habitation, it breeds freely in containers, cisterns, and other water storage units. As many as 140 eggs are laid singly at or near the waterline, and they can withstand desiccation for up to a year.

Aedes aegypti originated in Africa, whence it was widely distributed by the slave trade. It was transported to much of the world via water barrels in ships. With the mosquito went a *Flavivirus,* which causes yellow fever, a devastating disease that has wrought havoc wherever it has emerged. After establishing in the New World, *A. aegypti* caused many epidemics. For example, the British army lost 20,000 of 27,000 men who attempted to conquer Mexico in 1741; the French lost 29,000 of 33,000 men trying to acquire Haiti and the Mississippi Valley. It was largely because of the presence of yellow fever in Louisiana and parts north that France was willing to negotiate the Louisiana Purchase the following year. Many outbreaks hit coastal cities in the United States, such as Charleston, New Orleans, and Philadelphia, and gold-rush settlements in California. Yellow fever and malaria forced France to abandon the completion of the Panama Canal and prevented the job from being attempted again until William Gorgas developed a program of mosquito control in Havana and then applied it to Panama. Strangely, yellow fever has never established in Asia.

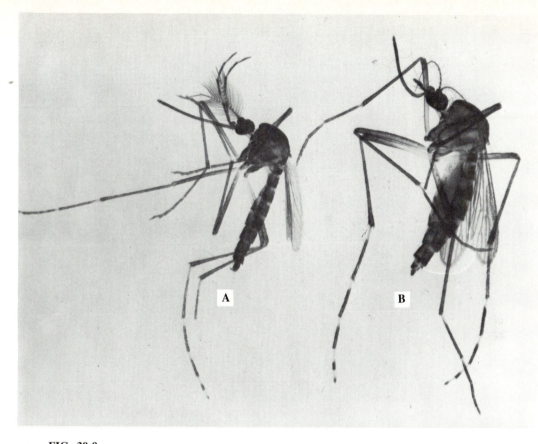

FIG. 39-9

Aedes aegypti, the yellow fever mosquito. **A,** Male. **B,** Female.

Photograph by Warren Buss.

Urban yellow fever is transmitted only by *A. aegypti,* but a sylvatic form existing in monkeys, both in Africa and South America, is transmitted by other *Aedes* and *Haemogogus* mosquitoes.

Another *Flavivirus* disease transmitted by *A. aegypti* is **dengue,** also called **breakbone fever** and **epidemic hemorrhagic fever.** The four distinct serotypes of dengue virus cannot be differentiated by symptoms. In uncomplicated cases the patient has fever, severe headaches, and pains in the muscles and joints. Weakness and temporary prostration are common, and recovery is rapid. A hemorrhagic complication occurs occasionally, especially in indigenous Asian children of 3 to 6 years old. This condition ranges from a rash and mottled skin to severe hemorrhaging in the lungs, digestive tract, and skin. The mortality rate is up to 7% in those who are hospitalized; unhospitalized cases must have a higher rate. Other *Aedes* spp. are known to be able to transmit the virus, but *A. aegypti* is the principal vector, being implicated in all serious epidemics. Dengue occurs from eastern Europe through most of Asia, North, Central and South America, and the Caribbean.

Genera *Mansonia* and *Coquillettidia*. These two genera are very similar; previous literature on *Mansonia* spp. often actually refers to *Coquillettidia.* The air tubes of the larvae of both are sharply pointed, enabling them to pierce stems of aquatic plants to obtain air. This has significance in control: simply coating the water with oil will not prevent these insects from obtaining oxygen. Species in this complex are important vectors of **brugian filariasis** (Chapter 31).

Genus *Culiseta*. Eight North American species and subspecies are in this genus. They are large, brownish mosquitoes, some of which are restricted to feeding on birds and other mammals. However, *C. inornata* and *C. melanura* are involved in the transmission of **western** and **eastern encephalitis viruses.** *Culiseta inornata* is the most widespread of the two, being found in southern Canada and the conterminous United

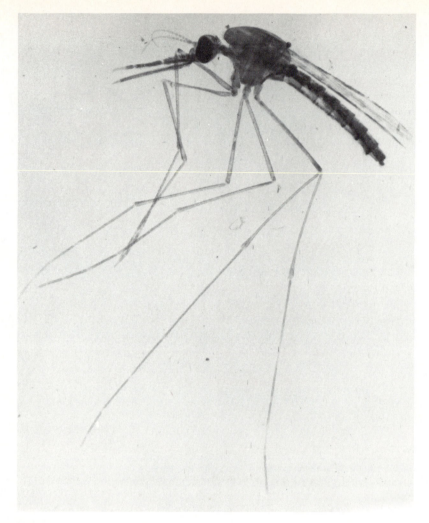

FIG. 39-10

Anopheles sp., female.

Photograph by Warren Buss

States, whereas *C. melanura* is restricted to the eastern and central United States.

Other genera of Culicinae are marginally important in the transmission of arboviruses.

■ Subfamily Anophelinae

Adult anophelines have a scutellum that is rounded or straight but never trilobed (except slightly in *Chagasia* spp.) in dorsal view. The abdominal sternites largely lack scales. The palpi of both sexes are almost as long as the proboscis (except in *Bironella* spp.). The larva lacks an air tube, and its dorsal surface bears branching hairs.

The subfamily contains three genera: *Bironella,* with seven species in New Guinea and Melanesia; *Chagasia,* with four species in tropical America; and *Anopheles,* with about 390 species, including 15 in North America. Resting and feeding postures are distinctive for *Anopheles* spp. When at rest, the head, proboscis, and abdomen are almost in a straight line; while feeding, the body is inclined at a sharp angle from the surface of the host. Because the genera *Bironella* and *Chagasia* are of no medical importance, we will consider only *Anopheles* spp. in this chapter.

Genus *Anopheles*. Female *Anopheles* (Fig. 39-10) lay up to a thousand eggs, depositing them singly on the water. The eggs have useful taxonomic characters, such as the presence or absence of lateral floats, which are characteristically marked, and a lateral frill. The eggs must remain in contact with water to survive. Usually they hatch within 2 to 6 days and develop through four larval instars in about 2 weeks, followed by a 3-day pupal stage. Development from egg to adult takes from 3 weeks to 1 month.

Preferred breeding sites vary tremendously among the species of *Anopheles*, a factor that must be understood before effective control of malaria vectors can be undertaken. Thus some species breed most efficiently in stagnant mangrove swamps; others, in sunny, partly shaded pools; and still others, along the edges of trickling streams. A few are tree-hole breeders.

Taxonomy of the genus *Anopheles* is complicated by the existence of several species complexes. For example, the species initially known as *A. maculipennis* is now known to consist of at least seven subspecies (or perhaps sibling species), differing slightly in host preferences and egg characteristics. American representatives of this complex are *A. quadrimaculatus, A. freeborni, A. aztecus, A. earlei,* and *A. occidentalis.* What was formerly known as *A. gambiae* in Africa is now known to comprise six species, including some freshwater and some saltwater breeders. Reproductive isolation and genetic barriers exist between all six species.[22]

Of all diseases transmitted to humans by insects, that caused by *Plasmodium falciparum* takes more lives and causes more suffering than the others put together (Chapter 9). This and the other malaria parasites of humans *(P. ovale, P. malariae,* and *P. vivax)* are all transmitted by species of *Anopheles*. Malaria is discussed extensively in Chapter 9, and the discussion will not be repeated here, except to say that historically the primary vectors of the disease in North America were *A. quadrimaculatus* and *A. freeborni.* Both are still common on that continent, as is *Aedes aegypti,* but like yellow fever, endemic malaria has been eradicated. The conquest of these two diseases in the United States stands among the greatest triumphs of this country. For a summary of mosquito-borne diseases and mosquito control, the reader is referred to the excellent treatment by Harwood and James.[10] Biological control with nematodes was discussed by Platzer.[25]

SUBORDER BRACHYCERA

In this group the antennae are reduced to three apparent segments, the terminal one being drawn into a sharp point, or **style.** A flagellum-like **arista** may be

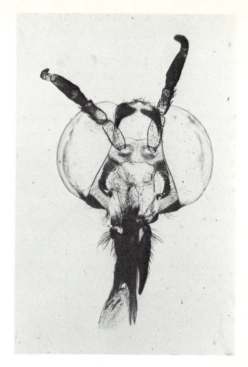

FIG. 39-11

Head of *Chrysops* sp., showing the tips of the mandibles.
Photograph by Warren Buss.

present. Wing venation is reduced. The larvae are active, usually predaceous, with an incomplete, retractable head. Life cycles are aquatic or semiaquatic. Only the Tabanidae and Rhagionidae among the Brachycera have bloodsucking habits.

Family Tabanidae

Horse flies and deer flies are widely distributed in the world, and their fierce questing for blood has earned them the animosity of all. They are large, powerful flies from 6 to 25 mm long. Tabanids are mainly daytime feeders. Only the female feeds on blood; the male lacks mandibles and eats only plant juices. The eyes are widely separated in females and are contiguous in males. About 3000 species are divided into 30 to 80 genera, depending on the authority.

The mouthparts of tabanids (Fig. 39-11) are of the horse fly subtype. They are similar to those of the Ceratopogonidae and Simuliidae but are stouter and stronger. The fascicle consists of six piercing organs: two flattened, bladelike mandibles with tooth-like serrations; two more narrow maxillae, also serrated; a median hypopharynx; and a median labrum-epipharynx. In biting, the mandibles cut in a scissors-like motion, whereas the maxillae pierce and rend the tissues, rupturing blood vessels. The fly feeds on the pool of blood that wells into the wound (telmophagy). The food canal is formed by the hypopharynx and labrum-epipharynx. Some fierce-looking species with long mouthparts, such as those in Pangoniinae, actually are not blood feeders.

Tabanids usually breed in aquatic or near aquatic environments, although some are known to complete larval development in soil. Generally, the female lays from a hundred to a thousand eggs at water's edge or on overhanging vegetation, rocks, and so on. At egg-laying time, such locations may be swarming with ovipositing flies. On hatching, the larva falls or crawls into the water or burrows into the mud. Many feed on organic debris, but others are voracious predators on insect larvae, worms, and other soft-bodied animals, including other horse fly larvae and even toads.

The larva has a small head, which is retractable and provided with powerful, sharp mandibles capable of inflicting a painful wound on the unwary. The body has 12 segments and a tracheal siphon that retracts into the posterior end of the body. In temperate zones the fly requires about a year to develop to pupation. At this time the larva crawls into drier earth and pupates. The pupa is obtect and requires from 5 days to 2 weeks to complete metamorphosis. The adult escapes the pupal case by cutting a T-shaped opening in the dorsal thorax; it then crawls out and makes it way to the surface. In the tropics two or more generations per year may occur.

Several species of horse flies are serious pests of humans and livestock in the United States. *Tabanus quinquevittatus* and *T. nigrovittatus* are the large, gray, "greenhead" horse flies of most of the United States. *Tabanus atratus* (Fig. 39-12) is a huge, uniformly black horse fly of eastern North America; *T. lineola* and *T. similis* are smaller, striped flies also in the eastern states. In the western states *T. punctifer* is a very large, black horse fly with a yellow thorax. Such species as *T. atratus* and *T. punctifer* seldom bite humans: they buzz so loudly when approaching that they seldom are allowed to land. *Haematopota americana, Hybomitra* spp., and *Silvius* spp. are also common western pests. *Diachlorus* spp. are common, aggressive pests in Central America, where they are called "doctor flies." Unlike most tabanids they freely enter houses.

The name deer fly is applied to the genus *Chrysops,* of which there are about 80 North American species. Deer flies (Fig. 39-13) usually are smaller than horse flies and have brown-spotted wings. Their flight is not so noisy as that of most horse flies, so they bite humans more commonly.

The medical importance of Tabanidae is two sided:

FIG. 39-12

Tabanus atratus.

Photograph by Warren Buss.

the annoyance and blood loss occasioned by the bite and infections transmitted mechanically and biologically by the flies.

Because of the large size of the mouthparts, tabanid bites are extremely painful. Most people have little or no allergic reaction to them, although such sequellae are known to occur. Their annoyance factor may seriously interfere with the use of recreational areas, and field and timber workers may have lowered productivity as a result of harassment by these flies.

A serious problem is blood loss and aggravated behavior in livestock. Beef and milk production in the United States were reduced an estimated $40 million in 1965 as a result of interrupted grazing, energy consumed in trying to escape the insects, and blood loss. Tethered or caged animals particularly suffer from these flies because they are unable to escape their tormentors, even briefly. One of us has seen a caged mule deer, simultaneously being fed on by a dozen or more *Tabanus punctifer,* with dozens of open, freely bleeding wounds covering the wretched animal's face.

FIG. 39-13

Deer flies. *Chrysops* sp.

Photograph by Warren Buss.

No tabanids are strictly host specific, but most have host preferences. Thus *Haematopota*, with at least 300 species, feeds mainly on cattle and antelope, with which it may have evolved. Birds are attacked uncommonly; amphibians and reptiles have been observed to provide nourishment for these parasites.

Certain adaptations of tabanids enhance their capabilities to transmit pathogens[15]: (1) **anautogeny,** the necessity of a blood meal for development of eggs, stimulating host-seeking behavior; (2) **telmophagy,** where blood-dwelling pathogens can enter the pools from which the fly sucks blood; (3) **relatively large blood meals,** enhancing the possibility that pathogens will be imbibed; (4) **long engorgement time,** enabling the pathogens to infect the fly's tissues; and (5) **intermittent feeding behavior,** increasing the chances for mechanical transmission of pathogens.

Tabanids are involved in the transmission of protozoan, helminthic, bacterial, and viral diseases of animals and humans. Among the diseases caused by protozoa, two species of *Trypanosoma* are transmitted mechanically by tabanids. *Trypanosoma evansi,* the causative agent of **surra** in many wild and domestic animals, is spread by species of *Tabanus* (Chapter 5). Other vectors, such as stable flies, other genera of horse flies, and vampire bats, also can be involved; but *Tabanus* spp. appear to be the most effective vectors of this trypanosome. *Trypanosoma theileri* is a cosmopolitan parasite of cattle and antelopes. Cyclopropagative development occurs in the insect gut, so

the tabanid species involved are actually true intermediate hosts. Examples of such species are *Haematopota pluvialis, Tabanus striatus,* and *T. glaucopis.*[15]

The African eye worm, *Loa loa,* is transmitted by species of *Chrysops* (Chapter 31). There appear to be two strains of this filariid, one in monkeys in the forest canopy and one in humans. The former is transmitted by night-feeding *Chrysops langi* and *C. centurionis* and the latter by diurnal *C. silaceus* and *C. dimidiatus.*

The arterial filariid *Elaeophora schneideri* lives in vessels of the head and neck of American species of deer, elk, moose, and domestic sheep in the western states. It is symptomless in deer but in other hosts causes much distress, such as blindness, nervous dysfunction, necrosis, and deformity of the head. Species of *Tabanus* and *Hybomitra* are the vectors of this worm.[12]

Bacterial infections known to be mechanically transmitted by tabanids are anaplasmosis and anthrax. Similarly, hog cholera and equine infectious anemia (swamp fever) viruses use horse flies as vectors.

Family Rhagionidae

Rhagionids, known as **snipe flies,** are predominantly nonbloodsucking, but the several species that feed on blood are considerably annoying. Dominant among the snipe flies that bite humans is the genus *Symphoromyia.* The behavior of its species is like that of *Chrysops* spp., landing suddenly and quickly piercing the skin. The mouthparts and mode of feeding are also like those of the Tabanidae.

The biology of Rhagionidae is poorly known. Larvae develop in boggy or mossy stream banks and similar places. It is not known if snipe flies transmit any diseases, but because *Symphoromyia* species and others take a considerable amount of blood at a meal and feed several times during their lives, they certainly must be regarded as potential vectors. As far as is known, the importance of rhagionids is primarily as pests whose biting proclivities may doom an otherwise pleasant outing.[30]

SUBORDER CYCLORRHAPHA

This is the most advanced group of flies, and except for the mosquitoes in Nematocera, they are the most important flies in veterinary and human medicine. The antennae are short and pendulous, with a conspicuous arista on the second segment. Usually, three prominent ocelli are arranged in a triangle on the vertex and frons. The compound eyes are very large, separated in females and close together in males. The arrangement of bristles on the head and thorax is important in the taxonomy of flies (**chaetotaxy**). At the anal angle of the wing is a prominent lobe, the **squama**, or **calypter**.

Larval cyclorrhaphids are **maggots**, with an elongate, simple body, usually tapered toward the head end. A true head is absent; the vertically biting mandibles are part of a conspicuous cephalopharyngeal skeleton. These sclerotized structures are important in the taxonomy of larvae. Two spiracles are found on the posterior end. Each species has distinctive markings on the spiracular plates that are useful in identification (see Figs. 39-15 and 39-16).

When fully developed, the third-stage larva undergoes **pupariation,** resulting in a pupa-like surrounding case (**puparium**) made of the hardened third-stage larval tegument. After internal reorganization, the adult emerges through a circular hole in the puparium. It pushes the operculum off the puparium by inflating a balloon-like organ, the **ptilinum,** in the head. The ptilinum extends from the frontal suture of the head, pushes off the operculum, and then is withdrawn back into the head, the frontal suture immediately healing.

Family Chloropidae

Looking very much like tiny house flies, the **eye gnats** are of considerable medical importance. Unlike those of the gnats discussed in the Nematocera, the antennae of this family are aristate. The antennae resemble those of Drosophilidae (fruit flies) except that the arista is nearly smooth, whereas it is distinctly feathered in fruit flies.

The most important genus in this family is *Hippolates*. These flies are very small, 1.5 to 2.5 mm long, and are acalypterate. They are called eye gnats because they are attracted to eye secretions, as well as to other body secretions and free blood. They do not bite but feed like house flies, sponging up liquid food. Also like house flies, they vomit liquid stomach contents onto their food, to be reeaten. For this and other reasons, eye gnats are important vectors of disease. In some species the labellum is provided with tiny spines that scarify the skin of a host, also leading to infection by pathogens. *Hippolates* spp. are very persistent when hungry and hence may become intensely irritating. They are strong flyers, able to fly against weak winds.

Hippolates spp. are found only in North America. *Hippolates pusio* and several other related species constitute the most important group.[26] *Siphunculina funicola* plays a similar role in Southeast Asia.

The life cycles all seem to be similar. About 50 eggs are laid on the surface of or slightly under the soil, which must be loose and have an abundance of well-aerated organic matter, either animal or vegetable. Larval development is completed in 7 to 12 days under optimal conditions; the pupal period requires about 6 days, and the adult ages 7 days before oviposition.[7]

Aside from the annoyance caused by these flies, which can be considerable, they are important vectors of disease. Although they are not biting insects, they congregate at wounds caused by others, such as tabanids and stable flies, thereby further contaminating the host.

Hippolates spp. probably are a factor in transmission of the bacillum causing human **pinkeye,** or **bacterial conjunctivitis,** although this has not been proved. Similarly, the spirochete *Treponema pertenue*, the etiological agent of **yaws,** most likely is transmitted mechanically by eye gnats, although this also has not been proved. **Bovine mastitis,** a bacterial disease, is known to be spread from cow to cow by *Hippolates* flies feeding at the tips of the teats.

Control of eye gnats is difficult. The best system in use so far is a combination of attractant baits with a pesticide and efficient soil management.

Family Muscidae

Members of this family are often **synanthropic;** that is, they live closely with humans. Many freely enter houses and readily avail themselves of whatever food and drink they may find there. They are small- to medium-sized flies, usually dull-colored, with well-developed squamae and mouthparts.

■ *Musca domestica*

The house fly is the most familiar of all flies, as well as being one of the most medically important. It is gray, 6 to 9 mm long, and has four conspicuous, dark, longitudinal stripes on the top of the thorax. Its distribution is nearly cosmopolitan but has changed markedly in advanced societies as sanitation has increased and dependence on the horse has decreased.

House flies breed in all types of organic wastes, except decaying flesh. Feces of any kind are preferred, although decaying milk around dairy barns, silage, slops around hog troughs, rotten fruit and vegetables, and so on are all stock in trade for house flies. Garbage cans are a favorite breeding place, and one of us has witnessed rotting garbage in the tropics apparently transformed into an equal mass of maggots, seemingly overnight.

Under ideal conditions, the egg can develop to adult in 10 days. One female deposits 120 to 150 eggs in each of at least six lots in its short lifetime. It is easily calculated that if all offspring of a pair of flies in April lived and reproduced, as did their succeeding generations, by August there would be 191,010,000,000,000,000,000 flies, which would cover the earth to a depth of 47 feet. This illustrates how a depleted population of flies can recover in a very short time.

The house fly is an efficient disease carrier for three primary reasons:

1. Its construction favors carrying bacteria. The multitude of tiny hairs covering most of the body readily collect bacteria, spores, and helminth eggs; and the mouthparts and six feet also have sticky pads that collect such matter.
2. It relishes human food and excrement alike. While walking on food and utensils, it not only leaves a trail of bacteria, but also while feeding, it defecates and vomits the remains of its last meal. Helminth eggs, protozoan cysts, and bacteria survive the intestinal tract of the fly and thus can be widely distributed from the site of their initial deposition.
3. Because of their synanthropy and powerful flight, house flies move about freely between indoor and outdoor attractions. Thus it would appear that house flies achieve the ideal of mechanical transmission of disease.

The list of diseases known to be transmitted by house flies is too long to be repeated here; the interested reader is referred to the excellent account by Harwood and James.[10] Briefly, most are enteric diseases occasioned by fecal contamination, such as typhoid fever, cholera, polio, hepatitis, shigellosis, salmonellosis, and other dysenteries. Others include

FIG. 39-14

Australian bush fly, *Musca vetustissima,* male.
Photograph by Warren Buss.

yaws, leprosy, anthrax, trachoma, tuberculosis, and diseases caused by various worms, such as *Ascaris.* Several diseases of domestic animals also are transmitted by these pests.

■ Other species of *Musca*

Twenty-six species have been placed in the genus *Musca.*[6] In Australia the bush fly, *M. vetustissima* (Fig. 39-14), occurs in incredible numbers. It is not so willing to enter human habitation as is *M. domestica.* Also, its importance as a vector is much less, probably partly because of the widely scattered human population in much of that continent. However, its propensity to walk on people's faces is a genuine nuisance. One of us has been photographed in the central Australian desert with at least a thousand of these flies on his back.

The face fly, *M. autumnalis,* is a native of Africa, Asia, and Europe and was introduced into the United States in 1950. It now is found from coast to coast and well into Canada. It is a little larger than the house fly. The sides of the abdomen of the female are black, and those of the male are orangish. Larval development is in cow dung. Adults feed on secretions around the head of cattle and other large ruminants. They will enter houses and can annoy people. They serve as vectors of the eye worm, *Thelazia* (Chapter 29). Annual losses in the United States are estimated at $60 million.

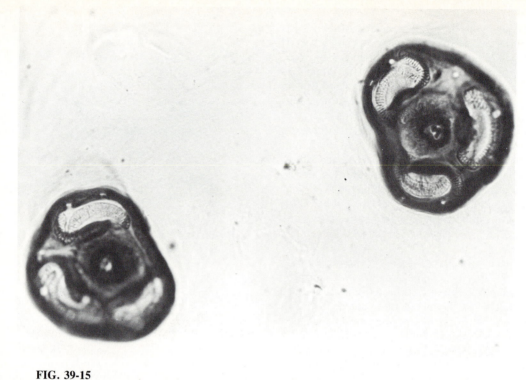

FIG. 39-15

Larval stigmata of *Stomoxys calcitrans*.

Photograph by Warren Buss.

■ **Genus** *Fannia*

About 200 species are known in this genus, two of which will be mentioned briefly. *Fannia canicularis*, the **lesser house fly,** looks much like a smaller version of *M. domestica*. It is more slender, dark, and has three brown longitudinal stripes on the thorax, rather than four as in the house fly. The biology of the lesser house fly parallels that of *M. domestica*, with the life cycle completed in 15 to 30 days. The larva is unlike that of *M. domestica*, since it is covered with long, slender projections. *Fannia canicularis* will enter houses freely; in fact it often is the dominant species at any given time. Unlike *M. domestica*, however, *F. canicularis* is not particularly attracted to food and so is not so efficient a vector as is the house fly.

The **latrine fly,** *F. scalaris*, is very similar in appearance to *F. canicularis* but seldom enters houses. It breeds in fresh dung, particularly that of swine. The larva also is similar, but its lateral processes are distinctly feathered.

Both species of *Fannia* are known to cause accidental myiasis of the rectum of humans, presumably when they lay eggs on the anus and the larvae crawl into the rectum and begin developing. No medical problem is caused by this beyond consternation in a person who discovers maggots in his stool. Myiasis is discussed at greater length later in this chapter.

■ *Stomoxys calcitrans*

The **stable fly** is a cosmopolitan species and the only species in the genus that occurs in North America. It is similar in appearance to *M. domestica* and is often mistaken for it. The gray abdomen is rather checked, and the long slender proboscis protrudes in front of the head. A valuable reference to this fly is Zumpt.[35]

Stable flies are daytime biters; both sexes feed on blood. The labella are equipped with rows of teeth that can readily pierce the skin and underlying tissues. The flies then sponge up blood that wells into the wound. Stable flies avidly bite humans and other animals, especially cattle and horses. They prefer to breed in decaying vegetation rather than manure but are adaptable. Larval stigmata are shown in Fig. 39-15.

When the insect occurs in great numbers, its attacks on humans are intolerable, affecting tourist industries in some areas. The fly is one of the most important pests of livestock, causing great losses annually. Loss of weight and milk production cost the United States $142 million in 1965. Hides can be damaged by the bites, and adult cattle can be killed if bitten enough times. It has been calculated that 25 flies a day per cow is the economic threshhold; more flies cause a recognizable loss.[27] A thousand flies have been observed on a cow at one time.

Several diseases are known or suspected to be transmitted by stable flies. Among these is *Trypanosoma evansi,* the agent of surra (Chapter 5). Mechanical transmission of the *T. brucei* complex also occurs. Other diseases transmitted by stable flies are epidemic relapsing fever, anthrax, brucellosis, swine erysipelas, equine swamp fever, African horse sickness, and fowl pox. Also, the stable fly is the intermediate host of the horse stomach worm, *Habronema microstoma,* which infects the horse when an infected fly is swallowed.

■ *Haematobia irritans*

The horn fly (also known as *Hydrotaea irritans)* is found in the Americas, Europe, Asia Minor, and Africa. It closely resembles the stable fly but is more slender. It feeds with its head toward the ground, whereas the stable fly feeds with its head up. Horn flies breed in fresh cow manure. They will bite humans but are not as active fliers as are stable flies. They have been implicated as vectors of bovine mastitis.[13]

The importance of horn flies is mainly veterinary, with loss to livestock approaching that caused by stable flies. However, no diseases have been proved to be transmitted by this insect, with the exception of the filarial nematode of cattle, *Stephanofilaria stilesi.* This skin parasite causes thickening and scabbing of the epidermis, especially around the naval, thus attracting more horn flies.

Other genera and species of muscoid flies are of some importance, but we do not have space to discuss them here. The reader is referred to Harwood and James.[10]

■ *Glossina* species

These are the infamous **tsetse flies** of Africa south of the Sahara. Although now restricted to that area, they were once widespread: four species have been found in the Oligocene shales of Colorado.

Tsetse (Fig. 4-4) are 7.5 to 14 mm long and brownish gray. When at rest, the wings are crossed like scissors. The palpi are almost as long as the proboscis, which protrudes from the front of the head. The mouthparts, and thus feeding habits, are much like those of stable flies. The base of the proboscis is swollen into a characteristic bulb. Tsetse are daytime feed-

ers and are visually attracted to moving objects. Both sexes feed exclusively on the blood of a wide variety of animals, including humans, and are particularly attracted to the Suidae.

Tsetse are larviparous and pupiparous, giving birth to a single, completely developed larva at intervals, producing from 8 to 20 in all. While in the oviduct, the larva feeds on secretions from specialized milk glands. Larvae are deposited on loose, dry soil, usually under shelter of some type. The larva has no locomotor structures but by contraction and extension buries itself under a few centimeters of loose soil. Hardening of the integument to form a puparium occurs within an hour of larviposition. The integument darkens to brownish black; it is barrel shaped and has two prominent posterior lobes. The adult emerges within 2 to 4 weeks. The biology and influence of tsetse are beautifully illustrated by Gerster.[5]

Twenty-two species are usually recognized and can be identified with the use of the key by Mulligan and Potts.[23] All but three have been found capable of transmitting trypanosomes of mammals. Six of these are of outstanding medical importance: *Glossina palpalis, G. fuscipes,* and *G. tachinoides* are found along rivers; *G. morsitans, G. swynnertoni,* and *G. pallidipes* are savannah species. The first three are the primary vectors of Gambian sleeping sickness; the last three principally transmit Rhodesian sleeping sickness. Probably all of these, as well as several other species, can transmit nagana to cattle. African trypanosomiases are discussed in detail in Chapter 4. The use of traps to control tsetse is discussed by Vale, et al.[31]

■ Myiasis

Myiasis is the name given to infection by fly maggots. Although families other than Calliphoridae cause myiasis, it seems appropriate to introduce the subject here, in a general way.

There are several categories of myiasis, as in other forms of parasitism. In **obligatory myiasis** the insect depends on a period of parasitism before it can complete its life cycle. In **facultative myiasis** a normally free-living maggot becomes parasitic when it accidentally gains entrance into a host. Other terms are useful in describing the location of the larva: **gastric, intestinal,** or **rectal myiasis,** for invasion of the digestive system; **nasopharyngeal** for nose, sinuses, and pharynx; **cutaneous,** either **creeping,** when the larva burrows through the skin, or **furuncular,** if it remains in a boil-like lesion; **urinary** or **urogenital; auricular** for the ears; and **ophthalmic** for the eyes. Some larvae are intermittent bloodsuckers and are loosely included under the term myiasis. **Accidental myiasis** is usually en-

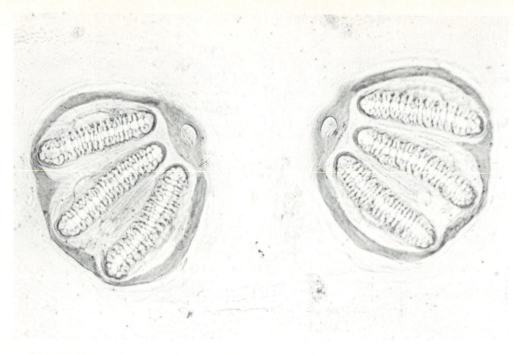

FIG. 39-16

Larval stigmata of *Calliphora vomitoria*.

Photograph by Warren Buss.

teric, when eggs or larvae are eaten with contaminated food or drink. Such cases also are called **pseudomyiasis.**[34] Pseudomyiasis usually involves muscoid flies, such as *M. domestica* and *Fannia* spp.

FAMILY CALLIPHORIDAE

This is the large family of blow flies, most of which are beneficial in helping destroy carcasses. A few, however, are of great importance in causing **myiasis.** The common blow flies are usually metallic green, blue, or copper, although some are nonmetallic.

Common species

Calliphora vomitoria (Fig. 39-16) is a large, metallic blue fly, probably the "blue-tailed fly" of song. It is conspicuous because of its large size and loud buzz as it flies. It has been recorded as a cause of pseudomyiasis. Other species of *Calliphora* are facultative parasites. *Phaenicia* and *Lucilia* spp. are metallic green with copper iridescence. *Phaenicia sericata* will breed in carrion, as well as excrement and garbage. It is important in **wool strike** in Australia. **Strike** is the term for the action of a fly laying its eggs or larvae on an animal. In wool strike the eggs are laid on the soiled wool on the rump of a sheep. The maggots feed

on feces and bacteria; their activities cause a great deal of irritation to the sheep, which leads to other complications. This is why the tail is docked from lambs soon after they are born. Laboratory-reared *P. sericata* were used in World War I to clean wounds in servicemen; some species of calliphorid blow flies limit infections in wounds.[3] It has been reported as a facultative parasite in the ear canal and in open wounds.

Other blow flies known to be facultative parasites are species of *Phormia*, *Lucilia*, *Cochliomyia*, and *Chrysomyia*.

■ *Cochliomyia hominivorax* (Fig. 39-17)

This is the **primary screwworm,** which is the most important cause of myiasis in the world. It is an obligate parasite, occurring throughout the Neotropical region. It causes dermal myiasis in nearly any mammal, as well as nasopharyngeal myiasis in humans.

The adult fly is a deep, greenish-blue metallic color with a yellow, orange, or reddish face and three dark stripes on the thorax. It is difficult to differentiate *C. hominivorax* from the **lesser screwworm,** *C. macellaria*, which is not an important myiasis-causing insect.

Screwworm larvae cannot penetrate intact skin, al-

FIG. 39-17

Life history stages of the primary screwworm, *Cochliomyia hominovorax*. **A,** Two egg clusters. **B,** Male. **C,** Female. **D,** Puparium. **E,** Larva.

though mucous membranes of the face and genitalia are susceptible to their attack. Usually a preexisting wound, however small, attracts the fly. Cuts from barbed wire or needle grass, castration and dehorning of calves, and insect and tick bites are all examples of sources of entry for screwworm maggots. Wounds from dog fights are commonly attacked. One of us helped remove more than a hundred screwworms from around the ear of a dog in Trinidad.

Screwworms in humans are not uncommon. Generally, the more abundant they are in livestock, the greater the changes of human infection. Infection in the head can be fatal, and urogenital infection can be grossly deforming.

The primary screwworm cannot survive winter in cold climates, but summer migrations have brought it as far north as Montana and Minnesota. Its normal range is from Mexico to Chile and Argentina. Severe epizootics have occurred in Texas cattle. More than 1.2 million cases were recorded in 1935 in that state alone.

The best control of this pest so far developed is to raise the flies in the laboratory, sterilize the males, and free them to mate with wild females. Since fe-males of this species usually mate only once, they thus cannot produce offspring after a sterile mating. *Cochliomyia hominivorax* has been eradicated from the United States and, in fact, is present in North America only in the Mexican state of Yucatan.[14]

The **Old World screwworm,** *Chrysomyia bezziana,* occupies the same niche in Africa, India, the Philippines, and the East Indies. It also attacks humans. Its biology and pathogenesis parallel those of *C. hominivorax.* It now is known to be endemic in South America.[17]

■ *Cordylobia anthropophaga*

The **tumbu fly** is an African calliphorid restricted to south of the Sahara. The adult is yellowish, as contrasted with the calliphorids we are accustomed to in the northern hemisphere. It is stimulated to lay its eggs on soil that has been contaminated with urine. When the first-stage larva contacts mammalian skin, it penetrates and begins to grow, causing furuncular myiasis. Reports of this parasite from other parts of the world probably reflect infection acquired in Africa and detected elsewhere. Many wild mammals are reservoirs.

■ *Auchmeromyia luteola*

The **Congo floor maggot** is a bloodsucking species found south of the Sahara. It is the only dipterous larva known to suck the blood of humans. Eggs are laid on floor mats, dry soil, or crevices in huts. The larvae are quite resistant to desiccation. They feed like bedbugs: when a person is asleep on the floor or on a mat, the maggots come out of hiding and pierce the skin with their powerful mouth hooks. Feeding is completed in 15 to 20 minutes, after which the larvae return to hiding. They are not not known to transmit any disease.

Bloodsucking maggots of birds are common. In the northern hemisphere *Protocalliphora* spp. sometimes destroy entire broods of young birds.

Family Sarcophagidae

These ubiquitous insects are known as **flesh flies.** They are closely related to Calliphoridae, but instead of being metallic, the abdomen is checkered gray and black. Most are parasites of invertebrates, including insects and snails, and some are carrion breeders, but others are parasitic in the skin of vertebrates, including humans.

Sarcophaga hemorrhoidalis is widespread in the northern hemisphere and well into the tropics. It looks like a large house fly, but the tip of the male abdomen is red. Most sarcophagids are larviparous and normally breed in carrion, but the female will deposit larvae in open wounds to become facultative parasites. Similarly, *Wohlfartia magnifica* is a facultative parasite of mammals in the warmer zones of the Palearctic region. Fatal human cases have been recorded.

Cutaneous furuncular myiasis is caused by *W. vigil* in Canada and the northern United States and by *W. opaca* in the western United States. The larvae are deposited on unbroken skin, which they quickly penetrate. Human infections usually occur in infants left unattended outdoors, although sleeping adults have been infected indoors. In the northern United States *W. vigil* is a serious pathogen of mink and fox kits in fur farms where newborns are often struck and soon die of the infection. Rodents and rabbits probably are reservoirs, as are carnivores.[4]

Family Hippoboscidae

The **louse flies** look neither like lice nor flies but rather like six-legged ticks. In most species the males are winged and the females wingless, although some, such as *Hippobosca* spp., are winged in both sexes. Both sexes are bloodsuckers, with some species parasitizing mammals and others, birds. This is another pupiparous family: the larvae are retained within the female, feeding on secretions from special glands, and when born are ready to pupariate.

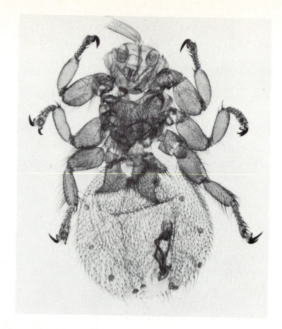

FIG. 39-18

Sheep ked, *Melophagus ovinus*.
Photograph by Warren Buss.

The **sheep ked,** *Melophagus ovinus* (Fig. 39-18), is distributed worldwide except in the tropics. Its puparium is glued to the wool of its host at any season of the year. Each female produces from 10 to 12 young. The entire life of the ked is spent on the host; when removed, it dies in about 4 days. A heavy infestation causes emaciation, anemia, and general unthriftiness of sheep. The skin may be so scarred by bites as to lose its market value. Hippoboscids are not loath to bite humans, and sheep shearers particularly are vulnerable to their attacks. The bite is said to be as painful as a yellow jacket wasp sting.

Other genera and species are found on mammals and birds in various parts of the world. *Olfersia coriacea* has been observed to bite humans in Panama. Its painless bite is a potential disease vector to humans.[9] In the western United States *Neolipoptena ferrisi* (Fig. 39-19) is common on mule deer. The **pigeon fly,** *Pseudolynchia canariensis,* is common on pigeons throughout most of the warm, temperate regions of the world. Both sexes are winged. Besides its importance as a bloodsucking parasite of pigeons, it also is the intermediate host of the malaria-like *Haemoproteus columbae,* discussed in Chapter 9. The pigeon fly also is willing to bite people, with painful results.

Family Gasterophilidae

This family comprises the **stomach bots** of equids, elephants, and rhinoceroses. The adult flies are similar

FIG. 39-19

Neolipoptena ferrisi, a louse fly of mule deer.

Photograph by Warren Buss.

to honeybees in size and appearance and are strong fliers. The ovipositor is long and protuberant. The larvae of these species cause true enteric myiasis, attaching to the mucosa of the host's stomach.

Three species have been introduced into the United States from the Old World. They are parasites of horses, asses, and mules.

Gasterophilus intestinalis is called the **horse bot fly.** It is very common in North America and throughout most of the world. The female attaches approximately a thousand eggs to the hair of the horse, mainly on the knees. When the animal licks its hair, the warmth and moisture stimulate hatching. The first-stage larva immediately penetrates the tongue epithelium and tunnels its way down to the stomach, where it emerges and attaches with powerful mouth hooks. Feeding on blood, it grows through two ecdyses to the third stage (Fig. 39-20). All instars have circles of strong spines on all but the last few segments. They remain attached until the following spring and early summer, when they detach and pass out with the feces. Pupation takes place in loose earth, and after 3 to 5 weeks the adults emerge. Migration in the oral cavity is well illustrated by Cogley et al.[1]

Gastrophilus nasalis, the **throat bot fly,** has a similar life cycle except that the eggs are attached to hairs under the jaw. The larvae hatch in 4 to 5 days without the need of moisture, crawl along the jaw, and enter between the lips.

The **nose bot fly,** *G. haemorrhoidalis,* strikes the horse on the lips. The remainder of its life cycle is similar to that of *G. intestinalis* except that the third instar larvae attach inside the anus for a short time before passing out.

Other genera are *Cobboldia, Platycobboldia,* and *Rodhainomyia* in elephants and *Gyrostigma* in rhinoceroses.

A few stomach bots cause little or no problems in horses, but a heavy infection may cause enough damage to the mucosa of the stomach and to the intestine during migration to kill the animal. Blockage of the pylorus also can occur.

Occasionally, first instars will penetrate human skin and cause creeping myiasis, but they cannot mature or move deeper into the tissues.

Family Hypodermatidae

Variously known as **cattle grubs, ox warbles,** and **heel flies,** these skin parasites are found over most of the northern hemisphere. They primarily infect cattle and Old World deer, including reindeer, but they have been known to parasitize horses and humans as well.

Hypoderma lineatum and *H. bovis* (Fig. 39-21) are the two species that infect cattle. The former is common in Asia, Europe, and the United States, whereas the latter is slightly more northern in its distribution. Both look much like small bumblebees, with light and dark bands on the bodies.

The life cycles of the two species are similar. Both flies strike the hair of cattle, mainly on the hind legs. Although this is painless, the cattle become agitated and even terrified and gallop back and forth to avoid them. This action is called **gadding** and gives rise to the term "gadfly," sometimes applied to people. On hatching within a week, the larva penetrates the skin and makes a remarkable migration, first to the front end of its relatively huge host, then back to the lumbar region, where it develops until pupation. All aspects of the migration are not yet known, but *H. bovis* reaches the spinal cord, usually in the neck, and burrows posteriad between the periosteum and the dura mater for a distance, then completes its journey through tissues to the back. *Hypoderma lineatum* rests for a time in the wall of the esophagus and appears not to invade the spinal cord. Both species, on arriving at the lumbar skin, cut a hole in it, reverse position, apply the spiracles to the hole, and begin to feed. When ready to pupate, the grub cuts its way out, falls to the ground, and buries itself. The entire life cycle requires about a year.

These flies cause considerable damage to their hosts. In 1965 the USDA cited a $192 million loss,

FIG. 39-20

Third-stage larvae of the horse stomach bot, *Gasterophilus intestinalis*.

Photograph by Warren Buss.

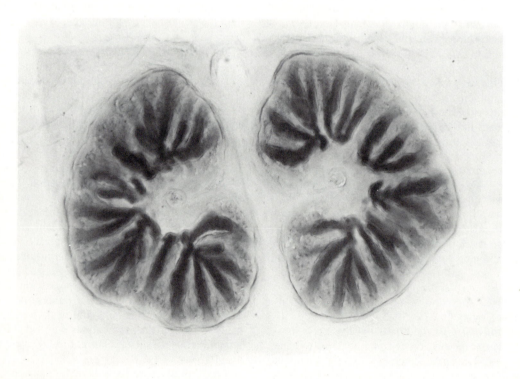

FIG. 39-21

Larval stigmata of *Hypderma bovis*.

Photograph by Warren Buss.

FIG. 39-22

Reindeer warbles, *Oedemagena tarandi*.

Photograph by Warren Buss.

resulting primarily from the loss of weight, reduced milk production, and damage to hides. Warble fly has been nearly eradicated in Britain. The methods used are reviewed by Tarry.[28]

Numerous cases of *Hypoderma* spp. in humans are recorded. Unlike those of *Gasterophilus* spp., the larvae of these flies can successfully migrate and develop in the human body. Usually they surface in the neck region, probably because of the upright position of humans. Results of the migration can be dire, including partial or total paralysis of the legs. Ocular myiasis can occur, with loss of an eye. Most people infected with this parasite have a close association with cattle.

Other species of warbles infect sheep and goats in Africa, Asia, and Europe. The **reindeer warble,** *Oedemagena tarandi* (Fig. 39-22), is distributed over the range of wild and domestic reindeer, causing considerable loss in young animals. Some Eskimo tribes consider the fresh, live grubs to be a delicacy to be eaten immediately on slaughter of a caribou.

Family Oestridae

The **head maggots** are about the same size and shape as honeybees and do not feed as adults. The larvae develop within the sinuses and nasal passages of hoofed animals.

The **sheep bot,** *Oestrus ovis,* is a cosmopolitan parasite of domestic sheep and goats and related wild

species. The female deposits active larvae in the nostrils of the host during summer or early autumn. The larvae rapidly crawl up into the sinuses, where they attach to the mucosa and feed. Often they are in great numbers, causing considerable damage and pain to the host. By spring the larvae are developed and crawl back down to the nostrils, where they fall or are sneezed out. Pupation in the soil lasts from 3 to 6 weeks. Heavy infections can be fatal, but usually the host is only tormented, showing evidence of great distress by sneezing, shaking of the head, loss of appetite, and a purulent discharge from the nose.

Other head maggots are *Rhinoestrus purpureus* in horses of Europe, Asia, and Africa; *Gedoelstia* spp. in African antelopes; and *Cephenemyia* spp. in Old and New World deer.

Ophthalmomyiasis occasionally occurs in humans, usually because of strike by *O. ovis* or *R. purpureus*. The larva cannot develop beyond first stage and usually does not last long. Inflammation and conjunctivitis may result. Head maggot strike in humans is most common in shepherds and others who work closely with sheep or horses.

Family Cuterebridae

The cuterebrids are the **skin bot flies.** The common genus *Cuterebra* is a large black or blue fly about the size of a bumblebee. Found from the north temperate

643

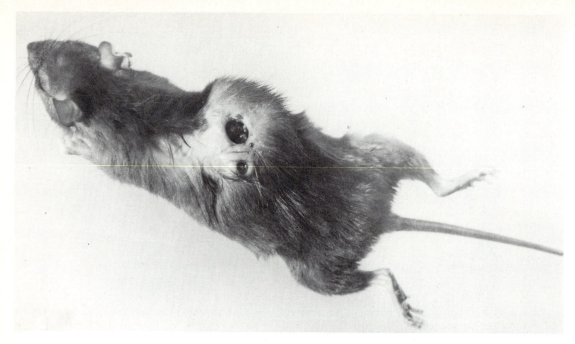

FIG. 39-23

Rice rat, *Oryzomys capito,* with larvae of *Cuterebra* sp. in the skin.

Photograph by C.O.R. Everard.

to tropical zones of the New World, species of this genus parasitize rodents, lagomorphs, and marsupials. Eggs are laid on or near natural orifices, and, after hatching, the larvae enter the body, tunnel under the skin, cut an air hole in it, and begin to feed. The larvae are densely covered with thick spines and in some cases grow as large as a man's thumb. Often the host is disproportionately small, and it seems incredible that it can survive parasitism by such a large bot (Fig. 39-23). When located in the scrotum, *Cuterebra* spp. often will castrate the rodent host. Human cases are rare, with entry made through the anus, nose, mouth, or eye.

Dermatobia hominis is the **human skin bot.** It is common from Mexico through most of South America. A forest-inhabiting fly, it develops in the skin of almost any warm-blooded animals, including birds. Adults resemble bluebottles. Unlike any other myiasis-causing fly, *D. hominis* does not lay its eggs directly on the host. Instead, it catches another insect, such as a mosquito, and glues its eggs to the side of it, with the operculated anterior end hanging down. At least 48 species of flies and one tick have been recorded as carriers.[8]

When the carrier insect lands on warm skin, the eggs immediately hatch, and the larva drops onto the new host, penetrating unbroken skin. It bores into the dermis and remains there without further wandering. Development to pupal stage requires about 6 weeks; pupation is in the soil.

This fly commonly parasitizes humans, in whom it causes painful lesions. Often a person has returned home to Europe or North America before the infection is noted and diagnosed. A small incision in the skin allows the larva to be expressed. It is readily recognized by two cauliflower-like projections at the posterior end. A valuable review of this fly is given by Guimarães and Papavero.[8]

Families Nycteribiidae and Streblidae

These two small, poorly known families are the **bat flies,** parasitic only on bats. Nycteribiids are called **bat spider flies** (Fig. 39-24) because of their superficial resemblance to spiders. They are wingless, with the head folded back into a groove in the dorsum of the thorax. Five species are found in North America; most species feed on Old World bats.

The streblids may be winged or wingless, or they may have reduced wings. Compound eyes are small or absent. Six species are found in North America; most are associated with New World tropical bats.

Both families are pupiparous.

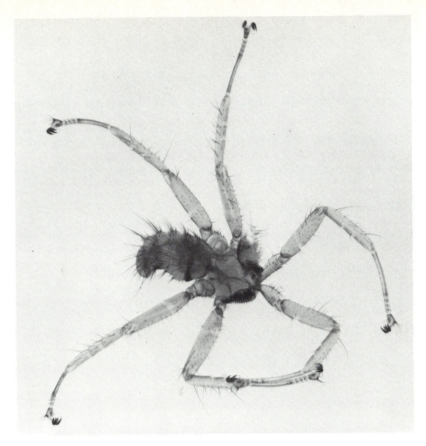

FIG. 39-24

Bat spider fly, family Nycteribiidae.

Photograph by Warren Buss.

REFERENCES

1. Cogley, T.P., J.R. Anderson, and L.J. Cogley. 1982. Migration of *Gasterophilus intestinalis* larvae (Diptera: Gasterophilidae) in the equine oral cavity. Int. J. Parasitol. 12:473-480.

2. Downes, J.A. 1971. The ecology of blood-sucking Diptera: an evolutionary perspective. In A.M. Fallis, editor. Ecology and physiology of parasites. University of Toronto Press, Toronto.

3. Erdmann, G.R. 1987. Antibacterial action of myiasis-causing flies. Parasitol. Today 3:214-216.

4. Eschle, J.L., and G.R. DeFoliart. 1965. Rearing and biology of *Wohlfartia vigil* (Diptera, Sarcophagidae). Ann. Entomol. Soc. Am. 58:849-855.

5. Gerster, G. 1986. Tsetse—the deadly fly. Natl. Geogr. 170:814-833.

6. Greenberg, B. 1971. Flies and disease, vol. 1. Ecology, classification, and biotic associations. Princeton University Press, Princeton, N.J.

7. Greenberg, B. 1973. Flies and disease, vol. 2. Biology and disease transmission. Princeton University Press, Princeton, N.J.

8. Guimarães, J.H., and N. Papavero. 1966. A tentative annotated bibliography of *Dermatobia hominis* (Linnaeus, 1781) (Diptera, Cuterebridae). Arq. Zool. 14:223-294.

9. Harlan, H.J., and B.N. Chaniotes. 1983. Report of *Olfersia coriacea* (Diptera: Hippoboscidae) feeding on a human in Panama. J. Parasitol. 69:1026.

10. Harwood, R.F., and M.T. James. 1979. Entomology in human and animal health, ed 7. Macmillan Publishing Co., Inc., New York.

11. Hawley, W.A., P. Reiter, R.S. Copeland, C.B. Pumpuni, and G.B. Craig, Jr. 1987. *Aedes albopictus* in North America: probable introduction in used tires from northern Asia. Science 236:1114-1116.

12. Hibler, C.P., and J.L. Adcock. 1971. Elaeophorosis. In Davis, J.W., and R.C. Anderson, editors. Parasitic diseases of wild mammals. Iowa State University Press, Ames.

13. Hillerton, J.E. 1987. Summer mastitis: vector transmission or not? Parasitol. Today 3:121-123.

14. Krafsur, E.S., C.J. Whitten, and J.E. Novy. 1987. Screwworm eradication in North and Central America. Parasitol. Today 3:131-137.

15. Krinsky, W.L. 1976. Animal disease agents transmitted by horse flies and deer flies (Diptera: Tabanidae). J. Med. Entomol. 13:225-275.

16. Lainson, R., and J.J. Shaw. 1974. The leishmanias and leish-

maniasis of the New World with particular reference to Brazil. Bol. Off. Sanit. Panama 76:93-114.

17. Lawrence, B.R. 1986. Old World blowflies in the New World. Parasitol. Today 2:77-79.

18. Lewis, D.J. 1974. The biology of Phlebotomidae in relation to leishmaniasis. Ann. Rev. Entomol. 19:363-384.

19. Lewis, D.J. 1975. Functional morphology of the mouthparts in New World phlebotomine sand flies (Diptera: Psychodidae). Trans. R. Soc. Entomol. 126:493-532.

20. Lewis, D.J., D.G. Young, G.B. Fairchild, and D.M. Minter. 1977. Proposal for a stable classification of the phlebotomine sandflies (Diptera: Psychodidae). Syst. Entomol. 2:319-332.

21. Lysenko, A.J. 1971. Distribution of leishmaniasis in the Old World. Bull. WHO 44:515-520.

22. Mattingly, P.F. 1977. Names for the *Anopheles gambiae* complex. Mosq. Syst. 9:323-328.

23. Mulligan, H.W., and W.H. Potts. 1970. The African trypanosomiases. Allen & Unwin Ltd., London.

24. Nielsen, B.O., and O. Christensen. 1975. A mass attack by the biting midge *Culicoides nubeculosus* (Mq.) (Diptera, Ceratopogonidae) on grazing cattle in Denmark. A new aspect of sewage discharge. Nord. Vet. Med. 27:365-372.

25. Platzer, E.G. 1981. Biological control of mosquitoes with mermithids. J. Nematol. 13:257-262.

26. Sabrosky, C.W. 1941. The *Hippolates* flies or eye gnats: preliminary notes. Can. Entomol. 73:23-27.

27. Steelman, C.D. 1976. Effects of external and internal arthropod parasites on domestic livestock production. Ann. Rev. Entomol. 21:155-178.

28. Tarry, D.W. 1986. Progress in warble fly eradication. Parasitol. Today 2:111-116.

29. Tod, M.E., D.E. Jacobs, and A.M. Dunn. 1971. Mechanisms for the dispersal of parasitic nematode larvae. I. Psychodid flies as transport hosts. Helminthology 45:133-137.

30. Turner, W.N. 1978. A case of severe human allergic reaction to bites of *Symphoromyia* (Diptera: Rhagionidae). J. Med. Entomol. 15:138-139.

31. Vale, G.A., E. Bursell, and J.W. Hargrove. 1985. Catching-out the tsetse fly. Parasitol. Today 1:106-110.

32. Wirth, W.W., and W.R. Atchley. 1973. A review of the North America *Leptoconops* (Diptera: Ceratopogonidae). Grad. Stud. Texas Tech Univ. 5:1-57.

33. Woodard, D.B., and H.C. Chapman. 1970. Hatching of floodwater mosquitoes in screened and unscreened enclosures exposed to natural flooding of Louisiana salt marshes. Mosq. News 30:545-550.

34. Zumpt, F. 1965. Myiasis in man and animals in the Old World. Butterworth & Co. (Publishers) Ltd., London.

35. Zumpt, F. 1973. The stomoxyine biting flies of the world. Gustav Fischer Verlag, Stuttgart.

CHAPTER 40

PARASITIC INSECTS: ORDERS STREPSIPTERA, HYMENOPTERA, AND OTHERS

Thus far we have considered the orders of insects containing members of medical or veterinary significance. Remaining are several orders that are not covered often in parasitology texts but that are of biological interest and, particularly in the case of the Hymenoptera, have considerable impact on human welfare. Half or more of hymenopterans are parasites of other insects, and many are extremely important natural controls of agricultural and forestry pests. Unchecked, the hosts of these hymenopteran parasites would undoubtedly destroy so much of our food supply that we would experience a worldwide famine. Although of less economic significance, all Strepsiptera are parasitic and demonstrate very interesting adaptations to parasitism. We will consider the Hymenoptera and Strepsiptera in some detail and briefly will mention some members of other orders that are parasitic.

Many of the insects discussed in this chapter, especially among the Hymenoptera, are parasitic as larvae, growing inside or outside the host and eventually killing it. Such insects would seem to fit the conventional definition of parasite when they are small, but because they kill their host during or at the completion of development, they would seem more akin to predators at this stage. Consequently, they are often referred to as **parasitoids,** rather than parasites. Other authors prefer to use the term **protelean** to describe all insects whose immature stages are parasitic and whose adults are free living. This designation would include some members of most orders covered in this chapter, as well as certain Diptera, for example, *Gasterophilus, Dermatobia,* and *Hypoderma.* The adults of protelean parasites are generally the dispersal mechanism and host finders, whereas the immature stages are largely rather helpless creatures whose sustenance the host provides.

Doutt[12] pointed out several ways in which parasitoids differ from typical parasites, including (1) that the host is usually of the same taxonomic class, that is, Insecta; (2) that the development of an individual

destroys its host; (3) that they are relatively large in size compared with their hosts; (4) that they are parasitic as larvae only, the adults being free living; and (5) that their trophic interaction resembles that of predators more than that of true parasites. Nonetheless, Doutt used the term *parasite* interchangeably with *parasitoid* and distinguished parasites from predators in that parasites (or parasitoids) consume merely a single host individual, and predators must devour several to reach maturity. Here we will adopt the broad definition of parasitoids as parasites.

ORDERS WITH FEW PARASITIC SPECIES
Order Dermaptera

The comparatively small order Dermaptera (1100 species) has hemimetabolous development and is composed primarily of the free-living earwigs. Only two genera (11 species) can be considered parasitic. One of these, *Hemimerus,* is considered to constitute a separate order by some investigators.[21] The abdomen of adult free-living earwigs terminates in a pair of unsegmented, pincer-like cerci, whereas the cerci of *Hemimerus* are short and threadlike. *Hemimerus* are parasites of pouched rats in tropical Africa. They are eyeless and wingless, and they feed on their host's epidermis and a fungus that grows thereon.[16,22]

The other genus of Dermaptera containing parasites is *Arixenia,* although whether its members actually should be considered parasitic is debatable. They are wingless, they have compound eyes, and their cerci are more heavily sclerotized and similar to those of typical earwigs. They have been reported from bat caves in Sarawak, Java, Malaya, and the Philippines. *Arixenia esau* feeds on the skin of dead or dying bats but also is known to feed on arthropods. Another species, *A. jacobsoni,* is found in bat caves but apparently is completely predatory on arthropods and has not been found on the bodies of the bats.

Order Neuroptera

Members of the Neuroptera are holometabolous, as are those of the remaining orders to be considered. The Neuroptera is an order of moderate size (5000 species), and only about 190 species can be considered protelean parasites. Most neuropterans are predators as larvae, and sometimes as adults, on other insects and mites. They are considered beneficial to humans because they help control numerous pests. Most larvae are terrestrial, although members of the Sisyridae have aquatic larvae. The larvae of *Sisyra* and *Climacia* spp. parasitize freshwater sponges, a food source apparently distasteful to most other animals. The first instar larvae float in the water until they encounter a sponge, enter ostia, and begin to feed. Larvae of the family Mantispidae parasitize the egg cocoons of several families of spiders. When fully developed, mantispid larvae are a contrast to the active, predaceous larvae of other families. They have a small head, large abdomen, and small legs, characteristics correlated with the fact that their food source is abundant and does not have to be caught and killed.

Order Lepidoptera

Members of the Lepidoptera are familiar to most persons as the butterflies and moths. The adults have generally rather large wings covered with tiny scales, and they have a long suctorial proboscis formed from parts of the maxillae, the mandibles being absent or rudimentary. They are adapted for feeding on nectar or other extruded plant juices. The polypodous larvae are different in form, with strong mandibles adapted for chewing. The order is large (120,000 species), and the adults and larvae are almost entirely phytophagous. A few species are ectoparasites of mammals as adults, and a few are protelean parasites of insects. Although small in number of species, the eye-feeding moths in Southeast Asia demonstrate an interesting progression from nectar feeding on plants to blood-sucking from animals. Members of several families occasionally visit the eyes of domestic ungulates, where they feed on lacrimation; those of a different group visit the eyes fairly frequently, feeding on lacrimation and fluid running down the host's cheeks; and those of another group (for example, *Arcyophora* and *Lobocraspis* [Fig. 2-2]) feed only at the eyelid on lacrimation and pus and sometimes penetrate the conjunctiva to feed on blood.[5] Finally, a species of noctuid, *Calpe eustrigata,* has developed the ability to pierce vertebrate skin and feed on blood[2] (Fig. 2-3).

Cyclotorna monocentra in Australia has a very curious life history.[11] Its first instar larva finds, attaches to, and feeds on but does not kill nymphs or adults of a species of leafhopper (Hemiptera, Homoptera).

When *Cyclotorna* molts to the second instar, it somehow induces an ant *(Irdomyrmex)* to pick it up and carry it back to its colony. Thereupon the caterpillar ingratiates itself by providing the ants with a sweet secretion and by running its mouthparts over the bodies of the ants—activities that the ants seem to enjoy. Payment for these services is exacted when the caterpillar feeds on ant larvae. Finally, the lepidopteran larva emerges from the ant colony and pupates on the trunk of a tree.

A few other Lepidoptera, such as *Fulgoraecia* spp., attach themselves to and become ectoparasites of homopterans.

Order Coleoptera

Judged by the number of described species, the order Coleoptera, or beetles, is the most successful order of animals on earth (300,000 species). Less than 2% are parasites, but both mammalian ectoparasites and protelean parasites of insects are represented. As to be expected in such a large and successful order, the Coleoptera exhibit a wide range of morphological form, habit, and adaptation. The mandibles are well developed for biting and chewing in most species, but in some they are adapted for piercing and sucking. Although most beetles can fly well, the forewings are hardened into sheathlike elytra. At rest the elytra cover the dorsum of the abdomen and the hindwings, which are folded beneath them. During flight the hindwings carry the burden, and the elytra are of little use, being held outstretched from the body. Because beetles are not as maneuverable in flight as are dipterans and hymenopterans, some entomologists have suggested that the adults cannot be as efficient in host finding as in the latter two orders. Consequently, most coleopterans that are protelean parasites are hypermetamorphic, leaving the host finding to the first instar larva. **Hypermetamorphosis** is a condition in which different larval instars have dissimilar forms, and it is found among the mantispid neuropterans, the Strepsiptera, and several families of Diptera and Hymenoptera, as well as in the Coleoptera. In all of these the first instar is rather heavily sclerotized and quite active. In the Neuroptera, Strepsiptera, and Coleoptera, the larva is a campodeiform oligopod and is called a **triungulin** or **triungulinid** (Fig. 40-1, *C*). The active dipteran and hymenopteran larvae are apodous, but they move about with the aid of thoracic and caudal setae; the larva is called a **planidium** (Fig. 40-1, *A* and *B*). Subsequent instars of both types are typically much less sclerotized and have a smaller head and much reduced means of locomotion. Planidia and triungulins are an interesting instance of convergent evolution.

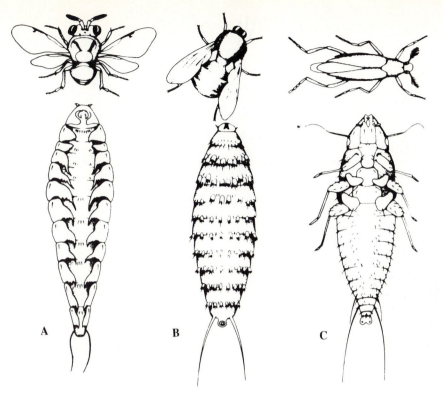

FIG. 40-1

First instar larvae with corresponding adults (*not* drawn to same scale). The larvae are active and must find the host. **A,** Perilampidae (Hymenoptera) and, **B,** Acroceridae (Diptera) exemplify the planidium type of larva. **C,** A triungulin larva (with well-developed legs) of Rhipiphoridae (Coleoptera).

From Askew, R.R. 1971. Parasitic insects. American Elsevier Publishing Co., Inc., New York.

Examples of protelean parasitic beetles include the aleocharine staphylinids, whose triungulins seek out the puparia of Diptera, penetrate them, and feed on the pupa within. The larvae of some Meloidae feed on grasshopper eggs, and others feed on the eggs of solitary bees and then on the honey stored in the cell in which the egg is laid. Triungulins of the latter may congregate on flowers, lying motionless until the flower is visited by a bee. They quickly attach themselves to the bee, hang on until the bee returns to its nest, and then drop off into the cell in which the bee deposits its egg.

A few beetles are symbiotic as adults. *Platypsyllus castoris* (Platypsyllidae) is an ectoparasite of beavers in both the Palearctic and Nearctic. It is a blind, obligate parasite and an ectoparasite in both the adult and larval states, feeding on skin debris. Some species of Staphylinidae are apparently mutuals of marsupials and certain rodents, feeding on fleas and mites.[13] The beetles are rather large (5 to 16 mm) but are well tolerated by the host, clinging to its hair in areas, such as the base of the tail, where a more noxious passenger would excite grooming attention.

ORDER STREPSIPTERA

In numbers of species (only about 300), the Strepsiptera is very small compared with the larger orders of insects. They are of great interest biologically, however, and demonstrate some of the most extreme adaptations to protelean parasitism of any insects. Commonly known as **stylops,** they show certain similarities to the Coleoptera and are considered by some investigators to be a superfamily (Stylopoidea) of the Coleoptera.[8] The two orders have important differences, including the fact that the last larval instar of the most primitive strepsipteran family, Mengeidae, has compound eyes. This is in striking contrast to the

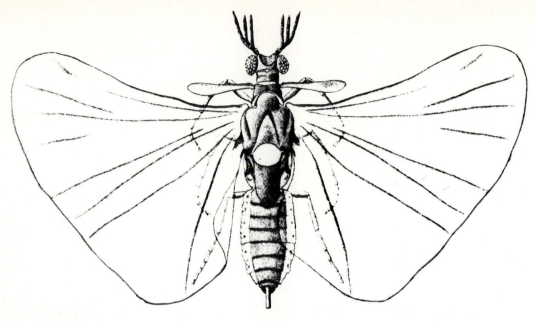

FIG. 40-2

Eoxenos laboulbenei, adult male (Strepsiptera). The mesothoracic wings have been reduced to halteres, and the large, membranous metathoracic wings are borne on the very large metathorax.

From Parker, H.L., and H.D. Smith, 1934. Ann. Entomol. Soc. Am. 27:468-477.

condition in all other endopterygote larvae, which have only simple, ocellus-like stemmata.

Morphology

The strepsipterans exhibit extreme sexual dimorphism. The males are small, robust, beetle-like insects about 1.5 to 4 mm long (Fig. 40-2). The forewings, believed by Crowson[8] to represent reduced elytra, are small halteres bearing numerous sensory endings. The hindwings are large, membranous, and fan shaped and are borne on a large, third thoracic somite. The insects are various shades of brown to black. The compound eyes are large and protuberant, and one or more segments of the antennae have lateral processes so that the antennae appear branched (Fig. 40-3). Mandibles are present but are simple and sickle shaped; other mouthparts are reduced or absent. These characteristics reflect the fact that the adult male spends its very short existence agitatedly seeking females to mate with. Thus it has little use for feeding appendages but great need for well-developed sense organs.

The adult female is parasitic in all stylops except in the Mengeidae. It is vermiform, up to 20 to 30 mm long according to species, and has no wings, eyes, legs or antennae (Fig. 40-4). The mouth and anus are tiny and nonfunctional, and the gut has no lumen. The head and thorax are fused to form a cephalothorax and

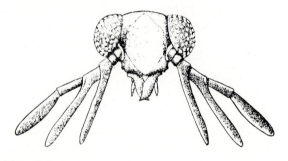

FIG. 40-3

Eoxenos laboulbenei, adult male, front view of head. The eyes are large and protuberant, and the processes of the antennal segments make the antennae appear branched. The simple mandibles and what are probably maxillary palps are the only mouthparts that can be distinguished.

From Parker, H.L., and H.D. Smith, 1934. Ann. Entomol. Soc. Am. 27:468-477.

are rather heavily sclerotized. The animal breathes through a spiracle on each side of the cephalothorax, which protrudes from the host between two sclerites, usually between tergites of the host abdomen. The abdomen of the female stylops is large and soft, lying within the host body and ensheathed in the cuticle of the last larval instar. The space between the stylops' abdomen and its larval cuticle forms a brood canal,

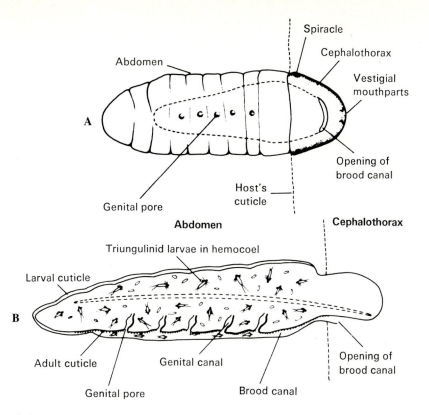

FIG. 40-4

Diagram of female strepsipteran. **A,** Ventral view. **B,** Longitudinal section. The more heavily sclerotized cephalothorax protrudes from the host body between abdominal tergites, whereas the soft abdomen lies within the host's hemocoel and is covered by the cuticle of the last larval instar. Initially closed, the brood canal is pierced by the male during copulation, and the triungulins subsequently exit through that opening.

From Askew, R.R. 1971. Parasitic insects. American Elsevier Publishing Co., Inc., New York.

which opens to the outside beneath the cephalothorax. Copulation is accomplished by insertion of the male's aedeagus into the brood canal. The sperm make their way through two to five genital openings in the female's abdomen and thence to her hemocoel. There the sperm fertilize the eggs, which are lying free in the hemocoel. This type of reproductive system is found in no other insects.

Development

The embryos develop and hatch within the hemocoel of the female to produce triungulin larvae, which exit through the genital pores and brood canal. Like other organisms with a high mortality in the progeny, stylops have a very high reproductive potential. One female can produce 2000 or more triungulins, and polyembryony has been reported. The triungulins are only about 0.2 mm long, but they have a well-sclerotized cuticle, well-developed legs, and one or two long caudal filaments (Fig. 40-5). They die in a short time if they do not reach a new host. Although the vast majority of triungulins will perish, they possess some adaptations to increase their chances of survival. For example, the triungulins of *Corioxenos antestiae* rest motionless with the anterior part of the body raised, resting on the hind legs and the central caudal bristles, which are bent forward beneath the abdomen.[18] A movement in their vicinity, particularly if the moving object is black and orange (the color of their host, a pentatomid bug),[17] stimulates the triungulin to leap up to 10 mm vertically and 25 mm horizontally. Adhesive pads on its front two pairs of tarsi help maintain contact if it hits a host (Fig. 40-5). Although it will attach to a wide variety of insects, it soon detaches unless the insect is the correct host. Other species are known, however, that will penetrate an un-

651

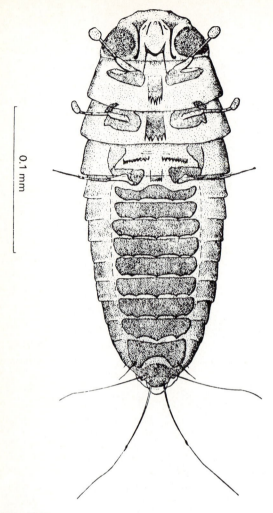

FIG. 40-5

First instar larva (triungulin) of the strepsipteran, *Coriox-enos antestiae*. The triungulin characteristically lies motionless, resting on its hind legs and the central caudal bristles. It is stimulated to leap in the air by movements of nearby black and orange objects, such as its host, a pentatomid bug. Adhesive pads on its forelegs help it stay on the host when it makes contact.

From Kirkpatrick, T.W. 1937. Trans. R. Entomol. Soc. Lond. 86:247-343.

suitable host and die therein. Although parasites of Hemiptera and those of social members of Hymenoptera reach their hosts directly, stylops parasitic in solitary Hymenoptera have special problems because the host larva is hidden in a cell, and they must be transported to that site. A bee infected with *Halictoxenos jonesi,* for example, drags her abdomen among the stamens of flowers, depositing triungulins.[3]

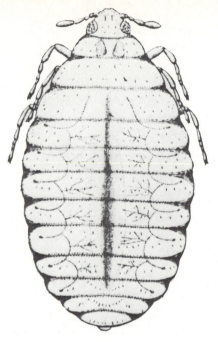

FIG. 40-6

Adult female of *Eoxenos laboulbenei,* a primitive strepsipteran (Mengeidae). In contrast to other strepsipteran females, members of this family emerge completely from their hosts, pupate, and are free-living. Although wingless, they have well-developed legs, eyes, antennae, and mandibles. The head is not fused with the thorax.

From Parker, H.L., and H.D. Smith. 1933. Ann. Entomol. Soc. Am. 26:217-233.

When another bee visits the flower, the triungulin attaches to it, is transported by the adult bee back to the cell containing its larva, and transfers to the larva while the adult bee is feeding its young.

Once on its host, the triungulin penetrates the hemocoel through an intersegmental membrane and begins to feed. The second instar, however, differs vastly from the triungulin. It is a grublike organism without mouthparts and legs, and it feeds by absorption of nutrient through its cuticle. There are normally six larval instars, including the triungulin.[14] The sixth instar regains its mandibles and chews its way through the intersegmental membrane between the sclerites of the host abdomen. If it is a male, it emerges completely and pupates within the cuticle of the last larval instar. If it is a female, only the cephalothorax emerges with no obvious pupation. Interestingly, females of the primitive Mengeidae, mentioned before, emerge completely, pupate, and become free living (Fig. 40-6). Although wingless, they have functional eyes, legs, and antennae. Mengeids are parasites of Thysanura, a very primitive order of insects, but they are thought not to have coevolved because the Strepsiptera are of more recent origin than are the Thysanura.[1]

Stylops differ from virtually all other protelean parasites of insects in that their host usually lives approximately a normal life span. Even so, the stylops may cause the reproductive death of the host[19] because parasitic castration often occurs. Although there is some variation, depending on whether the host is infected in an early or late instar, secondary sexual characteristics tend to take on an intersex appearance because effects on the gonads may be profound. Williams[28] reported that the aedeagus and parameres were often reduced or absent in leafhoppers *(Dicranotropis muiri)* infected with *Elenchus templetoni* in Mauritius, and the length of the ovipositors and sheath was reduced in female hosts. The gonads of hosts with advanced larvae or extruded forms were much reduced or could not be found at all.

ORDER HYMENOPTERA

The Hymenoptera is the largest order of insects after the Coleoptera, estimated to include more than 200,000 species; at least half of these are insect protelean parasites. Free-living hymenopterans are familiar to everyone as ants, bees, and wasps. Although feared and avoided by most people because of their sometimes potent sting, they are nevertheless extremely important pollinators of flowering plants. Most of our vegetable and fruit plants could ill dispense with the services of honeybees. At least as important, if not more so, is the role played by the parasitic Hymenoptera in population control of other insects.

Morphology

Hymenopterans usually have a fairly heavily sclerotized cuticle and show a vast range of body form, size, and color. They include the smallest insect known *(Alaptus,* 0.21 mm long), up to insects of 115 mm (some ichneumons, including the ovipositor). They have two pairs of membranous wings, of which the forewings are much larger (Fig. 40-7) except in the primitive Symphyta. The hindwings are attached to the forewings by a row of small hooks on the leading edge of the hindwings. Wing venation tends to be reduced and may be absent in some minute species. Some forms are wingless. The first abdominal segment is fused to the thorax and is called the **propodeum.** The second abdominal segment in most Hymenoptera is constricted to a waistlike **pedicel,** or **petiole.** The head is remarkably free, with a small neck

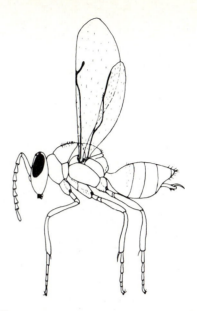

FIG. 40-7

Chalcidoid hymenopteran (Encyrtidae). Members of this family commonly are parasites of scale insects (Hemiptera, Homoptera), and several species have been used successfully in biological control of important pests.[24]

From An Introduction to the Study of Insects. Third Edition, by Donald J. Borror and Dwight M. DeLong. Copyright © 1964 and 1971 by Holt, Rinehart and Winston. Reprinted by permission of CBS College Publishing.

and large compound eyes. Antennae usually have 12 segments in males and 13 in females. The mandibles, maxillae, and labium are present and variously modified for different feeding habits. A **glossa** (considered the hypopharynx by some authors) is present, and it and some other mouthparts may be lengthened to form a tongue or proboscis for gathering nectar from flowers.

The **ovipositor** is modified to form a stinging organ in some species, but it is important to the biology of the parasitic forms as well. In common with many other insects, the ovipositor of hymenopterans has been derived from pairs of segmental appendages on the abdomen (Fig. 40-8). Its basal portions represent the coxae of abdominal segments eight and nine and are called the **valvifers.**[25] The coxal endites have been lengthened to become the body of the ovipositor and are called **valvulae.** Another process from the second valvifer (third valvula) has developed to become the ovipositor sheath. In Hymenoptera the pair of second valvulae have fused and have longitudinal ridges that interlock with grooves on the first valvulae. This forms the tube down which the egg must pass, and the functional unit is called the **terebra.** The much broader third valvulae form a sheath on each side of

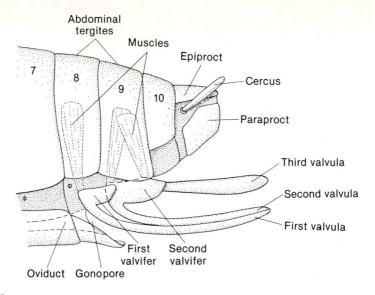

FIG. 40-8

Diagram of generalized appendiculate ovipositor of a pterygote insect. Some internal muscles and the position of the oviduct are shown. The valvulae and valvifers are all paired, although the second valvulae are fused in the Hymenoptera. They join with the first valvulae to form the ovipositor tube (terebra), and the third valvulae form the ovipositor sheath.

Drawing by William C. Ober.

the terebra; they are well endowed with sensory endings on their outer surfaces and are much less rigid than are the terebra. Cutting ridges on the end of the first and second valvulae enable the insect to drill through host cuticle, cells that contain the host, or even through hard vegetable matter, before oviposition. Since the lumen of the terebra has a diameter much smaller than that of the egg, the eggshell must be elastic enough to permit passage down the ovipositor to gain entrance to the host body.

Development

Parthenogenesis occurs throughout the order Hymenoptera. Three types are recognized: thelyotoky, deuterotoky, and arrhenotoky. In **thelyotoky** all individuals are uniparental (parthenogenetic), and virtually no males are produced. Some sawflies and parasitic Hymenoptera in several families are in this category. Some males are produced in **deuterotoky,** but all individuals are nevertheless uniparental. The most common condition is **arrhenotokous parthenogenesis,** in which only the males are uniparental and are haploid. The females come from fertilized eggs and are diploid. By some still obscure means, the mated, egg-laying female can influence whether a given egg will be fertilized. Needless to say, all of the foregoing types of parthenogenesis may lead to sex ratios that

diverge strongly from the 50/50 that we normally expect. An interesting discussion of parthenogenesis in Hymenoptera is given by Doutt.[12]

Larvae of the phytophagous sawflies have large heads and strong, chewing mouthparts. Many are polypodous and caterpillar-like, but the stem-boring forms are apodous. The larvae of the higher Hymenoptera reflect the fact that they must be placed in the midst of a plentiful food supply or be cared for by the adults. They are apodous and often headless and have a thin cuticle.

A few hymenopterans are hypermetamorphic and have a planidium larva. An example is *Perilampus hyalinus,* an American chalcid, which lays its eggs on foliage.[1] The planidia search for and penetrate caterpillars such as the fall webworm (*Hyphantria*). Once inside, however, their quest is not ended because they must find and penetrate the larvae of another parasite of *Hyphantria,* the tachinid dipteran, *Ernestia*. Thus they are hyperparasites, a common condition among Hymenoptera. Hypermetamorphosis is found among the Perilampidae, Echaritidae, and Ichneumonidae, but of course, not all species are hyperparasites.

Classification and examples

The Hymenoptera are divided into two suborders, the Symphyta and Apocrita, easily distinguished because the Symphyta do not have a pedicel. The latter are mostly phytophagous, but the small family Orussoidea parasitizes larvae of cerambycid and buprestid beetles.

Two divisions of the Apocrita, the Aculeata and the Parasitica, are commonly recognized; however, some of the members of the Aculeata are parasitic, and some of the Parasitica are not. In the Aculeata the eighth and ninth tergites are reduced and are retracted into the seventh, so that the ovipositor (sting) seems to issue from the apex of the abdomen. In the Parasitica the eighth segment is not retracted into the seventh, and the ovipositor is exposed almost to its base. Superfamilies in the Aculeata with parasitic forms are the Bethyloidea and Vespoidea (some); in the Parasitica, the Evanioidea, Trigonaloidea, Ichneumonoidea, Proctotrupoidea (Serphoidea), Chalcidoidea (many), and Cynipoidea (some).[1] We will discuss a few examples of the biology of this group.

The arbitrary nature of the distinction between parasitism and predation in hymenopterans is well illustrated in the Aculeata, which contains the ants, solitary and social wasps, and bees. A number of the wasps sting their prey (host) to paralyze it and then lay their eggs on the host, which provides food for the young. Some entomologists distinguish parasite from predator on the basis of whether the adult wasp constructs a cell to house the paralyzed host. An intermediate position is occupied by members of the Bethylidae, which drag the host to a sheltered position after paralyzing it, then oviposit on it. The female, who stands guard while the larvae develop as ectoparasites of the host, sometimes bites the host and feeds on its body fluids. Additional females may lay eggs on the same host individual, and the insects guard the young cooperatively. Such behavior may demonstrate an early stage in the evolution of the maternal care practiced by the social wasps. The various vespoid families exhibit the spectrum of maternal behavior, ranging from the typical parasitoid practice of laying an egg on the prey and then abandoning it, to the building of a cell to house the prey and wasp young, and finally to the maternal care of the social wasps.

The Ichneumonoidea comprises the Ichneumonidae, the Braconidae, and the Aphidiidae (sometimes considered a subfamily of braconids). The Ichneumonidae is a very large family, perhaps the largest of all insect families, with more than 3000 described species. They are all parasitic, sometimes on other ichneumonids, and most are endoparasitic. They tend to have a slender abdomen and often have a very long ovipositor that is permanently extruded. The largest ichneumonids in the United States may exceed 40 mm, and their ovipositors may be twice that length. Such in-

sects attack the larvae of horntails, wood wasps, and wood-boring beetles, somehow detecting the location of the host within its tunnel, which may be several centimeters below the surface. In Britain *Rhyssa persuasoria* is attracted by a substance produced by a fungus in the frass (feces and wood pulp) of the tunnels of the host, a wood wasp, *Sirex*. When the location of the host has been determined, apparently by the antennae, the ichneumonid raises its abdomen and places the tip of its ovipositor exactly in the location indicated by its antennae. Aided by sharp cutting ridges at the end of the ovipositor, the ichneumonid forces the ovipositor through the wood by pressure and twisting motions of its abdomen[1] (Fig. 40-9). It is astonishing that the apparently delicate ovipositor can penetrate the wood to reach the host. *Pseudorhyssa alpestris,* a parasite of the alder wood wasp, *Xiphydria camelus,* has solved the problem in another way. *Xiphydria* is also parasitized by *Rhysella curvipes,* which locates its host and oviposits much like *R. persuasoria. Pseudorhyssa,* however, simply locates the oviposition holes of *Rhysella,* inserts its ovipositor, and lays an egg in the same host individual. When the *Pseudorhyssa* hatches, it kills the larva of *Rhysella* and takes over the host![26]

Members of the Braconidae are similar to those of Ichneumonidae, although they are generally smaller (15 mm or less in length) and tend to be more heavy bodied. They also differ in that they pupate in silken cocoons on the outside surface of the host, rather than within its body (Fig. 40-10). A number of species of ichneumonid and braconid adults are known to feed on the host, as well as ovipositing within it. *Polysphincta* stings and paralyzes its spider host and then lays an egg on the spider's opisthosoma and feeds on its body fluids.[9] It is known that some species cannot mature their eggs or achieve full reproductive potential without first feeding on host body fluids,[4,20] and it is possible that many other species have this requirement. Larvae of the ichneumonid, *Hyposoter* spp., can parasitize the tussock moth only if the host has been previously parasitized by a braconid, *Cotesia melanoscela.*[15] The first parasite apparently modifies the host defense system in such a way that the second can survive.

Furthermore, the developing larvae may modify their host environment to the advantage of the parasite. No more than one larva of the aphidiid parasites of pea aphids, *Aphidius smithi,* can develop in a single host; if more than one egg is deposited, the supernumerary parasites are somehow eliminated. However, superparasitized pea aphids have a greater food incorporation efficiency and growth rate than singly parasitized hosts; thus the early presence of more than one

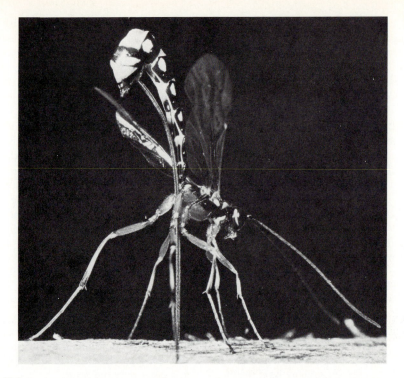

FIG. 40-9

Ichneumonid with the end of her abdomen raised to thrust her ovipositor through the wood to the host, a wood-boring, larval beetle. The coxae of the hind legs help steady and support the ovipositor as it is thrust into the wood.

Photograph by L.L. Rue III; from Hickman, C.P., Sr., C.P. Hickman, Jr., F.M. Hickman, and L.S. Roberts. 1979. Integrated principles of zoology, ed. 6. The C.V. Mosby Co., St. Louis.

FIG. 40-10

Larva of a tomato hornworm (Lepidoptera, Sphingidae) parasitized by the braconid, *Apanteles* sp. The white objects on the dorsum of the caterpillar are the pupal cocoons of the wasp.

Courtesy O.W. Olsen; from Hickman, C.P., Sr., C.P. Hickman, Jr., F.M. Hickman, and L.S. Roberts. 1979. Integrated principles of zoology, ed. 6. The C.V. Mosby Co., St. Louis.

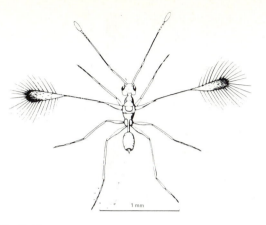

FIG. 40-11

Fairyfly, *Mymar pulchellus* female (Chalcidoidea, Mymaridae). This is one of the larger species in the family, commonly parasitic in the eggs of other insects.

From Askew, R.R. 1971. Parasitic insects. American Elsevier Publishing Co., Inc., New York.

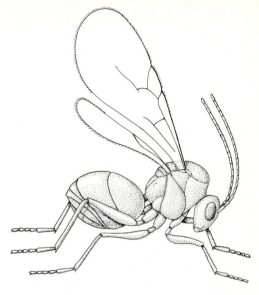

FIG. 40-12

Female *Charips victrix* (Cynipoidea, Cynipidae), a hyperparasite on a braconid primary parasite of aphids.

Drawing by William C. Ober.

parasite larva modifies the host physiology by some unknown mechanism.[7] Some investigators have proposed that the dominant egg releases substances that suppress the development of supernumerary parasites.[23] Many species of ichneumonoids are viewed as beneficial to humans, and there are numerous instances of their use as biological controls of pest species.

The Chalcidoidea is a large superfamily that contains 18 families. Its members are very small (less than 5 mm), metallic green or black wasps with few wing veins. Some of them are not parasitic: they feed on plant matter, and some induce the growth of plant galls. The parasitic chalcidoids attack a wide variety of hosts, the majority of which are in the Lepidoptera, Coleoptera, Hymenoptera, Diptera, and Hemiptera. A few are parasites of ticks, mites, and egg cocoons of spiders. The Trichogrammatidae and Mymaridae are parasites of insect eggs. The mymarids include the smallest insects known (0.2 mm); they are called "fairyflies" and are smaller than some protozoa (Fig. 40-11). Several Encyrtidae (Fig. 40-7) have been very valuable in control of scale insects (Hemiptera, Homoptera).

Most of the Cynipoidea are phytophagous, and they induce the formation of galls in the tissue on which they feed. However, a number are parasitic on various other insects. *Charips* is a hyperparasite of a braconid primary parasite of aphids (Fig. 40-12).

The Evanioidea, Trigonaloidea, and Procto-

trupoidea are less common parasites of a variety of insects. Trigonaloids parasitize larvae of social wasps (Vespoidea), and some are hyperparasites of ichneumonid and tachinid dipteran parasites of lepidopterans. The trigonalid female lays many eggs, not on its prospective host but on vegetation. The eggs do not hatch until they are eaten by a caterpillar, but they can remain viable for several months.[1] Once eaten by the caterpillar, they hatch, penetrate the gut to the hemocoel, and then must find a larval primary parasite in which to develop further. The Evanioidea contains several families, one of which (Evaniidae) contains parasites of cockroach eggs. The evaniid egg is laid within one of the cockroach eggs. After hatching, the evaniid larva consumes the roach egg and then may eat the other eggs in the egg case (ootheca) of the cockroach. The Proctotrupoidea is a somewhat larger group, including seven families.[1] They parasitize various insects, including homopterans, dipterans, neuropterans, and coleopterans. Pelecinidae (Fig. 40-13) have a long, attenuated abdomen, which may compensate for the short ovipositor when the female burrows through the ground in search of its host, the larvae (grubs) of May beetles.

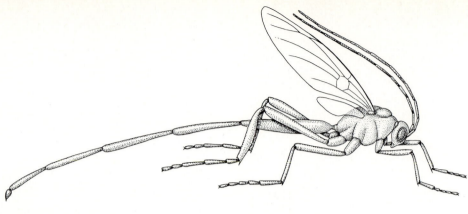

FIG. 40-13

Female *Pelecinus polyturator* (Proctotrupoidea, Pelecinidae). This striking insect is 2 inches or longer and shining black. The rare males are about an inch long and have a swollen abdomen. They are parasites of the larvae (grubs) of May beetles (Scarabaeidae), which burrow in the soil.
Drawing by William C. Ober.

BIOLOGICAL CONTROL

We will close this chapter with a few brief comments on biological control. Although biological control can include a number of other strategies besides use of insect protelean parasites, for example, viral, bacterial, nematode, and other pathogens and pheromones, our consideration will refer specifically to parasitoids. Strategies in using parasitoids as biological controls have been reviewed by Waage and Hassell.[27]

Since the development of modern insecticides, much food production in the world has become overwhelmingly dependent on the use of chemicals, both as fertilizers and as pesticides. For several reasons, the need for alternative means of pest control is becoming increasingly urgent. Not the least of these reasons is cost. Cost has always been an important factor in developing countries, but the enormous increase in the price of oil and hydrocarbon products in recent years has multiplied the relative cost of fuels and insecticides everywhere. Chemical insecticides must be applied repeatedly, often several times a year. In contrast, once a successful biological control agent has been introduced, no additional cost accrues except occasionally for reintroduction of the parasite. Furthermore, evolution works against chemical insecticides. When humans apply insecticides, they are applying a powerful selective pressure to the environment of the insect pest. The chemical rarely kills 100% of the organisms, since there are always some in the popula-

tion that are more resistant to the poison. All too commonly, resistance to specific insecticides develops in pest populations. Hence new chemicals must be developed continually to keep ahead of the unnatural selection caused by humans. Use of a good biological control agent, however, tends to keep the host population to a low but tolerable level in equilibrium. Evolution of greater resistance in the host population to the parasite is likely to be matched by adaptations in the parasite population to restore the equilibrium. Finally, chemical insecticides are toxic in some degree to other fauna, including humans; that is, they are nonspecific. In addition to being generalized hazards to humans and other animals, they are often quite toxic to beneficial species, such as natural predators of the pest insect.

Despite the long-term lower cost, however, the initial costs of developing biological controls for practical use may be higher than for chemical insecticides because extensive research is required. Careful ecological investigations of both the parasite and pest must be conducted before a foreign organism can be introduced into an area. Sibling species or different biological races of a parasite may have different host preferences or biological characteristics that determine success or failure. There is considerable danger in introducing an obligate hyperparasite in the mistaken belief that it is a primary parasite of the pest species. Parasites that are successful in biological control usually have the following qualities (modified from Askew)[1]:

1. They must have a high host-searching capacity.
2. They must have a very limited range of hosts but be able to use a few other host species in addition to the target; that is, when the pest population is reduced, the parasite should be able to maintain itself on alternative hosts. Thus a high enough parasite population is available to counteract surges in the pest population.
3. Their life cycle must be substantially shorter than that of the pest if the pest population consists of overlapping generations, or their life cycle must be synchronized with the pest life cycle if the pest population is composed of a single developmental stage at any time.
4. They must be able to survive in all habitats occupied by the pest.
5. They must be easily cultured in the laboratory so that large enough numbers are available for introductions.
6. They must control the pest population rapidly (some workers have suggested that control must occur within not more than 3 years from time of introduction).[6]

Despite these requirements and the research necessary before introductions can be made, the fact that numerous successful cases of pest control by parasites have been accomplished is encouraging.[24] DeBach[10] estimated that at least $110 million was saved in California from 1923 to 1959 as a result of biological control projects; since then the sum surely must be a great deal more.

REFERENCES

1. Askew, R.R. 1971. Parasitic insects. American Elsevier Publishing Co., Inc., New York.
2. Bänziger, H. 1968. Preliminary observations on a skin-piercing blood-sucking moth (Calyptra eustrigata Hmps.) (Lep., Noctuidae) in Malaya. Bull. Entomol. Res. 58:159-163.
3. Batra, S.W.T. 1965. Organisms associated with Lasioglossum zyphyrum (Hymenoptera: Halictidae). J. Kans. Entomol. Soc. 38:367-389.
4. Bracken, G.K. 1965. Effects of dietary components on fecundity of the parasitoid Exeristes comstockii (Cress.) (Hymenoptera: Ichneumonidae). Can. Entomol. 97:1037-1041.
5. Büttiker, W. 1967. Biological notes on eye-frequenting moths from N. Thailand. Mitt. schweiz. entomol. Ges. 39(1966):151-179.
6. Clausen, C.P. 1951. The time factor in biological control. J. Econ. Entomol. 44:1-9.
7. Cloutier, C., and M. Mackauer. 1980. The effect of superparasitism by Aphidius smithi (Hymenoptera: Aphidiidae) on the food budget of the pea aphid, Acyrthosiphon pisum (Homoptera: Aphidiidae). Can. J. Zool. 58:241-244.
8. Crowson, R.A. 1955. The natural classification of the families of Coleoptera. Nathaniel Lloyd, London.
9. Cushman, R.A. 1926. Some types of parasitism among the Ichneumonidae. Proc. Entomol. Soc. Wash. 28:5-6.
10. DeBach, P. 1964. Biological control of insect pests and weeds. Chapman & Hall Ltd., London.
11. Dodd, F.P. 1912. Some remarkable ant-friend Lepidoptera of Queensland. Trans. R. Entomol. Soc. Lond. 59(1911):577-590.
12. Doutt, R.L. 1959. The biology of parasitic Hymenoptera. Ann. Rev. Entomol. 4:161-182.
13. Durden, L.A. 1987. Predator-prey interactions between ectoparasites. Parasitol. Today 3:306-308.
14. Greathead, D.J. 1968. Further descriptions of Halictophagus pontilifex Fox and H. regina Fox (Strepsiptera: Halictophagidae) from Uganda. Proc. R. Entomol. Soc. Lond. [B]37:91-97.
15. Guzo, D., and D.B. Stoltz. 1985. Obligatory multiparasitism in the tussock moth, Orgyia leucostigma. Parasitology 90:1-10.
16. Jordan, K. 1910. Notes on the anatomy of Hemimerus talpoides. Novit. Zool. 16:327-330.
17. Kirkpatrick, T.W. 1937. Colour vision in the triungulin larva of a strepsipteron (Corioxenos antestiae Blair). Proc. R. Entomol. Soc. Lond. [A]12:40-44.
18. Kirkpatrick, T.W. 1937. Studies on the ecology of coffee plantations in East Africa. II. The autecology of Antestia spp. (Pentatomidae) with a particular account of a strepsipterous parasite. Trans. R. Entomol. Soc. Lond. 86:247-343.
19. Kuris, A.M. 1974. Trophic interactions: similarity of parasitic castrators to parasitoids. Q. Rev. Biol. 49:129-148.
20. Leius, K. 1961. Influence of food on fecundity and longevity of adults of Itoplectis conquisitor (Say) (Hymenoptera: Ichneumonidae). Can. Entomol. 93:771-800.
21. Popham, E.J. 1961. On the systematic position of Hemimerus Walker—a case for ordinal status. Proc. R. Entomol. Soc. Lond. [B]30:19-25.
22. Popham, E.J. 1962. The anatomy related to the feeding habits of Arixenia and Hemimerus (Dermaptera). Proc. Zool. Soc. Lond. 139:429-450.
23. Silvers, M.J., and A.J. Nappi. 1986. In vitro study of physiological suppression of supernumerary parasites by the endoparasitic wasp Leptopilina heterotoma. J. Parasitol. 72:405-409.
24. Simmonds, F.J., and F.D. Bennett. 1977. Biological control of agricultural pests. In White, D., editor. Proceedings of the Fifteenth International Congress of Entomology. Entomological Society of America, College Park, Md., pp. 464-472.
25. Snodgrass, R.E. 1935. Principles of insect morphology. McGraw-Hill Book Co., New York.
26. Thompson, G.H., and E.R. Skinner. 1961. The alder wood wasp and its insect enemies. Film, Commonwealth Forestry Institute. Quoted in Askew, R.R. 1971. Parasitic insects. American Elsevier Publishing Co., Inc., New York.
27. Waage, J.K., and M.P. Hassell. 1982. Parasitoids as biological control agents—a fundamental approach. In Anderson, R.M., and E.U. Canning, editors. Parasites as biological control agents. Symposia of the British Society for Parasitology, vol. 19. Parasitology 84(4):241-268.
28. Williams, J.R. 1957. The sugar-cane Delphacidae and their natural enemies in Mauritius. Trans. R. Entomol. Soc. Lond. 109:65-110.

CHAPTER 41

PARASITIC ARACHNIDS: ORDER ACARI, THE TICKS AND MITES

Ticks and mites are immensely important in human and veterinary medicine, many by causing diseases themselves and others by acting as vectors of serious pathogens. All ticks are epidermal parasites during their larval, nymphal, and adult instars; and many mites are parasites on or in the skin or in the respiratory system or other organs of their hosts. Some mites, although not actually parasites of vertebrates, stimulate allergic reactions when they or their remains come into contact with a susceptible individual. Ticks are found in nearly every country of the world; mites are even more ubiquitous, thriving on land, in fresh water, and in the oceans. Many millions of dollars are spent annually throughout the world in attempts to control these pests and the diseases they transmit.

The general morphology of the Acari, described in detail in Chapter 34, can be summarized as follows. Segmentation is reduced externally, having been obscured by fusion. Tagmatization has resulted in two body regions, an anterior **gnathosoma**, or **capitulum**, bearing the mouthparts, and a single **idiosoma**, containing most internal organs and bearing the legs. Most adult Acari have eight legs but some mites have only one to three pairs. The idiosoma is further divided into regions (Fig. 34-17) as follows: the portion bearing the legs is the **podosoma**, the first and second pairs of legs are on the **propodosoma**, and the third and fourth pairs are on the **metapodosoma**. The portion of the body posterior to the legs is the **opisthosoma**. The gnathosoma and propodosoma together comprise the **proterosoma**, and the metapodosoma and opisthosoma together are the **hysterosoma**. These terms may seem confusing at first, but they are very useful in describing, and therefore identifying, acarines.

The capitulum mainly is made up of feeding appendages surrounding the mouth. On each side of the mouth is a **chelicera**, which functions in piercing, tearing, or gripping host tissues. The form of the chelicerae varies greatly in different families; thus they are useful taxonomic features. Lateral to the chelicerae is a pair of segmental **pedipalps**, which also vary greatly in form and function related to feeding. Ventrally the coxae of the pedipalps are fused to form a **hypostome**, whereas a **rostrum**, or **tectum**, extends dorsally over the mouth. Some or all of these structures can be retracted in some acarines.

Ticks and mites, although basically similar, have distinct differences, as follows:

Ticks	Mites
Hypostome toothed, exposed	Hypostome unarmed, hidden
Large, easily macroscopic	Small, usually microscopic
Haller's organ (Fig. 41-1) present on first tarsi	Haller's organ absent
Peritreme absent	Peritreme present in Mesostigmata

The mouthparts of the Acari are modified for specialized feeding, with more variations found in the mites. In ticks the pedipalps grasp a fold of skin while the chelicerae cut through it. As cutting progresses, the hypostome is thrust into the wound, and its teeth help anchor the tick to its host. Blood and lymph from lacerated tissues well into the wound and are sucked up. Soft ticks feed rapidly, leaving their host after engorging, whereas hard ticks remain attached for several days. Some hard ticks, particularly those with short mouthparts, secrete a cementing substance that hardens, further securing them to the host.

Mites feed on vertebrates in much the same way, but because their mouthparts are so small, most feed on lymph or other secretions rather than on blood. There is no toothed hypostome, and the chelicerae vary from chelate cutters to hooks or stylets. Variations in feeding behavior of mites are discussed with the following respective groups.

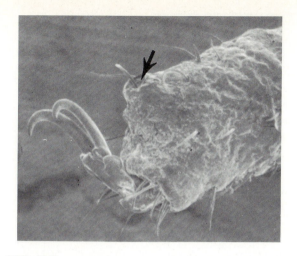

FIG. 41-1

Haller's organ *(arrow)*.

Photograph by Tyler Woolley.

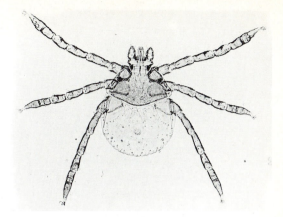

FIG. 41-2

Six-legged larva of *Ixodes* sp.

Photograph by Jay Georgi.

SUBORDER METASTIGMATA: THE TICKS

Because of their large size and pesky habits, ticks have been recognized for centuries. Both Homer and Aristotle referred to them in their writings, but Linnaeus in 1746 was the first to attempt to classify them among other animals. Their importance as agents and vectors of disease also has long been recognized. Pathogenesis attributable to these parasites appears in several ways:

1. **Anemia.** Blood loss in heavy infections can be considerable; as much as 200 pounds of blood can be lost from a single large host in one season.[17]
2. **Dermatosis.** Inflammation, swelling, ulcerations, and itching can result from a tick bite. These reactions often are caused by pieces of mouthparts remaining in the wound after the tick is forcibly removed, but probably constituents of the tick's saliva and secondary infection by bacteria are also involved.
3. **Paralysis.** A condition known as tick paralysis is common in humans, dogs, cattle, and other mammals when they are bitten near the base of the skull. This paralysis seems to result from toxic secretions by the tick and is quickly reversed when the parasite is removed.[45] Mechanisms were reviewed by Gothe, Kunze, and Hoogstraal.[10]
4. **Otoacariasis.** Infestation of the ear canal by ticks causes a serious irritation to the host, sometimes accompanied by severe infection.

5. **Infections.** In addition to common pyogenic infections, ticks are known to transmit viruses, bacteria, rickettsias, spirochetes, protozoa, and filariae. These will be discussed further with the appropriate vectors.

An indication of the economic importance of ticks to agriculture is suggested by the 1965 USDA report on the effects of arthropods on domestic livestock production. At that time the annual losses for cattle production caused by ticks were estimated at $60 million and for sheep production, at $4.7 million.[40] The loss to beef cattle in Australia is estimated at $25 million annually.[46] Much remains to be learned about the biology and taxonomy of these parasites before better control measures can be devised.[46]

Biology

All ticks undergo four basic stages in their life cycles: egg, larva, nymph, and adult, all of which might require from 6 weeks to 3 years to complete. Ixodids have only one nymphal instar, but argasids have as many as five. Copulation nearly always occurs on the host. The male tick produces a spermatophore, which it places under the genital operculum of the female. A blood meal is usually required for egg production, although exceptions are known. Also, some ticks are parthenogenetic. The engorged female drops to the ground, where she deposits her eggs in the soil or humus. A six-legged larva (Fig. 41-2) hatches from each of the 100 to 18,000 eggs and climbs onto low vegetation, where it quests for a host. On finding one, it

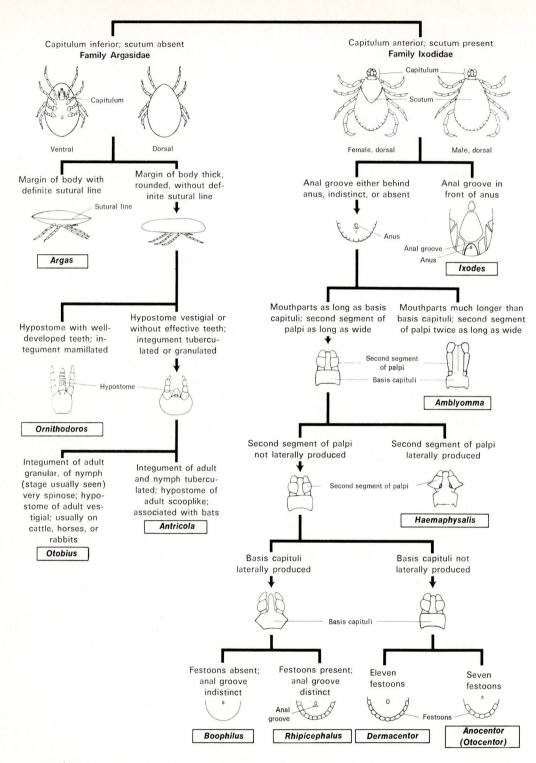

FIG. 41-3

Pictorial key to genera of adult ticks in the United States.

Courtesy U.S. Federal Security Agency. 1948. U.S. Public Health Service, Washington, D.C.

feeds and then molts to an eight-legged nymph. If molting through all instars occurs on the same host animal, the tick is called a **one-host tick,** such as *Boophilus.* If the nymph drops off, molts to adult, and attaches to another host, the tick is a **two-host tick.** Confinement of instars to one or two hosts is an adaptation to feeding or wide-ranging hosts.[16] Most ixodids are **three-host ticks,** whereas argasids, with their multiple nymphal stages, are **many-host ticks.** Clearly the use of a series of hosts increases the opportunities for transmission of pathogens.

Some ticks are rather host specific, but most are opportunists that will feed on a variety of hosts. Ticks are hardy and can withstand periods of starvation as long as 16 years. Some may have a life span of up to 21 years. Schlein and Gunders[32] discuss tick pheromones. (See key to genera, Fig. 41-3.)

Family Ixodidae

The family of **hard ticks** is divided into three subfamilies: Ixodinae, with the single genus *Ixodes;* Amblyominae, containing *Amblyomma, Haemaphysalis, Aponomma,* and *Dermacentor;* and Rhipicephalinae, with *Rhipicephalus, Anocentor, Hyalomma, Boophilus,* and *Margaropus.* We will discuss these genera individually.

Hard ticks are easily recognized as such, since the capitulum is terminal and can be seen in dorsal view. By contrast, the capitulum is subterminal in soft ticks and cannot be seen in dorsal view. Following are other characters of the Ixodidae: (1) a large anterodorsal sclerite, the **scutum,** is present; (2) eyes, when present, are on the scutum; (3) pedipalps are rigid, not leglike; (4) marked sexual dimorphism occurs in size and often coloration; (5) the female has **porose areas** on the basis capituli; (6) the coxae usually have spurs; (7) a pulvillus is present on the tarsus; (8) the stigmatal plates are behind the fourth pair of legs; (9) the posterior margin of the opisthosoma usually is subdivided into sclerites called **festoons;** and (10) there is only one nymphal instar.

■ Genus *Ixodes*

Ixodes is the largest genus of hard ticks, with about 250 species; nearly 40 species are known from North America. Most parasitize small mammals and are so small themselves that they are easily overlooked. Phylogenetically, *Ixodes* is a unique, highly specialized group; however, the absence of eyes is thought to be a primitive character. The pronounced sexual dimorphism of the mouthparts, which are longer in the female, is a condition unknown in any other genus. Fes-

FIG. 41-4

Ixodes scapularis, the black-legged tick.
Photograph by Jay Georgi.

toons are absent, and the anal groove is anterior to the anus. *Ixodes* spp. are three-host ticks.

Ixodes scapularis, the **blacklegged tick** (Fig. 41-4), is common in the eastern and south-central United States. It feeds on a wide variety of hosts and can be a major pest on dogs. It bites humans freely, commonly resulting in a strong reaction with pain at the site and generalized malaise for a short time. *Ixodes pacificus* is found along the west coast of California, Oregon, and Washington on deer, cattle, and other mammals. It also welcomes a human meal and is emerging as a vector of Lyme disease, caused by a spirochaete. Transovarial and transstadial passage of Lyme disease has been reported.[22] *Ixodes holocyclus* is the major cause of tick paralysis in Australia. Other species in the genus are also common agents of paralysis in several parts of the world.

Tick-borne encephalitis in Europe and Asia is known to be transmitted by the bite of *I. ricinus, I. persulcatus,* and *I. pavlovskyi.* Human infection with Lyme arthritis and the piroplasm *Babesia microti* is conveyed by the bite of the **deer tick,** *Ixodes dammini* on islands off the coast of Massachusetts, several northeastern states, and Long Island (Chapter 9). This tick has been reported as far west as Minnesota and Wisconsin. A novel method of control was proposed by Mather, Ribeiro, and Spielman.[26]

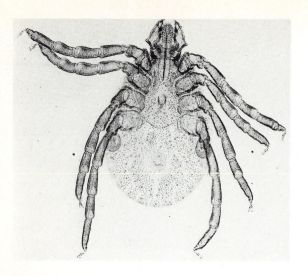

FIG. 41-5

Haemaphysalis leporispalustris, the rabbit tick.

Photograph by Jay Georgi.

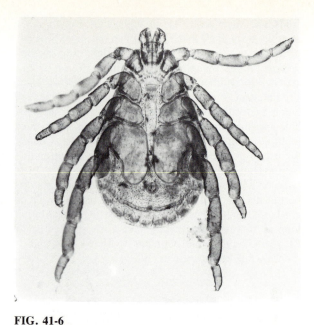

FIG. 41-6

Dermacentor andersoni, the Rocky Mountain wood tick.

Photograph by Jay Georgi.

■ Genus *Haemaphysalis*

Haemaphysalis spp. are easily recognized by the second segments of the pedipalps, which are produced laterally into spurs. These small ticks often are overlooked unless they are engorged. About 150 species are known worldwide, with only two found in North America. Most are three-host ticks. There is little sexual dimorphism, and both sexes have festoons.

Haemaphysalis leporispalustris, the **rabbit tick** (Fig. 41-5), is common on rabbits from Alaska to Argentina. It occasionally feeds on domestic animals but rarely bites humans. Its main importance is as a vector of tularemia and Rocky Mountain spotted fever among wild mammals. *Haemaphysalis cordeilis,* the **bird tick,** is common on turkeys, quail, pheasants, and related game birds. It may vector diseases among these hosts. It seldom is found on mammals and is not a pest on humans. Most species of *Haemaphysalis* are found in Asia, Africa, and the East Indies.

■ Genus *Dermacentor*

Among the most medically important of all the ticks, particularly in North America, this genus contains about 30 species. At least seven of these are found in the United States. These ticks are ornate, with punctations and colored markings, and usually show sexual dimorphism of color. Both sexes have festoons. The eyes are well developed, and the stigmatal plates have numerous shallow depressions called **goblets.** The sides of the basis capituli are par-allel. Most species are three-host ticks, but a few are one-host.

Dermacentor andersoni (Fig. 41-6) is the **Rocky Mountain wood tick.** It is distributed throughout most of the western United States west of the Great Plains, and is most prevalent in mountainous, brushy terrain. The larvae feed on small mammals, especially chipmunks, ground squirrels, and rabbits; nymphs also feed on these hosts, as well as on marmots, porcupines, and other medium-sized mammals. Adults may feed on any of these animals but seem to prefer larger hosts such as deer, sheep, cattle, and coyotes and are not reluctant to bite humans. When hosts are plentiful, the entire life cycle may take a year, but if long waits occur between meals, the life cycle can be extended up to 3 years. All stages can survive about a year without feeding. This species becomes most active in early spring, as soon as the snow cover is off, and actively continues to seek hosts until about the beginning of July, when the ticks become dormant.

Dermacentor andersoni is the agent, or vector, for several diseases that afflict humans; tick paralysis, Powassan encephalitis virus (mainly transmitted by *Ixodes spinipalpis),* Colorado tick fever virus, tularemia, Rocky Mountain spotted fever, and some viruses not known to be pathogenic. These organisms and others, such as anaplasmosis, are transmitted to other mammals by *D. andersoni.*

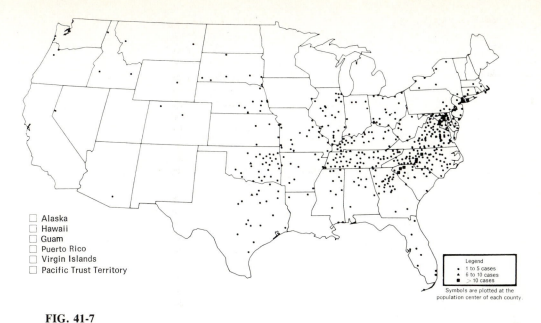

Alaska
Hawaii
Guam
Puerto Rico
Virgin Islands
Pacific Trust Territory

Legend
1 to 5 cases
6 to 10 cases
> 10 cases
Symbols are plotted at the
population center of each county.

FIG. 41-7

Distribution map of Rocky Mountain spotted fever, 1978.

Courtesy CDC Mortality and Morbidity Weekly Report.

Dermacentor variabilis, the **American dog tick,** is common throughout the eastern United States and is extending its range rapidly. One focus has been established in the Pacific Northwest, mainly along river valleys in Washington, Oregon, and Idaho. It prefers to feed on dogs but will attack horses and other mammals, including humans. This tick appears to occupy the same niche in the eastern states that *D. andersoni* occupies in the western states, although it is more urban. Isolated foci of *D. variabilis* are known in western Oregon.

The female lays 4000 to 6500 eggs on the ground, which hatch in about 35 days. The six-legged larva feeds on small rodents, especially voles and deer mice, and then drops off to molt. Nymphs feed again on these hosts for about a week, dropping off again to molt to adult. Adults prefer larger hosts. They often accumulate at the edges of trails and roadways where they have access to passersby.

Dermacentor variabilis is the principal vector of Rocky Mountain spotted fever in the central and eastern United States. Oddly enough, by far the majority of cases of this poorly named disease occur in the eastern half of the country (Fig. 41-7). This tick also causes paralysis in dogs and humans and transmits tularemia.

Dermacentor occidentalis, the **Pacific Coast tick,** is very similar in morphology and biology to *D. ander-*

soni. Adults are found on many species of large mammals, including humans, in California and Oregon. *Dermacentor occidentalis* is known to transmit Colorado tick fever virus, Rocky Mountain spotted fever, tularemia, anaplasmosis, and chlamydial abortion of cattle.

Dermacentor albipictus is called the **horse tick** or, because it does not feed during the summer months, **winter tick.** This one-host tick is widely distributed in northern United States and Canada, where it feeds on elk, moose, horse, and deer. It rarely attacks cattle or humans. The eggs are laid in the spring. The larvae hatch in 3 to 6 weeks and become dormant until the cold weather of autumn stimulates them to seek a host. Once aboard, the tick remains through its instars and subsequent mating, until it drops off in the spring. Infection can be so heavy as to kill the host. Because it is a one-host tick, *D. albipictus* is an inefficient vector for microorganisms.

Other species of *Dermacentor* throughout the world are vectors of several diseases caused by viruses and rickettsias.

■ Genus *Amblyomma*

The hundred or so species in this genus mostly are restricted to the tropics. The only exceptions are a few species in the southern United States. All appear to be one-host ticks and to have a 1-year life cycle. *Ambly-*

665

FIG. 41-8

Amblyomma americanum, the lone star tick.

Photograph by Warren Buss.

omma spp. are fairly long, highly ornate ticks with long mouthparts. The second segment of the pedipalp is longer than the others, resulting in the mouthparts, including the hypostome, being longer than the basis capituli.

Larvae and nymphs are common on birds, although adults usually feed only on large animals. Immature stages will feed on almost any terrestrial vertebrate and can also be found alongside adults on the final host. Because all three stages will readily bite humans, which is unusual among hard ticks, *Amblyomma* is exceptionally annoying.

Amblyomma americanum is the lone star tick (Fig. 41-8). Both sexes are dark brown with a bright silver spot at the posterior margin of the scutum. Males may have more than one spot. It ranges throughout much of the southern United States and well into Mexico.[34] It has a wide variety of hosts, including livestock and humans, and is a vector for Rocky Mountain spotted fever and tularemia.

The **Gulf Coast tick** *A. maculatum,* is very pestiferous, feeding in the ears of cattle. It ranges throughout the Gulf states.[33]

Amblyomma cajennense, the **Cayenne tick,** is found in South and Central America, the West Indies, Mexico, and Texas. All stages freely bite humans and many other animals. It is a vector of Rocky Mountain spotted fever.

■ Genus *Aponomma*

The few species of *Aponomma* parasitize reptiles, particularly in the Orient and Africa. *Aponomma elaphensis* is known from rat snakes in Texas, and four species have been described from South America and Haiti. They are not known to be of medical or economic importance, although it has been speculated that they may transmit haemogregarines among reptiles.[15] They are eyeless.

■ Genus *Rhipicephalus*

Continental Africa appears to be the place of origin and center of distribution of rhipicephalid ticks.[15] The approximately 60 species and subspecies in this genus usually are ecologically distributed, being restricted to forests, mountains, and semidesert regions or at least corresponding to certain limits of rainfall. Most species show little host specificity, biting many mammals and even reptiles and ground-dwelling birds. Both two-host and three-host species are known. The genus is easily recognized by the combination of festoons and a bilaterally pointed basis capituli. The only other

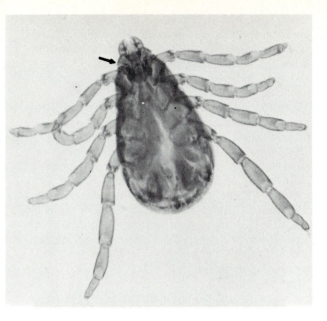

FIG. 41-9

Rhipicephalus sanguineus, the brown tick. Note the spurs on the basis capituli *(arrow).*

Photograph by Warren Buss.

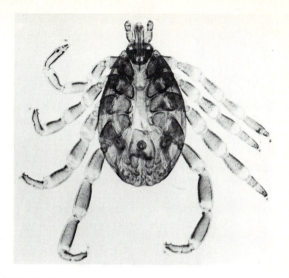

FIG. 41-10

Hyalomma sp.

Photograph by Jay Georgi.

genus with a similar basis capituli is *Boophilus,* which lacks festoons.

Rhipicephalus sanguineus, the **brown dog tick,** or **kennel tick** (Fig. 41-9), is the most widely distributed of all ticks, being found in practically all countries between 50° N and 35° S, including most of North America. It quickly becomes established whenever it is introduced.[16] *Rhipicephalus sanguineus* is a three-host tick, since it leaves the host to molt. All three stages feed mainly on dogs, mostly between their toes, in their ears, and behind the neck. The host range is very wide and occasionally includes human attacks, but it has a definite taste for dogs.

In some areas of Europe and Africa *R. sanguineus* is known to be a major vector of boutenneuse fever *(Rickettsia connorii),* which usually is acquired by crushing the tick against the skin. In Mexico it transmits Rocky Mountain spotted fever. Other diseases of animals are transmitted by this tick: *Borrelia theileri,* a spirochaete of sheep, goats, horses, and cattle that is carried by this arthropod in some areas; a highly fatal canine rickettsiosis, caused by *Rickettsia canis;* and malignant jaundice, caused by *Babesia canis.* The protozoan *Hepatozoon canis* infects dogs when they swallow infected ticks. For a list of successful and unsuccessful experimental transmissions of diseases by *R. sanguineus* see Hoogstraal.[15]

By far the most important disease transmitted by any species of *Rhipicephalus* is **East Coast fever,** a protozoan disease of the red blood cells of cattle (Chapter 9). This highly malignant infection mainly kills adult cattle, with mortality ranging from an average of 80% to 100% of infected animals. The causative agent, *Theileria parva,* is transmitted by several ticks, but the most important vector seems to be *R. appendiculatus.* An infected tick can transmit the disease only during the stadium following the infected meal; thus, the parasitic protozoan is transstadial, not transovarial.

■ Genus *Anocentor*

Only one species, *A. nitens,* is known in this genus. The **tropical horse tick** is found through South America up to Texas, Georgia, and Florida. It is very similar to *Dermacentor* but has 7 festoons rather than 11; its eyes are poorly developed, and it is inornate. It feeds primarily on horses and is not known to attack humans. It transmits *Babesia caballi,* a protozoan blood parasite in horses.

■ Genus *Hyalomma*

Few, if any, ticks are as difficult to identify as are species of *Hyalomma* (Fig. 41-10). This is caused in part by a natural genetic flexibility but also by a tendency toward hybridization. Furthermore, extrinsic factors, such as periods of starvation and climatic conditions, will cause morphological variations. This group probably originated in Iran or southern Soviet

Union[15] and has radiated into Asia, the Middle East, southern Europe, and Africa. Frequently, nymphs are carried from Africa to Europe by migrating birds.

These are fairly large ticks with no ornamentation. The legs are banded. Eyes are present, and the festoons are indistinct. They are very active, and it has been reported that in Africa and Arabian deserts these ticks will come rushing from beneath every shrub when persons or other animals stop by.[25]

Hyalommas must be the hardiest of ticks, since they are found in desert conditions where there is little shelter away from hosts, where small mammals available for larvae and nymphs to feed on are few, and where large mammals are undernourished and far ranging. They usually are the only ticks existing in such places.

Species of adult *Hyalomma* usually feed only on domestic animals. Occasionally they will bite humans, and because they transmit serious pathogens, they are among the most dangerous of ticks. Immature forms often feed on birds, rodents, and hares that are the reservoirs of viruses and rickettsias. For instance, **Crimean-Congo hemorrhagic fever** is carried between Africa and Europe by *H. marginatum* on migrating birds. Other viruses isolated from this species of tick are Dugbe virus and West Nile virus, and *H. anotolicum* harbors Thogoto virus and a swine poxvirus. Rickettsial diseases known to be transmitted by *Hyalomma* include Siberian tick typhus, boutonneuse fever, Q fever, and those that are caused by *Ehrlichia* spp. Malignant jaundice of dogs, caused by the protozoan *Babesia canis,* is transmitted by *H. marginatum* and *H. plumbeum* in Russia. Another protozoan, *Theileria annulata,* is transmitted to cattle by the bite of *H. anatolicum* in Eurasia.

■ Genus *Boophilus*

Boophilus ticks resemble *Rhipicephalus* spp. in that the basis capituli is bilaterally produced into points. They differ, however, in lacking festoons and an anal groove, which *Rhipicephalus* spp. have. Because unengorged specimens are quite small and easily overlooked, they have spread to many parts of the world when cattle are imported from an endemic zone. Two hypotheses on their place of origin have been suggested: either they came from the Indian subcontinent attached to Brahman cattle (zebu), or they originally were parasites of American bison or deer, they adapted to cattle, and they thence were exported to other places.[15]

Taxonomy of this genus has been confused, but at least three species are clearly recognized. *Boophilus annulatus* (Fig. 9-20) is the most widely distributed of these. It is often called the **American cattle tick,** since it was once widespread in the southern United States and is still common in Mexico, Central America, and some Caribbean islands. It is also known in Africa; the species known as *B. calcaratus,* from the Near East and Mediterranean region, is actually *B. annulatus.*[15]

This tick has been eradicated from the United States but appears sporadically along the Mexican-American border, as cattle and deer carry it across.

Boophilus microplus is similar in biology to *B. annulatus* and also has been eradicated from the United States. It still is found in Mexico and Africa, as well as Australia, Central and South America, Madagascar, and Taiwan. The high incidence of parthenogenesis in this species aids its survival when harsh conditions restrict the size of a population.[41] Cattle are the primary hosts of this tick; but sheep, goats, horses, and other animals may be infested.

The **blue tick,** *B. decoloratus,* occurs widely in continental Africa. Mainly it attacks cattle, but many other animals are bitten, including humans.

Boophilus spp. are all one-host ticks. Larval, nymphal, and adult stages are all spent on the same host animal, a rarity among ticks. Engorged females drop off and lay 2000 to 4000 eggs during the next 12 to 14 days. Newly hatched larvae are quite active, crawling to the tips of grasses and other plants, where they often accumulate in great numbers. After reaching a host, they remain until after breeding and feeding. Obviating the necessity of finding two or three hosts during its life has obvious survival value to these ticks.

Control, however, is greatly aided by the fact that all stages are to be found on the same animal. Dipping kills all parasitic stages at once. Unfed larvae die in about 65 days, so a pasture becomes tick free if cattle are dipped and kept off for this duration. Larvae, though, can be windblown for considerable distances.[24]

Species of *Boophilus* have been implicated as vectors of Crimean-Congo hemorrhagic fever and Ganjon viruses, as well as Bhanja virus in Nigeria and Thogoto virus in Kenya. The rickettsia *Anaplasma marginale* is transmitted to cattle by all three species of *Boophilus* in Africa. Mortality ranges from 30% to 50% in infected animals. Experimentally, *B. decoloratus* can transmit *Trypanosoma theileri* among cattle.[2]

By far the most important disease transmitted by a species of *Boophilus* is **Texas cattle fever,** also called **red-water fever.** The agent of this disease is a piroplasm, *Babesia bigemina,* discussed in detail in Chapter 9. To transmit most of the aforementioned diseases, ticks must change hosts, usually under crowded conditions in pens, railroad cars, and so on. However, because of transovarian transmission of *Babesia bigemina,* newly hatched ticks are already infected and capable of passing the disease on to cattle. Redwater fever was eradicated in the United States, along with the tick vectors, in 1939, but it persists in Central and South America, Africa, South Europe, Mexico, and the Philippines. Immunization of cattle against *B. microplus* shows considerable promise.[19]

■ Genus *Margaropus*

The four rare species in this genus, including *M. winthemi* and *M. reidi,* are found in East Africa and Sudan. They are parasites of giraffes and, occasionally, horses but are not known to be medically or otherwise economically important. For a review see Hoogstraal.[15]

Family Argasidae

The family of soft ticks includes only five genera, *Argas, Ornithodoros, Otobius, Nothoaspis,* and *Antricola,* with a total of about 140 species. *Antricola* and *Nothoaspis* infest cave-dwelling bats in North and Central America and will not be considered further here.

Soft ticks are easily distinguished from hard ticks by the following characters:

1. The capitulum in nymphs and adults (but not larvae) is subterminal and thus cannot be seen in dorsal view. It lies within a groove or depression called the **camerostome.** The dorsal wall of the camerostome, which extends over the capitulum, is called the **hood.**
2. There are no festoons or scutum.
3. Sexual dimorphism is slight.
4. The pedipalps are freely articulated and leglike.
5. There are no porose areas on the basis capituli.
6. Eyes are on the supracoxal fold.
7. The coxae lack spurs.
8. Pulvilli are absent on the tarsi.
9. The stigmatal plates are behind the third leg.
10. There may be two to eight nymphal stages.

In general, argasid ticks inhabit localities of extremely low relative humidity. Those that occur in wet climates seek dry microhabitats in which to live. Unlike ixodid ticks, most argasids feed repeatedly, resting away from the host between meals. This makes them difficult to collect, since they hide in loose soil, crevices, birds' nests, and the like. Examination, which includes sifting the soil or detritus of burrows, rodent nests, big game resting and rolling places, caves, and other animal lairs, is usually required to find them.

Adult females lay eggs in their hiding places several times between feedings. Even so, the total number of

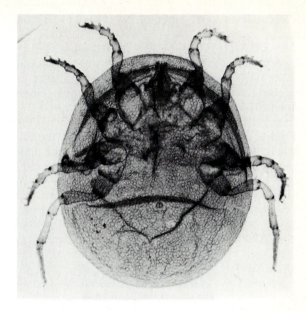

FIG. 41-11

Ornithodoros sp.

Photograph by Jay Georgi.

eggs produced is small, usually fewer than 500. Although larvae of some species remain dormant and molt to the first nymphal stage before feeding, most feed actively in this stage. Likewise, first-stage nymphs of a few species molt to second stage without feeding, although most feed first. Larvae usually remain on the host until molting, but nymphs, like the adult, leave the host. Exceptions to this are found in the genera *Otobius* and *Argas,* discussed later.

Hosts may be few and far between in desert habitats, and argasid ticks have adapted to potentially long periods without meals. They are capable of estivating for months or even many years without food. A blood meal, necessary for egg production, not only provides nutrition for developing eggs but also triggers a ganglion in the brain to release a hormone that instigates egg production in the ovaries.[36]

■ Genus *Ornithodoros*

Ornithodoros ticks are thick, leathery, and rounded (Fig. 41-11). The tegument in nonengorged specimens is densely wrinkled in fairly consistent patterns, allowing them great distention when feeding. The approximately 90 species parasitize mammals, including bats. It is unusual for them to feed on birds or reptiles, although *O. capensis* is found on marine birds in North America. Some species in this genus are very important in that they are vectors for relapsing fever spirochaetes, and the bite itself, of several species, is highly toxic and painful.

Ornithodoros hermsi is found throughout the Rocky Mountains and Pacific Coast states. It is an important vector of *Borrelia recurrentis,* the etiological agent of relapsing fever, which was first reported in North America from gold miners near Denver in 1915. Transmission of the spirochaete from tick to tick by hyperparasitism has been reported.[14] Basically the tick is a rodent parasite. Its life cycle is typical for most species of *Ornithodoros.* The female lays up to 200 eggs in crannies and crevices like those in which adults hide. The larvae actively seek hosts and feed for about 12 to 15 minutes. After molting, the two nymphal instars each feed again and then molt to adult. The life cycle under laboratory conditions takes about 4 months but may be greatly prolonged in the absence of food.

Ornithodoros cariaeceus occurs from Mexico to southern Oregon, hiding in the soil of bedding areas of large mammals, such as deer and cattle. It is greatly feared by many native peoples because of its painful bite and the venomous aftereffects of its attack. Like many bloodsucking arthropods, it is attracted to carbon dioxide and thus can be trapped using dry ice for bait.

Ornithodoros moubata is an eyeless argasid found in widely dispersed arid regions of Africa. Closely related species are *O. compactus* in South Africa, *O. aperatus* in mideastern Africa, and *O. porcinus* from middle to southern Africa. *Ornithodoros porcinus* is primarily a parasite of the burrowing warthog but readily invades human habitation. It feeds on many mammals and birds and can survive starvation for at least 5 years. The larvae of these species do not feed but molt directly into the first nymphal instar. Some populations are known to be parthenogenetic. An interesting physiological adaptation in this species complex is the absence of a passage between the midgut and hindgut, with the result that all waste matter must remain within the intestinal diverticuli during the life of the tick.[7,47] A great deal is known about the biology, control, and other aspects of this group,[15] all of which are important vectors of relapsing fever.

Ornithodoros savignyi is similar in appearance to *O. moubata,* except that it has eyes. It is found in arid regions of North, East, and southern Africa; the Near East; India; and Ceylon. Basically it is a parasite of camels but will bite practically any mammal and fowl. It does not invade human habitation, as does *O. porcinus,* but buries itself in shallow soil, awaiting its prey. It is quite bold, attacking across wide open areas, if the ground is not too hot, and feeds quickly. Its bite is

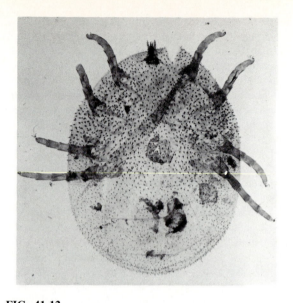

FIG. 41-12

Otobius megnini, the spinose ear tick: nymph.

Photograph by Warren Buss.

quite painful, but *O. savignyi* is not known to transmit diseases in nature.

■ Genus *Otobius*

These argasids are called **spinose ear ticks** because the nymphs have a spiny tegument and usually feed within the folds of the external auditory canal. *Otobius megnini* (Fig. 41-12) is widely distributed in the warmer parts of the United States and in British Columbia. It also has been introduced into India, South Africa, and South America. It feeds mainly on cattle but attacks many domestic and wild mammals, as well as humans.

Adult *O. megnini* do not feed. The adult capitulum is submarginal, but the hypostome is vestigial; its tegument is not spiny. The capitulum of the larva and nymph is marginal; the hypostome, well developed; and the tegument of the nymphs, especially the second stage, spiny. Eggs are laid on the soil. On hatching, the larvae contact a host, wander upward toward the head, and attach in the ear. There they remain through two molts, detaching and dropping to the ground for a final molt to the adult stage. Heavy infections can have serious, even fatal, effects on livestock.

Otobius lagophilus has a similar life cycle, except that the larvae and nymphs feed on the faces of rabbits in western North America.

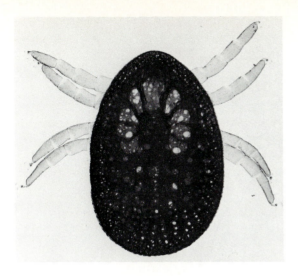

FIG. 41-13

Argas sp.

Photograph by Jay Georgi.

■ Genus *Argas*

Species of *Argas* are almost exclusively parasites of birds and bats. However, most of them have been known to bite humans, although they are not known to transmit diseases to them. *Argas* spp. (Fig. 41-13) are superficially similar to those of *Ornithodoros* but are flatter and retain a lateral ridge even when engorged. Furthermore, the peripheral tegument of the body is typically sculptured in *Argas* spp., whereas that of *Ornithodoros,* although minutely wrinkled, does not show an obvious pattern without very high magnification. Eyes are absent.

Argas ticks have bedbug-like habits, feeding briefly by night and hiding by day in crevices and under litter. Eggs are laid in these hiding places, and the hatched larvae eagerly seek a host. They usuallly remain attached, feeding for a few days before dropping off to molt to the first-stage nymph. The two nymphal stages feed in the same manner as do the adults, engorging in less than an hour, then leaving the host to hide and digest the meal.

Argas persicus, the **fowl tick,** is primarily an Old World species, although it does exist in the New World, along with the similar *A. miniatus, A. sanchezi,* and *A. radiatus,* all parasites of domestic fowl and other birds. Under favorable conditions, these ticks may build up a huge population in a henhouse, and their nocturnal depredations can exhaust a flock or even kill individuals. Vagabonds or others who intend to spend a night in a deserted chickenhouse are sometimes surprised by masses of ravenous

fowl ticks that literally come out of the woodwork to attack them. The bite is painful, often with toxic aftereffects, but such attacks on humans are rare.

The **pigeon tick,** *A. reflexus,* is a Near and Middle Eastern pest that has spread northward through Europe and Soviet Union and eastward to India and other Asian localities. It has been reported in North and South America but probably was misidentified. *Argas reflexus* mainly attacks domestic pigeons, but because these birds are closely associated with human habitation, this tick bites people more often than does *A. persicus.*

Argas cooleyi is commonly associated with cliff swallows and other birds in the United States. *Argas vespertillionis* is widely distributed among Old World bats and occasionally bites humans. The largest of all ticks is *A. brumpti,* a parasite inhabiting dens of the hyrax and some rodents in Africa. It is 15 to 20 mm long by 10 mm wide.

SUBORDER MESOSTIGMATA

The mesostigmatid mites (Fig. 41-14) have a pair of respiratory spiracles, the **stigmata,** which are located just behind and lateral to the third coxae. Usually extending anteriad from each stigma is a tracheal trunk, the **peritreme,** which makes it easy to recognize specimens belonging to this suborder. The gnathosoma forms a tube surrounding the mouthparts. A **tectum** is present above the mouth, and a ventral bristle-like organ, the **tritosternum,** usually is present immediately behind the gnathosoma. The palpal tarsus has a forked tine at its base. The dorsum of adults usually has one or two sclerites called **shields** or **dorsal plates.**

Family Laelaptidae

Laelaptid mites are worldwide in distribution. They are the most common ectoparasites of mammals, and some species parasitize invertebrates. Most species have pretarsi, caruncles, and claws on all legs. The dorsal shield is undivided. The second coxa has a toothlike projection from the anterior border.

The family Laelaptidae includes a large number of diverse genera. Taxonomy of the group is difficult because of the large number of species included and the lack of careful descriptions.

The **common rat mite,** *Echinolaelaps echidinus* (Fig. 41-15), transmits the protozoan *Hepatozoon muris* from rat to rat. Although laelaptids are not known to transmit diseases to humans, they are suspected of causing dermatitis. The virus of epidemic hemorrhagic fever has been demonstrated in several species collected in rodent burrows in the Far East.[4]

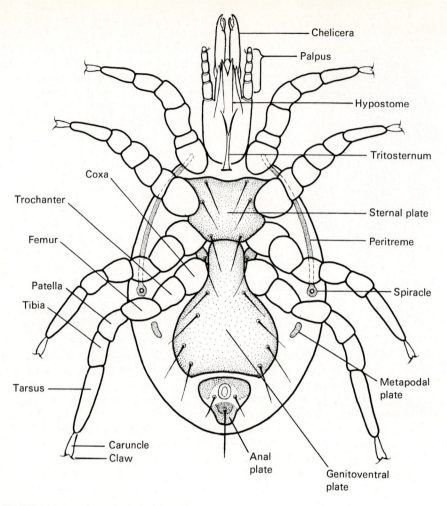

FIG. 41-14

Generalized mesostigmatid mite, ventral view.

Courtesy Communicable Disease Center, 1949, U.S. Public Health Service, Washington, D.C.

FIG. 41-15

Echinolaelaps echidinus, the common rat mite. Ventral view of female.

From Hirst, S. 1922. Br. Mus (Nat. Hist.) Econ. Ser. 13:1-107.

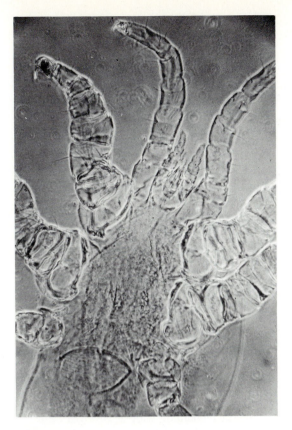

FIG. 41-16

Orthohalarachne attenuata from the nasal passages of a northern fur seal.

Photograph by Warren Buss.

FIG. 41-17

Lung of a baboon, *Papio cynocephalus*, with nodules caused by the mite, *Pneumonyssus* sp.

Photograph by R.E. Kuntz.

Family Halarachnidae

Closely related to the laelaptid mites, the halarachnids are parasites of the respiratory systems of mammals. They are easily recognized by a combination of morphological features: The dorsal shield is undivided and reduced; the sternal plate is reduced and has three pairs of setae (if much reduced, some setae may be displaced from the plate); the female genital sclerite, the epigynial plate, is rudimentary; and the male genital opening is in the anterior margin of the sternal plate. The tritosternum is absent, and the movable digit of the chelicera is more strongly developed than is the fixed digit.

Complete life cycles of halarachnids are unknown. The genus *Halarachne* is found only in the respiratory system of seals of the family Phocidae, and *Orthohalarachne attenuata* parasitizes other families of Pinnipedia (Fig. 41-16). Several species of *Pneumonyssus* are found in primates, although infection in hu-

mans is unknown. Respiratory problems caused by these mites in captive monkeys and baboons are common[44] (Fig. 41-17). *Pneumonyssus caninum* inhabits the nasal passages and sinuses of dogs and may cause central nervous system disorders.[3] *Raillietea auris* is the **cattle ear mite.** Not known to be pathogenic, apparently it feeds on secretions and dead cells of the external auditory meatus. This genus is considered by some authors to belong to a separate family, Raillietidae.

Family Dermanyssidae

Dermanyssids are parasites on vertebrates and are of considerable economic and medical importance. The dorsal plate in the female either is undivided or is divided with a very small posterior part. The sternal plate has three pairs of setae, and the metasternal plates are reduced and lateral to the genital plates. A tritosternum is present. The chelicerae may be normal, with reduced chelae, or they may be quite elongate and needle-like. All legs have pretarsi, caruncles, and claws.

673

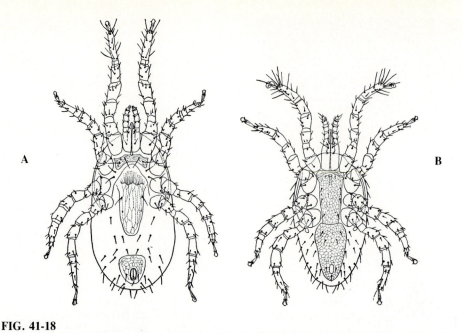

FIG. 41-18

Ventral views of *Dermanyssus gallinae,* the chicken mite. **A,** Female. **B,** Male.

From Hirst, S. 1922. Br. Mus. (Nat. Hist.) Econ. Ser. 13:1-107.

Dermanyssus gallinae, the **chicken mite** (Fig. 41-18), attacks domestic fowl, particularly chickens and pigeons, throughout the world. They hide by day in crevices near the roosting places of birds, emerging at night to feed. Their numbers may be so great as to kill the birds. Setting hens may be forced to abandon their nests, and young chicks may rapidly perish. This mite readily attacks humans, especially children, causing a severe dermatitis. Roosting pigeons may bring *D. gallinae* into proximity with human habitation, where wandering mites may encounter a mammalian meal.[35] They are attracted to warm objects and may accumulate in electric clocks and around fireplaces, water pipes, and so on.

The viruses of western and St. Louis equine encephalitis have been isolated from *D. gallinae.* It is unlikely that these mites play an important role in transmission of these diseases to mammals, but they may help keep up the reservoir of infection among birds.[38]

Natural and experimental transmissions of fowl poxvirus have been demonstrated.[37] Experiments have shown that *D. gallinae* also can transmit Q fever and fowl spirochaetosis.

Liponyssus sanguineus, the **house mouse mite,** prefers to feed on that host but will readily attack humans. It is important in that this mite can transmit the rickettsialpox pathogen to humans. Rickettsialpox is a mild, febrile condition with a vesicular rash commencing 3 to 4 days after the onset of fever. A scab develops at the site of the bite, and healing is slow. Besides fever, the patient has chills, sweating, backache, and muscle pains. Patients recover in 1 to 2 weeks; no fatalities are known. Q fever has been experimentally transmitted by this mite. The biology of the house mouse mite is summarized by Baker et al.[1]

The **tropical rat mite,** *Ornithonyssus bacoti* (Fig. 41-19), is found worldwide, in both temperate and tropical climates, where, as its name implies, it normally infests rats. It is a serious pathogen of laboratory mouse colonies, where it can retard growth and eventually kill young mice. When rats are killed or abandon their nests, the mites can migrate considerable distances to enter human habitation. They cause a sharp, itching pain at the time of their bite, and skin-sensitive persons may develop a severe dermatitis. The biology of the mite is summarized as follows: The egg produces a nonfeeding larva that rapidly molts to a bloodsucking **protonymph.** This molts to become a nonfeeding **deutonymph,** which in turn becomes a feeding adult. Parthenogenesis is common, producing only males. The minimal life cycle from egg to egg can be completed in 13 days. Adult females live about 60 days and produce approximately a hundred eggs.

These mites are not known to transmit any pathogens to humans, although experimentally they have

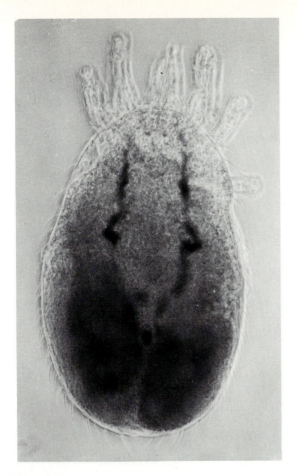

FIG. 41-19

Ornithonyssus bacoti, the tropical rat mite.

Photograph by Warren Buss.

been shown to transmit plague, rickettsialpox, Q fever, and murine typhus. *Ornithonyssus bacoti* is the intermediate host of the filarial nematode *Litomosoides carinii,* a parasite of the cotton rat, *Sigmodon hispidus.* Because rats, mites, and worms can easily be maintained in the laboratory, this system has been used as a model for studies on filariasis, including drug testing.[48]

The **northern fowl mite,** *Ornithonyssus sylviarum,* is widespread in northern temperate climes and has been reported from Australia and New Zealand. It will bite humans and can be a nuisance to egg processors. It does not appear to be particularly pathogenic to fowl.

Ornithonyssus bursa is the **tropical fowl mite.** It is ectoparasitic on chickens, turkeys, and some wild birds, including the English sparrow.[13] It can be pathogenic to poultry, causing them to be listless and

poorly developed. It bites humans but causes only a slight irritation.

Family Rhinonyssidae

This family is considered by some authorities to be a subfamily of Dermanyssidae. All members are parasitic in the respiratory tracts of birds. Rhinonyssids are oval in shape and have weakly sclerotized plates. All tarsi have pretarsi, caruncles, and claws. Stigmata are present with or without short dorsal peritremes. The tritosternum is absent.

These mites are viviparous, producing larvae in which the protonymph is already developed. Nearly every species of bird examined has nasal mites; many species have been described, and many more species undoubtedly are yet to be discovered. Because of their blood- or tissue-feeding habits, these mites may be regarded as significant disease agents in wild bird populations. The **canary lung mite,** *Sternostoma tracheacolum* (Fig. 41-20), can sicken and kill captive canaries and finches. Treatment of this disease is described by Jolivet.[20] Important taxonomic reviews of this family are provided by Pence.[27,28]

SUBORDER PROSTIGMATA

In this suborder the spiracles are located either between the chelicerae or on the dorsum of the hysterosoma. These mites usually are weakly sclerotized. The chelicerae vary from strongly chelate to reduced. Pedipalps are simple, fanglike, or clawed. There are phytophagous, terrestrial and aquatic free-living, and parasitic forms.

Family Cheyletidae

The cheyletid mites are small, measuring 0.2 to 0.8 mm long. Most are yellowish or reddish, oval, and plump, except for the feather-inhabiting species, which are elongate. The propodosoma and hysterosoma are clearly delineated and usually have one or more dorsal shields. Eyes are present or lacking. Strong peritremes, which usually surround the gnathostoma, are present. The chelicerae are short and stylet-like; the palpi are large and pincer-like.

In *Cheyletiella parasitivorax* the male genital opening is dorsal, a rare occurrence in arthropods. These are common denizens of fur coats, possibly predaceous on other mites. *Cheyletiella yasguri* of dogs and *C. blakei* of cats cause a mange dermatitis on their normal hosts. They also will feed on humans, although temporarily.

Family Pyemotidae

Pyemotid mites mainly parasitize insects that in turn infest cereal crops. They are brought into contact with

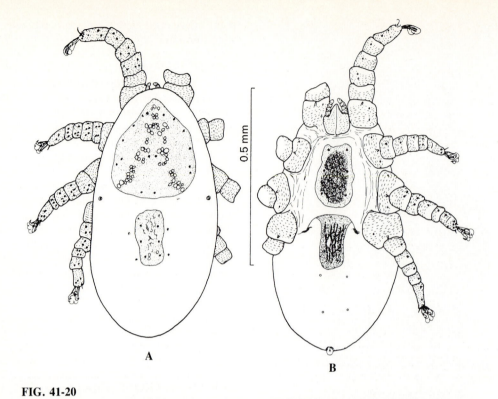

FIG. 41-20

Female *Sternostoma tracheacolum,* the canary lung mite. **A,** Dorsal view. **B,** Ventral view.

From Pence, D.B. 1975. Spec. Pub. Mus. Texas Tech. Univ. no. 8, p. 64.

humans when people harvest or work with stored grains or sleep on straw mattresses. When these mites bite, they leave a small, itching vesicle that rapidly may become inflamed and cause considerable discomfort. Itching, headache, nausea, and internal pains may accompany severe attacks.

These are soft-bodied mites with tiny chelicerae and pedipalps. A wide space occurs between the third and fourth pairs of legs. Sexual dimorphism is marked. Furthermore, the female becomes enormously swollen when gravid.

Pyemotes tritici normally is a parasite of various stored grain beetles. Known as the **straw itch mite,** it readily attacks humans. The male can be seen only with difficulty by the unaided eye, but the gravid female reaches nearly a millimeter in length. Her body contains 200 to 300 large eggs, which hatch internally. The developing mites complete all larval instars before being born. The few males emerge first and cluster around the genital pore of the mother, copulating with the females as they emerge. This species shows promise in controlling fire ants.

The **grain itch mite,** *Pyemotes ventricosus,* is similar in biology and pathogenicity. It normally infests-

boring beetle larvae, grain moths, and numerous other insects.

Family Psorergatidae

These are small to medium-sized mites that are unarmored and have striated skin and peritremes. Because they are soft, they are susceptible to desiccation and are less numerous during dry periods. The chelicerae are minute and stylet-like. The pedipalps are simple and minute and are not used for grasping. The first pair of legs is modified for grasping hairs.

Psorergates ovis is the itch mite of sheep and is a serious pest in sheep-raising countries, including the United States, New Zealand, and Australia. It causes skin injury and fleece derangement. *Psorergates simplex* is found on laboratory mice, and *P. bos* is known from cattle in the western United States.[18]

Family Demodicidae

These minute, cigar-shaped parasites are known as the **follicle mites.** They range in length from 100 to 400 μm and have short, stumpy, five-segmented legs on the anterior half of the body. The opisthosoma is transversely striated. Species of *Demodex* live in hair

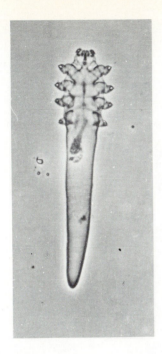

FIG. 41-21

Demodex folliculorum, the human follicle mite.
From Desch, C., and W.B. Nutting. 1972. J. Parasitol. 58:169-178.

follicles and sebaceous glands of many species of mammals. Although numerous species have been described, it is probable that many more exist, especially in wild mammals. There seems to be rigid host specificity.

Humans serve as hosts to two species. *Demodex folliculorum* (Fig. 41-21) lives in hair follicles, whereas *D. brevis,* a stubbier species, inhabits sebaceous glands.[5] Both exist mainly on the face, particularly around the nose and eyes. All life stages may be found in a single follicle. These mites may penetrate the skin and lodge in various internal organs, where they elicit a granulomatous response.

The incidence of these mites in humans is very high, from about 20% in persons 20 years of age or younger to nearly 100% in the aged. Infection usually is benign, although rarely there may be loss of eyelashes or granulomatous skin eruptions.[12] It is suspected that follicle mites may be involved in introducing acne-causing bacteria into the skin follicles of susceptible individuals. An easy means of diagnosing both species in humans is to examine microscopically some oil expressed from the side of the nose.

Much more pathogenic is the **dog follicle mite,** *De-modex canis.* This species, together with some form of the bacterium *Staphylococcus pyogenes,* causes **red mange,** or **canine demodectic mange.** Infection in young dogs can be serious, even fatal. There is hair loss on the muzzle, around the eyes, and on the forefeet. The skin develops reddish pimples and pustules, becoming hot, thickened, and covered with a foul-smelling reddish-yellow exudate. Exact diagnosis depends on demonstrating a mite in skin scrapings. Treatment is difficult, and severely infected puppies may have to be killed. Symptoms may disappear gradually, and older, although perhaps infected, dogs show no further signs of disease, probably because of acquired immunity.

Other demodicids of importance are the **cattle follicle mite,** *D. bovis,* the **horse follicle mite,** *D. equi,* and the **hog follicle mite,** *D. phylloides.* All three cause a pustular dermatitis with nodules and loss of hair. Holes in the skin caused by these mites may reduce the value of hides.

Family Trombiculidae

This family contains the infamous **chigger mites,** which are all too common in most tropical and temperate countries of the world. They are unique among mites that attack humans in that only the larval stage is parasitic; the nymphs and adults are predators on small terrestrial invertebrates or their eggs. Nearly 300 species have been described, many from the larva only; in numerous cases the nymphs and adults are unknown or have not been correlated with the larvae.

In a **generalized chigger life cycle** the egg hatches in about a week, exposing the developing larva, or **deutovum.** It rapidly completes differentiation into a six-legged larva (Fig. 41-22), which finds a host and feeds. The engorged larva leaves its host and becomes a quiescent **prenymph** or **nymphochrysalis.** Emerging nymphs feed on insect eggs or soft-bodied invertebrates and pass into a quiescent **imagochrysalis** before molting into an adult. Males deposit onto the substrate a stalked spermatophore, which the female inserts into her genital pore. Most chiggers show little host specificity, although species are known that appear to feed only on frogs.[6]

The taxonomy of trombiculids is based mainly on the larvae. The larval body is rounded and usually is red, although it may be colorless. It bears a dorsal plate, or scutum, at the level of the anterior two pairs of legs; usually two pairs of eyes are near the lateral margins of the scutum. The scutum bears a pair of sensillae and three to seven setae. The chelicerae have two segments; the basal segment is stout and muscular, whereas the distal segment is a curved blade with or without teeth. The pedipalps consist of five seg-

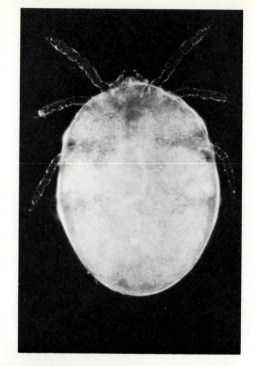

FIG. 41-22

A chigger larva.

Photograph by Mark Pope.

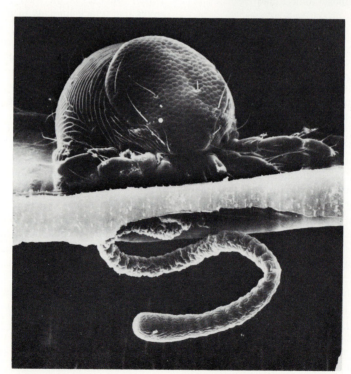

FIG. 41-23

Larval *Arrenurus* sp. feeding on a damselfly. Note the long stylostome penetrating the insect's body wall. The internal organs of the damselfly have been removed.

From Redmond, B.L., and J. Hochberg. 1981. J. Parasitol. 67:308-313.

ments. The fifth, or tarsus, bears several setae and opposes a tibial claw like a thumb. Distributed on the body are numerous plumose setae.

Adults are among the largest of mites, reaching a millimeter or more in length. There is a conspicuous constriction between the propodosoma and hysterosoma. Eyes may be present or absent. Both sexes are clothed with a dense covering of plumose setae, which gives them the appearance of velvet. Commonly they are bright red or yellow.

There are two medical aspects of chigger bite: **chigger dermatitis** and transmission of pathogens. These are considered separately.

Larval chiggers do not burrow into the skin, as is popularly thought. After the mouthparts penetrate the epidermis, the mite injects salivary secretions. These are proteolytic, killing and digesting host cells, which the parasite then sucks, along with interstitial fluids. Simultaneously, host cells harden under the influence of other salivary secretions to become a tube, the **stylostome** (Fig. 41-23). The mite retains its mouthparts in the stylostome, using it like a drinking straw, until engorged and then drops off. Not all chiggers cause an itching reaction; among those that do, the mite usually has detached before a host reaction begins. Some persons are immune to their bites, whereas others may incur a violent reaction.

Most chiggers of medical importance in North America are of the genus *Trombicula*. Of these, *T. alfreddugesi* is the most common species, ranging throughout the United States except in the western mountain states. It is most abundant in disturbed forest that has been overgrown with shrubs, vines, and similar second-growth vegetation. It feeds on most terrestrial vertebrates. *Trombicula splendens* is the most abundant chigger in the southeastern United States, especially in moist areas such as swamps and bogs. The two species overlap their ranges in many areas but are active in different seasons.

Other species, some representing other genera, are found in the United States and elsewhere. Many bite humans, and others are important pests of livestock. The turkey chigger, *Neoschoengastia americana*, causes discoloring of the skin and loss of feathers of turkeys and related birds, rendering them less fit for market.[8] *Euschoengastia latchmani* causes a mange-like dermatitis on horses in California.

Several species of *Leptotrombidium* are vectors of the rickettsial disease in humans called **scrub typhus.** The microorganism *Rickettsia tsutsugamushi* is transovarially transmitted among mites; wild rodents, particularly species of *Rattus*, are reservoirs. This disease was first described from Japan and now is known from Southeast Asia; adjacent islands of the Indian Ocean and the southwest Pacific; and coastal north Queensland, Australia. A primary lesion appears at the site of the chigger bite. It slowly enlarges to 8 to 12 mm and becomes necrotic in the center. By the fifth to eighth day, a red rash appears on the trunk and may spread to the extremities. Other symptoms are enlarged spleen, delirium and other nervous disturbances, prostration, and possibly deafness. Mortality rates range from 6% to 60%. Early treatment with broad-spectrum antibiotics usually is successful.

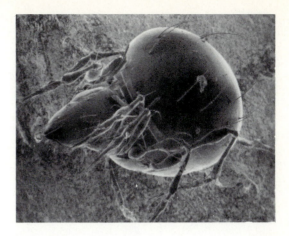

FIG. 41-24

Oppia coloradensis, a beetle mite.
Photograph by Tyler Woolley.

SUBORDER CRYPTOSTIGMATA

Cryptostigmatids are commonly called the **beetle mites,** not because they parasitize beetles, but because they have a superficial resemblance to tiny beetles (Fig. 41-24). The exoskeleton is strongly sclerotized or leathery and often deep brown. Stigmata and tracheae are usually present, opening into a **porose area.** The mouthparts are withdrawn into a tube, the **camerostome,** which may have a hoodlike sclerite over it.

A complex of families within this group is called the Oribatei. All feed on organic detritus and as such are among the dominant fauna of humus. They are of no direct medical importance, but many serve as intermediate hosts of *Moniezia expansa* and other anoplocephalid tapeworms. Members of this group are also intermediate hosts for *Bertiella studeri*, the primate cestode that sometimes infects humans (Chapter 22).

SUBORDER ASTIGMATA

Mites of the suborder Astigmata totally lack tracheal systems; they respire through the tegument, which is soft and thin. They lack claws, which are replaced with sucker-like structures on their pretarsi

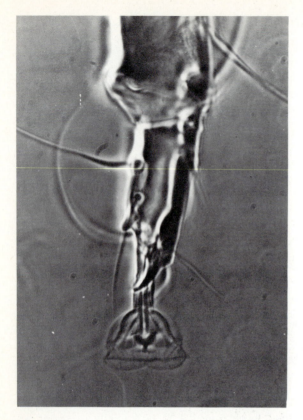

FIG. 41-25

Pretarsus of *Chorioptes* sp., showing sucker-like modification.

Photograph by Jay Georgi.

FIG. 41-26

Chorioptes sp., male.

Photograph by Jay Georgi.

(Fig. 41-25). Some of the most medically and economically important mites belong to this suborder.

Family Psoroptidae

Members of this family are very similar to those of the family Sarcoptidae and are easily confused with them. However, unlike the sarcoptids, they do not burrow into the skin; instead, psoroptids pierce the skin at the bases of hairs, causing an inflammation that can become severe. Furthermore, the Psoroptidae lack vertical setae on the propodosoma, which are present on the Sarcoptidae.

Chorioptic mange is a condition of domestic animals caused by mites of the genus *Chorioptes* (Fig. 41-26). Until recently, each host species of *Chorioptes* was considered to harbor a distinct species of parasite. However, careful laboratory experiments have shown them to represent a single species, *C. bovis*.[42] These mites usually inhabit the feet and lower hind legs of cattle and horses. In sheep, chorioptic mange of the scrotum is well known to cause seminal

degeneration. In fact surveys of sheep in the United States have shown *C. bovis* to be the most common arthropod parasite, although pathogenic results usually are rare.

Psoroptic mange is caused by several species of mites on several species of hosts. *Psoroptes* spp. are distinguished by long legs that extend beyond the body and by the pedicel of the caruncle, which is segmented. They pierce the skin and suck exudates. These fluids congeal to form scabs that provide a protective cover for the parasites; then the *Psoroptes* spp. can reproduce under ideal conditions, increasing their numbers into millions in only a few days.

The classification of *Psoroptes* mites parallels that of *Chorioptes* spp.: many nominal species probably are synonyms of a few real species.[43] Most domestic and a number of wild animals suffer from psoroptic mange. Wool production may be greatly inhibited by *P. ovis* in sheep. Unlike sarcoptic mange (following discussion), *P. ovis* infests parts of the body most densely covered with wool. The species was reviewed by Kirkwood.[21]

Mites very closely related to *Psoroptes* spp. are in the genus *Otodectes*. Fairly common in cats, dogs, foxes, and ferrets, *O. cynotis* (Fig. 41-27) is usually found in the ears, although other parts of the head

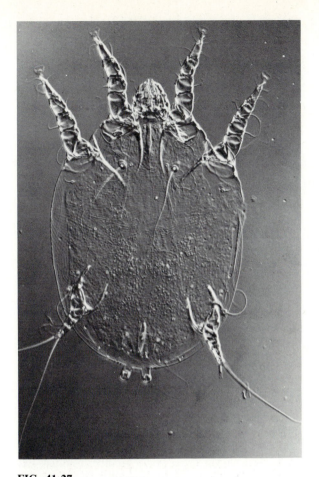

FIG. 41-27

Otodectes cynotis nymph.

Photograph by Jay Georgi.

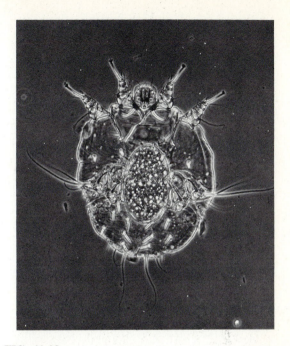

FIG. 41-28

Sarcoptes scabiei, the itch mite.

Photograph by Jay Georgi.

may be infested. Thousands of these mites swarming in the ears of a luckless host can cause desperate distress, with scabby, flowing ears and fitlike behavior.

Family Sarcoptidae

Sarcoptic mange, or **scabies,** may result from an infestation of the itch mite, *Sarcoptes scabiei* (Fig. 41-28). Although separate *Sarcoptes* spp. have been described from a wide variety of domestic animals, as well as humans, they are morphologically indistinguishable and probably represent physiological races or perhaps sibling species. Thus *S. scabiei* var. *equi* has a predilection for horses but will readily bite the rider as well.

The sarcoptids are skin parasites of homoiotherms. The body is rounded without a constriction separating the propodosoma from the hysterosoma. A propodosomal shield may be present or absent; either way, a pair

of vertical setae projects from the dorsal propodosoma. The tegument has fine striae arranged in fields interrupted by scales, spines, or setae. The legs are very short and may or may not have claws or caruncles.

Scabies mites mate on the skin of the host, the male inseminating immature females. The immature females move rapidly over the skin and probably are at this stage transmissible between hosts. Males do not burrow into the skin but remain on the surface, along with nymphs. The mature female uses the long bristles on the posterior legs to lift her back end up until she is nearly vertical. She cuts rapidly with her mouthparts and claws, becoming completely embedded in 2½ minutes. She remains within the horny layer of the skin, forming tortuous tunnels for about 2 months. Scattered along the burrows are eggs, hatched larvae, ecdysed cuticles, and excrement. Eggs hatch in 3 to 8 days, and the larvae and nymphs emerge to wander on the surface of the skin.

The tunneling and the secretory and excretory products produce an intense itching sensation in most infected persons. Usually a person does not notice any symptoms until the case is well advanced. A rash begins to show, and vesicles and crusts may begin to form in some cases. This disease has several names, such as **seven-year itch, Norwegian itch,** or simply

FIG. 41-29

Dermatophagoides sp., a house dust mite.

From Wharton. G.W. 1970. Science 167:1382-1383. Copyright 1970 by the American Association for the Advancement of Science.

scabies. The skin between the fingers, the breasts, the shoulder blades, the penis, and the creases of the knees and elbows are most often infected. Scratching can cause bleeding and secondary infection. Transmission occurs primarily by physical contact between persons. Scabies was well reviewed by Robinson.[31]

A 17- to 20-year cycle of resurgence of scabies infection in the world apparently occurs for unknown reasons, possibly because of a changing immunity in the human population.[39]

Sarcoptic mange in domestic and wild mammals is basically the same as that in humans. Hairless or short-haired regions of the body are most affected. Secondary infection by bacteria is more common in animals other than humans, resulting in severe weight loss or failure to gain, loss of hair, and pruritic dermatitis.[29] Infection is readily passed on to humans in contact with mangy animals.

Cats may develop mange caused by *Notoedres cati*. *Notoedres* is very similar in appearance to *Sarcoptes* but is smaller and more circular. It will affect rodents and dogs but apparently not humans. **Notoedric mange** usually begins at the tips of the ears and spreads down over the head, sometimes onto the body.

Family Knemidokoptidae

Knemidokoptid mites are very similar in morphology and biology to the Sarcoptidae, with which they have been aligned in the past. However, some author-

ities accord them distinct family status.[9] They all are parasites of birds.

Knemidokoptes mutans causes a condition known as **scaly leg** in chickens and other small domestic birds. The mites burrow into the skin and under the scales of the feet and lower legs, which become distorted and covered with thick, nodular, spongy crusts. Infection may be so severe as to kill the birds, either directly or by secondary sensitization, which can involve internal organs. Scaly leg is highly contagious.

Closely related to *K. mutans* is the **depluming mite,** *K. gallinae*. It embeds in the skin at the bases of the quills; infected birds pluck out their feather in an attempt to alleviate the itching, or the feathers may fall out by themselves. Usually, large patches of deplumation extend over the body.

Knemidokoptes jamaicensis is a parasite of canaries, whereas *K. pilae* infests the face and legs of budgerigars. For a discussion of control of these pests in cage birds see Rickards.[30]

Family Pyroglyphidae

Important mites in this family are species of *Dermatophagoides* (Fig. 41-29). Most are not parasitic, but *D. scheremetewskyi* can cause a severe dermatitis on the scalp, face, and ears of humans. Probably it is a normal parasite of sparrows, bats, or other house-inhabiting animals.

Of more importance, mites in this family are responsible for **house dust allergy.** Several species, especially those in *Dermatophagoides,* are abundant denizens of house dust. When whole mites or their parts or excrement are inhaled, as happens to everyone every day, they can stimulate an allergic reaction in sensitive individuals. Much literature exists on the subject.[23]

REFERENCES

1. Baker, E.W., et al. 1956. A manual of parasitic mites of medical and economic importance. National Pest Control Association, New York.
2. Burgdorfer, W., M.C. Schmidt, and H. Hoogstraal. 1973. Detection of *Trypanosoma theileri* in Ethiopian cattle ticks. Acta Trop. 30:340-346.
3. Christensen, H.A., and C. Rehbinder. 1971. *Pneumonyssus caninum,* a mite in the nasal cavities and sinuses of the dog. Nord. Vet. Med. 23:499-505.
4. Chumakov, M.P. 1957. Etiology, epidemiology and prophylaxis of hemorrhagic fevers. U.S. Public Health Service Monograph No. 50. U.S. Government Printing Office, Washington, D.C. pp. 19-25.
5. Desch, C., and W.B. Nutting. 1972. *Demodex folliculorum* (Simon) and *D. brevis* Akbulatova of man: redescription and reevaluation. J. Parasitol. 58:169-177.
6. Duszynski, D.W., and K.L. Jones. 1973. The occurrence of intradermal mites, *Hannemania* spp. (Acarina: Trombiculidae), in anurans in New Mexico with a histological description of the tissue capsule. J. Parasitol. 59:531-538.
7. Enigk, K., and I. Grittner, 1952. Die Excretion der Zecken. Z. Tropenmed. Parasitol. 4:77-94.
8. Everitt, R.E., M.A. Price, and S.E. Kunz. 1973. Biology of the chigger *Neoschöngastia americana* (Acarina: Trombiculidae). Ann. Entomol. Soc. Am. 66:429-435.
9. Fain, A., and P. Elsen. 1967. Mites of the family Knemidokoptidae that cause mange in birds (a revision with notes.) Acta Zool. Pathol. Antverp. no. 45.
10. Gothe, R., K. Kunze, and H. Hoogstraal. 1979. The mechanisms of pathogenicity in the tick paralyses. J. Med. Entomol. 16:357-369.
11. Gregson, J.D. 1949. Note on the longevity of certain ticks (Ixodidae). Proc. Entomol. Soc. Br. Col. 45:14.
12. Grosshans, E.M., M. Dremer, and J. Maleville. 1974. *Demodex folliculorum* and the histogenesis of granulomatous acne rosacea. Hautartz 25:166-177.
13. Hadani, A., K. Rauchbach, Y. Weissman, and R. Bock. 1975. The occurrence of the tropical fowl mite, *Ornithonyssus bursa* (Berlese, 1888), Dermanyssidae, on turkeys in Israel. Refuah Vet. 31:111-113.
14. Helmy, N., G.M. Khalil, and H. Hoogstraal. 1983. Hyperparasitism in *Ornithodoros erraticas.* J. Parasitol. 69:229-233.
15. Hoogstraal, H. 1956. African Ixodoidea, vol. 1. Ticks of the Sudan. U.S. Department of the Navy. U.S. Government Printing Office, Washington, D.C.
16. Hoogstraal, H. 1972. The influence of human activity on tick distribution, density, and diseases. Wiad. Parazytol. 18:501-511.
17. Hunter, W.D., and W.A. Hooker. 1907. Information concerning the North American fever tick. USDA Bur. Entomol. Bull. 72:1-87.
18. Johnston, D.E. 1964. *Psorergates bos,* a new mite parasite of domestic cattle (Acari, Psorergatidae). Ohio Agric. Exp. Sta. Res. Circ. 129:1-7.
19. Johnston, L.A.Y., D.H. Kemp, and R.D. Pearson. 1986. Immunization of cattle against *Boophilus microplus* using extracts derived from adult female ticks: effects of induced immunity on tick populations. Int. J. Parasitol. 16:27-34.
20. Jolivet, G. 1975. A new sickness of aviary birds in France. Respiratory acariasis due to *Sternostoma tracheolatum* Lawrence 1948. Rec. Med. Vet. 151:273-277.
21. Kirkwood, A.C. 1986. History, biology and control of sheep scab. Parasitol. Today 2:302-307.
22. Lane, R.S., and W. Burgdorfer. 1987. Transovarial and transtadial passage of *Borrelia burgdorferi* in the western black-legged tick, *Ixodes pacificus* (Acari: Ixodidae). Am. J. Trop. Med. Hyg. 37:188-192.
23. Lang, J.D., L.D. Charlet, and M.S. Mulla, 1976. Bibliography (1864 to 1974) of house-dust mites *Dermatophagoides* spp. (Acarina: Pyroglyphidae), and human allergy. Sci. Biol. J. 2:62-83.
24. Lewis, I.J. 1970. Observations on the dispersal of larvae of the cattle tick *Boophilus microplus* (Can.). Bull. Entomol. Res. 59:595-604.
25. Mann, W.M. 1915. A cursorial tick. Psyche 22:60.
26. Mather, T.N., J.M.C. Ribeiro, and A. Spielman. 1987. Lyme disease and babesiosis: acaricide focused on potentially infected ticks. Am. J. Trop. Med. Hyg. 36:609-614.
27. Pence, D.B. 1973. The nasal mites of birds from Louisiana. IX. Synopsis. J. Parasitol. 59:881-892.
28. Pence, D.B. 1975. Keys, species and host list, and bibliography for nasal mites of North American birds. (Acarina: Rhinonyssinae, Turbinoptinae, Speleognathinae, and Cytoditidae). Spec. Pub. Mus. Texas Tech Univ. 8:1-148.
29. Pence, D.B., et al. 1983. The epizootiology and pathology of sarcoptic mange in coyotes, *Canis latrans,* from south Texas. J. Parasitol. 69:1101-1115.
30. Rickards, D.A. 1975. Cnemidocoptic mange in parakeets. Vet. Med. Small Anim. Clin. 70:729-731.
31. Robinson, R. 1985. Fight the mite and ditch the itch. Parasitol. Today 1:140-142.
32. Schlein, Y., and A.E. Gunders. 1981. Pheromone of *Ornithodoros* spp. (Argasidae) in the coxal fluid of female ticks. Parasitology 82:467-471.
33. Semtner, P.J., and J.A. Hair. 1973. Distribution, seasonal abundance and hosts of the Gulf Coast tick in Oklahoma. Ann. Entomol. Soc. Am. 66:1264-1268.
34. Semtner, P.J., and J.A. Hair. 1973. The ecology and behavior of the lone star tick (Acarina: Ixodidae). V. Abundance and seasonal distribution in different habitat types. J. Med. Entomol. 10:618-628.
35. Sexton, D.J., and B. Hayes. 1975. Bird-mite infestation in a university hospital. Lancet 1(7904):445.
36. Shanbaky, N.M., and G.M. Khalil. 1975. The subgenus *Persicargus* (Ixodoidea: Argasidae: *Argus*). 22. The effect of feeding on hormonal control of egg development in Argas (*Persicargas*) *arboreus.* Exp. Parasitol. 37:361-366.
37. Shirinov, F.B., A.I. Ibragimova, and Z.G. Misirov. 1968. The

dissemination of the virus of fowl-pox by the mite *D. gallinae.* Veterinariia 4:48-49.

38. Smith, M.G., et al. 1948. Experiments on the role of the chicken mite *Dermanyssus gallinae* and the mosquito in the epidemiology of St. Louis encephalitis. J. Exp. Med. 87:119-138.

39. Sönnichsen, N., and H. Barthelmes. 1976. Epidemiological and immunological investigations on human scabies. Angew. Parasitol. 17:65-70.

40. Steelman, C.D. 1976. Effects of external and internal arthropod parasites on domestic livestock production. Ann. Rev. Entomol. 21:155-178.

41. Stone, B.F. 1963. Parthenogenesis in the cattle tick, *Boophilus microplus.* Nature 200:1233.

42. Sweatman, G.K. 1957. Life history, non-specificity, and revision of the genus *Chorioptes,* a parasitic mite of herbivores. Can. J. Zool. 35:641-689.

43. Sweatman, G.K. 1958. On the life history and validity of the species in *Psoroptes,* a genus of mange mites. Can. J. Zool. 36:905-929.

44. Testi, B., and F. de Michelis. 1972. An interesting parasitological discovery in monkeys imported into Italy for laboratory use. Zooprofilassi 27:353-369.

45. Viljoen, G.J., et al. 1986. Isolation of a neurotoxin from the salivary glands of female *Rhipicephalus evertsi evertsi.* J. Parasitol. 72:865-874.

46. Wharton, R.H., K.L.S. Harley, P.R. Wilkinson, K.B. Utech, and B.M. Kelly, 1969. A comparison of cattle tick control by pasture spelling, planned dipping, and tick-resistant cattle. Aust. J. Agric. Res. 20:783-797.

47. Wigglesworth, V.B. 1943. The fate of haemoglobin in *Rhodnius prolixus* (Hemiptera) and other blood-sucking arthropods. Proc. R. Soc. [B] 131:313-339.

48. Williams, R.W., and H.W. Brown. 1945. The development of *Litomosoides carinii* filariid parasite of the cotton rat in the tropical rat mite. Science 102:482-483.

SUGGESTED READINGS

Arthur, D.R. 1962. Ticks and disease. Pergamon Press, Oxford, Eng. (A standard resource on the subject.)

Baker, E.W., J.H. Camin, F. Cunliffe, T.A. Woolley, and C.E. Yunker. 1958. Guide to the families of mites. Inst. Acarol. Contrib. 3:1-242.

Baker, E.W., T.M. Evans, D.J. Gould, W.B. Hull, and H.L. Keegan. 1956. A manual of parasitic mites of medical or economic importance. Technical publication. National Pest Control Association, New York.

Baker, E.W., and G.W. Wharton. 1952. An introduction to acarology. The Macmillan Co., New York. (An excellent reference for the beginner and professional alike, although a bit out of date.)

Brennan, J.M., and E.K. Jones. 1959. Keys to the chiggers of North America with synonymic notes and descriptions of two new genera (Acarina: Trombiculidae). Ann. Entomol. Soc. Am. 52:7-16.

Brown, S.J. 1985. Immunology of acquired resistance to ticks. Parasitol. Today 1:166-171.

Krantz, G.W. 1978. A manual of acarology, ed. 2. Oregon State University Book Stores, Inc., Corvallis.

McDaniel, B. 1979. How to know the ticks and mites. William C. Brown Co., Dubuque, Iowa. (Useful, well-illustrated keys to genera and higher categories of ticks and mites in the United States.)

Piesman, J. 1987. Emerging tick-borne diseases in temperate climates. Parasitol. Today 3:197-199.

Strandtmann, R.W., and G.W. Wharton. 1958, Manual of mesostigmatid mites parasitic on vertebrates. Inst. Acarol. Contrib. 4.

Uilenberg, G. 1986. Highlights in recent research on tick-borne diseases of domestic animals. J. Parasitol. 72:485-491.

Woolley, T.A. 1961. A review of the phylogeny of mites. Ann. Rev. Entomol. 6:263-284.

GLOSSARY

abscess Tissue necrosis in a localized area with increase in hydrostatic pressure from pus accumulation.

acanthella Developing acanthocephalan larva between an acanthor and a cystacanth, in which the definitive organ systems are developed.

acanthor Acanthocephalan larva that hatches from the egg.

accidental myiasis Presence within a host of a fly not normally parasitic. Also called pseudomyiasis.

accidental parasite Parasite found in other than its normal host. Also called an incidental parasite.

acetabulum Sucker: the ventral sucker of a fluke; a sucker on the scolex of a tapeworm.

aclid organ Spined introvert near the anterior end of an acanthor.

acquired immunity Immunity arising from a specific immune response, either humoral or cell mediated, stimulated by antigen in the host's body (active) or in the body of another individual with the antibodies or lymphocytes transferred to the host (passive).

adoptive immunity Immune state conferred by inoculation of lymphocytes, not antibodies, from an immune animal rather than by exposure to the antigen itself.

aedeagus Copulatory organ or penis in insects and acarines.

agamete Germinative nucleus within an axial cell of a dicyemid mesozoan.

ala Term often applied to winglike structure on plants or animals: the lateral expansions of the branchiuran carapace to form respiratory alae, cuticular expansions of nematodes, and others.

allergic klendusity Disease-escaping ability produced by development of hypersensitivity to an antigen (allergen).

allograft Graft of a piece of tissue or organ from one individual to another of the same species.

amastigote Form of Trypanosomatidae that lacks a long flagellum; also called a Leishman-Donovan (L-D) body, as in *Leishmania*.

ameboma Granuloma containing active trophozoites, occasionally resulting from a chronic amebic ulcer; rare except in Central and South America.

amebula Daughter cell resulting from mitosis and cytokinesis of an encysted ameba.

amphid Sensory organ on each side of the "head" of nematodes.

amphistome Fluke with the ventral sucker located at the posterior end.

anamnestic response Immune response to a challenge or secondary antigen inoculation, marked by more rapid and stronger manifestation of the immune reaction (specifically, antibody titer) than after the primary immunizing dose.

anapolysis Detachment of a senile proglottid after it has shed its eggs.

anautogeny In some Diptera the necessity of a blood meal before eggs can develop within the female.

androgenic gland Gland located near the vas deferens in many Crustacea; its secretions are responsible for development of male secondary sexual characteristics.

anecdysis Ecdysis in which successive molts are separated by quite long intermolt phases; referred to as "terminal anecdysis" when maximal size is reached and no more ecdyses occur.

anisogametes Outwardly dissimilar male and female gametes.

antennae (second antennae of crustaceans) Second pair of appendages in Crustacea, with bases usually immediately posterior to antennules; primarily sensory but sometimes adapted for other functions; derived from appendages on primitive third preoral somite; no homologous appendage in insects.

antennules (first antennae) Anteriormost pair of appendages of Crustacea; primarily sensory but often adapted for additional or other functions in particular species; derived from appendages on primitive second preoral somite; homologous to *antennae* of insects.

anterior station Development of a protozoan in the middle or anterior intestinal portions of its insect host, such as the section *Salivaria* of Trypanosomatidae.

antibody Immunoglobulin protein, produced by B cells (or plasma cells derived from B cells), that binds with a specific antigen.

antibody titer Measure of the amount of antibody present, usually given in units per milliliter of serum.

antigen Any substance that will stimulate an immune response.

antigen challenge Dose or inoculation with an antigen given to an animal at some time after primary immunization with that antigen has been achieved.

antigenic determinant Area(s) on an antigen molecule that bind with antibody or specific receptor sites on the sensitized lymphocyte; they "determine" the specificity of the antibody or lymphocyte (see *epitope*).

apical organ Organ of unknown function at the apex of a cestode's scolex.

apodeme Spinelike inward projection of the cuticle in arthropods on which a muscle inserts; a ridgelike projection is an apophysis.

apodous larva Larva with no legs and with reduced head; usual in Hymenoptera, Diptera, some Coleoptera; requires maternal care or deposition in or on food source.

apolysis Disintegration or detachment of a gravid tapeworm segment; also, the detachment of the hypodermis from the old procuticle in arthropods before molting.

arista Flagellum-like appendage on the antenna of a fly of the suborder Brachycera and some members of the Nematocera.

arrhenotoky Type of reproduction in which females are diploid and come from fertilized eggs, whereas males are haploid and are produced parthenogenetically; the most common condition in Hymenoptera.

arthropodization Evolutionary development of the combination of characteristics associated with the Arthropoda, including a firm cuticular exoskeleton containing chitin.

ascaridine Protein of unknown function in the sperm of *Ascaris.*

ascaroside Glycoside found in *Ascaris,* made of the sugar *ascarylose* and a series of secondary monol and diol alcohols.

ascites Edema, or accumulation of tissue fluid, in the mesenteries and abdominal cavity.

autoinfection Reinfection by a parasite juvenile without its leaving the host.

autotrophic nutrition Feeding that does not require preformed organic molecules as nutritive substances.

axial cells Central cells of a dicyemid mesozoan.

axoneme Core of a cilium or flagellum, comprising microtubules.

axostyle Tubelike organelle in some flagellate protozoa, extending from the area of the kinetosomes to the posterior end, where it often protrudes.

B cell Type of lymphocyte that, on stimulation with an appropriate antigen, gives rise to plasma cells that liberate antibody to the antigen; so called because in birds they are processed through a lymphoid organ called the bursa of Fabricius; of primary importance in humoral immune response.

bacillary bands Lateral zones in the body wall of some nematodes, consisting of glandular and nonglandular cells of unknown function.

Baer's disc Large, ventral sucker of an aspidogastrean trematode.

ballonets Four inflated areas within the "head" of nematodes of the family Gnathostomatidae; each is connected to an internal cervical sac of unknown function.

basal body Centriole from which an axoneme arises; also called a kinetosome or blepharoplast.

basis (basipodite) Joint of a crustacean appendage from which the exopod and the endopod originate, that is, the joint between the coxa and the exopod and endopod.

basophil Least numerous of polymorphonuclear leukocytes, so called because it stains with basic stains.

bat fly Parasitic fly of the family Streblidae.

bat spider fly Parasitic dipteran of the family Nycteribiidae.

beetle mites Mites of the suborder Cryptostigmata.

bilharziasis Disease caused by *Schistosoma* spp. Also called schistosomiasis.

biological vector Vector in which disease organism lives or develops (contrast *mechanical vector)*.

biotic potential Reproductive potential of a species.

biramous appendage Appendage with two main branches from a common basal joint, characteristic of Crustacea, although not all appendages of a crustacean may be biramous.

biting midges Species in the family Ceratopogonidae.

black fly fever Combination of symptoms resulting from sensitization to bites by black flies (Simuliidae).

blackhead Disease of turkeys caused by the protozoan *Histomonas meleagridis.* Also called histomoniasis or infectious enterohepatitis.

blastocyst In cestodes, posterior portion of plerocercus metacestode into which the body can withdraw.

blastoderm "Primary epithelium" formed in early embryonic development of many arthropods, when the nuclei migrate to the periphery and undergo superficial cleavage; usually encloses the central yolk mass.

blepharoplast Centriole from which arises an axoneme. Also called a basal body or kinetosome.

bluetongue Virus disease of ruminants transmitted by biting midges (Ceratopogonidae).

bot fly Member of a family (Gasterophilidae) of flies whose maggots are parasitic in the stomachs of mammals, especially equids.

bothridium Muscular lappet on the dorsal or ventral side of the scolex of a tapeworm; bothridia are often highly specialized, with many types of adaptations for adhesion.

bothrium Dorsal or ventral groove, which may be variously modified, on the scolex of a cestode.

bradyzoite Small stage in various coccidia of the *Isospora* group that develops in a zoitocyst; similar to a merozoite.

breakbone fever Another name for dengue, a virus disease transmitted by mosquitoes.

bubo Swollen lymph node.

buccal cone Portion of the mouthparts of acarines composed of hypostome and labrum.

bulla Nonliving structure serving as an anchor to which the maxillae are permanently attached; secreted by head and maxillary glands of female copepods in the family Lernaeopodidae.

cadre Sclerotized mouth lining of a pentastomid.

calabar swelling Transient subcutaneous nodule, provoked by the filarial nematode *Loa loa.*

calotte "head" end of a dicyemid mesozoan.

calypter Squama or lobe in the anal angle of a dipteran wing.

camerostome Ventral groove in the propodosoma of soft ticks wherein lies the capitulum.

campestral Characteristic of rural locations, especially open country and grasslands.

capitulum Anterior of two basic body regions of a mite or tick. Also called a gnathosoma.

capsule In reference to eggshell of flatworms, that portion composed of sclerotin, precursors principally contributed by vitelline cells (contrast *coat).*

carapace Structure formed by posterior and lateral extension of dorsal sclerites of the head in many Crustacea, usually covering and/ or fusing with one or more thoracic somites; considered as arising from a fold of head exoskeleton. Also a dorsal sclerotized plate often covering the idiosoma of acarines.

Carrion's disease Bacterial disease transmitted by sand flies. See also *Oroya fever* and *verruga peruana.*

cell-mediated immunity (CMI) Immunity in which antigen is bound to receptor sites on the surface of sensitized T lymphocytes that have been produced in response to prior immunizing experience with that antigen and in which manifestation is through macrophage response with no intervention of antibody.

cellular immune response Binding of antigen with receptor sites on sensitized T lymphocytes to cause release of lymphokines that affect macrophages, a direct response with no intervention of antibody; also, the entire process by which the body responds to an antigen, resulting in a condition of cell-mediated immunity.

centrolecithal egg Type of egg found in many arthropods, in which the nucleus is located centrally in a small amount of nonyolky cytoplasm, surrounded by a large mass of yolk; after fertilization and some nuclear divisions, the nuclei migrate to the periphery to proceed with superficial cleavage, the yolk remaining central.

cephalogaster Contractile organ in adult epicaridean isopods that functions in sucking blood and perhaps in respiration.

cercaria Juvenile digenetic trematode, produced by asexual reproduction within a sporocyst or redia.

cerci Appendages on the eleventh abdominal somite of some insects; usually sensory.

cercomer Posterior, knoblike attachment on a procercoid or cysticercoid. It usually bears the hooks of the oncosphere.

chaetotaxy Taxonomic study of the location and arrangement of bristles on an insect. Especially important in the order Diptera.

Chagas' disease Disease of humans and other mammals caused by *Trypanosoma cruzi.*

chagoma Reddish nodule that forms at the site of entrance of *Trypanosoma cruzi* into the skin.

chalimus Specialized, parasitic copepodid, found in the copepod order Siphonostomatoida; attached to its host by an anterior "frontal filament" that is secreted by the frontal gland.

chelate Condition of an arthropod appendage in which the subterminal podomere bears a distal process to form a pincer with the terminal podomere; sometimes (incorrectly) used to describe the subchelate condition.

chelicerae Anteriormost pair of appendages in the chelicerate arthropods, which include spiders, ticks, and mites; generally the most important feeding appendages in these groups.

chigger Mite of the family Trombiculidae; also, sometimes applied to *Tunga penetrans,* the chigoe flea.

chitin High molecular weight polymer of N-acetyl glucosamine linked by $1,4$-β-glycosidic bonds.

choanomastigote Like a promastigote but with the flagellum emerging from a collar-like process, as in *Crithidia* spp.

chorioptic mange Disease caused by mites of the genus *Chorioptes.*

chromatoid bar Masses of RNA, visible with light microscopy, in young cysts of *Entamoeba* spp.

chyluria Lymph in the urine, characterized by a milky color.

ciliary organelles Organelles of specialized function formed by the fusion of cilia.

cirri Fused tufts of cilia in some protozoa, which function like tiny legs. Also, plural for cirrus.

cirrus Penis or copulatory organ of a flatworm.

clamp Complex set of sclerotized bars, forming a "pinching" organ on the opisthaptor of a monogenetic trematode.

claparedé organs See urstigmata.

coarctate pupa Pupa in which last larval cuticle is retained as puparium.

coat In reference to eggshell of many cestodes, the portion contributed by the outer envelope, derived from embryonic blastomeres.

coelozoic Living in the lumen of a hollow organ, such as the intestine.

coenurus Tapeworm metacestode in the family Taeniidae, in which several scolices bud from an internal germinative membrane, but none of which is enclosed in an internal secondary cyst.

colleterial glands Female accessory glands in insects that produce a substance to cement eggs together or material for an ootheca.

commensalism Kind of symbiosis in which one symbiont, the commensal, is benefited, whereas the other symbiont, the host, is neither harmed nor helped by the association.

complement Collective name for a series of proteins that bind in a complex series of reactions to antibody (either IgM or IgG) when the antibody is itself bound to an antigen; produces lysis of cells if the antibody is bound to antigens on the cell surface.

complement fixation test Immunological method used to detect presence of antibodies that bind (or fix) complement; a standard diagnostic test for many infections.

concomitant immunity Premunition.

condyles Bearing surfaces between arthropod joints, which provide the fulcra on which the joints move.

Congo floor maggot Bloodsucking African maggot, *Auchmeromyia luteola*

conoid Truncated cone of spiral fibrils located within the polar rings of the suborder Eimeriina.

contaminative antigen Antigen borne by the parasite that is common to both the host and the parasite, but which genetically is of host origin.

copepodid Juvenile stage(s) that succeed the naupliar stages in copepods, often quite similar in body form to the adult.

coracidium Larva with a ciliated epithelium, hatching from the egg of certain cestodes; a ciliated oncosphere.

costa Prominent striated rod in some flagellate protozoa that courses from one of the kinetosomes along the cell surface beneath the recurrent flagellum and undulating membrane.

coxa Most proximal podomere of an arthropod limb, sometimes called coxopodite in crustaceans.

creeping eruption Skin condition caused by hookworm larvae not able to mature in a given host.

crura Branches of intestine of a flatworm.

cryptoniscus Intermediate, free-swimming larval stage of the isopod suborder Epicaridea, developing after microniscus; attaches to definitive host.

cryptozoite Preerythrocytic schizont of *Plasmodium* spp.

ctenidium Series of stout, peglike spines on the head (genal ctenidium) and first thoracic tergite (pronotal ctenidium) of many fleas.

cypris Postnaupliar larva of barnacles (crustacean subclass Cirripedia) in which the carapace largely envelops the body; so called because of its resemblance to the ostracod genus *Cypris.*

cystacanth Juvenile acanthocephalan that is infective to its definitive host.

cysticercoid Metacestode developing from the oncosphere in most Cyclophyllidea. It usually has a "tail" and a well-formed scolex.

cysticercosis Infection with one or more cysticerci.

cystogenic cells Secretory cells in a cercaria that produce a metacercarial cyst.

cyton Cell body, contains nucleus and some other organelles, but excludes processes extending from cell; for example, the neurocyton is the nerve cell body excluding axon and dendrites.

cytophaneres Fibers radiating out from a zoitocyst into surrounding muscle; found in some species of Sarcocystidae.

dauer juvenile Nematode juvenile in which development is arrested during unsuitable conditions and resumes when conditions improve.

decacanth Ten-hooked larva that hatches from the egg of a cestodarian tapeworm; also called a lycophora.

definitive host Host in which a parasite achieves sexual maturity. If there is no sexual reproduction in the life of the parasite, the host most important to humans is the definitive host.

deirid Sensory papilla on each side near the anterior end of some nematodes.

delayed type hypersensitivity (DTH) Manifestation of cell-mediated immunity, distinguished from immediate hypersensitivity in that maximal response is reached about 24 hours or more after intradermal injection of the antigen; lesion site is infiltrated primarily by monocytes and macrophages.

dengue Virus disease transmitted by mosquitoes.

denticles (denticulate) Small, toothlike projections.

deuterotoky Type of parthenogenesis in which all individuals are uniparental but in which some males occur.

deutomerite Posterior half of a cephaline gregarine protozoan.

deutonymph In the life cycle of some mesostigmatid mites, a nonfeeding stage that molts into the adult.

deutovum Incompletely developed larva that hatches from the egg of a chigger mite.

diapause Quiescent phase in arthropods in which most physiological processes are suspended.

diapolar cells Ciliated somatodermal cells located between the parapolar and uropolar cells of a mesozoan.

diecdysis Condition in which ecdysis processes are going on continuously and one ecdysis cycle grades rapidly into another.

dioecious Separate sexes; males and females are different individuals.

diplokarya Having nuclei associated in pairs.

diplostomulum Strigeoid metacercaria in the family Diplostomatidae.

diporpa Larval stage in the life cycle of the monogenean *Diplozoon*.

direct development In arthropods, refers to development in which a juvenile hatches from the egg not distinctly different from adult except in size and maturity.

distal cytoplasm Distal cytoplasmic layer in the tegument of Monogenea, Digenea, and Cestoidea.

distome Fluke with two suckers: oral and ventral.

dorsal plate Dorsal plate on the body of a mesostigmatid mite.

dourine Disease of horses and other equids caused by *Trypanosoma equiperdum*.

dyspnea Difficult or labored breathing.

ecdysis Molting or discarding of inexpansible portions of cuticle, after which there is an increase in physical dimensions of the animal's body before the newly secreted cuticle hardens.

eclipsed antigen Antigen borne by the parasite that is common to both the host and the parasite but which genetically is of parasite origin.

ectocommensal Commensal symbiont that lives on the outer surface of its host.

ectoparasite Parasite that lives on the outer surface of its host.

ectopic Infection in a location other than normal or expected.

edema Accumulation of more than normal amounts of tissue fluid, or lymph, in the intercellular spaces, resulting in localized swelling of the area.

ELISA (enzyme-linked immunosorbent assay) Immunodiagnostic test designed to detect the presence of fixed antibody through linkage with an enzymatic reaction.

embryophore In reference to eggshell of many cestodes, that portion contributed by the inner envelope, derived from embryonic blastomeres.

endemic Normally present in a certain geographic area or part of an area.

endemicity Amount or severity of a disease in a particular geographic area.

endite Medial process from the protopod.

endocommensal Commensal symbiont that lives inside its host.

endocytosis Ingestion of particulate matter or fluid by phagocytosis or pinocytosis; that is, bringing material into a cell by invagination of its surface membrane, then pinching off the invaginated portion as a vacuole.

endodyogeny Same as endopolyogeny except that only two daughter cells are formed.

endoparasite Parasite that lives inside its host.

endopod (endopodite) Medial branch of a biramous appendage.

endopolyogeny Formation of daughter cells, each surrounded by its own membrane, while still in the mother cell.

endopterygote Condition of internal wing bud development in an insect; also, an insect in which the wing buds develop externally or any insect secondarily wingless but derived from such an ancestor; associated with holometabolous insects.

endosome Nucleolus-like organelle that does not disappear during mitosis.

eosinophil Type of polymorphonuclear leukocyte very important in many parasitic infections, so called because stains with acidic stains such as eosin.

eosinophilia Elevated eosinophil count in the circulating blood. Commonly associated with chronic parasite infections.

epicaridium First larval stage of the isopod suborder Epicaridea; attaches to a free-living copepod.

epicuticle Thin, outermost layer of arthropod cuticle; contains sclerotin but not chitin.

epidemic A sharp rise in the incidence of an infection or disease.

epidemic hemorrhagic fever Virus disease transmitted by mosquitoes. Also called dengue.

epidemiology Study concerned with all ecological aspects of disease to explain its transmission, distribution, prevalence, and incidence.

epimastigote Like a promastigote but with a short undulating membrane, such as in *Blastocrithidea*.

epipod (epipodite) Lateral process, from the protopod, usually with one or more joints; may be called an exite.

epitope Antigenic determinant, the portion of the antigen molecule displayed on the surface of an antigen-presenting cell (APC).

epizootic Massive infection rate among animals other than humans; identical to an epidemic in humans.

espundia Disease caused by *Leishmania braziliensis;* also called chiclero ulcer, uta, pian bois, or mucocutaneous leishmaniasis.

eukaryote Organism with membrane-bound nuclei in its cells.

eutely Cell or nuclear constancy; the adult has the same number of nuclei or cells as the first-stage juvenile; eutely may exist in tissues, organs, or entire animals.

exflagellation Rapid formation of microgametes from a microgametocyte of *Plasmodium* and related genera.

exite Lateral process or joint from the protopod, sometimes referred to as an epipod.

exopod (exopodite) Lateral branch of a biramous appendage.

exopterygote Condition of external wing bud development in an insect; also, any insect in which the wing buds develop externally; associated with hemimetabolous insects.

eye gnat Fly of the family Chloropidae.

facette Funnel-shaped opening through the inner membrane complex of the egg of a pentastomid; it receives the product of the dorsal organ.

facultative symbiont When facultative, a symbiont is an opportunist, establishing a relationship with a host only if the opportunity presents itself; it is not physiologically dependent on doing so.

fascicle Stylet bundle or combination of mouthparts used to pierce the skin in a blood-feeding arthropod; composition of a fascicle varies according to group.

femur Podomere of an insect or acarine leg fixed to the trochanter proximally and articulating with the tibia distally in insects and with the patella in acarines.

festoons Sclerites on the posterior margin of the opisthosoma of certain hard ticks.

flabellum Recurved process often found on the first two thoracic exopods of branchiuran crustaceans.

flesh fly Any member of the dipteran family Sarcophagidae.

genital atrium Cavity in the body wall of a flatworm into which male and female genital ducts open.

genitointestinal canal Duct connecting the oviduct and intestine of some polyopisthocotylean Monogenea.

gid Disorientation caused by cysticerci in the brain; usually manifested by staggering or whirling.

glial cells Nonnerve cells in a brain or ganglion; their function is obscure, but they may support the life processes of the neurons.

glossa Tonguelike mouthpart in Hymenoptera (considered a hypopharynx by some authors).

glycocalyx Finely filamentous layer containing carbohydrate, found on the outer surface of many cells, from 7.5 to 200 nm thick.

glycosomes Organelles found in *Trypanosoma* that contain enzymes of glycolysis and for oxidizing reduced NAD.

gnathopod Prehensile appendages of some Crustacea, for example, the second and third thoracic legs of Amphiphoda and the first thoracic legs of some Isopoda.

gnathosoma Anterior of two basic regions of the body of a mite or tick. Also called a capitulum.

goblets Markings on the stigmatal plates of certain hard ticks.

gonotyl Muscular sucker or other perigenital specialization surrounding or associated with the genital atrium of a digenetic trematode.

granuloma, granulomatous tissue Repaired area of body marked by fibrous connective tissue (fibrosis); also, fibrous connective tissue surrounding antigen source.

ground itch Skin rash caused by bacteria introduced by invasive hookworm larvae.

gynandry Maturation first of the female gonads within an individual, then of the male organs; also called protogyny.

gynocophoral canal Longitudinal groove in the ventral surface of a male schistosome fluke.

Haller's organ Depression on the first tarsi of ticks; functions as olfactory and humidity receptor.

haltere Vestigial wing on the metathorax of a fly of the order Diptera; necessary for balance during flight.

halzoun Disease resulting from blockage of the nasopharynx by a parasite.

hamuli Large hooks on the opisthaptor of a monogenetic trematode, referred to as anchors by American authors.

haptens Molecules of small molecular weight (usually) that are immunogenic only when attached to carrier molecules, usually proteins.

heel fly Fly maggot of the family Hypodermatidae. Also called a warble.

hemimetabolous metamorphosis Gradual metamorphosis in insects, in which the nymphs are generally similar in body form to the adults and become more like the adults with each instar.

hemocoel Main body cavity of arthropods, the embryonic development of which differs from that of a true coelom but which includes a vestige of a true coelom.

hemoglobinuria Bloody urine.

hemolymph Fluid within the hemocoel of arthropods; also, the pseudocoelomic fluid of nematodes.

hepatosplenomegaly Swollen liver and spleen.

hermaphroditism Possession of gonads of both sexes by a single (monoecious) individual.

heterogonic life cycle Life cycle involving alterations of parasitic and free-living generations.

heterophile reaction Antigen-antibody reaction, in which the antibody was not specifically elicited by the antigen to which it binds.

heteroxenous Living within more than one host during a parasite's life cycle.

hexacanth Oncosphere: a six-hooked larva hatching from the egg of a eucestode.

histozoic Dwelling within the tissues of a host.

holoblastic cleavage Each nuclear division in an early embryo that is accompanied or closely followed by complete cytokinesis, the nuclei being separated by cell membranes.

holometabolous metamorphosis Metamorphosis in an insect with a larva, pupa, and adult.

holophytic nutrition Formation of carbohydrates by chloroplasts.

holozoic nutrition Feeding by active ingestion of organisms or particles.

homogonic life cycle Life cycle in which all generations are parasitic or all are free living; there is no (or little) alternation of the two.

homothetogenic fission Mitotic fission across the rows of cilia of a protozoan.

hood Dorsal wall of the camerostome that extends over the capitulum.

host specificity The degree to which a parasite is able to mature in more than one host species.

humoral immune response Binding of antigen with soluble antibody in blood serum; also, the entire process by which the body responds to an antigen by producing antibody to that antigen.

hydatid cyst Metacestode of the cyclophyllidean cestode genus *Echinococcus,* with many protoscolices, some budding inside secondary brood cysts.

hydatid sand Free protoscolices forming sediment in a hydatid cyst.

hydrogenosomes Small organelles in certain anaerobic protozoa that produce molecular hydrogen as an end product of energy metabolism.

hyperapolysis Detachment of a tapeworm proglottid while still immature, before eggs are formed.

hyperendemic Condition in which a disease or infection has high transmission in a certain geographic area, usually seasonal.

hyperinfection Condition in *Strongyloides* infections in which filariform juveniles repenetrate mucosa of small intestine and proceed with migration.

hypermetamorphosis Type of metamorphic development in which different larval instars have markedly dissimilar body forms.

hyperparasitism Condition in which an organism is a parasite of another parasite.

hypnozoite Dormant exoerythrocytic form found in certain *Plasmodium* species.

hypopharynx Tonguelike lobe arising from floor of mouth in insects; variously modified for feeding in many groups.

hypostome Portion of the mouthparts of acarines; composed of fused coxae of pedipalps.

hysterosoma Combination of the metapodosoma and opisthosoma of the body of a tick or mite.

ick Serious disease of freshwater fish, caused by the ciliate protozoan *Ichthyophthirius multifiliis.*

icterus (jaundice) Yellowing of the skin and other organs because of bile pigments in the blood.

idiosoma Posterior of the two basic parts of the body of a mite or tick, bearing the legs and most internal organs.

imago Adult or final instar in development of an insect.

imagochrysalis Quiescent stage between the nymph and adult in the life cycle of a chigger mite.

immediate hypersensitivity Biological manifestation of an antigen-antibody reaction in which the maximal response is reached in a few minutes or hours; intradermal injection of antigen produces local swelling and redness with heavy infiltration of polymorphonuclear leukocytes; intravenous injection may produce anaphylactic shock and death.

immune cross-reaction Binding of an antibody or cell receptor site with an antigen other than the one that would provide an exact "fit," that is, an antigen-antibody reaction in which the antigen is not the same one that stimulated the production of that antibody.

immunity State in which a host is more or less resistant to an infective agent, preferably used in reference to resistance arising from tissues that are capable of recognizing and protecting the animal against "nonself."

immunoconglutinin Antibody against complexed antigen, antibody against that antigen, and complement.

immunogenic Refers to any substance that is antigenic, that is, stimulates production of antibody or cell-mediated immunity.

immunoglobulin Any one of five classes of proteins in blood serum that function as antibodies; abbreviated IgM, IgG, IgA, IgD, and IgE.

incidence In epidemiology, number of new cases of a disease per unit time, that is, a rate measurement (contrast *prevalence*).

incidental parasite Accidental parasite.

indirect development In arthropods, refers to development in which a larva or nymph hatches from egg, distinctly different in body form from the adult, that is, development with metamorphosis.

inflammation Defense process of body including congestion of blood vessels, escape of plasma to interstitial tissue space, swelling, and warmth.

infraciliature All cilia, basal bodies, and their associated fibrils in a ciliate protozoan.

infrapopulation of parasites All individuals of a single parasite species in one host.

infusoriform larva Ciliated larva produced by an infusorigen within a dicyemid mesozoan.

infusorigen Mass of reproductive cells within a rhombogen.

intermediate host Host in which a parasite develops to some extent but not to sexual maturity.

intermittent parasite Temporary parasite.

internuncial processes Cytoplasmic channels that connect one part of a cell to another, for example, the distal cytoplasm to the tegumental cytons in many flatworms.

intralecithal cleavage Cleavage in which the nuclei undergo several divisions within the yolk mass without concurrent cytokinesis; common in arthropods.

iodinophilous vacuole Vacuole within a protozoan that stains readily with iodine.

isogametes Outwardly similar male and female gametes.

jacket cells Ciliated somatoderm of an orthonectid mesozoan.

kala-azar Disease caused by *Leishmania donovani;* also called Dumdum fever or visceral leishmaniasis.

ked Louse fly of the family Hippoboscidae.

kentrogon Larva in the crustacean order Rhizocephala that is attached to its host crab; formed after the cypris larva molts and its appendages and carapace are discarded.

kinetid Axoneme of a cilium of flagellum together with its basal fibrils and organelles. Also called a mastigont.

kinetodesmose (kinetodesmata) Compound fiber joining cilia into rows.

kinetoplast Conspicuous part of a mitochondrion in a trypanosome, usually found near the kinetosome.

kinetosome Centriole from which an axoneme arises; also called a basal body or blepharoplast.

kinety Row of cilia basal bodies and their kinetodesmose; all kineties and kinetodesmata in the organism are its infraciliature.

Koch's blue bodies Schizonts of *Theileria parva* in circulating lymphocytes.

K-strategist Species of organism that uses a survival and reproductive "strategy" characterized by low fecundity, low mortality, longer life, and with populations approaching the carrying capacity of the environment, controlled by density-dependent factors.

Kupffer cells Phagocytic epithelial cells lining the sinusoids of the liver.

labium Mouthpart in insects composed of fused second maxillae; homologous to second maxillae of crustaceans.

labrum Sclerite forming the anterior closure of the mouth in arthropods, specifically, the free lobe overhanging the mouth.

lacunae Channels making up the lacunar system in Acanthocephala; also, in developing wings of insects, canals that contain nerves, tracheae, and hemolymph.

lacunar system System of canals in the body wall of an acanthocephalan, functioning as a circulatory system.

landscape epidemiology Approach to epidemiology that employs all ecological aspects of a nidus; by recognizing certain physical conditions, the epidemiologist can anticipate whether a disease can be expected to exist.

larva Progeny of any animal that is markedly different in body form from the adult.

larval stem nematogen Early stage in the development of a dicyemid mesozoan.

Laurer's canal Usually blind canal extending from the base of the seminal receptacle of a digenetic trematode; it probably represents a vestigial vagina.

Leishman-Donovan body Amastigote in the Trypanosmatidae; also known as L-D body.

leishmaniasis Infection by a species of *Leishmania*.

lemniscus Structure occurring in pairs attached to the inner, posterior margin of the neck of an acanthocephalan, extending into the trunk cavity. Its function is unknown.

loculi Shallow, sucker-like depressions in an adhesive organ of a flatworm.

louse fly Member of the dipteran family Hippoboscidae.

lumen Space within any hollow organ.

lunules Small, sucker-like discs on the anterior margin of some copepods in the family Caligidae, functioning as organs of adhesion.

lycophora Ten-hooked larva that hatches from the egg of a cestodarian tapeworm; also called a decacanth.

lymph varices Dilated lymph ducts.

lymphadenitis Inflamed lymph node.

lymphocyte Type of leukocyte vital in immune response, several different types known. (See *B cell* and *T cell*.)

lymphokine Any one of several kinds of effector molecules released by T lymphocytes when antigen to which the lymphocyte is sensitized binds to the cell surface.

lysosome An intracellular vacuole or vesicle containing digestive enzymes (lysozymes).

macrogamete Large, quiescent, "female" anisogamete.

macrogametocyte Cell giving rise to a macrogamete.

macrophage Important phagocytic cell and antigen-presenting cell, derived from monocyte.

macrophage migration inhibitory factor (MIF) Lymphokine released by sensitized lymphocytes that tends to inhibit migration of macrophages in the immediate vicinity, thus contributing to accumulation of larger numbers of macrophages close to the site of MIF release.

mamelon Ventral, serrated projection on the ventral surface of a male nematode of the family Syphaciidae; its function is unknown.

mandibles Third pair of appendages in Crustacea, second in Insecta; primarily function in feeding; derived from appendages on primitive fourth (first postoral) somite.

marginal bodies Sensory pits or short tentacles between the marginal loculi of the opisthaptor of an aspidogastrean trematode.

marrara Nasopharyngeal blockage by a parasite. Also called halzoun.

mast cell Type of cell in various tissues which releases pharmacologically active substances with a role in inflammation.

mastigont Axoneme of a cilium or flagellum together with its basal fibrils and organelles; also called a kinetid.

mastitis Infection of the udder of cattle.

Maurer's clefts Blotches on the surface of an erythrocyte infected with *Plasmodium falciparum.*

maxillae (second maxillae) Fifth pair of appendages in Crustacea, primarily feeding in function, derived from appendages on primitive sixth (third postoral) somite; homologous to *labium* in insects; maxillae of insects are third pair of head appendages, homologous to maxillules of Crustacea.

maxillipeds One or more pairs of head appendages originating posterior to maxillae in Crustacea; derived from appendages on somites that were primitively posterior to gnathocephalon; usually function in feeding but sometimes adapted for other functions, such as prehension, in parasitic forms.

maxillules (first maxillae) Fourth pair of appendages in Crustacea, primarily feeding in function; derived from appendages on primitive fifth (second postoral) somite; homologous to *maxillae* in insects.

mechanical vector Vector that transmits disease organism by mechanical means only (contrast *biological vector*).

megacolon Flabby distended colon caused by chronic Chagas' disease.

megaesophagus Distended esophagus caused by chronic Chagas' disease.

Mehlis' glands Unicellular mucous and serous glands surrounding the ootype of a flatworm.

membranelle Short, transverse rows of cilia, fused at their bases, serving to move food particles toward the oral groove of a protozoan.

merozoite Daughter cell resulting from schizogony.

mesocercaria Juvenile stage of the digenetic trematode *Alaria;* it is an unencysted form between the cercaria and the metacercaria.

metacercaria Stage between the cercaria and adult in the life cycle of most digenetic trematodes; usually encysted and quiescent.

metacestode Developmental stage of a cestode after metamorphosis of the oncosphere.

metacryptozoite Merozoite developed from a cryptozoite.

metacyclic Stage in the life cycle of a parasite that is infective to its definitive host.

metacyst Cystic stage of a parasite that is infective to a host.

metamere One of the segments in a metameric animal.

metamerism Division of the body along the anteroposterior axis into a serial succession of segments, each of which contains identical or similar representatives of all the organ systems of the body; primitively in arthropods, including, externally, a pair of appendages and, internally, a pair of nerve ganglia, a pair of nephridia, a pair of gonads, paired blood vessels and nerves, and a portion of the digestive and muscular systems.

metamorphosis Type of development in which one or more juvenile types differ markedly in body form from the adult; occurs in numerous animal phyla; also applies to the actual process of changing from larval to adult form.

metanauplius Later naupliar larvae of some crustaceans, that is, after several naupliar stages but before another larval type or preadult in developmental sequence.

metapodosoma Portion of the podosoma that bears the third and fourth pair of legs of a tick or mite.

metapolar cells Posterior tier of cells in the calotte of a dicyemid mesozoan.

metasome Portion of the body anterior to the major point of body flexion in many copepods; usually includes cephalothorax and several free thoracic segments.

metraterm Muscular, distended termination of the uterus of a digenetic trematode.

microfilaria First-stage juvenile of any filariid nematode that is ovoviviparous; usually found in the blood or tissue fluids of the definitive host.

microgamete Slender, active, "male" anisogamete.

microgametocyte Cell that gives rise to microgametes.

micronemes Slender, convoluted bodies that join a duct system with the rhoptries, opening at the tip of a sporozoite or merozoite.

microniscus Intermediate larval stages of the isopod suborder Epicaridea, parasitic on free-living copepods.

micropredator Temporary parasite.

micropyle A pore in the oocyst of some coccidia and in the egg of an insect.

microthrix (microtriches) Minute projections of the tegument of a cestode.

monoecious Hermaphroditic; an individual that contains reproductive systems of both sexes.

monostome Fluke that lacks a ventral sucker.

monoxenous Living within a single host during a parasite's life cycle.

monozoic Tapeworm whose "strobila" consists of a single unit.

moth fly Member of the dipteran subfamily Psychodinae, family Psychodidae.

mucron Apical anchoring device on an acephaline gregarine protozoan.

muscularis mucosae Smooth muscle fibers around the mucosa of the gut wall, surrounding the lamina propria and surrounded by the submucosa.

mutualism Type of symbiosis in which both host and symbiont benefit from the association.

mycetome Specialized organ in some insects that bears mutualistic bacteria.

myiasis Infection by fly maggots.

myzorhynchus Apical stalked, sucker-like organ on the scolex of some tetraphyllidean cestodes.

nagana Disease of ruminants caused by *Trypanosoma brucei brucei* or *T. congolense*.

nauplius Typically the earliest larval stage(s) of crustaceans; has only three pairs of appendages: antennules, antennae, and mandibles—all primarily of locomotive function.

neascus Strigeoid metacercaria with a spoon-shaped forebody.

necrosis Cell or tissue death.

neotenic plerocercoid Adult caryophyllidean cestode, except for *Archigetes,* which is a neotenic procercoid.

neotenic procercoid Adult *Archigetes,* a caryophyllidean cestode.

neutrophil Most abundant of polymorphonuclear leukocytes, important phagocyte, so called because it stains with both acidic and basic stains.

nidus Specific locality of a given disease; result of a unique combination of ecological factors that favors the maintenance and transmission of the disease organism.

no-see-um Biting midge of the family Ceratopogonidae.

nymphochrysalis Nonfeeding, prenymph stage in the life cycle of a chigger mite.

nymphs Juvenile instars in insects with hemimetabolous metamorphosis; also, juvenile instars of mites and ticks with a full complement of legs.

obligate symbiont Organism that is physiologically dependent on establishing a symbiotic relationship with another.

obtect pupa Pupa with wings and legs tightly appressed to body and covered by external cuticle.

oligopod larva Usual larva in Coleoptera and Neuroptera, with well-developed head and thoracic legs.

onchocercoma Subcutaneous nodule containing masses of the nematode *Onchocerca volvulus*.

oncomiracidium Ciliated larva of a monogenetic trematode.

oocyst Cystic form in the Apicomplexa, resulting from sporogony; the oocyst may be covered by a hard, resistant membrane (as in *Eimeria),* or it may not (as in *Plasmodium).*

oocyst residuum Cytoplasmic material not incorporated into the sporocyst within an oocyst; seen as an amorphous mass within an oocyst.

oogenotop Female genital complex of a flatworm, including oviduct, ootype, Mehlis' glands, common vitelline duct, and upper uterus.

ookinete Motile, elongate zygote of a *Plasmodium* or related organism.

oostegites Modified thoracic epipods in females of the crustacean superorder Peracarida; they form a pouch for brooding embryos.

ootheca Egg packet secreted by some insects; may be covered with sclerotin.

ootype Expansion of the flatworm female duct, surrounded by Mehlis' glands, where, in some flatworms, ducts from a seminal receptacle and vitelline reservoir join.

operculum Lidlike specialization of a parasite eggshell through which the larva escapes.

opisthaptor Posterior attachment organ of a monogenetic trematode.

opisthomastigote A form of Trypanosomatidae with the kinetoplast at the posterior end; the flagellum runs through a long reservoir to emerge at the anterior; no undulating membrane; for example, *Herpetomonas.*

opisthosoma Portion of the body posterior to the legs in a tick or mite.

opsonization Modification of the surface characteristics of an invading particle or organism by binding with antibody or a nonspecific molecule in such a manner as to facilitate phagocytosis by host cells.

orchitis Inflammation of the testis.

oriental sore Disease caused by *Leishmania tropica.* Also called Jericho boil, Delhi boil, Aleppo boil, or cutaneous leishmaniasis.

Oroya fever Clinical form of Carrion's disease, caused by the bacterium *Bartonella bacilliformis* and transmitted by sand flies.

otoacariasis Infestation of the external ear canal by ticks or mites.

overdispersion Ecological term meaning nonrandom dispersion of individuals in a habitat; for example, when a minority of host individuals will bear a majority of parasites.

ovicapt Sphincter on the oviduct of a flatworm.

ovipositor Structure on a female animal modified for deposition of eggs; in many insects, derived from segmental appendages of the abdomen.

ovisac External sac attached to the somite that bears openings of gonoducts in females of many Copepoda; fertilized eggs pass into the ovisacs for embryonation.

ovovitellarium Mixed mass of ova and vitelline cells; found in the monogenean genus *Gyrodactylus* and in a few tapeworms.

pandemic Very widely distributed epidemic.

pansporoblast Myxosporidean sporoblast that gives rise to more than one spore; also called a sporoblast mother cell.

Papatasi fever Virus disease transmitted by sand flies; also called sand fly fever.

parabasal body Golgi body located near the basal body of some flagellate protozoa, from which the parabasal filament runs to the basal body.

parabasal filament Fibril with periodicity visible in electron micrographs, that courses between the parabasal body and a kinetosome.

paramere Copulatory appendage in male cimicid bugs.

parapolar cells Cells making up the ciliated somatoderm immediately behind the calotte of a mesozoan.

parasite Raison d'être for parasitologists.

parasitic castration Condition in which a parasite causes retardation in development or atrophy of host gonads, often accompanied by failure of secondary sexual characteristics to develop.

parasitism Symbiosis in which the symbiont benefits from the association, whereas the host is harmed in some way.

parasitoid Organism that is a typical parasite early in its development but that finally kills the host during or at the completion of development; often used in reference to many insect parasites of other insects.

parasitologist Quaint person who seeks truth in strange places; he sits on one stool, while staring at another.

parasitophorous vacuole Vacuole within a host cell that contains a parasite.

paratenic host Host in which a parasite survives without undergoing further development; also known as a transport host.

pars prostatica Dilation of the ejaculatory duct of a flatworm, surrounded by unicellular prostate cells.

parthenogenesis Development of an unfertilized egg into a new individual.

paruterine organ Fibromuscular organ in some cestodes that replaces the uterus.

passive immunization Immune state in an animal created by inoculation with serum (containing antibodies) or lymphocytes from an immune animal, rather than by exposure to the antigen.

pathogenesis Production and development of disease.

pathogenicity Capability of an agent to produce disease.

pedicel (petiole) Slender, second abdominal segment that forms a "waist" in most Hymenoptera.

pedipalps Second pair of appendages in chelicerate arthropods, modified variously in different groups.

perikaryon (perikarya) Portion of the cell that contains the nucleus (karyon), sometimes called the cyton or cell body, used in reference to cells that have processes extending some distance away from the area of the nucleus, for example, nerve axons or tegumental cells of cestodes and trematodes.

peritreme Elongated sclerite extending forward from the stigma of certain mites, mainly in the suborder Mesostigmata.

peritrophic membrane Noncellular, delicate membrane lining an insect's midgut.

permanent parasite Parasite that lives its entire adult life within or on a host.

peroxisomes Small organelles containing enzymes of the glyoxylate cycle, catalase, and peroxidases.

Peyer's patches Lymphoid tissue in the wall of the intestine; not circumscribed by a tissue capsule.

phagocytosis Endocytosis of a particle by a cell.

phagosome Vacuole in a cell containing phagocytosed particle.

phasmid Sensory pit on each side near the end of the tail of nematodes of the class Phasmidea.

phoresis A form of symbiosis when the symbiont, the phoront, is mechanically carried about by its host. Neither is physiologically dependent on the other.

pigeon fly Family Hippoboscidae, a parasite of pigeons.

pinkeye Bacterial conjunctivitis, sometimes transmitted by flies of the genus *Hippolates*.

pipestem fibrosis Thickening of the walls of a bile duct as the result of the irritating presence of a parasite.

piroplasm Any of the class Piroplasmea, while in a circulating erythrocyte.

planidium First instar of hypermetamorphic, parasitic Diptera and Hymenoptera, which is apodous but moves actively by means of thoracic and caudal setae.

plasmotomy Division of a multinucleate cell into multinucleate daughter cells, without accompanying mitosis.

pleopods Abdominal appendages of Crustacea.

plerocercoid Metacestode that develops from a procercoid. It usually shows little differentiation.

plerocercus Tapeworm metacestode in the order Trypanorhyncha in which the posterior forms a bladder, the blastocyst, into which the rest of the body withdraws.

pleurite Lateral sclerite of a somite in an arthropod.

podomere More or less cylindrical segment of a limb of an arthropod, generally articulated at both ends.

podosoma Portion of the body of a tick or mite that bears the legs.

polar granule Refractile granule within a coccidian oocyst.

polar ring Electron-dense organelles of unknown function, located under the cell membrane at the anterior tip of sporozoites and merozoites.

polaroplast Organelle, apparently a vacuole, near the polar filament of a microsporidean.

polyembryony Development of a single zygote into more than one offspring.

polypod larva Caterpillar type of larva found in Lepidoptera and some Hymenoptera; has thoracic appendages and abdominal locomotory processes (prolegs); also called cruciform.

polyzoic Strobila, when consisting of more than one proglottid.

porose area Sunken areas on the basis capituli of certain mites and ticks.

posterior station Development of a protozoan in the hindgut or posterior midgut of its insect host, such as in the section Stercoraria of the Trypanosomatidae.

praniza Parasitic larva of the isopod suborder Gnathiidea; parasitizes fishes and feeds on blood.

predation Animal interaction in which the predator kills the prey outright; it does not subsist on the prey while the prey is alive.

premunition Resistance to reinfection or superinfection conferred by a still existing infection, that does not destroy the organisms of the infection already present.

prenymph Nonfeeding, quiescent stage in the life cycle of a chigger mite.

presoma The proboscis, neck and attached muscles and organs of an acanthocephalan.

prevalence In epidemiology, number of cases of a disease at a given time, that is, a static measurement (contrast *incidence*.)

primite Anterior member of a pair of gregarines in syzygy.

procercoid Cestode metacestode developing from a coracidium in some orders. It usually has a posterior cercomer.

procuticle Thicker layer beneath the epicuticle of arthropods that lends mass and strength to the cuticle; it contains chitin, sclerotin, and also inorganic salts in Crustacea; layers within procuticle vary in structure and composition.

proglottid One set of reproductive organs in a tapeworm strobila; it usually corresponds to a segment.

prohaptor Collective adhesive and feeding organs at the anterior end of a monogenetic trematode.

prokaryote Organism in which the chromosomes are not contained within membrane-bound nuclei.

promastigote Form of Trypanosomatidae with the free flagellum anterior and the kinetoplast anterior to the nucleus, as in *Leptomonas*.

propodeum First abdominal segment of hymenopterans, fused to the thorax.

propodosoma Portion of the podosoma that bears the first and second pairs of legs of a tick or mite.

propolar cells Anterior tier of cells in the calotte of a dicyemid mesozoan.

prosoma Anterior tagma of arachnids, consisting of cephalothorax; fused imperceptibly to opisthosoma in Acari.

protandry Maturation first of the male gonads, then of the female organs, within a hermaphroditic individual; also called androgyny.

protelean parasite Organism parasitic during its larval or juvenile stages and free living as an adult, usually changing form with each stage.

proterosoma Combination of the gnathosoma and propodosoma of the body of a tick or mite.

protomerite Anterior half of a cephaline gregarine protozoan.

protonymph Early, bloodsucking stage in the life cycle of some mesostigmatid mites.

protopod (protopodite) Coxa and basis together.

protopod larva Larva found in some parasitic Hymenoptera and Diptera; limbs rudimentary or absent; internal organs incompletely differentiated; requires highly nutritive and sheltered environment for further development.

protoscolex Juvenile scolex budded within a coenurus or a hydatid metacestode of a taeniid cestode.

pseudocyst Pocket of protozoa within a host cell but not surrounded by a cyst wall of parasite origin.

pseudolabia Bilateral lips around the mouth of many nematodes of the order Spirurata; they are not homologous to the lips of most other nematodes but develop from the inner wall of the buccal cavity.

pseudomyiasis Presence within a host of a fly not normally parasitic.

psoroptic mange Disease caused by mites of the genus *Psoroptes*.

ptilinum Balloon-like organ in the head of teneral dipterans that pushes off the operculum of the puparium.

punky Biting midge of the family Ceratopogonidae.

pupariation Formation of a puparium by third-stage larvae of certain families of Diptera.

puparium Pupal stage of certain families of Diptera.

pygidium Sensory organ on a posterior tergite of fleas, which apparently detects air currents.

quartan malaria Malaria with fevers recurring every 72 hours. Caused by *Plasmodium malariae*.

quotidian malaria Malaria with fevers recurring every 24 hours. Found in cases of overlapping infections.

rachis Central, longitudinal, supporting structure in the ovary of some nematodes.

red mange Disease caused by the dog follicle mite, *Demodex canis*.

redia A larval, digenetic trematode, produced by asexual reproduction within a miracidium, sporocyst, or mother redia.

reservoir Living or (rarely) nonliving means of maintaining infectious agent in nature that can serve as source of infection for humans or domestic animals.

reticuloendothelial system (RE) Total complement of fixed macrophages in the body, especially reticular connective tissue and the lining epithelium of the blood vascular system; some authorities also include the phagocytic white blood cells.

retrofection Process of reinfection, whereby juvenile nematodes hatch on the skin and reenter the body before molting to third-stage larvae.

rhombogen Stage in the life cycle of a dicyemid mesozoan.

rhoptries Elongate, electron-dense bodies extending within the polar rings of an apicomplexan.

Romaña's sign Symptoms of recent infection by *Trypanosoma cruzi*, consisting of edema of the orbit and swelling of the preauricular lymph node.

Romanovsky stain Complex stain, based on methylene blue and eosin, used to stain blood cells and hemoparasites; Wright's and Giemsa's stains are two common examples.

rostrum (tectum) Dorsal part of capitulum projecting over chelicerae in acarines.

r-strategist Species of organism that uses a survival and reproductive "strategy" characterized by high fecundity, high mortality, short longevity; populations controlled by density-independent factors.

ruffles Slender projections of the exterior surface of a dicyemid mesozoan.

Saefftigen's pouch Internal, muscular sac near the posterior end of a male acanthocephalan; it contains fluid that aids in manipulating the copulatory bursa.

salivarium Chamber in buccal cone of acarines into which salivary ducts open.

sand fly Member of the dipteran subfamily Phlebotominae, family Psychodidae; sometimes also applied to Simuliidae (New Zealand) and Ceratopogonidae (Caribbean).

sand fly fever Virus disease transmitted by sand flies.

saprophytic A plant living on dead organic matter.

saprozoic nutrition Nutrition of an animal by absorption of dissolved salts and simple organic nutrients from surrounding medium; also, refers to feeding on decaying organic matter.

sarcocystin Powerful toxin produced by zoitocysts of *Sarcocystis*.

sarcoptic mange Disease caused by mites of the genus *Sarcoptes*; also called scabies.

satellite Posterior member of a pair of gregarines in syzygy.

scabies Disease caused by mites of the genus *Sarcoptes*.

scaly leg Disease of birds, caused by mites of the genus *Knemidokoptes*.

schistosomule Juvenile stage of a blood fluke, between a cercaria and an adult; a migrating form taking the place of a metacercaria in the life cycle.

schizeckenosy System of waste elimination found in some mites with blindly ending midgut; lobe from ventriculus breaks free and is expelled through split in posterodorsal cuticle.

schizogony Form of asexual reproduction in which multiple mitoses take place, followed by simultaneous cytokineses, resulting in many daughter cells at once.

schizont Cell undergoing schizogony, in which nuclear divisions have occurred but cytokinesis is not completed, in its late phase, sometimes called a segmenter.

Schüffner's dots Small surface invaginations that appear as stippling on the membrane of an erythrocyte infected with *Plasmodium vivax* after Romanovsky staining.

sclerite Any well-defined, sclerotized area of arthropod cuticle limited by suture lines or flexible, membranous portions of cuticle.

sclerotin Highly resistant and insoluble protein occurring in the cuticle of arthropods; also thought to occur in structures secreted by various other animals, such as in the eggshells of some trematodes, in which stabilization of the protein is achieved by orthoquinone cross-links between free imino or amino groups of the protein molecules.

scolex "Head" or holdfast organ of a tapeworm.

scoliosis Lateral curvature of the spine.

screwworm Parasitic maggot of the species *Cochliomyia hominivorax*.

scrub typhus Rickettsial disease transmitted by certain chigger mites.

scutum Large, anteriodorsal sclerite on a tick or mite.

sheep bot *Oestrus ovis*, a fly maggot parasitic in the head sinuses of sheep and related animals.

skin bot fly Member of the family Cuterebridae.

sleeping sickness Name given to both African trypanosomiasis and mosquito-borne, virus-induced encephalitis.

slime ball Mass of mucus-covered cercariae of dicrocoeliid flukes, released from land snails.

snipe fly Fly of the family Rhagionidae.

solenophage Blood-feeding arthropod that introduces its mouthparts directly into a blood vessel to feed.

somite Body segment or metamere, a term usually used in reference to arthropods.

sparganum Cestode plerocercoid of unknown identity.

spermalege Organ that receives the sperm in the female cimicid bug during copulation.

spermatodactyl Modification of chelicera in some Acari, which functions in transfer of sperm from male's gonopore to copulatory receptacles between third and fourth coxae of female.

spermatophore Formed "container" or packet of sperm, which is placed in or on the body of a female, in contrast to copulation in which the sperm are conducted directly from male reproductive structures into the female's body.

spiracles Openings into the respiratory system in various arthropods.

spondylosis Degeneration of a vertebra.

sporadin A mature trophozoite of a gregarine protozoan.

sporoblast Cell mass that will differentiate into a sporocyst within an oocyst.

sporocyst Stage of development of a sporozoan protozoan, usually with an enclosing membrane, the oocyst; also, an asexual stage of development in some trematodes.

sporocyst residuum Cytoplasmic material "left over" within a sporocyst after sporozoite formation; seen as an amorphous mass.

sporogony Multiple fission of a zygote; such a cell also is called a sporont.

sporont Undifferentiated cell mass within an unsporulated oocyst.

sporoplasm Ameba-like portion of a microsporan or myxosporan cyst that is infective to the next host.

sporozoite Daughter cell resulting from sporogony.

spring dwindling Disease of honeybees caused by the microsporan protozoan *Nosema apis;* also called nosema disease, bee dysentery, bee sickness, and May sickness.

squama Prominent lobe in the anal angle of a dipteran wing.

sternite Main ventral sclerite of a somite of an arthropod.

stichosome Column of large, rectangular cells called stichocytes, supporting and secreting into most of the esophagus of nematodes of the family Trichuridae.

Stieda body Plug in the inner wall of one end of a coccidian oocyst.

stigma (pl. stigmata) Operculum-like area of an eggshell through which the miracidium of a schistosome fluke hatches; also, an arthropod spiracle.

strike Deposition of fly eggs or larvae on a living host.

strobilation Formation of a chain of zoids by budding, as in the strobila of a tapeworm.

strobilocercoid Cysticercoid that undergoes some strobilation; found only in *Schistotaenia.*

strobilocercus Simple cysticercus with some evident strobilation.

style Terminal segment of the antenna of a brachyceran dipteran. It is drawn into a sharp point.

stylostome Hardened, tubelike structure secreted by a feeding chigger mite.

subchelate Condition of an arthropod appendage in which the terminal podomere can fold back like a pincer against the subterminal podomere.

substiedal body Additional plug material underlying a Stieda body.

suprapopulation of parasites All individuals of a single parasite species at all stages in the life cycle in all hosts in an ecosystem.

surra Disease of large mammals caused by *Trypanosoma evansi.*

swarmer Daughter trophozoites resulting from multiple fissions of *Ichythophthirius multifiliis* and a few other protozoa.

sylvatic Existing normally in the wild, not in the human environment.

symbiology Study of symbioses.

symbiont Any organism involved in a symbiotic relationship with another organism, the host.

symbiosis Interaction in which one organism lives with, in, or on the body of another.

symmetrogenic fission Mitotic fission between the rows of flagella of protozoa.

synanthropism Habit of an organism to live in or around human dwellings.

syngamy Sexual reproduction by fusion of gametes.

syzygy Stage during sexual reproduction of some gregarines in which two or more sporadins connect end to end.

T cell Type of lymphocyte with vital regulatory role in immune response, subsets of T cells may be stimulatory or inhibitory; of primary importance in cell-mediated immunity; so called because they are processed through the thymus.

tachyzoite Small, merozoite-like stages of *Toxoplasma;* they develop in the host cells' parasitophorous vacuole by endodyogeny.

tagmatization Specialization of metameres in animals, particularly arthropods, into distinct body regions, each known as a tagma (pl., tagmata).

tarsus Most distal podomere of the insect or acarine limb; articulates proximally with the tibia and usually is subdivided into two to five subsegments in insects.

tectum Dorsal extension over the mouth of a crustacean or acarine; also called the rostrum.

telmophage Blood-feeding arthropod that cuts through skin and blood vessels to cause a small hemorrhage of blood from which it feeds.

temporary parasite Parasite that contacts its host only to feed and then leaves; also called an intermittent parasite or micropredator.

teneral Newly emerged adult arthropod that is soft and weak.

terebra Functional unit of hymenopteran ovipositor, formed from first and second valvulae.

tergite Main dorsal sclerite of a somite of an arthropod.

tertian malaria Malaria in which fevers recur every 48 hours. Caused by *Plasmodium vivax, P. ovale,* and *P. falciparum.*

tetracotyle Strigeoid metacercaria in the family Strigeidae.

tetrathyridium Only metacestode form known in the tapeworm cyclophyllidean genus *Mesocestoides;* a large, solid-bodied cysticercoid.

theileriosis Disease of cattle and other ruminants, caused by *Theileria parva;* also called East Coast fever.

thelyotoky Type of parthenogenesis in which all individuals are uniparental and essentially no males are produced.

three-day fever Virus disease transmitted by sand flies; also called sand fly fever.

thrombus Blood clot in blood vessel or in one of the cavities of the heart.

tibia Podomere of an insect or acarine leg that articulates proximally with the femur in insects and patella in acarines and distally with the tarsus in insects or with the metatarsus or tarsus in acarines.

titer Concentration of a substance in a solution as determined by titration.

trabecula In general anatomical usage, a septum extending from an envelope through enclosed substance, which, together with other trabeculae, forms part of the framework of various organs; here referring specifically to the cell processes connecting the perikarya of cestode and trematode tegumental cells with the distal cytoplasm; also called internuncial process.

tracheal system System of cuticle-lined tubes in many insects and acarines that functions in respiration; opens to outside through spiracles.

transport host Paratenic host.

tribocytic organ Glandular, padlike organ behind the acetabulum of a strigeoid trematode.

trickling filter fly *Psychoda alternata*, a common fly that breeds in sewage and similar habitats.

tritosternum Ventral, bristle-like sensory organ just behind the gnathosoma of a mesostigmatid mite.

triungulin (triungulinid) First instar larva of some parasitic, hypermetamorphic Neuroptera and Coleoptera and of the Strepsiptera, which is an active, campodeiform oligopod.

trochanter Podomere of insect or acarine leg that articulates basally with the coxa and distally with the femur (usually fixed to the femur in insects).

trophozoite Active, feeding stage of a protozoan, in contrast to a cyst; also called the vegetative stage.

trypomastigote Form of Trypanosomatidae with an undulating membrane and the kinetoplast located posterior to the nucleus; for example, *Trypanosoma*.

tsetse fly Bloodsucking fly of the genus *Glossina*.

tumbler Mosquito pupa.

Tumbu fly African Callophoridae, *Cordylobia anthropophaga*.

ulcer Area of inflammation that opens out to the skin or a mucous surface.

undulating membrane Name applied to two quite different structures in protozoa; in some Mastigophora it is a finlike ridge across the surface of a cell, with the axoneme of a flagellum near its surface; in some ciliates it is a line of cilia that are fused at their bases, usually beating to force food particles toward the gullet.

undulating ridges Undulatory waves in the surface of some protozoa, probably aided by subpellicular microtubules; the means of locomotion in some species.

uniramous appendage An arthropod appendage that is unbranched, characteristic of living arthropods other than Crustacea, although some crustacean appendages are uniramous.

urban Peculiar to the human environment, as contrasted with that found normally in wild animals.

urn A region near the center of an infusoriform larva of a dicyemid mesozoan.

uropolar cells Somatoderm cells at the posterior end of the trunk of a dicyemid mesozoan.

urosome Portion of the body posterior to the major point of body flexion in many copepods; usually includes one or more free thoracic segments and abdomen.

urstigmata Sense organs between the coxae of the first and second pairs of legs on some mites, apparently humidity receptors; also called Claparedé organs.

vagabond's disease Darkened, thickened skin caused by years of infestation with body lice, *Pediculus humanus humanus*.

valvifers Basal portions of the ovipositor in Hymenoptera, derived from coxae of segmental appendages.

valvulae Processes from the valvifers to form the body of the ovipositor (terebra) and the ovipositor sheath (third valvulae) in Hymenoptera.

variant antigen type (VAT) Applied to certain trypanosomes, any one of numerous antigenic types expressed on the surface of the organisms and which the immune system of the host "sees." (See *variant-specific surface glycoprotein.*)

variant-specific surface glycoprotein (VSG) The glycoprotein on the surface of certain trypanosomes recognized by the host's immune system; each VSG is responsible for one VAT.

vector Any agent, such as water, wind, or insect, that transmits a disease organism.

veins Blood vessels conducting blood toward the heart in any animal; also, more heavily sclerotized portions of wings of insects, which are remains of lacunae.

vermicle Infective stage of *Babesia* in a tick.

verminous intoxication Variable condition of systemic poisoning caused by absorbed metabolites produced by parasites.

verruga peruana Clinical form of Carrion's disease, caused by the bacterium *Bartonella bacilliformis* and transmitted by sand flies.

vesicular disease Any disease of the urinary bladder, such as vesicular schistosomiasis.

virulence Degree of pathogenicity of an agent; how much damage the agent can cause.

warble Fly maggot of the family Hypodermatidae; also called a heel fly.

whirling disease Disease of fishes, caused by the protozoan *Myxobolus cerebralis*.

wiggler Mosquito larva.

Winterbottom's sign Swollen lymph nodes at the base of the skull, symptomatic of African sleeping sickness.

xenodiagnosis Diagnosis of a disesae by infecting a test animal.

xenograft Graft of a piece of tissue or organ from one individual to another of a different species.

xenosomes Body or organelle living within a cell that contains its own DNA and is capable of reproducing itself, once having functioned as a free-living organism, for example, zooxanthellae and zoochlorellae.

xiphidiocercaria Cercaria with a stylet in the anterior rim of its oral sucker.

yaws Bacterial disease caused by the spirochete *Treponema pertenue*, often transmitted by flies.

yellow fever Virus disease transmitted by the mosquito *Aedes aegypti*.

zoid Member of a colonial organism

zoitocyst Tissue phase in some of the *Isospora* group of coccidia. They usually have internal septae and contain thousands of bradyzoites; also called a sarcocyst or Miescher's tubule.

zoonosis Disease of animals that is transmissible to humans; some authors subdivide the concept into zooanthroponosis, infections humans can acquire from animals, and anthropozoonosis, a disease of humans transmissible to other animals.

INDEX